BETTER DATA VISUALIZATIONS

BETTER DATA VISUALIZATIONS

A Guide for Scholars, Researchers, and Wonks

Jonathan Schwabish

COLUMBIA UNIVERSITY PRESS ▸ NEW YORK

Columbia University Press
Publishers Since 1893
New York   Chichester, West Sussex
cup.columbia.edu
Copyright © 2021 Columbia University Press
All rights reserved

Chapter 11, "Tables," based on Jonathan A. Schwabish, "Ten Guidelines for Better Tables,"
Journal of Benefit-Cost Analysis 11, no. 2 (2020): 151–178. Reprinted with permission.

Library of Congress Cataloging-in-Publication Data
Names: Schwabish, Jonathan A., author.
Title: Better data visualizations : a guide for scholars, researchers, and wonks /
    Jonathan Schwabish.
Description: New York : Columbia University Press, [2021] | Includes bibliographical
    references and index.
Identifiers: LCCN 2020017814 (print) | LCCN 2020017815 (ebook) | ISBN 9780231193108
    (hardback) | ISBN 9780231193115 (trade paperback) | ISBN 9780231550154 (ebook)
Subjects: LCSH: Information visualization. | Visual analytics.
Classification: LCC QA76.9.I52 S393 2021 (print) | LCC QA76.9.I52 (ebook) |
    DDC 001.4/226—dc23
LC record available at https://lccn.loc.gov/2020017814
LC ebook record available at https://lccn.loc.gov/2020017815

Columbia University Press books are printed on permanent and
durable acid-free paper.

Printed in the United States of America

For Aunt Vivi. Our Mendales. With love and Diet Coke.

CONTENTS

PART TWO: CHART TYPES

VISUAL PROCESSING AND PERCEPTUAL RANKINGS

Before we start creating our charts and graphs, we need to cover some basic theory of how the brain perceives visual stimuli. This will guide you as you decide what chart type is most appropriate to visualize your data.

When we consider how to visualize our data, we must ask ourselves how accurately the reader can perceive the data values. Are some graphs better equipped to guide the reader to the specific difference between, say, 2 percent and 2.3 percent? If so, how should we think about those differences as we create our visualizations?

There's a thread of research in the data visualization field that explores this very question. Based on original research over the past forty years or so, the image on the next page shows a spectrum of graphs—or more generally, types of data *encodings* like dots, lines, and bars—arrayed by how easily readers can estimate their value. The encodings that readers can most accurately estimate are arranged at the top, and those that enable more general estimates are at the bottom.

The rankings are unsurprising. It is easier to compare the data in line charts, bar charts, and area charts that have the same axis or baseline. Graphs on which the data are positioned on unaligned axes—think of a pair of bars that are offset from one another on different axes—are slightly harder for us to accurately discern the values.

Farther down the vertical axis are encodings based on angle, area, volume, and color. You intuitively know this: it's much easier to discern the exact data values and differences between values when reading a bar chart than when reading a map where countries are shaded with different colors.

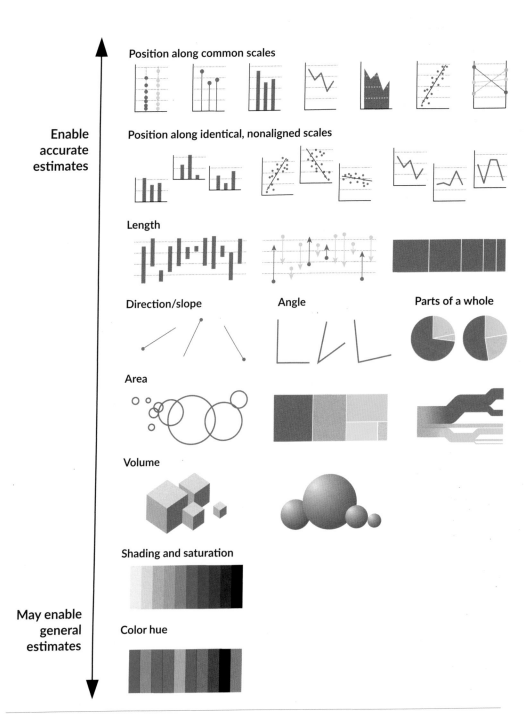

Perceptual ranking diagram. What kind of data visualization you choose to create will depend on your goals and your audience's needs, experiences, and expertise. This image is based on Alberto Cairo (2016) from research by Cleveland and McGill (1984), Heer, Bostock, and Ogievetsky (2010), and others.

Standard graphs, like bar and line charts, are so common because they are perceptually more accurate, familiar to people, and easy to create. Nonstandard graphs—those that use circles or curves, for instance—may not allow the reader to most accurately perceive the exact data values.

But perceptual accuracy is not always the goal. And sometimes it's not a goal at all.

Spurring readers to engage with a graph is sometimes just as important. Sometimes, it's more important. And nonstandard chart types may do just that. In some cases, nonstandard graphs may help show underlying patterns and trends in better ways than standard graphs. In other cases, the fact that these nonstandard graphs are different may make them more engaging, which we may sometimes need to first attract attention to the visualization.

This graphic from information designer Federica Fragapane shows the fifty most violent cities in the world in 2017. The vertical axis measures the population of each city and the horizontal axis captures the homicide rate per 100,000 people. The number of lines in each icon represents the number of homicides, and additional colors, shapes, and markers capture metrics like country of origin (the symbol in the middle of each), region (vertical dashed line), and change since 2016 (blue for decreases, red for increases). It could be a bar

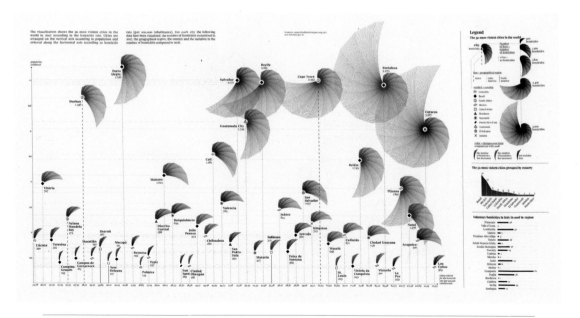

Graphic from Frederica Fragapane for La Lettura—Corriere della sera that shows the fifty most violent cities in the world. See the next page for a closer look at the legend.

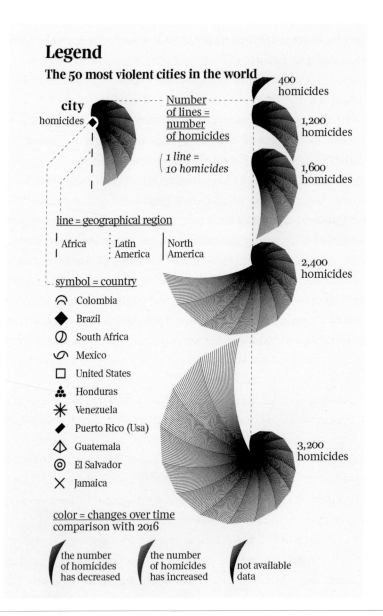

A zoom-in of the graphic from Frederica Fragapane. Notice all of the details and data elements included in each icon. It could be a bar chart or line chart, but would you then be inclined to zoom in and read it closely?

chart or a line chart or some other chart type. But if it were, would you be inclined to zoom in, read it closely, and examine it?

Data visualization is a mix of science and art. Sometimes we want to be closer to the science side of the spectrum—in other words, use visualizations that allow readers to more accurately perceive the absolute values of data and make comparisons. Other times we may want to be closer to the art side of the spectrum and create visuals that engage and excite the reader, even if they do not permit the most accurate comparisons.

Sometimes you must make your visuals interesting and engaging, even at the cost of absolute perceptual accuracy. Readers may not be as interested in the topic as we hope or may not have enough expertise to immediately grasp the content. As content creators, however, our job is to encourage people to read and use the graph, even if we "violate" perceptual rules that we know will hamper someone's ability to make the most accurate conclusions. Thinking about different audience types is not just about considering among decision makers, scholars, policymakers, and the general public—it also means thinking about different levels of interest or engagement with the visual itself. As historian Cecelia Watson writes in her book about the history and use of the semicolon, "What if we thought less about rules and more about communication, and considered it our obligation to one another to try to figure out what is really being communicated?"

We should not operate from the assumption that readers will pay attention to everything in our visual, even if we use a common, familiar chart type. Let's be honest: People see bar charts and line charts and pie charts all the time, and those charts are often boring. Boring graphs are forgettable. Different shapes and uncommon forms that move beyond the borders of our typical data visualization experience can draw readers in. Reading a graph is not like the spontaneous comprehension of seeing a photograph. Instead, reading a graph has more of the complex cognitive processes as reading a paragraph.

This isn't to say we should not concern ourselves with visual perception or allowing our readers to make the most accurate comparisons, but the goal of *engagement* can be worth a lot in its own right. Elijah Meeks, a data visualization engineer, wrote that, "Charts, like any other communication, need to be compelling to be convincing, and if your bar chart, as optimal as it may be, has been reduced to background noise by the constant hum of bar charts crossing a stakeholder's screen, then it's your responsibility to make it more compelling, even if it's not any more precise or accurate than a more simple form."

Introducing a new or different graph type can also introduce a hurdle to your reader. These can be big hurdles, like a completely new graph type or an exceptionally unusual

This graphic from an interactive visualization from the Organisation of Economic Co-Operation and Development (OECD) enables users to explore the different metrics and definitions of what it means to have a "better life." A more standard chart type, like a bar chart, might enable easier comparisons, but would it be as much fun?
Source: Organisation for Economic Co-Operation and Development

representation of the data. Or they can be small hurdles, graphs that rank lower on the perceptual-accuracy scale or graphs that people may have only seen a few times before. To overcome these hurdles, you may need to explain how to read the graph. But that might be worth it because sometimes different charts attract readers' attention and pique their curiosity.

When should you use a nonstandard graph? Likely not for many scholarly purposes, because they do not enable the most accurate perceptions of the data. For scholarly writing, accuracy is paramount. We want our reader to clearly and efficiently compare the values we're presenting. But in other cases—headline-style or standalone graphics, blog posts, shorter briefs or reports, or graphs for social media—creating something *different* may draw people in and hold their attention just long enough to convey your argument, data, or content.

This visualization from artist and journalist Jaime Serra Palou is a lovely example of this kind of nonstandard and creative data visualization. He plots his coffee consumption every day over the course of a year by using the stains from his coffee cups. You can immediately see those parts of the year when he needed an extra burst of caffeine. Yes, a line chart might convey the same data, but would you pause to spend an extra moment reading it?

Sometimes you can do both—a nonstandard, attention-grabbing graphic accompanied by a more familiar graph next to it. What you present and how you present it depends on your audience. The Serra piece might work as the lead graphic on a book or report about coffee consumption, but more detailed charts inside might take the form of standard charts and tables. Some academic research has shown that creating novel graphs, such as

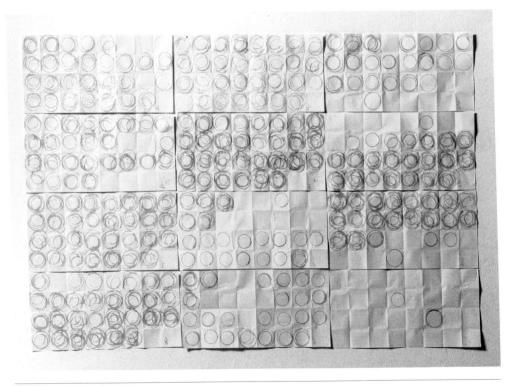

Artist and journalist Jaime Serra Palou plotted his coffee consumption every day for a year by using stains from his coffee cup.

those that enable the user to personalize the content (by inputting their own information) or are simply more aesthetically appealing, encourages readers to actively process the content.

ANSCOMBE'S QUARTET

The value of visualizing data is best illustrated by Anscombe's Quartet, published in 1973 by statistician Francis Anscombe. The Quartet demonstrates the power of graphs and how they, together with statistical calculations, can better communicate our data.

Examine the table below, which shows four pairs of data, an X and a Y.

We can make some basic observations about these data. We can see that the first three series of X's are all the same; the values of X's in the last series are all 8 except for the one 19; and the X's are all whole numbers while the Y's are not. We might even notice that the 12.7 value in the third column of Y is larger than the rest. In my experience, most people don't comment about the *relationship* between the different series, which, at the end of the day, is what we want to understand. It turns out that each of the four pairs yield the same standard information: the same average values of the X series and the Y series; the same variance for each; the same correlation between X and Y; and the same estimated regression equation.

Data set		1	1	2	2	3	3	4	4
Variable		x	y	x	y	x	y	x	y
Obs. No.	1 :	10	8.0	10	9.1	10	7.5	8	6.6
	2 :	8	7.0	8	8.1	8	6.8	8	5.8
	3 :	13	7.6	13	8.7	13	12.7	8	7.7
	4 :	9	8.8	9	8.8	9	7.1	8	8.8
	5 :	11	8.3	11	9.3	11	7.8	8	8.5
	6 :	14	10.0	14	8.1	14	8.8	8	7.0
	7 :	6	7.2	6	6.1	6	6.1	8	5.3
	8 :	4	4.3	4	3.1	4	5.4	19	12.5
	9 :	12	10.8	12	9.1	12	8.2	8	5.6
	10 :	7	4.8	7	7.3	7	6.4	8	7.9
	11 :	5	5.7	5	4.7	5	5.7	8	6.9
Mean		9.0	7.5	9.0	7.5	9.0	7.5	9.0	7.5
Variance		11.0	4.1	11.0	4.1	11.0	4.1	11.0	4.1
Correlation		0.816		0.816		0.816		0.817	
Regression line		y = 3 + 0.5x		y = 3 + 0.5x		y = 3 + 0.5x		y = 3 + 0.5x	

Source: Francis Anscombe

Known as Anscombe's Quartet, this example demonstrates how difficult it is for us to pull out basic patterns and summary statistics.

understand your data, content, and analysis. Ground your work in the lived experience of communities and people you care about and want to reach. Some audiences want a thirty-five-page PDF report. Others want a two-page brief. Others want an eight-hundred-word blog post. Still others want a more immersive, narrative experience like you might find on a major newspaper website. And some just want the data. An academic researcher, manager, practitioner, policymaker, and a reporter may need very different things. Your visualization should match the needs of your audience.

As you consider your audience's needs, you may need to balance the *accuracy* of your graph with how it *engages* your audience. One way to think of how to do this effectively is to be empathetic to your audience's needs. Take it from Alan Alda, in his book *If I Understood You, Would I Have This Look on My Face?*: "Developing empathy and learning to recognize what the other person is thinking are both essential to good communication."

PART TWO

CHART TYPES

4

COMPARING CATEGORIES

The graphs in this chapter are intended to help our readers compare values across categories. Bars, lines, and dots can all let our readers compare within and between groups. In some cases, we want our reader to see both levels *and* change, or some other variable combination; in other cases, we want to focus their attention on one comparison or another.

The challenge when comparing categorical data is deciding what we want the chart to convey. Is there a primary argument or story? Is there something you can identify as the most important comparison you want the reader to make? As chart creators, we need to prioritize what we want our charts to do. By putting *every* bar or dot in the graph, we can obscure the point we wish to convey.

This chapter starts with the bar chart. Like the line chart that will kick off the next chapter, the bar chart is familiar to most readers, which makes it a convenient choice to guide readers as they compare categories or view changes over time. It also sits at the top of the perceptual ranking diagram. It's not necessarily the case that we must *always* give our readers the exact values, but when we do, the bar chart is an excellent choice.

Graphs in this chapter are styled roughly following the guidelines published by Eurostat, the statistical office of the European Union. Eurostat's seventy-six-page style guide covers everything from color, typography, logos, tables, layout, and more elements of a comprehensive style guide that we will discuss in Chapter 12.

BAR CHARTS

One of the most familiar data visualizations, the lengths or heights of the rectangular bars in bar and column charts depict the value of your data. The rectangles can be arranged along the vertical axis so that the bars lie horizontally (often called a bar chart) or vertically on the horizontal axis (often called a column chart). For the sake of brevity, and the fact that whichever way you align them they are still bars, I call these bar charts throughout the book.

Bar charts sit at the top of the perceptual rankings list. With rectangles sitting on the same straight axis, it's easy to compare the values quickly and accurately. Bar charts are also easy to make, even with pen and paper. This one shows the total population in ten countries from around the world. It's easy to find the least (Italy) and most (Brazil) populous countries in the group, even when they are not labeled with the exact values.

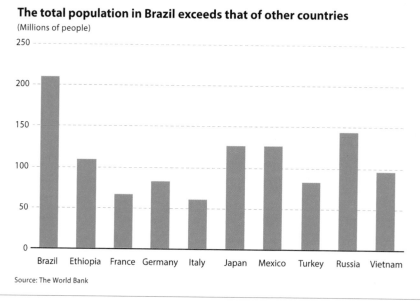

The bar chart is a familiar chart that's easy to read and make. It sits at the top of the perceptual ranking matrix.
Data Source: The World Bank.

It's even easier to see the highest and lowest values when the data are sorted according to their data values. This strategy doesn't always work, however. If, for example, I was showing population levels for sixty countries, I might sort the values alphabetically, so that readers

The total population in Brazil exceeds that of other countries
(Millions of people)

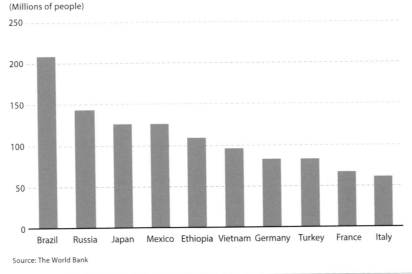

Source: The World Bank

When possible, sort the data in your bar charts. This makes it easier for your reader
to find the highest and lowest values.
Data Source: The World Bank.

could more easily find the bar for a specific country. But if I was making an argument about
the population level in a specific country or set of countries, I might sort the data so that the
country or countries of interest are at one end of the graph. Alternatively, I could simply use
a different color to highlight whichever bar or bars I want to set apart from the rest.

There are a few strategies to creating bar charts, many of which will apply to other charts
in this chapter as well.

START THE AXIS AT ZERO

Starting the axis of bar charts at zero is a rule of thumb upon which many data visualization
experts and authors agree. Because we perceive the values in the bar chart from the length of
the bars, starting the axis at something other than zero may overemphasize the differences
between the bars and bias our perception.

Take the bar chart of population. Because none are lower than fifty million, we might
be tempted to start the axis at fifty million. After all, this would emphasize the difference
between the values.

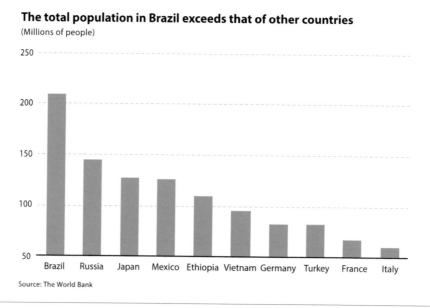

The total population in Brazil exceeds that of other countries
(Millions of people)

Source: The World Bank

Starting the vertical axis at 50 million overmphasizes the differences in values and skews our perception of the data.

But notice what happens when we do that. The differences in values are emphasized—in fact, they are *over*emphasized. Here, it looks as though Brazil is orders of magnitude larger than Italy, when, in fact, it is only about three-and-a-half times greater. This isn't a matter of moving from accurate perception to general perception—it's a matter of moving from accurate to inaccurate.

If you want to take a more extreme view of this, imagine starting the graph at a hundred million—and why not? If starting at fifty is OK, then we can pick any arbitrary number. Now at a glance it looks like nobody lives in half of these countries!

There is emerging research in this area that suggests that perhaps starting bar charts at something other than zero does not bias our perception of the data. In one recent study, participants were better able to assess the sensitivity of the results (e.g., no effect, small effect, medium effect, or big effect) and more accurate (e.g., the size of the effect) when the vertical axis was set at a range more consistent with the variation of the data. Until more research is conducted, however, my preference is to start the axis in bar charts at zero to avoid any confusion or possibility of visual bias.

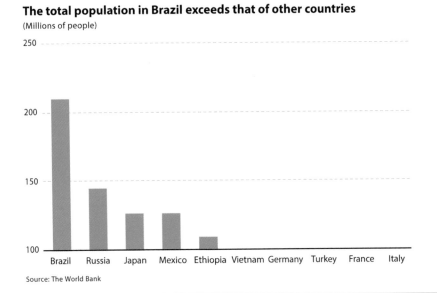

The total population in Brazil exceeds that of other countries
(Millions of people)

If starting the y-axis at fifty is OK, then why not one hundred?
Data Source: The World Bank.

DON'T BREAK THE BAR

Another cardinal sin of data visualization is what is called "breaking the bar"—that is, using a squiggly line or shape to show that you've cropped one or more of the bars. It's tempting to do this when you have an outlier (see Box on page 74), but it distorts the relative values between the bars.

Let's create a bar chart of population in the ten most populous countries of the world. In 2018, China and India were the most populous countries on the planet with 1.39 billion and 1.34 billion people, respectively, followed by the United States with 327 million people. We can see how dramatically larger China and India are relative to the rest of these countries in the top graph on the next page. If we wanted to make the differences between the less populous countries larger, we could break the bars, but this makes China and India look much less populous than they are. Chopping the lengths of the bars is completely arbitrary—I can place those squiggly lines wherever I like to zoom in on the other differences. But that's not being honest with the data.

If you run into a case where you have outliers but want to show the detailed differences between the smaller values, try using more graphs. You might think of this as a "zoom in" and "zoom out" approach—show all of your data so your reader can see the magnitude of the largest values, and then zoom in for a detailed look that omits the outliers. On the next page, I've highlighted the less populous countries to show the

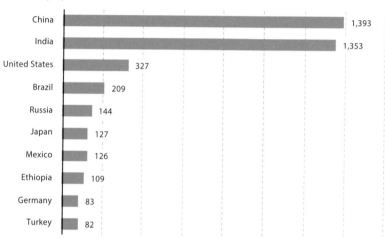

China and India are the most populous countries in the world
(Millions of people)

Source: The World Bank

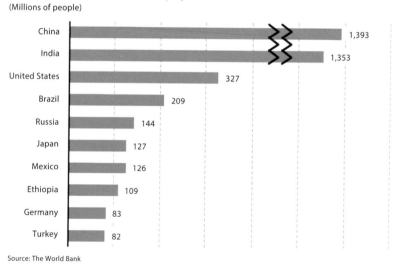

China and India are the most populous countries in the world
(Millions of people)

Source: The World Bank

Don't break the bar in your bar charts. The break can be arbitrarily set anywhere and distort our perception of the data.

differences between them, which we can't quite see in the main graph. Adding labels and an active title is another good way to communicate the differences between smaller values to the reader.

China and India are the most populous countries in the world
(Millions of people)

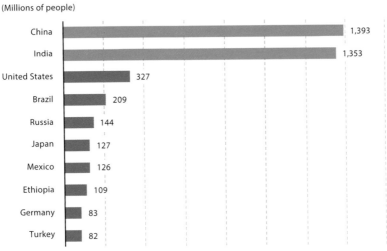

Source: The World Bank

Total population in these countries ranges from 82 million to 327 million
(Millions of people)

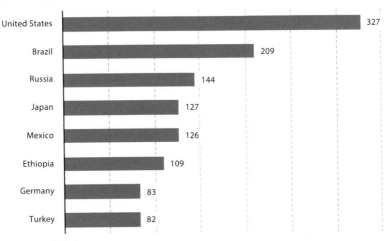

Source: The World Bank

In cases where you have large values or outliers but want to show the detailed differences between the smaller values, try using more graphs.

EXTREME VALUES OR OUTLIERS

An outlier is a data point that is far away from other observations in your data. It may be due to random variability in the data, measurement error, or an actual anomaly. Outliers are both an opportunity and a warning. They potentially give you something very interesting to talk about, or they may signal that something is wrong in the data.

In 2014, Buzzfeed teamed up with the website Pornhub to look at pornography viewing by state. Using geolocation data of people accessing their site, Pornhub calculated the number of page views per person in each state. People in Kansas, they found, watched far, far more pornography than any other state in the country: 194 page views per person. Nevada was second with 166 page views.

The data went into the scatterplot below, comparing blue-state and red-state porn consumption. You can clearly see Kansas as an outlier in page views. Do people in Kansas really watch that much more porn?

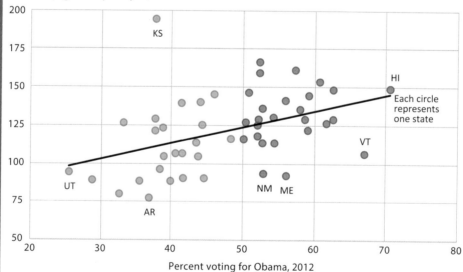

Presidential politics and porn per capita

(Pornhub pageviews per capita)

Source: Pornhub views from Buzzfeed; Voting percentages from *The Guardian* and NBC News.
Scatterplot originally created by Christopher Ingraham.

Turns out the answer is no. Apparently, Pornhub's methodology assigned missing geolocation data to the geographic center of the United States, which, as it turns out, is in Kansas.

Not all outliers are mistakes, however. As just one example, we can look at the rate of physical violence from a firearm across advanced countries. In 2017, more than 8 per 100,000 people were victims of firearm violence in the United States, compared with 0.90 per 100,000 people in Canada and 0.49 per 100,000 in Belgium. In some cases, outliers are truly outliers.

There are lots of ways to test for outliers in your data, some more complex than others. One way is to simply *look* at your data. Exploring your data does not need to start with complex math and statistics—you should always visually inspect your data.

But that approach is hardly mathematical. A standard method is to compare data values to 1.5 times the interquartile range (IQR). The IQR is a simple summary of your data and is the difference between the third and first quartiles (see the Box in Chapter 6 on percentiles).

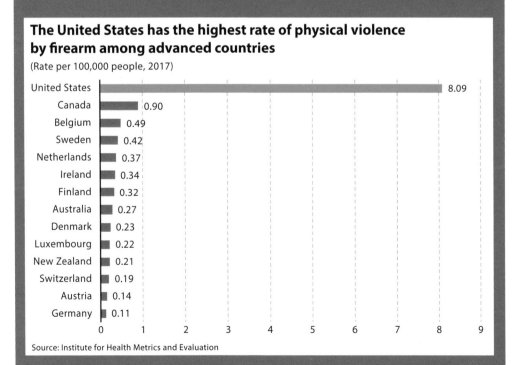

The United States has the highest rate of physical violence by firearm among advanced countries

(Rate per 100,000 people, 2017)

Country	Rate
United States	8.09
Canada	0.90
Belgium	0.49
Sweden	0.42
Netherlands	0.37
Ireland	0.34
Finland	0.32
Australia	0.27
Denmark	0.23
Luxembourg	0.22
New Zealand	0.21
Switzerland	0.19
Austria	0.14
Germany	0.11

Source: Institute for Health Metrics and Evaluation

USE TICK MARKS AND GRIDLINES JUDICIOUSLY

Bar charts don't need tick marks between the bars. White space is an effective separator and deleting the tick marks reduces clutter.

One exception is if you have a "major" category label that spans multiple bars. In such cases, larger tick marks can be helpful to group the labels (see the bottom chart on the next page).

Gridlines help the reader see the specific values for each bar and are especially useful for the bars farthest from the axis label. Because they serve as a visual guide, they can be rendered in a lighter color so the reader's eye stays on the data.

When it's important for the reader to know the *exact* values, you can add data labels to the chart. My preference is to forgo the gridlines and axis lines altogether in these cases.

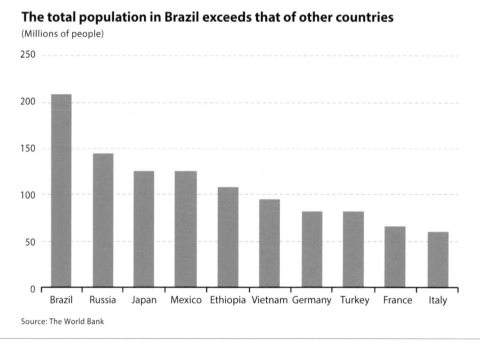

The total population in Brazil exceeds that of other countries
(Millions of people)

Source: The World Bank

In bar charts, tick marks are not necessary. The white space does
the job of separating the bars.

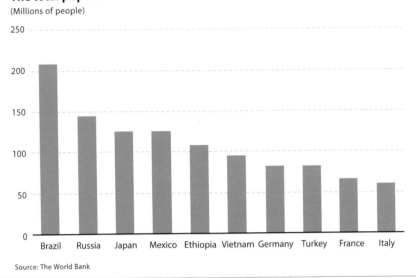

The total population in Brazil exceeds that of other countries

(Millions of people)

Source: The World Bank

Omitting tick marks is part of removing as many non-data elements as possible.

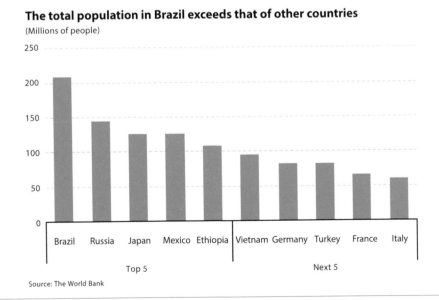

The total population in Brazil exceeds that of other countries

(Millions of people)

Source: The World Bank

Tick marks may be necessary when you have a "major" category.

Consider Italy in the next two graphs (highlighted in blue). Without the labels, the gridline helps us see that there are more than fifty million people living in the country; with the label, it is clear that it's sixty million people and thus the gridlines are probably not necessary.

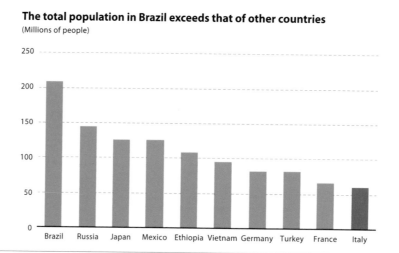

The total population in Brazil exceeds that of other countries
(Millions of people)

These horizontal gridlines help the reader see that, for example, there are more than fifty million people living in Italy.
Data Source: The World Bank.

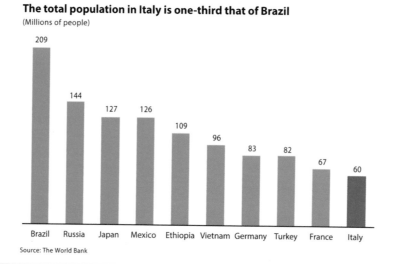

The total population in Italy is one-third that of Brazil
(Millions of people)

Source: The World Bank

Data labels make gridlines redundant and, by extension, the vertical axis.

The graph may look too cluttered with labels if I had fifty or maybe even twenty countries, so I might include a separate table or an appendix. Deleting the gridlines when including data labels is primarily an aesthetic choice and as you continue to work with data and make your own graphs, you will develop your own style for these graphic elements.

ROTATE LONG AXIS LABELS

The default solution for long horizontal axis labels is to run the text vertically, as on the spine of a book. But this approach forces your reader to turn their head to the side. One solution is to rotate them 45 degrees, but the reader still has to turn their head. Another approach is to shrink the font size so they are aligned horizontally—though this usually makes them too small.

The most elegant solution is to simply rotate the entire graph. This still uses the same pre-attentive attribute—the length of the bars—but the axis labels are now aligned horizontally; they are easy to read with no effect on data comprehension.

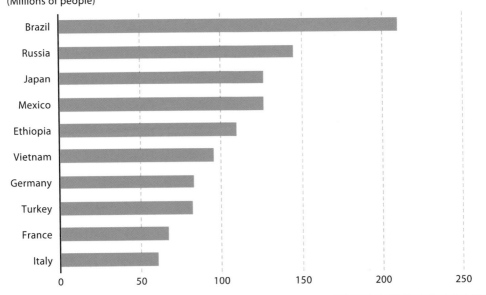

The total population in Brazil exceeds that of other countries
(Millions of people)

With long axis labels, consider rotating the chart to make the labels
horizontal and easier to read.
Data Source: The World Bank.

VARIATIONS ON THE BAR CHART

There are countless ways to modify the standard bar chart. One simple variation is to use other shapes in lieu of bars. The lollipop chart, for example, replaces the bar with a line and a dot at the end. This version lives a hair below the bar chart on our perceptual rankings, because it's not exactly clear which part of the circle encodes the value. But it removes a lot of ink from the page and gives you more white space to add labels or other annotation.

This is just one example of an alternative shape. Triangles, squares, and arrows are other options, as are bar-shaped images that reinforce your data. A chart showing data on urban growth may use building-shaped bars, and a chart on climate change may use trees for bars. Be careful with this approach, however, as readers may confuse the total *area* of the icons as a value indicator rather than just the height.

The total population in Brazil exceeds that of other countries
(Millions of people)

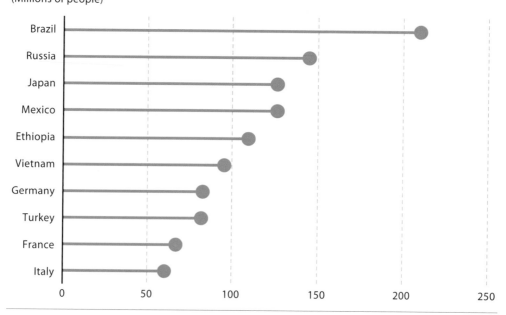

The lollipop chart replaces bars with a shape (usually a dot) and a line.
Data Source: The World Bank.

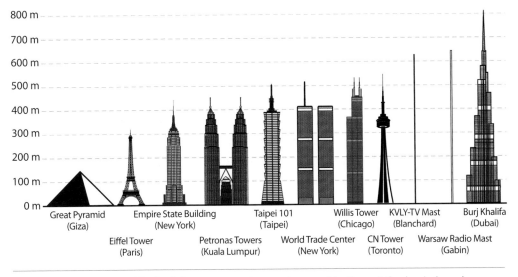

Alternative shapes, like buildings or people, can be used in lieu of the basic bar shape.

Source: Based on Wikimedia user BurjKhalifaHeight Petronas Towers

The total population in Brazil exceeds that of other countries
(Millions of people)

Change in Brazil's population from 2008 to 2018
(Millions of people)

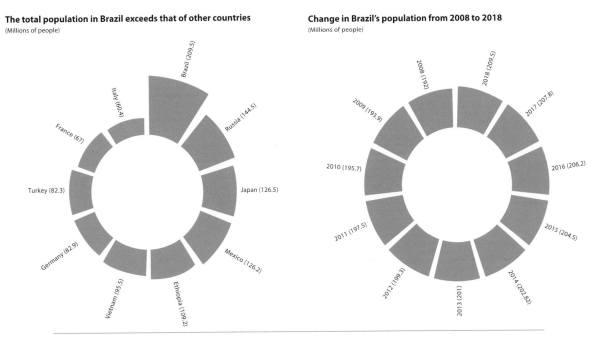

A radial bar chart wraps the standard bar chart around a circle. This chart type moves down the perceptual ranking list because it is harder to compare the heights of the bars.

Data Source: The World Bank.

Another approach to the basic bar chart is to abandon the usual grid and instead place the bars in a circle, called a radial layout. There are two common ways to do this: the radial bar chart and the circular bar chart.

The radial bar chart, also called the polar bar chart, arranges the bars to radiate outward from the center of a circle. This graph lies lower on the perceptual ranking list because it is harder to compare the heights of the bars arranged around a circle than when they are arranged along a single flat axis. But this layout does allow you to fit more values in a compact space, and makes the radial bar chart well-suited for showing more data, frequent changes (such as monthly or daily), or changes over a long period of time.

W. E. B. Du Bois used a circular bar chart in his famous *Exposition des Negres d'Amerique* at the 1900 Paris Exposition. He included this radial bar chart in his set of infographics for

Source: W. E. B. Du Bois, Assessed Value of Household and Kitchen Furniture Owned by Georgia Negroes (1900) via Library of Congress Prints and Photographs Division.

The Georgia Negro: A Social Study, which shows the dollar value of household and kitchen furniture held by African Americans in Georgia in six years (1875, 1880, 1885, 1890, 1895, and 1899). "The end result," wrote Whitney Battle-Baptiste and Britt Rusert in their book about Du Bois's graphics, "is simultaneously easy to read and hypnotic."

Perceptually speaking, the circular bar chart is problematic because it distorts our perception of the data—in this case, the lengths of the bars don't correspond to their actual value. Consider the case where the values of two bars are the same—the ends of the bars will line up in the same position, but the lengths of the bars are not actually the same because they lie along the circumference of two different circles. Author and data visualization expert Andy

Change in Brazil's population from 2008 to 2018
(Millions of people)

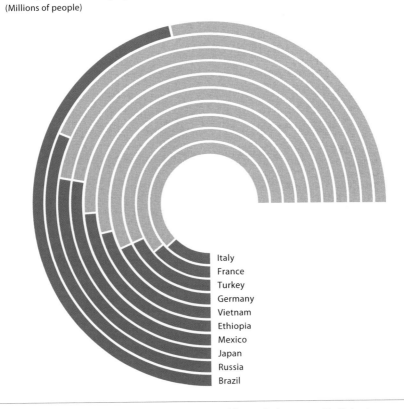

Italy
France
Turkey
Germany
Vietnam
Ethiopia
Mexico
Japan
Russia
Brazil

Perceptually speaking, the circular bar graph is problematic because it distorts our perception of the data. In this case, the lengths of the bars don't correspond to their actual value.
Data Source: The World Bank.

Kirk uses an Olympic footrace as a metaphor. Runners start at staggered positions on the track, one on the very inside of the track and another on the very outside of the track, but they all end up running the same distance. Here, the visualization doesn't move down the perceptual ranking, but off of it altogether because it distorts the data and for that reason, I recommend avoiding them altogether.

PAIRED BAR

A simple bar chart is perfect for making comparisons across categories, like comparing populations across countries. If I want to show comparisons not just across but also *within* countries, the paired bar chart is a good option. The paired bar chart will be familiar to most readers and is easy to read, and the shared baseline makes it easy to make comparisons.

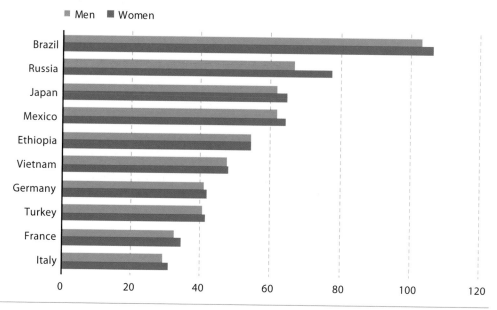

There are more women than men in each country except for Ethiopia
(Millions of people)

A simple paired bar chart is familiar to most readers and easy to read.
Data Source: The World Bank.

Difference between the number of women and men

(Millions of people)

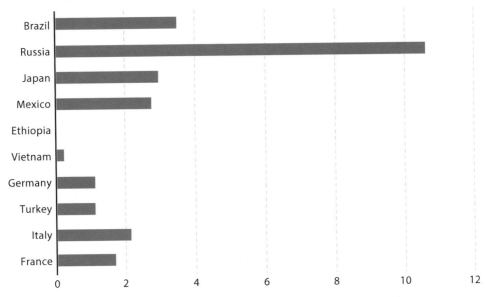

Instead of showing both data values, we could show the difference between them.
Data Source: The World Bank.

Say we want to show the number of men and women in each country in our sample. A paired bar chart allows us to do so.

Note that the paired bar chart directs the reader's attention not just to the levels, but also to the *difference*. If it's important that readers see both, this is a good option.

But if our goal is for the reader to focus only on the *difference* between the two values within each category, this isn't the most direct way to do so, because we are asking them to compare the difference in lengths. Instead, we could just show the difference between the two values in a single bar, like the one above.

In cases where you want the reader to see the level *and* the difference, you may need a different chart entirely. I prefer the parallel coordinates plot (see page 263), the slope chart (for data that vary over time; see page 150), and the dot plot (see page 97). Remember to ask yourself, What is the goal of this graph? That question will guide you to the best way to visualize your data.

Change in population from 2014 to 2018

(Millions of people)

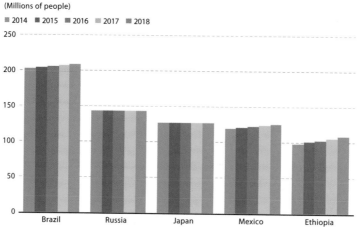

The paired bar chart can be used to show changes over time and can be used to examine changes within and between countries.

Data Source: The World Bank.

Another use for the paired bar chart is to show changes over time. And although I include the word *pair* in the title, these charts can have more than two values. The chart below, for example, shows the population in five of our countries from 2014 to 2018. This

Change in population from 2014 to 2018

(Millions of people)

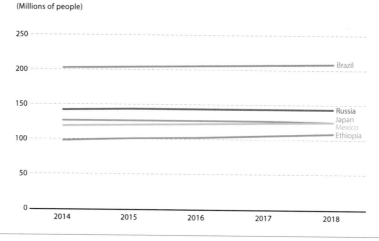

The line chart is a more familiar way to show changes over time.

Data Source: The World Bank.

allows the reader to examine the population change *within* countries and the differences *across* countries.

The patterns in your data can also drive your chart type selection. If the values decline evenly across the different years for all categories, a paired bar chart may look fine. But if the values move around over time, a line chart (as shown earlier) or cycle chart (see Chapter 5) may make for better comparisons over time within and across each group.

There are two instances when I prefer to use bar charts rather than line charts to show changes over time. First, when there are few data points—for example, only five years—the extra ink in the five bars gives the graph more visual weight. Second, when I have discrete time intervals (and few observations), such as the first quarter of the year.

Clutter is the main issue to keep in mind when assessing whether a paired bar chart is the right approach. With too many bars, and especially when there are more than two bars for each category, it can be difficult for the reader to see the patterns and determine whether the most important comparison is between or within the different categories.

When it comes to whether a paired bar chart is too cluttered, trust your eyes and your instincts. Put yourself in your readers' shoes—try to imagine where their eyes will go when they look at the graph for the first time. If there's too much going on, you may need to break up your data, use a different chart type, or try a small multiples approach.

STACKED BAR

Another variation on the bar chart is the stacked bar chart. While the paired bar chart shows two or more data values for each category, this chart subdivides the data within each category. The categories could sum to the same total, say, 100 percent, so that the total length of the bar is the same for every group. Or the totals may differ across the groups, in which case the total length of each bar may differ. Above, I've plotted the share of gross domestic product (GDP) each of ten countries spends on support for health care, old age, and other programs. The entire length of the bars shows how much each country spends on these programs as a share of GDP.

As with the bar charts we've looked at thus far, the stacked bar chart is familiar, easy to read, and easy to create. The biggest challenge, however, is that it can be difficult to compare

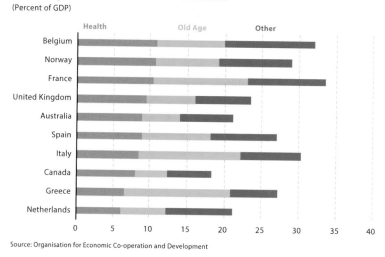

Social expenditures for 10 OECD countries
(Percent of GDP)

Source: Organisation for Economic Co-operation and Development

The stacked bar chart shows how different categories sum to a total. The interior series in the chart, however, are harder to compare with one another because they do not sit on the same baseline.
Data Source: Organisation for Economic Co-Operation and Development.

the different values of the segments *within* the chart. In the example above, it's easy to compare the values across the countries for the Health category, because the bar segments share the same vertical baseline. But that's harder to do with the two other series because they do not share a baseline. Which country spends more on old-age programs, Italy or Greece? You can quickly see that Italy spends more on health programs than Greece, because those segments are left-aligned on the vertical axis, but it's much harder to determine with the segments for the other categories.

One way to address the changing baseline is to break the graph apart so that each series sits on its own vertical baseline. This is a small multiples graph, arranged side by side. It's now easier to see that Greece spends more on Old Age programs than Italy. The tradeoff is that it is harder (if not impossible) to see the *total* values. But that too can be overcome: You can still break up the stacked graph and add a final segment that represents the total amount (this is not an issue when all of the series sum to 100 percent because the summed segments will all have the same length).

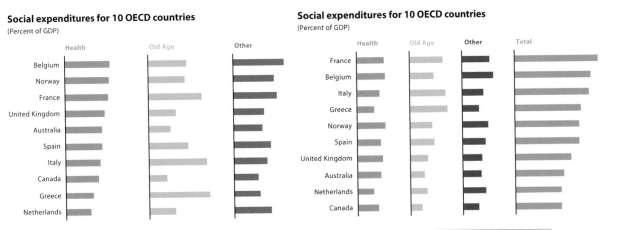

Social expenditures for 10 OECD countries
(Percent of GDP)

Social expenditures for 10 OECD countries
(Percent of GDP)

Instead of stacking all the data series together, we can break them up (either with or without the totals) to create a sort of small-multiples approach. Here, we move up to the top of the perceptual ranking list because each series sits on its own baseline.
Data Source: Organisation for Economic Co-Operation and Development.

In both versions, the horizontal spacing for each segment should be the same width, otherwise it might appear that a segment takes up a larger proportion of the space than it really does. In cases where you add the total, the width does not need to be the same as the other groups, but the increments along the axis should be the same. In other words, if the width of each segment above in which the data range from 0 percent to 50 percent is one inch wide, the total category that spans 0 percent to 100 percent should be two inches wide.

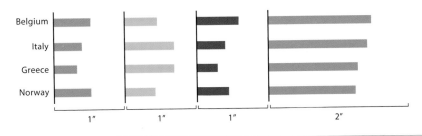

When creating these sorts of small multiples bar charts, be sure that each segment has the same width.

Even though different baselines in the standard stacked bar chart can make it more difficult to compare values, there are cases when the stacked bar chart is preferable. In this stacked bar chart, I've included more spending categories and divided them into shares of the total so the graph highlights the *distribution*. In this view, it becomes clear that around three-quarters of total government spending in these countries goes to programs for health care and old age programs. That observation is harder to see in the version on the right, where each category is placed on its own vertical baseline. Even though it is easier to compare differences in each category across countries, you don't see large differences between them.

As always, identify what you want to show and where you'd like to focus your readers' attention. In these examples, the *Health* category is emphasized because the data are sorted according to those values (shown as a percent of total spending on these specific programs) and it is situated along the vertical baseline. In this layout, the other segments become secondary in comparison to health spending.

There is one other stacked bar chart that you may have come across that shows a single set of data values and the gap between them and another value (often the total). The graph on the next page uses this approach to show the share of women elected to the U.S. House of Representatives from 1917 to 2018. The version on the left shows the raw percentages; the vertical axis ranges from 0 to 25 percent. Here, you see a dramatic increase in the share of

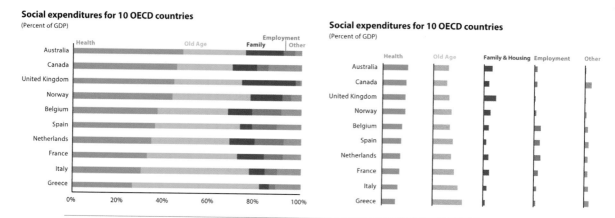

In these examples, we can see how our ability to compare different values within and across countries varies between these two views.
Data Source: Organisation for Economic Co-Operation and Development.

The 116th Congress represents the biggest jump in women members since the 1990s

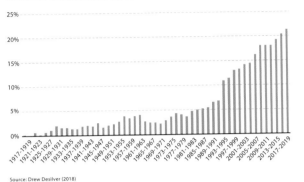

Source: Drew Desilver (2018)

The 116th Congress represents the biggest jump in women members since the 1990s

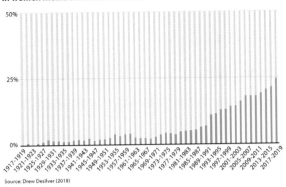

Source: Drew Desilver (2018)

A somewhat rare case where a stacked bar chart is used to focus attention on one series. This technique may be particularly valuable where the relative proportion is as important as the change.

women in Congress. The version on the right shows the same data, but stacks a gray series on top of the data values to 50 percent. In this version, we can emphasize that the share of women is still small even though that share is rising. It is in these cases—where the relative proportion is as important as the change—that this technique may be particularly valuable.

PERCENT CHANGE VS. PERCENTAGE POINT CHANGE

There is an important distinction between *percent change* and *percentage point change*, and it's a mistake that many often make.

Percent change compares an initial value *OLD* to a final value *NEW* according to this simple formula:

$$((NEW-OLD)/OLD) \times 100.$$

Positive percent changes (that is, *NEW>OLD*) mean there is a percent (or percentage) increase. Negative changes (*NEW<OLD*) mean there is a decline. You can calculate differences over time or between groups; all that really matters is that you follow the formula and know that you are comparing the change relative to the initial value of *OLD*.

Now, *percentage point change* is specific to looking at raw differences in percentages. The *percentage point change* is a simpler formula:

$$NEW-OLD$$

where both are already percentages.

These are very different things. Let's take a simple example. According to the U.S. Census Bureau, there were 40.6 million people in poverty in 2016 and 39.6 million people in poverty in 2017. The poverty rate (the number of people in poverty as a percent of the total population) was 12.7 percent in 2016 and 12.3 percent in 2017.

The number of people in poverty fell by 2.3 percent. The *percent change* was

$$[(39,698,000 - 40,616,000)/40,616,000] \times 100 = [-0.023] \times 100 = -2.3\%$$

But the poverty *rate* fell by 0.4 *percentage points* over the two years:

$$12.3\% - 12.7\% = -0.4 \text{ percentage points}$$

Obviously, those are two very different numbers, but people confuse them all the time. Clearly representing your data starts with clearly understanding your data, how they were collected, and how to calculate basic descriptive statistics.

DIVERGING BAR

A variation on the stacked bar chart is one in which the stacks diverge from a central baseline in opposite directions. These are often found in surveys where the responses are arrayed in ranges from, for instance, *strongly disagree* to *strongly agree*. These are often called "Likert Scales," named after the psychologist Rensis Likert, who invented the scales in the early 1930s.

This book is fun to read.

Strongly Disagree   Disagree   Undecided   Agree   Strongly Agree

In this example, drawing on data from the International Social Survey Programme, survey respondents were asked whether they believe it is the government's responsibility to reduce income inequality. By grouping the "disagrees" and the "agrees" together on either side of a central baseline, we can compare the *total* sentiment across the different countries.

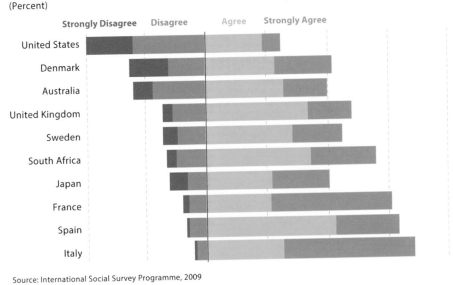

It is the responsibility of government to reduce differences in income between people with high & low incomes
(Percent)

Source: International Social Survey Programme, 2009

The diverging bar chart can show differences in opposing sentiments or groups, such as "agree/disagree" or "true/false."

One advantage of this chart is that the sentiments are clearly presented—the Disagrees jut out to the left (in what we might typically think of as a negative direction) and the Agrees out to the right. This works well if your audience is most interested in the *total* sentiment of each side and not necessarily comparisons between each individual component. If the individual comparisons are the primary focal point, then a paired bar chart could do the job just as well.

Why do we perceive these values to the left as negative? Throughout western history, the concept of left—and even left-handed people—has been plagued with negative connotations.

Consider the etymology of the word: *left* is derived from the Old English word *lyft*, which means "weak." In Latin the word *sinister* means the left or left-hand direction. The word *right* comes from the Old English *riht*, whose original meaning was "straight" and thus not bent or crooked. And this is why we have phrases such as "standing upright" or "do the right thing" or "the right answer," all of which connote goodness and correctness. You can also see this in other languages: In Spanish, for example, the word *derecha* means "right" and the closely-derived *derecho* means "straight."

As with the stacked bar chart, the challenge with visualizing these kinds of data is that we are comparing within *and* across the categories. Arranging the bars in opposite directions makes it difficult to compare the totals of the two groups. In other words, it's difficult to compare the *total share* of people who disagree with the *total share* of people who agree. That task is slightly easier in the paired bar chart, but then you lose the positive-negative connotation of the diverging chart. Depending on the patterns in your data and the number of categories and groups, you might find this chart looks cluttered and busy.

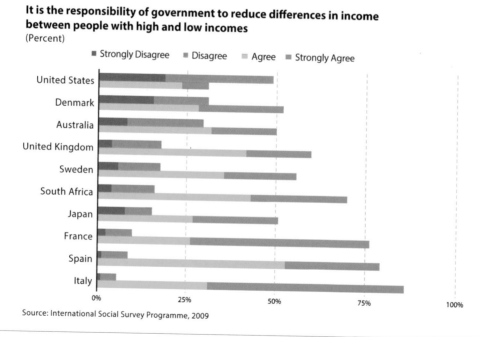

It is the responsibility of government to reduce differences in income between people with high and low incomes
(Percent)

Source: International Social Survey Programme, 2009

Taking the opposite sides of the diverging bar chart and placing them in a more standard paired bar chart approach can also work and allows us to more accurately compare the totals.

You must be especially careful using a diverging bar chart when you have a "neutral" category. By definition, the neutral survey response is neither agree nor disagree, and should therefore be grouped with neither category.

It is the responsibility of government to reduce differences in income between people with high & low incomes
(Percent)

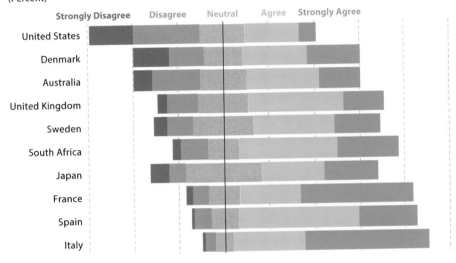

Source: International Social Survey Programme, 2009

Placing the *Neutral* category of a diverging bar chart in the middle wrongly implies that the neutral responses are split between the two sentiments.

Placing the neutral category in the middle of the chart along the vertical baseline creates a misalignment between the two groups and implies the neutral responses are split between the two sentiments. It also means that none of the segments sit on a vertical baseline. Placing it to the side of the chart is a better strategy because the disagree, agree, and neutral categories now all sit on their own vertical axes, even though the neutral category is somewhat emphasized as it sits to the side (see next page).

Another alternative—regardless of whether you have a neutral category—is the stacked bar chart as shown on the next page. In this view, the different categories sum to 100 percent, and one can more easily compare the totals between the countries. A good strategy is to mark specific aggregate values to guide the reader. Here, for example, I have marked the 50-percent position to make it clear for which countries the total "agree" and "disagree" sentiments are at least half of the total.

It is the responsibility of government to reduce differences in income between people with high & low incomes
(Percent)

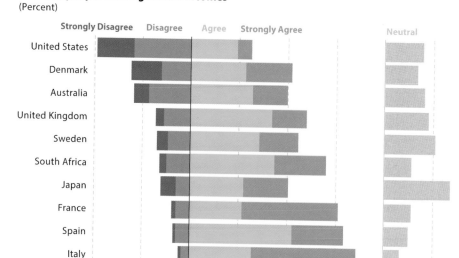

Strongly Disagree   Disagree   Agree   Strongly Agree   Neutral

United States
Denmark
Australia
United Kingdom
Sweden
South Africa
Japan
France
Spain
Italy

Source: International Social Survey Programme, 2009

A better approach is to place the *Neutral* category off to the side of the graph.

It is the responsibility of government to reduce differences in income between people with high & low incomes
(Percent)

Strongly Disagree   Disagree   Neutral   Agree   Strongly Agree

United States
Denmark
Australia
United Kingdom
Sweden
South Africa
Japan
France
Spain
Italy

Source: International Social Survey Programme, 2009

A stacked bar chart can be used to show these kinds of Likert scales.

It is the responsibility of government to reduce differences in income between people with high & low incomes
(Percent)

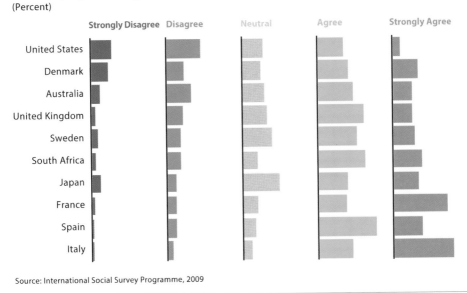

Source: International Social Survey Programme, 2009

The small multiples bar chart is yet another way to visualize these kinds of data.

You could take this further and break the chart into its components, as we discussed in the previous section. In general, as with many graphs, which variation you choose will depend on your goals.

DOT PLOT

The dot plot (sometimes called a dumbbell chart, barbell chart, or gap chart) is one of my favorite alternatives to a paired or stacked bar chart. Developed by William Cleveland, one of the early pioneers in data visualization research, the dot plot uses a symbol—often but not always a circle—corresponding to the data value, connected by a line or arrow. The data values correspond to one axis and the groups to the other, which do not necessarily need to be ordered in a specific way, though sorting can help.

The dot plot is an easy way to compare categories—especially many categories—when bars might add too much ink and clutter to the page. For this example, let's look at scholastic test

PISA scores for math and reading among 10 OECD countries

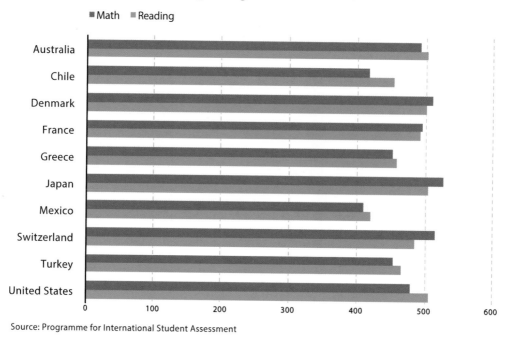

Source: Programme for International Student Assessment

This simple bar chart shows math and reading scores across multiple countries. Bar charts are often the default visual for these kind of data, but it looks heavy and dense.

scores around the world from the Programme for International Student Assessment (PISA), an international set of achievement tests taken by fifteen-year-old students in reading, mathematics, and science. We can easily plot the mathematics and reading scores for a set of countries using a simple bar chart, but the twenty bars make the chart heavy and dense.

By contrast, the dot plot shows the same data with a dot at each data value connected by a line to show the range or difference. The circles use less ink than the bars, which lightens the visual with more empty space. The country labels are placed close to the leftmost dot, though they could also be set off to the left along the vertical axis. If necessary, data values can be placed next to, above, or within each circle.

Dot plots are not restricted to two dots and a connecting line, nor are they restricted to simply comparing different categories. You can use dot plots to show a change between two years, for example. You could use different shapes or icons or arrows instead of lines to denote direction. You can also use more than two objects. For example, we could add science test scores to this plot, but we need to be sure to add sufficient labeling so our reader

PISA scores for math and reading among 10 OECD countries

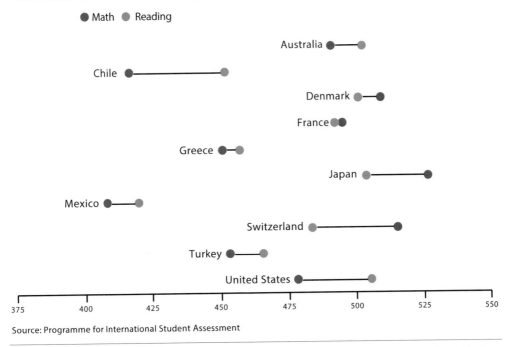

Source: Programme for International Student Assessment

The basic dot plot places a dot for each data point and connects them with a line. Notice how more white space lightens the visualization.

knows what each object on the graph represents. Axes and gridlines can be included or not, depending on how important it is for the reader to determine the exact values.

A few points of caution about the dot plot. First, it's not entirely obvious when the direction of the values change, as in the last chart. Did you notice that math scores were higher than reading scores in four of the countries in the dot plot above? That difference is not immediately evident unless the reader carefully examines the points and their coloring. In this and other cases, we should consider how sufficient annotation, clear labeling, and highlighting colors can help clarify different directions. The data are sorted by math scores in the dot plot on the next page, which helps organize the countries, but it is still not immediately clear that in only the first four countries are math scores higher than reading scores.

One approach is to split the graph into two groups, one for countries in which math scores are higher than reading scores and another for countries where the opposite occurred. In these versions (page 100), the groups are split and then sorted with larger, bold headers to distinguish them. We can also add data values—I will sometimes put them right inside the

PISA scores for math and reading among 10 OECD countries

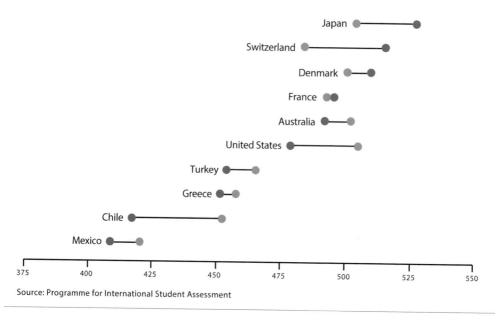

Source: Programme for International Student Assessment

As with the basic bar chart, sorting the data in a dot plot helps organize the space for the reader.

circle—but be careful because the labels can clutter the chart. An alternative is to include vertical gridlines, depending on how precisely we want to communicate the data to the reader.

When using a dot plot to show change over time, I prefer to change the linking lines to arrows, which helps make the direction clear.

Another word of caution for dot plots that show changes over time. The dot plot is, by definition, a summary chart. It does not show all of the data in the intervening years. If the data between the two dots generally move in the same direction, a dot plot is sufficient. But if the data contain sharp variations year by year, a dot plot will obscure that pattern (as it also does for bar charts). For example, if test scores decreased between 2015 and 2019 and then increased sharply between 2019 and 2020, the dot plot would only show an overall increase, masking the change in the intervening years. In some cases, you may not have a choice—if you are using data from the decennial U.S. Census, by definition you will only have data for every ten years. That's something you can't help, but if you are familiar enough with your content, you'll know whether showing only those points is enough to clearly and accurately make your point.

PISA scores for math and reading among 10 OECD countries

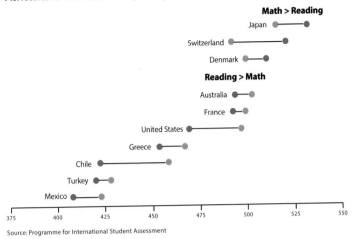

Labels and annotation can clarify differences in the relationships between the values. Grid-lines are not always necessary.

PISA math scores rose for 4 of 10 OECD countries between 2015 and 2018

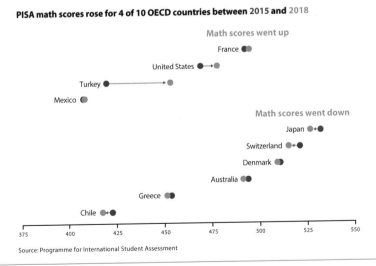

Dot plots can show changes over time. In these cases, I will often make the linking line an arrow to suggest the change over time.

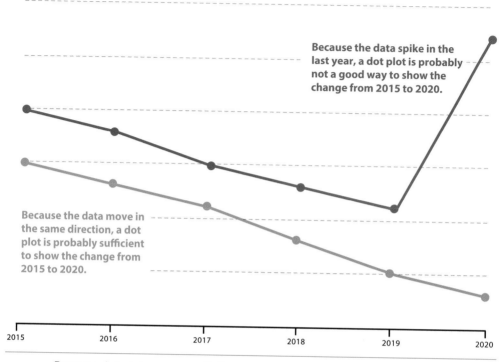

Because the data spike in the last year, a dot plot is probably not a good way to show the change from 2015 to 2020.

Because the data move in the same direction, a dot plot is probably sufficient to show the change from 2015 to 2020.

2015   2016   2017   2018   2019   2020

Because dot plots are essentially a summary plot, be wary of using highly variable data with dot plots.

MARIMEKKO AND MOSAIC CHARTS

Marimekko charts may look odd at first, but they are just an extension of the bar chart. This type of chart is useful when you want to make comparisons between two variables: one comparing categories and one showing how they sum to a total. The name of the chart comes from the Finnish design firm Marimekko, founded in 1951 by Armi Ratia and her husband, Viljo. Early Marimekko style featured straight, oversized, geometric patterns and bright colors.

In the standard vertical bar (or column) chart, the data are measured along the height of the vertical axis and the widths of the columns are identical. The Marimekko chart takes that standard column chart and expands the width of each bar according to another data value. The Marimekko chart is an easy way to add a second variable to your standard column or bar chart.

In this Marimekko chart, I show two variables for the ten most populous countries: the share of people with less than $5.20 per day and the share of the total population among these countries. The percent of people with less than $5.20 per day is plotted along the vertical axis

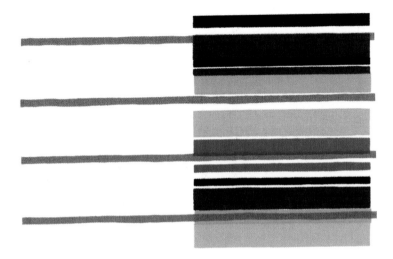

Early Marimekko fabric styles like this one featured straight, oversized, geometric patterns and bright colors.

as in a standard bar chart; the widths of each bar are then scaled according to the share of each country's population summing to 100 percent across these ten countries (an alternative is to show the raw counts rather than percentages). You can see that the most populous countries in this sample—the widest bars—and their distribution of poverty. You can also use color strategically here: if this graph were in an article about poverty in Brazil and China, we might shade all the bars the same color except for those two, as in the graph on the right.

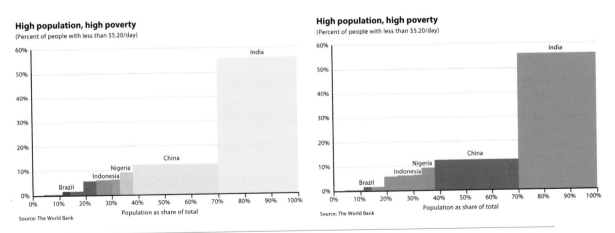

The Marimekko chart (sometimes called a Mekko chart) scales the widths of the bars in a bar chart corresponding to another variable. Color can be used to highlight specific values.

The two variables could also be plotted separately in two bar charts, and while these graphs are familiar and easy to read, they do not communicate the relationship between the two variables as well.

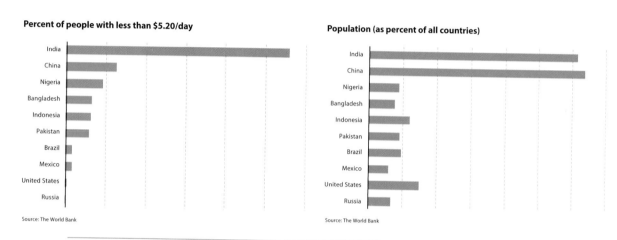

Instead of a Marimekko chart, the two variables could be plotted separately.

Putting two variables on a chart might get you thinking about the association or relationship between the two variables. If that's the case, you could visualize that relationship with other chart types. I've plotted the same data in this scatterplot (see also page 249). You can see how China and India are outliers, especially along the population axis, but the chart doesn't communicate the part-to-whole picture of population. The parallel coordinates plot on the right (see also page 263) similarly shows how many more people live in China and India, and how a greater share of the population in India lives on $5.20 per day. (One potential issue with the parallel coordinates plot is that the lines might suggest a change over time to some readers when instead, in this case, it is being used to compare the two variables.)

A variation on the Marimekko is to have *both* the heights and the widths of the bars sum to 100 percent. This is sometimes called a mosaic chart, though many people don't differentiate between these two charts and use the terms interchangeably. In this definition of the mosaic chart, you fill the entire graph space and can therefore provide a part-to-whole perspective of the data along both dimensions. In this way, the mosaic chart is also closely related to the treemap (see page 297), but does not necessarily show a hierarchical relationship.

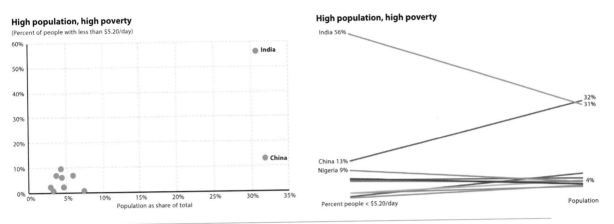

Other ways to plot two data series are a scatterplot (left) or parallel coordinates plot (right). Both are discussed later in this book.

In this example, population is still plotted along the horizontal axis, and the vertical axis now contains three categories for people with low levels of income: share of people with less than $1.90 per day, $3.20 per day, and $5.20 per day.

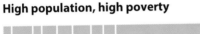

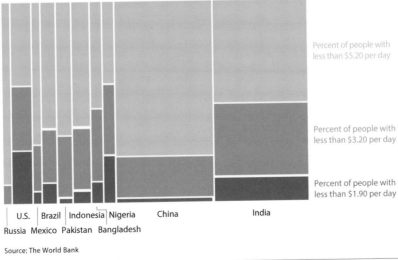

The mosaic chart is a variation on the Marimekko where both the heights and the widths of the bars sum to 100 percent.

Distribution of tax returns by size of tax change for the "Tax Cuts and Jobs Act"

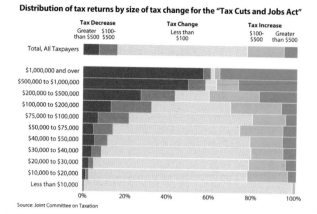

Source: Joint Committee on Taxation

Distribution of tax returns by size of tax change for the "Tax Cuts and Jobs Act"

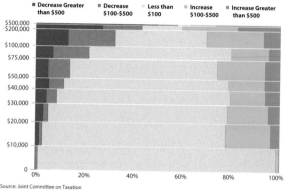

Source: Joint Committee on Taxation

Notice the difference between the stacked bar chart on the left (where all the bars are the same width) and the mosaic chart on the right. While the mosaic chart adds another variable, it is harder to see the details in the top category.

A mosaic chart can also serve as an extension of a stacked bar chart. These graphs show gains from the Tax Cuts and Jobs Act of 2017 for tax units at different points in the United States' income distribution. The stacked bar chart on the left shows five categories of tax gains across eleven income intervals, and each bar shares an equal width. If we scale the widths of the bars to the number of tax units in each income interval—so that the total vertical height of the chart sums to 100 percent—we can create the mosaic chart shown on the right. Notice that the mosaic chart gives a better sense of the distribution of the number of taxpayers in different groups, but because there are relatively few people in the top income group, it's harder to see those values.

UNIT, ISOTYPE, AND WAFFLE CHARTS

Unit charts show counts of a variable. Each symbol can represent an observation or a number of units. For example, if one symbol represents ten cars and there are ten car icons, the reader mentally multiplies the two for the total of one hundred cars. You can use unit charts to show percentages, dollars, or any other discrete amount. You can arrange them in different directions or break them down into subcategories by using colors or outlines.

Another advantage of these charts is that they can lend themselves to a more human connection. Bar charts, for example, are abstract and impersonal. They collapse all of the people reflected in that data point into a single shape. These charts, on the other hand, offer

Global out-of-school children of primary school age

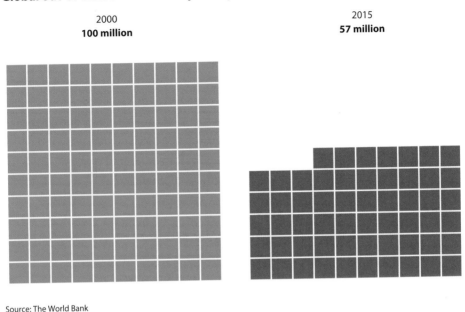

2000
100 million

2015
57 million

Source: The World Bank

Unit charts use symbols to show counts of a variable.

Global out-of-school children of primary school age

100 million 57 million
2000   2015

Source: The World Bank

BANs—or Big-Ass Numbers—are a way to just show the values.

an opportunity to connect with the subject by reminding readers of the number of people represented, particularly if each dot represents one person.

Another simple way to show these kinds of discrete counts is to just show the numbers. In *The Big Book of Dashboards*, authors Steve Wexler, Andy Cotgreave, and Jeff Shaffer call this the BAN approach: "Big-Ass Numbers." BANs might work best in a dashboard, infographic, social media post, or slide deck, but personally, I use them more sparingly in longer reports.

ISOTYPE CHARTS

Isotype charts are a subclass of unit charts that use images or icons instead of simple shapes. The term Isotype—International System of Typographic Picture Education—was coined by Austrian philosopher and political economist Otto Neurath, his wife Marie Neurath, and their collaborator Gerd Arntz in the 1920s. They used the Isotype system to visualize all kinds of data, from workers in different industries, to population density and distribution, to the number of machines used in specific factories. They believed that this kind of visual system would help people communicate demographic, economic, and environmental issues to a broader public regardless of people's educational attainment.

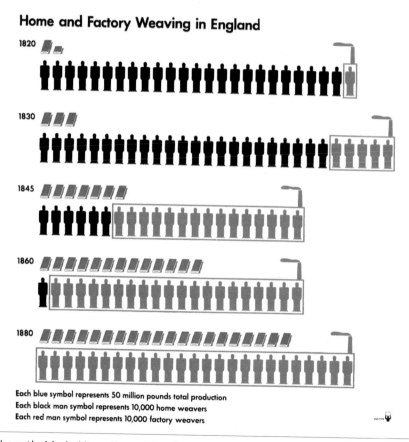

Otto Neurath, Marie Neurath, and Gerd Arntz developed the Isotype chart in the 1920s.

Extreme poverty rate in developing countries

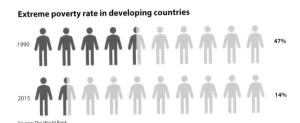

Millions of people in poverty

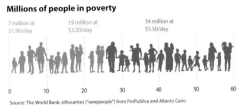

Two different ways to use icons in your data visualizations.

The graphic below is a classic example of their work. Each symbol represents a different count of workers (home or factory) and pounds of production. Aligned along a single vertical axis, it is easy to see how the values change over time.

We can take the same approach with the poverty data we've been using in this chapter. Notice that there's more than one way to use Isotype images in these two charts of extreme poverty rates. The version on the left uses individual icons to show each group of ten percentage points (the lighter icons could be included or not). The version on the right essentially orders the icons atop a bar chart. In either case, the icons connect the subject and content with an immediately recognizable visual image.

Extreme poverty rate in developing countries

47%
in 1990

14%
in 2015

Source: The World Bank

Icons can also be scaled according to their data values, but be careful because it's hard to know whether they are scaled according to the height, width, or area.

Instead of lining up the icons in rows or another arrangement, you can also scale them according to their value. But be careful, as sometimes it's hard to know whether the data are scaled according to the height, width, or area. That may not matter to every audience—it's clear here that the 47 percent is much larger than the 14 percent—but in cases where accuracy is paramount, this icon-scaling approach is inadvisable. Here, the vertical distance represents the data values, but the area of the icon on the left is about 10 times the size as the icon on the right.

These icon-driven graphs may look nice and engage readers, especially for few data values, but they can be difficult for your reader to count and compare. In his 1914 book, *Graphic Methods for Presenting Facts*, Willard Cope Brinton criticized this approach: "Charts of this kind with men represented in different sizes are usually so drawn that the data are represented by the height of the man. Such charts are misleading because the area of the pictured man increases more rapidly than his height." More recently, data visualization author and instructor Stephen Few wrote that unit charts force the reader "to either count, read the numbers, or do our best to compare the areas formed by each, which we can do poorly at best."

But sometimes the downsides of imprecision and slow comprehension may be offset by how memorable the chart is and how it engages readers, an issue that is borne out in recent research. Viewers in one study had clear preferences for graphs that included stacked icons rather than simple bars. They also found that images that sit in the background or are added to a chart but do not depict data are distracting to the reader, so if you choose to use these

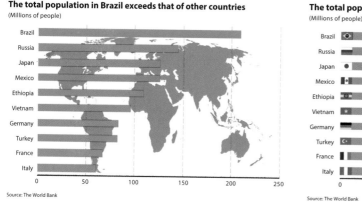

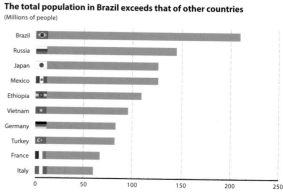

Both graphs show the population in ten countries. The graph on the left is cluttered by an unnecessary backdrop of the world, while the one on the right uses flag icons to add identifying detail.

kind of small unit icons or images in your work, be sure to use them only to encode your data and not for gratuitous decoration.

The graph on the left, which shows the population in ten countries, has a backdrop of a world map, a superfluous and distracting decoration. The graph on the right uses flag icons to add identifying detail and some visual engagement to a standard chart type.

Other research suggests that unit visualizations are intuitive and flexible, a way for readers to slowly wade into a visualization. Some have found that "unit visualizations are mostly useful when we want the readers to understand specific data item and encode its value (e.g., a unit, a person, a currency, a region, etc.)." Too much data and too many units, however, can create a visualization that is cluttered and obscures individual data points or the overall argument.

WAFFLE CHARTS

Waffle charts are another subclass of unit charts. They are especially good for visualizing part-to-whole relationships. Waffle charts are arranged in a 10 × 10 grid in which each colored cell represents one percentage point. You can use multiple waffle charts to show separate percentages—so the graph both shows part-to-whole relationships and lets your reader compare across the categories.

When creating unit or waffle charts, especially with icons, be mindful of your audience and how symbols may not appropriately represent your content. If you are visualizing child mortality rates in different countries, for example, an icon of a baby is not appropriate. Using icons of men to represent counts of people may ignore women in your data set. Alternatively,

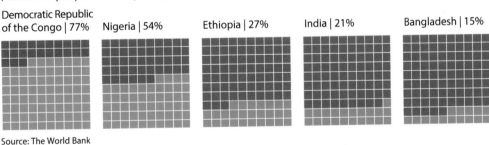

Overall poverty rate in five countries
(Percent of people at $1.90 per day)

Democratic Republic of the Congo | 77% Nigeria | 54% Ethiopia | 27% India | 21% Bangladesh | 15%

Source: The World Bank

Waffle charts are a subclass of unit charts and, in this case, arrange the squares in a 10 x 10 grid.

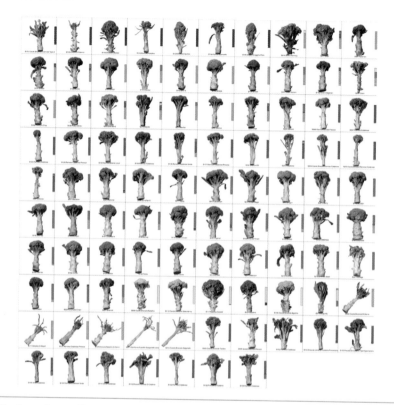

Zachary Stansell created this small multiples visualization of the different types of broccoli he grew in his garden.

if you want to measure the different types of broccoli in your garden, simply line up the pictures, as in this fun project from Zachary Stansell shown above.

HEATMAP

Heatmaps use colors and color saturations to represent data values. Simply put, a heatmap is a table with color-coded cells. They are often used to visualize high-frequency data or when seeing general patterns is more important than exact values.

These two heatmaps show the different components of total income for ten countries using data from the Luxembourg Income Study. The version on the left uses the same color scale for all six categories, where lighter colors encode smaller values and darker values

Composition of total income
(Percent of total income)

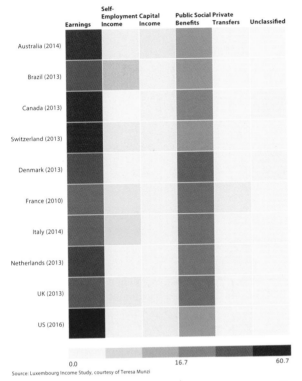

Composition of total income
(Percent of total income)

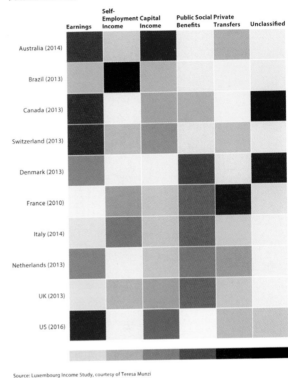

Source: Luxembourg Income Study, courtesy of Teresa Munzi

Heatmaps use colors and color saturations to represent data values and can focus the reader's attention along the columns or across the rows.

encode larger values. In this view, you can see that people's earnings account for the greatest share of their total income, and, in most countries, public social benefits appear to be the second-largest share. In the heatmap on the right, each category is assigned its own color scale. Here, you can more clearly see that public social benefits (in the fourth column) account for a smaller share of total income in Australia, Brazil, Switzerland, and the United States. Which one is better depends on your goals. Do you want your reader to compare across all of the values or within each category?

You can also use a heatmap to show changes over time. Imagine a spreadsheet that contains infection rates from the measles disease for every state in America from 1928 to 2008. If the spreadsheet had states ordered along the rows and years along the columns, your first instinct might be to create the line chart on the next page.

Measles incidence in the United States from 1928 to 2012

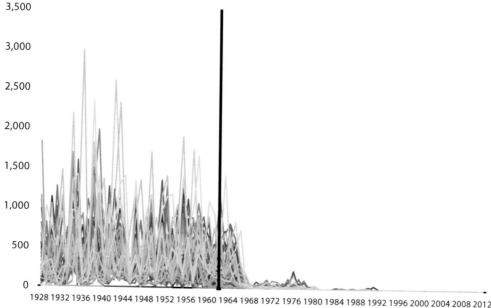

Source: Data from Project Tycho, https://www.tycho.pitt.edu/data

You can see basic patterns in measles infections across the United States in this dense line chart, but it's hard to pick out any specific values.

There's nothing inherently *wrong* with this next line chart—you can see the positive infection rate from 1928 to about 1963 (marked with the black vertical line), the year when the measles vaccine was introduced. Over the next five years or so, infections dropped quickly, and eventually reached around zero within about ten years. What you essentially get from this chart is that there were infections—going up, going down, in a tangle of lines—until about 1963.

The *Wall Street Journal* looked at the same spreadsheet and, instead of creating a dense line chart, they created a heatmap. I've created my own version here using a different color palette and discrete categories of the infection rate. You can see the darker blue cells (mostly above 16 infections per 100,000 people) before the introduction of the vaccine, again marked with a black line. After 1963, the colors quickly transition to lighter shades of blue, and ultimately to the lowest rates of infections (zero infections and fewer than 1 infection per 100,000 people).

This chart may not be inherently *better* than the line chart, but it does let you more easily examine each state or year far more easily than picking out a single line from the tangle in the line chart. Also remember that sometimes being different is good in itself. How many

Measles incidence in the United States from 1928 to 2012

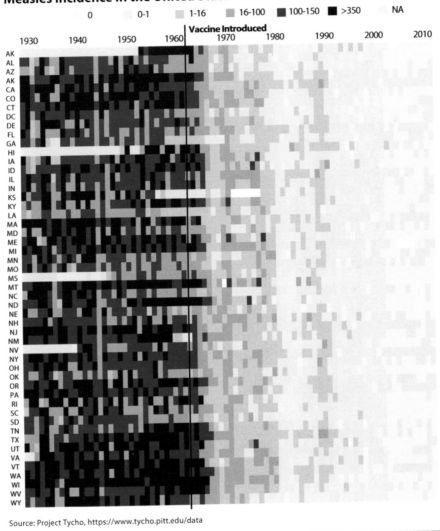

Source: Project Tycho, https://www.tycho.pitt.edu/data

This heatmap may not be inherently better than the line chart at showing measles infections rates, but it does let the reader more easily examine each state or year.

complex line charts have you seen and just immediately skipped over? The heatmap, with its different look and color, can draw readers in. As artist and data visualization expert Giorgia Lupi said, "Beauty is a very important entry point for readers to get interested about the visualization and be willing to explore more. Beauty cannot replace functionality, but beauty and functionality together achieve more."

Another way to use the heatmap is to modify the layout, for example, applying it to a calendar. In this example, vehicle fatalities in 2015 are plotted on heatmaps of months. Notice how easy it is to see the higher fatality rate on Fridays and Saturdays along the right edge of each column.

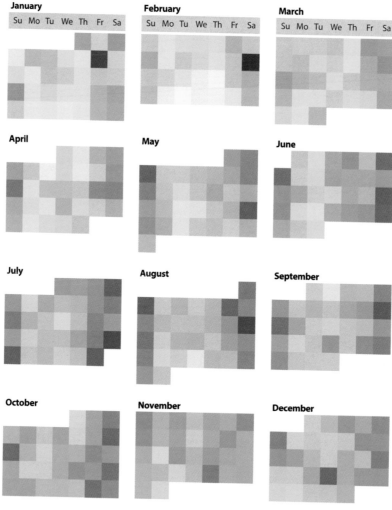

Vehicle fatalities in 2015

Source: National Highway Traffic Safety Administration
Note: Inspired by Nathan Yau at FlowingData.com

Another way to use a heatmap is to modify the layout, as in this version that shows vehicle fatalities in 2015.

Vehicle fatalities in 2015

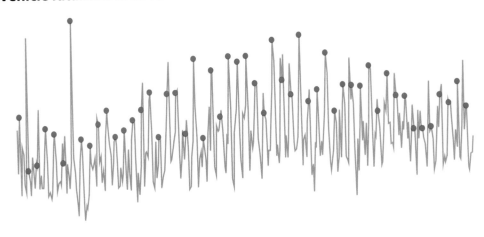

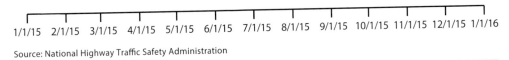

1/1/15 2/1/15 3/1/15 4/1/15 5/1/15 6/1/15 7/1/15 8/1/15 9/1/15 10/1/15 11/1/15 12/1/15 1/1/16

Source: National Highway Traffic Safety Administration

This line chart shows the same data as in the heatmap calendar, but it's more difficult to reach the same conclusion.

By comparison, consider plotting the same data in a line chart. Even with the additional blue circles used to mark Saturdays, it's difficult to reach the same conclusion about more fatalities on the weekends.

Unlike the measles example, where both charts had advantages and disadvantages, in this case the calendar heatmap is a better approach because it does a better job highlighting the important pattern of deaths on the weekends—and it's a more engaging graph placed in a familiar shape.

A final way to modify the heatmap is to change the rectangular layout altogether. On the next page, each of the fifty states is plotted along a radius of a circle, grouped into five geographic regions (separated by the black lines). Each ring represents a different (binary) data type, such as whether the state has right-to-work laws, an income tax, or a minimum wage.

This alternative to a set of six maps has some advantages and some disadvantages. On the one hand, it is a compact representation of six different data series that the reader can quickly and clearly see the categorical and part-to-whole data within and between states. On the other hand, it's not a familiar graph type, and that may turn off some readers. It's also

Employment rights in the United States

Source: National Employment Law Project, Wikipedia, Kaiser Family Foundation, National Conference of State Legislatures.

Heatmaps can be arranged in different ways. The radial layout has some advantages and disadvantages, but so does the six-pack of maps.

worth noting that the order of the rings can affect our perception of the data because the (red) squares on the outer ring are by definition larger than the (yellow) squares on the inner ring.

GAUGE AND BULLET CHARTS

The gauge chart (or gauge diagram) looks like the speedometer in your car's dashboard. Typically set up somewhere between a half-circle and a circle, it uses a pointer or needle to indicate where your data fall within a particular range. Sections of the gauge are shaded to illustrate sections such as poor, good, and excellent.

I see gauge diagrams most often in financial planning tools because they give an easy, familiar way to visualize targets or progress towards a goal. They also frequently show up in fundraising campaigns where the entire semicircle represents the goal, and the needle and filled area represent money raised so far. This can be a good example of using a familiar shape to support the metaphor of the visualization—everyone understands that "filling up" the gauge means the fundraising effort has reached its goal.

Gauge diagrams do, however, introduce perceptual challenges because, again, people are not very good at measuring and comparing angles. If you want to give your reader a general

PART THREE: DESIGNING AND REDESIGNING YOUR VISUAL

PART ONE

PRINCIPLES OF DATA VISUALIZATION

policy adoption. The advance of computing power, social media platforms, and the expanding media landscape made visual content more important, perhaps even necessary.

Today, I work with people in nonprofits, government agencies, private sector companies, and everything in between to improve how they create their graphs and communicate their content. I've worked with junior economists and analysts dealing with enormous data sets; health care workers trying to communicate results to patients, families, and hospital administrators; human resource representatives working with databases of job-seekers; advertisers and marketing executives selling products to clients; and many more.

I've seen hundreds of different kinds of data visualization challenges. The skills to meet them, unfortunately, are not yet regularly taught in schools or professional development programs. But these skills *can* be learned. We can learn how to read chart types we've never seen before, even if they are complex. And we can learn how to communicate our work in better and more effective ways.

Eventually, I discovered that one of the most important things I can show people is the extraordinarily wide array of graphs available to them. And that is precisely the content of this book, a survey of more than eighty types of data visualizations, from the familiar to the nonstandard.

But before we get to the library of graph types, we'll consider some of the science behind how we process visual information and some best practices and approaches to visualizing data.

That June, CBO's director sat in front the U.S. House Budget Committee to relay the results of our analysis. As the hearing played on a TV out in the hallway, I suddenly heard yells of, "Jon! Jon! Come out! Your infographic is on TV!"

And, sure enough, Congressman Chris Van Hollen was holding up the infographic on C-SPAN, covered with scribbles and notes. The visualization had captured and engaged the attention of one of the busiest people in America, and someone who could do something about the pressures facing the federal budget. That was the moment I knew that *how* we presented our data could matter as much as the data itself.

In 2014, I moved to the Urban Institute, a nonprofit research institution in Washington, DC, to spend half of my time conducting research and half of my time in the Communications department, helping colleagues present and visualize their data.

Since that time, I have conducted hundreds of workshops, delivered lectures around the globe, and published two books on data communication. The world, it seemed, had seen what I saw—better visual content and better presentations were the currency of research and

Maryland Congressman Chris Van Hollen holding up that Long-Term Budget Outlook infographic in a House Budget Committee hearing.
Source: C-SPAN2.

and strategically about our data visuals. From there, I started reading books on data visualization, design, color theory, and typography.

Working with our editorial department and designers, we began to improve the graphs in our basic reports and started creating new report and graph types. We made infographics—what was then a buzzword referring (sometimes derisively) to longer graphics that combine data, text, images, and more into a single visual. In 2012, we created this infographic to accompany and summarize *The Long-Term Budget Outlook*, a 109-page report.

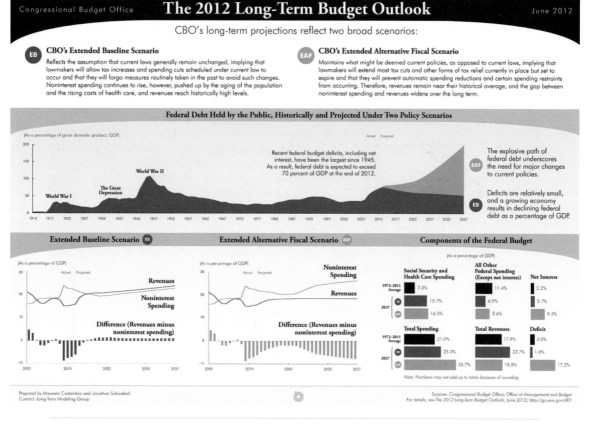

One-page infographic about the 2012 Long-Term Budget Outlook from the Congressional Budget Office.

Source: Congressional Budget Office.

You don't need to be a government economist to know that members of Congress are unlikely to read something that looks like a spreadsheet. There are too many rows, too many columns, too many numbers—too much information. It was right then that I first started thinking about better ways to present this information.

This was the result. We replaced some numbers with small area charts, which give the reader an immediate visual impression of each option—which ones increased the solvency of the program and which ones did not.

Table 2.

Changes to Social Security's Finances Under Various Options with Scheduled Benefits

(Percentage of GDP)

		2020	2040	2060	2080	Annual Finances	75-Year Present Value as a Percentage of GDP	Taxable Payroll
						Current Law[a]		
						Revenues and Outlays[b]		
	Revenues	4.9	4.9	4.9	5.0		5.2	14.4
	Outlays	5.2	6.2	6.0	6.3		5.8	16.0
	Balance	-0.3	-1.3	-1.1	-1.3		-0.6	-1.6
						Percentage-Point Change from Current Law[a]		
Change the Taxation of Earnings						**Change in Annual Balance**[c]		
1	Revenues	0.4	0.4	0.3	0.3		0.3	1.0
Increase the Payroll Tax Rate by 1 Percentage Point in 2012	Outlays[d]	*	*	*	*		*	*
	Balance	0.4	0.4	0.4	0.4		0.3	1.0
2	Revenues	0.3	0.7	0.7	0.7		0.5	1.6
Increase the Payroll Tax Rate by 2 Percentage Points Over 20 Years	Outlays[d]	*	*	*	*		*	*
	Balance	0.3	0.7	0.7	0.8		0.6	1.6
3	Revenues	0.2	0.5	0.8	1.0		0.5	1.5
Increase the Payroll Tax Rate by 3 Percentage Points Over 60 years	Outlays[d]	*	*	*	*		*	*
	Balance	0.2	0.5	0.9	1.1		0.5	1.4
4	Revenues	0.8	0.9	0.9	0.9		0.9	n.a.
Eliminate the Taxable Maximum[e]	Outlays	*	0.3	0.5	0.5		0.3	n.a.
	Balance	0.8	0.6	0.4	0.4		0.6	n.a.
5	Revenues	0.3	0.4	0.4	0.4		0.4	n.a.
Raise the Taxable Maximum to Cover 90% of Earnings[e]	Outlays	*	0.1	0.2	0.2		0.1	n.a.
	Balance	0.3	0.3	0.2	0.2		0.2	n.a.

Continued

Final version of that main exhibit in the Congressional Budget Office report on Social Security. Notice that there is less data and more graphs.
Source: Congressional Budget Office.

The report worked. We received good feedback from colleagues at CBO and other agencies, as well as readers on Capitol Hill and elsewhere, noting how easy it was to read and digest the graphs. It was maybe the first time I (and perhaps the agency) thought carefully

I moved to DC in 2005 to join the Congressional Budget Office (CBO). My job was to help work on the long-term microsimulation model that is used to examine the Social Security system and forecast the long-term finances of the federal budget. The spring of 2005 was an exciting time to work on Social Security—President George W. Bush had made Social Security a central component of his second term. In his 2005 State of the Union address, he said, "We must pass reforms that solve the financial problems of Social Security once and for all." Reform would stall later that year, but in the course of my first few months on the job, my group at CBO estimated and analyzed the effects of dozens of policy proposals.

Five years later, I had expanded my work to include issues around policies that affected disabled workers, immigration, and food stamps (now called the Supplemental Nutrition Assistance Program or SNAP). In 2010, three of my colleagues were drafting a special report on policy options for Social Security. In it, they would show the impact of thirty different options for reform. One of the central figures in the report would show changes in taxes received by the system, benefits paid out from the system, the balance between the two, and other measures of fiscal solvency for these thirty options. It looked something like this:

	Option Name		Revenues, Outlays, and Balances as a Percentage of GDP				75 Year Present Value as a Percentage of		Trust Fund Exhaustion Year
			Year				GDP	Taxable Payroll	
			2020	2040	2060	2080			
	Baseline[a]	Revenues[b]	4.9	4.9	4.9	5.0	5.2	14.4	
		Outlays[c]	5.2	6.2	6.0	6.3	5.8	16.0	20XX
		Balance[d]	-0.3	-1.3	-1.1	-1.3	-0.6	-1.6	

	Option Name		Changes in Revenues, Outlays, and Balances as a Percentage of GDP				Change in 75 Year Present Value as a Percentage of		Change in Trust Fund Exhaustion Year
			Year				GDP	Taxable Payroll	
			2020	2040	2060	2080			
1	Increase the Payroll Tax Rate by 1 Percentage Point in 2012	Revenues	0.4	0.4	0.3	0.3	0.3	1.0	XX
		Outlays	*	*	*	*	*	*	
		Balance	0.4	0.4	0.4	0.4	0.3	1.0	
2	Increase the Payroll Tax Rate by 2 Percentage Points over 20 Years	Revenues	0.3	0.7	0.7	0.7	0.5	1.6	YY
		Outlays	*	*	*	*	*	*	
		Balance	0.3	0.7	0.7	0.8	0.6	1.6	
3	Increase the Payroll Tax Rate by 3 Percentage Points over 60 Years	Revenues	0.2	0.5	0.8	1.0	0.5	1.5	ZZ
		Outlays	*	*	*	*	*	*	
		Balance	0.2	0.5	0.9	1.1	0.5	1.4	
4	Eliminate the Taxable Maximum	Revenues	0.8	0.9	0.9	0.9	0.9	n.a.	AA
		Outlays	*	0.3	0.5	0.5	0.3	n.a.	
		Balance	0.8	0.6	0.4	0.4	0.6	n.a.	
5	Raise the Taxable Maximum to Cover 90% of Earnings	Revenues	0.3	0.4	0.4	0.4	0.4	n.a.	BB
		Outlays	*	0.1	0.2	0.2	0.1	n.a.	
		Balance	0.3	0.3	0.2	0.2	0.2	n.a.	

Author's rendering of early draft of exhibit from the Congressional Budget Office.

forward creating more visuals and seeing their effect on your audience, you'll develop your own aesthetic and learn when to bend or break these guidelines.

Part 2 is the meat of the book. We will define and discuss more than eighty graphs, categorized into eight broad categories: Comparisons, Time, Distribution, Geospatial, Relationship, Part-to-Whole, Qualitative, and Tables. We will see how each graph works and the advantages and disadvantages of each.

Graphs overlap between these categories—a bar chart, for example, can be used to show changes over time or comparisons between groups. The categorizations here are based on a graph's primary purpose. But even that's not an objective truth, and your perspective and situation may differ. I do not discuss *every single possible graph*—there are many specialized graphs in fields like architecture, biology, and engineering that are excluded here. Instead, these chapters cover the most common and flexible graphs that can showcase the sorts of data most people will need to display.

I tie these chapters together in part 3 with a chapter on building a data visualization style guide and a chapter on how to pull the different lessons together in a series of graph redesigns. If you've ever written a research paper, or even a book report, you are probably aware of the array of writing style guides, from the *Chicago Manual of Style* to the *Modern Language Association*. These guides break down writing into component parts and prescribe their proper use. A data visualization style guide does the same for graphs—defines their parts and how to style and use them. In the final chapter, we apply the lessons to redesign a series of graphs to improve how they communicate data.

This book will guide you as you explore your data and how it might be visualized. Now more than ever, content must be visual if it is to travel far. Your clients and colleagues, and your audiences of policymakers, decisionmakers, and interested readers, are inundated with a flow of information. Visuals cut through that.

Anyone can improve the way they visualize and communicate their data—and you don't need a graduate degree in marketing or design or advertising. Take it from me, I started my career as an economist in the federal government.

HOW I LEARNED TO VISUALIZE MY DATA

Once I settled on declaring my economics major at the University of Wisconsin at Madison (there was an ill-fated attempt to also be a math major, but I hit a wall at Markov chains), I knew I wanted to end up in Washington, DC. I wanted to be near the center of public policy and politics. I wanted to explore the real problems of the day and help craft solutions.

Gauge charts are familiar and easy to read.

sense of the values, the gauge chart is a decent choice. But if enabling your reader to discern the specific values and compare those values to the ranges is of utmost importance, then it is not.

Given the familiarity with the gauge and the obvious metaphor it represents, it's perhaps no surprise that they show up in a variety of settings. As just one example, I once received

This series of gauge charts shows four real estate trends in my Northern Virginia neighborhood.
Source: MountJoy Properties, brokered by Keller Williams Realty.

a flyer from Mountjoy Properties (a residential real estate team that serves the DC Metro Area) that consisted of a series of gauge diagrams showing current real estate trends in my neighborhood of Northern Virginia. I could pretty quickly see the current state of the market, but adding more data or more detailed data might be more difficult.

BULLET CHARTS

Because of these perceptual challenges with the gauge chart, author Stephen Few invented the bullet chart, which is a linear, more compact way to show similar kinds of data. The basic bullet chart contains three different data elements:

1. First, there is the actual or *observed value*, shown here as the black horizontal bar. In this illustration, the bar represents an average customer satisfaction score of 4.0.
2. Second, there is a *target value*, shown here as the black vertical line. Here, we were aiming for a satisfaction score of 3.5.
3. Finally, there is the *background range*, which shows grades or bands of success, such as poor, good, and excellent. These sit behind the other two series so the reader can compare the actual and target values. Here, poor scores are 1.5 and below, good scores are from 1.5 to 3.0, and excellent scores are anything above 3.0.

The different components of the bullet charts can vary. There might be a scale of five ranges instead of three, or there may not be a target value. The scales can also show the underlying

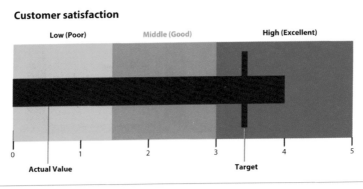

The bullet chart includes five separate data values.

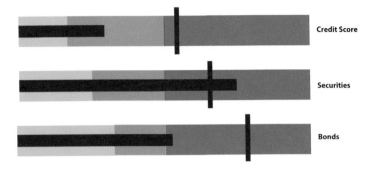

Combining different bullet charts is a compact way to let the reader make a series of different comparisons.

distribution of the data—for example, showing quartiles or ranges of percentiles (see Box on page 183 for more discussion of percentiles). Because the bullet chart is so compact, it's easy to create multiple versions and stack them together. The bullet charts above show three metrics you might find in a financial report, but they are more compact than gauge charts and, because the rectangles are aligned, it is easier to compare across the different categories.

BUBBLE COMPARISON AND NESTED BUBBLES

In the basic bubble graph, circles represent values. Like a bar chart, the purpose of these types of charts is to compare values between categories. But unlike the bar chart, humans are not very good at accurately comparing the sizes of circles (remember the perceptual ranking diagram from page 14). Still, circles may be more visually interesting, they can reinforce a visual or metaphor, and they are a good choice when discerning exact quantities is not paramount.

Another drawback of proportionally sized circles is that you cannot visualize negative values. While bars can go in both directions—typically right or upward for positive and left or downward for negative—that's much harder to visualize with circles.

In any case, we are not very good at making accurate estimates from the circles even when they are sized by area. Instead, what I think we try to do is make the comparison based on the diameter of the circle—like in a bar chart—which gives us incorrect conclusions. Take a look at the two graphs on the next page and try to guess the percent of people living on less than $1.90 per day in Pakistan. Do you think that task is easier in the bubble chart or in the bar chart?

This is not to say that you should never use circles. Remember, again, it's all about your audience. A bubble comparison chart, inserted in a short article off to the side with the

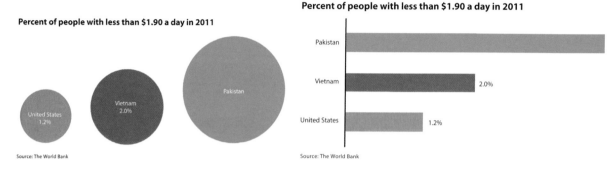

A bubble comparison chart, two versions

We are better at discerning differences from bars than by areas of circles. By the way, the percent of people with less than $1.90 a day in Pakistan is 4.0 percent.

numbers placed prominently in the middle of each circle may be more engaging than a standard bar chart. Too many circles, however, may make it difficult for your audience to discern any quantities or relationships. In this next example, yes, you can see that India, the Democratic Republic of the Congo, and Nigeria have the largest number of people in poverty, but it's difficult to quickly assess *how* different they are or the numbers of the next set of countries.

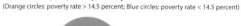

Number of people in poverty
(Orange circles: poverty rate > 14.5 percent; Blue circles: poverty rate < 14.5 percent)

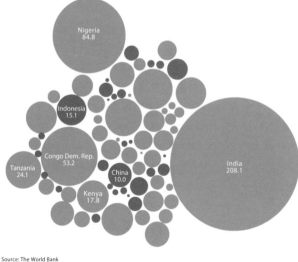

Source: The World Bank

A bubble comparison chart can be engaging and interesting, but it can also be hard to discern the values.

CALCULATING THE AREA OF A CIRCLE

Remember to size the circles by area, because using the radius or diameter generates circles that overemphasize differences (the radius or diameter scales in a linear way, but the area scales quadratically). The first black circle is sized relative to the gray circle using the diameter while the second black circle uses the area. As you can clearly see, using the diameter biases our perception and makes the difference between the two values look much larger.

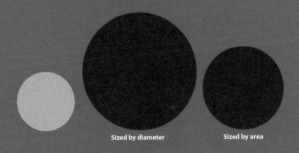

Sized by diameter   Sized by area

A simple example will demonstrate the importance of using the area rather than the radius/diameter for sizing these circles. In case you don't remember your middle-school math, the diameter is any straight line that passes through the middle of the circle. The radius (r) is half the diameter. And the area (A) is equal to the constant pi (π) times the radius-squared, or $A = \pi r^2$.

So, say the data value for the gray circle is 1 and for the black circle is 2. If we start with the gray circle and set the radius equal to 1, we can find the area is equal to: $A_O = \pi r^2 = \pi 1^2 = \pi$.

To find the size of the black circle the correct way, we say that the area of that circle is twice the size of the gray circle, corresponding to the difference in their data values. So, if we double the area of the black circle so that it is now 2π, we can then rearrange the formula and find the radius of the black circle to draw it: $r_B = \sqrt{A_B / \pi} = \sqrt{2\pi / \pi} = \sqrt{2}$.

Let's do this the wrong way and use the radius instead of the area. In this case, the radius of the gray circle is still 1, so let's make the radius of the black circle 2, again corresponding to the difference in their values. This makes the area of the black circle now $A_B = \pi r^2 = \pi(2)^2 = 4\pi$. In other words, the area of the black circle sized in this way is four times the size of the gray circle instead of twice the size, as the data suggest.

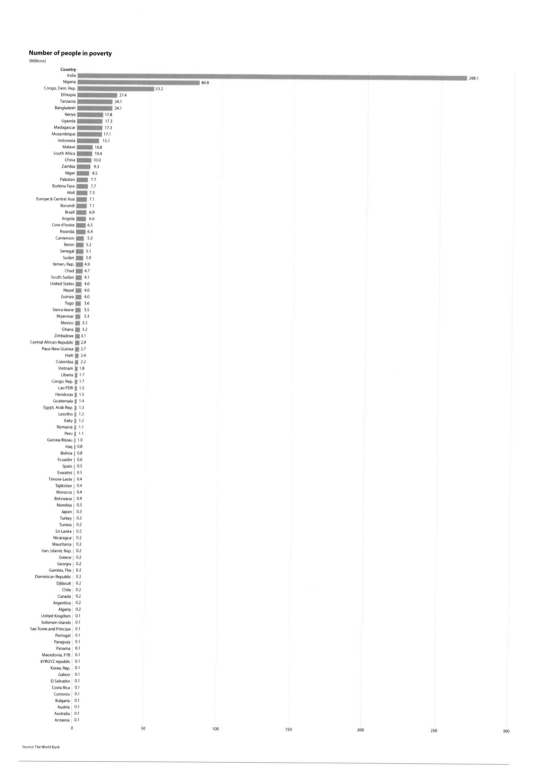

It's easier to pick out the countries with the most and least poor populations in this bar chart, but it takes up the entire page.

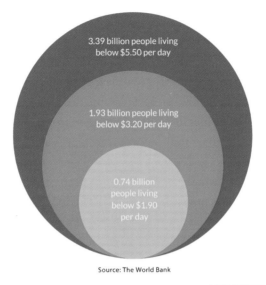

Source: The World Bank

The *nested bubble chart* can sometimes mask circles in the back—but it can also make for easier comparisons.

If we were to use a more perceptually accurate representation of these same data in, say, a bar chart, the visual becomes much larger as on the previous page. Each bar is labeled here, so a reader could find Madagascar, but is that important? Again, is the goal to show all of the countries or just a subset? As always, consider your goals and whether your reader needs a perceptually accurate view to understand your argument.

The bubble charts shown above are known as bubble comparison charts. Layering circles on top of each other, as in the graph above, are often called nested bubble charts. The nested bubble chart can sometimes mask circles in the back, but it can also make for easier comparisons.

You can use bubbles to demonstrate correlations (see the bubble chart in Chapter 8) or add bubbles to a map to encode another variable (see the point map in Chapter 7). In general, while there are perceptual issues with using circles and bubbles in data visualization, they can also be more engaging and enjoyable than yet one more bar or line chart. As Amanda Cox, Data Editor at the *New York Times* said, "There's a strand of the data viz world that argues that everything could be a bar chart. That's possibly true but also possibly a world without joy."

SANKEY DIAGRAM

Sankey diagrams—named for their creator Matthew Henry Phineas Riall Sankey in 1898—
are especially useful for showing how categories compare to one another and flow into other
states or categories. Arrows or lines display the transition from one state to another, and the
widths of the lines denote the magnitude of each transition. Changes can occur over time or
as comparisons between categories. For example, a Sankey diagram could be used show how
different companies in an industry have merged, broken apart, or failed in different years.

This Sankey diagram shows how fifty-two students tried to spell the word *camouflage*.
The first blue segment shows that all fifty-two students started with the letter "C", fifty
then went to "Cam", followed by thirty-seven with "Camof", and so on. Ten students
spelled the word correctly, shown in the orange segment near the top of the graph.

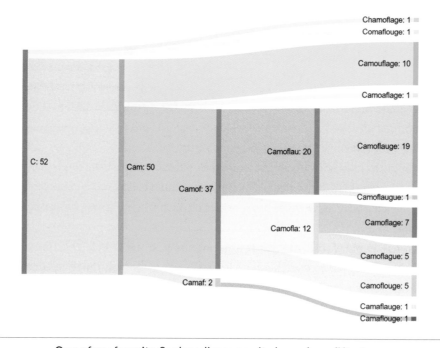

One of my favorite Sankey diagrams—it shows how fifty-two
students tried to spell the word *camouflage*. Graphic by Tim Bennett,
data collected by Reddit user iheartdna.

Financial support flows from Germany, the United States, and the United Kingdom to different areas of the world

(Percent of total support)

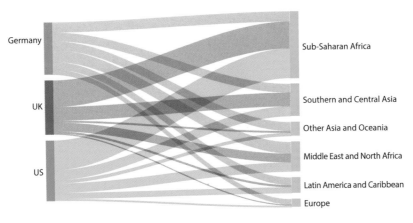

Source: Organisation for Economic Co-operation and Development

This Sankey diagram shows flows in federal aid from the United States, United Kingdom, and Germany to regions around the world.

More applicable to the sort of content we've looked at thus far in this chapter, the graph above shows flows in federal aid from the United States, United Kingdom, and Germany to regions around the world. You can see that countries in Sub-Saharan Africa and unspecified countries receive the bulk of the aid and that the United Kingdom and United States contribute more to Sub-Saharan African countries than does Germany.

Presenting these data as paired or stacked bar charts give us a different perspective. In the paired bar chart on the left, the obvious first comparison is across the funder countries for each region of the world. In the stacked bar chart, by comparison you're more likely to compare funding across the recipient regions—that, for example, the greatest share of spending on Sub-Saharan African countries is from the United States. The Sankey draws our attention in a different way, reading horizontally across the page in a way that mixes these two comparisons, privileging neither. Neither view is "right" or "wrong," but may serve different audiences differently, highlight different patterns, and answer different questions.

Sankey diagrams can be layered together with other chart types. In this example, the aid flows from the United States are presented on a world map. This provides a geographic view of the data, but also simplifies things—imagine how cluttered the visual would look if it included flows from the United States, United Kingdom, and Germany.

Financial support flows from Germany, the United States, and the United Kingdom to different areas of the world
(Percent of total support)

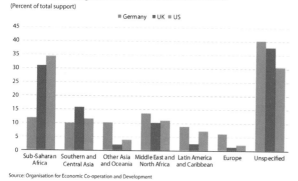

Source: Organisation for Economic Co-operation and Development

Financial support flows from Germany, the United States, and the United Kingdom to different areas of the world
(Percent of total support)

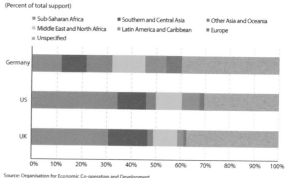

Source: Organisation for Economic Co-operation and Development

Presenting the financial flow data as a paired or stacked bar chart give us a different perspective than in the Sankey diagram.

Financial support flows from Germany, the United States, and the United Kingdom to different areas of the world

(Percent of total support)

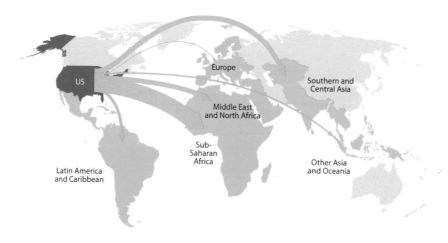

Source: Organisation for Economic Co-operation and Development

This flow map provides a geographic view of the financial flow data, but it also simplifies things by only showing flows from the United States.

The biggest problem with Sankey diagrams—and many charts for that matter—is plotting too many series, as in this version that includes more countries. With too many groups or too many crossings, the chart becomes difficult to navigate. If you find yourself with too

Financial support flows from Germany, the United States, and the United Kingdom to different areas of the world

(Percent of total support)

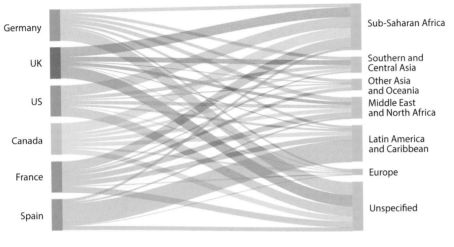

Source: Organisation for Economic Co-operation and Development

The biggest problem with Sankey diagrams—and many charts for that matter—is plotting too many series makes it difficult to identify any patterns or trends.

many lines or crossings, try simplifying your data, using multiple Sankey diagrams, or using a different chart type.

WATERFALL CHART

A waterfall chart shows a basic mathematical equation: adding or subtracting values from some initial value to produce a final amount. It is essentially a bar chart, but each subsequent bar starts where the previous one left off, showing how they accumulate across the graph. Typically, negative values are given a different color than positive values, and so are the totals at the beginning and end. Including lines that connect the bars can guide the reader through the visualization. Because the lines are guides and not actual data, they should be lighter and thinner than the other elements.

This next chart shows contributions to total gross income and total net income for Australia in 2016. The data are the same as those used in the heatmap example from earlier, but in that approach, I could fit data for ten countries in the same view. Imagine trying to do the

Income composition in Australia in 2016
(Percent of total gross income)

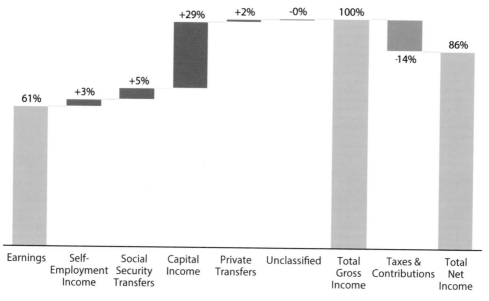

Source: Luxembourg Income Study, courtesy of Teresa Munzi

A waterfall chart shows a basic mathematical equation: adding or subtracting values from some initial value to produce a final amount.

same with a waterfall chart—we would need ten different graphs—something that might be useful under certain scenarios, but is certainly less compact than the heatmap.

Waterfall charts can also show changes over time. You might show, for example, contributions to total GDP from one year to the next, and how different values contribute to the change over the course of the year. Any data series that are added or subtracted to one another can be presented in this way, though, again, it is a nonstandard chart type and may require your reader more time to navigate it.

CONCLUSION

From single bars to groups of bars to stacks of bars, the bar chart is one of the most familiar data visualizations for showing categorical comparisons. It also ranks at the top of our perceptual ranking scale from earlier. But the bar chart also poses certain challenges: too many bars can make the visual seem overwhelming and cluttered, and stacking the series on top of one another makes it more difficult to compare series that are not aligned on the same axis.

The basic bar shape can be organized in many ways. They can sit next to each other or diverge from a central baseline. They can be stacked on top of one another on a horizontal or vertical dimension, or both as in a mosaic chart. They can also be arranged to show simple mathematical equations, as in a waterfall chart. We are generally good at discerning the data values from the lengths of the bars, so many of these chart types will make it easy for your reader to perceive the exact value.

There are other ways to let your reader make comparisons. I'm especially fond of dot plots because they remove a lot of the heavy ink from a standard bar chart and free up space to add annotation and labels. Using icons, squares, or other shapes can engage our audience in ways that standard charts may not, but may be less data dense.

While bar charts sit at the top of the perceptual ranking list, let's be honest: They can be very boring. We see bar charts every day. As chart creators, sometimes our challenge is to find ways to engage our audience, and deploying less common chart types from our data visualization toolbox can do just that. It's up to us to determine where we want to focus our readers' attention, on the level or the difference, the single or multiple comparison, or the relative or total values.

TIME

The graphs in this chapter show changes over time. Your readers will most likely be familiar with visuals like line, area, and stacked area charts. But others, like connected scatterplots and cycle charts, may need more labeling and annotation for the reader to navigate them successfully.

Many of the visuals in this chapter are variations on the line or area chart. Some let us include more data on the page than usual, while others allow us to combine changes over time with some other view of the data. With horizon charts and streamgraphs, for example, we can include more data in a single visual, but they are probably not best suited for detailed comparisons. Other graphs, like flow charts and timelines, may use qualitative data or narrative text and visual clues to guide the reader.

Graphs in this chapter are styled following the guidelines published by the Urban Institute, a nonprofit research institution based in Washington, DC. Urban's style guide outlines their color palette, fonts, and guidance for different chart types.

LINE CHART

The line chart and the bar chart may be the most common charts in the world. The line chart is easy to read, clear in its representation, and easily drawn with pen and paper. Data values are connected by lines to show values over a continuous period, tracking trends and patterns.

Total health care spending in the United States grew from 12.5 percent to 16.8 percent between 2000 and 2015
(Percent of GDP)

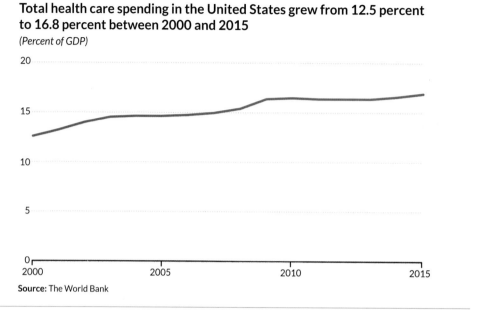

Source: The World Bank

The basic line chart.

This line chart shows the percent of gross domestic product (GDP) spent on health care in the United States over the sixteen-year period from 2000 to 2015.

As with the bar chart, the line chart sits near the top of the perceptual ranking scale. With lines relative to the same horizontal axis, it is easy to compare the values to each other and between different series.

As simple as the line chart can be to create and read, there are a number of considerations to take into account, some of which are aesthetic, and some of which are substantive.

THERE IS NO LIMIT TO THE NUMBER OF LINES YOU PLOT

There is no hard rule to dictate the number of series you can include in a single line graph. The key is not to worry about the sheer amount of data on the graph, but instead about the purpose of the graph and how you can focus your readers' attention to it. For example, in a line graph with many series, you can highlight or emphasize a subset of your data.

Say we were interested in showing the share of government spending on health care for the United States and Germany, but we also wanted to show them in relation to the other

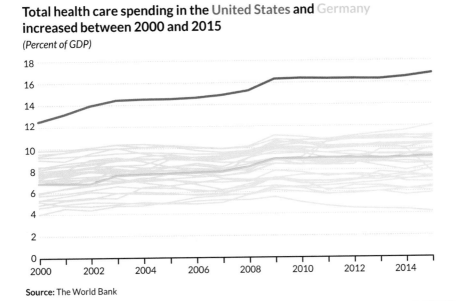

**Total health care spending in the United States and Germany
increased between 2000 and 2015**

(Percent of GDP)

Source: The World Bank

There is no hard rule to dictate the number of series you can include in a single line graph.

thirty-four countries that make up the Organisation of Economic Co-operation and Development (OECD). To do that, instead of giving each line the same color saturation and thickness, we might only color and thicken the lines for the United States and Germany and leave the lines for the other thirty-two countries gray and thin. The "Start with Gray" strategy from Chapter 2 comes in handy here.

Recall the section on preattentive processing: color and line width are two of the preattentive attributes (page 25); thus our attention is drawn to the thicker, colored lines. The advantage of the gray strategy is that the reader can appreciate the general pattern for the entire sample and yet focus on the two lines of interest.

We might also take the line graph and break it into multiple graphs, the small multiples approach. We might include just the line of interest in each small graph, or include all of the lines and use the gray strategy. The set of small multiple line graphs on the next page uses the former approach and shows spending on health care for nine of the thirty-four OECD countries. Instead of forcing all nine lines on a single graph, each country has its own panel. While we might lose some perspective of the *relative* values

Health care spending across major countries has largely increased since 2000

(Percent of GDP)

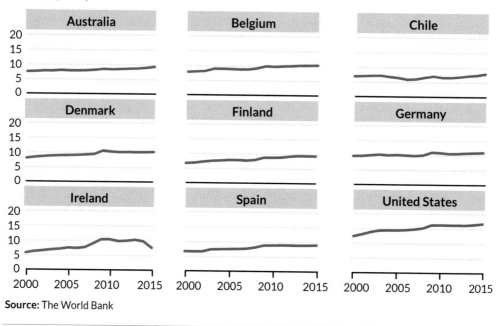

Source: The World Bank

The small-multiples approach breaks up a dense line chart into separate components.

of spending in each country, this layout provides more space for each country and thus the opportunity to provide more detail, labels, or other annotation.

YOU DON'T NEED TO START THE AXIS AT ZERO

One of the few rules of thumb of visualizing data is that bar chart axes must start at zero (see Chapter 4). Because we perceive the values in the bar from the length of the bars, starting the axis at something other than zero overemphasizes the differences in values.

This does not hold true for line charts. The axis of a line chart does not necessarily need to start at zero. As with many aspects of visualizating data, there are complications and different perspectives. If we say the axis does not need to start at zero, what is an appropriate range? Where should we start and end the axis?

To illustrate, let's look more closely at changes in health care spending in the United States. Each of the four charts below uses a different range in the vertical axis. As you can plainly see, those ranges affect our perception of the level and the change in spending. In the top-left graph, where the axis ranges from 0 to 20, we see a slight increase in spending. As you move clockwise through the graphs and the axis range gets smaller and smaller, the change in spending looks increasingly dramatic.

There is no right answer to the choice of the vertical axis; the answer depends on the data and your goal. If you need to demonstrate that the economy will falter if spending reaches 17 percent of GDP, then the bottom-right chart may be best. If you are telling a more general story, then one of the graphs in the top row might be preferable, because it still clearly shows the increase in spending over time. If you need to show a detailed examination of spending in each year, you might want to consider the graphs on the right.

Total health care spending in the United States grew from 12.5 percent to 16.8 percent between 2000 and 2015

(Percent of GDP)

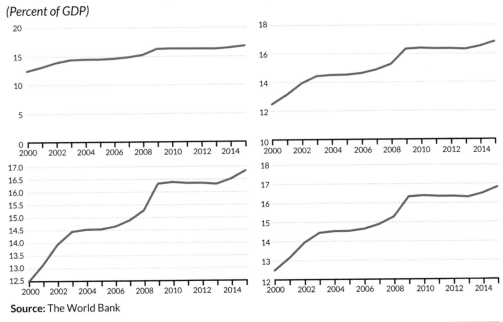

Source: The World Bank

There is no right answer to the choice of the vertical axis. The answer depends on the data and your goal.

Personally, I try to avoid the approach in the bottom-left graph where the axis does not start at zero *and* either the top or bottom value equals the minimum or maximum data value. In this case, the graph feels too "tight" to me and I think it can suggest the data can go no lower or higher than what's pictured, which is rarely the case.

Another way to think about this is how the data, context, content, and units all work together. A change in spending from 12 percent of GDP to 17 percent of GDP is a large change in the context of health care reform. But it's not as important that my kids can beat me seventeen times in a board game now instead of twelve times when they were a bit younger (for me—though it's pretty important for them!).

It's also worth noting that our perception of where zero lies in this space is affected by how we draw the vertical axis. Without looking carefully at a line chart, you are probably inclined to think the bottom of the vertical axis is zero, and in some cases— especially where the data series are both positive and negative—this can be especially important. Take these charts showing the year-to-year *percentage-point change* in health care spending instead of the level of spending as a portion of GDP. In the case on the left, it's not immediately clear that there are any spending declines over this period, because the zero baseline is not clearly delineated. By just darkening that axis line a bit more than the rest as in the graph on the right, it is more evident that there are three years when health care spending as a share of GDP declined year-over-year.

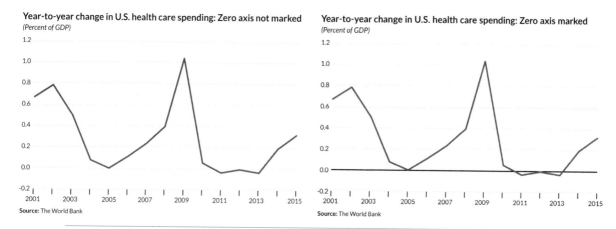

We are inclined to think the bottom of the vertical axis is zero, so using a different color or thickness for the axis makes it clear where zero lies.

BEWARE THE LINE-WIDTH ILLUSION (OR, BE CAREFUL OF THE AREA BETWEEN CURVES)

With line charts (and for other time-series charts, for that matter), we tend to *misestimate* the differences between two curves.

Take this graph from William Playfair, a Scottish engineer and political scientist who is often credited as the founder of graphic methods of statistics. In his chart from 1785, Playfair plots exports and imports (in millions of British pounds) between England and the East Indies from 1700 to 1780. The top line denotes imports, and the bottom line denotes exports. The vertical distance or gap between the two lines shows England's (positive) trade balance with the East Indies. Starting in 1700 on the left, you can see the balance grow over the first thirty years or so. Then, starting around 1730, the gap starts to shrink, reaching its narrowest point around 1755. The trade balance then appears to grow for a time and expands rapidly after around 1770.

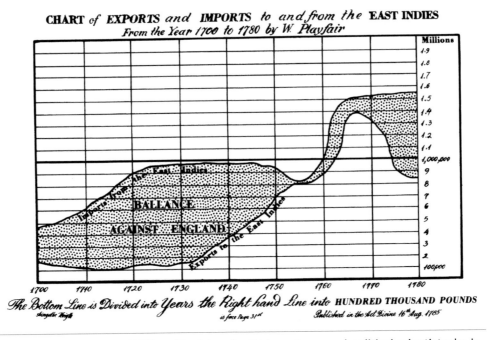

In his chart from 1785, William Playfair, a Scottish engineer and political scientist who is often credited as the founder of graphic methods of statistics plots, shows exports and imports between England and the East Indies.

Did you notice the hump in the trade balance after 1760? In the three-year period between 1762 and 1764, imports rose quickly while exports grew more slowly, creating a larger trade balance. Between 1764 and 1766 exports to the East Indies shoot up and brings the trade balance right back down. But the spike between 1762 and 1764 is hard to see in Playfair's original chart. Those changes are much easier to see in this line chart, which plots the gap between imports and exports.

This is the line-width illusion at work: we tend to assess the distance between curves at the closest point rather than the vertical distance. A variety of scholars have demonstrated this effect and have suggested alternative graph types, but the easiest solution may be to plot the *difference* whenever it's the metric of interest.

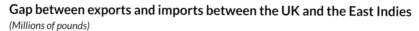

Gap between exports and imports between the UK and the East Indies
(Millions of pounds)

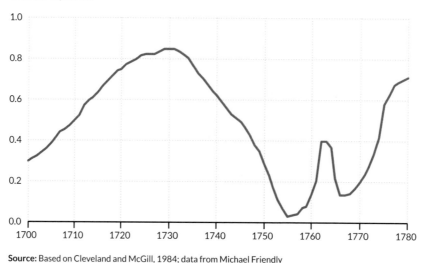

Source: Based on Cleveland and McGill, 1984; data from Michael Friendly

The line-width illusion at work here—the bump in the gap between exports and imports is easier to see here than in Playfair's original.

INCLUDE DATA MARKERS TO MARK SPECIFIC VALUES

Data markers are just what they sound like: symbols along the line to mark specific points in the series. There isn't a right answer to the question of when to deploy them. Personally, I include data markers when I have only few lines or data points, or for specific points I want to label or annotate. The data markers give the graph more visual weight.

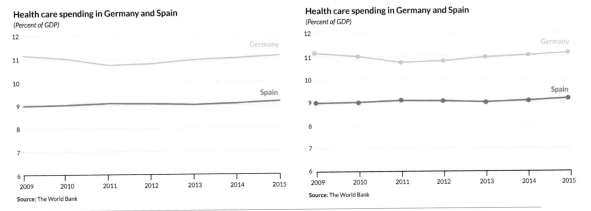

I like to add data markers to my line charts when I don't have a lot of data or when I want to highlight or label specific values.

These charts, for example, show health care spending as a share of GDP for Germany and Spain. There are so few data points and the changes in the series are so subtle that the addition of the circular data markers give the lines more visual heft.

I prefer to make my data markers circles rather than triangles, squares, or other shapes. This is partly an aesthetic preference, but there's also a logic to it. Circles are perfectly symmetrical, and so it never matters where the line intersects the circle. With other shapes, like triangles, the line might intersect the thinner top part or the thicker bottom part.

Other shapes may be necessary if you or your organization are required to comply with certain rules or laws that enable screen readers to differentiate between objects on a screen for people with vision disabilities. In the United States, federal government agencies are required to follow Federal Section 508 regulations that make websites accessible to people with disabilities (see Chapter 12 for more on data visualization accessibility). Even with different colors, most screen readers cannot differentiate between the different series if the shapes are all the same. In these cases, different data markers are a good choice.

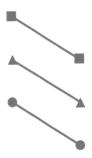

Circles are a symmetrical shape, which is why I prefer to use them as data markers.

USE VISUAL SIGNALS FOR MISSING DATA

At some level, there is *always* missing data. People change jobs every day, not just when unemployment numbers are published. Lots of things happen in the ten years between the publication of each U.S. Census. Most data are a snapshot in time, but we often treat them as continuous.

Missing data are *truly* missing when a regular series is interrupted because the data were not collected. In these cases, we should make it clear that the data are incomplete. In line charts, we can change the format of the line (for instance, with dashes) or not connect the points at all to signal that those data points are missing. We can also place a note on the chart or below the chart to explain that those data values are unavailable.

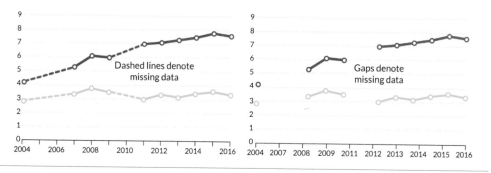

Here are two ways to signal missing data: a dashed line or annotation for the gaps.

What we should never do is ignore the missing values altogether and make it appear as though we have a continuous, uninterrupted series.

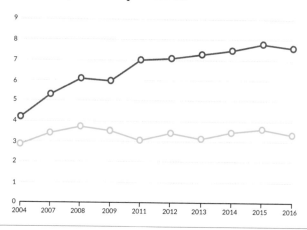

This chart ignores the missing data points and is misleading. It gives the false impression of a continuous, uninterrupted series.

AVOID DUAL-AXIS LINE CHARTS

You might be inclined to add another vertical axis to your line chart when comparing changes in two or more series of different units. Resist that urge. Consider this dual-axis line chart that shows the share of income devoted to paying for housing on the left axis and the quarterly unemployment rate on the right axis in the United States from 2000 to 2018. It's not immediately obvious that the unemployment rate is the blue line and associated with the right axis and housing debt is the yellow line plotted along the left axis. The purpose of this graph is to show that the economic climate for consumers in 2017 and 2018 was quite good—low unemployment rates and low housing debt.

But there are three problems with plotting the data like this.

First, they are often hard to read. Did you intuitively know which lines corresponded to which axis? I didn't. Even if the labels and axes were colored to match the lines (which many dual-axis charts don't include), it's hard to discern patterns in the data. They're extra work for the reader, especially when the labeling is not obvious.

Second, the gridlines may not match up. Notice how the horizontal gridlines in this graph are associated with the left axis, which leaves the numbers on the right axis floating in space. At the crossing point in 2009, it's hard to see that the value of the unemployment rate (the blue line) is just shy of 9 percent.

Third, and most importantly, the point where the lines cross becomes a focal point, even though it may have no real meaning. In this graph, the eye is drawn to the middle of

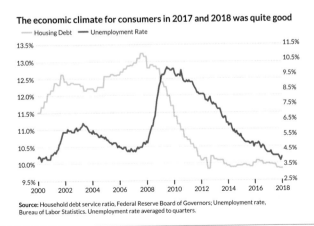

The economic climate for consumers in 2017 and 2018 was quite good

Housing Debt — Unemployment Rate

Source: Household debt service ratio, Federal Reserve Board of Governors; Unemployment rate, Bureau of Labor Statistics. Unemployment rate averaged to quarters.

The dual-axis chart introduces a series of perceptual issues, maybe the most important of which is that the eye is drawn to where the lines intersect, even though that may be meaningless.

the chart where the two lines intersect, because that's where the most interesting thing is happening. But there's nothing special about 2009; it's just a coincidence that they crossed at that time. The intended takeaway of the chart is how much the economic climate has improved since the 2007–2009 recession, but that's not what draws the eye.

The vertical axis in a line chart does not need to start at zero, so this chart—with the left axis starting at 9.5 percent and the right axis starting at 2.5 percent—is a perfectly reasonable way to plot the two series. By that logic, we could arbitrarily change the dimensions of each axis to make the lines cross wherever we like. And this is the problem with dual-axis line charts: the chart creator can deliberately mislead readers about the relationship between the series.

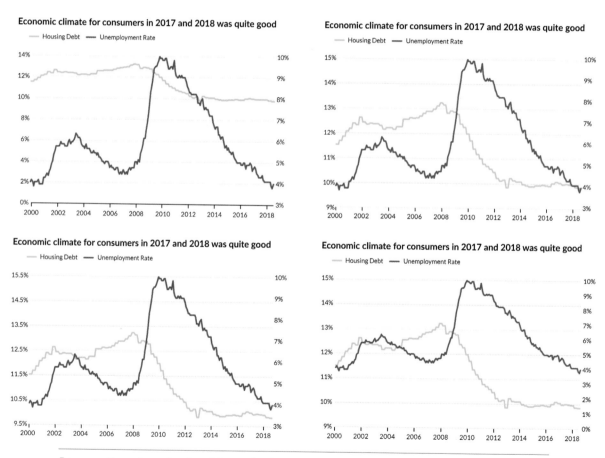

Because there is no hard and fast rule about how to set the dimensions of the vertical axis, we can arbitrarily change the dimensions to make the lines cross wherever we like.

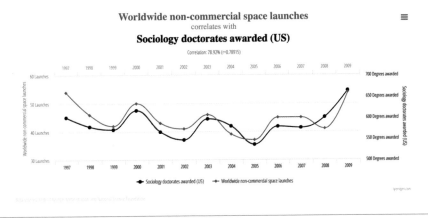

Tyler Vigen's (http://tylervigen.com/spurious-correlations) collection of dual line charts shows how we can imply correlation between seemingly independent data series simply by adjusting the vertical axes.

All of these four graphs are reasonable ways to set the vertical axes, and by manipulating those ranges, I can make the series look like they (a) are closely matched for a few years around 2010 and 2012; (b) cross in the middle and the end; (c) intersect around 2003 and then again a few years later; and (d) are closely related in the first half of the period but then diverge.

By arbitrarily choosing the axis range, we can make different data series look as correlated as we like. On his website, Spurious Correlations, Tyler Vigen shows all kinds of dual-axis charts in which arbitrary vertical axis scales create erroneous—and humorous—correlations.

Similar difficulties, though to a slightly different extent, exist in dual-axis charts that combine different graph types. On the next page you can see the same graph of the unemployment rate and housing debt, now with the unemployment rate plotted as an area chart. The right axis starts at zero so the gridlines on both sides match, but it's still not immediately obvious which variable goes with which axis. And though it is more obvious that there are two separate trends being visualized, the same perceptual pitfalls still exist, leading readers to see correlations that might not really be there.

There are a few solutions to the dual-axis chart challenge.

First, try setting the charts side by side. Remember, not everything needs to be packed in a single graph. We can break things up and use a small multiples approach. Although ideally side-by-side graphs should have the same vertical axis to facilitate easier comparisons, we've already determined that approach is impossible here, so splitting them up and using different axis ranges can work.

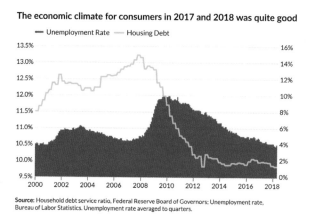

The issue with dual-axis charts is not resolved by combining area charts and line charts.

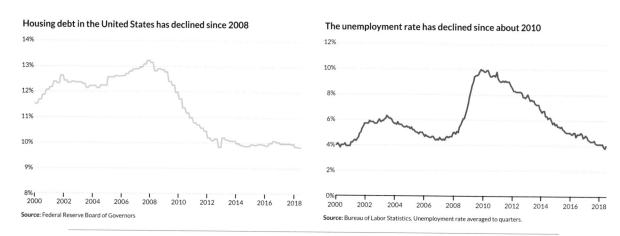

One alternative to the dual axis line chart is to use two, side-by-side charts.

If it's important to annotate a specific point on the horizontal axis, you could also vertically arrange the two and draw a line across both. This will change the rotation of the final graphic, but is an easier way to label a specific value or year.

Second, we might calculate an index or the percent change from some value or year (see page 148). This way the reader can see the change over time for both series and compare them along the same metric. In the data we've been looking at here, we calculate the difference between each year and 2000, the first year of the period (thus the percentage-point change). The obvious trade-off here is that we lose the *level* presentation of the data and instead present the *change*.

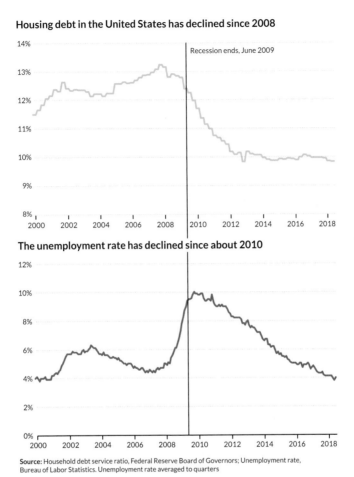

Housing debt in the United States has declined since 2008

Recession ends, June 2009

The unemployment rate has declined since about 2010

Source: Household debt service ratio, Federal Reserve Board of Governors; Unemployment rate, Bureau of Labor Statistics. Unemployment rate averaged to quarters

One alternative to the dual axis line chart is to use two charts aligned vertically, which makes it a little easier to mark a specific data point on both charts.

Third, try a different chart type. If showing the changes in the *associations* between the two series is important, try a connected scatterplot. The connected scatterplot—which has its own section at the end of this chapter—is like a scatterplot with a horizontal and vertical axis, but each point represents a different unit of time, such as a quarter or a year. As you can see on the next page, it's easier to see how the relationship has changed over time between these two metrics. You can also see how I have added more labels and annotation (along with different colors) to help the reader navigate the visual.

The economic climate for consumers in 2017 and 2018 was quite good
(Percent pointchange since 2000)

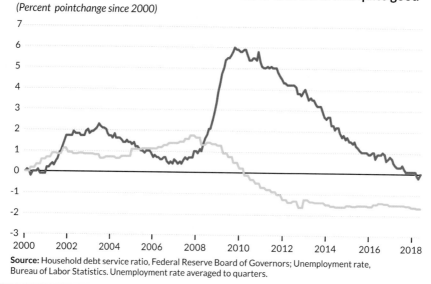

Source: Household debt service ratio, Federal Reserve Board of Governors; Unemployment rate, Bureau of Labor Statistics. Unemployment rate averaged to quarters.

Another alternative to the dual-axis chart is to normalize the data or calculate the percent change from some value.

The U.S. economy appears supportive of the consumer with low-unemployment rate and housing debt
(Household debt service ratio)

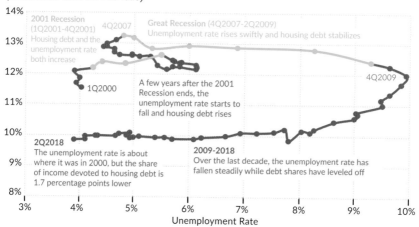

Source: Household debt service ratio, Federal Reserve Board of Governors; Unemployment rate, Bureau of Labor Statistics. Unemployment rate averaged to quarters.

Yet another alternative to the dual-axis chart is to use the connected scatterplot in which one data series corresponds to the horizontal axis and another to the vertical axis.

A possible exception to the "no dual axis chart" rule is if you are showing a translation of a single measure, for example Fahrenheit and Celsius temperatures. In these cases, we are not trying to track two different variables but showing how one maps directly onto another. In those cases, the usual pitfalls don't apply.

CIRCULAR LINE CHART

The radial bar chart and circular bar chart in Chapter 4 showed us how to take a bar chart and wrap it in a circle. The same can be done with lines showing changes over time. As before, using a circle may be less perceptually accurate for the reader, but it can be used, for example to improve a visual metaphor.

These two graphs show the percent of hospital emergency room visits for the flu in the United States for each week of the year from 2014 to 2017. Starting at the beginning of the flu season in October, the line chart on the left gives us the standard view: an increase in the flu during the winter months, which fades as we enter summer. The radial chart on the right shows the same data but with a different perspective—the "lean" toward three o'clock on the chart when more infections occur during the fall and winter, and fewer infections during the summer months on the left side of the circle. The radial chart is more compact than the standard line chart, but it is also harder to make precise comparisons because the lines do not sit on a single horizontal axis.

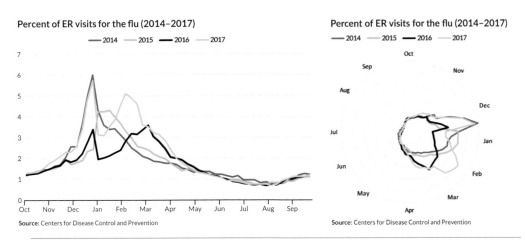

Two ways of showing the same time series data—as a standard line chart or by wrapping the lines around a circle.

SLOPE CHART

In some cases, it may not be necessary to show *all* of the data in your time series. In these cases, a slope chart—which is really just a simplified line chart—is a useful alternative.

The paired bar chart is a standard way to visualize two data points for multiple observations (also see page 84). As an example, consider these charts of changes in the unemployment rate for six states in the United States between 2000 and 2018. With this kind of visualization, we ask our reader to process the level *and* change in the unemployment rate *between and across* the six states. There is a lot of ink in the graph, and it asks the reader to do lots of mental math. We could, of course, just plot the change between the two time periods, but we often want to show both the level and the change.

The slope chart addresses this challenge by plotting each data point on a separate vertical axis and connecting the two with a line. In this example, the left vertical axis represents the first month of data (January 2000) and the right vertical axis represents the last month of data (January 2018). We can easily see the relative values of each data point. Here, for example, we can see—perhaps even more easily than in the paired bar chart—that Montana had the highest unemployment rate in the first month and Connecticut had the lowest. The line that connects the two data points visualizes the change over time. We can more easily see that the unemployment rate in Montana, Hawaii, and Idaho fell between 2000 and 2018 while it rose in the other three states.

There are many ways to style the slope chart. We can use two colors to denote increases and decreases. We can include or exclude labels for levels and changes. We can even adjust the

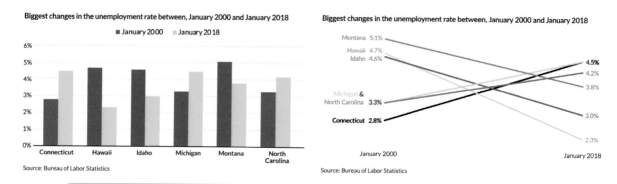

The paired bar chart asks the reader to make several comparisons on their own simultaneously, while the slope chart visualizes these comparisons for the reader.

thickness of the line to correspond to a third variable. We could also rely on the *Start with Gray* strategy and add more data to the basic slope chart. Here, I've included every state in the nation but highlighted and thickened the six states of interest and the national average.

In graphs like these, consider whether a taller chart will make it easier for the reader to see all of the detailed colors, labels, and annotations. As with the dot plot (see page 97), be careful about using the slope chart when a summary of the time series may mask changes in the intervening years. Of course, this is the same consideration when using the paired bar chart.

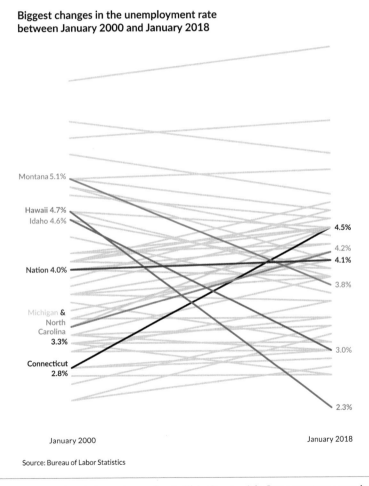

Biggest changes in the unemployment rate between January 2000 and January 2018

Montana 5.1%
Hawaii 4.7%
Idaho 4.6%
4.5%
4.2%
4.1%
Nation 4.0%
3.8%
Michigan & North Carolina 3.3%
3.0%
Connecticut 2.8%
2.3%

January 2000   January 2018

Source: Bureau of Labor Statistics

There are many ways to style a slope chart. This *Start with Gray* strategy can be especially useful here to show many observations while highlighting only a few.

SPARKLINES

There is a specific style of small multiples for line charts called sparklines. Invented by author and statistician Edward Tufte, sparklines are "small intense, simple, word-sized graphics with typographic resolution." They are typically used in data-rich tables and may appear at the end of a row or column. The purpose of sparklines is not necessarily to help the reader find *specific* values but instead to track *general* patterns and trends.

Let's use sparklines with the health spending data. The numbers in the two table columns show spending in 2000 and 2015, while the sparklines show the values for the entire sixteen-year period. This way, readers can see some specific values as well as the patterns over the entire period. Here, for example, you can quickly see that health spending rose for all of these countries except for Turkey, which I've also highlighted so it stands out.

Health care spending in selected countries

Country	2000	2015	2000-2015
Australia	7.6	9.4	
Canada	8.3	10.4	
Finland	6.8	9.4	
Japan	7.2	10.9	
Switzerland	9.3	12.1	
Turkey	4.6	4.1	
United Kingdom	6.0	9.9	
United States	12.5	16.8	

Source: The World Bank

Sparklines are a type of small multiples line chart typically used within data-rich tables.

BUMP CHART

A variation on the line chart is the bump chart, which is used for plotting changes in *ranks* over time, for example, political polling or positions in a golf tournament from hole to hole. When we want to show relative ranks rather than absolute values, the bump chart is a good choice.

A bump chart is, of course, a compromise. It does not show the raw values, which are often preferred, but it can be especially useful if your data have outliers. By plotting the ranks, we abstract from the large differences in magnitude.

These two bump charts show changes in health care spending across the ten countries that have the highest spending on health care as a share of their GDP in 2015. Those countries appear in the far right position of the horizontal axis, above the 2015 label. The difference between the two charts is that that the one on the left shows the patterns and ranks for only these ten countries for every year. You can see some gaps in certain years where other countries would appear in the rankings, but this chart only tracks those countries that end up in the top ten in 2015. The chart on the right, by comparison, includes every country in every year, which requires more labeling so the reader can understand why new countries (with different colors) suddenly appear in the chart. We could emphasize certain countries by changing line colors or the colors of the inside of the data marker circles, or even the thickness of the lines.

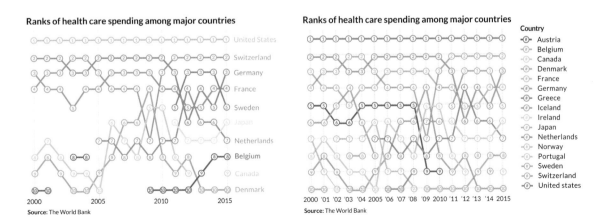

Bump charts plot changes in ranks over time.

Total health care spending in the United States and Germany increased between 2000 and 2015

(Percent of GDP)

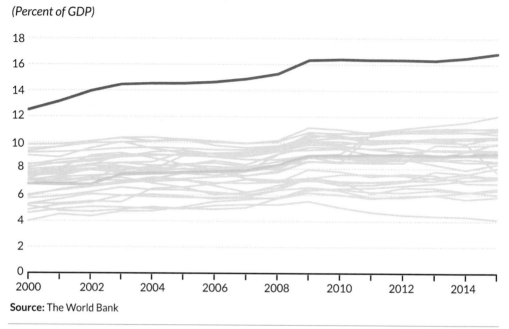

Source: The World Bank

Use color, data markers, or line thickness to highlight specific data series.

Compare these bump charts to the line chart above that shows all of the OECD countries in gray and highlights the United States and Germany with color. In this case, the United States stands far above the rest of the countries that here appear clumped together in a swirl of lines. As is often the case, there is a tradeoff between the bump chart and the line chart: In the standard line chart, you can see the relative differences between the series, but they're stacked together and hard to disentangle. In the bump chart, by comparison, we can't see the relative differences, but we can see relative ranks.

A modification on the bump chart is a *ribbon effect*. Here, in addition to rank, the widths of the ribbons are scaled according to the actual data values. Like the streamgraph, which we'll see later, this approach has a more organic, flowing look. This chart from the *Berliner Morgenpost* shows the rank, amount, and change in different sentiments around political problems in Germany.

Das sind die 15 wichtigsten politischen Probleme in Deutschland

Die Grafik zeigt, welche Themen die Deutschen bei dieser Bundestagswahl am meisten bewegen, und welche Bedeutung sie bei vergangenen Wahlen hatten.

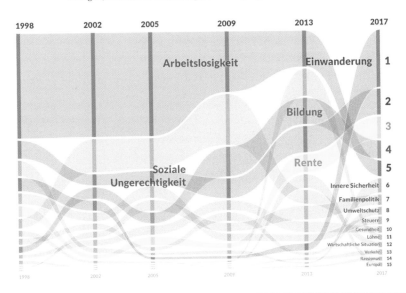

The ribbon effect is a modification on the standard bump chart. The title in this chart from the Berliner Morgenpost translates to "These are the 15 most important political problems in Germany". Arbeitslosigkeit translates to Unemployment; Einwanderung to Immigration; Bildung to education; and so on. It shows the changes in each ranks (and amounts) of these different sentiments.

CYCLE CHART

Cycle graphs typically compare small units of time, such as weeks or months, across a multiyear time frame. They are most commonly used to display strong seasonal trends. Here, we see the number of births in the United States in each month from 2007 to 2017. A yellow line marks each month's average value (a general but not necessary characteristic of the cycle chart). We can see the downward trend for births over the decade for every month and the higher birth rate during the summer months, July, August, and September. I've added a dot at the end of each line to mark the most recent year.

By comparison, this same data displayed as a standard line graph is less clear. We can see a spike in each year, but without more labeling, it's not clear in which month that spike occurs. Even though the cycle chart has more information on it—the average values shown in yellow and the point at the end of each line—it still feels less busy than the standard line chart.

A cycle graph can also split up a dense bar or line chart to give each series more space—something like a small multiples chart. Take this column chart of the unemployment rate for four groups in the United States. This kind of graph—with multiple years for different

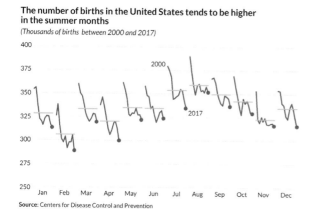

The number of births in the United States tends to be higher in the summer months
(Thousands of births between 2000 and 2017)

Source: Centers for Disease Control and Prevention

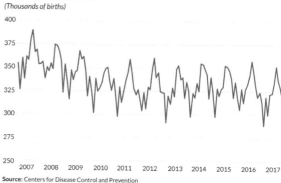

The number of births in the United States tends to be higher in the summer months
(Thousands of births)

Source: Centers for Disease Control and Prevention

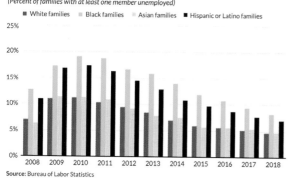

There are fewer families with at least one person unemployed than in the past
(Percent of families with at least one member unemployed)

■ White families   ■ Black families   ▨ Asian families   ■ Hispanic or Latino families

Source: Bureau of Labor Statistics

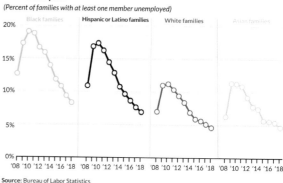

There are fewer families with at least one person unemployed than in the past
(Percent of families with at least one member unemployed)

Black families   Hispanic or Latino families   White families   Asian families

Source: Bureau of Labor Statistics

Cycle graphs compare small units of time, such as weeks or months, across a multiyear time frame.

groups pushed together—can be difficult for the reader to make comparisons within or across years. The cycle graph on the right separates each racial group into its own panel, sorted by the value in the most recent year. You could argue that the graph on the right is a small multiples line graph, but the organization and design make it more like a cycle graph.

AREA CHART

Area charts are line graphs with the area below the line filled in, giving the series more visual weight. The area chart on the left and the line chart on the right both show the number of people who died from prescription opioid overdoses in the United States between 2000 and 2016. You might think of the area chart as a bar chart where the bars have infinitely thin widths and thus, as we saw in the previous chapter, the vertical axis should always start at zero.

Placing two or more series in an area chart can be difficult because one series can hide (or "occlude") the other, an effect we will see more in the coming chapters. On the next page, the area chart on the left, for example, shows overdose deaths from cocaine and heroin, but the data series for heroin is hidden behind the series for cocaine. Even if the order of the data series are changed so that heroin overdose deaths are in front, now the heroin-deaths series blocks the cocaine-deaths series. Compounding that difficulty is that some readers might

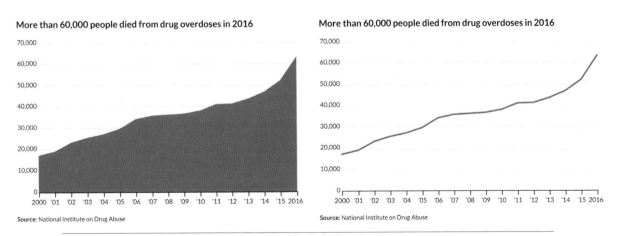

Area charts are line graphs with the area below the line filled in, giving the series more visual weight.

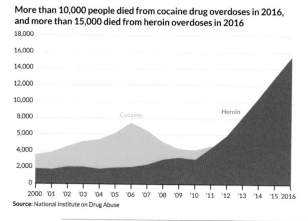

More than 10,000 people died from cocaine drug overdoses in 2016, and more than 15,000 died from heroin overdoses in 2016

Source: National Institute on Drug Abuse

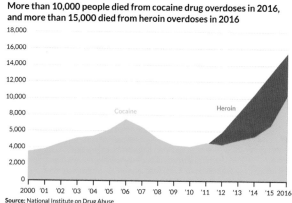

More than 10,000 people died from cocaine drug overdoses in 2016, and more than 15,000 died from heroin overdoses in 2016

Source: National Institute on Drug Abuse

Placing two or more series in an area chart can spell trouble, because one series can hide (or "occlude") the other.

mistake the two as summing to a total rather than as separate data series. Here, it's important to use the title and annotation to make it clear there are two distinct series.

One strategy to address this overlap is to add a transparency to the color of one (or both) series. But be careful: by adding a transparency to only one series, we deemphasize its importance. Another alternative is to use a line chart, as in the graph on the right.

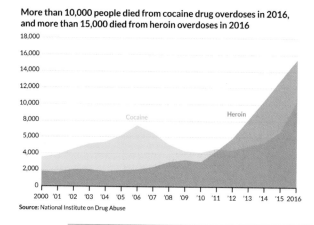

More than 10,000 people died from cocaine drug overdoses in 2016, and more than 15,000 died from heroin overdoses in 2016

Source: National Institute on Drug Abuse

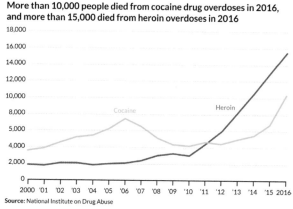

More than 10,000 people died from cocaine drug overdoses in 2016, and more than 15,000 died from heroin overdoses in 2016

Source: National Institute on Drug Abuse

One strategy to address the overlap between series on an area chart is to add a transparency to the color of one (or both) series. Another alternative is to use a line chart, as in the graph on the right.

STACKED AREA CHART

Stacked area charts build on the typical area chart by showing multiple data series simultaneously. Instead of sitting independently of one another as in the previous chart, the data in a stacked area chart sum to a total or a percentage.

The stacked area chart on the left shows the total number of drug overdose deaths between 1999 and 2016. The version on the right shows the same data, but presented as percentages that sum to 100 percent.

The reader will take away different conclusions from these two representations. In the graph on the left, the eye is drawn to the large increase in overall deaths over the period. In the version on the right, it is drawn to the changes in the distribution of deaths—a decline in overdoses from cocaine, but an increase in heroin, Benzodiazepines (drugs that are often used to treat anxiety, insomnia, and seizure disorders), and other drugs.

There are three disadvantages with the stacked area chart on the left. First, as earlier, we again see the line-width illusion—we tend to view steep changes as bigger than they actually are. Second, only the bottom series sits on a horizontal axis, so it is hard to accurately compare the changes over time for the other series. (Remember, this is the second row in the perceptual ranking table showed earlier.) Third, the ordering of the data series can affect

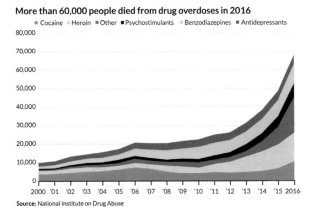

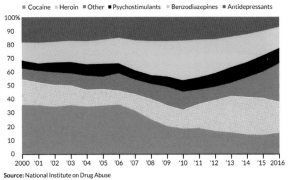

Stacked area charts build on the typical area chart by showing multiple data series simultaneously and sum to a total, often 100 percent.

More than 60,000 people died from drug overdoses in 2016

■ Cocaine ▪ Heroin ■ Other ■ Psychostimulants ▪ Benzodiazepines ■ Antidepressants

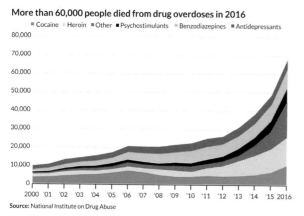

Source: National Institute on Drug Abuse

More than 60,000 people died from drug overdoses in 2016

■ Cocaine ▪ Heroin ■ Other ■ Psychostimulants ▪ Benzodiazepines ■ Antidepressants

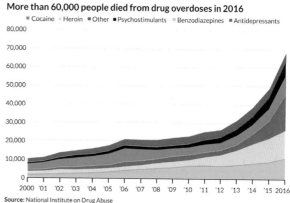

Source: National Institute on Drug Abuse

> Recall that the perceptual ranking list suggests we can better compare values when they sit on the same axis. That's why it's easiest to compare values for the bottom series.

our perception of the shares of the total and move the reader's attention around from one series to another.

To demonstrate, consider the two stacked area charts above. The version on the left is the same as before while the version on the right changes the order. Notice how in the new version, it is easier to compare changes in overdose deaths caused by Benzodiazepines (the yellow series), because the series sits along the same horizontal axis.

The share of people who died from overdoses from cocaine has declined since 2000

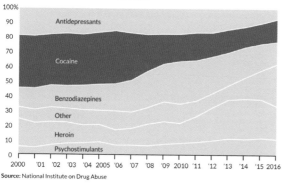

Source: National Institute on Drug Abuse

The share of people who died from overdoses from cocaine has declined since 2000

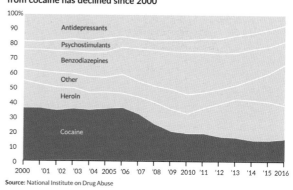

Source: National Institute on Drug Abuse

> There isn't necessarily a right way to stack the series in a stacked area chart, but how you decide to arrange them will influence how your reader perceives the data.

This isn't to say there is a "right" way to stack the data in an area chart, or that the most important data series should sit along the horizontal axis. If, say, you were telling a story about the declining share of overdose cocaine deaths over this period, you could keep it in the same position as above, but use the "start with gray" strategy and use color in just the cocaine series. Even with the line-width illusion, you can still see the share of deaths has declined. If it's important for your reader to see the exact change in the share, then putting that series along the horizontal axis is a better strategy. By placing the series in the middle of the chart, you can't compare the values to the horizontal baseline and thus less accurately perceive the values. (Also notice how I directly labeled the segments instead of using a legend so the reader can quickly and easily identify the different series.)

A small multiples approach (here, six different graphs) more clearly shows the exact patterns in each category, but does not as clearly show the relative shares of each. You make a couple of tradeoffs here. On the one hand, the stacked chart is more compact than the small multiples—you pack all of the information into a single visualization in which you can see the changes in shares. On the other hand, the small multiples gives you a more perceptually accurate view—because each series sits on its own horizontal baseline—but it is harder to compare across the different categories.

More than 60,000 people died from drug overdoses in 2016

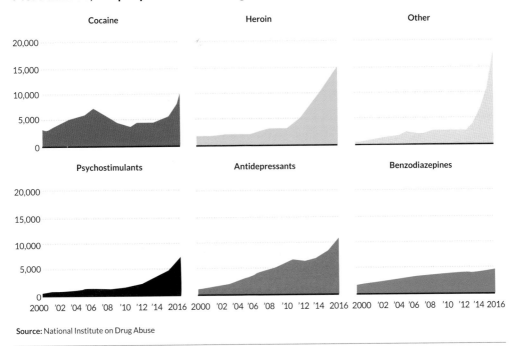

Source: National Institute on Drug Abuse

The small multiples approach more clearly shows the exact patterns in each series, but does not as clearly show the relative values of each.

Causes of death by age in the United States in 2017

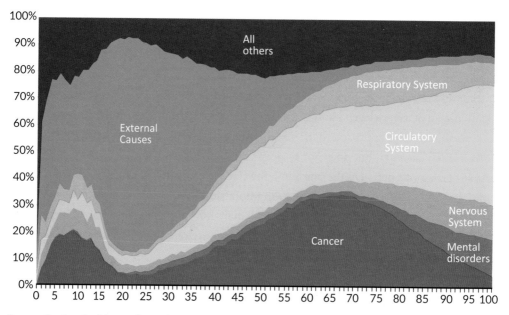

Source: Centers for Disease Control and Prevention

Stacked area charts can also show changes in the distribution of a data series.

Finally, the stacked area chart can also show changes in the distribution of some data series. This stacked area chart, for example, shows all of the different ways people from age zero to one hundred died in the United States in 2017. Instead of years or months along the horizontal axis, this graph shows the number of deaths for each single year of age, a different measure of time. Categorized into fifteen groups, most people who die around age 25 do so of "external causes," (the green series) such as falling or drowning, while many people who die around age sixty die of some form of cancer (blue). As before, we could modify the colors or the arrangement of the data to focus our readers' attention on specific patterns or trends.

STREAMGRAPH

Like the stacked area chart, a streamgraph also stacks the data series, but the central horizontal axis does not necessarily signal a zero value. Instead, data can be positive on both sides of the axis. Together, the streamgraph illustrates fluctuations in data over time in a flowing, organic shape. They are therefore best used for time series data when the series themselves have high volatility.

Streamgraphs are well suited for showing patterns that have peaks and troughs. Both the stacked area chart on the left and the streamgraph on the right show the *total number of deaths* rather than shares of deaths, as in the earlier stacked area chart. The streamgraph gives us a slightly different view of the data and may point us more toward overall increases rather than changes in specific series. The idea behind the streamgraph is to minimize the distortion in each layer's baseline that accumulates more rapidly with a stacked area chart.

Researchers are aware of how unusual the streamgraph looks, and how it may be more difficult for readers to understand. In a review of a streamgraph published by the *New York Times* in 2008, researchers noted that they "suspect that some of the aesthetically pleasing—or at least engaging—qualities may be in conflict with the need for legibility. The fact that the *New York Times* graph does not look like a standard statistical graphic may well be part of its appeal." Thus, while this kind of graph—or any different-looking graph—may at first confuse or confound readers, they may ultimately find the shapes, colors, and other attributes more interesting and engaging. It all depends on your audience.

On the next page, you can see a more recent example of a streamgraph. This visualization was published by the *Hindustan Times* in 2016 and shows the number and type of the highest civilian awards the Indian government confers. Additional streamgraphs in the original news story showed breakdowns by state, nationality, gender, and discipline.

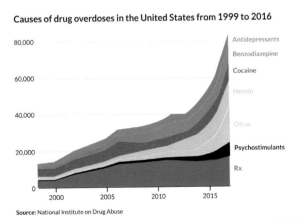

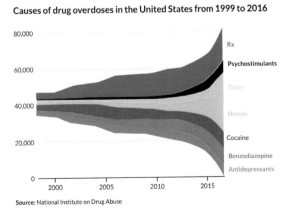

Streamgraphs are a variation on the area chart and are well suited for showing patterns that have peaks and troughs.

Delhiites have received the most awards, followed by Maharashtrians

State
- Delhi
- Maharashtra
- Tamil Nadu
- Uttar Pradesh
- West Bengal
- Karnataka
- Telangana
- Other

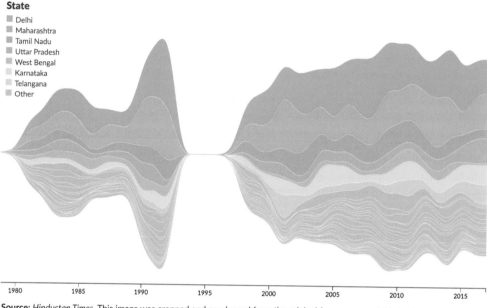

Source: *Hindustan Times*. This image was cropped and condensed from the original for purposes of this book.

This streamgraph from the *Hindustan Times* shows patterns in the number and type of the highest civilian awards conferred by the Indian government.

HORIZON CHART

A horizon chart is an area chart that is sliced into equal horizontal intervals and collapsed down into single bands, which makes the graph more compact and similar to a heatmap (page 112). The horizon chart is split into bands with positive numbers collapsed down and negative values flipped above the horizontal axis. Multiple horizon charts—which is how they are typically arranged—can condense a dense dataset into a single visualization. Horizon charts are especially useful when you are visualizing time series data that are so close in value so that the data marks in, for example, a line chart, would lie atop each other. Aligning the charts in this way allows us to include our data in a more compact space than in a series of area charts.

Color is the most important attribute in a horizon chart. Darker colors represent larger values and lighter colors smaller values. Like sparklines and to some extent heatmaps, the purpose of the horizon chart is not necessarily to enable readers to pick out specific values, but instead to easily spot general trends and identify extreme values.

This horizon chart uses the same data we've been using on changes in the percent of GDP spent on public health care. An area chart is built for each country, split and collapsed, and then arranged all together in rows. Notice how much data are packed into the single visualization (ten countries and fifteen years), and—recalling the importance of preattentive processing—see how your eye is drawn to the brighter and darker colors.

I'll use the series for Sweden to show how the horizon chart is built. The change in public health spending for Sweden (the third country from the bottom in this horizon chart) is shown on the next page as an area chart and sliced into equal increments (every 0.5 percentage points). Larger values have darker shades, and negative and positive values have different colors. The negatives are flipped above the horizontal axis and all are collapsed down to the first interval or band.

Color is the key here. The same visualization as a series of line charts does not have the same punch. The eye scans the entire visualization for important trends, but no particular region of the visual draws our attention. That could be modified by adding color to the

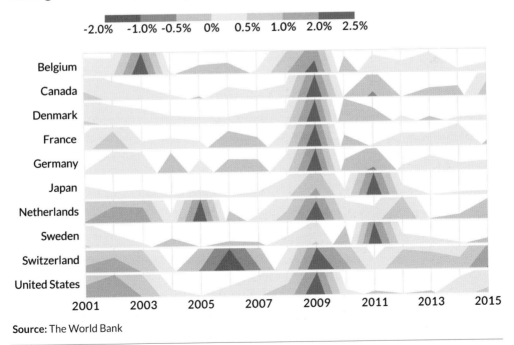

Change in health spending as a % of GDP

-2.0%   -1.0% -0.5%   0%   0.5%   1.0%   2.0%   2.5%

Source: The World Bank

The horizon chart is an area chart divided into equal intervals and collapsed into a band.

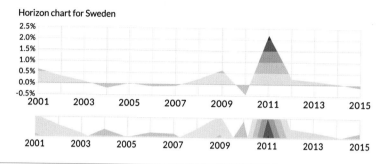

Horizon chart for Sweden

These charts show how the area chart for Sweden is divided and collapsed
to create the horizon chart.

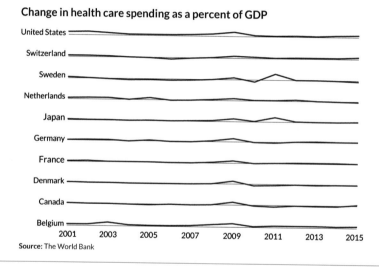

Change in health care spending as a percent of GDP

Source: The World Bank

A line chart is sufficient, but the use of color in the horizon chart attracts
and directs the reader's attention.

different lines to highlight extreme values, but the horizon chart does a much better job of
directing the eye by highlighting values through colors.

GANTT CHART

Another way to show changes over time is to use horizontal lines or bars to show the *duration* of different values or actions. Gantt charts are often used as schedule-tracking devices, for example, to track different phases of a project or budget. Invented by Henry Laurence

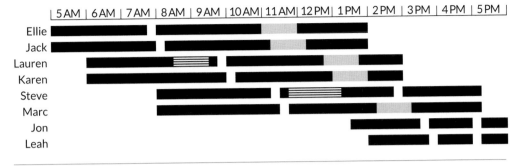

Gantt charts often show processes or schedules.

Gantt, an engineer working around the turn of the twentieth century, the charts were first used by production foremen and supervisors to track production schedules.

This Gantt chart shows staffing shifts at a coffee shop over the course of a day, denoting breaks with the white gaps, lunch breaks with the gray breaks, and other time away from the store with stripes.

Gantt charts can be extended by modifying the width of the bars to denote another variable. For example, this Gantt chart modifies the hypothetical chart above to scale the widths according to the pay of each employee.

Joseph Priestley, an eighteenth-century philosopher, chemist, and educator, published *A Chart of Biography* in 1765, showing the lifespans of approximately two thousand statesmen, poets, artists, and other notables who lived between 1200 BC and the mid-1700s. Often called a timeline, Priestley's chart looks more like a Gantt chart because of the use of horizontal bars/lines and the concrete beginnings (births) and ends (deaths).

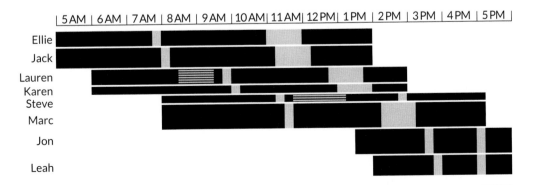

An extension to the Gantt chart is to adjust the widths of the bars to correspond to another data series.

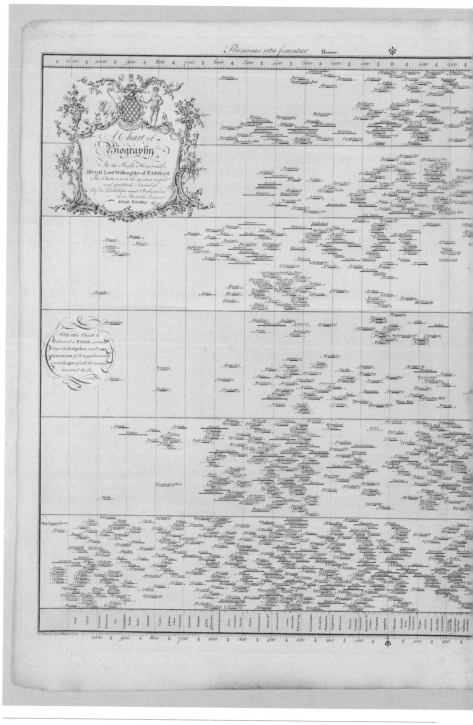

Joseph Priestley's *A Chart of Biography* (1765) shows the lifespans of about two thousand statesmen, poets, artists, and others. It covers an enormous time period,

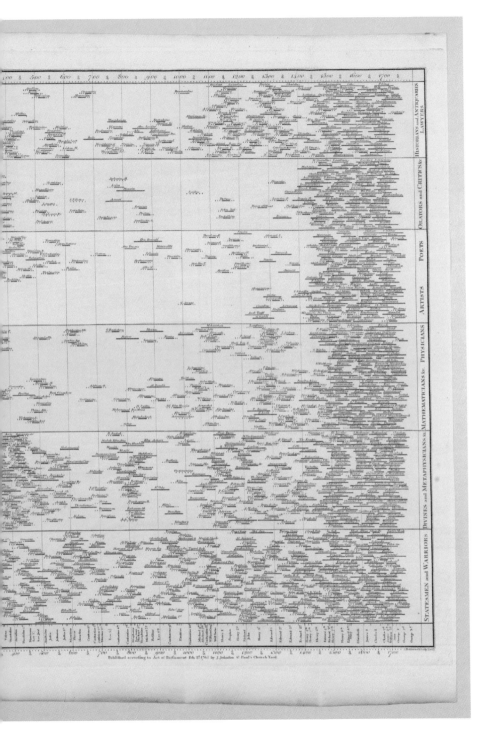

from 1200 BC to 1800 AD. A horizontal line for each person shows his or her life-span. Dots indicate wherever such there is uncertainty around those dates.

Source: Library Company of Philadelphia.

FLOW CHARTS AND TIMELINES

Flow charts and timelines are two examples of an array of visuals that can show changes over time or different kinds of processes, sequences, or hierarchies. This class of charts and diagrams can be explicitly tied to data or can be less quantitative and more illustrative, a way

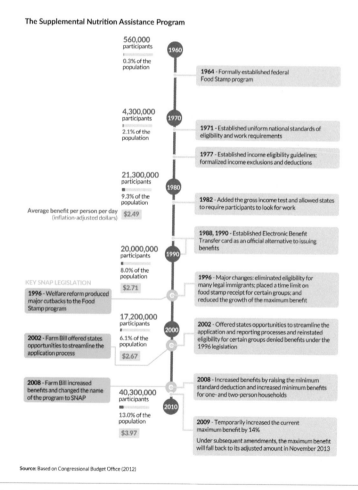

The Supplemental Nutrition Assistance Program

560,000 participants
0.3% of the population
1960

1964 - Formally established federal Food Stamp program

4,300,000 participants
2.1% of the population
1970

1971 - Established uniform national standards of eligibility and work requirements

1977 - Established income eligibility guidelines; formalized income exclusions and deductions

21,300,000 participants
9.3% of the population
1980

Average benefit per person per day (inflation-adjusted dollars) $2.49

1982 - Added the gross income test and allowed states to require participants to look for work

1988, 1990 - Established Electronic Benefit Transfer card as an official alternative to issuing benefits

20,000,000 participants
8.0% of the population
1990

$2.71

KEY SNAP LEGISLATION
1996 - Welfare reform produced major cutbacks to the Food Stamp program

1996 - Major changes: eliminated eligibility for many legal immigrants; placed a time limit on food stamp receipt for certain groups; and reduced the growth of the maximum benefit

17,200,000 participants
6.1% of the population
2000

2002 - Farm Bill offered states opportunities to streamline the application process

$2.67

2002 - Offered states opportunities to streamline the application and reporting processes and reinstated eligibility for certain groups denied benefits under the 1996 legislation

2008 - Farm Bill increased benefits and changed the name of the program to SNAP

40,300,000 participants
13.0% of the population
2010

$3.97

2008 - Increased benefits by raising the minimum standard deduction and increased minimum benefits for one- and two-person households

2009 - Temporarily increased the current maximum benefit by 14%

Under subsequent amendments, the maximum benefit will fall back to its adjusted amount in November 2013

Source: Based on Congressional Budget Office (2012)

This timeline, based on work I conducted at the Congressional Budget Office, shows major milestones and data for the Supplemental Nutrition Assistance Program (SNAP, formerly known as food stamps).

to demonstrate different structures or processes. In PowerPoint, for example, you can look through the "SmartArt" menu for a wide selection of layouts.

A timeline shows when certain events take place. It can be basic and flag events with a line, icon, or marker, or it can be more involved and include annotation, images, or even graphs. Though horizontal timelines are common, timelines can also be vertical or even a variety of different shapes. This timeline, based on work I conducted at the Congressional Budget Office, shows major milestones and data for Supplemental Nutrition Assistance Program (SNAP, formerly known as food stamps). The text in the gray boxes on the right gives details on specific legislation or program changes, and the information on the left presents changes in spending, number of program participants, and their share of the total population.

Flow charts are slightly different. They are not necessarily tied to time in the sense of days, months, and years, but instead they map a process, often step by step. Flow charts make it easier for readers to understand the paths of a process rather than reading through long passages of text or navigating a convoluted table. The flow chart on the next page shows the process by which people can apply and receive benefits through the U.S. Social Security Disability Insurance program (DI). Applicants start the program at what is called the "Disability Determination Services" stage. If their application is approved, they are "Allowed" onto the program; if not, they can either appeal the decision or exit the process altogether. The program is designed in such a way so applicants may appeal a denied request at each stage.

The shapes in a flow chart may carry different meanings, so we can use them strategically to denote different attributes of the system. For example, in a flow chart with rectangles, other shapes can denote choke points or decision points, and rounded rectangles might signal the beginning or end of a process. Adding different colors can help readers understand and differentiate the parts of the graph from each other.

If, for example, we wanted to highlight the different parts of the DI application system, we could use different colors and shapes, as in the version in the middle. Labels in a flow chart can sit alongside the lines and inside the boxes, but they should be large enough and have enough color contrast to be easily read. We could take this even further and scale aspects of the flow chart according to some data values; in the version on the right, for example, the branches are all scaled according to the respective shares at each stage, similar to a Sankey diagram (page 126).

I place the flow chart in the time chapter because these processes often occur over time, one by one. But that's not always the case. An *organizational chart* or *org chart*, for example,

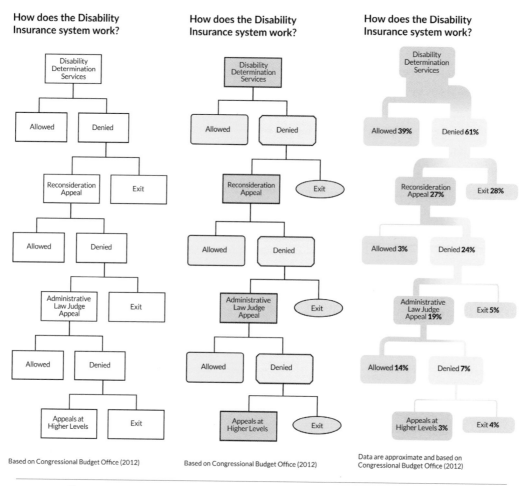

How does the Disability Insurance system work?

How does the Disability Insurance system work?

How does the Disability Insurance system work?

Based on Congressional Budget Office (2012)

Based on Congressional Budget Office (2012)

Data are approximate and based on Congressional Budget Office (2012)

Shapes color, and other elements can help readers understand the paths of a process in a flow chart or timeline. This chart is based on work from the Congressional Budget Office.

is a type of flow chart that shows the hierarchy or management structure of an organization, and how work flows from the top down. We'll see more examples of an org chart in Chapter 8.

Just as we saw with the line chart with many lines, it's not about the *amount* of content you place in the graph, but what meets the needs of your reader. The flow chart on page 174, for example, was prepared in 2010 by the Republican staff of the Joint Economic Committee in response to President Obama's Affordable Care Act proposal. The implicit purpose of this flow chart is to show how complex the proposal (and health care in general) is in the United States. In that sense, the chart does its job!

In these types of graphs, the amount of notes, text, icons, and other visual elements should, as always, meet the needs of the reader. Are specific details necessary at each point? Would an image anchor the moment in the reader's mind? Consider what information your reader needs most and provide it as engagingly as you can.

TOTALS VS. PER CAPITA

Totals can tell you a lot about a group, but they can also mislead. Take for example gross domestic product (GDP), a measure that shows up a lot in this book. In 2017, India and the United Kingdom had roughly the same total GDP at around $2.6 trillion. But their populations—and by consequence their *per capita* (or per person) GDP—are quite different.

In that same year, India's population was 1.3 billion people, more than *twenty times* that of the UK, which had a population of only 66 million. Thus, per capita GDP—total GDP divided by population—was $39,720 in the UK and $1,940 in India. If you treated GDP like a box of cash and gave out equal shares to everyone, each person in the UK would get roughly $38,000 more than each person in India.

These adjustments extend to what we call "normalizing" or "standardizing" metrics. We use this all the time when we drive our cars, for example—we drive sixty miles *per hour* and the price of gas is $2.75 *per gallon*. We can see this in all sorts of other areas and metrics, like mortality rates (deaths per 100,000 population) and wages (dollars per hour).

If you are working with totals in your data, consider whether per capita amounts or other adjustments may be a better and more informative measure. Knowing that India and the UK have similar total GDP doesn't tell you as much about their economies or the relative wealth of people living in those countries as does the per capita measure.

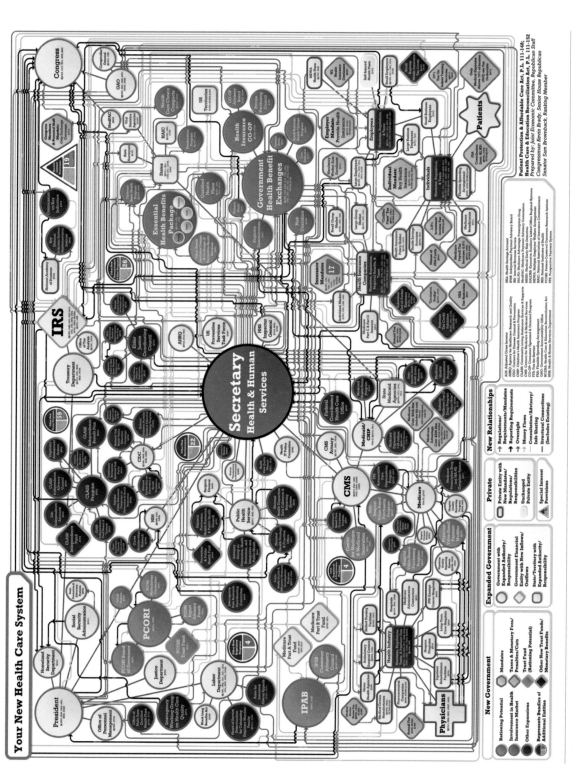

The implicit purpose of this flow chart from the United States Congress Joint Economic Committee (2010) is to demonstrate the complexity of the Affordable Care Act proposal.

CONNECTED SCATTERPLOT

Imagine showing two line charts side-by-side. You may be asking your reader to examine the relationship between the two. Do they move together? Do they diverge or converge? How are they related?

One way to bring two time series together *without* using a dual-axis chart is the connected scatterplot. A connected scatterplot shows two time series simultaneously—one each along horizontal and vertical axes—and are connected by a line to show relationships of the points over time.

As an example, the line chart on the left shows life expectancy in South Africa from 1996 to 2016. Over that twenty-year period, life expectancy followed a U-shape pattern, first falling from about sixty-three years old to fifty-three years old, and then increasing over the next decade, reaching about sixty-one years old in 2016. On the right, per capita GDP is plotted over that same period—that series was flat in the first few years, then increased until about 2008, when it dipped slightly before rising at a slightly slower rate.

With these two charts, we can make some basic visual comparisons—even as life expectancy fell, economic growth continued. When life expectancy started growing again, economic growth flattened out.

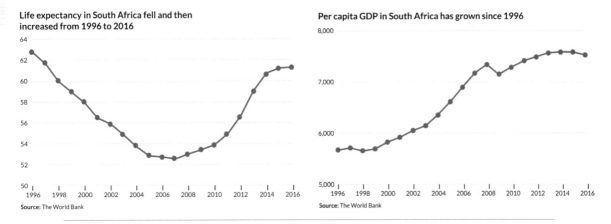

A common challenge is how to clearly show the association between two time series.

Now, notice what happens when the two lines are combined and plotted in a single graph. Here, life expectancy is shown along the horizontal axis and economic growth is plotted along the vertical axis.

Instead of jumping back and forth between the two charts, we can see that during the first half of the period life expectancy fell (moved to the left along the horizontal axis) and the economy grew (moved up along the vertical axis). Starting around 2006, life expectancy started to increase again (now moving to the right along the horizontal axis) while the economy grew, but at a slower rate (the slope of the line along this latter period is flatter than before).

Because this graph is different than standard graph types, your reader will need more time to understand how to read it. Annotation can help: Consider adding more axis labels, arrows and a label for the year on the first and last point. But when your reader understands this graph, it becomes part of their graphic toolbox, just like it has now become part of yours.

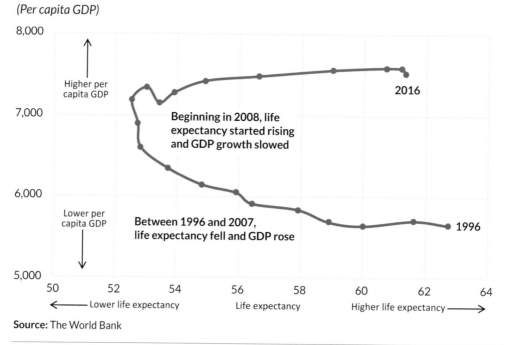

Life expectancy has turned around in South Africa
(Per capita GDP)

Source: The World Bank

The connected scatterplot is one way to show how two time series are related to one another. One series corresponds to the horizontal axis and the other to the vertical axis.

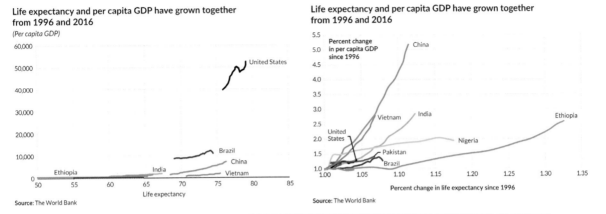

Life expectancy and per capita GDP have grown together from 1996 and 2016

Source: The World Bank

Life expectancy and per capita GDP have grown together from 1996 and 2016

Source: The World Bank

> Though less familiar to some readers, the connected scatterplots can be used to show more data series across two metrics.

You can also use connected scatterplots to show more groups. The connected scatterplot on the left shows the *levels* of economic growth and life expectancy for ten different countries. Here, the higher per capita GDP in the United States stands out, but patterns in the other countries are harder to see. The graph on the right shows percent *growth* in both variables since 1996. In this view, the United States is hardly visible, while the large gains in China and Ethiopia, for example, are clearly visible.

Which graph is better? As always, that depends on your audience, your argument, and what pattern, trend, or finding you want to bring to your readers' attention.

CONCLUSION

The graphs we covered in this chapter show changes over time. There are simple and familiar graphs, like line, area, and stacked area charts. But there are also more complex, less familiar, but equally useful chart types.

The line chart may be the most basic and familiar chart type to show changes over time. There is no limit to the number of lines you can plot, but if there are many lines, consider using color and line thickness to draw your readers' attention to the most important ones. Consider using data markers to add nuance to subtle or small data sets or ways to mark important points. As with many graphs we will explore, but perhaps even more so with line charts, use visual cues to signal missing data.

Alternative chart types are useful when you have too many data series to track in a single graph. Try sparklines, a small multiples approach, cycle charts, or horizon charts when you have a lot of data to visualize. For some of these approaches, enabling the reader to discern exact values is less important than showing them the overall trend or pattern.

Other graph types, like flow charts and timelines, have infinite varieties and styles. Horizontal layouts may work for some people, content, and platforms, while vertical layouts may be better online where they match the natural scrolling motion. Compact layouts are best for mobile platforms.

Whichever graph you use to plot your data, consider how much detail your reader needs and how you can guide them to the point you wish to convey. Many of these chart types are well-known and understood, so our challenge is to make them engaging and interesting without sacrificing accuracy.

DISTRIBUTION

This chapter covers visualizations of data distributions and statistical uncertainties. These may be inherently difficult for many readers because they may not be as familiar with the statistical terminology or the graphs themselves, which may look quite different from the standard graphs they are used to seeing.

Charts like the fan chart and the box-and-whisker plot show statistical measures like confidence intervals and percentiles. Violin plots, which depict entire distributions, may look so foreign that your reader will need detailed explanations to understand them. This doesn't mean that these charts are inherently *bad* at visualizing data—proper labeling and design can make even the most esoteric box-and-whisker plot interesting—but the hurdle of statistical literacy may make such graphs difficult for many readers.

Graphs in this chapter follow the guidelines published by the *Dallas Morning News* in 2005. The *News*'s guidelines include instructions for specific fonts and colors, as well as ways to design and style different graphs, tables, maps, icons, and a summary of the newsroom workflow. The guide uses two fonts, Gotham and Miller Deck, depending on the size and purpose; I use the Montserrat font, which is similar to Gotham.

HISTOGRAM

The histogram is the most basic graph type for visualizing a distribution. It is a specific kind of bar chart that presents the *tabulated frequency* of data over distinct intervals, called *bins*, that sum to the

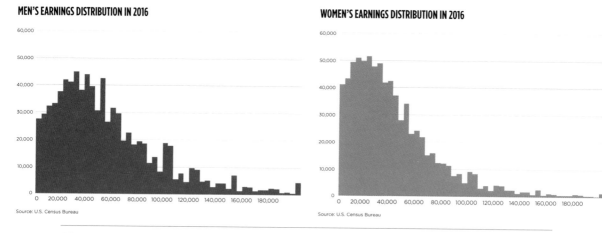

Histograms divide the entire sample into intervals (also called "bins"). The height of the bin shows the number of observations within it.

total distribution. The entire sample is divided into these bins, and the height of each bar shows the number of observations within each interval. Histograms can show where values are concentrated within a distribution, where extreme values are, and whether there are any gaps or unusual values.

We can layer histograms together to show how different distributions compare. The two histograms above depict the distribution of earnings for men and women working in the United States in 2016. We can make some general comparisons between them, but that task is made easier in the next two graphs, where the distributions are placed on top of each other.

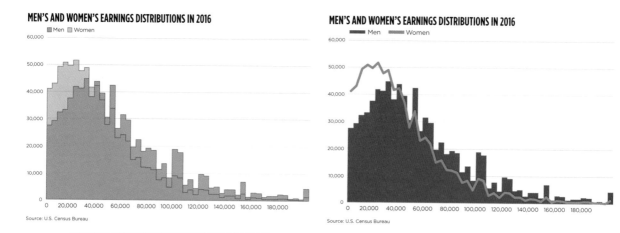

Histograms can be layered together by using saturated colors (left) or different encodings like bars and lines (right).

The graph on the left uses two column charts and transparent colors so both are visible. The graph on the right combines a column chart and line chart, which has the advantage of not using transparent colors, but the balance of how the two groups are presented is now unequal. You might also notice that the line intersects the *middle* of each bin as opposed to spanning the entire bin as the columns do—it's a minor difference but one that you may want to keep in mind.

A key consideration in creating a histogram is how wide to set the bins. Bins that are too wide may hide patterns in the distribution, while bins that are too narrow may obscure the overall shape of the distribution. While there is no "correct" number of bins, there are a number of statistical rules of thumb (using, for example, square roots, cube roots, or logarithms) that

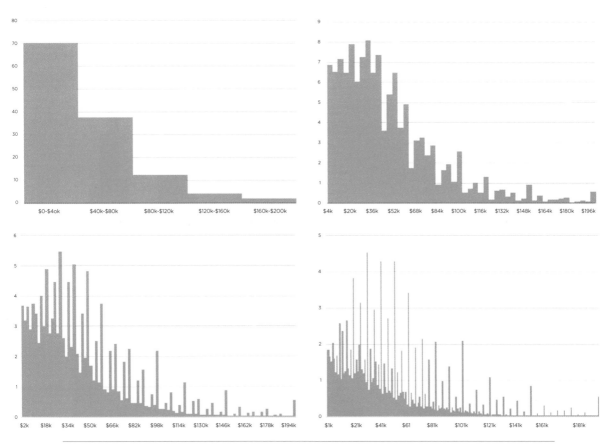

A data distribution will look different depending on the number bins, as it does here as the number of bins increases from 5 to 30 to 50 to 120.

can help determine the optimal bin width. In this example, notice how the distribution looks different as the number of bins increases from 5 to 30 to 50 to 120.

Histograms can help us understand whether our data *lean* to one side or another. A distribution in which more data are pushed to the left is known as *right-skewed*. A histogram with more observations to the right is called *left-skewed*. Distributions that have two peaks are called *bimodal* and distributions with multiple peaks are called *multimodal*. *Symmetric* distributions are those with a roughly equal number of observations on either side of a central value and *uniform* distributions in which the observations are roughly equally distributed.

When we understand the distribution of our data and its possible skew, we're better prepared to conduct more accurate statistical tests. Two completely different distributions may have the same mean and median, but if we don't understand the spread and structure of our data, our results may not paint a complete picture. This is where visualizing our data can be invaluable.

A modification to the basic histogram is the Pareto chart, named after the Italian engineer and economist Vilfredo Pareto. The Pareto chart (next page) consists of bars that represent

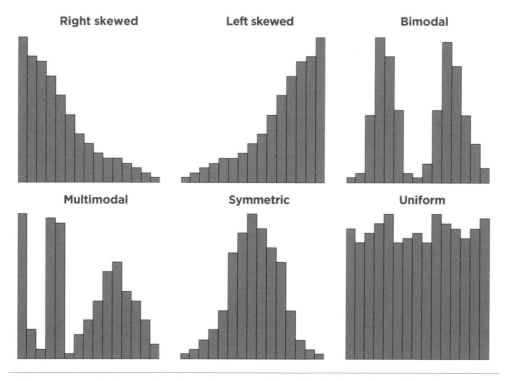

Histograms can help us understand the shape of the distribution of our data. Here we see six such forms of distribution.

TOTAL EARNINGS BY INDUSTRY

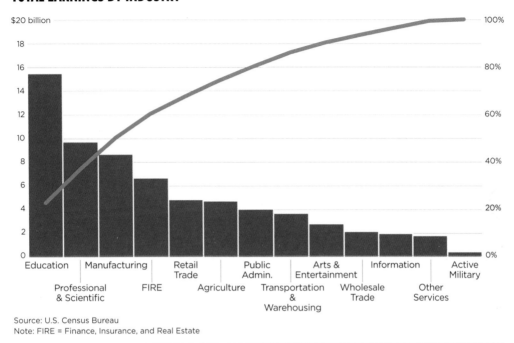

Source: U.S. Census Bureau
Note: FIRE = Finance, Insurance, and Real Estate

The Pareto chart shows values for each group (usually in bars) and the cumulative total as a line.

individual data points and a line that represents the cumulative total. The Pareto chart may be the exception to the rule against dual-axis charts because its *purpose* is to show the complementary distributions overlaid on each other. Of course, the two metrics are really not two different measures—it's the same metric, one as a marginal distribution (separate values of each group) and one as a cumulative distribution (where the values sum to a total).

This Pareto chart shows total earnings in thirteen different major industries in the United States—the bars show the total earnings in each industry and the line shows how the shares add up to total earnings in the economy.

UNDERSTANDING PERCENTILES

Imagine one hundred people standing on an auditorium stage. You stand in the audience as they line up from your left to right, arranged by their earnings. The person with the lowest earnings stands at the left side of the stage, and the person with

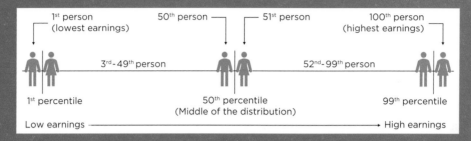

the highest earnings stands at the right. Together, they represent the entire earnings distribution.

The first person in the line has the lowest earnings. At their position, 99 percent of people—all those to their left—have higher earnings. They are said to be in the 1st percentile. Similarly, there is a worker on the other side of the stage in the 100th position. To their right stands 99 other people—99 percent of all people on the stage with lower earnings. They are in the 99th percentile of the distribution, in the top 1 percent. Finally, there is a point in the middle of the stage that splits everyone into two equal groups, 50 percent of the distribution on either side. That point (or more precisely, the *earnings* at that point) represents the 50th percentile or the median of the distribution.

The mathematics of increasing the number of people on stage from 100 to 200 to 1,000 to 150 million does not change—the middle of the ordered distribution is still the median and the person standing at the 10 percent position is at the 10th percentile. Because percentiles are independent of the population, you can compare them across any group such as country or industry.

While percentiles identify a specific location in the distribution, there are other metrics that will give you an overall measure of the distribution. The *mean* or *average* is equal to the sum of all values divided by the number of observations. Because we add up all the data, large values can generate a distorted picture of the true distribution. In the example above, the mean would change dramatically if we replaced one of the people on stage with someone who earned $100 million, but—take note—the median would *not* change because that person would still stand on the far right edge of the stage and the rest of the people in line would stay in the same position.

The *variance* is another metric of a distribution and measures how far each observation in a data set is spread out from the mean value. A large variance indicates the values in the data are far from the mean and from each other; a small variance, by contrast, indicates the opposite. A full decomposition of the variance and related formulas are beyond the scope of this book, but if you are working with data and creating data visualizations, it is certainly worth a bit of time and study to understand how they can be used to better understand the shape and scope of your data.

PYRAMID CHART

Most often used to show changes in population-based metrics such as birth rates, mortality rates, age, or overall population levels, pyramid charts put two groups on either side of a center vertical axis. The pyramid chart is a subcategory of the diverging bar chart (see page 92), but the name is reserved for comparing distributions, most often ages. As with the diverging bar chart, the layout may cause some confusion, because your reader may assume the leftward bars represent negative values, and the rightward bars represent positive values.

The advantage of the pyramid chart is that we can assess the overall shape of the distribution because both groups sit on the same vertical baseline. While many pyramid charts use different colors for the two groups, that's not a necessary characteristic.

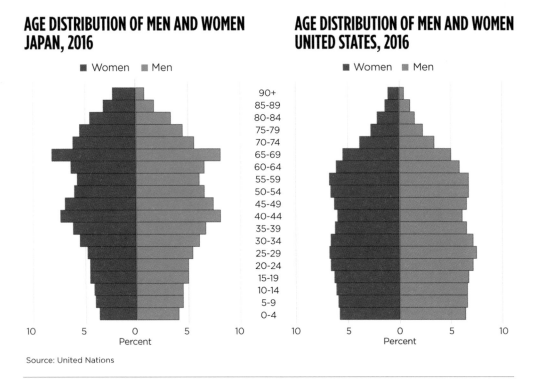

AGE DISTRIBUTION OF MEN AND WOMEN JAPAN, 2016

AGE DISTRIBUTION OF MEN AND WOMEN UNITED STATES, 2016

Source: United Nations

Pyramid charts are a type of diverging bar chart, typically used to show population-based metrics like birth rates, mortality rates, age, or overall population levels.

These pyramid charts show the distribution of ages in the United States and Japan in 2016. In both graphs, women are represented on the left branch of the vertical axis and men on the right. Each row represents a different age group: 0–4 years old, 5–9 years old, and so forth. The shape of the chart means we can immediately see that Japan has a greater share of older people, while there is a larger share of younger people in the United States.

Because the bars are not next to each other, it is difficult to compare the total shares of men and women. But again, whether that's a problem depends on the goal of your visual. If you want your reader to compare the shares of the two genders, then a different chart type—such as a paired bar chart or dot plot—would be a better choice. But if you want your reader to see the overall shape of the distribution, the pyramid chart is perfect.

A natural alternative to the pyramid chart is the dot plot or the lollipop chart. Dots for each gender are positioned along the horizontal axis corresponding to the data value and connected by a line. Or, as shown on the right, we could simply use a lollipop chart (also see page 80), replacing the bars with lines and dots. With either approach, we can still use different colors or just use a single color for the entire graph.

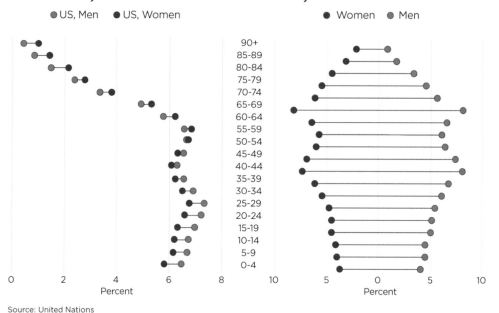

Alternatives to the basic pyramid chart are the dot plot or lollipop chart.

AGE DISTRIBUTION OF MEN AND WOMEN IN JAPAN AND THE UNITED STATES IN 2016

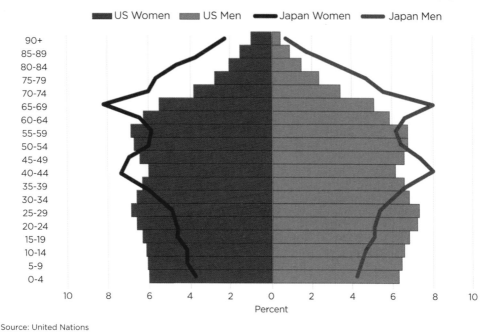

Source: United Nations

Combining distributions in one chart can be accomplished by adjust colors and combining different encodings.

One challenge with the pair of pyramid charts is to accurately compare the age distributions across the two countries. We can infer the main relationship, that Japan is, on average, older than the United States, but it's harder to make a more detailed comparison. Placing the charts atop each other—a technique we saw earlier with the histogram—makes that task easier. But be mindful that in using this approach the data for the two countries are shown with different encodings (bars and lines), which can risk emphasizing one set over the other.

VISUALIZING STATISTICAL UNCERTAINTY WITH CHARTS

There are lots of types of uncertainty in data and statistics. Before we learn the different ways to visualize uncertainty, it is worth pausing to understand what we mean by the term. Even if you're not a statistician or mathematician, it's important to understand how such uncertainty and measurement error can affect our results and, ultimately our

Source: Scott Adams

visualizations. Accounting for uncertainty—and making it clear you've done so—builds the reader's trust in your work.

We can think of the term uncertainty in two main ways. One is *uncertainty from randomness*, which applies to the statistical confidence in our statistical models and results. As an example, consider the standard margin of error built into political polling data: "Candidate Smith has a 54 percent approval rating with a margin of error of plus-or-minus 4 percentage points." Another kind is what we might call *uncertainty from unknowns*, where our data are inaccurate, untrusted, imprecise, or even unknown. A very simple example might be something like a data set that includes infants' ages in months instead of weeks. Using statistical and probabilistic models enables us to confront the first kind of uncertainty, which we can therefore visualize; the second kind of uncertainty concerns unknowns that can't necessarily be resolved through more data.

A thorough treatment of *uncertainty from randomness* (error margins, confidence intervals, and the like) is beyond the scope of this book. But *uncertainty from unknowns* is something that many readers can easily relate to. To illustrate, let's consider the data I use in this chapter: workers' earnings by industry and state. The data set used for this analysis is the 2016 U.S. Census Bureau's American Community Survey. The survey includes demographic and economic information for about 3.5 million people per year. For the data in this chapter, I examine individual earnings for more than a million people.

Now, imagine all the reasons why someone might tell the Census Bureau the wrong answer when they ask about their earnings. They might lie. They might round their earnings to the nearest dollar—or nearest hundred dollars, or nearest thousand dollars. Maybe they work side jobs they didn't mention. They may be asked about their spouse's or partner's earnings and have to hazard a guess. There are all sorts of reasons they may get it wrong, and recent economic research shows that reporting error (especially in government program participation) in some of the largest, most trusted government household surveys has been increasing over the past several years.

We must also recognize that for this survey, the Census Bureau only asks a *share* of Americans (so our calculations from these data also suffer from *uncertainty from randomness*). Consider all the reasons why that "sample" may not be truly representative. Maybe some people don't want to answer the survey. Maybe they moved and didn't get the form, or changed their phone number and didn't get the call.

Whenever we work with data, we should consider how these kinds of uncertainty may lead to some "error" around our final estimates. Not mistakes, but uncertainty. This error is fundamental to being careful with data and ultimately visualizing and explaining our results carefully. Line charts and bar charts suggest certainty with their sharp boundaries and crisp edges, but that certainty is rarely actually the case.

In his book *How Charts Lie*, Alberto Cairo remarks that, "Uncertainty confuses many people because they have the unreasonable expectation that science and statistics will unearth precise truths, when all they can yield is imperfect estimates that can always be subject to changes and updates." We should not expect flawless data, and we should be ready to explain those imperfections to our readers as best we can.

▶　▶　▶　▶　▶

We now move to the specific challenge of conveying uncertainty around data estimates or results from statistical models. This is a common problem: In a survey of ninety data visualization authors and developers, information visualization researcher Jessica Hullman found that graph creators did not include uncertainty in their work for four main reasons. First, they did not want to confuse or overwhelm viewers. Second, they did not have access to information about the uncertainty in their data. Third, they did not know how to calculate uncertainty. And fourth, they did not want to make the data appear questionable. Hullman argues that visualizing uncertainty is important because "a central problem is that authors often omit or downplay information, such that data are interpreted as being more credible than they are." More effectively conveying such uncertainty—especially when making statistical claims—builds trust and credibility.

This section will familiarize you with visual signals of uncertainty. There are many chart types that can be used to show uncertainty around a central estimate, and this section introduces a few of them: error bar charts, confidence interval charts, gradient charts, and fan charts.

ERROR BARS

Perhaps the simplest and most common way to visualize uncertainty is to use error bars: small markers that denote the error margin or confidence interval. Error bars are not really a visualization on their own, but are an addition to other charts, often bar or line charts. The ends of the error bars can correspond to any value you choose: percentiles, the standard error, the 95-percent confidence interval, or even a fixed number. And because error bars can convey these multiple statistical measures, recent research has shown that this can invite confusion on the part of the reader, making incorrect conclusions about the data. We must therefore clearly label the intervals, either in a chart note or, preferably, on the chart itself.

This bar chart shows average earnings in each of thirteen industries in 2016. Error bars denote the 25th and 75th percentiles.

AVERAGE EARNINGS IN U.S. INDUSTRIES IN 2016

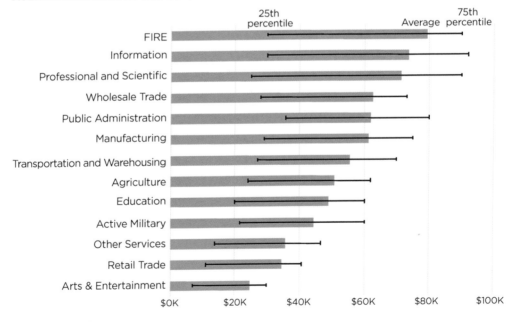

Source: U.S. Census Bureau

The simplest and most common way to visualize uncertainty or distributions is to use error bars, small markers that denote the margin of error or confidence interval.

Applying error bars to bar charts raises a potentially interesting complication: some research suggests that we tend to judge the points that fall *within* the bar as more likely than those *outside* the bar ("within-the-bar" bias). In the previous chart, that would mean a reader is more likely to assume that the salary of a worker in the Finance, Insurance, and Real Estate (FIRE) sector is less than $80,000 and not more than $80,000. Other research has found that we can better judge uncertainty and the distribution with other types of graphs, such as the violin plot, stripe plot, or gradient plot.

While it may be a familiar approach for many readers and requires less data than some of these other charts, existing research shows that we are not particularly good at assessing uncertainty through these kinds of visual approaches.

CONFIDENCE INTERVAL

A confidence interval chart typically uses lines or shaded areas to depict ranges or amounts of uncertainty, often over time. The basic confidence interval chart is literally a line chart with three lines: one for the central estimate, one for the upper confidence interval value, and another for the lower confidence interval value (these upper and lower lines can be confidence intervals, standard errors, or a fixed number). The lines may be solid, dashed, or colored, but if the central estimates are the primary numbers of interest, that line should be thicker or darker to highlight it against the confidence interval values.

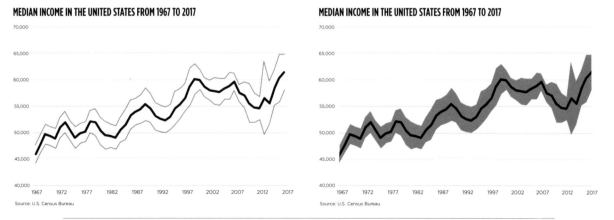

Lines or shaded areas around a central line can visualize ranges of uncertainty.

The two charts on the previous page show median earnings in the United States between 1967 and 2017. The standard error around those estimates is depicted by two separate lines in the left graph and by a shaded area in the right.

GRADIENT CHART

A gradient chart (sometimes called a stripe plot) shows distributions or differences in uncertainty. There are many ways to use the gradient plot, but the basic technique is to plot the primary number of importance and add a color gradient on one or both sides to visually demonstrate the measure of uncertainty around that single point. The plot is named not necessarily after the *shape* of the graph but after the use of the color gradient.

The gradient plot can show changes over time or, as in the graph here, the distribution around individual observations. This gradient plot shows the exact same data as the error

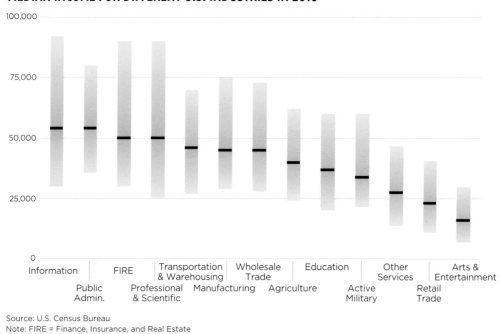

MEDIAN INCOME FOR DIFFERENT U.S. INDUSTRIES IN 2016

Source: U.S. Census Bureau
Note: FIRE = Finance, Insurance, and Real Estate

Gradient charts use a color gradient on one or both sides of the primary number of interest to show distributions or uncertainty.

bar chart above, but instead of bars with error lines jutting out in both directions, average earnings are encoded with the dark horizontal line and the 25th to 75th percentiles of the distribution are shown with the gradient. The color gradient might illustrate, for example, multiples of the standard error, which can be a signal to the reader that the outcomes are less certain the further they are from the central estimate line.

Stripe charts can also be an effective way to show changes over time. A strong example of this is the series of stripe charts created by Dr. Ed Hawkins, a climate scientist at the University of Reading, that showed temperature changes from 1850 to 2018. Each bar (stripe) showed a different temperature level, ranging from cooler blues to hotter reds. Together, readers could quickly and easily see the marked increase in temperatures around the world as a whole and in their specific region of the world by using an online tool. These stripe charts were published by a multitude of websites, television stations, and even became a cover of the *Economist* magazine.

In a 2019 interview on the *Data Stories* podcast, Hawkins said that he "was looking for a way to communicate to audiences that aren't used to seeing graphs, or axes, or labels—things that we see day-to-day, but are complicated to them. It may look too mathematical to them, so it turns them off straight away." In a later interview on that same podcast, Jennifer Christiansen, the senior graphics editor at *Scientific American*, described them as, "every region's version of the climate stripe pattern progresses from cool blue to a warm red. No labels

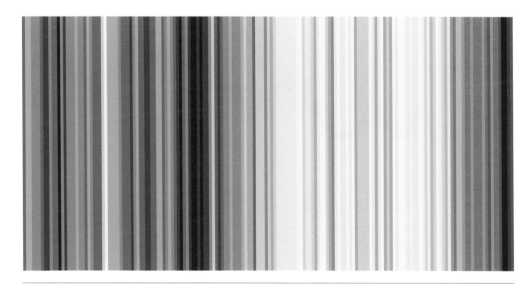

This stripe chart from ShowYourStripes.com shows global temperatures from 1850 to 2018. The simple colored stripes are easy to see and understand.

are needed; no caption is needed. It's a visceral and accessible nod to our warming planet with color representing annual temperature. And it prints legibly on everything from social media profiles, to pins, neckties, magazine covers, mugs, and concert screens."

FAN CHARTS

If the color or saturation of the shaded area between the confidence interval lines changes based on the value, it is often called a fan chart. Fan charts are like gradient plots for line charts, and they are great for visualizing changes in uncertainty over time. In the fan chart, values closest to the central estimate are the darkest and they lighten as the values move outward. The use of color distinguishes the move from higher levels of statistical confidence to lower levels. The advantage here is that it signals to the reader how the estimates become less certain the further they are from the central estimate.

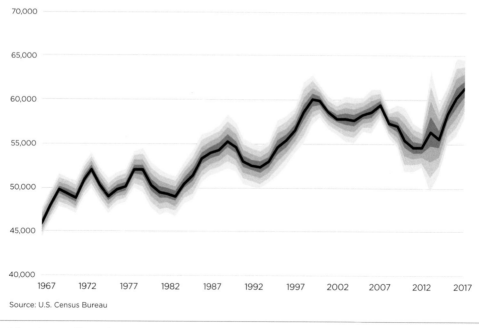

MEDIAN INCOME IN THE UNITED STATES FROM 1967 TO 2017

Source: U.S. Census Bureau

Like the gradient chart, the fan chart shows distributions (in this case, standard errors) around a central estimate.

This fan chart shows the change in median household income over the last fifty years. The color bands show the standard error divided into eight segments, though they could also show bands of percentiles or other measures. Similar to the gradient chart, the change in color saturation can show multiples of the standard error and signal less certainty as we move farther from the central estimate, which is here represented by the black line.

THE HAND-DRAWN LOOK

One last strategy to suggest uncertainty is not a visualization technique per se but a design technique. The hand-drawn, "sketchy," "gooey," or "painty" techniques can be used to add an

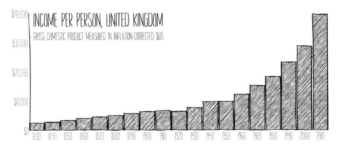

Hand-drawn, "sketchy," or "gooey" design effects use uneven edges to communicate a sense of uncertainty or imprecision.

Sources: Copyright Mona Chalabi (top) and Jo Wood, giCentre, City, University of London (bottom).

uneven edge or fuzziness to graph objects that will create a sense of uncertainty or imprecision. Research suggests that sketchy graphs generate more engagement in the graph and that
we can "tie sketchiness to uncertainty or significance values." These two examples, the first
from journalist Mona Chalabi at the *Guardian* and the second from Jo Wood at the University of London, both demonstrate these techniques in action.

BOX-AND-WHISKER PLOT

When you visualize the distribution of your data, you can show the entire distribution or just
specific points within it. The box-and-whisker plot (or boxplot), originally called a *schematic
plot* by its inventor John W. Tukey, uses a box and line markers to show specific percentile
values within a distribution. You can also add markers to show outliers or other interesting
data points or values. It is a compact summary of the data distribution, though it displays
less detail than a histogram or violin chart.

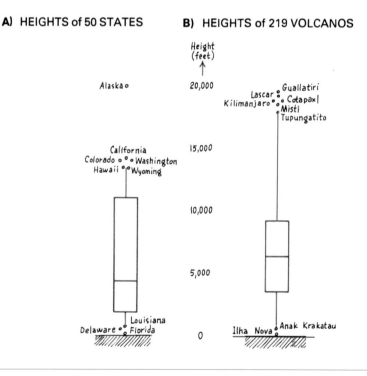

The original boxplot from Tukey 1977

The basic box-and-whisker plot consists of a rectangle (the *box*), two lines (the *whiskers*) that emanate from the top and bottom of the box, and dots for outliers or other specific data points. Most standard box-and-whisker plots have five major components:

1. The *median*, encoded by a single horizontal line inside the box.
2. Two *hinges*, which are the upper and lower edges of the box and typically correspond to the first quartile (or the 25th percentile) and third quartile (75th percentile). The difference between these two points is called the *Interquartile Range* or *IQR*.
3. The higher and lower extremes (sometimes the maximum and minimum) are placed at a position 1.5 times the *IQR* (recall the Box on page 74).
4. Two *whiskers* (the lines) connect the hinges to a specific observation or percentile.
5. *Outliers* are individual data points that are further away from the median than the edges of the whiskers.

Each of these components can vary depending on which parts of the distribution we wish to show. Some creators replace outliers with fixed quantiles such as the minimum and maximum values or the 1st and 99th percentiles. Some use the semi-interquartile range $(Q_3-Q_1)/2$, which can generate asymmetric whiskers. And some add other descriptive statistics like the mean or standard error. We can also vary the color, line thickness, and how and which parts of the chart are labeled.

As a practical example, the box-and-whisker plots on the next page show the distribution of earnings in our thirteen industries. The vertical line in the middle of each box represents

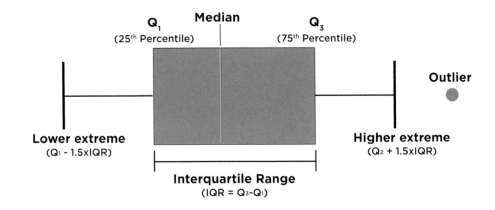

The basic box-and-whisker plot.

the median and the edges of the box represent the 25th and 75th percentiles. The ends of the lines (the whiskers) show the 10th and 90th percentiles.

The graph on the left sorts the industries alphabetically while the one on the right sorts them by the median value. I generally find it preferable to present data sorted by their values as opposed to alphabetical or some other arbitrary sorting. There may be cases when alphabetical sorting is best—for example, if we were presenting earnings across all fifty U.S. states, readers would find it easier to find individual states if they're sorted alphabetically. If, however, the goal of that graph is to discuss the high/low earnings of a particular state, we would sort the data by their values to make those comparisons easier.

As with all of these visualizations, showing percentiles and statistical or data uncertainty will often depend on the experience, interest, and expertise of our audience. In scientific or research applications, for example, communicating uncertainty is especially important to demonstrate whether a finding is statistically meaningful. But in other cases, where our data only have a single value for each observation—say, a single estimate of per capita GDP in the United States—we may not be able to visualize parts of the distribution.

In the case of the box-and-whisker plot, by plotting these specific percentile points, we are explicitly deciding *not* to visualize the entire distribution. This may not be entirely problematic, especially if other percentile points aren't particularly important, or if the data follow a fairly standard distribution. But we must always fully explore the data we're certain we're not hiding important patterns from our reader—or ourselves!

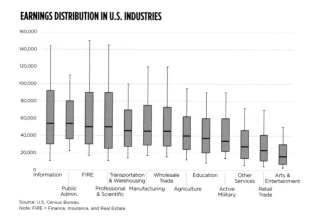

These charts show the distribution of earnings in thirteen industries either sorted alphabetically (left) or by median value (right). The edges of the box show the 25th and 75th percentiles and the whiskers show the 10th and 90th percentiles.

CANDLESTICK CHART

Candlestick or stock charts look like box-and-whisker plots, but they visualize different content. Whereas box-and-whisker plots visualize uncertainty or a distribution, candlestick charts visualize changes in the prices of stocks, bonds, securities, and commodities over time. Bars and lines show opening and closing prices and highs and lows in a day, plotted along a horizontal axis that measures time.

There are two elements of the candlestick chart. The central box—sometimes called the "real body"—shows the gap between the opening and closing prices. The lines that extend upward and downward from the real body—sometimes call the "wick"—show the low and high value for the day. Like the box-and-whisker chart, the candlestick chart includes specific points and does not show *all* of the activity during the day, such as price volatility.

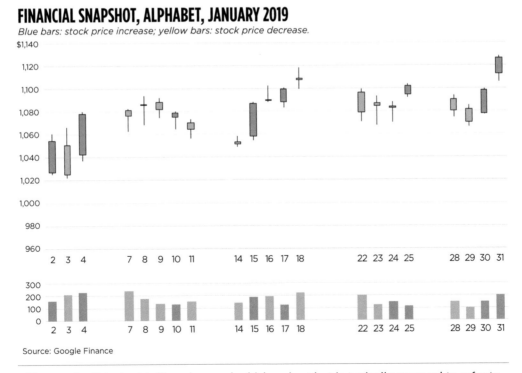

FINANCIAL SNAPSHOT, ALPHABET, JANUARY 2019

Blue bars: stock price increase; yellow bars: stock price decrease.

Source: Google Finance

The candlestick chart is like a box-and-whisker chart but is typically reserved to refer to prices of stocks, bonds, securities, and commodities over time.

Specific characteristics of the candlestick chart can vary in some obvious ways: color can be changed to differentiate between a drop in price during the day (i.e., the closing price is greater than or less than the opening price) and icons or other symbols can identify the high and low prices. Although I've placed this chart in this chapter, because of its relation to box-and-whisker plots, it could easily have also appeared in the *Time* chapter or even the *Comparing Categories* chapter.

The candlestick chart on the previous page shows overall daily trading patterns for shares of Alphabet, Inc.—the parent company of the Google search engine—from January to February 2018. The bar underneath shows trading volume. In both graphs, blue bars signal an increase in price over the day and yellow bars signal a decrease. Notice how the two graphs are stacked together as opposed to using a dual axis chart, which might be confusing or just plain cluttered.

VIOLIN CHART

Instead of showing selected percentiles from the distribution, the goal of the next set of charts is to show the *entire* distribution. Unlike the box-and-whisker plot, in which we choose specific points in the distribution, or the histogram in which values are grouped together into intervals, the violin chart shows the shape of the whole distribution.

These violin charts use the same data as above, the average earnings in thirteen industries in 2016. The thicker areas mean that there are more values in those sections while the thinner parts imply lower frequency of observations. I've added a dot in the middle to mark average earnings within each industry. Notice again the differences in the view when the chart is sorted alphabetically (on the left) versus by the mean income (on the right).

KERNEL DENSITY

One consideration in creating this chart type is that it requires estimating what is called the *kernel density* of each distribution. Kernel densities are a way to estimate the distribution of a variable—akin to a histogram—but can be smoothed or made to look more continuous using different algorithms. For the violin plot, those density estimates are plotted to mirror each other around an invisible central line.

Think of it this way: A histogram plots a summary view of a distribution along a single axis. The violin plot mirrors a smoothed version of the histogram on either side of that single

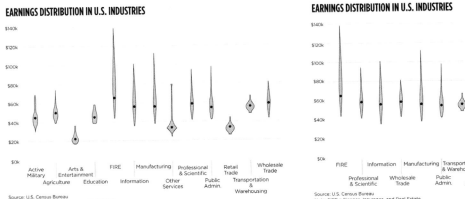

Instead of showing select points (percentiles) in a data distribution, the violin chart shows the estimated shape of the entire distribution using kernel densities.

axis. How that smoothing is accomplished will depend on what sort of kernel density estimator you choose, which can vary based on the data, underlying function, and more.

Violin charts, then, are richer than the box-and-whisker plot, but can also be more difficult to create and for our audience to understand. In modern versions of Excel, for example, the box-and-whisker plot is a default graph, while a violin plot requires manually calculating the probability densities and then finding a graphing solution.

RIDGELINE PLOT

The ridgeline plot is a series of histograms or density plots shown for different groups aligned along the same horizontal axis and presented with a slight overlap along the vertical axis. In a basic sense, the ridgeline plot is like a small multiples histogram or a horizon chart where the histograms are aligned in particular way.

The ridgeline plot on the next page shows the earnings distribution across the thirteen different industries. The horizontal axis is shared across all thirteen industries and the distributions sometimes overlap along the vertical dimension. Depending on the color scheme and density of the data, there may be more or less overlap between the series, but as we have seen in other graph types (for example, sparklines and horizon charts), sometimes showing the overall pattern is more important than the reader being able to pick out all of the specific

EARNINGS DISTRIBUTION IN U.S. INDUSTRIES

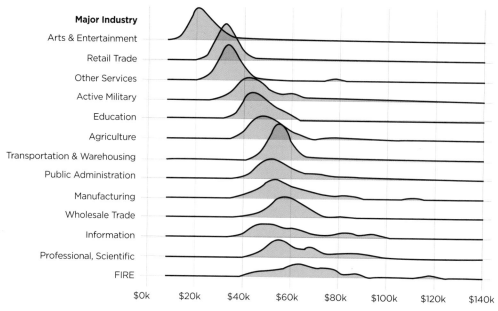

Source: U.S. Census Bureau

The ridgeline plot is a series of histograms shown for different groups aligned along the same horizontal axis and presented with a slight overlap along the vertical axis.

values. These overlaps can be a problem, but one that is mitigated by how quickly and easily readers see how the different distributions line up with one another along the same axis.

The most famous ridgeline plot is one that you didn't even know was a ridgeline plot: The album cover for *Unknown Pleasures*, the 1979 debut album of English post-punk band Joy Division, which had white lines on a black background with no band name, album title, or other identifiers.

In 2015, Jen Christiansen, the Senior Graphics Editor at *Scientific American*, tracked down the original image to the 1970 doctoral dissertation of Harold D. Craft, Jr., a radio astronomer at Cornell University. The original chart graphed the distribution of consecutive radio pulses emanating from a pulsar, a type of neutron star. The album cover designer, Peter Saville, called it "a wonderfully enigmatic symbol for a record cover."

Data that lend themselves well to a ridgeline plot are those in which the distributions differ from one category (row) to another so that the reader can see the shift up and down the page. With data in hand, variations on color, font, and layout can help engage and interest your reader. The ridgeline plot on the next page was published by the *Guardian* in 2018 and

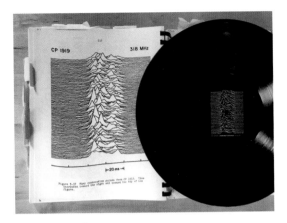

Source: Photograph by Jen Christiansen (featuring figure 5.37 from "Radio Observations of the Pulse Profiles and Dispersion Measures of Twelve Pulsars," by Harold D. Craft Jr., September 1970).

shows the distribution of the gap in pay between men and women in more than ten thousand companies and public bodies in the United Kingdom. Using different colors on either side of the vertical zero-percent line (perfect equality) and sorting the data from industries that have the highest pay gaps (e.g., Construction) to smallest pay gaps (e.g., Accommodation and Food Services) also helps direct the eye.

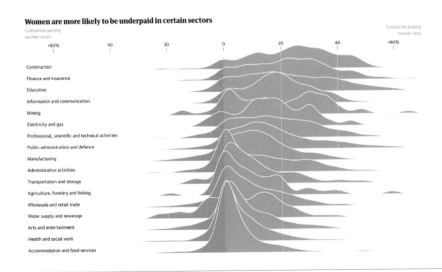

This ridgeline plot from the *Guardian* shows earnings distributions for men and women in different industries.

VISUALIZING UNCERTAINTY BY SHOWING THE DATA

The graphs presented in this chapter so far summarize data distributions with lines, points, bars, and color. To different degrees, this is what the histogram, violin, and ridgeline plots all do. Another way to visualize the distribution of your data is to just *show* the data.

STRIP PLOT

The basic way to show your data is with what is known as a *strip plot*. In this graph type, the data points are plotted along a single horizontal or vertical axis.

In this example, average earnings for each state are shown for each of the thirteen industries (the U.S. average is denoted with the vertical black line). We have already seen similar data presented in the box-and-whisker plot and violin chart, but here you can see individual

EARNINGS DISTRIBUTION IN U.S. INDUSTRIES

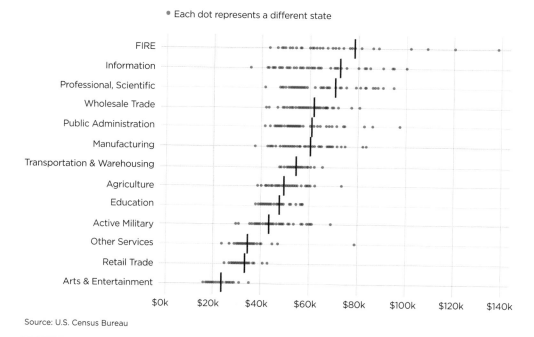

Source: U.S. Census Bureau

In a strip plot, data points are plotted along a single horizontal or vertical axis. This strip plot encodes the data with circles, but small lines are also often used.

points. Notice, however, that I'm showing average earnings for the fifty states, not for individual workers. Plotting earnings for everyone in my data (some 1.3 million people) would look like a single, dark line. There are too many values.

It's true that some of the data is obscured, but, especially by virtue of the overlapping transparent colors that make the patterns darker, it becomes clearer where the bulk of earnings lie in the distribution. There's no rule for how many data points are *too many*, but as you plot your data, you can always tell when you've passed that threshold.

This static image from an interactive strip plot from NPR is a good example of how this visualization can be richer than a standard bar chart or histogram. Here, they plot a point for every school district in every state. Darker orange dots (and below the black horizontal line) are districts in which spending per student is below the national average. Darker green dots are districts in which spending per student is above the national average. An interesting and useful design choice to make the dots transparent (with a solid border) lets us see where districts are close enough to overlap. Also notice that only Alabama, Florida, and Alaska are

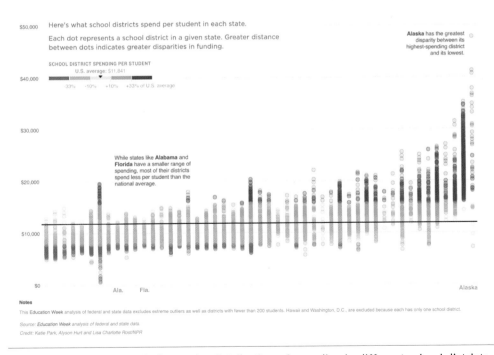

This strip plot from NPR shows the distribution of spending in different school districts around the United States.

labeled along the horizontal axis. Those three states are explicitly annotated in the chart. Other state labels only appear along the horizontal axis when, in the interactive version, the user hovers over a stack of circles.

BEESWARM PLOT

If we want to plot individual data points rather than distributions, one way to make the data more visible is to use a technique called "jittering." This is when we alter individual values slightly so that the data points don't lie on top of one another.

Consider the strip plot on the left, which leaves all of the data along the same horizontal axis. We can see clusters, but not all of the individual values. In the version on the right, the data are jittered along the horizontal and vertical axes to help make each point visible. There are different algorithms and approaches to jittering the data, and the most important consideration is to manipulate the values just enough to make them visible but not so much that it changes the overall view of the distribution. As with choosing a kernel density estimator for violin charts, the choice of jittering technique (for example, do we jitter both x and y variables and if so, do we jitter each variable independently?) will depend on the data and its underlying distribution.

Jittering doesn't work in every case, of course. We are limited by how many points we can plot. With too many data points, jittering would require so much movement that it would modify the underlying distribution. Plotting *everyone's* earnings in each industry, for example, creates a mass of dots that requires so much jittering that it modifies the presentation of the data by moving the points too far away from their true position. But showing average earnings in each of the fifty states across the thirteen industries is not as overwhelming, and you can see where the bulk of the distribution lies in each. Of course, if you're interested in just showing that there are a *lot* of points in your dataset and you can maintain the overall shape of the data, plotting many points may help make your exact argument.

To create what is called a *beeswarm plot*—because the clustering of the data points resembles a swarm of bees—we jitter the values so that they don't overlap and each point is visible. As with other charts in this chapter (and coming up in Chapter 8), there are different calculations we can use to arrange the dots—for example, arranging the points in increasing order or placing them in a square or hexagonal grid. Here, similar to the ridgeline plot, each industry shares the same horizontal axis so that we can easily compare across the different sectors.

Notice how I have added some simple annotation and labeling to some of the outlier values. These clearly stand out in the graph, and a curious reader will wonder what's going

EARNINGS DISTRIBUTION IN U.S. INDUSTRIES
(Major industries by state)

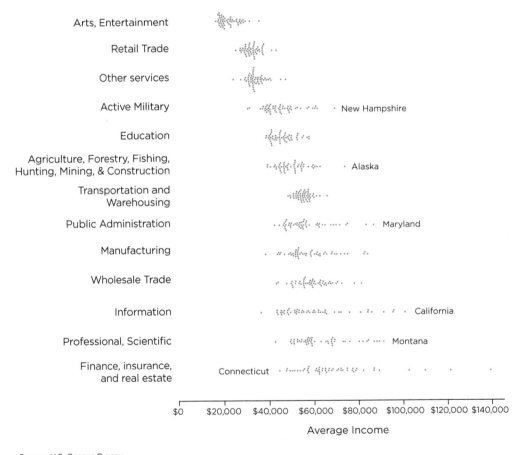

Source: U.S. Census Bureau

A beeswarm plot "jitters" all of the points in the data set so they don't overlap and each point is visible.

on with those points. Are they errors? If not, what is that state, and why are earnings so high relative to the rest of the country? I haven't labeled *every* outlier point, but enough to assure the reader I've thought about those values.

Beeswarm plots can also show changes over time. This beeswarm plot from Axios—really eight beeswarm plots aligned together—shows spending at properties owned by The Trump Organization before and after the 2016 election. The combination of color (spending origin), size (amount of spending), and the density of points (the time dimension) makes this an effective visualization to show the patterns around Election Day.

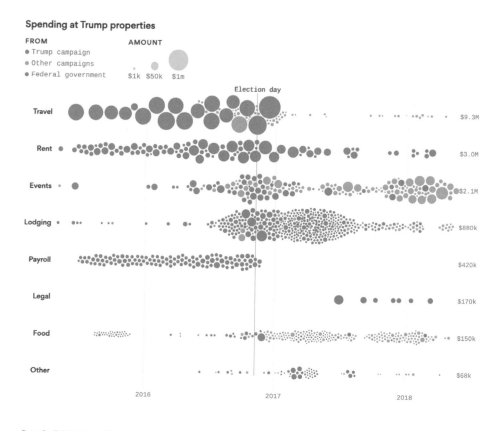

This beeswarm plot from Axios shows spending at properties owned by The Trump Organization before and after the 2016 election.

WILKINSON DOT PLOTS AND WHEAT PLOTS

The wheat plot, developed and named by Stephen Few, is a richer version of what is called a dot histogram or a Wilkinson Dot Plot (which is named after Leland Wilkinson author of the seminal data visualization book *The Grammar of Graphics*, though Wilkinson himself actually referred to these charts as *histodot plots*, a name that clearly did not stick). A Wilkinson Dot Plot is like a regular histogram except that instead of showing a single bar encoding all of the observations, data points are stacked on top of each other within their relative bins—something like combining a histogram and unit chart. With this approach, the points do not show their actual *value* (measured along the horizontal axis) because they are stacked in a single column. In other words, each dot represents an observation in each bin, not its actual value.

INCOME DISTRIBUTION

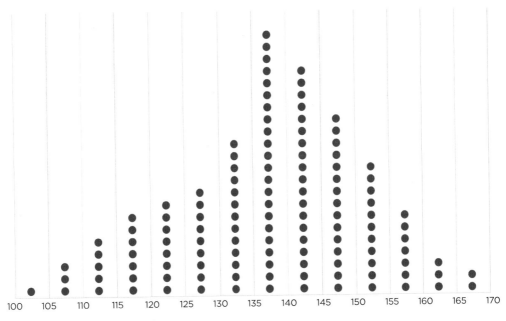

Note: Sample distribution of 200 workers

A dot histogram plots individual points within the bins of a histogram. Of course, it can only handle so many individual points.

The Wilkinson Dot Plot can be modified to create a wheat plot in which we show the *actual* data values, stacking them within their separate bins. The actual data values are plotted along the horizontal axis—still grouped into their bins—and stacked vertically to show the total number of observations. Stephen Few writes that, "The curved alignment of the dots is meaningful, for it graphically displays the distribution of values within each interval, based on their positions. Although this looks odd at first glance, it takes only a minute to understand and learn how to read." As with some of the previous distribution graphs, one of the limiting features of the wheat plot is that too many data values may lie on top of each other.

The wheat plot on the next page shows the distribution of earnings for a single industry for about two hundred workers. The histogram on the right is shown as a comparison—you can still see the relative share of observations in each part of the distribution, but not the actual data. There is an obvious tradeoff. On the one hand, the wheat plot shows more detail for the reader to explore and the graph can look more interesting and engaging. On the other hand, the histogram is likely more easily and immediately understandable to readers.

INCOME DISTRIBUTION

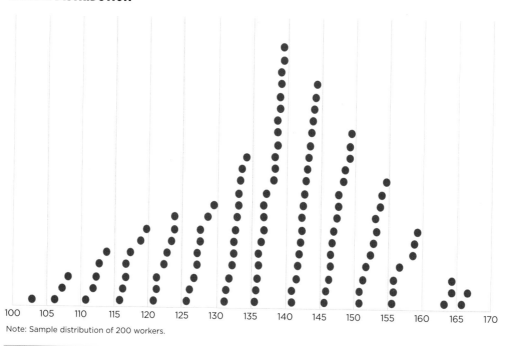

Note: Sample distribution of 200 workers.

The wheat plot, designed by Stephen Few, adjusts the dot histogram by showing the exact values, still within each bin.

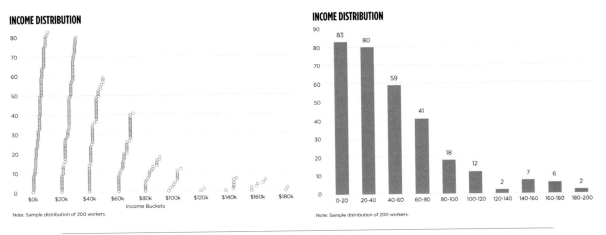

The tradeoff between wheat plots and simple histograms: the wheat plot has more detail but may be harder for people to understand.

We can see the difference between say, a wheat plot and a ridgeline plot in this visualization from the *Guardian*. Leading off the same article that has the ridgeline plot on page 201, this graph includes a dot for every company in their data set. It doesn't quite have the "lean" that a true wheat plot might have (because there are so many points), but you get a good sense of the overall distribution and the greater number of firms towards the right side of the graph. Plotting each individual company also lets the chart creators add labels to highlight

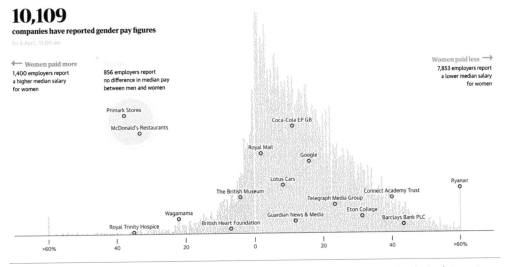

This wheat plot from the *Guardian* includes a dot for every company in their data set.

specific companies, which we couldn't do in a standard histogram or ridgeline plot. It is, however, important to realize that the selected labeled points are arbitrarily chosen along the vertical axis, which is simply stacking the points and not tied to any pay gap data.

RAINCLOUD PLOT

Sometimes it's useful to show *both* the distribution density of your data and the actual data points. The raincloud plot, perhaps first named by neuroscientist Micah Allen, shows the distribution (think violin chart) with the actual data plotted below. In this arrangement, it looks like a cloud raining data.

Raincloud plots show us a summary of the data and all the individual data points, so we can spot outliers and patterns. Again, the tradeoff is that this might require more work on the part of the reader to understand how to read the graph.

This raincloud plot shows the distribution of earnings across the fifty states with data values plotted below.

While the raincloud plot may seem like an esoteric chart—and, to be honest, right now it is— there are certainly scenarios and data for which this chart would be a useful choice.

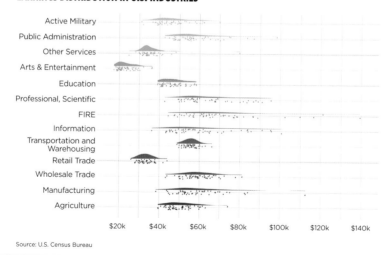

EARNINGS DISTRIBUTION IN U.S. INDUSTRIES

Source: U.S. Census Bureau

The raincloud plot shows the summary histogram with the actual data plotted below.

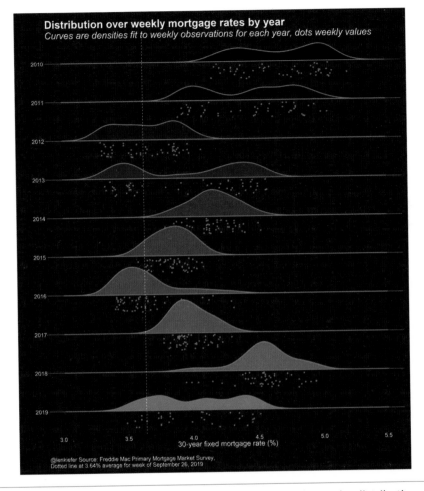

An example of a raincloud plot from Len Kiefer that shows the distribution
of weekly mortgage rates for different years. The visualization gives you
both an overall summary view of the data and shows the
specific data points.

This raincloud plot from Len Kiefer, the Deputy Chief Economist at Freddie Mac, shows
the distribution of mortgage rates from 2010 to 2019, with weekly observations shown in the
dots just below. This visualization both gives you an overall summary of the data and the
detailed values in the points below.

STEM-AND-LEAF PLOT

The stem-and-leaf plot is a table that shows the place values of each data value. Values are typically shown listed down the "stem" column with the first digit or digits. The rest of the table is reserved for the "leaf" that shows the last digit (or digits).

■池袋線 所沢 ◇池袋方面 平日 2018.03.10改正

時	分
5	00 08 18 24 28 34 41 45 50 54 59
6	02 06 10 14 17 19 24 27 30 33 37 42 45 48 51 54 57 59
7	02 05 08 11 14 16 20 23 26 28 31 35 38 41 44 46 50 53 56 58
8	01 04 07 10 13 17 19 22 25 29 32 35 39 42 45 50 55 58
9	02 05 10 15 19 23 26 31 35 39 44 48 55
10	04 09 12 15 19 23 25 30 35 39 44 49 52 55
11	00 04 09 12 19 23 25 30 35 39 44 49 52 55
12	00 04 09 12 19 23 25 30 35 39 44 49 52 55
13	00 04 09 12 19 23 25 30 35 39 44 49 52 55
14	00 04 09 12 19 23 25 30 35 39 44 49 52 55
15	00 04 09 12 19 23 25 30 35 39 44 49 52 55
16	00 04 09 12 18 20 23 25 30 35 40 44 48 52 55
17	00 04 08 12 16 20 22 24 30 34 38 42 46 52 54
18	00 05 08 12 16 20 22 25 30 35 38 42 46 52 55
19	00 05 08 12 16 20 24 27 29 32 37 40 42 46 52 55 59
20	02 07 10 12 16 20 22 25 28 32 37 40 46 52 55
21	02 07 10 16 22 25 32 37 40 47 52 55
22	02 07 10 17 22 25 32 37 40 47 52 55
23	02 08 14 19 25 28 32 36 42 50 58

種別　むさ＝特急むさし　ちち＝特急ちちぶ　S＝S-TRAIN　快速＝快速急行　F快＝快速急行東京メトロ線内急行　快★＝快速急行東京メトロ線内通勤急行
急行　通急＝通勤急行　快速　快○＝快速東京メトロ線内急行　快☆＝快速東京メトロ線内通勤急行　準急
準口＝準急東京メトロ線内普通　準★＝準急東京メトロ線内通勤急行　通準＝通勤準急　各☆＝各駅停車東京メトロ線内急行　各駅停車　快速(中)＝ひば
停車駅　特急＝池袋　S-TRAIN＝(保谷)、(石神井公園)、飯田橋、有楽町、豊洲　※（ ）は乗車専用です。　快速急行(中)＝ひば
ケ丘・石神井公園・練馬・新桜台・小竹向原　急行＝ひばりヶ丘・石神井公園・池袋　通勤急行＝東久留米・保谷・大泉学園・石神井公園・池袋・練馬・新桜台・小竹向原　快速＝ひばりヶ丘
までの各駅・石神井公園・練馬・池袋　準急＝石神井公園までの各駅・練馬・池袋
準急(中)☆＝石神井公園までの各駅・練馬・新桜台・小竹向原　通勤準急＝大泉学園までの各駅・練馬・池袋
行き先　武－武蔵小杉 ／横－横浜 ／中－元町・中華街 ／竹－小竹向原 ／木－新木場 ／洲－豊洲 ／無印－池袋 ／下線－当駅始発
野球開催日は一部変更になります。

Stem-and-leaf plots show the place values of each data value. They are sometimes used in transportation schedules, like this train schedule from the Tokorozawa station in Saitama, Japan.

As an example, take a simple dataset with just seven values: 4, 9, 12, 13, 18, 24, and 27. The data are arranged in downward-ascending order with the first digit on the left side and the second (tens) digit on the right. Obviously, for more detailed and complex data, the stem-and-leaf plot may not be a useful approach.

Stem-and-leaf plots are most useful as a reference tool, like a public transportation schedule, or to highlight basic distributions and outliers in a more limited set of data. The Japanese train schedule for the Tokorozawa station in Tokyo on the facing page shows the timing of train arrivals over the course of a day. The hour digit is shown in the far-left column, minutes are shown to the right. The first train begins running at 5:00 am, the next train leaves at 5:08 am, then 5:18 am, and so on. Because the stem-and-leaf plot is a table, it loses some of the advantages of a traditional data visualization, but the leaves illustrate some basic view of the distribution.

CONCLUSION

The collection of graphs in this chapter demonstrates how we can show the distribution of our data or uncertainty around specific values. Some of these charts show summary measures or specific values. We can aggregate the distribution into bins to visualize the distribution in a histogram. Or we might use specific percentiles to generate a box-and-whisker chart, for example, or show stock price variation in a candlestick chart.

With better data visualization tools and faster computers, we can show more data than ever before. Beeswarm charts, wheat plots, and raincloud plots include specific data points. While these visualization types are useful for presenting the full data set to our readers, they have their limits: We can only show so many data points before they begin to overlap and obscure one another.

The graphs in this chapter can introduce challenges to readers who are not familiar with statistical concepts and measures of dispersion. As always, the most important thing you can do when creating your graphs is to remember your audience. If you are a PhD economist presenting your work at your university lunchtime seminar, you don't need to explain the median or variance or even the 95-percent confidence interval. If you were to present the same results to a general audience, however, you would include definitions and annotation. This isn't to say you should avoid presenting these numbers or that you need to dumb things down, but rather that you may need to take time explaining concepts within the visual. The planning, testing, and conceptualizing of your visualization will pay off in the long run as you more effectively communicate your work with your audience.

GEOSPATIAL

There is an obvious advantage to plotting geographic data on a map—people can find themselves in the data. They can literally *see* themselves in the data, a connection with the subject matter that other visualizations cannot muster. Plotting geographic data may mean adding color to geographic areas likes states or countries, or adding circles, squares, lines, or other shapes on top of a geographic map.

Data-driven maps are not new. In 1922, the "Maps and Sales Visualization" on the next page shows the reader thirty-six different ways to place data on a map. The author writes:

> The use of maps has to do with the visual representation of space. Therefore, in all map work we start with an outline . . . The fact that the Earth is a globe makes visualization of maps a difficult process; arbitrary methods have to be used to get the surface of the ball represented in a flat picture.

This chapter begins with some of the basic challenges of visualizing geographic data and then presents some alternatives to the basic map, the modern version of this 1922 visualization.

The maps in this chapter use a color palette often used by the *Washington Post*. Maps of the U.S. political system in this chapter use the basic red-blue color palette used by many newsrooms.

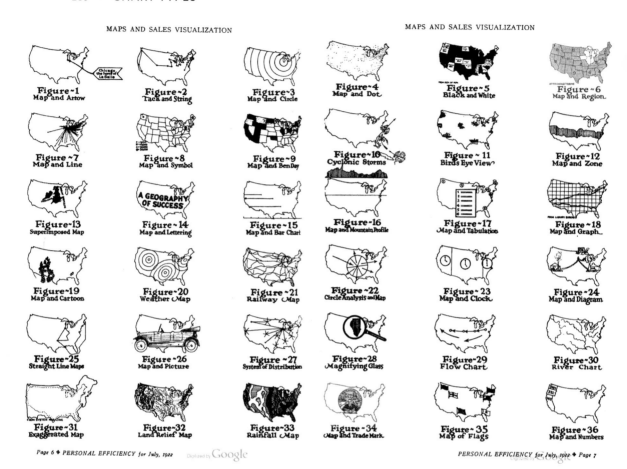

E.P. Hermann's (1922) "Maps and Sales Visualization" shows an early example of the multitude of ways to place data on a map.

THE CHALLENGES OF MAPS

Aaron Koblin's *Flight Patterns* is one of those maps that can entrance the reader. Koblin plots all flight paths in the skies above the United States in twenty-four hours. The static version of the map (there is an interactive version in which you can zoom into any area of the country) shows the entire country, major airports, and the activity of the skies above. It's not a visualization that ranks the biggest airports or tells you how to avoid delays, but it quickly shows the traffic patterns in American skies.

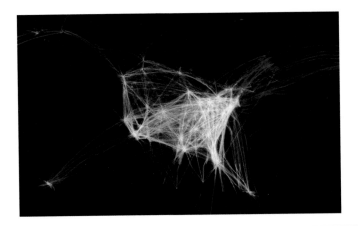

Aaron Koblin's *Flight Patterns* project, which consists of static and interactive maps, shows all flight paths in the skies above the United States over twenty-four hours.

There are some distinct challenges when creating maps. The biggest is that the size of a geographic area may not correspond to the importance of the data value. Russia is more than 6.6 million square miles, almost twice the size of Canada, and so it takes up a lot of space on a map. At 270,000 square miles, Texas is roughly the size of California and Colorado put together, but it's actually less than half the size of Alaska (665,000 square miles), which you might not know because most maps of the United States tend to distort it and arbitrarily position it out to sea, south of California. The point is that the data values for Russia, Texas, and Alaska may not correspond to their importance in the data, and the map can distort our perception of the important values being visualized.

I find that many people should approach their desire to create a map with more skepticism and critical thought. Is a map truly the best way to present geographic data? Or are you just showing where people live? Does it show the relationships we want to explore or are we simply relying on the fact that we have geographic identifiers? This chapter will explore the perceptual issues of maps and why they are not always the best medium through which to demonstrate our points. This isn't to say we should *never* make a map—in many cases we need to make a map to better understand our data—but, especially with maps, we should always take a step back and consider whether it is the right visualization choice.

The message is simple: there are many ways to present geographic data, and there are lots of objects, shapes, and colors you can add to maps. Which kind of map you use to visualize your data will depend on two questions: How important are the geographic patterns? And how important is it for your reader to see a familiar map?

CHOROPLETH MAP

Perhaps the most familiar data map has perhaps the most unfamiliar name: a choropleth map. Choropleth maps use colors, shades, or patterns on geographic units to show proportionate quantities and magnitudes. You likely already know how to read this choropleth map of per capita GDP around the world—it's a simple and recognizable shape that lets you quickly and easily find countries (and, by extension, yourself) in the visualization.

This color palette is also easy to understand—smaller numbers correspond to lighter colors and larger numbers to darker colors (and what is sometimes called a "color ramp"). More often than I care to count, map creators will use an incorrect color palette. For example, for this map, someone might use a diverging color palette in which colors progress outward from a central midpoint. Unless we are comparing per capita GDP above and below some midpoint number, such as the average GDP, the diverging color palette is a bad choice. Instead, we should follow what has become a simple standard of lighter colors to darker colors. We discuss color palettes in more detail in Chapter 12.

PER CAPITA GDP AROUND THE WORLD IN 2017

100,000
75,000
50,000
25,000

Source: The World Bank

The choropleth map is perhaps the most familiar data map. Colors correspond to data values and are assigned to the different geographic areas on the map.

It may be bit difficult to find smaller countries or countries that you don't already know, but the overall shape is familiar, well-known, and easy to understand. Did you know Luxembourg had the highest per capita GDP in 2017 at more than $104,000 per person (for reference, per capita GDP in the United States was $59,500)? It's a country of only one thousand square miles (for reference, France is almost 250,000 square miles) and difficult to find on this map, but it has the highest per capita income.

Maps like these introduce a geographic distortion—the size of the geographic area may not correspond to the importance of the data value. Even with this distortion, however, maps are an easy and familiar way to present geographic data to our readers.

There are a variety of alternative map types that correct this geographic distortion. Cartograms, for example, resize the geographic units according to their data values (see page 233) and a tile grid map uses a series of equal-sized squares (see page 238), not to mention the other chart types we can use to plot geographic data, such as a heat map (see page 112). The tradeoff with any of these alternative approaches is that the map is no longer as familiar to the reader as the standard map. But, as Kenneth Field, author of the data visualization and cartography book *Cartography*, once noted, "None of these maps are right and none of these maps are wrong. They are all just a different representation of the truth."

CHOOSING THE PROJECTION

One challenge with mapping data is the map *projection* the creator chooses. The world is a globe, but maps are flat. A mapmaker must choose a map projection to transform the world's sphere into a two-dimensional plane. All maps distort the surface of the planet to some degree and there is considerable debate about which projection does the best job depicting the earth in two dimensions.

The one you are likely most familiar with is the Mercator projection. This is the map used in early versions of Google Maps and is the default in many data visualization tools like Tableau and PowerBI. Developed in 1569 by Flemish geographer and cartographer Gerardus Mercator, it became the standard map projection for nautical purposes. A sailor could draw a straight line between two points and measure the angle between that line (called a *rhumb line*) and the meridian (the vertical line that runs between the north and south poles) to find their bearing. While it may be useful for sailing the seas, the Mercator projection distorts the size of objects as the latitude increases from the equator to the north and south poles. Thus, countries closer to the poles—like Greenland and Antarctica—appear much larger than they actually are. In the Mercator map shown on the next page (on the left), Greenland looks to

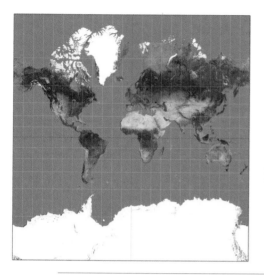

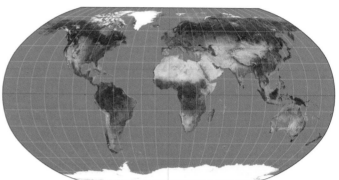

Map projections can influence our perception of the map. The Mercator projection on the left, for example, looks considerably different than the Robinson projection on the right.
Source: Wikimedia user Strebe.

be about the same size as South America, when it is in fact about one-eighth the actual size. In the Robinson projection on the right, country areas are closer to their true sizes.

There are three major categories of map projections:

CONIC

▶ Conic maps are as though a cone was placed over the Earth and unwrapped. Conic projections are best suited for mapping long east-west geographies, such as the United States and Russia, because the distortion is constant along common parallels. The Albers Equal Area Conic and the Lambert Conformal Conic are two of the more well-known projections.

CYLINDRICAL

▶ Cylindrical maps work like conic maps but use cylinders instead of cones. Like the Mercator projection, cylindrical maps inflate geographic areas farther from the center.

PLANAR (AZIMUTHAL)

▶ With this approach, the planet is projected on a flat surface. All points are at the same proportional distance from the center point, such as the north pole, but the distortion gets larger as you move further from that center point.

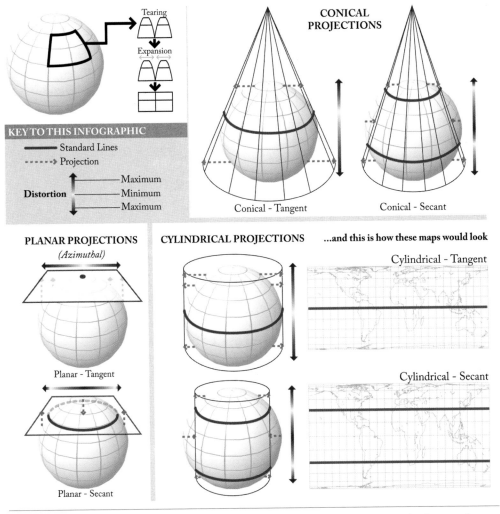

To some degree, all maps distort the surface of the planet. Alberto Cairo's diagram from *The Truthful Art* shows a selection of different types of map projections.

There isn't necessarily a right or wrong map projection, though I'm sure some cartographers would disagree! But each has tradeoffs, and serious mapmakers dig deep into the different properties of these projections and weigh the different options. Many people in the data visualization field shy away from the Mercator projection because of its obvious drawbacks, though it can work well for small areas. For the United States, you can tell a map is using the Mercator projection because the top left border of the country is a straight line. The Albers projection, by comparison, preserves the east-west perspective, which you can see by the slight curvature in the northern border.

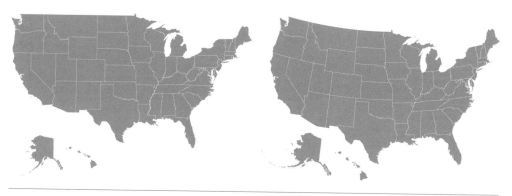

Notice how the northern border of the United States looks straight in the Mercator map on the left but is more curved in the Albers projection on the right.

CHOOSING THE BINS

As we start to add data to a choropleth map, our first consideration is the choice of intervals (or "bins") that will shade the geographic units. Placing data into discrete categories is, at its core, an aggregation problem. By combining several states or countries into a single bin, we don't know how different those units are from one another.

The map of the United States on the facing page, for example, shows median household income in 2018 in each state. How the states are placed into groups (the "bins"), the map shading, and ultimately our readers' perception of the data depends on our choices. Massachusetts ($86,345) and Maryland ($86,223) had the highest median household incomes in 2018 and fall into the highest bin with the darkest color; New Mexico ($48,283) and

Mississippi ($42,781) are at the other end of the distribution and are shown with the lightest shades.

There are four primary binning methods for creating maps.

NO BINS

This is essentially a continuous color palette (or "ramp") in which each data value receives its own unique color tone. On the one hand, this is easy because we don't need to think too much when we create the map—the colors ramp up from the lightest color for the lowest value to the darkest color for the highest value. On the other hand, the resulting color gradient may generate spatial patterns masked by subtle changes in color. In this example, it's hard to distinguish the differences between Iowa ($68,718), Nebraska ($67,515), and Wyoming ($62,539).

MEDIAN HOUSEHOLD INCOME IN THE UNITED STATES IN 2018

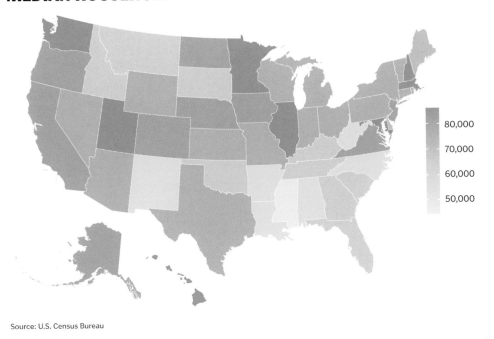

Source: U.S. Census Bureau

The continuous color palette (or "ramp") seamlessly goes from lighter colors (smaller values) to darker colors (larger values).

EQUAL INTERVAL BINS

In maps with a discrete number of bins, the default approach is to typically divide the data range into an equal number of groups. For example, in a map with four bins and a data range from 1 to 100, we end up with four equal groups (1–25, 26–50, 51–75, and 76–100).

This approach more clearly distinguishes geographic units (such as states) than the continuous (no bins) option, but it can mask the magnitudes of those changes by putting states in the same or different bins. In cases where the distributions are highly skewed, this approach may unevenly distribute the geographic units across the bins. In this map, the bins are split into equal units of $10,000, which results in five states in the bottom category, ten in the next, seventeen in the middle, fourteen in the next, and five in the top category.

MEDIAN HOUSEHOLD INCOME IN THE UNITED STATES IN 2018

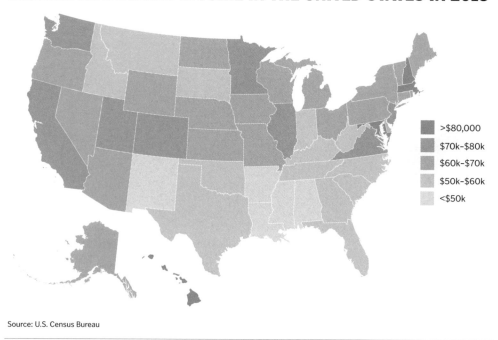

>$80,000
$70k-$80k
$60k-$70k
$50k-$60k
<$50k

Source: U.S. Census Bureau

Dividing the data into equal intervals—such as $10,000 gaps—is one way to add color to a data map.

DATA DISTRIBUTION BINS

We could also cut the data into different bins. For example, instead of having a bin at equal intervals, we could arrange the bins to hold the same *number of observations*, such as quartiles (four groups), quintiles (five groups), or deciles (ten groups). Or we could use other measures to collapse the data into groups, such as the variance or standard deviation.

The data distribution approach clearly shows differences between the geographic units, but the created cutoffs may not be numerically meaningful. In this map that divides the country into five equal groups (quintiles), Connecticut is just barely in the top group with a median household income of $72,812, while Minnesota is placed in the next lower bin, even though it has a very close income estimate of $71,817.

MEDIAN HOUSEHOLD INCOME IN THE UNITED STATES IN 2018

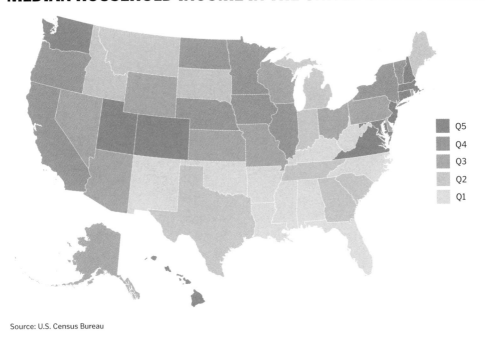

Source: U.S. Census Bureau

Another way to divide the geographic units is to divide the data into an equal number of observations, such as quartiles (four groups) or quintiles (five groups).

ARBITRARY BINS

In this approach, the map creator chooses the bin cutoffs based on round numbers, natural breaks, or some other arbitrary criterion. This method lets us avoid some of the odd breaks that might occur, such as the Connecticut-Minnesota example above, but it can also be misleading. A method in which the selected bins are based on larger groups or round numbers—even without looking at the data—might look like this:

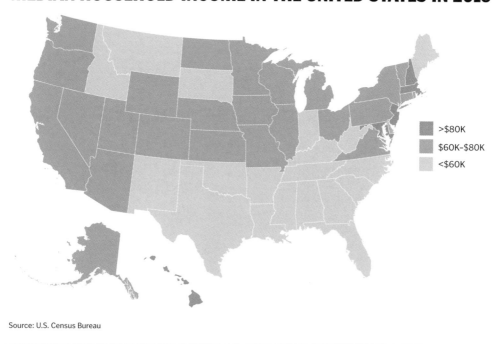

MEDIAN HOUSEHOLD INCOME IN THE UNITED STATES IN 2018

>$80K
$60K–$80K
<$60K

Source: U.S. Census Bureau

Depending on the goals of your data map, you can also divide the data into an arbitrary number of bins.

ALTERNATIVE OPTIONS

This isn't to say any of these maps are right or wrong, but they highlight the importance of binning decisions in choropleth maps. To make the best binning decision, we can draw on Mark Monmonier's 2018 book, *How to Lie with Maps.* Instead of using equally sized bins arbitrarily or letting the software tool decide which breaks to use, try considering—and showing—the actual distribution. If I add a column chart to the arbitrary bin method map, the bin breaks and the distinct differences between the values is apparent. Adding another graph takes up more space, but it gives readers a clear picture of the data.

Including the number of observations in each bin somewhere in the visualization can also reveal the distribution to your reader. In this version, the legend is converted to a small bar chart to show the number of observations in each bin.

MEDIAN HOUSEHOLD INCOME IN THE UNITED STATES IN 2018

>$80,000 2 states
$75k–$80k 5 states
$63k–$75k 16 states
$50k–$63k 22 states
<$50k 5 states

Source: U.S. Census Bureau

One way to help readers better understand the data in a map is to pair it with another visualization type, like a bar chart. This visualization has a small bar chart embedded within the legend to make clear how many states are in each group. This is not a necessary component, but one that can help the reader understand the distribution of the data on the map.

LABELING THE BINS

Another consideration when we create a map is how we *label* the bins. Let's use the visualization we just created of the map and bar chart of median household income.

In that map, the definition of the bins is arbitrary. The top bin is defined as ">$86,000" but because the maximum value in the next category $81,346 (New Hampshire), the top bin label could just as easily be ">$85,000," ">$82,000," or even ">$81,346."

The legend for this map can be defined several ways.

1. Instead of round numbers, we could use the actual income amounts, which still leaves us with arbitrary bins. In this case, it isn't clear whether $86,000 is in the fourth or fifth group. There are a few ways to do this, for instance with separate boxes or a single image with labels just below. (One possible solution is to explicitly note which bins are inclusive or exclusive of the upper and lower bounds.)

      <$50k $53k-$63k $63k-$77k $77k-$86k   >$86k

Note: Where data ranges appear to overlap, each range excludes its lower bound and includes its upper bound.

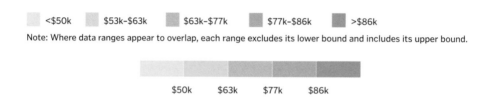

    $50k $63k $77k $86k

2. Another alternative is to define the bins on the actual data values.

This has the advantage of clearly showing the data values—for example, we can see the gap between $49,973 and $50,573 in the first two bins. The disadvantage is that the legend is overly detailed and complex. As the reader, you might wonder why all this precision is necessary and, depending on the content, you might wonder what's going on in the gap between the maximum of one bin and the minimum of the next.

  $42,781-$49,973 $50,573-$62,629 $63,938-$74,176 $77,067-$81,346 $86,223-$86,345

Note: Where data ranges appear to overlap, each range excludes its lower bound and includes its upper bound.

3. Alternatively, we could create a legend that includes these "gap" bins:

  $42,781- ☐ $49, 974- ■ $50,573- ☐ $62,630- ■ $63,938- ☐ $74,177- ■ $77,066- ☐ $81,347- ■ $86,223-
  $49,973 $50,572 $62,629 $63,937 $74,176 $77,065 $81,346 $86,222 $86,345

Although accurate and comprehensive, this legend feels busy and obscures the five original bins by adding four non-data bins that connect them.

There is no one-size-fits-all solution to this challenge, but here we have laid out the issues and tradeoffs of creating and labeling bins. Always consider the necessary level of precision (i.e., the number of decimal places), the overall number of bins, and the smoothness of the data (that is, if the data show no visible jumps, round-numbered bins may be fine).

SHOULD IT BE A MAP?

Before we start exploring other types of maps, let's first ask ourselves whether a map accomplishes our goals and best communicates our argument. Many maps are made simply because the creator has geographic data, not because the map is the best medium for that content.

Take this simple example. A 2016 *Washington Post* story examined the relationship between rates of suicide and gun ownership in the United States. The story explains:

> One 2006 study found that from the 1980s to the 2000s, every 10 percent decline in gun ownership in a census region accompanied a 2.5 percent drop in suicide rates. There are numerous other studies that show similar results.
>
> This pattern becomes clear when looking state by state. The states that have higher rates of gun ownership, where people have more access to guns, also have higher rates of suicide. Suicides are twice as common in states with high gun ownership than those with low gun ownership, even after controlling for rates of mental illness and other factors, according to a 2007 study.

Below these two paragraphs were two maps, which I have recreated on the next page as choropleth maps. Do these maps help you see the positive relationship between the gun ownership rate and the suicide rate? Can you pick out states that have the highest suicide rates and the highest gun ownership rates? I can't. Instead, I found myself jumping back and forth between the two maps trying to identify individual states and regions.

Instead, what if we placed the same data in a bubble plot? We could put the suicide rate on the vertical axis and the gun ownership rate on the horizontal axis, and scale the size of the circles by population. Adding color to differentiate between areas of the country makes it clearer that states in the upper-right area of the graph tend to be western and southern states, where gun ownership is higher and, it turns out, gun-control laws are weaker.

GUN OWNERSHIP RATE

Source: Kim Soffen via Miller et al 2007

SUICIDE RATE

Source: Kim Soffen via Miller et al 2007

Note: The original maps in the *Washington Post* story were actually tile grid maps—see page 238. Data were extracted from the *Washington Post* story and Miller et al. If you or someone you care for is in distress, suicide prevention and crisis resources are available at the National Suicide Prevention Lifeline in the United States at 1-800-273–8255. Many other countries have similar hotlines.

GUN OWNERSHIP AND SUICIDE RATES ARE POSITIVELY RELATED

(Suicide rate per 100,000)

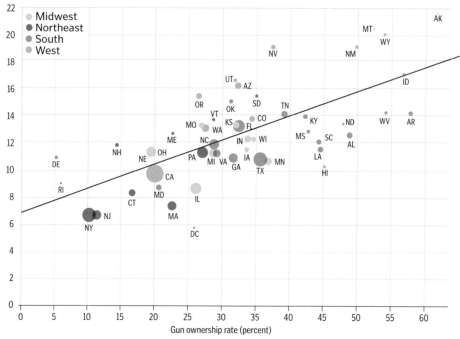

Source: Kim Soffen via Miller et al 2007 and the U.S. Census Bureau
Note: Circles sized by state population

A scatterplot can be an alternative to a pair of maps.

Bubble plots may be less familiar to readers than simple choropleth maps, but with a clear title and a little annotation, this chart can better demonstrate the relationship between gun ownership and suicide. As you prepare to create your next map, ask yourself, "Is a map the best visualization to communicate my argument?"

CARTOGRAM

One way to adjust for the geographic distortion of a typical choropleth map is with a *cartogram*, which reshapes geographic areas based on their values. There is an obvious tradeoff here: On one hand, these adjustments more accurately visualize the data because cartograms correlate the data and the geographic size. On the other hand, these graphs are not like the standard maps that we know and recognize. They are therefore not as intuitive for your reader. Your decision about whether to use a standard map or a cartogram will, as always, depend on your goals and your audience.

In his book *Cartography*, Kenneth Field summarizes the purpose of cartograms:

> The intent of most thematic maps is to provide the reader with a map from which comparisons can be made, and so geography is almost always inappropriate. This fact alone creates problems for perception and cognition. Accounting for these problems might be addressed in many ways such as manipulating the data itself. Alternatively, instead of changing the data and maintaining the geography, you can retain the data values but modify the geography to create a cartogram.

There are four primary types of cartograms: *contiguous*, *noncontiguous*, *graphical*, and *gridded*. One of the best ways to demonstrate the value of a cartogram is to examine the U.S. electoral college. In the U.S. election system, each state is assigned a number of electoral votes corresponding to its population, not its geographic size. Thus, states like Idaho, Montana, and Wyoming, which are very large in terms of square miles (325,412 square miles in total) but home to relatively few people, only have ten electoral votes between them. Massachusetts, by contrast, has eleven electoral votes and is 7,838 square miles, less than 2.5 percent of the size of those three states. In this choropleth map of the 2016 presidential election, Idaho, Montana, and Wyoming take up a disproportionate share of space on the map relative to their electoral votes.

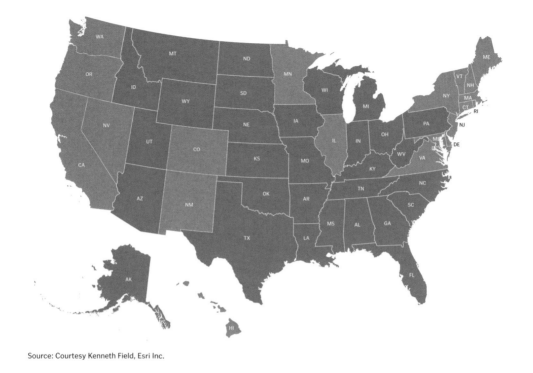

Source: Courtesy Kenneth Field, Esri Inc.

A standard choropleth map, this one shows the results of the 2016 U.S. Presidential election. Notice how much space large, but less populous states, like Idaho (ID), Montana (MT), and Wyoming (WY) in the northwest, are compared with smaller but more populous states, like Massachusetts (MA) and Rhode Island (RI) in the northeast.

CONTIGUOUS CARTOGRAM

The contiguous cartogram adjusts the size of each geographic unit according to the data. In the map at the top of the next page, for example, each state is sized to its number of electoral votes (or population, if you'd rather think of it that way). The version on the bottom-left of the next page uses squares to scale each state while retaining the original approximate geographic location and borders. The third map scales the counties in each state according to the vote share, which generates a more purple shade to the country, reflecting the split between the two political parties. Each of these approaches distorts the overall shape of the country as the data warps the geography, so they will look somewhat foreign to readers. The tradeoff becomes clear here—we can more accurately scale the states according to their data values, but the geography no longer looks as familiar.

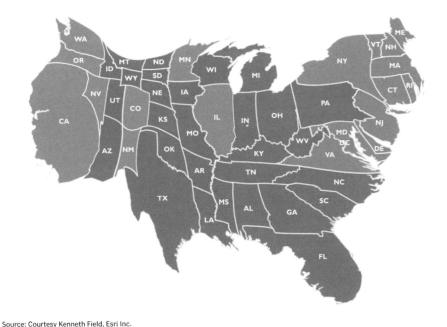

To address the fact that large but less populous states take up a disproportionate amount of space on the map, a cartogram scales the size of geographic units according to another data value. This map scales the states to their number of electoral votes.

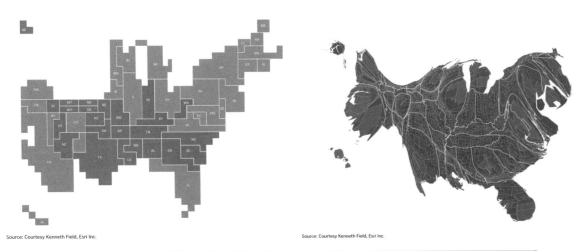

Other contiguous cartograms (using squares on the left or scaling counties on the right) are alternative ways to try to overcome the geographic distortions that occur in the standard choropleth map. But these maps are almost surely to be less familiar to readers than the standard map.

NONCONTIGUOUS CARTOGRAM

Now that you've seen a contiguous cartogram, you can probably guess what a noncontiguous cartogram looks like. In this approach, the size of the geographic units are based on the data value, such as population, but the units are broken apart and not kept adjacent to one another. In this way, we maintain the shape of the individual units but distort the overall view. One advantage of the noncontiguous cartogram is that we can build in more space for labels and annotation.

The map on the left scales each U.S. state according to its number of electoral votes and includes color to denote which candidate won those votes. The exact shape of each geographic unit also isn't necessary—the map on the right uses collections of squares for each state, here scaled to the number of electoral votes. In this version, we can again see how Idaho, Montana, and Wyoming look a lot smaller, while New York becomes much larger.

The noncontiguous cartogram was invented in the mid-1970s by Judy Olson, a geographer then at Boston University. "Probably one of the most interesting aspects of the noncontiguous cartogram," she wrote in her 1976 paper, "is that the empty area between units is meaningful. If the highest-density unit is used as the anchor [in other words, if the most dense geographic unit is used to scale all the other units], then the empty areas reflect the degree of discrepancy between the density in the most-crowded unit and the density in other units. The effect can be quite dramatic."

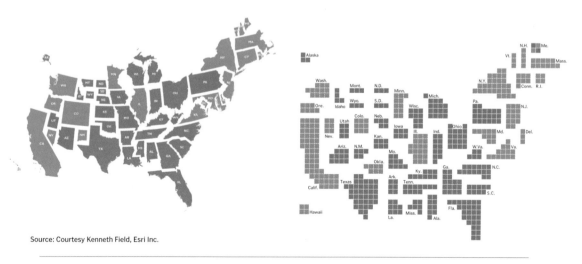

Source: Courtesy Kenneth Field, Esri Inc.

Noncontiguous cartograms like these break up the geographic areas. This helps for adding labels and annotation, but is a more unfamiliar chart type.

GRAPHICAL CARTOGRAM

Graphical cartograms do not maintain the original shape of the geographic units and instead use other shapes sized to the data values. Perhaps the most well-known graphical cartogram is the Dorling map—named for geographer Danny Dorling at the University of Oxford—which uses circles sized by area to the data.

A variant on the Dorling map is the DeMers Cartogram (or tilegram), which uses squares instead of circles. One advantage with the DeMers approach is that it minimizes the space between the geographic units. A disadvantage is that the entire geography becomes less recognizable. These two graphical cartograms again show the number of electoral votes in each state, and the colors again show which candidate won the state's electoral votes.

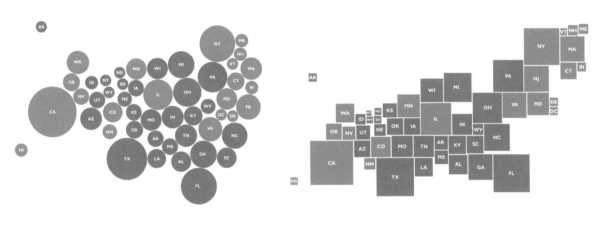

Source: Courtesy Kenneth Field, Esri Inc.   Source: Courtesy Kenneth Field, Esri Inc.

The Dorling (left) and Demers (right) cartograms move further away from the standard geographic maps and use shapes instead.

GRIDDED CARTOGRAMS

The fourth and final cartogram is the gridded cartogram, in which different shapes are scaled to the data and arranged so they maintain the general shape of the major geography. People most often use squares or hexagons to create these kinds of maps.

Take these hexagon grid maps, for example. The advantage of the hexagon over other shapes is that it offers us more flexibility to arrange the tiles closer to the true geography of the country. The map on the left shows one hexagon per state, shaded to encode median

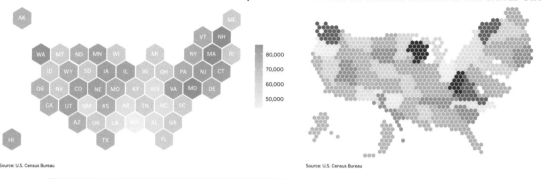

MEDIAN HOUSEHOLD INCOME IN THE UNITED STATES, 2018

Source: U.S. Census Bureau

MEDIAN HOUSEHOLD INCOME IN THE UNITED STATES, 2018

Source: U.S. Census Bureau

The hexagon grid map is named exactly for what it is: A gridded set of hexagons—either one for each geographic unit or scaled to its data value.

household income. The version on the right uses multiple hexagons per state with the color and number of hexagons corresponding to the data value.

Another popular gridded cartogram uses a single square for each geographic unit. This is often called a *tile grid map*. Here, median household income is divided into four groups (or quartiles).

MEDIAN HOUSEHOLD INCOME IN THE UNITED STATES, 2018

Lowest Quartile | 2nd Quartile | 3rd Quartile | Highest Quartile

Source: U.S. Census Bureau

The tile grid map uses a single square for each geographic unit.

As with all visualizations, there are tradeoffs. The advantage of the tile grid map is that each state is the same size, which abstracts from the geographic distortion. The disadvantage is that the geographic units are now not necessarily in the right place. In the tile grid map on the previous page, South Carolina is located east of North Carolina, California doesn't touch Arizona, and Wisconsin is north of Minnesota, all of which are not their real geographic relative locations. The arrangement of the tiles can be changed of course, but any decision is going to be arbitrary because we have moved away from true geography. But this map can also be easier to construct (it can be made in Excel with resized spreadsheet cells) than choropleths or cartograms.

Another advantage of the tile grid map is that it enables you to add more data in a consistent shape. In this tile grid map of the United States, small lines (or sparklines) are included in the square of each state showing the change in median household income between 2008 and 2018.

Another advantage of tile grid maps (and graphical cartograms, for that matter) is that we can add other shapes to the different geographic areas. Both of the tile grid maps on the next page use emojis to categorize the same household income estimates shown so far. Again, there's a clear tradeoff here: the emojis are a little more fun and a little more visual, but it has

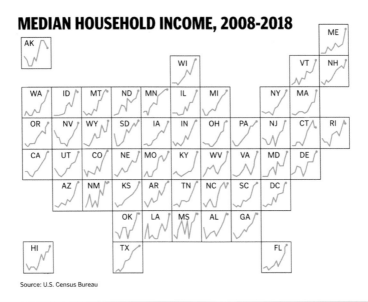

MEDIAN HOUSEHOLD INCOME, 2008-2018

Source: U.S. Census Bureau

One advantage of the tile grid map is because each state is the same size, we can add small lines, bars, or other graph types to each square.

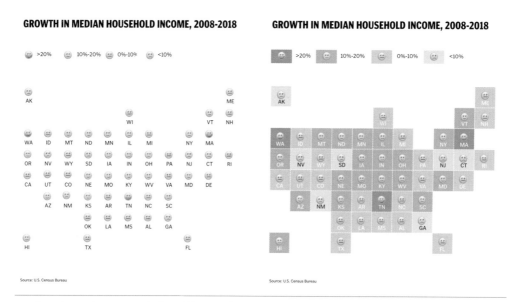

Another advantage of tile grid maps is that we can use other shapes, such as emojis.

a very different tone and makes it visually more difficult to immediately pick out values for groups of states or larger geographic patterns as in the original map. Adding some boundaries or shading (as in version on the right) merges the two approaches.

NON-AREA-BASED CARTOGRAM

There is one other class of cartograms worth mentioning. Maps do not always have to encode data, and they don't necessarily need to encode them accurately. A non-area based cartogram (or distance cartogram) distorts the physical geography by displaying relative time and distance. This version of the Washington, DC metro (subway) map, for example, shows relatively constant distances between stops, when in fact the distances vary considerably. Check out the western part of the Orange line that runs concurrently with the Silver line. The distance between my metro stop at East Falls Church and Ballston-MU, and between the Ballston-MU and Virginia Square-GMU stops are the same on the map, but the first trip is 2.7 miles long and the second trip is only half a mile long. Stretching the stops out to their actual distances is unnecessary here because the purpose of the map is to provide a compact view of the subway lines so riders can quickly and efficiently plan their trip.

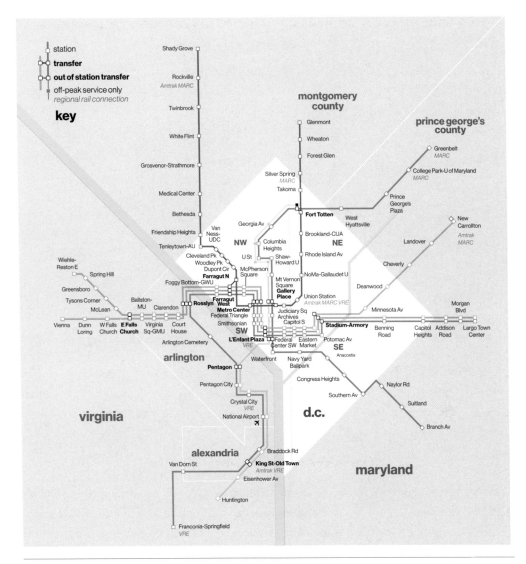

This version of the Washington, DC, metro map from designer Jacob Berman shows relatively constant distances between stops, even though that's not geographically the case.

A familiar map like this also gives us the opportunity to add data. As an example, the next map shows the number of passengers entering and exiting each station in the morning rush hour. Each station is turned into a pie chart in which the blue segment represents the share of people entering each station and the orange represents the share of people exiting. Even if

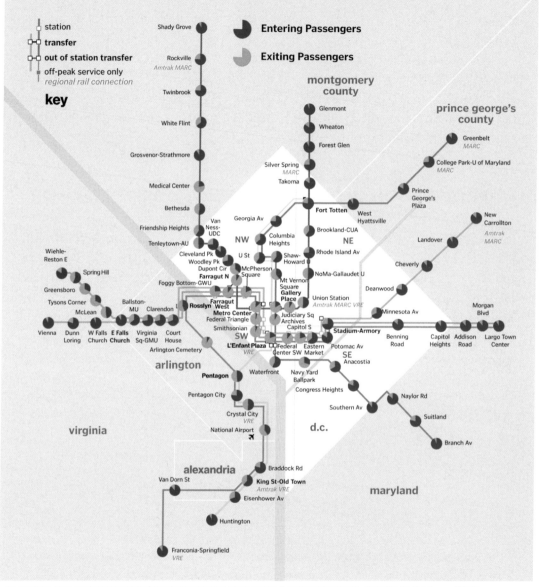

Source: Map from Jacob Berman; data from the Washington Metropolitan Area Transit Authority. Based on Matt Johnson (2012).

We can add other graphs to maps—pie charts here represent the share of people entering and exiting each metro stop in the morning. Even if you're not familiar with the DC metro, you can see how people tend to commute from the outer parts of the area (more blue areas in the pie charts) to downtown (more orange areas in the pie charts).

you're not familiar with this subway system, you can see the movement from the outer areas to the center city in the morning.

But let's make sure we remember our audience. These maps would be of interest to people who regularly use the DC metro system, but they would not be as valuable if I was an urban planner making an argument in Atlanta or Dallas or Berlin, because those audiences might not be familiar with the shape and arrangement. As always, we must consider the needs, expertise, and expectations of our audience.

PROPORTIONAL SYMBOL AND DOT DENSITY MAPS

Color and size are not the only encoding techniques we can use to visualize data on a map. Different shapes and objects—lines, arrows, points, circles, icons, even small compact bar graphs and pie charts—can all be placed on a map. These are known as *proportional symbol maps*, because the symbols are sized proportionate to the data. Be careful not to clutter the map or the reader will have difficulty identifying the most important information.

PER CAPITA GDP IN EUROPE, 2017

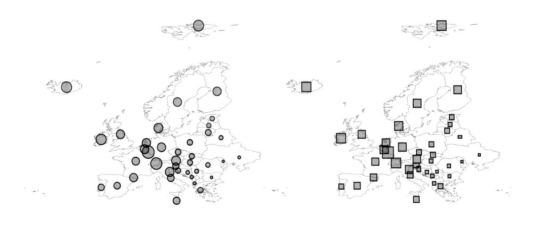

Source: The World Bank '

Different shapes and objects—lines, arrows, points, circles, and more—can all be placed on a map, sized according to the data value.

The two maps on the previous page show per capita GDP for European countries encoded with circles and squares, rather than using color to shade countries. Notice the importance of using a transparent color to make overlapping shapes visible, a technique we've seen in previous visualizations. Where there are dense clusters of areas, such as around Belgium and Netherlands, it can be difficult to find the shapes for individual countries. Aside from our inherent difficulty of discerning exact quantities from shapes like circles, the dense clusters can be a barrier to seeing specific geographic units.

A *dot density map* or *dot distribution map* takes the proportional symbol map in a slightly different direction by using dots or other symbols to show the presence of a data value. Symbols can represent either a single data value (one-to-one) or many values (one-to many). These kinds of maps can be data intensive, but they can also illuminate spatial patterns and clusters that would otherwise be difficult to visualize in a choropleth map or cartogram.

Dot density maps are valuable because they quickly and easily show geographic densities through the clustering of the symbols. The primary challenge, however, is that because these

Dot density maps include a dot or other symbol for a single (or many) data values. The "similarity" Gestalt principle helps us see the clusters of people around the country.
Source: Image Copyright, 2013, Weldon Cooper Center for Public Service, Rector and
Visitors of the University of Virginia (Dustin A. Cable, creator).

maps require exact geographic locations like addresses or longitude-latitude pairs, which are not usually available (or publishable), the symbols must be placed in a random or arbitrary position within a specific geographic area.

Consider the dot density map of the United States on the previous page. It uses data from the 2010 U.S. decennial census and places a dot for each of the country's 308 million residents in their Census block. Colors denote different racial and ethnic groups: blue for White people; green for Black people; red for Asian people; orange for Hispanic or Latino people; and brown for Native American people and people of multiple or other races. As with the map shown in Chapter 2, there is nothing in this map *except for the data*—no state borders, city markers, or other labels. We can still recognize it as the shape of the United States because people cluster in cities, and on borders and coasts.

FLOW MAP

Flow maps show movement between places. Arrows and lines denote the direction of the flow, and the width of the line can correspond to the data value. Flow maps can also encode qualitative data, but in those cases the width of the directional symbols may not be scaled to a data value. We saw one such example of a flow map on page 128 to show import and export trade flows between the United States and other areas of the world.

There are different types of flow maps. *Radial flow maps* (also called *origin-destination maps*) show flows from a single source to many destinations. A *distributive flow map* is similar, except that the flow from the single source can fork into many different lines. I like to think of these maps as those in the back pages of the airplane magazine, like this one in the back of the Delta Airlines magazine (see next page). This is a distributive flow map because it shows all of the various connections.

As a slight aside here, you might take a look at this map and think, "Wow, that's cluttered! How am I supposed to track my flight here?" But that's not the purpose of the map—instead, the intention of the map (or at least how I infer it) is to demonstrate Delta's domestic "unrivaled coverage"—all the many flights Delta offers around the United States. Showing the tangled web of *all* the flights communicates that point.

Perhaps the most famous flow map (at least in the data visualization field) is Charles-Joseph Minard's 1869 map of Napoleon's Russian Campaign of 1812 to 1813 and the connected, but lesser-known map of Hannibal's 218 BC march through the Alps to Rome. Minard was a French civil engineer who conducted in-depth studies over many decades to

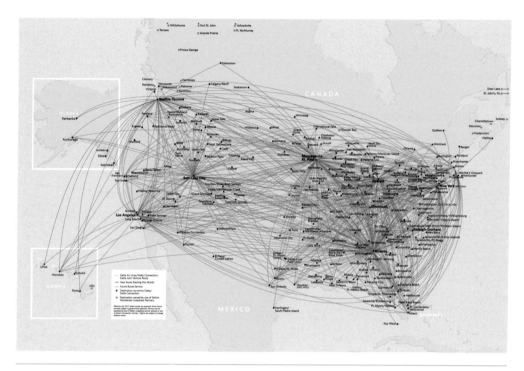

Flow maps show movement between places. Even though this flow map from Delta Airlines looks cluttered, it meets the airline's goal: demonstrating "unrivaled coverage."

create compelling visualizations in support of his research. Minard's Napoleon map—the bottom panel of the image on the next page—shows the "progressive losses in men" that Napoleon's army suffered as it marched into Russia and back.

Starting from the left, 420,000 men invaded Russia in June 1812. By the time the army reached Moscow four months later, only 100,000 troops were still alive. When Napoleon ordered a retreat in the fall, the army was forced to fight through blistering cold (dropping to a low of −30 degrees Celsius), and was ultimately reduced to 10,000 soldiers when it exited Russia in late 1813.

Minard's map integrates six different data values into a single view:

1. The number of troops (thickness of the lines);
2. The distance traveled (scale in the lower-right);
3. The temperature (line graph at the bottom);
4. The time (also included in the line graph at bottom);

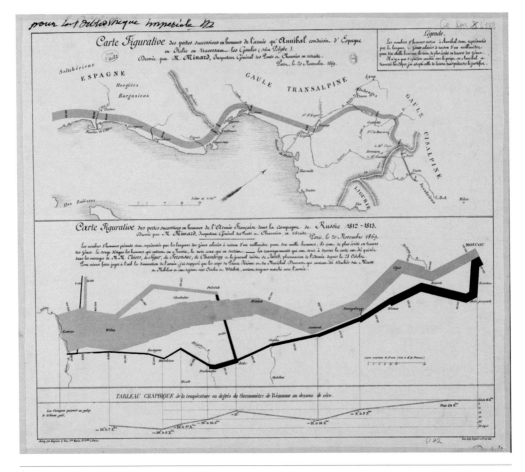

Charles Joseph Minard's Hannibal and Napoleon maps show the march of two different armies. Photo courtesy of Ecole nationale des ponts et chaussées.

5. The direction of travel (denoted by color—brown going eastward and black in retreat); and

6. Geography (cities, rivers, and battles—some but not all are included).

In her book on Minard's graphs, *The Minard System*, author Sandra Rendgen writes the following:

Minard created his visualization more than fifty years after the [Napoleon] campaign. It is a brilliant conceptual transfer: in applying the flow method to a military campaign, Minard shifts his entire focus to a single variable: the number of people in the flow. This variable sees only one type of variation—*a sharp and steady decline*. It seems to have been this potent and poignant message that made these two maps (and particularly the Napoleon one) so successful in telling a story about the cataclysm of war.

CONCLUSION

In this chapter we surveyed the promises and perils of visualizing geographic data. Sometimes the varying sizes of geographic units may distort the data. Other times sizing the geography to the data may make a familiar geography look foreign.

When working with geographic data, your instinct may be to create a map. But take a moment to consider: Is a map the best way to present your data? Does your reader need to see the exact differences between data values? If so, the aggregation problem inherent in many maps may make that difficult. Are there clear geographic patterns to be seen in the data? If not, then the map may not actually help the reader see your point.

If a data map *is* the right approach, carefully consider the map projection you use and whether the standard choropleth map is the best choice. Maybe some kind of cartogram—even with all its flaws—would be a better fit for your context and reader.

You may also determine that the best approach is to *combine* visualization types. Depending on your final publication type, you might use multiple visualizations, say, a map with a bar chart or table. This approach can help give your readers a familiar visualization type in which they can identify themselves and their location, but also help them gain a better, more detailed view of the actual data.

RELATIONSHIP

The charts in this chapter show relationships and correlations between two or more variables. Perhaps the most familiar chart type in this class is the scatterplot, a chart in which the data are encoded to a single horizontal and vertical axis. Other shapes and objects can also be used to visualize the relationship between two or more variables—a parallel coordinates plot uses lines, while a chord diagram uses arcs within a circle. These charts can show the reader correlations and even causal relationships.

For this chapter, I use the color palette and font from a famous bubble chart created by the Swedish academic Hans Rosling and his colleagues at Gapminder, a foundation dedicated to visualizing statistics. Rosling's Gapminder project didn't lay out a specific data visualization style guide, but the visualizations in this chapter use the basic colors and font (Bariol), with additional styles based on other visualizations from the Gapminder website.

SCATTERPLOT

The scatterplot is perhaps the most common visualization to illustrate correlations (or lack thereof) between two variables—one variable is plotted along a horizontal axis, and the other along a vertical axis. The specific observations are plotted in the created space. Unlike a bar chart, the scatterplot axes do not necessarily need to start at zero, especially if zero is not a possible value for the data series.

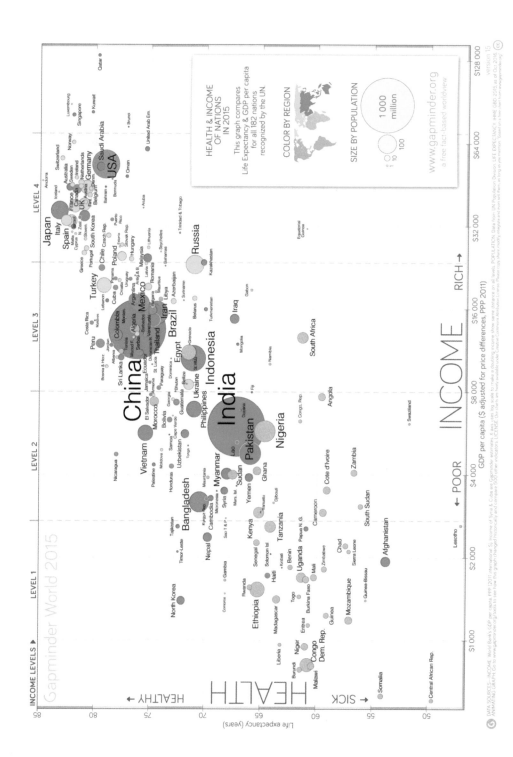

Source: Based on a free chart from www.gapminder.org

Source: XKCD

One of the most famous graphs that shows the relationship between two variables is the set of scatterplots created by Rosling and his colleagues at Gapminder. A physician by training who spent roughly two decades studying public health in rural areas across Africa, Rosling is perhaps best known for engaging presentations and data visualizations, and for promoting the use of data to explore issues around international development. In Rosling's 2014 TED Talk, he showed an animated scatterplot of the relationship between the fertility rate (number of births per woman) and life expectancy at birth from 1962 to 2003 for countries around the world.

As I've noted previously, there are many nonstandard graph types—scatterplots included—with which your reader may be unfamiliar. This doesn't mean you can't use these visualizations, but it does mean you should be mindful that your reader may be unfamiliar with them, and consider how to prepare them to understand your graphs.

Some readers may be familiar with scatterplots (or other nonstandard graph types for that matter) even if they are not familiar with reading *data* in scatterplots. *New York Magazine's* weekly *Approval Matrix* on the next page, for example, is their "deliberately oversimplified guide to who falls where on our taste hierarchies." Bits of text, images, and icons are plotted in a space defined by the "Highbrow-Lowbrow" vertical axis and "Despicable-Brilliant" horizontal axis. Though it's a light-hearted way to list popular news tidbits, it is, at its core, a scatterplot.

Moving to scatterplots that are a bit more, shall we say, data-driven, these two scatterplots on page 253 show net immigration (defined as the number of people migrating into a region divided by the total number of migrants moving in and out of a region) plotted along the horizontal axis and per capita gross domestic product (GDP) along the vertical axis. The version on the left uses a single color with a slight transparency (or "opacity") so the reader

THE APPROVAL MATRIX

Our deliberately oversimplified guide to who falls where on our taste hierarchies.

HIGHBROW

DESPICABLE — **BRILLIANT**

- Venice sinks.
- Mike Bloomberg's **curiously** timed apology for stop and frisk.
- President Obama's **side-eye** at "certain left-leaning Twitter feeds."
- Lydia Davis's *Essays One*: proof that books are still the **best** mobile devices.
- AI-generated art by Paris-based collective Obvious doesn't do well at auction. **Obviously.**
- Disney is **demolishing** an entire block of historic buildings in West Soho for its big, new, meh HQ.
- The **slow-cook** impeachment. (Sondland has the receipts.)
- The $1,250 Basquiat-**themed watch.**
- The **well-named** Brice Marden show at Gagosian: "It reminds me of something, and I don't know what it is."
- The decadent blitheness and high-minded **spectacle** of the Met's *Akhnaten*.
- Elon Musk's Starlink satellites are **blocking** astronomers' view of the stars.
- The editor of the Atlanta *Journal-Constitution* fact-checks the portrayal of the press in Clint Eastwood's *Richard Jewell*, says it "often differs from **reality**."
- The meticulous, beautiful, and **oh-so-very experimental** *History of Violence* at St. Ann's Warehouse.
- Stephen Miller revealed to be a … **white nationalist?** For real?
- Prince Andrew's attempt to **explain** his royal self on the BBC.
- Peter Dinklage's swashbuckling return to the stage in the **moody**, musical *Cyrano* at the Daryl Roth Theatre.
- Hannah Cullen's Young Dance Collective's **teen-angst** piece *Everything I Was Never Taught* at New York Live Arts.
- College students demand college journalists stop making **journalism.**
- Famous author Sarah Dessen and her famous-author pals drag a former Northern State University student for **snobbish** comments about her YA work.
- Flea's memoir and **child-rearing** guide, *Acid for the Children.*
- **Cruisy** holiday art gift book: *Peter Berlin: Icon, Artist, Photosexual.*
- Joe Biden is too worried that pot is a **gateway** drug to sign on to legalize it.
- WE VAPE! WE VOTE!
- TWA Hotel is opening a **jet-fuel-scented** runway ice-skating rink.
- Netflix **hires** Robert Towne to co-write a prequel series to *Chinatown.*
- Trying to watch the **brutal efficiency** of the Houston Rockets.
- Trump sours on e-cig ban for fear of losing the **vaper vote.**
- The new Milk Bar flagship at the Ace Hotel. **Mooooo!**
- Diana Vreeland's great-granddaughter Caroline sings, with **Pomeranians** in the audience, at the Top of the Standard.
- Pete Davidson is dating Cindy Crawford's **18-year-old** daughter, Kaia Gerber?
- The way to a single guy's heart is through **chicken parm**, the New York *Post* reports.
- Someone **dyes** the fountain in front of 1251 Sixth Avenue green …
- Two Arkansas chemistry professors are accused of cooking **meth** in a school lab …
- … Probably unrelated: Kermit the Frog pivots to **fashion** influencer.
- **Baby Yoda.**
- … And whoever thought of South Dakota's "**Meth.** We're on It" PSA definitely was.
- Now that your house is decluttered, Marie Kondo wants to **sell you** "joyful" objects via her online store.
- *Vogue* bleeps out Cardi B saying the word **fart** in a video interview. Censorship!
- METH. WE'RE ON IT.
- Soon on sale: Spot, the **terrifying** robo-dog! (Laser guns not included.)
- 23,000 NYCHA tenants were without **heat and hot water** during the cold weather.
- The **churro** vendor is now a public menace?
- The **red-panda** babies at Prospect Park Zoo!

LOWBROW

New York magazine's "Approval Matrix" is a scatterplot. Not truly data driven, but a scatterplot nonetheless.

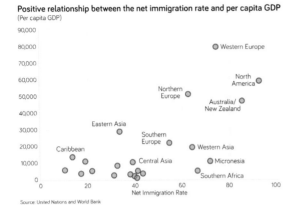

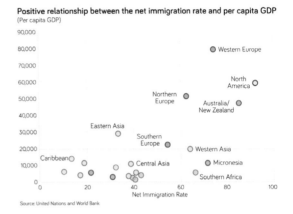

Positive relationship between the net immigration rate and per capita GDP
(Per capita GDP)

Positive relationship between the net immigration rate and per capita GDP
(Per capita GDP)

Source: United Nations and World Bank

Source: United Nations and World Bank

Both scatterplots show the association between net immigration and per capita GDP, using either a single transparent color (left) or different colors for regions of the world (right).

can see overlapping values. The same transparent effect is used in the scatterplot on the right, but this time colors capture the different regions of the world.

A scatterplot can help the reader see whether two variables are associated with one another. If the two variables move in the same direction—to the right along the horizontal axis and up along the vertical axis—they are said to be *positively correlated*. In other words, when both variables get bigger or smaller simultaneously, they are positively correlated. If they move in opposite directions, they are said to be *negatively correlated*. And if there is no apparent relationship, then they are not correlated (see Box on the next page). In the two scatterplots above, you get a visual sense that the two metrics are positively correlated—that net immigration is higher for regions with higher per capita GDP, in particular Western and Northern Europe, Australia/New Zealand, and North America.

One way to make the correlation even clearer is to add what statisticians call a *line of best fit* to the scatterplot. These are also called "regression lines" or "trendlines," and they show the general direction of the relationship. The statistical calculations to create lines of best fit are beyond the scope of this book, but the point is that you can make it even clearer to the reader in what direction (and to what magnitude) the two variables are correlated by calculating and including this line.

While the scatterplot is becoming a more common chart type, readers may still have difficulties reading and understanding it. A 2016 Pew Research Center survey showed that about 60 percent of people could correctly identify what they were seeing in a scatterplot.

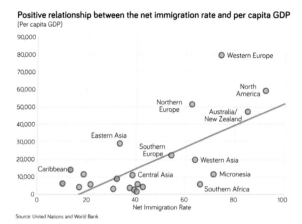

Positive relationship between the net immigration rate and per capita GDP
(Per capita GDP)

Source: United Nations and World Bank

A *line of best fit* visualizes the correlation between the two variables.

CORRELATIONS

You have probably heard the old adage that "Correlation does not imply causation." We hear this so often because people regularly assign a causal relationship between variables that is actually coincidental. People eat more ice cream when it's hot outside, but that doesn't mean that more ice cream consumption *causes* the temperature to rise. As you look at your data and visualize the relationship between the observations within, be careful to understand when something may be correlated and when it might be causal. The less we know, the more we observe correlations and not causation.

Correlation is a measure of the strength of the linear association between two quantitative variables. The most common measure of correlation is the *Pearson correlation coefficient*, which measures the linear association between variables and is typically denoted with the Greek letter rho (ρ).

The sign and value of a linear correlation coefficient describes the direction and magnitude of the relationship between the variables. The value of the correlation coefficient ranges between −1 and +1. Values of −1 signal a perfectly negative correlation, +1 signals a perfectly positive correlation, and 0 represents no linear correlation. Positive correlation coefficients denote a positive correlation, which means that if one variable gets bigger, the other variable also gets bigger; negative coefficients mean the variables move in opposite directions, one variable gets bigger as the other gets smaller.

This discussion is related to these linear associations (or relationships) and it is also possible for two variables to have a *nonlinear* relationship. A linear association is a statistical term that describes a straight-line relationship between one variable and another. A simple example is how we might calculate distance as rate times time. In this case, if we were driving sixty miles per hour for two hours, we would travel 120 miles. The driving speed does not change over time, so the relationship is linear.

A nonlinear association, by comparison, refers to patterns in data that curve or break from the straight linear trend. Consider, as an example, the profit a company makes from a new product. When it is first released, there is little competition and sales grow. As sales continue to rise, public awareness increases, and profits start to roll in. Competing companies then start making their own version of the product and prices for the original fall to keep up, so profits decline. The company then develops a new version, and the whole cycle starts again.

As you can see in the images below, the two data values lie on a single diagonal line when they are perfectly positively or negatively correlated. When the two values move in tandem, either up or down, they are said to be positively or negatively correlated. The data in the bottom-right graph demonstrates the impact outliers have on this measure of correlation—moving a single point to the top-left part of the graph reduces the correlation from +1.0 to +0.8.

These visuals reinforce the importance of looking at our data as we conduct our analysis. Data visualizations help us not only communicate our work to our readers, but also enable us to explore our data. They can reveal patterns and relationships we wouldn't otherwise see. It's important not to leave this to the end of the workflow.

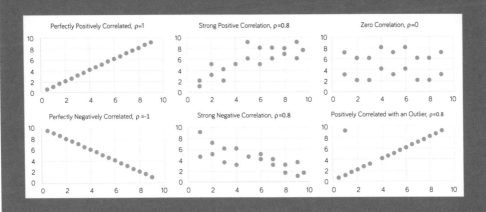

In 2016, University of Miami journalism professor Alberto Cairo drew a dinosaur with points in a scatterplot and dubbed it the "Datasaurus." His goal was to show the importance of visualizing your data in the exploratory phase. Imagine if you were teaching a data visualization class and asked your students to draw a scatterplot with 142 points, an average x value of 54.26, an average y value of 47.83, accompanying standard deviations, and a Pearson correlation of −0.06. Do you think anyone would draw a dinosaur?

In a 2017 paper, researchers Justin Matejka and George Fitzmaurice took the "Datasaurus" one step further and generated twelve alternatives that maintained the same summary statistics (mean, standard deviation, and correlation). The message of Cairo's "Datasaurus," Matejka and Fitzmaurice's paper, and Anscombe's quartet from Chapter 1, is that we should never rely on summary statistics alone but also on visuals of the data.

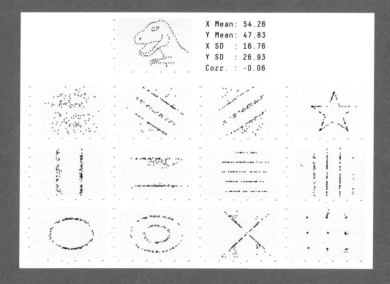

Source: Matejka and Fitzmaurice, 2017

BUBBLE PLOT

The scatterplot can be transformed into a bubble plot (or bubble scatterplot) by varying the sizes of the circles according to a third variable. The data points don't have to be circles; they can be any other shape that doesn't distort our perception of the data. As mentioned in the section on bubble charts, the circles should be sized by area, not radius (see page 123). Color can help group or highlight certain points or direct the reader's attention to different parts of the graph.

The circles in this bubble plot are scaled according to the population of each region. The same positive relationship is still evident, but you can now see the relative size of each circle.

Because it is a more uncommon graph, be especially mindful of how labeling and annotation can guide readers through the chart and its content. One strategy is to label each axis and the direction of the change along the axis. The bubble plot on the next page has a centrally located horizontal axis that reads "Net Immigration" and two other labels, "Higher Net Immigration" and "Lower Net Immigration."

To further guide the reader, we could add a 45-degree line, on which the values are equal. We could also highlight specific points with color or outlines, or we could add text to explain what a point or set of points means. Properly labeled, these elements can lead the reader through the graph and content. Even people who know how to read scatterplots can struggle for a moment to understand what is going on when there are lots of points.

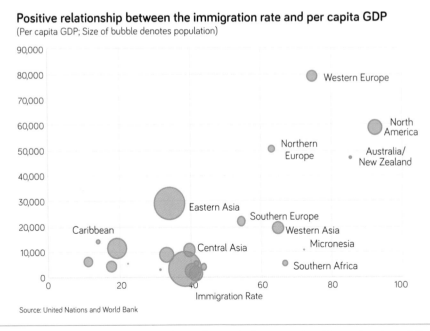

Positive relationship between the immigration rate and per capita GDP
(Per capita GDP; Size of bubble denotes population)

Source: United Nations and World Bank

A bubble chart adds a third variable to the typical scatterplot. Here, the size of the circles corresponds to the population in each region.

Positive relationship between the net immigration rate and per capita GDP
(Per capita GDP; Size of bubble denotes population)

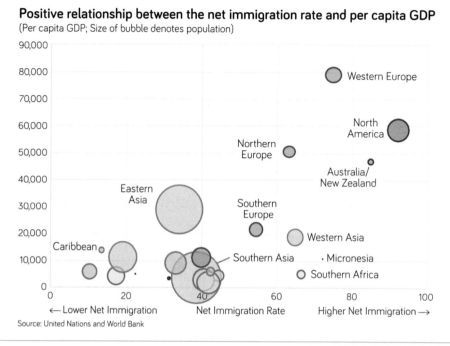

Source: United Nations and World Bank

As before, more colors can be added to denote another variable, such as region of the world.

Labeling certain points or groups of points—with text, color, or enclosing shapes—can help the reader navigate the chart and draw their attention. The plot above added color to denote regions of the world. The next two employ the same strategy and include the more than two hundred countries around the world. Using color like this lets the reader identify certain regions. If instead we want to highlight one specific region, we might use a single color for a region of interest and use gray to push the others to the background.

Two final points about scatterplots:

First, you will often see scatterplots that include labels for every single point, like the one on page 260. The end result is overwhelming clutter, with overlapping labels that are impossible to read. Luckily, we are far beyond the time where labels are the only way to convey information. If you believe there are readers who want to know the exact position of some of the points you didn't explicitly label, you can post a data file online, or create an interactive version using a tool like Tableau or PowerBI. Many academic researchers, for example, have an author page or webpage on their university website, as do many academic journals. These are great places to post the underlying data for your graphs.

Positive relationship between the immigration rate and per capita GDP
(Per capita GDP)

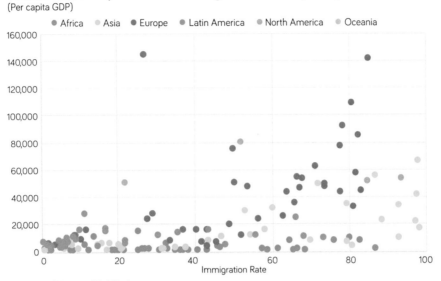

Source: United Nations and World Bank

European countries tend to have higher per capita GDP and immigration rates
(Per capita GDP)

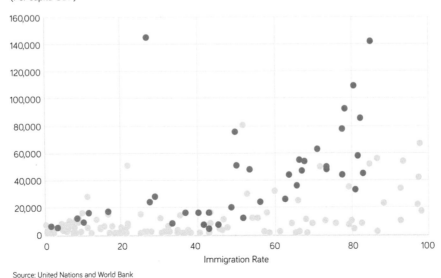

Source: United Nations and World Bank

As we've seen elsewhere, color can be used strategically to highlight different groups (for example, regions of the world as in the top graph) or to highlight a single group or data point (as in the bottom graph).

Positive relationship between the immigration rate and per capita GDP
(Per capita GDP)

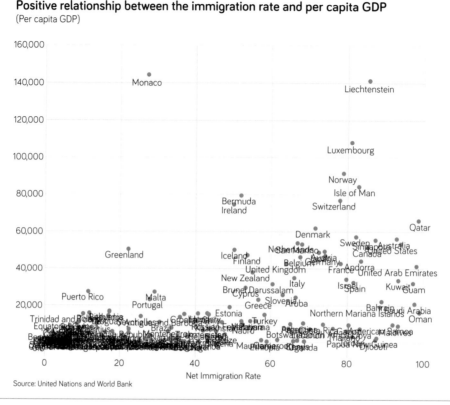

Source: United Nations and World Bank

There's rarely a case where labeling all of the data points is necessary. Reduce clutter like this so your readers can better see the data.

Second, there may be times when normalizing your data, calculating percent change, or taking the logarithm of your data may improve the visual clarity of your graph. This is especially true when our data are clustered too densely in a visual. The logarithm (or log for short), is, simply put, an exponent written in a different way. Using mathematical laws of exponents, the log transformation shows relative values instead of absolute ones. Visualizing log data can make highly skewed distributions appear less so.

In a log scale, the fact that 101 minus 100 and 2 minus 1 are the same doesn't matter. Instead, what matters is that going from 100 to 101 is a 1 percent increase and from 1 to 2 is a 100 percent increase. Thus, on a log scale, going from 100 to 101 is about 1 percent of the distance as going from 1 to 2.

Another way to understand the differences between absolute (level) and relative (log) values is to graph some sample data. The graph on the left shows a simple doubling of each

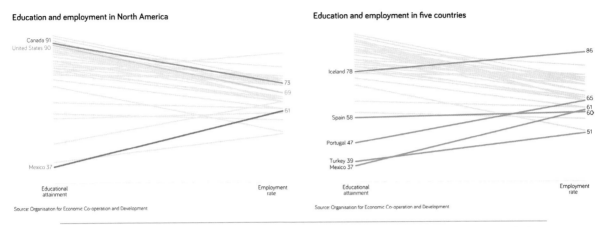

Education and employment in North America

Canada 91
United States 90

73
69

61

Mexico 37

Educational
attainment

Employment
rate

Source: Organisation for Economic Co-operation and Development

Education and employment in five countries

Iceland 78

Spain 58

Portugal 47

Turkey 39
Mexico 37

86

65
61
60

51

Educational
attainment

Employment
rate

Source: Organisation for Economic Co-operation and Development

As with the slope chart, you can use different colors, line thicknesses, or other visual elements to highlight areas or values.

As with the slope chart, you can use different colors, line thicknesses, or other visual or textual elements to highlight certain areas or values. You could, for example, highlight the North American countries (left graph) or maybe just those lines that are upward sloping (right graph).

Back to the parallel coordinates plot with all six variables shown at the top of the next page. Now that you understand how to read the chart, you can see the positive correlation between education and the employment rate in the first two axes. You can also see the positive correlation between the employment rate and life expectancy in the second and third axes. Our view of the data and the specific correlations we can most clearly identify are a function of how we organize the axes. The plot on the right changes the order of the vertical axes so we can now see the positive correlation between voter turnout and life expectancy in the first two axes, which we could not see in the original plot.

Placing all six metrics on the same vertical range also has the effect of suppressing the range (or variance) in some of these measures. For example, life expectancy varies only slightly, from 74.6 years to 83.9 years, while net migration varies more widely, from 8.6 percent to 93.9 percent. Allowing the ranges along the axes to fluctuate from one to the next (as in the middle chart on the next page) requires more labeling along each axis, but it also gives a better view of the data. The advantage of the two plots at the top is that you don't need to label every line. The disadvantage is that you suppress the variation within each variable. Notice, however, that this parallel coordinate plot—in which the range of each axis differs—looks more variable.

In sum, the challenge with many parallel coordinates plots is that they quickly become cluttered. With lots of observations (lines) and multiple axes, readers may have trouble finding the

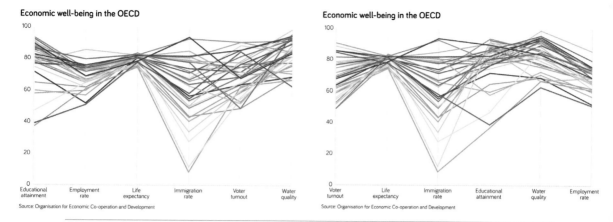

Parallel coordinates plots with too many observations quickly become cluttered.

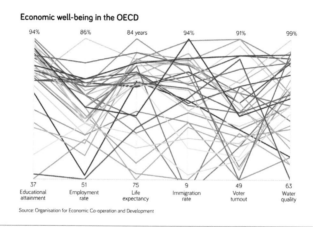

The axes in parallel coordinate plots can differ based on the minimum and maximum of the metric.

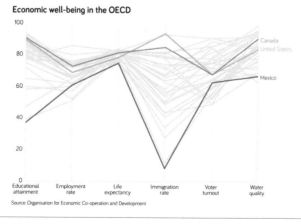

As we've seen previously, one way to simplify dense graphs like these is to use the "start with gray" strategy and add color to only a select number of observations.

correlations and picking out specific values. One way to alleviate this difficulty is to remember the "Start with Gray" guideline: Color a group of lines gray and highlight just a subset of the data.

RADAR CHARTS

Radar charts are like parallel coordinate plots, but the lines wrap around a circle instead of being arranged parallel to one another. These are also sometimes called *spider charts* or *star charts*, and they're a good way to show multiple comparisons within a relatively compact space. Data values are plotted along separate axes that radiate from the center (the axes themselves may or may not be shown) and are connected by lines or areas to show the relationships between the different variables.

The radar chart on the left shows the same six variables used above—the line for the United States and the gray area behind it the average for the thirty-two countries shown earlier. The version on the right shows the same variables for those six countries as well as the overall average in the gray. Both charts are compact and especially good at highlighting outliers. You can quickly and easily see the shape for Turkey (the pink line) is markedly

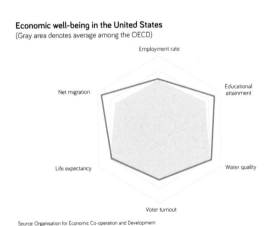

Economic well-being in the United States
(Gray area denotes average among the OECD)

Source: Organisation for Economic Co-operation and Development

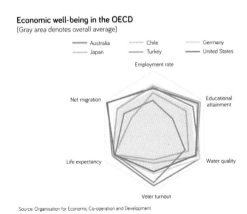

Economic well-being in the OECD
(Gray area denotes overall average)

Source: Organisation for Economic Co-operation and Development

Radar charts are like parallel coordinate plots, but the lines wrap around a circle instead of being arranged parallel to one another. The gray interior area represents the overall average for each metric.

Economic well-being in the OECD

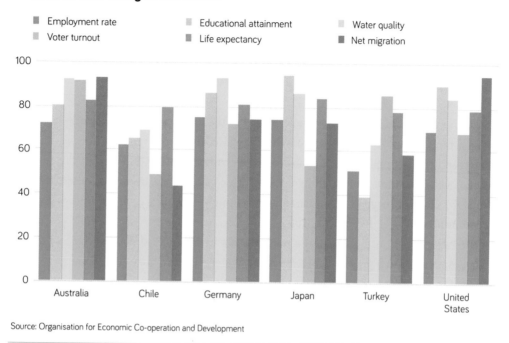

Source: Organisation for Economic Co-operation and Development

Too many bars in a bar chart like this make it hard to pick out specific observations or patterns.

different than the other countries. It's much harder to make that observation when the data are arrayed in the paired bar chart above.

As with many charts, the radar chart gets more complicated as more lines are added, and the crossing pattern around the circle can make perception even more difficult. As was the case with the parallel coordinates plot, plotting different metrics can also make perception difficult because it requires some normalizing or other modification of the data values—in the multi-country radar chart above, again notice how the values for life expectancy bunch together.

Another strategy is to use the small multiples approach, in which a separate radar chart is created for each country or group. In this case, the small multiples version takes up more space than the original and doesn't necessarily allow easy comparison across specific countries. But it is easier to see the values for each country relative to the overall average.

Economic well-being in the OECD
(Gray area denotes overall average)

Australia

Employment rate · Educational attainment · Water quality · Voter turnout · Life expectancy · Net migration

Chile

Employment rate · Educational attainment · Water quality · Voter turnout · Life expectancy · Net migration

Germany

Employment rate · Educational attainment · Water quality · Voter turnout · Life expectancy · Net migration

Japan

Employment rate · Educational attainment · Water quality · Voter turnout · Life expectancy · Net migration

Turkey

Employment rate · Educational attainment · Water quality · Voter turnout · Life expectancy · Net migration

United States

Employment rate · Educational attainment · Water quality · Voter turnout · Life expectancy · Net migration

Source: Organisation for Economic Co-operation and Development

A small-multiples approach lets us see the values for each country relative to the overall average.

CHORD DIAGRAM

Like the radar chart, the chord diagram is another way to show associations or relationships between observations arrayed in a circle. It is perhaps best used to show how observations have shared characteristics. In chord diagrams, observations (called *nodes*) are located around the circumference of the circle and connected by arcs within the circle to illustrate connections. The thickness of the arcs—often also differentiated by color or the transparency of the color—represent the degree of the connection between the different groups.

This chord diagram uses the same migration data used so far in this chapter to show migration flows between major regions of the world in 2017. Each region is placed along the circumference of the circle and the bands emanating from each correspond to the number of migrants entering or leaving each region. There are more than 110 values plotted in this

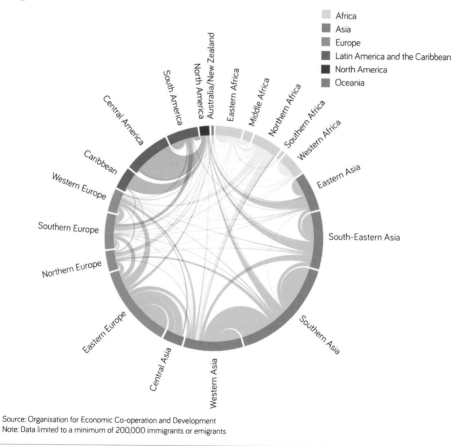

Migration around the world

Source: Organisation for Economic Co-operation and Development
Note: Data limited to a minimum of 200,000 immigrants or emigrants

In chord diagrams, the observations are located on the circumference of the circle and arcs within illustrate the connections.

single graph—though you could pick out specific values in a (very large) table, the chord diagram is clearly more visual and spatially efficient.

In the first chord diagram, you can see the large migrant flows within Asia (the red areas), and the movement between Central and North America (the thick green line to the blue segment near twelve o'clock in the circle). One danger is that the graph can quickly become cluttered and hard for the reader to easily see relationships. Again, we can use the strategy of highlighting specific groups with colors or lines. I've done that in this chord diagram in which the Asian region is red with all other regions in gray. The complexity of the chord diagram (and

Migration from Asia

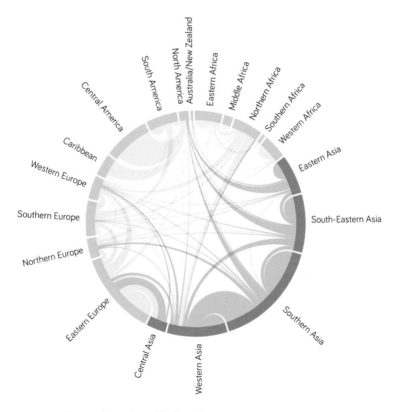

Source: Organisation for Economic Co-operation and Development
Note: Data limited to a minimum of 200,000 immigrants or emigrants

Using color strategically—especially with the color gray—can draw
attention to groups or points.

its relative compactness), make them visually intriguing and invites the reader to explore the data in more depth.

ARC CHART

Stretch a chord diagram out along a single horizontal axis and you have an arc chart. In this case, the nodes are placed along a line and are connected by arcing lines. The lines can vary in height, thickness, and color to illustrate the strength of the relationship or correlation. This arc chart shows the same migration flows between regions of the world as in the chord diagram.

A major consideration of the arc chart—and which also applies to many charts in this section—is that the order of the data can influence our perception of the results. Notice the high, wide green arcs stretching from North America to countries in the Caribbean, and Central and South America. If, by contrast, North America is placed to the far-right side of the graph and next to the other countries in the Western Hemisphere, the visualization is

Migration around the world

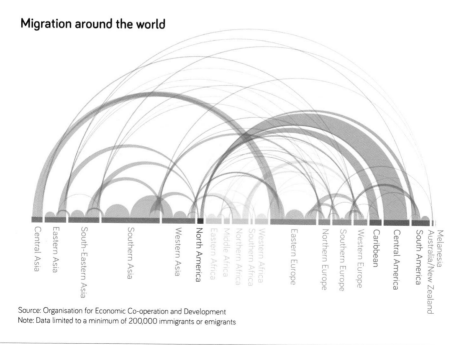

Source: Organisation for Economic Co-operation and Development
Note: Data limited to a minimum of 200,000 immigrants or emigrants

The arc chart is like a chord diagram stretched along a single horizontal axis.

Migration around the world

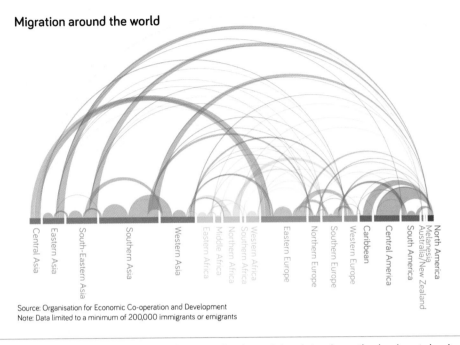

Source: Organisation for Economic Co-operation and Development
Note: Data limited to a minimum of 200,000 immigrants or emigrants

---

Like many graphs in this chapter, the organization of the data along the horizontal axis can affect our perception of the data. Compare this arc chart to the previous arc chart—same data, different shape.

substantially changed. There is no longer a tall green arc dominating the view, but a series of red bands that reach across the entire space between North America and countries in Asia. Some of this is obviously a function of the colors used, but the arrangement of the countries also matters. It is worth taking time to experiment with color and node placement to arrange the arc chart in the way that best communicates your argument.

A variation on the arc chart is an *arc-time chart* or *arc-connection chart*, in which connections over time are plotted in the same way. Instead of illustrating the correlation or relationship between two distinct variables, the nodes denote time. The arc-connection chart can also be thought of as an alternative to a timeline or flow chart, which we saw in Chapter 5. On the next page, the arc chart from Adam McCann shows the tenure of all Supreme Court justices in the United States since 1804. The origin (the left-most point) shows when each judge started his or her tenure, and the arc stretches to their retirement age. The height of each arc represents the age at the time of appointment (taller is younger) and the color represents their political party and year they were appointed (lighter shades are earlier years). An

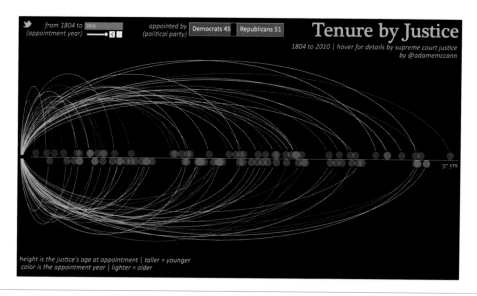

Adam McCann used an arc chart to show the tenure of all US Supreme Court justices since 1804.

Homeless relocations from New York City

The most popular US mainland destinations were two cities in the South: Orlando, Florida, and Atlanta, Georgia.

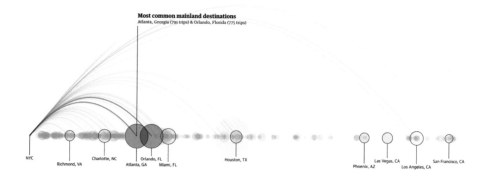

Most common mainland destinations
Atlanta, Georgia (791 trips) & Orlando, Florida (775 trips)

NYC · Richmond, VA · Charlotte, NC · Orlando, FL · Atlanta, GA · Miami, FL · Houston, TX · Phoenix, AZ · Las Vegas, CA · Los Angeles, CA · San Francisco, CA

Arc charts can also show geographic data, as in this one from the *Guardian* showing where New York City sends their homeless population.

alternative chart type, like a bar chart or a heatmap, could be used to show the same changes, but there is something arresting about the shapes in this view.

Another way to use the arc chart is to plot distances. The arc chart from the *Guardian* shows where New York City sends their homeless population. The cities are organized by distance from New York—Richmond, Virginia, on the left and San Francisco, California, on right. The height and thickness of the arcs and the size of the circles show how many people go to each city.

CORRELATION MATRIX

A correlation matrix is a table with the variables listed along the horizontal and vertical axes. Numbers in each cell represent the strength of that relationship, often as a Pearson's correlation coefficient (see Box on page 254).

The *correlation matrix graph* uses the same layout but instead of numbers it uses shapes—often circles—to show the strength of the correlation, and sometimes color and shades to organize the table. The correlation matrix is a cousin of the heatmap, and we can also think of it as a way to add a visual element to a standard table, a topic we will visit in Chapter 11.

World Migration

		Africa					Asia					Europe				Latin American and Caribbean			North America	Oceania			
		Eastern	Middle	Northern	Southern	Western	Central	Eastern	South-Eastern	Southern	Western	Eastern	Northern	Southern	Western	Caribbean	Central America	South America	North America	Australia/New Zealand	Melanesia	Micronesia	Polynesia
Africa	Eastern	49.0	10.0	7.1	0.6	0.1	0.0	0.1	0.6	0.0	0.1	0.0	0.1	0.1	1.1	0.0	0.0	0.0	0.0	0.0	0.0	0.0	0.0
	Middle	3.9	17.0	3.7	0.5	4.3	0.0	0.0	0.0	0.0	0.0	0.0	1.0	0.4	0.0	0.0	0.0	0.0	0.0	0.0	0.0	0.0	0.0
	Northern	7.2	1.1	3.2	0.0	0.4	0.0	0.1	0.1	0.4	9.0	0.2	0.2	0.3	0.8	0.0	0.0	0.0	0.2	0.0	0.0	0.0	0.0
	Southern	14.0	2.0	0.2	7.1	0.5	0.0	0.5	0.8	0.1	0.2	0.4	1.7	1.1	1.8	0.0	0.0	0.1	0.3	0.2	0.0	0.0	0.0
	Western	0.0	1.5	0.6	0.0	58.0	0.0	0.0	0.0	0.0	0.1	0.0	0.0	0.0	0.4	0.1	0.0	0.0	0.1	0.0	0.0	0.0	0.0
Asia	Central	0.0	0.0	0.0	0.0	0.0	4.9	1.0	0.1	0.0	1.6	44.0	0.1	0.1	2.4	0.0	0.0	0.0	0.0	0.0	0.0	0.0	0.0
	Eastern	0.0	0.0	0.0	0.0	0.1	0.3	53.0	1.8	12.0	0.0	0.2	0.6	0.1	0.3	0.0	0.0	0.0	0.9	0.4	0.0	0.0	0.0
	South-Eastern	0.0	0.0	0.0	0.0	0.0	0.0	9.2	13.0	68.0	0.2	0.0	0.6	0.1	0.3	0.0	0.0	0.0	3.6	2.2	0.4	0.0	0.0
	Southern	0.0	0.0	0.0	0.0	0.1	2.1	110.0	8.6	1.6	0.0	0.6	0.1	0.1	0.0	0.0	0.0	0.2	0.5	0.0	0.0	0.0	0.0
	Western	6.0	0.1	38.0	0.2	0.3	0.9	0.3	170.0	40.0	130.0	12.0	1.7	2.9	5.4	0.0	0.0	0.6	1.4	0.1	0.0	0.0	0.0
Europe	Eastern	0.0	0.0	0.2	0.0	0.1	56.0	1.5	0.4	1.0	21.0	100.0	4.7	2.5	3.9	0.0	0.0	0.1	0.5	0.0	0.0	0.0	0.0
	Northern	8.5	0.7	2.0	2.4	4.5	0.4	5.0	23.0	5.9	8.8	26.0	20.0	8.9	8.8	2.4	0.3	2.7	4.2	2.2	0.1	0.0	0.0
	Southern	2.0	2.4	15.0	0.3	5.3	0.5	4.1	6.2	2.3	3.2	32.0	5.8	31.0	16.0	3.8	1.7	25.0	1.9	0.6	0.0	0.0	0.0
	Western	4.2	4.0	34.0	0.5	6.5	12.0	4.8	8.7	8.5	31.0	60.0	7.7	49.0	29.0	2.0	0.7	7.3	3.6	0.6	0.0	0.0	0.0
Latin Amercian and Caribbean	Caribbean	0.0	0.0	0.0	0.0	0.0	0.0	0.1	0.1	0.1	0.0	0.0	0.2	0.3	1.4	7.1	0.2	1.0	2.6	0.0	0.0	0.0	0.0
	Central America	0.0	0.0	0.0	0.0	0.0	0.0	0.5	0.1	0.0	0.1	0.1	0.1	0.5	0.4	0.5	6.5	2.1	9.8	0.0	0.0	0.0	0.0
	South America	0.0	0.1	0.1	0.2	0.1	0.0	1.8	0.1	0.1	0.6	0.5	0.5	8.0	1.6	0.8	0.5	42.0	1.3	0.0	0.0	0.0	0.0
North America	North America	8.2	1.3	6.7	1.5	8.1	1.2	53.0	47.0	53.0	17.0	23.0	18.0	19.0	15.0	66.0	160.0	34.0	12.0	1.6	0.7	0.2	0.4
Oceania	Australia/New Zealand	1.6	0.1	0.8	2.5	0.2	0.0	8.8	8.3	10.0	3.1	1.9	18.0	6.8	3.7	0.1	0.2	1.3	2.1	7.4	1.4	0.0	1.5
	Melanesia	0.0	0.0	0.0	0.0	0.0	0.0	0.0	0.0	0.2	0.0	0.0	0.0	0.0	0.4	0.0	0.0	0.0	0.0	0.2	0.1	0.0	0.1
	Micronesia	0.0	0.0	0.0	0.0	0.0	0.0	0.1	0.0	0.5	0.0	0.0	0.0	0.0	0.0	0.0	0.0	0.0	0.2	0.0	0.0	0.2	0.0
	Polynesia	0.0	0.0	0.0	0.0	0.0	0.0	0.0	0.0	0.0	0.0	0.0	0.0	0.0	0.2	0.0	0.0	0.0	0.0	0.1	0.1	0.0	0.2

Source: Organisation for Economic Co-operation and Development

The basic correlation matrix is a table with numbers that show the strength of the relationship between observations.

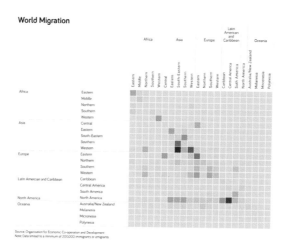

A heatmap approach to the matrix makes the patterns clear without (in this case) showing the numbers.

These two matrices show the relationship between immigrants (those entering each region) and emigrants (those leaving each region) across the world in 2017. The view on the left is the standard correlation matrix shown as a table. This gives us all the detail we would need to understand the exact correlation between different variables, but it's difficult to navigate. There are a lot of numbers, and the important values don't stand out. The matrix on the right is a heatmap (see page 112), which loses the detail of the table version in favor of highlighting the stronger (positive) correlations, especially within Asia. We could include both the colors and the numbers, but the view might end up looking cluttered and busy.

The next two visuals display the data as standard correlation matrix graphs. Circles represent the strength of the relationship, and color (in the version on the right) helps organize each area, though in this case there may be too many colors. It can be hard for the reader to clearly see differences because humans are not very good at assessing quantities from the sizes of circles. In both cases, the circles are sized to fit within each cell, but that doesn't necessarily need to be the case. We could make the circles larger to fill the entire space and use transparent colors when they overlap.

One final consideration with any correlation matrix or table is that the values along the diagonal are, by definition, equal to one. That is, migration between Eastern Africa and Eastern Africa is the same. This means they are often left out because they can visually dominate or clutter the visual.

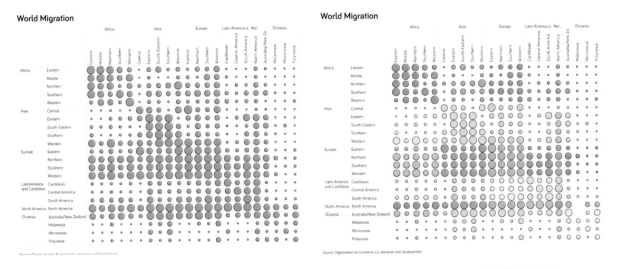

An alternative to the correlation matrix table is to use circles or other shapes, to which color can be added to visually organize the space.

NETWORK DIAGRAMS

We now enter a class of graphs for which I use the term *diagram* instead of graph, plot, or chart, largely because some of the decisions about layout and structure are not always determined by math or the data but by what looks best and is most clear. These diagrams are used to show hierarchies and connections within and across groups and systems. The thickness of the lines and size of the points can be sized according to data values to signal the strength of those relationships, and arrows can visualize movements inside groups and communities. Consider a family tree: the lines show links between parents, siblings, spouses, and children, but the connecting lines and the pictures or names of family members are not scaled according to a data value.

We start with the standard network diagram, which shows connections between people, groups, or other units. Generally speaking, the points in a network diagram (called *nodes* or *vertices*) denote the individual person or observation, and the lines (called *edges*) link them together and show the relationship. The position of the nodes and the length (and sometimes thickness) of the linking lines illustrates the strength of the relationship. While nodes are often depicted as circles, you could also use icons, symbols, or pictures.

The ultimate appearance and organization of a network diagram depends on the kind of network we want to visualize and the method with which we arrange the nodes and edges. When creating a network diagram, we must be careful about how edges cross and nodes overlap. In general, we want to achieve some kind of visual harmony in the visualization by finding a uniform and meaningful length of the edges and some symmetry for the entire graph.

To start, we can distinguish between four different kinds of network diagrams:

1. *Undirected and Unweighted*
   Lauren, Ara, and Katie are friends. Lauren is also friends with Deb and Kelly.
2. *Undirected and Weighted*
   Researchers in this diagram are connected if they published a paper together. The thickness of the line is the number of times they have published together.
3. *Directed and Unweighted*
   Jared follows Leah, Amelie, and Arielle on Twitter, but only Leah follows him back. Leah and Arielle follow each other, and Arielle follows Amelie. The connection is not weighted— they are either connected (in one or more directions) or not.

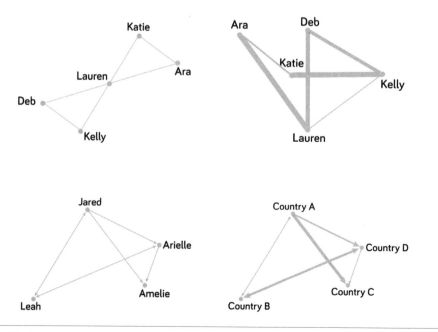

Four kinds of network diagrams, clockwise from top-left: Undirected and Unweighted; Undirected and Weighted; Directed and Unweighted; Directed and Weighted.

4. *Directed and Weighted*

People migrate from one country to another. The thickness of the line is the number of people migrating and the direction is the destination.

There are many algorithms we can choose from to lay out the nodes and edges in a network diagram. Usually, network algorithms try to minimize how often the edges cross one another and prevent overlap of the nodes. Generally, we want the edges in a network diagram to be of roughly uniform length and the vertices to be distributed evenly. Using example data

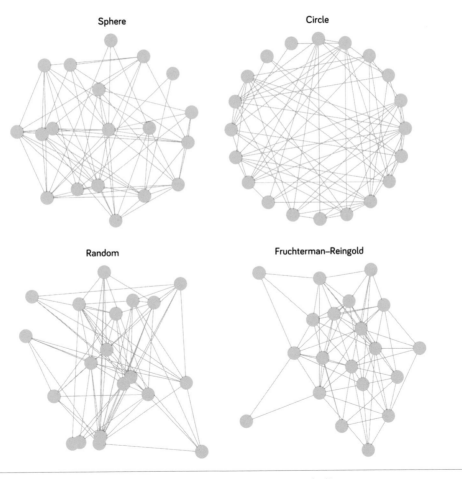

Four types of algorithms to create a network diagram.
Source: Based on the R Graph Gallery

around twenty points, these four network diagrams demonstrate how select organizing algorithms will generate different views of the network and the relationship between the points.

This network diagram shows the relationship of the seventy-five or so most populous
countries in the world (those with more than one million people) within their different
geographic regions. I'm not arguing that this network diagram is a better visualization than
a standard geographic map, but I show it here because you can easily understand the content

Regions and countries of the world

Source: United Nations

A simple network diagram that shows the arrangement of countries with more than one
million people.

and see how the diagram works. Imagine showing the links between people in your Twitter or Facebook network, grouped by family, friends, and coworkers.

Network diagrams are ideal for showing the structure and relationship between different agents in a system. In some cases, groupings or concentrations become clear as specific nodes cluster near one another. We can use color or other shapes to highlight specific groups within the larger network.

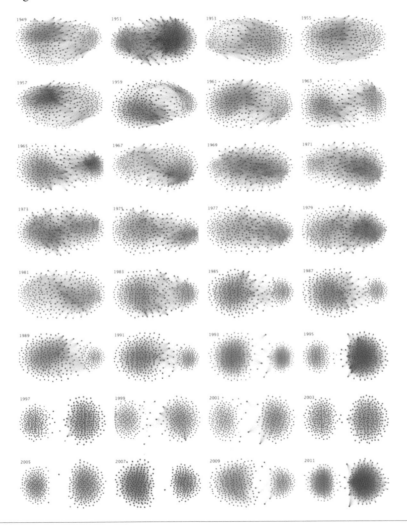

This small multiples set of network diagrams from Andris et al. (2015) shows voting behavior in the United States Congress.

As with all other charts, including too much data in a network diagram can clutter it and make it difficult to read. But unlike many charts, sometimes the goal of a network diagram *is* to show the dense clustering. The set of thirty-two network diagrams from Clio Andris and her collaborators shows the polarization of voting behavior in the United States Congress. The authors created network diagrams for each U.S. House of Representatives from 1949 (top-left) to 2011 (bottom-right). Republican members are represented by red dots, and Democrats by blue dots. The lines denote how often members of Congress voted with one another. Even though each network diagram is very dense, the use of these small multiples makes it clear that the two parties were much less likely to vote together in 2011 than in the past.

By comparison, the line chart below shows the measure of disagreement between the parties in a simple, straightforward way. Though it's immediately informative, it's not as visually stunning as the network view.

Average number of roll call vote disagreements

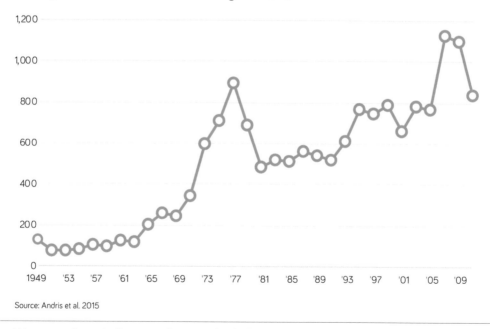

Source: Andris et al. 2015

We can make a similar case about voting behavior in the United States Congress using summary data from Andris et al. (2015), but the line chart probably doesn't grab you the way the small multiples network diagrams did.

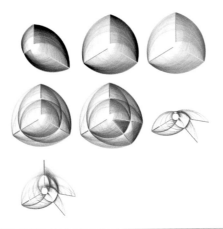

Martin Krzywinski's hive charts are another way to show networks.
Source: Canada's Michael Smith Genome Sciences Center.

Because network diagrams can look like hairballs with too many edges and nodes to make the visualization readable, some researchers have developed alternative visualization types. The *hive plot*, for example, first organizes the space along linear axes emanating outward from a single central point. Nodes are placed along three or more axes (possibly divided into

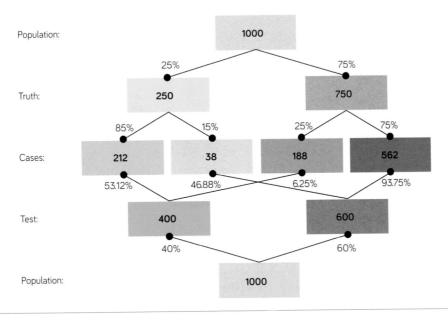

Some fields use network diagrams to visualize a process.

segments) and edges are drawn as curved links connecting the points. Martin Krzywinski, inventor of this visualization, writes that "The hive plot is itself founded on a layout algorithm. However, its output is not based on aesthetics but network structure. In this sense, the layout is rational—it depends on network features that you care about."

It's also worth noting that some fields refer to network diagrams in different ways. Instead of plotting how individual values correlate to one another, some network diagrams show flows or processes, similar to a flow chart or timeline. Examples might include how a computer network, a staff directory, or even a logic model of probabilities, as shown on the previous page.

TREE DIAGRAMS

Like the flow chart in Chapter 5, tree diagrams show levels of a hierarchy in a system or group. To imagine the basic tree diagram, think of a hierarchical organizational (or "org") chart. Nodes branch outward from an initial root connected by lines called *links*, *link lines*, or *branches*. The initial node is called the *root* and is the *parent* to all other nodes, some of which have child nodes of their own. Nodes who are not parent nodes are called *leaf nodes*.

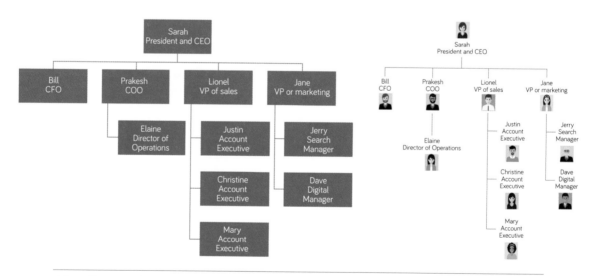

Tree diagrams describe hierarchies. These two org charts could be used for different purposes, depending on whether we think the reader would like a more designed version.

As with many of the visuals we have discussed so far, design is especially important here because there may not be much or even any data to define the elements in the diagram. Take these two imaginary org charts, for example. The chart on the left consists only of names, while the one on the right includes icons. Which one you would use depends entirely on your purpose. The one on the left might work well for the company board meeting or formal presentation; the one on the right might work better in a marketing campaign or website.

While the basic tree diagram will often branch downward, starting with the CEO or president at the top, there are lots of other ways to show these kinds of relationships. We could create a family tree that branches upward instead of downward, or a horizontal arrangement to show a different kind of hierarchy or taxonomy.

Another type of tree diagram is the word tree. Developed by Martin Wattenberg and Fernanda Viegas in 2007, the word tree is a visual representation of text in a book, article, or other passage (also see Chapter 10 on qualitative data visualization). The visualization is

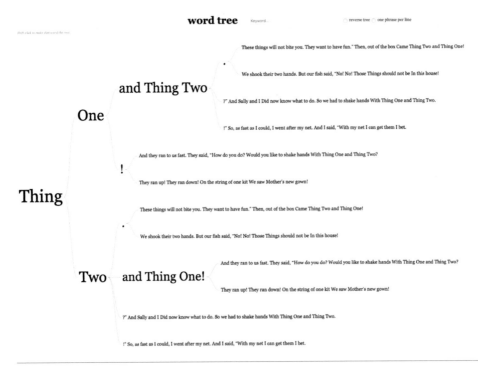

This word tree is a visual representation of the text of Dr. Seuss's book, *The Cat in the Hat*.
Source: Jason Davies

Regions and countries of the world

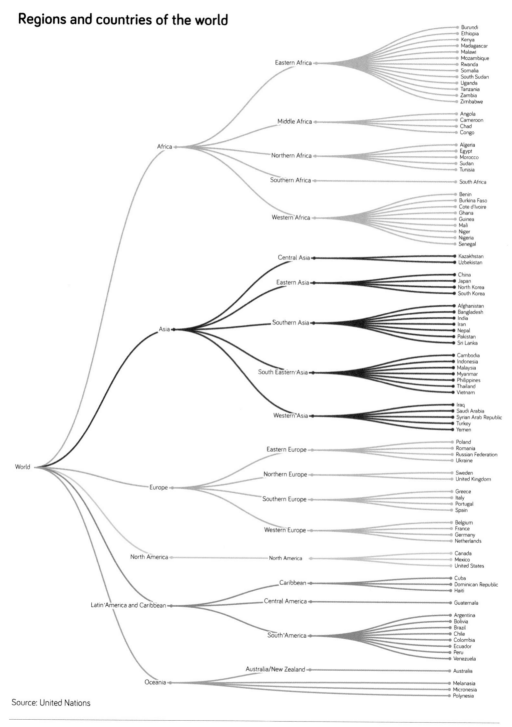

Source: United Nations

A simple tree diagram that shows the breakdown of regions into countries.

typically ordered horizontally with a word on the left or right that branches out to show the different contexts in which it appears. These contexts are arranged in a treelike structure so the reader can uncover themes and phrases. Individual words, which act here as the nodes, are often sized by the frequency in which they appear.

There are lots of different kinds of trees to visualize quantitative or qualitative data. They can show complex data, like the human genome, or simple data, like the breakdown of countries into regions. What kind of tree you create and the design touches you include will—as always—depend on your purpose and audience. The tree on the previous page, for example, shows the same data as the network diagram shown earlier. On the one hand, it's not particularly efficient, but on the other hand, it might be easier to navigate than the network diagram.

CONCLUSION

In this chapter, we surveyed charts and diagrams that visualize relationships between variables, individuals, or groups. We often want to understand how two or more things are related, but remember that because two variables might be correlated does not mean there is a causal relationship. Clearly understanding how elements in your data are related before presenting them to your reader or audience is of utmost importance.

This class of graphs uses different strategies and shapes to communicate these relationships, and there are advantages and disadvantages to each approach as they trade off between clarity, order, and compactness. Scatterplots have a single horizontal and vertical axis; bubble plots add a third variable. Parallel coordinate plots are defined by using two or more vertical axes. Radar charts pull the axes together and radiate outward from a center point of a circle while a chord diagram wraps everything around the circumference of a circle. An arc chart then stretches everything out along a single horizontal axis and a correlation matrix uses a square or rectangular format. Network and tree diagrams can be used to show relationships between individuals or groups or passages of text.

As with the graphs in previous chapters, some of the graphs in this chapter may be unfamiliar, even difficult for you or your readers to understand. This doesn't require you to dumb things down or leave things out, but it should prompt you to consider how to best communicate the content of nonstandard graphs. Use labels, annotations, active titles, and helpful pointers to teach them how to read the graph so that they can more easily understand the content.

PART-TO-WHOLE

This class of charts shows how the shares of some amount relate to the total. The most popular and familiar graph in this class is the pie chart, which introduces a variety of perceptual challenges, as we'll see. Other charts in this class like the treemap and sunburst have different perceptual issues and, as always, we must ask ourselves whether we must show *all* of the components and how they sum to the total. Graphs in this chapter can also be used to visualize hierarchical data—data that can be grouped into layers where natural groups exist—which we have already seen in cases such as the tree diagram.

Graphs in this chapter are based on the online style guide from the *Texas Tribune*, a digital-first publication based in Austin, Texas. In addition to the basic styles outlining colors and fonts, the *Tribune's* style guide also includes instructions for the online elements of their website.

PIE CHARTS

Disdain for pie charts pervades the entire field of data visualization. The most often-cited reason is that pie charts are a poor visualization choice because we have a hard time discerning exact quantities when they are visualized as slices of the pie. If we return to the perceptual ranking chart, we'll see that pie charts fall below the middle of the ranking. While the pie chart gets its fair share of complaints, it's also a very familiar chart type for many people, and familiarity can be useful. Research has also shown that people are more attracted to curves than to objects with sharp points, and author Manuel Lima has presented evidence that affinity for the circle shape dates back millennia in human evolution.

These pie charts show the distribution of imports (the dollar value of goods flowing into a country or region) in seven areas around the world. Notice that the version of the left shows the dollar amounts (in billions), while the right shows the percentages. Either approach works; just be sure to use correct labeling and note it in the title.

The most important rule for pie charts is that the slices must sum to 100 percent or at least some sort of total. You cannot leave segments out or—unfortunately more common—include segments that sum to more than 100 percent. When you arrange the slices of a pie chart, a good rule of thumb is to order them from largest to smallest beginning at the 12-o'clock position. This is often the best way, but sometime it isn't always possible or natural. For example, if you were visualizing shares of a total by age group, it would be better to start with the youngest group at 12 o'clock, then the next oldest group, and so on. In this case, it is more natural for our reader to comprehend the data when they are ordered by category rather than by value.

In situations where you might use more than one pie chart—something I would only recommend in very specific cases—always order the slices in the same way in all the charts. It becomes far too difficult to compare the slices in multiple pie charts when the slices are in different positions.

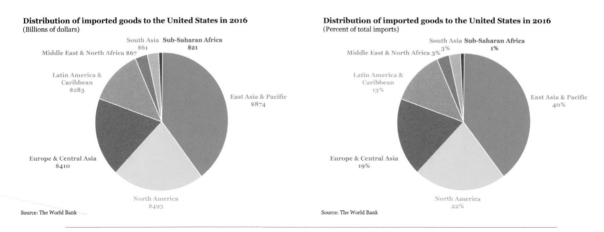

Pie charts show part-to-whole relationships. These two show the distribution of imported goods to the United States, in either dollars or percentages.

THE CASE AGAINST PIE CHARTS

The trouble with pie charts is that humans cannot easily compare differently sized slices. Here we have the same data arrayed as a pie chart and as a bar chart. See how it's easier to rank the values in the bar chart and how small differences are imperceptible in a pie chart?

Even if you may have correctly ranked the values, we can agree that it is easier and faster to rank them in the bar chart version. It's also easier to see how much the values differ from each other. The pie chart simply does not let us accurately discern the values, and if your goal is to help your reader make clear and accurate determinations about the data, the pie chart is not the best choice.

It's also worth noting that it's not really clear how we perceive the quantities in the pie chart. Is it the angle at the center? The area of each slice? The arc length? A pair of research papers in 2016 suggested that the angle of slice meeting at the center is not the main way we read pie charts and that area and arc length—the segment of the circumference—were better predictors of people's ability to read values than angle alone.

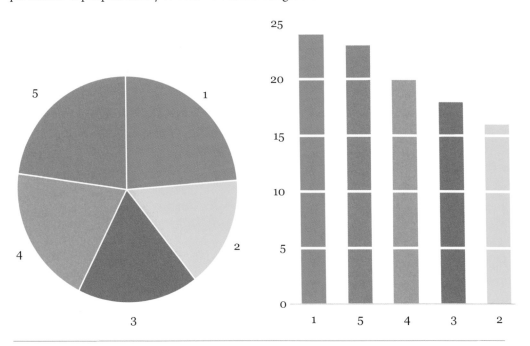

Comparing the values is easier to do in the bar chart on the right than in the pie chart on the left.

Distribution of imported goods to the United States in 2016
(Billions of dollars)

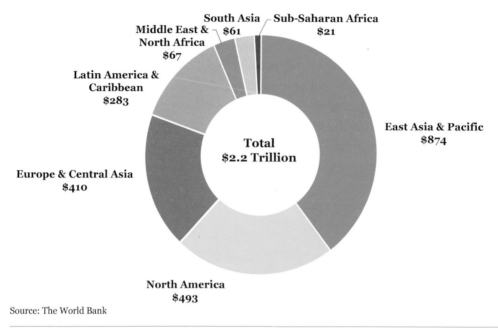

Source: The World Bank

The donut chart punches the middle out of a pie chart, which leaves
room for an additional title or text.

If we don't perceive the values in a pie chart from the angle at the center, it would mean that a donut chart—a pie chart with the center removed—is a viable, even preferable alternative. But if it's the case that we discern the quantities in the pie chart by the *angle* of the slices, then the donut chart is an even worse choice than the pie chart because the angles are held in the center, which is missing from the donut chart. No one has decisively proved it either way. Still, the primary advantage of the donut chart is that you can include a number or statement in the center of the chart.

I recommend avoiding using pairs of pie charts, even when they show only a few slices. The reader must look back and forth between the two pie charts to see whether the values of the different groups changed. That task is made faster and easier with a stacked bar chart or slope chart, though with the slope chart, we move away from the part-to-whole comparison and instead focus on the change over time.

Especially avoid pie charts that show too many slices. Five is already too many—more becomes incomprehensible. Also avoid the "breakout" pie chart that removes a single slice

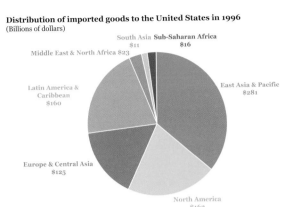

Distribution of imported goods to the United States in 1996
(Billions of dollars)

South Asia $11 Sub-Saharan Africa $16
Middle East & North Africa $23
East Asia & Pacific $281
Latin America & Caribbean $160
Europe & Central Asia $125
North America $162

Source: The World Bank

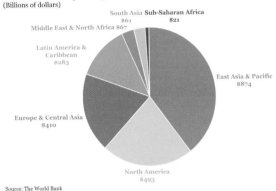

Distribution of imported goods to the United States in 2016
(Billions of dollars)

South Asia $61 Sub-Saharan Africa $21
Middle East & North Africa $67
Latin America & Caribbean $283
East Asia & Pacific $874
Europe & Central Asia $410
North America $493

Source: The World Bank

Distribution of imported goods to the United States in 1996 and 2016

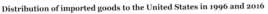
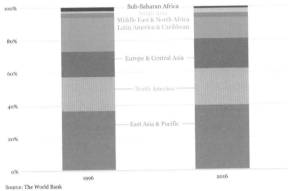

Sub-Saharan Africa
South Asia
Middle East & North Africa
Latin America & Caribbean
Europe & Central Asia
North America
East Asia & Pacific

Source: The World Bank

Distribution of imported goods to the United States in 1996 and 2016

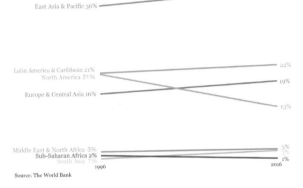

East Asia & Pacific 36% — 40%
Latin America & Caribbean 21% — 22%
North America 21% — 19%
Europe & Central Asia 16% — 13%
Middle East & North Africa 3% — 3%
Sub-Saharan Africa 2% — 2%
South Asia 1% — 1%
1996 2016

Source: The World Bank

Distribution of imported goods to the United States in 2016

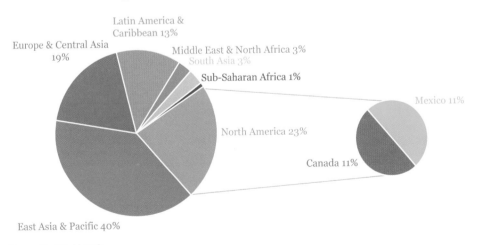

Latin America & Caribbean 13%
Europe & Central Asia 19%
Middle East & North Africa 3%
South Asia 3%
Sub-Saharan Africa 1%
North America 23%
East Asia & Pacific 40%
Mexico 11%
Canada 11%

Source: The World Bank

Pairs of pie charts are rarely useful to show changes over time. That task is made faster and easier with other visualization types, like a stacked bar chart or slope chart.

and breaks it down even further. These charts are hard to read and there are better ways to plot such data: in a bar chart or even a Sankey diagram (see page 126).

THE CASE FOR PIE CHARTS

Let me now make the case *for* using pie charts. These two pie charts show the same data but arranged differently. In the version on the left, the value of Group B (the purple slice) is not immediately clear. In the version on the right, the rotation of the chart creates a familiar right angle in the center of the chart, and now Group A's value (25 percent) is clear. A pie chart is therefore perhaps best used when the value(s) of the slice or slices sums to one of these round numbers (25 percent, 50 percent, and 75 percent), for which the angle is familiar. These cases let you you easily focus your readers' attention on three or maybe four slices.

Let's look at a practical example of when a pie chart can be useful. Imagine giving a presentation to report the fundraising efforts you and your colleagues have accomplished for your nonprofit organization. Let's say you have received one hundred donations, and a bit more than half of all money you have raised has come from a few, very large gifts of $1 million or more. You break down the distribution of gifts into seven categories and break out the top categories into a group of $1 million to $2 million donations and a group of $2 million donations and above. The perceptual rankings chart may guide you to use the most perceptually accurate graphs, like a bar chart or a scatterplot.

But for your audience, a pie chart might work best. They can read and understand a pie chart almost instantaneously, and because the share you want your audience to focus on

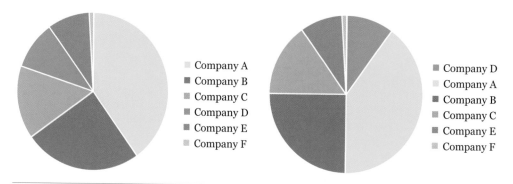

It's easier to guess the value of the purple slice for Company B when the chart is arranged with right angles.

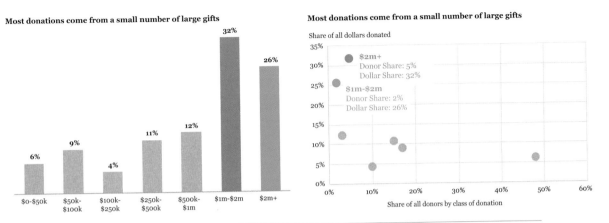

These two charts show that most donations to this nonprofit came from a small number of large gifts. But if the point is to show that slightly more than half came from these two groups, a pie chart is actually a good choice.

accounts for just more than half, it's easy for your audience to mentally draw the vertical line through the circle. Imagine explaining how a Marimekko chart or a scatterplot works to an audience who just wants the basic information as quickly as possible.

Notice how all seven values are plotted and labeled in the pie chart on the next page, but only the two slices of interest are colored, and together they are clearly larger than 50 percent. (Also notice how we are not following the start-at-12-o'clock rule, because arranging the slices by category value makes more sense here.) Because they are not the primary focus of the graph, the other slices are in gray and move to the background. For a presentation, we might even collapse our two groups to one and delete many of the labels to reduce the amount of information on the screen that would only distract the audience's attention from the speaker.

This example leads to another scenario in which a pie chart can be useful, and that is when you are focusing your readers' attention on a single value—the true part-to-whole relationship. In these cases, consider whether you actually need a chart at all. Do you need a visual to back up the statement that the U.S. poverty rate is 12.3 percent? Maybe on social media or on a presentation slide, but in a report or article you might do just as well leaving the number in the text.

A bad pie chart is still a bad chart. If you choose to use pie charts, do so strategically and thoughtfully. It will almost always be difficult for your reader to discern specific quantities or compare slices, but if you want them to understand a general difference in magnitude or focus on a single slice, the pie chart is appropriate.

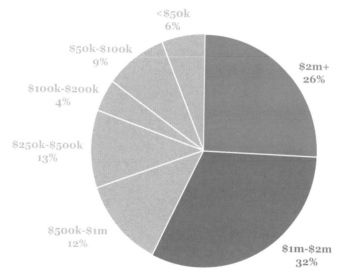

Most donations come from a small number of large gifts

<$50k
6%

$50k-$100k
9%

$100k-$200k
4%

$250k-$500k
13%

$500k-$1m
12%

$1m-$2m
32%

$2m+
26%

Source: The World Bank

This pie chart does a good job of highlighting the two groups of interest because (a) the groups sum to slightly more than 50 percent and (b) the other groups are gray.

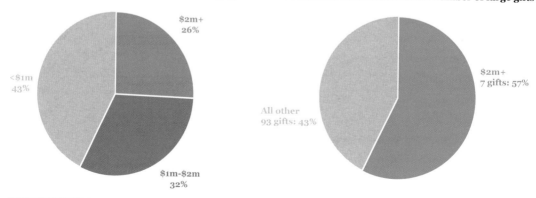

Most donations come from a small number of large gifts

$2m+
26%

<$1m
43%

$1m-$2m
32%

Most donations come from a small number of large gifts

$2m+
7 gifts: 57%

All other
93 gifts: 43%

Basic simplifications to these pie charts can focus attention on the groups of interest.

TREEMAP

Originally developed by Ben Shneiderman at the University of Maryland, the treemap divides sections of a square or rectangles into groups to illustrate a hierarchy or part-to-whole relationship. In other words, the treemap is a squarified version of a pie chart.

Distribution of imported goods to the United States in 2016

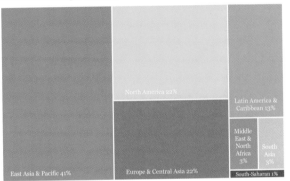

Source: The World Bank

Distribution of imported goods to the United States in 2016

Source: The World Bank

Distribution of imported goods to the United States in 2016

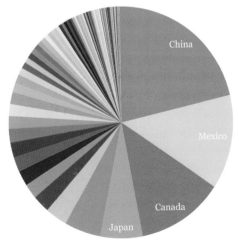

Source: The World Bank

Treemaps are an alternative to the pie chart and can show part-to-whole or hierarchical relationships. Notice how it is essentially impossible to read the pie chart.

The two treemaps on the previous page show the breakdown of total imports from specific countries to the United States in 2016. You may find this easier to read than a pie chart because rectangles are more easily compared. Or, because it's an unfamiliar graph type, you might find it slower to navigate or more difficult.

The treemap on the right breaks down the regional categories further into individual countries. Not every country in the world is labeled in this chart, but you do get a clear sense of which countries send the most goods to the United States.

Try to read the pie chart that shows the same data—it's impossible.

Shneiderman originally developed the treemap to concisely view the file directory on his computer. The compactness of the treemap is one of its biggest advantages because it can include many different groups and variables. Hierarchies are easy to visualize in the treemap because subsections can be labeled and embedded within the parent group. You can also add other encodings to the treemap—for example, this treemap shows the 2016 distribution

Distribution of imported goods to the United States in 2016
(Blue denotes increases between 1996 and 2016; red denotes decreases)

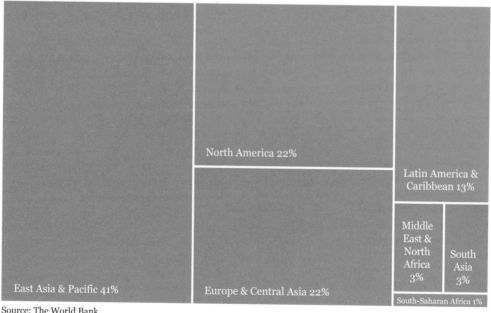

Source: The World Bank

Color can add another dimension to a treemap. In this case, the change in imports between 1996 and 2016.

of the value of imports to the United States and the colors denote the shares that increased (blue) and decreased (red) between 1996 and 2016.

SUNBURST DIAGRAM

If you want to show the proportions of parts to a whole at several levels in a hierarchy, you might use what is called a *sunburst diagram*. Like a treemap, the sunburst can show part-to-whole relationships or hierarchical relationships. This sunburst diagram, for example, shows the same data as the detailed treemap above, with an additional ring that breaks the major regions into their components.

Each ring in a sunburst corresponds to a different level in the hierarchy, and the slices of each ring—sometimes called *nodes*—refer to the different subgroups. The central ring shows

Distribution of imported goods to the United States in 2016

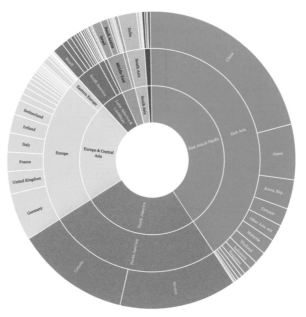

Source: The World Bank

The sunburst graph shows part-to-whole or hierarchical relationships.

the top level, sometimes referred to as the *root*. Outer rings show how the groups break into subgroups. You can use color to highlight different rings, groups, or hierarchies. The sunburst graph above uses color to differentiate the seven major areas of the world.

As with all circular visualizations, it's difficult to clearly and quickly see patterns. Sunburst diagrams with too many categories are likely too cluttered for a reader to pick out patterns even after a close look. Strategic use of color, labels, and annotation can guide the reader to the most important parts of the visualization.

NIGHTINGALE CHART

This type of chart is sometimes called the *coxcomb* or *rose diagram* but is most often known as the *Nightingale chart*. The first and most famous of these was created by Florence Nightingale to visualize soldier casualties during the Crimean War.

Nightingale was born in the early nineteenth century to wealthy parents who provided her with a comprehensive education in liberal arts and mathematics. Deciding early on that she wanted to dedicate her life to health care and helping the poor and needy, Nightingale became a nurse in the early 1850s. With experience in organizing hospital supplies, she eventually took care of British casualties at the Barrack Hospital of Scutari in Turkey during the Crimean War. For two years, she and her team of nurses helped care for the wounded and ill, all the while keeping careful records of their patients.

Convinced that cleanliness was a major reason for the high death rates at Scutari, Nightingale wrote hundreds of publications showing her data in dozens of visualizations. Ultimately, the British government created the Sanitary Commission to investigate the poor outcomes at the hospital and soon implemented improvements to sanitation, ventilation, and cleanliness.

Her most famous chart—and the one that bears her name—is drawn around a circle with values segmented into different time periods. Each slice represents the total deaths in each month from April 1854 (in the 9 o'clock position in the right diagram) to March 1856 (in the 9 o'clock position in the left diagram) during the Crimean War. Cause of death is broken down into three categories: deaths from wounds in battle (pink); deaths from "other causes" (black); and deaths from diseases (blue). The diagram on the right shows patterns in deaths over the first twelve months of the war. The one on the left shows the next twelve months after the Sanitation Commission implemented its reforms in March 1856.

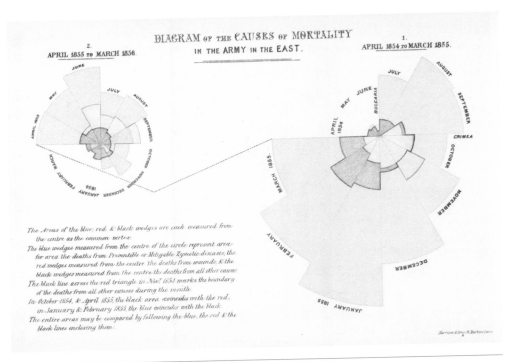

Florence Nightingale created her legendary chart to show the disproportionate share of deaths from disease like cholera, typhus, and dysentery. Image from the Wellcome Collection.

We can think of the Nightingale as a pie chart in which the slices have been expanded in different directions. The area of each slice represents its value relative to the whole, and the slices are arranged along the time dimension. The Nightingale chart therefore shows both changes over time *and* a part-to-whole relationship.

The visualization demonstrates two major points. First, deaths in battle were only a small portion of the total number deaths. Deaths from disease—in this case, cholera, typhus, and dysentery—made up a disproportionate share of deaths. Second, it illustrates how the Sanitary Commission, which started its work in the middle of the war, dramatically reduced the overall number of deaths.

Already in this book we've seen variations on the Nightingale chart. The circular column chart, for example, arranges columns around a central circle radiating outward. The radar chart also uses a circular layout to show relationships across variables aligned along the radii. As with many circular charts, the Nightingale can be difficult to read, and comparisons across the slices are hard to gauge. The major problem with the Nightingale chart is that the

Distribution of imported goods to the United States in 2016

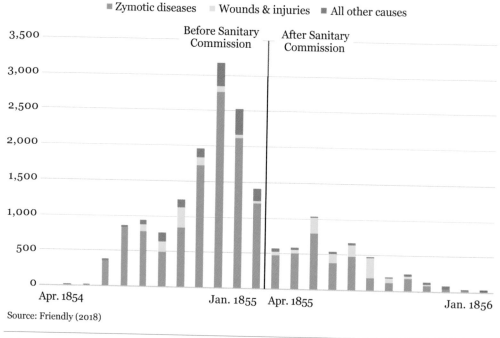

■ Zymotic diseases ▪ Wounds & injuries ■ All other causes

This stacked bar chart shows the same data as in the original Nightingale graph—but is it as memorable?

outer segments are necessarily larger and therefore are emphasized disproportionately. This distortion increases as we move farther from the center.

The stacked bar chart above uses the same Crimean War death data as the Nightingale chart on the previous page. I've added a vertical line between March and April 1855 to denote the break in the two circles. While the bar chart shows the totals more clearly, it does not make the break between the two years as clear as the original, nor perhaps is it necessarily as engaging.

The natural question, of course, is whether Nightingale's charts would be better as some kind of more standard chart, like this stacked bar chart. Data visualization expert RJ Andrews addressed this debate head-on:

Critics suggest that her mortality data is better shown in something more straightforward like a bar chart. But this is not true: Florence Nightingale made lots of bar charts. No one cares about them! Her roses gripped 1858 readers and they still hold our attention today.

In this case, as in many others, we must strike the balance between what is perceptually accurate and what will engage people and stick in their minds. Or, as author Alberto Cairo wrote about Nightingale, "I believe her goal wasn't just to inform but also to *persuade* with an intriguing, unusual, and beautiful picture. A bar graph conveys the same messages effectively, but it may not be as attractive to the eye."

Let's use the import data from early in this chapter and lay it out as a Nightingale chart. Again, this chart shows total imports to the United States from each of the seven regions

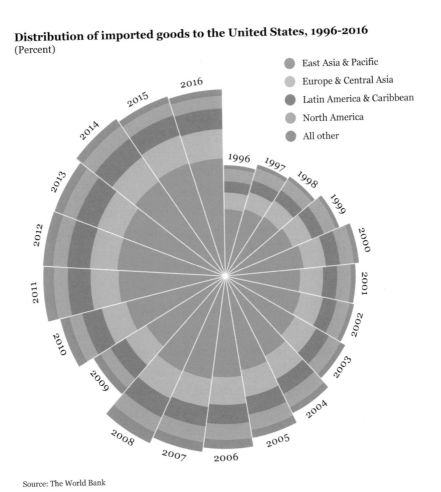

Distribution of imported goods to the United States, 1996-2016
(Percent)

East Asia & Pacific
Europe & Central Asia
Latin America & Caribbean
North America
All other

Source: The World Bank

Nightingale charts can visualize all sorts of data, though they introduce their own perceptual hurdles.

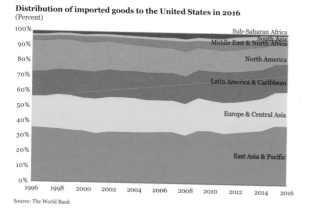

Distribution of imported goods to the United States in 2016
(Percent)

Source: The World Bank

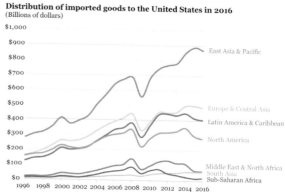

Distribution of imported goods to the United States in 2016
(Billions of dollars)

Source: The World Bank

A stacked area chart or a line chart are both simple alternatives to the Nightingale chart.

around the world from 1996 through 2016. Each region is stacked on top of each other, and the year data is arranged from clockwise from earliest to latest, beginning in 1996 at the top.

We could also use a stacked area chart or line chart to show these same data. Here, notice that the smaller values appear slightly larger in the Nightingale chart because, by definition, those areas become larger the further they are from the center of the circle. But as we've seen throughout, there might be cases where the Nightingale is useful over these standard chart types—it's more compact and looks very different from these basic chart types. And that may be a goal in itself.

VORONOI DIAGRAM

Named after Georgy Voronoi, a Russian mathematician who lived around the turn of the twentieth century, the *Voronoi diagram* divides a space of some number of points—called *sites*—into a corresponding number of shapes (polygons), which meet at boundaries that do not overlap. If we drop a new point in the space, it will be closer to the *site* in its area (polygon) than to any other in the space. Voronoi diagrams are sometimes used for geographic data, but I've placed them in this chapter because they can also show part-to-whole relationships. You can often find these diagrams in the biology literature (for cell structures), ecology (for the study of forest growth), and chemistry (for molecular positions).

The interesting property of Voronoi diagrams is that the border of each polygon is the same distance to the two nearest sites. In other words, each polygon is defined so that the distance from the edge of the polygon to its site is the shortest it can possibly be. When three borders meet, they result in a point—called a *vertex*—which is equidistant to the three nearest sites. There are a variety of algorithms to determine the shape and position of the various polygons.

That explanation is complex, so let's take a simple example. Say you live in a city with nine fire stations, and a fire breaks out at a building somewhere in the city. Which fire station should respond? The locations of the nine fire stations are the generating points, and the polygons that surround each tells you which fire station should respond to the fire on the principle that the closest fire station should respond. (Obviously, I'm ignoring issues of roads and bridges and other obstacles to actually *get* to the fire, but more complex algorithms can factor in these kinds of obstacles.) As you can see on the next page, the Voronoi diagram splits up the city into its different components, effectively letting us visualize a part-to-whole relationship.

Source: Map from the City Roads project by Andrei Kashcha

To demonstrate how a Voronoi diagram works, imagine there are nine fire stations located across the city.

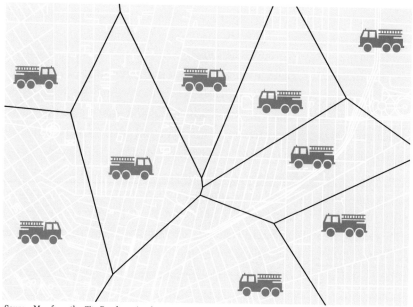

Source: Map from the City Roads project by Andrei Kashcha

Using a Voronoi map, we can divide the city into nine separate areas.

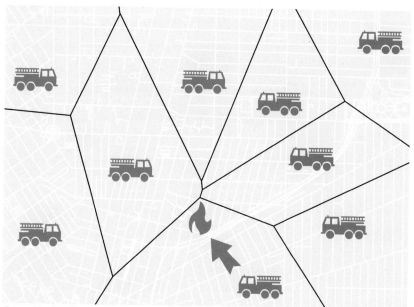

Source: Map from the City Roads project by Andrei Kashcha

If a fire breaks out, we can identify which fire station should respond based on its proximity.

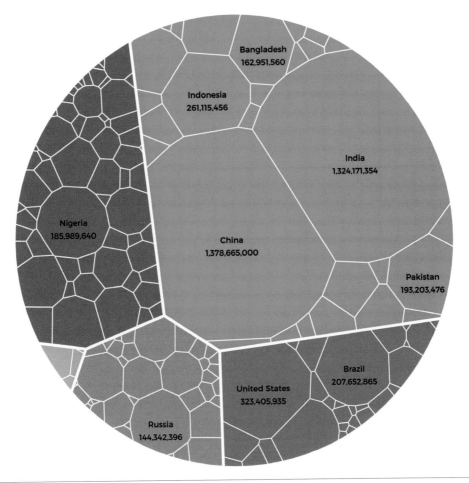

Maybe a more common use of the Voronoi chart is to show part-to-whole relationships, like this graph from Will Chase that shows the populations of countries around the world.

More generally, you may see Voronoi diagrams like this one, which shows the population of countries around the world categorized by region.

Many people probably don't know that one of the most famous maps in the history of data visualization is also a Voronoi diagram. In mid-1854, John Snow, an English physician and founder of the modern science of epidemiology, plotted each death from an outbreak of cholera as a small dash on a map of Soho in London. More than six hundred people died during the outbreak in a little more than a month.

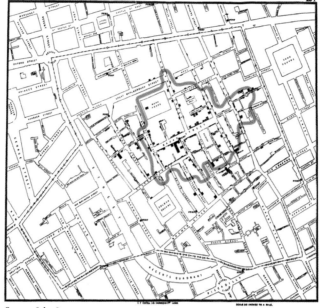

John Snow's famous cholera map is actually a Voronoi diagram. There is a clear clustering of deaths around the Broad Street pump—seen as a small dot in the center of the map just to the right of the thickest black bar. I've highlighted the dashed line that Snow included to show the distance from the Broad Street pump.

If we put a purple dot at each water pump in the map, we can create a Voronoi diagram from Snow's original map.

Snow's map showed a clustering around a single water pump on Broad Street. You can see a large number of dashes just to the left of the infected pump (the dot with the "PUMP" label directly in the center of the image) as well as other deaths up and down the street. In this version of the map (published slightly later than his original), a small dashed line is drawn around the affected area (which I've highlighted in purple). According to Snow, the dotted line "shows the various points which have been found by careful measurement to be at an equal distance by the nearest road from the pump in Broad Street and the surrounding pumps."

We can also apply a direct Voronoi approach by marking each water pump on the map as a *site* (purple dots in this map) and construct the polygons around each (purple lines on the map). Thus, the "Snow Map" is actually the "Snow Voronoi."

CONCLUSION

Pie charts are the default visualization of part-to-whole relationships because they're familiar, easy to make, and easy to read. But pie charts are rife with perceptual difficulties and should be used with care. Limit the number of slices in your pie charts and remember that right angles are familiar to our eyes and brains, so 25 percent increments are useful markers. It's difficult for readers to compare values across pairs of pie charts, but if we use more than one pie chart, we should at least keep the slices in the same order.

There are alternatives to the pie chart. Treemaps are essentially squarified pie charts, which can hold more annotation and show hierarchical relationships. Sunburst diagrams also show hierarchical relationships, though a sunburst diagram with many series can look complex and cluttered. Voronoi charts show part-to-whole relationships in a different layout. They can also be used to visualize geospatial data, cell structures, or ecological data.

Some of these other graph types may be less familiar to your readers, but they also help them more accurately discern the data values. Many of the graphs we have covered in earlier chapters—like the bar chart, stacked bar chart, and slope chart—can also be used to visualize part-to-whole relationships, but they often require explanatory text to make those relationships clear.

QUALITATIVE

Until now, we've explored charts that mostly communicate quantitative data. But there are also graphs that communicate *qualitative* data, non-numerical information collected through means of observations, interviews, focus groups, surveys, and other methods. The charts in this chapter primarily communicate words and phrases.

We can build narratives and tell stories around qualitative results in ways that can be more difficult with quantitative data. Downloading a big data set, running regressions, and creating tables may provide more generalizable results, but readers don't connect to them the way they connect to a story. Qualitative data can help tell those stories.

As with quantitative data, sometimes you may want to summarize your qualitative results. Images in this chapter are drawn from a variety of news and research organizations from around the world. Unlike previous chapters, they do not use the same style guide, which helps demonstrate some of the variety in layout, design, and approach. Some of the graphs in this chapter show an overall view of your data, while others ask the reader to review the detailed results, text, or quotations. Which you use will depend on your intended purpose and where you are publishing your results.

ICONS

Data visualization design can go a long way to deliver more content and captivate your reader. We've seen in previous chapters how color, layout, and different shapes can engage

Icon examples from the Noun Project. Modern versions of Microsoft Office have a built-in icon library, or you can use websites like The Noun Project and Flaticon where you can purchase or download free icons. There are also fonts that can be used like icons—each letter in the *StateFace* font, for example, is designed as a separate icon of a US state.

readers and bring them into a graphic. Icons, images, and photographs can similarly draw your reader in and categorize your qualitative data for their ease.

Iconography, especially, can help visualize your qualitative data. Icons can be purely decorative, they can represent data (as in a unit or Isotype chart, see page 106), or they can guide the reader from one phase of a visual to the next. Icons (including emojis) are themselves a visual language, so they can simplify and communicate ideas and feelings that are otherwise difficult to express. They may also help readers with certain intellectual or cognitive disabilities engage with your work. A body of research has shown that these kinds of visual-graph forms of communication have helped advance successful language development.

As an example of how icons can be used to support research and analysis, this short graphic from the Center on Budget and Policy Priorities uses icons to offset the five ways in which the Earned Income Tax Credit and the Child Tax Credit help families. The content is inherently qualitative, and the icons organize the text to make it easier for the reader.

WORD CLOUDS AND SPECIFIC WORDS

Word clouds are perhaps the most popular and familiar way to visualize qualitative data, but they are really a way to display quantitative data: the number of times a word appears in a text. In a word cloud, the size of each word is adjusted according to its frequency in a passage.

Working-Family Tax Credits Help at Every Stage of Life

The Earned Income Tax Credit (EITC) and Child Tax Credit (CTC) not only reward work and reduce poverty for low- and moderate-income working families with children, but a growing body of research shows that they help families at virtually every stage of life:

Improved infant and maternal health: Researchers have found links between increased EITCs and improvements in infant health indicators such as birth weight and premature birth. Research also suggests receiving an expanded EITC may improve maternal health.

Better school performance: Elementary and middle-school students whose families receive larger refundable credits (such as the EITC and CTC) tend to have higher test scores in the year of receipt.

Greater college enrollment: Young children in low-income families that benefit from expanded state or federal EITCs are more likely to go to college, research finds. Researchers attribute this to lasting academic gains from higher EITCs in middle school and earlier. Increased tax refunds also boost college attendance by making college more affordable for families with high-school seniors, research finds.

Increased work and earnings in the next generation: For each $3,000 a year in added income that children in a working-poor family receive before age 6, they work an average of 135 more hours a year between ages 25 and 37 and their average annual earnings increase by 17 percent, leading researchers have found.

Social Security retirement benefits: Research suggests that by boosting the employment and earnings of working-age women, the EITC boosts their Social Security retirement benefits, which should reduce poverty in old age. (Social Security benefits are based on how much one works and earns.)

Note: For further details on the research see Chuck Marr, Chye-Ching Huang, and Arloc Sherman, "Earned Income Tax Credit Promotes Work, Encourages Children's Success at School, Research Finds," CBPP

You can use icons to support research and analysis.
Source: Center on Budget and Policy Priorities.

These graphs are probably best used to visualize *overall* patterns or where a single value is obvious and stands out. It is less appropriate in cases where finding the specific values is especially important.

Word clouds are visually engaging, but they present two primary challenges. First, it is unclear what the *specific* frequency of each word is in the text. This word cloud uses the text from President Barack Obama's 2016 State of the Union address. You can see that he used the words *America*, *world*, and *people* frequently. But how much more frequently? It's hard to tell. If understanding the exact frequencies of words in the text is important, then this visual is insufficient.

What about the word *Americans* in this word cloud? It was also used frequently but its vertical orientation may have obscured it from your view. This is the second challenge with word clouds: some words may appear larger (and therefore more significant) than others because of their length, orientation, font, or color.

Remember, to create a word cloud you must calculate the frequency of each word. After you have quantified the text, you could use many different visuals—even a bar chart. It's much easier to see the most commonly used words in this bar chart than in the word cloud. It's worth noting that word clouds generally exclude the most common words, called *stop-words*, such as "the" and "at."

This word cloud shows the frequency of words spoken by President Barack
Obama in his 2016 State of the Union address.
Data Source: The White House.

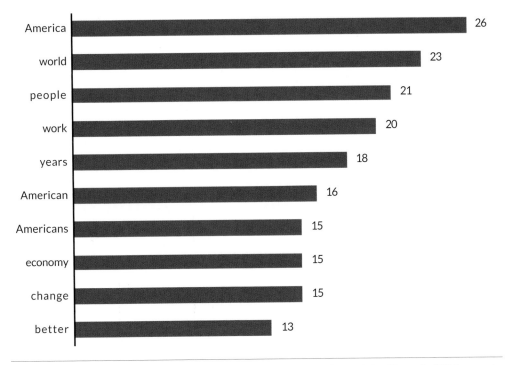

A standard bar chart that shows the ten most frequently used words in Obama's 2016 speech.
Data Source: The White House.

Another approach to creating effective word clouds may be to separate the overall body of text into *semantic* groupings—words grouped by their meaning. Following Obama's address, *USA Today* summarized his speech as follows:

> Obama defended the progress made over the last seven years and set out an agenda that will likely remain unfinished long after his presidency ends: turning back the effects of climate change, launching a "moonshot" to cure cancer, and a grassroots movement to demand changes in the political system.

Those important policy aspects of the speech don't come through in the basic word clouds, but organizing the text into semantic groups and then creating a set of smaller multiple word clouds might be a better way. In the next version, for example, you can better see the most important topics and words in the speech. This obviously requires

Instead of a single word cloud, Hearst et al. (2019) have suggested breaking up
the text into semantic groups.
Data Source: The White House.

more work on the part of the analyst to determine the groups and clusters, but it also results in a better visualization.

WORD TREES

Another way to visualize text data is to examine the contexts in which different words appear. *Word trees*, a version of which was developed by Martin Wattenberg and Fernanda Viegas in 2007, show all of the ways in which specific words are used within a text. The tree structure shows the combinations in which each word appears within the text, and the words are sized according to their frequency.

Word trees show all of the ways in which specific words are used within a text or, in this case, the Beatles' song "Come Together."
Source: AnyChart.

We've already seen one word tree in Chapter 8 that depicted how to show hierarchical relationships. Similarly, this word tree of the lyrics from the Beatles' song "Come Together" shows the elements of the song laid out in a hierarchy.

Trees can be used in other ways to show qualitative data. The next tree is based on interviews with fifty science bloggers and classifies their interview responses into their goals and

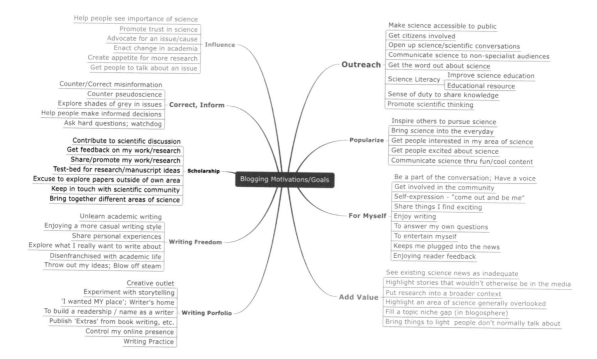

Paige Jarreau used a word tree to classify interview responses into different groups.

motivations. You can see the branching from each of nine different categories into specific comments and quotes. Word trees suffer the same pitfalls as word clouds: it is difficult to see the exact frequency of the words. But they do provide an engaging, interesting view of the text within its context.

SPECIFIC WORDS

Another way to visualize qualitative text data is to combine individual words with a quantitative metric. This 2019 visualization from Matt Daniels at the digital publication *The Pudding* is a histogram (recall page 179) of the number of unique words used by individual rap artists. Instead of a set of bars, the author plotted the artist's name, using color to distinguish the decade of album release. We see an overall view of the data, for example, that the most

# of Unique Words Used Within Artist's First 35,000 lyrics

BY ERA[1]

1980s | 1990s | 2000s | 2010s

<2,675 unique words: Lil Uzi Vert, NF

2,675–3,050 unique words: DMX, 21 Savage, A Boogie wit..., Lil Baby, Lil Durk, Wiz Khalifa, YG, YoungBoy Nev...

3,050–3,425 unique words: Bone Thugs-n..., 50 Cent, Juicy J, Drake, Future, Kid Cudi, Kid Ink, Kodak Black, Lil Yachty, Logic, Migos, Travis Scott, Young Thug

3,425–3,800 unique words: Foxy Brown, Juvenile, Master P, Salt-n-Pepa, Snoop Dogg, Eve, Gucci Mane, Kanye West, Lil Wayne, Missy Elliot, Trick Daddy, Trina, Young Jeezy, Big Sean, BoB, Childish Gam..., G-Eazy, Machine Gun..., Meek Mill, Nicki Minaj, Russ

3,800–4,175 unique words: Run-D.M.C., 2Pac, Big L, Insane Clown..., MC Lyte, Scarface, Three 6 Mafia, UGK, Dizzee Rascal, Jadakiss, Kano, Lil' Kim, Nelly, Rick Ross, T.I., 2 Chainz, A$AP Ferg, Big KRIT, Brockhampton, Cupcakke, Hopsin, Jay Rock, Kendrick Lamar, Mac Miller, ScHoolboy Q, Tyga, Vince Staples

4,175–4,550 unique words: Biz Markie, Ice T, Rakim, Brand Nubian, Geto Boys, Ice Cube, Jay-Z, Mobb Deep, Outkast, Public Enemy, Cam'ron, Eminem, The Game, Joe Budden, Kevin Gates, Royce da 5'9, Tech n9ne, Twista, Ab-Soul, A$AP Rocky, Danny Brown, Death Grips, Denzel Curry, $uicideboy$, Tyler the Cr..., Wale

4,550–4,925 unique words: Beastie Boys, Big Daddy Kane, LL Cool J, Busta Rhymes, Cypress Hill, De La Soul, Fat Joe, Gang Starr, KRS-One, Method Man, A Tribe Call..., Atmosphere, Ludacris, Lupe Fiasco, Mos Def, Murs, Talib Kweli, Xzibit, Flatbush Zom..., Joey BadA$$, Rittz

4,925–5,300 unique words: Common, Das EFX, E-40, Goodie Mob, Nas, Redman, Brother Ali, Action Bronson, KAAN

5,300–5,675 unique words: Kool G Rap, Kool Keith, Raekwon, CunninLynguists, Sage Francis, Watsky

5,675–6,050 unique words: Del the Funk..., The Roots, Blackalicious, Canibus, Ghostface Ki..., Immortal Tec..., Jean Grae, Killah Priest, RZA

6,050–6,425 unique words: GZA, Wu-Tang Clan, Jedi Mind Tr..., MF DOOM

6,425+ unique words: Aesop Rock, Busdriver

Just like specific data points can be shown in a scatterplot, specific words can be included in data visualizations. Matt Daniels from *The Pudding* used such an approach in his visualization of "The Largest Vocabulary in Hip Hop."

common number of words is around 4,000. We see a shift over time toward fewer words (more names in red to the left side of the graph). And we can focus on the details of the names and eras of specific artists if we want to.

As with other graph forms we have explored—for example, the beeswarm chart (page 206) and network diagram (page 277)—there are a many ways to visualize individual words in your data. The key is to organize that information logically, so the reader can sense the pattern immediately and engage with the graph.

QUOTES

If context is important, visualizing individual words is not the most useful way to show your qualitative data. Sometimes you need to show full quotations.

Following parliamentary elections in 2016, the *Berliner Morgenpost* published a story about how far some of the elected officials lived from their constituencies. Along with a series of maps that showed the residence of each of six officials relative to the

Warum kandidieren Sie so weit entfernt?

Erol Özkaraca
SPD

Der Sozialdemokrat engagiert sich nach wie vor in Neukölln, obwohl er schon vor Jahren nach Frohnau gezogen ist. Die Familie habe ein Haus mit Garten gewollt, das gebe es im Neuköllner Norden nun mal nicht. „Ich bin da eigentlich nur zum Schlafen, kenne mich dort kaum aus." Ihm sei aber klar, dass das durchaus Sozialneid im Wahlkreis auslösen kann.

Wolfgang Albers
DIE LINKE

Der Chirurg aus Wannsee ist seit 2006 im Abgeordnetenhaus · und überzeugter Experte für Gesundheits- und Hochschulpolitik. „Das Entscheidende ist, dass man gute Sachpolitik macht." Im Lichtenberger Wahlkreis ist er regelmäßig und hält Kontakt über sein Büro. „Das Einzige was nervt, ist der 34 Kilometer lange Weg im Stadtverkehr."

Iris Spranger
SPD

Die stellvertretende SPD-Landesvorsitzende hat lange in Marzahn-Hellersdorf gewohnt, ist dort 1994 in die Partei eingetreten „Es ist ja kein Geheimnis, ich habe meinen Mann kennengelernt und bin vor drei Jahren mit ihm zusammengezogen · in Frohnau. Ich habe aber keine Minute überlegt, ob ich politisch nach Reinickendorf gehe."

Anja-Beate Hertel
SPD

Ihre politische Heimat war lange Reinickendorf, wo sie auch wohnt. Nach einem internen Streit verlegte sie den Schwerpunkt ihrer Parteiarbeit nach Neukölln-Buckow. „Als Innenpolitikerin stehe ich den Positionen von Heinz Buschkowsky und der Neuköllner SPD in der Innen- und Integrationspolitik nahe, die in anderen Bezirken lange nicht mehrheitsfähig waren."

Holger Krestel
FDP

Der Liberale ist schon lange mit Tempelhof-Schöneberg verbunden. Hier hat er 1974 seinen Schulabschluss gemacht. „Bis heute bin ich im Bezirk aktiv". Von 2010 bis 2013 hat er ihn im Bundestag vertreten, war zuvor auch im Abgeordnetenhaus. „In Spandau wohne ich nicht zuletzt, um mich mit meiner Frau um meine 86-jährige Mutter zu kümmern."

André Lefeber
PIRATEN

Der Kandidat der Piraten wohnt noch in Lichterfelde, doch es zieht ihn immer wieder in den Südosten der Stadt. „Da viele meiner Bekannten und Freunde im Bezirk Treptow-Köpenick wohnen, finden viele meiner Freizeitaktivitäten dort statt, wo ich kandidiere." Er plane sogar einen Umzug in die Gegend. „Doch dies scheiterte bisher an den Mieten."

To communicate qualitative data, sometimes you need to just show the entire quote. The title of this story from the Berliner Morgenpost translates to "Why are you running for election from far away?" The quote from Ero Özkaraca, the image on the left, translates to, "The Social Democrat is still involved in Neukölln, although he moved to Frohnau years ago. The family had wanted a house with a garden, but that was not available in the north of Neukölln. 'I'm only there to sleep, I hardly know my way around.' But he is aware that this can certainly trigger social envy in the constituency."

district they represent, the newspaper also presented a series of direct quotes. Readers could see the crux of the story from the maps and find more detail by reading the quotes.

As an analyst working with survey data, this might be a difficult task. You may not have the exact names or pictures of your survey respondents, and even if you did, you might not be allowed to publish them because of concerns about privacy and security. But it is worth considering whether a generic quotation with a picture of a person would be sufficient in making the content more visual and personalized. As the example from an Urban Institute research project on unplanned pregnancies shows on the facing page, sometimes the quotes alone are sufficient to help communicate the content.

Women have diverse reasons for wanting to avoid unplanned pregnancies

Nearly all focus group participants said they did not want to become pregnant in the next year. Many had already experienced an unplanned pregnancy. They told us that having a child, or another child, would pose significant financial challenges and strain their relationships with their partner, parents, and other children.

I'M NOT CAPABLE OF HAVING A CHILD AND TAKING CARE OF THEM AT THE MOMENT. IT'S NOT THAT I WOULDN'T WANT TO BE A MOTHER, BUT I DON'T HAVE THE FUNDS TO DO IT.

I live with my boyfriend's mom and dad now, but we don't have enough space. The baby has his own room, but he doesn't use it... if we had another one, we'd have two babies in the bed.

Again, using specific quotes or phrases can be a powerful way to communicate your message.
Source: The Urban Institute

COLORING PHRASES

Passages of text build up from words to phrases to sentences and then to paragraphs. Depending on the goal of your visual, highlighting specific portions of text may be useful to summarize your analysis. You can do so by highlighting with color or boldface to make important passages visible to your reader.

The next image is from a *Bloomberg News* story that annotated and highlighted the transcripts from the Democratic primary debates in 2019. Each color represents a different content area, which helps the reader easily navigate the piece and quickly see which topics were discussed most frequently.

The Preservation of Favoured Traces project on the next page depicts the sequence of edits in Charles Darwin's *On the Origin of Species*. Darwin's final manuscript was actually the fifth draft, so by color-coding each word in the final text by the edition in which it first appeared, we can see the evolution of Darwin's writing and thinking over several years. An interactive version enables the user to zoom in, search, and explore the text in more detail.

This qualitative data visualization approach gives us a bird's-eye view of the data, but it isn't quantified. You can also use this technique to highlight quotes or passages in text to tag them as, say, "positive" or "negative," which might be subjective but also potentially enlightening. Exciting advances in natural-language processing, text analytics, and machine learning algorithms over the past few years now let researchers more accurately measure tone and semantics.

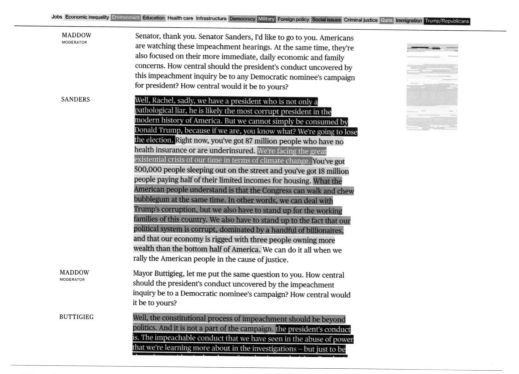

Bloomberg News highlighted phrases in words in a news story about the Democratic primary debates in 2019.

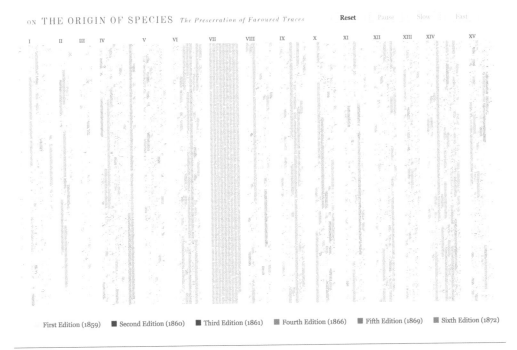

First Edition (1859)   ■ Second Edition (1860)   ■ Third Edition (1861)   ■ Fourth Edition (1866)   ■ Fifth Edition (1869)   ■ Sixth Edition (1872)

Especially in cases where there is a lot of text, highlighting specific phrases draws the reader's attention. Ben Fry used this very approach in his "Highlighting words in Darwin" project.

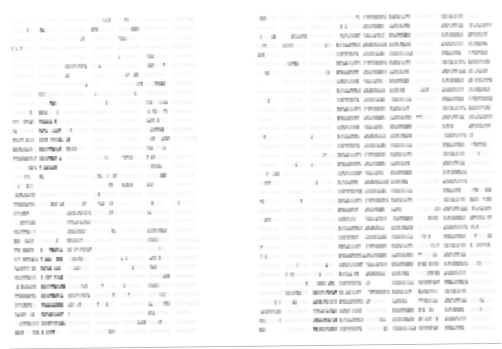

A closer look at "Highlighting words in Darwin."

MATRICES AND LISTS

If showing an entire qualitative data set is impossible, we can simplify and organize the data into groups or categories. Essentially, this approach creates a table of qualitative data so readers can more easily see the arguments and narratives. With this method, we can use lists to make it easier for our reader to navigate through what might be a very dense data set.

This graphic from the *Texas Tribune* organized the major issues in the platforms of the Texas Republican and Democratic parties in 2018. Combining the qualitative information

Private school subsidies

What Texans said in a June 2017 University of Texas/Texas Tribune Poll:

Redirect state tax revenue to help parents pay private school costs

48%	10%	42%
Strongly or somewhat oppose	Don't know	Strongly or somewhat support

Numbers may not add up to 100 due to rounding. Margin of error ± 2.83 percentage points.

What Democrats think

"Texas Democrats oppose the misnamed 'school choice' schemes of using public tax money for the support of private and sectarian schools."

Read the full Democratic platform

What Republicans think

"Texas families should be empowered to choose from public, private, charter, or homeschool options for their children's education, using tax credits or exemptions without government constraints or intrusion."

Read the full Republican platform

OUR TAKE

Democrats have long opposed subsidizing private school education but added language to this year's platform to argue that such subsidies would particularly affect access to special education services for Texas students with disabilities. The GOP modified its platform, which has long supported school choice, to assert that "no child should be forced to attend a failing school," and to "reject the intrusion of government in private, parochial, or homeschools."

The *Texas Tribune* combined quotes, text, icons, and quantitative data about poll results.
Source: *Texas Tribune*, texastribune.org

from the published platforms with quantitative data from other surveys, the reporters provide a broader view of six major policy issues. In this example for school subsidies, we can see how each section begins with a set of three icons, a sentence describing the issue, and a bar chart of survey results from a 2017 poll. Below, separate quotes from each party's platform documents and a summary from the reporters bring the entire issue together.

While in previous qualitative graph types, the analyst might need to make a judgment about semantics or meaning, in this case, the decision is to choose into which category the person falls. In other words, visualizing qualitative data doesn't need to be a complex process but simply a task of organizing the text into ways our reader can easily navigate and process.

CONCLUSION

Presenting qualitative data succinctly is a challenge. Taking a series of interviews and summarizing the results so that our reader can quickly and easily understand our argument is no small feat. But qualitative data can engage readers in ways that quantitative data may not. With good design and organization, we can hook our reader with qualitative data and encourage them to explore the words, quotes, and phrases.

Word clouds are one of the more common ways to visualize qualitative data, but they introduce perceptual issues depending on the length of the words, font, and layout. They are engaging, however, so they are useful in certain scenarios. Using recent research, we might consider ways to split our text into semantic groups and create a small-multiple version of the word cloud, which assuages some of these perceptual issues.

There is an array of alternatives to word clouds. Showing specific words or phrases, pairing quotes with photographs, or simply using icons in and around text are all ways to illuminate your qualitative data. As we've seen in the examples in this chapter, highlighting those specific passages and phrase often rely on specific design, color, and layout choices. But, as with many other chart types, practice and experimentation will lead to better visualizations of our qualitative data.

TABLES

Yes, tables are a form of data visualization. If you want to show the exact amount of every value in your data, a table might be your best solution. They are not the best solution if you want to show a lot of data or if you want to show the data in a compact space—but still, a well-designed table can help your reader find specific numbers and discover patterns and outliers.

As we've seen with other graphs, gridlines, tick marks, and other clutter can crowd a visualization and obscure takeaways. Tables are especially susceptible to clutter. The same guiding principles of creating effective visualizations apply here as well—clearly *show the data* so that our reader can find the most important patterns, trends, or values; *reduce the clutter* of gridlines, extra spacing, and uneven alignment; and *integrate the table and the text* by using concise active titles and subtitles, and including unit labels like percentage signs and dollar signs with care.

In this chapter, we'll cover ten steps for making better tables.

THE PROPER ANATOMY OF A TABLE

We must first understand the components of a table before we can understand how and when to adjust them to improve our data presentation. This diagram shows the ten primary components of a table. Many of these parallel the parts of a chart that you will see in the Style Guides section in Chapter 12. As with chart style, some of the style decisions you choose for

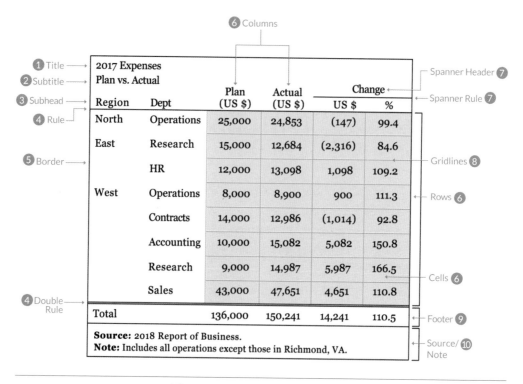

The components of a data table.

your tables will be subjective and depend on nothing more than your preferences for shading colors, font size, and line width.

1. **Title.** Use concise, active titles. "Table 1. Regression Results" is not particularly informative. Instead, guide your reader to the conclusion with a title like "A one year increase in work experience increases annual earnings by 2.8 percent." Left-aligning the title and subtitle will line it flush with the rest of the table, creating a grid, which is easier to navigate.

2. **Subtitle.** This sits below the title, often in a smaller font size or set in a different color to differentiate it. The subtitle should specify the units of the data in the table (like "Percent" or "Thousands of dollars") or make a secondary point (such as "The experience effect is greater for men than for women").

3. **Stubheads or Column Headers.** These are the titles of your columns. Differentiate these from the rest of the table cells with boldface or separate them with a line, also called a "rule."

4. **Rules.** The lines that separate the parts of the table from one another. At minimum, place rules below the stubheads and between the bottom row and any sources or notes.

5. **Border.** The set of lines that surround the table. Including a border around the whole table depends on how the table is arranged in the rest of the document. Sometimes you need to add a visual differentiator to set the table apart, and in those cases a border is useful. If, however, too many lines and borders clutter the document, omit the border altogether.

6. **Columns, Rows, and Cells.** Columns run vertically, and rows run horizontally. The intersecting areas are called cells.

7. **Spanner Header and Spanner Rule.** The text and line that span multiple columns. Text is usually centered over the multiple columns even if the specific column headers are left- or right-aligned.

8. **Gridlines.** The intersecting lines within the table that separate the cells. As with charts, take a light touch with your gridlines—heavy gridlines clutter the table.

9. **Footer.** The bottom area of a table where you might include a row for the total or average. As with the stubhead, we should differentiate this row from the rest of the table. We can do so by bolding the numbers, separating them with a line, or color shading the cells.

10. **Sources and Notes.** The text below a table containing the citation or additional details or notes to the table. Modern Language Association (MLA) style, for example, suggests putting the sources first and the notes second.

THE TEN GUIDELINES OF BETTER TABLES

These guidelines will take us from tables that have too much color, lines, and clutter to ones in which readers can easily see the important numbers and patterns. On the next page, we can see how these guidelines move us from the table on the left to the much more clear and readable table on the right.

RULE 1. OFFSET THE HEADERS FROM BODY

Make your column titles clear. Try using boldface text or lines to offset them from the numbers and text in the body of the table. It should be clear that the headers are not data values

Role	Name	ID	Start Date	Quarterly Profit	Percent Change
Operations	Waylon Dalton	A1873	May-11	5692.88	34.1
Operations	Justine Henderson	B56	Jan-10	4905.02	43.522
Operations	Abdullah Lang	J5867	Jun-14	4919.53	38
Operations	Marcu Cruz	B395	Dec-13	9877.52	37.1
Research	Thalia Cobb	C346	Apr-13	3179.49	-9
Research	Mathias Little	D401	Mar-11	5080.26	3.2
Research	Eddie Randolph	A576	Jul-18	7218.24	43.1
Contracts	Angela Walker	B31	Feb-18	6207.53	-1.788
Contracts	Lia Shelton	C840	Jan-16	1070.61	4.31
Contracts	Hadassah Hartman	D411	Nov-15	3735.96	3.01

Role	Name	ID	Start Date	Quarterly Profit	Percent Change
Operations	Waylon Dalton	A1873	May-11	$5,693	34.1
	Justine Henderson	B56	Jan-10	4,905	43.5
	Abdullah Lang	J5867	Jun-14	4,920	38.0
	Marcu Cruz	B395	Dec-13	9,878	37.1
Research	Thalia Cobb	C346	Apr-13	3,179	-9.0
	Mathias Little	D401	Mar-11	5,080	3.2
	Eddie Randolph	A576	Jul-18	7,218	43.1
Contracts	Angela Walker	B31	Feb-18	6,208	-1.8
	Lia Shelton	C840	Jan-16	1,071	4.3
	Hadassah Hartman	D411	Nov-15	3,736	3.0

Inspired by DarkHorse Analytics

but categories or headers. In this example, which uses growth in per capita GDP, the column labels are boldface and separated from the data with a single line.

Country	2013	2014	2015	2016
China	7.23	6.76	6.36	6.12
India	5.10	6.14	6.90	5.89
United States	0.96	1.80	2.09	0.74
Indonesia	4.24	3.73	3.65	3.85
Mexico	-0.06	1.45	1.90	1.68
Pakistan	2.21	2.51	2.61	3.44

Country	**2013**	**2014**	**2015**	**2016**
China	7.23	6.76	6.36	6.12
India	5.10	6.14	6.90	5.89
United States	0.96	1.80	2.09	0.74
Indonesia	4.24	3.73	3.65	3.85
Mexico	-0.06	1.45	1.90	1.68
Pakistan	2.21	2.51	2.61	3.44

Rule 1. Offset the headers from body.

RULE 2. USE SUBTLE DIVIDERS INSTEAD OF HEAVY GRIDLINES

As with the basic principle to reduce clutter for graphs, you can lighten or even remove much of the heavy borders and dividers in your tables. There is rarely a case when every single cell border is necessary. For series that show the total, use shading, boldface, or subtle line breakers to distinguish these.

Notice in the table on the left how the two columns that show the average (between 2007–2011 and 2012–2016) blend in with the other columns. At a quick glance you don't even

Country	2007	2008	2009	2010	2011	Avg.	2012	2013	2014	2015	2016	Avg.
China	13.64	9.09	8.86	10.10	6.36	10.74	7.33	7.23	6.76	6.36	6.12	6.76
India	8.15	2.38	6.95	8.76	6.90	6.30	4.13	5.10	6.14	6.90	5.89	5.63
United States	0.82	-1.23	-3.62	1.68	2.09	-0.30	1.46	0.96	1.80	2.09	0.74	1.41
Indonesia	4.91	4.59	3.24	4.83	3.65	4.47	4.68	4.24	3.73	3.65	3.85	4.03
Mexico	0.70	-0.48	-6.80	3.49	1.90	-0.19	2.15	-0.06	1.45	1.90	1.68	1.41
Pakistan	2.72	-0.36	0.74	-0.48	2.61	0.64	1.34	2.21	2.51	2.61	3.44	2.42
Average	5.15	2.33	1.56	4.73	3.92	3.51	3.52	3.28	3.73	3.92	3.60	3.61

Country	2007	2008	2009	2010	2011	Avg.	2012	2013	2014	2015	2016	Avg.
China	13.64	9.09	8.86	10.10	6.36	10.74	7.33	7.23	6.76	6.36	6.12	6.76
India	8.15	2.38	6.95	8.76	6.90	6.30	4.13	5.10	6.14	6.90	5.89	5.63
United States	0.82	-1.23	-3.62	1.68	2.09	-0.30	1.46	0.96	1.80	2.09	0.74	1.41
Indonesia	4.91	4.59	3.24	4.83	3.65	4.47	4.68	4.24	3.73	3.65	3.85	4.03
Mexico	0.70	-0.48	-6.80	3.49	1.90	-0.19	2.15	-0.06	1.45	1.90	1.68	1.41
Pakistan	2.72	-0.36	0.74	-0.48	2.61	0.64	1.34	2.21	2.51	2.61	3.44	2.42
Average	5.15	2.33	1.56	4.73	3.92	3.51	3.52	3.28	3.73	3.92	3.60	3.61

Rule 2. Use subtle dividers instead of heavy gridlines.

notice that there is a break in the annual series. In the version on the right, a light shade in those columns sets them apart.

RULE 3. RIGHT-ALIGN NUMBERS AND HEADERS

Right-align numbers along the decimal place or comma. You might need to add zeros to maintain the alignment, but it's worth it so the numbers are easier to read and scan. Here, for example, it is much easier to compare the values in the far-right column where the numbers are right-aligned than in either of the other two columns. To maintain the grid layout, the column header is right-aligned with the numbers as well.

Along these lines, choose the fonts in your tables carefully. Some fonts use what are called "oldstyle figures," in which some numbers drop below the horizontal baseline, the same way the letters *p* or *g* or *q* do. This is fine for cases where numbers are not a matter of data—like the numbering of chapters in a novel. But in data tables, they can be distracting and more difficult to read. Always use fonts that have "lining numbers," where all the numerals hit the baseline, and none drop below it.

Notice how the commas and decimal points in the table on the next page don't line up with custom fonts like Karla and Cabin. When choosing a font, be mindful that the numerals

	2016	2016	2016
China	6,894.40	6,894.40	6,894.40
India	1,862.43	1,862.43	1,862.43
United States	52,319.10	52,319.10	52,319.10
Indonesia	3,974.73	3,974.73	3,974.73
Mexico	9,871.67	9,871.67	9,871.67
Pakistan	1,179.41	1,179.41	1,179.41
Average	**12,683.62**	**12,683.62**	**12,683.62**

Rule 3. Right-align numbers and headers.

	Calibri	Karla	Cabin	Georgia
China	6,894.40	6,894.40	6,894.40	6,894.40
India	1,862.43	1,862.43	1,862.43	1,862.43
United States	52,319.10	52,319.10	52,319.10	52,319.10
Indonesia	3,974.73	3,974.73	3,974.73	3,974.73
Mexico	9,871.67	9,871.67	9,871.67	9,871.67
Pakistan	1,179.41	1,179.41	1,179.41	1,179.41
Average	**12,683.62**	**12,683.62**	**12,683.62**	**12,683.62**

Be aware of how numbers appear in different fonts.

are not always the same size. Also be aware that oldstyle figures of Georgia drop some of the digits below the horizontal baseline (I've added an underline in each cell to make this clear).

RULE 4. LEFT-ALIGN TEXT AND HEADER

Once we've right-aligned the numbers, we should left-align the text. The English language is read from left to right, so lining up the entries in that way generates an even, vertical border and is natural for the reader. Notice how much easier it is to read the country names in the far-right column than in the other two columns.

Right-aligned and hard to read	Centered and even harder to read	Left-aligned and easiest to read
British Virgin Islands	British Virgin Islands	British Virgin Islands
Cayman Islands	Cayman Islands	Cayman Islands
Democratic Republic of Korea	Democratic Republic of Korea	Democratic Republic of Korea
Luxembourg	Luxembourg	Luxembourg
United States	United States	United States
Germany	Germany	Germany
New Zealand	New Zealand	New Zealand
Costa Rica	Costa Rica	Costa Rica
Peru	Peru	Peru

Rule 4. Left-align text and headers.

RULE 5. SELECT THE APPROPRIATE LEVEL OF PRECISION

Precision to the fifth-decimal place is almost never necessary. Strike a balance between necessary precision and a clean, spare table. The per capita GDP growth rate, for example, is never

Country	Too many decimals	Too few decimals	About right
China	6.12380	6	6.1
India	5.88984	6	5.9
United States	0.74279	1	0.7
Indonesia	3.84530	4	3.8
Mexico	1.58236	2	1.6
Pakistan	3.43865	3	3.4
Average	**2.63104**	3	2.6

Rule 5. Select the appropriate level of precision.

reported to five decimals—that would be unnecessary and suggest a level of precision that is not supported by the data. This can also go the other way: Don't report too few digits. Showing per capita GDP growth as whole numbers masks important variation across countries.

RULE 6. GUIDE YOUR READER WITH SPACE BETWEEN ROWS AND COLUMNS

Your use of space in and around the table can influence the direction in which your reader reads the data. In the table on the left, for example, there is more space between the columns than between the rows, so your eye is drawn to read the table top-to-bottom rather than left-to-right. By comparison, the table on the right has more space between the rows

Country	2014	2015	2016
China	6.76	6.36	6.12
India	6.14	6.90	5.89
United States	1.80	2.09	0.74
Indonesia	3.73	3.65	3.85
Mexico	-0.38	-4.37	-4.25
Pakistan	2.51	2.61	3.44
Average	**3.43**	**2.87**	**2.63**

Country	2014	2015	2016
China	6.76	6.36	6.12
India	6.14	6.90	5.89
United States	1.80	2.09	0.74
Indonesia	3.73	3.65	3.85
Mexico	-0.38	-4.37	-4.25
Pakistan	2.51	2.61	3.44
Average	**3.43**	**2.87**	**2.63**

Rule 6. Guide your reader with space between rows and columns.

than between the columns, so your eye is more likely to track horizontally rather than vertically. Use spacing strategically to match the order in which you want your reader to take in the table.

RULE 7. REMOVE UNIT REPETITION

Your reader knows that the values in your table are dollars because you told them in the title or subtitle. Repeating the symbol throughout the table is overkill and cluttering. Use the title or column title area to define the units, or place them in the first row only (remembering to align the numbers along the decimal). If you are mixing units within the table, be sure to make your labels clear.

Country	2014	2015	2016
China	6.76%	6.36%	6.12%
India	6.14%	6.90%	5.89%
United States	1.80%	2.09%	0.74%
Indonesia	3.73%	3.65%	3.85%
Mexico	-0.38%	-4.37%	-4.25%
Pakistan	2.51%	2.61%	3.44%
Average	**3.43%**	**2.87%**	**2.63%**

Country	2014	2015	2016
China	6.76%	6.36%	6.12%
India	6.14	6.90	5.89
United States	1.80	2.09	0.74
Indonesia	3.73	3.65	3.85
Mexico	-0.38	-4.37	-4.25
Pakistan	2.51	2.61	3.44
Average	**3.43**	**2.87**	**2.63**

Rule 7. Remove unit repetition

RULE 8. HIGHLIGHT OUTLIERS

Instead of showing just six countries and three years as in the previous example, what if we need to show twenty countries and ten years of data? In this case, we might want to highlight outlier values by making the text boldface, shading it with color, or even shading the entire cell. Some readers will wade through all of the numbers in the table because they need specific information, but many readers are more likely to look for only the most important values. Guiding them to those important numbers lets them answer their own questions about the data or better comprehend your argument.

	2010	2011	2012	2013	2014	2015	2016
China	10.10	9.01	7.33	7.23	6.76	6.36	6.12
India	8.76	5.25	4.13	5.10	6.14	6.90	5.89
United States	1.68	0.85	1.46	0.96	1.80	2.09	0.74
Indonesia	4.83	4.79	4.68	4.24	3.73	3.65	3.85
Brazil	6.50	3.00	0.98	2.07	-0.38	-4.37	-4.25
Pakistan	-0.48	0.61	1.34	2.21	2.51	2.61	3.44
Nigeria	5.00	2.12	1.52	2.61	3.52	-0.02	-4.16
Bangladesh	4.40	5.25	5.28	4.77	4.84	5.37	5.96
Russia	4.46	5.20	3.48	1.57	-1.04	-3.04	-0.41
Mexico	3.49	2.12	2.15	-0.06	1.45	1.90	1.58

	2010	2011	2012	2013	2014	2015	2016
China	10.10	9.01	7.33	7.23	6.76	6.36	6.12
India	8.76	5.25	4.13	5.10	6.14	6.90	5.89
United States	1.68	0.85	1.46	0.96	1.80	2.09	0.74
Indonesia	4.83	4.79	4.68	4.24	3.73	3.65	3.85
Brazil	6.50	3.00	0.98	2.07	**-0.38**	**-4.37**	**-4.25**
Pakistan	**-0.48**	0.61	1.34	2.21	2.51	2.61	3.44
Nigeria	5.00	2.12	1.52	2.61	3.52	**-0.02**	**-4.16**
Bangladesh	4.40	5.25	5.28	4.77	4.84	5.37	5.96
Russia	4.46	5.20	3.48	1.57	**-1.04**	**-3.04**	**-0.41**
Mexico	3.49	2.12	2.15	**-0.06**	1.45	1.90	1.58

Rule 8. Highlight outliers.

RULE 9. GROUP SIMILAR DATA AND INCREASE WHITE SPACE

Reduce repetition by grouping similar data or labels. Similar to eliminating dollar signs on every number value, we can reduce some of the clutter in our tables by grouping like terms or labels. In this next example, grouping the names of the country regions reduces the amount of repetitive information in the first column. You can also use spanner headers and rules to combine the same entry and reduce unnecessary repetition. Here, besides grouping the country names, I've also applied some of the guidelines discussed so far such as left-aligning text, right-aligning numbers, and using boldface headers and footers.

Region	Country	Per Capita GDP		Percent Change
		2015	2016	
Asia	China	6496.62	6894	6.1238
Asia	India	1758.84	1862	5.8898
North America	United States	51933.40	52319	0.7428
Asia	Indonesia	3827.55	3975	3.8453
South America	Brazil	11351.57	10869	-4.2541
Asia	Pakistan	1140.21	1179	3.4387
Africa	Nigeria	2562.52	2456	-4.1601
Asia	Bangladesh	971.64	1030	5.9627
North America	Mexico	9717.90	9872	1.5824
Asia	Japan	47163.49	47661	1.0546
Africa	Ethiopia	487.29	511	4.9041
Middle East	Egypt	2665.35	2726	2.2633
Europe	Germany	45412.56	45923	1.1240
Middle East	Iran	6007.00	6734	12.1010
Middle East	Turkey	13898.75	14117	1.5734
Europe	France	41642.31	41969	0.7845
Average		15440	15631	2.6860

Region	Country	Per Capita GDP		Percent Change
		2015	2016	
Africa	Ethiopia	487	511	4.90
	Nigeria	2,563	2,456	-4.16
Asia	Bangladesh	972	1,030	5.96
	China	6,497	6,894	6.12
	India	1,759	1,862	5.89
	Indonesia	3,838	3,975	3.85
	Japan	47,163	47,661	1.05
	Pakistan	1,140	1,179	3.44
Europe	France	41,642	41,969	0.78
	Germany	45,413	45,923	1.12
Middle East	Egypt	2,665	2,726	2.26
	Iran	6,007	6,734	12.10
	Turkey	13,899	14,117	1.57
North America	Mexico	9,718	9,872	1.58
	United States	51,933	52,319	0.74
South America	Brazil	11,352	10,869	-4.25
Average		15,440	15,631	2.69

Rule 9. Group similar data and increase white space.

While grouping like elements does help reduce the amount of clutter on the page, be aware that posting tables to the internet may require some concessions in this regard. If you post tables to websites as images, users will be unable to copy and paste the data from the table to another tool, and screen readers—which literally step through the table and read the values out loud (see Chapter 12)—will be unable to read the data values. Instead, because of current constraints in web programming languages and formats, you might need to forgo spanner headers and other special formatting decisions, depending on the tools you use to post it to the internet.

RULE 10. ADD VISUALIZATIONS WHEN APPROPRIATE

We can make larger changes to our tables by adding small visualizations. Just like highlighting outliers with color or boldface, you might add sparklines (see page 152) to visualize some data rather than showing every number. Or you can use small bar charts to visually illustrate a series of numbers. Or you could use a heatmap (see page 112) and leave the numbers in the table or hide them, which can help the reader focus on the overall patterns and ignore the details.

We can also embed a chart-type structure right into our table. If you want a full chart embedded within the table, a dot plot (see Chapter 4) is succinct and can line up well within the linear structure of a table. You can also use a modification on the standard dot plot to place the numbers in their relative positions directly in a table.

Country	2007	2016	2007-2016
China	13.64	6.12	
India	8.15	5.89	
United States	0.82	0.74	
Indonesia	4.91	3.85	
Mexico	0.70	1.58	
Pakistan	2.72	3.44	
Average	**5.15**	**3.60**	

Country	2016	
China	6.12	
India	5.89	
United States	0.74	
Indonesia	3.85	
Mexico	1.58	
Pakistan	3.44	
Average	**3.60**	

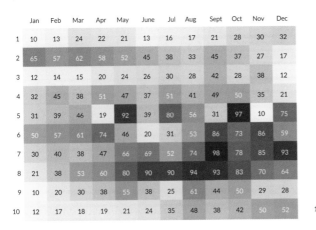

	Jan	Feb	Mar	Apr	May	June	Jul	Aug	Sept	Oct	Nov	Dec
1	10	13	24	22	21	13	16	17	21	28	30	32
2	65	57	62	58	52	45	38	33	45	37	27	17
3	12	14	15	20	24	26	30	28	42	28	38	12
4	32	45	38	51	47	37	51	41	49	50	35	21
5	31	39	46	19	92	39	80	56	31	97	10	75
6	50	57	61	74	46	20	31	53	86	73	86	59
7	30	40	38	47	66	69	52	74	98	78	85	93
8	21	38	53	60	80	90	90	94	93	83	70	64
9	10	20	30	38	55	38	25	61	44	50	29	28
10	12	17	18	19	21	24	35	48	38	42	50	52

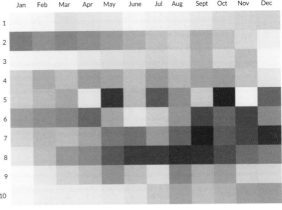

Country	2007	2016
China	———— 13.64	——— 6.12
India	——— 8.15	—— 5.89
United States	– 0.82	– 0.74
Indonesia	—— 4.91	—— 3.85
Mexico	– 0.7	—— 1.58
Pakistan	— 2.72	—— 3.44
Average	—— 5.86	— 2.63

Rule 10. Add visualizations when appropriate, for example, sparklines,
bar charts, heatmaps or dot plots.

DEMONSTRATION: A BASIC DATA TABLE REDESIGN

This table from the U.S. Department of Agriculture Food and Nutrition Service shows the number of people who participate in the Food Distribution Programs on Indian Reservations. The table presents participation estimates for twenty-four states over fiscal years 2013 through 2016, plus preliminary estimates for fiscal year 2017. Note the very dark, thick

FOOD DISTRIBUTION PROGRAM ON INDIAN RESERVATIONS: PERSONS PARTICIPATING					
(Data as of March 9, 2018)					
State	FY 2013	FY 2014	FY 2015	FY 2016	FY 2017
					Preliminary
Alaska	204	347	479	650	724
Arizona	10,835	11,556	11,880	11,887	11,235
California	5,593	5,495	5,159	4,795	4,463
Colorado	419	454	402	442	353
Idaho	1,440	1,566	1,688	1,706	1,530
Kansas	416	551	569	592	613
Michigan	1,299	1,846	1,971	2,061	1,960
Minnesota	2,297	2,756	2,645	2,600	2,487
Mississippi	701	863	958	1,056	1,169
Montana	2,375	3,144	3,149	3,313	3,271
Nebraska	1,010	1,229	1,339	1,396	1,267
Nevada	1,373	1,611	1,508	1,468	1,328
New Mexico	2,533	2,853	2,966	2,890	2,809
New York	380	384	369	452	350
North Carolina	584	736	743	700	671
North Dakota	3,840	4,800	4,976	5,661	5,569
Oklahoma	25,678	29,012	31,042	33,588	32,795
Oregon	678	871	800	785	687
South Dakota	7,457	8,123	8,208	8,505	8,525
Texas	117	131	142	124	114
Utah	117	167	217	421	384
Washington	3,164	3,185	3,284	3,410	3,221
Wisconsin	2,441	2,978	3,240	3,442	3,367
Wyoming	657	742	881	1,096	1,190
TOTAL	75,608	85,397	88,615	93,038	90,083

FDPIR is an alternative to the Supplemental Nutrition Assistance Program for Indian tribal organizations which prefer food distribution. Participation numbers are 12-month averages. Data are subject to revision.

This table from the U.S. Department of Agriculture Food and Nutrition Service
is cluttered and difficult to read.

10,835	11,556	11,880
5,593	5,495	5,159
419	454	402
1,440	1,566	1,688
416	551	569
1,299	1,846	1,971

Notice those heavy gridlines in the USDA table.

Source: US Department of Agriculture

gridlines, which make the table cluttered and difficult to read. As we zoom in, we can see that the numbers are top-aligned in each cell, which cuts them off ever-so-slightly.

There is a better way. Instead of including all of the gridlines—and making them dark and thick—we can remove them and keep only the line below the column header row. The

Number of People Participating in Food Distribution Programs on Indian Reservations
(Data as of March 9, 2018)

State	FY 2013	FY 2014	FY 2015	FY 2016	Preliminary, FY 2017
Alaska	204	347	479	650	724
Arizona	10,835	11,556	11,880	11,887	11,235
California	5,593	5,495	5,159	4,795	4,463
Colorado	419	454	402	442	353
Idaho	1,440	1,566	1,688	1,706	1,530
Kansas	416	551	569	592	613
Michigan	1,299	1,846	1,971	2,061	1,960
Minnesota	2,297	2,756	2,645	2,600	2,487
Mississippi	701	863	958	1,056	1,169
Montana	2,375	3,144	3,149	3,313	3,271
Nebraska	1,010	1,229	1,339	1,396	1,267
Nevada	1,373	1,611	1,508	1,468	1,328
New Mexico	2,533	2,853	2,966	2,890	2,809
New York	380	384	369	452	350
North Carolina	584	736	743	700	671
North Dakota	3,840	4,800	4,976	5,661	5,569
Oklahoma	25,678	29,012	31,042	33,588	32,795
Oregon	678	871	800	785	687
South Dakota	7,457	8,123	8,208	8,505	8,525
Texas	117	131	142	124	114
Utah	117	167	217	421	384
Washington	3,164	3,185	3,284	3,410	3,221
Wisconsin	2,441	2,978	3,240	3,442	3,367
Wyoming	657	742	881	1,096	1,190
Total	75,608	85,397	88,615	93,038	90,083

Note: FDPIR is an alternative to the Supplemental Nutrition Assistance Program for Indian tribal organizations which prefer food distribution. Participation numbers are 12-month averages. Data are subject to revision.

A simple redesign of the USDA table removes the clutter and lightens the view.

column header text is now bold to distinguish it from the numbers in the table. A line at the bottom of the table separates it from the note, and the *Total* row is bolded to set it apart from the body of the table.

Let's take this a step further by adding some visuals and color to the table.

Number of People Participating in Food Distribution Programs on Indian Reservations
(Data as of March 9, 2018)

State	FY 2013	FY 2014	FY 2015	FY 2016	Preliminary, FY 2017
Alaska	204	347	479	650	724
Arizona	10,835	11,556	11,880	11,887	11,235
California	5,593	5,495	5,159	4,795	4,463
Colorado	419	454	402	442	353
Idaho	1,440	1,566	1,688	1,706	1,530
Kansas	416	551	569	592	613
Michigan	1,299	1,846	1,971	2,061	1,960
Minnesota	2,297	2,756	2,645	2,600	2,487
Mississippi	701	863	958	1,056	1,169
Montana	2,375	3,144	3,149	3,313	3,271
Nebraska	1,010	1,229	1,339	1,396	1,267
Nevada	1,373	1,611	1,508	1,468	1,328
New Mexico	2,533	2,853	2,966	2,890	2,809
New York	380	384	369	452	350
North Carolina	584	736	743	700	671
North Dakota	3,840	4,800	4,976	5,661	5,569
Oklahoma	25,678	29,012	31,042	33,588	32,795
Oregon	678	871	800	785	687
South Dakota	7,457	8,123	8,208	8,505	8,525
Texas	117	131	142	124	114
Utah	117	167	217	421	384
Washington	3,164	3,185	3,284	3,410	3,221
Wisconsin	2,441	2,978	3,240	3,442	3,367
Wyoming	657	742	881	1,096	1,190
Total	**75,608**	**85,397**	**88,615**	**93,038**	**90,083**

Note: FDPIR is an alternative to the Supplemental Nutrition Assistance Program for Indian tribal organizations which prefer food distribution. Participation numbers are 12-month average. Data are subject to revision.

Adding a little color to the USDA table—such as a simple heatmap—makes it easier and faster for our reader to pick out specific values or patterns.

Number of People Participating in Food Distribution Programs on Indian Reservations
(Data as of March 9, 2018)

State	FY 2013	FY 2014	FY 2015	FY 2016	FY 2017	Average FY 2013-FY 2017
Alaska	204	347	479	650	724	481
Arizona	10,835	11,556	11,880	11,887	11,235	11,479
California	5,593	5,495	5,159	4,795	4,463	5,101
Colorado	419	454	402	442	353	414
Idaho	1,440	1,566	1,688	1,706	1,530	1,586
Kansas	416	551	569	592	613	548
Michigan	1,299	1,846	1,971	2,061	1,960	1,827
Minnesota	2,297	2,756	2,645	2,600	2,487	2,557
Mississippi	701	863	958	1,056	1,169	949
Montana	2,375	3,144	3,149	3,313	3,271	3,050
Nebraska	1,010	1,229	1,339	1,396	1,267	1,248
Nevada	1,373	1,611	1,508	1,468	1,328	1,458
New Mexico	2,533	2,853	2,966	2,890	2,809	2,810
New York	380	384	369	452	350	387
North Carolina	584	736	743	700	671	687
North Dakota	3,840	4,800	4,976	5,661	5,569	4,969
Oklahoma	25,678	29,012	31,042	33,588	32,795	30,423
Oregon	678	871	800	785	687	764
South Dakota	7,457	8,123	8,208	8,505	8,525	8,164
Texas	117	131	142	124	114	126
Utah	117	167	217	421	384	261
Washington	3,164	3,185	3,284	3,410	3,221	3,253
Wisconsin	2,441	2,978	3,240	3,442	3,367	3,094
Wyoming	657	742	881	1,096	1,190	913
Total	75,608	85,397	88,615	93,038	90,083	86,548

Note: FDPIR is an alternative to the Supplemental Nutrition Assistance Program for Indian tribal organizations which prefer food distribution. Participation numbers are 12-month average. Data are subject to revision.

Number of People Participating in Food Distribution Programs on Indian Reservations
(Data as of March 9, 2018)

State	FY 2013	FY 2014	FY 2015	FY 2016	FY 2017	Percent Change FY 2013-FY 2017
Alaska	204	347	479	650	724	254.9 ▲
Arizona	10,835	11,556	11,880	11,887	11,235	3.7 ▲
California	5,593	5,495	5,159	4,795	4,463	-20.2 ▼
Colorado	419	454	402	442	353	-15.8 ▼
Idaho	1,440	1,566	1,688	1,706	1,530	6.3 ▲
Kansas	416	551	569	592	613	47.4 ▲
Michigan	1,299	1,846	1,971	2,061	1,960	50.9 ▲
Minnesota	2,297	2,756	2,645	2,600	2,487	8.3 ▲
Mississippi	701	863	958	1,056	1,169	66.8 ▲
Montana	2,375	3,144	3,149	3,313	3,271	37.7 ▲
Nebraska	1,010	1,229	1,339	1,396	1,267	25.4 ▲
Nevada	1,373	1,611	1,508	1,468	1,328	-3.3 ▼
New Mexico	2,533	2,853	2,966	2,890	2,809	10.9 ▲
New York	380	384	369	452	350	-7.9 ▼
North Carolina	584	736	743	700	671	14.9 ▲
North Dakota	3,840	4,800	4,976	5,661	5,569	45.0 ▲
Oklahoma	25,678	29,012	31,042	33,588	32,795	27.7 ▲
Oregon	678	871	800	785	687	1.3 ▲
South Dakota	7,457	8,123	8,208	8,505	8,525	14.3 ▲
Texas	117	131	142	124	114	-2.6 ▼
Utah	117	167	217	421	384	228.2 ▲
Washington	3,164	3,185	3,284	3,410	3,221	1.8 ▲
Wisconsin	2,441	2,978	3,240	3,442	3,367	37.9 ▲
Wyoming	657	742	881	1,096	1,190	81.1 ▲
Total	75,608	85,397	88,615	93,038	90,083	19.1

Note: FDPIR is an alternative to the Supplemental Nutrition Assistance Program for Indian tribal organizations which prefer food distribution. Participation numbers are 12-month average. Data are subject to revision.

Adding other visualizations—bar charts or icons denoting change—are other ways to add visual elements to your tables.

The first example is a heatmap. Until I made this, I didn't realize by how much program participation in Oklahoma exceeded the rest of the states. It was only after the row appeared in dark blue that the magnitude became clear.

Another approach is to maintain the core look of the original table but add additional visual elements. In the tables above, the version on the left in the above pair adds a new data point—the average between fiscal year 2013 and 2017—and a bar chart to its right. This small graphic element gives the table a visual anchor and directs the eye to the states with more participation. The chart on the right adds the percentage change between 2013 and 2017 and a small up- or down-arrow to signal the change.

DEMONSTRATION: A REGRESSION TABLE REDESIGN

A typical regression table contains point estimates, standard errors, and some symbol (usually asterisks) to denote the level of statistical significance, such as 1 percent, 5 percent, and 10 percent. Such basic tables are especially useful when readers need the detailed numbers.

	Model 1	Model 2	Model 3
r_age	0.0509***	0.0119***	0.0207***
	(0.0062)	(0.0044)	(0.0026)
gndr	0.0442***	0.0616***	0.0630***
	(0.0057)	(0.0037)	(0.0043)
_educ	0.0027***	0.0052***	0.0157***
	(0.0087)	(0.0050)	(0.0072)
hrswkd	0.0397***	0.0075***	0.0211***
	(0.0053)	(0.0025)	(0.0029)
expr	0.0003***	0.0043***	0.0030***
	(0.0051)	(0.0026)	(0.0024)
marstat	0.0191***	0.0066***	0.0069***
	(0.0053)	(0.0025)	(0.0027)

* $p < 0.05$, ** $p < 0.01$, *** $p < 0.001$

	Model 1	Model 2	Model 3
Age	0.0509***	0.0119***	0.0207***
	(0.0062)	(0.0044)	(0.0026)
Gender	0.0442***	0.0616***	-0.0630***
	(0.0057)	(0.0037)	(0.0043)
Education	0.0027	0.0052	0.0157**
	(0.0087)	(0.0050)	(0.0072)
Hours Worked	0.0397***	0.0075*	0.0211***
	(0.0053)	(0.0044)	(0.0029)
Experience	0.0003	0.0043*	0.0030
	(0.0051)	(0.0026)	(0.0024)
Married	0.0191***	0.0066***	0.0069*
	(0.0053)	(0.0025)	(0.0041)

* $p < 0.05$, ** $p < 0.01$, *** $p < 0.001$

A table of basic regression results can be improved following the ten rules
shown earlier.

We can make a table of regression estimates clearer and more visually engaging by follow-ing the ten table rules and the visualization strategies from earlier in this chapter. We might also consider putting the dense table in an appendix (in the paper itself or maybe online) and using a graph in the main body of the paper instead.

Consider this relatively simple regression table that includes the coefficient estimates with asterisks, standard errors with parentheses, and unreadable variable names in the first col-umn. Don't use variable names to list your results! Your reader—even a reader of an aca-demic journal article—does not know what "_educ" or "expr" means. For our final table, let's use real words like "Education" and "Experience," and use the rules above to make the table cleaner and easier to read.

We can also convert these kinds of tables into data visualizations. A standard way is to use a bar chart with error bars, though, as noted in Chapter 6, some research has shown that we tend to discount the end of the error bar that sits within the bar itself.

Or we could try a dot plot approach (with or without error bars) and maybe use color to further signify statistical significance. In the graph on the right, solid circles contain esti-mates that are statistically significant and empty circles are estimates that are not statistically significant.

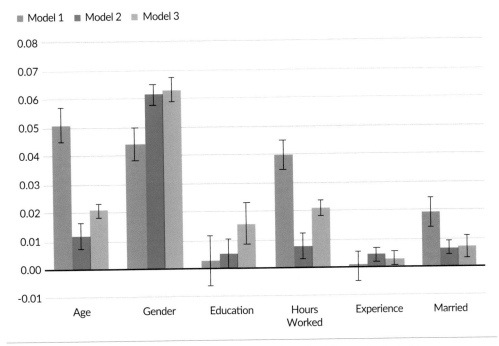

Regression results can be shown as a bar chart instead of a table.

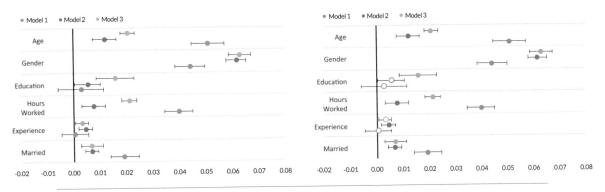

A dot plot is another way to visualize regression results.

Ultimately, if you decide such visual elements are unnecessary or insufficient, stick with the ten rules to make the table clearer and easier to read. Remember, the goal of our tables is to let the reader to more easily find the important numbers and patterns in the data and not ask them to wade through clutter.

CONCLUSION

Tables are themselves a form of data visualization, and the same rules apply. Many researchers and scholars rely heavily on tables, likely because they don't require much creative thinking—filled with text and numbers, columns and rows intersect, and details are left for the reader to navigate and decipher. And while tables are valuable and have their place, we can use these ten strategies to elevate our tables and make them clearer and easier to read.

Rule 1. Offset the headers from body
Rule 2. Use subtle dividers instead of heavy gridlines
Rule 3. Right-align numbers and headers
Rule 4. Left-align text and header
Rule 5. Select the appropriate level of precision
Rule 6. Guide your reader with space between rows and columns
Rule 7. Remove unit repetition
Rule 8. Highlight outliers
Rule 9. Group similar data and increase white space
Rule 10. Add visualizations when appropriate

PART THREE

DESIGNING AND REDESIGNING
YOUR VISUAL

DEVELOPING A DATA VISUALIZATION STYLE GUIDE

A data visualization style guide does for graphs what the *Chicago Manual of Style* does for English grammar. It defines the components of a graph and their proper, consistent use. Like a writing style guide, a comprehensive data visualization style guide breaks down the parts of graphs, charts, and tables to demonstrate best practices and strategies to design and style your charts. Elements like font and color, the widths of lines and style gridlines, and the use of tick marks are all choices that determine whether a graph is clear, engaging, and consistent—or whether it isn't.

The difference between a grammar guide and a data visualization guide is that many of our data style decisions are subjective. While the word *their* is objectively different than *they're*, and the use of one in a particular case is either correct or incorrect, there is no objectively correct or incorrect line thickness for a chart. There are, however, certain principles to consider, many of which we have covered so far. But for the most part, the styles you choose will reflect you and your organization's preferences.

THE ELEMENTS OF A DATA VISUALIZATION GUIDE

In organizations, a data visualization style guide serves three purposes.

First, it provides team members with the detailed styles and expectations about what should and should not be included in a visualization. Where should the title go? How large should it be? What font? What color?

Second, it guides those who may not be familiar with (or care about) all the styling and branding guidelines the organization may value. Instead of asking researchers and analysts to compile the data, create the graph, and then worry about which colors and fonts to use, a style guide makes those decisions easier. Building these styles into software tools streamlines the process and automates the application of graph styles.

Finally, a style guide sets the tone and expectations for people in the organization that the style, look, and details about data visualization are as important as other branding materials.

Even if you're an individual working with data, a style guide can be worthwhile. A custom style guide will make your work more consistent and efficient, and it will build your individual brand so your work stands out. A good style guide handles the basic style decisions for you, so you can focus on more important aspects of creating data visualizations.

As you build your style guide, test the components to make sure you or your team members can use and implement them. The style needs of your charts may differ from those of other branding materials. Colors that might look great in a logo may not work in a line chart or bar chart. Also remember to treat your data visualization style guide as a living document, just as you would a style guide for text or design. The guide should change as your personal or organizational aesthetic changes and evolve alongside changes in publication types and software tools.

Consider these Marimekko charts from the *Economist* and the *Financial Times*. Both publications have a distinct look and feel, and even those who are not regular readers may recognize the style. This branding is an important aspect of the organization's identity.

There can be many sections to a detailed style guide, which we'll cover in detail in this chapter, but here are the basics that any data visualization style guide should cover.

1. **Graph Anatomy.** Where should labels, titles, and other elements be placed? What is the proper size of charts and should this size differ for different types of output?
2. **Color Palette.** What colors should be used across graph and data types? Does the color palette vary across graph types? Does it vary for print and digital products?
3. **Font.** What font should be used and how should its size, boldness, and position vary? Should there be one font style for the title and another for the text in the body of the graph?

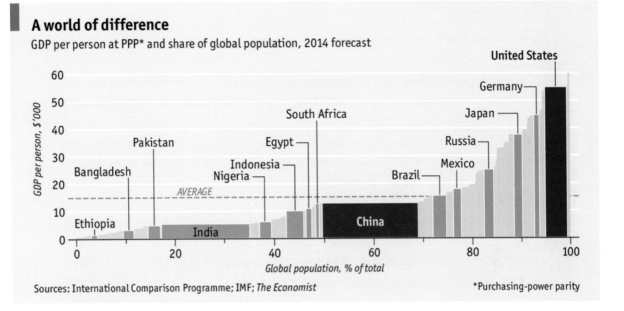

A world of difference

GDP per person at PPP* and share of global population, 2014 forecast

United States

Germany

Japan

South Africa

Russia

Pakistan

Egypt

Mexico

Indonesia
Nigeria

Brazil

Bangladesh

AVERAGE

Ethiopia

India

China

GDP per person, $'000: 0, 10, 20, 30, 40, 50, 60

Global population, % of total: 0, 20, 40, 60, 80, 100

Sources: International Comparison Programme; IMF; *The Economist*   *Purchasing-power parity

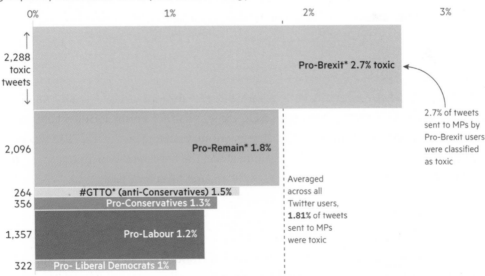

Self-described pro-Brexit tweeters directed the most toxicity at MPs

Percentage and number of tweets sent to MPs that were classified as toxic, grouped by the senders' use of political terminology

0%   1%   2%   3%

2,288 toxic tweets

Pro-Brexit* 2.7% toxic

2.7% of tweets sent to MPs by Pro-Brexit users were classified as toxic

2,096

Pro-Remain* 1.8%

264   #GTTO* (anti-Conservatives) 1.5%
356   Pro-Conservatives 1.3%

Averaged across all Twitter users, **1.81%** of tweets sent to MPs were toxic

1,357   Pro-Labour 1.2%

322   Pro- Liberal Democrats 1%

*Pro-Brexit tweeters are those using any of the following terms in their Twitter bio: Brexit party, #Brexit, #StandUp4Brexit, #GetBrexitDone, Pro-Brexit, Brexiteer. Pro-Remain tweeters used any of: #FBPE, Pro-EU, #RevokeA50, #Remain, #PeoplesVote, #StopBrexit, Revoke.
#FBPE = 'Follow back pro-European; #GTTO = 'Get the Tories out'
Source: FT research
© FT

The color, fonts, and overall style of these Marimekko charts from the *Economist* (top) and the *Financial Times* (bottom) make them easily identifiable.

4. **Graph Types.** Are there special considerations for specific graphs? For example, are pie charts forbidden in all situations? Is there a maximum number of series allowed in a line chart?

5. **Exporting Images.** How should team members move graphs from their software tool to the final report or website? Should they use PNG, JPEG, or some other image format? How should people create those image formats if they are not native to their software tool?

6. **Accessibility, Diversity, and Inclusion.** What steps do you and your organization need to take to make your graphs accessible to people with vision impairments or intellectual or other disabilities? Are you being mindful of how you're presenting results for different races, genders, and other groups?

THE ANATOMY OF A GRAPH

To set graph styles, we should first define each part of a graph. To illustrate how this can be done in practice, we can use the basic template in the style guide published by the Urban Institute, a nonprofit research institution in Washington, DC.

1. OVERALL DIMENSIONS

Specify the overall size of the chart. This may differ for different product types—an online graph, for example, is often measured in pixels, while a chart for a print document is typically measured in inches or centimeters. The dimensions may depend on the tools and workflow your organization uses. In the Urban Institute guide, the horizontal dimensions for a print graphic are specified at the top (6.5 inches) to fit on an 8.5 × 11 inch page.

2. FIGURE NUMBER

Figures and tables can be numbered, lettered, or left alone. You can place it above the chart title, centered or left-aligned, in a different font size and color. You could also leave it in-line with the chart title like, "Figure 1. Chart Title." As you can see in the image, the Urban style is to put the figure number above the chart title in all capital letters in its standard blue color. The note includes details for font (Lato Regular), font size (9 pt), capitalization (Uppercase), and color (RGB: 22 150 210).

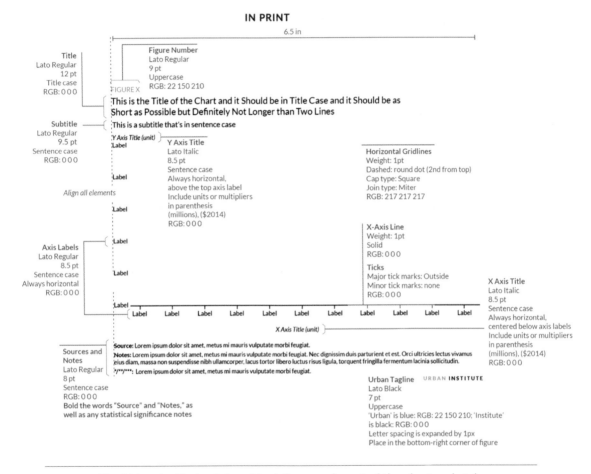

The Urban Institute style guide defines each part of the chart and style choices that apply to them.

3. TITLE

Will the title be aligned with the left side of the chart or centered over the plot space? Left-aligning the title (and other text objects) has the advantage of creating a vertical grid along the left edge, which makes it appear more organized and therefore easier to read. You should also lay out a basic style for the tone of the titles. Will they be purely descriptive or more active, summarizing the important points of the chart? Will you use sentence case or title case? The Urban style is to use title case, as detailed in the sample title. Notice also how the use of the title is a direction to the user—it should be title case, short as possible, and no more than two lines.

4. SUBTITLE

If there are going to be subtitles, how will they be used in the chart? Is this a place to insert a more active statement or is it a place to list the units in the chart? The subtitle is a good place to include the vertical axis title because, when left-aligned, it's located close to the top of the axis. To offset it from the title, you might place it in parentheses, make the size smaller, or even change the color. In the Urban example, the subtitle is written out in sentence case with a smaller size and black color.

5. AXIS TITLES

Where will the vertical and horizontal axis titles go? In many software tools, the vertical axis title is rotated and placed alongside the vertical axis. A better position is to have it horizontally oriented and positioned above the vertical axis, aligned with the title and subtitle (or, as mentioned, it might be the subtitle). For the horizontal axis title, you might need to decide how far below the axis labels it will sit. There are cases—such as months or years—where the units are obvious and a horizontal axis title can be omitted. Axis titles can be differentiated by using smaller text or different colors. You should also decide whether to spell out and capitalize units like "dollar" or "percent," or to use a symbol. The Urban style is to place the vertical (y-) axis title above the axis and to have units in parentheses; the horizontal axis label sits below the axis in 8.5 pt Lato Italic font, horizontal, and centered.

6. AXIS LABELS

How these should be formatted? Boldface, italics, different font size? The vertical axis labels (as distinct from the title) typically sit to the left of the chart, though they can also be added to the right side if the chart is very wide. For the horizontal axis labels, are there specific formats for certain units? For example, when using years along the axis, would a series like 2000, '01, '02 . . . be acceptable or should each number be written out in full?

7. AXIS LINES AND TICK MARKS

What color and thickness will you use for the axis lines? Will the tick marks be inside or outside the chart? Some organizations leave out the vertical axis line altogether, but the horizontal axis line is typically included to give the chart a consistent anchor. I prefer to make the zero-axis line slightly darker than the other gridlines because it acts as a baseline. This is especially true in cases with negative values: We want to make it clear that the zero-axis line is not at the bottom of the chart. Tick marks are likely not needed in the space between the bars in a bar chart, but may be necessary in a line chart. In the Urban example, there is no vertical axis line, but the horizontal axis line is a 1 pt black line with major tick marks that are outside the chart.

8. GRIDLINES

Many charts include horizontal gridlines, though the exact formatting varies. Will they be solid, dashed, or dotted? How thick will they be? And what color? At what increments will they be added? Many charts do not include vertical gridlines, though the occasional scatterplot will include them to create a visible grid.

9. SOURCES AND NOTES

Data sources should be documented and note any important modeling or modifications. A box for sources and notes is typically found at the bottom of the chart, left-aligned with the vertical axis labels, title, and subtitle. In many cases, the word *Source* and *Note* are boldfaced. The Chicago Manual of Style (section 3.20), for example, suggests placing the source line above the note line. In the Urban style, the words *Source* and *Notes* are in bold face and ordered in that way.

10. LOGO

If you want to include a logo on the graph, decide where it will go and what size it will be (and be sure to use a high-resolution image). Logos are often placed in the bottom-right corner, but sometimes in other places. The advantage of placing it in the bottom-right corner is that it is out of the way of the title/subtitle and sources/notes areas. Urban adds one of its logo formats to the bottom-right area of the graph, with specific instructions for color and spacing.

11. LEGEND

Will a legend be used and if so, where will it go, what size will it be, and what markers will be used? It is not labeled on this image, but the Urban style guide includes a separate section that specifies font sizes for other elements of graphs, including the legend.

12. DATA MARKERS

Will graphs, especially line graphs, include data markers, like circles or squares? Will the markers be filled or hollow? When will data values be labeled? You may want to set rules about using data markers for graphs with some number of values.

13. DATA LABELS

Determine when data points should be labeled and how they should be placed and formatted. The Urban guide has a separate table of font sizes that describes how these labels should appear.

14. DATA SERIES

This will vary by chart type—thickness of lines, space between bars and columns, colors for each element. You may need a separate section of the style guide to address issues of specific chart types, depending on the complexity of the charts your organization uses.

▶ ▶ ▶ ▶ ▶

As just two other examples of how published style guides define different parts of a chart, the first image is from the data visualization style guide published by the London Datastore, an effort by the city of London to open and share its data and resources. Similarly, the Sunlight Foundation, a nonpartisan organization that advocates for open government, has a style guide that defines each part of their charts to reflect their styling preferences.

CHART DESIGN
STYLING & LAYOUT

Consistent layout, labels & lines

For ease of comprehension, it's important that your charts are presented consistently, and are as clean and uncluttered as possible.

A descriptive chart title in 18pt Bold #353d42
More detailed explanation matching the document body copy in 14pt

Remember, your charts should feel part of the document they are displayed within, and so text sizes should relate to the document text hierarchy and all chart text should be consistent across the document.

Shown are typical text and line weight settings derived from the London Datastore's body copy. Depending on your tools, device / document context, and resolution, you may want to change the specific settings, but the relative relationships between lines & type settings should be similar.

A descriptive chart title in 18pt Bold #FFFFFF
More detailed explanation matching the document body copy in 14pt

CITY INTELLIGENCE

Source: Mike Brondbjerg of the Greater London Authority, reproduced under the Open Government License

Basic Structure

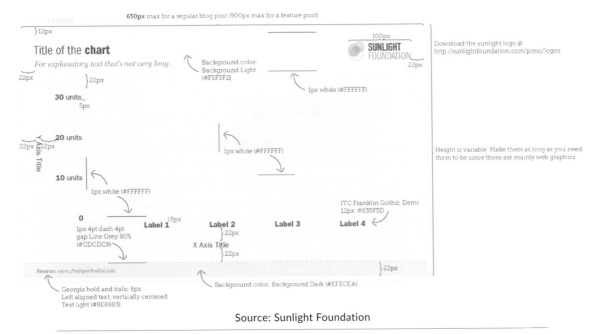

Source: Sunlight Foundation

A data visualization style guide should lay out specific chart fonts, styles, colors, and sizes.

COLOR PALETTES

Color has unmistakable power in our visualizations. It may be the first thing people notice about our graphs. Color can evoke emotions and draw attention. As Vincent van Gogh wrote to his brother in 1885, "Color expresses something in itself. One can't do without it; one must make use of it."

Successful brands have recognizable color palettes for everything from their logo and letterhead to their data visualizations. But a palette that works for company letterhead or website may not necessarily work for a line chart with five lines. There are a number of free online color tools to develop color palettes: Adobe Color, Color Brewer, Colour Lovers, and Design Seeds are a few examples, and the Appendix contains a longer list. Besides the basic colors, we will also need different shades and tints for each color in the palette.

A style guide should contain different schemes to guide chart creators in their color choices. The easier you make it for an analyst to apply the branding and design elements, the more time they have to work with the data and develop the best graph for their purpose.

There are five primary color schemes you can apply to your data visualizations.

Binary. Nominal differences divided into two (binary) categories: urban-rural, Democrat-Republican, agree-disagree.

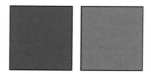

Sequential. Data values that are logically arranged from low to high should use sequential color schemes. Low values are usually represented by light colors, and high values by

Low   High

dark colors. Choropleth maps that show poverty rates or population, for example, would use sequential color palettes.

Diverging. In this scheme, the colors progress outward, growing darker from a central midpoint. A diverging color palette will share sequential schemes on two different colors and diverge from a shared, lighter color, for example, deviations from zero or a central number.

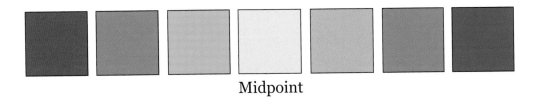

Midpoint

Categorical. Color schemes that use separate colors to represent nominal differences, for example, different race or gender groups.

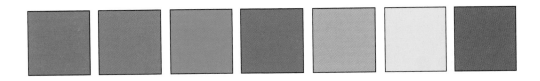

Highlighting. This is a special case of the categorical color scheme. These color schemes highlight a certain value or group within the visualization. For example, we could use this palette to emphasize a single or small group of points in a scatterplot.

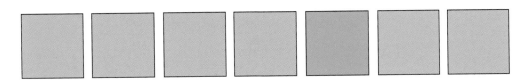

Transparency. Not so much a color scheme but a technique for using color, transparency in graph objects lets us (and our reader) see the object behind it. We've seen this technique a few times already (see the area chart section on page 157 as an example). You can use transparent colors—with or without a solid border—to make overlapping objects visible.

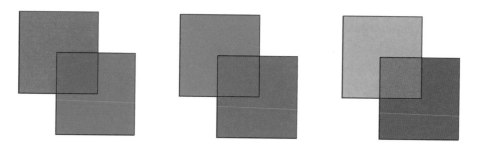

As an example of how color palettes are defined in practice, this section of the National Cancer Institute style guide shows primary and secondary color palettes along with a full list of tints and shades. The Consumer Finance Protection Bureau (CFPB) style guide includes sets of colors to "maintain CFPB brand cohesion."

We should also be mindful of readers with color vision deficiency (CVD) or color blindness. About 300 million people around the world have some form of CVD, most of them

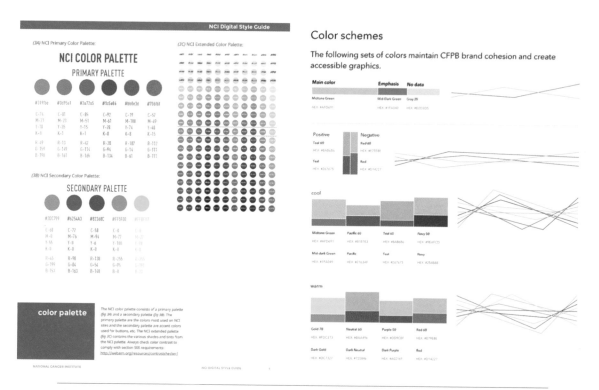

There are many ways to define branded colors and styles. The National Cancer Institute (left) and the Consumer Finance Protection Bureau (right) are demonstrations of two such ways.

men, and most of them have difficulty discerning between reds and greens, though there are other forms as well. There are a variety of online color and color contrast checking tools such as Vischeck.com and WebAim that can be used to test colors.

AVOID THE RAINBOW

When choosing a color palette, avoid the rainbow color palette. In most cases, the rainbow palette is a poor choice for visualizing your data for at least three reasons. First, while a color ramp from light blue (small data values) to dark blue (large data values) makes logical sense, it isn't really logical to say that "purple" means more than "orange." Second, and more importantly, the rainbow color palette does not map to our number system. Notice how wide the green area is in the rainbow palette below compared with the thin light blue area. If we were to show a unit change from, say, 1 to 2, we might not see a change in greens; but the same unit change from, say, 9 to 10, in the blue spectrum might shift all the way from teal to

Avoid the rainbow color palette. It doesn't map to our number system, isn't logical when mapped to data, is not comprehensible for people with color vision deficiency, and does not translate to grayscale.

navy. Finally, the rainbow palette is not consistent for people with CVD (the middle image) or when printed in black and white (the last image).

COLORS AND CULTURE

Finally, be mindful that colors can reinforce stereotypes or hold different meanings in different cultures. For many years, pink and blue colors were used to differentiate data values for women and men. But in modern-day western cultures, these colors come with gendered stereotype baggage: pink suggests weakness and blue suggests strength. Interestingly, this was not always the case—up until about the mid-twentieth century, it was the opposite. In her book, *The Secret Lives of Color*, Kassia St. Clair writes, "Pink is, after all, just faded red, which in the era of scarlet-jacketed soldiers and red-robed cardinals was the most masculine color, while blue was the signature hue of the Virgin Mary." Instead of the standard pink-blue pairing, consider using other color combinations such as purples and greens (as in the *Telegraph* newspaper) or blues and oranges (as in the *Guardian*).

More generally, also consider how different cultures use and perceive different colors. In western cultures, for example, the color red may evoke emotions of passion and excitement and has both positive and negative associations. In eastern cultures, however, the color red represents happiness, joy, and celebration. In India, red relates to purity, and in Japan it is associated with life, anger, and danger.

DEFINING FONTS FOR THE STYLE GUIDE

A data visualization style guide should define typefaces (or fonts) for each part of the chart. You probably don't need more than two different fonts, and one will usually suffice. Also remember that you can vary the look of a single font by changing its thickness (thin, bold), angle (italics), and color.

BE WARY OF CUSTOM FONTS

A custom font can differentiate your style from the standard fonts found in, for instance, the Microsoft Office package. But beware: using custom fonts requires that they be installed on any machine that shows the graph. While custom fonts make your graphs stand out,

Preferred Typefonts

Here is a list of preferred typefonts.

Primary	Optional Primary
Helvetica Light	Arial Regular
ABCDefgh1234	ABCDefgh1234
Helvetica Light Italic	Arial Regular Italic
ABCDefgh1234	ABCDefgh1234
Helvetica Regular	Arial Bold
ABCDefgh1234	ABCDefgh1234
Helvetica Regular Italic	Arial Bold Italic
ABCDefgh1234	ABCDefgh1234
Helvetica Bold	
ABCDefgh1234	
Helvetica Bold Italic	
ABCDefgh1234	

Secondary	
Times Roman Regular	
ABCDefgh1234	
Times Roman Regular Italic	
ABCDefgh1234	
Times Roman Bold	
ABCDefgh1234	
Times Roman Bold Italic	
ABCDefgh1234	

BBC AUDIENCES

CONSISTENCY STYLE GUIDE BETA

TYPOGRAPHY
CONSISTENT
USER
EXPERIENCES

FONTS
Your dashboard should have **Reith Sans** throughout.

Your default font in the style guide is sized to 10pts, but be aware that font **sizes below 12 points** will be difficult to read.

CHART TITLES & LEGENDS
Use the chart title to convey what's being displayed in the graph. This is also a good place to position your key metric (which means you can remove it from the axis).
Use the subtitle space as a colour legend when applicable.

Chart Title (12pts)
Chart Subtitle (10pts)
Thing 1 | Thing 2 (10pts Bold)

TYPE DYNAMICS
You can acheive a lot by using variations in text size on your dashboard.

MORE INK = MORE EMPHASIS
Good for KPIs and callouts

and vice versa
Good for de-emphasising information

AXIS FONTS
The emphasis on your axes will vary, but where possible de-emphasise axes text by reducing size (10pts) and / or reduce shading (to #898989).

ALIGNMENT
As a rule of thumb, align your text so that it is closer to the data (especially useful for bar charts and tables).

Contact the team

A data visualization style guide should also define appropriate fonts to use and when to use them. The BBC (left) and US Department of Agriculture (right) are just two examples of how to provide this guidance.

they can also raise trouble when sharing files or presenting from a laptop different from your own.

Default fonts like Century Gothic, Tahoma, Trebuchet MS, and Verdana are examples of effective fonts for data visualization that are available on most operating systems but are less commonly used and therefore appear more novel.

The BBC style guide for data visualization in Tableau (on the left) includes a typography section that demonstrates which fonts to use, where, and how to align them to the broader chart space. Their Reith Sans font is not a default font type, and they must therefore make sure that everyone in the organization has that font installed on their computers. The *Visuals Standards Guide* from the U.S. Department of Agriculture (on the right) features a broad set of fonts that are used across their publication types and at least two of the three (Arial and Times New Roman) are typically default fonts.

GUIDANCE FOR SPECIFIC GRAPH TYPES

Another section you might want to include in your style guide is a set of examples or instructions about specific chart types. Your organization may want to specify certain styling or data visualization best practices that differ chart to chart. You might also include examples of less common chart types to broaden your organization's data visualization toolbox, just like you did while reading this book.

Start by constructing guidelines for the most common chart types your organization uses. For example, you might specify that dual-axis line charts should never be used (see page 143) or that pie charts should have some upper limit of series (see page 289). There are also more granular specifications, such as where exactly labels should sit in a stacked bar chart or whether data points on a line chart sit on or between the tick marks. Or you might specify never to include tick marks on bar charts, or that whenever data labels are included, gridlines and tick marks must be omitted.

Another issue you might address in the specific chart area is how to manage the color palette among different data series. If the main colors in your palette are blue, red, and orange, the order of those colors may change if you have two or three series or may vary for, say, a paired bar chart versus a stacked bar chart.

TIPS FROM THE URBAN INSTITUTE STYLE GUIDE

▸ All of Urban's charts will be full-width (685px), so it is important to keep the data density as high as possible. Always include a text reference to your figure to give the data context to the content of the report/brief/blog post. If your chart has only two or three values, consider a couple of sentences of text to explain the figure.

▸ If you find your explanatory sentences do a better job of distilling the information, you might want to consider going without a chart.

▸ Title: Keep it short and simple. Try to explain the chart in a few words. If you need to add qualifiers (e.g., years, dollars) or further clarification, use a subtitle

▸ Source and Notes: This is where the technical information about methodology can go. Try to avoid putting this information in the title, labels, or on the chart.

▸ Legends: Stretch legends across the top of the chart, or to the right. Order them in a logical way, mirroring the order of the data in the charts.

Source: The Urban Institute Style Guide, accessed January 2020.

You might also include data visualization tips and tricks. These five tips are listed at the very top of the Urban Institute's guide. You can develop your own rules and tips or borrow those published by other groups and organizations.

EXPORTING IMAGES

Once a visualization is ready for external consumption, the chart creator must export it to a usable file format. This is another opportunity for things to go wrong: improper exporting might compress the resolution and pixelate the whole image. You can see the difference in resolution between these two versions of the same chart from Chapter 4. You can spend all the time you want creating a great, effective graph with clear colors and fonts, so don't waste all that effort with a blurry, hard-to-read final image.

Choosing the right file format for your visualizations is key. There are many file formats from which to choose, each with its own advantages and disadvantages. The biggest difference between image file formats is whether they are bitmap or vector. Images in bitmap format (also called raster) are stored as a series of squares (called pixels), each assigned a specific color. When you take a bitmap image and stretch it out, the pixels get larger and the resolution falls. You may have seen something like this if you put a photograph in a document and then tried to make it bigger—each pixel is now larger, and the crispness of the image deteriorates.

The other image format is vector. As opposed to bitmap images, vector images contain information about the actual shape in the image. A vector image is recreated when you

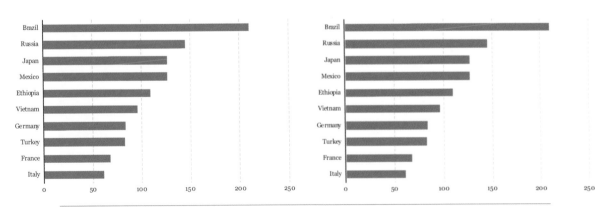

When moving images from your data visualization tool to your final product—either a report or web image—be sure the image has a sufficiently high resolution. The key is to test the image before you publish or post it.

Type	Acronym	Name	Application
vector	pdf	Portable Document Format	general purpose
	eps	Encapsulated PostScript	general purpose
	svg	Scalable Vector Graphics	online
bitmap	png	Portable Network Graphics	optimized for line drawings
	jpeg	Joint Photographic Experts Group	optimized for photographic images
	tiff	Tagged Image File Format	print production; better color reproduction
	gif	Graphics Interchange Format	typically used for animations

Source: Adapted from Claus O. Wilke, *Fundamentals of Data Visualization*

stretch it, so it won't lose resolution the way bitmap images do. Vector images are also called "resolution-independent" because they can be stretched forever without ever losing sharpness or detail. It might not surprise you to learn, then, that one of the biggest downsides of vector images is that the file size can be surprisingly large.

How you and your organization guide the export of graphs from software tool to the final product depends on a variety of factors, including the primary data visualization software tool, operating system, and where the final output will appear: Will it be in a PDF report? A standalone image on a website? Embedded in a tweet? The best strategy is to try a variety of approaches, but double-check the final product to make sure you have the sharpest, clearest image possible.

ACCESSIBILITY, DIVERSITY, AND INCLUSION

Many people with vision impairments rely on screen readers to navigate the internet. A screen reader reads the content on a screen aloud to the user, so if you post a graph with the filename "Image1.png," that is what the user will hear. People with other physical, cognitive, or intellectual disabilities may have difficulty reading your work or using your website if you have not taken into account their accessibility needs. Accessibility also extends to whether people can access the internet and the speed of their Internet connection. It is worth considering how your content (and website more generally) can be made more accessible by users who may require different levels of assistance.

To create accessible content, you might follow the guidelines laid out in Section 508 of the Rehabilitation Act of 1973. Section 508 requires U.S. federal government agencies to develop, procure, maintain, and use information and communications technology (ICT) that is accessible to people with disabilities. This means that federal agencies that fall under Section 508 compliance rules must make their ICT—such as online training and websites—accessible for everyone.

One 508 standard for images that we can all apply is to use "alternative text" (commonly called "alt text") in our images. Alt text succinctly describes the content in the image. For data visualizations, this might be text communicating the general conclusion or message of the chart. In other words, what is the single concise sentence that summarizes your chart?

You can find some basic issues to consider by following the recommendations of the Web Content Accessibility Guidelines (WCAG), an international group charged with leading "the Web to its full potential." With respect to accessibility, WCAG defines four main areas:

1. **Perceivable.** Information must be presented in ways that users can perceive them. This might mean making non-text content available to other forms people need such as speech, symbols, or large print. Text should have sufficient color contrast with the background and images should have information ("alt tags") that will make them readable by screen readers and other assistive technology.
2. **Operable.** Make all functionality available from a keyboard. This means, for example, that using the tab, enter, and space bar keys enables the user to navigate the page and each interaction can be triggered.
3. **Understandable.** Text content should be readable and understandable, and web pages should operate in predictable ways. For example, significantly rearranging the content on the page can make content more difficult to read and understand.
4. **Robust.** Online content should be robust enough to be compatible with current and future users as well as assistive technologies. This might mean, for example, developing the website in such ways that screen readers and other assistive technologies can accurately interpret the content.

To date, there are no concrete rules about how to make a website completely accessible, though there are existing threads of research exploring how to use different assistive technologies to make visual content more accessible. Computer operating systems, browsers, and programming languages change and evolve, and any accessibility guide would be trying to hit a moving target. What we *can* do, however, is to consider how people with different abilities can or cannot access our content. A lot of these strategies are just good practices we can use to more effectively communicate our work with text and explanations. Considering better accessibility then leads to better usability for everyone.

Another issue to keep in mind in data visualization is how you refer to different groups. You may have considered this when using terms like "Black," "African American," or "Hispanic" in your writing, tables, or graphs. Use the phrasing accepted and recognized by your audience and the communities you are studying. Consider the lived experiences of the people and groups you study and write about. Also consider using "people-first" language, such as "people with disabilities" instead of "disabled people." It is important to remember that data are a reflection of the lives of real people.

This also applies to the layout of your graphs and the language you use. How do you order the bars or lines in your tables and graphs? Is it alphabetical, based on sample size, or is it based on some unknown, arbitrary decision? Again, there are not many answers to these questions, but it is worth taking some time to consider approaches and strategies to make your work more accessible and inclusive of different groups.

PUTTING IT ALL TOGETHER

There are not necessarily right or wrong answers to some of these questions and style decisions. Whether the thickness of your gridlines is 1 pt or 2 pt, one shade of gray or another—these are primarily style decisions, but they are also functional decisions. As you saw in the first chapter, the goal is to emphasize the data over the gridlines, tick marks, and markers.

An effective, comprehensive data visualization style guide is best developed at the organizational level. If possible, bring your design and data teams together to determine branding guidelines that meet the needs of your organization, including data visualization. If your organization does not have these divisions, or if you are working to develop your own individual style guide, you might reach out to experts or refer to other published style guides to develop branding guidelines and styles.

Remember to treat your data visualization style guide as a living document. Revisit the guide as technologies and trends change. And remember to be flexible to the different needs, tools, and skills in your organization. Creating an instructive and clear guide that can be accessed and implemented by everyone can serve you, your organization, and your reader.

REDESIGNS

By this chapter, your data visualization toolbox contains much more than it did when you began this book. We've seen dozens of graphs, many of which may have been new to you. As you develop your own eye for data visualization, you'll find places where these new graph types may be especially useful.

In this chapter, we'll cover a handful of data visualization redesigns. The graphs I choose to redesign here are not all especially *bad* graphs. Some are simply chosen because I believe there are more effective ways to plot the data. My goal is not to criticize these chart creators or their efforts but to demonstrate how the lessons we have learned can be applied to making data visualizations cleaner, clearer, and more effective.

The changes made here are by no means the only ways to modify these graphs, but each redesign follows the guidelines discussed throughout this book. In general, there is no "right" or "wrong" approach, just different ways of making improvements. As you develop an eye for better data visualization design, you will develop your own aesthetic and preferences.

PAIRED BAR CHART: ACREAGE FOR MAJOR FIELD CROPS

Take a moment and examine this bar chart from the U.S. Department of Agriculture that shows the number of harvested acres for five major crops in the United States for six different years. What do you see first?

My guess is you saw what I first saw: The acreage for all five crops increased over time. Your second observation, which quickly follows the first, is that cotton acreage (the second

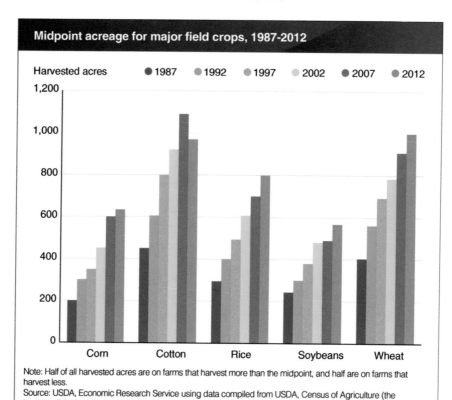

Basic bar chart from the U.S. Department of Agriculture.

group) fell in the last year. Unlike the other groups, the last bar for cotton (the green bar) is shorter than the bar for the preceding year. But it doesn't jump out at you because there is so much ink and color in the graph.

If the goal with this chart is to show relative trends in acreage among five crops, a bar chart is a poor choice. The paired bar chart is good at showing exact values, but the relative trends are not clear or immediately evident.

We could redesign this as a simple line chart.

Here, the drop in acreage for cotton is very clear, as are the relative sizes of the five crops. In the bar chart, I couldn't see immediately that rice acreage sits right in the middle of the five crops, but here I can see that right away. I didn't use a legend here, as might be the default approach, but instead added the labels at the end of each line, using color to link them with the lines.

Midpoint acreage for major field crops, 1987-2012
(Midpoint acreages more than doubled for all five major field crops)

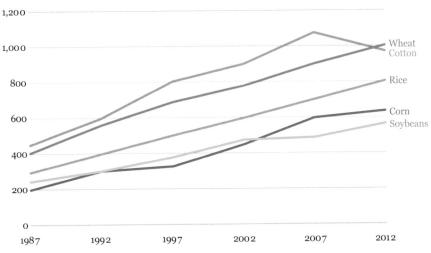

Wheat
Cotton
Rice
Corn
Soybeans

1987   1992   1997   2002   2007   2012

Source: U.S. Department of Agriculture

Midpoint acreage for major field crops, 1987-2012
(Midpoint acreages more than doubled for all five major field crops)

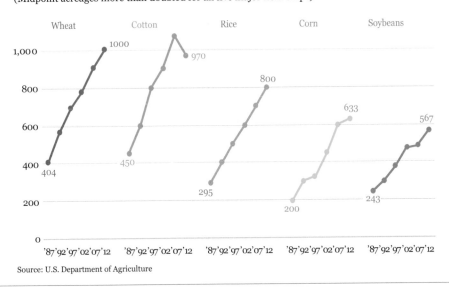

Wheat   Cotton   Rice   Corn   Soybeans

Source: U.S. Department of Agriculture

Two ways to redesign the USDA bar chart: A line chart or a cycle chart.

Another approach is a cycle chart. Instead of putting the lines together, this cycle chart is essentially a small multiples line chart where each crop gets its own panel. The advantage is that there's more space for the graph and it's perhaps a little more engaging because it's different. The disadvantage is that relative patterns are slightly less clear than in the line chart.

STACKED BAR CHART: SERVICE DELIVERY

Let's go back to page 14 in Chapter 1 and consider the perceptual rankings diagram. At the very top are graphs positioned along common scales—the bar chart or line chart with a single horizontal axis, for example. One step below are those graphs that are not positioned along common scales graphs. It is slightly harder to accurately assess the values in these.

This graph contains data from both sections of the ranking diagram. We can clearly discern the differences between the values of the blue series (Functional assignment) because they all sit on the same vertical baseline. We are not as well equipped, however, to similarly

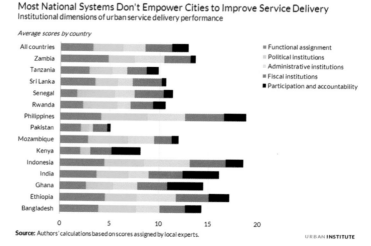

We can barely see that the value for Political Institutions is larger for *All countries* than *Zambia*.
Source: Roth and Malik, 2016

Most national systems don't empower cities to improve service delivery
(Institutional dimensions of urban service delivery performance, Average scores by country)

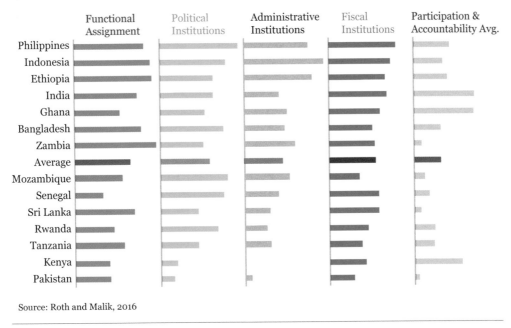

Source: Roth and Malik, 2016

One way to redesign the stacked bar chart is to break them up
and use a small multiples approach.

assess the values for the other series, because they don't share the same baseline. You can test this yourself: Is the value for Political Institutions (the yellow series) larger for "All Countries" or Zambia (the first two series)?

Instead of packing all of the data onto a single chart, we can break it into five separate charts. In this case, each series is given its own vertical baseline, so it's easier to make comparisons across countries within each series. The important point with making a graph like this is that the horizontal space for each series is the same. If we shrank the space for "Fiscal Institutions," for example, it might look like those values are larger than others.

This approach, however, doesn't tell you much about the *overall* values between countries. There's nothing wrong with adding a series for the overall *Total*, again, as long as we use the same horizontal spacing. In other words, the space between each gridline is the same. This approach works even better in cases where the values sum to the same total or to 100 percent, because the total length will be the same for all of the bars and thus a *Total* segment is unnecessary.

Most national systems don't empower cities to improve service delivery
(Institutional dimensions of urban service delivery performance, Average scores by country)

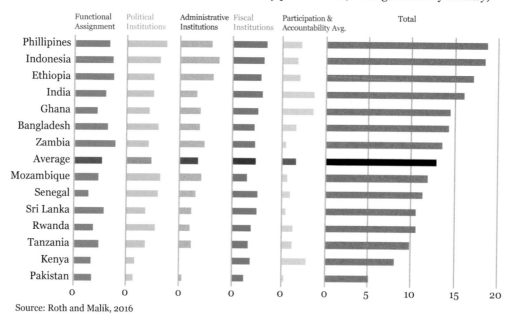

Source: Roth and Malik, 2016

When breaking up stacked bar charts, it is sometimes important to include the totals.

LINE CHART: THE SOCIAL SECURITY TRUSTEES

Each year, the Board of Trustees of the Federal Old-Age and Survivors Insurance and Federal Disability Insurance Trust Funds report on the current and projected status of the U.S. Social Security program. The Trustees are responsible for estimating the current and future financial picture of the program to communicate to the public and policymakers the challenges the program faces. The Social Security Technical Panel is an independent expert panel responsible for reviewing the work of the Trustees, including the methodological details, economic and demographic assumptions, and the Trustees communication efforts.

The 2019 Technical Panel placed an emphasis on this latter category: "The Panel believes that trust in public institutions is enhanced by greater understanding . . . In this context, we believe it is paramount for the Trustees to communicate clearly and effectively with the general public about its finances." The panel emphasized clear, plain language, a focus on the core message, and better data visualizations in the Trustee's work.

Let's look, then, at two of their data visualizations.

A CLEANUP

The first example is a relatively simple clean-up rather than a wholesale redesign. This line chart—which has appeared in virtually every Trustees Report—shows the time series of the basic finances of the Social Security system. System income (taxes paid into the system) are set next to system costs (benefits paid to beneficiaries) for a short historical period (here from 2000 to 2018) and out in the longer projection period (here from 2018 through 2092). Two sets of costs are shown: one that shows how many benefits are *scheduled* to be paid (the dashed line) and one that shows how many benefits can *actually* be paid (the solid bold line).

The existing graph has annotation and labels to help the reader better understand the content and the concepts. A small table near the bottom of the graph lists benefit shares in

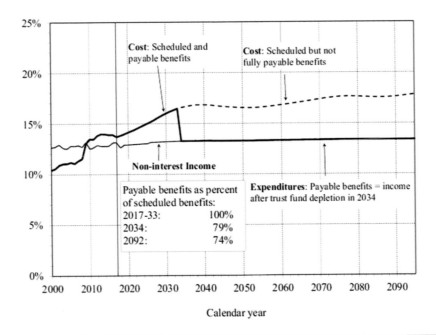

The Social Security Administration (2019) has published this graph showing the basic finances of the Social Security system for many years.

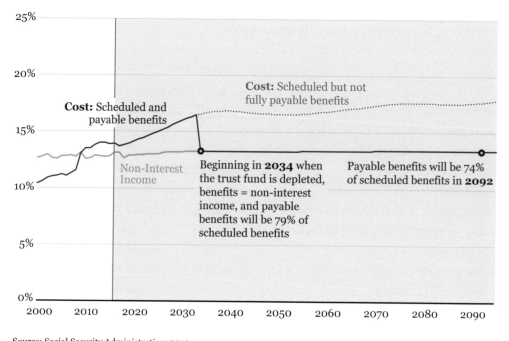

Source: Social Security Administration, 2019

Some basic cleanup and annotation improves the clarity of the Social Security finances chart.

specific years. But there is also a lot of ink used on extraneous details: horizontal and vertical gridlines and tick marks for every percentage point and year.

Let's take a simple approach to redesigning this graph by removing some of these extraneous details and markers. Here, I've removed the vertical gridlines and all of the tick marks. I deleted the small table and instead directly labeled the years those numbers referenced. I used some slight color here—which is consistent with black-and-white printing—and added a gray box to the projection period (after 2018) to draw attention to the imbalance.

A BETTER DOT PLOT

The next graph appeared in the 2011 Technical Panel Report. It shows the sensitivity of different assumptions of the Social Security model. Most of the six Technical Panel Reports published since 1999 include this information as a series of tables. But in 2011, the Panel

differences in school achievement scores for Black and white students, arranged by scores of Black students.

Let's focus on the bars on the far-left side of the graph. There are three numbers here: 28 percent, 32 percent, and 60 percent. The numbers in the green boxes show test scores for white (60 percent) and Black (28 percent) students, and the middle number shows the gap between the two groups (32 percent). But the green boxes make it appear as if the 28 percent represents a range of numbers, from, say, 22 percent to 28 percent. By using rectangles instead of points or markers, it resembles a stacked chart rather than the dot plot, which was likely the intention.

Figure 8. Percentage of Black and White students who were National School Lunch Program (NSLP) eligible and percentage who had a parent with more than a high school education, by Black student density category: 2011

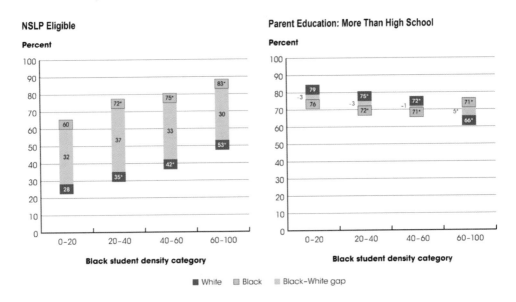

* Significantly different (*p* < .05) from the 0 percent to 20 percent density category.

NOTE: The measures displayed in this figure are percentages of students within each Black student density category.

SOURCE: U.S. Department of Education, Institute of Education Sciences, National Center for Education Statistics, National Assessment of Educational Progress (NAEP), 2011 Mathematics Grade 8 Assessment.

This graph from the National Center for Education Statistics shows the percentage of students eligible for the National School Lunch Program.

Percentage of students eligible for the National School Lunch Program

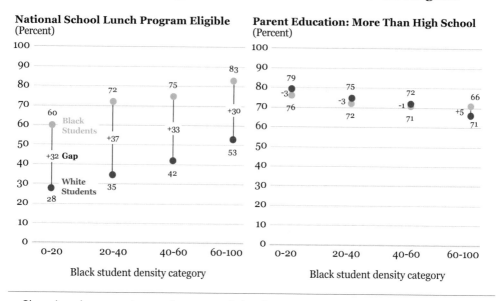

Changing shapes and removing some of the clutter makes the graph from the National Center for Education Statistics easier to read.

As an alternative, let's make it a true (vertical) dot plot. We can replace the green boxes with green circles and connect them with a gray, vertical line. Now we perceive the green circles as specific points rather than ranges or a stacked set of values.

You might also notice that I deleted the legend and labeled the three series directly on the left chart. I didn't repeat the labeling in the chart on the right for two reasons: First, because the gaps are smaller, there is less space for the labels. And second, the reader doesn't need to be reminded of the definition of each dot and line at every single occurrence on the page.

DOT PLOT: GDP GROWTH IN THE UNITED STATES

Every quarter, the U.S. Bureau of Economic Analysis (BEA)—the federal agency responsible for producing some of the most important measures of the U.S. economy—releases their

TABLE: FIRM ENGAGEMENT

As we saw in Chapter 11, there are many ways to make our tables more visual. We can add color, icons, bars, or other elements to highlight the important values for our reader instead of asking them to sift through all the data values.

This table uses different shapes and shades of gray to show the share of firms that engage in different business activities like design and market research. As the reader, we must understand which shapes correspond to which percentages and then figure out the different

Figure 1: Proportion of firm-years engaging in each intangible acitivty, by industry

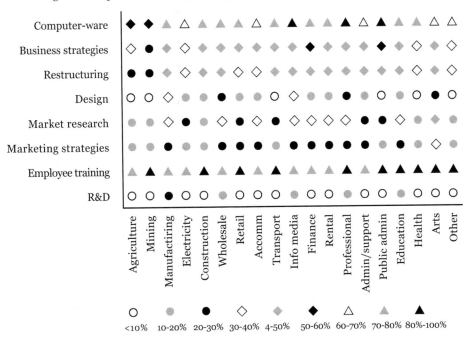

Source: Author's rendering of original chart by Chappell and Jaffe, 2018
Note: Data based on a visual inspection of the original graphic.

Author's rendering of an original chart by Chappell and Jaffe (2018), which could be improved by changing how the values are displayed.

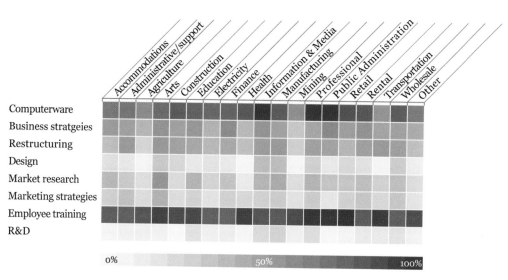

Source: Chappell and Jaffe, 2018
Note: Data based on a visual inspection of the original graphic.

A heatmap is one alternative to the Chappell and Jaffe (2018) chart.

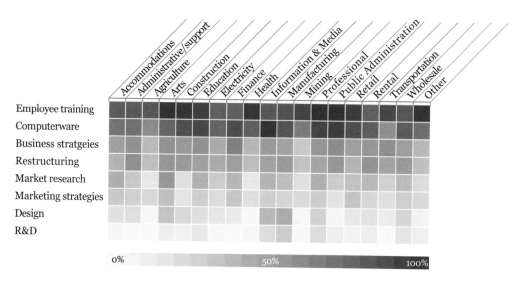

Source: Chappell and Jaffe, 2018
Note: Data based on a visual inspection of the original graphic.

This heatmap alternative to the Chappell and Jaffe (2018) chart sorts the data.

shading styles. Of course, triangles don't necessarily mean "more" of something than circles, so the rank-ordering of the values is hard to interpret.

Instead, what if we use a monochromatic color ramp moving from a light blue for the lower percentages to darker blues for the higher values? In this heatmap approach, it's much easier to see that there is a lot of time spent on *Employee training*, the dark blue row towards the bottom of the table.

We can take this a small step forward and sort the data, which will naturally focus the reader's visual attention. The longer labels require us to use rotated labels, as I've done here, or perhaps to rotate the entire graph and change the spacing or cell size so all of the text can fit.

CONCLUSION

With more graphs in your data visualization toolbox to choose from, and having seen more graphs and best practices, I'm confident that you're ready to improve upon your own graphs. Finding and redesigning even the simplest graph—I find that mining the academic peer-reviewed literature a good place to start—can help you refine your skills and develop your own data visualization aesthetic. Like any other skill, practice makes better.

Two important caveats. First, if you critique a graph publicly, keep in mind that some-one made that graph and that even your well-intentioned efforts to redesign it may not be appreciated. The chart creator may have had time pressures, software limitations, or organi-zational demands of which you are not aware. Reaching out to the person who created the original graph may be worth your effort. Second, try to identify the central goal of the chart and the possible challenges of the data series. This will help lead you to the best chart type for the task at hand.

CONCLUSION

Now, at the end of this book, your data visualization toolbox has expanded considerably. Instead of using whatever the default graphs are available in your favorite software tool, you can now draw upon more examples to visualize your data in the ways that best serve your reader, user, and audience.

The graphs presented in these chapters have been tested over and over again. Analysts, researchers, reporters, and scholars have used them with different data sets and layers of text, annotation, color, font, purpose, and platform. But the set of graphs available to you is infinite. The bar chart did not exist before someone invented it. Maybe it's you who will invent the next great graph type.

Sources of inspiration for new and innovative visualizations are all around us, from public and private organizations, the media, data scientists, designers, and artists. The images on the next few pages are from some popular data visualization projects and stories. They use different shapes, layouts, approaches, and techniques to visualize data in ways that are unique to the field and form.

Freelance data designer Maarten Lambrechts maintains a special project website, Xenographics, with the tagline, "Weird but (sometimes) useful charts." It is a repository of "novel, innovative and experimental visualizations to inspire you, to fight xenographphobia and popularize new chart types." There, you can find the different, unique, and strange graphs that can be used to show data in different ways. By experimenting with such forms, we can move the field forward and communicate data and information in new and better ways.

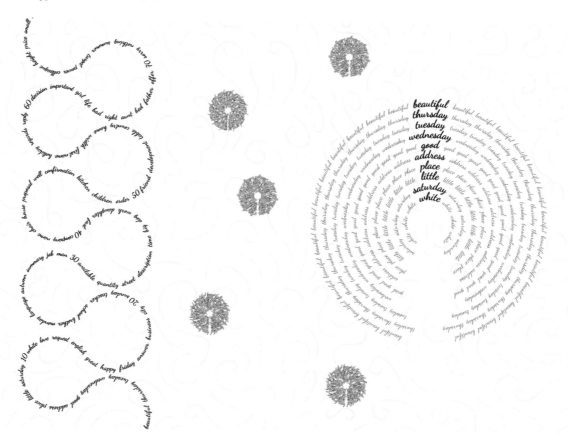

Source: Nadieh Bremer, Visual Cinnamon

As you go forth and create your visualizations, continue to explore. There are yet undiscovered options, forms, shapes, and graphic types that can best show your data. Even if you come up with a great idea but don't know how to create it on the computer, you may find colleagues or partners who can put your ideas in practice.

LESSONS LEARNED

Over the course of this book, we have covered lots of rules, principles, and guidelines. If I were to sum these up into the absolute universal principles of data visualization, I would leave you with just these four.

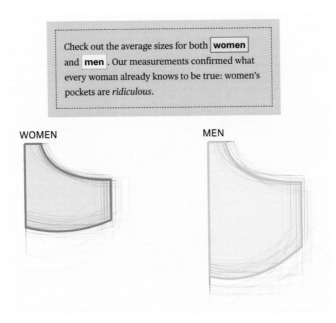

Source: Jan Diehm and Amber Thomas, "Women's Pockets are Inferior," The Pudding

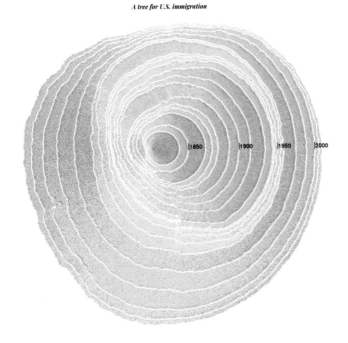

Source: Cruz, Wihbey, Ghael, and Shibuya, 2018

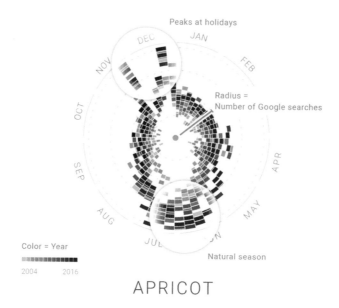

Peaks at holidays

Radius =
Number of Google searches

Color = Year

2004   2016

Natural season

APRICOT

Source: Moritz Stefaner

SHOW YOUR DATA

People are reading your graph to learn something, and they do that best by seeing the data. This doesn't mean you need to show them *all* of the data, but that you should always highlight the most important data.

REDUCE CLUTTER

Reduce and remove all of the clutter that distracts your audience from the data or distorts the representation. Make it as easy as possible for your reader to see the most important points in your graph.

INTEGRATE THE GRAPHICS WITH THE TEXT

Directly label your data, remove legends, use active titles, and employ good labels and annotation. You may need to guide your reader through the graph before you help them

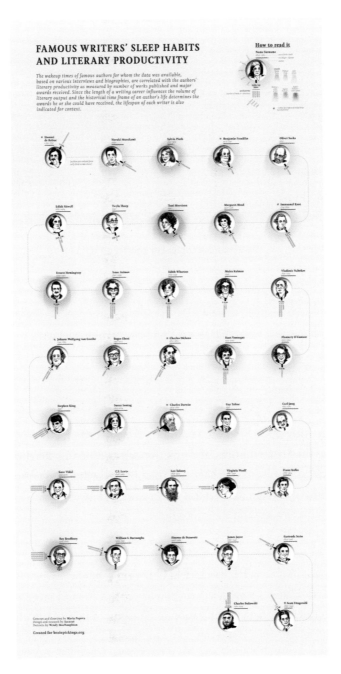

Source: Accurat. Portraits by Wendy MacNaughton; Design and research by Giorgia Lupi, Simone Quadri, Gabriele Rossi, Davide Ciuffi, Federica Fragapane, Francesco Majno. 2013.

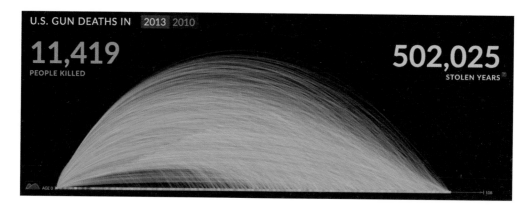

Source: Periscopic: Do good with data

understand the content, but even that lesson can be accomplished with simple, smart labels and callouts.

CONSIDER YOUR AUDIENCE

Always remember with whom you are communicating. Academic researchers are looking for different things than practitioners, and practitioners are looking for different things than managers and policymakers. Try to identify your likely audience—try to talk to them if possible—and design your graphs to meet their needs. That way you can help them find insights, make discoveries, and do their jobs better.

FINAL THOUGHTS

I first became interested in data visualization after seeing much of my and my colleagues' work go unnoticed and unused. I did not come to the field with a degree in design or computer science or data science. And because I did it, I believe you can too. In fact, anyone can effectively communicate their data by thinking critically about their own work and the needs of their audience, readers, and users.

APPENDIX 1
DATA VISUALIZATION TOOLS

This book is tool agnostic. My goal is not to show you how to create each of the eighty-plus graphs we explored. There are far too many tools you can use and far too many approaches within each tool. What matters is not which tool you use, but that it helps you create the graphs you need to best serve the needs of your audience.

There are many, many data visualization tools available, for use on different platforms and with different purchasing and subscription options. The number, types, and capabilities of these tools are constantly changing to reflect updated underlying technologies and coding languages. Which tool you use will depend on your personal preferences and skills, as well as those of your colleagues and support within your organization.

Data visualization tools live along a spectrum. On one end are click or drag-and-drop tools like Excel, which allow the user to click and insert a chart. On the other end are programming languages like R and JavaScript that require written code to create data visuals. The barrier to entry is much lower in Excel—virtually anyone can create a line chart or bar chart in seconds. It's much different on the other end, where programming languages require an understanding of how to write code and the different syntax across the different languages. Yet programming languages give you substantially more flexibility, while tools like Excel can box you in within a small subset of graphs.

How "difficult" these tools are depends on the person and sometimes the organization. You may have an affinity for computer programming languages, in which case Java-Script, Python, or R may be appropriate, but for a variety of reasons, your organization may not allow you to use these open source tools. Many social science researchers use

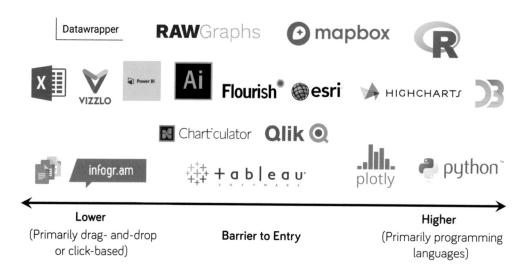

statistical languages like SAS, SPSS, and Stata, but, in my opinion, the graphing capabilities of these tools lag behind those of other languages. Some of the drag-and-drop tools like Excel and Tableau are easier to start with but designing more bespoke visualizations either is impossible or may require some coding (for example, Calculated Fields in Tableau).

I used a variety of tools to create the graphs in this book. Many, if not all, could probably be created in programming language tools like R or JavaScript; a sizable proportion could be created in drag-and-drop tools like Excel and Tableau. By utilizing a variety of tools, I found that some were easier to create the graph, but harder to style the way I liked and some were more difficult to learn while others were more intuitive. This list of tools is not exhaustive (by far) and I based my decision on which tools to include on my experience in the field, not on any formal survey or data set. Before you invest substantial time or money to learn any of these, you should explore the landscape of available tools and products.

PRIMARILY DRAG-AND-DROP OR CLICK-BASED

Adobe Illustrator. This is primarily a design tool. Adobe Illustrator, and the rest of the Adobe Creative Suite like Photoshop and InDesign, are the workhorse tools for designers. The graphing library is actually pretty poor in Illustrator, but you can insert graphs made in other tools to add more styling, labels, and annotation. The Adobe Creative Cloud is now primarily subscription-based, but is not a particularly cheap tool to purchase.

Charticulator. Launched in 2018, Microsoft's Charticulator is an online tool that enables you to create a custom chart layout. It differs from some other tools in that you don't select from predefined charts but instead the creators have transformed chart specifications into mathematical layout parameters, such as marks (e.g., rectangle, line, or text) and axes (e.g., property of one direction in a plot). Charticulator charts are primarily static, but can be integrated with Microsoft's PowerBI tool to create interactive visualizations. At the time of this writing, Charticulator is free to use.

Datawrapper. An online tool from a team based in Germany where you can upload your data, select a graph template, refine and style, and publish or download. Created charts can be embedded in websites and can be made interactive. Datawrapper is free for most purposes, and paid versions allow for custom themes and additional exporting options. There are many tools in this same vein (Flourish and RAW below are two other examples)—some are better than others with different default options and user experience.

Excel. Likely the primary data and data visualization tool for many people around the world. As part of the Microsoft Office suite, Excel is not free, but it is your basic click-based tool. At the time of this writing, you can create more than 16 basic chart types in Excel, with variations within most of them. There is a basic library of charts that you can expand with clever "hacking" or additional coding using the Visual Basic for Applications (VBA) programming language that sits behind all Microsoft Office tools.

Flourish. Launched in 2016, Flourish is an online tool primarily aimed at newsrooms to help journalists create both static and interactive data visualizations in a drag-and-drop framework. There are options to customize and further develop Flourish graphs using the underlying JavaScript framework. Pricing options range from the free Public version to the paid Personal version (which gives you additional features) and the paid Business version (targeted towards large teams and organizations). Through a partnership with the Google News Lab, Flourish provides newsrooms free premium accounts.

Google Sheets. Akin to Excel, Google Sheets is part of the Google suite of tools. It works very similarly to Excel, though without some of the more sophisticated options. Because it is based online, the sharing capabilities are somewhat better than Excel (with the side effect being you need to have internet access to use it).

PowerBI. Microsoft's business intelligence tool that allows you to create interactive dashboards and visualizations. It directly links with the rest of the Microsoft Office suite (especially Excel) and can be modified and customized in ways similar to Tableau. There is the free Power BI Desktop version, the paid Power BI Pro, and the Power BI Premium package for organizations.

RAW. Created by DensityDesign Research Lab in Italy in 2013, RAW was an early project designed to help creators link spreadsheet tools like Excel to graphic editing tools like Adobe Illustrator. It is an open source tool, which means you can download the code to further customize the visualization options. There is also an online platform in which, like other tools, you upload your data, select the graph, and customize. RAW has a variety of options for certain nonstandard graphs (like streamgraphs and bump charts) that are not typically available in other tools. RAW is free to use.

Tableau. Perhaps the most popular business intelligence dashboarding tool, Tableau's drag-and-drop interface enables you to create interactive dashboards and visualizations. Like Excel, users have customized their Tableau work to create an array of visualizations outside the basic graph menu. There are a number of versions of Tableau, from the free Tableau Public (but which means you save your work to the Tableau website) to paid versions like Tableau Desktop and Tableau Server (for large organizations).

ONLINE TOOLS (CLICK-BASED)

Infogram, Venngage, and Vizzlo. These are just three of the many click-based online tools that are aimed more for people who want to quickly create infographics and reports. In my experience, these tools sometimes have more graph options than other online tools, but they are not always based on best practices. Pricing varies from free packages that usually mean your data and visualizations can be viewed by anyone, to enterprise packages for large teams and organizations.

PROGRAMMING LANGUAGES

D3. We first need to define JavaScript. JavaScript is a programming language that allows you to implement information onto a webpage. Every time a web page does something, like display updates, animate graphics, or play videos, JavaScript is probably involved. D3 is a JavaScript library for manipulating objects based on data and was developed by Mike Bostock along with Jeff Heer and Vadim Ogievetsky at Stanford University in the early 2010s. Most of the interactive data visualizations we currently see on the web are run on D3—virtually every interactive graph you play with on the *New York Times*, *Washington Post*, and *Guardian* websites is built with D3. Like other programming languages, there is a steep learning curve to using D3, but you are basically unlimited in the kinds of visualizations you can create. D3 is an open source language, which means it is free to use.

Highcharts. Launched in 2009 by a team in Norway, Highcharts—and its cousin tools Highstock, Highmaps, Highcharts Cloud, and Highslide—is a suite of interactive data visualization tools rooted in JavaScript. You do need to know a bit of coding to use Highcharts, but templates and libraries help you create the basics of a graph and then add additional styling and formats. Highcharts is free for personal use and nonprofit organizations; pricing then varies by number of licenses and package.

Python. Python is used in a wide number of fields and industries from basic and complex data analysis to web application to artificial intelligence. Like D3 and R, the language is open source, which means there are a lot of open, free libraries to help you create data visualizations including, for example, Matplotlib, Seaborn, Bokeh, and ggplot.

R. Conceived in 1992 and initially released in 1995, R is a free, open source programming language for statistical computing and graphics. R is becoming more and more widely used for data visualization, especially since the launch of the "ggplot2" package by Hadley Wickham in 2005 (based on the "Grammar of Graphics" by Leland Wilkinson). R allows you to conduct statistical analyses and create customizable data visualizations; additional tools and packages can enable you to create interactive visualizations.

APPENDIX 2
FURTHER READINGS AND RESOURCES

GENERAL DATA VISUALIZATION BOOKS

SCOTT BERINATO. Berinato's book *Good Charts: The HBR Guide to Making Smarter, More Persuasive Data Visualizations* focuses on graphs for the business community. The book is wide-ranging and discusses more of the differences between explanatory and exploratory charts than some of the other leading data visualization books. He has a follow-up workbook that provides hands-on examples and tutorials.

ALBERTO CAIRO. Author of several books on data visualization including *The Functional Art*, *The Truthful Art*, and *How Charts Lie*. Cairo is a journalism professor, so his books focus primarily on creating data visualizations for telling stories to a wide audience. His books provide fundamental overviews of data, data visualization, introductory statistics, and how to create visualizations. His most recent book *How Charts Lie* helps readers spot lies in deceptive graphs and how to become better consumers of data visualization.

JORGE CAMÕES. His book *Data at Work* covers a wide range of data visualization principles and strategies, ranging from rules of visual perception to design considerations to data preparation and visualization.

STEPHEN FEW. Author of several books on data visualization, his *Show Me the Numbers* and *Now You See It: Simple Visualization Techniques for Quantitative Analysis* are comprehensive overviews of how to present data effectively and strategically.

ANDY KIRK. Author of two books on data visualization, his *Data Visualisation: A Handbook for Data Driven Design* provides readers with a system to conceptualize and develop data visualizations, and a process to help readers make design choices that result in clear and effective visualizations.

JUSSO KOPONEN AND JONATAN HILDÉN. Their *Data Visualization Handbook* (translated from Finnish) is a practical guide to data visualization and contains lots of examples of data visualizations and information graphics.

COLE NUSSBAUMER KNAFLIC. Knaflic's book *Storytelling with Data*, and blog of the same name, provides an introductory treatment of data visualization, and how to pair text with graphs to tell effective, compelling stories. Her follow-up book, *Storytelling with Data: Let's Practice!*, takes the reader through a series of hands-on exercises to practice their data visualization skills.

ISABEL MEIRELLES. A professor of design, Meirelles' *Design for Information* surveys current examples of data visualizations for both elements of content and design. Numerous examples provide a library of visualization types and approaches.

TAMARA MUNZNER. Her *Visualization Analysis and Design* book approaches data visualization from a more systematic academic-based perspective. It features a unified approach to reading and creating visualizations, all rooted in the academic literature. Munzner's book is more along the lines of a true data visualization textbook.

STEVE WEXLER, JEFF SHAFFER, AND ANDY COTGREAVE. Their *Big Book of Dashboards* is one of the more comprehensive treatments of creating interactive dashboards. Mostly rooted in the Tableau software tool, the authors review nearly thirty examples to help the reader determine good design and interactive experiences.

CLAUS O. WILKE. Wilke's book *Fundamentals of Data Visualization: A Primer on Making Informative and Compelling Figures*, takes a fundamental and practical approach to data visualization. Wilke presents basic principles of good data visualization strategies and practices and also shows how to create the visuals in his book using the R programming language. Like this book, it is one of the few to include a wider array of graphs than the typical line, bar, pie, and map.

DONA WONG. Wong's *Guide to Information Graphics* dedicates individual pages to specific graph types, how and why to choose the best chart to fit the data, the most effective way to communicate data, and what to include and not include in different graphs.

HISTORICAL DATA VISUALIZATION BOOKS

RJ ANDREWS. Andrews' *Info We Trust: How to Inspire the World with Data* takes a different perspective on data visualization techniques, drawing heavily on historical visualizations and design approaches by luminaries in the field over the past century.

WHITNEY BATTLE-BAPTISTE AND BRITT RUSERT. The first full exploration of W.E.B. Du Bois's data visualizations from the 1900 Paris Exposition, *W.E.B. Du Bois's Data Portraits: Visualizing Black America*, is an image-by-image description of these early data visualizations.

BRUCE BERKOWITZ. His biography of William Playfair shows the real life of the man who invented "statistical graphics." Berkowitz goes beyond the statistics and graphics that Playfair is best known for to tell a detailed story of his life.

MANUEL LIMA. In his two books, *The Book of Trees: Visualizing Branches of Knowledge* and *The Book of Circles: Visualizing Spheres of Knowledge*, Lima explores the long history of the tree diagram and circular information design. Both take you on a tour of the long history of both kinds of visualization methods.

SANDRA RENDGEN. Similar to the Du Bois book, in Rendgen's *The Minard System: The Complete Statistical Graphics of Charles-Joseph Minard*, she explores the career and story behind Minard's graphs, maps, and tables.

BOOKS ON DATA VISUALIZATION TOOLS

The list in Appendix 1 includes just a selection of tools I've used for this book. Which tool is best for you depends on your existing expertise, your needs, and the needs of your audience. In my experience, these books have the best tool-specific data visualization tutorials. There are countless online blogs and resources you can also explore.

KIERAN HEALY. Maybe more of an introduction to basic core principles of data visualization, Healy's *Data Visualization: A Practical Introduction* mixes in a how-to in R. There is an online companion with code snippets.

NORMAN MATLOFF. At more than 400 pages, Matloff's *The Art of R Programming* is a beast—and that's because it covers everything in R, an essential reference book in your library.

ERIC MATTHES. Every Python programmer I know loves Matthes' *Python Crash Course: A Hands-On, Project-Based Introduction to Programming*. Beginners and more experienced coders are really fond of how he leads the reader through the language and the useful exercises in each chapter.

RYAN SLEEPER. One of the more recent (and better) books on creating calculations and custom visualizations in the Tableau software tool, Sleeper's *Practical Tableau* is a great start to using Tableau.

AMELIA WATTENBERGER. One of the few books about the D3 JavaScript library, likely because coding examples and snippets are really important, *Fullstack D3 and Data Visualization* (and the companion website) is a comprehensive introduction to D3. This is the most recent book about D3 (and perhaps the best) because it walks the reader through the entire process and includes a large amount of code snippets and examples on the book's website.

HADLEY WICKHAM AND GARRETT GROLEMUND. Probably my go-to book to learn R, *R for Data Science*, takes a comprehensive view at how to work with and visualize data in R. It is especially strong on the side of wrangling and visualizing data.

DATA VISUALIZATION LIBRARIES

There is no one-to-one mapping between data types and graph types. A column chart, for example, can be used to show changes over time or to compare differences between categories. There are a variety of resources you can use to help you select a graph, but the ultimate decision will rest with you, your data, your audience, and your creativity.

THE CHARTMAKER DIRECTORY. A crowd-sourced interactive matrix that consists of 50 graph types and 40 different tools. Users can post links to tutorials on how to make the graphs in the different tools. http://chartmaker.visualisingdata.com/

CHART SUGGESTIONS—A THOUGHT STARTER. A relatively smaller, static project with about 20 graphs broken into four categories. One of the first of these charting libraries. http://extremepresentation.typepad.com/blog/2006/09/choosing_a_good.html

THE DATA VISUALISATION CATALOGUE. Online graph resource with descriptions, anatomy, and some video guides to each chart type. https://datavizcatalogue.com/

THE DATA VIZ PROJECT. An interactive website with more than 100 graph types including select examples of each. http://datavizproject.com/

THE GRAPHIC CONTINUUM. A poster, smaller sheet, flash cards, and card game I produced with a friend, *The Graphic Continuum* shows more than 90 different graph types grouped into six categories. https://policyviz.com/product/graphic-continuum-poster/

INTERACTIVE CHART CHOOSER. An interactive chart chooser that lets you sort and filter more than 30 graph types based on the data you are trying to visualize. Links to real examples help you see these chart types in action. https://depictdatastudio.com/charts/

THE R GRAPH GALLERY AND PYTHON GRAPH GALLERY. A pair of webpages that contain hundreds of charts made with the R and Python programming languages. Each chart contains reproducible code. https://www.r-graph-gallery.com/ and https://python -graph-gallery.com/

TEXT VISUALIZATION BROWSER. Nearly 450 examples of ways to visualize text data. https://textvis.lnu.se/

THE VISUAL VOCABULARY. A poster project from the graphics desk at the *Financial Times*, the Visual Vocabulary shows more than 70 different graph types in 9 categories. https:// github.com/ft-interactive/chart-doctor/blob/master/visual-vocabulary/Visual-vocabulary.pdf

XENO.GRAPHICS. A collection of unusual charts and maps to help you expand your visualization repertoire even further. https://xeno.graphics/

WHERE TO PRACTICE

Maybe the best way to improve and refine your data visualization technique is to create visualizations. Exploring different data sets, visualization types, and tools can help you refine your aesthetic, and play with different techniques and forms. There are a few community projects that you may want to explore to help you on your way.

MAKEOVER MONDAY. A weekly learning project in which participants work with a sample data set to try to create better, more effective visualizations. It tends to be focused in the

Tableau community, but visualizations can be created in any tool. Visualizations are created and posted publicly with review and feedback on social media and a mid-week webinar to enable people to give feedback and refine their ideas.

OBSERVE, COLLECT, DRAW. Primarily shown on Instagram, *ObserveCollectDraw* is based on the book by designers Giorgia Lupi and Stefanie Posavec that encourages people to collect their own personal data and draw their visualizations. This project—and their original book *Dear Data*—shows people how to collect and record their personal experiences in analog ways.

STORYTELLING WITH DATA. The Storytelling with Data Community is a place to practice your data visualization skills, get and give feedback, and discuss topics related to effectively communicating data. There is a monthly "SWDchallenge," which asks participants to address a specific data visualization challenge and a periodic "SWDexercise" that is based on real-world scenarios with actual data.

TIDY TUESDAY. A weekly project aimed at R users, project managers post a dataset and a published chart or article related to the dataset. Participants are asked to explore the data and publish their results (and code) for others to use and adapt.

OTHERS. There are other similar projects, but tend to have dedicated audiences in different ways. The Data Visualization Society, a community of data visualization practitioners, hosts Slack channels revolving around practice and critique. There are also various forums on HelpMeViz, Reddit, Stack Exchange, Twitter, and other social media platforms where you can practice and publish your work.

ACKNOWLEDGMENTS

I n 2014, I wrote an introductory article on data visualization for the *Journal of Economic Perspectives (JEP)*. I had always wanted to publish in the *JEP*, but never imagined that data visualization—rather than public policy or economics—would be the topic of interest. Shortly after the article was published, I received a call from Bridget Flannery-McCoy at Columbia University Press asking to see if I was interested in writing a full book on data visualization. At that time, I wasn't ready to write this book—there were a lot of books just out or coming out and to do it right, I would need to dedicate a lot more time and resources than I had available. Bridget's call proved fruitful nonetheless, ultimately resulting in my first book, *Better Presentations*.

Fast-forward five years and Bridget called again to see if I was ready for a follow-up book. Now, having changed jobs and focusing more on data communication, I found myself with more to say and an idea for a book that would place it in a unique space in the field.

I wrote the first draft of the entire manuscript on two train rides between Washington, DC and New York City. But it would take another two-and-a-half years to fully develop it, reorganize it, expand it, and create the graphs. The result is the book you are now holding.

For getting me to this point, my gratitude belongs to several important people: Brittany Fong, Ajjit Narayanan, Jon Peltier, Anthea Piong, and Aaron Williams helped me with various Tableau, Excel, R, and JavaScript challenges and tasks. RJ Andrews, John Burn-Murdoch, Alberto Cairo, Jennifer Christiansen, Alice Feng, John Grimwade, Steve Haroz, Robert Kosara, and Severino Ribecca were generous in their time to have discussions, provide feedback, and assist with a number of requests to find research and images, both modern and historical.

More thanks are due to the people and organizations who have granted me permission to include their work in this book. Their work is truly special and I am grateful they allowed me to include it in these pages.

Additional thanks to Ken Skaggs for helping to manage the PolicyViz Podcast and to the more than 200 guests who have appeared on the show.

A special thanks to Alberto Cairo, Nigel Holmes, Jessica Hullman, David Napoli, and Chad Skelton for reviewing parts or all of the manuscript and providing invaluable feedback.

Very special thanks are due to Kenneth Field for creating many of the maps and for keeping up with my "feature creep." Special thanks are also due to Hiram Henriquez for creating original illustrations and making all of the graphs ready for print.

This book could not have come together without the help of my Columbia University Press editor, Stephen Wesley, who kept up with my incessant emails and questions. Additional thanks to Christian Winting, Ben Kolstad, and the rest of the Columbia University Press team for helping to bring this project to life.

I am grateful for my Urban Institute colleagues whose dedication to fact-based research has the power to improve public policy and practice, strengthen communities, and transform people's lives for the better. They have helped create a special place that values not only in-depth scholarship, but also innovative ways to communicate that work.

I am also indebted to the many, many friends and strangers with whom I have discussed and debated aspects of data visualization and data communication over the years. The data visualization community is a special place—the warm welcome it gave me about 10 years ago changed my career in unimaginable ways, and I hope to pass it forward to another crop of data communicators. I am also especially thankful to people in the field whose creativity continues to inspire.

Writing a third book is simultaneously easier and harder than writing the first and second. I could not have summoned the courage to go through another project without the love and support of friends and family who have cheered me on and taken interest in even the most mundane details of my work. Special thanks are due to the ongoing support and love of my entire family.

Finally, my most special thanks are due to the three most important people in my life: my wife and kids. My kids, Ellie and Jack, are my favorite sources of happiness, pride, and fun. They have watched their dad struggle with code, complain about color palettes and fonts, and watched me stay up late into the night, reading and writing. Through it all, they supplied me with steady encouragement, love, and support. As a dad, you know you're doing something right when they lean over your shoulder and say, "How's it coming with that heatmap?"

My deepest thanks goes to my wife, Lauren, who has edited every page of this book, cutting out repetitive text and keeping the language clear and descriptive, always working with my reader in mind. She keeps all of us moving forward, especially when I'm writing, speaking, or traveling demands take me far and wide. She routinely reminds me she's the best thing that's ever happened to me. A truer statement has never been spoken.

REFERENCES

Ahmed, Naema, Cassi Pollock, and Alex Samuels. "How the Texas Democratic and Republic Party Platforms Compare." *Texas Tribune*, July 5, 2018, https://apps.texastribune.org/features/2018/party-platforms/?_ga=2.129478090.770685496.1576106215–1729798919.1576004948.

Alda, Alan. *If I Understood You, Would I Have This Look on My Face?: My Adventures in the Art and Science of Relating and Communicating*. New York: Random House, 2018.

Andrews, RJ. "Florence Nightingale Is a Design Hero." *Medium*, July 15, 2019, https://medium.com/nightingale/florence-nightingale-is-a-design-hero-8bf6e5f2147.

Andrews, RJ. *Info We Trust: How to Inspire the World with Data*. Hoboken, NJ: Wiley, 2019.

Andris, Clio, David Lee, Marcus J. Hamilton, Mauro Martino, Christian E. Gunning, and John Armistead Selden. "The Rise of Partisanship and Super-Cooperators in the US House of Representatives." *PloS one* 10, no. 4 (2015): e0123507.

Anscombe, Francis J. "Graphs in Statistical Analysis." *The American Statistician* 27, no. 1 (1973): 17–21.

Arellano, Cristina, Juan Carlos Conesa, and Timothy J. Kehoe. "Chronic Sovereign Debt Crises in the Eurozone, 2010–2012." Federal Reserve Bank of Minneapolis Economic Policy paper 12–4, May 2012, https://www.minneapolisfed.org/~/media/files/pubs/eppapers/12-4/epp_12-4_chronic_sovereign_debt_crisis_eurozone.pdf.

Avery, Beth. "Ban the Box: U.S. Cities, Counties, and States Adopt Fair Hiring Policies." National Employment Law Project. July 1, 2019. https://www.nelp.org/publication/ban-the-box-fair-chance-hiring-state-and-local-guide/.

AxisMaps. "Cartography Guide." https://www.axismaps.com/guide/general/map-projections/, accessed November 2019.

Bard Graduate Gallery. "Marimekko: Fabrics, Fashion, Architecture." https://www.bgc.bard.edu/gallery/exhibitions/43/marimekko.

Battle-Baptiste, Whitney, and Britt Rusert. "WEB Du Bois's Data Portraits: Visualizing Black America: The Color Line at the Turn of the Twentieth Century." Amherst, Massachusetts: WEB Du Bois Center at the University of Massachusetts Amherst (2018).

BBC. "BBC Audiences Tableau Style Guide." https://public.tableau.com/profile/bbc.audiences#!/vizhome /BBCAudiencesTableauStyleGuide/Hello

Béland, Antoine, and Thomas Hurtut. "Unit Visualizations for Visual Storytelling." OSF Preprints, https://osf.io/bshpc/.

Bennet, Tim. "How 52 Ninth-Graders Spell 'Camouflage', Sankey Diagram." https://www.reddit .com/r/dataisbeautiful/comments/6a4pb8/how_52_ninthgraders_spell_camouflage_sankey/?st =J2NBTEoQ&sh=ddd5c5e1

Berinato, Scott. "The Power of Visualizations' 'Aha!' Moments." *Harvard Business Review*, March 19, 2013, https://hbr.org/2013/03/power-of-visualizations-aha-moment.

Berman, Jacob. "Washington Metro Map Print Original Art." Etsy listing, 2019, https://www.etsy.com /listing/647623499/washington-metro-map-print-original-art

Bontemps, Xtophe. "Why You Should Never Use Radar Plots." data.visualisation.free.fr, March 2017, https://rpubs.com/Xtophe/268920.

Borkin, Michelle A., Zoya Bylinskii, Nam Wook Kim, Constance May Bainbridge, Chelsea S. Yeh, Daniel Borkin, Hanspeter Pfister, and Aude Oliva. "Beyond Memorability: Visualization Recognition and Recall." *IEEE Transactions on Visualization and Computer Graphics* 22, no. 1 (2015): 519–528.

Börner, Katy, Andreas Bueckle, and Michael Ginda. "Data Visualization Literacy: Definitions, Conceptual Frameworks, Exercises, and Assessments." *Proceedings of the National Academy of Sciences* 116, no. 6 (2019): 1857–1864.

Bortins, I., Demers S., & Clarke, K. (2002). Cartogram Types. http://www.ncgia.ucsb.edu/projects /Cartogram_Central/types.html

Bostock, Michael, Vadim Ogievetsky, and Jeffrey Heer. "D³ Data-Driven Documents." *IEEE Transactions on Visualization and Computer Graphics* 17, no. 12 (2011): 2301–2309.

Brehmer, Matt. "Visualizing Ranges Over Time on Mobile Phones." Medium, November 3, 2018, https://medium.com/multiple-views-visualization-research-explained/ranges-900a04d7d32a.

Brehmer, Matthew, Bongshin Lee, Petra Isenberg, and Eun Kyoung Choe. "Visualizing ranges over time on mobile phones: a task-based crowdsourced evaluation." *IEEE Transactions on Visualization and Computer Graphics* 25, no. 1 (2018): 619–629.

Bremer, Nadieh. "More Fun Data Visualizations with the Gooey Effect." Visual Cinnamon (blog). June 20, 2016. https://www.visualcinnamon.com/2016/06/fun-data-visualizations-svg-gooey-effect.html.

Bremer, Nadieh. "Techniques for Data Visualization on Both Mobile & Desktop." Visual Cinnamon (blog). April 17, 2019. https://www.visualcinnamon.com/2019/04/mobile-vs-desktop-dataviz#recap.

Brinton, Willard Cope. *Graphic Methods for Presenting Facts*. Engineering Magazine Company, 1914.

Broderick, Ryan. "Who Watches More Porn: Republicans or Democrats?" BuzzFeed News. April 2014. https://www.buzzfeednews.com/article/ryanhatesthis/who-watches-more-porn-republicans-or-democrats.

Bureau of Labor Statistics. Unemployment rate. https://www.bls.gov/cps/

Burn-Murdoch, John. "Episode #155: John Burn-Murdoch." The PolicyViz Podcast, June 18, 2019, https://policyviz.com/podcast/episode-155-john-burn-murdoch/.

Burn-Murdoch, John. Twitter post. October 28, 2019, 4:04 a.m., https://twitter.com/jburnmurdoch/status/1188728193945100288.

Bush, George W. "Address Before a Joint Session of the Congress on the State of the Union." The American Presidency Project, February 2, 2005, https://www.presidency.ucsb.edu/documents/address-before-joint-session-the-congress-the-state-the-union-14.

Byron, Lee, and Martin Wattenberg. "Stacked Graphs–Geometry and Aesthetics." *IEEE Transactions on Visualization and Computer Graphics* 14, no. 6 (2008): 1245–1252.

Cairo, Alberto. "Annotation, Narrative, and Storytelling in Infographics and Visualization." The Functional Art (blog), April 16, 2014, http://www.thefunctionalart.com/2014/04/annotation-narrative-and-storytelling.html.

Cairo, Alberto. "Download the Datasaurus: Never Trust Summary Statistics Alone; Always Visualize Your Data." The Functional Art (blog), August 29, 2016, http://www.thefunctionalart.com/2016/08/download-datasaurus-never-trust-summary.html.

Cairo, Alberto. *The Functional Art: An Introduction to Information Graphics and Visualization.* Berkeley, CA: New Riders, 2012.

Cairo, Alberto. *How Charts Lie: Getting Smarter About Visual Information.* New York: Norton, 2019.

Cairo, Alberto. "Visual Storytelling w/ Alberto Cairo and Robert Kosara." The Data Stories Podcast, Episode #35, April 16, 2014, https://datastori.es/data-stories-35-visual-storytelling-w-alberto-cairo-and-robert-kosara/.

Camões, Jorge. "Perception: Gestalt Laws." excelcharts.com (blog), January 10, 2012, https://excelcharts.com/data-visualization-excel-users/gestalt-laws/.

Carpenter, Patricia A., and Priti Shah. "A Model of the Perceptual and Conceptual Processes in Graph Comprehension." *Journal of Experimental Psychology: Applied* 4, no. 2 (1998): 75.

Center on Budget and Policy Priorities. "Working-Family Tax Credits Help at Every Stage of Life." Graphic, https://www.cbpp.org/working-family-tax-credits-help-at-every-stage-of-life-0.

Centers for Disease Control and Prevention. "CDC Wonder." https://wonder.cdc.gov/.

Centers for Disease Control and Prevention. "Community Mitigation Guidelines to Prevent Pandemic Influenza—United States, 2017." *Morbidity and Mortality Weekly Report* 66, No. 1, April 21, 2017. https://stacks.cdc.gov/view/cdc/45220.

Centers for Disease Control and Prevention. "National, Regional, and State Level Outpatient Illness and Viral Surveillance." https://gis.cdc.gov/grasp/fluview/fluportaldashboard.html.

Cesal, Amy. "The Sunlight Foundation's Data Visualization Style Guidelines." Sunlight Foundation, March 12, 2014, https://sunlightfoundation.com/2014/03/12/datavizguide/.

Chalabi, Mona. Presentation, Tapestry Conference, University of Miami, November 29, 2018.

Chang, Kenneth. "A Different Way to Chart the Spread of Coronavirus." *New York Times*, March 20, 2020. https://www.nytimes.com/2020/03/20/health/coronavirus-data-logarithm-chart.html.

Chappell, Nathan and Adam B. Jaffe. "Intangible Investment and Firm Performance." *National Bureau of Economic Research Working Paper No. 24363*, March 2018, https://www.nber.org/papers/w24363.

Chase, Will. "Voronoi Treemap" (website). May 2, 2019, https://observablehq.com/@will-r-chase /voronoi-treemap.

Chase, Will. "Why I'm Not Making COVID19 Visualizations, and Why You (Probably) Shouldn't Either" (blog post). March 31, 2020, https://www.williamrchase.com/post/why-i-m-not-making -covid19-visualizations-and-why-you-probably-shouldn-t-either/.

Cherdarchuk, Joey. "Clear off the Table." Dark Horse Analytics (blog). March 27, 2014, https://www .darkhorseanalytics.com/blog/clear-off-the-table.

Choi, Jinho, Sanghun Jung, Deok Gun Park, Jaegul Choo, and Niklas Elmqvist. "Visualizing for the Non-Visual: Enabling the Visually Impaired to Use Visualization." *Computer Graphics Forum* 38, no. 3 (2019): 249–260.

Christiansen, Jen. "Pop Culture Pulsar: Origin Story of Joy Division's Unknown Pleasures Album Cover." *Scientific American*, February 18, 2015, https://blogs.scientificamerican.com/sa-visual /pop-culture-pulsar-origin-story-of-joy-division-s-unknown-pleasures-album-cover-video/.

Cleveland, William S. *The Elements of Graphing Data*. 2nd ed. Summit, NJ: Hobart Press, 1994.

Cleveland, William S. *Visualizing Data*. Summit, NJ: Hobart Press, 1993.

Cleveland, William S., and Robert McGill. "Graphical Perception: Theory, Experimentation, and Application to the Development of Graphical Methods." *Journal of the American Statistical Association* 79, no. 387 (1984): 531–554.

Colour Blindness Awareness. "Colour Blindness." http://www.colourblindawareness.org/colour-blindness /, accessed November 2019.

Congressional Budget Office. "CBO's August 2018 report An Update to the Economic Outlook: 2018 to 2028." www.cbo.gov/publication/54318.

Congressional Budget Office. "Changes in the Distribution of Workers' Annual Earnings Between 1979 and 2007." Congressional Budget Office, October 2009, https://www.cbo.gov/sites/default /files/111th-congress-2009-2010/reports/10-02-workers.pdf.

Congressional Budget Office. "Social Security Policy Options." Congressional Budget Office, July 2010, https://www.cbo.gov/sites/default/files/111th-congress-2009-2010/reports/07-01-ssoptions _forweb.pdf

Congressional Budget Office. "The 2012 Long-Term Budget Outlook | House Budget Committee." Congressional Budget Office, September 12, 2014, https://www.youtube.com/watch?v=dqhqZYYGNnA

Congressional Budget Office. "The Budget and Economic Outlook: Fiscal Years 2012 to 2022." Congressional Budget Office." January 31, 2012, https://www.cbo.gov/publication/42905?index=12699

Congressional Budget Office. "The Budget and Economic Outlook: Infographic." Congressional Budget Office." June 5, 2012, https://www.cbo.gov/publication/43289

Congressional Budget Office. "The Social Security Disability Insurance Program." Infographic. https://www.cbo.gov/publication/43432

Congressional Budget Office. "The Supplemental Nutrition Assistance Program." Infographic. https://www.cbo.gov/publication/43174.

Cook, Albert M., and Janice Miller Polgar. *Assistive Technologies-E-Book: Principles and Practice*. St. Louis, MO: Elsevier, 2014.

Correll, Michael, and Michael Gleicher. "Error Bars Considered Harmful: Exploring Alternate Encodings for Mean and Error." *IEEE Transactions on Visualization and Computer Graphics* 20, no. 12 (2014): 2142–2151.

Correll, Michael, Enrico Bertini, and Steven Franconeri. "Truncating the Y-Axis: Threat or Menace?" In *Proceedings of the 2020 CHI Conference on Human Factors in Computing Systems*, ed. Regina Bernhaupt et al., 1–12. New York: Association for Computing Machinery. 2020.

Correll, Michael. "Ethical Dimensions of Visualization Research." Tableau Research, CHI 2019, 2019.

Correll, Michael. "Visualization Design Principles for the Pandemic." Medium, April 3, 2020, https://medium.com/@mcorrell/visualization-design-principles-for-the-pandemic-e65388280d16.

Cousins, Carrie. "Color and Cultural Design Considerations." WebDesignerDepot (blog), June 11, 2012, https://www.webdesignerdepot.com/2012/06/color-and-cultural-design-considerations/.

Cox, Amanda. "Amanda Cox—Eyeo Festival 2011." https://vimeo.com/29391942

Cruz, Pedro de, John Wihbey, and Felipe Shibuya. *Simulated Dendrochronology of the United States*. 2018.

D'Ignazio, Catherine and Lauren F. Klein. *Data Feminism*. Boston, MA: MIT Press. 2020.

Dallas Morning News. *Dallas Morning News Graphics Stylebook*. https://knightcenter.utexas.edu/mooc/file/tdmn_graphics.pdf.

Daniels, Matt. "The Largest Vocabulary In Hip Hop." The Pudding, January 21, 2019, https://pudding.cool/projects/vocabulary/index.html.

DeBelius, Danny. "Let's Tesselate: Hexagons For Tile Grid Maps." NPR visuals team, May 11, 2015, https://blog.apps.npr.org/2015/05/11/hex-tile-maps.html.

DeBold, Tynan and Dov Friedman. "Battling Infectious Diseases in the 20th Century: The Impact of Vaccines." *Wall Street Journal*, February 11, 2015, http://graphics.wsj.com/infectious-diseases-and-vaccines/.

Desilver, Drew. "A Record Number of Women Will Be Serving in the New Congress." Pew Research Center, December 18, 2018. https://www.pewresearch.org/fact-tank/2018/12/18/record-number-women-in-congress/

Dorling, Daniel. "Area cartograms: their use and creation." *The map reader: Theories of mapping practice and cartographic representation* (2011): 252–260.

Du Bois, William Edward Burghardt, ed. *The College-bred Negro: Report of a Social Study Made Under the Direction of Atlanta University; Together with the Proceedings of the Fifth Conference for the Study of the Negro Problems, Held at Atlanta University, May 29–30, 1900*. No. 5. Atlanta University Press, 1900.

Economist. "The Climate Issue." https://www.economist.com/leaders/2019/09/19/the-climate-issue.

Economist. "The Dragon Takes Wing." May 1, 2014, https://www.economist.com/news/finance-and -economics/21601568-new-data-suggest-chinese-economy-bigger-previously-thought-dragon.

Economist. "Flattening the Curve: Covid-19 Is Now in 50 Countries, and Things Will Get Worse." *The Economist*, February 29, 2020. https://www.economist.com/briefing/2020/02/29/covid-19-is-now -in-50-countries-and-things-will-get-worse.

Eurostat, "Graphical Style Guide—A Practical Layout Guide for Eurostat Publications—2016 Edition." *Eurostat News*, February 25, 2016, https://ec.europa.eu/eurostat/web/products-eurostat-news /-/STYLE-GUIDE_2016.

Federal Home Loan Bank Board. Home Owners' Loan Corporation. 1933-7/1/1939. "City of Richmond, Virginia and Environs." U.S. National Archives and Records Administration. https://catalog .archives.gov/id/85713737

Federal Reserve Board of Governors. "Household Debt Services and Financial Obligations Ratios." Accessed January 2020. https://www.federalreserve.gov/releases/housedebt/

Few, Stephen. "Bullet Graph Design Specification." Perceptual Edge, Visual Business Intelligence Newsletter, October 10, 2013, https://www.perceptualedge.com/articles/misc/Bullet_Graph_Design _Spec.pdf.

Few, Stephen. "The DataVis Jitterbug: Let's Improve an Old Dance." Perceptual Edge, Visual Business Intelligence Newsletter, April/May/June 2017, https://www.perceptualedge.com/articles/visual _business_intelligence/the_datavis_jitterbug.pdf.

Few, Stephen. *Show Me the Numbers: Designing Tables and Graphs to Enlighten*. Oakland, CA: Analytics Press, 2004.

Few, Stephen. "Unit Charts Are For Kids." Perceptual Edge, Visual Business Intelligence Newsletter, October, November, December 2010, https://www.perceptualedge.com/articles/visual_business _intelligence/unit_charts_are_for_kids.pdf.

Field, Kenneth. *Cartography*. Redlands, CA: Esri, 2018.

FiveThirtyEight. "50 Years Of World Cup Doppelgangers." July 15, 2018, https://projects.fivethirtyeight .com/world-cup-comparisons/romelu-lukaku-2018/

Florence, Philip Sargant. *Only an Ocean Between*. New York: Essential, 1946.

Fontenot, Kayla, Jessica Semega, and Melissa Kollar. "Income and Poverty in the United States: 2017." U.S. Bureau of the Census, September 2008, https://www.census.gov/content/dam/Census/library /publications/2018/demo/p60-263.pdf.

Friendly, Michael. "Playfair Balance of Trade Data." September 13, 2018, http://www.datavis.ca /courses/RGraphics/R/playfair-east-indies.html.

Fruchterman, Thomas MJ, and Edward M. Reingold. "Graph Drawing by Force-Directed Placement." *Software: Practice and experience* 21, no. 11 (1991): 1129–1164.

Fry, Ben. "Tracing the Origin of Species." https://fathom.info/traces/.

Gallo, Carmine. *The Presentation Secrets of Steve Jobs: How to Be Insanely Great in Front of Any Audience.* Upper Saddle River, NJ: Prentice Hall, 2010.

Gambino, Megan. "Do Our Brains Find Certain Shapes More Attractive Than Others?" Smithsonian Magazine. November 14, 2013. https://www.smithsonianmag.com/science-nature/do-our-brains-find -certain-shapes-more-attractive-than-others-180947692/.

Gantt, Henry Laurence. "A Graphical Daily Balance in Manufacture." No. 1002 in American Society of Mechanical Engineers. *Transactions of the American Society of Mechanical Engineers.* New York City: The Society, 1880. https://babel.hathitrust.org/cgi/pt?id=mdp.39015023119541&view=1up& seq=1363

Gee, Alastair, Julia Carrie Wong, Paul Lewis, Adithya Sambamurthy, Charlotte Simmonds, Nadieh Bremer, and Shirley Wu. "Bussed Out: How America Moves Its Homeless." *The Guardian*, December 20, 2017, https://www.theguardian.com/us-news/ng-interactive/2017/dec/20/bussed-out-america -moves-homeless-people-country-study

Google Finance. http://finance.google.com. Accessed January 2020.

Gramacki, Artur. *Nonparametric Kernel Density Estimation and Its Computational Aspects.* Cham, Switzerland: Springer International, 2018.

Groeger, Lena V. "A Big Article About Wee Things." ProPublica (blog), September 25, 2014, https:// www.propublica.org/nerds/a-big-article-about-wee-things.

Guardian. "EU Referendum: Full Results and Analysis." *The Guardian*, https://www.theguardian.com /politics/ng-interactive/2016/jun/23/eu-referendum-live-results-and-analysis.

Guardian. "Full US 2012 Election County-Level Results to Download." https://www.theguardian.com /news/datablog/2012/nov/07/us-2012-election-county-results-download#data.

Hansen, Wallace R. *Suggestions to Authors of the Reports of the United States Geological Survey*, US GPO, January 1991.

Haroz, Steve, Robert Kosara, and Steven L. Franconeri. "Isotype Visualization: Working Memory, Performance, and Engagement with Pictographs." In *Proceedings of the 33rd annual ACM conference on human factors in computing systems*, 1191–1200. ACM, 2015.

Hawkins, Ed. "147 | Iconic Climate Visuals with Ed Hawkins." Data Stories podcast, September 10, 2019, https://datastori.es/147-iconic-climate-visuals-with-ed-hawkins/.

Hawkins, Ed. ShowYourStripes.com. https://showyourstripes.info/

Healey, Christopher, and James Enns. "Attention and Visual Memory in Visualization and Computer Graphics." *IEEE Transactions on Visualization and Computer Graphics* 18, no. 7 (2011): 1170–1188.

Hearst, Marti, Emily Pedersen, Lekha Priya Patil, Elsie Lee, Paul Laskowski, and Steven Franconeri. "An Evaluation of Semantically Grouped Word Cloud Designs." *IEEE Transactions on Visualization and Computer Graphics* (2019).

Heer, Jeffrey, Michael Bostock, and Vadim Ogievetsky. "A Tour Through the Visualization Zoo." *Commun. Acm* 53, no. 6 (2010): 59–67.

Hermann, E.P. "Maps and Sales Visualization" *Personal Efficiency: The How and Why Magazine*, Published by the LaSalle Extension University, US Department of Education, July 1922, pages 6–7.

Hofman, Jake M., Daniel G. Goldstein, and Jessica Hullman. "How visualizing inferential uncertainty can mislead readers about treatment effects in scientific results." *Proceedings of the 2020 CHI Conference on Human Factors in Computing Systems*.

Hofmann, Heike, and Marie Vendettuoli. "Common Angle Plots as Perception-True Visualizations of Categorical Associations." *IEEE Transactions on Visualization and Computer Graphics* 19.12 (2013): 2297–2305.

Horowitz, Juliana Mesace, Kim Parker, and Renee Stepler. "Wide Partisan Gaps in U.S. Over How Far the Country Has Come on Gender Equality." Pew Research Center, October 18, 2017, https://www.pewsocialtrends.org/2017/10/18/wide-partisan-gaps-in-u-s-over-how-far-the-country-has-come-on-gender-equality/.

Hullman, Jessica and Matthew Kay. "Uncertainty + Visualization, Explained." Medium, Visualization Research Explained (blog), https://medium.com/multiple-views-visualization-research-explained/uncertainty-visualization-explained-67e7a73f031b.

Hullman, Jessica, Eytan Adar, and Priti Shah. "Benefitting Infovis with Visual Difficulties." *IEEE Transactions on Visualization and Computer Graphics* 17, no. 12 (2011): 2213–2222.

Hullman, Jessica, Paul Resnick, and Eytan Adar. "Hypothetical Outcome Plots Outperform Error Bars and Violin Plots for Inferences About Reliability of Variable Ordering." *PloS one* 10, no. 11 (2015).

Hullman, Jessica. "Leading with the Unknowns in COVID-19 Models." *Scientific American*, April 11, 2020. https://blogs.scientificamerican.com/observations/leading-with-the-unknowns-in-covid-19-models/.

Ingraham, Christopher. "Kansas Is the Nation's Porn Capital, According to Pornhub." Wonkviz, no date, https://wonkviz.tumblr.com/post/82488570278/kansas-is-the-nations-porn-capital-according-to

Institute for Health Metrics and Evaluation. "Causes of Death (COD) Visualization | Viz Hub." Accessed January 2020. https://vizhub.healthdata.org/cod/.

International Social Survey Programme (ISSP). "ISSP 2009 'Social Inequality IV'—ZA No. 5400." Accessed January 2020. https://www.gesis.org/issp/modules/issp-modules-by-topic/social-inequality/2009/

Irvin-Erickson, Yasemin, Jonathan Schwabish, and Nicole Weissman. "What We Know About Gun Violence in the United States: Who's Affected?" *Urban Wire Urban Institute*, October 4, 2016, https://www.urban.org/urban-wire/what-we-know-about-gun-violence-united-states-whos-affected.

Japanese: https://seibu.ekitan.com/pdf/20180310/235-16-1-0.pdf

Jarreau, Paige. "#MySciBlog Interviewee Motivations to Blog about Science." Figshare.com (blog), March 21, 2015, https://figshare.com/articles/_MySciBlog_Interviewee_Motivations_to_Blog_about _Science/1345026/2.

Johnson, Brian, and Ben Shneiderman. "Tree-Maps: A Space-Filling Approach to the Visualization of Hierarchical Information Structures." *VIS '91: Proceedings of the 2nd conference on Visualization '91.* IEEE, 1991.

Johnson, Matt. "Which Metro Stations Are the Most Balanced?" Greater Greater Washington, November 29, 2012, https://ggwash.org/view/29468/which-metro-stations-are-the-most-balanced.

Joint Committee on Taxation. "A Distribution of Returns by the Size of the Tax Change for the 'Tax Cuts and Jobs Act,' As Ordered Reported by the Committee on Finance on November 16, 2017." November 27, 2017.

Joint Committee on Taxation. "Distributional Effects of Public Law 115–97." Scheduled for a Public Hearing Before the House Committee on Ways and Means on March 27, 2019, March 25, 2019, JCX-10-19.

Kaiser Family Foundation. "Status of State Action on the Medicaid Expansion Decision." Accessed January 2020. https://www.kff.org/health-reform/state-indicator/state-activity-around-expanding -medicaid-under-the-affordable-care-act/.

Kastellec, Jonathan P., and Eduardo L. Leoni. "Using Graphs Instead of Tables in Political Science." *Perspectives on Politics* 5, no. 4 (2007): 755–771.

Katz, Josh. "Who Will Be President?" *New York Times*, November 8, 2016, https://www.nytimes.com /interactive/2016/upshot/presidential-polls-forecast.html

Kiefer, Len. Twitter post. September 26, 2019, 4:44 p.m., https://twitter.com/lenkiefer/status /1177323018143580160.

Kight, Stef W. and Harry Stevens. "By the Numbers: How Trump Properties Profited from His Presidency." *Axios*, June 28, 2018, https://www.axios.com/donald-trump-properties-taxpayer-campaigns -presidency-91e3755d-23cd-42d1-897d-c0c81ac509dd.html.

Kirk, Andy. "The Problems with B'Arc Charts." Visualisingdata.com, September 1, 2017, https://www .visualisingdata.com/2017/09/problems-barc-charts/.

Klein, Scott. "Infographics in the Time of Cholera." *Pro Publica*, March 16, 2016, https://www.propublica .org/nerds/infographics-in-the-time-of-cholera.

Koblin, Aaron. "Flight Patterns." http://www.aaronkoblin.com/work/flightpatterns/.

Koblin, Aaron. "Visualizing Ourselves . . . with Crowd-Sourced data." TED2011, March 2011, https:// www.ted.com/talks/aaron_koblin_visualizing_ourselves_with_crowd_sourced_data.

Kommenda, Niko, Caelainn Barr, and Josh Holder. "Gender Pay Gap: What We Learned and How to Fix It." *The Guardian*, April 5, 2018, https://www.theguardian.com/news/ng-interactive/2018 /apr/05/women-are-paid-less-than-men-heres-how-to-fix-it.

Krzywinski, Martin, Inanc Birol, Steven JM Jones, and Marco A. Marra. "Hive plots—rational approach to visualizing networks." *Briefings in bioinformatics* 13, no. 5 (2011): 627–644.

Krzywinski, Martin. "Hive plots—rational approach to visualizing networks." http://www.hiveplot .com/.

Lekovic, Jovan. "How We Made the BBC Audiences Tableau Style Guide." Medium, August 29, 2018, https://medium.com/bbc-data-science/how-we-made-the-bbc-audiences-tableau-style-guide -4f0a6b7525ce.

Likert, Rensis. "A Technique for the Measurement of Attitudes." *Archives of Psychology* 22 (1932): 15–55.

Lima, Manuel. "Why Humans Love Pie Charts." Medium (blog). July 23, 2018. https://blog.usejournal .com/why-humans-love-pie-charts-9cd346000bdc?.

Lima, Manuel. *The Book of Circles: Visualizing Spheres of Knowledge.* New York: Princeton Architectural Press, 2017.

Liu, Yang, and Jeffrey Heer. "Somewhere Over the Rainbow: An Empirical Assessment of Quantitative Colormaps." In *Proceedings of the 2018 CHI Conference on Human Factors in Computing Systems*, pp. 1–12. 2018.

Luxembourg Income Study. https://www.lisdatacenter.org/.

Malkulec, Amanda. "Leading with the Unknowns in COVID-19 Models." Medium, March 11, 2020, https:// medium.com/nightingale/ten-considerations-before-you-create-another-chart-about-covid -19-27d3bd691be8.

Manski, Charles F., and Francesca Molinari. "Rounding Probabilistic Expectations in Surveys." *Journal of Business & Economic Statistics* 28, no. 2 (2010): 219–231.

Matejka, Justin, and George Fitzmaurice. "Same Stats, Different Graphs: Generating Datasets with Varied Appearance and Identical Statistics Through Simulated Annealing." In *Proceedings of the 2017 CHI Conference on Human Factors in Computing Systems*, pp. 1290–1294. ACM, 2017.

Mayr, Eva and Günther Schreder. "Isotype Visualizations: A Chance for Participation and Civic Education." *Journal of Democracy* 6(2): 136–150, 2014.

McCann, Adam. "Arc Chart in Tableau." August 27, 2015, http://duelingdata.blogspot.com/2015/08 /arc-chart-in-tableau.html.

McKee, Robert. *Substance, Structure, Style, and the Principles of Screenwriting.* New York: HarperCollins, 1997.

Medina, John. *Brain Rules: 12 Principles for Surviving and Thriving at Work, Home, and School.* ReadHowYouWant.com, 2011.

Meeks, Elijah. "Exploratory Design in Data Visualization: Understanding and leveraging chart similarity." *Medium*, January 15, 2019, https://medium.com/@Elijah_Meeks/exploratory-design -in-data-visualization-87bc60ce7f04.

Meeks, Elijah. "Sketchy Data Visualization in Semiotic." Medium (blog), September 11, 2017, https:// medium.com/@Elijah_Meeks/sketchy-data-visualization-in-semiotic-5811a52f59bc.

Meyer, Bruce D., Nikolas Mittag, and Robert M. George. *Errors in Survey Reporting and Imputation and their Effects on Estimates of Food Stamp Program Participation*. No. w25143. National Bureau of Economic Research, 2018.

Meyer, Bruce D., Wallace K.C. Mok, and James X. Sullivan. "Household Surveys in Crisis." *Journal of Economic Perspectives* 29, no. 4 (2015): 199–226.

Miller, Matthew, Steven J. Lippmann, Deborah Azrael, and David Hemenway. "Household firearm ownership and rates of suicide across the 50 United States." *Journal of Trauma and Acute Care Surgery* 62, no. 4 (2007): 1029–1035.

Minard, Charles-Joseph. "*Carte figurative des pertes successives en hommes de l'Armée Française dans la campagne de Russie 1812–1813*." 1861, *Photo courtesy of Ecole nationale des ponts et chaussées*

Monmonier, Mark. *How to Lie with Maps*. Chicago: University of Chicago Press, 2018.

Munzner, Tamara. *Visualization Analysis and Design*. AK Peters, 2014.

National Conference of State Legislatures. "State Family and Medical Leave Laws." July 2016. http://www.ncsl.org/research/labor-and-employment/state-family-and-medical-leave-laws.aspx

National Conference of State Legislatures. "State Minimum Wages | 2020 Minimum Wage by State." January 2020. http://www.ncsl.org/research/labor-and-employment/state-minimum-wage-chart.aspx#Table.

National Highway Traffic Safety Administration. Fatality Analysis Reporting System (FARS). https://one.nhtsa.gov/Data/Fatality-Analysis-Reporting-System-(FARS).

National Institute on Drug Abuse. "Overdose Death Rates." January 2019. https://www.drugabuse.gov/related-topics/trends-statistics/overdose-death-rates.

NBC News. "Presidential Election Results." http://elections.nbcnews.com/ns/politics/2012/all/president/#.XiZR4lNKhTZ.

Nediger, Midori. "How to Use an Icon Story to Take Your Infographic to the Next Level." Venngage (blog), April 27, 2018, https://venngage.com/blog/icon-story/.

Neurath, O. *International Picture Language. The First Rules of Isotype*. London: Kegan Paul, 1936.

Neurath, O. *Modern Man in the Making*. New York: Knopf, 1939.

New York Magazine. "Approval Matrix." provided via email.

New York Times. Paths to the White House.

New York Times. Twitter post. August 7, 2016, 10:52 p.m. https://twitter.com/nytgraphics/status/762481520565030919?ref_src=twsrc%5Etfw

Newman, George E., and Brian J. Scholl. "Bar Graphs Depicting Averages are Perceptually Misinterpreted: the Within-the-Bar Bias." *Psychonomic Bulletin and Review* 19, no. 4 (2012): 601–607.

Noble, Safiya Umoja. *Algorithms of Oppression: How Search Engines Reinforce Racism*. New York: New York University Press. 2018.

O'Brady, Kevin. "Fight to Repeal." Joint Economic Committee Republicans, 2010, https://kevinbrady.house.gov/obamacare/fighting-to-repeal-obamacare.htm.

O'Neill, Catherine. *Weapons of Math Destruction: How Big Data Increases Inequality and Threatens Democracy*. New York: Broadway, 2016.

Obama, Barack. "Remarks of President Barack Obama—State of the Union Address As Delivered." State of the Union Address, Washington, DC, January 13, 2016, https://obamawhitehouse.archives .gov/the-press-office/2016/01/12/remarks-president-barack-obama-%E2%80%93-prepared-delivery -state-union-address.

Olson, J. M. "Noncontiguous Area Cartograms." *The Professional Geographer*, 28, no. 4 (1976): 371–380.

Ondov, Brian, Nicole Jardine, Niklas Elmqvist, and Steven Franconeri. "Face to Face: Evaluating Visual Comparison." *IEEE Transactions on Visualization and Computer Graphics* 25, no. 1 (2018): 861–871.

Organization for Economic Cooperation and Development (OECD). "Better Life Index." Accessed December 2019. http://www.oecdbetterlifeindex.org/.

Organization for Economic Cooperation and Development (OECD). Program for International Student Assessment (PISA), 2015 Reading, Mathematics and Science Assessment.

Organization for Economic Cooperation and Development (OECD). Social Expenditure Database. Accessed January 2019. https://www.oecd.org/social/expenditure.htm.

Padilla, Lace, Matthew Kay, and Jessica Hullman. "Uncertainty Visualization." PsyArXiv. April 27, 2020. doi:10.31234/osf.io/ebd6r.

Pätzold, André, Julius Tröger, Joachim Fahrun, Christopher Möller, David Wendler and Marie-Louise Timcke. "These Candidates Live the Furthest Away from Their Voters." *Berliner Morgenpost*, September 8, 2016, https://interaktiv.morgenpost.de/waehlernaehe-berlin/.

Perez, Caroline Criado. *Invisible Women: Exposing Data Bias in a World Designed for Men*. New York: Random House, 2019.

Playfair, William. *Playfair's Commercial and Political Atlas and Statistical Breviary*. Cambridge: Cambridge University Press, 2005.

Porter, Mark and Rhys Blakely. 2020. "Coronavirus: How Safe Are You? All Your Health Questions Answered." *Australian*, March 31, 2020, https://www.theaustralian.com.au/world/the-times /coronavirus-how-safe-are-you-all-your-health-questions-answered/news-story/c40ec5841763c6d 57f986c3ea3e19e4f

Presentitude. "A Guide to 44 Safe Fonts & 74 Safe Combos for PowerPoint." Presentitude (blog) no date, access January 2020, http://presentitude.com/fonts/.

Priestley, Joseph. "A Specimen of a Chart of Biography." 1765.

Project Tycho. https://www.tycho.pitt.edu/data.

Rendgen, Sandra. *The Minard System: The Complete Statistical Graphics of Charles-Joseph Minard*. New York: Princeton Architectural Press, 2018.

R Graph Gallery. "Network Graph." https://www.r-graph-gallery.com/network.html.

Wexler, Steve, Jeffrey Shaffer, and Andy Cotgreave. *The Big Book of Dashboards: Visualizing Your Data Using Real-World Business Scenarios*. Hoboken, NJ: Wiley, 2017.

The White House. "President Obama's Final State of the Union." February 12, 2016. https://obama whitehouse.archives.gov/sotu.

Wickham, Hadley, and Lisa Stryjewski. "40 Years of Boxplots." *American Statistician* (2011).

Wikipedia. "Right-to-Work Law." Accessed January 2020. https://en.wikipedia.org/wiki/Right-to-work _law.

Wikipedia. "State Income Tax." Accessed January 2020. https://en.wikipedia.org/wiki/State_income _tax.

Wilke, Claus O. *Fundamentals of Data Visualization: A Primer on Making Informative and Compelling Figures*. Sebastopol, CA: O'Reilly, 2019.

Wilkinson, Krista M., and William J. McIlvane. "Perceptual Factors Influence Visual Search for Meaningful Symbols in Individuals with Intellectual Disabilities and Down Syndrome or Autism Spectrum Disorders." *American Journal on Intellectual and Developmental Disabilities* 118, no. 5 (2013): 353–364.

Wilkinson, Leland. "Dot Plots." *American Statistician* 53, no. 3 (1999): 276–281.

Wilkinson, Leland. "The Grammar of Graphics." In *Handbook of Computational Statistics*, 375–414. Berlin: Springer, 2012.

Wilkinson, Leland. *The Grammar of Graphics*. Chicago: Springer Science & Business Media, 2013.

Witt, Jessica. "Graph Construction: An Empirical Investigation on Setting the Range of the Y-Axis." http://amplab.colostate.edu/reprints/Witt_Graphs_YaxisRange.pdf.

Wood, Jo, Petra Isenberg, Tobias Isenberg, Jason Dykes, Nadia Boukhelifa, and Aidan Slingsby. "Sketchy rendering for information visualization." *IEEE Transactions on Visualization and Computer Graphics* 18, no. 12 (2012): 2749–2758.

World Bank. "World Development Indicators." The World Bank Group. Accessed January 2020. https://datacatalog.worldbank.org/dataset/world-development-indicators.

XKCD. "Correlation." https://xkcd.com/552/.

Yau, Nathan. "Vehicles Involved in Fatal Crashes." FlowingData blog, January 11, 2012, https://flowing data.com/2012/01/11/vehicles-involved-in-fatal-crashes/.

INDEX

FOURTH
EDITION

Single-Variable
Calculus
with Analytic Geometry
Early Transcendentals Version

C. H. Edwards, Jr.
The University of Georgia, Athens

David E. Penney
The University of Georgia, Athens

PRENTICE HALL Inc.
Englewood Cliffs, New Jersey 07632

Acquisitions Editor: Jacqueline Wood
Developmental Editor: Maurice Esses
Production Editor: Kelly Dickson
Cover Design: Carole Giguère
Cover Image: Michael Portman
Page Layout: Andrew Zutis, Karen Noferi
Text Composition: Interactive Composition Corporation

MATLAB® is a registered trademark of:
The MathWorks/24 Prime Park Way/ Natick MA 01760

ISBN 0-13-300591-7

Prentice-Hall Inc., Englewood Cliffs, New Jersey
Prentice-Hall International (UK) Limited, London
Prentice-Hall of Australia Pty. Limited, Sydney
Prentice-Hall Hispanoamericana, S.A., Mexico
Prentice-Hall of India Private Limited, New Delhi
Prentice-Hall of Japan, Inc., Tokyo
Simon & Schuster Asia Private Limited., Singapore
Editora Prentice-Hall do Brasil, Ltda., Rio de Janeiro

Contents

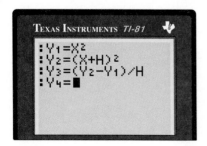
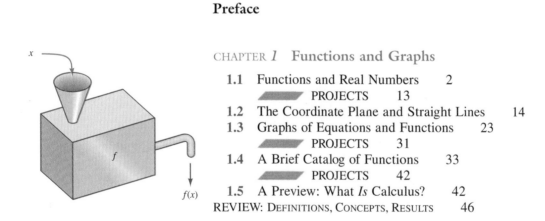

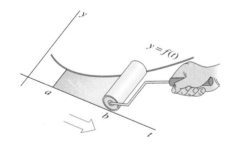

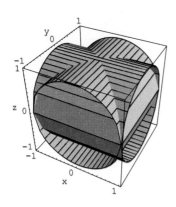

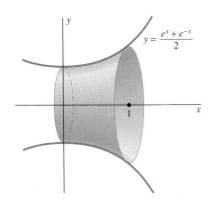

$y = \dfrac{e^x + e^{-x}}{2}$

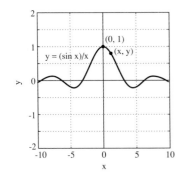

$y = (\sin x)/x$

Contents   V

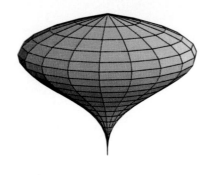

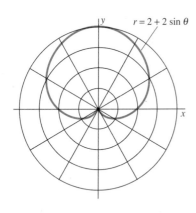

$r = 2 + 2\sin\theta$

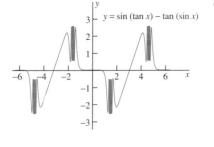

$y = \sin(\tan x) - \tan(\sin x)$

Appendices A-1

About the Authors

C. Henry Edwards, University of Georgia, received his Ph.D. from the University of Tennessee in 1960. He then taught at the University of Wisconsin for three years and spent a year at the Institute for Advanced Study (Princeton) as an Alfred P. Sloan Research Fellow. Professor Edwards has just completed his thirty-fifth year of teaching (including teaching calculus almost every year) and has received numerous university-wide teaching awards. His scholarly career has ranged from research and the direction of dissertations in topology to the history of mathematics to applied mathematics to computers and technology in mathematics (his focus in recent years). In addition to his calculus, advanced calculus, linear algebra, and differential equations textbooks, he is well known to calculus instructors as the author of *The Historical Development of the Calculus* (Springer-Verlag, 1979). He has served as a principal investigator on three recent NSF-supported projects: (1) A project to introduce technology throughout the mathematics curricula in two northeastern Georgia public school systems (including *Maple* for beginning algebra students); (2) a *Calculus with Mathematica* pilot program at the University of Georgia; and (3) a *MATLAB*-based computer lab project for upper-division numerical analysis and applied mathematics students.

David E. Penney, University of Georgia, completed his Ph.D. at Tulane University in 1965 while teaching at the University of New Orleans. Earlier he had worked in experimental biophysics at both Tulane University and the Veteran's Administration Hospital in New Orleans. He actually began teaching calculus in 1957 and has taught the course almost every term since then. He joined the mathematics department at Georgia in 1966 and has since received numerous university-wide teaching awards. He is the author of several research papers in number theory and topology and is author or co-author of books on linear algebra, differential equations, and calculus.

Preface

The role and practice of mathematics in the world at large is now undergoing a revolution that is driven largely by computational technology. Calculators and computer systems provide students and teachers with mathematical power that no previous generation could have imagined. We read even in daily newspapers of stunning current events like the recently announced proof of Fermat's last theorem. Surely *today* is the most exciting time in all history to be mathematically alive! So in preparing this new edition of **CALCULUS with Analytic Geometry**, we wanted first of all to bring at least some sense of this excitement to the students who will use it.

We also realize that the calculus course is a principal gateway to technical and professional careers for a still increasing number of students in an ever widening range of curricula. Wherever we look—in business and government, in science and technology—almost every aspect of professional work in the world involves mathematics. We therefore have re-thought once again the goal of providing calculus students the solid foundation for their subsequent work that they deserve to get from their calculus textbook.

For the first time since the original version was published in 1982, the text for this fourth edition has been reworked from start to finish. Discussions and explanations have been rewritten throughout in language that (we hope) today's students will find more lively and accessible. Seldom-covered topics have been trimmed to accommodate a leaner calculus course. Historical and biographical notes have been added to show students the human face of calculus. Graphics calculator and computer lab projects (with Derive, Maple, and *Mathematica* options) for key sections throughout the text have been added. Indeed, a new spirit and flavor reflecting the prevalent interest in graphics calculators and computer systems will be discernible throughout this edition. Consistent with the graphical emphasis of the current calculus reform movement, the number of figures in the text has been almost doubled, with must new computer-generated artwork added. Many of these additional figures serve to illustrate a more deliberative and exploratory approach to problem-solving. Our own teaching experience suggests that use of contemporary technology can make calculus more concrete and accessible to many students.

Fourth Edition Features

In preparing this edition, we have benefited from many valuable comments and suggestions from users of the first three editions. This revision was so pervasive that the individual changes are too numerous to be detailed in a preface, but the following paragraphs summarize those that may be of widest interest.

ADDITIONAL PROBLEMS   The number of problems has steadily increased since the first edition, and now totals well over 6000. In the third edition we inserted many additional practice exercises near the beginnings of problem sets to insure that students gain sufficient confidence and computational skill before moving on to the more conceptual problems that constitute the real goal of calculus. In this edition we have added graphics-based problems that emphasize conceptual understanding and accommodate student use of graphics calculators.

NEW EXAMPLES AND COMPUTATIONAL DETAILS   In many sections throughout this edition, we have inserted a simpler first example or have replaced existing examples with ones that are computationally simpler. Moreover, we have inserted an additional line or two of computational detail in many of the worked-out examples to make them easier for student readers to follow. The purpose of these computational changes is to make the computations themselves less of a barrier to conceptual understanding.

PROJECT MATERIAL   Several supplementary projects have been inserted in each chapter—a total of four dozen in all. Each project employs some aspect of modern computational technology to illustrate the principal ideas of the preceding section, and typically contains additional problems intended for solution with the use of a graphics calculator or computer. Figures and data illustrate the use of graphics calculators and computer systems such as Derive, Maple, and *Mathematica.* This project material is suitable for use in a computer/calculator lab that is conducted in association with a standard calculus course, perhaps meeting weekly. It can also be used as a basis for graphics calculator or computer assignments that students will complete outside of class, or for individual study.

COMPUTER GRAPHICS   Now that graphics calculators and computers are here to stay, an increased emphasis on graphical visualization along with numeric and symbolic work is possible as well as desirable. About 300 new MATLAB-generated figures illustrate the kind of figures students using graphics calculators can produce for themselves. Many of these are included with new graphical problem material. *Mathematica*-generated color graphics are included to highlight all sections involving 3-dimensional material.

HISTORICAL MATERIAL   Historical and biographical chapter openings have been inserted to give students a sense of the development of our subject by real, live human beings. Both authors are fond of the history of mathematics, and believe that it can favorably influence our teaching of mathematics. For this reason, numerous historical comments appear in the text itself.

INTRODUCTORY CHAPTERS   Chapters 1 and 2 have been streamlined for a leaner and quicker start on calculus. Chapter 1 concentrates on functions and graphs. It now includes a section cataloging the elementary functions of calculus, and provides a foundation for an earlier emphasis on transcendental functions. Chapter 1 now concludes with a section addressing the question "What *is* calculus?" Chapter 2 on limits begins with a section on tangent lines to motivate the official introduction of limits in Section 2.2. In contrast with the third edition, trigonometric limits now are treated throughout Chapter 2, in order to encourage a richer and more visual introduction to the limit concept.

DIFFERENTIATION CHAPTERS   The sequence of topics in Chapters 3 and 4 varies a bit from the most traditional order. We attempt to build student confidence by introducing topics more nearly in order of increasing difficulty. The chain rule appears quite early (in Section 3.3) and we cover the basic techniques for differentiating algebraic functions before discussing maxima and minima in Sections 3.5 and 3.6. The appearance of inverse functions is now delayed until Chapter 7. Section 3.7 now treats the derivatives of all six trigonometric functions. Implicit differentiation and related rates are combined in a single section (Section 3.8). The mean value theorem and its applications are deferred to Chapter 4. Sections 4.4 on the first derivative test and 4.6 on higher derivatives and concavity have been simplified and streamlined. A great deal of new graphic material has been added in the curve-sketching sections that conclude Chapter 4.

INTEGRATION CHAPTERS   New and simpler examples have been inserted throughout Chapters 5 and 6. Antiderivatives (formerly at the end of Chapter 4) now begin Chapter 5. Section 5.4 (Riemann sums) has been simplified greatly, with upper and lower sums eliminated and endpoint and midpoint sums emphasized instead. Many instructors now believe that first applications of integration ought not be confined to the standard area and volume computations; Section 6.5 is an optional section that introduces separable differential equations. To eliminate redundancy, the material on centroids and the theorems of Pappus is delayed to Chapter 15 (Multiple Integrals) where it can be treated in a more natural context.

EARLY TRANSCENDENTAL FUNCTIONS OPTIONS   In the "regular version" of this book the appearance of exponential and logarithmic functions is delayed until Chapter 7 (after integration). In the present "early transcendental version" these functions are introduced at the earliest practical opportunity in differential calculus—in Section 3.8, immediately following the differentiation of trigonometric functions in Section 3.7. Section 3.8 begins with an intuitive approach to exponential functions regarded as variable powers of a constant base, followed by the elementary idea of a logarithm as "the power to which the base a must be raised to get the number x". On this basis, the section includes a low-key review of the laws of exponents and of logarithms, and investigates somewhat informally the differentiation of exponential and logarithmic functions. Consequently, a diverse collection of transcendental functions is available for use in examples and problems throughout the balance of differential calculus (Chapters 3 and 4) and in the study of integral calculus (Chapters 5 and 6). Chapter 7 returns to exponential and logarithmic functions, offering a more complete and rigorous treatment plus further applications. But Section 7.2—based on the formal definition of the logarithm as an integral—actually can be covered anytime after the integral has been defined early in Chapter 5 (along with as much of the remainder of Chapter 7 as the instructor desires). Thus this version of the text is designed to support a course syllabus that includes exponential functions early in differential calculus, as well as logarithmic functions (defined as integrals) early in integral calculus.

The remaining transcendental functions—inverse trigonometric and hyperbolic—are now treated in Chapter 8. This newly organized chapter now includes also indeterminate forms and l'Hopital's rule (much earlier than in the third edition).

STREAMLINING TECHNIQUES OF INTEGRATION Chapter 9 is organized to accommodate those instructors who feel that methods of formal integration now require less emphasis, in view of modern techniques for both numerical and symbolic integration. Presumably everyone will want to cover the first four sections of the chapter (through integration by parts in Section 9.4.). The method of partial fractions appears in Section 9.5, and trigonometric substitutions and integrals involving quadratic polynomials follow in Section 9.6 and 9.7. Improper integrals now appear in Section 9.8, and the more specialized rationalizing substitutions have been relegated to the Chapter 9 miscellaneous problems. This rearrangement of Chapter 9 makes it more convenient to stop wherever the instructor desires.

INFINITE SERIES After the usual introduction to convergence of infinite sequences and series in Section 11.2 and 11.3, a combined treatment of Taylor polynomials and Taylor series appears in Section 11.4. This makes it possible for the instructor to experiment with a much briefer treatment of infinite series, but still offer exposure to the Taylor series that are so important for applications.

DIFFERENTIAL EQUATIONS Many calculus instructors now believe that differential equations should be seen as early and as often as possible. The very simplest differential equations (of the form $y' = f(x)$) appear in a subsection at the end of Section 5.2 (Antiderivatives). Section 6.5 illustrates applications of integration to the solution of separable differential equations. Section 9.5 includes applications of the method of partial fractions to population problems and the logistic equation. In such ways we have distributed enough of the spirit and flavor of differential equations throughout the text that it seemed expeditious to eliminate the (former) final chapter devoted solely to differential equations. However, those who so desire can arrange with the publisher to obtain for supplemental use appropriate sections of Edwards and Penney, *Elementary Differential Applications with Boundary Value Problems,* third edition (Englewood Cliffs, N.J.: Prentice-Hall, 1993).

Maintaining Traditional Strengths

While many new features have been added, five related objectives remained in constant view: concreteness, readability, motivation, applicability, and accuracy.

CONCRETENESS

The power of calculus is impressive in its precise answers to realistic questions and problems. In the necessary conceptual development of the subject, we keep in sight the central question: How does one actually *compute* it? We place special emphasis on concrete examples, applications, and problems that serve both to highlight the development of the theory and to demonstrate the remarkable versatility of calculus in the investigation of important scientific questions.

READABILITY

Difficulties in learning mathematics often are complicated by language difficulties. Our writing style stems from the belief that crisp exposition, both intuitive and precise, makes mathematics more accessible—and hence more readily learned—with no loss of rigor. We hope our language is clear and attractive to students and that they can and actually will read it, thereby enabling the instructor to concentrate class time on the less routine aspects of teaching calculus.

MOTIVATION

Our exposition is centered around examples of the use of calculus to solve real problems of interest to real people. In selecting such problems for examples and exercises, we took the view that stimulating interest and motivating effective study go hand in hand. We attempt to make it clear to students how the knowledge gained with each new concept or technique will be worth the effort expended. In theoretical discussions, especially, we try to provide an intuitive picture of the goal before we set off in pursuit of it.

APPLICATIONS

Its diverse applications are what attract many students to calculus, and realistic applications provide valuable motivation and reinforcement for all students. This book is well-known for the broad range of applications that we include, but it is neither necessary nor desirable that the course cover all the applications in the book. Each section or subsection that may be omitted without loss of continuity is marked with an asterisk. This provides flexibility for each instructor to determine his or her own flavor and emphasis.

ACCURACY

Our coverage of calculus is complete (though we hope it is somewhat less than encyclopedic). Still more than its predecessors, this edition was subjected to a comprehensive reviewing process to help ensure accuracy. For example, essentially every problem answer appearing in the Answers Section at the back of the book in this edition has been verified using *Mathematica*. With regard to the selection and sequence of mathematical topics, our approach is traditional. However, close examination of the treatment of standard topics may betray our own participation in the current movement to revitalize the teaching of calculus. We continue to favor an intuitive approach that emphasizes both conceptual understanding and care in the formulation of definitions and key concepts of calculus. Some proofs that may be omitted at the discretion of the instructor are placed at the ends of sections, and others are deferred to the book's appendices. In this way we leave ample room for variation in seeking the proper balance between rigor and intuition.

Supplementary Material

Answers to most of the odd-numbered problems appear in the back of the book. Solutions to most problems (other than those odd-numbered ones for which an answer alone is sufficient) are available in the *Instructor's Solutions Manual*. A subset of this manual, containing solutions to problems numbered 1, 4, 7, 10, · · · is availabe as a *Student's Solutions Manual*. A collection of some 1700 additional problems suitable for use as test questions, the *Calculus Test Item File,* is available (in both electronic and hard copy form) for use by instructors. Finally, an *Instructor's Edition* including section-by-section teaching outlines and suggestions is available to those who are using this book to teach calculus.

STUDENT LAB MANUALS

A variety of additional supplements are provided by the publisher, including author-written project manuals keyed to specific calculators and computer systems. These include

> Calculus Projects Using Derive
> Calculus Projects Using HP Graphics Calculators
> Calculus Projects Using Maple
> Calculus Projects Using *Mathematica*
> Calculus Projects Using MATLAB
> Calculus Projects Using TI Graphics Calculators
> Calculus Projects Using X(PLORE)

Each of these manuals consists of versions of the text's 48 projects that have been expanded where necessary to take advantage of specific computational technology in the teaching of calculus. Each project is designed to provide the basis for an outside class assignment that will engage students for a period of several days (or perhaps longer). The Derive, Maple, *Mathematica,* and MATLAB manuals are accompanied by command-specific diskettes that will relieve students of much of the burden of typing (by providing "templates" for the principal commands used in each project). In many cases, the diskettes also contain additional discussion and examples.

Acknowledgments

All experienced textbook authors know the value of critical reviewing during the preparation and revision of a manuscript. In our work on various editions of this book, we have benefited greatly from the advice (and frequently the consent) of the following exceptionally able reviewers:

Leon E. Arnold, Delaware County Community College
H. L. Bentley, University of Toledo
Michael L. Berry, West Virginia Wesleyan College
William Blair, Northern Illinois University
George Cain, Georgia Institute of Technology
Wil Clarke, Atlantic Union College
Peter Colwell, Iowa State University
James W. Daniel, University of Texas at Austin
Robert Devaney, Boston University
Dan Drucker, Wayne State University
William B. Francis, Michigan Technological University
Dianne H. Haber, Westfield State College
John C. Higgins, Brigham Young University
W. Cary Huffman, Loyola University of Chicago
Calvin Jongsma, Dordt College
Louise E. Knouse, Le Tourneau College
Morris Kalka, Tulane University
Catherine Lilly, Westfield State College
Joyce Longman, Villanova University
E. D. McCune, Stephen F. Austin State University
Arthur L. Moser, Illinois Central College
Barbara Moses, Bowling Green University
Barbara L. Osofsky, Rutgers University at New Brunswick
John Petro, Western Michigan University
Wayne B. Powell, Oklahoma State University
James P. Qualey, Jr., University of Colorado
Thomas Roe, South Dakota State University
Lawrence Runyan, Shoreline Community College
William L. Siegmann, Rensselaer Polytechnic Institute
John Spellman, Southwest Texas State University
Virginia Taylor, University of Lowell
Samuel A. Truitt, Jr., Middle Tennessee State University
Robert Urbanski, Middlesex County College
Robert Whiting, Villanova University
Cathleen M. Zucco, LeMoyne College

Many of the best improvements that have been made must be credited to colleagues and users of the first three editions throughout the United States, Canada, and abroad. We are grateful to all those, especially students, who have written to us, and hope they will continue to do so. We thank Terri Bittner of Laurel Tutoring (San Carlos, CA) who with her staff checked the accuracy of every example solution and odd-numbered answer. We also believe that the quality of the finished book itself is adequate testimony of the skill, diligence, and talent of an exceptional staff at Prentice-Hall; we owe special thanks to George Lobell, mathematics editor; Karen Karlin, development editor, Ed Thomas, production editor; Andrew Zutis, designer; and Network Graphics who did the illustrations. Finally, we again are unable to thank Alice Fitzgerald Edwards and Carol Wilson Penney adequately for their continued assistance, encouragement, support, and patience.

C. H. E., Jr.
D. E. P.

Athens, Georgia

Functions and Graphs

René Descartes (1596–1650)

❑ The seventeenth-century French scholar René Descartes is perhaps better remembered today as a philosopher than as a mathematician. But most of us are familiar with the "Cartesian plane" in which the location of a point P is specified by its coordinates (x, y).

❑ As a schoolboy Descartes was often permitted to sleep late because of allegedly poor health. He claimed he always thought most clearly about philosophy, science, and mathematics while he was lying comfortably in bed on cold mornings. After graduating from college, where he studied law (apparently without much enthusiasm), Descartes traveled with various armies for a number of years, but more as a gentleman soldier than as a professional military man.

❑ After finally settling down (in Holland), Descartes published in 1637 his famous philosophical treatise *Discourse on the Method* (of Reasoning Well and Seeking Truth in the Sciences). One of three appendices to this work set forth his new "analytic" approach to geometry. His principal idea (set forth almost simultaneously by his countryman Pierre de Fermat) was the correspondence between an *equation* and its *graph*, generally a curve in the plane. The equation could be used to study the curve, or vice versa.

❑ Suppose that we want to solve the equation $f(x) = 0$. Its solutions are the intersection points of the graph $y = f(x)$ with the x-axis, so an accurate picture of the curve shows the number and approximate locations of the solutions of the equation. For instance, the graph

$$y = x^3 - 3x^2 + 1$$

has three x-intercepts, showing that the equation

$$x^3 - 3x^2 + 1 = 0$$

has three real solutions—one between -1 and 0, one between 0 and 1, and one between 2 and 3. A modern graphics calculator or computer graphing program can approximate these solutions more accurately by magnifying the regions in which they are located. For instance, the magnified center region shows that the corresponding solution is $x \approx 0.65$.

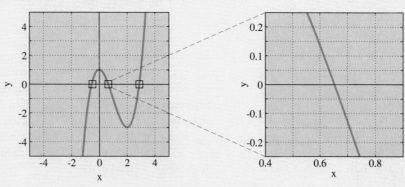

The graph $y = x^3 - 3x^2 + 1$

1.1

Functions and Real Numbers

Calculus is one of the supreme accomplishments of the human intellect. This mathematical discipline stems largely from the seventeenth-century investigations of Isaac Newton (1642–1727) and Gottfried Wilhelm Leibniz (1646–1716). Yet some of its ideas date back to the time of Archimedes (287–212 B.C.) and originated in cultures as diverse as those of Greece, Egypt, Babylonia, India, China, and Japan. Many of the scientific discoveries that have shaped our civilization during the past three centuries would have been impossible without the use of calculus.

The principal objective of calculus is the analysis of problems of change and motion. These problems are fundamental because we live in a world of ceaseless change, filled with bodies in motion and with phenomena of ebb and flow. Consequently, calculus remains a vibrant subject, and today this body of computational technique continues to serve as the principal quantitative language of science and technology.

Much of calculus involves the use of real numbers or variables to describe changing quantities and the use of functions to describe relationships between different variables. In this initial section we first review briefly the notation and terminology of real numbers and then discuss functions in more detail.

REAL NUMBERS

The **real numbers** are already familiar to you. They are just those numbers ordinarily used in most measurements. The mass, speed, temperature, and charge of a body are measured by real numbers. Real numbers can be represented by **terminating** or by **nonterminating** decimal expansions; in fact, every real number has a nonterminating decimal expansion because a terminating expansion can be followed by infinitely many zeros:

$$\tfrac{3}{8} = 0.375 = 0.375000000\ldots.$$

Any **repeating** decimal, such as

$$\tfrac{7}{22} = 0.31818\,181818\,8\ldots,$$

represents a **rational** number, one that is the ratio of two integers. Conversely, every rational number is represented by a repeating decimal expansion like the ones displayed here. The decimal expansion of an **irrational** number (a number that is not rational), such as

$$\sqrt{2} = 1.41421\,3562\ldots \quad \text{or} \quad \pi = 3.14159\,26535\,89793\ldots,$$

is both nonterminating and nonrepeating.

The geometric interpretation of real numbers as points on the **real line** (or *real number line*) **R** should also be familiar to you. Each real number is represented by precisely one point of **R**, and each point of **R** represents precisely one real number. By convention, the positive numbers lie to the right of zero and the negative numbers to the left, as in Fig. 1.1.1.

The following properties of inequalities of real numbers are fundamental and are often used:

Fig. 1.1.1 The real line **R**

If $a < b$ and $b < c$, then $a < c$.

If $a < b$, then $a + c < b + c$.

If $a < b$ and $c > 0$, then $ac < bc$.

$\qquad(1)$

If $a < b$ and $c < 0$, then $ac > bc$.

The last two statements mean that an inequality is preserved when its members are multiplied by a *positive* number but is *reversed* when they are multiplied by a *negative* number.

ABSOLUTE VALUE

The (nonnegative) distance along the real line between zero and the real number a is the **absolute value** of a, written $|a|$. Equivalently,

$$|a| = \begin{cases} a & \text{if } a \geqq 0; \\ -a & \text{if } a < 0. \end{cases} \qquad(2)$$

The notation $a \geqq 0$ means that a is *either* greater than zero *or* equal to zero. Equation (2) implies that $|a| \geqq 0$ for every real number a and that $|a| = 0$ if and only if $a = 0$.

Fig. 1.1.2 The absolute value of a real number is simply its distance from zero (Example 1).

EXAMPLE 1 As Fig. 1.1.2 shows,

$$|4| = 4 \quad \text{and} \quad |-3| = 3.$$

Moreover, $|0| = 0$ and $|\sqrt{2} - 2| = 2 - \sqrt{2}$, the latter being true because $2 > \sqrt{2}$. Thus, $\sqrt{2} - 2 < 0$, and hence

$$|\sqrt{2} - 2| = -(\sqrt{2} - 2) = 2 - \sqrt{2}.$$

The following properties of absolute values are frequently used:

$$|a| = |-a| = \sqrt{a^2} \geqq 0,$$
$$|ab| = |a||b|,$$
$$-|a| \leqq a \leqq |a|, \quad \text{and} \qquad(3)$$
$$|a| < b \quad \text{if and only if} \quad -b < a < b.$$

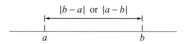

Fig. 1.1.3 The distance between a and b

The **distance** between the real numbers a and b is defined to be $|a - b|$ (or $|b - a|$; there's no difference). This distance is simply the length of the line segment of the real line $\boldsymbol{R}$ with endpoints a and b (Fig. 1.1.3).

The properties of inequalities and of absolute values in Eqs. (1) through (3) imply the following important theorem.

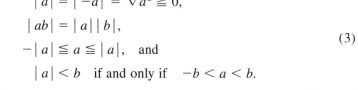

Triangle Inequality
For all real numbers a and b,

$$|a + b| \leqq |a| + |b|. \qquad(4)$$

Proof There are several cases to consider, depending upon whether the numbers a and b are positive or negative and which has the larger absolute value. If both are positive, then so is $a + b$; hence

$$|a + b| = a + b = |a| + |b|. \tag{5}$$

If $a > 0$ but $b < 0$ with $|b| < |a|$, then

$$0 < a + b < a,$$

so

$$|a + b| = a + b < a = |a| < |a| + |b|, \tag{6}$$

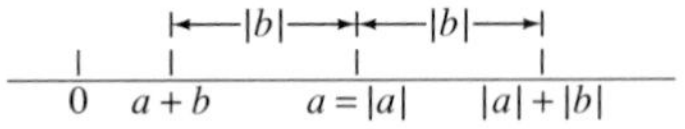

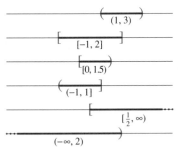

Fig. 1.1.4 The triangle inequality with $a > 0$, $b < 0$, and $|b| < |a|$

as illustrated in Fig. 1.1.4. The other cases are similar. In particular, we see that the triangle inequality is actually an equality [as in Eq. (5)] unless a and b have different signs, in which case it is a strict inequality [as in Eq. (6)]. ❑

INTERVALS

Suppose that S is a set (collection) of real numbers. It is common to describe S by the notation

$$S = \{x : \text{condition}\},$$

where the "condition" is true for those numbers x in S and false for those numbers x not in S. The most important sets of real numbers in calculus are **intervals**. If $a < b$, then the **open interval** (a, b) is defined to be the set

$$(a, b) = \{x : a < x < b\}$$

of real numbers, and the **closed interval** $[a, b]$ is

$$[a, b] = \{x : a \leq x \leq b\}.$$

Thus a closed interval contains its endpoints, whereas an open interval does not. We also use the **half-open intervals**

$$[a, b) = \{x : a \leq x < b\} \quad \text{and} \quad (a, b] = \{x : a < x \leq b\}.$$

Thus the open interval $(1, 3)$ is the set of those real numbers x such that $1 < x < 3$, the closed interval $[-1, 2]$ is the set of those real numbers x such that $-1 \leq x \leq 2$, and the half-open interval $(-1, 2]$ is the set of those real numbers x such that $-1 < x \leq 2$. In Fig. 1.1.5 we show examples of such intervals as well as some **unbounded** intervals, which have forms such as

$$[a, \infty) = \{x : x \geq a\},$$

$$(-\infty, a] = \{x : x \leq a\},$$

$$(a, \infty) = \{x : x > a\}, \quad \text{and}$$

$$(-\infty, a) = \{x : x < a\}.$$

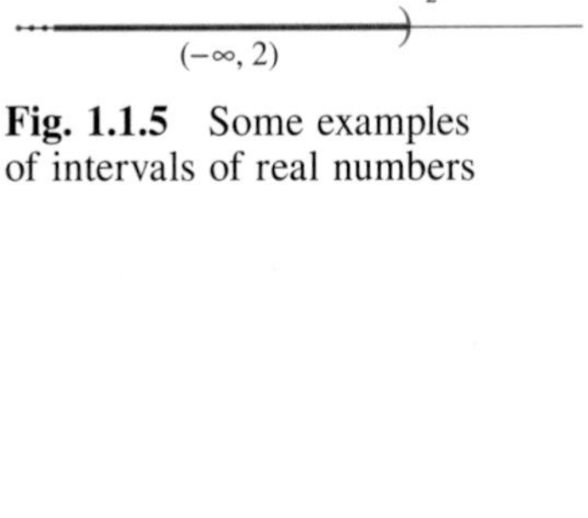

Fig. 1.1.5 Some examples of intervals of real numbers

The symbol ∞, denoting infinity, is merely a notational convenience and does *not* represent a real number—the real line $\boldsymbol{R}$ does *not* have "endpoints at infinity." The use of this symbol is motivated by the brief and natural descriptions $[\pi, \infty)$ and $(-\infty, 2)$ for the sets

$$\{x : x \geq \pi\} \quad \text{and} \quad \{x : x < 2\}$$

of all real numbers x such that $x \geq \pi$ and $x < 2$, respectively.

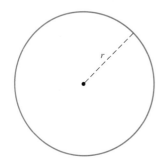

Fig. 1.1.6 Circle: area $A = \pi r^2$, circumference $C = 2\pi r$

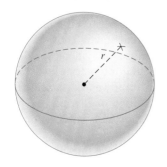

Fig. 1.1.7 Sphere: volume $V = \frac{4}{3}\pi r^3$, surface area $S = 4\pi r^2$

FUNCTIONS

The key to the mathematical analysis of a geometric or scientific situation is typically the recognition of relationships among the variables that describe the situation. Such a relationship may be a formula that expresses one variable as a *function* of another. For example, the area A of a circle of radius r is given by $A = \pi r^2$ (Fig. 1.1.6). The volume V and surface area S of a sphere of radius r are given by

$$V = \tfrac{4}{3}\pi r^3 \quad \text{and} \quad S = 4\pi r^2,$$

respectively (Fig. 1.1.7). After t seconds (s) a body that has been dropped from rest has fallen a distance

$$s = \tfrac{1}{2}gt^2$$

feet (ft) and has speed $v = gt$ feet per second (ft/s), where $g \approx 32$ ft/s² is gravitational acceleration. The volume V (in liters, L) of 3 grams (g) of carbon dioxide (CO_2) at 27°C is given in terms of its pressure p in atmospheres (atm) by $V = 1.68/p$. These are all examples of real-valued functions of a real variable.

Definition of Function
A real-valued **function** f defined on a set D of real numbers is a rule that assigns to each number x in D exactly one real number, denoted by $f(x)$.

The set D of all numbers for which $f(x)$ is defined is called the **domain** or **domain of definition** of the function f. The number $f(x)$, read "f of x," is called the **value** of f at the number (or point) x. The set of all values $y = f(x)$ is called the **range** of f. That is, the range of f is the set

$$\{y : y = f(x) \quad \text{for some } x \text{ in } D\}.$$

In this section we will be concerned more with the domain of a function than with its range.

Often a function is described by means of a formula that specifies how to compute the number $f(x)$ in terms of the number x. The symbol $f(\ )$ may be regarded as an operation that is to be performed whenever a number or expression is inserted between the parentheses.

EXAMPLE 2 The formula

$$f(x) = x^2 + x - 3 \tag{7}$$

is the rule of a function f whose domain is the entire real line $\mathbf{R}$. Some typical values of f are $f(-2) = -1, f(0) = -3$, and $f(3) = 9$. Some other values of the function f are

$$f(4) = (4)^2 + 4 - 3 = 17,$$

$$f(c) = c^2 + c - 3,$$

$$f(2 + h) = (2 + h)^2 + (2 + h) - 3$$

$$= (4 + 4h + h^2) + (2 + h) - 3 = 3 + 5h + h^2, \quad \text{and}$$

$$f(-t^2) = (-t^2)^2 + (-t^2) - 3 = t^4 - t^2 - 3.$$

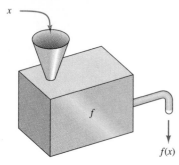

x

f

$f(x)$

Fig. 1.1.8 A "function machine"

When we describe the function f by writing a formula $y = f(x)$, we call x the **independent variable** and y the **dependent variable** because the value of y depends—through f—upon the choice of x. As x changes, or varies, then so does y. The way that y varies with x is determined by the rule of the function f. For example, if f is the function of Eq. (7), then $y = -1$ when $x = -2$, $y = -3$ when $x = 0$, and $y = 9$ when $x = 3$.

You may find it useful to visualize the dependence of the value $y = f(x)$ on x by thinking of a function as a kind of machine that accepts as input a number x and then produces as output the number $f(x)$, perhaps displayed or printed (Fig. 1.1.8).

One such machine is the familiar pocket calculator with a square root key. When a nonnegative number x is entered and this key is pressed, the calculator displays (an approximation to) the number $\sqrt{x}$. Note that the domain of this *square root function* $f(x) = \sqrt{x}$ is the set of all nonnegative real numbers, because no negative number has a real square root. Its range is also the set of all nonnegative real numbers, because the symbol $\sqrt{x}$ always denotes the *nonnegative* square root of x. The calculator illustrates its knowledge of the domain by displaying an adverse reaction if we ask it to calculate the square root of a negative number (unless it's one of the more sophisticated calculators, such as the TI-85 or HP-48S, which handle complex numbers).

Not every function has a rule expressible as a simple one-part formula such as $f(x) = \sqrt{x}$. For example, if we write

$$f(x) = \begin{cases} x^2 & \text{if } x \geqq 0, \\ -x & \text{if } x < 0, \end{cases}$$

then we have defined a perfectly good function with domain $\boldsymbol{R}$. Some of its values are $f(-3) = 3$, $f(0) = 0$, and $f(2) = 4$. The function in Example 3 is defined initially by means of a verbal description rather than by means of formulas.

EXAMPLE 3 For each real number x, let $f(x)$ denote the greatest integer that is less than or equal to x. For instance, $f(2.5) = 2$, $f(0) = 0$, $f(-3.5) = -4$, and $f(\pi) = 3$. If n is an integer, then $f(x) = n$ for every number x in the half-open interval $[n, n + 1)$. This function f is called the **greatest integer function** and is often denoted by

$$f(x) = [\![x]\!]. \tag{8}$$

Thus $[\![2.5]\!] = 2$, $[\![0]\!] = 0$, $[\![-3.5]\!] = -4$, and $[\![\pi]\!] = 3$. Note that although $[\![x]\!]$ is defined for all real x, the range of the greatest integer function consists of the set of integers only.

What is the domain of a function supposed to be when it is not specified? This is a common situation and occurs when we give a function f by writing only its formula $y = f(x)$. If no domain is given, we agree for convenience that the domain D is the set of all real numbers x for which the expression $f(x)$ produces a real number. For example, the domain of $f(x) = 1/x$ is the set of all nonzero real numbers (because $1/x$ is defined exactly when $x \neq 0$).

Ch. 1 / Functions and Graphs

EXAMPLE 4 Find the domain of the function g with formula

$$g(x) = \frac{1}{\sqrt{2x + 4}}.$$

Solution For the square root $\sqrt{2x + 4}$ to be defined, it is necessary that $2x + 4 \geq 0$. This holds when $2x \geq -4$ and thus when $x \geq -2$. For the reciprocal $1/\sqrt{2x + 4}$ to be defined, we also require that $\sqrt{2x + 4} \neq 0$ and thus that $x \neq -2$. Hence the domain of g is the interval $D = (-2, \infty)$.

FUNCTIONS AND APPLICATIONS

The investigation of an applied problem often hinges on the definition of a function that captures the essence of a geometrical or physical situation. Examples 5 and 6 illustrate this process.

EXAMPLE 5 A rectangular box with a square base has volume 125. Express its total surface area A as a function of the edge length x of its base.

Solution The first step is to draw a sketch and to label the relevant dimensions. Figure 1.1.9 shows a rectangular box with square base of edge length x and with height y. We are given that the volume of the box is

$$V = x^2y = 125. \tag{9}$$

Both the top and bottom of the box have area x^2, and each of its four vertical sides has area xy, so its total surface area is

$$A = 2x^2 + 4xy. \tag{10}$$

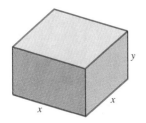

Fig. 1.1.9 The box of Example 5

But this is a formula for A in terms of the *two* variables x and y rather than a function of the *single* variable x. To eliminate y and thereby obtain A in terms of x alone, we solve Eq. (9) for $y = 125/x^2$ and then substitute this result in Eq. (10) to obtain

$$A = 2x^2 + 4x \cdot \frac{125}{x^2} = 2x^2 + \frac{500}{x}.$$

Thus the surface area is given as a function of the edge length x by

$$A(x) = 2x^2 + \frac{500}{x}, \qquad 0 < x < \infty. \tag{11}$$

It is necessary to specify the domain, because negative values of x make sense in the *formula* in Eq. (11) but do not belong in the domain of the *function A*. Because every $x > 0$ determines such a box, the domain does, in fact, include all positive numbers.

COMMENT In Example 5 our goal was to express the dependent variable A as a *function* of the independent variable x. Initially, the geometric situation provided us instead with

1. The *formula* in Eq. (10) expressing A in terms of both x and the additional variable y, and

2. The *relation* in Eq. (9) between x and y, which we used to eliminate y and thereby obtain A as a function of x alone.

We will see that this is a common pattern in many different applied problems, such as the one that follows.

THE ANIMAL PEN PROBLEM   You must build a rectangular holding pen for animals. To save material, you will use an existing wall as one of its four sides. The fence for the other three sides costs \$5/ft, and you must spend \$1/ft to paint the portion of the wall that forms the fourth side of the pen. If you have a total of \$180 to spend, what dimensions will maximize the area of the pen you can build?

Figure 1.1.10 shows the animal pen and its dimensions x and y, along with the cost per foot of each of its four sides. When we are confronted with a verbally stated applied problem such as this, our first question is, How on earth do we get started on it? The function concept is the key to getting a handle on such a situation. If we can express the quantity to be maximized—the dependent variable—as a function of some independent variable, then we have something tangible to do: Find the maximum value attained by this function.

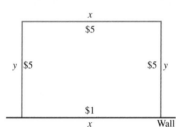

Fig. 1.1.10   The animal pen

EXAMPLE 6   In connection with the animal pen problem, express the area A of the pen as a function of the length x of its wall side.

Solution   The area A of the rectangular pen of length x and width y is

$$A = xy. \tag{12}$$

When we multiply the length of each side in Fig. 1.1.10 by its cost per foot and then add the results, we find that the total cost C of the pen is given by

$$C = x + 5y + 5x + 5y.$$

So

$$6x + 10y = 180, \tag{13}$$

because we are given $C = 180$. Choosing x to be the independent variable, we use the relation in Eq. (13) to eliminate the additional variable y from the area formula in Eq. (12). We solve Eq. (13) for y and substitute the result

$$y = \tfrac{1}{10}(180 - 6x) = \tfrac{3}{5}(30 - x) \tag{14}$$

in Eq. (12). Thus we obtain the desired function

$$A(x) = \tfrac{3}{5}(30x - x^2)$$

that expresses the area A as a function of the length x.

In addition to this formula for the function A, we must also specify its domain. Only if $x > 0$ will actual rectangles be produced, but we find it convenient to include the value $x = 0$ as well. This value of x corresponds to a "degenerate rectangle" of base zero and height

$$y = \tfrac{3}{5} \cdot 30 = 18,$$

x	$A(x)$
0	0
5	75
10	120
15	135 ←
20	120
25	75
30	0

Fig. 1.1.11 Area $A(x)$ of a pen with side of length x

x	$A(x)$
10	120
11	125.4
12	129.6
13	132.6
14	134.4
15	135 ←
16	134.4
17	132.6
18	129.6
19	125.4
20	120

Fig. 1.1.12 Further indication that $x = 15$ yields maximal area $A = 135$

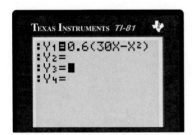

Fig. 1.1.13 A calculator programmed to evaluate $A(x) = 0.6 \cdot (30x - x^2)$

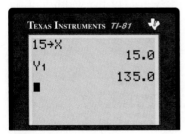

Fig. 1.1.14 Calculation of the apparent maximum value $A(15) = 135$

a consequence of Eq. (14). For similar reasons, we have the restriction $y \geqq 0$. Because

$$y = \tfrac{3}{5}(30 - x),$$

it follows that $x \leqq 30$. Thus the complete definition of the area function is

$$A(x) = \tfrac{3}{5}(30x - x^2), \qquad 0 \leqq x \leqq 30. \tag{15}$$

Example 6 illustrates an important part of the solution of a typical problem involving applications of mathematics. The domain of a function is a necessary part of its definition, and for each function we must specify the domain of values of the independent variable. In applications, we use the values of the independent variable that are relevant to the problem at hand.

NUMERICAL INVESTIGATION

Armed with the result of Example 6, we might attack the animal pen problem by calculating a table of values of the area function $A(x)$ in Eq. (15). Such a table is shown in Fig. 1.1.11. The data in this table suggest strongly that the maximum area is $A = 135$ ft^2, attained with side length $x = 15$ ft, in which case Eq. (14) yields $y = 9$ ft. This conjecture appears to be corroborated by the more refined data shown in Fig. 1.1.12.

Thus it seems that the animal pen with maximal area (costing \$180) is $x = 15$ ft long and $y = 9$ ft wide. The tables in Figs. 1.1.11 and 1.1.12 show only *integral* values of x, however, and the possibility remains that the length x of the pen of maximal area is *not* an integer. Consequently, numerical tables alone do not settle the matter. A new mathematical idea is needed in order to *prove* that $A(15) = 135$ is the maximum value of

$$A(x) = \tfrac{3}{5}(30x - x^2), \qquad 0 \leqq x \leqq 30$$

for *all* x in its domain. We attack this problem again in Section 1.3 after a review of rectangular coordinates in Section 1.2.

NOTE Many scientific calculators allow the user to program a given function for repeated evaluation and thereby to compute painlessly tables like those in Figs. 1.1.11 and 1.1.12. For example, Fig. 1.1.13 shows the display of a TI–81 calculator programmed to evaluate the dependent variable

$$\mathrm{Y1} = A(x) = (0.6)(30x - x^2)$$

when the desired value $\mathrm{x} = x$ of the independent variable is stored and the name of the dependent variable is entered (Fig. 1.1.14).

TABULATION OF FUNCTIONS

The use of a calculator to tabulate values of a function (as in Figs. 1.1.11 and 1.1.12) is a simple technique with surprisingly many applications. Here we illustrate a method of solving an equation of the form $f(x) = 0$ by *repeated tabulation* of values $f(x)$ of the function f.

To give a specific example, suppose that we ask what value of x in Eq. (15) yields an animal pen of area $A = 100$. Then we need to solve the equation

$$A(x) = \tfrac{3}{5}(30x - x^2) = 100,$$

which is equivalent to the equation

$$f(x) = \tfrac{3}{5}(30x - x^2) - 100 = 0. \qquad (16)$$

This is a quadratic equation that could be solved using the quadratic formula of basic algebra, but we want to take a more direct, numerical approach. The reason is that the numerical approach is applicable even when no simple formula (such as the quadratic formula) is available.

The data in Fig. 1.1.11 suggest that one value of x for which $A(x) = 100$ lies somewhere between $x = 5$ and $x = 10$ and that a second such value lies between $x = 20$ and $x = 25$. Indeed, substitution in Eq. (16) yields

$$f(5) = -25 < 0 \quad \text{and} \quad f(10) = 20 > 0.$$

The fact that $f(x)$ is *negative* at one endpoint of the interval $[5, 10]$ but *positive* at the other endpoint suggests that $f(x)$ is *zero* somewhere between $x = 5$ and $x = 10$.

To see *where,* we tabulate values of $f(x)$ on $[5, 10]$. In the table of Fig. 1.1.15 we see that

$$f(7) = -3.4 < 0 \quad \text{and} \quad f(8) = 5.6 > 0,$$

so we focus next on the interval $[7, 8]$.

Tabulation of $f(x)$ on $[7, 8]$ gives the table of Fig. 1.1.16, where we see that

$$f(7.3) = -0.574 < 0 \quad \text{and} \quad f(7.4) = 0.344 > 0.$$

We therefore tabulate $f(x)$ once more, this time on the interval $[7.3, 7.4]$. In Fig. 1.1.17 we see that

$$f(7.36) \approx -0.02 \quad \text{and} \quad f(7.37) \approx 0.07.$$

Because $f(7.36)$ is considerably closer to zero than is $f(7.37)$, we conclude that the desired solution of Eq. (16) is given approximately by $x \approx 7.36$, accurate to two decimal places. If greater accuracy were needed, we could continue to tabulate $f(x)$ on smaller and smaller intervals.

If we were to begin with the interval $[20, 25]$ and proceed similarly, we would find the second value $x \approx 22.64$ such that $f(x) = 0$. (You should do this for practice.)

Finally, let's calculate the corresponding values of the width y of the animal pen such that $A = xy = 100$:

❏ If $x \approx 7.36$, then $y \approx 13.59$.
❏ If $x \approx 22.64$, then $y \approx 4.42$.

Thus, under the cost constraint of the animal pen problem, we can construct either a 7.36-ft by 13.59-ft or a 22.64-ft by 4.42-ft rectangle, both of area 100 ft^2.

This *method of repeated tabulation* can be applied to a wide range of equations of the form $f(x) = 0$. If the interval $[a, b]$ contains a solution and

x	$f(x)$
5	-25.0
6	-13.6
7	-3.4
8	5.6
9	13.4
10	20.0

Fig. 1.1.15   Values of $f(x)$ on $[5, 10]$

x	$f(x)$
7.0	-3.400
7.1	-2.446
7.2	-1.504
7.3	-0.574
7.4	0.334
7.5	1.250
7.6	2.144
7.7	3.026
7.8	3.896
7.9	4.754
8.0	5.600

Fig. 1.1.16   Values of $f(x)$ on $[7, 8]$

x	$f(x)$
7.30	-0.5740
7.31	-0.4817
7.32	-0.3894
7.33	-0.2973
7.34	-0.2054
7.35	-0.1135
7.36	-0.0218
7.37	0.0699
7.38	0.1614
7.39	0.2527
7.40	0.3440

Fig. 1.1.17   Values of $f(x)$ on $[7.3, 7.4]$

the endpoint values $f(a)$ and $f(b)$ differ in sign, then we can approximate this solution by tabulating values on successively smaller subintervals. Problems 61 through 70 and the projects at the end of this section are applications of this concrete numerical method for the approximate solution of equations.

1.1 Problems

Simplify each expression in Problems 1 through 10 by writing it without using absolute value symbols.

1. $|3 - 17|$

2. $|-3| - |17|$

3. $|-0.25 - \frac{1}{4}|$

4. $|5| - |-7|$

5. $|(-5)(4 - 9)|$

6. $\dfrac{|-6|}{|4| + |-2|}$

7. $|(-3)^3|$

8. $|3 - \sqrt{3}|$

9. $|\pi - \frac{22}{7}|$

10. $-|7 - 4|$

In Problems 11 through 14, find and simplify each of the following values: (a) $f(-a)$; (b) $f(a^{-1})$; (c) $f(\sqrt{a})$; (d) $f(a^2)$.

11. $f(x) = \dfrac{1}{x}$

12. $f(x) = x^2 + 5$

13. $f(x) = \dfrac{1}{x^2 + 5}$

14. $f(x) = \sqrt{1 + x^2 + x^4}$

In Problems 15 through 20, find all values of a such that $g(a) = 5$.

15. $g(x) = 3x + 4$

16. $g(x) = \dfrac{1}{2x - 1}$

17. $g(x) = \sqrt{x^2 + 16}$

18. $g(x) = x^3 - 3$

19. $g(x) = \sqrt[3]{x + 25}$

20. $g(x) = 2x^2 - x + 4$

In Problems 21 through 26, compute and then simplify the quantity $f(a + h) - f(a)$.

21. $f(x) = 3x - 2$

22. $f(x) = 1 - 2x$

23. $f(x) = x^2$

24. $f(x) = x^2 + 2x$

25. $f(x) = \dfrac{1}{x}$

26. $f(x) = \dfrac{2}{x + 1}$

In Problems 27 through 30, find the range of values of the given function.

27. $f(x) = \begin{cases} \dfrac{x}{|x|} & \text{if } x \neq 0; \\ 0 & \text{if } x = 0 \end{cases}$

28. $f(x) = [\![3x]\!]$ (Recall that $[\![x]\!]$ is the largest integer not exceeding x.)

29. $f(x) = (-1)^{[\![x]\!]}$

30. $f(x)$ is the first-class postage (in cents) for a letter mailed in the United States and weighing x ounces,

$0 < x < 12$. In 1993 the postage rate for such a letter was 29¢ for the first ounce plus 23¢ for each additional ounce or fraction thereof.

In Problems 31 through 45, find the largest domain (of real numbers) on which the given formula determines a (real-valued) function.

31. $f(x) = 10 - x^2$

32. $f(x) = x^3 + 5$

33. $f(t) = \sqrt{t^2}$

34. $g(t) = (\sqrt{t})^2$

35. $f(x) = \sqrt{3x - 5}$

36. $g(t) = \sqrt[3]{t + 4}$

37. $f(t) = \sqrt{1 - 2t}$

38. $g(x) = \dfrac{1}{(x + 2)^2}$

39. $f(x) = \dfrac{2}{3 - x}$

40. $g(t) = \left(\dfrac{2}{3 - t}\right)^{1/2}$

41. $f(x) = \sqrt{x^2 + 9}$

42. $h(z) = \dfrac{1}{\sqrt{4 - z^2}}$

43. $f(x) = \sqrt{4 - \sqrt{x}}$

44. $f(x) = \sqrt{\dfrac{x + 1}{x - 1}}$

45. $g(t) = \dfrac{t}{|t|}$

46. Express the area A of a square as a function of its perimeter P.

47. Express the circumference C of a circle as a function of its area A.

48. Express the volume V of a sphere as a function of its surface area S.

49. Given: 0°C is the same as 32°F, and a temperature change of 1°C is the same as a change of 1.8°F. Express the Celsius temperature C as a function of the Fahrenheit temperature F.

50. Show that if a rectangle has base x and perimeter 100 (Fig. 1.1.18), then its area A is given by the function

$$A(x) = x(50 - x), \qquad 0 \leq x \leq 50.$$

Fig. 1.1.18 $A = xy$ (Problem 50)

51. A rectangle with base of length x is inscribed in a circle of radius 2 (Fig. 1.1.19). Express the area A of the rectangle as a function of x.

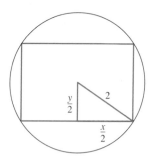

Fig. 1.1.19 $A = xy$ (Problem 51)

52. An oil field containing 20 wells has been producing 4000 barrels of oil daily. For each new well that is drilled, the daily production of each well decreases by 5 barrels. Write the total daily production of the oil field as a function of the number x of new wells drilled.

53. Suppose that a rectangular box has volume 324 cm³ and a square base of edge length x centimeters. The material for the base of the box costs 2¢/cm², and the material for its top and four sides costs 1¢/cm². Express the total cost of the box as a function of x. See Fig. 1.1.20.

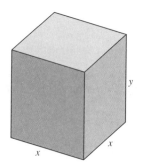

Fig. 1.1.20 $V = x^2 y$ (Problem 53)

54. A rectangle of fixed perimeter 36 is rotated about one of its sides S to generate a right circular cylinder. Express the volume V of this cylinder as a function of the length x of the side S. See Fig. 1.1.21.

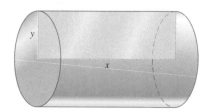

Fig. 1.1.21 $V = \pi x y^2$ (Problem 54)

55. A right circular cylinder has volume 1000 in.³, and the radius of its base is r inches. Express the total surface area A of the cylinder as a function of r. See Fig. 1.1.22.

Fig. 1.1.22 $V = \pi r^2 h$ (Problem 55)

56. A rectangular box has total surface area 600 cm² and a square base with edge length x centimeters. Express the volume V of the box as a function of x.

57. An open-topped box is to be made from a square piece of cardboard of edge length 50 in. First, four small squares, each of edge length x inches, are cut from the four corners of the cardboard (Fig. 1.1.23). Then the four resulting flaps are turned up—folded along the dotted lines—to form the four sides of the box, which will thus have a square base and a depth of x inches. Express its volume V as a function of x.

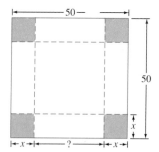

Fig. 1.1.23 Fold the edges up to make a box (Problem 57).

58. Continue Problem 50 by numerically investigating the area of a rectangle of perimeter 100. What dimensions (length and width) would appear to maximize the area of such a rectangle?

59. Determine numerically the number of new oil wells that should be drilled to maximize the total daily production of the oil field of Problem 52.

60. Investigate numerically the total surface area A of the rectangular box of Example 5. Assuming that both $x \geqq 1$ and $y \geqq 1$, what dimensions x and y would appear to minimize A?

In Problems 61 through 70, a quadratic equation $ax^2 + bx + c = 0$ and an interval $[p, q]$ containing one of its solutions are given. *Use the method of repeated tabulation to approximate this solution with two digits correct or correctly rounded to the right of the decimal. Check that your result agrees with one of the two solutions given by the quadratic formula,*

$$x = \frac{-b \pm \sqrt{b^2 - 4ac}}{2a}.$$

61. $x^2 - 3x + 1 = 0$, $[0, 1]$
62. $x^2 - 3x + 1 = 0$, $[2, 3]$

63. $x^2 + 2x - 4 = 0$, $[1, 2]$
64. $x^2 + 2x - 4 = 0$, $[-4, -3]$
65. $2x^2 - 7x + 4 = 0$, $[0, 1]$
66. $2x^2 - 7x + 4 = 0$, $[2, 3]$
67. $x^2 - 11x + 25 = 0$, $[3, 4]$
68. $x^2 - 11x + 25 = 0$, $[7, 8]$
69. $3x^2 + 23x - 45 = 0$, $[1, 2]$
70. $3x^2 + 23x - 45 = 0$, $[-10, -9]$

1.1 Projects

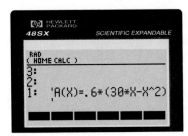

Fig. 1.1.24 A calculator prepared to define the function $A(x) = \frac{3}{5}(30x - x^2)$

In the projects described here, you are to apply the method of repeated tabulation using a scientific calculator or a computer. Figure 1.1.24 shows a HP–48 calculator prepared to define the function $A(x) = \frac{3}{5}(30x - x^2)$ as you press the [DEFINE] key. Then a value $A(x)$ can be computed simply by entering the desired value of x and pressing the [A] key.

Some computer systems have "one-liners" for tabulation of functions. The table in Fig. 1.1.25 lists commands for several common systems that can be used to tabulate values of the function $f(x)$ on the interval $[a, b]$ with subintervals of length h. The formula defining $f(x)$ and the numbers a, b, and h are entered as in the typical command

```
table(0.6*(30*x − x^2), x=7 to x=8 step 0.1)
```

to tabulate the function $f(x) = \frac{3}{5}(30x - x^2)$ on the interval $[7, 8]$.

Fig. 1.1.25 Commands for tabulating the function $f(x)$

BASIC	`FOR x = a to b STEP h : PRINT x,f(x) : NEXT`
Derive	`[VECTOR(x,x,a,b,h) , VECTOR(f(x) ,x,a,b,h)]`'
Maple	`for x from a by h to b do print (x,f(x)) od`
Mathematica	`MatrixForm[ Table[ {x,f[x]} , {x,a,b,h} ] ]`
(X)PLORE	`table( f(x), x = a to x = b step h )`

PROJECT A Suppose that you need to find the (positive) square root of 2 accurate to three decimal places, but your calculator has no square root key, only keys for the four arithmetic operations $+$, $-$, $\times$, and $\div$. Could you nevertheless approximate $\sqrt{2}$ accurately by using such a simple calculator?

You are simply looking for a number x such that $x^2 = 2$; that is, such that $x^2 - 2 = 0$. Thus $x = \sqrt{2}$ is a solution of the equation

$$f(x) = x^2 - 2 = 0. \qquad (17)$$

Even with your simple four-function calculator, you can easily calculate any desired value of $f(x)$: Merely multiply x by x and then subtract 2. Consequently, you can readily tabulate values of the function $f(x) = x^2 - 2$.

Apply the method of repeated tabulation to approximate $\sqrt{2}$ accurate to three decimal places. Use whatever calculator or computer is available to you, but do not use its square root function.

Similarly, you could apply the method of repeated tabulation to find three-place approximations to such roots as

❑ $\sqrt{17}$ as a solution of $x^2 = 17$,
❑ $\sqrt[3]{25}$ as a solution of $x^3 = 25$, or
❑ $\sqrt[5]{100}$ as a solution of $x^5 = 100$.

PROJECT B Figure 1.1.26 shows a 50-ft by 100-ft rectangular plot that you plan to enclose by a sidewalk of width x costing 25¢/ft². If you have $250 to pay for the sidewalk, determine the value of x accurate to 0.01 ft. First express the area of the sidewalk as the difference of the areas of the outer and inner rectangles in Fig. 1.1.26. Then show that

$$(2x + 100)(2x + 50) - 5000 = 1000. \tag{18}$$

Finally, approximate x by repeated tabulation.

For some more interesting possibilities, you might replace the rectangular plot with

❑ An L-shaped plot,
❑ A plot shaped like an isosceles right triangle, an equilateral triangle, or a 3-4-5 right triangle,
❑ A plot shaped like a Norman window (a rectangle beneath a semicircle), or
❑ A regular hexagon.

For comparable results, let the plot in each case have a perimeter of approximately 300 ft.

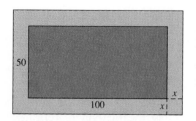

Fig. 1.1.26 Rectangular plot with sidewalk

1.2
The Coordinate Plane and Straight Lines

Imagine the flat, featureless, two-dimensional plane of Euclid's geometry. Install a copy of the real number line R, with the line horizontal and the positive numbers to the right. Add another copy of R perpendicular to the first, with the two lines crossing where zero is located on each. The vertical line should have the positive numbers above the horizontal line, as in Fig. 1.2.1; the negative numbers should be below it. The horizontal line is called the **x-axis**, and the vertical line is the **y-axis**.

With these added features, we call the plane the **coordinate plane,**

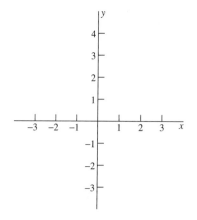

Fig. 1.2.1 The coordinate plane

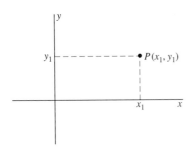

Fig. 1.2.2 The point P has rectangular coordinates (x_1, y_1).

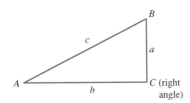

Fig. 1.2.3 The Pythagorean theorem

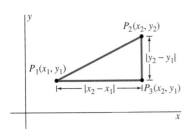

Fig. 1.2.4 Use this triangle to deduce the distance formula.

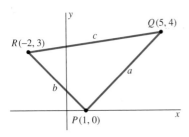

Fig. 1.2.5 Is this a right triangle (Example 1)?

because it's now possible to locate any point there by a pair of numbers called the *coordinates of the point.* Here's how: If P is a point in the plane, draw perpendiculars from P to the **coordinate axes**, as shown in Fig. 1.2.2. One perpendicular meets the x-axis at the **x-coordinate** (or **abscissa**) of P, labeled x_1 in Fig. 1.2.2. The other meets the y-axis in the **y-coordinate** (or **ordinate**) y_1 of P. The pair of numbers (x_1, y_1), in that order, is called the **coordinate pair** for P, or simply the **coordinates** of P. To be concise, we speak of "the point $P(x_1, y_1)$."

This coordinate system is called the **rectangular coordinate system**, or the **Cartesian coordinate system** (because its use in geometry was popularized, beginning in the 1630s, by the French mathematician and philosopher René Descartes [1596–1650]). The plane, thus coordinatized, is denoted by R^2 because two copies of R are used; it is known also as the **Cartesian plane**.

Rectangular coordinates are easy to use, because $P(x_1, y_1)$ and $Q(x_2, y_2)$ denote the same point if and only if $x_1 = x_2$ and $y_1 = y_2$. Thus when you know that P and Q are different points, you may conclude that P and Q have different abscissas, different ordinates, or both.

The point of symmetry $(0, 0)$ where the coordinate axes cross is called the **origin**. All points on the x-axis have coordinates of the form $(x, 0)$. Although the *real number* x is not the same as the geometric point $(x, 0)$, there are situations in which it is useful to think of the two as the same. Similar remarks apply to points $(0, y)$ on the y-axis.

The concept of distance in the coordinate plane is based on the **Pythagorean theorem**: If ABC is a right triangle with its right angle at the point C and with hypotenuse c, as in Fig. 1.2.3, then

$$c^2 = a^2 + b^2. \tag{1}$$

The converse of the Pythagorean theorem is also true: If the three sides of a given triangle satisfy the Pythagorean relation in Eq. (1), then the angle opposite side c must be a right angle.

The *distance* $d(P_1, P_2)$ between the points P_1 and P_2 is, by definition, the length of the straight-line segment joining P_1 and P_2. The following formula gives $d(P_1, P_2)$ in terms of the coordinates of the two points.

> **Distance Formula**
> The **distance** between the two points $P_1(x_1, y_1)$ and $P_2(x_2, y_2)$ is
> $$d(P_1, P_2) = \sqrt{(x_2 - x_1)^2 + (y_2 - y_1)^2}. \tag{2}$$

Proof   If $x_1 \neq x_2$ and $y_1 \neq y_2$, then the formula in Eq. (2) follows from the Pythagorean theorem. Use the right triangle with vertices P_1, P_2, and $P_3(x_2, y_1)$ shown in Fig. 1.2.4.

If $x_1 = x_2$, then P_1 and P_2 lie in a vertical line. In this case

$$d(P_1, P_2) = |y_2 - y_1| = \sqrt{(y_2 - y_1)^2}.$$

This agrees with the formula in Eq. (2) because $x_1 = x_2$. The remaining case, in which $y_1 = y_2$, is similar.   ❑

EXAMPLE 1   Show that the triangle PQR with vertices $P(1, 0)$, $Q(5, 4)$, and $R(-2, 3)$ is a right triangle (Fig. 1.2.5).

Solution The distance formula gives

$$a^2 = [d(P, Q)]^2 = (5 - 1)^2 + (4 - 0)^2 = 32,$$

$$b^2 = [d(P, R)]^2 = (-2 - 1)^2 + (3 - 0)^2 = 18, \quad \text{and}$$

$$c^2 = [d(Q, R)]^2 = (-2 - 5)^2 + (3 - 4)^2 = 50.$$

Because $a^2 + b^2 = c^2$, the *converse* of the Pythagorean theorem implies that RPQ is a right angle. The right angle is at P because P is the vertex opposite the longest side, QR.

Another application of the distance formula is an expression for the coordinates of the midpoint M of the line segment P_1P_2 with endpoints P_1 and P_2 (Fig. 1.2.6). Recall from geometry that M is the one (and only) point of the line segment P_1P_2 that is equally distant from P_1 and P_2. The following formula tells us that the coordinates of M are the *averages* of the corresponding coordinates of P_1 and P_2.

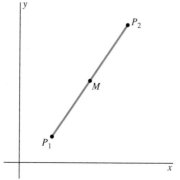

Fig. 1.2.6 The midpoint M

> **Midpoint Formula**
> The **midpoint** of the line segment with endpoints $P_1(x_1, y_1)$ and $P_2(x_2, y_2)$ is the point $M(\bar{x}, \bar{y})$ with coordinates
> $$\bar{x} = \tfrac{1}{2}(x_1 + x_2), \qquad \bar{y} = \tfrac{1}{2}(y_1 + y_2). \tag{3}$$

Proof If you substitute the coordinates of P_1, M, and P_2 in the distance formula, you find that $d(P_1, M) = d(P_2, M)$. All that remains is to show that M lies on the line segment P_1P_2. We ask you to do this, and thus complete this proof, in Problem 31. ❑

STRAIGHT LINES AND SLOPE

We want to define the *slope* of a straight line, a measure of its rate of rise or fall from left to right. Given a nonvertical line L in the coordinate plane, choose two points $P_1(x_1, y_1)$ and $P_2(x_2, y_2)$ on L. Consider the **increments** Δx and Δy (read "delta x" and "delta y") in the x- and y-coordinates from P_1 to P_2. These are defined to be

$$\Delta x = x_2 - x_1 \quad \text{and} \quad \Delta y = y_2 - y_1. \tag{4}$$

Engineers (and others) call Δx the **run** from P_1 to P_2 and Δy the **rise** from P_1 to P_2, as in Fig. 1.2.7. The **slope** m of the nonvertical line L is then the ratio of the rise to the run:

$$m = \frac{\Delta y}{\Delta x} = \frac{y_2 - y_1}{x_2 - x_1}. \tag{5}$$

This, too, is the definition in civil engineering and elsewhere (including calculus). In a surveying text you are likely to find the memory aid

$$\text{"slope} = \frac{\text{rise}}{\text{run}} \text{,"}$$

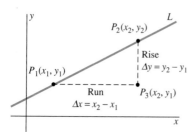

Fig. 1.2.7 The slope of a straight line

Ch. 1 / Functions and Graphs

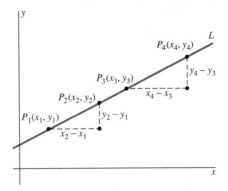

Fig. 1.2.8 The result of the slope computation does not depend on which two points of L are used.

Recall that corresponding sides of similar (that is, equal-angled) triangles have equal ratios. Hence, if $P_3(x_3, y_3)$ and $P_4(x_4, y_4)$ are two other points on L, then the similarity of the triangles in Fig. 1.2.8 implies that

$$\frac{y_4 - y_3}{x_4 - x_3} = \frac{y_2 - y_1}{x_2 - x_1}.$$

Therefore, the slope m as defined in Eq. (5) does *not* depend on the particular choice of P_1 and P_2.

If the line L is horizontal, then $\Delta y = 0$. In this case Eq. (5) gives $m = 0$. If L is vertical, then $\Delta x = 0$ and the slope of L is *not defined*. Thus we have the following statements:

Horizontal lines have slope zero.

Vertical lines have no defined slope.

EXAMPLE 2 (a) The slope of the line through the points $(3, -2)$ and $(-1, 4)$ is

$$m = \frac{4 - (-2)}{(-1) - 3} = \frac{6}{-4} = -\frac{3}{2}.$$

(b) The points $(3, -2)$ and $(7, -2)$ have the same y-coordinate. Therefore, the line through them is horizontal and thus has slope zero.

(c) The points $(3, -2)$ and $(3, 4)$ have the same x-coordinate. Thus the line through them is vertical, and so its slope is undefined.

EQUATIONS OF STRAIGHT LINES

Our immediate goal is to be able to write equations of given straight lines. That is, if L is a straight line in the coordinate plane, we wish to construct a mathematical sentence—an equation—about points (x, y) in the plane. We want this equation to be *true* when (x, y) is a point on L and *false* when (x, y) is not a point on L. Clearly this equation will involve x and y and some numerical constants determined by L itself. For us to write this equation, the concept of the slope of L is essential.

Suppose, then, that $P(x_0, y_0)$ is a fixed point on the nonvertical line L of slope m. Let $P(x, y)$ be any *other* point on L. We apply Eq. (5) with P and P_0 in place of P_1 and P_2 to find that

$$m = \frac{y - y_0}{x - x_0};$$

that is,

$$y - y_0 = m(x - x_0). \tag{6}$$

Because the point (x_0, y_0) satisfies Eq. (6), as does every other point of L, and because no other point of the plane can do so, Eq. (6) is indeed an equation for the given line L. In summary, we have the following result.

The Point-Slope Equation

The point $P(x, y)$ lies on the line with slope m through the fixed point (x_0, y_0) if and only if its coordinates satisfy the equation

$$y - y_0 = m(x - x_0). \tag{6}$$

Equation (6) is called the *point-slope* equation of L, partly because the coordinates of the point (x_0, y_0) and the slope m of L may be read directly from the equation.

EXAMPLE 3 Write an equation for the straight line L through the points $P_1(1, -1)$ and $P_2(3, 5)$.

Solution The slope m of L may be obtained from the two given points:

$$m = \frac{5 - (-1)}{3 - 1} = 3.$$

Either P_1 or P_2 will do for the fixed point. We use $P_1(1, -1)$. Then, with the aid of Eq. (6), the point-slope equation of L is

$$y + 1 = 3(x - 1).$$

If simplification is appropriate, we write $3x - y = 4$.

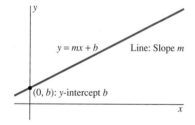

Fig. 1.2.9 The straight line with equation $y = mx + b$ has slope m and y-intercept b.

Equation (6) can be written in the form

$$y = mx + b, \tag{7}$$

where $b = y_0 - mx_0$ is a constant. Because $y = b$ when $x = 0$, the **y-intercept** of L is the point $(0, b)$ shown in Fig. 1.2.9. Equations (6) and (7) are different forms of the equation of a straight line.

The Slope-Intercept Equation

The point $P(x, y)$ lies on the line with slope m and y-intercept b if and only if the coordinates of P satisfy the equation

$$y = mx + b. \tag{7}$$

Perhaps you noticed that both Eq. (6) and Eq. (7) can be written in the form of the general linear equation

$$Ax + By = C, \tag{8}$$

where A, B, and C are constants. Conversely, if $B \neq 0$, then Eq. (8) can be written in the form of Eq. (7) if we divide each term by B. Therefore, Eq. (8) represents a straight line with its slope being the coefficient of x *after* solution of the equation for y. If $B = 0$, then Eq. (8) reduces to the equation of a vertical line: $x = K$ (where K is a constant). If $A = 0$, then Eq. (8) reduces to the equation of a horizontal line: $y = H$ (H a constant). Thus we see that Eq. (8) is always an equation of a straight line unless $A = B = 0$. Conversely, every straight line in the coordinate plane—even a vertical one—has an equation of the form in (8).

PARALLEL LINES AND PERPENDICULAR LINES

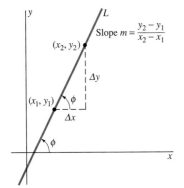

If the line L is not horizontal, it must cross the x-axis. Then its **angle of inclination** is the angle ϕ measured counterclockwise from the positive x-axis to L. It follows that $0° < \phi < 180°$ if ϕ is measured in degrees. Figure 1.2.10 makes it clear that this angle ϕ and the slope m of a nonvertical line are related by the equation

$$m = \frac{\Delta y}{\Delta x} = \tan \phi. \tag{9}$$

This is true because if ϕ is an acute angle in a right triangle, then $\tan \phi$ is the ratio of the leg opposite ϕ to the leg adjacent to ϕ.

Your intuition correctly assures you that two lines are parallel if and only if they have the same angle of inclination. So it follows from Eq. (9) that two parallel nonvertical lines have the same slope and that two lines with the same slope must be parallel. This completes the proof of Theorem 1.

Fig. 1.2.10 How is the angle of inclination ϕ related to the slope m?

> **Theorem 1   *Slopes of Parallel Lines***
> Two nonvertical lines are parallel if and only if they have the same slope.

Theorem 1 can also be proved without the use of the tangent function. The two lines shown in Fig. 1.2.11 are parallel if and only if the two right triangles are similar, which is equivalent to the slopes of the lines being equal.

EXAMPLE 4   Write an equation of the line L that passes through the point $P(3, -2)$ and is parallel to the line L' with the equation $x + 2y = 6$.

Solution   When we solve the equation of L' for y, we get $y = -\frac{1}{2}x + 3$. So L' has slope $m = -\frac{1}{2}$. Because L has the same slope, its point-slope equation is then

$$y + 2 = -\tfrac{1}{2}(x - 3),$$

or, if you prefer, $x + 2y = -1$.

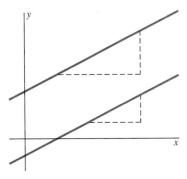

Fig. 1.2.11 Two parallel lines

> **Theorem 2   *Slopes of Perpendicular Lines***
> Two lines L_1 and L_2 with slopes m_1 and m_2, respectively, are perpendicular if and only if
> $$m_1 m_2 = -1. \tag{10}$$
> That is, the slope of each is the *negative reciprocal* of the slope of the other.

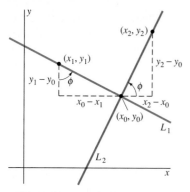

Fig. 1.2.12 Illustration of the proof of Theorem 2

Proof If the two lines L_1 and L_2 are perpendicular and the slope of each exists, then neither is horizontal or vertical. Thus the situation resembles that shown in Fig. 1.2.12, in which the two lines meet at the point (x_0, y_0). It is easy to see that the two right triangles are similar, so equality of ratios of corresponding sides yields

$$m_2 = \frac{y_2 - y_0}{x_2 - x_0} = \frac{x_0 - x_1}{y_1 - y_0} = -\frac{x_1 - x_0}{y_1 - y_0} = -\frac{1}{m_1}.$$

Thus Eq. (10) holds if the two lines are perpendicular. This argument can be reversed to prove the converse—that the lines are perpendicular if $m_1 m_2 = -1$. ☐

EXAMPLE 5 Write an equation of the line L through the point $P(3, -2)$ that is perpendicular to the line L' with the equation $x + 2y = 6$.

Solution As we saw in Example 4, the slope of L' is $m' = -\frac{1}{2}$. By Theorem 2, the slope of L is $m = -1/m' = 2$. Thus L has the point-slope equation

$$y + 2 = 2(x - 3);$$

equivalently, $2x - y = 8$.

You will find it helpful to remember that the *sign* of the slope m of the line L indicates whether L runs upward or downward as your eyes move from left to right. If $m > 0$, then the angle of inclination ϕ of L must be an acute angle, because $m = \tan \phi$. In this case, L "runs upward" to the right. If $m < 0$, then ϕ is obtuse, so L "runs downward." Figure 1.2.13 shows the geometry behind these observations.

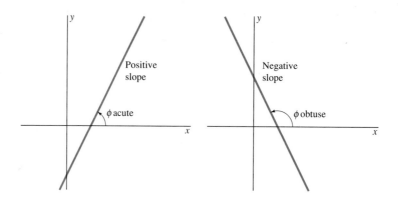

Fig. 1.2.13 Positive and negative slope; effect on ϕ

GRAPHICAL INVESTIGATION

Many mathematical problems require the simultaneous solution of a pair of linear equations of the form

$$\begin{align} a_1 x + b_1 y &= c_1 \\ a_2 x + b_2 y &= c_2. \end{align} \tag{11}$$

The graphs of these two equations are a pair of straight lines in the xy-plane. If these two lines are not parallel, then they must intersect at a single point P whose coordinates (x_0, y_0) constitute the solution of (11). That is, $x = x_0$ and

Ch. 1 / Functions and Graphs

$y = y_0$ are the (only) values of x and y for which both equations in (11) are true.

In elementary algebra you studied various elimination and substitution methods for solving linear systems such as the one in (11). Example 6 illustrates an alternative *graphical method* that is sometimes useful when a graphing utility—a graphics calculator or a computer with a graphing program—is available.

EXAMPLE 6 We want to investigate the simultaneous solution of the linear equations

$$10x - 8y = 17$$
$$15x + 18y = 67. \tag{12}$$

With many graphics calculators, it is necessary first to solve each equation for y:

$$y = (17 - 10x)/(-8)$$
$$y = (67 - 15x)/18. \tag{13}$$

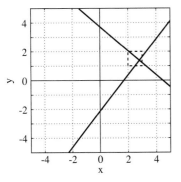

Fig. 1.2.14 A calculator prepared to graph the lines in Eq. (12) (Example 6)

Figure 1.2.14 shows a calculator prepared to graph the two lines represented by the equations in (12), and Fig. 1.2.15 shows the result in the *viewing window* $-5 \leq x \leq 5$, $-5 \leq y \leq 5$.

Before proceeding, note that in Fig. 1.2.15 the two lines *appear* to be perpendicular. But their slopes, $(-10)/(-8) = \frac{5}{4}$ and $(-15)/18 = -\frac{5}{6}$, are *not* negative reciprocals of one another. It follows from Theorem 2 that the two lines are *not* perpendicular.

Figures 1.2.16, 1.2.17, and 1.2.18 show successive magnifications produced by "zooming in" on the point of intersection of the two lines. The dashed-line box in each figure is the viewing window for the next figure. Looking at Fig. 1.2.18, we see that the intersection point is given by the approximations

$$x \approx 2.807, \quad y \approx 1.383, \tag{14}$$

rounded to three decimal places.

The result in (14) can be checked by equating the right-hand sides in (13) and solving for x. This gives $x = 421/150 \approx 2.8067$. Substitution of this value into either equation in (13) then yields $y = 83/60 \approx 1.3833$.

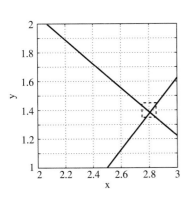

Fig. 1.2.15 $-5 \leq x \leq 5$, $-5 \leq y \leq 5$ (Example 6)

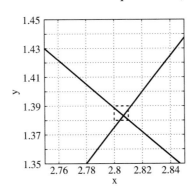

Fig. 1.2.16 $2 \leq x \leq 3$, $1 \leq y \leq 2$ (Example 6)

Fig. 1.2.17 $2.75 \leq x \leq 2.85$, $1.35 \leq y \leq 1.45$ (Example 6)

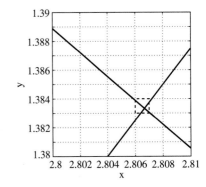

Fig. 1.2.18 $2.80 \leq x \leq 2.81$, $1.38 \leq y \leq 1.39$ (Example 6)

The graphical method illustrated by Example 6 typically produces approximate solutions that are sufficiently accurate for practical purposes. But the method is especially useful for *nonlinear* equations, for which exact algebraic techniques of solution may not be available.

1.2 Problems

Three points A , B, and C lie on a single straight line if and only if the slope of AB is equal to the slope of BC. In Problems 1 through 4, plot the three given points and then determine whether or not they lie on a single line.

1. $A(-1, -2)$, $B(2, 1)$, $C(4, 3)$
2. $A(-2, 5)$, $B(2, 3)$, $C(8, 0)$
3. $A(-1, 6)$, $B(1, 2)$, $C(4, -2)$
4. $A(-3, 2)$, $B(1, 6)$, $C(8, 14)$

In Problems 5 and 6, use the concept of slope to show that the four points given are the vertices of a parallelogram.

5. $A(-1, 3)$, $B(5, 0)$, $C(7, 4)$, $D(1, 7)$
6. $A(7, -1)$, $B(-2, 2)$, $C(1, 4)$, $D(10, 1)$

In Problems 7 and 8, show that the three given points are the vertices of a right triangle.

7. $A(-2, -1)$, $B(2, 7)$, $C(4, -4)$
8. $A(6, -1)$, $B(2, 3)$, $C(-3, -2)$

In Problems 9 through 13, find the slope m and y-intercept b of the line with the given equation. Then sketch the line.

9. $2x = 3y$
10. $x + y = 1$
11. $2x - y + 3 = 0$
12. $3x + 4y = 6$
13. $2x = 3 - 5y$

In Problems 14 through 23, write an equation of the straight line L described.

14. L is vertical and has x-intercept 7.
15. L is horizontal and passes through $(3, -5)$.
16. L has x-intercept 2 and y-intercept -3.
17. L passes through $(2, -3)$ and $(5, 3)$.
18. L passes through $(-1, -4)$ and has slope $\frac{1}{2}$.
19. L passes through $(4, 2)$ and has angle of inclination $135°$.
20. L has slope 6 and y-intercept 7.
21. L passes through $(1, 5)$ and is parallel to the line with equation $2x + y = 10$.
22. L passes through $(-2, 4)$ and is perpendicular to the line with equation $x + 2y = 17$.

23. L is the perpendicular bisector of the line segment that has endpoints $(-1, 2)$ and $(3, 10)$.

24. Find the perpendicular distance from the point $(2, 1)$ to the line with equation $y = x + 1$.

25. Find the perpendicular distance between the parallel lines $y = 5x + 1$ and $y = 5x + 9$.

26. The points $A(-1, 6)$, $B(0, 0)$, and $C(3, 1)$ are three consecutive vertices of a parallelogram. Find the fourth vertex. (What happens if the word *consecutive* is omitted?)

27. Prove that the diagonals of the parallelogram of Problem 26 bisect each other.

28. Show that the points $A(-1, 2)$, $B(3, -1)$, $C(6, 3)$, and $D(2, 6)$ are the vertices of a *rhombus*—a parallelogram with all sides of equal length. Then prove that the diagonals of this rhombus are perpendicular to each other.

29. The points $A(2, 1)$, $B(3, 5)$, and $C(7, 3)$ are the vertices of a triangle. Prove that the line joining the midpoints of AB and BC is parallel to AC.

30. A **median** of a triangle is a line joining a vertex to the midpoint of the opposite side. Prove that the three medians of the triangle of Problem 29 intersect in a single point.

31. Complete the proof of the midpoint formula in Eq. (3). It is necessary to show that the point M lies on the segment P_1P_2. One way to do this is to show that the slope of P_1M is equal to the slope of MP_2.

32. Let $P(x_0, y_0)$ be a point of the circle with center $C(0, 0)$ and radius r. Recall that the line tangent to the circle at P is perpendicular to the radius CP. Prove that the equation of this tangent line is $x_0x + y_0y = r^2$.

33. The Fahrenheit temperature F and the absolute temperature K satisfy a linear equation. Moreover, $K = 273.16$ when $F = 32$, and $K = 373.16$ when $F = 212$. Express K in terms of F. What is the value of F when $K = 0$?

34. The length L (in centimeters) of a copper rod is a linear function of its Celsius temperature C. If $L = 124.942$ when $C = 20$ and $L = 125.134$ when $C = 110$, express L in terms of C.

35. The owner of a grocery store finds that she can sell 980 gal of milk each week at \$1.69/gal and 1220 gal of milk each week at \$1.49/gal. Assume a linear relationship

between price and sales. How many gallons would she then expect to sell each week at $1.56/gal?

36. Figure 1.2.19 shows the graphs of the equations

$$17x - 10y = 57$$

$$25x - 15y = 17.$$

Are these two lines parallel? If not, find their point of intersection. If you have a graphing utility, find the solution by graphical approximation as well as by exact algebraic methods.

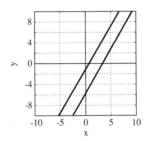

Fig. 1.2.19 The lines of Problem 36

In Problems 37 through 46, use a graphics calculator or computer to approximate graphically (with three digits to the right of the decimal correct or correctly rounded) the solution of the given linear system. Then check your approximate solution by solving the system by an exact algebraic method.

37. $2x + 3y = 5$
$2x + 5y = 12$

38. $6x + 4y = 5$
$8x - 6y = 13$

39. $3x + 3y = 17$
$3x + 5y = 16$

40. $2x + 3y = 17$
$2x + 5y = 20$

41. $4x + 3y = 17$
$5x + 5y = 21$

42. $4x + 3y = 15$
$5x + 5y = 29$

43. $5x + 6y = 16$
$7x + 10y = 29$

44. $5x + 11y = 21$
$4x + 10y = 19$

45. $6x + 6y = 31$
$9x + 11y = 37$

46. $7x + 6y = 31$
$11x + 11y = 47$

47. Justify the phrase "no other point of the plane can do so" that follows the first appearance of Eq. (6).

48. The discussion of the linear equation $Ax + By = C$ in Eq. (8) does not include a description of the graph of this equation if $A = B = 0$. What is the graph in this case?

1.3

Graphs of Equations and Functions

We saw in Section 1.2 that the points (x, y) satisfying the linear equation $Ax + By = C$ form a very simple set: a straight line (if A and B are not both zero). In contrast, the set of points (x, y) that satisfy the equation

$$x^4 - 4x^3 + 3x^2 + 2x^2y^2 = y^2 + 4xy^2 - y^4$$

is the exotic curve shown in Fig. 1.3.1—though this is certainly not obvious! But both the straight line and the complicated curve are examples of *graphs*.

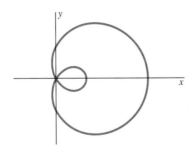

Fig. 1.3.1 The graph of the equation
$x^4 - 4x^3 + 3x^2 + 2x^2y^2$
$= y^2 + 4xy^2 - y^4$

> **Definition** *Graph of an Equation*
> The **graph** of an equation in two variables x and y is the set of all points (x, y) in the plane that satisfy the equation.

For example, the distance formula tells us that the graph of the equation

$$x^2 + y^2 = r^2 \tag{1}$$

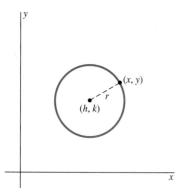

Fig. 1.3.2   A translated circle

is the circle of radius r centered at the origin $(0, 0)$. More generally, the graph of the equation

$$(x - h)^2 + (y - k)^2 = r^2 \tag{2}$$

is the circle of radius r with center (h, k). This follows from the distance formula, because the distance between the points (x, y) and (h, k) in Fig. 1.3.2 is r.

EXAMPLE 1   The equation of the circle with center $(3, 4)$ and radius 10 is

$$(x - 3)^2 + (y - 4)^2 = 100,$$

which may also be written in the form

$$x^2 + y^2 - 6x - 8y - 75 = 0.$$

We may regard the general circle, Eq. (2), as a *translate* of the origin-centered circle of Eq. (1): Each point of the former is obtained by translating (moving) each point of the xy-plane h units to the right and k units upward. (A negative value of h corresponds to a translation $|h|$ units to the left; a negative value of k means a downward translation.) You see that Eq. (2) of the translated circle with center (h, k) is obtained from Eq. (1) by replacing x with $x - h$ and y with $y - k$. We will see that this principle applies to arbitrary curves:

When the graph of an equation is translated h units to the right and k units upward, the equation of the translated curve is obtained from the original equation by replacement of x with $x - h$ and of y with $y - k$.

Observe that we can write the equation of a translated circle in Eq. (2) in the general form

$$x^2 + y^2 + ax + by = c. \tag{3}$$

What, then, can we do when we encounter an equation already of the form in Eq. (3)? We first recognize that it is an equation of a circle. Next, we can discover its center and radius by the technique of *completing the square*. To do so, we note that

$$x^2 + ax = \left(x + \frac{a}{2}\right)^2 - \frac{a^2}{4},$$

which shows that $x^2 + ax$ can be made into a perfect square by adding to it the square of *half* the coefficient of x.

EXAMPLE 2   Find the center and radius of the circle that has the equation

$$x^2 + y^2 - 4x + 6y = 12.$$

Solution   We complete the square separately for both variables x and y. This gives

$$(x^2 - 4x + 4) + (y^2 + 6y + 9) = 12 + 4 + 9;$$

$$(x - 2)^2 + (y + 3)^2 = 25.$$

Hence the circle—shown in Fig. 1.3.3—has center $(2, -3)$ and radius 5. Solution of the last equation for y gives

$$y = -3 \pm \sqrt{25 - (x - 2)^2}.$$

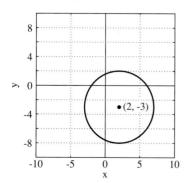

Fig. 1.3.3   The circle of Example 2

24

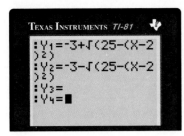

Fig. 1.3.4 A graphics calculator prepared to graph the circle of Example 2

Thus the circle consists of the graphs of the *two* functions

$$y_1(x) = -3 + \sqrt{25 - (x - 2)^2}$$

and

$$y_2(x) = -3 - \sqrt{25 - (x - 2)^2}$$

that describe its upper and lower semicircles. Figure 1.3.4 shows a graphics calculator prepared to graph this circle.

GRAPHS OF FUNCTIONS

The graph of a function is a special case of the graph of an equation.

> **Definition** *Graph of a Function*
> The **graph** of the function f is the graph of the equation $y = f(x)$.

Thus the graph of the function f is the set of all points in the plane that have the form $(x, f(x))$, where x is in the domain of f. Because the second coordinate of such a point is uniquely determined by its first coordinate, we obtain the following useful principle:

No *vertical* line can intersect the graph of a function in more than one point.

Alternatively,

Each vertical line through a point in the domain of a function meets its graph in exactly one point.

If you examine Fig. 1.3.1, you will see from these remarks that the graph shown there cannot be the graph of a *function*, although it *is* the graph of an equation.

EXAMPLE 3 Construct the graph of the absolute value function $f(x) = |x|$.

Solution Recall that

$$|x| = \begin{cases} x & \text{if } x \geqq 0; \\ -x & \text{if } x < 0. \end{cases}$$

So the graph of $y = |x|$ consists of the right half of the line $y = x$ together with the left half of the line $y = -x$, as shown in Fig. 1.3.5.

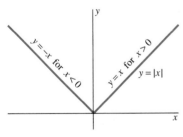

Fig. 1.3.5 The graph of the absolute value function $y = |x|$ of Example 3

EXAMPLE 4 Sketch the graph of the reciprocal function

$$f(x) = \frac{1}{x}.$$

Solution Let's examine four natural cases.

❑ When x is positive and numerically large, $f(x)$ is small and positive.
❑ When x is positive and near zero, $f(x)$ is large and positive.

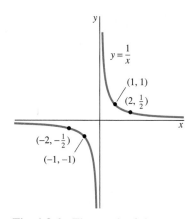

Fig. 1.3.6 The graph of the reciprocal function $y = 1/x$ of Example 4

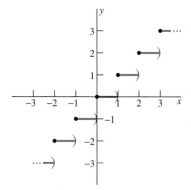

Fig. 1.3.7 The graph of the greatest integer ("staircase") function $f(x) = [\![x]\!]$ of Example 5

□ When x is negative and numerically small (negative and close to zero), $f(x)$ is large and negative.

□ When x is large and negative (x is negative but $|x|$ is large), $f(x)$ is small and negative (negative and close to zero).

To get started with the graph, we can plot a few points, such as $(1, 1)$, $(-1, -1)$, $(10, 0.1)$, $(0.1, 10)$, $(-10, -0.1)$, and $(-0.1, -10)$. The result of the information displayed here suggests that the actual graph is much like the one shown in Fig. 1.3.6.

Figure 1.3.6 exhibits a "gap," or "discontinuity," in the graph of $y = 1/x$ at $x = 0$. Indeed, the gap is called an *infinite discontinuity* because y increases without bound as x approaches zero from the right, whereas y decreases without bound as x approaches zero from the left. This phenomenon generally is signaled by the presence of denominators that are zero at certain values of x, as in the case of the functions

$$f(x) = \frac{1}{1 - x} \quad \text{and} \quad f(x) = \frac{1}{x^2},$$

which we ask you to graph in the problems.

EXAMPLE 5 Figure 1.3.7 shows the graph of the greatest integer function $f(x) = [\![x]\!]$ of Example 3 of Section 1.1. Note the "jumps" that occur at integral values of x. On calculators, the greatest integer function is typically denoted by $\boxed{\text{INT}}$; in some programming languages, it is FLOOR.

EXAMPLE 6 Graph the function with the formula

$$f(x) = x - [\![x]\!] - \tfrac{1}{2}.$$

Solution Recall that $[\![x]\!] = n$, where n is the greatest integer not exceeding x: $n \leqq x < n + 1$. Hence if n is an integer, then

$$f(n) = n - n - \tfrac{1}{2} = -\tfrac{1}{2}.$$

This implies that the point $(n, -\tfrac{1}{2})$ lies on the graph for each integer n. Next, if $n \leqq x < n + 1$ (where, again, n is an integer), then

$$f(x) = x - n - \tfrac{1}{2}.$$

Because $y = x - n - \tfrac{1}{2}$ has as its graph a straight line of slope 1, it follows that the graph of f takes the form shown in Fig. 1.3.8. This *sawtooth function* is another example of a discontinuous function. The values of x where the value of $f(x)$ makes a jump are called **points of discontinuity** of the function f. Thus the points of discontinuity of the sawtooth function are the integers. As x approaches the integer n from the left, the value of $f(x)$ approaches $+\tfrac{1}{2}$,

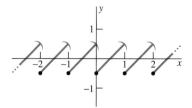

Fig. 1.3.8 The graph of the sawtooth function $f(x) = x - [\![x]\!] - \tfrac{1}{2}$ of Example 6

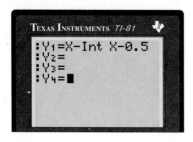

Fig. 1.3.9 A graphics calculator prepared to graph the sawtooth function of Example 6

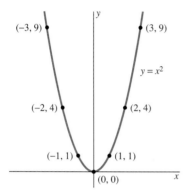

Fig. 1.3.10 The graph of the parabola $y = x^2$ of Example 7

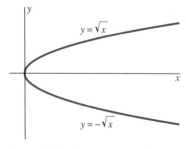

Fig. 1.3.11 The graph of the parabola $x = y^2$ of Example 8

but $f(x)$ abruptly jumps to the value $-\frac{1}{2}$ when $x = n$. A precise definition of continuity and discontinuity for functions appears in Section 2.4. Figure 1.3.9 shows a graphics calculator prepared to graph the sawtooth function.

PARABOLAS

The graph of a *quadratic* function of the form

$$f(x) = ax^2 + bx + c \qquad (a \neq 0) \tag{4}$$

is a *parabola* whose shape resembles that of the particular parabola in Example 7.

EXAMPLE 7 Construct the graph of the parabola $y = x^2$.

Solution We plot some points in a short table of values.

x	-3	-2	-1	0	1	2	3
$y = x^2$	9	4	1	0	1	4	9

When we draw a smooth curve through these points, we obtain the curve shown in Fig. 1.3.10.

The parabola $y = -x^2$ would look similar to the one in Fig. 1.3.10 but would open downward instead of upward. More generally, the graph of the equation

$$y = ax^2 \tag{5}$$

is a parabola with its *vertex* at the origin, provided that $a \neq 0$. This parabola opens upward if $a > 0$ and downward if $a < 0$. [For the time being, we may regard the vertex of a parabola as the point at which it "changes direction." The vertex of a parabola of the form $y = ax^2$ ($a \neq 0$) is always at the origin. A precise definition of the *vertex* of a parabola appears in Chapter 10.]

EXAMPLE 8 Construct the graphs of the functions $f(x) = \sqrt{x}$ and $g(x) = -\sqrt{x}$.

Solution After plotting and connecting points as in Example 7, we obtain the parabola $y^2 = x$ shown in Fig. 1.3.11. This parabola opens to the right. The upper half is the graph of $f(x) = \sqrt{x}$, whereas the lower half is the graph of $g(x) = -\sqrt{x}$. Thus the union of the graphs of these two functions is the graph of the *single* equation $y^2 = x$. (Compare this with the circle of Example 2.) More generally, the graph of the equation

$$x = by^2 \tag{6}$$

is a parabola with its vertex at the origin, provided that $b \neq 0$. This parabola opens to the right if $b > 0$ (as in Fig. 1.3.11) but to the left if $b < 0$.

The *size* of the coefficient a in Eq. (5) [or of b in Eq. (6)] determines the "width" of the parabola; its *sign* determines the direction in which the

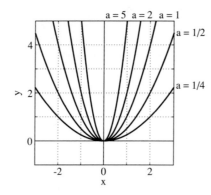
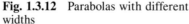

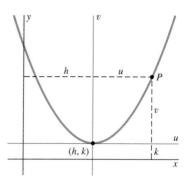

Fig. 1.3.12 Parabolas with different widths

Fig. 1.3.13 A translated parabola

parabola opens. Specifically, the larger $a > 0$ is, the steeper the curve rises and hence the narrower the parabola is (Fig. 1.3.12).

The parabola in Fig. 1.3.13 has the shape of the "standard parabola" of Example 7, but its vertex is located at the point (h, k). In the indicated uv-coordinate system, the equation of this parabola is $v = u^2$, in analogy to Eq. (5) with $a = 1$. But the uv-coordinates and xy-coordinates are related as follows:

$$u = x - h, \qquad v = y - k.$$

Hence the xy-coordinate equation of this parabola is

$$y - k = (x - h)^2. \tag{7}$$

Thus when the parabola $y = x^2$ is translated h units to the right and k units upward, the equation in (7) of the translated parabola is obtained by replacing x with $x - h$ and y with $y - k$. This is another instance of the *translation principle* that we observed in connection with circles.

More generally, the graph of any equation of the form

$$y = ax^2 + bx + c \qquad (a \neq 0) \tag{8}$$

can be recognized as a translated parabola by first completing the square in x to obtain an equation of the form

$$y - k = a(x - h)^2. \tag{9}$$

The graph of this equation is a parabola with its vertex at (h, k).

EXAMPLE 9 Determine the shape of the graph of the equation

$$y = 2x^2 - 4x - 1. \tag{10}$$

Solution If we complete the square in x, Eq. (10) takes the form

$$y = 2(x^2 - 2x + 1) - 3;$$

$$y + 3 = 2(x - 1)^2.$$

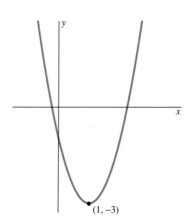

Fig. 1.3.14 The parabola $y = 2x^2 - 4x - 1$ of Example 9

Hence the graph of Eq. (10) is the parabola shown in Fig. 1.3.14. It opens upward, and its vertex is at $(1, -3)$.

APPLICATIONS OF QUADRATIC FUNCTIONS

In Section 1.1 we saw that a certain type of applied problem may call for us to find the maximum or minimum value attained by a certain function f. If the function f is a quadratic function as in Eq. (4), then the graph of $y = f(x)$ is a parabola. In this case the maximum (or minimum) value of $f(x)$ corresponds to the highest (or lowest) point of the parabola. We can therefore find this maximum (or minimum) value graphically—at least approximately—by zooming in on the vertex of the parabola.

For instance, recall the animal pen problem of Section 1.1. In Example 6 we saw that the area A of the pen (see Fig. 1.3.15) is given as a function of its base length x by

$$A(x) = \tfrac{3}{5}(30x - x^2), \qquad 0 \leq x \leq 30. \tag{11}$$

Figure 1.3.16 shows the graph $y = A(x)$, and Figs. 1.3.17, 1.3.18, and 1.3.19 show successive magnifications of the region near the high point (vertex) of the parabola. The dashed rectangle in each figure is the viewing window for the next. Figure 1.3.19 makes it *seem* that the maximum area of the pen is $A(15) = 135$. It is clear from the figure that the maximum value of $f(x)$ is within 0.001 of $A = 135$.

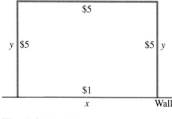

Fig. 1.3.15 The animal pen

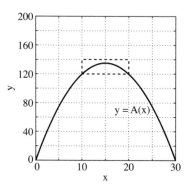

Fig. 1.3.16 The graph $y = A(x)$

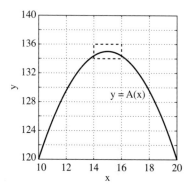

Fig. 1.3.17 The first zoom

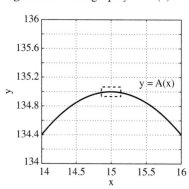

Fig. 1.3.18 The second zoom

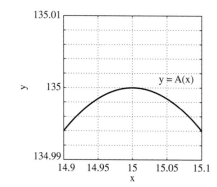

Fig. 1.3.19 The third zoom

We can verify by completing the square as in Example 9 that the maximum value is *precisely* $f(15) = 135$:

$$A = -\tfrac{3}{5}(x^2 - 30x) = -\tfrac{3}{5}(x^2 - 30x + 225 - 225)$$

$$= -\tfrac{3}{5}(x^2 - 30x + 225) + 135;$$

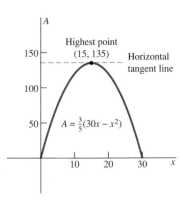

Fig. 1.3.20 The graph of $A(x) = \frac{3}{5}(30x - x^2)$ for $0 \le x \le 30$

that is,

$$A - 135 = -\frac{3}{5}(x - 15)^2. \qquad (12)$$

It follows from Eq. (12) that the graph of Eq. (11) is the parabola shown in Fig. 1.3.20, which opens downward from its vertex (15, 135). This *proves* that the maximum value of $A(x)$ on the interval $[0, 30]$ is the value $A(15) = 135$, as both our numerical investigation in Section 1.1 and our graphical investigation here suggest. And when we glance at Eq. (12) in the form

$$A(x) = 135 - \frac{3}{5}(x - 15)^2,$$

it's clear and unarguable that the maximum possible value of $135 - \frac{3}{5}u^2$ is 135 when $u = x - 15 = 0$—that is, when $x = 15$.

The technique of completing the square is quite limited: It can be used to find maximum or minimum values only of *quadratic* functions. One of our goals in calculus is to develop a more general technique that can be applied to a far wider variety of functions.

The basis of this more general technique lies in the following observation. Visual inspection of the graph of

$$A(x) = \frac{3}{5}(30x - x^2)$$

in Fig. 1.3.20 suggests that the line tangent to the curve at its highest point is horizontal. If we *knew* that the line tangent to a graph at its highest point must be horizontal, then our problem would reduce to showing that (15, 135) is the only point of the graph of $y = A(x)$ at which the tangent line is horizontal.

But what do we mean by the *tangent line* to an arbitrary curve? We pursue this question in Section 2.1. The answer will open the door to the possibility of finding maximum and minimum values of virtually arbitrary functions.

1.3 Problems

Sketch each of the translated circles in Problems 1 through 6. Indicate the center and radius of each.

1. $x^2 + y^2 = 4x$

2. $x^2 + y^2 + 6y = 0$

3. $x^2 + y^2 + 2x + 2y = 2$

4. $x^2 + y^2 + 10x - 20y + 100 = 0$

5. $2x^2 + 2y^2 + 2x - 2y = 1$

6. $9x^2 + 9y^2 - 6x - 12y = 11$

Sketch each of the translated parabolas in Problems 7 through 12. Indicate the vertex of each.

7. $y = x^2 - 6x + 9$ **8.** $y = 16 - x^2$

9. $y = x^2 + 2x + 4$ **10.** $2y = x^2 - 4x + 8$

11. $y = 5x^2 + 20x + 23$ **12.** $y = x - x^2$

The graph of the equation $(x - h)^2 + (y - k)^2 = C$ is a circle if $C > 0$, is the single point (h, k) if $C = 0$, and contains no points if $C < 0$. (Why?) Identify the graphs of

the equations in Problems 13 through 16. If the graph is a circle, give its center and radius.

13. $x^2 + y^2 - 6x + 8y = 0$

14. $x^2 + y^2 - 2x + 2y + 2 = 0$

15. $x^2 + y^2 + 2x + 6y + 20 = 0$

16. $2x^2 + 2y^2 - 2x + 6y + 5 = 0$

Sketch the graphs of the functions in Problems 17 through 40. Take into account the domain of definition of each function, and plot points as necessary.

17. $f(x) = 2 - 5x, \quad -1 \le x \le 1$

18. $f(x) = 2 - 5x, \quad 0 \le x < 2$

19. $f(x) = 10 - x^2$ **20.** $f(x) = 1 + 2x^2$

21. $f(x) = x^3$ **22.** $f(x) = x^4$

23. $f(x) = \sqrt{4 - x^2}$ **24.** $f(x) = -\sqrt{9 - x^2}$

25. $f(x) = \sqrt{x^2 - 9}$ **26.** $f(x) = \dfrac{1}{1 - x}$

27. $f(x) = \dfrac{1}{x + 2}$ **28.** $f(x) = \dfrac{1}{x^2}$

29. $f(x) = \dfrac{1}{(x - 1)^2}$ **30.** $f(x) = \dfrac{|x|}{x}$

31. $f(x) = \dfrac{1}{2x + 3}$ **32.** $f(x) = \dfrac{1}{(2x + 3)^2}$

33. $f(x) = \sqrt{1 - x}$ **34.** $f(x) = \dfrac{1}{\sqrt{1 - x}}$

35. $f(x) = \dfrac{1}{\sqrt{2x + 3}}$ **36.** $f(x) = |2x - 2|$

37. $f(x) = |x| + x$ **38.** $f(x) = |x - 3|$

39. $f(x) = |2x + 5|$ **40.** $f(x) = \begin{cases} |x| & \text{if } x < 0 \\ x^2 & \text{if } x \geq 0 \end{cases}$

Graph the functions given in Problems 41 through 46. Indicate any points of discontinuity.

41. $f(x) = \begin{cases} 0 & \text{if } x < 0 \\ 1 & \text{if } x \geq 0 \end{cases}$

42. $f(x) = 1$ if x is an integer; otherwise, $f(x) = 0$.

43. $f(x) = [\![2x]\!]$ **44.** $f(x) = \dfrac{x - 1}{|x - 1|}$

45. $f(x) = [\![x]\!] - x$

46. $f(x) = [\![x]\!] + [\![-x]\!] + 1$

In Problems 47 through 54, use a graphics calculator or computer to find (by zooming) the highest or lowest (as appropriate) point P on the given parabola. Determine the

coordinates of P with two digits to the right of the decimal correct or correctly rounded. Then verify your result by completing the square to find the actual vertex of the parabola.

47. $y = 2x^2 - 6x + 7$

48. $y = 2x^2 - 10x + 11$

49. $y = 4x^2 - 18x + 22$

50. $y = 5x^2 - 32x + 49$

51. $y = -32 + 36x - 8x^2$

52. $y = -53 - 34x - 5x^2$

53. $y = 3 - 8x - 3x^2$

54. $y = -28 + 34x - 9x^2$

In Problems 55 through 58, use the method of completing the square to graph the appropriate function and thereby determine the maximum or minimum value requested.

55. If a ball is thrown straight upward with initial velocity 96 ft/s, then its height t seconds later is $y = 96t - 16t^2$ (ft). Determine the maximum height that the ball attains.

56. Find the maximum possible area of the rectangle described in Problem 50 of Section 1.1.

57. Find the maximum possible value of the product of two positive numbers whose sum is 50.

58. In Problem 52 of Section 1.1, you were asked to express the daily production of a specific oil field as a function $P = f(x)$ of the number x of new oil wells drilled. Construct the graph of f, and use it to find the value of x that maximizes P.

1.3 Projects

The projects here require the use of a graphics calculator or a computer with a graphing program.

    With the typical graphics calculator, you can carry out successive magnifications simply by pressing the ZOOM key. Figure 1.3.21 lists graphing commands for common computer systems. Using such a system, you ordinarily must define each successive viewing window explicitly, as in the typical command

```
plot(x^2 - 2, x = -5..5, y = -10..10)
```

to graph $f(x) = x^2 - 2$ in the window $-5 \leq x \leq 5$, $-10 \leq y \leq 10$.

Derive	Author f(x) , then use Plot and Zoom commands
Maple	plot(f(x) , x = a..b, y = c..d)
Mathematica	Plot[f[x] , {x,a,b} , PlotRange -> {c,d}]
(X)PLORE	window(a,b, c,d) graph(f(x) , x)

Fig. 1.3.21 Commands for graphing the function $f(x)$

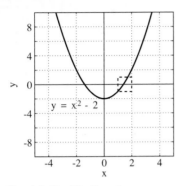

Fig. 1.3.22 Finding the (positive) square root of 2

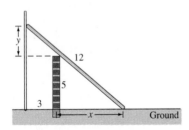

Fig. 1.3.23 The line and parabola in the sidewalk problem

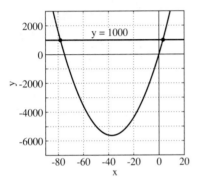

Fig. 1.3.24 The ladder of Project C

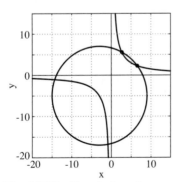

Fig. 1.3.25 The circle and hyperbola in the ladder problem

PROJECT A Project A of Section 1.1 discusses the number $\sqrt{2}$ as the positive solution of the equation

$$x^2 - 2 = 0,$$

which is the same as the intersection of the parabola

$$y = x^2 - 2$$

and the positive x-axis. Beginning with Fig. 1.3.22, approximate $\sqrt{2}$ (accurate to four decimal places) graphically—that is, by the **method of successive zooms**.

Similarly, you might apply the method of successive zooms to find four-place approximations to such roots as $\sqrt{17}$, $\sqrt[3]{25}$, and $\sqrt[5]{100}$.

PROJECT B Project B of Section 1.1 deals with a sidewalk of width x feet that borders a 50-ft by 100-ft rectangular plot. Conclude from the discussion there that if the total area of the sidewalk alone is to be 1000 ft^2, then x must be the x-coordinate of a point of intersection of the line $y = 1000$ and the parabola

$$y = (2x + 100)(2x + 50) - 5000.$$

Then apply the method of successive zooms to find (accurate to four decimal places) both of the two intersection points shown in Fig. 1.3.23. Are there actually two possible values of x?

PROJECT C Figure 1.3.24 shows a 12-ft ladder leaning across a 5-ft fence and touching a high wall located 3 ft behind the fence. We want to find the distance x from the base of this ladder to the bottom of the fence.

Carefully write up the following discussion. Application of the Pythagorean theorem to the large right triangle in Fig. 1.3.24 yields

$$(x + 3)^2 + (y + 5)^2 = 144, \tag{13}$$

the equation of a circle with center $(-3, -5)$ and radius 12. Then note that the two small triangles in the figure are similar. Hence

$$\frac{y}{3} = \frac{5}{x},$$

so

$$y = \frac{15}{x}. \tag{14}$$

The graph of Eq. (14) is essentially the same as that in Example 4 (see Fig. 1.3.6), with an infinite discontinuity at $x = 0$.

The graphs of Eqs. (13) and (14) are shown together in Fig. 1.3.25. The two indicated first-quadrant intersection points yield *two* physically possible positions of the ladder. Apply the method of successive zooms to find them, with x and y accurate to four decimal places. Sketch, roughly to scale, these two possibilities. Why do the two third-quadrant intersection points not yield two additional physically possible positions of the ladder?

1.4
A Brief Catalog of Functions

In this section we briefly survey a variety of functions that are used in applications of calculus to describe changing phenomena in the world around us. Our viewpoint here is largely graphical. The objective is for you to attain a general understanding of major differences between different types of functions. In later chapters we use calculus to investigate further the graphs presented here.

COMBINATIONS OF FUNCTIONS

At first we concentrate on simple functions, because many varied and complex functions can be assembled out of simple "building-block functions." Here we discuss some of the ways of combining functions to obtain new ones.

Suppose that f and g are functions and that c is a fixed real number. The **(scalar) multiple cf**, the **sum $f + g$**, the **difference $f - g$**, the **product $f \cdot g$**, and the **quotient f/g** are the new functions determined by these formulas:

$$(cf)(x) = cf(x), \tag{1}$$

$$(f + g)(x) = f(x) + g(x), \tag{2}$$

$$(f - g)(x) = f(x) - g(x), \tag{3}$$

$$(f \cdot g)(x) = f(x) \cdot g(x), \quad \text{and} \tag{4}$$

$$\left(\frac{f}{g}\right)(x) = \frac{f(x)}{g(x)}. \tag{5}$$

The combinations in Eqs. (2) through (4) are defined for every number x that lies both in the domain of f and in the domain of g. In Eq. (5) we must require that $g(x) \neq 0$.

EXAMPLE 1   Let $f(x) = x^2 + 1$ and $g(x) = x - 1$. Then

$$(3f)(x) = 3(x^2 + 1),$$

$$(f + g)(x) = (x^2 + 1) + (x - 1) = x^2 + x,$$

$$(f - g)(x) = (x^2 + 1) - (x - 1) = x^2 - x + 2,$$

$$(f \cdot g)(x) = (x^2 + 1)(x - 1) = x^3 - x^2 + x - 1, \quad \text{and}$$

$$\left(\frac{f}{g}\right)(x) = \frac{x^2 + 1}{x - 1} \quad (x \neq 1).$$

EXAMPLE 2   If $f(x) = \sqrt{1 - x}$ for $x \leq 1$ and $g(x) = \sqrt{1 + x}$ for $x \geq -1$, then the sum and product of f and g are defined where *both* f and g are defined. Thus the domain of both

$$f(x) + g(x) = \sqrt{1 - x} + \sqrt{1 + x}$$

and

$$f(x) \cdot g(x) = \sqrt{1 - x}\sqrt{1 + x} = \sqrt{1 - x^2}$$

is the closed interval $[-1, 1]$. But the domain of the quotient

$$\frac{f(x)}{g(x)} = \frac{\sqrt{1 - x}}{\sqrt{1 + x}} = \sqrt{\frac{1 - x}{1 + x}}$$

is the half-open interval $(-1, 1]$, because $g(-1) = 0$.

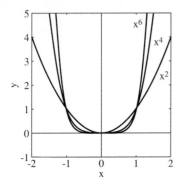

Fig. 1.4.1 Graphs of power functions of even degree (Example 3)

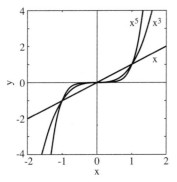

Fig. 1.4.2 Graphs of power functions of odd degree (Example 3)

POLYNOMIALS

A **polynomial** of degree n is a function of the form

$$p(x) = a_n x^n + a_{n-1} x^{n-1} + \cdots + a_2 x^2 + a_1 x + a_0 \qquad (6)$$

where the coefficients $a_0, a_1, \ldots, a_n$ are fixed real numbers. Thus an nth-degree polynomial is a sum of constant multiples of the **power functions**

$$1, x, x^2, \ldots, x^{n-1}, x^n. \qquad (7)$$

EXAMPLE 3 The graphs of the even-degree power functions x^2, x^4, $x^6, \ldots$ all "cup" upward, as shown in Fig. 1.4.1. But the graphs of the odd-degree power functions $x^1, x^3, x^5, \ldots$ all go from southwest to northeast, as shown in Fig. 1.4.2. In both cases the larger the exponent n, the "flatter" the graph $y = x^n$ near the origin.

A first-degree polynomial is simply a *linear function* $a_1 x + a_0$ whose graph is a straight line (see Section 1.2). A second-degree polynomial is a *quadratic function* whose graph $y = a_2 x^2 + a_1 x + a_0$ is a parabola (see Section 1.3).

Recall that a **zero** of the function f is a solution of the equation

$$f(x) = 0. \qquad (8)$$

Is it obvious to you that *the zeros of $f(x)$ are precisely the x-intercepts of the graph*

$$y = f(x)? \qquad (9)$$

Indeed, a major reason that we are interested in the graph of a function is to see the number and approximate locations of its zeros.

A key to understanding graphs of higher-degree polynomials is the *fundamental theorem of algebra*. It states that every nth-degree polynomial has n zeros (possibly complex, possibly repeated). It follows that an nth-degree polynomial has *no more than n* distinct real zeros.

EXAMPLE 4 Figures 1.4.3 and 1.4.4 exhibit polynomials that both have the maximum number of real zeros allowed by the fundamental theorem of

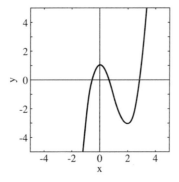

Fig. 1.4.3 $f(x) = x^3 - 3x^2 + 1$ has three real zeros (Example 4).

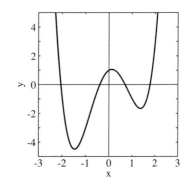

Fig. 1.4.4 $f(x) = x^4 - 4x^2 + x + 1$ has four real zeros (Example 4).

Ch. 1 / Functions and Graphs

algebra. But Figs. 1.4.1 and 1.4.2 show high-degree polynomials with only a single real zero. And the quadratic function

$$f(x) = x^2 + 4x + 13 = (x + 2)^2 + 9$$

has no real zeros at all. (Why not?) In fact, in case n is even, an nth-degree polynomial can have any number of real zeros from 0 to n (1 to n if n is odd).

In Chapter 4 we will use calculus to investigate graphs of polynomials (and other functions) more fully. There we will see that every polynomial graph shares a number of qualitative features with those in Figs. 1.4.1 through 1.4.4.

If $p(x)$ is a polynomial of even degree, then $y = p(x)$ goes in the same direction (either to $+\infty$ or to $-\infty$) as x goes to $+\infty$ and to $-\infty$. If $p(x)$ is a polynomial of odd degree, then y goes in opposite directions as x goes to $+\infty$ and to $-\infty$. Moreover, "between the extremes" to the right and left, an nth-degree polynomial has at most $n - 1$ "bends." Thus the two bends in Fig. 1.4.3 are the most a third-degree (cubic) polynomial graph can have, and the three bends in Fig. 1.4.4 are the most a fourth-degree (quartic) polynomial graph can have. (One task in Chapter 4 will be to make precise the notion of a "bend" in a curve.)

RATIONAL FUNCTIONS

Just as a rational number is a quotient of two integers, a **rational function** is a quotient

$$f(x) = \frac{p(x)}{q(x)} \tag{10}$$

of two polynomials $p(x)$ and $q(x)$. Graphs of rational functions and polynomials have several features in common. For instance, a rational function has only a finite number of zeros, because $f(x)$ in Eq. (10) can be zero only when the numerator polynomial $p(x)$ is zero. Similarly, the graph of a rational function has only a finite number of bends.

But the denominator polynomial in Eq. (10) may have a zero at a point $x = a$ where the numerator is nonzero. In this case the value of $f(x)$ will be extremely large when x is very close to a. This observation implies that the graph of a rational function may have a feature that no polynomial graph can have—an *asymptote*.

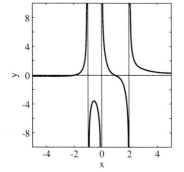

Fig. 1.4.5 The graph of the rational function in Eq. (11) (Example 5)

EXAMPLE 5 Figure 1.4.5 shows the graph of the rational function

$$f(x) = \frac{(x + 2)(x - 1)}{x(x + 1)(x - 2)}. \tag{11}$$

Note the x-intercepts $x = -2$ and $x = 1$, corresponding to the zeros of the numerator $(x + 2)(x - 1)$. The vertical lines $x = -1$, $x = 0$, and $x = 2$ shown in the graph correspond to the zeros of the denominator $x(x + 1)(x - 2)$. These vertical lines are *asymptotes* of the graph of f.

EXAMPLE 6 Figure 1.4.6 shows the graph of the rational function

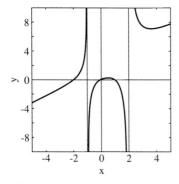

Fig. 1.4.6 The graph of the rational function in Eq. (12) (Example 6)

$$f(x) = \frac{x(x + 2)(x - 1)}{(x + 1)(x - 2)}. \tag{12}$$

The x-intercepts $x = -2$, $x = 0$, and $x = 1$ correspond to the zeros of the numerator, whereas the asymptotes $x = -1$ and $x = 2$ correspond to the zeros of the denominator.

It should be clear that—by counting x-intercepts and asymptotes—you could match the rational functions in Eqs. (11) and (12) with their graphs in Figs. 1.4.5 and 1.4.6 without knowing in advance which was which.

TRIGONOMETRIC FUNCTIONS

A review of trigonometry is included in Appendix A. In elementary trigonometry a trigonometric function such as sin A, cos A, or tan A ordinarily is first defined as a function of an *angle* A in a right triangle. But here a trigonometric function of a *number* corresponds to that function of an angle measuring x radians. Thus

$$\sin \frac{\pi}{6} = \frac{1}{2}, \qquad \cos \frac{\pi}{6} = \frac{\sqrt{3}}{2}, \quad \text{and} \quad \tan \frac{\pi}{6} = \frac{\sin \dfrac{\pi}{6}}{\cos \dfrac{\pi}{6}} = \frac{1}{\sqrt{3}}$$

because $\pi/6$ is the radian measure of an angle of $30°$.

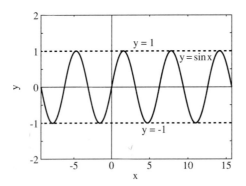

Fig. 1.4.7 $y = \sin x$

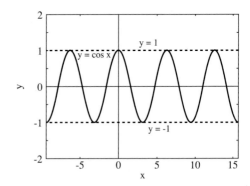

Fig. 1.4.8 $y = \cos x$

Figures 1.4.7 and 1.4.8 show the graphs $y = \sin x$ and $y = \cos x$ of the sine and cosine functions, respectively. The value of each oscillates between $+1$ and -1, exhibiting the characteristic *periodicity* of the trigonometric functions:

$$\sin(x + 2\pi) = \sin x, \qquad \cos(x + 2\pi) = \cos x. \tag{13}$$

If we translate the graph $y = \cos x$ by $\pi/2$ units to the right, we get the graph $y = \sin x$. This observation corresponds to the familiar relation

$$\cos\left(x - \frac{\pi}{2}\right) = \cos\left(\frac{\pi}{2} - x\right) = \sin x. \tag{14}$$

Figure 1.4.9 shows the translated sine curve obtained by translating the origin to the point $(1, 2)$. Its equation is obtained upon replacing x and y in $y = \sin x$ with $x - 1$ and $y - 2$, respectively:

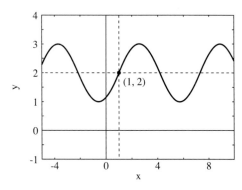

Fig. 1.4.9 The translated sine curve $y - 2 = \sin(x - 1)$

$$y - 2 = \sin(x - 1); \qquad \text{that is,}$$

$$y = 2 + \sin(x - 1). \tag{15}$$

The world around us is full of quantities that oscillate like the trigonometric functions. Think of the alternation of day and night, the endless repetition of the seasons, the monthly cycle of the moon, the rise and fall of the tides, the beat of your heart.

EXAMPLE 7 Figure 1.4.10 shows the cosine-like behavior of temperatures in Athens, Georgia. The average temperature T (in °F) on a day t months after July 15 is given approximately by

$$T = 61.3 + 17.9 \cos(0.5236t).$$

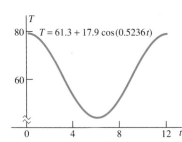

Fig. 1.4.10 Average daily temperature in Athens, Georgia, t months after July 15 (Example 7)

The periodicity and oscillatory behavior of the trigonometric functions make them quite unlike polynomial functions. Because

$$\sin n\pi = 0 \quad \text{and} \quad \cos\left(\frac{2n + 1}{2}\pi\right) = 0 \tag{16}$$

for $n = 0, 1, 2, 3, \ldots$, we see that the simple trigonometric equations

$$\sin x = 0 \quad \text{and} \quad \cos x = 0 \tag{17}$$

have *infinitely many solutions*. In contrast, a polynomial equation can have only a finite number of solutions.

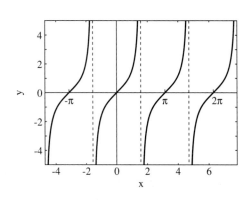

Fig. 1.4.11 $y = \tan x$

Figure 1.4.11 shows the graph of $y = \tan x$. The x-intercepts correspond to the zeros of the numerator $\sin x$ in

$$\tan x = \frac{\sin x}{\cos x}, \tag{18}$$

whereas the asymptotes correspond to the zeros of the denominator cos x. Observe the "infinite gaps" in the graph $y = \tan x$ at these odd-integral multiples of $\pi/2$. We call these gaps *discontinuities*, phenomena we discuss further in Chapter 2.

EXPONENTIAL FUNCTIONS AND LOGARITHMIC FUNCTIONS

An **exponential function** is a function of the form

$$f(x) = a^x, \tag{19}$$

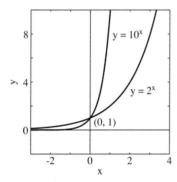

Fig. 1.4.12 Increasing exponential functions $y = 2^x$ and $y = 10^x$

where the **base** a is a fixed real number—a constant. Note the difference between an exponential function and a power function. In the power function x^n, the *variable x* is raised to a *constant* power; in the exponential function a^x, a *constant* is raised to a *variable* power.

Many computers and programmable calculators use $a^\wedge x$ to denote the exponential a^x. If $a > 1$, then the graph $y = a^x$ looks much like those in Fig. 1.4.12, which depicts $y = 2^x$ and $y = 10^x$. The graph of an exponential function with base a, $a > 1$, increases steadily from left to right. Therefore, such a graph is nothing like the graph of a polynomial or trigonometric function. The larger the base a, the more rapid the rate at which the curve $y = a^x$ rises (for $x > 0$). Thus $y = 10^x$ climbs more steeply than $y = 2^x$.

If we replace x in Eq. (19) with $-x$, we get the function a^{-x}. Its graph $y = a^{-x}$ *decreases* from left to right if $a > 1$. Figure 1.4.13 shows the graphs $y = 3^{-x}$ and $y = 7^{-x}$.

Whereas trigonometric functions are used to describe periodic phenomena of ebb and flow, exponential functions are used to describe natural quantities that always increase or always decrease.

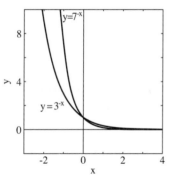

Fig. 1.4.13 Decreasing exponential functions $y = 3^{-x}$ and $y = 7^{-x}$

EXAMPLE 8 Let $P(t)$ denote the number of rodents after t months in a certain prolific population that doubles every month. If there are $P(0) = 10$ rodents initially, then there are

❑ $P(1) = 10 \cdot 2^1 = 20$ rodents after 1 month,
❑ $P(2) = 10 \cdot 2^2 = 40$ rodents after 2 months,
❑ $P(3) = 10 \cdot 2^3 = 80$ rodents after 3 months,

and so forth. Thus the rodent population after t months is given by the exponential function

$$P(t) = 10 \cdot 2^t \tag{20}$$

if t is a nonnegative integer. Under appropriate conditions, Eq. (20) gives an accurate approximation to the rodent population even when t is not an integer.

EXAMPLE 9 Suppose that you invest $5000 in a money-market account that pays 8% annually. This means that the amount in the account is multiplied by 1.08 at the end of each year. Let $A(t)$ denote the amount in your account at the end of t years. Then:

Ch. 1 / Functions and Graphs

❑ $A(1) = 5000 \cdot 1.08^1$ ($5400) after 1 yr,

❑ $A(2) = 5000 \cdot 1.08^2$ ($5832) after 2 yr,

❑ $A(3) = 5000 \cdot 1.08^3$ ($6298.56) after 3 yr,

and so on. Thus after t years (t a nonnegative integer), the amount in your account is given by the exponential function

$$A(t) = 5000 \cdot 1.08^t. \tag{21}$$

In the case of a slightly lower interest rate (about 7.696%) compounded *continuously* (rather than annually), this formula holds also when t is not an integer. Figure 1.4.14 shows the graph $A(t) = 5000 \cdot 1.08^t$ as well as the horizontal line $A = 10,000$. From this graph we see, for instance, that the amount in the account has doubled (to $10,000) after approximately $t = 9$ yr. We could approximate the "doubling time" t more accurately by magnifying the graph near the intersection of the horizontal line and the rising curve.

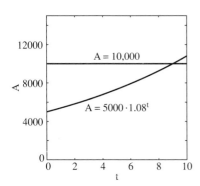

Fig. 1.4.14 The graph for Example 9

In analogy with the inverse trigonometric functions that you may have seen in trigonometry, logarithms are "inverse" to exponential functions. The **base a logarithm** of the positive number x is the power to which a must be raised to get x. That is,

$$y = \log_a x \quad \text{if} \quad a^y = x. \tag{22}$$

The [LOG] key on most calculators gives the base 10 (*common*) *logarithm* $\log_{10} x$. The [LN] key gives the *natural logarithm*

$$\ln x = \log_e x, \tag{23}$$

where e is a special irrational number:

$$e = 2.71828\ 18284\ 59045\ 23536\ \dots. \tag{24}$$

You'll see the significance of this base in Chapter 7.

Figure 1.4.15 shows the graphs $y = \ln x$ and $y = \log_{10} x$. Both graphs pass through the point $(1, 0)$ and rise steadily (though slowly) from left to right. Because exponential functions never take on zero or negative values, negative numbers are not in the domain of any logarithm function; nor is the number zero.

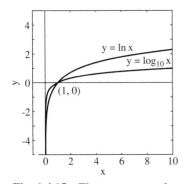

Fig. 1.4.15 The common and natural logarithm functions

TRANSCENDENTAL EQUATIONS

The trigonometric, exponential, and logarithmic functions are typically called *transcendental* functions. As we saw in Eq. (17), an equation that includes transcendental functions can have infinitely many solutions. But it also may have only a finite number of solutions; which of these two possibilities is the case can be difficult to determine. One approach is to write the given equation in the form

$$f(x) = g(x), \tag{25}$$

where both the functions f and g are readily graphed. Then the solutions of Eq. (25) correspond to the intersections of the two graphs $y = f(x)$ and $y = g(x)$.

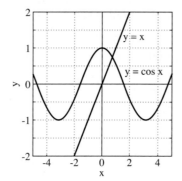

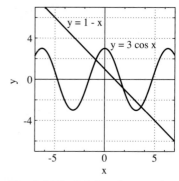

Fig. 1.4.16 Solving the equation $x = \cos x$ of Example 10

Fig. 1.4.17 Solving the equation $1 - x = 3 \cos x$ of Example 11

EXAMPLE 10 The single point of intersection of the graphs $y = x$ and $y = \cos x$, shown in Fig. 1.4.16, indicates that the equation

$$x = \cos x \qquad (26)$$

has only a single solution. Moreover, from the graph you can glean the additional information that the solution lies in the interval $(0, 1)$.

EXAMPLE 11 The graphs of $y = 1 - x$ and $y = 3 \cos x$ are shown in Fig. 1.4.17. In contrast with Example 10, there are three points of intersection of the graphs. This makes it clear that the equation

$$1 - x = 3 \cos x \qquad (27)$$

has one negative solution and two positive solutions. They could be approximated by (separately) zooming in on the three intersection points.

1.4 Problems

In Problems 1 through 10, find $f + g, f \cdot g$, and f/g, and give the domain of definition of each of these new functions.

1. $f(x) = x + 1, \quad g(x) = x^2 + 2x - 3$

2. $f(x) = \dfrac{1}{x - 1}, \quad g(x) = \dfrac{1}{2x + 1}$

3. $f(x) = \sqrt{x}, \quad g(x) = \sqrt{x - 2}$

4. $f(x) = \sqrt{x + 1}, \quad g(x) = \sqrt{5 - x}$

5. $f(x) = \sqrt{x^2 + 1}, \quad g(x) = \dfrac{1}{\sqrt{4 - x^2}}$

6. $f(x) = \dfrac{x - 1}{x - 2}, \quad g(x) = \dfrac{x + 1}{x + 2}$

7. $f(x) = x, \quad g(x) = \sin x$

8. $f(x) = \sqrt{x}, \quad g(x) = \cos x^2$

9. $f(x) = \sqrt{x^2 + 1}, \quad g(x) = \tan x$

10. $f(x) = \cos x, \quad g(x) = \tan x$

In Problems 11 through 14, match the given polynomial with its graph among those shown in Figs. 1.4.18 through 1.4.21.

11. $f(x) = x^3 - 3x + 1$

12. $f(x) = 1 + 4x - x^3$

13. $f(x) = x^4 - 5x^3 + 13x + 1$

14. $f(x) = 2x^5 - 10x^3 + 6x - 1$

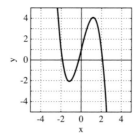

Fig. 1.4.18

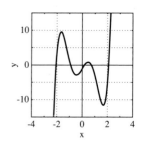

Fig. 1.4.19

40

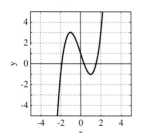

Fig. 1.4.20

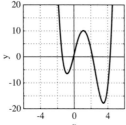

Fig. 1.4.21

Fig. 1.4.26

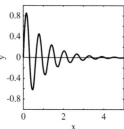

Fig. 1.4.27

In Problems 15 through 18, match the given rational function with its graph among those shown in Figs. 1.4.22 through 1.4.25.

15. $f(x) = \dfrac{1}{(x + 1)(x - 2)}$

16. $f(x) = \dfrac{x}{x^2 - 9}$

17. $f(x) = \dfrac{3}{x^2 + 1}$

18. $f(x) = \dfrac{x^2 + 1}{x^3 - 1}$

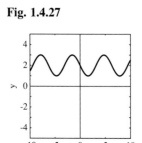

Fig. 1.4.28

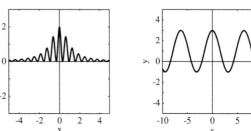

Fig. 1.4.29

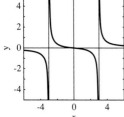

Fig. 1.4.22

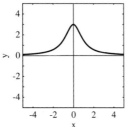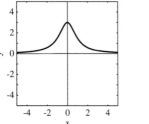

Fig. 1.4.23

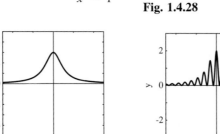

Fig. 1.4.30

Fig. 1.4.31

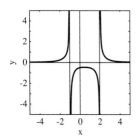

Fig. 1.4.24

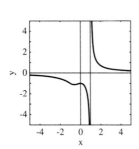

Fig. 1.4.25

Use a graphics calculator (or computer graphing program) to determine by visual inspection of graphs the number of real solutions of the equations in Problems 25 through 36.

25. $x^3 - 3x + 1 = 0$

26. $x^3 - 3x + 2 = 0$

27. $x^3 - 3x + 3 = 0$

28. $x^4 - 3x^3 + 5x - 4 = 0$

29. $x^4 - 3x^3 + 5x + 4 = 0$

30. $3 \cos x = x + 1$

31. $3 \cos x = x - 1$

32. $x = 5 \cos x$

33. $x = 7 \cos x$

34. $\ln x = \cos x \quad (x > 0)$

35. $\ln x = 2 \cos x \quad (x > 0)$

36. $\dfrac{x}{5} = \cos x + \ln x \quad (x > 0)$

In Problems 19 through 24, match the given function with its graph among those shown in Figs. 1.4.26 through 1.4.31.

19. $f(x) = x^3 - 2x^2 + 2$

20. $f(x) = 1 + 2 \cos x$

21. $f(x) = 2 - \sin x$

22. $f(x) = e^x - 1$

23. $f(x) = \dfrac{1 + \cos 10x}{1 + x^2}$

24. $f(x) = e^{-x} \sin 10x$

These projects require the use of a graphics calculator or a computer with a graphing program (as discussed in the Section 1.3 projects). Each calls for the graphical solution of one or more equations by the method of successive magnifications (or zooms). Determine each desired solution accurate to three decimal places.

1. Find all real solutions of the cubic equations
(a) $x^3 - 3x^2 + 1 = 0$ (Fig. 1.4.3);
(b) $x^3 - 3x^2 - 2 = 0$.

2. Find all real solutions of the quartic equation

$$x^4 - 4x^2 + x + 1 = 0 \qquad \text{(Fig. 1.4.4).}$$

3. (a) Suppose that you invest \$5000 in an account that pays interest compounded continuously, with an annual interest rate of 7.696%, so that the amount on deposit at time t (in years) is given by

$$A(t) = 5000 \cdot 1.08^t.$$

Beginning with Fig. 1.4.14, determine graphically how long it takes—to the nearest day—for your initial investment of \$5000 to double. (b) If the interest rate were instead 9.531%, but still compounded continuously, then the amount you would have on deposit after t years would be

$$A(t) = 5000 \cdot (1.10)^t.$$

Find graphically how long it would take for your investment to *triple*.

4. Suppose that a population is described by an exponential function, as in Example 8. (a) If this population doubles in six months, how long does it take to triple? (b) If this population triples in six months, how long does it take to double?

5. Find all real solutions of the equations
(a) $x = \cos x$ (Fig. 1.4.16);
(b) $x^2 = \cos x$;
(c) $1 - x = 3 \cos x$ (Fig. 1.4.17).

6. Find all *positive* solutions of the equations
(a) $2^x = 3 \cos x$;
(b) $2^x = 3 \cos 4x$.

1.5
A Preview: What *Is* Calculus?

Surely this question is on your mind as you begin a study of calculus that may extend over two or three terms. Following our review of functions and graphs in Sections 1.1 through 1.4, we can preview here at least the next several chapters, where the central concepts of calculus are developed.

THE TWO FUNDAMENTAL PROBLEMS

The body of computational technique that constitutes "the calculus" revolves around two fundamental geometric problems that people have been investi-

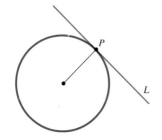

Fig. 1.5.1 The tangent line L touching the circle at the point P

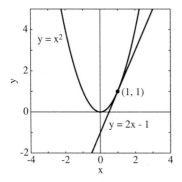

Fig. 1.5.2 The line tangent to the parabola $y = x^2$ at the point (1, 1)

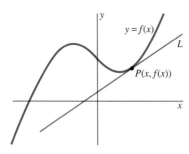

Fig. 1.5.3 What is the slope of the line L tangent to the graph $y = f(x)$ at the point $P(x, f(x))$?

gating for more than 2000 years. Each problem involves the graph $y = f(x)$ of a given function.

The first fundamental problem is this: What do we mean by the *line tangent* to the curve $y = f(x)$ at a given point? The word *tangent* stems from the Latin word *tangens*, for "touching." Thus a line tangent to a curve is one that "just touches" the curve. Lines tangent to circles (Fig. 1.5.1) are well known from elementary geometry. Figure 1.5.2 shows the line tangent to the parabola $y = x^2$ at the point (1, 1). We see in Section 2.1 that this particular tangent line has slope 2, so its point-slope equation is

$$y - 1 = 2 \cdot (x - 1); \quad \text{that is,} \quad y = 2x - 1.$$

Our first problem is how to find tangent lines in more general cases.

The Tangent Problem Given a point $P(x, f(x))$ on the curve $y = f(x)$, how do we calculate the slope of the tangent line at P (Fig. 1.5.3)?

We begin to explore the answer to this question in Chapter 2. If we denote by $m(x)$ the slope of the tangent line at $P(x, f(x))$, then m is a *new function*. It might informally be called a *slope-predictor function* for the curve $y = f(x)$. In calculus this slope-predictor function is called the **derivative** of the function f. In Chapter 3 we learn to calculate derivatives of a variety of functions, and in both Chapter 3 and Chapter 4 we see numerous applications of derivatives in solving real-world problems. This gives an introduction to the part of calculus called *differential calculus.*

The tangent problem is a geometric problem—thus, a purely mathematical question. But its answer (in the form of derivatives) is the key to the solution of diverse applied problems in many scientific and technical areas. Examples 1 and 2 may suggest to you the *connections* that are the key to the pivotal role of calculus in science and technology.

EXAMPLE 1 Suppose that you're driving a car along a long, straight road (Fig. 1.5.4). If $f(t)$ denotes the *distance* (in miles) the car has traveled at time t (in hours), then the slope of the line tangent to the curve $y = f(t)$ at the point $(t, f(t))$ (Fig. 1.5.5) is the *velocity* (in miles per hour) of the car at time t.

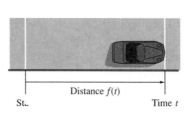

Fig. 1.5.4 A car on a straight roads (Example 1)

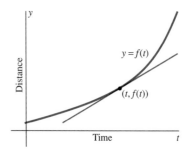

Fig. 1.5.5 The slope of the tangent line at the point $(t, f(t))$ is the velocity at time t (Example 1).

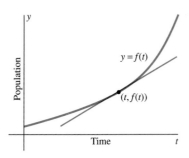

Fig. 1.5.6 The rate of growth of $f(t)$ at time t is the slope of the tangent line at the point $(t, f(t))$ (Example 2).

EXAMPLE 2 Suppose that $f(t)$ denotes the number of people in the United States who have a certain serious disease at time t (measured in days from the beginning of the year). Then the slope of the line tangent to the curve $y = f(t)$ at the point $(t, f(t))$ (Fig. 1.5.6) is the *rate of growth* (the number of persons newly afflicted per day) of the diseased population at time t.

NOTE The truth of the statements made in these two examples is *not* supposed to be obvious to you. To learn such things is one reason you study calculus! We return to the concepts of velocity and rate of change at the beginning of Chapter 3.

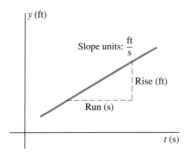

Fig. 1.5.7 Here slope has the dimensions of velocity (ft/s).

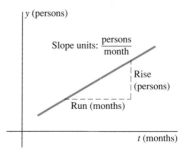

Fig. 1.5.8 Here slope has the dimensions of rate of change of population.

Here we will be content with the observation that the tangent lines in Example 1 and 2 at least have the correct *units*. If in the time-distance plane of Example 1 we measure time t (on the horizontal axis) in seconds and distance y (on the vertical axis) in feet (or meters), then the slope ("rise/run") of a straight line has the dimensions of feet (or meters) per second—the proper units for velocity (Fig. 1.5.7). Similarly, if in the ty-plane of Example 2 time t is measured in months and y is measured in persons, then the slope of a straight line has the proper units of persons per month for measuring the rate of growth of the afflicted population (Fig. 1.5.8).

The second fundamental problem of calculus is the problem of area. Given the graph $y = f(x)$, what is the area between the graph of f and the x-axis, for instance, over an interval?

The Area Problem If $f(x) \geqq 0$ for x in the interval $[a, b]$, how do we calculate the area A of the plane region that lies under the curve $y = f(x)$ and above this interval (Fig. 1.5.9)?

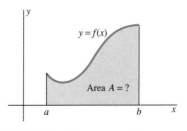

Fig. 1.5.9 The area problem

We begin to explore the answer to this second question in Chapter 5. In calculus the area A is called an *integral* of the function f. Chapters 5 and 6 are devoted to the calculation and application of integrals. This gives an introduction to the other part of calculus, which is called *integral calculus*.

Like the tangent problem, the area problem is a purely mathematical geometric question, but its answer (in the form of integrals) has extensive ramifications of practical importance. Examples 3 and 4 have an obvious kinship with Examples 1 and 2.

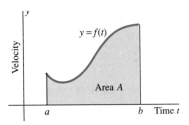

Fig. 1.5.10 The area A under the velocity curve is equal to the distance traveled during the time interval $a \leq t \leq b$ (Example 3).

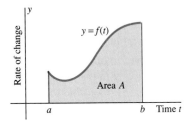

Fig. 1.5.11 The area A under the rate-of-change curve is equal to the net change in the population from time $t = a$ to time $t = b$ (Example 4).

EXAMPLE 3 If $f(t)$ denotes the *velocity* of a car at time t, then the area under the curve $y = f(t)$ over the interval $[a, b]$ is equal to the *distance* traveled by the car between time $t = a$ and time $t = b$ (Fig. 1.5.10).

EXAMPLE 4 If $f(t)$ denotes the *rate of growth* of a diseased population at time t, then the area under the curve $y = f(t)$ over the interval $[a, b]$ is equal to the net change in the *size* of this population between time $t = a$ and time $t = b$ (Fig. 1.5.11).

When we discuss integrals in Chapter 5, you will learn why the statements in Examples 3 and 4 are true.

THE FUNDAMENTAL RELATIONSHIP

Examples 1 and 3 are two sides of a certain coin: There is an "inverse relationship" between the *distance* traveled and the *velocity* of a moving car. Examples 2 and 4 exhibit a similar relationship between the *size* of a population and its *rate of change*.

Both the distance/velocity relationship and the size/rate-of-change relationship illustrated by Examples 1 through 4 are consequences of a deep and fundamental relationship between the tangent problem and the area problem. This more general relationship is described by the *fundamental theorem of calculus,* which we discuss in Section 5.6. It was discovered in 1666 by Isaac Newton at the age of 23 while he was still a student at Cambridge University. A few years later it was discovered independently by Gottfried Wilhelm Leibniz, who was then a German diplomat in Paris who studied mathematics privately. Although the tangent problem and the area problem had, even then, been around for almost 2000 years, and much progress on separate solutions had been made by predecessors of Newton and Leibniz, their joint discovery of the fundamental relationship between the area and tangent problems made them famous as "the inventors of the calculus."

So calculus centers around the computation *and application* of derivatives and integrals—that is, of tangent line slopes and areas under graphs. Throughout this textbook, you will see concrete applications of calculus to different areas of science and technology. The following list of a dozen such applications gives just a brief indication of the extraordinary range and real-world power of calculus.

❏ Suppose that you make and sell tents. How can you make the biggest tent from a given amount of cloth and thereby maximize your profit? (Section 3.6)

❏ You throw into a lake a cork ball that has one-fourth the density of water. How deep will it sink in the water? (Section 3.9)

❏ A driver involved in an accident claims he was going only 25 mi/h. Can you determine from his skid marks the actual speed of his car at the time of the accident? (Section 6.6)

❏ The great pyramid of Khufu at Gizeh, Egypt, was built well over 4000 years ago. No personnel records from the construction remain, but nevertheless we can calculate the approximate number of laborers involved. (Section 6.6)

❑ If the earth's population continues to grow at its present rate, when will there be "standing room only"? (Section 7.5)

❑ The factories polluting Lake Erie are forced to cease dumping wastes into the lake immediately. How long will it take for natural processes to restore the lake to an acceptable level of purity? (Section 7.6)

❑ In 1845 the Belgian demographer Verhulst used calculus to predict accurately the course of U.S. population growth (to within 1%) well into the twentieth century, long after his death. How? (Section 9.5)

❑ Suppose that you win the Florida lottery and decide to use part of your winnings to purchase a "perpetual annuity" that will pay you and your heirs (and theirs, *ad infinitum*) $10,000 per year. What is a fair price for an insurance company to charge you for such an annuity? (Section 9.8)

❑ What explains the fact that a well-positioned reporter can eavesdrop on a quiet conversation between two diplomats 50 feet away in the Whispering Gallery of the U.S. Senate, even if this conversation is inaudible to others in the same room? (Section 10.5)

❑ Suppose that Paul and Mary alternately toss a fair six-sided die in turn until one wins the pot by getting the first "six." How advantageous is it to be the one who tosses first? (Section 11.3)

❑ How can a submarine traveling in darkness beneath a polar icecap keep accurate track of its position without being in radio contact with the rest of the world? (Section 12.4)

❑ Suppose that your club is designing an unpowered race car for the annual downhill derby. You have a choice of solid wheels, bicycle wheels with thin spokes, or even solid spherical wheels (like giant ball bearings). Can you apply calculus to determine (without time-consuming experimentation) which will make the race car go the fastest? (Section 15.5)

Chapter 1 Review: DEFINITIONS, CONCEPTS, RESULTS

Use this list as a guide to concepts that you may need to review.

1. Rational and irrational numbers
2. The real number line
3. Properties of inequalities
4. Absolute value of a real number
5. Properties of the absolute value function
6. The triangle inequality
7. Open and closed intervals
8. The definition of function
9. The domain of a function
10. Independent and dependent variables
11. The coordinate plane
12. The Pythagorean theorem

13. The distance formula
14. The midpoint formula
15. The slope of a straight line
16. The point-slope equation of a line
17. The slope-intercept equation of a line
18. The slope relationship between parallel lines
19. The slope relationship between perpendicular lines
20. The graph of an equation
21. The graph of a function
22. Equations and translates of circles
23. Parabolas and graphs of quadratic functions
24. Qualitative differences between the graphs of polynomials, rational functions, trigonometric functions, and exponential and logarithmic functions

Ch. 1 / Functions and Graphs

for the difference of the x-coordinates of the points P and Q. (The notation Δx is as old as calculus itself, but it means now what it did 300 years ago: an **increment,** or *change,* in the value of x. As we discussed in Section 1.2 when we introduced the concept of the slope of a line, Δx isn't the product of Δ and x but should be thought of as a single symbol, like h.) Then the coordinates of Q are given by the formulas

$$b = a + h = a + \Delta x \quad \text{and} \quad b^2 = (a + h)^2 = (a + \Delta x)^2.$$

Hence the difference in the y-coordinates of P and Q is

$$\Delta y = b^2 - a^2 = (a + h)^2 - a^2.$$

Because $Q \neq P$, we can use the definition of slope to compute the slope of the secant line K through P and Q. We denote this slope by the function notation $m(h)$, because the slope m is a function of h: If you change the value of h, you change the line K and thereby change its slope. Therefore,

$$m(h) = \frac{\Delta y}{\Delta x} = \frac{b^2 - a^2}{b - a} = \frac{(a + h)^2 - a^2}{(a + h) - a} = \frac{2ah + h^2}{h}. \tag{1}$$

Because h is nonzero, we may cancel it in the last fraction. Thus we find that

$$m(h) = 2a + h. \tag{2}$$

Now imagine what happens as you move the point Q closer and closer to the point P. (This situation corresponds to h approaching zero.) The line K still passes through P and Q, but it pivots around the fixed point P. As h approaches zero, the secant line K comes closer to coinciding with the tangent line L. This phenomenon is suggested in Fig. 2.1.3, which shows the secant line K as almost indistinguishable from the tangent line L.

The tangent line L must, by definition, lie in the *limiting position* of the secant line K. To see precisely what this means, examine what happens to the slope of K as K pivots into coincidence with L:

> As h approaches zero,
> Q approaches P, and so
> K approaches L; meanwhile,
> *the slope of K approaches the slope of L.*

But we found in Eq. (2) that the slope of the secant line K is

$$m(h) = 2a + h.$$

If we think of values of h that become closer and closer to zero (such as $h = 0.01$, $h = 0.0001$, and $h = 0.000001$), then it is clear from the table in Fig. 2.1.4 that

$$m(h) = 2a + h \quad \text{approaches } 2a \text{ as } h \text{ approaches zero.}$$

Consequently, we *must* define the tangent line L to be the straight line through P that has slope

$$m = 2a.$$

In this circumstance we say that $m = 2a$ is the **limit of $m(h)$** as h approaches zero, and we write

$$m = \lim_{h \to 0} m(h) = \lim_{h \to 0} (2a + h) = 2a. \tag{3}$$

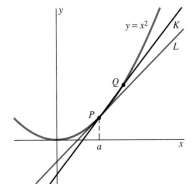

Fig. 2.1.3 As $h \to 0$, Q approaches P, and K moves into coincidence with the tangent line L (Example 1).

h	$2a + h$
0.1	$2a + 0.1$
0.01	$2a + 0.01$
0.001	$2a + 0.001$
$\vdots$	$\vdots$
$\downarrow$	$\downarrow$
0	$2a$

Fig. 2.1.4 As h approaches zero, $2a + h$ approaches $2a$ (Example 1).

The phrase $h \to 0$ is read "h approaches zero." An alternative to the form in Eq. (3) is

$$2a + h \to 2a \quad \text{as} \quad h \to 0.$$

The result $m = 2a$ in Eq. (3) is a "slope predictor" for lines tangent to the parabola $y = x^2$. You can use it to write the *equation* of the line tangent to the parabola $y = x^2$ at any desired point (a, a^2) of the parabola.

EXAMPLE 2   To find the equation of the line tangent to the parabola $y = x^2$ at the point $(3, 9)$, we take $a = 3$ in Eq. (3). We find that the slope of the tangent line is $m = 6$. The point-slope equation of the tangent line is therefore

$$y - 9 = 6(x - 3);$$

its slope-intercept form is $y = 6x - 9$.

The general case $y = f(x)$ is scarcely more complicated than the special case $y = x^2$. Suppose that $y = f(x)$ is given and that we want to find the slope of the tangent line L to its graph at the point $(a, f(a))$. As shown in Fig. 2.1.5, let K be the secant line passing through the point $P(a, f(a))$ and a nearby point $Q(a + h, f(a + h))$ on the graph. The slope of this secant line K is the **difference quotient**

$$m(h) = \frac{\Delta y}{\Delta x} = \frac{f(a + h) - f(a)}{h} \qquad \text{(for } h \neq 0\text{).} \qquad (4)$$

We now force Q to approach P along the graph of f by making h approach zero. Suppose that $m(h)$ approaches the number m as h gets closer and closer to zero. Then the *tangent line L* to the curve $y = f(x)$ at the point $P(a, f(a))$ is, *by definition*, the line through P that has slope m.

To describe the fact that $m(h)$ approaches m as h approaches zero, we call m the **limit of $m(h)$** as h approaches zero, and we write

$$m = \lim_{h \to 0} m(h) = \lim_{h \to 0} \frac{f(a + h) - f(a)}{h}. \qquad (5)$$

The slope m in Eq. (5) depends both on the function f and on the number a. We indicate this by writing

$$m = f'(a) = \lim_{h \to 0} \frac{f(a + h) - f(a)}{h}. \qquad (6)$$

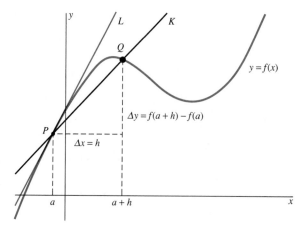

Fig. 2.1.5   As $h \to 0$, $Q \to P$, and the slope of K approaches the slope of the tangent line L.

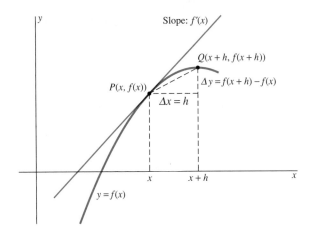

Fig. 2.1.6 The slope of the tangent line at $(x, f(x))$ is $f'(x)$.

In Eq. (6), f' (read "f prime") is a *function* with independent variable a. To return to the standard symbol x for the independent variable, we simply replace a by x. (We assumed nothing about the particular value of a, so we change nothing by writing x in place of a.) Thus we obtain the definition of the new function f':

$$f'(x) = \lim_{h \to 0} \frac{f(x + h) - f(x)}{h}. \tag{7}$$

This new function f' is defined for all values of x for which the limit in Eq. (7) exists. Because f' is "derived" from f, f' is called the **derivative** of the original function f, and the process of computing the formula for f' is called **differentiation** of f.

Our discussion leading up to Eq. (7) shows that the derivative of f has the following important geometric interpretation:

The slope of the line tangent to the graph of the curve $y = f(x)$ at the point $(x, f(x))$ is $f'(x)$.

This interpretation is illustrated in Fig. 2.1.6.

The relationship between the curve and its tangent line is such that if we "zoom in" on the point of tangency, successive magnifications show less and less of a difference between the curve and the tangent line. This phenomenon is illustrated in Figs. 2.1.7 through 2.1.10.

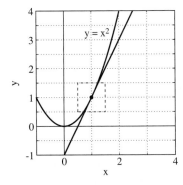

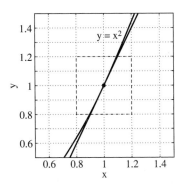

Fig. 2.1.7 The parabola $y = x^2$ and its tangent line at $P(1, 1)$

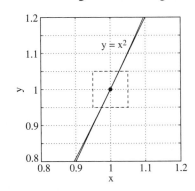

Fig. 2.1.8 First magnification

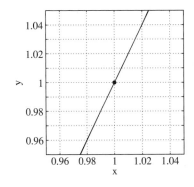

Fig. 2.1.9 Second magnification

Fig. 2.1.10 Can you see the difference?

EXAMPLE 3 Upon replacing a by x in Eq. (3), we see that the slope of the line tangent to the parabola $y = x^2$ at the point (x, x^2) is $m = 2x$. In the language of derivatives,

$$\text{If } f(x) = x^2, \quad \text{then} \quad f'(x) = 2x. \tag{8}$$

What's important is the *idea* that a formula such as Eq. (8) expresses, not the particular notation used. Thus it follows immediately from Eq. (8) that if $g(t) = t^2$, then $g'(t) = 2t$. In short, the derivative of the "squaring function" is the "doubling function." In any case, Eq. (8) implies that the slope of the line tangent to the parabola $y = x^2$ at the point $(1, 1)$ is

$$f'(1) = 2 \cdot 1 = 2,$$

and the slope of the tangent line at the point $(-2, 4)$ is

$$f'(-2) = 2 \cdot (-2) = -4.$$

For us to find the derivative of a given function f, the definition

$$f'(x) = \lim_{h \to 0} \frac{f(x + h) - f(x)}{h}$$

in Eq. (7) calls for us to carry out the following four steps:

1. Write the definition of $f'(x)$.
2. Substitute into this definition the formula of the given function f.
3. Make algebraic substitutions until Step 4 can be carried out.
4. Determine the value of the limit as $h \to 0$.

Remember that x may be thought of as a *constant* throughout this computation—it is h that is the variable in this four-step process.

EXAMPLE 4 The derivative of the function

$$f(x) = 3x + 2$$

must be the constant function $f'(x) = 3$, because $y = 3x + 2$ is the equation of a straight line of slope $m = 3$, and that line *must* be its own tangent line at every point. To check this reasoning by using the four-step process just given, we find that

$$f'(x) = \lim_{h \to 0} \frac{f(x + h) - f(x)}{h}$$

$$= \lim_{h \to 0} \frac{[3(x + h) + 2] - [3x + 2]}{h}$$

$$= \lim_{h \to 0} \frac{3h}{h} = \lim_{h \to 0} 3 = 3.$$

Therefore,

$$f'(x) \equiv 3$$

(a constant function). We conclude in the final step that the limit is 3 because the constant 3 does not change as $h \to 0$.

The symbol $\lim_{h \to 0}$ signifies an operation to be performed, so we must continue to write it until the final step, when the operation of taking the limit *is* performed.

With computations such as those in Examples 1 and 4, it is just as easy to find—once and for all—the derivative of an arbitrary *quadratic function* $f(x) = ax^2 + bx + c$ (where a, b, and c are constants).

> **Principle** *Differentiation of Quadratic Functions*
> If $f(x) = ax^2 + bx + c$, then
> $$f'(x) = 2ax + b. \tag{9}$$

Proof   The four-step process for applying the definition of the derivative yields

$$f'(x) = \lim_{h \to 0} \frac{f(x + h) - f(x)}{h}$$

$$= \lim_{h \to 0} \frac{[a(x + h)^2 + b(x + h) + c] - [ax^2 + bx + c]}{h}$$

$$= \lim_{h \to 0} \frac{2axh + bh + ah^2}{h} = \lim_{h \to 0} (2ax + b + ah).$$

Thus we find that

$$f'(x) = 2ax + b,$$

because the value of ah approaches zero as $h \to 0$. ❏

In geometric terms, this theorem tells us that the slope-prediction formula for curves of the form $y = ax^2 + bx + c$ is

$$m = 2ax + b. \tag{10}$$

NORMAL LINES

How would you find the point $P(c, c^2)$ that lies on the parabola $y = x^2$ and is closest to the point $(3, 0)$? Intuitively, the line segment N with endpoints $(3, 0)$ and (c, c^2) should be perpendicular, or *normal,* to the graph at (c, c^2) (Fig. 2.1.11). This warrants a precise definition: The line N through the point P on the curve $y = f(x)$ is said to be **normal** to the curve at P provided that N is perpendicular to the tangent line at P, as in Fig. 2.1.12. By Theorem 2 in Section 1.2, the slope of N is $-1/f'(c)$ provided that $f'(c) \neq 0$. Thus with the aid of Eq. (10), you can write equations of normal lines at points of quadratic curves as easily as you can write equations of tangent lines.

REMARK   Now that the derivative formula in Eq. (9) is available, you can apply it to differentiate specific quadratic functions immediately, instead of retracing the steps in the proof of the theorem. For example, if

$$f(x) = 2x^2 - 3x + 5,$$

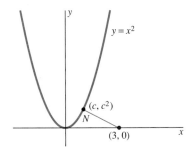

Fig. 2.1.11   The normal line N from the point $(3, 0)$ to the point (c, c^2) on the parabola $y = x^2$

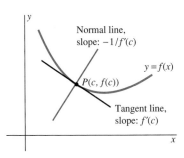

Fig. 2.1.12   The tangent line and normal line to a curve at a point

then you can immediately write

$$f'(x) = 2 \cdot 2x - 3 = 4x - 3.$$

EXAMPLE 5 Write equations for both the line tangent to and the line normal to the parabola $y = f(x) = 2x^2 - 3x + 5$ at the point $P\,(-1, 10)$.

Solution We use the derivative just computed. The slope of the tangent line at $(-1, 10)$ is

$$f'(-1) = 4 \cdot (-1) - 3 = -7.$$

Hence the point-slope equation of the desired tangent line is

$$y - 10 = -7(x + 1).$$

The normal line has slope $m = -1/(-7)$, so its point-slope equation is

$$y - 10 = \tfrac{1}{7}(x + 1).$$

EXAMPLE 6 Differentiate the function

$$f(x) = \frac{1}{x + 1}.$$

Solution We have yet to establish a formula for the derivative of f, so we must carry out the four steps that use the definition of the derivative. We get

$$f'(x) = \lim_{h \to 0} \frac{f(x + h) - f(x)}{h} = \lim_{h \to 0} \frac{1}{h}\left(\frac{1}{x + h + 1} - \frac{1}{x + 1}\right)$$

$$= \lim_{h \to 0} \frac{1}{h}\left(\frac{(x + 1) - (x + h + 1)}{(x + h + 1)(x + 1)}\right) = \lim_{h \to 0} \frac{1}{h} \cdot \frac{-h}{(x + h + 1)(x + 1)}$$

$$= \lim_{h \to 0} \frac{-1}{(x + h + 1)(x + 1)}.$$

It is apparent that $\lim_{h \to 0}(x + h + 1) = x + 1$. Hence we finally obtain the derivative of $f(x)$,

$$f'(x) = -\frac{1}{(x + 1)^2}.$$

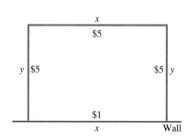

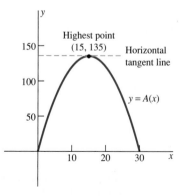

Fig. 2.1.13 The animal pen

THE ANIMAL PEN PROBLEM COMPLETED In conclusion, we apply the derivative to wrap up our continuing discussion of the animal pen problem of Section 1.1. In Example 6 there, we found that the area A of the pen (see Fig. 2.1.13) is given as a function of its base length x by

$$A(x) = \tfrac{3}{5}(30x - x^2) = -\tfrac{3}{5}x^2 + 18x, \qquad 0 \leq x \leq 30. \qquad (11)$$

Our problem is to find the maximum value of $A(x)$ for x in the closed interval $[0, 30]$.

Let us accept as intuitively obvious (we shall give a rigorous proof in Chapter 3) the fact that the maximum value of $A(x)$ occurs at a point where the line tangent to the graph of $y = A(x)$ is horizontal, as shown in Fig. 2.1.14. Then we can apply the result on differentiation of quadratic functions [Eq. (9)] to find this maximum point. The slope of the tangent line at an arbitrary point $(x, A(x))$ is given by

$$m = A'(x) = -\tfrac{3}{5} \cdot 2x + 18 = -\tfrac{6}{5}x + 18.$$

Fig. 2.1.14 The graph of $y = A(x), 0 \leq x \leq 30$

Ch. 2 / Prelude to Calculus

We ask when $m = 0$ and find that this happens when

$$-\tfrac{6}{5}x + 18 = 0,$$

and thus when $x = 15$. In agreement with the result obtained by algebraic methods in Section 1.3, we find that the maximum possible area of the pen is

$$A(15) = \tfrac{3}{5}(30 \cdot 15 - 15^2) = 135 \quad (\text{ft}^2).$$

NUMERICAL INVESTIGATION OF DERIVATIVES

Suppose that you are given a function f and a specific numerical value of x. You can use a calculator to investigate the value

$$f'(x) = \lim_{h \to 0} \frac{f(x+h) - f(x)}{h} \tag{12}$$

of the derivative of f by computing the values

$$m(x, h) = \frac{f(x+h) - f(x)}{h} \tag{13}$$

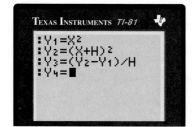

Fig. 2.1.15 A calculator prepared to compute $\dfrac{(x+h)^2 - x^2}{h}$

of the "h-quotient at x" with successively smaller values of h. Figure 2.1.15 shows a TI calculator prepared to compute Eq. (13) with $f(x) = x^2$, as indicated in Fig. 2.1.16. Figure 2.1.17 shows an HP calculator prepared to define the same quotient; then evaluation of the expression `'M(2, 0.001)'` yields the value $m(2, 0.001) = 4.001$. In this way we get the table shown in Fig. 2.1.18, which corroborates the fact that $f'(2) = 4$.

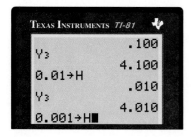

Fig. 2.1.16 Approximating $f'(2)$

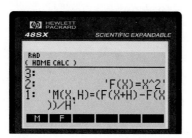

Fig. 2.1.17 A calculator prepared to compute $\dfrac{f(x+h) - f(x)}{h}$

h	$m(2, h)$
0.1	4.1
0.01	4.01
0.001	4.001
⋮	⋮
↓	↓
0	4

Fig. 2.1.18 Numerical investigation of the limit in Eq. (12) with $f(x) = x^2$, $x = 2$

2.1 Problems

In Problems 1 through 14, apply the formula for differentiation of quadratic functions [Eq. (9)] to write the derivative $f'(x)$ of the given function f.

1. $f(x) = 5$

2. $f(x) = x$

3. $f(x) = x^2$

4. $f(x) = 1 - 2x^2$

5. $f(x) = 4x - 5$

6. $f(x) = 7 - 3x$

7. $f(x) = 2x^2 - 3x + 4$

8. $f(x) = 5 - 3x - x^2$

9. $f(x) = 2x(x + 3)$

10. $f(x) = 3x(5 - x)$

11. $f(x) = 2x - \left(\dfrac{x}{10}\right)^2$ **12.** $f(x) = 4 - (3x + 2)^2$

13. $f(x) = (2x + 1)^2 - 4x$

14. $f(x) = (2x + 3)^2 - (2x - 3)^2$

In Problems 15 through 20, find all points of the curve $y = f(x)$ at which the tangent line is horizontal.

15. $y = 10 - x^2$ **16.** $y = 10x - x^2$

17. $y = x^2 - 2x + 1$ **18.** $y = x^2 + x - 2$

19. $y = x - \left(\dfrac{x}{10}\right)^2$ **20.** $y = x(100 - x)$

Apply the definition of the deriviative—that is, follow the four-step process of this section—to find $f'(x)$ for the functions in Problems 21 through 32. Then write an equation for the straight line that is tangent to the graph of the curve $y = f(x)$ at the point $(2, f(2))$.

21. $f(x) = 3x - 1$ **22.** $f(x) = x^2 - x - 2$

23. $f(x) = 2x^2 - 3x + 5$ **24.** $f(x) = 70x - x^2$

25. $f(x) = (x - 1)^2$ **26.** $f(x) = x^3$

27. $f(x) = \dfrac{1}{x}$ **28.** $f(x) = x^4$

29. $f(x) = \dfrac{1}{x^2}$ **30.** $f(x) = x^2 + \dfrac{3}{x}$

31. $f(x) = \dfrac{2}{x - 1}$ **32.** $f(x) = \dfrac{x}{x - 1}$

In Problems 33 through 43, use Eq. (9) to find the derivative of a quadratic function if necessary. In Problems 33 through 35, write equations for the tangent line and for the normal line to the curve $y = f(x)$ at the point P.

33. $y = x^2;$ $P(-2, 4)$

34. $y = 5 - x - 2x^2;$ $P(-1, 4)$

35. $y = 2x^2 + 3x - 5;$ $P(2, 9)$

36. Prove that the line tangent to the parabola $y = x^2$ at the point (x_0, y_0) intersects the x-axis at the point $(x_0/2, 0)$. See Fig. 2.1.19.

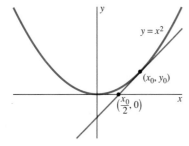

Fig. 2.1.19 The parabola and tangent line of Problem 36

37. If a ball is thrown straight upward with initial velocity 96 ft/s, then its height t seconds later is $y(t) = 96t - 16t^2$ feet. Determine the maximum height the ball attains by finding the time t when $y'(t) = 0$.

38. Find the maximum possible area of the rectangle with perimeter 100 described in Problem 50 of Section 1.1.

39. Find the maximum possible value of the product of two positive numbers whose sum is 50.

40. Suppose that a projectile is fired at an angle of $45°$ from the horizontal. Its initial position is the origin in the xy-plane, and its initial velocity is $100\sqrt{2}$ ft/s (Fig. 2.1.20). Then its trajectory will be the part of the parabola $y = x - (x/25)^2$ for which $y \geqq 0$. (a) How far does the projectile travel (horizontally) before it hits the ground? (b) What is the maximum height above the ground that the projectile attains?

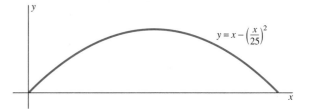

Fig. 2.1.20 The trajectory of the projectile of Problem 40

41. One of the two lines that pass through the point $(3, 0)$ and are tangent to the parabola $y = x^2$ is the x-axis. Find an equation for the *other* line. [*Suggestion:* First find the value of the number a shown in Fig. 2.1.21.]

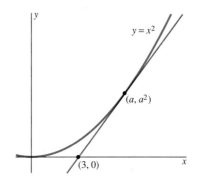

Fig. 2.1.21 Two lines tangent to the parabola of Problem 41

42. Write equations for the two straight lines that pass through the point $(2, 5)$ and are tangent to the parabola $y = 4x - x^2$. [*Suggestion:* Draw a figure like Fig. 2.1.21.]

43. Between Examples 4 and 5 we raised—but did not answer—the question of how to locate the point on the graph of $y = x^2$ closest to the point $(3, 0)$. It's now time for you to find that point. [*Suggestion:* Draw a figure like Fig. 2.1.21. The cubic equation you obtain has one solution apparent by inspection.]

In Problems 44 through 50, use a calculator to investigate (as in the Numerical Investigation of Derivatives at the end of this section) the value of f′ at the number a.

44. $f(x) = x^2$; $a = -1$
45. $f(x) = x^3$; $a = 2$
46. $f(x) = x^3$; $a = -1$
47. $f(x) = \sqrt{x}$; $a = 1$
48. $f(x) = \sqrt{x}$; $a = 4$
49. $f(x) = \dfrac{1}{x}$; $a = 1$
50. $f(x) = \dfrac{1}{x}$; $a = -\frac{1}{2}$

2.1 Project

The following problems are motivated by Figs. 2.1.7 through 2.1.10, which illustrate the fact that a curve and its tangent line are virtually indistinguishable at a sufficiently high level of magnification. They require a graphics utility (calculator or computer) with a *trace facility* that enables you to move the cross hairs (cursor) and to read the xy-coordinates of selected points on a graph. For example, with some TI graphics calculators, you can simply press the [TRACE] key and then move the cursor along the graph. With some computer graphing programs, the xy-coordinates of the current position of the cursor are automatically displayed at the bottom of the monitor screen.

Each problem lists a curve $y = f(x)$ and a point $P(a, f(a))$ at which the slope $f'(a)$ of its tangent line is to be approximated graphically. Zoom in on P (magnifying at least six times) until the graph $y = f(x)$ looks (in the resulting viewing window) precisely like a straight line. At each zoom, record the coordinates (x_1, y_1) and (x_2, y_2) of two points located on either side of P (as in Fig. 2.1.22). Then let

$$m_k = \frac{\Delta y}{\Delta x} = \frac{y_2 - y_1}{x_2 - x_1}$$

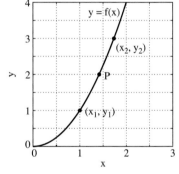

Fig. 2.1.22 Points on either side of P

denote the approximate slope that results at the kth zoom. Is it clear what limiting value the approximate slopes $m_1, m_2, m_3, \ldots$ are approaching? [In each case $f'(a)$ should be an integer or an integral multiple of $\frac{1}{2}$, $\frac{1}{4}$, or $\frac{1}{8}$.]

1. $f(x) = x^2$, $P = P(-2, 4)$; $f'(-2) = ?$
2. $f(x) = \sqrt{x}$, $P = P(1, 1)$; $f'(1) = ?$
3. $f(x) = \dfrac{1}{x}$, $P = P(2, \frac{1}{2})$; $f'(2) = ?$
4. $f(x) = \dfrac{12}{x^2}$, $P = P(-4, \frac{3}{4})$; $f'(-4) = ?$
5. $f(x) = \sqrt{x^2 - 9}$, $P = P(5, 4)$; $f'(5) = ?$

2.2
The Limit Concept

In Section 2.1 we defined the derivative $f'(a)$ of the function f at the number a to be

$$f'(a) = \lim_{h \to 0} \frac{f(a + h) - f(a)}{h}. \tag{1}$$

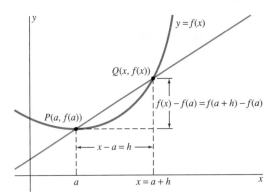

Fig. 2.2.1 The derivative can be defined in this way:

$$f'(a) = \lim_{x \to a} \frac{f(x) - f(a)}{x - a}.$$

The graph that motivated this definition is repeated in Fig. 2.2.1, with $a + h$ relabeled as x (so that $h = x - a$). We see that x approaches a as h approaches zero, so Eq. (1) can be rewritten as

$$f'(a) = \lim_{x \to a} \frac{f(x) - f(a)}{x - a}. \tag{2}$$

Thus the computation of $f'(a)$ amounts to the determination of the limit, as x approaches a, of the function

$$G(x) = \frac{f(x) - f(a)}{x - a}. \tag{3}$$

To prepare for the differentiation of more complicated functions than we could handle in Section 2.1, we now turn to the meaning of the statement

$$\lim_{x \to a} F(x) = L. \tag{4}$$

This is read "the limit of $F(x)$ as x approaches a is L." We shall sometimes write Eq. (4) in the concise form

$$F(x) \to L \quad \text{as} \quad x \to a.$$

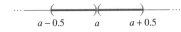

Fig. 2.2.2 A deleted neighborhood of a: here, the open interval $(a - 0.5, a + 0.5)$ with the point a removed. Thus this deleted neighborhood of a consists of the two open intervals $(a - 0.5, a)$ and $(a, a + 0.5)$.

The function F need not be defined at a for us to discuss the limit of F at a. Neither its value there nor the possibility that it has no assigned value at a is important. We require only that F be defined in some **deleted neighborhood** of a: a set obtained by deleting the single point a from some open interval containing a. For example, it would suffice for F to be defined on the open intervals $(a - 0.5, a)$ and $(a, a + 0.5)$ (see Fig. 2.2.2) but not necessarily at the point a itself. This is exactly the situation for the function in Eq. (3), which is defined *except* at a (where the denominator is zero).

The following statement rephrases Eq. (4) in intuitive language.

> **Idea of the Limit**
>
> We say that the number L is the *limit* of $F(x)$ as x approaches a provided that we can make the number $F(x)$ as close to L as we please merely by choosing x sufficiently near, though not equal to, the number a.

What this means, roughly, is that $F(x)$ tends to get closer and closer to L as x gets closer and closer to a. Once we decide how close to L we want $F(x)$

Ch. 2 / Prelude to Calculus

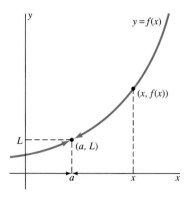

Fig. 2.2.3 Graphical interpretation of the limit concept

to be, it is necessary that $F(x)$ be that close to L for *all* x sufficiently close to (but not equal to) a.

Figure 2.2.3 shows a graphical interpretation of the limit concept. As x approaches a (from either side), the point $(x, f(x))$ on the graph $y = f(x)$ must approach the point (a, L).

In this section we explore the idea of the limit, mainly through the investigation of specific examples. A precise statement of the definition of the limit appears at the end of this section.

EXAMPLE 1 Evaluate $\lim\limits_{x \to 2} x^2$.

Investigation We call this an investigation rather than a solution. The reason is that the technique used here is useful for gathering preliminary information about what the value of a limit might be. This technique may reinforce our intuitive guess as to what the value of a limit is likely to be, but it does not constitute a rigorous proof.

We form a table of values of x^2, using values of x that approach $a = 2$. Such a table is shown in Fig. 2.2.4. We use regularly changing values of x because that makes the behavior exhibited in the table easier to understand. We use values of x that are nicely expressible in the decimal system for the same reason.

x	x^2	x	x^2
1.9	3.6100	2.1	4.4100
1.99	3.9601	2.01	4.0401
1.999	3.9960	2.001	4.0040
1.9999	3.9996	2.0001	4.0004
1.99999	4.0000	2.00001	4.0000
⋮	⋮	⋮	⋮
↓	↓	↓	↓
2	4	2	4

Fig. 2.2.4 What happens to x^2 as x approaches 2 (Example 1)?

Note that we consider—in separate columns—values of x that approach from the left (through values of x less than 2) as well as values of x that approach 2 from the right (through values of x greater than 2).

Now examine the table—read down the columns for x, because *down* is the table's direction for "approaches"—to see what happens to the corresponding values of x^2. Even though Fig. 2.2.4 uses only a few special values of x, it provides evidence that

$$\lim_{x \to 2} x^2 = 4.$$

IMPORTANT *Note that we did not substitute the value $x = 2$ into the function $F(x) = x^2$ to obtain the value 4 of the limit.* Although such substitution produces the correct answer in this particular case, in many limits it produces either an incorrect answer or no answer at all. (See Examples 4 and 6 and, among many others, Problems 9, 10, and 51.)

EXAMPLE 2 Evaluate $\lim\limits_{x \to 3} \dfrac{x - 1}{x + 2}$.

x	$\dfrac{x-1}{x+2}$ (Values rounded)
3.1	0.41176
3.01	0.40120
3.001	0.40012
3.0001	0.40001
3.00001	0.40000
3.000001	0.40000
$\vdots$	$\vdots$
$\downarrow$	$\downarrow$
3	0.4

Fig. 2.2.5 Investigating the limit in Example 2

Investigation   A natural guess is that the value of the limit is plainly $\frac{2}{5} = 0.4$. The data shown in Fig. 2.2.5 reinforce this guess. If you experiment with other values of x that approach 3, the results will support the guess that the limit is indeed 0.4.

EXAMPLE 3   Evaluate $\lim\limits_{x \to -4} \sqrt{x^2 + 9}$.

Investigation   The evidence in Fig. 2.2.6 certainly suggests that

$$\lim_{x \to -4} \sqrt{x^2 + 9} = 5.$$

EXAMPLE 4   Evaluate $\lim\limits_{x \to 0} \dfrac{\sqrt{x + 25} - 5}{x}$.

Investigation   Here we cannot make a preliminary guess by substituting $x = 0$, because the fraction is meaningless when $x = 0$. But Fig. 2.2.7 indicates that

$$\lim_{x \to 0} \frac{\sqrt{x + 25} - 5}{x} = 0.1.$$

x	$\sqrt{x^2 + 9}$ (Values rounded)
-4.1	5.080354
-4.01	5.008004
-4.001	5.000800
-4.0001	5.000080
-4.00001	5.000008
-4.000001	5.000001
-4.0000001	5.000000
$\vdots$	$\vdots$
$\downarrow$	$\downarrow$
-4	5

Fig. 2.2.6   The behavior of $\sqrt{x^2 + 9}$ as $x \to -4$ (Example 3)

x	$\dfrac{\sqrt{x + 25} - 5}{x}$ (Values rounded)
10	0.09161
1	0.09902
0.1	0.09990
0.01	0.09999
0.001	0.10000
0.0001	0.10000
$\vdots$	$\vdots$
$\downarrow$	$\downarrow$
0	0.1

Fig. 2.2.7   Numerical data for Example 4

The numerical investigations in Examples 2 through 4 are incomplete in that each of the associated tables shows values of the function $F(x)$ on only one side of the point $x = a$. But in order that $\lim_{x \to a} F(x) = L$, it is necessary for $F(x)$ to approach L *both* as x approaches a from the left *and* as x approaches a from the right. If $F(x)$ approaches different values as x approaches a from different sides, then $\lim_{x \to a} F(x)$ does not exist. In Section 2.3 we discuss *one-sided* limits in more detail.

EXAMPLE 5   Investigate $\lim\limits_{x \to 0} F(x)$, given

$$F(x) = \frac{x}{|x|} = \begin{cases} 1 & \text{if } x > 0; \\ -1 & \text{if } x < 0. \end{cases}$$

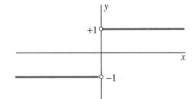

Fig. 2.2.8   The graph of $f(x) = \dfrac{x}{|x|}$ (Example 5)

Solution   From the graph of F shown in Fig. 2.2.8, it is apparent that $F(x) \to 1$ as $x \to 0$ from the right and that $F(x) \to -1$ as $x \to 0$ from the left. In particular, there are positive values of x as close to zero as we please such that $F(x) = 1$ and negative values of x equally close to zero such that $F(x) = -1$. Hence we cannot make $F(x)$ as close as we please to any *single* value of L *merely* by choosing x sufficiently close to zero. Therefore,

$$\lim_{x \to 0} \frac{x}{|x|} \quad \text{does not exist.}$$

In Example 6 the value obtained by substituting $x = a$ in $F(x)$ to find $\lim_{x \to a} F(x)$ is incorrect.

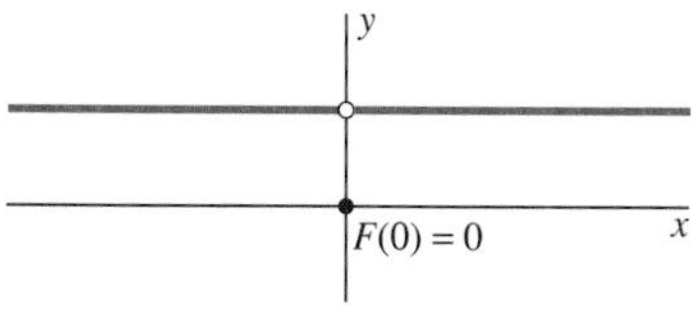

EXAMPLE 6 Evaluate $\lim_{x \to 0} F(x)$ where

$$F(x) = \begin{cases} 1 & \text{if } x \neq 0; \\ 0 & \text{if } x = 0. \end{cases}$$

The graph of F is shown in Fig. 2.2.9.

Fig. 2.2.9 The graph of the function F of Example 6

Solution The fact that $F(x) = 1$ for *every* value of x in any deleted neighborhood of zero implies that

$$\lim_{x \to 0} F(x) = 1.$$

But note that the value of the limit at $x = 0$ is *not* equal to the functional value $F(0) = 0$ there.

Numerical investigations such as those in Examples 1 through 4 provide us with an intuitive feeling for limits and typically suggest the correct value of a limit. But most limit computations are based neither on merely suggestive (and imprecise) numerical estimates nor on direct (but difficult) applications of the definition of limit. Instead, such computations are performed most easily and naturally with the aid of the *limit laws* that we give next. These "laws" actually are *theorems,* whose proofs (based on the precise definition of the limit) are included in Appendix B.

THE LIMIT LAWS

Constant Law
If $F(x) \equiv C$, where C is a constant [so $F(x)$ is a **constant function**], then

$$\lim_{x \to a} F(x) = \lim_{x \to a} C = C. \tag{5}$$

Sum Law
If both limits

$$\lim_{x \to a} F(x) = L \quad \text{and} \quad \lim_{x \to a} G(x) = M$$

exist, then

$$\lim_{x \to a} [F(x) \pm G(x)] = \lim_{x \to a} F(x) \pm \lim_{x \to a} G(x) = L \pm M. \tag{6}$$

(The limit of a sum is the sum of the limits; the limit of a difference is the difference of the limits.)

Product Law

If both limits

$$\lim_{x \to a} F(x) = L \quad \text{and} \quad \lim_{x \to a} G(x) = M$$

exist, then

$$\lim_{x \to a} [F(x)G(x)] = \left[\lim_{x \to a} F(x)\right]\left[\lim_{x \to a} G(x)\right] = LM. \tag{7}$$

(The limit of a product is the product of the limits.)

Quotient Law

If both limits

$$\lim_{x \to a} F(x) = L \quad \text{and} \quad \lim_{x \to a} G(x) = M$$

exist *and* if $M \neq 0$, then

$$\lim_{x \to a} \frac{F(x)}{G(x)} = \frac{\lim_{x \to a} F(x)}{\lim_{x \to a} G(x)} = \frac{L}{M}. \tag{8}$$

(The limit of a quotient is the quotient of the limits, provided that the limit of the denominator is not zero.)

Root Law

If n is a positive integer and if $a > 0$ for even values of n, then

$$\lim_{x \to a} \sqrt[n]{x} = \sqrt[n]{a}. \tag{9}$$

The case $n = 1$ of the root law is obvious:

$$\lim_{x \to a} x = a. \tag{10}$$

Examples 7 and 8 show how the limit laws can be used to evaluate limits of polynomials and of rational functions.

EXAMPLE 7

$$\begin{aligned}
\lim_{x \to 3} (x^2 + 2x + 4) &= \lim_{x \to 3} x^2 + \lim_{x \to 3} 2x + \lim_{x \to 3} 4 \\
&= \left(\lim_{x \to 3} x\right)^2 + 2\left(\lim_{x \to 3} x\right) + \lim_{x \to 3} 4 \\
&= 3^2 + 2 \cdot 3 + 4 = 19.
\end{aligned}$$

EXAMPLE 8

$$\begin{aligned}
\lim_{x \to 3} \frac{2x + 5}{x^2 + 2x + 4} &= \frac{\lim_{x \to 3} (2x + 5)}{\lim_{x \to 3} (x^2 + 2x + 4)} \\
&= \frac{2 \cdot 3 + 5}{3^2 + 2 \cdot 3 + 4} = \frac{11}{19}.
\end{aligned}$$

Ch. 2 / Prelude to Calculus

NOTE   In Examples 7 and 8, we systematically applied the limit laws until we could simply substitute 3 for $\lim_{x \to 3} x$ at the final step. To determine the limit of a quotient of polynomials, we must verify before this final step that the limit of the denominator is not zero. If the denominator limit is zero, the limit *may* fail to exist.

EXAMPLE 9   Investigate $\lim_{x \to 1} \dfrac{1}{(x-1)^2}$.

Solution   Because $\lim_{x \to 1} (x-1)^2 = 0$, we cannot apply the quotient law. Moreover, we can make $1/(x-1)^2$ arbitrarily large by choosing x sufficiently close to 1. Hence $1/(x-1)^2$ cannot approach any (finite) number L as x approaches 1. Therefore, the limit in this example does not exist. You can see the geometric reason if you examine the graph of $y = 1/(x-1)^2$ in Fig. 2.2.10. As x approaches 1, the corresponding point on the curve must leave the indicated strip between the two horizontal lines $y = L - \epsilon$ and $y = L + \epsilon$ that bracket the proposed limit L and hence cannot approach the point $(1, L)$.

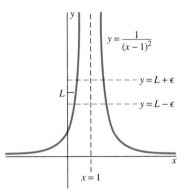

Fig. 2.2.10   The graph of $y = \dfrac{1}{(x-1)^2}$ (Example 9)

EXAMPLE 10   Investigate $\lim_{x \to 2} \dfrac{x^2 - 4}{x^2 + x - 6}$.

Solution   We cannot immediately apply the quotient law (as we did in Example 8) because the denominator approaches zero as x approaches 2. If the numerator were approaching some number *other* than zero as $x \to 2$, then the limit would fail to exist (as in Example 9). But here the numerator *does* approach zero, so there is a possibility that a factor of the numerator can be canceled with a factor of the denominator, thus removing the zero-denominator problem. Indeed,

$$\lim_{x \to 2} \frac{x^2 - 4}{x^2 + x - 6} = \lim_{x \to 2} \frac{(x-2)(x+2)}{(x-2)(x+3)}$$
$$= \lim_{x \to 2} \frac{x+2}{x+3} = \frac{4}{5}.$$

We can cancel the factor $x - 2$ because it is nonzero: $x \neq 2$ when we evaluate the limit as x approaches 2.

SUBSTITUTION OF LIMITS

Let us reconsider the limit of Example 3. It is tempting to write

$$\lim_{x \to -4} \sqrt{x^2 + 9} = \sqrt{\lim_{x \to -4} (x^2 + 9)}$$
$$= \sqrt{(-4)^2 + 9} = \sqrt{25} = 5. \qquad (11)$$

But can we simply "move the limit inside the radical" in Eq. (11)? To analyze this question, let us write

$$f(x) = \sqrt{x} \quad \text{and} \quad g(x) = x^2 + 9.$$

Then the function that appears in Eq. (11) is a *function of a function*:

$$f(g(x)) = \sqrt{g(x)} = \sqrt{x^2 + 9}.$$

(The left-hand expression in this equation is read "f of g of x.") Hence our question is whether or not

$$\lim_{x \to a} f(g(x)) = f\left(\lim_{x \to a} g(x)\right).$$

The next limit law answers this question in the affirmative, provided that the "outside function" f meets a certain condition; if so, then the limit of the combination function $f(g(x))$ as $x \to a$ may be found by substituting into the function f the limit of $g(x)$ as $x \to a$.

Substitution Law

Suppose that

$$\lim_{x \to a} g(x) = L \quad \text{and that} \quad \lim_{x \to L} f(x) = f(L).$$

Then

$$\lim_{x \to a} f(g(x)) = f\left(\lim_{x \to a} g(x)\right) = f(L). \tag{12}$$

Thus the condition under which Eq. (12) holds is that the limit of the *outer* function f not only exists at $x = L$, but also is equal to the "expected" value of f—namely, $f(L)$. In particular, because

$$\lim_{x \to -4} (x^2 + 9) = 25 \quad \text{and} \quad \lim_{x \to 25} \sqrt{x} = \sqrt{25} = 5,$$

this condition is satisfied in Eq. (11). Hence the computations shown there are valid.

In this section we use only the following special case of the substitution law. With $f(x) = x^{1/n}$, where n is a positive integer, Eq. (12) takes the form

$$\lim_{x \to a} \sqrt[n]{g(x)} = \sqrt[n]{\lim_{x \to a} g(x)}, \tag{13}$$

under the assumption that the limit of $g(x)$ exists as $x \to a$ (and is positive if n is even). With $g(x) = x^m$, where m is a positive integer, Eq. (13) in turn yields

$$\lim_{x \to a} x^{m/n} = a^{m/n}, \tag{14}$$

with the condition that $a > 0$ if n is even. Equations (13) and (14) may be regarded as generalized root laws. Example 11 illustrates the use of these special cases of the substitution law.

EXAMPLE 11

$$\lim_{x \to 4} \left(3x^{3/2} + 20\sqrt{x}\right)^{1/3} = \left(\lim_{x \to 4}\left(3x^{3/2} + 20\sqrt{x}\right)\right)^{1/3} \quad \text{[using Eq. (13)]}$$

$$= \left(\lim_{x \to 4} 3x^{3/2} + \lim_{x \to 4} 20\sqrt{x}\right)^{1/3} \quad \text{[using the sum law]}$$

$$= \left[3 \cdot 4^{3/2} + 20\sqrt{4}\right]^{1/3} \quad \text{[using Eq. (14)]}$$

$$= (24 + 40)^{1/3} = \sqrt[3]{64} = 4.$$

Our discussion of limits began with the derivative. If we write the definition of the derivative as

$$f'(a) = \lim_{x \to a} \frac{f(x) - f(a)}{x - a},$$

then the limit involved is similar to those in the preceding examples in that x is changing while other numbers (such as a) are held fixed. But when we let $h = x - a$ and use the definition of f' in the form

$$f'(x) = \lim_{h \to 0} \frac{f(x + h) - f(x)}{h},$$

then x plays the role of a *constant*, and h is the variable approaching zero.

Example 12 illustrates an algebraic device often used in "preparing" functions before taking limits. This device can be applied when roots are present and resembles the simple computation

$$\frac{1}{\sqrt{3} - \sqrt{2}} = \frac{1}{\sqrt{3} - \sqrt{2}} \cdot \frac{\sqrt{3} + \sqrt{2}}{\sqrt{3} + \sqrt{2}}$$

$$= \frac{\sqrt{3} + \sqrt{2}}{3 - 2} = \sqrt{3} + \sqrt{2}.$$

EXAMPLE 12   Differentiate $f(x) = \sqrt{x}$.

Solution

$$f'(x) = \lim_{h \to 0} \frac{\sqrt{x + h} - \sqrt{x}}{h}. \tag{15}$$

To prepare the fraction for evaluation of the limit, we first multiply the numerator and denominator by the *conjugate* $\sqrt{x + h} + \sqrt{x}$ of the numerator:

$$f'(x) = \lim_{h \to 0} \frac{\sqrt{x + h} - \sqrt{x}}{h} \cdot \frac{\sqrt{x + h} + \sqrt{x}}{\sqrt{x + h} + \sqrt{x}}$$

$$= \lim_{h \to 0} \frac{(x + h) - x}{h\left(\sqrt{x + h} + \sqrt{x}\right)}$$

$$= \lim_{h \to 0} \frac{1}{\sqrt{x + h} + \sqrt{x}}.$$

Thus

$$f'(x) = \frac{1}{2\sqrt{x}}. \tag{16}$$

(In the final step we used the sum, quotient, and root laws—we did not simply substitute 0 for h.)

Note that if we equate the right-hand sides of Eqs. (15) and (16) and take $x = 25$, then we get the limit of Example 4:

$$\lim_{h \to 0} \frac{\sqrt{25 + h} - 5}{h} = \frac{1}{10}.$$

THE DEFINITION OF THE LIMIT

There are few people for whom a full understanding of the limit concept comes quickly or easily. Indeed, the precise meaning of the statement "$F(x)$ approaches L as x approaches a" was debated vigorously (and sometimes acrimoniously) for hundreds of years—until late in the nineteenth century. Then the German mathematician Karl Weierstrass (1815–1897) finally formulated the rigorous definition of the limit that is accepted today.

Definition of the Limit

The number L is the *limit* of $F(x)$ as x approaches a provided that, given any number $\epsilon > 0$, there exists a number $\delta > 0$ such that

$$|F(x) - L| < \epsilon$$

for all x such that

$$0 < |x - a| < \delta.$$

Figure 2.2.11 illustrates this definition. The points on the graph of $y = F(x)$ that satisfy the inequality $|F(x) - L| < \epsilon$ are those points that lie between the two horizontal lines $y = L - \epsilon$ and $y = L + \epsilon$. The points on this graph that satisfy the inequality $|x - a| < \delta$ are those points that lie between the two vertical lines $x = a - \delta$ and $x = a + \delta$. Consequently, the definition implies that $\lim_{x \to a} F(x) = L$ if and only if the following is true:

> Suppose that the two horizontal lines $y = L - \epsilon$ and $y = L + \epsilon$ (with $\epsilon > 0$) are given. Then it is possible to choose two vertical lines $x = a - \delta$ and $x = a + \delta$ (with $\delta > 0$) with the following property: Every point on the graph of $y = F(x)$ (with $x \neq a$) that lies between the two vertical lines must also lie between the two horizontal lines.

Figure 2.2.11 suggests that the closer together the two horizontal lines, the closer together the two vertical lines will have to be. This is what we mean by "making $F(x)$ closer to L by making x closer to a."

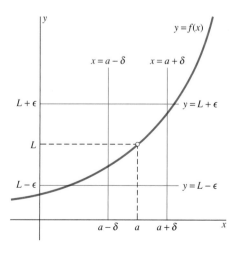

Fig. 2.2.11 Geometric illustration of the limit definition

Ch. 2 / Prelude to Calculus

In some cases in which $\lim_{x \to a} F(x) = L$, $F(x)$ does not get *steadily* closer to L as x moves steadily toward a. This situation is perfectly permissible, as you can see either by carefully examining the definition of limit or by studying Fig. 2.2.11.

2.2 Problems

In Problems 1 through 30, use the limit laws of this section to evaluate those limits that exist.

1. $\lim_{x \to 0} (3x^2 + 7x - 12)$ **2.** $\lim_{x \to 2} (4x^2 - x + 5)$

3. $\lim_{x \to -1} (2x - x^5)$ **4.** $\lim_{x \to -2} (x^2 - 2)^5$

5. $\lim_{x \to 3} \dfrac{(x - 1)^7}{(2x - 5)^4}$ **6.** $\lim_{x \to 1} \dfrac{x + 1}{x^2 - x - 2}$

7. $\lim_{x \to -1} \dfrac{x + 1}{x^2 - x - 2}$ **8.** $\lim_{t \to 2} \dfrac{t^2 + 2t - 5}{t^3 - 2t}$

9. $\lim_{t \to 3} \dfrac{t^2 - 9}{t - 3}$ **10.** $\lim_{y \to 3} \dfrac{\frac{1}{y} - \frac{1}{3}}{y - 3}$

11. $\lim_{x \to 3} (x^2 - 1)^{3/2}$ **12.** $\lim_{t \to -4} \sqrt{\dfrac{t + 8}{25 - t^2}}$

13. $\lim_{z \to 8} \dfrac{z^{2/3}}{z - \sqrt{2z}}$ **14.** $\lim_{t \to 2} \sqrt[3]{3t^3 + 4t - 5}$

15. $\lim_{x \to 0} \dfrac{\sqrt{x + 4} - 2}{x}$ **16.** $\lim_{h \to 0} \dfrac{\frac{1}{2 + h} - \frac{1}{2}}{h}$

17. $\lim_{h \to 0} \dfrac{\frac{1}{\sqrt{9 + h}} - \frac{1}{3}}{h}$ **18.** $\lim_{x \to 2} \dfrac{(x - 2)^2}{x^4 - 16}$

19. $\lim_{x \to 3} |1 - x|$ **20.** $\lim_{x \to -5} |3x - 2|$

21. $\lim_{x \to 4} \dfrac{x - 4}{\sqrt{x} - 2}$ **22.** $\lim_{x \to 9} \dfrac{3 - \sqrt{x}}{9 - x}$

23. $\lim_{x \to 1} \dfrac{x^2 + x - 2}{x^2 - 4x + 3}$ **24.** $\lim_{x \to -1/2} \dfrac{4x^2 - 1}{4x^2 + 8x + 3}$

25. $\lim_{x \to 4} \dfrac{x^2 - 16}{2 - \sqrt{x}}$ **26.** $\lim_{x \to 5} \left(\dfrac{2x^2 + 2x + 4}{6x - 3} \right)^{1/3}$

27. $\lim_{x \to 1} \sqrt{(x - 2)^2}$ **28.** $\lim_{x \to -4} \sqrt[3]{(x + 1)^6}$

29. $\lim_{x \to -2} \sqrt[3]{\dfrac{x + 2}{(x - 2)^2}}$ **30.** $\lim_{x \to 0} \dfrac{\sqrt{1 + x} - \sqrt{1 - x}}{x}$

Use a calculator to investigate numerically (as in Examples 1 through 4) the limits in Problems 31 through 40.

31. $\lim_{x \to -3} x^2$ **32.** $\lim_{x \to -2} x^3$

33. $\lim_{x \to 4} \sqrt{x}$ **34.** $\lim_{x \to -8} \sqrt[3]{x}$

35. $\lim_{x \to -2} \sqrt{x^2}$ **36.** $\lim_{x \to -3} \sqrt[3]{x^3}$

37. $\lim_{x \to 12} \sqrt{x^2 + 25}$ **38.** $\lim_{x \to -5} \sqrt[3]{x^2 - 33}$

39. $\lim_{x \to 1} \dfrac{x + 2}{x - 1}$ **40.** $\lim_{x \to -2} \dfrac{x - 1}{x + 2}$

In Problems 41 through 46, use the method of Example 12—where appropriate—to find the derivative $f'(x)$ of $f(x)$.

41. $f(x) = \dfrac{1}{\sqrt{x}}$ **42.** $f(x) = \dfrac{1}{2x + 3}$

43. $f(x) = \dfrac{x}{2x + 1}$ **44.** $f(x) = \sqrt{3x + 1}$

45. $f(x) = \dfrac{x^2}{x + 1}$ **46.** $f(x) = \dfrac{1}{\sqrt{x + 4}}$

47. Apply the product law to prove that $\lim_{x \to a} x^n = a^n$ if n is a positive integer.

48. Suppose that $p(x) = b_n x^n + \cdots + b_1 x + b_0$ is a polynomial. Prove that $\lim_{x \to a} p(x) = p(a)$.

49. Let $f(x) = [\![x]\!]$ be the greatest integer function. For what values of a does $\lim_{x \to a} f(x)$ exist?

50. Let $r(x) = p(x)/q(x)$, where $p(x)$ and $q(x)$ are polynomials. Suppose that $p(a) \neq 0$ and that $q(a) = 0$. Prove that $\lim_{x \to a} r(x)$ does not exist. [*Suggestion:* Consider $\lim_{x \to a} q(x) r(x)$.]

51. Let $g(x) = 1 + [\![x]\!] + [\![-x]\!]$. Show that the limit of $g(x)$ as $x \to 3$ cannot be obtained by substitution of 3 for x in the formula for g.

52. Let $h(x) = x - (0.01)[\![100x]\!]$. Show that $h(x) \to 0$ as $x \to 0$ but that $h(x)$ does not approach zero "steadily" as $x \to 0$. First decide what the phrase "approaches steadily" ought to mean.

53. Repeat Problem 52 with

$$h(x) = \dfrac{1}{[\![1/x]\!]} - x.$$

The object of this project is the systematic use of a calculator or computer for the numerical investigation of limits. Suppose that we want to investigate the value (if any) of the limit

$$\lim_{x \to a} f(x)$$

of a given function f at $x = a$. We shall begin with a fixed increment h and then calculate (as efficiently as possible) the values of f at the points

$$a + \frac{h}{5}, a + \frac{h}{5^2}, \ldots, a + \frac{h}{5^n}, \ldots,$$

which approach the number a as n increases. Fig. 2.2.12 shows a simple TI calculator program for this purpose.

Program steps	Comments
`Prgm1:LIMIT`	Name of program
`:Disp "A"`	
`:Input A`	Input limiting value a of x
`:1→H`	Initial increment $h = 1$
`:0→N`	Initial counter value $n = 0$
`:Lbl 1`	Label to begin loop
`:H/5→H`	Divide h by 5
`:A+H→X`	Set $x = a + h$
`:Disp X`	Display new value of x
`:Disp Y1`	Display new value of $f(x)$
`:N+1→N`	Increment counter
`:Pause`	Press ENTER to continue
`:If N<8`	If $n < 8$ go through loop again
`:Goto 1`	
`:End`	Stop otherwise

Fig. 2.2.12 TI-81 program for investigating $\lim_{x \to a} f(x)$

To investigate a limit as $x \to 0$, we might take $a = 0$ and $h = 1$ and then calculate the numerical value $f(x)$ at the points

$$0.2, 0.04, 0.008, \ldots, (0.2)^n, \ldots,$$

which approach zero as n increases. With the function $f(x) = (\sqrt{x + 25} - 5)/x$ of Example 4, for instance, we get the data shown in Fig. 2.2.13. These numerical results suggest that

$$\lim_{x \to 0} \frac{\sqrt{x + 25} - 5}{x} = \frac{1}{10}.$$

Some computer systems have "one-liners" for function tabulation that can be used to construct tables such as Fig. 2.2.13. The table in Fig. 2.2.14 lists commands in several common systems for tabulating numerical values of $f(a + h/5^n)$.

Investigate the numerical values (and existence) of the limits given in Problems 1 through 10. You should use several different values of h (both positive and negative) in each problem.

x	$\dfrac{\sqrt{x + 25} - 5}{x}$ (Values rounded)
0.2	0.09980
0.04	0.09996
0.008	0.09999
0.0016	0.10000
0.00032	0.10000
0.000064	0.10000
$\vdots$	$\vdots$
$\downarrow$	$\downarrow$
0	0.1

Fig. 2.2.13 Investigating $\lim_{x \to 0} \dfrac{\sqrt{x + 25} - 5}{x}$

BASIC	`FOR n=1 to 8 : PRINT a+h/5^n, f(a+h/5^n) : NEXT`
Derive	`[VECTOR(a+h/5^n,n,1,8), VECTOR(f(a+h/5^n),n,1,8)]´`
Maple	`for n from 1 to 8 do print (a+h/5^n,f(a+h/5^n)) od`
Mathematica	`MatrixForm[ Table[ {a+h/5^n,f[a+h/5^n]}, {n,1,8} ]`
(X)PLORE	`table( f(a+h/5^n), n = 1 to n = 8)`

Fig. 2.2.14 Commands for tabulating values of $f(a + h/5^n)$

1. $\displaystyle\lim_{x \to 0} \frac{(1 + x)^2 - 1}{x}$

2. $\displaystyle\lim_{x \to 1} \frac{x^4 - 1}{x - 1}$

3. $\displaystyle\lim_{x \to 0} \frac{\sqrt{x + 9} - 3}{x}$

4. $\displaystyle\lim_{x \to 4} \frac{x^{3/2} - 8}{x - 4}$

5. $\displaystyle\lim_{x \to 0} \left(\frac{1}{x + 5} - \frac{1}{5} \right)$

6. $\displaystyle\lim_{x \to 8} \frac{x^{2/3} - 4}{x - 8}$

7. $\displaystyle\lim_{x \to 0} \frac{\sin x}{x}$

8. $\displaystyle\lim_{x \to 0} \frac{2^x - 1}{x}$

9. $\displaystyle\lim_{x \to 0} \frac{10^x - 1}{x}$

10. $\displaystyle\lim_{x \to 0} \left(1 + \frac{1}{x} \right)^x$

2.3
More About Limits

To investigate limits of trigonometric functions, we begin with Fig. 2.3.1, which shows an angle θ with its vertex at the origin, its initial side along the positive x-axis, and its terminal side intersecting the unit circle at the point P. By the definition of the sine and cosine functions, the coordinates of P are $P(\cos \theta, \sin \theta)$. From geometry we see that, as $\theta \to 0$, the point $P(\cos \theta, \sin \theta)$ approaches the point $R(1, 0)$. Hence $\cos \theta \to 1$ and $\sin \theta \to 0$ as $\theta \to 0^+$. A similar picture gives the same results as $\theta \to 0^-$, so we see that

$$\lim_{\theta \to 0} \cos \theta = 1 \quad \text{and} \quad \lim_{\theta \to 0} \sin \theta = 0. \tag{1}$$

Equation (1) says simply that the *limits* of the functions $\cos \theta$ and $\sin \theta$ as $\theta \to 0$ are equal to their *values* at $\theta = 0$: $\cos 0 = 1$ and $\sin 0 = 0$.

The limit of the quotient $(\sin \theta)/\theta$ as $\theta \to 0$ plays a special role in the calculus of trigonometric functions: It is essential for computing the derivatives of the sine and cosine functions. Note that the value of the quotient $(\sin \theta)/\theta$ at $\theta = 0$ is not defined. (Why not?) But a calculator set in *radian*

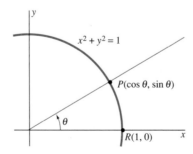

Fig. 2.3.1 An angle θ

θ	$\dfrac{\sin\theta}{\theta}$
± 1.0	0.84147
± 0.5	0.95885
± 0.1	0.99833
± 0.05	0.99958
± 0.01	0.99998
± 0.005	1.00000
± 0.001	1.00000
$\vdots$	$\vdots$
$\downarrow$	$\downarrow$
0	1

Fig. 2.3.2 The numerical data suggest that

$$\lim_{\theta \to 0} \frac{\sin\theta}{\theta} = 1.$$

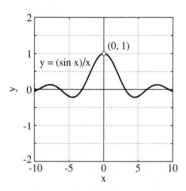

Fig. 2.3.3 $y = \dfrac{\sin x}{x}$ for $x \neq 0$

mode provides us with the numerical evidence shown in Fig. 2.3.2. This table strongly suggests that the limit of $(\sin\theta)/\theta$ is 1 as $\theta \to 0$. This conclusion is supported by the graph of $y = (\sin x)/x$ shown in Fig. 2.3.3, where it appears that the point (x, y) on the curve is near $(0, 1)$ when x is near zero. At the end of this section we provide a geometric proof of the following result.

Theorem 1   The Basic Trigonometric Limit

$$\lim_{x \to 0} \frac{\sin x}{x} = 1. \tag{2}$$

As in Examples 1 and 2, many other trigonometric limits can be reduced to the one in Theorem 1.

EXAMPLE 1   Show that

$$\lim_{x \to 0} \frac{1 - \cos x}{x} = 0. \tag{3}$$

Solution   We multiply the numerator and denominator in Eq. (3) by the *conjugate* $1 + \cos x$ of the numerator $1 - \cos x$. Then we apply the fundamental identity $1 - \cos^2 x = \sin^2 x$. This gives

$$\lim_{x \to 0} \frac{1 - \cos x}{x} = \lim_{x \to 0} \frac{1 - \cos x}{x} \cdot \frac{1 + \cos x}{1 + \cos x} = \lim_{x \to 0} \frac{\sin^2 x}{x(1 + \cos x)}$$

$$= \left(\lim_{x \to 0} \frac{\sin x}{x} \right)\left(\lim_{x \to 0} \frac{\sin x}{1 + \cos x} \right) = 1 \cdot \frac{0}{1 + 1} = 0.$$

In the last step we used *all* the limits in Eqs. (1) and (2).

EXAMPLE 2   Evaluate $\displaystyle\lim_{x \to 0} \frac{\tan 3x}{x}$.

Solution

$$\lim_{x \to 0} \frac{\tan 3x}{x} = 3\left(\lim_{x \to 0} \frac{\tan 3x}{3x} \right) = 3\left(\lim_{\theta \to 0} \frac{\tan \theta}{\theta} \right) \qquad (\theta = 3x)$$

$$= 3\left(\lim_{\theta \to 0} \frac{\sin \theta}{\theta \cos \theta} \right) \qquad \left(\text{because } \tan \theta = \frac{\sin \theta}{\cos \theta} \right)$$

$$= 3\left(\lim_{\theta \to 0} \frac{\sin \theta}{\theta} \right)\left(\lim_{\theta \to 0} \frac{1}{\cos \theta} \right) \qquad \begin{array}{l}\text{(by the product law of}\\ \text{limits)}\end{array}$$

$$= 3 \cdot 1 \cdot \frac{1}{1} = 3.$$

We used the fact that $\tan \theta = (\sin \theta)/(\cos \theta)$ as well as some of the limits in Eqs. (1) and (2).

Example 3 constitutes a *warning:* The results of numerical investigation can be misleading unless they are interpreted with care.

x	$\sin \dfrac{\pi}{x}$
1	0
0.5	0
0.1	0
0.05	0
0.01	0
0.005	0
0.001	0

Fig. 2.3.4 Do you think that $\displaystyle\lim_{x \to 0} \sin \dfrac{\pi}{x} = 0$ (Example 3)?

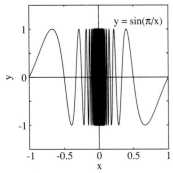

Fig. 2.3.5 The graph of $y = \sin \dfrac{\pi}{x}$ shows infinite oscillation as $x \to 0$ (Example 3).

sx	$\sin \dfrac{\pi}{x}$
$\frac{2}{9}$	$+1$
$\frac{2}{11}$	-1
$\frac{2}{101}$	$+1$
$\frac{2}{103}$	-1
$\frac{2}{1001}$	$+1$
$\frac{2}{1003}$	-1

Fig. 2.3.6 Verify the entries in the second column (Example 3).

EXAMPLE 3 The numerical data shown in the table of Fig. 2.3.4 suggest that the limit

$$\lim_{x \to 0} \sin \frac{\pi}{x} \qquad (4)$$

has the value zero. But it appears in the graph of $y = \sin(\pi/x)$ (for $x \neq 0$), shown in Fig. 2.3.5, that the value of $\sin(\pi/x)$ oscillates infinitely often between $+1$ and -1 as $x \to 0$. Indeed, this fact follows from the periodicity of the sine function, because π/x increases without bound as $x \to 0$. Hence $\sin(\pi/x)$ cannot approach zero (or any other number) as $x \to 0$. Therefore, the limit in (4) *does not exist*.

We can explain the potentially misleading results tabulated in Fig. 2.3.4 as follows: Each value of x shown there just happens to be of the form $1/n$, the reciprocal of an integer. Therefore,

$$\sin \frac{\pi}{x} = \sin \frac{\pi}{1/n} = \sin n\pi = 0$$

for every nonzero integer n. But with a different selection of "trial values" of x, we might have obtained the results shown in Fig. 2.3.6, which immediately suggest the nonexistence of the limit in (4).

THE SQUEEZE LAW OF LIMITS

A final property of limits that we will need is the *squeeze law*. It is related to the fact that taking limits preserves inequalities among functions.

> **Squeeze Law**
> Suppose that $f(x) \leqq g(x) \leqq h(x)$ for all x in some deleted neighborhood of a and also that
> $$\lim_{x \to a} f(x) = L = \lim_{x \to a} h(x).$$
> Then
> $$\lim_{x \to a} g(x) = L$$
> as well.

Figure 2.3.7 illustrates how and why the squeeze law works and how it got its name. The idea is that $g(x)$ is trapped between $f(x)$ and $h(x)$ near a; both $f(x)$ and $h(x)$ approach the same limit L, so $g(x)$ must approach L as well.

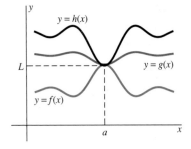

Fig. 2.3.7 How the squeeze law works

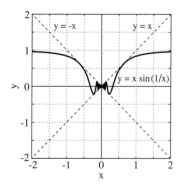

Fig. 2.3.8 The graph of
$g(x) = x \sin \dfrac{1}{x}$ for $x \neq 0$
(Example 4)

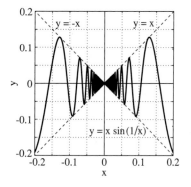

Fig. 2.3.9 The graph magnified near the origin (Example 4)

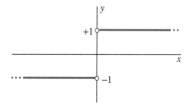

Fig. 2.3.10 The graph of
$f(x) = \dfrac{x}{|x|}$ again

EXAMPLE 4 Figures 2.3.8 and 2.3.9 show two views of the graph of the function g defined for $x \neq 0$ by

$$g(x) = x \sin \frac{1}{x}.$$

As in Example 3, $\sin(1/x)$ oscillates infinitely often between $+1$ and -1 as $x \to 0$. Therefore, $g(x)$ oscillates infinitely often between the values $+x$ and $-x$. Because $|\sin(1/x)| \leq 1$ for all $x \neq 0$,

$$-|x| \leq x \sin \frac{1}{x} \leq +|x|$$

for all $x \neq 0$. Moreover, $\pm|x| \to 0$ as $x \to 0$, so with $f(x) = -|x|$ and $h(x) = +|x|$, it follows from the squeeze law of limits that

$$\lim_{x \to 0} x \sin \frac{1}{x} = 0. \tag{5}$$

QUESTION Why *doesn't* the limit in Eq. (5) follow from the product law of limits with $f(x) = x$ and $g(x) = \sin(1/x)$ (Problem 66)?

ONE-SIDED LIMITS

In Example 5 of Section 2.2 we examined the function

$$f(x) = \frac{x}{|x|} = \begin{cases} 1 & \text{if } x > 0; \\ -1 & \text{if } x < 0. \end{cases}$$

The graph of $y = f(x)$ is shown in Fig. 2.3.10. We argued that the limit of $f(x) = x/|x|$ as $x \to 0$ does not exist because $f(x)$ approaches $+1$ as x approaches zero from the right, whereas $f(x) \to -1$ as x approaches zero from the left. A natural way of describing this situation is to say that at $x = 0$ the *right-hand limit* of $f(x)$ is $+1$ and the *left-hand limit* of $f(x)$ is -1.

Here we define and investigate these one-sided limits. Their definitions will be stated initially in the informal language we used in Section 2.2 to describe the "idea of the limit." To define the right-hand limit of $f(x)$ at $x = a$, we must assume that f is defined on an open interval immediately to the right of a. To define the left-hand limit, we must assume that f is defined on an open interval just to the left of a.

The Right-Hand Limit of a Function

Suppose that f is defined on the open interval (a, c) immediately to the right of a. Then we say that the number L is the **right-hand limit** of $f(x)$ as x approaches a, and we write

$$\lim_{x \to a^+} f(x) = L, \tag{6}$$

provided that we can make the number $f(x)$ as close to L as we please merely by choosing the point x in (a, c) sufficiently close to a.

Ch. 2 / Prelude to Calculus

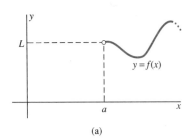

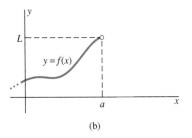

Fig. 2.3.11 (a) The right-hand limit of $f(x)$ is L. (b) The left-hand limit of $f(x)$ is L.

We may describe the right-hand limit in Eq. (6) by saying that $f(x) \to L$ as $x \to a^+$, or as x approaches a from the right. (The symbol a^+ denotes the right-hand, or positive, side of a.) More precisely, Eq. (6) means that, given $\epsilon > 0$, there exists $\delta > 0$ such that

$$|f(x) - L| < \epsilon \tag{7}$$

for all x such that

$$a < x < a + \delta. \tag{8}$$

See Fig. 2.3.11 for a geometric interpretation of one-sided limits.

The Left-Hand Limit of a Function

Suppose that f is defined on an open interval (c, a) immediately to the left of a. Then we say that the number L is the **left-hand limit** of $f(x)$ as x approaches a, and we write

$$\lim_{x \to a^-} f(x) = L, \tag{9}$$

provided that we can make the number $f(x)$ as close to L as we please merely by choosing the point x in (c, a) sufficiently near the number a.

A consequence of these definitions is that the value of $f(a)$ itself is not relevant to the existence or value of the one-sided limits, just as it is not relevant to the existence or value of the (regular) two-sided limit.

We may describe the left-hand limit in Eq. (9) by saying that $f(x) \to L$ as $x \to a^-$, or as x approaches a from the left. (The symbol a^- denotes the left-hand, or negative, side of a.) We get a precise definition of the left-hand limit in Eq. (9) merely by changing the open interval in (8) to the interval $a - \delta < x < a$.

Our preliminary discussion of the function $f(x) = x/|x|$ amounts to saying that this function's one-sided limits at $x = 0$ are

$$\lim_{x \to 0^+} \frac{x}{|x|} = 1 \quad \text{and} \quad \lim_{x \to 0^-} \frac{x}{|x|} = -1.$$

We argued further in Example 5 of Section 2.2 that, because these two limits are not equal, *the* two-sided limit of $x/|x|$ as $x \to 0$ does not exist. The following theorem is intuitively obvious and can be proved by using the precise definitions of all the limits involved.

Theorem 2 *One-Sided Limits and Two-Sided Limits*

Suppose that the function f is defined on a deleted neighborhood of the point a. Then the limit $\lim_{x \to a} f(x)$ exists and is equal to the number L if and only if the one-sided limits $\lim_{x \to a^+} f(x)$ and $\lim_{x \to a^-} f(x)$ both exist and are equal to L.

Theorem 2 is particularly useful in showing that certain (two-sided) limits do *not* exist, usually by showing that the left-hand and right-hand limits are not equal to each other.

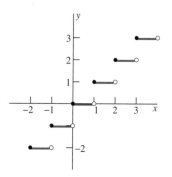

Fig. 2.3.12 The graph of the greatest integer function $f(x) = [\![x]\!]$ (Example 5)

EXAMPLE 5 The graph of the greatest integer function $f(x) = [\![x]\!]$ is shown in Fig. 2.3.12. It should be apparent that if a is not an integer, then

$$\lim_{x \to a^+} [\![x]\!] = \lim_{x \to a^-} [\![x]\!] = \lim_{x \to a} [\![x]\!] = [\![a]\!] .$$

But if $a = n$, an integer, then

$$\lim_{x \to n^-} [\![x]\!] = n - 1 \quad \text{and} \quad \lim_{x \to n^+} [\![x]\!] = n.$$

Because these left-hand and right-hand limits are not equal, it follows from Theorem 2 that the limit of $f(x) = [\![x]\!]$ does not exist as x approaches an integer n.

EXAMPLE 6 According to the root law in Section 2.2,

$$\lim_{x \to a} \sqrt{x} = \sqrt{a} \quad \text{if } a > 0.$$

But the limit of $f(x) = \sqrt{x}$ as $x \to 0^-$ is not defined because the square root of a negative number is undefined. Hence f is undefined on every deleted neighborhood of zero. What we can say about $a = 0$ is that

$$\lim_{x \to 0^+} \sqrt{x} = 0,$$

although the left-hand limit $\lim_{x \to 0^-} \sqrt{x}$ does not exist.

To each of the limit laws stated in Section 2.2 there correspond two *one-sided limit laws*—a right-hand version and a left-hand version. You may apply these one-sided limit laws in the same way you apply the two-sided limit laws in evaluation of limits.

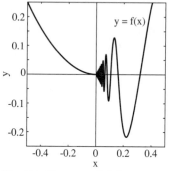

Fig. 2.3.13 $y = f(x)$ (Example 7)

EXAMPLE 7 Figure 2.3.13 shows the graph of the function f defined by

$$f(x) = \begin{cases} x^2 & \text{if } x \leq 0; \\ x \sin \dfrac{1}{x} & \text{if } x > 0. \end{cases}$$

Clearly

$$\lim_{x \to 0^-} f(x) = 0 \quad \text{and} \quad \lim_{x \to 0^+} f(x) = 0$$

by a one-sided version of the squeeze law (as in Example 4). It therefore follows from Theorem 2 that

$$\lim_{x \to 0} f(x) = 0.$$

EXAMPLE 8 Upon applying the appropriate one-sided limit laws, we find that

$$\lim_{x \to 3^-} \left(\frac{x^2}{x^2 + 1} + \sqrt{9 - x^2} \right) = \frac{\lim\limits_{x \to 3^-} x^2}{\lim\limits_{x \to 3^-} (x^2 + 1)} + \sqrt{\lim\limits_{x \to 3^-} (9 - x^2)}$$

$$= \frac{9}{9 + 1} + \sqrt{0} = \frac{9}{10}.$$

Note that the two-sided limit as $x \to 3$ is not defined because $\sqrt{9 - x^2}$ is not defined when $x > 3$.

76

Recall that the derivative $f'(x)$ of the function $f(x)$ is defined by

$$f'(x) = \lim_{h \to 0} \frac{f(x + h) - f(x)}{h} \tag{10}$$

provided that this (two-sided) limit exists, in which case we say that the function f is **differentiable** at the point x. Example 9 shows that it is possible for a function to be defined everywhere but fail to be differentiable at certain points of its domain.

EXAMPLE 9 Show that the function $f(x) = |x|$ is *not* differentiable at $x = 0$.

Solution When $x = 0$, we have

$$\frac{f(x + h) - f(x)}{h} = \frac{|h|}{h} = \begin{cases} -1 & \text{if } h < 0; \\ +1 & \text{if } h > 0. \end{cases}$$

Hence the left-hand limit of the quotient is -1, whereas the right-hand limit is $+1$. Therefore, the *two-sided* limit in (10) does not exist, so $f(x) = |x|$ is not differentiable at $x = 0$. To put it another way, $f'(0)$ does not exist.

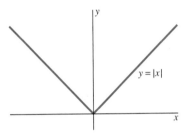

Fig. 2.3.14 The graph of $f(x) = |x|$ has a corner point at $(0, 0)$.

Figure 2.3.14 shows the graph of the function $f(x) = |x|$ of Example 9. The graph has a sharp corner at the point $(0, 0)$, and this explains why there can be no tangent line there—no single straight line is a good approximation to the shape of the graph at $(0, 0)$. But the figure makes it evident that $f'(x)$ exists if $x \neq 0$. Indeed,

$$f'(x) = \begin{cases} -1 & \text{if } x < 0; \\ +1 & \text{if } x > 0. \end{cases}$$

In Problem 64 we ask you to derive this result directly from the definition of the derivative, without appeal to Fig. 2.3.14.

INFINITE LIMITS

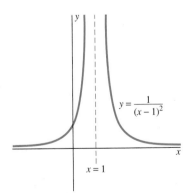

Fig. 2.3.15 The graph of the function $f(x) = 1/(x - 1)^2$

In Example 9 of Section 2.2, we investigated the function $f(x) = 1/(x - 1)^2$; the graph of f is shown in Fig. 2.3.15. The value of $f(x)$ *increases without bound* (that is, eventually exceeds any preassigned number) as x approaches 1 either from the right or from the left. This situation can be described by writing

$$\lim_{x \to 1^-} \frac{1}{(x - 1)^2} = +\infty = \lim_{x \to 1^+} \frac{1}{(x - 1)^2}, \tag{11}$$

and we say that each of these one-sided limits is equal to "plus infinity."

CAUTION The expression

$$\lim_{x \to 1^+} \frac{1}{(x - 1)^2} = +\infty \tag{12}$$

does not mean that there exists an "infinite real number" denoted by $+\infty$—there does not! Neither does it mean that the limit on the left-hand side in Eq.

(12) exists—it does not! To the contrary, the expression in Eq. (12) is just a convenient way of saying *why* the right-hand limit in Eq. (12) does not exist: because the quantity $1/(x - 1)^2$ increases without bound as $x \rightarrow 1^+$.

With similar provisos we may write

$$\lim_{x \to 1} \frac{1}{(x - 1)^2} = +\infty \tag{13}$$

despite the fact that the (two-sided) limit in Eq. (13) does not exist. The expression in Eq. (13) is merely a convenient way of saying that the limit in Eq. (13) does not exist because $1/(x - 1)^2$ increases without bound as $x \rightarrow 1$ from either side.

Now consider the function $f = 1/x$; its graph is shown in Fig. 2.3.16. This function increases without bound as x approaches zero from the right but decreases without bound—it becomes less than any preassigned negative number—as x approaches zero from the left. We therefore write

$$\lim_{x \to 0^-} \frac{1}{x} = -\infty \quad \text{and} \quad \lim_{x \to 0^+} \frac{1}{x} = +\infty. \tag{14}$$

There is no shorthand for the two-sided limit in this case. We may say only that

$$\lim_{x \to 0} \frac{1}{x} \quad \text{does not exist.}$$

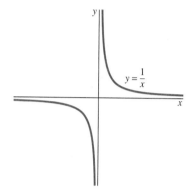

Fig. 2.3.16 The graph of the function $f(x) = 1/x$

EXAMPLE 10 Investigate the behavior of the function

$$f(x) = \frac{2x + 1}{x - 1}$$

near the point $x = 1$, where its limit does not exist.

Solution First we look at the behavior of $f(x)$ just to the right of the number 1. If x is greater than 1 but close to 1, then $2x + 1$ is close to 3 and $x - 1$ is a small *positive* number. In this case the quotient $(2x + 1)/(x - 1)$ is a large positive number, and the closer x is to 1, the larger this positive quotient will be. For such x, $f(x)$ increases without bound as x approaches 1 from the right. That is,

$$\lim_{x \to 1^+} \frac{2x + 1}{x - 1} = +\infty, \tag{15}$$

as the data in Fig. 2.3.17 suggest.

Fig. 2.3.17 The behavior of

$$f(x) = \frac{2x + 1}{x - 1}$$

for x near 1 (Example 10)

x	$\dfrac{2x + 1}{x - 1}$	x	$\dfrac{2x + 1}{x - 1}$
1.1	32	0.9	−28
1.01	302	0.99	−298
1.001	3002	0.999	−2998
1.0001	30002	0.9999	−29998
⋮	⋮	⋮	⋮
↓	↓	↓	↓
1	$+\infty$	1	$-\infty$

If x is instead less than 1 but close to 1, then $2x + 1$ is still close to 3, but now $x - 1$ is a *negative* number close to zero. In this case the quotient $(2x + 1)/(x - 1)$ is a (numerically) large negative number and decreases without bound as $x \to 1^-$. Hence we conclude that

$$\lim_{x \to 1^-} \frac{2x + 1}{x - 1} = -\infty. \tag{16}$$

The results in Eqs. (15) and (16) provide a concise description of the behavior of $f(x) = (2x + 1)/(x - 1)$ near the point $x = 1$. Finally, to remain consistent with Theorem 2 on one-sided and two-sided limits, we say in this case that

$$\lim_{x \to 1} \frac{2x + 1}{x - 1} \quad \text{does not exist.}$$

THE BASIC TRIGONOMETRIC LIMIT

We conclude this section with a geometric proof that

$$\lim_{\theta \to 0} \frac{\sin \theta}{\theta} = 1. \tag{17}$$

Proof Figure 2.3.18 shows the angle θ, the triangles OPQ and ORS, and the circular sector OPR that contains the triangle OPQ and is contained in the triangle ORS. Hence

$$\text{area}(\triangle OPQ) < \text{area}(\text{sector } OPR) < \text{area}(\triangle ORS).$$

In terms of θ, this means that

$$\frac{1}{2} \sin \theta \cos \theta < \frac{1}{2} \theta < \frac{1}{2} \tan \theta = \frac{\sin \theta}{2 \cos \theta}.$$

Here we use the standard formula for the area of a triangle to obtain the area of $\triangle OPQ$ and $\triangle ORS$. We also use the fact that the area of a circular sector in a circle of radius r is $A = \frac{1}{2} r^2 \theta$ if the sector is subtended by an angle of θ radians; here $r = 1$. If $0 < \theta < \pi/2$, we can divide each member of the last inequality by $\frac{1}{2} \sin \theta$ to obtain

$$\cos \theta < \frac{\theta}{\sin \theta} < \frac{1}{\cos \theta}.$$

We take reciprocals, which reverses the inequalities:

$$\cos \theta < \frac{\sin \theta}{\theta} < \frac{1}{\cos \theta}.$$

Now we apply the squeeze law of limits with

$$f(\theta) = \cos \theta, \quad g(\theta) = \frac{\sin \theta}{\theta}, \quad \text{and} \quad h(\theta) = \frac{1}{\cos \theta}.$$

Because it is clear from Eq. (1) at the beginning of this section that $f(\theta)$ and $h(\theta)$ both approach 1 as $\theta \to 0^+$, so does $g(\theta) = (\sin \theta)/\theta$. This geometric argument shows that $(\sin \theta)/\theta \to 1$ for *positive* values of θ that approach zero. But the same result follows for negative values of θ, because $\sin(-\theta) = -\sin \theta$. So we have proved Eq. (17). ❑

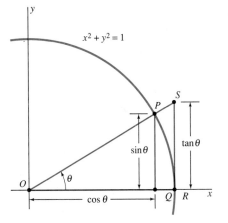

Fig. 2.3.18 Aid to the proof of the basic trigonometric limit

2.3 Problems

Find the trigonometric limits in Problems 1 through 25. If you have a graphics calculator or a computer with graphing facility, verify that graphical evidence supports your answer.

1. $\lim\limits_{\theta \to 0} \dfrac{\theta^2}{\sin \theta}$

2. $\lim\limits_{\theta \to 0} \dfrac{\sin^2 \theta}{\theta^2}$

3. $\lim\limits_{\theta \to 0} \dfrac{1 - \cos \theta}{\theta^2}$

4. $\lim\limits_{\theta \to 0} \dfrac{\tan \theta}{\theta}$

5. $\lim\limits_{x \to 0} \dfrac{2x}{(\sin x) - x}$

6. $\lim\limits_{\theta \to 0} \dfrac{\sin (2\theta^2)}{\theta^2}$

7. $\lim\limits_{x \to 0} \dfrac{\sin 5x}{x}$

8. $\lim\limits_{x \to 0} \dfrac{\sin 2x}{x \cos 3x}$

9. $\lim\limits_{x \to 0} \dfrac{\sin x}{\sqrt{x}}$

10. $\lim\limits_{x \to 0} \dfrac{1 - \cos 2x}{x}$

11. $\lim\limits_{x \to 0} \dfrac{1}{x} \sin \dfrac{x}{3}$

12. $\lim\limits_{x \to 0} \dfrac{(\sin 3x)^2}{x^2 \cos x}$

13. $\lim\limits_{x \to 0} \dfrac{1 - \cos x}{\sin x}$

14. $\lim\limits_{x \to 0} \dfrac{\tan 3x}{\tan 5x}$

15. $\lim\limits_{x \to 0} x \sec x \csc x$

16. $\lim\limits_{\theta \to 0} \dfrac{\sin 2\theta}{\theta}$

17. $\lim\limits_{\theta \to 0} \dfrac{1 - \cos \theta}{\theta \sin \theta}$

18. $\lim\limits_{\theta \to 0} \dfrac{\sin^2 \theta}{\theta}$

19. $\lim\limits_{x \to 0} \dfrac{\tan x}{x}$

20. $\lim\limits_{x \to 0} \dfrac{\tan 2x}{3x}$

21. $\lim\limits_{x \to 0} x \cot 3x$

22. $\lim\limits_{x \to 0} \dfrac{x - \tan x}{\sin x}$

23. $\lim\limits_{x \to 0} \dfrac{1}{x^2} \sin^2\!\left(\dfrac{x}{2}\right)$

24. $\lim\limits_{x \to 0} \dfrac{\sin 2x}{\sin 5x}$

25. $\lim\limits_{x \to 0} x^2 \csc 2x \cot 2x$

Use the squeeze law of limits to find the limits in Problems 26 through 28.

26. $\lim\limits_{x \to 0} x^2 \sin \dfrac{1}{x^2}$

27. $\lim\limits_{x \to 0} x^2 \cos \dfrac{1}{\sqrt[3]{x}}$

28. $\lim\limits_{x \to 0} \sqrt[3]{x} \sin \dfrac{1}{x}$

Use one-sided limit laws to find the limits in Problems 29 through 48 or to determine that they do not exist.

29. $\lim\limits_{x \to 0^+} (3 - \sqrt{x})$

30. $\lim\limits_{x \to 0^+} (4 + 3x^{3/2})$

31. $\lim\limits_{x \to 1^-} \sqrt{x - 1}$

32. $\lim\limits_{x \to 4^-} \sqrt{4 - x}$

33. $\lim\limits_{x \to 2^+} \sqrt{x^2 - 4}$

34. $\lim\limits_{x \to 3^+} \sqrt{9 - x^2}$

35. $\lim\limits_{x \to 5^-} \sqrt{x(5 - x)}$

36. $\lim\limits_{x \to 2^-} x \sqrt{4 - x^2}$

37. $\lim\limits_{x \to 4^+} \sqrt{\dfrac{4x}{x - 4}}$

38. $\lim\limits_{x \to -3^+} \sqrt{6 - x - x^2}$

39. $\lim\limits_{x \to 5^-} \dfrac{x - 5}{|x - 5|}$

40. $\lim\limits_{x \to -4^+} \dfrac{16 - x^2}{\sqrt{16 - x^2}}$

41. $\lim\limits_{x \to 3^+} \dfrac{\sqrt{x^2 - 6x + 9}}{x - 3}$

42. $\lim\limits_{x \to 2^+} \dfrac{x - 2}{x^2 - 5x + 6}$

43. $\lim\limits_{x \to 2^+} \dfrac{2 - x}{|x - 2|}$

44. $\lim\limits_{x \to 7^-} \dfrac{7 - x}{|x - 7|}$

45. $\lim\limits_{x \to 1^+} \dfrac{1 - x^2}{1 - x}$

46. $\lim\limits_{x \to 0^-} \dfrac{x}{x - |x|}$

47. $\lim\limits_{x \to 5^+} \dfrac{\sqrt{(5 - x)^2}}{5 - x}$

48. $\lim\limits_{x \to -4^-} \dfrac{4 + x}{\sqrt{(4 + x)^2}}$

For each of the functions in Problems 49 through 58, there is exactly one point a where both the right-hand and left-hand limits of $f(x)$ fail to exist. Describe (as in Example 10) the behavior of $f(x)$ for x near a.

49. $f(x) = \dfrac{1}{x - 1}$

50. $f(x) = \dfrac{2}{3 - x}$

51. $f(x) = \dfrac{x - 1}{x + 1}$

52. $f(x) = \dfrac{2x - 5}{5 - x}$

53. $f(x) = \dfrac{1 - x^2}{x + 2}$

54. $f(x) = \dfrac{1}{(x - 5)^2}$

55. $f(x) = \dfrac{|1 - x|}{(1 - x)^2}$

56. $f(x) = \dfrac{x + 1}{x^2 + 6x + 9}$

57. $f(x) = \dfrac{x - 2}{4 - x^2}$

58. $f(x) = \dfrac{x - 1}{x^2 - 3x + 2}$

In Problems 59 through 63, $[\![x]\!]$ denotes the greatest integer not exceeding x. First find the limits $\lim\limits_{x \to n^-} f(x)$ and $\lim\limits_{x \to n^+} f(x)$ for each integer n (the answer will normally be in terms of n); then sketch the graph of f.

59. $f(x) = \begin{cases} 2 & \text{if } x \text{ is not an integer;} \\ 2 + (-1)^x & \text{if } x \text{ is an integer.} \end{cases}$

60. $f(x) = \begin{cases} x & \text{if } x \text{ is not an integer;} \\ 0 & \text{if } x \text{ is an integer.} \end{cases}$

61. $f(x) = (-1)^{[\![x]\!]}$

62. $f(x) = \left[\!\left[\dfrac{x}{2} \right]\!\right]$

63. $f(x) = 1 + [\![x]\!] - x$

64. Given $f(x) = |x|$, use the definition of the derivative to find $f'(x)$ in the two separate cases $x > 0$ and $x < 0$.

65. Suppose that there exists a number M such that $|f(x)| \leqq M$ for all x. Apply the squeeze law to show that

$$\lim_{x \to 0} x f(x) = 0.$$

66. Find the flaw in the following argument. Be precise and concise. Point out *exactly* where the error occurs and what the error is.

$$\lim_{x \to 0} x \sin \frac{1}{x} = \left(\lim_{x \to 0} x \right) \left(\lim_{x \to 0} \sin \frac{1}{x} \right)$$

$$= 0 \cdot \left(\lim_{x \to 0} \sin \frac{1}{x} \right) = 0. \qquad \textbf{(Wrong!)}$$

2.4
The Concept of Continuity

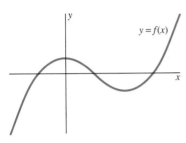

Fig. 2.4.1 A continuous graph

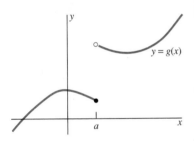

Fig. 2.4.2 A graph that is not continuous

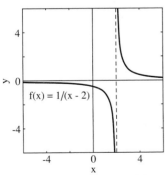

Fig. 2.4.3 The function $f(x) = 1/(x - 2)$ has an infinite discontinuity at $x = 2$ (Example 1).

Anyone can see a drastic difference between the graphs in Figs. 2.4.1 and 2.4.2. Figure 2.4.1 is intended to suggest that the graph $y = f(x)$ can be traced with a continuous motion—without any jumps—of the pen from left to right. But in Fig. 2.4.2 the pen must make a sudden jump at $x = a$.

The concept of continuity isolates the property that the function f of f Fig. 2.4.1 has but that the function g of Fig. 2.4.2 lacks. We first define *continuity* of a function at a single point.

Definition of Continuity at a Point

Suppose that the function f is defined in a neighborhod of a. We say that f is **continuous at a** provided that $\lim_{x \to a} f(x)$ exists and, moreover, that the value of this limit is $f(a)$. In other words, f is continuous at a provided that

$$\lim_{x \to a} f(x) = f(a). \qquad (1)$$

Briefly, continuity of f at a means this:

The limit of f at a is equal to the value of f there.

Another way to put it is this: The limit of f at a is the "expected" value—the value that you would assign if you knew the values of f in a deleted neighborhood of a and you knew f to be "predictable." Alternatively, continuity of f at a means this: When x is close to a, then $f(x)$ is close to $f(a)$.

Analysis of the definition of continuity shows us that to be continuous at the point a, a function must satisfy the following three conditions:

1. The function f must be defined at a [so that $f(a)$ exists].

2. The limit of $f(x)$ as x approaches a must exist.

3. The numbers in conditions 1 and 2 must be equal:

$$\lim_{x \to a} f(x) = f(a).$$

If any one of these conditions is not satisfied, then f is not continuous at a. Examples 1 through 3 illustrate these three possibilities for *discontinuity* at a point. If the function f is *not* continuous at a, then we say that it is **discontinuous** there, or that a is a **discontinuity** of f. Intuitively, a discontinuity of f is a point where the graph of f has a "gap," or "jump," of some sort.

EXAMPLE 1 Figure 2.4.3 shows the graph of the function f defined by

$$f(x) = \frac{1}{x - 2} \qquad \text{for } x \neq 2.$$

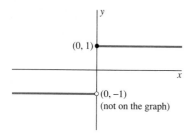

Fig. 2.4.4 The function g has a finite jump discontinuity at $x = 0$ (Example 2).

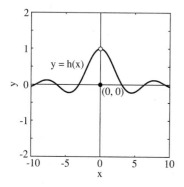

Fig. 2.4.5 The point $(0, 0)$ is on the graph; the point $(0, 1)$ is not (Example 3).

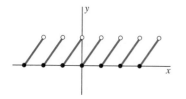

Fig. 2.4.6 The "sawtooth function" of Example 4

Because f is not defined at the point $x = 2$, it is not continuous there. Moreover, f has what might be called an *infinite discontinuity* at $x = 2$.

EXAMPLE 2 Figure 2.4.4 shows the graph of the function g defined by

$$g(x) = \text{sign}(x) = \begin{cases} +1 & \text{if } x \geqq 0; \\ -1 & \text{if } x < 0. \end{cases}$$

Its left-hand and right-hand limits at $x = 0$ are unequal, so $g(x)$ has no limit as $x \to 0$. Consequently, the function g is not continuous at $x = 0$; it has what might be called a *finite jump discontinuity* there.

EXAMPLE 3 Figure 2.4.5 shows the graph of the function h defined by

$$h(x) = \begin{cases} \dfrac{\sin x}{x} & \text{if } x \neq 0; \\ 0 & \text{if } x = 0. \end{cases}$$

Because we saw in Section 2.3 that

$$\lim_{x \to 0} h(x) = \lim_{x \to 0} \frac{\sin x}{x} = 1,$$

whereas $h(0) = 0$, we see that the limit and the value of h at $x = 0$ are not equal. Thus the function h is not continuous there. As x moves from negative values through $x = 0$ to positive values, the value of $h(x)$ jumps from "near 1" to 0 and back again.

EXAMPLE 4 Figure 2.4.6 shows the graph of the function f defined by

$$f(x) = x - [\![x]\!].$$

As before, $[\![x]\!]$ denotes the largest integer no greater than x. If $x = n$, an integer, then $[\![n]\!] = n$, so $f(n) = 0$. On the open interval $(n, n + 1)$, the graph of f is linear and has slope 1. It should be clear that f is

❏ Continuous at x if x is *not* an integer;
❏ Discontinuous at each integer point on the x-axis.

COMBINATIONS OF CONTINUOUS FUNCTIONS

Frequently we are most interested in functions that *are* continuous. Suppose that the function f is defined on an open interval or a union of open intervals. Then we say simply that f is **continuous** if it is continuous at each point of its domain of definition.

It follows readily from the limit laws in Section 2.2 that *any sum or product of continuous functions is continuous.* That is, if the functions f and g are continuous at $x = a$, then so are $f + g$ and $f \cdot g$. For instance, if f and g are continuous at $x = a$, then

$$\lim_{x \to a} [f(x) + g(x)] = \left(\lim_{x \to a} f(x) \right) + \left(\lim_{x \to a} g(x) \right) = f(a) + g(a).$$

EXAMPLE 5 Because $f(x) = x$ and constant-valued functions are clearly continuous everywhere, it follows that the cubic polynomial function

$$f(x) = x^3 - 3x^2 + 1 = x \cdot x \cdot x + (-3) \cdot x \cdot x + 1$$

is continuous everywhere.

More generally, it follows in a similar way that *every* **polynomial function**

$$p(x) = b_n x^n + b_{n-1} x^{n-1} + \cdots + b_1 x + b_0$$

is continuous at each point of the real line. In short, every polynomial is continuous everywhere.

If $p(x)$ and $q(x)$ are polynomials, then the quotient law for limits and the continuity of polynomials imply that

$$\lim_{x \to a} \frac{p(x)}{q(x)} = \frac{\lim_{x \to a} p(x)}{\lim_{x \to a} q(x)} = \frac{p(a)}{q(a)}$$

provided that $q(a) \neq 0$. Thus every **rational function**

$$f(x) = \frac{p(x)}{q(x)} \tag{2}$$

is continuous wherever it is defined—that is, wherever the denominator polynomial is nonzero. More generally, the quotient of any two continuous functions is continuous at every point where the denominator is nonzero.

At a point $x = a$ where the denominator in Eq. (2) *is* zero, $q(a) = 0$, there are two possibilities:

❏ If $p(a) \neq 0$, then f has an infinite discontinuity (as in Figs. 2.4.3 and 2.4.7) at $x = a$.
❏ Otherwise, f *may* have a *removable discontinuity* at $x = a$.

The point $x = a$ where the function f is discontinuous is called a **removable discontinuity** of f provided that there exists a function F such that

❏ $F(x) = f(x)$ for all $x \neq a$ in the domain of f, and
❏ This new function F is continuous at $x = a$.

Thus, by adjoining the single point $(a, F(a))$ to the graph of f, we "remove" the discontinuity, obtaining the graph of the function F that is continuous at $x = a$.

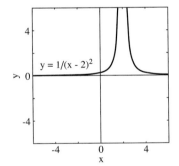

Fig. 2.4.7 The function $f(x) = 1/(x - 2)^2$ has an infinite discontinuity at $x = 2$.

EXAMPLE 6 Suppose that

$$f(x) = \frac{x - 2}{x^2 - 3x + 2}. \tag{3}$$

We factor the denominator: $x^2 - 3x + 2 = (x - 1)(x - 2)$. This shows that f is not defined either at $x = 1$ or at $x = 2$. Thus the rational function defined in Eq. (3) is continuous except at these two points. Because cancellation gives

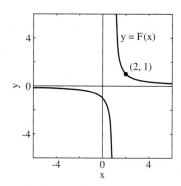

Fig. 2.4.8 In Example 6, the graph $y = F(x)$ consists of the graph $y = f(x)$ with the single point $(2, 1)$ adjoined.

$$f(x) = \frac{x - 2}{(x - 1)(x - 2)} = \frac{1}{x - 1}$$

except at the single point $x = 2$, the new function

$$F(x) = \frac{1}{x - 1} \tag{4}$$

agrees with $f(x)$ if $x \neq 2$ but is continuous at $x = 2$ also, where $F(2) = 1$. See Fig. 2.4.8.

CONTINUITY OF TRIGONOMETRIC FUNCTIONS

At the beginning of Section 2.3 we noted that

$$\lim_{x \to 0} \cos x = 1 \quad \text{and} \quad \lim_{x \to 0} \sin x = 0. \tag{5}$$

Because $\cos 0 = 1$ and $\sin 0 = 0$, the sine and cosine functions must be continuous at $x = 0$. But this fact implies that they are continuous everywhere.

Theorem 1 *Continuity of Sine and Cosine*
The functions $f(x) = \sin x$ and $g(x) = \cos x$ are continuous functions of x on the whole real line.

Proof   We give the proof only for $\sin x$; the proof for $\cos x$ is similar (see Problem 69). We want to show that $\lim_{x \to a} \sin x = \sin a$ for every real number a. If we write $x = a + h$, so that $h = x - a$, then $h \to 0$ as $x \to a$. Thus we need only show that

$$\lim_{h \to 0} \sin(a + h) = \sin a.$$

But the addition formula for the sine function yields

$$\lim_{h \to 0} \sin(a + h) = \lim_{h \to 0} (\sin a \cos h + \cos a \sin h)$$

$$= (\sin a)\left(\lim_{h \to 0} \cos h \right) + (\cos a)\left(\lim_{h \to 0} \sin h \right) = \sin a$$

as desired; we used the limits in Eq. (5) in the last step.   ❏

It now follows that the function

$$\tan x = \frac{\sin x}{\cos x}$$

is continuous except where $\cos x = 0$—that is, except when x is an *odd* integral multiple of $\pi/2$. As illustrated in Fig. 2.4.9, $\tan x$ has an infinite discontinuity at each such point.

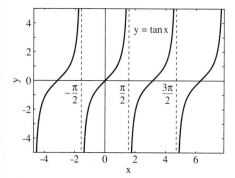

Fig. 2.4.9 The function $\tan x$ has infinite discontinuities at $x = \pm\pi/2, \pm3\pi/2, \ldots$.

COMPOSITION OF CONTINUOUS FUNCTIONS

Many varied and complex functions can be "put together" by using quite simple "building-block" functions. In addition to adding, subtracting, multi-

84

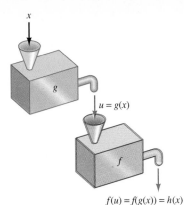

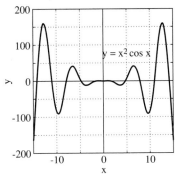

Fig. 2.4.10 The composition of f and g

Fig. 2.4.11 $y = x^2 \cos x$ (Example 8)

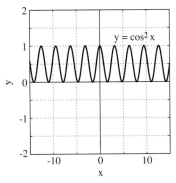

Fig. 2.4.12 $y = \cos^2 x$ (Example 8)

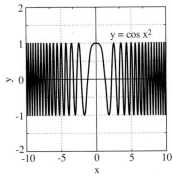

Fig. 2.4.13 $y = \cos x^2$ (Example 8)

plying, or dividing two given functions, we can also combine functions by letting one function act on the output of the other.

> **Definition** *Composition of Functions*
>
> The **composition** of the two functions f and g is the function $h = f \circ g$ defined by
> $$h(x) = f(g(x)) \tag{6}$$
> for all x in the domain of g such that $u = g(x)$ is in the domain of f. [The right-hand side in Eq. (6) is read "f of g of x."]

Thus the output $u = g(x)$ of the function g is used as the input to the function f (Fig. 2.4.10). We sometimes refer to g as the *inner function* and to f as the *outer function* in Eq. (6).

EXAMPLE 7 If $f(x) = \sqrt{x}$ and $g(x) = 1 - x^2$, then
$$f(g(x)) = \sqrt{1 - x^2} \qquad \text{for } |x| \leq 1,$$
whereas
$$g(f(x)) = 1 - \left(\sqrt{x}\right)^2 = 1 - x \qquad \text{for } x \geq 0.$$

The $f(g(x))$ notation for compositions is most commonly used in ordinary computations, but the $f \circ g$ notation emphasizes that the composition may be regarded as a new kind of combination of the functions f and g. But Example 7 shows that $f \circ g$ is quite unlike the product fg of the two functions f and g:
$$f \circ g \neq g \circ f,$$
whereas $fg = gf$ [because $f(x) \cdot g(x) = g(x) \cdot f(x)$ whenever $f(x)$ and $g(x)$ are defined]. So, remember that composition is quite different in character from ordinary multiplication of functions.

EXAMPLE 8 If
$$f(x) = x^2 \quad \text{and} \quad g(x) = \cos x,$$
then the functions
$$f(x)g(x) = x^2 \cos x,$$
$$f(g(x)) = \cos^2 x = (\cos x)^2, \quad \text{and}$$
$$g(f(x)) = \cos x^2 = \cos(x^2)$$
are defined for all x. Figures 2.4.11 through 2.4.13 illustrate vividly how different these three functions are.

EXAMPLE 9 Given the function $h(x) = (x^2 + 4)^{3/2}$, find two functions f and g such that $h(x) = f(g(x))$.

Solution It is technically correct—but useless—simply to let $g(x) = x$ and $f(u) = (u^2 + 4)^{3/2}$. We seek a nontrivial answer here. To calculate

85

$(x^2 + 4)^{3/2}$, we must first calculate $x^2 + 4$. So we choose $g(x) = x^2 + 4$ as the inner function. The last step is to raise $u = g(x)$ to the power $\frac{3}{2}$, so we take $f(u) = u^{3/2}$ as the outer function. Thus if

$$f(x) = x^{3/2} \quad \text{and} \quad g(x) = x^2 + 4,$$

then $f(g(x)) = f(x^2 + 4) = (x^2 + 4)^{3/2} = h(x)$.

Theorem 2 implies that functions built by forming compositions of continuous functions are themselves continuous.

> **Theorem 2** *Continuity of Compositions*
> The composition of two continuous functions is continuous. More pre-cisely, if g is continuous at a and f is continuous at $g(a)$, then $f(g)$ is continuous at a.

Proof The continuity of g at a means that $g(x) \to g(a)$ as $x \to a$, and the continuity of f at $g(a)$ implies that $f(g(x)) \to f(g(a))$ as $g(x) \to g(a)$. Hence the substitution law for limits (Section 2.2) yields

$$\lim_{x \to a} f(g(x)) = f\left(\lim_{x \to a} g(x)\right) = f(g(a)),$$

as desired. ❑

Recall from the root law in Section 2.2 that

$$\lim_{x \to a} \sqrt[n]{x} = \sqrt[n]{a}$$

under the conditions that n is an integer and that $a > 0$ if n is even. Thus the nth-root function $f(x) = \sqrt[n]{x}$ is continuous everywhere if n is odd and con-tinuous for $x > 0$ if n is even.

We may combine this result with Theorem 2. Then we see that a root of a continuous function is continuous wherever it is defined. That is, the composition

$$g(x) = [f(x)]^{1/n}$$

is continuous at a if f is, assuming that $f(a) > 0$ if n is even (so that $\sqrt[n]{f(a)}$ is defined).

EXAMPLE 10 Show that the function

$$f(x) = \left(\frac{x - 7}{x^2 + 2x + 2}\right)^{2/3}$$

is continuous on the whole real line.

Solution Note first that the denominator

$$x^2 + 2x + 2 = (x + 1)^2 + 1$$

is never zero. Hence the rational function

$$r(x) = \frac{x - 7}{x^2 + 2x + 2}$$

is defined and continuous everywhere. It then follows from Theorem 2 and the continuity of the cube root function that

$$f(x) = ([r(x)]^2)^{1/3}$$

is continuous everywhere. Hence, for example,

$$\lim_{x \to -1} \left(\frac{x - 7}{x^2 + 2x + 2} \right)^{2/3} = f(-1) = (-8)^{2/3} = 4.$$

CONTINUOUS FUNCTIONS ON CLOSED INTERVALS

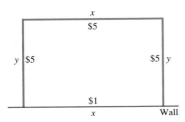

An applied problem typically involves a function whose domain is a *closed interval*. For example, in the animal pen problem of Section 1.1, we found that the area A of the rectangular pen in Fig. 2.4.14 was expressed as a function of its base length x by

$$A = f(x) = \tfrac{3}{5}x(30 - x).$$

Fig. 2.4.14   The animal pen

Although this formula for f is meaningful for all x, only values in the closed interval $[0, 30]$ correspond to actual rectangles, so only such values are pertinent to the animal pen problem.

The function f defined on the closed interval $[a, b]$ is said to be **continuous on $[a, b]$** provided that it is continuous at each point of the open interval (a, b) *and* that

$$\lim_{x \to a^+} f(x) = f(a) \quad \text{and} \quad \lim_{x \to b^-} f(x) = f(b).$$

The last two conditions mean that at each endpoint, the value of the function is equal to its limit from within the interval. For instance, every polynomial is continuous on every closed interval. The square root function $f(x) = \sqrt{x}$ is continuous on the closed interval $[0, 1]$ even though f is not defined for $x < 0$.

Continuous functions defined on closed intervals have very special properties. For example, every such function has the *intermediate value property* of Theorem 3. (A proof of this theorem is given in Appendix C.) We suggested earlier that continuity of a function is related to the possibility of tracing its graph without lifting the pen from the paper. Theorem 3 expresses this fact precisely.

> **Theorem 3**   *Intermediate Value Property*
>
> Suppose that the function f is continuous on the closed interval $[a, b]$. Then $f(x)$ assumes every intermediate value between $f(a)$ and $f(b)$. That is, if K is any number between $f(a)$ and $f(b)$, then there exists at least one number c in (a, b) such that $f(c) = K$.

Figure 2.4.15 shows the graph of a typical continuous function f whose domain is the closed interval $[a, b]$. The number K is located on the y-axis, somewhere between $f(a)$ and $f(b)$. In the figure $f(a) < f(b)$, but this is not important. The horizontal line through K must cross the graph of f somewhere, and the x-coordinate of the point where the graph and the line meet yields the value of c. The number c is the one whose existence is guaranteed by the intermediate value property of the continuous function f.

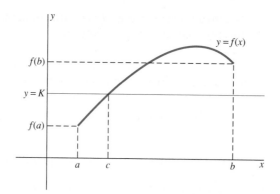

Fig. 2.4.15 The continuous function f attains the intermediate value K at $x = c$.

Thus the intermediate value property implies that each horizontal line meeting the y-axis between $f(a)$ and $f(b)$ must cross the graph of the continuous function f somewhere. This is a way of saying that the graph has no gaps or jumps, suggesting that the idea of being able to trace such a graph without lifting the pen from the paper is accurate.

EXAMPLE 11 The discontinuous function defined on $[-1, 1]$ as

$$f(x) = \begin{cases} 0 & \text{if } x < 0, \\ 1 & \text{if } x \geqq 0 \end{cases}$$

does *not* attain the intermediate value 0.5. See Fig. 2.4.16.

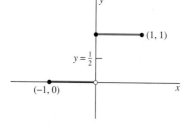

Fig. 2.4.16 This discontinuous function does not have the intermediate value property (Example 11).

EXISTENCE OF SOLUTIONS OF EQUATIONS

An important application of the intermediate value property is the verification of the existence of solutions of equations written in the form

$$f(x) = 0. \tag{7}$$

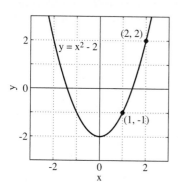

Fig. 2.4.17 The graph of $f(x) = x^2 - 2$ (Example 12)

EXAMPLE 12 Part A of Project 1.3 asked you to approximate the number $\sqrt{2}$ by zooming in on the intersection of the parabola $y = x^2 - 2$ with the positive x-axis. (See Fig. 2.4.17.) The x-coordinate of the intersection yields the solution of the equation

$$f(x) = x^2 - 2 = 0. \tag{8}$$

Perhaps it makes no sense to zoom in on this point unless we know that it's "really there." But we can see from Eq. (8) that

$$f(1) = -1 < 0, \quad \text{whereas} \quad f(2) = 2 > 0.$$

We note that the function f is continuous on $[1, 2]$ (it is continuous everywhere) and that $K = 0$ is an intermediate value of f on the interval $[1, 2]$. Therefore, it follows from Theorem 3 that $f(c) = 0$ for some number c in $[1, 2]$—that is, that

$$c^2 = 2.$$

This number c is the desired square root of 2. Thus it is the intermediate value property of continuous functions that guarantees the existence of the number $\sqrt{2}$: There *is* a real number whose square is 2.

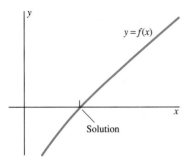

Fig. 2.4.18 A root of the equation $f(x) = 0$

As indicated in Fig. 2.4.18, the solutions of Eq. (7) are simply the points where the graph $y = f(x)$ crosses the x-axis. Suppose that f is continuous and that we can find a closed interval $[a, b]$ (such as the interval $[1, 2]$ of Example 12) such that the value of f is positive at one endpoint of $[a, b]$ and negative at the other. That is, suppose that $f(x)$ *changes sign* on the closed interval $[a, b]$. Then the intermediate value property ensures that $f(x) = 0$ at some point of $[a, b]$.

EXAMPLE 13 Show that the equation

$$x^3 + x - 1 = 0$$

has a solution between $x = 0$ and $x = 1$.

Solution The function $f(x) = x^3 + x - 1$ is continuous on $[0, 1]$ because it is a polynomial. Next, $f(0) = -1$ and $f(1) = +1$. So every number between -1 and $+1$ is a value of f on $(0, 1)$. In particular,

$$f(0) < 0 < f(1),$$

so the intermediate value property of f implies that f attains the value 0 at some number c between $x = 0$ and $x = 1$. That is, $f(c) = 0$ for some number c in $(0, 1)$. In other words, $c^3 + c - 1 = 0$. Therefore, the equation $x^3 + x - 1 = 0$ has a solution in $(0, 1)$.

CONTINUITY AND DIFFERENTIABILITY

We have seen that a wide variety of functions are continuous. Although there are continuous functions so exotic that they are nowhere differentiable, Theorem 4 tells us that every differentiable function is continuous.

> **Theorem 4** *Differentiability Implies Continuity*
> Suppose that the function f is defined in a neighborhood of a. If f is differentiable at a, then f is continuous at a.

Proof Because $f'(a)$ exists, the product law for limits yields

$$\lim_{x \to a} [f(x) - f(a)] = \lim_{x \to a} (x - a) \cdot \frac{f(x) - f(a)}{x - a}$$

$$= \left(\lim_{x \to a} (x - a) \right) \left(\lim_{x \to a} \frac{f(x) - f(a)}{x - a} \right)$$

$$= 0 \cdot f'(a) = 0.$$

Thus $\lim_{x \to a} f(x) = f(a)$, so f is continuous at a. ❑

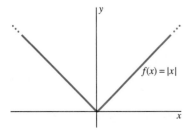

Fig. 2.4.19 The graph of $f(x) = |x|$

Although is is complicated to exhibit a nowhere-differentiable but everywhere-continuous function, the absolute value function $f(x) = |x|$ provides us with a simple example of an everywhere-continuous function that is not differentiable at the one point $x = 0$ (as we saw in Example 9 of Section 2.3). The "corner point" on the graph evident in Fig. 2.4.19 indicates that $y = |x|$ has no tangent line at $(0, 0)$ and thus that $f(x) = |x|$ cannot be differentiable there.

2.4 Problems

In Problems 1 through 10, find $f(g(x))$ and $g(f(x))$.

1. $f(x) = 1 - x^2$, $g(x) = 2x + 3$

2. $f(x) = -17$, $g(x) = |x|$

3. $f(x) = \sqrt{x^2 - 3}$, $g(x) = x^2 + 3$

4. $f(x) = x^2 + 1$, $g(x) = \dfrac{1}{x^2 + 1}$

5. $f(x) = x^3 - 4$, $g(x) = \sqrt[3]{x + 4}$

6. $f(x) = \sqrt{x}$, $g(x) = \cos x$

7. $f(x) = \sin x$, $g(x) = x^3$

8. $f(x) = \sin x$, $g(x) = \cos x$

9. $f(x) = 1 + x^2$, $g(x) = \tan x$

10. $f(x) = 1 - x^2$, $g(x) = \sin x$

In Problems 11 through 20, find a function of the form $f(x) = x^k$ (you must specify k) and a function g such that $f(g(x)) = h(x)$.

11. $h(x) = (2 + 3x)^2$ **12.** $h(x) = (4 - x)^3$

13. $h(x) = \sqrt{2x - x^2}$ **14.** $h(x) = (1 + x^4)^{17}$

15. $h(x) = (5 - x^2)^{3/2}$ **16.** $h(x) = \sqrt[3]{(4x - 6)^4}$

17. $h(x) = \dfrac{1}{x + 1}$ **18.** $h(x) = \dfrac{1}{1 + x^2}$

19. $h(x) = \dfrac{1}{\sqrt{x + 10}}$ **20.** $h(x) = \dfrac{1}{(1 + x + x^2)^3}$

In Problems 21 through 42, tell where the function with the given formula is continuous. Recall that when the domain of a function is not specified, it is the set of all real numbers for which the formula of the function is meaningful.

21. $f(x) = 2x + \sqrt[3]{x}$ **22.** $f(x) = x^2 + \dfrac{1}{x}$

23. $f(x) = \dfrac{1}{x + 3}$ **24.** $f(x) = \dfrac{5}{5 - x}$

25. $f(x) = \dfrac{1}{x^2 + 1}$ **26.** $f(x) = \dfrac{1}{x^2 - 1}$

27. $f(x) = \dfrac{x - 5}{|x - 5|}$ **28.** $f(x) = \dfrac{x^2 + x + 1}{x^2 + 1}$

29. $f(x) = \dfrac{x^2 + 4}{x - 2}$ **30.** $f(x) = \sqrt[4]{4 + x^4}$

31. $f(x) = \sqrt[3]{\dfrac{x + 1}{x - 1}}$ **32.** $f(x) = \sqrt[3]{3 - x^3}$

33. $f(x) = \dfrac{3}{x^2 - x}$ **34.** $f(x) = \sqrt{9 - x^2}$

35. $f(x) = \dfrac{x}{\sqrt{4 - x^2}}$ **36.** $f(x) = \sqrt{\dfrac{1 - x^2}{4 - x^2}}$

37. $f(x) = \dfrac{\sin x}{x^2}$ **38.** $f(x) = \dfrac{x}{\cos x}$

39. $f(x) = \dfrac{1}{\sin 2x}$ **40.** $f(x) = \sqrt{\sin x}$

41. $f(x) = \sin |x|$ **42.** $f(x) = \dfrac{1}{\sqrt{1 + \cos x}}$

In Problems 43 through 52, find the points where the given function is not defined and is therefore not continuous. For each such point a, tell whether this discontinuity is removable.

43. $f(x) = \dfrac{x}{(x + 3)^3}$ **44.** $f(x) = \dfrac{x}{x^2 - 1}$

45. $f(x) = \dfrac{x - 2}{x^2 - 4}$ **46.** $f(x) = \dfrac{x + 1}{x^2 - x - 6}$

47. $f(x) = \dfrac{1}{1 - |x|}$ **48.** $f(x) = \dfrac{|x - 1|}{(x - 1)^3}$

49. $f(x) = \dfrac{x - 17}{|x - 17|}$ **50.** $f(x) = \dfrac{x^2 + 5x + 6}{x + 2}$

51. $f(x) = \begin{cases} -x & \text{if } x < 0; \\ x^2 & \text{if } x > 0 \end{cases}$

52. $f(x) = \begin{cases} x + 1 & \text{if } x < 1; \\ 3 - x & \text{if } x > 1 \end{cases}$

In Problems 53 through 58, apply the intermediate value property of continuous functions to show that the given equation has a solution in the given interval.

53. $x^2 - 5 = 0$ on $[2, 3]$

54. $x^3 + x + 1 = 0$ on $[-1, 0]$

55. $x^3 - 3x^2 + 1 = 0$ on $[0, 1]$

56. $x^3 = 5$ on $[1, 2]$

57. $x^4 + 2x - 1 = 0$ on $[0, 1]$

58. $x^5 - 5x^3 + 3 = 0$ on $[-3, -2]$

In Problems 59 and 60, show that the given equation has three distinct roots by calculating the values of the left-hand side at $x = -3, -2, -1, 0, 1, 2,$ and 3 and then applying the intermediate value property of continuous functions on appropriate closed intervals.

59. $x^3 - 4x + 1 = 0$ **60.** $x^3 - 3x^2 + 1 = 0$

61. Apply the intermediate value property of continuous functions to show that every positive number a has a square root. That is, given $a > 0$, prove that there exists a number r such that $r^2 = a$.

62. Apply the intermediate value property to prove that every real number has a cube root.

63. Determine where the function $f(x) = [\![x/3]\!]$ is continuous.

64. Determine where the function $f(x) = x + [[x]]$ is continuous.

65. Suppose that $f(x) = 0$ if x is a rational number, whereas $f(x) = 1$ if x is irrational. Prove that f is discontinuous at every real number.

66. Suppose that $f(x) = 0$ if x is a rational number, whereas $f(x) = x^2$ if x is irrational. Prove that f is continuous only at the single point $x = 0$.

67. Show that the function $f(x) = x \sin(1/x)$ of Example 4 in Section 2.3 is *not* differentiable at $x = 0$. [*Suggestion:*

Show that whether $z = 1$ or $z = -1$, there are arbitrarily small values of h such that $[f(h) - f(0)]/h = z$.]

68. Let $f(x) = x^2 \sin(1/x)$ for $x \neq 0$; $f(0) = 0$. (The graph of f appears in Figs. 2.4.20 and 2.4.21.) Apply the definition of the derivative to show that f is differentiable at $x = 0$ and that $f'(0) = 0$.

69. Show that the cosine function is continuous on the set of all real numbers. [*Suggestion:* Alter the proof in Theorem 1 of the continuity of the sine function.]

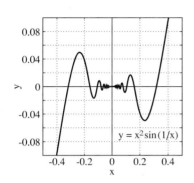

Fig. 2.4.20 The graph of $y = x^2 \sin \dfrac{1}{x}$ (Problem 68)

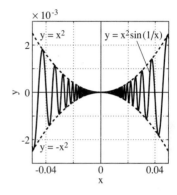

Fig. 2.4.21 The graph of Fig. 2.4.20 magnified (Problem 68)

2.4 Projects

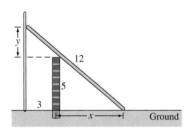

Fig. 2.4.22 The leaning ladder

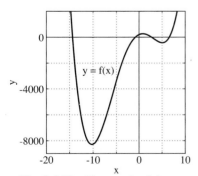

Fig. 2.4.23 The graph of the leaning-ladder equation

These projects require a calculator that readily tabulates functions or, preferably, a graphics calculator or computer with zoom capability.

PROJECT A Figure 2.4.22 shows the leaning ladder of part C of Project 1.3. There you were asked to show that the indicated lengths x and y satisfy the equations

$$(x + 3)^2 + (y + 5)^2 = 144,$$
$$xy = 15.$$

Now eliminate y to obtain the single equation

$$f(x) = x^4 + 6x^3 - 110x^2 + 150x + 225 = 0. \tag{9}$$

Figure 2.4.23 shows the graph of f. Apply the intermediate value property of continuous functions to *prove* that Eq. (9) has four real solutions, and locate them in intervals whose endpoints are consecutive integers. Then use repeated tabulation or successive zooms to approximate each of these roots accurate to at least three places. What are the physically possible values of the distance x from the bottom of the fence to the foot of the ladder?

PROJECT B A 100-ft tree stands 20 ft from a 10-ft fence. Then the tree is "broken" at a height of x feet (Fig. 2.4.24). The tree falls so that its trunk

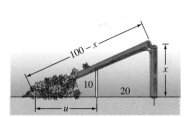

Fig. 2.4.24 The bent tree

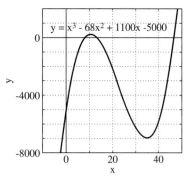

Fig. 2.4.25 The graph of the bent-tree equation

barely touches the top of the fence when the tip of the tree strikes the ground on the other side of the fence. Use similar triangles and the Pythagorean theorem to show that x satisfies the equation

$$f(x) = x^3 - 68x^2 + 1100x - 5000 = 0. \qquad (10)$$

Figure 2.4.25 shows the graph of f. Apply the intermediate value property of continuous functions to *prove* that Eq. (10) has three real solutions, and locate them in intervals whose endpoints are consecutive integers. Then use repeated tabulation or successive zooms to approximate each of the roots accurate to at least three decimal places. What are the physical possibilities for the height x?

Chapter 2 Review: DEFINITIONS, CONCEPTS, RESULTS

Use the following list as a guide to concepts that you may need to review.

1. The limit of $f(x)$ as x approaches a
2. Limits laws: constant, addition, product, quotient, root, substitution, and squeeze
3. Definition of the derivative of a function
4. The four-step process for finding the derivative
5. The derivative of $f(x) = ax^2 + bx + c$
6. Tangent and normal lines to the graph of a function
7. The basic trigonometric limit
8. Evaluation of limits of trigonometric functions
9. Right-hand and left-hand limits
10. The relation between one-sided and two-sided limits
11. Infinite limits
12. Composition of functions
13. Continuity of a function at a point
14. Continuity of polynomials and of rational functions
15. Continuity of trigonometric functions
16. Continuity of compositions of functions
17. Continuity of a function on a closed interval
18. The intermediate value property of continuous functions
19. Differentiability implies continuity

Chapter 2 Miscellaneous Problems

Apply the limit laws to evaluate the limits in Problems 1 through 40 or to show that the indicated limit does not exist, as appropriate.

1. $\lim\limits_{x \to 0} (x^2 - 3x + 4)$

2. $\lim\limits_{x \to -1} (3 - x + x^3)$

3. $\lim\limits_{x \to 2} (4 - x^2)^{10}$

4. $\lim\limits_{x \to 1} (x^2 + x - 1)^{17}$

5. $\lim\limits_{x \to 2} \dfrac{1 + x^2}{1 - x^2}$

6. $\lim\limits_{x \to 3} \dfrac{2x}{x^2 - x - 3}$

7. $\lim\limits_{x \to 1} \dfrac{x^2 - 1}{1 - x}$

8. $\lim\limits_{x \to -2} \dfrac{x + 2}{x^2 + x - 2}$

9. $\lim\limits_{t \to -3} \dfrac{t^2 + 6t + 9}{9 - t^2}$

10. $\lim\limits_{x \to 0} \dfrac{4x - x^3}{3x + x^2}$

11. $\lim\limits_{x \to 3} (x^2 - 1)^{2/3}$

12. $\lim\limits_{x \to 2} \sqrt{\dfrac{2x^2 + 1}{2x}}$

13. $\lim\limits_{x \to 3} \left(\dfrac{5x + 1}{x^2 - 8}\right)^{3/4}$

14. $\lim\limits_{x \to 1} \dfrac{x^4 - 1}{x^2 + 2x - 3}$

15. $\lim\limits_{x \to 7} \dfrac{\sqrt{x + 2} - 3}{x - 7}$

16. $\lim\limits_{x \to 1^+} (x - \sqrt{x^2 - 1})$

17. $\lim\limits_{x \to -4} \dfrac{\dfrac{1}{\sqrt{13 + x}} - \dfrac{1}{3}}{x + 4}$

18. $\lim\limits_{x \to 1^+} \dfrac{1 - x}{|1 - x|}$

19. $\lim\limits_{x \to 2^+} \dfrac{2 - x}{\sqrt{4 - 4x + x^2}}$

20. $\lim\limits_{x \to -2^-} \dfrac{x + 2}{|x + 2|}$

21. $\lim\limits_{x \to 4^+} \dfrac{x - 4}{|x - 4|}$

22. $\lim\limits_{x \to 3^-} \sqrt{x^2 - 9}$

23. $\lim\limits_{x \to 2^+} \sqrt{4 - x^2}$

24. $\lim\limits_{x \to -3} \dfrac{x}{(x + 3)^2}$

25. $\lim\limits_{x \to 2} \dfrac{x + 2}{(x - 2)^2}$

26. $\lim\limits_{x \to 1^-} \dfrac{x}{x - 1}$

27. $\lim\limits_{x \to 3^+} \dfrac{x}{x - 3}$

28. $\lim\limits_{x \to 1^-} \dfrac{x - 2}{x^2 - 3x + 2}$

29. $\lim\limits_{x \to 1^-} \dfrac{x + 1}{(x - 1)^3}$

30. $\lim\limits_{x \to 5^+} \dfrac{25 - x^2}{x^2 - 10x + 25}$

31. $\lim\limits_{x \to 0} \dfrac{\sin 3x}{x}$

32. $\lim\limits_{x \to 0} \dfrac{\tan 5x}{x}$

33. $\lim\limits_{x \to 0} \dfrac{\sin 3x}{\sin 2x}$

34. $\lim\limits_{x \to 0} \dfrac{\tan 2x}{\tan 3x}$

35. $\lim\limits_{x \to 0^+} \dfrac{x}{\sin \sqrt{x}}$

36. $\lim\limits_{x \to 0} \dfrac{1 - \cos 3x}{2x}$

37. $\lim\limits_{x \to 0} \dfrac{1 - \cos 3x}{2x^2}$

38. $\lim\limits_{x \to 0} x^3 \cot x \csc x$

39. $\lim\limits_{x \to 0} \dfrac{\sec 2x \tan 2x}{x}$

40. $\lim\limits_{x \to 0} x^2 \cot^2 3x$

Apply the formula for the derivative of $f(x) = ax^2 + bx + c$ to differentiate the functions in Problems 41 through 46. Then write an equation for the line tangent to the curve $y = f(x)$ at the point $(1, f(1))$.

41. $f(x) = 3 + 2x^2$

42. $f(x) = x - 5x^2$

43. $f(x) = 3x^2 + 4x - 5$

44. $f(x) = 1 - 2x - 3x^2$

45. $f(x) = (x - 1)(2x - 1)$

46. $f(x) = \dfrac{x}{3} - \left(\dfrac{x}{4}\right)^2$

In Problems 47 through 53, apply the definition of the derivative to find $f'(x)$.

47. $f(x) = 2x^2 + 3x$

48. $f(x) = x - x^3$

49. $f(x) = \dfrac{1}{3 - x}$

50. $f(x) = \dfrac{1}{2x + 1}$

51. $f(x) = x - \dfrac{1}{x}$

52. $f(x) = \dfrac{x}{x + 1}$

53. $f(x) = \dfrac{x + 1}{x - 1}$

54. Find the derivative of
$$f(x) = 3x - x^2 + |2x + 3|$$
at the points where it is differentiable. Find the point where f is *not* differentiable. Sketch the graph of f.

55. Write equations of the two lines through $(3, 4)$ that are tangent to the parabola $y = x^2$. [*Suggestion:* Let (a, a^2) denote either point of tangency; first solve for a.]

56. Write an equation of the circle with center $(2, 3)$ that is tangent to the line with equation $x + y + 3 = 0$.

57. Suppose that $f(x) = 1 + x^2$. Find g such that
$$f(g(x)) = 1 + x^2 - 2x^3 + x^4.$$

58. Suppose that $g(x) = 1 + \sqrt{x}$. Find f such that
$$f(g(x)) = 3 + 2\sqrt{x} + x.$$

In Problems 59 through 62, find a function g such that $f(g(x)) = h(x)$.

59. $f(x) = 2x + 3$, $h(x) = 2x + 5$

60. $f(x) = x + 1$, $h(x) = x^3$

61. $f(x) = x^2$, $h(x) = x^4 + 1$

62. $f(x) = \dfrac{1}{x}$, $h(x) = x^5$

In Problems 63 through 66, explain why each function is continuous wherever it is defined by the given formula. For each point a where f is not defined by the formula, tell whether a value can be assigned to $f(a)$ in such a way to make f continuous at a.

63. $f(x) = \dfrac{1 - x}{1 - x^2}$

64. $f(x) = \dfrac{1 - x}{(2 - x)^2}$

65. $f(x) = \dfrac{x^2 + x - 2}{x^2 + 2x - 3}$

66. $f(x) = \dfrac{|x^2 - 1|}{x^2 - 1}$

67. Apply the intermediate value property of continuous functions to prove that the equation $x^5 + x = 1$ has a solution.

68. Apply the intermediate value property of continuous functions to prove that the equation $x^5 - 3x^2 + 1 = 0$ has three different solutions.

69. Show that there is a number x between 0 and $\pi/2$ such that $x = \cos x$.

70. Show that there is a number x between $\pi/2$ and π such that $\tan x = -x$. [*Suggestion:* First sketch the graphs of $y = \tan x$ and $y = -x$.]

The Derivative

Isaac Newton (1642–1727)

❏ Isaac Newton was born in a rural English farming village on Christmas Day in 1642, three months after his father's death. When the boy was three, his mother remarried and left him with his grandmother. Nothing known about his childhood and early schooling hinted that his life and work would constitute a turning point in the history of humanity.

❏ But due to the influence of an uncle who suspected hidden potential in young Isaac, Newton was able to enter Cambridge University in 1661. During the years 1665 and 1666, when Cambridge closed because of the bubonic plague then sweeping Europe, he returned to his country home and there laid the foundations for the three towering achievements of his scientific career—the invention of the cal-

culus, the discovery of the spectrum of colors in light, and the theory of gravitation. Of these two years he later wrote that "in those days I was in the prime of my age of invention and minded mathematics and philosophy more than at any time since." Indeed, his thirties were devoted more to smoky chemical (and even alchemical) experiments than to serious mathematical investigations.

❏ In his forties while a mathematics professor (and apparently a quite unsuccessful teacher) at Cambridge, Newton wrote the *Principia Mathematica* (1687), perhaps the single most influential scientific treatise ever published. In it he applied the concepts of the calculus to explore the workings of the universe, including the motions of the earth, moon, and planets about the sun. A student is said to have remarked, "there goes the man that wrote a book that neither he nor anyone else understands." But it established for Newton such fame that upon his death in 1727 he was buried alongside his country's greats in Westminster Abbey with such pomp that the French philospher Voltaire remarked, "I have seen a professor of mathematics . . . buried like a king who had done good to his subjects."

❏ Shortly after his Cambridge graduation in 1665, Newton discovered a new method for solving an equation of the form $f(x) = 0$. Unlike special methods such as the quadratic formula that apply only to equations of

special form, *Newton's method* can be used to approximate numerical solutions of virtually any equation. In Section 3.10 we present an iterative formulation of Newton's method that is especially adaptable to calculators and computers. There we describe how the combination of Newton's method with modern computer graphics has led to the generation of striking fractal images associated with the science of *chaos*. The pictures result from the application of a complex-number version of Newton's method to the simple equation $x^3 + 1 = 0$.

3.1

The Derivative and Rates of Change

Our preliminary investigation of tangent lines and derivatives in Section 2.1 led to the study of limits in the remainder of Chapter 2. Armed with this knowledge of limits, we are now prepared to study derivatives more extensively.

In Section 2.1 we introduced the derivative $f'(x)$ as the slope of the line tangent to the graph of the function f at the point $(x, f(x))$. More precisely, we were motivated by the geometry to *define* the line tangent to the graph at the point $P(a, f(a))$ to be the straight line through P with slope

$$m = \lim_{h \to 0} \frac{f(a + h) - f(a)}{h}, \tag{1}$$

as indicated in Fig. 3.1.1. If we replace the arbitrary number a in Eq. (1) with the independent variable x, we get a new function f', the *derivative* of the original function f.

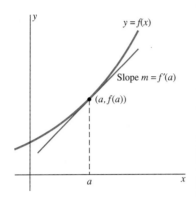

Fig. 3.1.1 The geometric motivation for the definition of the derivative

---

Definition of the Derivative

The **derivative** of the function f is the function f' defined by

$$f'(x) = \lim_{h \to 0} \frac{f(x + h) - f(x)}{h} \tag{2}$$

for all x for which this limit exists.

The last phrase of the definition requires, in particular, that f must be defined in a neighborhood of x for $f'(x)$ to exist.

We emphasized in Section 2.1 that we hold x *fixed* in Eq. (2) while h approaches zero. When we are specifically interested in the value of the derivative f' at $x = a$, we sometimes rewrite Eq. (2) in the form

$$f'(a) = \lim_{h \to 0} \frac{f(a + h) - f(a)}{h} = \lim_{x \to a} \frac{f(x) - f(a)}{x - a}. \tag{3}$$

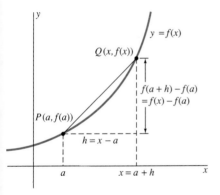

Fig. 3.1.2 The notation in Eq. (3)

The second limit in Eq. (3) is obtained from the first by writing $x = a + h$, $h = x - a$, and by noting that $x \to a$ as $h \to 0$ (Fig. 3.1.2). The statement that these equivalent limits exist can be abbreviated as "$f'(x)$ exists." In this case we say that the function f is **differentiable** at $x = a$. The process of finding the derivative f' is called **differentiation of** f.

We used several examples in Sections 2.1 and 2.2 to illustrate the process of differentiating a given function f by direct evaluation of the limit in Eq. (2). This involves carrying out four steps:

1. Write the definition in Eq. (2) of the derivative.
2. Substitute the expressions $f(x + h)$ and $f(x)$ as determined by the particular function f.
3. Simplify the result by algebraic methods to make Step 4 possible.
4. Evaluate the limit—typically, by application of some of the limit laws.

EXAMPLE 1 Differentiate $f(x) = \dfrac{x}{x + 3}$.

Solution

$$f'(x) = \lim_{h \to 0} \frac{f(x+h) - f(x)}{h} = \lim_{h \to 0} \frac{1}{h} \cdot \left(\frac{x+h}{x+h+3} - \frac{x}{x+3} \right)$$

$$= \lim_{h \to 0} \frac{(x+h)(x+3) - x(x+h+3)}{h(x+h+3)(x+3)}$$

$$= \lim_{h \to 0} \frac{3h}{h(x+h+3)(x+3)} = \lim_{h \to 0} \frac{3}{(x+h+3)(x+3)}$$

$$= \frac{3}{\left(\lim_{h \to 0}(x+h+3) \right)\left(\lim_{h \to 0}(x+3) \right)}.$$

Therefore,

$$f'(x) = \frac{3}{(x+3)^2}.$$

Even when the function f is rather simple, this four-step process for computing f' directly from the definition of the derivative can be tedious. Also, Step 3 may require considerable ingenuity. Moreover, it would be very repetitious to continue to rely on this process. To avoid tedium, we want a fast, easy, short method for computing $f'(x)$.

That new method is one focus of this chapter: the development of systematic methods ("rules") for differentiating those functions that occur most frequently. Such functions include polynomials, rational functions, the trigonometric functions $\sin x$ and $\cos x$, and combinations of such functions. Once we establish these general differentiation rules, we can apply them formally, almost mechanically, to compute derivatives. Only rarely should we need to return to the definition of the derivative.

An example of a "differentiation rule" is the theorem of Section 2.1 on differentiation of quadratic functions:

$$\text{If} \quad f(x) = ax^2 + bx + c, \quad \text{then} \quad f'(x) = 2ax + b. \tag{4}$$

Once we know this rule, we never again need apply the definition of the derivative to differentiate a quadratic function. For example, if $f(x) = 3x^2 - 4x + 5$, we can apply Eq. (4) to write the following immediately, without having to go through the four-step process:

$$f'(x) = 2 \cdot (3x) + (-4) = 6x - 4.$$

Similarly, if $g(t) = 2t - 5t^2$, then

$$g'(t) = (2) + 2 \cdot (-5t) = 2 - 10t.$$

It makes no difference what the name for the function is or whether we write x or t for the independent variable. This flexibility is valuable—in general, it is such adaptability that makes mathematics applicable to virtually every other branch of human knowledge. In any case, you should learn every differentiation rule in a form independent of the notation used to state it.

We develop additional differentiation rules in Sections 3.2 through 3.4. First, however, we must introduce new notation and a new interpretation of the derivative.

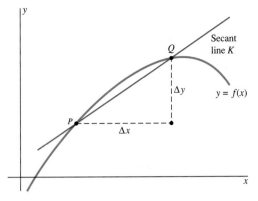

Fig. 3.1.3 Origin of the dy/dx notation

DIFFERENTIAL NOTATION

An important alternative notation for the derivative originates from the early custom of writing Δx in place of h (because $h = \Delta x$ is an increment in x) and

$$\Delta y = f(x + \Delta x) - f(x)$$

for the resulting change (or increment) in y. The slope of the secant line K of Fig. 3.1.3 is then

$$m_{\text{sec}} = \frac{\Delta y}{\Delta x} = \frac{f(x + \Delta x) - f(x)}{\Delta x},$$

and the slope of the tangent line is

$$m = \frac{dy}{dx} = \lim_{\Delta x \to 0} \frac{\Delta y}{\Delta x}. \tag{5}$$

Hence, if $y = f(x)$, we often write

$$\frac{dy}{dx} = f'(x). \tag{6}$$

(The so-called *differentials dy* and *dx* are discussed carefully in Chapter 4.) The symbols $f'(x)$ and dy/dx for the derivative of the function $y = f(x)$ are used almost interchangeably in mathematics and its applications, so you need to be familiar with both versions of the notation. You also need to know that dy/dx is a single symbol representing the derivative; it is *not* the quotient of two separate quantities dy and dx.

EXAMPLE 2 If $y = ax^2 + bx + c$, then the derivative in Eq. (4) in differential notation takes the form

$$\frac{dy}{dx} = 2ax + b.$$

Consequently,

$$\text{if} \quad y = 3x^2 - 4x + 5, \quad \text{then} \quad \frac{dy}{dx} = 6x - 4;$$

$$\text{if} \quad z = 2t - 5t^2, \quad \text{then} \quad \frac{dz}{dt} = 2 - 10t.$$

The letter d in the notation dy/dx stands for the word "differential." Whether we write dy/dx or dz/dt, the dependent variable appears "upstairs" and the independent variable "downstairs."

RATES OF CHANGE

In Section 2.1 we introduced the derivative of a function as the slope of the line tangent to its graph. Here we introduce the equally important interpretation of the derivative of a function as the rate of change of that function with respect to the independent variable.

We begin with the *instantaneous rate of change* of a function whose independent variable is time t. Suppose that Q is a quantity that varies with time t, and write $Q = f(t)$ for the value of Q at time t. For example, Q might be

❑ The size of a population (such as kangaroos, people, or bacteria);
❑ The number of dollars in a bank account;
❑ The volume of a balloon being inflated;
❑ The amount of water in a reservoir with variable inflow and outflow;
❑ The amount of a chemical product produced in a reaction; or
❑ The distance traveled t hours after the beginning of a journey.

The change in Q from time t to time $t + \Delta t$ is the **increment**

$$\Delta Q = f(t + \Delta t) - f(t).$$

The **average rate of change** of Q (per unit of time) is, by definition, the ratio of the change ΔQ in Q to the change Δt in t. Thus it is the quotient

$$\frac{\Delta Q}{\Delta t} = \frac{f(t + \Delta t) - f(t)}{\Delta t} \tag{7}$$

illustrated in Fig. 3.1.4.

We define the **instantaneous rate of change** of Q (per unit of time) to be the limit of this average rate as $\Delta t \to 0$. That is, the instantaneous rate of change of Q is

$$\lim_{\Delta t \to 0} \frac{\Delta Q}{\Delta t} = \lim_{\Delta t \to 0} \frac{f(t + \Delta t) - f(t)}{\Delta t}. \tag{8}$$

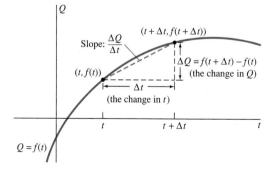

Fig. 3.1.4 Average rate of change as a slope

But the right-hand limit in Eq. (8) is simply the derivative $f'(t)$. So we see that the instantaneous rate of change of $Q = f(t)$ is the derivative

$$\frac{dQ}{dt} = f'(t). \tag{9}$$

To interpret intuitively the concept of instantaneous rate of change, think of the point $P(t, f(t))$ moving along the graph of the function $Q = f(t)$. As Q changes with time t, the point P moves along the curve. But if suddenly, at the instant t, the point P begins to follow a straight line path, then the new path of P would appear as in Fig. 3.1.5. The dashed curve in the figure corresponds to the "originally planned" behavior of Q (before P decided to fly off along the straight line path). But the straight line path of P (of constant slope) corresponds to the quantity Q "changing at a constant rate." Because the straight line is tangent to the graph $Q = f(t)$, we can interpret dQ/dt as the instantaneous rate of change of the quantity Q at the instant t:

The instantaneous rate of change of $Q = f(t)$ at time t is equal to the slope of the line tangent to the curve $Q = f(t)$ at the point $(t, f(t))$.

We can draw additional important conclusions. Because a positive slope corresponds to a rising tangent line and a negative slope corresponds to a falling tangent line (as in Figs. 3.1.6 and 3.1.7), we say that

$$Q \text{ is increasing at time } t \text{ if } \frac{dQ}{dt} > 0;$$

$$Q \text{ is decreasing at time } t \text{ if } \frac{dQ}{dt} < 0. \tag{10}$$

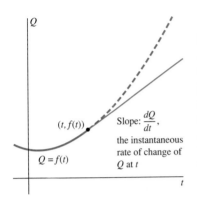

Fig. 3.1.5 The relation between the tangent line at $(t, f(t))$ and the instantaneous rate of change of f at t

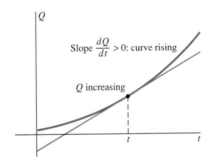

Fig. 3.1.6 Quantity increasing—derivative positive

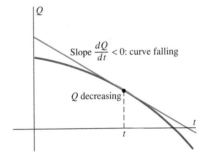

Fig. 3.1.7 Quantity decreasing—derivative negative

NOTE   The meaning of the phrase "$Q = f(t)$ is increasing *over* (or *during*) *the time interval from* $t = a$ to $t = b$" should be intuitively clear. The expressions in (10) give us a way to make precise what we mean by "$Q = f(t)$ is increasing *at time* t"—that is, at the instant t.

EXAMPLE 3   The cylindrical tank sketched in Fig. 3.1.8 has a vertical axis and is initially filled with 600 gal of water. This tank takes 60 min to empty after a drain in its bottom is opened. Suppose that the drain is opened at time $t = 0$. Suppose also that the volume V of water remaining in the tank after t minutes is

$$V(t) = \tfrac{1}{6}(60 - t)^2 = 600 - 20t + \tfrac{1}{6}t^2$$

Fig. 3.1.8 The draining tank of Example 3

gallons. Find the instantaneous rate at which the water is flowing out of the tank at time $t = 15$ (min) *and* at time $t = 45$ (min). Also find the average rate at which water flows out of the tank during the half hour from $t = 15$ to $t = 45$.

Solution   The instantaneous rate of change of the volume $V(t)$ of water in the tank is given by the derivative

$$\frac{dV}{dt} = -20 + \tfrac{1}{3}t.$$

At the instants $t = 15$ and $t = 45$ we obtain

$$V'(15) = -20 + \tfrac{1}{3} \cdot 15 = -15$$

and

$$V'(45) = -20 + \tfrac{1}{3} \cdot 45 = -5.$$

The units here are gallons per minute (gal/min). The fact that $V'(15)$ and $V'(45)$ are negative is consistent with the observation that V is a decreasing function of t (as t increases, V decreases). One way to indicate this is to say that after 15 min, the water is flowing *out* of the tank at 15 gal/min; after 45 min, the water is flowing *out* at 5 gal/min. The instantaneous rate of change of V at $t = 15$ is -15 gal/min, and the instantaneous rate of change of V at $t = 45$ is -5 gal/min. We could have predicted the units, because $\Delta V / \Delta t$ is a ratio of gallons to minutes, and therefore its limit $V'(t) = dV/dt$ must be expressed in the same units.

During the time interval of length $\Delta t = 30$ min from time $t = 15$ to $t = 45$, the *average* rate of change in the volume $V(t)$ is

$$\frac{\Delta V}{\Delta t} = \frac{V(45) - V(15)}{45 - 15}$$

$$= \frac{\tfrac{1}{6}(60 - 45)^2 - \tfrac{1}{6}(60 - 15)^2}{45 - 15} = \frac{-300}{30}.$$

Each numerator in the last equation is measured in gallons—this is especially apparent when you examine the second numerator—and each denominator is measured in minutes. Hence the ratio in the last fraction is a ratio of gallons to minutes, so the average rate of change of the volume V of water *in* the tank is -10 gal/min. Thus the average rate of flow of water *out* of the tank during this half-hour interval is 10 gal/min.

Our examples of functions up to this point have been restricted to those with formulas or with verbal descriptions. Scientists and engineers often work with tables of values obtained from observations or experiments. Example 4 shows how the instantaneous rate of change of such a tabulated function can be estimated.

EXAMPLE 4   The table in Fig. 3.1.9 gives the U.S. population P (in millions) in the nineteenth century at 10-yr intervals. Estimate the instantaneous rate of population growth in 1850.

Solution   We take $t = 0$ (yr) in 1800, so $t = 50$ corresponds to the year 1850. In Fig. 3.1.10 we have plotted the data and added a freehand sketch of a smooth curve that fits these data.

t	Year	U.S. population (millions)
0	1800	5.3
10	1810	7.2
20	1820	9.6
30	1830	12.9
40	1840	17.1
50	1850	23.2
60	1860	31.4
70	1870	38.6
80	1880	50.2
90	1890	62.9
100	1900	76.0

Fig. 3.1.9   Data for Example 4

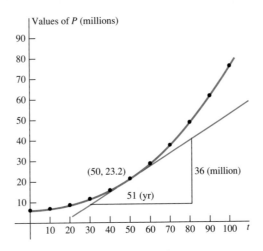

Fig. 3.1.10 A smooth curve that fits the data of Fig. 3.1.9 well (Example 4)

However we obtain it, a curve that fits the data should be a good approximation to the true graph of the unknown function $P = f(t)$. The instantaneous rate of change dP/dt in 1850 is the slope of the tangent line at the point $(50, 23.2)$. We draw the tangent as accurately as we can by visual inspection and then measure the base and height of the triangle in Fig. 3.1.10. In this way we approximate the slope of the tangent at $t = 50$ as

$$\frac{dP}{dt} \approx \frac{36}{51} \approx 0.71$$

millions of people per year (in 1850). Although there was no national census in 1851, we would expect the U.S. population then to have been approximately $23.2 + 0.7 = 23.9$ million.

VELOCITY AND ACCELERATION

Fig. 3.1.11 The particle in motion is at the point $x = f(t)$ at time t.

Suppose that a particle moves along a horizontal straight line, with its location x at time t given by its **position function** $x = f(t)$. Thus we make the line of motion a coordinate axis with an origin and a positive direction; $f(t)$ is merely the x-coordinate of the moving particle at time t (Fig. 3.1.11).

Think of the time interval from t to $t + \Delta t$. The particle moves from position $f(t)$ to position $f(t + \Delta t)$ during this interval. Its displacement is then the increment

$$\Delta f = f(t + \Delta t) - f(t).$$

We calculate the *average velocity* of the particle during this time interval exactly as we would calculate average speed on a long motor trip: We divide the distance by the time to obtain an average speed in miles per hour. In this case we divide the displacement of the particle by the elapsed time to obtain the **average velocity**

$$\bar{v} = \frac{\Delta x}{\Delta t} = \frac{f(t + \Delta t) - f(t)}{\Delta t}. \tag{11}$$

(The overbar is a standard symbol that usually connotes an average of some sort.) We define the **instantaneous velocity** v of the particle at the time t to be the limit of the average velocity $\bar{v}$ as $\Delta t \to 0$. That is,

$$v = \lim_{\Delta t \to 0} \frac{\Delta x}{\Delta t} = \lim_{\Delta t \to 0} \frac{f(t + \Delta t) - f(t)}{\Delta t}. \qquad (12)$$

We recognize the limit on the right in Eq. (12)—it is the definition of the derivative of f at time t. Therefore, the velocity of the moving particle at time t is simply

$$v = \frac{dx}{dt} = f'(t). \qquad (13)$$

Thus *velocity is instantaneous rate of change of position.* The velocity of a moving particle may be positive or negative, depending on whether the particle is moving in the positive or negative direction along the line of motion. We define the **speed** of the particle to be the *absolute value* of the velocity: $|v|$.

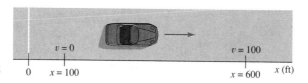

Fig. 3.1.12 The car of Example 5

EXAMPLE 5 Figure 3.1.12 shows a car moving along the (horizontal) x-axis. Suppose that its position (in feet) at time t (in seconds) is given by

$$x(t) = 5t^2 + 100.$$

Then its velocity at time t is

$$v(t) = x'(t) = 10t.$$

Because $x(0) = 100$ and $v(0) = 0$, the car starts at time $t = 0$ from rest—$v(0) = 0$—at the point $x = 100$. Substituting $t = 10$, we see that $x(10) = 600$ and $v(10) = 100$, so after 10 s the car has traveled 500 ft (from its starting point $x = 100$), and its speed then is 100 ft/s.

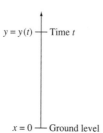

Fig. 3.1.13 Vertical motion with position function $y(t)$

VERTICAL MOTION

In the case of vertical motion—such as that of a ball thrown straight upward—it is common to denote the position function by $y(t)$ rather than $x(t)$. Typically, $y(t)$ denotes the height above the ground at time t, as in Fig. 3.1.13. But velocity is still the derivative of position:

$$v(t) = \frac{dy}{dt}.$$

Upward motion with y increasing corresponds to *positive velocity*, $v > 0$ (Fig. 3.1.14). *Downward motion* with y decreasing corresponds to *negative velocity*, $v < 0$.

The case of vertical motion under the influence of constant gravity is of special interest. If a particle is projected straight upward from an initial height y_0 (ft) above the ground at time $t = 0$ (s) and with initial velocity v_0 (ft/s) *and* if air resistance is negligible, then its height y (in feet above the ground) at time t is given by a formula known from physics,

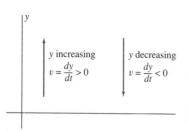

Fig. 3.1.14 Upward motion and downward motion

$$y(t) = -\tfrac{1}{2}gt^2 + v_0 t + y_0. \tag{14}$$

Here g denotes the *acceleration* due to the force of gravity. Near the surface of the earth, g is nearly constant, so we assume that it is exactly constant, and at the surface of the earth $g \approx 32$ ft/s^2, or $g \approx 9.8$ m/s^2.

If we differentiate y with respect to time t, we obtain the velocity of the particle at time t:

$$v = \frac{dy}{dt} = -gt + v_0. \tag{15}$$

The **acceleration** of the particle is defined to be the instantaneous time rate of change (derivative) of its velocity:

$$a = \frac{dv}{dt} = -g. \tag{16}$$

Your intuition should tell you that a body projected upward in this way will reach its maximum height at the instant that its velocity becomes zero—when $v(t) = 0$. (We shall see in Section 3.5 why this is true.)

EXAMPLE 6   Find the maximum height attained by a ball thrown straight upward from the ground with initial velocity $v_0 = +96$ ft/s. Also find the velocity with which it hits the ground upon its return.

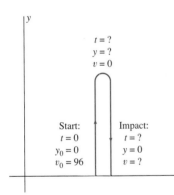

y

t = ?
y = ?
v = 0

Start:
t = 0
y₀ = 0
v₀ = 96

Impact:
t = ?
y = 0
v = ?

Fig. 3.1.15   Data for the ball of Example 6

Solution   To begin the solution of a motion problem such as this, we sketch a diagram like Fig. 3.1.15, indicating both the given data and the data that are unknown at the time instants in question. Here we focus on the time $t = 0$ when the ball leaves the ground ($y = 0$), the unknown time when it reaches its maximum height with velocity $v = 0$, and the unknown time when it returns to the ground.

We begin with Eq. (15), with $v_0 = 96$ and $g = 32$. We find the velocity of the ball at time t to be

$$v(t) = -32t + 96$$

while it remains aloft. The ball attains its maximum height when $v(t) = 0$—that is, when

$$-32t + 96 = 0.$$

This occurs at time $t = 3$ (s). Upon substituting this value of t into the altitude function in Eq. (14), taking $y_0 = 0$, we find that the maximum height of the ball is

$$y_{\max} = y(3) = (-16)(3^2) + (96)(3) = 144 \quad \text{(ft)}.$$

The ball returns to the ground when $y(t) = 0$. The equation

$$y(t) = -16t^2 + 96t = -16t(t - 6) = 0$$

has the two solutions $t = 0$ and $t = 6$. Thus the ball returns to the ground at time $t = 6$. The velocity with which it strikes the ground is

$$v(6) = (-32)(6) + 96 = -96 \quad \text{(ft/s)}.$$

The derivative of any function—not merely a function of time—may be interpreted as its instantaneous rate of change with respect to the independent variable. If $y = f(x)$, then the **average rate of change of y** (per unit change in x) on the interval $[x, x + \Delta x]$ is the quotient

$$\frac{\Delta y}{\Delta x} = \frac{f(x + \Delta x) - f(x)}{\Delta x}.$$

The **instantaneous rate of change of y with respect to x** is the limit, as $\Delta x \to 0$, of the average rate of change. Thus the instantaneous rate of change of y with respect to x is

$$\lim_{\Delta x \to 0} \frac{\Delta y}{\Delta x} = \frac{dy}{dx} = f'(x). \tag{17}$$

Example 7 illustrates the fact that a dependent variable may sometimes be expressed as two different functions of two different independent variables. The derivatives of these functions are then rates of change of the dependent variable with respect to the two different independent variables.

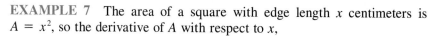

EXAMPLE 7 The area of a square with edge length x centimeters is $A = x^2$, so the derivative of A with respect to x,

$$\frac{dA}{dx} = 2x, \tag{18}$$

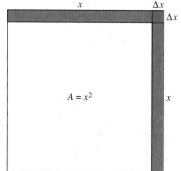

$A = x^2$

is the rate of change of its area A with respect to x. (See the computations in Fig. 3.1.16.) The units of dA/dx are square centimeters *per centimeter*. Now suppose that the edge length of the square is increasing with time: $x = 5t$, with time t in seconds. Then the area of the square at time t is

$$A = (5t)^2 = 25t^2.$$

Fig. 3.1.16 The square of Example 7:

$A + \Delta A = (x + \Delta x)^2;$
$\Delta A = 2x\,\Delta x + (\Delta x)^2;$
$\dfrac{\Delta A}{\Delta x} = 2x + \Delta x;$
$\dfrac{dA}{dx} = 2x.$

The derivative of A with respect to t is

$$\frac{dA}{dt} = (2)(25t) = 50t; \tag{19}$$

this is the rate of change of A with respect to time t, with units of square centimeters *per second*. For instance, when $t = 10$ (so $x = 50$), the values of the two derivatives of A in Eqs. (18) and (19) are

$$\left.\frac{dA}{dx}\right|_{x=50} = (2)(50) = 100 \quad (\text{cm}^2/\text{cm})$$

and

$$\left.\frac{dA}{dt}\right|_{t=10} = (50)(10) = 500 \quad (\text{cm}^2/\text{s}).$$

The notation dA/dt for the derivative suffers from the minor inconvenience of not providing a "place" to substitute a particular value of t, such as $t = 10$. The last lines of Example 7 illustrate one way around this difficulty.

3.1 Problems

In Problems 1 through 10, find the indicated derivative by using the differentiation rule in (4):

$$\text{If } f(x) = ax^2 + bx + c, \quad \text{then} \quad f'(x) = 2ax + b.$$

1. $f(x) = 4x - 5$; find $f'(x)$.

2. $g(t) = 100 - 16t^2$; find $g'(t)$.

3. $h(z) = z(25 - z)$; find $h'(z)$.

4. $f(x) = 16 - 49x$; find $f'(x)$.

5. $y = 2x^2 + 3x - 17$; find dy/dx.

6. $x = 16t - 100t^2$; find dx/dt.

7. $z = 5u^2 - 3u$; find dz/du.

8. $v = 5y(100 - y)$; find dv/dy.

9. $x = -5y^2 + 17y + 300$; find dx/dy.

10. $u = 7t^2 + 13t$; find du/dt.

In Problems 11 through 20, apply the definition of the derivative (as in Example 1) to find $f'(x)$.

11. $f(x) = 2x - 1$ **12.** $f(x) = 2 - 3x$

13. $f(x) = x^2 + 5$ **14.** $f(x) = 3 - 2x^2$

15. $f(x) = \dfrac{1}{2x + 1}$ **16.** $f(x) = \dfrac{1}{3 - x}$

17. $f(x) = \sqrt{2x + 1}$ **18.** $f(x) = \dfrac{1}{\sqrt{x + 1}}$

19. $f(x) = \dfrac{x}{1 - 2x}$ **20.** $f(x) = \dfrac{x + 1}{x - 1}$

In Problems 21 through 25, the position function $x = f(t)$ of a particle moving in a horizontal straight line is given. Find its location x when its velocity v is zero.

21. $x = 100 - 16t^2$

22. $x = -16t^2 + 160t + 25$

23. $x = -16t^2 + 80t - 1$

24. $x = 100t^2 + 50$

25. $x = 100 - 20t - 5t^2$

In Problems 26 through 29, the height $y(t)$ (in feet at time t seconds) of a ball thrown vertically upward is given. Find the maximum height that the ball attains.

26. $y = -16t^2 + 160t$

27. $y = -16t^2 + 64t$

28. $y = -16t^2 + 128t + 25$

29. $y = -16t^2 + 96t + 50$

30. The Celsius temperature C is given in terms of the Fahrenheit temperature F by $C = \frac{5}{9}(F - 32)$. Find the rate of change of C with respect to F and the rate of change of F with respect to C.

31. Find the rate of change of the area A of a circle with respect to its circumference C.

32. A stone dropped into a pond at time $t = 0$ s causes a circular ripple that travels out from the point of impact at 5 m/s. At what rate (in square meters/s) is the area within the circle increasing when $t = 10$?

33. A car is traveling at 100 ft/s when the driver suddenly applies the brakes ($x = 0, t = 0$). The position function of the skidding car is $x(t) = 100t - 5t^2$. How far and for how long does the car skid before it comes to a stop?

34. A water bucket containing 10 gal of water develops a leak at time $t = 0$, and the volume V of water in the bucket t seconds later is given by

$$V(t) = 10\left(1 - \frac{t}{100}\right)^2$$

until the bucket is empty at time $t = 100$. (a) At what rate is water leaking from the bucket after exactly 1 min has passed? (b) When is the instantaneous rate of change of V equal to the average rate of change of V from $t = 0$ to $t = 100$?

35. A population of chipmunks moves into a new region at time $t = 0$. At time t (in months), the population numbers

$$P(t) = 100[1 + (0.3)t + (0.04)t^2].$$

(a) How long does it take for this population to double its initial size $P(0)$? (b) What is the rate of growth of the population when $P = 200$?

36. The following data describe the growth of the population P (in thousands) of Gotham City during a 10-year period. Use the graphical method of Example 4 to estimate its rate of growth in 1989.

Year	1984	1986	1988	1990	1992	1994
P	265	293	324	358	395	437

37. The following data give the distance x in feet traveled by an accelerating car (that starts from rest at time $t = 0$) in the first t seconds. Use the graphical method of Example 4 to estimate its speed (in miles/hour) when $t = 20$ and again when $t = 40$.

t	0	10	20	30	40	50	60
x	0	224	810	1655	2686	3850	5109

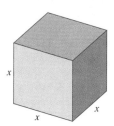

Fig. 3.1.17 The cube of Problem 38: volume $V = x^3$, surface area $S = 6x^2$.

Fig. 3.1.18 The sphere of Problem 39: volume $V = \frac{4}{3}\pi r^3$, surface area $S = 4\pi r^2$.

Fig. 3.1.19 The cylinder of Problem 40: volume $V = \pi r^2 h$, surface area $S = 2\pi r^2 + 2\pi rh$.

In Problems 38 through 43, use the fact (proved in Section 3.2) that the derivative of $y = ax^3 + bx^2 + cx + d$ is $dy/dx = 3ax^2 + 2bx + c$.

38. Prove that the rate of change of the volume V of a cube with respect to its edge length x is equal to half the surface area A of the cube (Fig. 3.1.17).

39. Show that the rate of change of the volume V of a sphere with respect to its radius r is equal to its surface area S (Fig. 3.1.18).

40. The height h of a cylinder whose height changes is always twice its radius r. Show that the rate of change of its volume V with respect to r is equal to its total surface area S (Fig. 3.1.19).

41. A spherical balloon with an initial radius r of 5 in. begins to leak at time $t = 0$, and its radius t seconds later is $r = (60 - t)/12$ in. At what rate (in cubic inches/ second) is air leaking from the balloon when $t = 30$?

42. The volume V (in liters) of 3 g of CO_2 at 27°C is given in terms of its pressure p (in atmospheres) by the formula $V = 1.68/p$. What is the rate of change of V with respect to p when $p = 2$ (atm)? [Use the fact that the derivative of

$f(x) = c/x$ is $f'(x) = -c/x^2$ if c is a constant; you can establish this by using the definition of the derivative.]

43. As a snowball with an initial radius of 12 cm melts, its radius decreases at a constant rate. It begins to melt when $t = 0$ (h) and takes 12 h to disappear. (a) What is its rate of change of volume when $t = 6$? (b) What is its average rate of change of volume from $t = 3$ to $t = 9$?

44. A ball thrown vertically upward at time $t = 0$ (s) with initial velocity 96 ft/s and with initial height 112 ft has height function $y(t) = -16t^2 + 96t + 112$. (a) What is the maximum height attained by the ball? (b) When and with what impact speed does the ball hit the ground?

45. A spaceship approaching touchdown on the planet Gzyx has height y (meters) at time t (seconds) given by $y = 100 - 100t + 25t^2$. When and with what speed does it hit the ground?

46. The population (in thousands) of the city Metropolis is given by

$$P(t) = 100[1 + (0.04)t + (0.003)t^2],$$

with t in years and with $t = 0$ corresponding to 1980. (a) What was the rate of change of P in 1986? (b) What was the average rate of change of P from 1983 to 1988?

3.1 Project

This project involves a graphical analysis of the population growth of a small city during the decade of the 1990s. You will need a graphics calculator or computer with plotting facility.

For your own small city, choose a positive integer k with $k \leq 9$ (perhaps the last nonzero digit of your student I.D. number). Then suppose that the population P of the city t years after 1990 is given (in thousands) by

$$P(t) = 10 + t - (0.1)t^2 + (0.001)(k + 5)t^3.$$

Investigate the following questions.

1. Does the graph of $P(t)$ indicate that the population is increasing throughout the 1990s? Explain your answer.

2. Does the graph of the derivative $P'(t)$ confirm that $P(t)$ is increasing throughout the 1990s? What property of this graph is pertinent to the question?

3. What points on the graph of $P'(t)$ correspond to the time (or times) at which the instantaneous rate of change of P is equal to its average rate of change between the years 1990 and 2000? Apply the method of successive magnification to find each such time (accurate to two decimal places).

4. What points on the graph of the derivative $P'(t)$ correspond to the time (or times) at which the population $P(t)$ is increasing the slowest? The fastest? Apply the method of successive magnification to find each such time (accurate to two decimal places).

3.2

Basic Differentiation Rules

Here we begin our development of formal rules for finding the derivative f' of the function f:

$$f'(x) = \lim_{h \to 0} \frac{f(x + h) - f(x)}{h}. \tag{1}$$

Some alternative notation for derivatives will be helpful.

When we interpreted the derivative in Section 3.1 as a rate of change, we found it useful to employ the dependent-independent variable notation

$$y = f(x), \quad \Delta x = h, \quad \Delta y = f(x + \Delta x) - f(x). \tag{2}$$

This led to the "differential notation"

$$\frac{dy}{dx} = \lim_{\Delta x \to 0} \frac{\Delta y}{\Delta x} = \lim_{\Delta x \to 0} \frac{f(x + \Delta x) - f(x)}{\Delta x} \tag{3}$$

for the derivative. When you use this notation, remember that the symbol dy/dx is simply another notation for the derivative $f'(x)$; it is *not* the quotient of two separate entities dy and dx.

A third notation is sometimes used for the derivative $f'(x)$; it is $Df(x)$. Here, think of D as a "machine" that operates on the function f to produce its derivative Df (Fig. 3.2.1). Thus we can write the derivative of $y = f(x) = x^2$ in any of three ways:

$$f'(x) = \frac{dy}{dx} = Dx^2 = 2x.$$

These three notations for the derivative—the function notation $f'(x)$, the differential notation dy/dx, and the operator notation $Df(x)$—are used interchangeably in mathematical and scientific writing, so you need to be familiar with each.

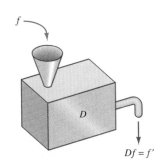

Fig. 3.2.1 The "differentiation machine" D

DERIVATIVES OF POLYNOMIALS

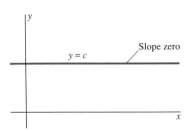

Fig. 3.2.2 The derivative of a constant-valued function is zero (Theorem 1).

Our first differentiation rule says that *the derivative of a constant function is identically zero.* Geometry makes this obvious, because the graph of a constant function is a horizontal straight line that is its own tangent line, with slope zero at every point (Fig. 3.2.2).

> **Theorem 1** *Derivative of a Constant*
> If $f(x) = c$ (a constant) for all x, then $f'(x) = 0$ for all x. That is,
> $$\frac{dc}{dx} = Dc = 0. \qquad (4)$$

Proof   Because $f(x + h) = f(x) = c$, we see that

$$f'(x) = \lim_{h \to 0} \frac{f(x + h) - f(x)}{h} = \lim_{h \to 0} \frac{c - c}{h} = \lim_{h \to 0} \frac{0}{h} = 0. \quad \square$$

As motivation for the next rule, consider the following list of derivatives, all computed in Chapter 2.

$$Dx = 1 \qquad \text{(Problem 2, Section 2.1)}$$
$$Dx^2 = 2x \qquad \text{(Example 3, Section 2.1)}$$
$$Dx^3 = 3x^2 \qquad \text{(Problem 26, Section 2.1)}$$
$$Dx^4 = 4x^3 \qquad \text{(Problem 28, Section 2.1)}$$
$$Dx^{-1} = -x^{-2} \qquad \text{(Problem 27, Section 2.1)}$$
$$Dx^{-2} = -2x^{-3} \qquad \text{(Problem 29, Section 2.1)}$$
$$Dx^{1/2} = \tfrac{1}{2}x^{-1/2} \qquad \text{(Example 12, Section 2.2)}$$
$$Dx^{-1/2} = -\tfrac{1}{2}x^{-3/2} \qquad \text{(Problem 41, Section 2.2)}$$

Each of these formulas fits the simple pattern

$$Dx^n = nx^{n-1}. \qquad (5)$$

Equation (5), inferred from the preceding list of derivatives, as yet is only a conjecture. But many discoveries in mathematics are made by detecting such patterns and then proving that they hold universally.

Eventually, we shall see that the formula in Eq. (5), called the **power rule,** is valid for all real numbers n. At this time we give a proof only for the case in which the exponent n is a *positive integer*. We need the *binomial formula* from high-school algebra:

$$(a + b)^n = a^n + na^{n-1}b + \frac{n(n - 1)}{2 \cdot 1}a^{n-2}b^2 + \cdots$$

$$+ \frac{n(n - 1) \cdots (n - k + 1)}{k(k - 1) \cdots 3 \cdot 2 \cdot 1}a^{n-k}b^k + \cdots + nab^{n-1} + b^n \quad (6)$$

if n is a positive integer. The cases $n = 2$ and $n = 3$ are the familiar formulas

$$(a + b)^2 = a^2 + 2ab + b^2$$

and

$$(a + b)^3 = a^3 + 3a^2b + 3ab^2 + b^3.$$

There are $n + 1$ terms on the right-hand side in Eq. (6), but all we need to know to prove the power rule is the exact form of the first two terms and the fact that all the other terms include b^2 as a factor. For instance, with $a = x$, $b = h$, and $n = 5$ in Eq. (6), we obtain

$$(x + h)^5 = x^5 + 5x^4h + 10x^3h^2 + 10x^2h^3 + 5xh^4 + h^5$$
$$= x^5 + 5x^4h + h^2(10x^3 + 10x^2h + 5xh^2 + h^3)$$
$$= x^5 + 5x^4h + h^2 \cdot P(h).$$

Here $P(h)$ is a polynomial of degree 3 in the variable h, with coefficients involving x. With the general exponent n in place of 5, Eq. (6) similarly yields the result

$$(x + h)^n = x^n + nx^{n-1}h + h^2 P(h), \tag{7}$$

where $P(h)$ is a polynomial of degree $n - 2$ in the variable h. Now we are ready for Theorem 2.

Theorem 2   *Power Rule for a Positive Integer n*
If n is a positive integer and $f(x) = x^n$, then $f'(x) = nx^{n-1}$.

Proof

$$f'(x) = \lim_{h \to 0} \frac{f(x + h) - f(x)}{h} = \lim_{h \to 0} \frac{(x + h)^n - x^n}{h}$$

$$= \lim_{h \to 0} \frac{nx^{n-1}h + h^2 P(h)}{h} \qquad \text{[by Eq. (7)]}$$

$$= \lim_{h \to 0} [nx^{n-1} + h \cdot P(h)] = nx^{n-1} + 0 \cdot P(0) \quad \text{(by limit laws and continuity of } P\text{)};$$

$$f'(x) = nx^{n-1}. \qquad \square$$

We need not always use the same symbols x and n for the independent variable and the constant exponent in the power rule. For instance,

$$Dt^m = mt^{m-1} \quad \text{and} \quad Dz^k = kz^{k-1},$$

where D denotes differentiation with respect to the indicated independent variable. If it is not perfectly clear which is the independent variable, indicate it by affixing it as a subscript to the operator D:

$$D_t t^m = mt^{m-1} \quad \text{and} \quad D_z z^k = kz^{k-1}.$$

EXAMPLE 1 $Dx^7 = 7x^6, \qquad Dt^{17} = 17t^{16}, \qquad Dz^{100} = 100z^{99}.$

To use the power rule to differentiate polynomials, we need to know how to differentiate *linear combinations*. A **linear combination** of the functions f and g is a function of the form $af + bg$, where a and b are constants. It follows from the sum and product laws for limits that

$$\lim_{x \to c} [af(x) + bg(x)] = a\left(\lim_{x \to c} f(x)\right) + b\left(\lim_{x \to c} g(x)\right) \tag{8}$$

provided that the two limits on the right in Eq. (8) both exist. The formula in Eq. (8) is called the **linearity property** of the limit operation. It implies an analogous linearity property of differentiation.

Theorem 3   *Derivative of a Linear Combination*

If f and g are differentiable functions and a and b are fixed real numbers, then

$$D[af(x) + bg(x)] = aDf(x) + bDg(x). \tag{9}$$

With $u = f(x)$ and $v = g(x)$, this takes the form

$$\frac{d(au + bv)}{dx} = a\frac{du}{dx} + b\frac{dv}{dx}. \tag{9'}$$

Proof   The linearity property of limits immediately gives

$$D[af(x) + bg(x)] = \lim_{h \to 0} \frac{[af(x + h) + bg(x + h)] - [af(x) + bg(x)]}{h}$$

$$= a\left(\lim_{h \to 0} \frac{f(x + h) - f(x)}{h}\right) + b\left(\lim_{h \to 0} \frac{g(x + h) - g(x)}{h}\right)$$

$$= aDf(x) + bDg(x). \quad \square$$

Now take $a = c$ and $b = 0$ in Eq. (9). The result is

$$D[cf(x)] = cDf(x); \tag{10}$$

alternatively,

$$\frac{d(cu)}{dx} = c\frac{du}{dx}. \tag{10'}$$

Thus *the derivative of a constant multiple of a function is the same constant multiple of its derivative.*

EXAMPLE 2 $D(16x^6) = 16 \cdot 6x^5 = 96x^5.$

Next, take $a = b = 1$ in Eq. (9). We find that

$$D[f(x) + g(x)] = Df(x) + Dg(x). \tag{11}$$

In differential notation,

$$\frac{d(u + v)}{dx} = \frac{du}{dx} + \frac{dv}{dx}. \tag{11'}$$

Thus *the derivative of the sum of two functions is the sum of their derivatives.* Similarly, for differences we have

$$\frac{d(u - v)}{dx} = \frac{du}{dx} - \frac{dv}{dx}. \tag{12}$$

It's easy to see that these rules generalize to sums and differences of more than two functions. For example, repeated application of Eq. (11) to the sum of a finite number of differentiable functions gives

$$\frac{d(u_1 + u_2 + \cdots + u_n)}{dx} = \frac{du_1}{dx} + \frac{du_2}{dx} + \cdots + \frac{du_n}{dx}. \qquad (13)$$

EXAMPLE 3

$$D(36 + 26x + 7x^5 - 5x^9) = 0 + 26 \cdot 1 + 7 \cdot 5x^4 - 5 \cdot 9x^8$$
$$= 26 + 35x^4 - 45x^8.$$

When we apply Eqs. (10) and (13) and the power rule to the polynomial

$$p(x) = a_n x^n + a_{n-1} x^{n-1} + \cdots + a_2 x^2 + a_1 x + a_0,$$

we find the derivative *as fast as we can write it*:

$$p'(x) = n a_n x^{n-1} + (n-1) a_{n-1} x^{n-2} + \cdots + 3 a_3 x^2 + 2 a_2 x + a_1. \quad (14)$$

With this result, it becomes a routine matter to write an equation for a line tangent to the graph of a polynomial.

EXAMPLE 4   Write an equation for the straight line that is tangent to the graph of $y = 2x^3 - 7x^2 + 3x + 4$ at the point $(1, 2)$.

Solution   We compute the derivative as in Eq. (14):

$$\frac{dy}{dx} = 2 \cdot 3x^2 - 7 \cdot 2x + 3 = 6x^2 - 14x + 3.$$

We substitute $x = 1$ in the derivative and find that the slope of the tangent line at $(1, 2)$ is $m = -5$. So the point-slope equation of the tangent line is

$$y - 2 = -5(x - 1).$$

EXAMPLE 5   The volume V (in cubic centimeters) of a quantity of water varies with changing temperature T. For T between 0°C and 30°C, the relationship is almost exactly

$$V = V_0(1 + \alpha T + \beta T^2 + \gamma T^3),$$

where V_0 is the volume at 0°C and the three constants have the values

$$\alpha = -0.06427 \times 10^{-3}, \quad \beta = 8.5053 \times 10^{-6},$$
$$\text{and} \quad \gamma = -6.7900 \times 10^{-8}.$$

The rate of change of volume with respect to temperature is

$$\frac{dV}{dT} = V_0(\alpha + 2\beta T + 3\gamma T^2).$$

Suppose that $V_0 = 10^5$ cm³ and that $T = 20$°C. Substitution of these numerical data into the formulas for V and dV/dT yields $V \approx 100{,}157$ cm³ and $dV/dT \approx 19.4$ cm³/°C. We may conclude that, at $T = 20$°C, this volume of water should increase by approximately 19.4 cm³ for each 1°C rise in temperature. For comparison, direct substitution into the formula for V shown here yields

$$V(20.5) - V(19.5) \approx 19.4445 \quad \text{(cm³)}.$$

THE PRODUCT RULE AND QUOTIENT RULE

It might be natural to conjecture that the derivative of a product $f(x)g(x)$ is the product of the derivatives. This is *false*! For example, if $f(x) = g(x) = x$, then

$$D[f(x)g(x)] = Dx^2 = 2x.$$

But

$$[Df(x)] \cdot [Dg(x)] = (Dx) \cdot (Dx) = 1 \cdot 1 = 1.$$

In general, the derivative of a product is *not* merely the product of the derivatives. Theorem 4 tells us what it *is*.

Theorem 4 *The Product Rule*

If f and g are differentiable at x, then fg is differentiable at x, and

$$D[f(x)g(x)] = f'(x)g(x) + f(x)g'(x). \tag{15}$$

With $u = f(x)$ and $v = g(x)$, this **product rule** takes the form

$$\frac{d(uv)}{dx} = u\frac{dv}{dx} + v\frac{du}{dx}. \tag{15'}$$

When it is clear what the independent variable is, we can make the product rule even briefer:

$$(uv)' = u'v + uv'. \tag{15''}$$

Proof   We use an "add and subtract" device.

$$D[f(x)g(x)] = \lim_{h \to 0} \frac{f(x+h)g(x+h) - f(x)g(x)}{h}$$

$$= \lim_{h \to 0} \frac{f(x+h)g(x+h) - f(x)g(x+h) + f(x)g(x+h) - f(x)g(x)}{h}$$

$$= \lim_{h \to 0} \frac{f(x+h)g(x+h) - f(x)g(x+h)}{h}$$

$$+ \lim_{h \to 0} \frac{f(x)g(x+h) - f(x)g(x)}{h}$$

$$= \left(\lim_{h \to 0} \frac{f(x+h) - f(x)}{h} \right)\left(\lim_{h \to 0} g(x+h) \right)$$

$$+ f(x)\left(\lim_{h \to 0} \frac{g(x+h) - g(x)}{h} \right)$$

$$= f'(x)g(x) + f(x)g'(x).$$

In this proof we used the sum law and product law for limits, the definitions of $f'(x)$ and $g'(x)$, and the fact that

$$\lim_{h \to 0} g(x+h) = g(x).$$

This last equation holds because g is differentiable and therefore continuous at x, by Theorem 4 in Section 2.4.   ❏

In words, the product rule says that *the derivative of the product of two functions is formed by multiplying the derivative of each by the other and then adding the results.*

EXAMPLE 6 Find the derivative of

$$f(x) = (1 - 4x^3)(3x^2 - 5x + 2)$$

without first multiplying out the two factors.

Solution
$$D[(1 - 4x^3)(3x^2 - 5x + 2)]$$
$$= [D(1 - 4x^3)](3x^2 - 5x + 2) + (1 - 4x^3)[D(3x^2 - 5x + 2)]$$
$$= (-12x^2)(3x^2 - 5x + 2) + (1 - 4x^3)(6x - 5)$$
$$= -60x^4 + 80x^3 - 24x^2 + 6x - 5.$$

We can apply the product rule repeatedly to find the derivative of a product of three or more differentiable functions $u_1, u_2, \ldots, u_n$ of x. For example,

$$D[u_1 u_2 u_3] = [D(u_1 u_2)] \cdot u_3 + (u_1 u_2) \cdot Du_3$$
$$= [(Du_1) \cdot u_2 + u_1 \cdot (Du_2)] \cdot u_3 + (u_1 u_2) \cdot Du_3$$
$$= (Du_1)u_2 u_3 + u_1(Du_2)u_3 + u_1 u_2(Du_3).$$

Note that the derivative of each factor in the original product is multiplied by the other two factors, and then the three resulting products are added. This is, indeed, the general result:

$$D(u_1 u_2 \cdots u_n) = (Du_1)u_2 \cdots u_n + u_1(Du_2)u_3 \cdots u_n$$
$$+ \cdots + u_1 u_2 \cdots u_{n-1}(Du_n), \tag{16}$$

where the sum in Eq. (16) has one term corresponding to each of the n factors in the product $u_1 u_2 \cdots u_n$. It is easy to establish this **extended product rule** (see Problem 62) one step at a time—next with $n = 4$, then with $n = 5$, and so forth.

Our next result tells us how to find the derivative of the reciprocal of a function if we know the derivative of the function itself.

The Reciprocal Rule

If f is differentiable at x and $f(x) \neq 0$, then

$$D\frac{1}{f(x)} = -\frac{f'(x)}{[f(x)]^2}. \tag{17}$$

With $u = f(x)$, the reciprocal rule takes the form

$$\frac{d}{dx}\left(\frac{1}{u}\right) = -\frac{1}{u^2} \cdot \frac{du}{dx} \tag{17'}$$

If there can be no doubt what the independent variable is, we can write

$$\left(\frac{1}{u}\right)' = -\frac{u'}{u^2}. \tag{17''}$$

Proof As in the proof of Theorem 4, we use the limit laws, the definition of the derivative, and the fact that a function is continuous wherever it is differentiable (by Theorem 4 of Section 2.4). Moreover, note that $f(x + h) \neq 0$ for h near zero because $f(x) \neq 0$ and f is continuous at x (see Problem 16 in Appendix B). Therefore,

$$D\frac{1}{f(x)} = \lim_{h \to 0} \frac{1}{h}\left(\frac{1}{f(x + h)} - \frac{1}{f(x)}\right) = \lim_{h \to 0} \frac{f(x) - f(x + h)}{hf(x + h)f(x)}$$

$$= -\left(\lim_{h \to 0} \frac{1}{f(x + h)f(x)}\right)\left(\lim_{h \to 0} \frac{f(x + h) - f(x)}{h}\right) = -\frac{f'(x)}{[f(x)]^2}. \quad \Box$$

EXAMPLE 7 With $f(x) = x^2 + 1$ in Eq. (17), we get

$$D\frac{1}{x^2 + 1} = -\frac{D(x^2 + 1)}{(x^2 + 1)^2} = -\frac{2x}{(x^2 + 1)^2}.$$

We now combine the reciprocal rule with the power rule for positive integral exponents to establish the power rule for negative integral exponents.

> **Theorem 5 Power Rule for a Negative Integer n**
> If n is a negative integer, then $Dx^n = nx^{n-1}.$

Proof Let $m = -n$, so that m is a positive integer. Then

$$Dx^n = D\frac{1}{x^m} = -\frac{D(x^m)}{(x^m)^2} = -\frac{mx^{m-1}}{x^{2m}} = (-m)x^{(-m)-1} = nx^{n-1}. \quad \Box$$

This proof also shows that the rule of Theorem 5 holds exactly when the function being differentiated is defined: when $x \neq 0$.

EXAMPLE 8

$$D\frac{5x^4 - 6x + 7}{2x^2} = D(\tfrac{5}{2}x^2 - 3x^{-1} + \tfrac{7}{2}x^{-2})$$

$$= \tfrac{5}{2}(2x) - 3(-x^{-2}) + \tfrac{7}{2}(-2x^{-3}) = 5x + \frac{3}{x^2} - \frac{7}{x^3}.$$

The key here was to "divide out" before differentiating.

Now we apply the product rule and reciprocal rule to get a rule for differentiation of the quotient of two functions.

> **Theorem 6 The Quotient Rule**
> If f and g are differentiable at x and $g(x) \neq 0$, then f/g is differentiable at x and
>
> $$D\frac{f(x)}{g(x)} = \frac{f'(x)g(x) - f(x)g'(x)}{[g(x)]^2}. \qquad (18)$$

With $u = f(x)$ and $v = g(x)$, this rule takes the form

$$\frac{d}{dx}\left(\frac{u}{v}\right) = \frac{v\dfrac{du}{dx} - u\dfrac{dv}{dx}}{v^2}. \tag{18'}$$

If it is clear what the independent variable is, we can write the quotient rule as

$$\left(\frac{u}{v}\right)' = \frac{u'v - uv'}{v^2}. \tag{18''}$$

Proof We apply the product rule to the factorization

$$\frac{f(x)}{g(x)} = f(x) \cdot \frac{1}{g(x)}.$$

This gives

$$D\frac{f(x)}{g(x)} = [Df(x)] \cdot \frac{1}{g(x)} + f(x) \cdot D\frac{1}{g(x)}$$

$$= \frac{f'(x)}{g(x)} + f(x) \cdot \left(-\frac{g'(x)}{[g(x)]^2}\right) = \frac{f'(x)g(x) - f(x)g'(x)}{[g(x)]^2}. \quad \square$$

Note that the numerator in Eq. (18) is *not* the derivative of the product of f and g. And the minus sign means that the *order* of terms in the numerator is important.

EXAMPLE 9 Find dz/dt, given

$$z = \frac{1 - t^3}{1 + t^4}.$$

Solution Primes denote derivatives with respect to t. With t (rather than x) as the independent variable, the quotient rule gives

$$\frac{dz}{dt} = \frac{(1 - t^3)'(1 + t^4) - (1 - t^3)(1 + t^4)'}{(1 + t^4)^2}$$

$$= \frac{(-3t^2)(1 + t^4) - (1 - t^3)(4t^3)}{(1 + t^4)^2} = \frac{t^6 - 4t^3 - 3t^2}{(1 + t^4)^2}.$$

3.2 Problems

Apply the differentiation rules of this section to find the derivatives of the functions in Problems 1 through 40.

1. $f(x) = 3x^2 - x + 5$

2. $g(t) = 1 - 3t^2 - 2t^4$

3. $f(x) = (2x + 3)(3x - 2)$

4. $g(x) = (2x^2 - 1)(x^3 + 2)$

5. $h(x) = (x + 1)^3$

6. $g(t) = (4t - 7)^2$

7. $f(y) = y(2y - 1)(2y + 1)$

8. $f(x) = 4x^4 - \dfrac{1}{x^2}$

9. $g(x) = \dfrac{1}{x + 1} - \dfrac{1}{x - 1}$

10. $f(t) = \dfrac{1}{4 - t^2}$

11. $h(x) = \dfrac{3}{x^2 + x + 1}$

12. $f(x) = \dfrac{1}{1 - \dfrac{2}{x}}$

13. $g(t) = (t^2 + 1)(t^3 + t^2 + 1)$

14. $f(x) = (2x^3 - 3)(17x^4 - 6x + 2)$

15. $g(z) = \dfrac{1}{2z} - \dfrac{1}{3z^2}$

16. $f(x) = \dfrac{2x^3 - 3x^2 + 4x - 5}{x^2}$

17. $g(y) = 2y(3y^2 - 1)(y^2 + 2y + 3)$

18. $f(x) = \dfrac{x^2 - 4}{x^2 + 4}$ **19.** $g(t) = \dfrac{t - 1}{t^2 + 2t + 1}$

20. $u(x) = \dfrac{1}{(x + 2)^2}$ **21.** $v(t) = \dfrac{1}{(t - 1)^3}$

22. $h(x) = \dfrac{2x^3 + x^2 - 3x + 17}{2x - 5}$

23. $g(x) = \dfrac{3x}{x^3 + 7x - 5}$ **24.** $f(t) = \dfrac{1}{\left(t + \dfrac{1}{t}\right)^2}$

25. $g(x) = \dfrac{\dfrac{1}{x} - \dfrac{2}{x^2}}{\dfrac{2}{x^3} - \dfrac{3}{x^4}}$ **26.** $f(x) = \dfrac{x^3 - \dfrac{1}{x^2 + 1}}{x^4 + \dfrac{1}{x^2 + 1}}$

27. $y = x^3 - 6x^5 + \frac{3}{2}x^{-4} + 12$

28. $y = \dfrac{3}{x} - \dfrac{4}{x^2} - 5$ **29.** $y = \dfrac{5 - 4x^2 + x^5}{x^3}$

30. $y = \dfrac{2x - 3x^2 + 2x^4}{5x^2}$ **31.** $y = 3x - \dfrac{1}{4x^2}$

32. $y = \dfrac{1}{x(x^2 + 2x + 2)}$ **33.** $y = \dfrac{x}{x - 1} + \dfrac{x + 1}{3x}$

34. $y = \dfrac{1}{1 - 4x^{-2}}$ **35.** $y = \dfrac{x^3 - 4x + 5}{x^2 + 9}$

36. $y = x^2\left(2x^3 - \dfrac{3}{4x^4}\right)$ **37.** $y = \dfrac{2x^2}{3x - \dfrac{4}{5x^4}}$

38. $y = \dfrac{4}{(x^2 - 3)^2}$ **39.** $y = \dfrac{x^2}{x + 1}$

40. $y = \dfrac{x + 10}{x^2}$

In Problems 41 through 50, write an equation of the line tangent to the curve $y = f(x)$ at the given point P on the curve. Express the answer in the form $ax + by = c$.

41. $y = x^3$; $P(2, 8)$

42. $y = 3x^2 - 4$; $P(1, -1)$

43. $y = \dfrac{1}{x - 1}$; $P(2, 1)$

44. $y = 2x - \dfrac{1}{x}$; $P(0.5, -1)$

45. $y = x^3 + 3x^2 - 4x - 5$; $P(1, -5)$

46. $y = \left(\dfrac{1}{x} - \dfrac{1}{x^2}\right)^{-1}$; $P(2, 4)$

47. $y = \dfrac{3}{x^2} - \dfrac{4}{x^3}$; $P(-1, 7)$

48. $y = \dfrac{3x - 2}{3x + 2}$; $P(2, 0.5)$

49. $y = \dfrac{3x^2}{x^2 + x + 1}$; $P(-1, 3)$

50. $y = \dfrac{6}{1 - x^2}$; $P(2, -2)$

51. Apply the formula of Example 5 to answer the following two questions. (a) If 1000 cm³ of water at 0°C is heated, does it initially expand or contract? (b) What is the rate (in cm³/°C) at which it initially contracts or expands?

52. Susan's weight in pounds is given by the formula $W = (2 \times 10^9)/R^2$, where R is her distance in miles from the center of the earth. What is the rate of change of W with respect to R when $R = 3960$ mi? If Susan climbs a mountain, beginning at sea level, at what rate in ounces per (vertical) mile does her weight initially decrease?

53. The conical tank shown in Fig. 3.2.3 has radius 160 cm and height 800 cm. Water is running out a small hole in the bottom of the tank. When the height h of water in the tank is 600 cm, what is the rate of change of its volume V with respect to h?

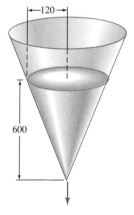

Fig. 3.2.3 The leaky tank of Problem 53

54. Find the x- and y-intercepts of the straight line that is tangent to the curve $y = x^3 + x^2 + x$ at the point $(1, 3)$ (Fig. 3.2.4).

55. Find an equation for the straight line that passes through the point $(1, 5)$ and is tangent to the curve $y = x^3$. [*Suggestion:* Denote by (a, a^3) the point of tangency, as indicated in Fig. 3.2.5. Find by inspection small integral solutions of the resulting cubic equation in a.]

116

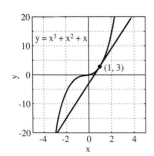 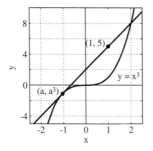 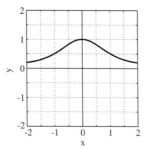 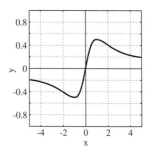

Fig. 3.2.4 The tangent line of Problem 54

Fig. 3.2.5 The tangent line of Problem 55

Fig. 3.2.6 The graph of $y = \dfrac{1}{1 + x^2}$

Fig. 3.2.7 The graph of $y = \dfrac{x}{1 + x^2}$

56. Find *two* lines through the point $(2, 8)$ that are tangent to the curve $y = x^3$. [See the suggestion for Problem 55.]

57. Prove that no straight line can be tangent to the curve $y = x^2$ at two different points.

58. Find the two straight lines of slope -2 that are tangent to the curve $y = 1/x$.

59. Let $n \geq 2$ be a fixed but unspecified integer. Find the x-intercept of the line that is tangent to the curve $y = x^n$ at the point $P(x_0, y_0)$.

60. Prove that the curve $y = x^5 + 2x$ has no horizontal tangents. What is the smallest slope that a line tangent to this curve can have?

61. Apply Eq. (16) with $n = 3$ and $u_1 = u_2 = u_3 = f(x)$ to show that

$$D([f(x)]^3) = 3[f(x)]^2 f'(x).$$

62. (a) First write $u_1 u_2 u_3 u_4 = (u_1 u_2 u_3)u_4$ to verify Eq. (16) for $n = 4$. (b) Then write $u_1 u_2 u_3 u_4 u_5 = (u_1 u_2 u_3 u_4)u_5$ and apply the result in part (a) to verify Eq. (16) for $n = 5$.

63. Apply Eq. (16) to show that

$$D([f(x)]^n) = n[f(x)]^{n-1} f'(x)$$

if n is a positive integer and $f'(x)$ exists.

64. Use the result of Problem 63 to compute $D(x^2 + x + 1)^{100}$.

65. Find $g'(x)$, given $g(x) = (x^3 - 17x + 35)^{17}$.

66. Find constants a, b, c, and d such that the curve $y = ax^3 + bx^2 + cx + d$ has horizontal tangent lines at the points $(0, 1)$ and $(1, 0)$.

In connection with Problems 67 through 71, Figs. 3.2.6 through 3.2.9 show the curves

$$y = \frac{x^n}{1 + x^2}$$

for n = 0, 1, 2, 3.

67. Show that for $n = 0$ and $n = 2$, the curve has only a single point where the tangent line is horizontal (Figs. 3.2.6 and 3.2.8).

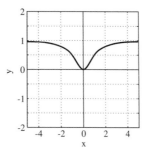 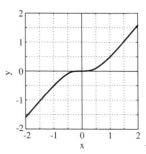

Fig. 3.2.8 The graph of $y = \dfrac{x^2}{1 + x^2}$

Fig. 3.2.9 The graph of $y = \dfrac{x^3}{1 + x^2}$

68. When $n = 1$, there are two points on the curve where the tangent line is horizontal (Fig. 3.2.7). Find them.

69. Show that for $n \geq 3$, $(0, 0)$ is the only point on the graph of

$$y = \frac{x^n}{1 + x^2}$$

at which the tangent line is horizontal (Fig. 3.2.9).

70. Figure 3.2.10 shows the graph of the derivative $f'(x)$ of the function

$$f(x) = \frac{x^3}{1 + x^2}.$$

There appear to be two points on the graph of $y = f(x)$ where the tangent line has slope 1. Find them.

71. It appears in Fig. 3.2.10 that there are three points on the curve $y = f'(x)$ where the tangent line is horizontal. Find them.

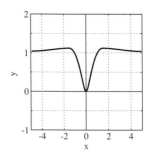

Fig. 3.2.10 The graph of $y = D_x\left(\dfrac{x^3}{1 + x^2}\right)$ of Problems 70 and 71

3.3
The Chain Rule

We saw in Section 3.2 how to differentiate polynomials and rational functions. But we often need to differentiate *powers* of such functions. For instance, if

$$y = [f(x)]^3, \tag{1}$$

then the extended product rule [Eq. (16) in Section 3.2] yields

$$\frac{dy}{dx} = D[f(x) \cdot f(x) \cdot f(x)]$$

$$= f'(x) \cdot f(x) \cdot f(x) + f(x) \cdot f'(x) \cdot f(x) + f(x) \cdot f(x) \cdot f'(x),$$

so collecting terms gives

$$\frac{dy}{dx} = 3[f(x)]^2 f'(x). \tag{2}$$

Is it a surprise that the derivative of $[f(x)]^3$ is not simply the quantity $3[f(x)]^2$, which you might expect in analogy with the (correct) formula $D_x x^3 = 3x^2$? There is an additional factor $f'(x)$ whose origin can be explained by writing $y = [f(x)]^3$ in the form

$$y = u^3 \quad \text{with} \quad u = f(x).$$

Then

$$\frac{dy}{dx} = D[f(x)]^3,$$

$$\frac{dy}{du} = 3u^2 = 3[f(x)]^2, \quad \text{and} \tag{3}$$

$$\frac{du}{dx} = f'(x),$$

so the derivative formula in Eq. (2) takes the form

$$\frac{dy}{dx} = \frac{dy}{du} \cdot \frac{du}{dx}. \tag{4}$$

Equation (4), the **chain rule,** holds for *any* two differentiable functions $y = g(u)$ and $u = f(x)$. The formula in Eq. (2) is simply the special case of Eq. (4) with $g(u) = u^3$.

EXAMPLE 1 If

$$y = (3x + 5)^{17},$$

it would be impractical to write the binomial expansion of the seventeenth power of $3x + 5$: The result would be a polynomial with 18 terms, and some of the coefficients would have 14 digits! But if we write

$$y = u^{17} \quad \text{with} \quad u = 3x + 5,$$

then

$$\frac{dy}{du} = 17u^{16} \quad \text{and} \quad \frac{du}{dx} = 3.$$

118

Hence the chain rule yields

$$D(3x + 5)^{17} = \frac{dy}{dx} = \frac{dy}{du} \cdot \frac{du}{dx} = (17u^{16})(3)$$

$$= 17(3x + 5)^{16} \cdot 3 = 51(3x + 5)^{16}.$$

The formula in Eq. (4) is one that, once learned, is unlikely to be forgotten. Although dy/du and du/dx are *not* fractions—they are merely symbols representing the derivatives $g'(u)$ and $f'(x)$—it is just as though they *were* fractions, with the du in the first factor "canceling" the du in the second. Of course, such "cancellation" no more proves the chain rule than canceling d's proves that

$$\frac{dy}{dx} = \frac{y}{x} \qquad \text{(an absurdity).}$$

It is nevertheless an excellent way to *remember* the chain rule. Such manipulations with differentials are so suggestive (even when invalid) that they played a substantial role in the early development of calculus in the seventeenth and eighteenth centuries. Many formulas were thereby produced that were later proved valid (as were some formulas that were incorrect).

EXAMPLE 2 For a physical interpretation of the chain rule, imagine an oil refinery that first makes u liters of gasoline from x barrels of crude oil. Then, in a second process, the refinery makes y grams of a marketable petrochemical from the u liters of gasoline. (The two processes are illustrated in Fig. 3.3.1.) Then y is a function of u and u is a function of x, so the final output y is a function also of the input x. Consider the *units* in which the *derivatives* of these functions are measured.

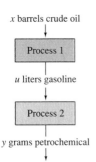

x barrels crude oil

Process 1

u liters gasoline

Process 2

y grams petrochemical

Fig. 3.3.1 The two-process oil refinery (Example 2)

$$\frac{dy}{du} : \frac{\text{g}}{\text{L}} \qquad \text{(grams of petrochemical per liter of gasoline)}$$

$$\frac{du}{dx} : \frac{\text{L}}{\text{barrel}} \qquad \text{(liters of gasoline per barrel of oil)}$$

$$\frac{dy}{dx} : \frac{\text{g}}{\text{barrel}} \qquad \text{(grams of petrochemical per barrel of oil)}$$

When we include these units in Eq. (4), we get

$$\frac{dy}{dx} \frac{\text{g}}{\text{barrel}} = \left(\frac{dy}{du} \frac{\text{g}}{\cancel{\text{L}}} \right) \cdot \left(\frac{du}{dx} \frac{\cancel{\text{L}}}{\text{barrel}} \right)$$

$$= \left(\frac{dy}{du} \cdot \frac{du}{dx} \right) \frac{\text{g}}{\text{barrel}}.$$

The handy cancellation of units seems to confirm the validity of the chain rule (at least in this application). For example, if we get 3 g of petrochemical per liter of gasoline and 75 L of gasoline per barrel of oil, how could we fail to get $225 = 3 \cdot 75$ g of petrochemical per barrel of oil?

Although Eq. (4) is a memorable statement of the chain rule in differential notation, it has the disadvantage of not specifying the values of the variables at which the derivatives are evaluated. This problem is solved by the use of function notation for the derivatives. Let us write

$$y = g(u), \qquad u = f(x), \qquad y = h(x) = g(f(x)). \tag{5}$$

Then

$$\frac{du}{dx} = f'(x), \qquad \frac{dy}{dx} = h'(x),$$

and

$$\frac{dy}{du} = g'(u) = g'(f(x)). \tag{6}$$

Substitution of these derivatives into Eq. (4) recasts the chain rule in the form

$$h'(x) = g'(f(x)) \cdot f'(x). \tag{7}$$

This version of the chain rule gives the derivative of the *composition* $h = g \circ f$ of two functions g and f in terms of *their* derivatives.

Theorem 1   *The Chain Rule*
Suppose that f is differentiable at x and that g is differentiable at $f(x)$. Then the composition $h = g \circ f$ defined by $h(x) = g(f(x))$ is differentiable at x, and its derivative is

$$h'(x) = D[g(f(x))] = g'(f(x)) \cdot f'(x). \tag{8}$$

NOTE   Although the derivative of $h = g \circ f$ is a product of the derivatives of f and g, these two derivatives are evaluated at *different* points. The derivative g' is evaluated at $f(x)$, whereas f' is evaluated at x. For a particular number $x = a$, Eq. (7) tells us that

$$h'(a) = g'(b) \cdot f'(a), \qquad \text{where } b = f(a). \tag{9}$$

EXAMPLE 3   If $h(x) = g(f(x))$, where f and g are differentiable functions, and

$$f(2) = 17, \qquad f'(2) = -3, \quad \text{and} \quad g'(17) = 5,$$

then the chain rule gives

$$h'(2) = g'(f(2)) \cdot f'(2) = g'(17) \cdot f'(2) = (5)(-3) = -15.$$

OUTLINE OF THE PROOF OF THE CHAIN RULE

To *outline* a proof of the chain rule, suppose that we are given differentiable functions $y = g(u)$ and $u = f(x)$ and want to compute the derivative

$$\frac{dy}{dx} = \lim_{\Delta x \to 0} \frac{\Delta y}{\Delta x} = \lim_{\Delta x \to 0} \frac{g(f(x + \Delta x)) - g(f(x))}{\Delta x}. \tag{10}$$

The differential form of the chain rule suggests the factorization

$$\frac{\Delta y}{\Delta x} = \frac{\Delta y}{\Delta u} \frac{\Delta u}{\Delta x}. \qquad (11)$$

The product law of limits then gives

$$\frac{dy}{dx} = \lim_{\Delta x \to 0} \frac{\Delta y}{\Delta u} \frac{\Delta u}{\Delta x} = \left(\lim_{\Delta u \to 0} \frac{\Delta y}{\Delta u} \right) \left(\lim_{\Delta x \to 0} \frac{\Delta u}{\Delta x} \right) = \frac{dy}{du} \frac{du}{dx}. \qquad (12)$$

This will suffice to prove the chain rule *provided that*

$$\Delta u = f(x + \Delta x) - f(x) \qquad (13)$$

is a *nonzero* quantity that approaches zero as $\Delta x \to 0$. Certainly $\Delta u \to 0$ as $\Delta x \to 0$, because f is differentiable and therefore continuous. But it is still quite possible that Δu is zero for some—even all—nonzero values of Δx. In such a case, the factorization in Eq. (11) would include the *invalid* step of division by zero. Thus our proof is incomplete. A complete proof of the chain rule is given in Appendix D.

THE GENERALIZED POWER RULE

If we substitute $f(x) = u$ and $f'(x) = du/dx$ into Eq. (8), we get the hybrid form

$$D_x g(u) = g'(u) \frac{du}{dx} \qquad (14)$$

of the chain rule that typically is the most useful form for purely computational purposes. Recall that the subscript x in D_x specifies that $g(u)$ is being differentiated with respect to x rather than with respect to u.

Let us set $g(u) = u^n$ in Eq. (14), where n is an integer. Because $g'(u) = nu^{n-1}$, we thereby obtain

$$D_x u^n = nu^{n-1} \frac{du}{dx}, \qquad (15)$$

the *chain rule version* of the power rule. If $u = f(x)$ is a differentiable function, then Eq. (15) says that

$$D_x[f(x)]^n = n[f(x)]^{n-1} f'(x). \qquad (16)$$

[If $n - 1 < 0$, we must add the proviso that $f(x) \neq 0$ in order for the right-hand side in Eq. (16) to be meaningful.] We refer to this chain rule version of the power rule as the **generalized power rule.**

EXAMPLE 4   To differentiate

$$y = \frac{1}{(2x^3 - x + 7)^2},$$

we first write

$$y = (2x^3 - x + 7)^{-2}$$

in order to apply the generalized power rule, Eq. (16), with $n = -2$. This gives

$$\frac{dy}{dx} = (-2)(2x^3 - x + 7)^{-3} D(2x^3 - x + 7)$$

$$= (-2)(2x^3 - x + 7)^{-3}(6x^2 - 1) = \frac{2(1 - 6x^2)}{(2x^3 - x + 7)^3}.$$

EXAMPLE 5 Find the derivative of the function

$$h(z) = \left(\frac{z - 1}{z + 1}\right)^5.$$

Solution The key to applying the generalized power rule is observing *what* the given function is a power *of*. Here,

$$h(z) = u^5, \quad \text{where} \quad u = \frac{z - 1}{z + 1},$$

and z, not x, is the independent variable. Hence we apply first Eq. (15) and then the quotient rule to get

$$h'(z) = 5u^4 \frac{du}{dz} = 5\left(\frac{z - 1}{z + 1}\right)^4 D_z\left(\frac{z - 1}{z + 1}\right)$$

$$= 5\left(\frac{z - 1}{z + 1}\right)^4 \cdot \frac{(1)(z + 1) - (z - 1)(1)}{(z + 1)^2}$$

$$= 5\left(\frac{z - 1}{z + 1}\right)^4 \cdot \frac{2}{(z + 1)^2} = \frac{10(z - 1)^4}{(z + 1)^6}.$$

The importance of the chain rule goes far beyond the power function differentiations illustrated in Examples 1, 4, and 5. We shall learn in later sections how to differentiate exponential, logarithmic, and trigonometric functions. Each time we learn a new differentiation formula—for the derivative $g'(x)$ of a new function $g(x)$—the formula in Eq. (14) immediately provides us with the chain rule version of that formula,

$$D_x g(u) = g'(u) D_x u.$$

The step from the power rule $Dx^n = nx^{n-1}$ to the generalized power rule $D_x u^n = nu^{n-1} D_x u$ is our first instance of this general phenomenon.

RATE-OF-CHANGE APPLICATIONS

Suppose that the physical or geometric quantity p depends on the quantity q, which in turn depends on time t. Then the *dependent* variable p is a function both of the *intermediate* variable q and of the *independent* variable t. Hence the derivatives that appear in the chain rule formula

$$\frac{dp}{dt} = \frac{dp}{dq} \frac{dq}{dt}$$

are rates of change (as in Section 3.1) of these variables with respect to one another. For instance, suppose that a spherical balloon is being inflated or deflated. Then its volume V and its radius r are changing with time t, and

$$\frac{dV}{dt} = \frac{dV}{dr} \cdot \frac{dr}{dt}.$$

Remember that a positive derivative signals an increasing quantity and that a negative derivative signals a decreasing quantity.

EXAMPLE 6   A spherical balloon is being inflated (Fig. 3.3.2). The radius r of the balloon is increasing at the rate of 0.2 cm/s when $r = 5$ cm. At what rate is the volume V of the balloon increasing at that instant?

Solution   Given $dr/dt = 0.2$ cm/s when $r = 5$ cm, we want to find dV/dt at that instant. Because the volume of the balloon is

$$V = \tfrac{4}{3}\pi r^3,$$

we see that $dV/dr = 4\pi r^2$. So the chain rule gives

$$\frac{dV}{dt} = \frac{dV}{dr} \cdot \frac{dr}{dt} = 4\pi r^2 \frac{dr}{dt} = 4\pi (5)^2 (0.2) \approx 62.83 \quad \text{(cm/s)}$$

at the instant when $r = 5$ cm.

Fig. 3.3.2   The spherical balloon with volume $V = \tfrac{4}{3}\pi r^3$ (Example 6)

In Example 6 we did not need to know r explicitly as a function of t. But suppose we are told that after t seconds the radius (in centimeters) of an inflating balloon is $r = 3 + (0.2)t$ (until the balloon bursts). Then the volume of this balloon is

$$V = \frac{4}{3}\pi r^3 = \frac{4}{3}\pi \left(3 + \frac{t}{5}\right)^3,$$

so dV/dt is given explicitly as a function of t by

$$\frac{dV}{dt} = \frac{4}{3}\pi(3)\left(3 + \frac{t}{5}\right)^2\left(\frac{1}{5}\right) = \frac{4}{5}\pi\left(3 + \frac{t}{5}\right)^2.$$

EXAMPLE 7   Imagine a spherical raindrop that is falling through water vapor in the air. Suppose that the vapor adheres to the surface of the raindrop in such a way that the time rate of increase of the mass M of the droplet is proportional to the surface area S of the droplet. If the initial radius of the droplet is, in effect, zero and the radius is $r = 1$ mm after 20 s, when is the radius 3 mm?

Solution   We are given

$$\frac{dM}{dt} = kS,$$

where k is some constant that depends upon atmospheric conditions. Now

$$M = \tfrac{4}{3}\pi\rho r^3 \quad \text{and} \quad S = 4\pi r^2,$$

where ρ is the density of water. Hence the chain rule gives

$$4\pi k r^2 = kS = \frac{dM}{dt} = \frac{dM}{dr} \cdot \frac{dr}{dt};$$

that is,

$$4\pi k r^2 = 4\pi \rho r^2 \frac{dr}{dt}.$$

This implies that

$$\frac{dr}{dt} = \frac{k}{\rho},$$

a constant. So the radius of the droplet grows at a *constant* rate. Thus if it takes 20 s for r to grow to 1 mm, it will take 1 min for r to grow to 3 mm.

3.3 Problems

Find dy/dx in Problems 1 through 12.

1. $y = (3x + 4)^5$

2. $y = (2 - 5x)^3$

3. $y = \dfrac{1}{3x - 2}$

4. $y = \dfrac{1}{(2x + 1)^3}$

5. $y = (x^2 + 3x + 4)^3$

6. $y = (7 - 2x^3)^{-4}$

7. $y = (2 - x)^4(3 + x)^7$

8. $y = (x + x^2)^5(1 + x^3)^2$

9. $y = \dfrac{x + 2}{(3x - 4)^3}$

10. $y = \dfrac{(1 - x^2)^3}{(4 + 5x + 6x^2)^2}$

11. $y = [1 + (1 + x)^3]^4$

12. $y = [x + (x + x^2)^{-3}]^{-5}$

In Problems 13 through 20, express the derivative dy/dx in terms of x.

13. $y = (u + 1)^3$ and $u = \dfrac{1}{x^2}$

14. $y = \dfrac{1}{2u} - \dfrac{1}{3u^2}$ and $u = 2x + 1$

15. $y = (1 + u^2)^3$ and $u = (4x - 1)^2$

16. $y = u^5$ and $u = \dfrac{1}{3x - 2}$

17. $y = u(1 - u)^3$ and $u = \dfrac{1}{x^4}$

18. $y = \dfrac{u}{u + 1}$ and $u = \dfrac{x}{x + 1}$

19. $y = u^2(u - u^4)^3$ and $u = \dfrac{1}{x^2}$

20. $y = \dfrac{u}{(2u + 1)^4}$ and $u = x - \dfrac{2}{x}$

In Problems 21 through 26, identify a function u of x and an integer $n \neq 1$ such that $f(x) = u^n$. Then compute $f'(x)$.

21. $f(x) = (2x - x^2)^3$

22. $f(x) = \dfrac{1}{2 + 5x^3}$

23. $f(x) = \dfrac{1}{(1 - x^2)^4}$

24. $f(x) = (x^2 - 4x + 1)^3$

25. $f(x) = \left(\dfrac{x + 1}{x - 1}\right)^7$

26. $f(x) = \dfrac{(x^2 + x + 1)^7}{(x + 1)^4}$

Differentiate the functions given in Problems 27 through 36.

27. $g(y) = y + (2y - 3)^5$

28. $h(z) = z^2(z^2 + 4)^3$

29. $F(s) = \left(s - \dfrac{1}{s^2}\right)^3$

30. $G(t) = \left[t^2 + \left(1 + \dfrac{1}{t}\right)\right]^2$

31. $f(u) = (1 + u)^3(1 + u^2)^4$

32. $g(w) = (w^2 - 3w + 4)(w + 4)^5$

33. $h(v) = \left[v - \left(1 - \dfrac{1}{v}\right)^{-1}\right]^{-2}$

34. $p(t) = \left(\dfrac{1}{t} + \dfrac{1}{t^2} + \dfrac{1}{t^3}\right)^{-4}$

35. $F(z) = \dfrac{1}{(3 - 4z + 5z^5)^{10}}$

36. $G(x) = \{1 + [x + (x^2 + x^3)^4]^5\}^6$

In Problems 37 through 44, dy/dx can be found in two ways—one way using the chain rule, the other way without using it. Use both techniques to find dy/dx and then compare the answers. (They should agree!)

37. $y = (x^3)^4 = x^{12}$

38. $y = x = \left(\dfrac{1}{x}\right)^{-1}$

39. $y = (x^2 - 1)^2 = x^4 - 2x^2 + 1$

40. $y = (1 - x)^3 = 1 - 3x + 3x^2 - x^3$

41. $y = (x + 1)^4 = x^4 + 4x^3 + 6x^2 + 4x + 1$

42. $y = (x + 1)^{-2} = \dfrac{1}{x^2 + 2x + 1}$

43. $y = (x^2 + 1)^{-1} = \dfrac{1}{x^2 + 1}$

44. $y = (x^2 + 1)^2 = (x^2 + 1)(x^2 + 1)$

We shall see in Section 3.7 that $D_x \sin x = \cos x$ (provided that x is in radian measure). Use this fact and the chain rule to find the derivatives of the functions in Problems 45 through 48.

45. $f(x) = \sin(x^3)$ **46.** $g(t) = (\sin t)^3$

47. $g(z) = (\sin 2z)^3$ **48.** $k(u) = \sin(1 + \sin u)$

49. The radius of a circle is increasing at the rate of 2 in./s. At what rate is its area increasing when its radius is 10 in.?

50. The area of a circle is decreasing at the rate of 2π cm^2/s. At what rate is the radius of the circle decreasing when its area is 75π cm^2?

51. Each edge x of a square is increasing at the rate of 2 in./s. At what rate is the area A of the square increasing when each edge is 10 in.?

52. Each edge of an equilateral triangle is increasing at 2 cm/s (Fig. 3.3.3). At what rate is the area of the triangle increasing when each edge is 10 cm?

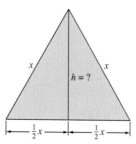

Fig. 3.3.3 The triangle of Problem 52 with area $A = \frac{1}{2}xh$

53. A cubical block of ice is melting in such a way that each edge decreases steadily by 2 in. every hour. At what

rate is its volume decreasing when each edge is 10 in. long?

54. Find $f'(-1)$, given $f(y) = h(g(y))$, $h(2) = 55$, $g(-1) = 2$, $h'(2) = -1$, and $g'(-1) = 7$.

55. Given: $G(t) = f(h(t))$, $h(1) = 4$, $f'(4) = 3$, and $h'(1) = -6$. Find $G'(1)$.

56. Suppose that $f(0) = 0$ and that $f'(0) = 1$. Calculate the value of the derivative of $f(f(f(x)))$ at $x = 0$.

57. Air is being pumped into a spherical balloon in such a way that its radius r is increasing at the rate of $dr/dt = 1$ cm/s. What is the time rate of increase, in cubic centimeters per second, of the balloon's volume when $r = 10$ cm?

58. Suppose that the air is being pumped into the balloon of Problem 57 at the constant rate of 200π cm^3/s. What is the time rate of increase of the radius r when $r = 5$ cm?

59. Air is escaping from a spherical balloon at the constant rate of 300π cm^3/s. What is the radius of the balloon when its radius is decreasing at the rate of 3 cm/s?

60. A spherical hailstone is losing mass by melting uniformly over its surface as it falls. At a certain time, its radius is 2 cm, and its volume is decreasing at the rate of 0.1 cm^3/s. How fast is its radius decreasing at that time?

61. A spherical snowball is melting in such a way that the rate of decrease of its volume is proportional to its surface area. At 10 A.M. its volume is 500 in.^{3}, and at 11 A.M. its volume is 250 in.^{3} When does the snowball finish melting? (See Example 7.)

62. A cubical block of ice with edges 20 in. long begins to melt at 8 A.M. Each edge decreases steadily thereafter and is 8 in. long at 4 P.M. What was the rate of change of the block's volume at noon?

63. Suppose that u is a function of v, that v is a function of w, that w is a function of x, and that all these functions are differentiable. Explain why it follows from the chain rule that

$$\frac{du}{dx} = \frac{du}{dv} \cdot \frac{dv}{dw} \cdot \frac{dw}{dx}.$$

3.4

Derivatives of Algebraic Functions

We saw in Section 3.3 that the chain rule yields the differentiation formula

$$D_x u^n = nu^{n-1} \frac{du}{dx} \tag{1}$$

if $u = f(x)$ is a differentiable function and the exponent n is an integer. We shall see in Theorem 1 of this section that this **generalized power rule** holds not only when the exponent is an integer, but also when it is a rational number $r = p/q$ (where p and q are integers and $q \neq 0$). Recall that rational powers are defined in terms of integral roots and powers as follows:

$$u^{p/q} = \sqrt[q]{u^p} = \left(\sqrt[q]{u}\right)^p.$$

We first consider the case of a rational power of the independent variable x:

$$y = x^{p/q}, \tag{2}$$

where p and q are integers with q positive. We show independently in Section 3.8 that $g(x) = x^{p/q}$ is differentiable wherever its derivative does not involve division by zero or an even root of a negative number. Assuming this fact, let us take the qth power of each side in Eq. (2) to obtain

$$y^q = x^p \tag{3}$$

[because $(x^{p/q})^q = x^p$]. Note that Eq. (3) is an identity—the functions y^q and x^p of x are identical where defined. Therefore, their derivatives with respect to x must also be identical. That is,

$$D_x(y^q) = D_x(x^p);$$

$$qy^{q-1}\frac{dy}{dx} = px^{p-1}.$$

To differentiate the left-hand side, we use Eq. (1) with $u = y$ and $n = q$. Finally, we solve for the derivative:

$$\frac{dy}{dx} = \frac{px^{p-1}}{qy^{q-1}} = \frac{p}{q}x^{p-1}y^{1-q} = \frac{p}{q}x^{p-1}(x^{p/q})^{1-q}$$

$$= \frac{p}{q}x^{p-1}x^{p/q}x^{-p} = \frac{p}{q}x^{(p/q)-1}.$$

Thus we have shown that

$$D_x x^{p/q} = \frac{p}{q}x^{(p/q)-1},$$

that is, that the power rule

$$D_x x^r = rx^{r-1} \tag{4}$$

holds if the exponent $r = p/q$ is a rational number (subject to the provisos previously mentioned).

Using Eq. (4) we can differentiate a simple "radical" (or "root") function by first rewriting it as a power with a fractional exponent.

EXAMPLE 1

$$D\sqrt{x} = Dx^{1/2} = \frac{1}{2}x^{-1/2} = \frac{1}{2\sqrt{x}};$$

$$D\sqrt{x^3} = Dx^{3/2} = \frac{3}{2}x^{1/2} = \frac{3}{2}\sqrt{x};$$

$$D\frac{1}{\sqrt[3]{x^2}} = Dx^{-2/3} = -\frac{2}{3}x^{-5/3} = -\frac{2}{3\sqrt[3]{x^5}}.$$

For the more general form of the power rule, let

$$y = u^r,$$

where u is a differentiable function of x and $r = p/q$ is rational. Then

$$\frac{dy}{du} = ru^{r-1}$$

by Eq. (4), so the chain rule gives

$$\frac{dy}{dx} = \frac{dy}{du}\frac{du}{dx} = ru^{r-1}\frac{du}{dx}.$$

Thus

$$D_x u^r = ru^{r-1}\frac{du}{dx}, \tag{5}$$

which is the generalized power rule for rational exponents.

> **Theorem 1   Generalized Power Rule**
> If r is a rational number, then
> $$D_x[f(x)]^r = r[f(x)]^{r-1}f'(x) \tag{6}$$
> wherever the function f is differentiable and the right-hand side is defined.

For the right-hand side in Eq. (6) to be "defined" means that $f'(x)$ exists, there is no division by zero, and no even root of a negative number appears.

EXAMPLE 2

$$D\sqrt{4 - x^2} = D(4 - x^2)^{1/2} = \tfrac{1}{2}(4 - x^2)^{-1/2}D(4 - x^2)$$
$$= \tfrac{1}{2}(4 - x^2)^{-1/2}(-2x);$$

$$D\sqrt{4 - x^2} = -\frac{x}{\sqrt{4 - x^2}} \tag{7}$$

except where $x = \pm 2$ (division by zero) or where $|x| > 2$ (square root of a negative number). Thus Eq. (7) holds if $-2 < x < 2$. In writing derivatives of algebraic functions, we ordinarily omit such disclaimers unless they are pertinent to some specific purpose at hand.

A template for the application of the generalized power rule is

$$D_x[\ast\ast\ast]^n = n[\ast\ast\ast]^{n-1}D_x[\ast\ast\ast],$$

where $\ast\ast\ast$ represents a function of x and (as we now know) n can be either an integer or a fraction (a quotient of integers).

But to differentiate a *power of a function,* we must first recognize *what function* it is a power *of.* So to differentiate a function involving roots (or radicals), we first "prepare" it for an application of the generalized power rule by rewriting it as a power function with fractional exponent. Examples 3 through 6 illustrate this technique.

EXAMPLE 3   If $y = 5\sqrt{x^3} - \dfrac{2}{\sqrt[3]{x}}$, then

$$y = 5x^{3/2} - 2x^{-1/3}, \quad \text{so}$$

$$\frac{dy}{dx} = 5 \cdot \left(\frac{3}{2}x^{1/2}\right) - 2 \cdot \left(-\frac{1}{3}x^{-4/3}\right)$$

$$= \frac{15}{2}x^{1/2} + \frac{2}{3}x^{-4/3} = \frac{15}{2}\sqrt{x} + \frac{2}{3\sqrt[3]{x^4}}.$$

EXAMPLE 4   With $f(x) = 3 - 5x$ and $r = 7$, the generalized power rule yields

$$D(3 - 5x)^7 = 7(3 - 5x)^6 D(3 - 5x)$$

$$= 7(3 - 5x)^6(-5) = -35(3 - 5x)^6.$$

EXAMPLE 5   With $f(x) = 2x^2 - 3x + 5$ and $r = \frac{1}{2}$, the generalized power rule yields

$$D\sqrt{2x^2 - 3x + 5} = D(2x^2 - 3x + 5)^{1/2}$$

$$= \frac{1}{2}(2x^2 - 3x + 5)^{-1/2} D(2x^2 - 3x + 5)$$

$$= \frac{4x - 3}{2\sqrt{2x^2 - 3x + 5}}.$$

EXAMPLE 6   If $y = [5x + \sqrt[3]{(3x - 1)^4}]^{10}$, then Eq. (5) with $u = 5x + (3x - 1)^{4/3}$ gives

$$\frac{dy}{dx} = 10u^9 \frac{du}{dx} = 10[5x + (3x - 1)^{4/3}]^9 D[5x + (3x - 1)^{4/3}]$$

$$= 10[5x + (3x - 1)^{4/3}]^9[D(5x) + D(3x - 1)^{4/3}]$$

$$= 10[5x + (3x - 1)^{4/3}]^9[5 + \tfrac{4}{3}(3x - 1)^{1/3}(3)]$$

$$= 10[5x + (3x - 1)^{4/3}]^9[5 + 4\sqrt[3]{3x - 1}].$$

VERTICAL TANGENT LINES

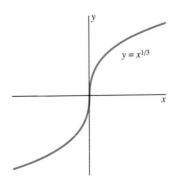

Fig. 3.4.1 The graph of the cube root function

Figure 3.4.1 shows the graph of the function

$$y = \sqrt[3]{x} = x^{1/3}.$$

Its derivative,

$$\frac{dy}{dx} = \frac{1}{3}x^{-2/3} = \frac{1}{3\sqrt[3]{x^2}},$$

increases without bound as $x \to 0$ but does not exist at $x = 0$. Therefore, the definition of the tangent line in Section 2.1 does not apply to this graph at $(0, 0)$. Nevertheless, from the figure it seems appropriate to regard the vertical line $x = 0$ as the line tangent to the curve $y = x^{1/3}$ at the point $(0, 0)$. Observations such as this motivate the following definition.

Ch. 3 / The Derivative

Definition *Vertical Tangent Line*

The curve $y = f(x)$ has a **vertical tangent line** at the point $(a, f(a))$ provided that f is continuous at a and that

$$| f'(x) | \rightarrow +\infty \quad \text{as} \quad x \rightarrow a. \tag{8}$$

The requirement that f be continuous at $x = a$ implies that $f(a)$ must be defined. Thus it would be pointless to ask about a line (vertical or not) tangent to the curve $y = 1/x$ where $x = 0$.

If f is defined (and differentiable) on only one side of $x = a$, we mean in Eq. (8) that $| f'(x) | \rightarrow +\infty$ as x approaches a from that side.

EXAMPLE 7   Find the points on the curve

$$y = f(x) = x\sqrt{1 - x^2}, \qquad -1 \leq x \leq 1,$$

at which the tangent line is either horizontal or vertical.

Solution   We differentiate using first the product rule and then the chain rule:

$$f'(x) = (1 - x^2)^{1/2} + \frac{x}{2}(1 - x^2)^{-1/2}(-2x)$$

$$= (1 - x^2)^{-1/2}[(1 - x^2) - x^2] = \frac{1 - 2x^2}{\sqrt{1 - x^2}}.$$

Thus $f'(x) = 0$ only when the numerator $1 - 2x^2$ is zero—that is, when $x = \pm 1/\sqrt{2}$. Because $f(\pm 1/\sqrt{2}) = \pm 0.5$, the curve has a horizontal tangent line at each of the two points $(1/\sqrt{2}, 0.5)$ and $(-1/\sqrt{2}, -0.5)$.

We also observe that the denominator $\sqrt{1 - x^2}$ approaches zero as $x \rightarrow -1^+$ and as $x \rightarrow +1^-$. Because $f(\pm 1) = 0$, we see that the curve has a vertical tangent line at each of the two points $(1, 0)$ and $(-1, 0)$. The graph of f is shown in Fig. 3.4.2.

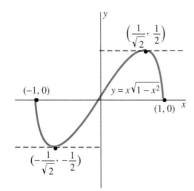

Fig. 3.4.2   The graph of $f(x) = x\sqrt{1 - x^2}, -1 \leq x \leq 1$ (Example 7)

3.4 Problems

Differentiate the functions given in Problems 1 through 44.

1. $f(x) = 4\sqrt{x^5} + \dfrac{2}{\sqrt{x}}$

2. $f(x) = 9\sqrt[3]{x^4} - \dfrac{3}{\sqrt[3]{x}}$

3. $f(x) = \sqrt{2x + 1}$

4. $f(x) = \dfrac{1}{\sqrt[3]{7 - 6x}}$

5. $f(x) = \dfrac{6 - x^2}{\sqrt{x}}$

6. $f(x) = \dfrac{7 + 2x - 3x^4}{\sqrt[3]{x^2}}$

7. $f(x) = (2x + 3)^{3/2}$

8. $f(x) = (3x + 4)^{4/3}$

9. $f(x) = (3 - 2x^2)^{-3/2}$

10. $f(x) = (4 - 3x^3)^{-2/3}$

11. $f(x) = \sqrt{x^3 + 1}$

12. $f(x) = \dfrac{1}{(x^4 + 3)^2}$

13. $f(x) = \sqrt{2x^2 + 1}$

14. $f(x) = \dfrac{x}{\sqrt{1 + x^4}}$

15. $f(t) = \sqrt{2t^3}$

16. $g(t) = \sqrt{\dfrac{1}{3t^5}}$

17. $f(x) = (2x^2 - x + 7)^{3/2}$

18. $g(z) = (3z^2 - 4)^{97}$

19. $g(x) = \dfrac{1}{(x - 2x^3)^{4/3}}$

20. $f(t) = [t^2 + (1 + t)^4]^5$

21. $f(x) = x\sqrt{1 - x^2}$

22. $g(x) = \sqrt{\dfrac{2x + 1}{x - 1}}$

23. $f(t) = \sqrt{\dfrac{t^2 + 1}{t^2 - 1}}$

24. $h(y) = \left(\dfrac{y + 1}{y - 1}\right)^{17}$

25. $f(x) = \left(x - \dfrac{1}{x}\right)^3$

26. $g(z) = \dfrac{z^2}{\sqrt{1 + z^2}}$

27. $f(v) = \dfrac{\sqrt{v + 1}}{v}$

28. $h(x) = \left(\dfrac{x}{1 + x^2}\right)^{5/3}$

29. $f(x) = \sqrt[3]{1 - x^2}$　　**30.** $g(x) = \sqrt{x + \sqrt{x}}$

31. $f(x) = x(3 - 4x)^{1/2}$　**32.** $g(t) = \dfrac{1}{t^2}[t - (1 + t^2)^{1/2}]$

33. $f(x) = (1 - x^2)(2x + 4)^{4/3}$

34. $f(x) = (1 - x)^{1/2}(2 - x)^{1/3}$

35. $g(t) = \left(1 + \dfrac{1}{t}\right)^2 (3t^2 + 1)^{1/2}$

36. $f(x) = x(1 + 2x + 3x^2)^{10}$

37. $f(x) = \dfrac{2x - 1}{(3x + 4)^5}$　　**38.** $h(z) = (z - 1)^4(z + 1)^6$

39. $f(x) = \dfrac{(2x + 1)^{1/2}}{(3x + 4)^{1/3}}$

40. $f(x) = (1 - 3x^4)^5(4 - x)^{1/3}$

41. $h(y) = \dfrac{\sqrt{1 + y} + \sqrt{1 - y}}{\sqrt[3]{y^5}}$

42. $f(x) = \sqrt{1 - \sqrt[3]{x}}$　**43.** $g(t) = \sqrt{t + \sqrt{t + \sqrt{t}}}$

44. $f(x) = x^3 \sqrt{1 - \dfrac{1}{x^2 + 1}}$

For each curve given in Problems 45 through 50, find all points on the graph where the tangent line is either horizontal or vertical.

45. $y = x^{2/3}$　　　　**46.** $y = x\sqrt{4 - x^2}$

47. $y = x^{1/2} - x^{3/2}$　**48.** $y = \dfrac{1}{\sqrt{9 - x^2}}$

49. $y = \dfrac{x}{\sqrt{1 - x^2}}$　**50.** $y = \sqrt{(1 - x^2)(4 - x^2)}$

51. The period of oscillation P (in seconds) of a simple pendulum of length L (in feet) is given by $P = 2\pi \sqrt{L/g}$, where $g = 32$ ft/s^2. Find the rate of change of P with respect to L when $P = 2$.

52. Find the rate of change of the volume $V = \frac{4}{3}\pi r^3$ of a sphere of radius r with respect to its surface area $A = 4\pi r^2$ when $r = 10$.

53. Find the two points on the circle $x^2 + y^2 = 1$ at which the slope of the tangent line is -2 (Fig. 3.4.3).

54. Find the two points on the circle $x^2 + y^2 = 1$ at which the slope of the tangent line is 3.

55. Find a line through the point $P(18, 0)$ that is normal to the tangent line to the parabola $y = x^2$ at some point $Q(a, a^2)$ (Fig. 3.4.4). [*Suggestion:* You will obtain a cubic equation in the unknown a. Find by inspection a small integral root r. The cubic polynomial is then the product of $a - r$ and a quadratic polynomial; you can find the quadratic by dividing $a - r$ into the cubic.]

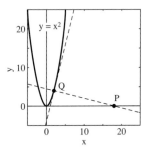

Fig. 3.4.4 The tangent and normal of Problem 55

56. Find three distinct lines through the point $P(3, 10)$ that are normal to the parabola $y = x^2$ (Fig. 3.4.5). [See the suggestion for Problem 55. This problem will require a fair amount of calculator-aided computation.]

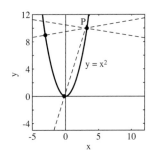

Fig. 3.4.5 The three normal lines of Problem 56

57. Find two distinct lines through the point $P(0, 2.5)$ that are normal to the curve $y = x^{2/3}$ (Fig. 3.4.6).

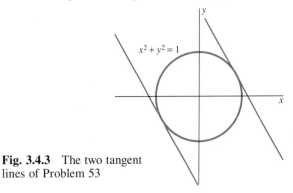

Fig. 3.4.3 The two tangent lines of Problem 53

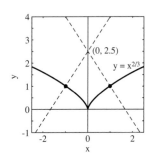

Fig. 3.4.6 The two normal lines of Problem 57

58. Verify that the tangent line at a point P of the curve $x^2 + y^2 = a^2$ is perpendicular to the radial line segment OP (Fig. 3.4.7).

59. Consider the cubic equation $x^3 = 3x + 8$. If we differentiate each side with respect to x, we obtain $3x^2 = 3$, which has the two solutions $x = 1$ and $x = -1$. But neither of these is a solution of the original equation. What went wrong? After all, in several examples and theorems of this section, we *appeared* to differentiate both sides of an equation. Explain carefully why differentiation of both sides in Eq. (3) is valid and why the differentiation in this problem is not.

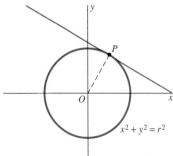

Fig. 3.4.7 The circle, radius, and tangent line of Problem 58

3.5

Maxima and Minima of Functions on Closed Intervals

In applications we often need to find the maximum (largest) or minimum (smallest) value that a specified quantity can attain. The animal pen problem posed in Section 1.1 is a simple yet typical example of an applied maximum-minimum problem. There we investigated the animal pen shown in Fig. 3.5.1, with the indicated dollar-per-foot cost figures for its four sides. We showed that if \$100 is allocated for materials to construct this pen, then its area $A = f(x)$ is given as a function of its base x by

$$f(x) = \tfrac{3}{5} x(30 - x), \qquad 0 \leq x \leq 30. \tag{1}$$

Hence the question of the largest possible area of the animal pen is equivalent to the purely mathematical problem of finding the maximum value attained by the function $f(x) = \tfrac{3}{5} x(30 - x)$ on the closed interval $[0, 30]$.

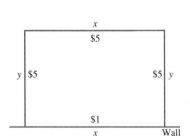

Fig. 3.5.1 The animal pen

Definition *Minimum and Maximum Values*

If c is in the closed interval $[a, b]$, then $f(c)$ is called the **minimum value** of $f(x)$ on $[a, b]$ if $f(c) \leq f(x)$ for all x in $[a, b]$. Similarly, if d is in $[a, b]$, then $f(d)$ is called the **maximum value** of $f(x)$ on $[a, b]$ if $f(d) \geq f(x)$ for all x in $[a, b]$.

Thus if $f(c)$ is the minimum value and $f(d)$ the maximum value of $f(x)$ on $[a, b]$, then

$$f(c) \leq f(x) \leq f(d) \tag{2}$$

for all x in $[a, b]$, and hence $f(x)$ attains no value smaller than $f(c)$ or larger than $f(d)$. In geometric terms, $(c, f(c))$ is a *low point* and $(d, f(d))$ is a *high point* on the curve $y = f(x)$, $a \leq x \leq b$, as illustrated in Figs. 3.5.2 and 3.5.3.

Theorem 1 (proved in Appendix C) says that a continuous function f on a closed interval $[a, b]$ attains a minimum value $f(c)$ and a maximum value $f(d)$, so the inequalities in (2) hold: The curve $y = f(x)$ over $[a, b]$ has both a lowest point and a highest point.

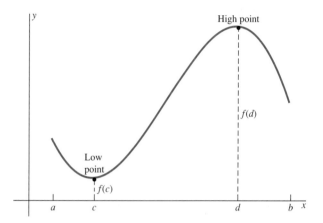

Fig. 3.5.2 $f(c)$ is the minimum value and $f(d)$ is the maximum value of $f(x)$ on $[a, b]$.

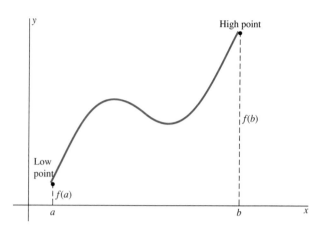

Fig. 3.5.3 Maximum and minimum values can occur at the endpoints of an interval. Here $f(a)$ is the minimum value and $f(b)$ is the maximum value of $f(x)$ on $[a, b]$.

Theorem 1   *Minimum and Maximum Value Property*

If the function f is continuous on the closed interval $[a, b]$, then there exist numbers c and d in $[a, b]$ such that $f(c)$ is the minimum value, and $f(d)$ is the maximum value, of f on $[a, b]$.

In short, a continuous function on a closed and bounded interval attains both a minimum value and a maximum value at points of the interval. Hence we see it is the *continuity* of the function

$$f(x) = \tfrac{3}{5}x(30 - x)$$

on the *closed* interval $[0, 30]$ that guarantees that the maximum value of f exists and is attained at some point of the interval $[0, 30]$.

Suppose that the function f is defined on the interval I. Examples 1 and 2 show that if *either f is not continuous or I is not closed*, then f may fail to attain maximum and minimum values at points of I. Thus both hypotheses in Theorem 1 are necessary.

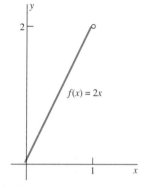

Fig. 3.5.4 The graph of the function of Example 1

EXAMPLE 1   Let the continuous function $f(x) = 2x$ be defined only for $0 \leq x < 1$, so that its domain of definition is a half-open interval rather than a closed interval. From the graph shown in Fig. 3.5.4, it is clear that f attains its minimum value 0 at $x = 0$. But $f(x) = 2x$ attains *no* maximum value at any point of $[0, 1)$. The only possible candidate for a maximum value would be the value 2 at $x = 1$, but $f(1)$ is not defined.

EXAMPLE 2   The function f defined on the closed interval $[0, 1]$ with the formula

$$f(x) = \begin{cases} \dfrac{1}{x} & \text{if } 0 < x \leq 1 \\ 1 & \text{if } x = 0 \end{cases}$$

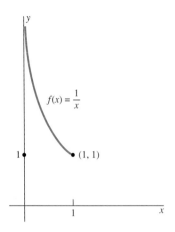

Fig. 3.5.5 The graph of the function of Example 2

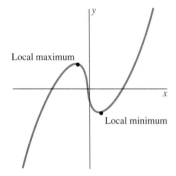

Fig. 3.5.6 Local extrema

is not continuous on [0, 1], because $\lim_{x \to 0^+} (1/x)$ does not exist (Fig. 3.5.5). This function does attain its minimum value of 1 at $x = 0$ and also at $x = 1$. But it attains no maximum value on [0, 1], because $1/x$ can be made arbitrarily large by choosing x positive and very close to zero.

For a variation on Example 2, the function $g(x) = 1/x$ with domain the *open* interval (0, 1) attains neither a minimum nor a maximum value there.

LOCAL MAXIMA AND MINIMA

Once we know that the continuous function f *does* attain minimum and maximum values on the closed interval [a, b], the remaining question is this: Exactly *where* are these values located? We solved the animal pen problem in Section 2.1 on the basis of the following assumption, motivated by geometry: The function $f(x) = \frac{3}{5}x(30 - x)$ attains its maximum value on [0, 30] at an interior point of that interval, a point at which the tangent line is horizontal. Theorems 2 and 3 of this section provide a rigorous basis for the method we used there.

We say that the value $f(c)$ is a **local maximum value** of the function f if $f(x) \leqq f(c)$ for all x sufficiently near c. More precisely, if this inequality holds for all x that are simultaneously in the domain of f and in some open interval containing c, then $f(c)$ is a local maximum of f. Similarly, we say that the value $f(c)$ is a **local minimum value** of f if $f(x) \geqq f(c)$ for all x sufficiently near c.

As Fig. 3.5.6 shows, a local maximum is a point such that no nearby points on the graph are higher, and a local minimum is one such that no nearby points on the graph are lower. A **local extremum** of f is a value of f that is either a local maximum or a local minimum.

> **Theorem 2** *Local Maxima and Minima*
>
> If f is differentiable at c and is defined on an open interval containing c and if $f(c)$ is either a local maximum value or a local minimum value of f, then $f'(c) = 0$.

Thus a local extremum of a *differentiable* function on an *open* interval can occur only at a point where the derivative is zero and, therefore, where the line tangent to the graph is horizontal.

Proof of Theorem 2 Suppose, for instance, that $f(c)$ is a local maximum value of f. The assumption that $f'(c)$ exists means that that the right-hand and left-hand limits

$$\lim_{h \to 0^+} \frac{f(c + h) - f(c)}{h} \quad \text{and} \quad \lim_{h \to 0^-} \frac{f(c + h) - f(c)}{h}$$

both exist and are equal to $f'(c)$.
If $h > 0$, then

$$\frac{f(c + h) - f(c)}{h} \leqq 0,$$

because $f(c) \geqq f(c + h)$ for all small positive values of h. Hence, by a one-sided version of the squeeze law for limits (in Section 2.2), this inequality is preserved when we take the limit as $h \to 0$. We thus find that

$$f'(c) = \lim_{h \to 0^+} \frac{f(c + h) - f(c)}{h} \leqq \lim_{h \to 0^+} 0 = 0.$$

Similarly, in the case $h < 0$, we find that

$$\frac{f(c + h) - f(c)}{h} \geqq 0.$$

Therefore,

$$f'(c) = \lim_{h \to 0^-} \frac{f(c + h) - f(c)}{h} \geqq \lim_{h \to 0^-} 0 = 0.$$

Because both $f'(c) \leqq 0$ and $f'(c) \geqq 0$, we conclude that $f'(c) = 0$. This establishes Theorem 2. □

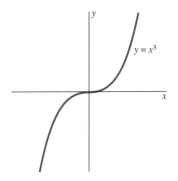

Fig. 3.5.7 No extremum at $x = 0$ even though the derivative is zero there

BEWARE   The converse of Theorem 3 is false. That is, the fact that $f'(c) = 0$ is *not enough* to imply that $f(c)$ is a local extremum. For example, consider the function $f(x) = x^3$. Its derivative $f'(x) = 3x^2$ is zero at $x = 0$. But a glance at its graph (Fig. 3.5.7) shows us that $f(0)$ is *not* a local extremum of f.

Thus the equation $f'(c) = 0$ is a *necessary* condition for $f(c)$ to be a local maximum or minimum value for a function f that is differentiable on an open interval. It is not a *sufficient* condition. The reason: $f'(x)$ may well be zero at points other than local maxima and minima. We give sufficient conditions for local maxima and minima in Chapter 4.

ABSOLUTE MAXIMA AND MINIMA

In most types of optimization problems, we are less interested in the local extrema (as such) than in the *absolute,* or *global,* maximum and minimum values attained by a given continuous function. If f is a function with domain D, we call $f(c)$ the **absolute maximum value,** or the **global maximum value,** of f on D provided that $f(c) \geqq f(x)$ for *all* x in D. Briefly, $f(c)$ is the largest value of f on D. It should be clear how the global minimum of f is to be defined. Figure 3.5.8 illustrates some local and global extrema. On the one hand, every global extremum is, of course, local as well. On the other hand, the graph shows local extrema that are not global.

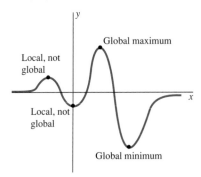

Fig. 3.5.8   Some extrema are global; others are merely local.

Theorem 3 tells us that the absolute maximum and absolute minimum values of the continuous function f on the closed interval $[a, b]$ occur either at one of the endpoints a or b or at a *critical point* of f. The number c in the domain of f is called a **critical point** of f if either

- $f'(c) = 0$, or
- $f'(c)$ does not exist.

> **Theorem 3   Absolute Maxima and Minima**
> Suppose that $f(c)$ is the absolute maximum (or absolute minimum) value of the continuous function f on the closed interval $[a, b]$. Then c is either a critical point of f or one of the endpoints a and b.

Proof   This result follows almost immediately from Theorem 2. If c is not an endpoint of $[a, b]$, then $f(c)$ is a local extremum of f on the open interval (a, b). In this case Theorem 2 implies that $f'(c) = 0$, provided that f is differentiable at c.   ❑

As a consequence of Theorem 3, we can find the (absolute) maximum and minimum values of the function f on the closed interval $[a, b]$ as follows:

1. *Locate* the critical points of f: those points where $f'(x) = 0$ and those points where $f'(x)$ does not exist.
2. *List* the values of x that yield *possible* extrema of f: the two endpoints a and b and those critical points that lie in $[a, b]$.
3. *Evaluate* $f(x)$ at each point in this list of possible locations of extrema.
4. *Inspect* these values of $f(x)$ to see which is the smallest and which is the largest.

The largest of the values in Step 4 is the absolute maximum value of f; the smallest, the absolute minimum. We call this procedure the **closed-interval maximum-minimum method**.

EXAMPLE 3   For our final discussion of the animal pen problem, let us apply the closed-interval maximum-minimum method to find the maximum and minimum values of the differentiable function

$$f(x) = \tfrac{3}{5}x(30 - x) = \tfrac{3}{5}(30x - x^2)$$

on the closed interval $[0, 30]$.

Solution   The derivative of f is

$$f'(x) = \tfrac{3}{5}(30 - 2x),$$

which is zero only at the point $x = 15$ in $[0, 30]$. Including the two endpoints, our list of the only values of x that can yield extrema of f consists of 0, 15, and 30. We evaluate the function at each:

$$f(0) = 0, \quad \leftarrow \quad \text{absolute minimum}$$
$$f(15) = 135, \quad \leftarrow \quad \text{absolute maximum}$$
$$f(30) = 0. \quad \leftarrow \quad \text{absolute minimum}$$

Thus the maximum value of $f(x)$ on $[0, 30]$ is 135 (attained at $x = 15$), and the minimum value is 0 (attained both at $x = 0$ and at $x = 30$).

EXAMPLE 4 Find the maximum and minimum values of

$$f(x) = 2x^3 - 3x^2 - 12x + 15$$

on the closed interval $[0, 3]$.

Solution The derivative of f is

$$f'(x) = 6x^2 - 6x - 12 = 6(x - 2)(x + 1).$$

So the critical points of f are the solutions of the equation

$$6(x - 2)(x + 1) = 0$$

and the numbers c for which $f'(c)$ does not exist. There are none of the latter, so the critical points of f occur at $x = -1$ and at $x = 2$. The first of these is not in the domain of f; we discard it, and thus the only critical point of f in $[0, 3]$ is $x = 2$. Including the two endpoints, our list of values of x that yield a possible maximum or minimum value of f consists of 0, 2, and 3. We evaluate the function f at each:

$$f(0) = 15, \quad \leftarrow \quad \text{absolute maximum}$$

$$f(2) = -5, \quad \leftarrow \quad \text{absolute minimum}$$

$$f(3) = 6.$$

Therefore, the maximum value of f on $[0, 3]$ is $f(0) = 15$, and the minimum value is $f(2) = -5$.

If in Example 4 we had asked for the maximum and minimum values of $f(x)$ on the interval $[-2, 3]$ (instead of the interval $[0, 3]$), then we would have included *both* critical points $x = -1$ and $x = 2$ in our list of possibilities. The resulting values of f would have been

$$f(-2) = 11,$$

$$f(-1) = 22, \quad \leftarrow \quad \text{absolute maximum}$$

$$f(2) = -5, \quad \leftarrow \quad \text{absolute minimum}$$

$$f(3) = 6.$$

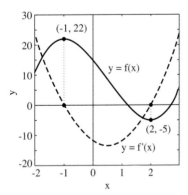

Fig. 3.5.9 The critical points of $f(x)$ are the zeros of $f'(x)$.

Figure 3.5.9 shows both the curve $y = f(x)$ and the graph of its derivative. Note the vertical line segments joining high and low points on $y = f(x)$ with x-intercepts of $dy/dx = f'(x)$. Thus the figure illustrates the following fact:

The critical points of a differentiable function $f(x)$ are the zeros of its derivative $f'(x)$.

On the basis of this principle, we can approximate a critical point of f graphically by "zooming in" on a zero of f'.

In Example 4 the function f was differentiable everywhere. Examples 5 and 6 illustrate the case of an extremum at a critical point where the function is not differentiable.

EXAMPLE 5 Find the maximum and minimum values of the function $f(x) = 3 - |x - 2|$ on the interval $[1, 4]$.

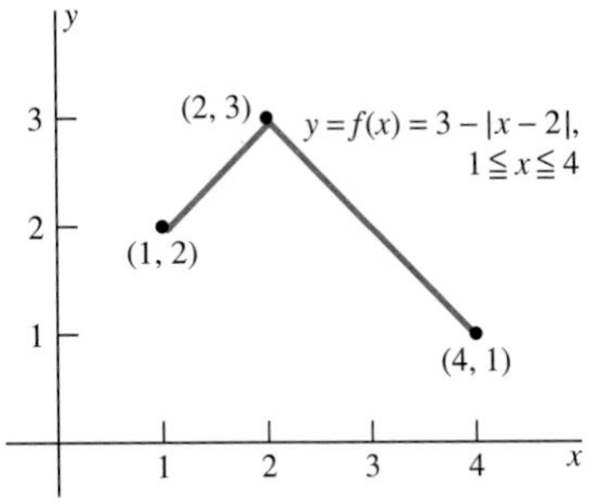

Solution   If $x \leqq 2$, then $x - 2 \leqq 0$, so

$$f(x) = 3 - (2 - x) = x + 1.$$

If $x \geqq 2$, then $x - 2 \geqq 0$, so

$$f(x) = 3 - (x - 2) = 5 - x.$$

Consequently, the graph of f looks like the one shown in Fig. 3.5.10. The only critical point of f in $[1, 4]$ is the point $x = 2$, because $f'(x)$ takes on only the two values $+1$ and -1 (and so is never zero), and $f'(2)$ does not exist. (Why not?) Evaluation of f at this critical point and at the two endpoints yields

$$f(1) = 2,$$
$$f(2) = 3, \quad \leftarrow \quad \text{absolute maximum}$$
$$f(4) = 1. \quad \leftarrow \quad \text{absolute minimum}$$

Fig. 3.5.10 Graph of the function of Example 5

EXAMPLE 6 Find the maximum and minimum values of

$$f(x) = 5x^{2/3} - x^{5/3}$$

on the closed interval $[-1, 4]$.

Solution   Differentiation of f yields

$$f'(x) = \frac{10}{3}x^{-1/3} - \frac{5}{3}x^{2/3} = \frac{5}{3}x^{-1/3}(2 - x) = \frac{5(2 - x)}{3x^{1/3}}.$$

Hence f has two critical points in the interval: $x = 2$, where $f'(x) = 0$, and $x = 0$, where f' does not exist [the graph of f has a vertical tangent line at $(0, 0)$]. When we evaluate f at these two critical points and at the two endpoints, we get

$$f(-1) = 6, \quad \leftarrow \quad \text{absolute maximum}$$
$$f(0) = 0, \quad \leftarrow \quad \text{absolute minimum}$$
$$f(2) = (5)(2^{2/3}) - 2^{5/3} \approx 4.76,$$
$$f(4) = (5)(4^{2/3}) - 4^{5/3} \approx 2.52.$$

Thus the maximum value $f(-1) = 6$ occurs at an endpoint. The minimum value $f(0) = 0$ occurs at a point where f is not differentiable.

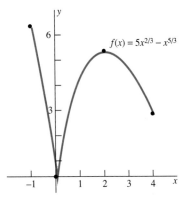

Fig. 3.5.11 Graph of the function of Example 6

By using a graphics calculator or computer with graphics capabilities, you can verify that the graph of the function f of Example 6 is that shown in Fig. 3.5.11. But in the usual case of a continuous function that has only finitely many critical points in a given closed interval, the closed-interval maximum-minimum method suffices to determine its maximum and minimum values without requiring any detailed knowledge of the graph of the function.

3.5 Problems

In Problems 1 through 10, state whether the given function attains a maximum value or a minimum value (or both) on the given interval. [Suggestion: Begin by sketching a graph of the function.]

1. $f(x) = 1 - x;$ $[-1, 1)$

2. $f(x) = 2x + 1;$ $[-1, 1)$

3. $f(x) = |x|;$ $(-1, 1)$

4. $f(x) = \dfrac{1}{\sqrt{x}};$ $(0, 1]$

5. $f(x) = |x - 2|;$ $(1, 4]$

6. $f(x) = 5 - x^2;$ $[-1, 2)$

7. $f(x) = x^3 + 1;$ $[-1, 1]$

8. $f(x) = \dfrac{1}{x^2 + 1};$ $(-\infty, \infty)$

9. $f(x) = \dfrac{1}{x(1 - x)};$ $[2, 3]$

10. $f(x) = \dfrac{1}{x(1 - x)};$ $(0, 1)$

In Problems 11 through 40, find the maximum and minimum values attained by the given function on the indicated closed interval.

11. $f(x) = 3x - 2;$ $[-2, 3]$

12. $g(x) = 4 - 3x;$ $[-1, 5]$

13. $h(x) = 4 - x^2;$ $[1, 3]$

14. $f(x) = x^2 + 3;$ $[0, 5]$

15. $g(x) = (x - 1)^2;$ $[-1, 4]$

16. $h(x) = x^2 + 4x + 7;$ $[-3, 0]$

17. $f(x) = x^3 - 3x;$ $[-2, 4]$

18. $g(x) = 2x^3 - 9x^2 + 12x;$ $[0, 4]$

19. $h(x) = x + \dfrac{4}{x};$ $[1, 4]$

20. $f(x) = x^2 + \dfrac{16}{x};$ $[1, 3]$

21. $f(x) = 3 - 2x;$ $[-1, 1]$

22. $f(x) = x^2 - 4x + 3;$ $[0, 2]$

23. $f(x) = 5 - 12x - 9x^2;$ $[-1, 1]$

24. $f(x) = 2x^2 - 4x + 7;$ $[0, 2]$

25. $f(x) = x^3 - 3x^2 - 9x + 5;$ $[-2, 4]$

26. $f(x) = x^3 + x;$ $[-1, 2]$

27. $f(x) = 3x^5 - 5x^3;$ $[-2, 2]$

28. $f(x) = |2x - 3|;$ $[1, 2]$

29. $f(x) = 5 + |7 - 3x|;$ $[1, 5]$

30. $f(x) = |x + 1| + |x - 1|;$ $[-2, 2]$

31. $f(x) = 50x^3 - 105x^2 + 72x;$ $[0, 1]$

32. $f(x) = 2x + \dfrac{1}{2x};$ $[1, 4]$

33. $f(x) = \dfrac{x}{x + 1};$ $[0, 3]$

34. $f(x) = \dfrac{x}{x^2 + 1};$ $[0, 3]$

35. $f(x) = \dfrac{1 - x}{x^2 + 3};$ $[-2, 5]$

36. $f(x) = 2 - \sqrt[3]{x};$ $[-1, 8]$

37. $f(x) = x\sqrt{1 - x^2};$ $[-1, 1]$

38. $f(x) = x\sqrt{4 - x^2};$ $[0, 2]$

39. $f(x) = x(2 - x)^{1/3};$ $[1, 3]$

40. $f(x) = x^{1/2} - x^{3/2};$ $[0, 4]$

41. Suppose that $f(x) = Ax + B$ is a linear function. Explain why the maximum and minimum values of f on a closed interval $[a, b]$ must occur at the endpoints of the interval.

42. Suppose that f is continuous on $[a, b]$ and differentiable on (a, b) and that $f'(x)$ is never zero at any point of (a, b). Explain why the maximum and minimum values of f must occur at the endpoints of the interval $[a, b]$.

43. Explain why every real number is a critical point of the greatest integer function $f(x) = [\![x]\!]$.

44. Prove that every quadratic function

$$f(x) = ax^2 + b + c \qquad (a \neq 0)$$

has exactly one critical point on the real line.

45. Explain why the cubic polynomial function

$$f(x) = ax^3 + bx^2 + cx + d \quad (a \neq 0)$$

can have either two, one, or no critical points on the real line. Produce examples that illustrate each of the three cases.

46. Define $f(x)$ to be the distance from x to the nearest integer. What are the critical points of f?

In Problems 47 through 52, match the given graph of the function f with the graph of its derivative f' from those in Fig. 3.5.12, parts (a) through (f).

47. Fig. 3.5.13

48. Fig. 3.5.14

49. Fig. 3.5.15

50. Fig. 3.5.16

51. Fig. 3.5.17

52. Fig. 3.5.18

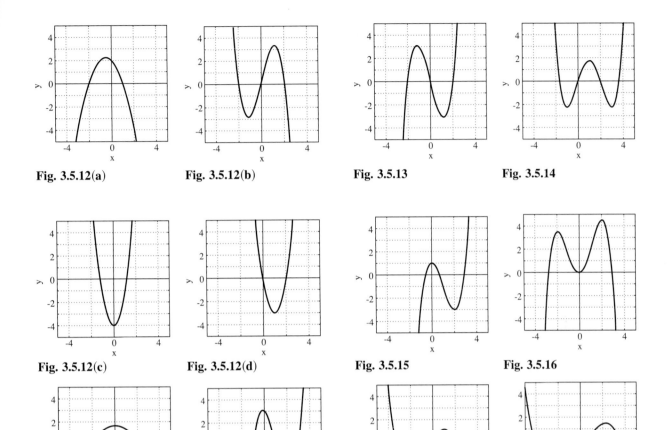

Fig. 3.5.12(a)   Fig. 3.5.12(b)   Fig. 3.5.13   Fig. 3.5.14

Fig. 3.5.12(c)   Fig. 3.5.12(d)   Fig. 3.5.15   Fig. 3.5.16

Fig. 3.5.12(e)   Fig. 3.5.12(f)   Fig. 3.5.17   Fig. 3.5.18

3.5 Project

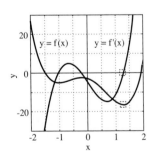

Fig. 3.5.19 The graphs of $y = f(x)$ and $y = f'(x)$

This project requires a graphics calculator or computer with a graphing utility.

Figure 3.5.19 shows the graphs of the function

$$f(x) = 4x^4 - 11x^2 - 5x - 3$$

and its derivative

$$f'(x) = 16x^3 - 22x - 5$$

on the interval $[-2, 2]$. The maximum value of $f(x)$ on $[-2, 2]$ is $f(-2) = 27$ at the left endpoint. The lowest point on the curve $y = f(x)$ and the corresponding zero of the derivative $dy/dx = f'(x)$ lie within the small boxes.

If we attempt to zoom in on the lowest point—without changing the "range factors" or "aspect ratio" of the viewing windows—we get the picture

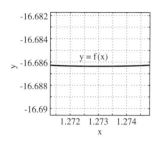

Fig. 3.5.20 Zooming in on the minimum of Fig. 3.5.19

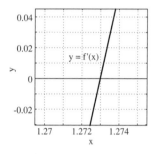

Fig. 3.5.21 Zooming in instead on the zero of $f'(x)$ of Fig. 3.5.19

in Fig. 3.5.20, where it is difficult to locate the lowest point precisely. The reason is this: After sufficient magnification, the graph is indistinguishable from its tangent line, which is horizontal at a local maximum or minimum point.

Consequently, it is much better to zoom in on the corresponding zero of the derivative $f'(x)$. We can then locate the indicated critical point with much greater precision (Fig. 3.5.21). Here it is clear that the minimum value attained by $f(x)$ on $[-2, 2]$ is approximately $f(1.273) \approx -16.686$.

In Problems 1 through 8, find the maximum and minimum values of the given function on the indicated closed interval by zooming in on the zeros of the derivative.

1. $f(x) = x^3 + 3x^2 - 7x + 10;$ $[-2, 2]$
2. $f(x) = x^3 + 3x^2 - 7x + 10;$ $[-4, 2]$
3. $f(x) = x^4 - 3x^3 + 7x - 5;$ $[-3, 3]$
4. $f(x) = x^4 - 5x^3 + 17x - 5;$ $[-3, 3]$
5. $f(x) = x^4 - 5x^3 + 17x - 5;$ $[0, 2]$
6. $f(x) = x^5 - 5x^4 - 15x^3 + 17x^2 + 23x;$ $[-1, 1]$
7. $f(x) = x^5 - 5x^4 - 15x^3 + 17x^2 + 23x;$ $[-3, 3]$
8. $f(x) = x^5 - 5x^4 - 15x^3 + 17x^2 + 23x;$ $[0, 10]$

3.6

Applied Maximum-Minimum Problems

This section is devoted to applied maximum-minimum problems (like the animal pen problem of Section 1.1) for which the closed-interval maximum-minimum method of Section 3.5 can be used. When we confront such a problem, there is an important first step: We must determine the quantity to be maximized or minimized. This quantity will be the dependent variable in our analysis of the problem.

This dependent variable must then be expressed as a function of an independent variable, one that "controls" the values of the dependent variable. If the domain of values of the independent variable—those that are pertinent to the applied problem—is a closed interval, then we may proceed with the closed-interval maximum-minimum method. This plan of attack can be summarized in the following steps:

1. *Find the quantity to be maximized or minimized.* This quantity, which you should describe with a word or short phrase and label with a letter, will be the dependent variable. Because it is a *dependent* variable, it depends on something else; that something will be the independent variable. We call the independent variable x here.

2. *Express the dependent variable as a function of the independent variable.* Use the information in the problem to write the dependent variable as a function of x. Always draw a figure and *label the variables;* this is generally the best way to find the relationship between the dependent and independent variables. Use auxiliary variables if they help, but do not use too many, for you must eventually eliminate them. You *must* express the dependent variable as a function of the *single* independent variable x and various constants before you can compute any derivatives. Find the domain of this function as well as its formula. Force the domain to be a closed and bounded interval if possible—if the natural domain is an open interval, adjoin the endpoints if you can.

3. *Apply calculus to find the critical points.* Compute the derivative f' of the function f that you found in Step 2. Use the derivative to find the critical points—where $f'(x) = 0$ and where $f'(x)$ does not exist. If f is differentiable everywhere, then its only critical points occur where

$$f'(x) = 0.$$

4. *Identify the extrema.* Evaluate f at each critical point in its domain *and* at the two endpoints. The values you obtain will tell you which is the absolute maximum and which is the absolute minimum. Of course, either or both of these may occur at more than one point.

5. *Answer the question posed in the problem.* In other words, interpret your results. The answer to the original problem may be something other than merely the largest (or smallest) value of f. Give a precise answer to the specific question originally asked.

Observe how we follow this five-step process in Example 1.

EXAMPLE 1   A farmer has 200 yd of fence with which to construct three sides of a rectangular pen; an existing long, straight wall will form the fourth side. What dimensions will maximize the area of the pen?

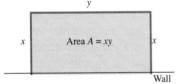

Fig. 3.6.1 The rectangular pen of Example 1

Solution   We want to maximize the area A of the pen shown in Fig. 3.6.1. To get a formula for the *dependent* variable A, we observe that the area of a rectangle is the product of its base and its height. So we let x denote the length of each of the two sides of the pen perpendicular to the wall. We also let y denote the length of the side parallel to the wall. Then the area of the rectangle is given by the *formula*

$$A = xy.$$

Now we need to write A as a *function* of either x or y. Because all 200 yd of fence are to be used,

$$2x + y = 200, \quad \text{so} \quad y = 200 - 2x. \tag{1}$$

(We choose to express y in terms of x merely because the algebra is slightly simpler.) Next, we substitute this value of y into the formula $A = xy$ to obtain

$$A(x) = x(200 - 2x) = 200x - 2x^2. \tag{2}$$

This equation expresses the dependent variable A as a function of the independent variable x.

Before proceeding, we must find the domain of the function A. It is clear from Fig. 3.6.2 that $0 < x < 100$. But to apply the closed-interval maximum-minimum method, we need a closed interval. In this example, we may adjoin the endpoints to $(0, 100)$ to get the *closed* interval $[0, 100]$. The values $x = 0$ and $x = 100$ correspond to "degenerate" pens of area zero. Because zero cannot be the maximum value of A, there is no harm in thus enlarging the domain of the function A.

Fig. 3.6.2 The relation in Eq. (1) between x and y (Example 1)

Now we compute the derivative of the function A in Eq. (2):

$$\frac{dA}{dx} = 200 - 4x.$$

Because A is differentiable, its only critical points occur when

$$\frac{dA}{dx} = 0,$$

that is, when

$$200 - 4x = 0.$$

So $x = 50$ is the only interior critical point. Including the endpoints, the extrema of A can occur only at $x = 0$, 50, and 100. We evaluate A at each:

$$A(0) = 0,$$

$$A(50) = 5000, \quad \leftarrow \quad \text{absolute maximum}$$

$$A(100) = 0.$$

Thus the maximal area is $A(50) = 5000$ yd². From Eq. (1) we find that $y = 100$ when $x = 50$. Therefore, for the pen to have maximal area, each of the two sides perpendicular to the wall should be 50 yd long, and the side parallel to the wall should be 100 yd long (Fig. 3.6.3).

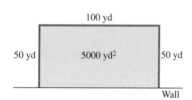

100 yd

50 yd | 5000 yd² | 50 yd

Wall

Fig. 3.6.3 The pen with maximal area of Example 1

EXAMPLE 2 A piece of sheet metal is rectangular, 5 ft wide and 8 ft long. Congruent squares are to be cut from its four corners. The resulting piece of metal is to be folded and welded to form an open-topped box (Fig. 3.6.4). How should this be done to get a box of largest possible volume?

Solution The quantity to be maximized—the dependent variable—is the volume V of the box to be constructed. The shape and thus the volume of the box are determined by the length x of the edge of each corner square removed. Hence x is a natural choice for the independent variable.

To write the volume V as a function of x, note that the finished box will have height x, and its base will measure $8 - 2x$ ft by $5 - 2x$ ft. Hence its volume is given by

$$V(x) = x(5 - 2x)(8 - 2x) = 4x^3 - 26x^2 + 40x.$$

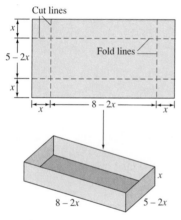

Cut lines

Fold lines

x | $8 - 2x$ | x

$8 - 2x$ $5 - 2x$ x

Fig. 3.6.4 Making the box of Example 2

x x

5

Fig. 3.6.5 The 5-ft width of the metal sheet (Example 2)

The procedure described in this example will produce an actual box only if $0 < x < 2.5$ (Fig. 3.6.5). But we make the domain the *closed* interval [0, 2.5] to ensure that a maximum of $V(x)$ exists and to use the closed-interval maximum-minimum method. The values $x = 0$ and $x = 2.5$ correspond to "degenerate" boxes of zero volume, so adjoining these points to (0, 2.5) will affect neither the location of the absolute maximum nor its value.

Now we compute the derivative of V:

$$V'(x) = 12x^2 - 52x + 40 = 4(3x - 10)(x - 1).$$

The only critical points of the differentiable function V occur where

$$V'(x) = 0,$$

that is, where

$$4(3x - 10)(x - 1) = 0.$$

The solutions of this equation are $x = 1$ and $x = \frac{10}{3}$. We discard the latter because it does not lie in the domain [0, 2.5] of V. So we examine these values of V:

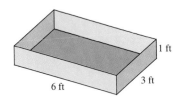

Fig. 3.6.6 The box with maximal volume of Example 2

$$V(0) = 0,$$

$$V(1) = 18, \quad \leftarrow \quad \text{absolute maximum}$$

$$V(2.5) = 0.$$

Thus the maximum value of $V(x)$ on $[0, 2.5]$ is $V(1) = 18$. The answer to the question posed is this: The squares cut from the corners should be of edge length 1 ft each. The resulting box will measure 6 ft by 3 ft by 1 ft, and its volume will be 18 ft^3 (Fig. 3.6.6).

For our next application of the closed-interval maximum-minimum method, let us consider a typical problem in business management. Suppose that x units of computer diskettes are to be manufactured at a total cost of $C(x)$ dollars. We make the simple (but not always valid) assumption that the cost function $C(x)$ is the sum of two terms:

❑ A constant term a representing the fixed cost of acquiring and maintaining production facilities (overhead), and

❑ A variable term representing the additional cost of making x units at, for example, b dollars each.

Then $C(x)$ is given by

$$C(x) = a + bx. \tag{3}$$

We assume also that the number of units that can be sold is a linear function of the selling price p, so that $x = m - np$, where m and n are positive constants. The minus sign indicates that an increase in selling price will result in a decrease in sales. If we solve this last equation for p, we get the so-called price function

$$p = p(x) = A - Bx \tag{4}$$

(A and B are also constants).

The quantity to be maximized is profit, given here by the profit function $P(x)$, which is equal to the sales revenue minus the production costs. Thus

$$P = P(x) = xp(x) - C(x). \tag{5}$$

EXAMPLE 3 Suppose that the cost of publishing a small book is $10,000 to set up the (annual) press run plus $8 for each book printed. The publisher sold 7000 copies last year at $13 each, but sales dropped to 5000 copies this year when the price was raised to $15 per copy. Assume that up to 10,000 copies can be printed in a single press run. How many copies should be printed, and what should be the selling price of each copy, to maximize the year's profit on this book?

Solution The dependent variable to be maximized is the profit P. As independent variable we choose the number x of copies to be printed; also, $0 \leq x \leq 10,000$. The given cost information then implies that

$$C(x) = 10,000 + 8x.$$

Now we substitute into Eq. (4) the data $x = 7000$ when $p = 13$ as well as the data $x = 5000$ when $p = 15$. We obtain the equations

$$A - 7000B = 13, \quad A - 5000B = 15.$$

When we solve these equations simultaneously, we find that $A = 20$ and $B = 0.001$. Hence the price function is

$$p = p(x) = 20 - \frac{x}{1000},$$

and thus the profit function is

$$P(x) = x\left(20 - \frac{x}{1000}\right) - (10{,}000 + 8x).$$

We expand and collect terms to obtain

$$P(x) = 12x - \frac{x^2}{1000} - 10{,}000, \qquad 0 \leqq x \leqq 10{,}000.$$

Now

$$\frac{dP}{dx} = 12 - \frac{x}{500},$$

and the only critical points of the differentiable function P occur when

$$\frac{dP}{dx} = 0,$$

that is, when

$$12 - \frac{x}{500} = 0; \qquad x = 12 \cdot 500 = 6000.$$

We check P at this value of x as well as the values of $P(x)$ at the endpoints to find the maximum profit:

$$P(0) = -10{,}000,$$
$$P(6000) = 26{,}000, \quad \leftarrow \quad \text{absolute maximum}$$
$$P(10{,}000) = 10{,}000.$$

Therefore, the maximum possible annual profit of \$26,000 results from printing 6000 copies of the book. Each copy should be sold for \$14, because

$$p = 20 - \frac{6000}{1000} = 14.$$

EXAMPLE 4 We need to design a cylindrical can with radius r and height h. The top and bottom must be made of copper, which will cost 2¢/in.² The curved side is to be made of aluminum, which will cost 1¢/in.² We seek the dimensions that will maximize the volume of the can. The only constraint is that the total cost of the can is to be 300π ¢.

Solution We need to maximize the volume V of the can, which we can compute if we know its radius r and its height h (Fig. 3.6.7). With these dimensions, we find that

$$V = \pi r^2 h, \tag{6}$$

but we need to express V as a function of r alone (or as a function of h alone).

Both the circular top and bottom of the can have area πr^2 in.², so the area of copper to be used is $2\pi r^2$, and its cost is $4\pi r^2$ cents. The area of the

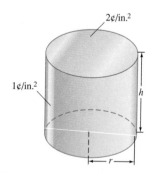

Fig. 3.6.7 The cylindrical can of Example 4

144

curved side of the can is $2\pi rh$ in.², so the area of aluminum used is the same, and the aluminum costs $2\pi rh$ cents.

We obtain the total cost of the can by adding the cost of the copper to the cost of the aluminum. This sum must be 300π ¢, and therefore

$$4\pi r^2 + 2\pi rh = 300\pi. \tag{7}$$

We eliminate h in Eq. (6) by solving Eq. (7) for h:

$$h = \frac{300\pi - 4\pi r^2}{2\pi r} = \frac{1}{r}(150 - 2r^2). \tag{8}$$

Hence

$$V = V(r) = (\pi r^2)\frac{1}{r}(150 - 2r^2) = 2\pi(75r - r^3). \tag{9}$$

To determine the domain of definition of V, we note from Eq. (7) that $4\pi r^2 < 300\pi$, so $r < \sqrt{75}$ for the desired can; with $r = \sqrt{75} = 5\sqrt{3}$, we get a degenerate can with height $h = 0$. With $r = 0$, we obtain *no* value of h in Eq. (8) and therefore no can, but $V(r)$ is nevertheless continuous at $r = 0$. Consequently, we can take the closed interval $[0, 5\sqrt{3}]$ to be the domain of V.

Calculating the derivative yields

$$V'(r) = 2\pi(75 - 3r^2) = 6\pi(25 - r^2).$$

Because $V(r)$ is a polynomial, $V'(r)$ exists for all values of r, so we obtain all critical points by solving the equation

$$V'(r) = 0;$$

that is,

$$6\pi(25 - r^2) = 0.$$

We discard the solution -5, as it does not lie in the domain of V. Thus we obtain only the single critical point $r = 5$ in $[0, 5\sqrt{3}]$. Now

$$V(0) = 0,$$
$$V(5) = 500\pi, \quad \leftarrow \quad \text{absolute maximum}$$
$$V(5\sqrt{3}) = 0.$$

Thus the can of maximum volume has radius $r = 5$ in., and Eq. (8) yields its height to be $h = 20$ in. Figure 3.6.8 shows such a can.

Fig. 3.6.8 The can of maximal volume in Example 4

EXAMPLE 5 *(A Sawmill Problem)*   Suppose that you need to cut a beam with maximal rectangular cross section from a circular log of radius 1 ft. (This is the geometric problem of finding the rectangle of greatest area that can be inscribed in a circle of radius 1.) What are the shape and cross-sectional area of such a beam?

Solution   Let x and y denote half the base and half the height, respectively, of the inscribed rectangle (Fig. 3.6.9). Apply the Pythagorean theorem to the small right triangle in the figure. This yields the equation

$$x^2 + y^2 = 1, \quad \text{so} \quad y = \sqrt{1 - x^2}.$$

The area of the inscribed rectangle is $A = (2x)(2y) = 4xy$. You may now express A as a function of x alone:

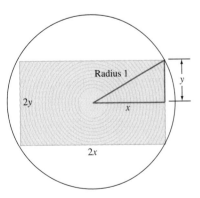

Fig. 3.6.9 A sawmill problem—Example 5

$$A(x) = 4x\sqrt{1 - x^2}.$$

The practical domain of definition of A is $(0, 1)$, and there is no harm (and much advantage) in adjoining the endpoints, so you take $[0, 1]$ to be the domain. Next,

$$\frac{dA}{dx} = 4 \cdot (1 - x^2)^{1/2} + 2x(1 - x^2)^{-1/2}(-2x) = \frac{4 - 8x^2}{(1 - x^2)^{1/2}}.$$

You observe that $A'(1)$ does not exist, but this causes no trouble, because differentiability at the endpoints is not assumed in Theorem 3 of Section 3.5. Hence you need only solve the equation

$$A'(x) = 0;$$

that is,

$$\frac{4 - 8x^2}{\sqrt{1 - x^2}} = 0.$$

A fraction can be zero only when its numerator is zero and its denominator is *not*, so $A'(x) = 0$ when $4 - 8x^2 = 0$. Thus you find the only critical point of A in the open interval $(0, 1)$ to be $x = \frac{1}{2}\sqrt{2}$ (and $2x = 2y = \sqrt{2}$). You evaluate A here and at the two endpoints to find that

$$A(0) = 0,$$
$$A(\tfrac{1}{2}\sqrt{2}) = 2, \quad \leftarrow \quad \text{absolute maximum}$$
$$A(1) = 0.$$

Therefore, the beam with rectangular cross section of maximal area is square, with edges $\sqrt{2}$ ft long and with cross-sectional area 2 ft².

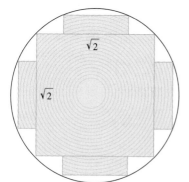

Fig. 3.6.10 Cut four more beams after cutting one large beam.

In Problem 43 we ask you to maximize the total cross-sectional area of the four planks that can be cut from the four pieces of log that remain after cutting the square beam (Fig. 3.6.10).

PLAUSIBILITY

You should always check your answers for plausibility. In Example 5, the cross-sectional area of the log from which the beam is to be cut is $\pi \approx 3.14$ ft². The beam of maximal cross-sectional area 2 ft² thus uses a little less than 64% of the log. This *is* plausible. Had the fraction been an extremely inefficient 3% or a wildly optimistic 99%, you should have searched for an error in arithmetic, algebra, calculus, or logic (as you would had the fraction been -15% or 150%). Check the results of Examples 1 through 4 for plausibility.

DIMENSIONS

Another way to check answers is to use *dimensional analysis*. Work the problem with unspecified constants in place of the actual numbers. In Example 5, it would be good practice to find the beam of maximal rectangular cross section that can be cut from a circular log of radius R rather than radius 1 ft.

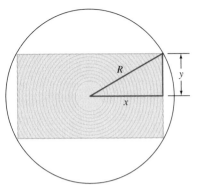

Fig. 3.6.11 The log with radius R

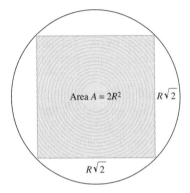

Area $A = 2R^2$

$R\sqrt{2}$

$R\sqrt{2}$

Fig. 3.6.12 The inscribed square beam with maximal cross-sectional area

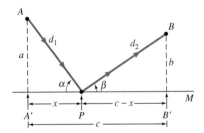

Fig. 3.6.13 Reflection at P of a light ray by a mirror M (Example 6)

You can always substitute the given value $R = 1$ at the conclusion of the problem. A brief solution to this problem might go as follows:

Dimensions of beam: base $2x$, height $2y$.

Area of beam: $A = 4xy$.

Draw a radius of the log from its center to one corner of the rectangular beam, as in Fig. 3.6.11. This radius has length R, so the Pythagorean theorem gives

$$x^2 + y^2 = R^2; \qquad y = \sqrt{R^2 - x^2}.$$

Area of beam:

$$A = A(x) = 4x\sqrt{R^2 - x^2}, \qquad 0 \leq x \leq R.$$

$$A'(x) = 4(R^2 - x^2)^{1/2} + 2x(R^2 - x^2)^{-1/2}(-2x) = \frac{4R^2 - 8x^2}{\sqrt{R^2 - x^2}}.$$

$A'(x)$ does not exist when $x = R$, but that's an endpoint; we'll check it separately.

$A'(x) = 0$ when $x = \frac{1}{2}R\sqrt{2}$ (ignore the negative root; it's not in the domain of A).

$$A(0) = 0,$$

$$A\left(\tfrac{1}{2}R\sqrt{2}\right) = 2R^2, \quad \leftarrow \quad \text{absolute maximum}$$

$$A(R) = 0.$$

Figure 3.6.12 shows the dimensions of the inscribed rectangle of maximal area.

Now you can check the results for dimensional accuracy. The value of x that maximizes A is a length (R) multiplied by a pure numerical constant ($\frac{1}{2}\sqrt{2}$), so x has the dimensions of length—that's correct; had it been anything else, you would need to search for the error. Moreover, the maximum cross-sectional area of the beam is $2R^2$, the product of a pure (dimensionless) number and the square of a length, thus having the dimensions of area. This, too, is correct.

EXAMPLE 6 We consider the reflection of a ray of light by a mirror M as in Fig. 3.6.13, which shows a ray traveling from point A to point B via reflection off M at the point P. We assume that the location of the point of reflection is such that the total distance $d_1 + d_2$ traveled by the light ray will be minimized. This is an application of *Fermat's principle of least time* for the propagation of light. The problem is to find P.

Solution Drop perpendiculars from A and B to the plane of the mirror M. Denote the feet of these perpendiculars by A' and B' (Fig. 3.6.13). Let a, b, c, and x denote the lengths of the segments AA', BB', $A'B'$, and $A'P$, respectively. Then $c - x$ is the length of the segment PB'. By the Pythagorean theorem, the distance to be minimized is then

$$d_1 + d_2 = f(x) = \sqrt{a^2 + x^2} + \sqrt{b^2 + (c - x)^2}. \qquad (10)$$

We may choose as the domain of f the interval $[0, c]$, because the minimum of f must occur somewhere within that interval. (To see why, examine the picture you get if x is *not* in that interval.)

Then

$$f'(x) = \frac{x}{\sqrt{a^2 + x^2}} + \frac{(c - x)(-1)}{\sqrt{b^2 + (c - x)^2}}. \qquad (11)$$

Because

$$f'(x) = \frac{x}{d_1} - \frac{c - x}{d_2}, \qquad (12)$$

we find that any horizontal tangent to the graph of f must occur over the point x determined by the equation

$$\frac{x}{d_1} = \frac{c - x}{d_2}. \qquad (13)$$

At such a point, $\cos \alpha = \cos \beta$, where α is the angle of the incident light ray and β is the angle of the reflected ray (Fig. 3.6.13). Both α and β lie between 0 and $\pi/2$, and thus we find that $\alpha = \beta$. In other words, the angle of incidence is equal to the angle of reflection, a familiar principle from physics.

The computation in Example 6 has an alternative interpretation that is interesting, if somewhat whimsical. Figure 3.6.14 shows a feedlot 200 ft long with a water trough along one edge and a feed bin located on an adjacent edge. A cow enters the gate at the point A, 90 ft from the water trough. She walks straight to point P, gets a drink from the trough, and then walks straight to the feed bin at point B, 60 ft from the trough. If the cow knew calculus, what point P along the water trough would she select to minimize the total distance she walks?

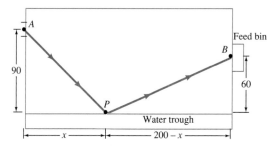

Fig. 3.6.14 The feedlot

In comparing Figs. 3.6.13 and 3.6.14, we see that the cow's problem is to minimize the distance function f in Eq. (10) with the numerical values $a = 90$, $b = 60$, and $c = 200$. When we substitute these values and

$$d_1 = \sqrt{a^2 + x^2} \quad \text{and} \quad d_2 = \sqrt{b^2 + (c - x)^2}$$

into Eq. (13), we get

$$\frac{x}{\sqrt{8100 + x^2}} = \frac{200 - x}{\sqrt{3600 + (200 - x)^2}}.$$

We square both sides, clear the equation of fractions, and simplify. The result is

$$x^2[3600 + (200 - x)^2] = (200 - x)^2(8100 + x^2);$$

$$3600x^2 = 8100\,(200 - x)^2; \quad \text{(Why?)}$$

$$60x = 90(200 - x);$$

$$150x = 18{,}000;$$

$$x = 120.$$

Thus the cow should proceed directly to the point P located 120 ft along the water trough.

These examples indicate that the closed-interval maximum-minimum method is applicable to a wide range of problems. Indeed, applied optimization problems that seem as different as light rays and cows may have essentially identical mathematical models. This is only one illustration of the power of generality that calculus exploits so effectively.

3.6 Problems

1. Find two positive real numbers x and y such that their sum is 50 and their product is as large as possible.

2. Find the maximum possible area of a rectangle of perimeter 200 m.

3. A rectangle with sides parallel to the coordinate axes has one vertex at the origin, one on the positive x-axis, one on the positive y-axis, and its fourth vertex in the first quadrant on the line with equation $2x + y = 100$ (Fig. 3.6.15). What is the maximum possible area of such a rectangle?

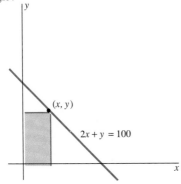

Fig. 3.6.15 The rectangle of Problem 3

4. A farmer has 600 m of fencing with which to enclose a rectangular pen adjacent to a long existing wall. He will use the wall for one side of the pen and the available fencing for the remaining three sides. What is the maximum area that can be enclosed in this way?

5. A rectangular box has a square base with edges at least 1 in. long. It has no top, and the total area of its five sides is 300 in.^{2} (Fig. 3.6.16). What is the maximum possible volume of such a box?

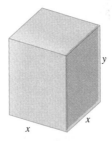

Fig. 3.6.16 A box with square base and volume $V = x^2 y$ (Problems 5, 17, and 20)

6. If x is in the interval $[0, 1]$, then $x - x^2$ is not negative. What is the maximum value that $x - x^2$ can have on that interval? In other words, what is the greatest amount by which a real number can exceed its square?

7. The sum of two positive numbers is 48. What is the smallest possible value of the sum of their squares?

8. A rectangle of fixed perimeter 36 is rotated about one of its sides, thus sweeping out a figure in the shape of a right circular cylinder (Fig. 3.6.17). What is the maximum possible volume of that cylinder?

Fig. 3.6.17 The rectangle and cylinder of Problem 8

9. The sum of two nonnegative numbers is 10. Find the minimum possible value of the sum of their cubes.

10. Suppose that the strength of a rectangular beam is proportional to the product of the width and the *square* of the height of its cross section. What shape beam should be cut from a cylindrical log of radius r to achieve the greatest possible strength?

11. A farmer has 600 yd of fencing with which to build a rectangular corral. Some of the fencing will be used to construct two internal divider fences, both parallel to the same two sides of the corral (Fig. 3.6.18). What is the maximum possible total area of such a corral?

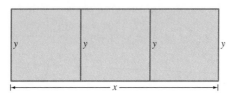

Fig. 3.6.18   The divided corral of Problem 11

12. Find the maximum possible volume of a right circular cylinder if its total surface area—including both circular ends—is 150π.

13. Find the maximum possible area of a rectangle with diagonals of length 16.

14. A rectangle has a line of fixed length L reaching from one vertex to the midpoint of one of the far sides (Fig. 3.6.19). What is the maximum possible area of such a rectangle?

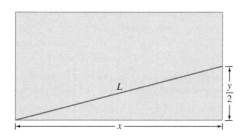

Fig. 3.6.19   The rectangle of Problem 14

15. The volume V (in cubic centimeters) of 1 kg of water at temperature T between 0°C and 30°C is very closely approximated by

$$V = 999.87 - (0.06426)T$$

$$+ (0.0085043)T^2 - (0.0000679)T^3.$$

At what temperature does water have its maximum density?

16. What is the maximum possible area of a rectangle with a base that lies on the x-axis and with two upper vertices that lie on the graph of the equation $y = 4 - x^2$ (Fig. 3.6.20)?

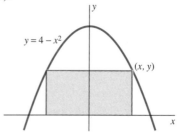

Fig. 3.6.20   The rectangle of Problem 16

17. A rectangular box has a square base with edges at least 1 cm long. Its total surface area is 600 cm². What is the largest possible volume that such a box can have?

18. You must make a cylindrical can with a bottom but no top from 300π in.² of sheet metal. No sheet metal will be wasted; you are allowed to order a circular piece of any size for its base and any appropriate rectangular piece to make into its curved side so long as the given conditions are met. What is the greatest possible volume of such a can?

19. Three large squares of tin, each with edges 1 m long, have four small, equal squares cut from their corners. All twelve resulting small squares are to be the same size (Fig. 3.6.21). The three large cross-shaped pieces are then folded and welded to make boxes with no tops, and the twelve small squares are used to make two small cubes. How should this be done to maximize the total volume of all five boxes?

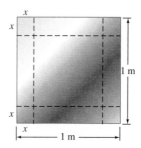

Fig. 3.6.21   One of the three tin 1-m squares of Problem 19

20. Suppose that you are to make a rectangular box with a square base from two different materials. The material for the top and four sides of the box costs $1/ft²; the material for the base costs $2/ft². Find the dimensions of the box of greatest possible volume if you are allowed to spend $144 for the material to make it.

21. A piece of wire 80 in. long is cut into at most two pieces. Each piece is bent into the shape of a square. How should this be done to minimize the sum of the area(s) of the square(s)? To maximize it?

22. A wire of length 100 cm is cut into two pieces. One piece is bent into a circle, the other into a square. Where should the cut be made to maximize the sum of the areas of the square and the circle? To minimize that sum?

23. A farmer has 600 m of fencing with which she plans to enclose a rectangular pasture adjacent to a long existing wall. She plans to build one fence parallel to the wall, two to form the ends of the enclosure, and a fourth (parallel to the two ends of the enclosure) to divide it equally. What is the maximum area that can be enclosed?

24. A zookeeper needs to add a rectangular outdoor pen to an animal house with a corner notch, as shown in Fig. 3.6.22. If 85 m of new fence is available, what dimensions of the pen will maximize its area? No fence will be used along the walls of the animal house.

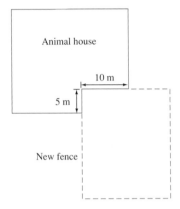

Fig. 3.6.22 The rectangular pen of Problem 24

25. Suppose that a post office can accept a package for mailing only if the sum of its length and its girth (the circumference of its cross section) is at most 100 in. What is the maximum volume of a rectangular box with square cross section that can be mailed?

26. Repeat Problem 25, except use a cylindrical package; its cross section is circular.

27. A printing company has eight presses, each of which can print 3600 copies per hour. It costs $5.00 to set up each press for a run and $10 + 6n$ dollars to run n presses for 1 h. How many presses should be used to print 50,000 copies of a poster most profitably?

28. A farmer wants to hire workers to pick 900 bushels of beans. Each worker can pick 5 bushels per hour and is paid $1.00 per bushel. The farmer must also pay a supervisor $10 per hour while the picking is in progress, and he has additional miscellaneous expenses of $8 per worker. How many workers should he hire to minimize the total cost? What will then be the cost per bushel picked?

29. The heating and cooling costs for a certain uninsulated house are $500/yr, but with $x \leq 10$ in. of insulation, the costs are $1000/(2 + x)$ dollars/yr. It costs $150 for each

inch (thickness) of insulation installed. How many inches of insulation should be installed to minimize the *total* (initial plus annual) costs over a 10-yr period? What will then be the annual savings resulting from this optimal insulation?

30. A concessionaire had been selling 5000 burritos each game night at 50¢ each. When she raised the price to 70¢ each, sales dropped to 4000 per night. Assume a linear relationship between price and sales. If she has fixed costs of $1000 per night and each burrito costs her 25¢, what price will maximize her nightly profit?

31. A commuter train carries 600 passengers each day from a suburb to a city. It costs $1.50 per person to ride the train. Market research reveals that 40 fewer people would ride the train for each 5¢ increase in the fare, 40 more for each 5¢ decrease. What fare should be charged to get the largest possible revenue?

32. Find the shape of the cylinder of maximal volume that can be inscribed in a sphere of radius R (Fig. 3.6.23). Show that the ratio of the height of the cylinder to its radius is $\sqrt{2}$ and that the ratio of the volume of the sphere to that of the maximal cylinder is $\sqrt{3}$.

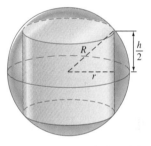

Fig. 3.6.23 The sphere and cylinder of Problem 32

33. Find the dimensions of the right circular cylinder of greatest volume that can be inscribed in a right circular cone of radius R and height H (Fig. 3.6.24).

Fig. 3.6.24 The cone and cylinder of Problem 33

34. Figure 3.6.25 shows a circle of radius 1 in which a trapezoid is inscribed. The longer of the two parallel sides of the trapezoid coincides with a diameter of the circle. What is the maximum possible area of such a trapezoid? [*Suggestion:* A positive quantity is maximized when its square is maximized.]

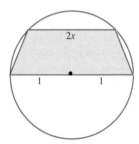

Fig. 3.6.25 The circle and trapezoid of Problem 34

35. Show that the rectangle of maximal perimeter that can be inscribed in a circle is a square.

36. Find the dimensions of the rectangle (with sides parallel to the coordinate axes) of maximal area that can be inscribed in the ellipse with equation

$$\frac{x^2}{25} + \frac{y^2}{9} = 1$$

(Fig. 3.6.26).

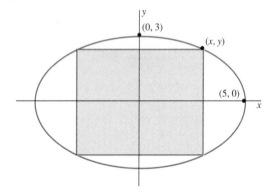

Fig. 3.6.26 The ellipse and rectangle of Problem 36

37. A right circular cone of radius r and height h has slant height $L = \sqrt{r^2 + h^2}$. What is the maximum possible volume of a cone with slant height 10?

38. Two vertical poles 10 ft apart are both 10 ft tall. Find the length of the shortest rope that can reach from the top of one pole to a point on the ground between them and then to the top of the other pole.

39. The sum of two nonnegative numbers is 16. Find the maximum possible value and the minimum possible value of the sum of their cube roots.

40. A straight wire 60 cm long is bent into the shape of an L. What is the shortest possible distance between the two ends of the bent wire?

41. What is the shortest possible distance from a point on the parabola $y = x^2$ to the point $(0, 1)$?

42. Given: There is exactly one point on the graph of $y = \sqrt[3]{3x - 4}$ that is closest to the origin. Find it. [*Suggestion:* See Fig. 3.6.27, and solve the equation you obtain by inspection.]

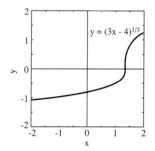

Fig. 3.6.27 The curve of Problem 42

43. Find the dimensions that maximize the cross-sectional area of the four planks that can be cut from the four pieces of the circular log of Example 5—the pieces that remain after a square beam has been cut from the log (Fig. 3.6.10).

44. Find the maximal area of a rectangle inscribed in an equilateral triangle with edges of length 1 (Fig. 3.6.28).

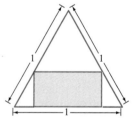

Fig. 3.6.28 The rectangle and equilateral triangle of Problem 44

45. A small island is 2 km off shore in a large lake. A woman on the island can row her boat 10 km/h and can run at a speed of 20 km/h. If she rows to the closest point of the straight shore, she will land 6 km from a village on the shore. Where should she land to reach the village most quickly by a combination of rowing and running?

46. A factory is located on one bank of a straight river that is 2000 m wide. On the opposite bank but 4500 m downstream is a power station from which the factory draws its electricity. Assume that it costs three times as much per meter to lay an underwater cable as to lay an aboveground cable. What path should a cable connecting the power station to the factory take to minimize the cost of laying the cable?

47. A company has plants that are located (in an appropriate coordinate system) at the points $A(0, 1)$, $B(0, -1)$, and $C(3, 0)$ (Fig. 3.6.29). The company plans to construct a distribution center at the point $P(x, 0)$. What value of x would minimize the sum of the distances from P to A, B, and C?

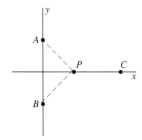

Fig. 3.6.29   The locations in Problem 47

48. Light travels at speed c in air and at a slower speed v in water. (The constant c is approximately 3×10^{10} cm/s; the ratio $n = c/v$, known as the **index of refraction,** depends on the color of the light but is approximately 1.33 for water.) Figure 3.6.30 shows the path of a light ray traveling from point A in air to point B in water, with what appears to be a sudden change in direction as the ray moves through the air-water interface. (a) Write the time T required for the ray to travel from A to B in terms of the variable x and the constants a, b, c, s, and v, all of which have been defined or are shown in the figure. (b) Show that the equation $T'(x) = 0$ for minimizing T is equivalent to the condition

$$\frac{\sin \alpha}{\sin \beta} = \frac{c}{v} = n.$$

This is **Snell's law:** The ratio of the sines of the angles of incidence and refraction is equal to the index of refraction.

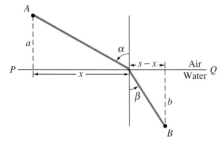

Fig. 3.6.30   Snell's law gives the path of refracted light (Problem 48).

Refraction of light at an air – water interface

49. The mathematics of Snell's law (Problem 48) is applicable to situations other than the refraction of light. Fig. 3.6.31 shows a horizontal geologic fault that separates two towns at points A and B. Assume that A is a miles north of the fault, that B is b miles south of the fault, and that B is L miles east of A. We want to build a road from A to B. Because of differences in terrain, the cost of construction is C_1 (in millions of dollars per mile) north of the fault and C_2 south of it. Where should the point P be placed to minimize the total cost of road construction? (a) Using the notation in the figure, show that the cost is minimized when $C_1 \sin \theta_1 = C_2 \sin \theta_2$. (b) Take $a = b = C_1 = 1$, $C_2 = 2$, and $L = 4$. Show that the equation in part (a) is equivalent to

$$f(x) = 3x^4 - 24x^3 + 51x^2 - 32x + 64 = 0.$$

To approximate the desired solution of this equation, calculate $f(0), f(1), f(2), f(3)$, and $f(4)$. You should find that $f(3) > 0 > f(4)$. Interpolate between $f(3)$ and $f(4)$ to approximate the desired root of this equation.

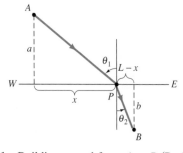

Fig. 3.6.31   Building a road from A to B (Problem 49)

50. The sum of the volumes of two cubes is 2000 in.³ What should their edges x and y be to maximize the sum of their surface areas? To minimize it?

51. The sum of the surface areas of a cube and a sphere is 1000 in.² What should their dimensions be to minimize the sum of their volumes? To maximize it?

52. Your younger brother has six pieces of wood with which to make the kite frame shown in Fig. 3.6.32. The four outer pieces with the indicated lengths have already been cut. How long should the lengths of the inner struts be to maximize the area of the kite?

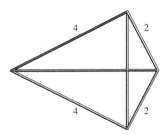

Fig. 3.6.32 The kite frame (Problem 52)

3.6 Projects

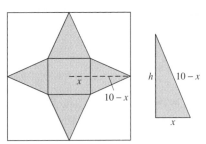

Fig. 3.6.33 The canvas square—first attempt

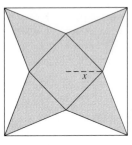

Fig. 3.6.34 The canvas square—second attempt

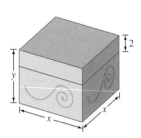

Fig. 3.6.35 The square-based hatbox with lid

These projects require the use of a graphics calculator or a computer with graphing utility.

PROJECT A The following problems deal with alternative methods of constructing a tent.

1. Figure 3.6.33 shows a 20- by 20-ft square of canvas tent material. Girl Scout Troop A must cut pieces from its four corners as indicated, so that the four remaining triangular flaps can be turned up to form a tent in the shape of a pyramid with a square base. How should this be done to maximize the volume of the tent?

Let A denote the area of the base of the tent and h its height. With x as indicated in the figure, show that the volume $V = \frac{1}{3}Ah$ of the tent is given by

$$V(x) = \tfrac{4}{3}x^2 \sqrt{100 - 20x}, \quad 0 \leqq x \leqq 5.$$

Maximize V by graphing $V(x)$ and $V'(x)$ and zooming in on the zero of $V'(x)$.

2. Girl Scout Troop B must make a tent in the shape of a pyramid with a square base from a similar 20- by 20-ft square of canvas but in the manner indicated in Fig. 3.6.34. With x as indicated in the figure, show that the volume of the tent is given by

$$V(x) = \tfrac{2}{3}x^2 \sqrt{200 - 20x}, \qquad 0 \leqq x \leqq 10.$$

Maximize V graphically as in Problem 1.

3. Solve Problems 1 and 2 analytically to verify that the maximal volume in Problem 2 is exactly $2\sqrt{2}$ times the maximal volume in Problem 1. It pays to think before making a tent!

PROJECT B

1. Hattie the milliner needs a hatbox in the shape of a rectangular box with a square base and no top. The volume of the hatbox is to be 1000 in.³, and the edge x of its base must be at least 8 in. long (Fig. 3.6.35). Hattie must also make a square lid with a 2-in. rim. Thus the hatbox-with-lid is, in effect, two open-topped boxes—the first box with height $y \geqq 2$ in. and the lid with height 2 in. What should the dimensions x and y be to minimize the *total area* A of the two open-topped boxes?

Solve Hattie's problem. Begin by expressing the total area as a function A of x. Don't forget to find the domain of A. Show that the equation $A'(x) = 0$ simplifies to the cubic equation

$$x^3 + 2x^2 - 1000 = 0.$$

Instead of attempting to solve this equation exactly, graph $A(x)$ and $A'(x)$ on the same set of coordinate axes and zoom in on the zero of $A'(x)$.

2. Repeat Problem 1 for the case of a cylindrical hatbox, as shown in Fig. 3.6.36. The base and lid are each to have radius $r \geqq 4$ in., the height h of the base must be at least 2 in., the rim of the lid must be exactly 2 in. tall, and the volume of the hatbox is to be 1000 in.³

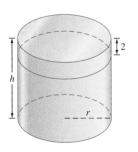

Fig. 3.6.36 The cylindrical hatbox with lid

3.7
Derivatives of Trigonometric Functions

In this section we begin our study of the calculus of trigonometric functions, focusing first on the sine and cosine functions. The definitions and the elementary properties of trigonometric functions are reviewed in Appendix A.

When we write $\sin \theta$ (or $\cos \theta$) in calculus, we mean the sine (or cosine) of an angle of θ radians (rad). Recall the fundamental relation between radian measure and degree measure of angles:

$$\text{There are } \pi \text{ radians in 180 degrees.} \qquad (1)$$

Figure 3.7.1 shows radian-degree conversions for some frequently occurring angles.

The derivatives of the sine and cosine functions depend on the limits

$$\lim_{\theta \to 0} \frac{\sin \theta}{\theta} = 1, \qquad \lim_{\theta \to 0} \frac{1 - \cos \theta}{\theta} = 0 \qquad (2)$$

that we established in Section 2.3. The addition formulas

$$\cos(x + y) = \cos x \cos y - \sin x \sin y,$$
$$\sin(x + y) = \sin x \cos y + \cos x \sin y \qquad (3)$$

are needed as well.

Radians	Degrees
0	0
$\pi/6$	30
$\pi/4$	45
$\pi/3$	60
$\pi/2$	90
$2\pi/3$	120
$3\pi/4$	135
$5\pi/6$	150
π	180
$3\pi/2$	270
2π	360
4π	720

Fig. 3.7.1 Some radian-degree conversions

Theorem 1 *Derivatives of Sines and Cosines*

The functions $f(x) = \sin x$ and $g(x) = \cos x$ are differentiable for all x, and

$$D \sin x = \cos x, \qquad (4)$$

$$D \cos x = -\sin x. \qquad (5)$$

Proof To differentiate $f(x) = \sin x$, we begin with the definition of the derivative,

$$f'(x) = \lim_{h \to 0} \frac{f(x + h) - f(x)}{h} = \lim_{h \to 0} \frac{\sin(x + h) - \sin x}{h}.$$

Next we apply the addition formula for the sine and the limit laws to get

$$f'(x) = \lim_{h \to 0} \frac{(\sin x \cos h + \sin h \cos x) - \sin x}{h}$$

$$= \lim_{h \to 0} \left[(\cos x) \frac{\sin h}{h} - (\sin x) \frac{1 - \cos h}{h} \right]$$

$$= (\cos x) \left(\lim_{h \to 0} \frac{\sin h}{h} \right) - (\sin x) \left(\lim_{h \to 0} \frac{1 - \cos h}{h} \right).$$

The limits in Eq. (2) now yield

$$f'(x) = (\cos x)(1) - (\sin x)(0) = \cos x,$$

which proves Eq. (4). The proof of Eq. (5) is quite similar (see Problem 62). ❑

Examples 1 through 4 illustrate the applications of Eqs. (4) and (5) in conjunction with the general differentiation formulas of Sections 3.2, 3.3, and 3.4 to differentiate various combinations of trigonometric and other functions.

EXAMPLE 1

$$D_x(x^2 \sin x) = (D\, x^2)(\sin x) + (x^2)(D \sin x)$$
$$= 2x \sin x + x^2 \cos x.$$

EXAMPLE 2

$$D_t(\cos^3 t) = D[(\cos t)^3] = 3(\cos t)^2\, D(\cos t)$$
$$= 3(\cos^2 t)(-\sin t) = -3 \cos^2 t \sin t.$$

EXAMPLE 3   If $y = \dfrac{\cos x}{1 - \sin x}$, then

$$\frac{dy}{dx} = \frac{(D\cos x)(1 - \sin x) - (\cos x)[D(1 - \sin x)]}{(1 - \sin x)^2}$$

$$= \frac{(-\sin x)(1 - \sin x) - (\cos x)(-\cos x)}{(1 - \sin x)^2}$$

$$= \frac{-\sin x + \sin^2 x + \cos^2 x}{(1 - \sin x)^2} = \frac{-\sin x + 1}{(1 - \sin x)^2};$$

$$\frac{dy}{dx} = \frac{1}{1 - \sin x}.$$

EXAMPLE 4   If $g(t) = (2 - 3 \cos t)^{3/2}$, then

$$g'(t) = \tfrac{3}{2}(2 - 3 \cos t)^{1/2}\, D(2 - 3 \cos t) = \tfrac{3}{2}(2 - 3 \cos t)^{1/2}(3 \sin t)$$
$$= \tfrac{9}{2}(2 - 3 \cos t)^{1/2} \sin t.$$

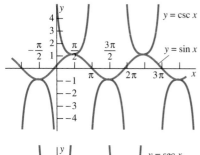

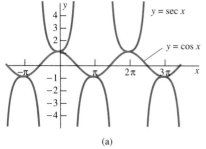

(a)

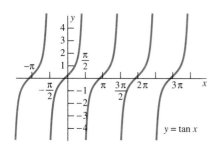

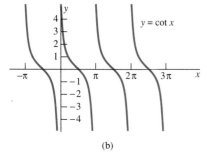

(b)

Fig. 3.7.2 Graphs of the six trigonometric functions

THE REMAINING TRIGONOMETRIC FUNCTIONS

It is easy to differentiate the other four trigonometric functions, because they can be expressed in terms of the sine and cosine functions:

$$\tan x = \frac{\sin x}{\cos x}, \qquad \cot x = \frac{\cos x}{\sin x},$$

$$\sec x = \frac{1}{\cos x}, \qquad \csc x = \frac{1}{\sin x}. \tag{6}$$

Each of these formulas is valid except where its denominator is zero. Thus $\tan x$ and $\sec x$ are undefined when x is an odd integral multiple of $\pi/2$, whereas $\cot x$ and $\csc x$ are undefined when x is an integral multiple of π. The graphs of the six trigonometric functions appear in Fig. 3.7.2. There we show the sine and its reciprocal, the cosecant, in the same coordinate plane; we also pair the cosine with the secant but show the tangent and the cotangent functions separately.

The functions in Eq. (6) can be differentiated by using the quotient rule and the derivatives of the sine and cosine functions. For example,

$$\tan x = \frac{\sin x}{\cos x},$$

so

$$D \tan x = \frac{(D \sin x)(\cos x) - (\sin x)(D \cos x)}{(\cos x)^2}$$

$$= \frac{(\cos x)(\cos x) - (\sin x)(-\sin x)}{\cos^2 x} = \frac{\cos^2 x + \sin^2 x}{\cos^2 x} = \frac{1}{\cos^2 x};$$

$$D \tan x = \sec^2 x.$$

As an exercise, you should derive in similar fashion the differentiation formulas in Eqs. (8) through (10) of Theorem 2.

> **Theorem 2   Derivatives of Trigonometric Functions**
> The functions $f(x) = \tan x$, $g(x) = \cot x$, $p(x) = \sec x$, and $q(x) = \csc x$ are differentiable wherever they are defined, and
>
> $$D \tan x = \sec^2 x, \tag{7}$$
>
> $$D \cot x = -\csc^2 x, \tag{8}$$
>
> $$D \sec x = \sec x \tan x, \tag{9}$$
>
> $$D \csc x = -\csc x \cot x. \tag{10}$$

The patterns in the formulas of Theorem 2 and in Eqs. (4) and (5) make them easy to remember. The formulas in Eqs. (5), (8), and (10) are the "cofunction analogues" of those in Eqs. (4), (7), and (9), respectively. Note that the derivative formulas for the three cofunctions are those involving minus signs.

EXAMPLE 5

$$D_x(x \tan x) = (Dx)(\tan x) + (x)(D \tan x)$$
$$= (1)(\tan x) + (x)(\sec^2 x) = \tan x + x \sec^2 x.$$
$$D_x(\cot^3 x) = D(\cot x)^3 = 3(\cot x)^2 D \cot x$$
$$= 3(\cot x)^2(-\csc^2 x) = -3 \csc^2 x \cot^2 x.$$
$$D_x\left(\frac{\sec x}{\sqrt{x}}\right) = \frac{(D \sec x)(\sqrt{x}) - (\sec x)(D\sqrt{x})}{(\sqrt{x})^2}$$
$$= \frac{(\sec x \tan x)(\sqrt{x}) - (\sec x)(\frac{1}{2}x^{-1/2})}{x}$$
$$= \frac{1}{2}x^{-3/2}(2x \tan x - 1) \sec x.$$

CHAIN RULE FORMULAS

Recall from Eq. (14) in Section 3.3 that the chain rule gives

$$D_x g(u) = g'(u)\frac{du}{dx} \tag{11}$$

for the derivative of the composition $g(u(x))$ of two differentiable functions g and u. This formula yields a *chain rule version* of each new differentiation formula that we learn.

If we apply Eq. (11) first with $g(u) = \sin u$, then with $g(u) = \cos u$, and so on, we get the chain rule versions of the trigonometric differentiation formulas:

$$D_x \sin u = (\cos u)\frac{du}{dx}, \tag{12}$$

$$D_x \cos u = (-\sin u)\frac{du}{dx}, \tag{13}$$

$$D_x \tan u = (\sec^2 u)\frac{du}{dx}, \tag{14}$$

$$D_x \cot u = (-\csc^2 u)\frac{du}{dx}, \tag{15}$$

$$D_x \sec u = (\sec u \tan u)\frac{du}{dx}, \tag{16}$$

$$D_x \csc u = (-\csc u \cot u)\frac{du}{dx}. \tag{17}$$

The cases in which $u = kx$ (where k is a constant) are worth mentioning. For example,

$$D_x \sin kx = k \cos kx \quad \text{and} \quad D_x \cos kx = -k \sin kx. \tag{18}$$

The formulas in Eq. (18) provide an explanation of why radian measure is more appropriate than degree measure. Because it follows from Eq. (1) that an angle of degree measure x has radian measure $\pi x/180$, the "sine of an angle of x degrees" is a *new* and *different* function with the formula

$$\sin x° = \sin \frac{\pi x}{180},$$

expressed on the right-hand side in terms of the standard (radian-measure) sine function. Hence the first formula in Eq. (18) yields

$$D \sin x° = \frac{\pi}{180} \cos \frac{\pi x}{180},$$

so

$$D \sin x° \approx (0.01745) \cos x°.$$

The necessity of using the approximate value 0.01745 here—and indeed its very presence—is one reason why radians instead of degrees are used in the calculus of trigonometric functions: When we work with radians, we don't need such approximations.

EXAMPLE 6   If $y = 2 \sin 10t + 3 \cos \pi t$, then

$$\frac{dy}{dt} = 20 \cos 10t - 3\pi \sin \pi t.$$

EXAMPLE 7 $D_x(\sin^2 3x \cos^4 5x)$

$= [D(\sin 3x)^2](\cos^4 5x) + (\sin^2 3x)[D (\cos 5x)^4]$

$= 2(\sin 3x)(D \sin 3x) \cdot (\cos^4 5x) + (\sin^2 3x) \cdot 4(\cos 5x)^3(D \cos 5x)$

$= 2(\sin 3x)(3 \cos 3x)(\cos^4 5x) + (\sin^2 3x)(4 \cos^3 5x)(-5 \sin 5x)$

$= 6 \sin 3x \cos 3x \cos^4 5x - 20 \sin^2 3x \sin 5x \cos^3 5x.$

EXAMPLE 8   Differentiate $f(x) = \cos \sqrt{x}$.

Solution   If $u = \sqrt{x}$, then $du/dx = 1/(2\sqrt{x})$, so Eq. (13) yields

$$D_x \cos \sqrt{x} = D_x \cos u = (-\sin u) \frac{du}{dx}$$

$$= -(\sin \sqrt{x}) \frac{1}{2\sqrt{x}} = -\frac{\sin \sqrt{x}}{2\sqrt{x}}.$$

Alternatively, we can carry out this computation without introducing the auxiliary variable u:

$$D \cos \sqrt{x} = (-\sin \sqrt{x})D(\sqrt{x}) = -\frac{\sin \sqrt{x}}{2\sqrt{x}}.$$

EXAMPLE 9   Differentiate

$$y = \sin^2(2x - 1)^{3/2} = [\sin(2x - 1)^{3/2}]^2.$$

Solution   Here, $y = u^2$, where $u = \sin(2x - 1)^{3/2}$, so

$$\frac{dy}{dx} = 2u\,\frac{du}{dx} = 2[\sin(2x - 1)^{3/2}]D_x[\sin(2x - 1)^{3/2}]$$

$$= 2[\sin(2x - 1)^{3/2}][\cos(2x - 1)^{3/2}]D(2x - 1)^{3/2}$$

$$= 2[\sin(2x - 1)^{3/2}][\cos(2x - 1)^{3/2}]\tfrac{3}{2}(2x - 1)^{1/2}\cdot 2$$

$$= 6(2x - 1)^{1/2}[\sin(2x - 1)^{3/2}][\cos(2x - 1)^{3/2}].$$

EXAMPLE 10

$$D_x \tan 2x^3 = (\sec^2 2x^3)D(2x^3) = 6x^2 \sec^2 2x^3.$$

$$D_x \cot^3 2x = D(\cot 2x)^3 = 3(\cot 2x)^2\,D(\cot 2x)$$

$$= (3\,\cot^2 2x)(-\csc^2 2x)D(2x)$$

$$= -6\,\csc^2 2x\,\cot^2 2x.$$

$$D_x \sec\sqrt{x} = (\sec\sqrt{x}\,\tan\sqrt{x})D\sqrt{x} = \frac{\sec\sqrt{x}\,\tan\sqrt{x}}{2\sqrt{x}}.$$

$$D_x \sqrt{\csc x} = D(\csc x)^{1/2} = \tfrac{1}{2}(\csc x)^{-1/2}\,D(\csc x)$$

$$= \tfrac{1}{2}(\csc x)^{-1/2}(-\csc x\,\cot x) = -\tfrac{1}{2}(\cot x)\sqrt{\csc x}.$$

Examples 11 and 12 illustrate the applications of trigonometric functions to rate-of-change and maximum-minimum problems.

EXAMPLE 11   A rocket is launched vertically and is tracked by a radar station located on the ground 5 mi from the launch pad. Suppose that the elevation angle θ of the line of sight to the rocket is increasing at $3°$ per second when $\theta = 60°$. What is the velocity of the rocket at this instant?

Solution   First we convert the given data from degrees into radians. Because there are $\pi/180$ rad in $1°$, the rate of increase of θ becomes

$$\frac{3\pi}{180} = \frac{\pi}{60} \quad \text{(rad/s)}$$

at the instant when

$$\theta = \frac{60\pi}{180} = \frac{\pi}{3} \quad \text{(rad)}.$$

From Fig. 3.7.3 we see that the height y (in miles) of the rocket is

$$y = 5 \tan \theta.$$

Hence its velocity is

$$\frac{dy}{dt} = \frac{dy}{d\theta}\cdot\frac{d\theta}{dt} = 5(\sec^2 \theta)\,\frac{d\theta}{dt}.$$

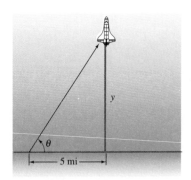

Fig. 3.7.3 Tracking an ascending rocket (Example 11)

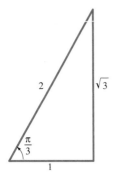

Fig. 3.7.4 $\sec \dfrac{\pi}{3} = 2$
(Example 11)

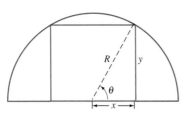

Fig. 3.7.5 The rectangle of Example 12

Because $\sec(\pi/3) = 2$ (Fig. 3.7.4), the velocity of the rocket is

$$\frac{dy}{dt} = 5 \cdot 2^2 \cdot \frac{\pi}{60} = \frac{\pi}{3} \quad \text{(mi/s)},$$

about 3770 mi/h, at the instant when $\theta = 60°$.

EXAMPLE 12 A rectangle is inscribed in a semicircle of radius R (Fig. 3.7.5). What is the maximum possible area of such a rectangle?

Solution   If we denote the length of *half* the base of the rectangle by x and its height by y, then its area is $A = 2xy$. We see in Fig. 3.7.5 that the right triangle has hypotenuse R, the radius of the circle. So

$$x = R \cos \theta \quad \text{and} \quad y = R \sin \theta. \tag{19}$$

Each value of θ between 0 and $\pi/2$ corresponds to a possible inscribed rectangle. The values $\theta = 0$ and $\theta = \pi/2$ yield degenerate rectangles.

We substitute the data in Eq. (19) into the formula $A = 2xy$ to obtain the area

$$A = A(\theta) = 2(R \cos \theta)(R \sin \theta)$$
$$= 2R^2 \cos \theta \sin \theta \tag{20}$$

as a function of θ on the closed interval $[0, \pi/2]$. To find the critical points, we differentiate:

$$\frac{dA}{d\theta} = 2R^2(-\sin \theta \sin \theta + \cos \theta \cos \theta) = 2R^2(\cos^2 \theta - \sin^2 \theta).$$

Because $dA/d\theta$ always exists, we have critical points only if

$$\cos^2 \theta - \sin^2 \theta = 0; \qquad \sin^2 \theta = \cos^2 \theta;$$
$$\tan^2 \theta = 1; \qquad \tan \theta = \pm 1.$$

The only value of θ in $[0, \pi/2]$ such that $\tan \theta = \pm 1$ is $\theta = \pi/4$.

Upon evaluation of $A(\theta)$ at each of the possible values $\theta = 0$, $\theta = \pi/4$, and $\pi/2$ (the endpoints and the critical point), we find that

$$A(0) = 0,$$

$$A\left(\frac{\pi}{4}\right) = 2R^2\left(\frac{1}{\sqrt{2}}\right)\left(\frac{1}{\sqrt{2}}\right) = R^2, \quad \longleftarrow \quad \text{absolute maximum}$$

$$A\left(\frac{\pi}{2}\right) = 0.$$

Thus the largest inscribed rectangle has area R^2, and its dimensions are $2x = R\sqrt{2}$ and $y = R/\sqrt{2}$.

3.7 Problems

Differentiate the functions given in Problems 1 through 20.

1. $f(x) = 3 \sin^2 x$

2. $f(x) = 2 \cos^4 x$

3. $f(x) = x \cos x$

4. $f(x) = \sqrt{x} \sin x$

5. $f(x) = \dfrac{\sin x}{x}$

6. $f(x) = \dfrac{\cos x}{\sqrt{x}}$

7. $f(x) = \sin x \cos^2 x$

8. $f(x) = \cos^3 x \sin^2 x$

9. $g(t) = (1 + \sin t)^4$

10. $g(t) = (2 - \cos^2 t)^3$

11. $g(t) = \dfrac{1}{\sin t + \cos t}$ **12.** $g(t) = \dfrac{\sin t}{1 + \cos t}$

13. $f(x) = 2x \sin x - 3x^2 \cos x$

14. $f(x) = x^{1/2} \cos x - x^{-1/2} \sin x$

15. $f(x) = \cos 2x \sin 3x$ **16.** $f(x) = \cos 5x \sin 7x$

17. $g(t) = t^3 \sin^2 2t$ **18.** $g(t) = \sqrt{t} \cos^3 3t$

19. $g(t) = (\cos 3t + \cos 5t)^{5/2}$

20. $g(t) = \dfrac{1}{\sqrt{\sin^2 t + \sin^2 3t}}$

Find dy/dx in Problems 21 through 60.

21. $y = \sin^2 \sqrt{x}$ **22.** $y = \dfrac{\cos 2x}{x}$

23. $y = x^2 \cos(3x^2 - 1)$ **24.** $y = \sin^3 x^4$

25. $y = \sin 2x \cos 3x$ **26.** $y = \dfrac{x}{\sin 3x}$

27. $y = \dfrac{\cos 3x}{\sin 5x}$ **28.** $y = \sqrt{\cos \sqrt{x}}$

29. $y = \sin^2 x^2$ **30.** $y = \cos^3 x^3$

31. $y = \sin 2\sqrt{x}$ **32.** $y = \cos 3\sqrt[3]{x}$

33. $y = x \sin x^2$ **34.** $y = x^2 \cos\left(\dfrac{1}{x}\right)$

35. $y = \sqrt{x} \sin \sqrt{x}$ **36.** $y = (\sin x - \cos x)^2$

37. $y = \sqrt{x}(x - \cos x)^3$ **38.** $y = \sqrt{x} \sin \sqrt{x} + \sqrt{x}$

39. $y = \cos(\sin x^2)$ **40.** $y = \sin(1 + \sqrt{\sin x})$

41. $y = \tan x^7$ **42.** $y = \sec x^7$

43. $y = (\tan x)^7$ **44.** $y = (\sec 2x)^7$

45. $y = x^7 \tan 5x$ **46.** $y = \dfrac{\sec x^5}{x}$

47. $y = \sqrt{x} \sec \sqrt{x}$ **48.** $y = \sec \sqrt{x} \tan \sqrt{x}$

49. $y = \csc\left(\dfrac{1}{x^2}\right)$ **50.** $y = \cot\left(\dfrac{1}{\sqrt{x}}\right)$

51. $y = \dfrac{\sec 5x}{\tan 3x}$ **52.** $y = \sec^2 x - \tan^2 x$

53. $y = x \sec x \csc x$ **54.** $y = x^3 \tan^3 x^3$

55. $y = \sec(\sin x)$ **56.** $y = \cot(\sec 7x)$

57. $y = \dfrac{\sin x}{\sec x}$ **58.** $y = \dfrac{\sec x}{1 + \tan x}$

59. $y = \sqrt{1 + \cot 5x}$ **60.** $y = \sqrt{\csc \sqrt{x}}$

61. Derive the differentiation formulas in Eqs. (8) through (10).

62. Use the definition of the derivative to show directly that $g'(x) = -\sin x$ if $g(x) = \cos x$.

63. If a projectile is fired from ground level with initial velocity v_0 and inclination angle α and if air resistance can be ignored, then its range—the horizontal distance it travels—is

$$R = \tfrac{1}{16} v_0^2 \sin \alpha \cos \alpha$$

(Fig. 3.7.6). What value of α maximizes R?

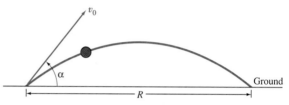

Fig. 3.7.6 The projectile of Problem 63

64. A weather balloon that is rising vertically is observed from a point on the ground 300 ft from the spot directly beneath the balloon (Fig. 3.7.7). At what rate is the balloon rising when the angle between the ground and the observer's line of sight is 45° and is increasing at 1° per second?

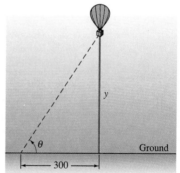

Fig. 3.7.7 The weather balloon of Problem 64

65. A rocket is launched vertically upward from a point 2 mi west of an observer on the ground. What is the speed of the rocket when the angle of elevation (from the horizontal) of the observer's line of sight to the rocket is 50° and is increasing at 5° per second?

66. A plane flying at an altitude of 25,000 ft has a defective airspeed indicator. To determine her speed, the pilot sights a fixed point on the ground. At the moment the angle of depression (from the horizontal) of her line of sight is 65°, she notes that this angle is increasing at 1.5° per second (Fig. 3.7.8). What is the speed of the plane?

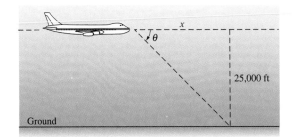

Fig. 3.7.8 The airplane of Problem 66

67. An observer on the ground sights an approaching plane flying at constant speed and at an altitude of 20,000 ft. From his point of view, the plane's angle of elevation is increasing at 0.5° per second when the angle is 60°. What is the speed of the plane?

68. Find the largest possible area A of a rectangle inscribed in the unit circle $x^2 + y^2 = 1$ by maximizing A as a function of the angle θ indicated in Fig. 3.7.9.

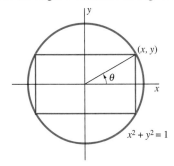

Fig. 3.7.9 A rectangle inscribed in the unit circle (Problem 68)

69. A water trough is to be made from a long strip of tin 6 ft wide by bending up at an angle θ a 2-ft strip on each side (Fig. 3.7.10). What angle θ would maximize the cross-sectional area, and thus the volume, of the trough?

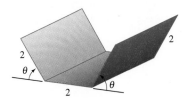

Fig. 3.7.10 The water trough of Problem 69

70. A circular patch of grass of radius 20 m is surrounded by a walkway, and a light is placed atop a lamppost at the circle's center. At what height should the light be placed to illuminate the walkway most strongly? The intensity of illumination I of a surface is given by $I = (k \sin \theta)/D^2$, where D is the distance from the light source to the sur-

face, θ is the angle at which light strikes the surface, and k is a positive constant.

71. Find the minimum possible volume V of a cone in which a sphere of given radius R is inscribed. Minimize V as a function of the angle θ indicated in Fig. 3.7.11.

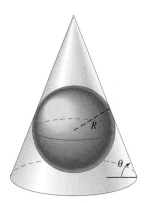

Fig. 3.7.11 Finding the smallest cone containing a fixed sphere (Problem 71)

72. A very long rectangular piece of paper is 20 cm wide. The bottom right-hand corner is folded along the crease shown in Fig. 3.7.12, so that the corner just touches the left-hand side of the page. How should this be done so that the crease is as short as possible?

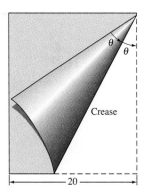

Fig. 3.7.12 Fold a piece of paper; make the crease of minimal length (Problem 72).

73. Find the maximum possible area A of a trapezoid inscribed in a semicircle of radius 1, as shown in Fig. 3.7.13.

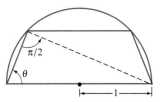

Fig. 3.7.13 A trapezoid inscribed in a circle (Problem 73)

Begin by expressing A as a function of the angle θ shown there.

74. A logger must cut a six-sided beam from a circular log of diameter 30 cm so that its cross section is as shown in Fig. 3.7.14. The beam is symmetrical, with only two different internal angles α and β. Show that the area of the cross section is maximal when the cross section is a regular hexagon, with equal sides and angles (corresponding to $\alpha = \beta = 2\pi/3$). Note that $\alpha + 2\beta = 2\pi$.   (Why?)

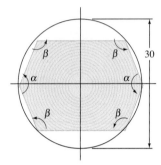

Fig. 3.7.14   A hexagonal beam cut from a circular log (Problem 74)

75. Consider a circular arc of length s with its endpoints on the x-axis (Fig. 3.7.15). Show that the area A bounded by this arc and the x-axis is maximal when the circular arc is in the shape of a semicircle. [*Suggestion:* Express A in terms of the angle θ subtended by the arc at the center of the circle, as shown in Fig. 3.7.15. Show that A is maximal when $\theta = \pi$.]

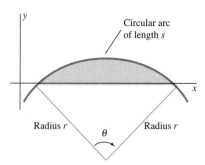

Fig. 3.7.15   Finding the maximum area bounded by a circular arc and its chord (Problem 75)

76. A hiker starting at a point P on a straight road wants to reach a forest cabin that is 2 km from a point Q 3 km down the road from P (Fig. 3.7.16). She can walk 8 km/h along the road but only 3 km/h through the forest. She wants to minimize the time required to reach the cabin. How far down the road should she walk before setting off through the forest straight for the cabin? [*Suggestion:* Use the angle θ between the road and the path she takes through the forest as the independent variable.]

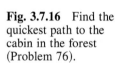

Fig. 3.7.16   Find the quickest path to the cabin in the forest (Problem 76).

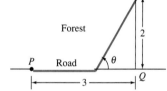

3.8

Exponential and Logarithmic Functions

Our study of calculus has, to this point, been concentrated on algebraic and trigonometric functions. Exponential and logarithmic functions complete the list of the most elementary functions that are important in applications. This introductory section gives a brief and somewhat informal overview of these nonalgebraic functions, partly from the perspective of precalculus mathematics. In addition to reviewing the laws of exponents and the laws of logarithms, we provide an intuitive foundation for the systematic treatment of exponential and logarithmic functions in Chapter 7.

EXPONENTIAL FUNCTIONS

An *exponential function* is a function of the form

$$f(x) = a^x \tag{1}$$

where $a > 0$. Note that x is the variable; the number a, called the *base*, is a constant. Thus

- ❑ An exponential function $f(x) = a^x$ is a constant raised to a variable power, whereas
- ❑ The power function $p(x) = x^k$ is a variable raised to a constant power.

In elementary algebra a *rational* power of the positive real number a is defined in terms of integral roots and powers. There we learn that if p and q are integers (with $q > 0$), then

$$a^{p/q} = \sqrt[q]{a^p} = \left(\sqrt[q]{a}\right)^p.$$

The following **laws of exponents** are then established for all *rational* exponents r and s:

$$a^{r+s} = a^r \cdot a^s, \qquad (a^r)^s = a^{r \cdot s},$$

$$a^{-r} = \frac{1}{a^r}, \qquad (ab)^r = a^r \cdot b^r. \tag{2}$$

Moreover, recall that

$$a^0 = 1 \tag{3}$$

for every positive real number a.

EXAMPLE 1 A typical calculation using the laws of exponents is

$$\frac{(2^2)^3(3)^{-4}}{(2)^{-2}} = \frac{2^6 \cdot 2^2}{3^4} = \frac{2^8}{3^4} = \frac{256}{81}.$$

Some fractional powers of 2 are

$$2^{3/2} = \sqrt{2^3} = \sqrt{8}, \qquad 2^{5/3} = \sqrt[3]{2^5} = \sqrt[3]{32},$$

and

$$2^{0.7} = 2^{7/10} = \sqrt[10]{2^7} = \sqrt[10]{128} \approx 1.6245.$$

EXAMPLE 2 Applications often call for irrational exponents as well as rational ones. Consider a bacteria population $P(t)$ that begins (at time $t = 0$) with initial population

$$P(0) = 1 \quad \text{(million)}$$

and doubles every hour thereafter, increasing by the same factor during any two time intervals of the same length. In 3 h, P will increase by a factor of $2 \cdot 2 \cdot 2 = 2^3$, so

$$P(3) = 2^3 = 8 \quad \text{(million)}.$$

If k is the factor by which P increases in $\frac{1}{3}$ h, then in 1 h P will increase by a factor of $k \cdot k \cdot k = k^3 = 2$, so

$$k = 2^{1/3} \quad \text{and} \quad P(\tfrac{1}{3}) = 2^{1/3} \quad \text{(million)}.$$

More generally, if p and q are positive integers, then $P(1/q) = 2^{1/q}$. In p/q hours, P will increase by a factor of $2^{1/q}$ each of p times, so

$$P\left(\frac{p}{q}\right) = (2^{1/q})^p = 2^{p/q}.$$

Thus the bacteria population after t hours is given (in millions) by

$$P(t) = 2^t \tag{4}$$

if t is a rational number. But because time is not restricted to rational values alone, we surely ought to conclude that $P(t) = 2^t$ for *all* $t > 0$.

But what do we mean by an expression involving an irrational exponent, such as $2^{\sqrt{2}}$ or 2^π? To find the value of 2^π, we might work with (rational) finite

t	2^t
3.1	8.5742
3.14	8.8152
3.141	8.8214
3.1415	8.8244
3.14159	8.8250
3.141592	8.8250
3.1415926	8.8250
↓	↓
π	2^π

Fig. 3.8.1 Investigating 2^π (Example 2)

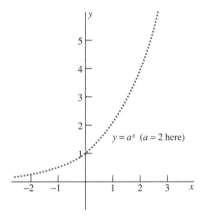

Fig. 3.8.2 The graph of $y = a^x$ has "holes" if only rational values of x are used.

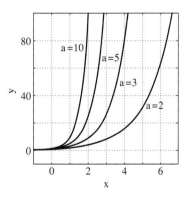

Fig. 3.8.3 $y = a^x$ for $a = 2, 3, 5, 10$

decimal approximations to the irrational number $\pi = 3.14159\,26\ldots$. For example, a calculator gives

$$2^{3.1} = 2^{31/10} = \left(\sqrt[10]{2}\right)^{31} \approx 8.5742.$$

The approximate values shown in the table in Fig. 3.8.1 indicate that the bacteria population after π hours is

$$P(\pi) = 2^\pi \approx 8.8250 \quad \text{(million)}.$$

Example 2 suggests the practical importance of the exponential function 2^x with base 2. In Chapter 7 we'll see that for any fixed base $a > 0$, the exponential function $f(x) = a^x$ can be defined for *all* x so that

❑ The laws of exponents in Eq. (2) hold for irrational exponents as well as for rational exponents;

❑ The function $f(x) = a^x$ is continuous, indeed differentiable.

Let us begin by investigating the function $f(x) = a^x$ in the case $a > 1$. We first note that $a^r > 1$ if r is a positive rational number. If $r < s$, another rational number, then the laws of exponents give

$$a^r < a^r \cdot a^{s-r} = a^s,$$

so $f(r) < f(s)$ whenever $r < s$. That is, $f(x) = a^x$ is an *increasing* function of the rational exponent x. If we plot points on the graph of $y = a^x$ for rational values x by using a typical fixed value of a such as $a = 2$, we obtain a graph like the one in Fig. 3.8.2. This graph is shown with a dotted curve to suggest that it is densely filled with "holes" corresponding to the missing points (x, y) when x is irrational. We shall show in Section 7.4 that the holes in this graph can be filled to obtain the graph of an increasing and continuous function f that is defined for all real x and for which $f(r) = a^r$ for every rational number r. We therefore write $f(x) = a^x$ for all x and call f the **exponential function with base a.**

As illustrated in Fig. 3.8.3, the function $f(x) = a^x$ increases rapidly as $x > 0$ increases, and the graphs of $y = a^x$ look qualitatively similar for different values of $a > 1$ as the base.

DERIVATIVES OF EXPONENTIAL FUNCTIONS

To compute the derivative of the exponential function $f(x) = a^x$, we begin with the definition of the derivative and then use the first law of exponents in Eq. (2) to simplify. This gives

$$f'(x) = D_x a^x = \lim_{h \to 0} \frac{f(x + h) - f(x)}{h} = \lim_{h \to 0} \frac{a^{x+h} - a^x}{h}$$

$$= \lim_{h \to 0} \frac{a^x a^h - a^x}{h} \qquad \text{(by the laws of exponents)}$$

$$= a^x \left(\lim_{h \to 0} \frac{a^h - 1}{h} \right) \qquad \begin{array}{l}\text{(because } a^x \text{ is "constant"}\\ \text{with respect to } h\text{).}\end{array}$$

Under the assumption that $f(x) = a^x$ is differentiable, it follows that the limit

$$m(a) = \lim_{h \to 0} \frac{a^h - 1}{h} \tag{5}$$

exists. Although its value $m(a)$ depends on a, the limit is a constant as far as x is concerned. Thus we find that the derivative of a^x is a *constant multiple* of a^x itself:

$$D_x a^x = m(a) \cdot a^x. \tag{6}$$

Because $a^0 = 1$, we see from Eq. (6) that the constant $m(a)$ is the slope of the line tangent to the curve $y = a^x$ at the point $(0, 1)$, where $x = 0$.

The numerical data shown in Fig. 3.8.4 suggest that $m(2) \approx 0.693$ and that $m(3) \approx 1.099$. The tangent lines with these slopes are shown in Fig. 3.8.5. Thus it appears that

$$D_x 2^x \approx (0.693) \cdot 2^x \quad \text{and} \quad D_x 3^x \approx (1.099) \cdot 3^x. \tag{7}$$

h	$\dfrac{2^h - 1}{h}$	$\dfrac{3^h - 1}{h}$
0.1	0.718	1.161
0.01	0.696	1.105
0.001	0.693	1.099
0.0001	0.693	1.099

Fig. 3.8.4 Investigating the values of $m(2)$ and $m(3)$

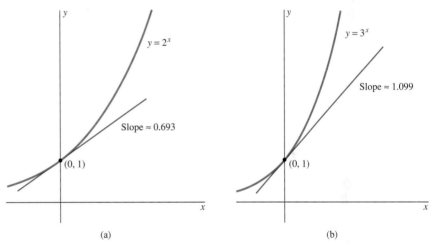

(a) (b)

Fig. 3.8.5 The graphs (a) $y = 2^x$ and (b) $y = 3^x$

We would like somehow to avoid awkward numerical factors like those in Eq. (7). It seems plausible that the value $m(a)$ defined in Eq. (5) is a continuous function of a. If so, then because $m(2) < 1$ and $m(3) > 1$, the intermediate value property implies that $m(e) = 1$ (exactly) for some number e between 2 and 3. If we use this particular number e as the base, then it follows from Eq. (6) that the derivative of the resulting exponential function $f(x) = e^x$ is

$$D_x e^x = e^x. \tag{8}$$

Thus the function e^x is its own derivative. We call $f(x) = e^x$ the **natural exponential function.** Its graph is shown in Fig. 3.8.6.

We will see in Section 7.3 that the number e is given by the limit

$$e = \lim_{n \to \infty} \left(1 + \frac{1}{n} \right)^n. \tag{9}$$

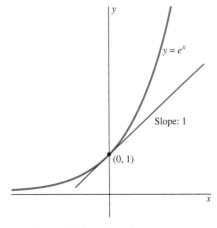

Fig. 3.8.6 The graph $y = e^x$

n	$\left(1 + \dfrac{1}{n}\right)^n$
10	2.594
100	2.705
1,000	2.717
10,000	2.718
100,000	2.718

Fig. 3.8.7 Numerical estimate of the number e

Let us investigate this limit numerically. With a calculator we obtain the values in the table of Fig. 3.8.7. The evidence suggests (but does not prove) that $e \approx 2.718$ to three places. This number e is one of the most important special numbers in mathematics. It is known to be irrational; its value to 20 places is

$$e \approx 2.71828\ 18284\ 59045\ 23536.$$

EXAMPLE 3

(a) If $y = x^2 e^x$, then the fact that $D_x e^x = e^x$ and the product rule yield

$$\frac{dy}{dx} = (D_x x^2)e^x + x^2(D_x e^x) = (2x)(e^x) + (x^2)(e^x) = (2x + x^2)e^x.$$

(b) If

$$y = \frac{e^x}{x + 1},$$

then the quotient rule gives

$$\frac{dy}{dx} = \frac{(D_x e^x)(x + 1) - (e^x)D_x(x + 1)}{(x + 1)^2} = \frac{(e^x)(x + 1) - (e^x)(1)}{(x + 1)^2} = \frac{xe^x}{(x + 1)^2}.$$

INVERSE FUNCTIONS

In precalculus courses, the **base a logarithm function $\log_a x$** is introduced as the "opposite" of the exponential function $f(x) = a^x$ with base $a > 1$. That is, $\log_a x$ is the power to which a must be raised to get x. Thus

$$y = \log_a x \quad \text{if and only if} \quad a^y = x. \tag{10}$$

With $a = 10$, this is the base 10 *common logarithm* $\log_{10} x$.

EXAMPLE 4

$$\log_{10} 1000 = 3 \quad \text{because} \quad 1000 = 10^3;$$

$$\log_{10}(0.1) = -1 \quad \text{because} \quad 0.1 = 10^{-1};$$

$$\log_2 16 = 4 \quad \text{because} \quad 16 = 2^4;$$

$$\log_3 9 = 2 \quad \text{because} \quad 9 = 3^2.$$

If $y = \log_a x$, then $a^y = x > 0$. Hence it follows that

$$a^{\log_a x} = x \tag{11a}$$

and

$$\log_a(a^y) = y. \tag{11b}$$

Thus the base a exponential and logarithmic functions are natural opposites, in the sense that each undoes the result of applying the other. Apply both in

EXAMPLE 8 To find the derivative of $h(x) = \ln x^2$, we can apply a law of logarithms to simplify before differentiation. That is,

$$h(x) = \ln x^2 = 2 \ln x,$$

and hence

$$h'(x) = 2 \cdot D_x(\ln x) = \frac{2}{x}.$$

In the preliminary discussion of this section, we introduced in an informal manner

❑ The number $e \approx 2.71828$,
❑ The natural exponential function e^x, and
❑ The natural logarithm function $\ln x$.

Our investigation of the derivatives of e^x and $\ln x$—given in Eqs. (8) and (18)—should be regarded as provisional, pending a more complete discussion of these new functions in Sections 7.2 and 7.3. This brief preview of the connections between exponential functions and logarithmic functions should help you better understand their systematic development in Chapter 7. We shall see there that the functions e^x and $\ln x$ play a vital role in the quantitative analysis of a wide range of natural phenomena—including population growth, radioactive decay, the spread of epidemics, growth of investments, diffusion of pollutants, and motion with the effect of resistance taken into account.

3.8 Problems

Use the laws of exponents to simplify the expressions in Problems 1 through 10. Then write each answer as an integer.

1. $2^3 \cdot 2^4$ **2.** $3^2 \cdot 3^3$

3. $(2^2)^3$ **4.** $2^{(2^3)}$

5. $3^5 \cdot 3^{-5}$ **6.** $10^{10} \cdot 10^{-10}$

7. $(2^{12})^{1/3}$ **8.** $(3^6)^{1/2}$

9. $4^5 \cdot 2^{-6}$ **10.** $6^5 \cdot 3^{-5}$

Note that $\log_{10} 100 = \log_{10} 10^2 = 2 \log_{10} 10 = 2$. Use a similar technique to simplify and evaluate (without the use of a calculator) the expressions in Problems 11 through 16.

11. $\log_2 16$ **12.** $\log_3 27$

13. $\log_5 125$ **14.** $\log_7 49$

15. $\log_{10} 1000$ **16.** $\log_{12} 144$

Use the laws of logarithms to express the natural logarithms in Problems 17 through 26 in terms of the three numbers $\ln 2$, $\ln 3$, and $\ln 5$.

17. $\ln 8$ **18.** $\ln 9$

19. $\ln 6$ **20.** $\ln 15$

21. $\ln 72$ **22.** $\ln 200$

23. $\ln \frac{8}{27}$ **24.** $\ln \frac{12}{25}$

25. $\ln \frac{27}{40}$ **26.** $\ln \frac{1}{90}$

27. Which is larger, $2^{(3^4)}$ or $(2^3)^4$?

28. Evaluate $\log_{0.5} 16$.

29. By inspection, find two values of x such that $x^2 = 2^x$.

30. Show that the number $\log_2 3$ is irrational. [*Suggestion:* Assume to the contrary that $\log_2 3 = p/q$, where p and q are positive integers, and then express the consequence of this assumption in exponential form. Under what circumstances can an integral power of 2 equal an integral power of 3?]

In Problems 31 through 40, solve for x without using a calculator.

31. $2^x = 64$ **32.** $10^{-x} = 0.001$

33. $10^{-x} = 100$ **34.** $(3^x)^2 = 81$

35. $x^x = x^2$ (Find *all* solutions.)

36. $\log_x 16 = 2$

37. $\log_3 x = 4$ **38.** $e^{5x} = 7$

39. $3e^x = 3$ **40.** $2e^{-7x} = 5$

In Problems 41 through 54, find dy/dx.

41. $y = xe^x$

42. $y = x^3 e^x$

43. $y = \sqrt{x}\, e^x$

44. $y = \dfrac{1}{x}\, e^x$

45. $y = \dfrac{e^x}{x^2}$

46. $y = \dfrac{e^x}{\sqrt{x}}$

47. $y = x \ln x$

48. $y = x^2 \ln x$

49. $y = \sqrt{x} \ln x$

50. $y = \dfrac{\ln x}{\sqrt{x}}$

51. $y = \dfrac{x}{e^x}$

52. $y = e^x \ln x$

53. $y = \ln x^3$ [This means $\ln(x^3)$.]

54. $y = \ln \sqrt[3]{x}$

55. Use the chain rule to deduce from $D_x e^x = e^x$ that $D_x(e^{kx}) = ke^{kx}$ if k is a constant.

Apply the formula in Problem 55 to find the derivatives of the functions in Problems 56 through 58.

56. $f(x) = e^{3x}$

57. $f(x) = e^{x/10}$

58. $f(x) = e^{-10x}$

59. Substitute $2 = e^{\ln 2}$ into $P(t) = 2^t$; then apply the chain rule to show that $P'(t) = 2^t \ln 2$.

Use the method of Problem 59 to find the derivatives of the functions in Problems 60 through 63.

60. $P(t) = 10^t$

61. $P(t) = 3^t$

62. $P(t) = 2^{3t}$

63. $P(t) = 2^{-t}$

64. Suppose that a population $P(t)$ of bacteria (in millions) at time t (in hours) is initially (at time $t = 0$) 1 (million). If the population doubles every hour, then the population at time $t \geqq 0$ is given by $P(t) = 2^t$. Use the result of Problem 59 to find the instantaneous rate of growth (in millions of bacteria per hour) of this population (a) at time $t = 0$; (b) after 4 h (that is, at time $t = 4$). [*Note:* The natural logarithm of 2 is $\ln 2 \approx 0.69315$.]

65. Repeat Problem 64 under the assumption that the bacteria population $P(t)$ *triples* every hour, so $P(t) = 3^t$ (millions) at time $t \geqq 0$. [*Note:* The natural logarithm of 3 is $\ln 3 \approx 1.09861$.]

3.8 Project

You can investigate numerically the limit

$$m(a) = \lim_{h \to 0} \frac{a^h - 1}{h}$$

in Eq. (5) by using a calculator or computer with which the function

$$\phi(h) = \frac{a^h - 1}{h}$$

(with a fixed) can be defined and rapidly calculated for successively smaller values of h.

For instance, calculate $\phi(h)$ with $a = 2$ and with $a = 3$ for $h = 0.1$, $0.01, 0.001, \ldots$, and conclude that

$$m(2) \approx 0.6931 < 1,$$

whereas

$$m(3) \approx 1.0986 > 1.$$

It follows that the mysterious number e for which

$$m(e) = 1$$

lies somewhere between 2 and 3.

Indeed, interpolation between the values $m(2) \approx 0.6931$ and $m(3) \approx 1.0986$ suggests that $e \approx 2.7$ or $e \approx 2.8$. Hence investigate the values $m(2.7)$ and $m(2.8)$ to verify the entries shown in Fig. 3.8.10.

Continue in this way to "close in" on the number e. Don't quit until you're convinced that $e \approx 2.718$ accurate to three decimal places.

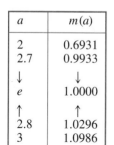

a	$m(a)$
2	0.6931
2.7	0.9933
$\downarrow$	$\downarrow$
e	1.0000
$\uparrow$	$\uparrow$
2.8	1.0296
3	1.0986

Fig. 3.8.10 Closing in on the number e

3.9

Implicit Differentiation and Related Rates

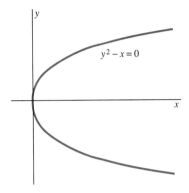

Fig. 3.9.1 The parabola $y^2 - x = 0$

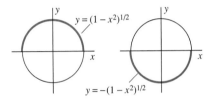

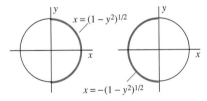

Fig. 3.9.2 Continuous functions defined implicitly by $x^2 + y^2 = 1$

An equation in two variables x and y may have one or more solutions for y in terms of x or for x in terms of y. These solutions are functions that are said to be **implicitly defined** by the equation. Here we discuss the differentiation of such functions and the use of their derivatives in solving specific rate-of-change problems.

For example, the equation $y^2 - x = 0$ implicitly defines two continuous functions of x:

$$y = \sqrt{x} \quad \text{and} \quad y = -\sqrt{x}.$$

Each has domain the half-line $x \geqq 0$. The graphs of these two functions are the upper and lower branches of the parabola shown in Fig. 3.9.1. The whole parabola is not the graph of a function of x because no vertical line can meet the graph of such a function at more than one point.

The equation of the unit circle, $x^2 + y^2 = 1$, implicitly defines four functions (among others):

$$y = +\sqrt{1 - x^2} \quad \text{for } x \text{ in } [-1, 1],$$
$$y = -\sqrt{1 - x^2} \quad \text{for } x \text{ in } [-1, 1],$$
$$x = +\sqrt{1 - y^2} \quad \text{for } y \text{ in } [-1, 1], \quad \text{and}$$
$$x = -\sqrt{1 - y^2} \quad \text{for } y \text{ in } [-1, 1].$$

The four graphs are highlighted against four copies of the unit circle shown in Fig. 3.9.2.

The equation $2x^2 + 2y^2 + 3 = 0$ implicitly defines *no* function, because this equation has no real solution (x, y). (Clearly, $2x^2 + 2y^2 + 3 > 0$ for all x and y.)

In advanced calculus we can study conditions that will guarantee when an implicitly defined function is differentiable. Here we will proceed on the assumption that our implicitly defined functions are differentiable at almost all points in their domains. (The functions with the graphs shown in Fig. 3.9.2 are *not* differentiable at the endpoints of their domains.)

When we assume differentiability, we can use the chain rule to differentiate the given equation, thinking of x as the independent variable. We can then solve the resulting equation for the derivative $dy/dx = f'(x)$ of the implicitly defined function f. This process is called **implicit differentiation.**

EXAMPLE 1 Use implicit differentiation to find the derivative of a differentiable function $y = f(x)$ implicitly defined by the equation

$$x^2 + y^2 = 1.$$

Solution The equation $x^2 + y^2 = 1$ is to be thought of as an *identity* that implicitly defines $y = y(x)$ as a function of x. Because $x^2 + [y(x)]^2$ is then a function of x, it has the same derivative as the constant function 1 on the other side of the identity. Thus we may differentiate both sides of $x^2 + y^2 = 1$ with respect to x and equate the results. We obtain

$$2x + 2y \frac{dy}{dx} = 0.$$

In this step, it is essential to remember that y is a function of x, so that the chain rule yields $D_x(y^2) = 2y\, D_x y$.

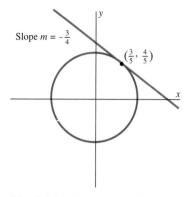

Slope $m = -\frac{3}{4}$

$(\frac{3}{5}, \frac{4}{5})$

Fig. 3.9.3 The tangent line of Example 1

Then we solve for

$$\frac{dy}{dx} = -\frac{x}{y}. \tag{1}$$

It may be surprising to see a formula for dy/dx containing both x and y, but such a formula can be just as useful as one containing only x. For example, the formula in Eq. (1) tells us that the slope of the line tangent to the circle $x^2 + y^2 = 1$ at the point $(\frac{3}{5}, \frac{4}{5})$ is

$$\frac{dy}{dx}\bigg|_{(\frac{3}{5}, \frac{4}{5})} = -\frac{0.6}{0.8} = -0.75.$$

The circle and this line are shown in Fig. 3.9.3.

If we solve for $y = \pm\sqrt{1 - x^2}$ in Example 1, then

$$\frac{dy}{dx} = \frac{-x}{\pm\sqrt{1 - x^2}} = -\frac{x}{y},$$

in agreement with Eq. (1). Thus Eq. (1) simultaneously gives us the derivatives of both the functions $y = +\sqrt{1 - x^2}$ and $y = -\sqrt{1 - x^2}$ implicitly defined by the equation $x^2 + y^2 = 1$.

EXAMPLE 2 The *folium of Descartes* is the graph of the equation

$$x^3 + y^3 = 3xy. \tag{2}$$

This curve was first proposed by René Descartes as a challenge to Pierre de Fermat (1601–1665) to find its tangent line at an arbitrary point. The graph of the folium appears in Fig. 3.9.4. The graph very nearly coincides with the straight line $x + y + 1 = 0$ when $|x|$ and $|y|$ are large and produces a loop shaped like a laurel leaf (thus the name *folium*) in the first quadrant. We indicate in Problem 24 of Section 12.1 how this graph can be constructed. Here we want to find the slope of its tangent line.

Solution We use implicit differentiation (rather than the *ad hoc* methods with which Fermat met Descartes' challenge). We differentiate both sides in Eq. (2) and find that

$$3x^2 + 3y^2\frac{dy}{dx} = 3y + 3x\frac{dy}{dx}.$$

Then we solve for the derivative:

$$\frac{dy}{dx} = \frac{y - x^2}{y^2 - x}. \tag{3}$$

For instance, at the point $(\frac{3}{2}, \frac{3}{2})$ of the folium, the slope of the tangent line is

$$\frac{dy}{dx}\bigg|_{(\frac{3}{2}, \frac{3}{2})} = \frac{\frac{3}{2} - (\frac{3}{2})^2}{(\frac{3}{2})^2 - \frac{3}{2}} = -1,$$

and this result agrees with our intuition about the figure. At the point $(\frac{2}{3}, \frac{4}{3})$, Eq. (3) gives

$$\frac{dy}{dx}\bigg|_{(\frac{2}{3}, \frac{4}{3})} = \frac{\frac{4}{3} - (\frac{2}{3})^2}{(\frac{4}{3})^2 - \frac{2}{3}} = \frac{4}{5}.$$

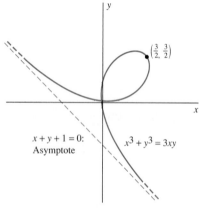

$x + y + 1 = 0$:
Asymptote

$(\frac{3}{2}, \frac{3}{2})$

$x^3 + y^3 = 3xy$

Fig. 3.9.4 The folium of Descartes; Example 2
2

176

It follows that

$$2z = 3x,$$

and implicit differentiation gives

$$2\frac{dz}{dt} = 3\frac{dx}{dt}.$$

We substitute $dx/dt = 8$ and find that

$$\frac{dz}{dt} = \frac{3}{2}\cdot\frac{dx}{dt} = \frac{3}{2}\cdot 8 = 12.$$

So the tip of the man's shadow is moving at 12 ft/s.

Example 5 is somewhat unusual in that the answer is independent of the man's distance from the light pole—the given value $x = 100$ is superfluous. Example 6 is a related-rates problem with two relationships between the variables, which is not quite so unusual.

EXAMPLE 6 Two radar stations at A and B, with B 6 km east of A, are tracking a ship. At a certain instant, the ship is 5 km from A, and this distance is increasing at the rate of 28 km/h. At the same instant, the ship is also 5 km from B, but this distance is increasing at only 4 km/h. Where is the ship, how fast is it moving, and in what direction is it moving?

Solution With the distances indicated in Fig. 3.9.7, we find—again with the aid of the Pythagorean theorem—that

$$x^2 + y^2 = u^2 \quad \text{and} \quad (6 - x)^2 + y^2 = v^2. \tag{6}$$

We are given these data: $u = v = 5$, $du/dt = 28$, and $dv/dt = 4$ at the instant in question. Because the ship is equally distant from A and B, it is clear that $x = 3$. Thus $y = 4$. Hence the ship is 3 km east and 4 km north of A.

We differentiate implicitly the two equations in (6), and we obtain

$$2x\frac{dx}{dt} + 2y\frac{dy}{dt} = 2u\frac{du}{dt}$$

and

$$-2(6 - x)\frac{dx}{dt} + 2y\frac{dy}{dt} = 2v\frac{dv}{dt}.$$

When we substitute the numerical data given and data deduced, we find that

$$3\frac{dx}{dt} + 4\frac{dy}{dt} = 140 \quad \text{and} \quad -3\frac{dx}{dt} + 4\frac{dy}{dt} = 20.$$

These equations are easy to solve: $dx/dt = dy/dt = 20$. Therefore, the ship is sailing northeast at a speed of

$$\sqrt{20^2 + 20^2} = 20\sqrt{2} \quad \text{(km/h)}$$

—*if* the figure is correct! A mirror along the line AB will reflect *another* ship, 3 km east and 4 km *south* of A, sailing *southeast* at a speed of $20\sqrt{2}$ km/h.

The lesson? Figures are important, helpful, often essential—but potentially misleading. Avoid taking anything for granted when you draw a figure. In this example there would be no real problem, for each radar station could determine whether the ship was generally to the north or to the south.

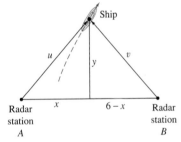

Radar
station
A

Radar
station
B

Fig. 3.9.7 Radar stations tracking a ship (Example 6)

In Problems 1 through 10, find dy/dx by implicit differentiation.

1. $x^2 - y^2 = 1$

2. $xy = 1$

3. $16x^2 + 25y^2 = 400$

4. $x^3 + y^3 = 1$

5. $\sqrt{x} + \sqrt{y} = 1$

6. $x^2 + xy + y^2 = 9$

7. $x^{2/3} + y^{2/3} = 1$

8. $(x - 1)y^2 = x + 1$

9. $x^2(x - y) = y^2(x + y)$

10. $(x^2 + y^2)^2 = 4xy$

In Problems 11 through 20, first find dy/dx by implicit differentiation. Then write an equation of the line tangent to the graph of the equation at the given point.

11. $x^2 + y^2 = 25$; $(3, -4)$

12. $xy = -8$; $(4, -2)$

13. $x^2 y = x + 2$; $(2, 1)$

14. $x^{1/4} + y^{1/4} = 4$; $(16, 16)$

15. $xy^2 + x^2 y = 2$; $(1, -2)$

16. $\dfrac{1}{x + 1} + \dfrac{1}{y + 1} = 1$; $(1, 1)$

17. $12(x^2 + y^2) = 25xy$; $(3, 4)$

18. $x^2 + xy + y^2 = 7$; $(3, -2)$

19. $\dfrac{1}{x^3} + \dfrac{1}{y^3} = 2$; $(1, 1)$

20. $(x^2 + y^2)^3 = 8x^2 y^2$; $(1, -1)$

21. Find dy/dx, given $xy^3 - x^5 y^2 = 4$. Then find the slope of the line tangent to the graph of the given equation at the point $(1, 2)$.

22. Show that the graph of $xy^5 + x^5 y = 1$ has no horizontal tangents.

23. Show that there are no points on the graph of the equation $x^3 + y^3 = 3xy - 1$ at which the tangent line is horizontal.

24. Find all points on the graph of the equation

$$x^4 + y^4 + 2 = 4xy^3$$

at which the tangent line is horizontal.

25. Find all points on the graph of $x^2 + y^2 = 4x + 4y$ at which the tangent line is horizontal.

26. Find the first-quadrant points of the folium of Example 2 at which the tangent line is either horizontal $(dy/dx = 0)$ or vertical (where $dx/dy = 1/(dy/dx) = 0$).

27. The graph of the equation $x^2 - xy + y^2 = 9$ is the rotated ellipse shown in Fig. 3.9.8. Find the lines tangent to this curve at the two points where it intersects the *x*-axis, and show that these lines are parallel.

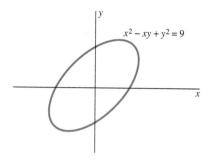

Fig. 3.9.8 The rotated ellipse of Problem 27

28. Find the points on the curve of Problem 27 where the tangent line is horizontal $(dy/dx = 0)$ and those where it is vertical $(dx/dy = 0)$.

29. The graph in Fig. 3.9.9 is a *lemniscate* with equation $(x^2 + y^2)^2 = x^2 - y^2$. Find by implicit differentiation the four points on the lemniscate where the tangent line is horizontal. Then find the two points where the tangent line is vertical—that is, where $dx/dy = 1/(dy/dx) = 0$.

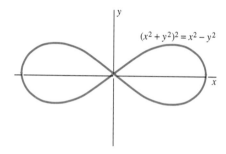

Fig. 3.9.9 The lemniscate of Problem 29

30. Water is being collected from a block of ice with a square base (Fig. 3.9.10). The water is produced because the ice is melting in such a way that each edge of the base of the block is decreasing at 2 in./h while the height of the block is decreasing at 3 in./h. What is the rate of flow of water into the collecting pan when the base has edge length 20 in. and the height of the block is 15 in.? Make the simplifying assumption that the water and ice have the same density.

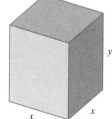

Fig. 3.9.10 The ice block of Problem 30

31. Figure 3.9.11 shows sand being emptied from a hopper at the rate of 10 ft³/s. The sand forms a conical pile whose height is always twice its radius. At what rate is the radius of the pile increasing when its height is 5 ft?

Fig. 3.9.11 The conical sand pile of Problem 31 with volume $V = \frac{1}{3}\pi r^2 h$

32. Suppose that water is being emptied from a spherical tank of radius 10 ft (Fig. 3.9.12). If the depth of water in the tank is 5 ft and is decreasing at the rate of 3 ft/s, at what rate is the radius r of the top surface of the water decreasing?

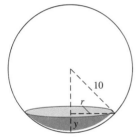

Fig. 3.9.12 The spherical tank of Problem 32

33. A circular oil slick of uniform thickness is caused by a spill of 1 m³ of oil. The thickness of the oil slick is decreasing at the rate of 0.1 cm/h. At what rate is the radius of the slick increasing when the radius is 8 m?

34. Suppose that an ostrich 5 ft tall is walking at a speed of 4 ft/s directly toward a street light 10 ft high. How fast is the tip of the ostrich's shadow moving along the ground? At what rate is the ostrich's shadow decreasing in length?

35. The width of a rectangle is half its length. At what rate is its area increasing if its width is 10 cm and is increasing at 0.5 cm/s?

36. At what rate is the area of an equilateral triangle increasing if its base is 10 cm long and is increasing at 0.5 cm/s?

37. A gas balloon is being filled at the rate of 100π cm³ of gas per second. At what rate is the radius of the balloon increasing when the radius is 10 cm?

38. The volume V (in cubic inches) and pressure p (in pounds per square inch) of a certain gas sample satisfy the equation $pV = 1000$. At what rate is the volume of the sample changing if the pressure is 100 lb/in.² and is increasing at the rate of 2 lb/in.² per second?

39. Figure 3.9.13 shows a kite in the air at an altitude of 400 ft. The kite is being blown horizontally at the rate of 10 ft/s away from the person holding the kite string at ground level. At what rate is the string being payed out when 500 ft of string is already out?

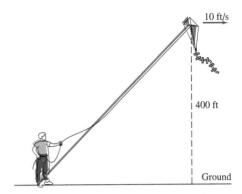

Fig. 3.9.13 The kite of Problem 39

40. A weather balloon that is rising vertically is being observed from a point on the ground 300 ft from the spot directly beneath the balloon. At what rate is the balloon rising when the angle between the ground and the observer's line of sight is 45° and is increasing at 1° per second?

41. An airplane flying horizontally at an altitude of 3 mi and at a speed of 480 mi/h passes directly above an observer on the ground. How fast is the distance from the observer to the airplane increasing 30 s later?

42. Figure 3.9.14 shows a spherical tank of radius a partly filled with water. The maximum depth of water in the tank is y. A formula for the volume V of water in the tank—a formula you can derive after you study Chapter 5—is $V = \frac{1}{3}\pi y^2(3a - y)$. Suppose that water is being drained from a spherical tank of radius 5 ft at the rate of 100 gal/min. Find the rate at which the depth y of water is decreasing when (a) $y = 7$ (ft); (b) $y = 3$ (ft). [*Note:* One gallon of water occupies a volume of approximately 0.1337 ft³.]

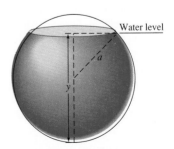

Fig. 3.9.14 The spherical water tank of Problem 42

Sec. 3.9 / Implicit Differentiation and Related Rates

43. Repeat Problem 42, but use a tank that is hemispherical, flat side on top, with radius 10 ft.

44. A swimming pool is 50 ft long and 20 ft wide. Its depth varies uniformly from 2 ft at the shallow end to 12 ft at the deep end (Fig. 3.9.15). Suppose that the pool is being filled at the rate of 1000 gal/min. At what rate is the depth of water at the deep end increasing when the depth is 6 ft? [*Note:* One gallon of water occupies a volume of approximately 0.1337 ft³.]

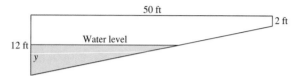

Fig. 3.9.15 Cross section of the swimming pool of Problem 44

45. A ladder 41 ft long that was leaning against a vertical wall begins to slip. Its top slides down the wall while its bottom moves along the level ground at a constant speed of 10 ft/s. How fast is the top of the ladder moving when it is 9 ft above the ground?

46. The base of a rectangle is increasing at 4 cm/s while its height is decreasing at 3 cm/s. At what rate is its area changing when its base is 20 cm and its height is 12 cm?

47. The height of a cone is decreasing at 3 cm/s while its radius is increasing at 2 cm/s. When the radius is 4 cm and the height is 6 cm, is the volume of the cone increasing or decreasing? At what rate is the volume changing then?

48. A square is expanding. When each edge is 10 in., its area is increasing at 120 in.²/s. At what rate is the length of each edge changing then?

49. A rocket that is launched vertically is tracked by a radar station located on the ground 4 mi from the launch site. What is the vertical speed of the rocket at the instant its distance from the radar station is 5 mi and this distance is increasing at the rate of 3600 mi/h?

50. Two straight roads intersect at right angles. At 10 A.M a car passes through the intersection headed due east at 30 mi/h. At 11 A.M. a truck heading due north at 40 mi/h passes through the intersection. Assume that the two vehicles maintain the given speeds and directions. At what rate are they separating at 1 P.M.?

51. A 10-ft ladder is leaning against a wall. The bottom of the ladder begins to slide away from the wall at a speed of 1 mi/h. Find the rate at which the top of the ladder is moving when it is (a) 4 ft above the ground; (b) 1 in. above the ground.

52. Two ships are sailing toward a very small island. One ship, the Pinta, is east of the island and is sailing due west

at 15 mi/h. The other ship, the Niña, is north of the island and is sailing due south at 20 mi/h. At a certain time the Pinta is 30 mi from the island and the Niña is 40 mi from it. At what rate are the two ships drawing closer together at that time?

53. At time $t = 0$, a single-engine military jet is flying due east at 12 mi/min. At the same altitude and 208 mi directly ahead of the military jet, still at time $t = 0$, a commercial jet is flying due north at 8 mi/min. When are the two planes closest to each other? What is the minimum distance between them?

54. A ship with a long anchor chain is anchored in 11 fathoms of water. The anchor chain is being wound in at the rate of 10 fathoms/min, causing the ship to move toward the spot directly above the anchor resting on the seabed. The hawsehole—the point of contact between ship and chain—is located 1 fathom above the waterline. At what speed is the ship moving when there are exactly 13 fathoms of chain still out?

55. A water tank is in the shape of a cone with vertical axis and vertex downward. The tank has radius 3 ft and is 5 ft high. At first the tank is full of water, but at time $t = 0$ (in seconds), a small hole at the vertex is opened and the water begins to drain. When the height of water in the tank has dropped to 3 ft, the water is flowing out at 2 ft³/s. At what rate, in feet per second, is the water level dropping then?

56. A spherical tank of radius 10 ft is being filled with water at the rate of 200 gal/min. How fast is the water level rising when the maximum depth of water in the tank is 5 ft? [*Suggestion:* See Problem 42 for a useful formula and a helpful note.]

57. A water bucket is shaped like the frustum of a cone with height 2 ft, base radius 6 in., and top radius 12 in. Water is leaking from the bucket at 10 in.³/min. At what rate is the water level falling when the depth of water in the bucket is 1 ft? [*Note:* The volume V of a conical frustum with height h and base radii a and b is

$$V = \frac{\pi h}{3}(a^2 + ab + b^2).$$

Such a frustum is shown in Fig. 3.9.16.]

Fig. 3.9.16 The conical frustum whose volume is given in Problem 57

Because $f'(x) = 2x$, Eq. (6) gives the iterative formula

$$x_{n+1} = x_n - \frac{x_n^2 - A}{2x_n} = \frac{1}{2}\left(x_n + \frac{A}{x_n}\right). \qquad (8)$$

Thus we have derived the Babylonian iterative formula as a special case of Newton's method. The use of Eq. (8) with $A = 2$ therefore yields exactly the values of $x_1, x_2, x_3,$ and x_4 that we computed in Example 1, and after performing another iteration we find that

$$x_5 = \frac{1}{2}\left(x_4 + \frac{2}{x_4}\right) \approx 1.41421\,3562,$$

which agrees with x_4 to nine decimal places. The very rapid convergence here is an important characteristic of Newton's method. As a general rule (with some exceptions), each iteration doubles the number of decimal places of accuracy.

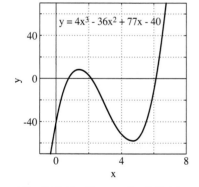

Fig. 3.10.4 The tray of Example 3

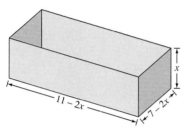

Fig. 3.10.5 The graph of $f(x)$ in Eq. (10) of Example 3

EXAMPLE 3 Figure 3.10.4 shows an open-topped tray constructed by the method of Example 2 in Section 3.6. We begin with a 7-in. by 11-in. rectangle of sheet metal. We cut a square with edge length x from each of its four corners and then fold up the resulting flaps to obtain a rectangular tray with volume

$$V(x) = x(7 - 2x)(11 - 2x)$$
$$= 4x^3 - 36x^2 + 77x, \qquad 0 \leqq x \leqq 3.5. \qquad (9)$$

In Section 3.6 we inquired about the maximum possible volume of such a tray. Here we want to find the value(s) of x that will yield a tray with volume 40 in.³; we will find x by solving the equation

$$V(x) = 4x^3 - 36x^2 + 77x = 40.$$

To solve this equation for x, first we write an equation of the form in Eq. (7):

$$f(x) = 4x^3 - 36x^2 + 77x - 40 = 0. \qquad (10)$$

Figure 3.10.5 shows the graph of f. We see three solutions: a root r_1 between 0 and 1, a root r_2 slightly greater than 2, and a root r_3 slightly larger than 6. Because

$$f'(x) = 12x^2 - 72x + 77,$$

Newton's iterative formula in Eq. (6) takes the form

$$x_{n+1} = x_n - \frac{f(x_n)}{f'(x_n)}$$
$$= x_n - \frac{4x_n^3 - 36x_n^2 + 77x_n - 40}{12x_n^2 - 72x_n + 77}. \qquad (11)$$

Beginning with the initial guess $x_0 = 1$ (because it's reasonably close to r_1), Eq. (11) gives

$$x_1 = 1 - \frac{4 \cdot 1^3 - 36 \cdot 1^2 + 77 \cdot 1 - 40}{12 \cdot 1^2 - 72 \cdot 1 + 77} \approx 0.7059,$$

$$x_2 \approx 0.7736,$$

$$x_3 \approx 0.7780,$$

$$x_4 \approx 0.7780.$$

Thus we obtain the root $r_1 \approx 0.7780$, retaining only four decimal places.

If we had begun with a different initial guess, the sequence of Newton iterates might well have converged to a different root of the equation $f(x) = 0$. The approximate solution obtained therefore depends upon the initial guess. For example, with $x_0 = 2$ and, later, with $x_0 = 6$, the iteration in Eq. (11) produces the two sequences

$$\begin{array}{ll} x_0 = 2 & x_0 = 6 \\ x_1 \approx 2.1053 & x_1 \approx 6.1299 \\ x_2 \approx 2.0993 & x_2 \approx 6.1228 \\ x_3 \approx 2.0992 & x_3 \approx 6.1227 \\ x_4 \approx 2.0992 & x_4 \approx 6.1227 \end{array}$$

Thus the other two roots of Eq. (10) are $r_2 \approx 2.0992$ and $r_3 \approx 6.1227$.

With $x = r_1 \approx 0.7780$, the tray in Fig. 3.10.4 has the approximate dimensions 9.4440 in. by 5.4440 in. by 0.7780 in. With $x = r_2 \approx 2.0992$, its approximate dimensions are 6.8015 in. by 2.8015 in. by 2.0992 in. But the third root $r_3 \approx 6.1227$ would *not* lead to a tray that is physically possible. (Why not?) Thus the *two* values of x that yield trays with volume 40 in. are $x \approx 0.7780$ and $x \approx 2.0992$.

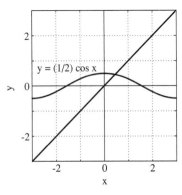

Fig. 3.10.6 Solving the equation $x = \frac{1}{2} \cos x$ (Example 4)

EXAMPLE 4 Figure 3.10.6 indicates that the equation

$$x = \tfrac{1}{2} \cos x \qquad (12)$$

has a solution r near 0.5. To apply Newton's method to approximate r, we rewrite Eq. (12) in the form

$$f(x) = 2x - \cos x = 0.$$

Because $f'(x) = 2 + \sin x$, the iterative formula of Newton's method is

$$x_{n+1} = x_n - \frac{2x_n - \cos x_n}{2 + \sin x_n}.$$

Beginning with $x_0 = 0.5$ and retaining five decimal places, this formula yields

$$x_1 \approx 0.45063, \quad x_2 \approx 0.45018, \quad x_3 \approx 0.45018.$$

Thus the root is 0.45018 to five decimal places.

EXAMPLE 5 Figure 3.10.7 indicates that the equation

$$3 \sin x = \ln x$$

has either five or six positive solutions. To approximate the smallest solution $r \approx 3$, we apply Newton's method with

$$f(x) = 3 \sin x - \ln x, \qquad f'(x) = 3 \cos x - \frac{1}{x}.$$

Then the iterative formula of Newton's method is

$$x_{n+1} = x_n - \frac{3 \sin x_n - \ln x_n}{3 \cos x_n - (1/x_n)}.$$

Fig. 3.10.7 $y = 3 \sin x$ and $y = \ln x$

When we begin with $x_0 = 3$ and retain five decimal places, this formula gives

$$x_1 \approx 2.79558, \qquad x_2 \approx 2.79225, \qquad x_3 \approx 2.79225.$$

Thus $r \approx 2.79225$ to five decimal places. In Problem 40 we ask you to find the remaining solutions indicated in Fig. 3.10.7.

Newton's method is one for which "the proof is in the pudding." If it works, it's obvious that it does, and everything's fine. When Newton's method fails, it may do so spectacularly. For example, suppose that we want to solve the equation

$$f(x) = x^{1/3} = 0.$$

Here $r = 0$ is the only solution. The iterative formula in Eq. (6) becomes

$$x_{n+1} = x_n - \frac{(x_n)^{1/3}}{\frac{1}{3}(x_n)^{-2/3}} = x_n - 3x_n = -2x_n.$$

If we begin with $x_0 = 1$, Newton's method yields $x_1 = -2$, $x_2 = +4$, $x_3 = -8$, and so on. Figure 3.10.8 indicates why our "approximations" are not converging.

When Newton's method fails, a graph will typically indicate the reason why. Then the use of an alternative method such as repeated tabulation or successive magnification is appropriate.

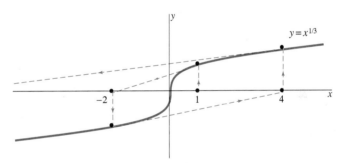

Fig. 3.10.8 A failure of Newton's method

NEWTON'S METHOD AND COMPUTER GRAPHICS

Newton's method and similar iterative techniques are often used to generate vividly colored "fractal patterns," in which the same or similar structures are replicated on smaller and smaller scales at successively higher levels of magnification. To describe how this is done, we replace the real numbers in our Newton's method computations with *complex* numbers. We illustrate this idea with the cubic equation

$$f(x) = x^3 - 3x^2 + 1 = 0. \tag{13}$$

In Project A we ask you to approximate the three solutions

$$r_1 \approx -0.53, \quad r_2 \approx 0.65, \quad r_3 \approx 2.88$$

of this equation.

First, recall that a *complex number* is a number of the form $a + bi$, where $i = \sqrt{-1}$, so $i^2 = -1$. The real numbers a and b are called the *real part* and the *imaginary part,* respectively, of $a + bi$. You add, multiply, and divide complex numbers as if they were binomials, with real and imaginary parts "collected" as in the computations

$$(3 + 4i) + (5 - 7i) = (3 + 5) + (4 - 7)i = 8 - 3i,$$

$$(2 + 5i)(3 - 4i) = 2(3 - 4i) + 5i(3 - 4i)$$
$$= 6 - 8i + 15i - 20i^2 = 26 + 7i,$$

and

$$\frac{2 + 5i}{3 + 4i} = \frac{2 + 5i}{3 + 4i} \cdot \frac{3 - 4i}{3 - 4i} = \frac{26 + 7i}{9 - 16i^2} = \frac{26 + 7i}{25} = 1.04 + 0.28i.$$

The use of the *conjugate* $3 - 4i$ of the denominator $3 + 4i$ in the last computation is a very common technique for writing a complex fraction in the standard form $a + bi$. (The **conjugate** of $x + yi$ is $x - yi$; it follows that the conjugate of $x - yi$ is $x + yi$.)

Now let us substitute the complex number $z = x + iy$ into the cubic polynomial

$$f(z) = z^3 - 3z^2 + 1$$

of Eq. (13) and into its derivative $f'(z) = 3z^2 - 6z$. We find that

$$f(z) = (x + iy)^3 - 3(x + iy)^2 + 1$$
$$= (x^3 - 3xy^2 - 3x^2 + 3y^2 + 1) + (3x^2y - y^3 + 6xy)i \quad (14)$$

and

$$f'(z) = 3(x + iy)^2 - 6(x + iy)$$
$$= (3x^2 - 3y^2 - 6x) + (6xy - 6y)i. \quad (15)$$

Consequently, there is nothing to prevent us from applying Newton's method to Eq. (13) with complex numbers. Beginning with a *complex* initial guess $z_0 = x_0 + iy_0$, we can substitute Eqs. (14) and (15) into Newton's iterative formula

$$z_{n+1} = z_n - \frac{f(z_n)}{f'(z_n)} \quad (16)$$

to generate the complex sequence $\{z_n\}$, which may yet converge to a (real) solution of Eq. (13).

With this preparation, we can now explain how Fig. 3.10.9 was generated: A computer was programmed to carry out Newton's iteration repeatedly, beginning with many thousands of initial guesses $z_0 = x_0 + iy_0$ that "fill" the rectangle $-2 \le x \le 4$, $-2.25 \le y \le 2.25$ in the complex plane. The initial point $z_0 = (x_0, y_0)$ was then color-coded according to the root (if any) to which the corresponding sequence $\{z_n\}$ converged:

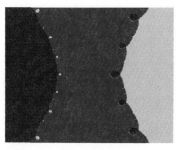

Fig. 3.10.9 $-2 \le x \le 4$, $-2.25 \le y \le 2.25$

Color z_0 green if $\{z_n\}$ converges to the root $r_1 \approx -0.53$;

Color z_0 red   if $\{z_n\}$ converges to the root $r_2 \approx \ \ 0.65$;

Color z_0 yellow if $\{z_n\}$ converges to the root $r_3 \approx \ \ 2.88$.

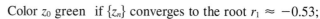

Thus we use different colors to distinguish different "*Newton basins* of attraction" for the equation we are investigating. It is not surprising that a red region containing the root r_2 appears in the middle of Fig. 3.10.9, separating a green region to the left that contains r_1 and a yellow region to the right that contains r_3. But why would yellow lobes protrude from the green region into the red region and green lobes protrude from the yellow region into the red one? To see what's happening near these lobes, we generated some blowups.

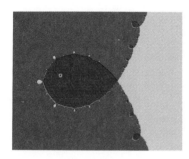

Fig. 3.10.10 $1.6 \leqq x \leqq 2.4$, $-0.3 \leqq y \leqq 0.3$

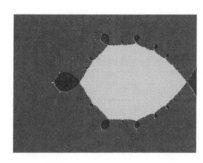

Fig. 3.10.11 $1.64 \leqq x \leqq 1.68$, $-0.015 \leqq y \leqq 0.015$

Fig. 3.10.12 $1.648 \leqq x \leqq 1.650$, $-0.00075 \leqq y \leqq 0.00075$

Figure 3.10.10 shows a blowup of the rectangle $1.6 \leqq x \leqq 2.4$, $-0.3 \leqq y \leqq 0.3$ containing the green lobe indicated in Fig. 3.10.9. Figure 3.10.11 ($1.64 \leqq x \leqq 1.68$, $-0.015 \leqq y \leqq 0.015$) and Fig. 3.10.12 ($1.648 \leqq x \leqq 1.650$, $-0.00075 \leqq y \leqq 0.00075$) are further magnifications. The rectangle shown in Fig. 3.10.12 corresponds to less than one millionth of a square inch of Fig. 3.10.9.

At every level of magnification, each green lobe has smaller yellow lobes protruding into the surrounding red region, and each of these yellow lobes has still smaller green lobes protruding from it, and so on ad infinitum (just like the proverbial little fleas that are bitten by still smaller fleas, and so on ad infinitum).

Fig. 3.10.13 Newton basins for the twelfth-degree polynomial

Figure 3.10.13 shows the Newton basins picture for the twelfth-degree polynomial equation

$$f(x) = x^{12} - 14x^{10} + 183x^8 - 612x^6$$
$$- 2209x^4 - 35,374x^2 + 38,025 = 0, \qquad (17)$$

which has as its solutions the twelve complex numbers

$$1, \quad 1 \pm 2i, \quad -1, \quad -1 \pm 2i,$$
$$3, \quad 3 \pm 2i, \quad -3, \quad -3 \pm 2i.$$

Twelve different colors are used to distinguish the Newton basins of these twelve solutions of Eq. (17).

Fig. 3.10.14 The flower at the center of Fig. 3.10.13

Fig. 3.10.15 A bud on a petal of the flower in Fig. 3.10.14

Where the fractal common boundary appears to separate basins of different colors, it is studded with "flowers" like the one at the center of Fig. 3.10.13, which is magnified in Fig. 3.10.14. Each of these flowers has ten "leaves" (in the remaining ten colors). Each of these leaves has "buds" like the one shown in Fig. 3.10.15. Each of these buds is encircled with flowers that have leaves that have buds that are encircled with flowers—and so on ad infinitum.

3.10 Problems

In Problems 1 through 20, use Newton's method to find the solution of the given equation $f(x) = 0$ in the indicated interval $[a, b]$ accurate to four decimal places. You may choose the initial guess x_0 either on the basis of a calculator graph or by interpolation between the values $f(a)$ and $f(b)$.

1. $x^2 - 5 = 0$; $[2, 3]$ (to find the positive square root of 5)

2. $x^3 - 2 = 0$; $[1, 2]$ (to find the cube root of 2)

3. $x^5 - 100 = 0$; $[2, 3]$ (to find the fifth root of 100)

4. $x^{3/2} - 10 = 0$; $[4, 5]$ (to find $10^{2/3}$)

5. $x^2 + 3x - 1 = 0$; $[0, 1]$

6. $x^3 + 4x - 1 = 0$; $[0, 1]$

7. $x^6 + 7x^2 - 4 = 0$; $[-1, 0]$

8. $x^3 + 3x^2 + 2x = 10$; $[1, 2]$

9. $x - \cos x = 0$; $[0, 2]$

10. $x^2 - \sin x = 0$; $[0.5, 1.0]$

11. $4x - \sin x = 4$; $[1, 2]$

12. $5x + \cos x = 5$; $[0, 1]$

13. $x^5 + x^4 = 100$; $[2, 3]$

14. $x^5 + 2x^4 + 4x = 5$; $[0, 1]$

15. $x + \tan x = 0$; $[2, 3]$

16. $x + \tan x = 0$; $[11, 12]$

17. $x - e^{-x} = 0$; $[0, 1]$

18. $x^3 - 2x - 5 = 0$; $[2, 3]$ (Newton's own example)

19. $e^x + x - 2 = 0$; $[0, 1]$

20. $e^{-x} - \ln x = 0$; $[1, 2]$

21. (a) Show that Newton's method applied to the equation $x^3 - a = 0$ yields the iteration

$$x_{n+1} = \frac{1}{3}\left(2x_n + \frac{a}{x_n^2}\right)$$

for approximating the cube root of a. (b) Use this iteration to find $\sqrt[3]{2}$ accurate to five decimal places.

22. (a) Show that Newton's method yields the iteration

$$x_{n+1} = \frac{1}{k}\left[(k-1)x_n + \frac{a}{(x_n)^{k-1}}\right]$$

for approximating the kth root of the positive number a. (b) Use this iteration to find $\sqrt[10]{100}$ accurate to five decimal places.

23. Equation (12) has the special form $x = G(x)$, where $G(x) = \frac{1}{2}\cos x$. For an equation of this form, the iterative formula $x_{n+1} = G(x_n)$ *sometimes* produces a sequence $x_1, x_2, x_3, \ldots$ of approximations that converge to a root. In the case of Eq. (12), this *repeated substitution* formula is simply $x_{n+1} = \frac{1}{2}\cos x_n$. Begin with $x_0 = 0.5$ as in Example 4 and retain five decimal places in your computation of the solution of Eq. (12). [*Check:* You should find that $x_8 \approx 0.45018$.]

24. The equation $x^4 = x + 1$ has a solution between $x = 1$ and $x = 2$. Use the initial guess $x_0 = 1.5$ and the method of repeated substitution (see Problem 23) to discover that one of the solutions of this equation is approximately 1.220744. Iterate using the formula

$$x_{n+1} = (x_n + 1)^{1/4}.$$

Then compare the result with what happens when you iterate using the formula

$$x_{n+1} = (x_n)^4 - 1.$$

25. The equation $x^3 - 3x^2 + 1 = 0$ has a solution between $x = 0$ and $x = 1$. To apply the method of repeated substitution (see Problem 23) to this equation, you may write it either in the form

$$x = 3 - \frac{1}{x^2}$$

or in the form

$$x = (3x^2 - 1)^{1/3}.$$

If you begin with $x_0 = 0.5$ in the hope of finding the nearby solution (approximately 0.6527) of the original equation by using each of the preceding iterative formulas, you will observe some of the drawbacks of the method. Describe what goes wrong.

26. Show that Newton's method applied to the equation

$$\frac{1}{x} - a = 0$$

yields the iterative formula

$$x_{n+1} = 2x_n - a(x_n)^2$$

and thus provides a method for approximating the reciprocal $1/a$ without performing any divisions. Such a method is useful because, in most high-speed computers, the operation of division is more time consuming than even several additions and multiplications.

27. Prove that the equation $x^5 + x = 1$ has exactly one real solution. Then use Newton's method to find it with three places correct to the right of the decimal.

In Problems 28 through 30, use Newton's method to find all real roots of the given equation with two digits correct to the right of the decimal. [Suggestion: In order to determine the number of roots and their approximate locations, graph the left-hand and right-hand sides of each equation and observe where the graphs cross.]

28. $x^2 = \cos x$

29. $x = 2 \sin x$

30. $\cos x = -\frac{1}{5}x$ (There are exactly three solutions, as indicated in Fig. 3.10.16.)

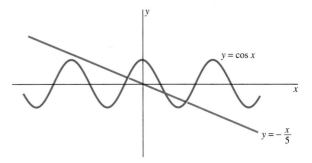

Fig. 3.10.16   Solving the equation of Problem 30

31. Prove that the equation $x^7 - 3x^3 + 1 = 0$ has at least one solution. Then use Newton's method to find one solution.

32. Use Newton's method to approximate $\sqrt[3]{5}$ to three-place accuracy.

33. Use Newton's method to find the value of x for which $x^3 = \cos x$.

34. Use Newton's method to find the smallest positive value of x for which $x = \tan x$.

35. In Problem 49 of Section 3.6, we dealt with the problem of minimizing the cost of building a road to two points on opposite sides of a geologic fault. This problem led to the equation

$$f(x) = 3x^4 - 24x^3 + 51x^2 - 32x + 64 = 0.$$

Use Newton's method to find, to four-place accuracy, the root of this equation that lies in the interval [3, 4].

36. The moon of Planet Gzyx has an elliptical orbit with eccentricity 0.5, and its period of revolution about the planet is 100 days. If the moon is at the position $(a, 0)$ when $t = 0$, then (Fig. 3.10.17) the central angle after t days is given by *Kepler's equation*

$$\frac{2\pi t}{100} = \theta - \frac{1}{2} \sin \theta.$$

Use Newton's method to solve for θ when $t = 17$ (days). Take $\theta_0 = 1.5$ (rad), and calculate the first two approximations θ_1 and θ_2. Express θ_2 in degrees as well.

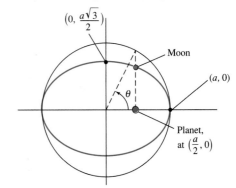

Fig. 3.10.17   The elliptical orbit of Problem 36

37. A great problem of Archimedes was that of using a plane to cut a sphere into two segments with volumes in a given (preassigned) ratio. Archimedes showed that the volume of a segment of height h of a sphere of radius a is $V = \frac{1}{3}\pi h^2(3a - h)$. If a plane at distance x from the center of a sphere of radius 1 cuts the sphere into two segments, one with twice the volume of the other, show that $3x^3 - 9x + 2 = 0$. Then use Newton's method to find x accurate to four decimal places.

38. The equation $f(x) = x^3 - 4x + 1 = 0$ has three distinct real roots. Locate them by calculating the values of f for $x = -3, -2, -1, 0, 1, 2,$ and 3. Then use Newton's method to approximate each of the three roots to four-place accuracy.

39. The equation $x + \tan x = 0$ is important in a variety of applications—for example, in the study of the diffusion of heat. It has a sequence $\alpha_1, \alpha_2, \alpha_3, \ldots$ of positive roots, with the nth one slightly larger than $(n - 0.5)\pi$. Use Newton's method to compute α_1 and α_2 to three-place accuracy.

40. Find the remaining positive solutions of the equation $3 \sin x - \ln x = 0$ of Example 5 (Fig. 3.10.7). Do whatever is necessary to determine if there is or is not a solution near $x = 20$.

3.10 Projects

These projects require the use of a calculator or computer with which Newton's method can be implemented efficiently.

PROJECT A Figure 3.10.18 shows a large cork ball of radius 1 ft floating in water. If its density is one-fourth that of water, then Archimedes' law of buoyancy implies that the ball floats in water with one-fourth its total volume submerged. Because the volume of the cork ball is $4\pi/3$, the volume of the part of the ball beneath the waterline is $V = \pi/3$.

The volume of a spherical segment of radius r and height $h = x$ (as in Fig. 3.10.18) is given by the formula

$$V = \frac{\pi x}{6}(3r^2 + x^2).$$

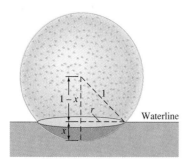

Fig. 3.10.18 The floating cork ball

This formula was derived by Archimedes.

Proceed as follows to find the depth x to which the ball sinks in the water. Equate the two previous expressions for V, and then use the right triangle in Fig. 3.10.18 to eliminate r. You should find that x must be a solution of the cubic equation

$$f(x) = x^3 - 3x^2 + 1 = 0. \qquad (1)$$

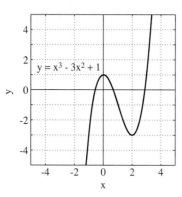

Fig. 3.10.19 Graph for the cork-ball equation

As the graph of f in Fig. 3.10.19 indicates, this equation has three real solutions—one in $[-1, 0]$, one in $[0, 1]$, and one in $[2, 3]$. The solution between 0 and 1 gives the actual depth x to which the ball sinks. But use Newton's method to find all three solutions accurate to four decimal places.

NEWTON'S METHOD WITH CALCULATORS AND COMPUTERS

With calculators and computers that permit user-defined functions, Newton's method is very easy to set up and apply repeatedly. It is helpful to interpret Newton's iteration

$$x_{n+1} = x_n - \frac{f(x_n)}{f'(x_n)} \qquad (2)$$

as follows. Having first defined the functions f and f', we then define the "iteration function"

$$g(x) = x - \frac{f(x)}{f'(x)}. \tag{3}$$

Newton's method is then equivalent to the following procedure. Begin with an initial estimate x_0 of the solution of the equation

$$f(x) = 0.$$

Calculate successive approximations $x_1, x_2, x_3, \ldots$ to the exact solution by means of the iteration

$$x_{n+1} = g(x_n). \tag{4}$$

That is, apply the function g to each approximation to get the next.

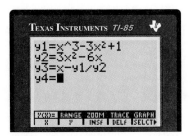

Fig. 3.10.20 Preparing to solve the cork-ball equation

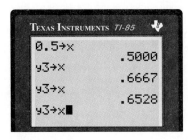

Fig. 3.10.21 Solving the cork-ball equation

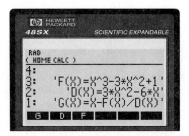

Fig. 3.10.22 Preparing to solve the cork-ball equation

Figure 3.10.20 shows a TI graphics calculator prepared to solve the cork ball equation in Eq. (1). Then we need only store the initial guess, $0.5 \rightarrow X$, and enter repeatedly the command $Y3 \rightarrow X$, as indicated in Fig. 3.10.21.

Figure 3.10.22 shows an HP calculator prepared to carry out the same iteration. The functions $F(X)$, $D(X)$ (for $f'(x)$), and $G(X)$ are each defined by pressing the DEFINE key. Then it is necessary only to ENTER the initial guess x_0 and press the G key repeatedly to generate the desired successive approximations.

With *Maple* or *Mathematica* you can define the functions f and g and then repeatedly enter the command $x = g(x)$, as shown in Fig. 3.10.23. The implementation of Newton's method using *Derive* or *X(plore)* is similar.

Mathematica command	*Maple* command	Result
f[x_] := x^3 − 3 x^2 + 1	f:= x -> x^3 − 3*x^2 + 1;	
g[x_] := x − f[x]/f'[x]	g := x -> x − f(x)/D(f)(x);	
x = 0.5	x := 0.5;	0.500000
x = g[x]	x := g(x);	0.666667
x = g[x]	x := g(x);	0.652778
x = g[x]	x := g(x);	0.652704
x = g[x]	x := g(x);	0.652704

Fig. 3.10.23 *Mathematica* and *Maple* implementations of Newton's method

PROJECT B Investigate the cubic equation

$$4x^3 - 42x^2 - 19x - 28 = 0.$$

Perhaps you can see graphically that it has only a single real solution. Find it (accurate to four decimal places). First try the initial guess $x_0 = 0$; be prepared for at least 25 iterations. Then try initial guesses $x_0 = 10$ and $x_0 = 100$.

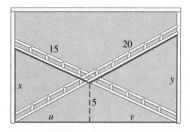

Fig. 3.10.24 The crossing ladders of Project C

PROJECT C   A 15-ft ladder and a 20-ft ladder lean in opposite directions against the vertical walls of a hall (Fig. 3.10.24). The ladders cross at a height of 5 ft. You must find the width w of the hall. First, let x and y denote the heights of the tops of the ladders on the walls and u and v denote the lengths shown in the figure, so that $w = u + v$. Use similar triangles to show that

$$x = 5\left(1 + \frac{u}{v}\right), \qquad y = 5\left(1 + \frac{v}{u}\right).$$

Then apply the Pythagorean theorem to show that $t = u/v$ satisfies the equation

$$t^4 + 2t^3 - 7t^2 - 2t - 1 = 0.$$

Finally, use Newton's method to find first the possible values of t and then those of w, accurate to four decimal places.

Chapter 3 Review: FORMULAS, CONCEPTS, DEFINITIONS

DIFFERENTIATION FORMULAS

$$D_x(cu) = c\frac{du}{dx}$$

$$D_x(u + v) = \frac{du}{dx} + \frac{dv}{dx}$$

$$D_x(uv) = u\frac{dv}{dx} + v\frac{du}{dx}$$

$$D_x\frac{u}{v} = \frac{v\dfrac{du}{dx} - u\dfrac{dv}{dx}}{v^2}$$

$$D_x\, g(u) = g'(u)\frac{du}{dx}$$

$$D_x(u^r) = ru^{r-1}\frac{du}{dx}$$

$$D_x \sin u = (\cos u)\frac{du}{dx}, \quad D_x \cos u = (-\sin u)\frac{du}{dx}$$

$$D_x \ln u = \frac{1}{u}\frac{du}{dx}, \quad D_x\, e^u = e^u\frac{du}{dx}$$

Use the list below as a guide to concepts that you may need to review.

1. Definition of the derivative
2. Average rate of change of a function
3. Instantaneous rate of change of a function

4. Position function; velocity and acceleration
5. Differential, function, and operator notation for derivatives
6. The power rule
7. The binomial formula
8. Linearity of differentiation
9. The product rule
10. The reciprocal rule
11. The quotient rule
12. The chain rule
13. The generalized power rule
14. Vertical tangent lines
15. Local maxima and minima
16. $f'(c) = 0$ as a necessary condition for local extrema
17. Absolute (or global) extrema
18. Critical points
19. The closed-interval maximum-minimum method
20. Steps in the solution of applied maximum-minimum problems
21. Derivatives of the sine and cosine functions
22. Derivatives of the other four trigonometric functions
23. Derivatives of e^x and $\ln x$
24. Implicit differentiation
25. Solving related-rates problems
26. Newton's method

Chapter 3 Miscellaneous Problems

Find dy/dx in Problems 1 through 35.

1. $y = x^2 + \dfrac{3}{x^2}$

2. $y^2 = x^2$

3. $y = \sqrt{x} + \dfrac{1}{\sqrt[3]{x}}$

4. $y = (x^2 + 4x)^{5/2}$

5. $y = (x - 1)^7(3x + 2)^9$

6. $y = \dfrac{x^4 + x^2}{x^2 + x + 1}$

7. $y = \left(3x - \dfrac{1}{2x^2}\right)^4$

8. $y = x^{10} \sin 10x$

9. $xy = 9$

10. $y = \sqrt{\dfrac{1}{5x^6}}$

11. $y = \dfrac{1}{\sqrt{(x^3 - x)^3}}$

12. $y = \sqrt[3]{2x + 1}\ \sqrt[5]{3x - 2}$

13. $y = \dfrac{1}{1 + u^2}$, where $u = \dfrac{1}{1 + x^2}$

14. $x^3 = \sin^2 y$

15. $y = \left(\sqrt{x} + \sqrt[3]{2x}\right)^{7/3}$

16. $y = \sqrt{3x^5 - 4x^2}$

17. $y = \dfrac{u + 1}{u - 1}$, where $u = \sqrt{x + 1}$

18. $y = \sin(2 \cos 3x)$

19. $x^2y^2 = x + y$

20. $y = \sqrt{1 + \sin \sqrt{3}}$

21. $y = \sqrt{x + \sqrt{2x + \sqrt{3x}}}$

22. $y = \dfrac{x + \sin x}{x^2 + \cos x}$

23. $\sqrt[3]{x} + \sqrt[3]{y} = 4$

24. $x^3 + y^3 = xy$

25. $y = (1 + 2u)^3$, where $u = \dfrac{1}{(1 + x)^3}$

26. $y = \cos^2(\sin^2 x)$

27. $y = \sqrt{\dfrac{\sin^2 x}{1 + \cos x}}$

28. $y = \left(1 + \sqrt{x}\right)^3\left(1 - 2\sqrt[3]{x}\right)^4$

29. $y = \dfrac{\cos 2x}{\sqrt{\sin 3x}}$

30. $x^3 - x^2y + xy^2 - y^3 = 4$

31. $y = e^x \cos x$

32. $y = e^{-2x} \sin 3x$

33. $y = [1 + (2 + 3e^x)^{-3/2}]^{2/3}$

34. $y = (e^x + e^{-x})^5$ 　　　**35.** $y = \cos^3\left(\sqrt[3]{1 + \ln x}\right)$

In Problems 36 through 39, find the line tangent to the given curve at the indicated point.

36. $y = \dfrac{x + 1}{x - 1};\quad (0, -1)$

37. $x = \sin 2y;\quad (1, \pi/4)$

38. $x^2 - 3xy + 2y^2 = 0;\quad (2, 1)$

39. $y^3 = x^2 + x;\quad (0, 0)$

40. If a hemispherical bowl with radius 1 ft is filled with water to a depth of x in., the volume of water in the bowl is

$$V = \dfrac{\pi}{3}(36x^2 - x^3) \quad (\text{in.}^3).$$

If the water flows out a hole at the bottom of the bowl at the rate of 36π in.³/s, how fast is x decreasing when $x = 6$ in.?

41. Falling sand forms a conical sandpile. Its height h always remains twice its radius r while both are increasing. If sand is falling onto the pile at the rate of 25π ft³/min, how fast is r increasing when $r = 5$ ft?

Find the limits in Problems 42 through 47.

42. $\lim\limits_{x \to 0} \dfrac{x - \tan x}{\sin x}$ 　　　**43.** $\lim\limits_{x \to 0} x \cot 3x$

44. $\lim\limits_{x \to 0} \dfrac{\sin 2x}{\sin 5x}$ 　　　**45.** $\lim\limits_{x \to 0} x^2 \csc 2x \cot 2x$

46. $\lim\limits_{x \to 0} x^2 \sin \dfrac{1}{x^2}$ 　　　**47.** $\lim\limits_{x \to 0^+} \sqrt{x} \sin \dfrac{1}{x}$

In Problems 48 through 53, identify two functions f and g such that $h(x) = f(g(x))$. Then apply the chain rule to find $h'(x)$.

48. $h(x) = \sqrt[3]{x + x^4}$ 　　　**49.** $h(x) = \dfrac{1}{\sqrt{x^2 + 25}}$

50. $h(x) = \sqrt{\dfrac{x}{x^2 + 1}}$ 　　　**51.** $h(x) = \sqrt[3]{(x - 1)^5}$

52. $h(x) = \dfrac{(x + 1)^{10}}{(x - 1)^{10}}$ 　　　**53.** $h(x) = \cos(x^2 + 1)$

54. The period T of oscillation (in seconds) of a simple pendulum of length L (in feet) is given by $T = 2\pi\sqrt{L/32}$. What is the rate of change of T with respect to L when $L = 4$ ft?

55. What is the rate of change of the volume $V = 4\pi r^3/3$ of a sphere with respect to its surface area $S = 4\pi r^2$?

56. What is an equation for the straight line through $(1, 0)$ that is tangent to the graph of

$$h(x) = x + \frac{1}{x}$$

at a point in the first quadrant?

57. A rocket is launched vertically upward from a point 2 mi west of an observer on the ground. What is the speed of the rocket when the angle of elevation (from the horizontal) of the observer's line of sight to the rocket is $50°$ and is increasing at $5°$ per second?

58. An oil field containing 20 wells has been producing 4000 barrels of oil daily. For each new well drilled, the daily production of each well decreases by 5 barrels. How many new wells should be drilled to maximize the total daily production of the oil field?

59. A triangle is inscribed in a circle of radius R. One side of the triangle coincides with a diameter of the circle. In terms of R, what is the maximum possible area of such a triangle?

60. Five rectangular pieces of sheet metal measure 210 cm by 336 cm each. Equal squares are to be cut from all their corners, and the resulting five cross-shaped pieces of metal are to be folded and welded to form five boxes without tops. The 20 little squares that remain are to be assembled in groups of four into five larger squares, and these five larger squares are to be assembled into a cubical box with no top. What is the maximum possible total volume of the six boxes that are constructed in this way?

61. A mass of clay of volume V is formed into two spheres. For what distribution of clay is the total surface area of the two spheres a maximum? A minimum?

62. A right triangle has legs of lengths 3 m and 4 m. What is the maximum possible area of a rectangle inscribed in the triangle in the "obvious" way—with one corner at the triangle's right angle, two adjacent sides of the rectangle lying on the triangle's legs, and the opposite corner on the hypotenuse?

63. What is the maximum possible volume of a right circular cone inscribed in a sphere of radius R?

64. A farmer has 400 ft of fencing with which to build a rectangular corral. He will use some or even all of an existing straight wall 100 ft long as part of the perimeter of the corral. What is the maximum area that can be enclosed?

65. In one simple model of the spread of a contagious disease among members of a population of M people, the incidence of the disease, measured as the number of new cases per day, is given in terms of the number x of individuals already infected by

$$R(x) = kx(M - x) = kMx - kx^2,$$

where k is some positive constant. How many individuals in the population are infected when the incidence R is the greatest?

66. Three sides of a trapezoid have length L, a constant. What should be the length of the fourth side if the trapezoid is to have maximum area?

67. A box with no top must have a base twice as long as it is wide, and the total surface area of the box is to be 54 ft². What is the maximum possible volume of such a box?

68. A small right circular cone is inscribed in a larger one (Fig. 3.MP.1). The larger cone has fixed radius R and fixed altitude H. What is the largest fraction of the volume of the larger cone that the smaller one can occupy?

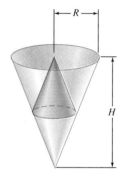

Fig. 3.MP.1 A small cone inscribed in a larger one (Problem 68)

69. Two vertices of a trapezoid are at $(-2, 0)$ and $(2, 0)$, and the other two lie on the semicircle $x^2 + y^2 = 4$, $y \geqq 0$. What is the maximum possible area of the trapezoid? [*Note:* The area of a trapezoid with bases b_1 and b_2 and height h is $A = h(b_1 + b_2)/2$.]

70. Suppose that f is a differentiable function defined on the whole real number line $\mathbf{R}$ and that the graph of f contains a point $Q(x, y)$ closest to the point $P(x_0, y_0)$ not on the graph. Show that

$$f'(x) = -\frac{x - x_0}{y - y_0}$$

at Q. Conclude that the segment PQ is perpendicular to the line tangent to the curve at Q. [*Suggestion:* Minimize the square of the distance PQ.]

71. Use the result of Problem 70 to show that the minimum distance from the point (x_0, y_0) to a point of the straight line $Ax + By + C = 0$ is

$$\frac{|Ax_0 + By_0 + C|}{\sqrt{A^2 + B^2}}.$$

72. A race track is to be built in the shape of two parallel and equal straightaways connected by semicircles on each end (Fig. 3.MP.2). The length of the track, one lap, is to be exactly 5 km. What should its design be to maximize the rectangular area within it?

Fig. 3.MP.2 Design the race track to maximize the shaded area (Problem 71).

73. Two towns are located on the straight shore of a lake. Their nearest distances to points on the shore are 1 mi and 2 mi, respectively, and these points on the shore are 6 mi apart. Where should a fishing pier be located to minimize the total amount of paving necessary to build a straight road from each town to the pier?

74. A hiker finds herself in a forest 2 km from a long straight road. She wants to walk to her cabin, which is 10 km away in the forest and also 2 km from the road (Fig. 3.MP.3). She can walk at a rate of 8 km/h along the road but only 3 km/h through the forest. So she decides to walk first to the road, then along the road, and finally through the forest to the cabin. What angle θ (shown in the figure) would minimize the total time required for the hiker to reach her cabin? How much time is saved in comparison with the straight route through the forest?

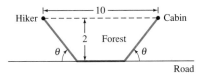

Fig. 3.MP.3 The hiker's quickest path to the cabin (Problem 74)

75. When an arrow is shot from the origin with initial velocity v and initial angle of inclination α (from the horizontal x-axis, which represents the ground), then its trajectory is the curve

$$y = mx - \frac{16}{v^2}(1 + m^2)x^2,$$

where $m = \tan \alpha$. (a) Find the maximum height reached by the arrow in terms of m and v. (b) For what value of m (and hence, for what α) does the arrow travel the greatest horizontal distance?

76. A projectile is fired with initial velocity v and angle of elevation θ from the base of a plane inclined at 45° from

the horizontal (Fig. 3.MP.4). The range of the projectile, as measured up this slope, is given by

$$R = \frac{v^2\sqrt{2}}{16}(\cos \theta \sin \theta - \cos^2 \theta).$$

What value of θ maximizes R?

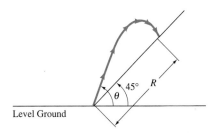

Fig. 3.MP.4 A projectile fired uphill (Problem 76)

In Problems 77 through 88, use Newton's method to find the solution of the given equation $f(x) = 0$ in the indicated interval $[a, b]$ accurate to four decimal places.

77. $x^2 - 7 = 0$; $[2, 3]$ (to find the positive square root of 7)

78. $x^3 - 3 = 0$; $[1, 2]$ (to find the cube root of 3)

79. $x^5 - 75 = 0$; $[2, 3]$ (to find the fifth root of 75)

80. $x^{4/3} - 10 = 0$; $[5, 6]$ (to approximate $10^{3/4}$)

81. $x^3 - 3x - 1 = 0$; $[-1, 0]$

82. $x^3 - 4x - 1 = 0$; $[-1, 0]$

83. $x^6 + 7x^2 - 4 = 0$; $[0, 1]$

84. $x^3 - 3x^2 + 2x + 10 = 0$; $[-2, -1]$

85. $x + \cos x = 0$; $[-2, 0]$

86. $x^2 + \sin x = 0$; $[-1.0, -0.5]$

87. $4x - \sin x + 4 = 0$; $[-2, -1]$

88. $5x - \cos x + 5 = 0$; $[-1, 0]$

89. Find the depth to which a wooden ball with radius 2 ft sinks in water if its density is one-third that of water. A useful formula appears in Problem 37 of Section 3.10.

90. The equation $x^2 + 1 = 0$ has no real solutions. Try finding a solution by using Newton's method, and report what happens. Use the initial estimate $x_0 = 2$.

91. At the beginning of Section 3.10 we mentioned the fifth-degree equation $x^5 - 3x^3 + x^2 - 23x + 19 = 0$; its graph appears in Fig. 3.10.1. The graph makes it clear that this equation has exactly three real solutions. Find all of them, to four-place accuracy, using Newton's method.

92. The equation

$$\tan x = \frac{1}{x}$$

has a sequence α_1, α_2, α_3, ... of positive roots, with α_n slightly larger than $(n - 1)\pi$. Use Newton's method to approximate α_1 and α_2 to three-place accuracy.

93. Criticize the following "proof" that $3 = 2$. Begin by writing

$$x^3 = x \cdot x^2 = x^2 + x^2 + \cdots + x^2 \qquad (x \text{ summands}).$$

Differentiate to obtain

$$3x^2 = 2x + 2x + \cdots + 2x \qquad (\text{still } x \text{ summands}).$$

Thus $3x^2 = 2x^2$, and "therefore" $3 = 2$.

If we substitute $z = x + h$ into the definition of the derivative, the result is

$$f'(x) = \lim_{z \to x} \frac{f(z) - f(x)}{z - x}.$$

Use this formula in Problems 94 and 95, together with the formula

$$a^3 - b^3 = (a - b)(a^2 + ab + b^2)$$

for factoring the difference of two cubes.

94. Show that

$$D x^{3/2} = \lim_{z \to x} \frac{z^{3/2} - x^{3/2}}{z - x} = \frac{3}{2} x^{1/2}.$$

[*Suggestion*: Factor the numerator as a difference of cubes and the denominator as a difference of squares.]

95. Prove that

$$D x^{2/3} = \lim_{z \to x} \frac{z^{2/3} - x^{2/3}}{z - x} = \frac{2}{3} x^{-1/3}.$$

[*Suggestion*: Factor the numerator as a difference of squares and the denominator as a difference of cubes.]

96. A rectangular block with square base is being squeezed in such a way that its height y is decreasing at the rate of 2 cm/min while its volume remains constant. At what rate is the edge x of its base increasing when $x = 30$ cm and $y = 20$ cm?

97. Air is being pumped into a spherical balloon at the constant rate of 10 in.^{3}/s. At what rate is the surface area of the balloon increasing when its radius is 5 in.?

98. A ladder 10 ft long is leaning against a wall. If the bottom of the ladder slides away from the wall at the constant rate of 1 mi/h, how fast (in miles per hour) is the top of the ladder moving when it is 0.01 ft above the ground?

99. A water tank in the shape of an inverted cone, axis vertical and vertex downward, has a top radius of 5 ft and height 10 ft. The water is flowing out of the tank through a hole at the vertex at the rate of 50 ft^3/min. What is the time rate of change of the water depth at the instant when the water is 6 ft deep?

100. Plane A is flying west toward an airport at an altitude of 2 mi. Plane B is flying south toward the same airport at an altitude of 3 mi. When both planes are 2 mi (ground distance) from the airport, the speed of plane A is 500 mi/h and the distance between the two planes is decreasing at 600 mi/h. What is the speed of plane B then?

101. A water tank is shaped such that the volume of water in the tank is $V = 2y^{3/2}$ in.^{3} when its depth is y inches. If water flows out a hole at the bottom at the rate of $3\sqrt{y}$ in.^{3}/min, at what rate does the water level in the tank fall? Can you think of a practical application for such a water tank?

102. Water is being poured into the conical tank of Problem 99 at the rate of 50 ft^3/min and is draining out the hole at the bottom at the rate of $10\sqrt{y}$ ft^3/min, where y is the depth of water in the tank. (a) At what rate is the water level rising when the water is 5 ft deep? (b) Suppose that the tank is initially empty, water is poured in at 25 ft^3/min, and water continues to drain at $10\sqrt{y}$ ft^3/min. What is the maximum depth attained by the water?

103. Let L be a straight line passing through the fixed point $P(x_0, y_0)$ and tangent to the parabola $y = x^2$ at the point $Q(a, a^2)$. (a) Show that $a^2 - 2ax_0 + y_0 = 0$. (b) Apply the quadratic formula to show that if $y_0 < (x_0)^2$ (that is, if P lies below the parabola), then there are two possible values for a and thus two lines through P that are tangent to the parabola. (c) Similarly, show that if $y_0 > (x_0)^2$ (P lies above the parabola), then no line through P is also tangent to the parabola.

Additional Applications of the Derivative

G. W. Leibniz (1646–1716)

❑ Gottfried Wilhelm Leibniz entered the University of Leipzig when he was 15, studied philosophy and law, graduated at 17, and received his doctorate in philosophy at 21. Upon completion of his academic work, Leibniz entered the political and governmental service of the Elector of Mainz (Germany). His serious study of mathematics did not begin until 1672 (when he was 26) when he was sent to Paris on a diplomatic mission. During the next four years there he conceived the principal features of calculus. For this work he is remembered (with Newton) as a co-discoverer of the subject. Newton's discoveries had come slightly earlier (in the late 1660s), but Leibniz's were the first to be published, beginning in 1684. Despite an unfortunate

priority dispute between supporters of Newton and supporters of Leibniz that raged for more than a century, it is clear now that the discoveries were made independently.

❑ Throughout his life, Leibniz sought a universal language incorporating notation and terminology that would provide *all* educated people with the powers of clear and correct reasoning in all subjects. But only in mathematics did he largely accomplish this goal. His differential notation for calculus is arguably the best example of a system of notation chosen so as to mirror perfectly the basic operations and processes of the subject. Indeed, it can be said that Leibniz's notation for calculus brings within the range of ordinary students problems that once required the ingenuity of an Archimedes or a Newton. For this reason, Leibniz's approach to calculus dominated during the eighteenth century, even though Newton's somewhat different approach may have been closer to our modern understanding of the subject.

❑ The origin of differential notation was an infinitesimal right triangle with legs dx and dy and with hypotenuse a tiny segment of the curve $y = f(x)$. Leibniz later described the moment he first visualized this "characteristic

triangle" as a burst of light that was the inception of his calculus. Indeed, he sometimes referred to his calculus as "my method of the Characteristic Triangle."

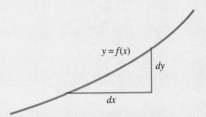

Leibniz's characteristic triangle

❑ The following excerpt shows the opening paragraphs of Leibniz' first published article (in the 1684 *Acta Eruditorum*) in which differential notation initially appeared. In the fifth line of the second paragraph, the product rule for differentiation is expressed as

$$d(xv) = x\,dv + v\,dx.$$

MENSIS OCTOBRIS A. M DC LXXXIV. 467
*NOVA METHODVS PRO MAXIMIS ET Mi-
nimis, itemque tangentibus, quæ nec fractas, nec irrati-
onales quantitates moratur, & singulare pro
illis calculi genus, per G.G.L.*

Sit axis AX, & curvæ plures, ut VV, WW, YY, ZZ, quarum ordi-
natæ, ad axem normales, VX, WX, YX, ZX, quæ vocentur respectivè, v, vv, y, z, & ipsæ AX abscissæ ab axe, vocetur x. Tangentes sint
VB, WC, YD, ZE axi occurrentes respectivè in punctis B, C, D, E.
Jam recta aliqua pro arbitrio assumta vocetur dx, & recta quæ sit ad
dx, ut v (vel vv, vel y, vel z) est ad VB (vel WC, vel YD, vel ZE) vocetur dv (vel dv, vel dy vel dz) sive differentia ipsarum v (vel ipsarum vv, aut y, aut z) His positis calculi regulæ erunt tales:

Sit a quantitas data constans, erit da æqualis o, & d ax erit æqu-
a dx: si sit y æqu. v (seu ordinata quævis curvæ YY, æqualis cuivis or-
dinatæ respondenti curvæ VV) erit dy æqu. dv. Jam *Additio & Sub-
tractio*: si sit z−y +̇ vv +̇ x æqu. v, erit d z−y +̇ vv +̇ x seu dv, æqu.
dz −dy +̇ dvv +̇ dx. *Multiplicatio*, d x v æqu. x dv +̇ v dx, seu posito
y æqu. xv, fiet dy æqu. x dv +̇ v dx. In arbitrio enim est vel formulam,
ut xv, vel compendio pro ea literam, ut y, adhibere. Notandum & x
& dx eodem modo in hoc calculo tractari, ut y & dy, vel aliam literam
indeterminatam cum sua differentiali. Notandum etiam non dari
semper regressum a differentiali Æquatione, nisi cum quadam cautio-

4.1
Introduction

We learned in Chapter 3 how to differentiate a wide variety of algebraic, trigonometric, exponential, and logarithmic functions. We saw that derivatives have such diverse applications as maximum-minimum problems, related-rates problems, and the solution of equations by Newton's method. The further applications of differentiation that we discuss in this chapter all depend ultimately upon a single fundamental question. Suppose that $y = f(x)$ is a differentiable function defined on the closed interval $[a, b]$ of length $\Delta x = b - a$. Then the *increment* Δy in the value of $f(x)$ as x changes from $x = a$ to $x = b = a + \Delta x$ is

$$\Delta y = f(b) - f(a). \tag{1}$$

The question is this: How is the increment Δy related to the derivative—the rate of change—of the function f at the points of the interval $[a, b]$?

An *approximate* answer is given in Section 4.2. If the function continued throughout the interval with the same rate of change $f'(a)$ that it had at $x = a$, then the change in its value would be $f'(a)(b - a) = f'(a) \Delta x$. This observation motivates the tentative approximation

$$\Delta y \approx f'(a) \Delta x. \tag{2}$$

A precise answer to the preceding question is provided by the mean value theorem of Section 4.3. This theorem implies that the exact increment is given by

$$\Delta y = f'(c) \Delta x \tag{3}$$

for some number c in (a, b). The mean value theorem is the central theoretical result of differential calculus and is also the key to many of the more advanced applications of derivatives.

4.2
Increments, Differentials, and Linear Approximation

Sometimes we need a quick and simple estimate of the change in $f(x)$ that results from a given change in x. We write y for $f(x)$ and suppose first that the change in the independent variable is the *increment* Δx, so that x changes from its original value to the new value $x + \Delta x$. The change in the value of y is the **increment** Δy, computed by subtracting the old value of y from its new value:

$$\Delta y = f(x + \Delta x) - f(x). \tag{1}$$

The increments Δx and Δy are represented geometrically in Fig. 4.2.1.

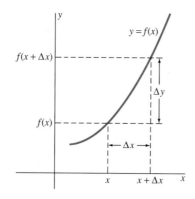

Fig. 4.2.1   The increments Δx and Δy

Ch. 4 / Additional Applications of the Derivative

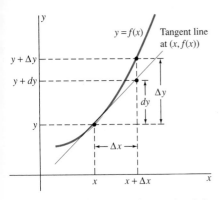

Fig. 4.2.2 The estimate dy of the actual increment Δy

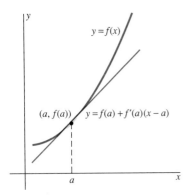

Fig. 4.2.3 The linear approximation $y = f(a) + f'(a)(x - a)$

Now we compare the actual increment Δy with the change that *would* occur in the value of y *if* it continued to change at the *fixed* rate $f'(x)$ while the value of the independent variable changes from x to $x + \Delta x$. This hypothetical change in y is the **differential**

$$dy = f'(x)\,\Delta x. \tag{2}$$

As Fig. 4.2.2 shows, dy is the change in height of a point that moves along the tangent line at the point $(x, f(x))$ rather than along the curve $y = f(x)$.

Think of x as fixed. Then Eq. (2) shows that the differential dy is a *linear* function of the increment Δx. For this reason, dy is called the **linear approximation** to the increment Δy. We can approximate $f(x + \Delta x)$ by substituting dy for Δy:

$$f(x + \Delta x) = y + \Delta y \approx y + dy.$$

Because $y = f(x)$ and $dy = f'(x)\,\Delta x$, this gives the **linear approximation formula**

$$f(x + \Delta x) \approx f(x) + f'(x)\,\Delta x. \tag{3}$$

The point is that this approximation is a "good" one, at least when Δx is relatively small. If we combine Eqs. (1), (2), and (3), we see that

$$\Delta y \approx f'(x)\,\Delta x = dy. \tag{4}$$

Thus the differential $dy = f'(x)\,\Delta x$ is a good approximation to the increment $\Delta y = f(x + \Delta x) - f(x)$.

If we replace x with a in Eq. (3), we get the approximation

$$f(a + \Delta x) \approx f(a) + f'(a)\,\Delta x. \tag{5}$$

If we now write $\Delta x = x - a$, so that $x = a + \Delta x$, the result is

$$f(x) \approx f(a) + f'(a)(x - a). \tag{6}$$

Because the right-hand side in Eq. (6) is a linear function of x, we call the right-hand side the **linear approximation to the function f near the point $x = a$** (Fig. 4.2.3).

EXAMPLE 1 Find the linear approximation to the function $f(x) = \sqrt{1 + x}$ near the point $a = 0$.

Solution Note that $f(0) = 1$ and that

$$f'(x) = \frac{1}{2}(1 + x)^{-1/2} = \frac{1}{2\sqrt{1 + x}},$$

so $f'(0) = \frac{1}{2}$. Hence Eq. (6) with $a = 0$ yields

$$f(x) \approx f(0) + f'(0)(x - 0) = 1 + \tfrac{1}{2}x;$$

that is,

$$\sqrt{1 + x} \approx 1 + \tfrac{1}{2}x. \tag{7}$$

IMPORTANT   The linear approximation in Eq. (7) is likely to be accurate only if x is close to zero. For instance, the approximations

$$\sqrt{1.1} \approx 1 + \tfrac{1}{2}(0.1) = 1.05 \quad \text{and} \quad \sqrt{1.03} \approx 1 + \tfrac{1}{2}(0.03) = 1.015,$$

using $x = 0.1$ and $x = 0.03$, are accurate to two and three decimal places (rounded), respectively. But

$$\sqrt{3} \approx 1 + \tfrac{1}{2} \cdot 2 = 2,$$

using $x = 2$, is a very poor approximation to $\sqrt{3} \approx 1.732$.

The approximation $\sqrt{1 + x} \approx 1 + \tfrac{1}{2}x$ is a special case of the approximation

$$(1 + x)^k \approx 1 + kx \qquad (8)$$

(k is a constant, x is near zero), an approximation with numerous applications. The derivation of Eq. (8) is similar to Example 1 (see Problem 39).

EXAMPLE 2 Use the linear approximation formula to approximate $(122)^{2/3}$. Note that

$$(125)^{2/3} = [(125)^{1/3}]^2 = 5^2 = 25.$$

Solution   We need to approximate a particular value of $x^{2/3}$, so our strategy is to apply Eq. (6) with $f(x) = x^{2/3}$. We first note that $f'(x) = \tfrac{2}{3}x^{-1/3}$. We choose $a = 125$, because we know the *exact* values

$$f(125) = (125)^{2/3} = 25 \quad \text{and} \quad f'(125) = \tfrac{2}{3}(125)^{-1/3} = \tfrac{2}{15}$$

and because 125 is relatively close to 122. Then the linear approximation in (6) to $f(x) = x^{2/3}$ near $a = 125$ takes the form

$$f(x) \approx f(125) + f'(125)(x - 125);$$

that is,

$$x^{2/3} \approx 25 + \tfrac{2}{15}(x - 125).$$

With $x = 122$ we get

$$(122)^{2/3} \approx 25 + \tfrac{2}{15}(-3) = 24.6.$$

Thus $(122)^{2/3}$ is approximately 24.6. The actual value of $(122)^{2/3}$ is about 24.5984, so the formula in (6) gives a relatively good approximation in this case.

EXAMPLE 3 Use the linear approximation formula to approximate $\sqrt[10]{e}$.

Solution   We take $f(x) = e^x$, so $f'(x) = e^x$ as well. Then

$$f(0) = f'(0) = e^0 = 1,$$

so the linear approximation in (6) to $f(x) = e^x$ near $a = 0$ takes the form

$$f(x) \approx f(0) + f'(0)(x - 0);$$

that is,

$$e^x \approx 1 + x.$$

With $x = 0.1$ we obtain

$$e^{0.1} \approx 1 + 0.1 = 1.1.$$

(To five decimal places, the correct value is 1.10517.)

EXAMPLE 4 A hemispherical bowl of radius 10 in. is filled with water to a depth of x inches. The volume V of water in the bowl (in cubic inches) is given by the formula

$$V = \frac{\pi}{3}(30x^2 - x^3) \qquad (9)$$

Fig. 4.2.4 The bowl of Example 4

(Fig. 4.2.4). (You will be able to derive this formula after you study Chapter 6.) Suppose that you *measure* the depth of water in the bowl to be 5 in. with a maximum possible measured error of $\frac{1}{16}$ in. Estimate the maximum error in the calculated volume of water in the bowl.

Solution The error in the calculated volume $V(5)$ is the difference

$$\Delta V = V(x) - V(5)$$

between the actual volume $V(x)$ and the calculated volume. We do not know the depth x of water in the bowl. We are given only that the difference

$$\Delta x = x - 5$$

between the actual and the measured depths is numerically at most $\frac{1}{16}$ in.: $|\Delta x| \leq \frac{1}{16}$. Because Eq. (9) yields

$$V'(x) = \frac{\pi}{3}(60x - 3x^2) = \pi(20x - x^2),$$

the linear approximation

$$\Delta V \approx dV = V'(5)\,\Delta x$$

at $x = 5$ gives

$$\Delta V \approx \pi(20 \cdot 5 - 5^2)\,\Delta x = 75\pi\,\Delta x.$$

With the common practice in science of writing $\Delta x = \pm\frac{1}{16}$ to signify that $-\frac{1}{16} \leq \Delta x \leq \frac{1}{16}$, this gives

$$\Delta V \approx (75\pi)(\pm\tfrac{1}{16}) \approx \pm 14.73 \quad (\text{in.}^3).$$

The formula in Eq. (9) gives the calculated volume $V(5) \approx 654.50$ in.^{3}, but we now see that this may be in error by almost 15 in.^{3} in either direction.

The **absolute error** in a measured or approximated value is defined to be the remainder when the approximate value is subtracted from the true value. The **relative error** is the ratio of the absolute error to the true value. Thus in Example 3, a relative error in the measured depth x of

$$\frac{\Delta x}{x} = \frac{\frac{1}{16}}{5} = 0.0125 = 1.25\%$$

leads to a relative error in the estimated volume of

$$\frac{dV}{V} \approx \frac{14.73}{654.50} \approx 0.0225 = 2.25\%.$$

The relationship between these two relative errors is of some interest. The formulas for dV and V in Example 4 give

$$\frac{dV}{V} = \frac{\pi(20x - x^2)\,\Delta x}{\frac{1}{3}\pi(30x^2 - x^3)} = \frac{3(20 - x)}{30 - x} \cdot \frac{\Delta x}{x}.$$

When $x = 5$, this gives

$$\frac{dV}{V} = (1.80)\,\frac{\Delta x}{x}.$$

Hence, to approximate the volume of water in the bowl with a relative error of at most 0.5%, for instance, we would need to measure the depth with a relative error of at most (0.5%)/1.8, thus with a relative error of less than 0.3%.

THE ERROR IN LINEAR APPROXIMATION

Now we consider briefly the question of how closely the differential dy approximates the increment Δy. It is apparent in Fig. 4.2.2 that the smaller Δx is, the closer are the corresponding points on the curve $y = f(x)$ and its tangent line. Because the difference in the heights of two such points is the value of $\Delta y - dy$ determined by a particular choice of Δx, we conclude that $\Delta y - dy$ approaches zero as $\Delta x \to 0$.

But even more is true: As $\Delta x \to 0$, the difference $\Delta y - dy$ is small *even in comparison with* Δx. For

$$\frac{\Delta y - dy}{\Delta x} = \frac{f(x + \Delta x) - f(x) - f'(x)\,\Delta x}{\Delta x} = \frac{f(x + \Delta x) - f(x)}{\Delta x} - f'(x).$$

That is,

$$\frac{\Delta y - dy}{\Delta x} = \epsilon, \tag{10}$$

where, by the definition of the derivative $f'(x)$, we see that $\epsilon = \epsilon(\Delta x)$ is a function of Δx that approaches zero as $\Delta x \to 0$. If Δx is "very small," so that ϵ is also "very small," we might well describe the product $\epsilon \cdot \Delta x = \Delta y - dy$ as "very *very* small." These concepts and quantities are illustrated in Fig. 4.2.5.

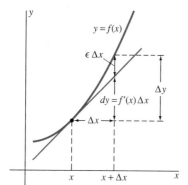

Fig. 4.2.5 The error $\epsilon \cdot \Delta x$ in the linear approximation $\Delta y \approx f'(x)\,\Delta x$

DIFFERENTIALS

The linear approximation formula in (3) is often written with dx in place of Δx:

$$f(x + dx) \approx f(x) + f'(x)\,dx. \tag{11}$$

In this case dx is an independent variable, called the **differential** of x, and x is fixed. Thus the differentials of x and y are defined to be

$$dx = \Delta x \quad \text{and} \quad dy = f'(x)\,\Delta x = f'(x)\,dx. \tag{12}$$

From this definition it follows immediately that

$$\frac{dy}{dx} = \frac{f'(x)\,dx}{dx} = f'(x),$$

in accord with the notation we have been using. Indeed, Leibniz originated differential notation by visualizing "infinitesimal" increments dx and dy (Fig. 4.2.6), with their ratio dy/dx being the slope of the tangent line. The key to Leibniz's independent discovery of differential calculus in the 1670s was his

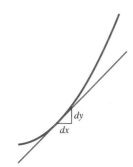

Fig. 4.2.6 The slope of the tangent line as the ratio of the infinitesimals dy and dx

Ch. 4 / Additional Applications of the Derivative

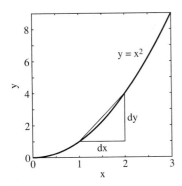

Fig. 4.2.7 $dx = 1$

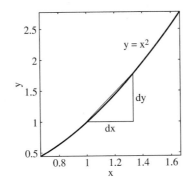

Fig. 4.2.8 $dx = \frac{1}{3}$

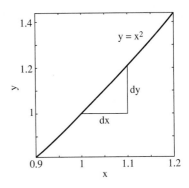

Fig. 4.2.9 $dx = \frac{1}{10}$

insight that if dx and dy are sufficiently small, then the segment of the curve $y = f(x)$ and the straight line segment joining (x, y) and $(x + dx, y + dy)$ are virtually indistinguishable. This insight is illustrated by the successive magnifications in Figs. 4.2.7 through 4.2.9 of the curve $y = x^2$ near the point $(1, 1)$.

Differential notation provides us with a convenient way to write derivative formulas. Suppose that $z = g(u)$, so $dz = g'(u)\, du$. For particular choices of the function g, we get the formulas

$$d(u^n) = nu^{n-1}\, du, \tag{13}$$

$$d(\sin u) = (\cos u)\, du, \tag{14}$$

and so on. Thus we can write differentiation rules in differential form without having to identify the independent variable. The sum, product, and quotient rules take the respective forms

$$d(u + v) = du + dv, \tag{15}$$

$$d(uv) = u\, dv + v\, du, \quad \text{and} \tag{16}$$

$$d\left(\frac{u}{v}\right) = \frac{v\, du - u\, dv}{v^2}. \tag{17}$$

If $u = f(x)$ and $z = g(u)$, we may substitute $du = f'(x)\, dx$ into the formula $dz = g'(u)\, du$. This gives

$$dz = g'(f(x)) \cdot f'(x)\, dx. \tag{18}$$

This is the differential form of the chain rule

$$D_x g(f(x)) = g'(f(x)) \cdot f'(x).$$

Thus the chain rule appears here as though it were the result of mechanical manipulations of the differential notation. This compatibility with the chain rule is one reason for the extraordinary usefulness of differential notation in calculus.

4.2 Problems

In Problems 1 through 16, write dy in terms of x and dx.

1. $y = 3x^2 - \dfrac{4}{x^2}$

2. $y = 2\sqrt{x} - \dfrac{3}{\sqrt[3]{x}}$

3. $y = x - \sqrt{4 - x^3}$

4. $y = \dfrac{1}{x - \sqrt{x}}$

5. $y = 3x^2(x - 3)^{3/2}$

6. $y = \dfrac{x}{x^2 - 4}$

7. $y = x(x^2 + 25)^{1/4}$

8. $y = \dfrac{1}{(x^2 - 1)^{4/3}}$

9. $y = \cos\sqrt{x}$

10. $y = x^2 \sin x$

11. $y = \sin 2x \cos 2x$

12. $y = \cos^3 3x$

13. $y = \dfrac{\sin 2x}{3x}$

14. $y = xe^{2x}$

15. $y = \dfrac{1}{1 - x \sin x}$

16. $y = x^2 \ln x$

In Problems 17 through 24, find—as in Example 1—the linear approximation to the given function f near the point a = 0.

17. $f(x) = \dfrac{1}{1 - x}$

18. $f(x) = \dfrac{1}{\sqrt{1 + x}}$

19. $f(x) = (1 + x)^2$

20. $f(x) = (1 - x)^3$

21. $f(x) = (1 - 2x)^{3/2}$

22. $y = \ln(1 + x)$

23. $f(x) = \sin x$

24. $f(x) = \cos x$

In Problems 25 through 34, use—as in Example 2—a linear approximation to an appropriate function, with an appropriate value of a, to estimate the given number.

25. $\sqrt[3]{25}$

26. $\sqrt{102}$

27. $\sqrt[4]{15}$

28. $\sqrt{80}$

29. $65^{-2/3}$

30. $80^{3/4}$

31. $\cos 43°$

32. $\sin 32°$

33. $\sin 88°$

34. $\ln 1.1$

In Problems 35 through 38, compute the differential of each side of the given equation, regarding x and y as dependent variables (as if both were functions of some third, unspecified variable). Then solve for dy/dx.

35. $x^2 + y^2 = 1$

36. $xe^y = 1$

37. $x^3 + y^3 = 3xy$

38. $x \ln y = 1$

39. Assuming that $D_x x^k = kx^{k-1}$ for any real constant k (which we shall establish in Chapter 7), derive the linear approximation formula $(1 + x)^k \approx 1 + kx$ for x near zero.

In Problems 40 through 47, use linear approximations to estimate the change in the given quantity.

40. The circumference of a circle, if its radius is increased from 10 in. to 10.5 in.

41. The area of a square, if its edge length is decreased from 10 in. to 9.8 in.

42. The surface area of a sphere, if its radius is increased from 5 in. to 5.2 in. (Fig. 4.2.10).

43. The volume of a cylinder, if both its height and its radius are decreased from 15 cm to 14.7 cm (Fig. 4.2.11).

Fig. 4.2.10 The sphere of Problem 42: Area $A = 4\pi r^2$, volume $V = \frac{4}{3}\pi r^3$

Fig. 4.2.11 The cylinder of Problem 43: Volume $V = \pi r^2 h$

44. The volume of the conical sandpile of Fig. 4.2.12, if its radius is 14 in. and its height is increased from 7 in. to 7.1 in.

Fig. 4.2.12 The conical sandpile of Problem 44: Volume $V = \frac{1}{3}\pi r^2 h$

45. The range $R = \frac{1}{16}v^2 \sin 2\theta$ of a shell fired at inclination angle $\theta = 45°$, if its initial velocity v is increased from 80 ft/s to 81 ft/s.

46. The range $R = \frac{1}{16}v^2 \sin 2\theta$ of a projectile fired with initial velocity $v = 80$ ft/s, if its initial inclination angle θ is increased from 45° to 46°.

47. The wattage $W = RI^2$ of a floodlight with resistance $R = 10$ ohms, if the current I is increased from 3 amperes to 3.1 amperes.

48. The equatorial radius of the earth is approximately 3960 mi. Suppose that a wire is wrapped tightly around the earth at the equator. Approximately how much must this wire be lengthened if it is to be strung all the way around the earth on poles 10 ft above the ground? Use the linear approximation formula!

49. The radius of a spherical ball is measured as 10 in., with a maximum error of $\frac{1}{16}$ in. What is the maximum resulting error in its calculated volume?

50. With what accuracy must the radius of the ball of Problem 49 be measured to ensure an error of at most 1 in.³ in its calculated volume?

51. The radius of a hemispherical dome is measured as 100 m with a maximum error of 1 cm (Fig. 4.2.13). What is the maximum resulting error in its calculated surface area?

Fig. 4.2.13 The hemisphere of Problem 51: Curved surface area $A = 2\pi r^2$

52. With what accuracy must the radius of a hemispherical dome be measured to ensure an error of at most 0.01% in its calculated surface area?

4.3

Increasing and Decreasing Functions and the Mean Value Theorem

The significance of the *sign* of the derivative of a function is simple but crucial:

$f(x)$ is increasing on an interval where $f'(x) > 0$;

$f(x)$ is decreasing on an interval where $f'(x) < 0$.

Geometrically, this means that where $f'(x) > 0$, the graph of $y = f(x)$ is rising as you scan it from left to right. Where $f'(x) < 0$, the graph is falling. We can clarify the terms *increasing* and *decreasing* as follows.

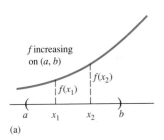

(a)

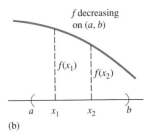

(b)

Fig. 4.3.1 (a) An increasing function and (b) a decreasing function

Definition *Increasing and Decreasing Functions*
The function f is **increasing** on the interval $I = (a, b)$ provided that

$$f(x_1) < f(x_2)$$

for all pairs of numbers x_1 and x_2 in I for which $x_1 < x_2$. The function f is **decreasing** on I provided that

$$f(x_1) > f(x_2)$$

for all pairs of numbers x_1 and x_2 in I for which $x_1 < x_2$.

Figure 4.3.1 illustrates this definition. In short, the function f is increasing on $I = (a, b)$ if the values of $f(x)$ increase as x increases [Fig. 4.3.1(a)]; f is decreasing on I if the values of $f(x)$ decrease as x increases [Fig. 4.3.1(b)].

We speak of a function as increasing or decreasing *on an interval*, not at a single point. Nevertheless, if we consider the sign of f', the *derivative* of f, at a single point, we get a useful intuitive picture of the significance of the sign of the derivative. This is because the derivative $f'(x)$ is the slope of the tangent line at the point $(x, f(x))$ on the graph of f. If $f'(x) > 0$, then the tangent line has positive slope. Therefore, it rises as you scan from left to right. Intuitively, a rising tangent would seem to correspond to a rising graph

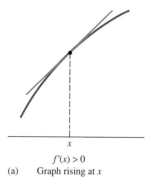

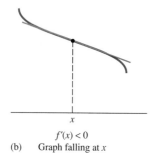

$f'(x) > 0$

(a)   Graph rising at x

$f'(x) < 0$

(b)   Graph falling at x

Fig. 4.3.2   (a) A graph rising at x and (b) a graph falling at x

and thus to an increasing function. Similarly, we expect to see a falling graph where $f'(x)$ is negative (Fig. 4.3.2). One caution: In order to determine whether a function f is increasing or decreasing, we must examine the sign of f' on a whole interval, not merely at a single point (see Problem 55).

THE MEAN VALUE THEOREM

Although pictures of rising and falling graphs are suggestive, they provide no actual *proof* of the significance of the sign of the derivative. To establish rigorously the connection between a graph's rising and falling and the sign of the derivative of the graphed function, we need the *mean value theorem*, stated later in this section. This theorem is the principal theoretical tool of differential calculus, and we shall see that it has many important applications.

As an introduction to the mean value theorem, we pose the following question. Suppose that P and Q are two points on the surface of the sea, with Q lying generally to the east of P (Fig. 4.3.3). Is it possible to sail a boat from P to Q, always sailing roughly east, without *ever* (even for an instant) sailing in the exact direction from P to Q? That is, can we sail from P to Q without our instantaneous line of motion ever being parallel to the line PQ?

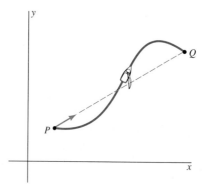

Fig. 4.3.3   Can you sail from P to Q without ever sailing—even for an instant—in the direction PQ (the direction of the arrow)?

Fig. 4.3.4   The sailboat problem in mathematical terminology

The mean value theorem answers this question: No. There will always be at least one instant when we are sailing parallel to the line PQ, no matter which path we choose.

To paraphrase: Let the path of the sailboat be the graph of a differentiable function $y = f(x)$ with endpoints $P(a, f(a))$ and $Q(b, f(b))$. Then we say that there must be some point on this graph where the tangent line (corresponding to the instantaneous line of motion of the boat) to the curve is parallel to the line PQ that joins the curve's endpoints. This is a *geometric interpretation* of the mean value theorem.

But the slope of the tangent line at the point $(c, f(c))$ (Fig. 4.3.4) is $f'(c)$, whereas the slope of the line PQ is

$$\frac{f(b) - f(a)}{b - a}.$$

210

We may think of this last quotient as the average (or *mean*) value of the slope of the curve $y = f(x)$ over the interval $[a, b]$. The mean value theorem guarantees that there is a point c in (a, b) for which the tangent line at $(c, f(c))$ is indeed parallel to the line PQ. In the language of algebra, there's a number c in (a, b) such that

$$f'(c) = \frac{f(b) - f(a)}{b - a}. \tag{1}$$

We first give a preliminary result, a lemma to expedite the proof of the mean value theorem. This theorem is called *Rolle's theorem,* after Michel Rolle (1652–1719), who discovered it in 1690. In his youth Rolle studied the emerging subject of calculus but later renounced it. He argued that the subject was based on logical fallacies, and he is remembered today only for the single theorem that bears his name. It is ironic that his theorem plays an important role in the rigorous proofs of several calculus theorems.

Rolle's Theorem
Suppose that the function f is continuous on the closed interval $[a, b]$ and is differentiable in its interior (a, b). If $f(a) = 0 = f(b)$, then there exists some number c in (a, b) such that $f'(c) = 0$.

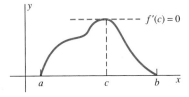

Fig. 4.3.5 The idea of the proof of Rolle's theorem

Figure 4.3.5 illustrates the first case in the following proof of Rolle's theorem. The idea of the proof is this: Suppose that the smooth graph $y = f(x)$ starts $(x = a)$ at height zero and ends $(x = b)$ at height zero. Then if it goes up, it must come back down. But where it stops going up and starts coming back down, its tangent line must be horizontal. Therefore, the derivative is zero at that point.

Proof of Rolle's Theorem   Because f is continuous on $[a, b]$, it must attain both a maximum and a minimum value on $[a, b]$ (by the maximum value property of Section 3.5). If f has any positive values, consider its maximum value $f(c)$. Now c is not an endpoint of $[a, b]$ because $f(a) = 0$ and $f(b) = 0$. Therefore, c is a point of (a, b). But we know that f is differentiable at c. So it follows from Theorem 2 of Section 3.5 that $f'(c) = 0$.

Similarly, if f has any negative values, we can consider its minimum value $f(c)$ and conclude that $f'(c) = 0$.

If f has neither positive nor negative values, then f is identically zero on $[a, b]$, and it follows that $f'(c) = 0$ for *every* c in (a, b).

Thus we see that the conclusion of Rolle's theorem is justified in every case.   ❑

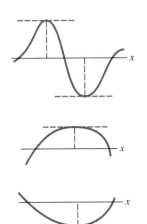

Fig. 4.3.6 The existence of the horizontal tangent is a consequence of Rolle's theorem.

An important consequence of Rolle's theorem is that between each pair of zeros of a differentiable function, there is *at least one* point at which the tangent line is horizontal. Some possible pictures of the situation are indicated in Fig. 4.3.6.

EXAMPLE 1 Suppose that $f(x) = x^{1/2} - x^{3/2}$ on $[0, 1]$. Find a number c that satisfies the conclusion of Rolle's theorem.

Solution Note that f is continuous on $[0, 1]$ and differentiable on $(0, 1)$. Because the term $x^{1/2}$ is present, f is *not* differentiable at $x = 0$, but this is irrelevant. Also, $f(0) = 0 = f(1)$, so all the hypotheses of Rolle's theorem are satisfied. Finally,

$$f'(x) = \tfrac{1}{2}x^{-1/2} - \tfrac{3}{2}x^{1/2} = \tfrac{1}{2}x^{-1/2}(1 - 3x),$$

so we see that $f'(c) = 0$ for $c = \tfrac{1}{3}$. An accurate graph of f on $[0, 1]$, including c and the horizontal tangent line, is shown in Fig. 4.3.7.

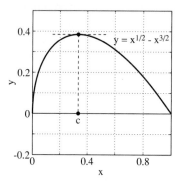

Fig. 4.3.7 The number c of Example 1

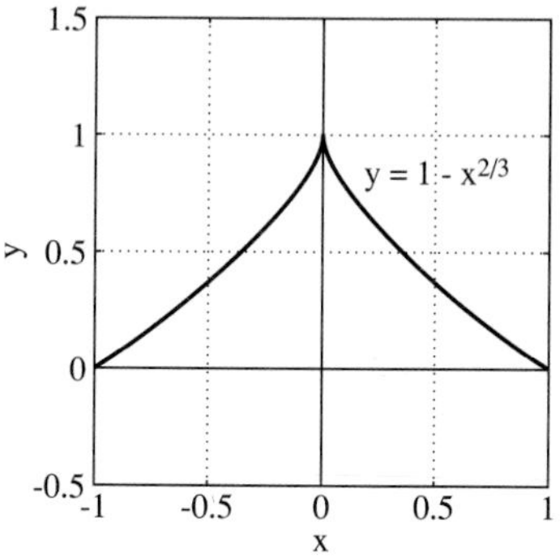

Fig. 4.3.8 The function $f(x) = 1 - x^{2/3}$ of Example 2

EXAMPLE 2 Suppose that $f(x) = 1 - x^{2/3}$ on $[-1, 1]$. Then f satisfies the hypotheses of Rolle's theorem *except* for the fact that $f'(0)$ does not exist. It is clear from the graph of f that there is *no* point where the tangent line is horizontal (Fig. 4.3.8). Indeed,

$$f'(x) = -\frac{2}{3}x^{-1/3} = -\frac{2}{3\sqrt[3]{x}},$$

so $f'(x) \neq 0$ for $x \neq 0$, and we see that $|f'(x)| \to \infty$ as $x \to 0$. Hence the graph of f has a vertical tangent line—rather than a horizontal one—at the point $(0, 1)$.

Now we are ready to state formally and prove the mean value theorem.

The Mean Value Theorem
Suppose that the function f is continuous on the closed interval $[a, b]$ and differentiable on the open interval (a, b). Then

$$f(b) - f(a) = f'(c)(b - a) \tag{2}$$

for some number c in (a, b).

COMMENT Because Eq. (2) is equivalent to Eq. (1), the conclusion of the mean value theorem is that there must be at least one point on the curve $y = f(x)$ at which the tangent line is parallel to the line joining its endpoints $P(a, f(a))$ and $Q(b, f(b))$.

Ch. 4 / Additional Applications of the Derivative

Motivation for the Proof of the Mean Value Theorem   We consider the auxiliary function ϕ suggested by Fig. 4.3.9. The value of $\phi(x)$ is, by definition, the vertical height difference over x of the point $(x, f(x))$ on the curve and the corresponding point on the line PQ. It appears that a point on the curve $y = f(x)$ where the tangent line is parallel to PQ corresponds to a maximum or minimum of ϕ. It's also clear that $\phi(a) = 0 = \phi(b)$, so Rolle's theorem can be applied to the function ϕ on $[a, b]$. So our plan for proving the mean value theorem is this: First, we obtain a formula for the function ϕ. Second, we locate the point c such that $\phi'(c) = 0$. Finally, we show that this number c is exactly the number needed to satisfy the conclusion of the theorem in Eq. (2).

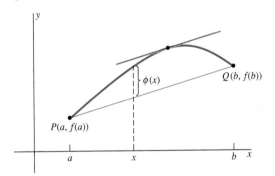

Fig. 4.3.9   The construction of the auxiliary function ϕ

Proof of the Mean Value Theorem   Because the line PQ passes through $P(a, f(a))$ and has slope

$$m = \frac{f(b) - f(a)}{b - a},$$

the point-slope formula for the equation of a straight line gives us the following equation for PQ:

$$y = y_{\text{line}} = f(a) + m(x - a).$$

Thus

$$\phi(x) = y_{\text{curve}} - y_{\text{line}} = f(x) - f(a) - m(x - a).$$

You may verify by direct substitution that $\phi(a) = 0 = \phi(b)$. And, because ϕ is continuous on $[a, b]$ and differentiable in (a, b), we may apply Rolle's theorem to it. Thus there is a point c somewhere in the open interval (a, b) at which $\phi'(c) = 0$. But

$$\phi'(x) = f'(x) - m = f'(x) - \frac{f(b) - f(a)}{b - a}.$$

Because $\phi'(c) = 0$, we conclude that

$$0 = f'(c) - \frac{f(b) - f(a)}{b - a}.$$

That is,

$$f(b) - f(a) = f'(c)(b - a). \qquad \square$$

The proof of the mean value theorem is an application of Rolle's theorem, whereas Rolle's theorem is the special case of the mean value theorem in which $f(a) = 0 = f(b)$.

EXAMPLE 3 Suppose that we drive from Kristiansand, Norway, to Oslo—a road distance of almost exactly 350 km—in exactly 4 h, from time $t = 0$ to time $t = 4$. Let $f(t)$ denote the distance we have traveled at time t, and assume that f is a differentiable function. Then the mean value theorem implies that

$$350 = f(4) - f(0) = f'(c)(4 - 0) = 4f'(c)$$

and thus that

$$f'(c) = \tfrac{350}{4} = 87.5$$

at some instant c in $(0, 4)$. But $f'(c)$ is our *instantaneous* velocity at time $t = c$, and 87.5 km/h is our *average* velocity for the trip. Thus the mean value theorem implies that we must have an instantaneous velocity of exactly 87.5 km/h at least once during the trip.

The argument in Example 3 is quite general—during any trip, the instantaneous velocity must at *some* instant equal the average velocity for the whole trip. For instance, it follows that if two toll stations are 60 mi apart and you drive between the two in exactly 1 h, then at some instant you must have been speeding in excess of the posted limit of 55 mi/h. (Thus the mean value theorem precludes some courtroom defenses from being plausible.)

CONSEQUENCES OF THE MEAN VALUE THEOREM

The first of these consequences is the *non*trivial converse of the trivial fact that the derivative of a constant function is identically zero. That is, we prove that there can be *no* unknown exotic function that is nonconstant but has a derivative that is identically zero. In Corollaries 1 through 3 we assume that f and g are continuous on the closed interval $[a, b]$ and differentiable on (a, b).

Corollary 1   Functions with Zero Derivative
If $f'(x) \equiv 0$ on (a, b) [that is, $f'(x) = 0$ for all x in (a, b)], then f is a constant function on $[a, b]$. In other words, there exists a constant C such that $f(x) \equiv C$.

Proof   Apply the mean value theorem to the function f on the interval $[a, x]$, where x is a fixed but arbitrary point of the interval (a, b). We find that

$$f(x) - f(a) = f'(c)(x - a)$$

for some number c between a and x. But $f'(x)$ is always zero on the interval (a, b), so $f'(c) = 0$. Thus $f(x) - f(a) = 0$, and so $f(x) = f(a)$.

But this last equation holds for *all* x in (a, b). Therefore, $f(x) = f(a)$ for all x in $(a, b]$ and, indeed, for all x in $[a, b]$. That is, $f(x)$ has the fixed value $C = f(a)$. This establishes Corollary 1.   ❑

Corollary 1 is usually applied in a different but equivalent form, which we state and prove next.

Corollary 2 *Functions with Equal Derivatives*

Suppose that $f'(x) = g'(x)$ for all x in the open interval (a, b). Then f and g differ by a constant on $[a, b]$. That is, there exists a constant K such that

$$f(x) = g(x) + K$$

for all x in $[a, b]$.

Proof Given the hypotheses, let $h(x) = f(x) - g(x)$. Then

$$h'(x) = f'(x) - g'(x) = 0$$

for all x in (a, b). So, by Corollary 1, $h(x)$ is a constant K on $[a, b]$. That is, $f(x) - g(x) = K$ for all x in $[a, b]$; therefore,

$$f(x) = g(x) + K$$

for all x in $[a, b]$. This establishes Corollary 2. ❏

EXAMPLE 4 If $f'(x) = 2 \cos x$ and $f(0) = 5$, what is the function $f(x)$?

Solution From our knowledge of the derivatives of trigonometric functions, we know that one explicit function with derivative $2 \cos x$ is

$$g(x) = 2 \sin x.$$

Hence Corollary 2 implies that there exists a constant K such that

$$f(x) = g(x) + K = 2 \sin x + K$$

on any given interval $[a, b]$ containing zero. But we can find the value of K by substituting $x = 0$:

$$f(0) = 2 \sin 0 + K;$$

$$5 = 2 \cdot 0 + K;$$

so $K = 5$. Thus the function f is

$$f(x) = 2 \sin x + 5.$$

The following consequence of the mean value theorem verifies the remarks about increasing and decreasing functions with which we opened this section.

Corollary 3 *Increasing and Decreasing Functions*

If $f'(x) > 0$ for all x in (a, b), then f is an increasing function on $[a, b]$.
If $f'(x) < 0$ for all x in (a, b), then f is a decreasing function on $[a, b]$.

Proof Suppose, for example, that $f'(x) > 0$ for all x in (a, b). We need to show the following: If u and v are points of $[a, b]$ with $u < v$, then $f(u) < f(v)$. We apply the mean value theorem to f, but on the closed interval $[u, v]$. This is legitimate because $[u, v]$ is contained in $[a, b]$, so f satisfies the hypotheses of the mean value theorem on $[u, v]$ as well as on $[a, b]$. The result is that

$$f(v) - f(u) = f'(c)(v - u)$$

for some number c in (u, v). Because $v > u$ and because, by hypothesis, $f'(c) > 0$, it follows that

$$f(v) - f(u) > 0; \quad \text{that is,} \quad f(u) < f(v),$$

as we wanted to show. The proof is similar in the case that $f'(x)$ is negative on (a, b). $\square$

The meaning of Corollary 3 is summarized in Fig. 4.3.10. Figure 4.3.11 shows a graph $y = f(x)$ labeled in accord with this correspondence between the sign of the derivative $f'(x)$ and the increasing or decreasing behavior of the function $f(x)$.

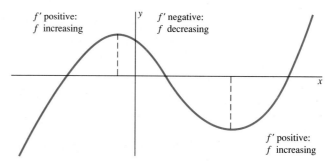

$f'(x)$	$f(x)$
Negative	Decreasing
Positive	Increasing

Fig. 4.3.10 Corollary 3

Fig. 4.3.11 The significance of the sign of $f'(x)$

EXAMPLE 5 Because

$$D_x e^x = e^x > 0 \qquad \text{for all } x \quad \text{and}$$

$$D_x \ln x = \frac{1}{x} > 0 \qquad \text{for all } x > 0,$$

Corollary 3 implies that the natural exponential and logarithmic functions are increasing wherever they are defined.

EXAMPLE 6 Where is the function $f(x) = x^2 - 4x + 5$ increasing, and where is it decreasing?

Solution The derivative of f is $f'(x) = 2x - 4$. Clearly $f'(x) > 0$ if $x > 2$, whereas $f'(x) < 0$ if $x < 2$. Hence f is decreasing on $(-\infty, 2)$ and increasing on $(2, +\infty)$, as we see in Fig. 4.3.12.

EXAMPLE 7 Show that the equation $x^3 + x - 1 = 0$ has exactly one (real) solution.

Solution We need a *function* in order to apply the tools of this section, so we let $f(x) = x^3 + x - 1$. Because $f(0) = -1 < 0, f(1) = +1 > 0$, and f is continuous (everywhere), the intermediate value property guarantees that

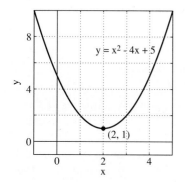

Fig. 4.3.12 The parabola of Example 6

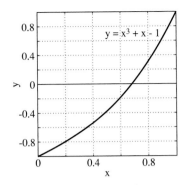

Fig. 4.3.13 The increasing function of Example 7

$f(x)$ has at *least* one real zero in $[0, 1]$ (Fig. 4.3.13). But it might have more; because we cannot plot $y = f(x)$ on its entire domain $(-\infty, \infty)$, we do not *know* that the equation $f(x) = 0$ has no additional real solutions. We need conclusive evidence that this is true.

But

$$f'(x) = 3x^2 + 1,$$

and it is evident that $f'(x) > 0$ for all x. So Corollary 3 implies that the graph of $y = f(x)$ is increasing on the whole real number line. Therefore, the equation $f(x) = 0$ can have at most one solution. (Why?)

In summary, the equation $f(x) = 0$ has both at *least* one solution and at *most* one solution. Therefore, it must have exactly one (real) solution.

EXAMPLE 8 Determine the open intervals of the x-axis on which the function

$$f(x) = 3x^4 - 4x^3 - 12x^2 + 5$$

is increasing and those on which it is decreasing.

Solution   The derivative of f is

$$f'(x) = 12x^3 - 12x^2 - 24x$$
$$= 12x(x^2 - x - 2) = 12x(x + 1)(x - 2). \qquad (3)$$

Fig. 4.3.14 The signs of $x + 1$ and $x - 2$ (Example 8)

The critical points $x = -1, 0, 2$ separate the x-axis into the four open intervals $(-\infty, -1)$, $(-1, 0)$, $(0, 2)$, and $(2, +\infty)$ (Fig. 4.3.14). The derivative $f'(x)$ does not change sign within any of these intervals, because

❑ The factor $x + 1$ in Eq. (3) changes sign only at $x = -1$,
❑ The factor $12x$ changes sign only at $x = 0$, and
❑ The factor $x - 2$ changes sign only at $x = 2$.

Let's illustrate two methods of determining the sign of $f'(x)$ on each of the four intervals (Fig. 4.3.14).

METHOD 1   The second, third, and fourth columns of the next table record the signs of the factors in Eq. (3) on each of the four intervals listed in the first column. The signs of $f'(x)$ shown in the fifth column are then obtained by multiplication. The sixth column lists the resulting increasing or decreasing behavior of f on the four intervals.

Interval	$x + 1$	$12x$	$x - 2$	$f'(x)$	f
$(-\infty, -1)$	Neg.	Neg.	Neg.	Neg.	Decreasing
$(-1, 0)$	Pos.	Neg.	Neg.	Pos.	Increasing
$(0, 2)$	Pos.	Pos.	Neg.	Neg.	Decreasing
$(2, +\infty)$	Pos.	Pos.	Pos.	Pos.	Increasing

METHOD 2   Because the derivative $f'(x)$ does not change sign within any of the four intervals, we need to calculate its value at only a single point in each interval. Whatever the sign at that point may be, it is the sign of $f'(x)$ throughout that interval.

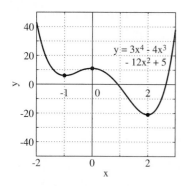

Fig. 4.3.15 The critical points of the polynomial of Example 8

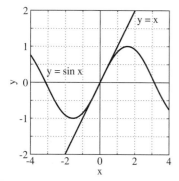

Fig. 4.3.16 x and $\sin x$ (Example 9)

In $(-\infty, -1)$: $f'(-2) = -96 < 0$; f is decreasing.

In $(-1, 0)$: $f'(-0.5) = 7.5 > 0$; f is increasing.

In $(0, 2)$: $f'(1) = -24 < 0$; f is decreasing.

In $(2, +\infty)$: $f'(3) = 144 > 0$; f is increasing.

The second method is especially convenient if the derivative is complicated but an appropriate calculator for computation of its values is available.

Finally, note that the results we have obtained in each method are consistent with the graph of $y = f(x)$ shown in Fig. 4.3.15.

EXAMPLE 9   The graphs in Fig. 4.3.16 suggest that $\sin x < x$ for $x > 0$. To show that this is so, we consider the function

$$f(x) = x - \sin x.$$

Because $|\sin x| \leqq 1$ for *all* x, $\sin x < x$ if $x > 1$, so we need to show only that $\sin x < x$ for x in the interval $(0, 1]$. But

$$f'(x) = 1 - \cos x > 0$$

for such x, because $\cos x < 1$ if $0 < x < 2\pi$, and so $\cos x < 1$ for $0 < x \leqq 1$. Hence $f(x)$ is an increasing function on $(0, 1]$. But $f(0) = 0$, so it follows that

$$f(x) = x - \sin x > 0$$

for all x in $(0, 1]$. Consequently $\sin x < x$ for x in $(0, 1]$. By our earlier remarks, it now follows that $\sin x < x$ for all $x > 0$.

4.3 Problems

For the functions in Problems 1 through 6, first determine (as in Example 8) the open intervals on the x-axis on which each function is increasing and those where it is decreasing. Then use this information to match the function to its graph, one of the six shown in Fig. 4.3.17.

1. $f(x) = 4 - x^2$ **2.** $f(x) = x^2 - 2x - 1$

3. $f(x) = x^2 + 4x + 1$ **4.** $f(x) = \frac{1}{4}x^3 - 3x$

5. $f(x) = \frac{1}{3}x^3 - \frac{1}{2}x^2 - 2x + 1$

6. $f(x) = 2x - \frac{1}{6}x^2 - \frac{1}{9}x^3$

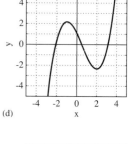

(c)   (d)

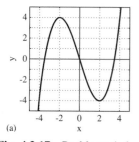

(a)   (b)

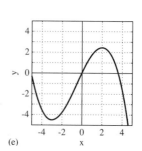

(e)   (f)

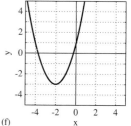

Fig. 4.3.17   Problems 1 through 6

Ch. 4 / Additional Applications of the Derivative

In Problems 7 through 10, the derivative $f'(x)$ and the value $f(0)$ are given. Use the method of Example 4 to find the function $f(x)$.

7. $f'(x) = 4x$; $f(0) = 5$

8. $f'(x) = 3\sqrt{x}$; $f(0) = 4$

9. $f'(x) = \dfrac{1}{x^2}$; $f(1) = 1$

10. $f'(x) = \dfrac{2}{\sqrt{x}}$; $f(0) = 3$

In Problems 11 through 24, determine (as in Example 8) the open intervals on the x-axis on which the function is increasing as well as those on which it is decreasing. If you have a graphics calculator or computer, plot the graph $y = f(x)$ to see whether it agrees with your results.

11. $f(x) = 3x + 2$

12. $f(x) = 4 - 5x$

13. $f(x) = 8 - 2x^2$

14. $f(x) = 4x^2 + 8x + 13$

15. $f(x) = 6x - 2x^2$

16. $f(x) = x^3 - 12x + 17$

17. $f(x) = x^4 - 2x^2 + 1$

18. $f(x) = \dfrac{x}{x + 1}$ [Note: $f'(x)$ can change sign at $x = -1$. Why?]

19. $f(x) = 3x^4 + 4x^3 - 12x^2$

20. $f(x) = x\sqrt{x^2 + 1}$

21. $f(x) = 8x^{1/3} - x^{4/3}$

22. $f(x) = 2x^3 + 3x^2 - 12x + 5$

23. $f(x) = \dfrac{(x - 1)^2}{x^2 - 3}$ (See the note for Problem 18.)

24. $f(x) = x^2 + \dfrac{16}{x^2}$ (See the note for Problem 18.)

In Problems 25 through 28, show that the given function satisfies the hypotheses of Rolle's theorem on the indicated interval [a, b], and find all numbers c in (a, b) that satisfy the conclusion of the theorem.

25. $f(x) = x^2 - 2x$; $[0, 2]$

26. $f(x) = 9x^2 - x^4$; $[-3, 3]$

27. $f(x) = \dfrac{1 - x^2}{1 + x^2}$; $[-1, 1]$

28. $f(x) = 5x^{2/3} - x^{5/3}$; $[0, 5]$

In Problems 29 through 31, show that the given function f does not satisfy the conclusion of Rolle's theorem on the indicated interval. Which of the hypotheses does it fail to satisfy?

29. $f(x) = 1 - |x|$; $[-1, 1]$

30. $f(x) = 1 - (2 - x)^{2/3}$; $[1, 3]$

31. $f(x) = x^4 + x^2$; $[0, 1]$

In Problems 32 through 36, show that the given function f satisfies the hypotheses of the mean value theorem on the indicated interval, and find all numbers c in (a, b) that satisfy the conclusion of that theorem.

32. $f(x) = x^3$; $[-1, 1]$

33. $f(x) = 3x^2 + 6x - 5$; $[-2, 1]$

34. $f(x) = \sqrt{x - 1}$; $[2, 5]$

35. $f(x) = (x - 1)^{2/3}$; $[1, 2]$

36. $f(x) = x + \dfrac{1}{x}$; $[1, 2]$

In Problems 37 through 40, show that the given function f satisfies neither the hypotheses nor the conclusion of the mean value theorem on the indicated interval.

37. $f(x) = |x - 2|$; $[1, 4]$

38. $f(x) = 1 + |x - 1|$; $[0, 3]$

39. $f(x) = [\![x]\!]$ (the greatest integer function); $[-1, 1]$

40. $f(x) = 3x^{2/3}$; $[-1, 1]$

In Problems 41 through 43, show that the given equation has exactly one solution in the indicated interval.

41. $x^5 + 2x - 3 = 0$; $[0, 1]$

42. $x^{10} = 1000$; $[1, 2]$

43. $x^4 - 3x = 20$; $[2, 3]$

44. Show that the function $f(x) = x^{2/3}$ does not satisfy the hypotheses of the mean value theorem on $[-1, 27]$ but that nevertheless there is a number c in $(-1, 27)$ such that

$$f'(c) = \frac{f(27) - f(-1)}{27 - (-1)}.$$

45. Prove that the function

$$f(x) = (1 + x)^{3/2} - \tfrac{3}{2}x - 1$$

is increasing on $(0, +\infty)$. Explain carefully how you could conclude that

$$(1 + x)^{3/2} > 1 + \tfrac{3}{2}x$$

for all $x > 0$.

46. Suppose that f' is a constant function on the interval $[a, b]$. Prove that f must be a linear function (a function whose graph is a straight line).

47. Suppose that $f'(x)$ is a polynomial of degree $n - 1$ on the interval $[a, b]$. Prove that $f(x)$ must be a polynomial of degree n on $[a, b]$.

48. Suppose that there are k different points of $[a, b]$ at which the differentiable function f vanishes (is zero). Prove that f' must vanish on at least $k - 1$ points of $[a, b]$.

49. (a) Apply the mean value theorem to $f(x) = \sqrt{x}$ on $[100, 101]$ to show that

$$\sqrt{101} = 10 + \frac{1}{2\sqrt{c}}$$

for some number c in $(100, 101)$. (b) Show that if $100 < c < 101$, then $10 < \sqrt{c} < 10.5$, and use this fact to conclude from part (a) that $10.0475 < \sqrt{101} < 10.0500$.

50. Prove that the equation $x^7 + x^5 + x^3 + 1 = 0$ has exactly one real solution.

51. (a) Show that $D \tan^2 x = D \sec^2 x$ on the open interval $(-\pi/2, \pi/2)$. (b) Conclude that there exists a constant C such that $\tan^2 x = \sec^2 x + C$ for all x in $(-\pi/2, \pi/2)$. Then evaluate C.

52. Explain why the mean value theorem does not apply to the function $f(x) = |x|$ on the interval $[-1, 2]$.

53. Suppose that the function f is differentiable on the interval $[-1, 2]$, with $f(-1) = -1$ and $f(2) = 5$. Prove that there is a point on the graph of f at which the tangent line is parallel to the line with the equation $y = 2x$.

54. Let $f(x) = x^4 - x^3 + 7x^2 + 3x - 11$. Prove that the graph of f has at least one horizontal tangent line.

55. Let the function g be defined as follows:

$$g(x) = \frac{x}{2} + x^2 \sin \frac{1}{x}$$

for $x \neq 0$; $g(0) = 0$. (a) Show that $g'(0) = \frac{1}{2} > 0$. (b) Sketch the graph of g near $x = 0$. Is g increasing on any open interval containing $x = 0$? [*Answer:* No.]

56. Suppose that f is increasing on every closed interval $[a, b]$ provided that $2 \le a < b$. Prove that f is increasing on the unbounded open interval $(2, +\infty)$. Note that the principle you discover was used implicitly in Example 6 of this section.

Approximations

Problems 57 through 59 illustrate the use of the mean value theorem to approximate numerical values of functions.

57. Use the method of Example 9 with $f(x) = \cos x$ and $g(x) = 1 - \frac{1}{2}x^2$ to show that

$$\cos x > 1 - \tfrac{1}{2}x^2$$

for all $x > 0$ (Fig. 4.3.18).

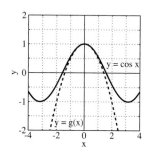 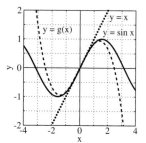

Fig. 4.3.18 $\cos x$ and $g(x) = x - \frac{1}{2}x^2$ (Problem 57)

Fig. 4.3.19 x, $\sin x$, and $g(x) = x - \frac{1}{6}x^3$ (Problem 58)

58. (a) Use the method of Example 9 and the result of Problem 57 to show that

$$\sin x > x - \tfrac{1}{6}x^3$$

for all $x > 0$ (Fig. 4.3.19). (b) Use the results of Example 9 and part (a) to calculate the sine of a 5° angle accurate to three decimal places.

59. (a) Use the result of Problem 58(a) to show that

$$\cos x < 1 - \tfrac{1}{2}x^2 + \tfrac{1}{24}x^4$$

for all $x > 0$. (b) Use the results of Problem 57 and part (a) to calculate the cosine of a 10° angle accurate to three decimal places.

4.4

The First Derivative Test

In Section 3.5 we discussed absolute maximum and minimum values of a function defined on a closed and bounded interval $[a, b]$. Now we consider extreme values of functions defined on more general domains, including open or unbounded intervals as well as closed and bounded intervals.

Recall that if c is a point of the domain D of the function f such that $f(c) \ge f(x)$ for every x in D, then $f(c)$ is called the **absolute maximum value** of $f(x)$ on D. If $f(c) \le f(x)$ for every x in D, then $f(c)$ is the **absolute minimum value** of $f(x)$ on D. If the domain of definition of f is clear—such as when f is defined on the whole real line—we may say simply that $f(c)$ is the absolute maximum (or minimum) value of $f(x)$.

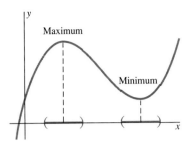

Fig. 4.4.1 Local extrema are absolute extrema on sufficiently small intervals.

The word *global* often is used in place of the word *absolute* in this context. The distinction between global and local extreme values is important. Recall from Section 3.5 that $f(c)$ is a **local maximum value** of $f(x)$ if $f(c)$ is the global maximum value of $f(x)$ on some open interval containing c; similarly, a **local minimum value** $f(c)$ is a global minimum value of $f(x)$ on some open interval containing c. Figure 4.4.1 shows a typical example of a function that has neither a global maximum nor a global minimum. But each of the two local extrema pictured there is a global extreme value on a sufficiently small open interval.

Theorem 2 of Section 3.5 tells us that any extremum of the *differentiable* function f on an open interval I must occur at a critical point where the derivative vanishes:

$$f'(x) = 0.$$

But the mere fact that $f'(c) = 0$ does *not*, by itself, imply that the critical value $f(c)$ is an extreme value of f. Figures 4.4.2 through 4.4.5 illustrate different possibilities for the nature of $f(c)$: whether it is a local or global maximum or minimum value, or neither.

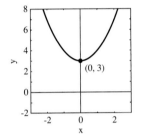

Fig. 4.4.2 The graph of $f(x) = x^2 + 3$. The local minimum value $f(0) = 3$ is also the global minimum value of $f(x)$.

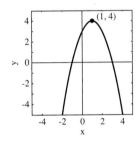

Fig. 4.4.3 The graph of $f(x) = 4 - (x - 1)^2$. The local maximum value $f(1) = 4$ is also the global maximum value of $f(x)$.

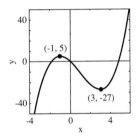

Fig. 4.4.4 The graph of $f(x) = x^3 - 3x^2 - 9x$. The local minimum value $f(3) = -27$ clearly is not the global minimum value. Similarly, the local maximum value $f(-1) = 5$ is not the global maximum value.

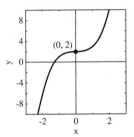

Fig. 4.4.5 The graph of $f(x) = x^3 + 2$. The critical value $f(0) = 2$ is neither a global nor a local extreme value of $f(x)$.

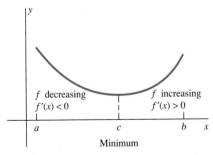 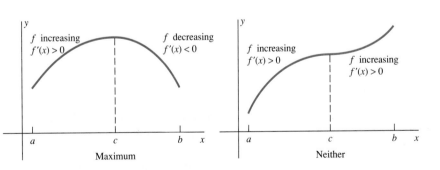

Fig. 4.4.6 The first derivative test

What we need is a way to test whether, at the critical point $x = c$, the value $f(c)$ is actually a maximum or a minimum value of $f(x)$, either local or global. Figure 4.4.6 shows how such a test might be developed. Suppose that the function f is continuous at c and that c is an **interior point** of the domain of f—that is, f is defined on some open interval that contains c. If f is decreasing immediately to the left of c and increasing immediately to the right, then $f(c)$ should be a local minimum value of $f(x)$. But if f is increasing immediately to the left of c and decreasing immediately to its right, then $f(c)$ should be a local maximum. If f is increasing on both sides or decreasing on both sides, then $f(c)$ should be neither a maximum value nor a minimum value of $f(x)$.

Moreover, we know from Corollary 3 in Section 4.3 that the *sign* of the derivative $f'(x)$ determines where $f(x)$ is decreasing and where it is increasing:

❑ $f(x)$ is decreasing where $f'(x) < 0$;
❑ $f(x)$ is increasing where $f'(x) > 0$.

In the following test for local extrema, we say that

❑ $f'(x) < 0$ *to the left of* c if $f'(x) < 0$ on some interval (a, c) of numbers immediately to the left of c, and that
❑ $f'(x) > 0$ *to the right of* c if $f'(x) > 0$ on some interval (c, b) of numbers immediately to the right of c,

and so forth (see Fig. 4.4.7). Theorem 1 tells us how to use the *signs* of $f'(x)$ to the left and right of the point c to determine whether $f(x)$ has a local maximum or local minimum value at $x = c$.

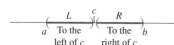

Fig. 4.4.7 Open intervals to the left and right of the point c

---

Theorem 1 *The First Derivative Test for Local Extrema*

Suppose that the function f is continuous on the interval I and also is differentiable there except possibly at the interior point c of I.

1. If $f'(x) < 0$ to the left of c and $f'(x) > 0$ to the right of c, then $f(c)$ is a *local minimum value* of $f(x)$ on I.

2. If $f'(x) > 0$ to the left of c and $f'(x) < 0$ to the right of c, then $f(c)$ is a *local maximum value* of $f(x)$ on I.

3. If $f'(x) > 0$ both to the left of c and to the right of c, *or if* $f'(x) < 0$ both to the left of c and to the right of c, then $f(c)$ is *neither* a maximum nor a minimum value of $f(x)$.

COMMENT   Thus $f(c)$ is a local extremum if the first derivative $f'(x)$ *changes sign* as x increases through c, and the direction of this sign change determines whether $f(c)$ is a local maximum or a local minimum. A good way to remember the first derivative test for local extrema is simply to visualize Fig. 4.4.6.

Proof of Theorem 1   We will prove only part 1; the other two parts have similar proofs. Suppose that the hypotheses of Theorem 1 hold: that f is continuous on the interval I, that c is an interior point of I, and that f is differentiable on I except possibly at $x = c$. Suppose also that $f'(x) < 0$ to the left of c and that $f'(x) > 0$ to the right of c. Then there exist two intervals (a, c) and (c, b), each wholly contained in I, such that $f'(x) < 0$ on (a, c) and $f'(x) > 0$ on (c, b).

Ch. 4 / Additional Applications of the Derivative

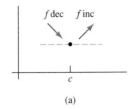

(a)

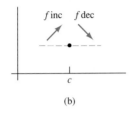

(b)

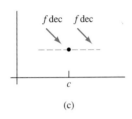

(c)

Fig. 4.4.8 The three cases in the first derivative test

Suppose that x is in (a, b). Then there are three cases to consider. First, if $x < c$, then x is in (a, c) and f is decreasing on $(a, c]$, so $f(x) > f(c)$. Second, if $x > c$, then x is in (c, b) and f is increasing on $[c, b)$, so again $f(x) > f(c)$. Finally, if $x = c$, then $f(x) \geqq f(c)$. Thus, for each x in (a, b), $f(x) \geqq f(c)$. Therefore, by definition, $f(c)$ is a local minimum value of $f(x)$. ☐

The idea of this proof is illustrated in Fig. 4.4.8. Part (a) shows f decreasing to the left of c and increasing to the right, so there must be a local minimum at $x = c$. Part (b) shows f increasing to the left of c and decreasing to the right, so $f(c)$ is a local maximum value of $f(x)$. In part (c), the derivative has the same sign on each side of c, and so there can be no extremum of any sort at $x = c$.

REMARK Figures 4.4.9 through 4.4.13 illustrate cases in which Theorem 1 applies, where the interval I is the entire real number line R. In Figs. 4.4.9 through 4.4.11, the origin $c = 0$ is a critical point because $f'(0) = 0$. In Figs. 4.4.12 and 4.4.13, $c = 0$ is a critical point because $f'(0)$ does not exist.

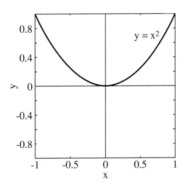

Fig. 4.4.9 $f(x) = x^2, f'(x) = 2x$: a local minimum at $x = 0$

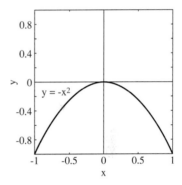

Fig. 4.4.10 $f(x) = -x^2$, $f'(x) = -2x$: a local maximum at $x = 0$

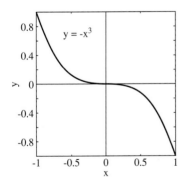

Fig. 4.4.11 $f(x) = -x^3$, $f'(x) = -3x^2$: no extremum at $x = 0$

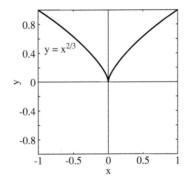

Fig. 4.4.12 $f(x) = x^{2/3}, f'(x) = \frac{2}{3}x^{-1/3}$: a local minimum at $x = 0$

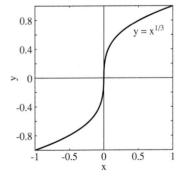

Fig. 4.4.13 $f(x) = x^{1/3}, f'(x) = \frac{1}{3}x^{-2/3}$: no extremum at $x = 0$

CLASSIFICATION OF CRITICAL POINTS

Suppose that we have found the critical points of a function. Then we can attempt to classify them—as local maxima, local minima, or neither—by applying the first derivative test at each point in turn. Example 1 illustrates a procedure that can be used.

EXAMPLE 1 Find and classify the critical points of the function

$$f(x) = 2x^3 - 3x^2 - 36x + 7.$$

Solution   The derivative is

$$f'(x) = 6x^2 - 6x - 36 = 6(x + 2)(x - 3), \tag{1}$$

so the critical points [where $f'(x) = 0$] are $x = -2$ and $x = 3$. These two points separate the x-axis into the three open intervals $(-\infty, -2)$, $(-2, 3)$, and $(3, +\infty)$. The derivative $f'(x)$ cannot change sign within any of these intervals. One reason is that the factor $x + 2$ in Eq. (1) changes sign only at -2, whereas the factor $x - 3$ changes sign only at 3 (Fig. 4.4.14). As in Example 7 of Section 4.3, we illustrate here two methods of determining the signs of $f'(x)$ on the intervals $(-\infty, -2)$, $(-2, 3)$, and $(3, +\infty)$.

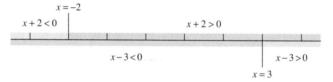

Fig. 4.4.14 The signs of $x + 2$ and $x - 3$ (Example 1)

METHOD 1   The second and third columns of the following table record (from Fig. 4.4.14) the signs of the factors $x + 2$ and $x - 3$ in Eq. (1) on the three intervals listed in the first column. The signs of $f'(x)$ in the fourth column are then obtained by multiplication.

Interval	$x + 2$	$x - 3$	$f'(x)$
$(-\infty, -2)$	Neg.	Neg.	Pos.
$(-2, 3)$	Pos.	Neg.	Neg.
$(3, +\infty)$	Pos.	Pos.	Pos.

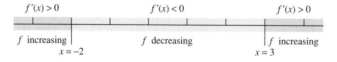

Fig. 4.4.15 The three intervals of Example 1

METHOD 2   Because the derivative $f'(x)$ does not change sign within any of the three intervals, we need to calculate its value only at a single point in each interval:

$$\text{In } (-\infty, -2): \quad f'(-3) = 36 > 0; \quad \text{positive;}$$

$$\text{In } (-2, 3): \quad f'(0) = -36 < 0; \quad \text{negative;}$$

$$\text{In } (3, +\infty): \quad f'(4) = 36 > 0; \quad \text{positive.}$$

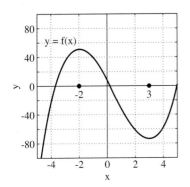

Fig. 4.4.16 $y = f(x)$
(Example 1)

Figure 4.4.15 summarizes our information about the signs of $f'(x)$. Because $f'(x)$ is positive to the left and negative to the right of the critical point $x = -2$, the first derivative test implies that $f(-2) = 51$ is a local maximum value. Because $f'(x)$ is negative to the left and positive to the right of $x = 3$, it follows that $f(3) = -74$ is a local minimum value. The graph of $y = f(x)$ in Fig. 4.4.16 confirms this classification of the critical points -2 and 3.

OPEN-INTERVAL MAXIMUM-MINIMUM PROBLEMS

In Section 3.6 we discussed applied maximum-minimum problems in which the values of the dependent variable are given by a function defined on a closed and bounded interval. Sometimes, though, the function f describing the variable to be maximized is defined on an *open* interval (a, b), possibly an *unbounded* open interval such as $(1, +\infty)$ or $(-\infty, +\infty)$, and we cannot "close" the interval by adjoining endpoints. Typically, the reason is that $|f(x)| \to \infty$ as x approaches a or b. But if f has only a single critical point in (a, b), then the first derivative test can tell us that $f(c)$ is the desired extremum and can even determine whether it is a maximum or a minimum value of $f(x)$.

EXAMPLE 2 Figure 4.4.17 shows the graph of the function

$$f(x) = \frac{2 \ln x}{x},$$

which is defined on the open interval $(0, +\infty)$. Because

$$f'(x) = \frac{2}{x} \cdot \frac{1}{x} - \frac{2 \ln x}{x^2} = \frac{2}{x^2}(1 - \ln x),$$

there is a lone critical point $x = e$. Note that

❏ If $x < e$, then $\ln ex < 1$, so $f'(x) > 0$ if $x < e$;
❏ If $x > e$, then $\ln ex > 1$, so $f'(x) < 0$ if $x > e$.

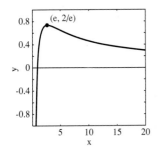

Fig. 4.4.17 $y = \dfrac{2 \ln x}{x}$

Therefore the first derivative test implies that $f(e) = 2/e$ is a local maximum value of f. Indeed, because f is increasing if $0 < x < e$ and decreasing if $x > e$, it follows that $2/e$ is the absolute maximum value of f.

EXAMPLE 3 Find the (absolute) minimum value of

$$f(x) = x + \frac{4}{x} \qquad \text{for } 0 < x < +\infty.$$

Solution The derivative is

$$f'(x) = 1 - \frac{4}{x^2} = \frac{x^2 - 4}{x^2}. \tag{2}$$

The roots of the equation

$$f'(x) = \frac{x^2 - 4}{x^2} = 0$$

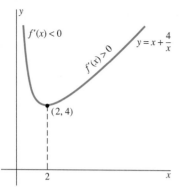

Fig. 4.4.18 The graph of the function of Example 3

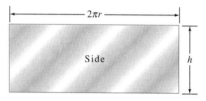

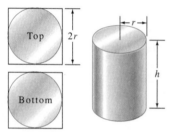

Fig. 4.4.19 The parts to make the cylindrical can of Example 4

are $x = -2$ and $x = 2$. But $x = -2$ is not in the open interval $(0, +\infty)$, so we have only the critical point $x = 2$ to consider.

We see immediately from Eq. (2) that

❑ $f'(x) < 0$ to the left of $x = 2$ (because $x^2 < 4$ there), whereas
❑ $f'(x) > 0$ to the right of $x = 2$ (because $x^2 > 4$ there).

Therefore, the first derivative test implies that $f(2) = 4$ is a local minimum value. We note also that $f(x) \to +\infty$ as either $x \to 0^+$ or $x \to +\infty$. Hence the graph of f must resemble Fig. 4.4.18, and we see that $f(2) = 4$ is the absolute minimum value of $f(x)$ on the entire interval $(0, +\infty)$.

EXAMPLE 4 We must make a cylindrical can with volume 125 in.³ (about 2 L) by cutting its top and bottom from squares of metal and forming its curved side by bending a rectangular sheet of metal to match its ends. What radius r and height h of the can will minimize the total amount of material required for the rectangle and the two squares?

Solution We assume that the corners cut from the two squares, shown in Fig. 4.4.19, are wasted but that there is no other waste. As the figure shows, the area of the total amount of sheet metal required is

$$A = 8r^2 + 2\pi rh.$$

The volume of the resulting can is then

$$V = \pi r^2 h = 125,$$

so $h = 125/(\pi r^2)$. Hence A is given as a function of r by

$$A(r) = 8r^2 + 2\pi r \cdot \frac{125}{\pi r^2} = 8r^2 + \frac{250}{r}, \qquad 0 < r < +\infty.$$

The domain of A is the unbounded open interval $(0, +\infty)$ because r can have any positive value, so $A(r)$ is defined for every number r in $(0, +\infty)$. But $A(r) \to +\infty$ as $r \to 0^+$ and as $r \to +\infty$. So we cannot use the closed-interval maximum-minimum method. But we *can* use the first derivative test.

The derivative of $A(r)$ is

$$\frac{dA}{dr} = 16r - \frac{250}{r^2} = \frac{16}{r^2}\left(r^3 - \frac{125}{8}\right). \tag{3}$$

Thus the only critical point in $(0, +\infty)$ is where $r^3 = \frac{125}{8}$; that is,

$$r = \sqrt[3]{\tfrac{125}{8}} = \tfrac{5}{2} = 2.5.$$

We see immediately from Eq. (3) that

❑ $dA/dr < 0$ to the left of $r = \frac{5}{2}$, because $r^3 < \frac{125}{8}$ there, whereas
❑ $dA/dr > 0$ to the right, where $r^3 > \frac{125}{8}$.

Therefore, the first derivative test implies that a local minimum value of $A(r)$ on $(0, +\infty)$ is

$$A(2.5) = 8 \cdot (2.5)^2 + \frac{250}{2.5} = 150.$$

226

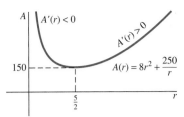

Fig. 4.4.20 Graph of the function of Example 4

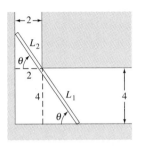

Fig. 4.4.21 Carrying a rod around a corner (Example 5)

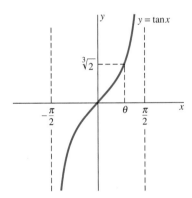

Fig. 4.4.22 $y = \tan x$ (Example 5)

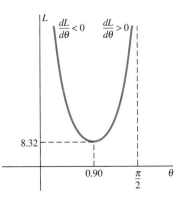

Fig. 4.4.23 The graph of $L(\theta)$ (Example 5)

Considering that $A(r) \to +\infty$ as $x \to 0^+$ and as $x \to +\infty$, we see that the graph of $A(r)$ on $(0, +\infty)$ looks like Fig. 4.4.20. This clinches the fact that $A(2.5) = 150$ is the *absolute* minimum value of $A(r)$. Therefore, we minimize the amount of material required by making a can with radius $r = 2.5$ in. and height

$$h = \frac{125}{\pi(2.5)^2} = \frac{20}{\pi} \approx 6.37 \quad (\text{in.}).$$

The total amount of material used is 150 in.^{2}

EXAMPLE 5 Find the length of the longest rod that can be carried horizontally around the corner from a hall 2 m wide into one that is 4 m wide.

Solution The desired length is the *minimum* length $L = L_1 + L_2$ of the rod being carried around the corner in Fig. 4.4.21. We see from the two similar triangles in the figure that

$$\frac{4}{L_1} = \sin \theta \quad \text{and} \quad \frac{2}{L_2} = \cos \theta,$$

so

$$L_1 = 4 \csc \theta \quad \text{and} \quad L_2 = 2 \sec \theta.$$

Therefore, the length $L = L_1 + L_2$ of the rod is given as a function of θ by

$$L(\theta) = 4 \csc \theta + 2 \sec \theta$$

on the open interval $(0, \pi/2)$. Note that $L \to +\infty$ as either $\theta \to 0^+$ or $\theta \to (\pi/2)^-$. (Why?)

The derivative of $L(\theta)$ is

$$\frac{dL}{d\theta} = -4 \csc \theta \cot \theta + 2 \sec \theta \tan \theta$$

$$= -\frac{4 \cos \theta}{\sin^2 \theta} + \frac{2 \sin \theta}{\cos^2 \theta} = \frac{2 \sin^3 \theta - 4 \cos^3 \theta}{\sin^2 \theta \cos^2 \theta}$$

$$= \frac{(2 \cos \theta)(\tan^3 \theta - 2)}{\sin^2 \theta}. \tag{4}$$

Hence $dL/d\theta = 0$ exactly when

$$\tan \theta = \sqrt[3]{2}, \quad \text{so} \quad \theta \approx 0.90 \quad (\text{rad}).$$

We now see from Eq. (4) and from the graph of the tangent function (Fig. 4.4.22) that

❑ $dL/d\theta < 0$ to the left of $\theta = 0.90$, where $\tan \theta < \sqrt[3]{2}$, so $\tan^3 \theta < 2$, and
❑ $dL/d\theta > 0$ to the right, where $\tan^3 \theta > 2$.

Hence the graph of L resembles Fig. 4.4.23. This means that the absolute minimum value of L—and therefore the maximum length of the rod in question—is about

$$L(0.90) = 4 \csc(0.90) + 2 \sec(0.90),$$

approximately 8.32 m.

The method we used in Examples 2 through 5 to establish absolute extrema illustrates the following global version of the first derivative test.

Theorem 2 *The First Derivative Test for Global Extrema*

Suppose that f is defined on an open interval I, either bounded or unbounded, and that f is differentiable at each point of I except possibly at the critical point c where f is continuous.

1. If $f'(x) < 0$ for all x in I with $x < c$ and $f'(x) > 0$ for all x in I with $x > c$, then $f(c)$ is the absolute minimum value of $f(x)$ on I.
2. If $f'(x) > 0$ for all x in I with $x < c$ and $f'(x) < 0$ for all x in I with $x > c$, then $f(c)$ is the absolute maximum value of $f(x)$ on I.

The proof of this theorem is essentially the same as that of Theorem 1.

REMARK When the function $f(x)$ has only one critical point c in an open interval I, Theorem 2 may apply to tell us either that $f(c)$ is the absolute minimum or that it is the absolute maximum of $f(x)$ on I. But it is good practice to verify your conclusion by sketching the graph as we did in Examples 2 through 5.

4.4 Problems

Apply the first derivative test to classify each of the critical points of the functions in Problems 1 through 16 (local or global, maximum or minimum, or not an extremum). If you have a graphics calculator or computer, plot $y = f(x)$ to see whether the appearance of the graph corresponds to your classification of the critical points.

1. $f(x) = x^2 - 4x + 5$ **2.** $f(x) = 6x - x^2$
3. $f(x) = x^3 - 3x^2 + 5$ **4.** $f(x) = x^3 - 3x + 5$
5. $f(x) = x^3 - 3x^2 + 3x + 5$
6. $f(x) = 2x^3 + 3x^2 - 36x + 17$
7. $f(x) = 10 + 60x + 9x^2 - 2x^3$
8. $f(x) = 27 - x^3$ **9.** $f(x) = x^4 - 2x^2$
10. $f(x) = 3x^5 - 5x^3$ **11.** $f(x) = x + \dfrac{1}{x}$
12. $f(x) = x + \dfrac{9}{x}$ **13.** $f(x) = x^2 + \dfrac{2}{x}$
14. $f(x) = xe^{-x}$ **15.** $f(x) = x^2 e^{-x}$
16. $f(x) = x^2 e^{-x^2}$

In Problems 17 through 26, find and classify the critical points of the given function in the indicated open interval. You may find it useful to construct a table of signs as in Example 1.

17. $f(x) = \sin^2 x$; $(0, 3)$
18. $f(x) = \cos^2 x$; $(-1, 3)$
19. $f(x) = \sin^3 x$; $(-3, 3)$
20. $f(x) = \cos^4 x$; $(0, 4)$
21. $f(x) = \tan^2 x$; $(-1, 1)$
22. $f(x) = \tan^3 x$; $(-1, 1)$
23. $f(x) = \sin x - x \cos x$; $(-5, 5)$
24. $f(x) = \cos x + x \sin x$; $(-3, 3)$
25. $f(x) = \dfrac{\ln x}{x^2}$; $(0, 5)$
26. $f(x) = \dfrac{1 - \ln x}{x}$; $(5, 10)$

In Problems 27 through 50, which are applied maximum-minimum problems, use the first derivative test to verify your answers.

27. Determine two real numbers with difference 20 and minimum possible product.

28. A long, rectangular sheet of metal is to be made into a rain gutter by turning up two sides at right angles to the remaining center strip (Fig. 4.4.24). The rectangular cross

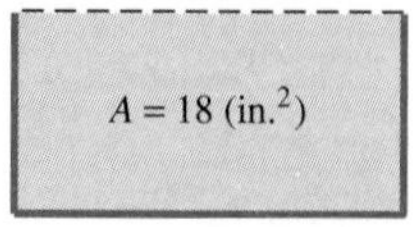

$A = 18$ (in.^{2})

Fig. 4.4.24 The rectangular cross section of Problem 28

section of the gutter is to have area 18 in.^{2} Find the minimum possible width of the strip.

29. Find the point (x, y) on the line $2x + y = 3$ that is closest to the point $(3, 2)$.

30. You must construct a closed rectangular box with volume 576 in.^{3} and with its bottom twice as long as it is wide (Fig. 4.4.25). Find the dimensions of the box that will minimize its total surface area.

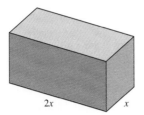

Fig. 4.4.25 The box of Problem 30

31. Repeat Problem 30, but use an open-topped box with volume 972 in.^{3}

32. An open-topped cylindrical pot is to have volume 125 in.^{3} What dimensions will minimize the total amount of material used in making this pot (Fig. 4.4.26)? Neglect the thickness of the material and possible wastage.

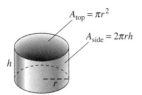

Fig. 4.4.26 The cylinder of Problems 32, 33, 38, and 39

33. An open-topped cylindrical pot is to have volume 250 cm^3 (Fig. 4.4.26). The material for the bottom of the pot costs 4¢/cm^2; that for its curved side costs 2¢/cm^2. What dimensions will minimize the total cost of this pot?

34. Find the point (x, y) on the parabola $y = 4 - x^2$ that is closest to the point $(3, 4)$. [*Suggestion:* The cubic equation that you should obtain has a small integer as one of its roots. *Suggestion:* Minimize the *square* of the distance.]

35. Show that the rectangle with area 100 and minimum perimeter is a square.

36. Show that the rectangular solid with a square base, volume 1000, and minimum total surface area is a cube.

37. A box with a square base and an open top is to have volume 62.5 in.^{3} Neglect the thickness of the material used to make the box, and find the dimensions that will minimize the amount of material used.

38. You need a tin can in the shape of a right circular cylinder of volume 16π cm^3 (Fig. 4.4.26). What radius r

and height h would minimize its total surface area (including top and bottom)?

39. The metal used to make the top and bottom of a cylindrical can (Fig. 4.4.26) costs 4¢/in.^{2}; the metal used for the sides costs 2¢/in.^{2} The volume of the can must be exactly 100 in.^{3} What dimensions of the can would minimize its total cost?

40. Each page of a book will contain 30 in.^{2} of print, and each page must have 2-in. margins at top and bottom and 1-in. margins at each side. What is the minimum possible area of such a page?

41. What point or points on the curve $y = x^2$ are nearest the point $(0, 2)$? [*Suggestion:* The square of a distance is minimized exactly when the distance is itself minimized.]

42. What is the length of the shortest line segment lying wholly in the first quadrant with its endpoints on the coordinate axes and tangent to the graph of $y = 1/x$?

43. A rectangle has area 64 cm^2. A straight line is to be drawn from one corner of the rectangle to the midpoint of one of the two more distant sides. What is the minimum possible length of such a line?

44. An oil can is to have volume 1000 in.^{3} and is to be shaped like a cylinder with a flat bottom but capped by a hemisphere (Fig. 4.4.27). Neglect the thickness of the material of the can, and find the dimensions that will minimize the amount of material needed to construct it.

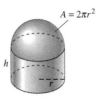

Fig. 4.4.27 The oil can of Problem 44

45. Find the length L of the longest rod that can be carried horizontally around a corner from a corridor 2 m wide into one 4 m wide. Do this by *minimizing* the length of the rod in Fig. 4.4.28 by minimizing the square of that length as a function of x.

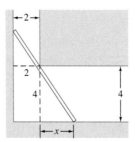

Fig. 4.4.28 Carrying a rod around a corner (Problem 45)

46. Find the length of the shortest ladder that will reach from the ground, over a wall 8 ft high, to the side of a building 1 ft behind the wall. That is, minimize the length $L = L_1 + L_2$ shown in Fig. 4.4.29.

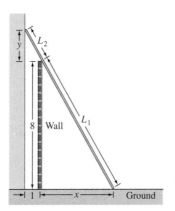

Fig. 4.4.29 The ladder of Problem 46

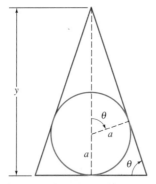

Fig. 4.4.30 Cross section through the centers of the sphere and pyramid of Problem 47.

47. A sphere with fixed radius a is inscribed in a pyramid with a square base so that the sphere touches the base of the pyramid and also each of its four sides. Show that the

minimum possible volume of the pyramid is $8/\pi$ times the volume of the sphere. [*Suggestion:* Use the two right triangles in Fig. 4.4.30 to show that the volume of the pyramid is

$$V = V(y) = \frac{4a^2y^2}{3(y - 2a)}.$$

This can be done easily with the aid of the angle θ and *without* the formula for $\tan(\theta/2)$.] Don't forget the domain of $V(y)$.

48. Two noisy discothèques, one four times as noisy as the other, are located on opposite ends of a block 1000 ft long. What is the quietest point on the block between the two discos? The intensity of noise at a point away from its source is proportional to the noisiness and inversely proportional to the square of the distance from the source.

49. A floored tent with fixed volume V is to be shaped like a pyramid with a square base and congruent sides (Fig. 4.4.31). What height y and base edge $2x$ would minimize its total surface area (including its floor)?

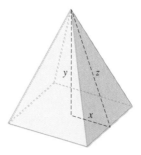

Fig. 4.4.31 The tent of Problem 49

50. Suppose that the distance from the building to the wall in Problem 46 is a and that the height of the wall is b. Show that the minimal length of the ladder is

$$L_{\min} = (a^{2/3} + b^{2/3})^{3/2}.$$

4.4 Project

The problems in this project require the use of a graphics calculator or computer with a graphing utility.

Figure 4.4.32 shows a rectangular box with square base. Suppose that its volume $V = x^2y$ is to be 1000 in.³ We want to make this box at minimal cost. Each of the six faces costs a cents per square inch, and gluing each of the 12 edges costs b cents per inch of edge length.

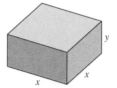

Fig. 4.4.32 A box with square base

1. Show first that the total cost C is given as a function of x by

$$C(x) = 2ax^2 + 8bx + \frac{4000a}{x} + \frac{4000b}{x^2}. \tag{5}$$

2. Suppose that $a = b = 1$. Find the dimensions of the box of minimal cost by zooming in on the appropriate solution of $C'(x) = 0$.

3. Repeat Problem 2 with a and b being the last two nonzero digits of your student I.D. number.

4. After doing Problems 2 and 3, you should smell a rat. Is it possible that the shape of the optimal box is independent of the values of a and b? Show that the equation $C'(x) = 0$ simplifies to the fourth-degree equation

$$ax^4 + 2bx^3 - 1000ax - 2000b = 0. \tag{6}$$

Then solve this equation (for x) by using a symbolic algebra system such as *Derive, Maple,* or *Mathematica,* if one is available. If not, you may still be able to solve Eq. (6) by hand—if you begin by factoring x^3 from the first two terms and 1000 from the last two.

5. Suppose that the top and bottom of the box in Fig. 4.4.32 cost p cents per square inch and that the four sides cost q cents per square inch (the 12 edges still cost b cents per inch). For instance, let p, q, and b be the last three nonzero digits of your student I.D. number. Then determine graphically the dimensions of the box with volume 1000 in.^{3} and minimal cost.

4.5
Simple Curve Sketching

We can construct a reasonably accurate graph of the polynomial function

$$f(x) = a_nx^n + a_{n-1}x^{n-1} + \cdots + a_2x^2 + a_1x + a_0 \tag{1}$$

by assembling the following information.

1. **The critical points of f**—that is, the points on the graph where the tangent line is horizontal, so that $f'(x) = 0$.
2. **The increasing/decreasing behavior of f**—that is, the intervals on which f is increasing and those on which it is decreasing.
3. **The behavior of f "at" infinity**—that is, the behavior of $f(x)$ as $x \to +\infty$ and as $x \to -\infty$.

The same information often is the key to understanding the structure of a graph that has been plotted with a calculator or computer.

To carry out the task in item 3, we write $f(x)$ in the form

$$f(x) = x^n\left(a_n + \frac{a_{n-1}}{x} + \cdots + \frac{a_1}{x^{n-1}} + \frac{a_0}{x^n}\right).$$

Thus we conclude that the behavior of $f(x)$ as $x \to \pm \infty$ is much the same as that of its *leading term a_nx^n,* because all the terms that have powers of x in the denominator approach zero as $x \to \pm\infty$. In particular, if $a_n > 0$, then

$$\lim_{x \to \infty} f(x) = +\infty, \tag{2}$$

meaning that $f(x)$ increases without bound as $x \to +\infty$. Also,

$$\lim_{x \to -\infty} f(x) = \begin{cases} +\infty & \text{if } n \text{ is even;} \\ -\infty & \text{if } n \text{ is odd.} \end{cases} \tag{3}$$

If $a_n < 0$, simply reverse the signs on the right-hand sides in Eqs. (2) and (3). Every polynomial, such as $f(x)$ in Eq. (1), is differentiable everywhere.

So the critical points of $f(x)$ are the roots of the polynomial equation $f'(x) = 0$—that is, solutions of

$$na_nx^{n-1} + (n - 1)a_{n-1}x^{n-2} + \cdots + 2a_2x + a_1 = 0. \qquad (4)$$

Sometimes we can find all (real) solutions of such an equation by factoring, but other times we must resort to numerical methods aided by calculator or computer.

But suppose that we have somehow found *all* the (real) solutions c_1, $c_2, \ldots, c_k$ of Eq. (4). Then these solutions are the critical points of f. If they are arranged in increasing order, as in Fig. 4.5.1, they separate the x-axis into the finite number of open intervals

$$(-\infty, c_1), (c_1, c_2), (c_2, c_3), \ldots, (c_{k-1}, c_k), (c_k, +\infty)$$

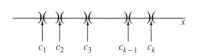

Fig. 4.5.1 The zeros of $f'(x)$ divide the x-axis into intervals on which $f'(x)$ does not change sign.

that also appear in the figure. The intermediate value property applied to $f'(x)$ tells us that $f'(x)$ can change sign only at the critical points of f, so $f'(x)$ has only one sign on each of these open intervals. It is typical for $f'(x)$ to be negative on some intervals and positive on others. Moreover, it's easy to find the sign of $f'(x)$ on any such interval I: We need only substitute *any* convenient number in I into $f'(x)$.

Once we know the sign of $f'(x)$ on each of these intervals, we know where f is increasing and where it is decreasing. We then apply the first derivative test to find which of the critical values are local maxima, which are local minima, and which are neither—merely places where the tangent line is horizontal. With this information, the knowledge of the behavior of f as $x \to \pm\infty$, and the fact that f is continuous, we can sketch its graph. We plot the critical points $(c_i, f(c_i))$ and connect them with a smooth curve that is consistent with our other data.

It may also be helpful to plot the y-intercept $(0, f(0))$ and also any x-intercepts that are easy to find. But we recommend (until inflection points are introduced in Section 4.6) that you plot *only* these points—critical points and intercepts—and rely otherwise on the increasing and decreasing behavior of f.

EXAMPLE 1 Sketch the graph of $f(x) = x^3 - 27x$.

Solution Because the leading term is x^3, we see that

$$\lim_{x\to+\infty} f(x) = +\infty \quad \text{and} \quad \lim_{x\to-\infty} f(x) = -\infty.$$

Moreover, because

$$f'(x) = 3x^2 - 27 = 3(x + 3)(x - 3), \qquad (5)$$

we see that the critical points where $f'(x) = 0$ are $x = -3$ and $x = 3$. The corresponding points on the graph of f are $(-3, 54)$ and $(3, -54)$. The critical points separate the x-axis into the three open intervals $(-\infty, -3)$, $(-3, 3)$, and $(3, +\infty)$ (Fig. 4.5.2). To determine the increasing or decreasing

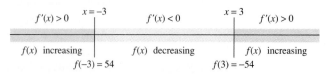

Fig. 4.5.2 The three open intervals of Example 1

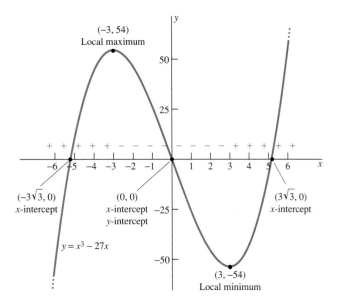

(−3, 54)
Local maximum

(−3√3, 0)
x-intercept

(0, 0)
x-intercept
y-intercept

(3√3, 0)
x-intercept

$y = x^3 - 27x$

(3, −54)
Local minimum

Fig. 4.5.3 Graph of the function of Example 1

behavior of f on these intervals, let's substitute a number in each interval into the derivative in Eq. (5):

On $(-\infty, -3)$: $f'(-4) = (3)(-1)(-7) = 21 > 0;$ f is increasing;

On $(-3, 3)$: $f'(0) = (3)(3)(-3) = -27 < 0;$ f is decreasing;

On $(3, +\infty)$: $f'(4) = (3)(7)(1) = +21 > 0;$ f is increasing.

We plot the critical points and the intercepts $(0, 0)$, $(3\sqrt{3}, 0)$, and $(-3\sqrt{3}, 0)$. Then we use this information about where f is increasing or decreasing to connect them with a smooth curve. Remembering that there are horizontal tangents at the two critical points, we obtain the graph shown in Fig. 4.5.3.

In the figure we use plus and minus signs to mark the sign of $f'(x)$ in each interval. This makes it clear that $(-3, 54)$ is a local maximum and that $(3, -54)$ is a local minimum. The limits we found at the outset show that neither is global.

EXAMPLE 2   Sketch the graph of $f(x) = 8x^5 - 5x^4 - 20x^3$.

Solution   Because

$$f'(x) = 40x^4 - 20x^3 - 60x^2 = 20x^2(x + 1)(2x - 3), \tag{6}$$

the critical points where $f'(x) = 0$ are $x = -1$, $x = 0$, and $x = \frac{3}{2}$. These three critical points separate the x-axis into the four open intervals shown in Fig. 4.5.4. This time, let's determine the increasing or decreasing behavior of

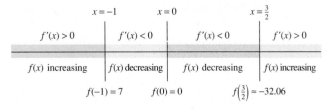

Fig. 4.5.4   The four open intervals of Example 2

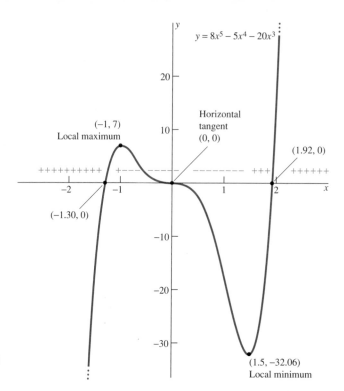

Fig. 4.5.5 Graph of the function of Example 2

f by recording the signs of the factors in Eq. (6) on each of the subintervals shown in Fig. 4.5.4. In this way we get the following table:

Interval	$x + 1$	$20x^2$	$2x - 3$	$f'(x)$	f
$(-\infty, -1)$	Neg.	Pos.	Neg.	Pos.	Increasing
$(-1, 0)$	Pos.	Pos.	Neg.	Neg.	Decreasing
$(0, \frac{3}{2})$	Pos.	Pos.	Neg.	Neg.	Decreasing
$(\frac{3}{2}, +\infty)$	Pos.	Pos.	Pos.	Pos.	Increasing

The points on the graph that correspond to the critical points are $(-1, 7)$, $(0, 0)$, and $(1.5, -32.0625)$.

We write $f(x)$ in the form

$$f(x) = x^3(8x^2 - 5x - 20)$$

in order to use the quadratic formula to find the intercepts. They turn out to be $(-1.30, 0)$, $(1.92, 0)$ (the abscissas are given only approximately), and the origin $(0, 0)$. The latter is also the y-intercept. We apply the first derivative test to the increasing or decreasing behavior shown in the table. It follows that $(-1, 7)$ is a local maximum, $(1.5, -32.0625)$ is a local minimum, and $(0, 0)$ is neither. The graph looks like the one shown in Fig. 4.5.5.

In Example 3, the function is not a polynomial. Nevertheless, the methods of this section suffice for sketching its graph.

EXAMPLE 3 Sketch the graph of

$$f(x) = x^{2/3}(x^2 - 2x - 6) = x^{8/3} - 2x^{5/3} - 6x^{2/3}.$$

Solution The derivative of f is

$$f'(x) = \frac{8}{3}x^{5/3} - \frac{10}{3}x^{2/3} - \frac{12}{3}x^{-1/3} = \frac{2}{3}x^{-1/3}(4x^2 - 5x - 6)$$

$$= \frac{2(4x + 3)(x - 2)}{3x^{1/3}}. \tag{7}$$

The tangent line is horizontal at the two critical points $x = -\frac{3}{4}$ and $x = 2$, where the numerator in the last fraction of Eq. (7) is zero (and the denominator is not). Moreover, because of the presence of the term $x^{1/3}$ in the denominator, $|f'(x)| \to +\infty$ as $x \to 0$. Thus $x = 0$ (a critical point because f is not differentiable there) is a point where the tangent line is vertical. These three critical points separate the x-axis into the four open intervals shown in Fig. 4.5.6. We determine the increasing or decreasing behavior of f by substituting a number from each interval in Eq. (7).

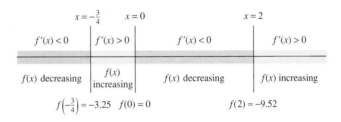

Fig. 4.5.6 The four open intervals of Example 3

$$\text{On } (-\infty, -\tfrac{3}{4}): \quad f'(-1) = \frac{2 \cdot (-1)(-3)}{3 \cdot (-1)} < 0; \quad f \text{ is decreasing;}$$

$$\text{On } (-\tfrac{3}{4}, 0): \quad f'(-\tfrac{1}{2}) = \frac{2 \cdot (+1)(-\tfrac{5}{2})}{3 \cdot (-\tfrac{1}{2})^{1/3}} > 0; \quad f \text{ is increasing;}$$

$$\text{On } (0, 2): \quad f'(1) = \frac{2 \cdot (+7)(-1)}{3 \cdot (+1)} < 0; \quad f \text{ is decreasing;}$$

$$\text{On } (2, +\infty): \quad f'(3) = \frac{2 \cdot (+15)(+1)}{3 \cdot (+3)^{1/3}} > 0; \quad f \text{ is increasing.}$$

The three critical points $x = -\frac{3}{4}$, $x = 0$, and $x = 2$ give the points $(-0.75, -3.25)$, $(0, 0)$, and $(2, -9.52)$ on the graph (using approximations where appropriate).

The first derivative test now shows local minima at $(-0.75, -3.25)$ and $(2, -9.52)$ and a local maximum at $(0, 0)$. Although $f'(0)$ does not exist, f is continuous at $x = 0$, so it is continuous everywhere.

We use the quadratic formula to find the x-intercepts. In addition to the origin, they occur where $x^2 - 2x - 6 = 0$, and thus they are located at $(1 - \sqrt{7}, 0)$ and at $(1 + \sqrt{7}, 0)$. We then plot the approximations $(-1.65, 0)$ and $(3.65, 0)$. Finally, we note that $f(x) \to +\infty$ as $x \to \pm\infty$. So the graph has the shape shown in Fig. 4.5.7.

Fig. 4.5.7 The technique is effective for nonpolynomial functions, as in Example 3.

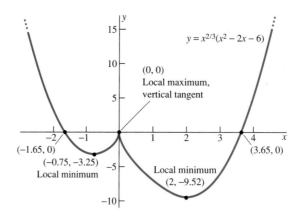

$y = x^{2/3}(x^2 - 2x - 6)$

(0, 0)
Local maximum,
vertical tangent

(−1.65, 0)
(3.65, 0)

(−0.75, −3.25)
Local minimum

Local minimum
(2, −9.52)

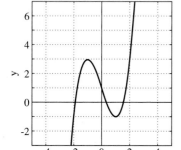

Fig. 4.5.8 $y = x^3 - 3x + 1$

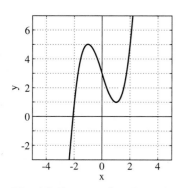

Fig. 4.5.9 $y = x^3 - 3x + 2$

CURVE SKETCHING AND SOLUTION OF EQUATIONS

An important application of curve-sketching techniques is the solution of an equation of the form

$$f(x) = 0. \qquad (8)$$

The real (as opposed to complex) solutions of this equation are simply the x-intercepts of the graph of $y = f(x)$. Hence by sketching this graph with reasonable accuracy—either "by hand" or with a calculator or computer—we can glean information about the number of real solutions of Eq. (8) as well as their approproximate locations.

For example, Figs. 4.5.8 through 4.5.10 show the graphs of the cubic polynomials on the left-hand sides of the equations

$$x^3 - 3x + 1 = 0, \qquad (9)$$

$$x^3 - 3x + 2 = 0, \qquad (10)$$

$$x^3 - 3x + 3 = 0. \qquad (11)$$

Note that the polynomials differ only in their constant terms.

It is clear from Fig. 4.5.8 that Eq. (9) has three real solutions, one in each of the intervals $[-2, -1]$, $[0, 1]$, and $[1, 2]$. These solutions could be approximated graphically by successive magnification or analytically by Newton's method. (There are even formulas—*Cardan's formulas*—for the exact solution of an arbitrary cubic equation, but they are unwieldy and are seldom used.)

It appears in Fig. 4.5.9 that Eq. (10) has the two real solutions $x = 1$ and $x = -2$. Once we verify that $x = 1$ is a solution, then it follows from the *factor theorem* of algebra that $x - 1$ is a factor of $x^3 - 3x + 2$. The other factor can be found by division:

$$
\begin{array}{r}
x^2 + x - 2 \\
x - 1 \overline{)\, x^3 - 3x + 2} \\
\underline{x^3 - x^2 } \\
x^2 - 3x \\
\underline{x^2 - x } \\
-2x + 2 \\
\underline{-2x + 2} \\
0
\end{array}
$$

Fig. 4.5.10 $y = x^3 - 3x + 3$

236

Thus we see that

$$x^3 - 3x + 2 = (x - 1)(x^2 + x - 2) = (x - 1)^2(x + 2).$$

Hence $x = 1$ is a "double root" and $x = -2$ is a "single root" of Eq. (10), thereby accounting for the three solutions that a cubic equation "ought to have."

We see in Fig. 4.5.10 that Eq. (11) has only one real solution. It is given approximately by $x \approx -2.1038$. Problem 45 asks you to divide $x + 2.1038$ into $x^3 - 3x + 3$ to obtain a factorization of the form

$$x^3 - 3x + 3 \approx (x + 2.1038)(x^2 + bx + c). \tag{12}$$

The quadratic equation $x^2 + bx + c = 0$ has two complex conjugate solutions, which are the other two solutions of Eq. (11).

4.5 Problems

In Problems 1 through 4, use behavior "at infinity" to match the given function with its graph in Fig. 4.5.11.

1. $f(x) = x^3 - 5x + 2$

2. $f(x) = x^4 - 3x^2 + x - 2$

3. $f(x) = -\frac{1}{3}x^5 - 3x^2 + 3x + 2$

4. $f(x) = -\frac{1}{3}x^6 + 2x^5 - 3x^4 + \frac{1}{2}x + 5$

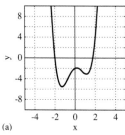
(a)

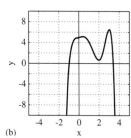
(b)

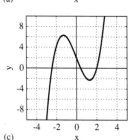
(c)

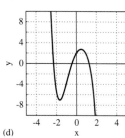
(d)

Fig. 4.5.11 Problems 1 through 4

In Problems 5 through 38, find the intervals on which the function f is increasing and those on which it is decreasing. Sketch the graph of $y = f(x)$, and label the local maxima and minima.

5. $f(x) = 3x^2 - 6x + 5$

6. $f(x) = 5 - 8x - 2x^2$

7. $f(x) = x^3 - 12x$

8. $f(x) = 2x^3 + 3x^2 - 12x$

9. $f(x) = x^3 - 6x^2 + 9x$

10. $f(x) = x^3 + 6x^2 + 9x$

11. $f(x) = x^3 + 3x^2 + 9x$

12. $f(x) = x^3 - 27x$

13. $f(x) = (x - 1)^2(x + 2)^2$

14. $f(x) = (x - 2)^2(2x + 3)^2$

15. $f(x) = 3\sqrt{x} - x\sqrt{x}$

16. $f(x) = x^{2/3}(5 - x)$

17. $f(x) = 3x^5 - 5x^3$

18. $f(x) = x^4 + 4x^3$

19. $f(x) = x^4 - 8x^2 + 7$

20. $f(x) = \dfrac{1}{x}$

21. $f(x) = 2x^2 - 3x - 9$

22. $f(x) = 6 - 5x - 6x^2$

23. $f(x) = 2x^3 + 3x^2 - 12x$

24. $f(x) = x^3 + 4x$

25. $f(x) = 50x^3 - 105x^2 + 72x$

26. $f(x) = x^3 - 3x^2 + 3x - 1$

27. $f(x) = 3x^4 - 4x^3 - 12x^2 + 8$

28. $f(x) = x^4 - 2x^2 + 1$

29. $f(x) = 3x^5 - 20x^3$

30. $f(x) = 3x^5 - 25x^3 + 60x$

31. $f(x) = 2x^3 + 3x^2 + 6x$

32. $f(x) = x^4 - 4x^3$

33. $f(x) = 8x^4 - x^8$

34. $f(x) = 1 - x^{1/3}$

35. $f(x) = x^{1/3}(4 - x)$

36. $f(x) = x^{2/3}(x^2 - 16)$

37. $f(x) = x(x - 1)^{2/3}$

38. $f(x) = x^{1/3}(2 - x)^{2/3}$

In Problems 39 through 44, the values of the function $f(x)$ at its critical points are given, together with the graph $y = f'(x)$ of its derivative. Use this information to construct a rough sketch of the graph $y = f(x)$ of the function.

39. $f(-3) = 78, f(2) = -47$; Fig. 4.5.12

40. $f(-2) = 106, f(4) = -110$; Fig. 4.5.13

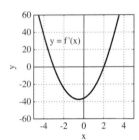
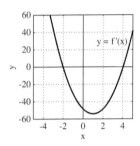

Fig. 4.5.12 $y = f'(x)$ of Problem 39

Fig. 4.5.13 $y = f'(x)$ of Problem 40

41. $f(-3) = -66, f(2) = 59$; Fig. 4.5.14

42. $f(-3) = -130, f(0) = 5, f(1) = -2$; Fig. 4.5.15

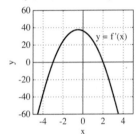
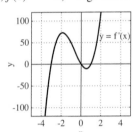

Fig. 4.5.14 $y = f'(x)$ of Problem 41

Fig. 4.5.15 $y = f'(x)$ of Problem 42

43. $f(-2) = -107, f(1) = 82, f(3) = 18$; Fig. 4.5.16

44. $f(-3) = 5336, f(0) = 17, f(2) = 961, f(4) = -495$; Fig. 4.5.17

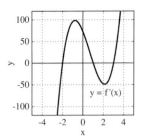
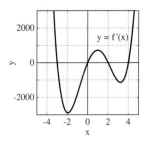

Fig. 4.5.16 $y = f'(x)$ of Problem 43

Fig. 4.5.17 $y = f'(x)$ of Problem 44

45. (a) Verify the approximate solution $x \approx -2.1038$ of Eq. (11). (b) Divide $x^3 - 3x + 3$ by $x + 2.1038$ to obtain the factorization in Eq. (12). (c) Use the quotient in part (b) to find (approximately) the complex conjugate pair of solutions of Eq. (11).

46. Explain why Figs. 4.5.8 and 4.5.9 imply that the cubic equation $x^3 - 3x + q = 0$ has exactly one real solution if $|q| > 2$ but has three distinct real solutions if $|q| < 2$. What is the situation if $q = -2$?

47. The computer-generated graph in Fig. 4.5.18 shows how the curve $y = [x(x - 1)(2x - 1)]^2$ looks on any "reasonable" scale with integral units of measurement on the y-axis. Use the methods of this section to show that the graph really has the appearance shown in Fig. 4.5.19 (the values on the y-axis are in thousandths), with critical points at $0, \frac{1}{2}, \frac{1}{6}(3 \pm \sqrt{3})$, and 1.

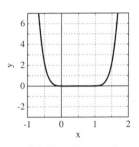
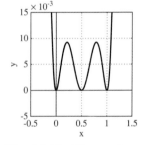

Fig. 4.5.18 The graph $y = [x(x - 1)(2x - 1)]^2$ on a "reasonable" scale (Problem 47)

Fig. 4.5.19 The graph $y = [x(x - 1)(2x - 1)]^2$ on a finer scale: $-0.005 \leqq y \leqq 0.015$ (Problem 47)

4.5 Projects

These projects require the use of a graphics calculator or a computer with a graphing utility.

PROJECT A First show that, on a "reasonable" scale with integral units of measurement on the y-axis, the graph of the polynomial

$$f(x) = [\tfrac{1}{6}x(9x - 5)(x - 1)]^4$$

looks much like Fig. 4.5.18, with an apparent flat section. Then produce a plot that reveals the true structure of the graph, as in Fig. 4.5.19. Finally, find (graphically or otherwise) the approximate coordinates of the local maximum and minimum points on the graph of $y = f(x)$.

PROJECT B   The quartic (fourth-degree) equations

$$f(x) = x^4 - 55x^3 + 500x^2 + 11{,}000x - 110{,}000 = 0 \qquad (13)$$

and

$$g(x) = x^4 - 55x^3 + 550x^2 + 11{,}000x - 110{,}000 = 0 \qquad (14)$$

differ only in a single digit in the coefficient of x^2. However small this difference may seem, show graphically that Eq. (13) has four distinct real solutions but Eq. (14) has only two. Find (approximately) the local maximum and minimum points on each graph.

4.6
Higher Derivatives and Concavity

We saw in Section 4.3 that the sign of the first derivative f' indicates whether the graph of the function f is rising or falling. Here we shall see that the sign of the *second* derivative of f, the derivative of f', indicates which way the curve $y = f(x)$ is *bending*, upward or downward.

HIGHER DERIVATIVES

The **second derivative** of f is denoted by f'', and its value at x is

$$f''(x) = D(f'(x)) = D(Df(x)) = D^2 f(x).$$

(The superscript 2 is not an exponent but only an indication that the derivative is the second one.) The derivative of f'' is the **third derivative** f''' of f, with

$$f'''(x) = D(f''(x)) = D(D^2 f(x)) = D^3 f(x).$$

The third derivative is also denoted by $f^{(3)}$. More generally, the result of beginning with the function f and differentiating n times in succession is the **nth derivative** $f^{(n)}$ of f, with $f^{(n)}(x) = D^n f(x)$.

If $y = f(x)$, then the first n derivatives are written in operator notation as

$$D_x y, \; D_x{}^2 y, \; D_x{}^3 y, \; \ldots , \; D_x{}^n y,$$

in dependent-independent variable notation as

$$y'(x), \; y''(x), \; y'''(x), \; \ldots , \; y^{(n)}(x),$$

or in differential notation as

$$\frac{dy}{dx}, \frac{d^2 y}{dx^2}, \frac{d^3 y}{dx^3}, \ldots , \frac{d^n y}{dx^n}.$$

The history of the curious use of superscripts in differential notation for higher derivatives involves the metamorphosis

$$\frac{d}{dx}\left(\frac{dy}{dx}\right) \to \frac{d}{dx}\frac{dy}{dx} \to \frac{(d)^2 y}{(dx)^2} \to \frac{d^2 y}{dx^2}.$$

EXAMPLE 1 Find the first four derivatives of

$$f(x) = 2x^3 + \frac{1}{x^2} + 16x^{7/2}.$$

Solution   Write

$$f(x) = 2x^3 + x^{-2} + 16x^{7/2}. \qquad \text{Then}$$

$$f'(x) = 6x^2 - 2x^{-3} + 56x^{5/2} = 6x^2 - \frac{2}{x^3} + 56x^{5/2},$$

$$f''(x) = 12x + 6x^{-4} + 140x^{3/2} = 12x + \frac{6}{x^4} + 140x^{3/2},$$

$$f'''(x) = 12 - 24x^{-5} + 210x^{1/2} = 12 - \frac{24}{x^5} + 210\sqrt{x}, \quad \text{and}$$

$$f^{(4)}(x) = 120x^{-6} + 105x^{-1/2} = \frac{120}{x^6} + \frac{105}{\sqrt{x}}.$$

Example 2 shows how to find higher derivatives of implicitly defined functions.

EXAMPLE 2   Find the second derivative $y''(x)$ of the function $y = y(x)$ that is defined implicitly by the equation

$$x^2 - xy + y^2 = 9.$$

Solution   A first implicit differentiation of the given equation *with respect to x* gives

$$2x - y - x\frac{dy}{dx} + 2y\frac{dy}{dx} = 0,$$

so

$$\frac{dy}{dx} = \frac{y - 2x}{2y - x}.$$

We obtain d^2y/dx^2 by differentiating implicitly, again with respect to x, using the quotient rule. After that, we substitute the expression we just found for dy/dx:

$$\frac{d^2y}{dx^2} = D_x\left(\frac{y - 2x}{2y - x}\right) = \frac{\left(\dfrac{dy}{dx} - 2\right)(2y - x) - (y - 2x)\left(2\dfrac{dy}{dx} - 1\right)}{(2y - x)^2}$$

$$= \frac{3x\dfrac{dy}{dx} - 3y}{(2y - x)^2} = \frac{3x\dfrac{y - 2x}{2y - x} - 3y}{(2y - x)^2}.$$

Thus

$$\frac{d^2y}{dx^2} = -\frac{6(x^2 - xy + y^2)}{(2y - x)^3}.$$

We now substitute the original equation, $x^2 - xy + y^2 = 9$, for one final simplification:

$$\frac{d^2y}{dx^2} = -\frac{54}{(2y - x)^3}.$$

THE SECOND DERIVATIVE TEST

Now we shall investigate the significance of the *sign* of the second derivative. If $f''(x) > 0$ on the interval I, then the first derivative f' is an increasing function on I, because *its* derivative $f''(x)$ is positive. Thus, as we scan the graph $y = f(x)$ from left to right, we see the tangent line turning counterclockwise (Fig. 4.6.1). We describe this situation by saying that the curve $y = f(x)$ is **bending upward.** Note that a curve can bend upward without rising, as in Fig. 4.6.2.

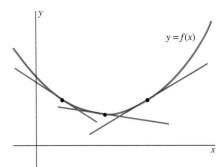

Fig. 4.6.1 The graph is bending upward (concave upward).

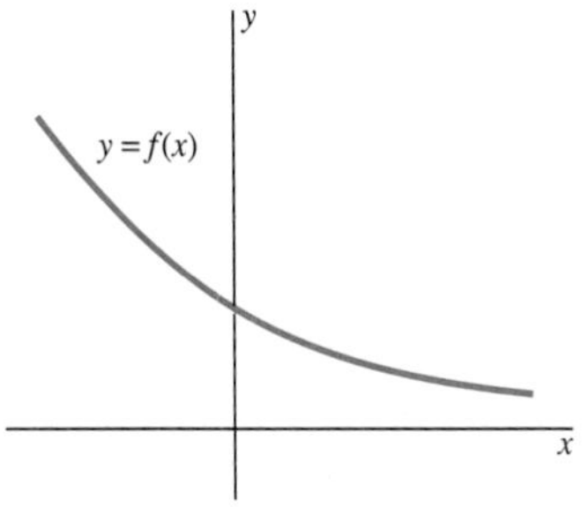

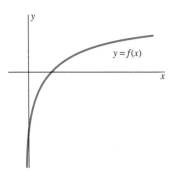

Fig. 4.6.2 Another graph bending upward (concave upward)

If $f''(x) < 0$ on the interval I, then the first derivative f' is decreasing on I, so the tangent line turns clockwise as x increases. We say in this case that the curve $y = f(x)$ is **bending downward.** Figures 4.6.3 and 4.6.4 show two ways this can happen.

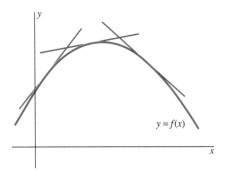

$f''(x)$	$y = f(x)$
Negative	Bending downward
Positive	Bending upward

Fig. 4.6.5 Significance of the sign of $f''(x)$ on an interval

Fig. 4.6.3 A graph bending downward (concave downward)

Fig. 4.6.4 Another graph bending downward (concave downward)

The two cases are summarized in the brief table in Fig. 4.6.5.

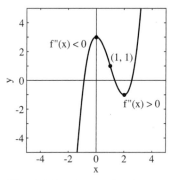

Fig. 4.6.6 The graph of $y = x^3 - 3x^2 + 3$ (Example 3)

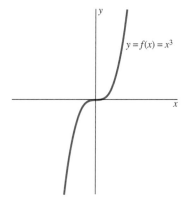

Fig. 4.6.7 Although $f'(0) = 0$, $f(0)$ is not an extremum.

EXAMPLE 3 Figure 4.6.6 shows the graph of the function

$$f(x) = x^3 - 3x^2 + 3.$$

Because

$$f'(x) = 3x^2 - 6x \quad \text{and} \quad f''(x) = 6x - 6 = 6(x - 1),$$

we see that

$$f''(x) < 0 \quad \text{for} \quad x < 1,$$
$$f''(x) > 0 \quad \text{for} \quad x > 1.$$

Observe in the figure that the curve bends downward on $(-\infty, 1)$ but bends upward on $(1, \infty)$, consistent with the correspondences in Fig. 4.6.5.

We know from Section 3.5 that a local extremum of a differentiable function f can occur only at a critical point c where $f'(c) = 0$, so the tangent line at the point $(c, f(c))$ on the curve $y = f(x)$ is horizontal. But the example $f(x) = x^3$, for which $x = 0$ is a critical point but not an extremum (Fig. 4.6.7), shows that the *necessary condition* $f'(c) = 0$ is *not* a sufficient condition by which to conclude that $f(c)$ is an extreme value of the function f.

Now suppose not only that $f'(c) = 0$, but also that the curve $y = f(x)$ is bending upward on some open interval that contains the critical point $x = c$. It is apparent from Fig. 4.6.8(a) that $f(c)$ is a local minimum value. Similarly, $f(c)$ is a local maximum value if $f'(c) = 0$ while $y = f(x)$ is bending downward on some open interval about c [Fig. 4.6.8(b)]. But the *sign* of the second derivative $f''(x)$ tells us whether $y = f(x)$ is bending upward or downward and therefore provides us with a *sufficient* condition for a local extremum.

Theorem 1 *Second Derivative Test*

Suppose that the function f is twice differentiable on the open interval I containing the critical point c at which $f'(c) = 0$. Then

1. If $f''(x) > 0$ on I, then $f(c)$ is the minimum value of $f(x)$ on I.
2. If $f''(x) < 0$ on I, then $f(c)$ is the maximum value of $f(x)$ on I.

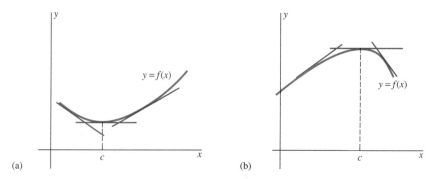

Fig. 4.6.8 The second derivative test (Theorem 1). (a) $f''(x) > 0$; tangent turning counterclockwise; graph concave upward; local minimum at c. (b) $f''(x) < 0$; tangent turning clockwise; graph concave downward; local maximum at c

Proof We will prove only part 1. If $f''(x) > 0$ on I, then it follows that the first derivative f' is an increasing function on I. Because $f'(c) = 0$, we conclude that $f'(x) < 0$ for $x < c$ in I and that $f'(x) > 0$ for $x > c$ in I. Consequently, the first derivative test of Section 4.4 implies that $f(c)$ is the minimum value of $f(x)$ on I. ❑

$f''(x)$	$f(c)$
Positive	Minimum
Negative	Maximum

Fig. 4.6.9 Significance of the sign of $f''(x)$ on an interval containing the critical point c

REMARK 1 Rather than memorizing verbatim the conditions in parts 1 and 2 of Theorem 1 (summarized in Fig. 4.6.9), it is easier and more reliable to remember the second derivative test by visualizing continuously turning tangent lines (Fig. 4.6.8).

REMARK 2 Theorem 1 implies that the function f has a local minimum at the critical point c if $f''(c) > 0$ on some open interval about c but a local maximum if $f''(x) < 0$ near c. But the hypothesis on $f''(x)$ in Theorem 1 is *global* in that $f''(x)$ is assumed to have the same sign at *every* point of the open interval I that contains the critical point c. There is a strictly *local* version of the second derivative test that involves only the sign of $f''(c)$ at the critical point c (rather than on a whole open interval). According to Problem 80, if $f'(c) = 0$, then $f(c)$ is a local minimum value of f if $f''(c) > 0$ but a local maximum if $f''(c) < 0$.

REMARK 3 The second derivative test says *nothing* about what happens if $f''(c) = 0$ at the critical point c. Consider the three functions $f(x) = x^4$, $f(x) = -x^4$, and $f(x) = x^3$. For each, $f'(0) = 0$ and $f''(0) = 0$. But their graphs, shown in Fig. 4.6.10, demonstrate that *anything* can happen at such a point.

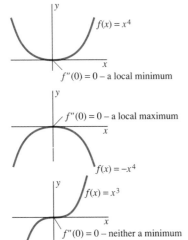

$f(x) = x^4$

$f''(0) = 0$ – a local minimum

$f''(0) = 0$ – a local maximum

$f(x) = -x^4$

$f(x) = x^3$

$f''(0) = 0$ – neither a minimum nor a maximum

Fig. 4.6.10 No conclusion is possible if $f'(c) = 0 = f''(c)$.

REMARK 4 Suppose that we want to maximize or minimize the function f on the open interval I, and we find that f has only one critical point in I, a number c at which $f'(c) = 0$. If $f''(x)$ has the same sign at all points of I, then Theorem 1 implies that $f(c)$ is an *absolute* extremum of f on I—a minimum if $f''(x) > 0$ and a maximum if $f''(x) < 0$. This absolute interpretation of the second derivative test is useful in applied open-interval maximum-minimum problems.

EXAMPLE 3 continued Consider again the function $f(x) = x^3 - 3x^2 + 3$, for which

$$f'(x) = 3x(x - 2) \quad \text{and} \quad f''(x) = 6(x - 1).$$

Then f has the two critical points, $x = 0$ and $x = 2$, as marked in Fig. 4.6.6. Because $f''(x) < 0$ for x near zero, the second derivative test implies that $f(0) = 3$ is a local maximum value of f. And because $f''(x) > 0$ for x near 2, it follows that $f(2) = -1$ is a local minimum value.

EXAMPLE 4 An open-topped rectangular box with square base has volume 500 cm³. Find the dimensions that minimize the total area A of its base and four sides.

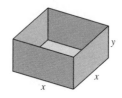

Fig. 4.6.11 The open-topped box of Example 4

Solution We denote by x the edge length of the square base and by y the height of the box (Fig. 4.6.11). The volume of the box is

$$V = x^2 y = 500, \tag{1}$$

and the total area of its base and four sides is

$$A = x^2 + 4xy. \tag{2}$$

When we solve Eq. (1) for $y = 500/x^2$ and substitute this into Eq. (2), we get the area function

$$A(x) = x^2 + \frac{2000}{x}, \qquad 0 < x < +\infty.$$

The domain of A is the open and unbounded interval $(0, +\infty)$ because x can take on any positive value; to make the box volume 500, simply choose $y = 500/x^2$. But x cannot be zero or negative.

The first derivative of $A(x)$ is

$$A'(x) = 2x - \frac{2000}{x^2} = \frac{2(x^3 - 1000)}{x^2}. \tag{3}$$

The equation $A'(x) = 0$ yields $x^3 = 1000$, so the only critical point of A in $(0, +\infty)$ is $x = 10$. To investigate this critical point, we calculate the second derivative,

$$A''(x) = 2 + \frac{4000}{x^3}. \tag{4}$$

Because it is clear that $A''(x) > 0$ on $(0, +\infty)$, it follows from the second derivative test and Remark 4 that $A(10) = 300$ is the absolute minimum value of $A(x)$ on $(0, +\infty)$. Finally, because $y = 500/x^2$, $y = 5$ when $x = 10$. Therefore, this absolute minimum corresponds to a box with base 10 cm by 10 cm and height 5 cm.

CONCAVITY AND CURVE SKETCHING

A comparison of Fig. 4.6.1 with Fig. 4.6.3 suggests that the question of whether the curve $y = f(x)$ is bending upward or downward is closely related to the question of whether it lies above or below its tangent lines. The latter question refers to the important property of *concavity*.

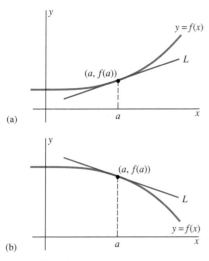

(a)

(b)

Fig. 4.6.12 (a) At $x = a$, f is concave upward. (b) At $x = a$, f is concave downward.

Definition *Concavity*

Suppose that the function f is differentiable at the point a and that L is the line tangent to the graph $y = f(x)$ at the point $(a, f(a))$. Then the function f (or its graph) is said to be

1. **Concave upward** at a if, on some open interval containing a, the graph of f lies *above* L;
2. **Concave downward** at a if, on some open interval containing a, the graph of f lies *below* L.

Figure 4.6.12(a) shows a graph that is concave upward at $(a, f(a))$. Figure 4.6.12(b) shows a graph that is concave downward at $(a, f(a))$.

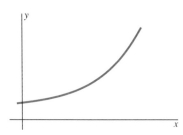

Fig. 4.6.13 $f'(x) > 0$, f increasing; $f''(x) > 0$, f concave upward

Fig. 4.6.14 $f'(x) > 0$, f increasing; $f''(x) < 0$, f concave downward

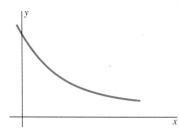

Fig. 4.6.15 $f'(x) < 0$, f decreasing; $f''(x) > 0$, f concave upward

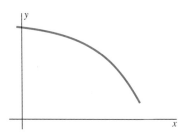

Fig. 4.6.16 $f'(x) < 0$, f decreasing; $f''(x) < 0$, f concave downward

Theorem 2 establishes the connection between concavity and the sign of the second derivative. That connection is the one suggested by our discussion of bending.

Theorem 2   Test for Concavity
Suppose that the function f is twice differentiable on the open interval I.

1. If $f''(x) > 0$ on I, then f is concave upward at each point of I.
2. If $f''(x) < 0$ on I, then f is concave downward at each point of I.

A proof of Theorem 2 based on the second derivative test is given at the end of this section.

NOTE   The significance of the sign of the *first* derivative must not be confused with the significance of the sign of the *second* derivative. The possibilities illustrated in Figs. 4.6.13 through 4.6.16 show that the signs of f' and f'' are independent of each other.

EXAMPLE 3 continued again   For the function $f(x) = x^3 - 3x^2 + 3$, the second derivative
$$f''(x) = 6(x - 1)$$
changes sign from positive to negative at the point $x = 1$. Observe in Fig. 4.6.6 that the corresponding point $(1, 1)$ on the graph of f is where the curve changes from bending downward to bending upward.

Observe that the test for concavity in Theorem 2 says nothing about the case in which $f''(x) = 0$. A point where the second derivative is zero *may or may not* be a point where the function changes from concave upward on one side to concave downward on the other. But a point like $(1, 1)$ in Fig. 4.6.6, where the concavity *does* change in this manner, is called an *inflection point* of the graph of f. More precisely, the point $x = a$ where f is continuous is an **inflection point** of the function f provided that f is concave upward on one side of $x = a$ and concave downward on the other side. We also refer to $(a, f(a))$ as an inflection point on the graph of f.

Theorem 3   Inflection Point Test
Suppose that the function f is continuous on an open interval containing the point a. Then a is an inflection point of f provided that $f''(x) < 0$ on one side of a and $f''(x) > 0$ on the other side.

The fact that a point where the second derivative changes sign is an inflection point follows from Theorem 2 and the definition of an inflection point.

REMARK   At the inflection point itself, either

❏ $f''(a) = 0$,   or
❏ $f''(a)$ does not exist.

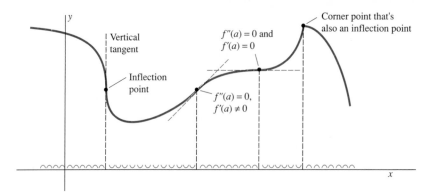

Fig. 4.6.17   Some inflection points

Thus we find *inflection points of f* by examining the *critical points of f'*. Some of the possibilities are indicated in Fig. 4.6.17. We mark the intervals of upward concavity and downward concavity by small cups opening upward and downward, respectively.

EXAMPLE 5   Figure 4.6.18 shows the graph of $f(x) = x^2 e^{-x}$. Two evident inflection points are marked. Find their coordinates.

Solution   We calculate

$$f'(x) = 2xe^{-x} - x^2 e^{-x} = (2x - x^2)e^{-x}$$

and

$$f''(x) = (2 - 2x)e^{-x} - (2x - x^2)e^{-x} = (x^2 - 4x + 2)e^{-x}.$$

Because e^{-x} is never zero, it follows that $f''(x) = 0$ only when

$$x^2 - 4x + 2 = 0$$

—that is, when

$$x = 2 - \sqrt{2} \approx 0.586 \quad \text{or} \quad x = 2 + \sqrt{2} \approx 3.414$$

(with the aid of the quadratic formula). Only at these two points can $f''(x)$ change sign. But

$$f''(0) \approx 2 > 0,$$
$$f''(2) \approx -0.271 < 0, \quad \text{and}$$
$$f''(4) \approx 0.037 > 0.$$

So it follows that

$$f''(x) > 0 \quad \text{if} \quad x < 2 - \sqrt{2},$$
$$f''(x) < 0 \quad \text{if} \quad 2 - \sqrt{2} < x < 2 + \sqrt{2}, \quad \text{and}$$
$$f''(x) > 0 \quad \text{if} \quad 2 + \sqrt{2} < x.$$

Thus $f(x) = x^2 e^{-x}$ has inflection points at $x = 2 \pm \sqrt{2}$. The corresponding points marked on the graph in Fig. 4.6.18 are approximately $(0.586, 0.191)$ and $(3.414, 0.384)$.

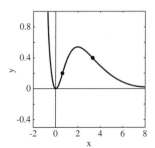

Fig. 4.6.18 $y = x^2 e^{-x}$

246

Let the function f be twice differentiable for all x. Just as the critical points where $f'(x) = 0$ separate the x-axis into open intervals on which $f'(x)$ does not change sign, the *possible* inflection points where $f''(x) = 0$ separate the x-axis into open intervals on which $f''(x)$ does not change sign. On each of these intervals, the curve $y = f(x)$ either is bending downward $[f''(x) < 0]$ or is bending upward $[f''(x) > 0]$. We can determine the sign of $f''(x)$ in each of these intervals in either of two ways:

1. Evaluation of $f''(x)$ at a typical point of each interval. The sign of $f''(x)$ at that particular point is the sign of $f''(x)$ throughout that interval.
2. Construction of a table of signs of the factors of $f''(x)$. Then the sign of $f''(x)$ on each interval can be deduced from the table.

These are the same two methods we used in Sections 4.4 and 4.5 to determine the sign of $f'(x)$. We use the first method in Example 6 and the second in Example 7.

EXAMPLE 6   Sketch the graph of $f(x) = 8x^5 - 5x^4 - 20x^3$, indicating local extrema, inflection points, and concave structure.

Solution   We sketched this curve in Example 2 of Section 4.5; see Fig. 4.5.5 for the graph. In that example we found the first derivative to be

$$f'(x) = 40x^4 - 20x^3 - 60x^2 = 20x^2(x + 1)(2x - 3),$$

so the critical points are $x = -1$, $x = 0$, and $x = \frac{3}{2}$. The second derivative is

$$f''(x) = 160x^3 - 60x^2 - 120x = 20x(8x^2 - 3x - 6).$$

When we compute $f''(x)$ at each critical point, we find that

$$f''(-1) = -100 < 0, \qquad f''(0) = 0, \quad \text{and} \quad f''(\tfrac{3}{2}) = 225 > 0.$$

Continuity of f'' ensures that $f''(x) < 0$ near the critical point $x = -1$ and that $f''(x) > 0$ near the critical point $x = \frac{3}{2}$. The second derivative test therefore tells us that f has a local maximum at $x = -1$ and a local minimum at $x = \frac{3}{2}$. We cannot determine from the second derivative test the behavior of f at $x = 0$.

Because $f''(x)$ exists everywhere, the possible inflection points are the solutions of the equation

$$f''(x) = 0; \quad \text{that is,} \quad 20x(8x^2 - 3x - 6) = 0.$$

Clearly, one solution is $x = 0$. To find the other two, we use the quadratic formula to solve the equation

$$8x^2 - 3x - 6 = 0.$$

This gives

$$x = \tfrac{1}{16}(3 \pm \sqrt{201}),$$

so $x \approx 1.07$ and $x \approx -0.70$ are possible inflection points along with $x = 0$.

These three possible inflection points separate the x-axis into the intervals indicated in Fig. 4.6.19. We check the sign of $f''(x)$ on each.

On $(-\infty, -0.70)$: $\quad f''(-1) = -100 < 0$; $\quad f$ is concave downward;

On $(-0.70, 0)$: $\quad f''(-\frac{1}{2}) = 25 > 0$; $\quad f$ is concave upward;

On $(0, 1.07)$: $\quad f''(1) = -20 < 0$; $\quad f$ is concave downward;

On $(1.07, +\infty)$: $\quad f''(2) = 800 > 0$; $\quad f$ is concave upward.

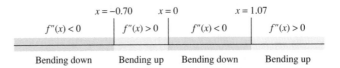

Fig. 4.6.19 Intervals of concavity of Example 6

Thus we see that the direction of concavity of f changes at each of the points $x \approx -0.70$, $x = 0$, and $x \approx 1.07$. These three points are indeed inflection points. This information is shown in the graph of f sketched in Fig. 4.6.20.

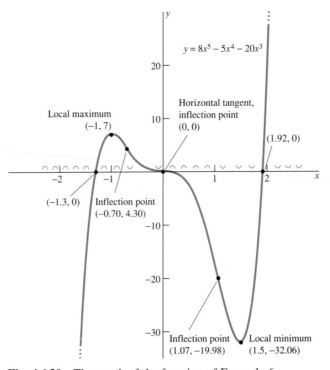

Fig. 4.6.20 The graph of the function of Example 6

EXAMPLE 7 Sketch the graph of $f(x) = 4x^{1/3} + x^{4/3}$. Indicate local extrema, inflection points, and concave structure.

Solution First,

$$f'(x) = \frac{4}{3}x^{-2/3} + \frac{4}{3}x^{1/3} = \frac{4(x + 1)}{3x^{2/3}},$$

248

so the critical points are $x = -1$ (where the tangent line is horizontal) and $x = 0$ (where it is vertical). Next,

$$f''(x) = -\frac{8}{9}x^{-5/3} + \frac{4}{9}x^{-2/3} = \frac{4(x-2)}{9x^{5/3}},$$

so the possible inflection points are $x = 2$ (where $f''(x) = 0$) and $x = 0$ (where $f''(x)$ does not exist).

To determine where f is increasing and where it is decreasing, we construct the following table.

Interval	$x + 1$	$x^{2/3}$	$f'(x)$	f
$(-\infty, -1)$	Neg.	Pos.	Neg.	Decreasing
$(-1, 0)$	Pos.	Pos.	Pos.	Increasing
$(0, +\infty)$	Pos.	Pos.	Pos.	Increasing

Thus f is decreasing when $x < -1$ and increasing when $x > -1$ (Fig. 4.6.21).

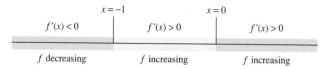

$x = -1$ $x = 0$

$f'(x) < 0$ $f'(x) > 0$ $f'(x) > 0$

f decreasing f increasing f increasing

Fig. 4.6.21 Increasing and decreasing intervals of Example 7

To determine the concavity of f, we construct a table to find the sign of $f''(x)$ on each of the intervals separated by its zeros.

Interval	$x^{5/3}$	$x - 2$	$f'(x)$	f
$(-\infty, 0)$	Neg.	Neg.	Pos.	Concave upward
$(0, 2)$	Pos.	Neg.	Neg.	Concave downward
$(2, +\infty)$	Pos.	Pos.	Pos.	Concave upward

The table shows that f is concave downward on $(0, 2)$ and concave upward for $x < 0$ and for $x > 2$ (Fig. 4.6.22).

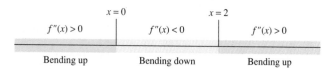

$x = 0$ $x = 2$

$f''(x) > 0$ $f''(x) < 0$ $f''(x) > 0$

Bending up   Bending down   Bending up

Fig. 4.6.22 Intervals of concavity of Example 7

We note that $f(x) \to +\infty$ as $x \to \pm\infty$, and we mark with plus signs the intervals on the x-axis where f is increasing, minus signs where it is decreasing, cups opening upward where f is concave upward, and cups opening downward where f is concave downward. We plot (at least approximately) the points on the graph of f that correspond to the zeros and discontinuities of f'

and f''; these are $(-1, -3)$, $(0, 0)$, and $(2, 6\sqrt[3]{2})$. Finally, we use all this information to draw the smooth curve shown in Fig. 4.6.23.

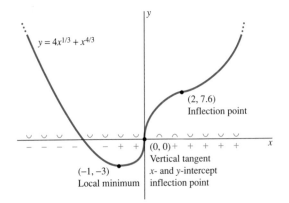

Fig. 4.6.23 The graph of the function of Example 7

Proof of Theorem 2 We will prove only part 1—the proof of part 2 is similar. Given a fixed point a of the open interval I where $f''(x) > 0$, we want to show that the graph $y = f(x)$ lies above the tangent line at $(a, f(a))$. The tangent line in question has the equation

$$y = T(x) = f(a) + f'(a)(x - a). \tag{5}$$

Consider the auxiliary function

$$g(x) = f(x) - T(x) \tag{6}$$

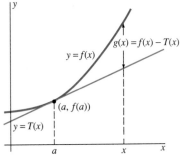

Fig. 4.6.24 Illustrating the proof of Theorem 2

illustrated in Fig. 4.6.24. Note first that $g(a) = g'(a) = 0$, so $x = a$ is a critical point of g. Moreover, Eq. (5) implies that $T'(x) \equiv f'(a)$ and that $T''(x) \equiv 0$, so

$$g''(x) = f''(x) - T''(x) = f''(x) > 0$$

at each point of I. Therefore, the second derivative test implies that $g(a) = 0$ is the minimum value of $g(x) = f(x) - T(x)$ on I. It follows that the curve $y = f(x)$ lies above the tangent line $y = T(x)$. ❑

4.6 Problems

Calculate the first three derivatives of the functions given in Problems 1 through 15.

1. $f(x) = 2x^4 - 3x^3 + 6x - 17$

2. $f(x) = 2x^5 + x^{3/2} - \dfrac{1}{2x}$

3. $f(x) = \dfrac{2}{(2x - 1)^2}$

4. $g(t) = t^2 + \sqrt{t + 1}$

5. $g(t) = (3t - 2)^{4/3}$

6. $f(x) = x\sqrt{x + 1}$

7. $h(y) = \dfrac{y}{y + 1}$

8. $f(x) = (1 + \sqrt{x})^3$

9. $g(t) = t^2 \ln t$

10. $h(z) = \dfrac{e^z}{\sqrt{z}}$

11. $f(x) = \sin 3x$

12. $f(x) = \cos^2 2x$

13. $f(x) = \sin x \cos x$

14. $f(x) = x^2 \cos x$

15. $f(x) = \dfrac{\sin x}{x}$

In Problems 16 through 22, calculate dy/dx and d^2y/dx^2, assuming that y is defined implicitly as a function of x by the given equation.

16. $x^2 + y^2 = 4$

17. $x^2 + xy + y^2 = 3$

18. $x^{1/3} + y^{1/3} = 1$

19. $y^3 + x^2 + x = 5$

20. $\dfrac{1}{x} + \dfrac{1}{y} = 1$ **21.** $\sin y = xy$

22. $xe^y = y + 1$

In Problems 23 through 30, find the exact *coordinates of the inflection points and critical points marked on the given graph.*

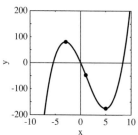

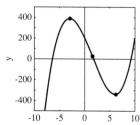

Fig. 4.6.25 The graph of $f(x) = x^3 - 3x^2 - 45x$ (Problem 23)

Fig. 4.6.26 The graph of $f(x) = 2x^3 - 9x^2 - 108x + 200$ (Problem 24)

23. The graph of $f(x) = x^3 - 3x^2 - 45x$ (Fig. 4.6.25)
24. The graph of $f(x) = 2x^3 - 9x^2 - 108x + 200$ (Fig. 4.6.26)
25. The graph of $f(x) = 4x^3 - 6x^2 - 189x + 137$ (Fig. 4.6.27)
26. The graph of $f(x) = -40x^3 - 171x^2 + 2550x + 4150$ (Fig. 4.6.28)

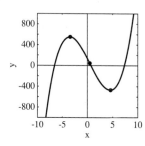

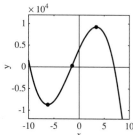

Fig. 4.6.27 The graph of $f(x) = 4x^3 - 6x^2 - 189x + 137$ (Problem 25)

Fig. 4.6.28 The graph of $f(x) = -40x^3 - 171x^2 + 2550x + 4150$

27. The graph of $f(x) = x^4 - 54x^2 + 237$ (Fig. 4.6.29)
28. The graph of $f(x) = x^4 - 10x^3 - 250$ (Fig. 4.6.30)
29. The graph of $f(x) = 3x^5 - 20x^4 + 1000$ (Fig. 4.6.31)
30. The graph of $f(x) = 3x^5 - 160x^3$ (Fig. 4.6.32)

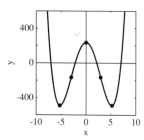

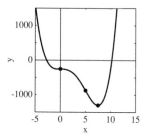

Fig. 4.6.29 The graph of $f(x) = x^4 - 54x^2 + 237$ (Problem 27)

Fig. 4.6.30 The graph of $f(x) = x^4 - 10x^3 - 250$ (Problem 28)

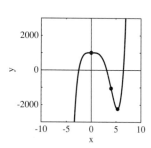

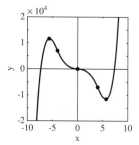

Fig. 4.6.31 The graph of $f(x) = 3x^5 - 20x^4 + 1000$ (Problem 29)

Fig. 4.6.32 The graph of $f(x) = 3x^5 - 160x^3$ (Problem 30)

Apply the second derivative test to find the local maxima and local minima of the functions given in Problems 31 through 50, and apply the inflection point test to find all inflection points.

31. $f(x) = x^2 - 4x + 3$ **32.** $f(x) = 5 - 6x - x^2$
33. $f(x) = x^3 - 3x + 1$ **34.** $f(x) = x^3 - 3x^2$
35. $f(x) = x^3$ **36.** $f(x) = x^4$
37. $f(x) = x^5 + 2x$ **38.** $f(x) = x^4 - 8x^2$
39. $f(x) = x^2(x - 1)^2$ **40.** $f(x) = x^3(x + 2)^2$
41. $f(x) = \sin x$ on $(0, 2\pi)$
42. $f(x) = \cos x$ on $(-\pi/2, 3\pi/2)$
43. $f(x) = \tan x$ on $(-\pi/2, \pi/2)$
44. $f(x) = \sin^2 x$ on $(0, \pi)$
45. $f(x) = \cos^2 x$ on $(-\pi/2, 3\pi/2)$
46. $f(x) = \sin^3 x$ on $(-\pi, \pi)$
47. $f(x) = xe^{-x}$ **48.** $f(x) = x^3 e^{-x}$
49. $f(x) = \dfrac{\ln x}{x}$ **50.** $f(x) = e^{-x}\sin x$ on $(0, 2\pi)$

In Problems 51 through 62, rework the indicated problem from Section 4.4, now using the second derivative test to verify that you have found the desired absolute maximum or minimum value.

51. Problem 27   **52.** Problem 28
53. Problem 29   **54.** Problem 30

55. Problem 31   **56.** Problem 32

57. Problem 33   **58.** Problem 36

59. Problem 37   **60.** Problem 38

61. Problem 39   **62.** Problem 40

Sketch the graphs of the functions in Problems 63 through 76, indicating all critical points and inflection points. Apply the second derivative test at each critical point. Show the correct concave structure in your sketches, and indicate the behavior of $f(x)$ as $x \to \pm\infty$.

63. $f(x) = 2x^3 - 3x^2 - 12x + 3$

64. $f(x) = 3x^4 - 4x^3 - 5$

65. $f(x) = 6 + 8x^2 - x^4$ **66.** $f(x) = 3x^5 - 5x^3$

67. $f(x) = 3x^4 - 4x^3 - 12x^2 - 1$

68. $f(x) = 3x^5 - 25x^3 + 60x$

69. $f(x) = x^3(1 - x)^4$ **70.** $f(x) = (x - 1)^2(x + 2)^3$

71. $f(x) = 1 + x^{1/3}$ **72.** $f(x) = 2 - (x - 3)^{1/3}$

73. $f(x) = (x + 3)\sqrt{x}$ **74.** $f(x) = x^{2/3}(5 - 2x)$

75. $f(x) = (4 - x)\sqrt[3]{x}$ **76.** $f(x) = x^{1/3}(6 - x)^{2/3}$

In Problems 77 through 82, the graph of a function $f(x)$ is shown. Match it with the graph of its second derivative $f''(x)$ in Fig. 4.6.33.

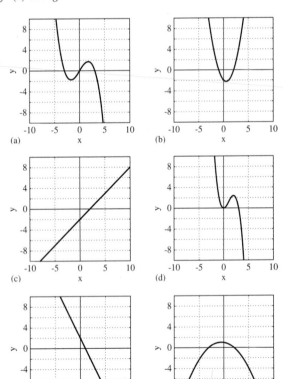

(a)   (b)   (c)   (d)   (e)   (f)

Fig. 4.6.33

77. See Fig. 4.6.34.

78. See Fig. 4.6.35.

79. See Fig. 4.6.36.

80. See Fig. 4.6.37.

81. See Fig. 4.6.38.

82. See Fig. 4.6.39.

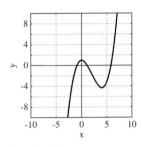

Fig. 4.6.34

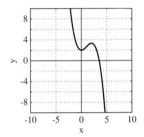

Fig. 4.6.35

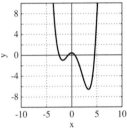

Fig. 4.6.36

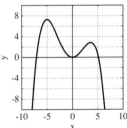

Fig 4.6.37

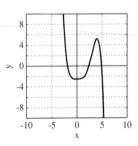

Fig. 4.6.38

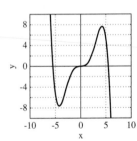

Fig. 4.6.39

83. (a) Show first that the nth derivative of $f(x) = x^n$ is $f^{(n)}(x) \equiv n! = n \cdot (n - 1) \cdot (n - 2) \cdots 3 \cdot 2 \cdot 1$. (b) Conclude that if $f(x)$ is a polynomial of degree n, then $f^{(k)}(x) \equiv 0$ if $k > n$.

84. (a) Calculate the first four derivatives of $f(x) = \sin x$. (b) Conclude that $D^{n+4} \sin x = D^n \sin x$ if n is a positive integer.

85. Suppose that $z = g(y)$ and that $y = f(x)$. Show that

$$\frac{d^2z}{dx^2} = \frac{d^2z}{dy^2}\left(\frac{dy}{dx}\right)^2 + \frac{dz}{dy} \cdot \frac{d^2y}{dx^2}.$$

86. Prove that the graph of a quadratic polynomial has no inflection points.

Ch. 4 / Additional Applications of the Derivative

87. Prove that the graph of a cubic polynomial has exactly one inflection point.

88. Prove that the graph of a polynomial function of degree 4 has either no inflection point or exactly two inflection points.

89. Suppose that the pressure p (in atmospheres), volume V (in cubic centimeters), and temperature T (in kelvins) of n moles of carbon dioxide (CO_2) satisfies van der Waals' equation

$$\left(p + \frac{n^2 a}{V^2}\right)(V - nb) = nRT,$$

where a, b, and R are empirical constants. The following experiment was carried out to find the values of these constants.

One mole of CO_2 was compressed at the constant temperature $T = 304$ K. The measured pressure-volume (pV) data were then plotted as in Fig. 4.6.40, with the pV curve showing a *horizontal* inflection point at $V = 128.1$, $p = 72.8$. Use this information to calculate a, b, and R. [*Suggestion:* Solve van der Waals' equation for p and then calculate dp/dV and d^2p/dV^2.]

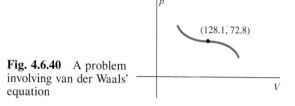

Fig. 4.6.40 A problem involving van der Waals' equation

90. Suppose that the function f is differentiable on an open interval containing the point c at which $f'(c) = 0$ and that the second derivative

$$f''(c) = \lim_{h \to 0} \frac{f'(c + h) - f'(c)}{h} = \lim_{h \to 0} \frac{f'(c + h)}{h}$$

exists. (a) First assume that $f''(c) > 0$. Reason that $f'(c + h)$ and h have the same sign if $h \neq 0$ is sufficiently small. Hence apply the first derivative test to show in this case that $f(c)$ is a local minimum value of f. (b) Show similarly that $f(c)$ is a local maximum value of f if $f''(c) < 0$.

4.6 Projects

Projects A and B require the use of a graphics calculator or a computer with a graphing utility. For each project, choose in advance an integer n between 0 and 9. For instance, you could let n be the last digit of your student ID number. Project C requires also a computer with a symbolic algebra program.

PROJECT A   If the coefficients a, b, and c are defined to be

$$a = 30{,}011 + 2n,$$

$$b = 30{,}022 + 4n, \quad \text{and}$$

$$c = 10{,}010 + 2n,$$

then the curve

$$y = 10{,}000x^3 - ax^2 + bx + c$$

has two good "wiggles," as a cubic should. Find them. In particular, find the local maximum and minimum points *and* the inflection point (or points) on this curve. Give the coordinates of each of these points accurate to five decimal places. You can use the [SOLVE] key on your calculator or a computer command such as

```
N[ Solve[ f[x] == 0, x ] ]
```

for the numerical solution of an equation. Produce a graph that plainly exhibits these points—you can mark and label the points by hand. As you zoom in, you'll need to control carefully the successive viewing windows.

PROJECT B   Your task is to analyze the structure of the curve

$$y = x^7 + 5x^6 - 11x^5 - 21x^4 + 31x^3 - 57x^2$$

$$- (101 + 2n)x + (89 - 3n).$$

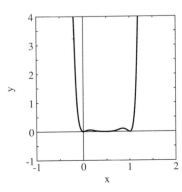

Fig. 4.6.41 The graph of $y = f(x)$ of Project C

Provide the same information that was specified in Project A. You will probably need to produce separate plots with different scales, showing different parts of this curve. In the end, use all the information accumulated to produce a careful hand sketch (not to scale) displaying all the maxima, minima, and inflection points on the curve with their (approximate) coordinates labeled.

PROJECT C   Let

$$f(x) = [x(1 - x)(2x - 1)(4 - 9x)]^2.$$

The graph of f is shown in Fig. 4.6.41. We promise that f has at least four local minima, three local maxima, and six inflection points in $[0, 1]$. Find the approximate coordinates of all 13 points, and show the graph of f on a scale that makes all these points evident.

4.7

Curve Sketching and Asymptotes

We now want to extend the limit concept to include infinite limits and limits at infinity. This extension will add a powerful weapon to our arsenal of curve-sketching techniques, the notion of an *asymptote* to a curve—a straight line that the curve approaches arbitrarily closely in a sense we soon make precise.

Recall from Section 2.3 that $f(x)$ is said to **increase without bound,** or **become infinite,** as x approaches a, and we write

$$\lim_{x \to a} f(x) = +\infty, \tag{1}$$

provided that $f(x)$ can be made arbitrarily large by choosing x sufficiently close (but not equal) to a. The statement that $f(x)$ **decreases without bound,** or **becomes negatively infinite,** as $x \to a$, written

$$\lim_{x \to a} f(x) = -\infty, \tag{2}$$

has an analogous definition.

EXAMPLE 1   It is apparent that

$$\lim_{x \to -2} \frac{1}{(x + 2)^2} = +\infty$$

because, as $x \to -2$, $(x + 2)^2$ is positive and approaches zero. By contrast,

$$\lim_{x \to -2} \frac{x}{(x + 2)^2} = -\infty$$

because, as $x \to -2$, the denominator $(x + 2)^2$ is still positive and approaches zero, but the numerator x is negative. Thus when x is very close to -2, we have in $x/(x + 2)^2$ a negative number close to -2 divided by a very small positive number. Hence the quotient becomes a negative number of large magnitude.

One-sided versions of Eqs. (1) and (2) are valid also. For instance, if n is an *odd* positive integer, then it is apparent that

$$\lim_{x \to 2^-} \frac{1}{(x - 2)^n} = -\infty \quad \text{and that} \quad \lim_{x \to 2^+} \frac{1}{(x - 2)^n} = +\infty,$$

because $(x - 2)^n$ is negative when x is to the left of 2 and positive when x is to the right of 2.

VERTICAL ASYMPTOTES

The line $x = a$ is a **vertical asymptote** of the curve $y = f(x)$ provided that *either*

$$\lim_{x \to a^-} f(x) = \pm\infty \qquad (3a)$$

or

$$\lim_{x \to a^+} f(x) = \pm\infty \qquad (3b)$$

or both. It is usually the case that both one-sided limits, rather than only one, are infinite. If so, we write

$$\lim_{x \to a} f(x) = \pm\infty. \qquad (3c)$$

The geometry of a vertical asymptote is illustrated by the graphs of $y = 1/(x - 1)$ and $y = 1/(x - 1)^2$ (Figs. 4.7.1 and 4.7.2). In both cases, as $x \to 1$ and $f(x) \to \pm\infty$, the point $(x, f(x))$ on the curve approaches the vertical asymptote $x = 1$ and the shape and direction of the curve are better and better approximated by the asymptote.

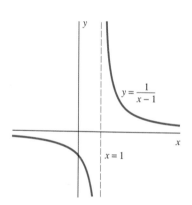

Fig. 4.7.1 The graph of $y = 1/(x - 1)$

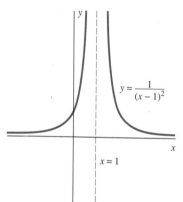

Fig. 4.7.2 The graph of $y = 1/(x - 1)^2$

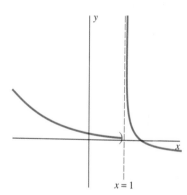

Fig. 4.7.3 A "right-hand-only" vertical asymptote

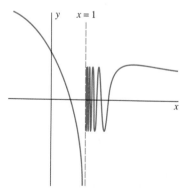

Fig. 4.7.4 The behavior of the graph to its *left* produces the vertical asymptote.

Figure 4.7.3 shows the graph of a function whose left-hand limit is zero at $x = 1$. But the right-hand limit there is $+\infty$, which explains why the line $x = 1$ is also a vertical asymptote for this graph. The right-hand limit in Fig. 4.7.4 does not even exist, but because the left-hand limit at $x = 1$ is $-\infty$, the vertical line at $x = 1$ is again a vertical asymptote.

A vertical asymptote typically appears in the case of a rational function $f(x) = p(x)/q(x)$ at a point $x = a$ where $q(a) = 0$ but $p(a) \neq 0$. (See Examples 4, 5, and 6.)

LIMITS AT INFINITY

In Section 4.5 we mentioned infinite limits at infinity in connection with the behavior of a polynomial as $x \to \pm\infty$. There is also such a thing as a *finite* limit at infinity. We say that **$f(x)$ approaches the number L as x increases without bound** and write

$$\lim_{x \to +\infty} f(x) = L \qquad (4)$$

provided that $|f(x) - L|$ can be made arbitrarily small (close to zero) merely by choosing x sufficiently large. That is, given $\epsilon > 0$, there exists $M > 0$ such that

$$x > M \quad \text{implies that} \quad |f(x) - L| < \epsilon. \qquad (5)$$

The statement that

$$\lim_{x \to -\infty} f(x) = L$$

has a definition of similar form—merely replace the condition $x > M$ by the condition $x < -M$.

The analogues for limits at infinity of the limit laws of Section 2.2 all hold, including, in particular, the sum, product, and quotient laws. In addition, it is not difficult to show that if

$$\lim_{x \to +\infty} f(x) = L \quad \text{and} \quad \lim_{x \to +\infty} g(x) = \pm\infty,$$

then

$$\lim_{x \to +\infty} \frac{f(x)}{g(x)} = 0.$$

It follows from this result that

$$\lim_{x \to +\infty} \frac{1}{x^k} = 0 \qquad (6)$$

for any choice of the positive rational number k.

Using Eq. (6) and the limit laws, we can easily evaluate limits at infinity of rational functions. The general method is this: First divide each term in both the numerator and the denominator by the highest power of x that appears in any of the terms. Then apply the limit laws.

EXAMPLE 2 Find

$$\lim_{x \to +\infty} f(x) \quad \text{if} \quad f(x) = \frac{3x^3 - x}{2x^3 + 7x^2 - 4}.$$

Solution We begin by dividing each term in the numerator and denominator by x^3:

$$\lim_{x \to +\infty} \frac{3x^3 - x}{2x^3 + 7x^2 - 4} = \lim_{x \to +\infty} \frac{3 - \dfrac{1}{x^2}}{2 + \dfrac{7}{x} - \dfrac{4}{x^3}}$$

$$= \frac{\lim\limits_{x \to +\infty} \left(3 - \dfrac{1}{x^2}\right)}{\lim\limits_{x \to +\infty} \left(2 + \dfrac{7}{x} - \dfrac{4}{x^3}\right)} = \frac{3 - 0}{2 + 0 - 0} = \frac{3}{2}.$$

The same computation, but with $x \to -\infty$, also gives the result

$$\lim_{x \to -\infty} f(x) = \frac{3}{2}.$$

Ch. 4 / Additional Applications of the Derivative

EXAMPLE 3 Find $\lim\limits_{x \to +\infty} (\sqrt{x + a} - \sqrt{x})$.

Solution We use the familiar "divide and multiply" technique:

$$\lim_{x \to +\infty} (\sqrt{x + a} - \sqrt{x}) = \lim_{x \to +\infty} (\sqrt{x + a} - \sqrt{x}) \cdot \frac{\sqrt{x + a} + \sqrt{x}}{\sqrt{x + a} + \sqrt{x}}$$

$$= \lim_{x \to +\infty} \frac{a}{\sqrt{x + a} + \sqrt{x}} = 0.$$

HORIZONTAL ASYMPTOTES

In geometric terms, the statement

$$\lim_{x \to +\infty} f(x) = L$$

means that the point $(x, f(x))$ on the curve $y = f(x)$ approaches the horizontal line $y = L$ as $x \to +\infty$. In particular, with the numbers M and ϵ of the condition in Eq. (5), the part of the curve for which $x > M$ lies between the horizontal lines $y = L - \epsilon$ and $y = L + \epsilon$ (Fig. 4.7.5). Therefore, we say that the line $y = L$ is a **horizontal asymptote** of the curve $y = f(x)$ if either

$$\lim_{x \to +\infty} f(x) = L \quad \text{or} \quad \lim_{x \to -\infty} f(x) = L.$$

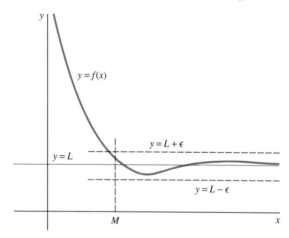

Fig. 4.7.5 Geometry for the definition of horizontal asymptote

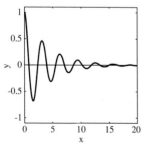

Fig. 4.7.6 $y = e^{-x/4} \cos 2x$

EXAMPLE 4 Figure 4.7.6 shows the graph of $f(x) = e^{-x/4} \cos 2x$ for $x > 0$. Because $|\cos 2x| \leq 1$ for all x and

$$e^{-x/4} = \frac{1}{e^{x/4}} \to 0$$

as $x \to +\infty$, the squeeze law of limits implies that $f(x) \to 0$ as $x \to +\infty$. Thus the x-axis, $y = 0$, is a horizontal asymptote of the curve $y = e^{-x/4} \cos 2x$.

EXAMPLE 5 Sketch the graph of $f(x) = x/(x - 2)$. Indicate any horizontal or vertical asymptotes.

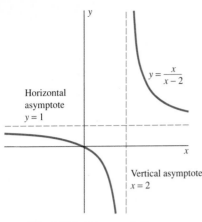

Horizontal asymptote
y = 1

$y = \dfrac{x}{x-2}$

Vertical asymptote
x = 2

Fig. 4.7.7 The graph for Example 5

Solution First we note that $x = 2$ is a vertical asymptote because $|f(x)| \to +\infty$ as $x \to 2$. Also,

$$\lim_{x \to \pm\infty} \frac{x}{x-2} = \lim_{x \to \pm\infty} \frac{1}{1 - \dfrac{2}{x}} = \frac{1}{1-0} = 1.$$

So the line $y = 1$ is a horizontal asymptote. The first two derivatives of f are

$$f'(x) = -\frac{2}{(x-2)^2} \quad \text{and} \quad f''(x) = \frac{4}{(x-2)^3}.$$

Neither $f'(x)$ nor $f''(x)$ is zero anywhere, so the function f has no critical points and no inflection points. Because $f'(x) < 0$ for $x \ne 2$, we see that $f(x)$ is decreasing on the open intervals $(-\infty, 2)$ and $(2, +\infty)$. And because $f''(x) < 0$ for $x < 2$, whereas $f''(x) > 0$ for $x > 2$, the graph of f is concave downward on $(-\infty, 2)$ and concave upward on $(2, +\infty)$. The graph of f appears in Fig. 4.7.7.

The curve-sketching techniques of Sections 4.5 and 4.6, together with those of this section, can be summarized as a list of steps. If you follow these steps, loosely rather than rigidly, you will obtain a qualitatively accurate sketch of the graph of a given function f.

1. Solve the equation $f'(x) = 0$ and also find where $f'(x)$ does not exist. This gives the critical points of f. Note whether the tangent line is vertical, horizontal, or nonexistent at each critical point.

2. Determine the intervals on which f is increasing and those on which it is decreasing.

3. Solve the equation $f''(x) = 0$ and also find where $f''(x)$ does not exist. These points are the *possible* inflection points of the graph.

4. Determine the intervals on which the graph of f is concave upward and those on which it is concave downward.

5. Find the y-intercept and the x-intercepts (if any) of the graph.

6. Plot and label the critical points, possible inflection points, and intercepts.

7. Determine the asymptotes (if any), discontinuities (if any), and *especially* the behavior of $f(x)$ and $f'(x)$ near discontinuities of f. Also determine the behavior of $f(x)$ as $x \to +\infty$ and as $x \to -\infty$.

8. Finally, join the plotted points with a curve that is consistent with the information you have amassed. Remember that corner points are rare and that straight sections of graph are even rarer.

You may follow these steps in any convenient order and omit any that present formidable computational difficulties. Many problems require fewer than all eight steps; see Example 5. But Example 6 requires them all.

EXAMPLE 6 Sketch the graph of

$$f(x) = \frac{2 + x - x^2}{(x-1)^2}.$$

Solution   We notice immediately that

$$\lim_{x \to 1} f(x) = +\infty,$$

because the numerator approaches 2 as $x \to 1$, whereas the denominator approaches zero through *positive* values. So the line $x = 1$ is a vertical asymptote. Also,

$$\lim_{x \to \pm\infty} \frac{2 + x - x^2}{(x - 1)^2} = \lim_{x \to \pm\infty} \frac{\dfrac{2}{x^2} + \dfrac{1}{x} - 1}{\left(1 - \dfrac{1}{x}\right)^2} = -1,$$

so the line $y = -1$ is a horizontal asymptote.

Next we apply the quotient rule and simplify to find that

$$f'(x) = \frac{x - 5}{(x - 1)^3}.$$

So the only critical point in the domain of f is $x = 5$, and we plot the point $(5, f(5)) = (5, -1.125)$ on a convenient coordinate plane and mark the horizontal tangent there. To determine the increasing or decreasing behavior of f, we use both the critical point $x = 5$ and the point $x = 1$ (where f' is not defined) to separate the x-axis into open intervals. Here are the results.

Interval	$(x - 1)^3$	$x - 5$	$f'(x)$	f
$(-\infty, 1)$	Neg.	Neg.	Pos.	Increasing
$(1, 5)$	Pos.	Neg.	Neg.	Decreasing
$(5, +\infty)$	Pos.	Pos.	Pos.	Increasing

After some simplifications, we find the second derivative to be

$$f''(x) = \frac{2(7 - x)}{(x - 1)^4}.$$

The only possible inflection point is at $x = 7$, corresponding to the point $(7, -\frac{10}{9})$ on the graph. We use both $x = 7$ and $x = 1$ (where f'' is not defined) to separate the x-axis into open intervals. The concave structure of the graph can be deduced with the aid of the next table.

Interval	$(x - 1)^4$	$7 - x$	$f''(x)$	f
$(-\infty, 1)$	Pos.	Pos.	Pos.	Concave upward
$(1, 7)$	Pos.	Pos.	Pos.	Concave upward
$(7, \infty)$	Pos.	Neg.	Neg.	Concave downward

The y-intercept of f is $(0, 2)$, and the equation $2 + x - x^2 = 0$ readily yields the x-intercepts $(-1, 0)$ and $(2, 0)$. We plot these intercepts, sketch the asymptotes, and finally sketch the graph with the aid of the two tables; their information now appears along the x-axis in Fig. 4.7.8.

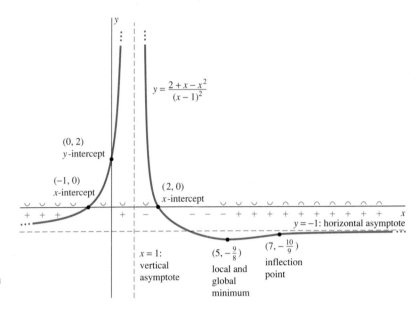

Fig. 4.7.8 Graphing the function of Example 6

Some asymptotes are inclined: Not all asymptotes are horizontal or vertical. The nonvertical line $y = mx + b$ is an **asymptote** for the curve $y = f(x)$ provided that either

$$\lim_{x \to +\infty} [f(x) - (mx + b)] = 0 \qquad (7a)$$

or

$$\lim_{x \to -\infty} [f(x) - (mx + b)] = 0 \qquad (7b)$$

(or both). These conditions mean that as $x \to +\infty$ or as $x \to -\infty$ (or both), the vertical distance between the point $(x, f(x))$ on the curve and the point $(x, mx + b)$ on the line approaches zero.

Suppose that $f(x) = p(x)/q(x)$ is a rational function for which the degree of p is greater by 1 than the degree of q. Then, by long division of $q(x)$ into $p(x)$, we find that $f(x)$ has the form

$$f(x) = mx + b + g(x),$$

where

$$\lim_{x \to \pm\infty} g(x) = 0.$$

Thus the nonvertical line $y = mx + b$ is an asymptote of the graph of $y = f(x)$. Such an asymptote is called an **oblique** asymptote.

EXAMPLE 7 Sketch the graph of

$$f(x) = \frac{x^2 + x - 1}{x - 1}.$$

Solution The long division suggested previously takes the form

Ch. 4 / Additional Applications of the Derivative

$$\begin{array}{r}
x + 2 \\
x - 1 \overline{)\smash{x^2 + x - 1}} \\
\underline{x^2 - x\phantom{{}-1}} \\
2x - 1 \\
\underline{2x - 2} \\
1
\end{array}$$

Thus

$$f(x) = x + 2 + \frac{1}{x - 1}.$$

So $y = x + 2$ is an asymptote of the curve. Also,

$$\lim_{x \to 1} |f(x)| = +\infty,$$

so $x = 1$ is a vertical asymptote. The first two derivatives of f are

$$f'(x) = 1 - \frac{1}{(x - 1)^2} = \frac{x(x - 2)}{(x - 1)^2}$$

and

$$f''(x) = \frac{2}{(x - 1)^3}.$$

It follows that f has critical points at $x = 0$ and at $x = 2$ but no inflection points. The sign of f' tells us that f is increasing on $(-\infty, 0)$ and on $(2, +\infty)$, decreasing on $(0, 1)$ and on $(1, 2)$. Examination of $f''(x)$ reveals that f is concave downward on $(-\infty, 1)$ and concave upward on $(1, +\infty)$. In particular, $f(0) = 1$ is a local maximum value, and $f(2) = 5$ is a local minimum value. The graph of f looks much like the one in Fig. 4.7.9.

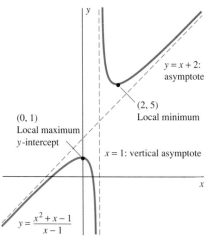

Fig. 4.7.9 A function with asymptote $y = x + 2$ (Example 7)

4.7 Problems

Investigate the limits in Problems 1 through 16.

1. $\displaystyle\lim_{x \to +\infty} \frac{x}{x + 1}$

2. $\displaystyle\lim_{x \to -\infty} \frac{x^2 + 1}{x^2 - 1}$

3. $\displaystyle\lim_{x \to 1} \frac{x^2 + x - 2}{x - 1}$

4. $\displaystyle\lim_{x \to 1} \frac{x^2 - x - 2}{x - 1}$

5. $\displaystyle\lim_{x \to +\infty} \frac{2x^2 - 1}{x^2 - 3x}$

6. $\displaystyle\lim_{x \to -\infty} \frac{x^2 + 3x}{x^3 - 5}$

7. $\displaystyle\lim_{x \to -1} \frac{x^2 + 2x + 1}{(x + 1)^2}$

8. $\displaystyle\lim_{x \to +\infty} \frac{5x^3 - 2x + 1}{7x^3 + 4x^2 - 2}$

9. $\displaystyle\lim_{x \to 4} \frac{x - 4}{\sqrt{x} - 2}$

10. $\displaystyle\lim_{x \to +\infty} \frac{2x + 1}{x - x\sqrt{x}}$

11. $\displaystyle\lim_{x \to -\infty} \frac{8 - \sqrt[3]{x}}{2 + x}$

12. $\displaystyle\lim_{x \to +\infty} \frac{2x^2 - 17}{x^3 - 2x + 27}$

13. $\displaystyle\lim_{x \to +\infty} \sqrt{\frac{4x^2 - x}{x^2 + 9}}$

14. $\displaystyle\lim_{x \to -\infty} \frac{\sqrt[3]{x^3 - 8x + 1}}{3x - 4}$

15. $\displaystyle\lim_{x \to -\infty} (\sqrt{x^2 + 2x} - x)$

16. $\displaystyle\lim_{x \to -\infty} (2x - \sqrt{4x^2 - 5x})$

Apply your knowledge of limits and asymptotes to match each function in Problems 17 through 28 with its graph-with-asymptotes in one of the twelve parts of Fig. 4.7.10.

17. $f(x) = \dfrac{1}{x - 1}$

18. $f(x) = \dfrac{1}{1 - x}$

19. $f(x) = \dfrac{1}{(x - 1)^2}$

20. $f(x) = -\dfrac{1}{(1 - x)^2}$

21. $f(x) = \dfrac{1}{x^2 - 1}$

22. $f(x) = \dfrac{1}{1 - x^2}$

23. $f(x) = \dfrac{x}{x^2 - 1}$

24. $f(x) = \dfrac{x}{1 - x^2}$

25. $f(x) = \dfrac{x}{x - 1}$

26. $f(x) = \dfrac{x^2}{x^2 - 1}$

27. $f(x) = \dfrac{x^2}{x - 1}$

28. $f(x) = \dfrac{x^3}{x^2 - 1}$

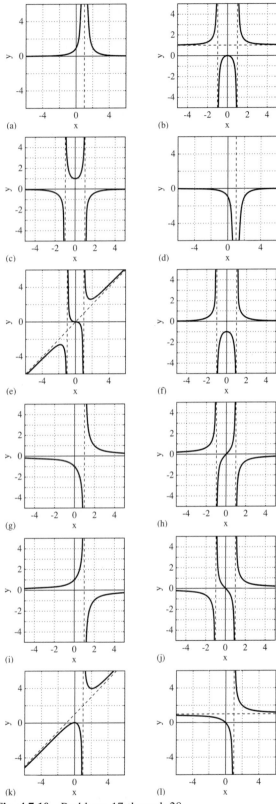

(a)

(b)

(c)

(d)

(e)

(f)

(g)

(h)

(i)

(j)

(k)

(l)

Fig. 4.7.10 Problems 17 through 28

Sketch the graph of each function in Problems 29 through 54. Identify and label all extrema, inflection points, intercepts, and asymptotes. Show the concave structure clearly as well as the behavior of the graph for |x| large and for x near any discontinuities of the function.

29. $f(x) = \dfrac{2}{x - 3}$ **30.** $f(x) = \dfrac{4}{5 - x}$

31. $f(x) = \dfrac{3}{(x + 2)^2}$ **32.** $f(x) = -\dfrac{4}{(3 - x)^2}$

33. $f(x) = \dfrac{1}{(2x - 3)^3}$ **34.** $f(x) = \dfrac{x + 1}{x - 1}$

35. $f(x) = \dfrac{x^2}{x^2 + 1}$ **36.** $f(x) = \dfrac{2x}{x^2 + 1}$

37. $f(x) = \dfrac{1}{x^2 - 9}$ **38.** $f(x) = \dfrac{x}{4 - x^2}$

39. $f(x) = \dfrac{1}{x^2 + x - 6}$ **40.** $f(x) = \dfrac{2x^2 + 1}{x^2 - 2x}$

41. $f(x) = x + \dfrac{1}{x}$ **42.** $f(x) = 2x + \dfrac{1}{x^2}$

43. $f(x) = \dfrac{x^2}{x - 1}$ **44.** $f(x) = \dfrac{2x^3 - 5x^2 + 4x}{x^2 - 2x + 1}$

45. $f(x) = \dfrac{1}{(x - 1)^2}$ **46.** $f(x) = \dfrac{1}{x^2 - 4}$

47. $f(x) = \dfrac{x}{x + 1}$ **48.** $f(x) = \dfrac{1}{(x + 1)^3}$

49. $f(x) = \dfrac{1}{x^2 - x - 2}$

50. $f(x) = \dfrac{1}{(x - 1)(x + 1)^2}$

51. $f(x) = \dfrac{x^2 - 4}{x}$ **52.** $f(x) = \dfrac{x}{x^2 - 1}$

53. $f(x) = \dfrac{x^3 - 4}{x^2}$ **54.** $f(x) = \dfrac{x^2 + 1}{x - 2}$

55. Suppose that

$$f(x) = x^2 + \frac{2}{x}.$$

Note that

$$\lim_{x \to \pm\infty} [f(x) - x^2] = 0,$$

so the curve $y = f(x)$ approaches the parabola $y = x^2$ as $x \to \pm\infty$. Use this observation to make an accurate sketch of the graph of f.

56. Use the method of Problem 55 to make an accurate sketch of the graph of

$$f(x) = x^3 - \frac{12}{x - 1}.$$

Chapter 4 Review: DEFINITIONS, CONCEPTS, RESULTS

Use the list below as a guide to concepts that you may need to review.

1. Increment Δy
2. Differential dy
3. Linear approximation formula
4. Differentiation rules in differential form
5. Increasing functions and decreasing functions
6. Significance of the sign of the first derivative
7. Rolle's theorem
8. The mean value theorem
9. Consequences of the mean value theorem
10. First derivative test
11. Open-interval maximum-minimum problems
12. Graphs of polynomials
13. Calculation of higher derivatives
14. Concave-upward functions and concave-downward functions
15. Test for concavity
16. Second derivative test
17. Inflection points
18. Inflection point test
19. Infinite limits
20. Vertical asymptotes
21. Limits as $x \to \pm\infty$
22. Horizontal asymptotes
23. Oblique asymptotes
24. Curve-sketching techniques

Chapter 4 Miscellaneous Problems

In Problems 1 through 6, write dy in terms of x and dx.

1. $y = (4x - x^2)^{3/2}$
2. $y = 8x^3 \sqrt{x^2 + 9}$
3. $y = \dfrac{x + 1}{x - 1}$
4. $y = \sin x^2$
5. $y = x^2 \cos \sqrt{x}$
6. $y = \dfrac{x}{\sin 2x}$

In Problems 7 through 16, estimate the indicated number by linear approximation.

7. $\sqrt{6401}$ (Note that $80^2 = 6400$.)
8. $\dfrac{1}{1.000007}$
9. $(2.0003)^{10}$ (Note that $2^{10} = 1024$.)
10. $\sqrt[3]{999}$ (Note that $10^3 = 1000$.)
11. $\sqrt[3]{1005}$
12. $\sqrt[3]{62}$
13. $26^{3/2}$
14. $\sqrt[5]{30}$
15. $\sqrt[4]{17}$
16. $\sqrt[10]{1000}$

In Problems 17 through 22, estimate by linear approximation the change in the indicated quantity.

17. The volume $V = s^3$ of a cube, if its side length s is increased from 5 in. to 5.1 in.

18. The area $A = \pi r^2$ of a circle, if its radius r is decreased from 10 cm to 9.8 cm.

19. The volume $V = \frac{4}{3}\pi r^3$ of a sphere, if its radius r is increased from 5 cm to 5.1 cm.

20. The volume $V = 1000/p$ in.^{3} of a gas, if the pressure p is decreased from 100 lb/in.^{2} to 99 lb/in.^{2}

21. The period of oscillation $T = 2\pi \sqrt{L/32}$ of a pendulum, if its length L is increased from 2 ft to 25 in. (Time T is in seconds.)

22. The lifetime $L = 10^{30}/E^{13}$ hours of a light bulb with applied voltage E volts (V), if the voltage is increased from 110 V to 111 V. Compare your result with the actual decrease of the lifetime.

If the mean value theorem applies to the function f on the interval [a, b], it ensures the existence of a solution c in the interval (a, b) of the equation

$$f'(c) = \frac{f(b) - f(a)}{b - a}.$$

In Problems 23 through 28, a function f and an interval [a, b] are given. Verify that the hypotheses of the mean value theorem are satisfied for f on [a, b]. Then use the given equation to find the value of the number c.

23. $f(x) = x - \dfrac{1}{x}$; $\quad [1, 3]$
24. $f(x) = x^3 + x - 4$; $\quad [-2, 3]$
25. $f(x) = x^3$; $\quad [-1, 2]$
26. $f(x) = x^3$; $\quad [-2, 1]$
27. $f(x) = \frac{11}{5} x^5$; $\quad [-1, 2]$
28. $f(x) = \sqrt{x}$; $\quad [0, 4]$

Sketch the graphs of the functions in Problems 29 through 33. Indicate the local maxima and minima of each function and the intervals on which the function is increasing or decreasing. Show the concave structure of the graph, and identify all inflection points.

29. $f(x) = x^2 - 6x + 4$
30. $f(x) = 2x^3 - 3x^2 - 36x$

31. $f(x) = 3x^5 - 5x^3 + 60x$

32. $f(x) = (3 - x)\sqrt{x}$

33. $f(x) = (1 - x)\sqrt[3]{x}$

34. Show that the equation $x^5 + x = 5$ has exactly one real solution.

Calculate the first three derivatives of the functions in Problems 35 through 44.

35. $f(x) = x^3 - 2x$ **36.** $f(x) = (x + 1)^{100}$

37. $g(t) = \dfrac{1}{t} - \dfrac{1}{2t + 1}$ **38.** $h(y) = \sqrt{3y - 1}$

39. $f(t) = 2t^{3/2} - 3t^{4/3}$ **40.** $g(x) = \dfrac{1}{x^2 + 9}$

41. $h(t) = \dfrac{t + 2}{t - 2}$ **42.** $f(z) = \sqrt[3]{z} + \dfrac{3}{\sqrt[5]{z}}$

43. $g(x) = \sqrt[3]{5 - 4x}$ **44.** $g(t) = \dfrac{8}{(3 - t)^{3/2}}$

In Problems 45 through 52, calculate dy/dx and d^2y/dx^2 under the assumption that y is defined implicitly as a function of x by the given equation.

45. $x^{1/3} + y^{1/3} = 1$ **46.** $2x^2 - 3xy + 5y^2 = 25$

47. $y^5 - 4y + 1 = \sqrt{x}$ **48.** $\sin xy = xy$

49. $x^2 + y^2 = 5xy + 5$ **50.** $x^5 + xy^4 = 1$

51. $y^3 - y = x^2 y$ **52.** $(x^2 - y^2)^2 = 4xy$

Sketch the graphs of the functions in Problems 53 through 72, indicating all critical points, inflection points, and asymptotes. Show the concave structure clearly.

53. $f(x) = x^4 - 32x$ **54.** $f(x) = 18x^2 - x^4$

55. $f(x) = x^6 - 2x^4$ **56.** $f(x) = x\sqrt{x - 3}$

57. $f(x) = x\sqrt[3]{4 - x}$ **58.** $f(x) = \dfrac{x - 1}{x + 2}$

59. $f(x) = \dfrac{x^2 + 1}{x^2 - 4}$ **60.** $f(x) = \dfrac{x}{x^2 - x - 2}$

61. $f(x) = \dfrac{2x^2}{x^2 - x - 2}$ **62.** $f(x) = \dfrac{x^3}{x^2 - 1}$

63. $f(x) = 3x^4 - 4x^3$ **64.** $f(x) = x^4 - 2x^2$

65. $f(x) = \dfrac{x^2}{x^2 - 1}$ **66.** $f(x) = x^3 - 12x$

67. $f(x) = -10 + 6x^2 - x^3$

68. $f(x) = \dfrac{x}{1 + x^2}$;   note that

$$f'(x) = -\frac{(x - 1)(x + 1)}{(x^2 + 1)^2}$$

and that

$$f''(x) = \frac{2x(x^2 - 3)}{(x^2 + 1)^3}.$$

69. $f(x) = x^3 - 3x$

70. $f(x) = x^4 - 12x^2$

71. $f(x) = x^3 + x^2 - 5x + 3$

72. $f(x) = \dfrac{1}{x} + \dfrac{1}{x^2}$

73. The function

$$f(x) = \frac{1}{x^2 + 2x + 2}$$

has a maximum value, and only one. Find it.

74. You need to manufacture a cylindrical pot, without a top, with a volume of 1 ft³. The cylindrical part of the pot is to be made of aluminum, the bottom of copper. Copper is five times as expensive as aluminum. What dimensions would minimize the total cost of the pot?

75. An open-topped rectangular box is to have a volume of 4500 cm³. If its bottom is a rectangle whose length is twice its width, what dimensions would minimize the total area of the bottom and four sides of the box?

76. A small rectangular box must be made with a volume of 324 in.³ Its bottom is square and costs twice as much (per square inch) as its top and four sides. What dimensions would minimize the total cost of the material needed to make this box?

77. You must make a small rectangular box with a volume of 400 in.³ Its bottom is a rectangle whose length is twice its width. The bottom costs 7¢/in.²; the top and four sides of the box cost 5¢/in.² What dimensions would minimize the cost of the box?

78. Suppose that $f(x)$ is a cubic polynomial with exactly three distinct real zeros. Prove that the two zeros of $f'(x)$ are real and distinct.

79. Suppose that it costs $1 + (0.0003)v^{3/2}$ dollars per mile to operate a truck at v miles per hour. If there are additional costs (such as the driver's pay) of $10/h, what speed would minimize the total cost of a 1000-mi trip?

80. The numbers $a_1, a_2, \ldots, a_n$ are fixed. Find a simple formula for the number x such that the sum of the squares of the distances of x from the n fixed numbers is as small as possible.

81. Sketch the curve $y^2 = x(x - 1)(x - 2)$, indicating that it consists of two pieces—one bounded and the other unbounded—and has two horizontal tangent lines, three vertical tangent lines, and two inflection points. [*Suggestion:* Note that the curve is symmetric about the x-axis and begin by determining the intervals on which the product $x(x - 1)(x - 2)$ is positive. Compute dy/dx and d^2y/dx^2 by implicit differentiation.]

82. Farmer Rogers wants to fence in a rectangular plot of area 2400 ft². She wants also to use additional fencing to build an internal divider fence parallel to two of the boundary sections (Fig. 4.MP.1). What is the minimum total length of fencing that this project will require? Verify that your answer yields the global minimum.

Fig. 4.MP.1   The fencing of Problem 82

83. Farmer Simmons wants to fence in a rectangular plot of area 1800 ft². He wants also to use additional fencing to build two internal divider fences, both parallel to the same two outer boundary sections (Fig. 4.MP.2). What is the minimum total length of fencing that this project will require? Verify that your answer yields the global minimum.

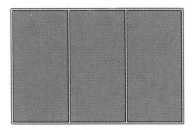

Fig. 4.MP.2   The fencing of Problem 83

84. Farmer Taylor wants to fence in a rectangular plot of area 2250 m². She wants also to use additional fencing to build three internal divider fences, all parallel to the same two outer boundary sections. What is the minimum total length of fencing that this project will require? Verify that your answer yields the global minimum.

85. Farmer Upshaw wants to fence in a rectangular plot of area A ft². He wants also to use additional fencing to build n (a fixed but unspecified positive integer) internal divider fences, all parallel to the same two outer boundary sections. What is the minimum total length of fencing that this project will require? Verify that your answer yields the global minimum.

86. What is the length of the shortest line segment that lies in the first quadrant with its endpoints on the coordinate axes and is tangent to the graph of $y = 1/x^2$? Verify that your answer yields the global minimum.

87. A right triangle is formed in the first quadrant by a line segment that is tangent to the graph of $y = 1/x^2$ and whose endpoints are on the coordinate axes. Is there a maximum possible area of such a triangle? Is there a minimum? Explain your answers.

88. A right triangle is formed in the first quadrant by a line segment that is tangent to the graph of $y = 1/x$ and whose endpoints are on the coordinate axes. Is there a maximum possible area of such a triangle? Is there a minimum? Explain your answers.

89. A rectangular box (with a top) is to have volume 288 in.³, and its base is to be exactly three times as long as it is wide. What is the minimum possible surface area of such a box? Verify that your answer yields the global minimum.

90. A rectangular box (with a top) is to have volume 800 in.³, and its base is to be exactly four times as long as it is wide. What is the minimum possible surface area of such a box? Verify that your answer yields the global minimum.

91. A rectangular box (with a top) is to have volume 225 cm³, and its base is to be exactly five times as long as it is wide. What is the minimum possible surface area of such a box? Verify that your answer yields the global minimum.

92. A rectangular box (with a top) is to have volume V, and its base is to be exactly n times as long as it is wide (n is a fixed positive integer). What is the minimum possible surface area of such a box? Verify that your answer yields the global minimum.

93. The graph of $f(x) = x^{1/3}(1 - x)^{2/3}$ is shown in Fig. 4.MP.3. Recall from Section 4.7 that this graph has an oblique asymptote with equation $y = mx + b$ provided that

$$\lim_{x \to +\infty} [f(x) - (mx + b)] = 0 \qquad \text{or that}$$
$$\lim_{x \to -\infty} [f(x) - (mx + b)] = 0.$$

(The values of m and b may be different in the two cases $x \to +\infty$ and $x \to -\infty$.) The graph here appears to have such an asymptote as $x \to +\infty$. Find m by evaluating

$$\lim_{x \to +\infty} \frac{f(x)}{x}.$$

Then find b by evaluating

$$\lim_{x \to +\infty} [f(x) - mx].$$

Finally, find m and b for the case $x \to -\infty$.

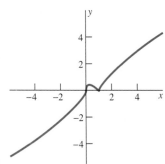

Fig. 4.MP.3   The graph of $y = f(x)$ of Problem 93

The Integral

Archimedes (287–212 B.C.)

❑ Archimedes of Syracuse was the greatest mathematician of the ancient era from the fifth century B.C. to the second century A.D., when the seeds of modern mathematics sprouted in Greek communities located mainly on the shores of the Mediterranean Sea. He was famous in his own time for mechanical inventions—the so-called Archimedean screw for pumping water, lever-and-pulley devices ("give me a place to stand and I can move the earth"), a planetarium that duplicated the motions of heavenly bodies so accurately as to show eclipses of the sun and moon, and machines of war that terrified Roman soldiers in the siege of Syracuse, during which Archimedes was killed. But it is said that for Archimedes himself

these inventions were merely the "diversions of geometry at play," and his writings are devoted to mathematical investigations.

❑ Archimedes carried out many area and volume computations that now use integral calculus—ranging from areas of circles, spheres, and segments of conic sections to volumes of cones, spheres, ellipsoids, and paraboloids. It had been proved earlier in Euclid's *Elements* that the area A of a circle is proportional to the square of its radius r, so $A = \pi r^2$ for some proportionality constant π. But it was Archimedes who accurately approximated the numerical value of π, showing that it lies between the value $3\frac{1}{7}$ memorized by elementary school children and the lower bound $3\frac{10}{71}$. Euclid had also proved that the volume V of a sphere of radius r is given by $V = \mu r^3$ (μ constant), but it was Archimedes who discovered (and proved) that $\mu = 4\pi/3$. He also discovered the now-familiar volume formulas $V = \pi r^2 h$ and $V = \frac{1}{3}\pi r^2 h$ for a cylinder and a cone, respectively, of base radius r and height h respectively.

❑ It was long suspected that Archimedes had not originally discovered his area and volume formulas by means of the limit-based arguments he used to establish them rigorously. In 1906 an Archimedean treatise entitled *The Method* was rediscovered virtually by accident after having

been lost since ancient times. In it he described a "method of discovery" based on using infinitesimals much as they were employed during the invention and exploration of calculus in the seventeenth and eighteenth centuries.

❑ To commemorate his sphere and cylinder formulas, Archimedes requested that on his tombstone be carved a sphere inscribed in a circular cylinder. If the height of the cylinder is $h = 2r$, can you verify that the total surface areas A_C and A_S of the cylinder and sphere, and their volumes V_C and V_S, are related by Archimedes' formulas

$$A_S = \tfrac{2}{3}A_C \quad \text{and} \quad V_S = \tfrac{2}{3}V_C\,?$$

Thus the volumes and surface areas of the sphere and cylinder have the same $2:3$ ratio.

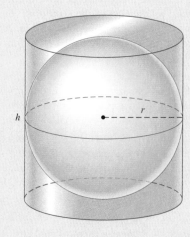

5.1
Introduction

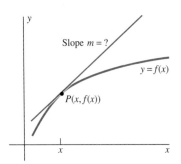

Fig. 5.1.1 The tangent-line problem motivates differential calculus.

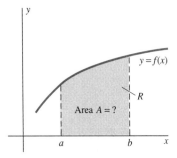

Fig. 5.1.2 The area problem motivates integral calculus.

Chapters 1 through 4 dealt with **differential calculus**, which is one of two closely related parts of *the* calculus. Differential calculus is centered on the concept of the *derivative*. Recall that the original motivation for the derivative was the problem of defining tangent lines to graphs of functions and calculating the slopes of such lines (Fig. 5.1.1). By contrast, the importance of the derivative stems from its applications to diverse problems that may, upon initial inspection, seem to have little connection with tangent lines to graphs.

Integral calculus is based on the concept of the *integral*. The definition of the integral is motivated by the problem of defining and calculating the area of the region that lies between the graph of a positive-valued function f and the x-axis over a closed interval $[a, b]$. The area of the region R of Fig. 5.1.2 is given by the *integral* of f from a to b, denoted by the symbol

$$\int_a^b f(x)\ dx. \tag{1}$$

But the integral, like the derivative, is important due to its applications in many problems that may appear unrelated to its original motivation—problems involving motion and velocity, population growth, volume, arc length, surface area, and center of gravity, among others.

The principal theorem of this chapter is the *fundamental theorem of calculus* in Section 5.6. It provides a vital connection between the operations of differentiation and integration, one that provides an effective method for computing values of integrals. It turns out that instead of finding the derivative of the function $f(x)$ in Eq. (1), we need instead to find a new function $F(x)$ whose derivative is $f(x)$:

$$F'(x) = f(x). \tag{2}$$

Thus we need to do "differentiation in reverse." We therefore begin in Section 5.2 with an investigation of *antidifferentiation*.

5.2
Antiderivatives and Initial Value Problems

The language of change is the natural language for the statement of most scientific laws and principles. For example, Newton's law of cooling says that the *rate of change* of the temperature T of a body is proportional to the difference between T and the temperature of the surrounding medium (Fig. 5.2.1). That is,

$$\frac{dT}{dt} = -k(T - A), \tag{1}$$

where k is a positive constant and A, normally assumed to be constant, is the surrounding temperature. Similarly, the *rate of change* of a population P with

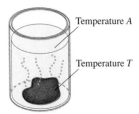

Temperature A

Temperature T

Fig. 5.2.1 Newton's law of cooling, Eq. (1), describes the cooling of a hot rock in cold water.

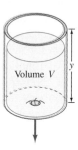

Fig. 5.2.2 Torricelli's law of draining, Eq. (3), describes the draining of a water tank.

Function $f(x)$	Antiderivative $F(x)$
1	x
$2x$	x^2
x^3	$\frac{1}{4}x^4$
$\cos x$	$\sin x$
$\sin 2x$	$-\frac{1}{2}\cos 2x$

Fig. 5.2.3 Some antiderivatives

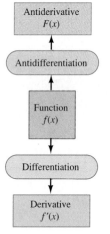

Fig. 5.2.4 Differentiation and antidifferentiation are opposites.

constant birth and death rates is proportional to the size of the population:

$$\frac{dP}{dt} = kP \qquad (k \text{ constant}). \qquad (2)$$

Torricelli's law of draining (Fig. 5.2.2) implies that the *rate of change* of the volume V of water in a draining tank is proportional to the square root of the depth y of the water; that is,

$$\frac{dV}{dt} = -k\sqrt{y} \qquad (k \text{ constant}). \qquad (3)$$

Mathematical models of real-world situations frequently involve equations that contain *derivatives* of unknown functions. Such equations, including Eqs. (1) through (3), are called **differential equations.**

The simplest kind of differential equation has the form

$$\frac{dy}{dx} = f(x), \qquad (4)$$

where f is a given (known) function and the function $y(x)$ is unknown. The process of finding a function from its derivative is the opposite of differentiation and is therefore called **antidifferentiation**. If we can find a function $y(x)$ whose derivative is $f(x)$,

$$y'(x) = f(x),$$

then we call $y(x)$ an *antiderivative* of $f(x)$.

Definition *Antiderivative*

An **antiderivative** of the function f is a function F such that

$$F'(x) = f(x)$$

wherever $f(x)$ is defined.

The table in Fig. 5.2.3 shows some examples of functions, each paired with one of its antiderivatives. Figure 5.2.4 illustrates the operations of differentiation and antidifferentiation, beginning with the same function $f(x)$ and going in opposite directions. Figure 5.2.5 illustrates differentiation "undoing" the result of antidifferentiation—the derivative of the antiderivative of $f(x)$ is the original function $f(x)$.

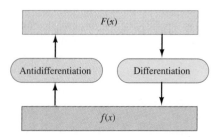

Fig. 5.2.5 Differentiation undoes the result of antidifferentiation.

EXAMPLE 1 Given: The function $f(x) = 3x^2$. Then $F(x) = x^3$ is an antiderivative of $f(x) = 3x^2$, as are the functions

$$G(x) = x^3 + 17, \qquad H(x) = x^3 + \pi, \quad \text{and} \quad K(x) = x^3 - \sqrt{2}.$$

Indeed, $J(x) = x^3 + C$ is an antiderivative of $f(x) = 3x^2$ for *any* choice of the *constant C.*

Thus a single function has *many* antiderivatives, whereas a function can have *only one* derivative. If $F(x)$ is an antiderivative of $f(x)$, then so is $F(x) + C$ for any choice of the constant C. The converse of this statement is more subtle: If $F(x)$ is one antiderivative of $f(x)$ *on the interval I*, then *every* antiderivative of $f(x)$ on I is of the form $F(x) + C$. This follows directly from Corollary 2 of the mean value theorem in Section 4.3, according to which two functions with the same derivative on an interval differ only by a constant on that interval.

Thus the graphs of any two antiderivatives $F(x) + C_1$ and $F(x) + C_2$ of the same function $f(x)$ on the same interval I are "parallel" in the sense illustrated in Figs. 5.2.6 through 5.2.8. There we see that the constant C is the vertical distance between the curves $y = F(x)$ and $y = F(x) + C$ for each x in I. This is the geometric interpretation of Theorem 1.

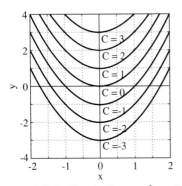

Fig. 5.2.6 Graph of $y = x^2 + C$ for various values of C

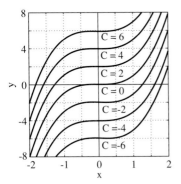

Fig. 5.2.7 Graph of $y = x^3 + C$ for various values of C

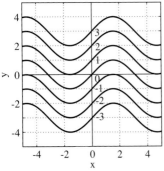

Fig. 5.2.8 Graph of $y = \sin x + C$ for various values of C

Theorem 1   *The Most General Antiderivative*

If $F'(x) = f(x)$ at each point of the open interval I, then every antiderivative G of f on I has the form

$$G(x) = F(x) + C, \qquad (5)$$

where C is a constant.

Thus if F is any single antiderivative of f on the interval I, then the *most general* antiderivative of f on I has the form $F(x) + C$, as given in Eq. (5). The collection of *all* antiderivatives of the function $f(x)$ is called the **indefinite integral** of f with respect to x and is denoted by

$$\int f(x)\,dx.$$

On the basis of Theorem 1, we write

$$\int f(x)\,dx = F(x) + C, \qquad (6)$$

where $F(x)$ is one particular antiderivative of $f(x)$. Therefore,

$$\int f(x)\,dx = F(x) + C \quad \text{if and only if} \quad F'(x) = f(x).$$

The integral symbol $\int$ is made like an elongated capital S. It is, in fact, a medieval S, used by Leibniz as an abbreviation for the Latin word *summa* ("sum"). We think of the combination $\int \cdots dx$ as a single symbol; we fill in the "blank" with the formula of the function whose antiderivative we seek. We may regard the differential dx as specifying the independent variable x both in the function $f(x)$ and in its antiderivatives.

EXAMPLE 2 The entries in Fig. 5.2.3 yield the indefinite integrals

$$\int 1 \, dx = x + C,$$

$$\int 2x \, dx = x^2 + C,$$

$$\int x^3 \, dx = \tfrac{1}{4}x^4 + C,$$

$$\int \cos x \, dx = \sin x + C, \quad \text{and}$$

$$\int \sin 2x \, dx = -\tfrac{1}{2}\cos 2x + C.$$

You can verify each such formula by differentiating the right-hand side. Indeed, this is the *surefire* way to check any antidifferentiation: To verify that $F(x)$ is an antiderivative of $f(x)$, compute $F'(x)$ to see whether or not you obtain $f(x)$. For instance, the differentiation

$$D_x(-\tfrac{1}{2}\cos 2x + C) = -\tfrac{1}{2}(-2\sin 2x) + 0 = \sin 2x$$

is sufficient to verify the fifth formula of this example.

The differential dx in Eq. (6) specifies that the independent variable is x. But we can describe a specific antidifferentiation in terms of *any* independent variable that is convenient. For example, the indefinite integrals

$$\int 3t^2 \, dt = t^3 + C, \qquad \int 3y^2 \, dy = y^3 + C, \quad \text{and} \quad \int 3u^2 \, du = u^3 + C$$

mean exactly the same thing as

$$\int 3x^2 \, dx = x^3 + C.$$

Every differentiation formula yields immediately—by "reversal" of the differentiation—a corresponding indefinite integral formula. The now-familiar derivatives of power functions and trigonometric functions yield the integral formulas stated in Theorem 2.

> **Theorem 2   Some Integral Formulas**
>
> $$\int x^k \, dx = \frac{x^{k+1}}{k+1} + C \qquad (\text{if} \quad k \neq -1), \qquad (7)$$
>
> $$\int \cos kx \, dx = \frac{1}{k}\sin kx + C, \qquad (8)$$
>
> $$\int \sin kx \, dx = -\frac{1}{k}\cos kx + C, \qquad (9)$$

$$\int \sec^2 kx \, dx = \frac{1}{k} \tan kx + C, \tag{10}$$

$$\int \csc^2 kx \, dx = -\frac{1}{k} \cot kx + C, \tag{11}$$

$$\int \sec kx \tan kx \, dx = \frac{1}{k} \sec kx + C, \quad \text{and} \tag{12}$$

$$\int \csc kx \cot kx \, dx = -\frac{1}{k} \csc kx + C. \tag{13}$$

REMARK   Be sure you see why there is a minus sign in Eq. (9) but none in Eq. (8)!

Furthermore, the differentiation formulas $D_x e^{kx}$ and $D_x \ln x = 1/x$ of Section 3.8 yield the integral formulas

$$\int e^{kx} \, dx = \frac{1}{k} e^{kx} + C$$

and

$$\int \frac{1}{x} \, dx = \ln x + C \qquad (x > 0).$$

Recall that the operation of differentiation is *linear,* meaning that

$$D_x[cF(x)] = cF'(x) \quad \text{(where } c \text{ is a constant)}$$

and

$$D_x[F(x) \pm G(x)] = F'(x) \pm G'(x).$$

It follows in the notation of antidifferentiation that

$$\int cf(x) \, dx = c \int f(x) \, dx \qquad (c \text{ is a constant}) \tag{14}$$

and

$$\int [f(x) \pm g(x)] \, dx = \int f(x) \, dx \pm \int g(x) \, dx. \tag{15}$$

We can summarize these two equations by saying that antidifferentiation is **linear.** In essence, then, we antidifferentiate a sum of functions by antidifferentiating each function individually. This is *termwise* (or *term-by-term*) antidifferentiation. Moreover, a constant coefficient in any such term is merely "carried through" the antidifferentiation.

EXAMPLE 3   Find

$$\int \left(x^3 + 3\sqrt{x} - \frac{4}{x^2} \right) dx.$$

Solution   Just as in differentiation, we prepare for antidifferentiation by writing roots and reciprocals as powers with fractional or negative exponents. Thus

$$\int \left(x^3 + 3\sqrt{x} - \frac{4}{x^2} \right) dx = \int (x^3 + 3x^{1/2} - 4x^{-2})\, dx$$

$$= \int x^3\, dx + 3 \int x^{1/2}\, dx - 4 \int x^{-2}\, dx \qquad \text{[using Eqs. (14) and (15)]}$$

$$= \frac{x^4}{4} + 3 \cdot \frac{x^{3/2}}{\frac{3}{2}} - 4 \cdot \frac{x^{-1}}{-1} + C \qquad \text{[using Eq. (7)]}$$

$$= \frac{1}{4}x^4 + 2x\sqrt{x} + \frac{4}{x} + C.$$

There's only one "$+C$" because the surefire check verifies that $\frac{1}{4}x^4 + 2x^{2/3} + 4x^{-1}$ is a particular antiderivative. Hence any other anti-derivative differs from this one by only a (single) constant C.

EXAMPLE 4

$$\int (2 \cos 3t + 5 \sin 4t + 3e^{7t})\, dt$$

$$= 2 \int \cos 3t\, dt + 5 \int \sin 4t\, dt + 3 \int e^{7t}\, dt$$

$$= 2(\tfrac{1}{3} \sin 3t) + 5(-\tfrac{1}{4} \cos 4t) + 3(\tfrac{1}{7}e^{7t}) + C$$

$$= \tfrac{2}{3} \sin 3t - \tfrac{5}{4} \cos 4t + \tfrac{3}{7}e^{7t} + C.$$

Equation (7) is the power rule "in reverse." The generalized power rule in reverse is

$$\int u^k\, du = \frac{u^{k+1}}{k+1} + C \qquad (\text{if } k \neq -1), \qquad (16)$$

where

$$u = g(x) \qquad \text{and} \qquad du = g'(x)dx.$$

EXAMPLE 5 With $u = x + 5$ (so that $du = dx$), Eq. (16) yields

$$\int (x + 5)^{10}\, dx = \frac{1}{11}(x + 5)^{11} + C.$$

EXAMPLE 6 We want to find

$$\int \frac{20}{(4 - 5x)^3}\, dx.$$

We plan to use Eq. (16) with $u = 4 - 5x$. But we must get the differential $du = -5\, dx$ into the act. The "constant multiplier rule" of Eq. (14) permits us to do this:

$$\int \frac{20}{(4 - 5x)^3}\, dx = 20 \int (4 - 5x)^{-3}\, dx$$

$$= \frac{20}{-5} \int (4 - 5x)^{-3}(-5\, dx) \qquad (17)$$

and Q have sufficiently many sides, all very short, then it would appear that their areas $a(P)$ and $a(Q)$ closely approximate the area of the region R. Moreover, error control is possible: We see that

$$a(P) < a(R) < a(Q) \tag{1}$$

because R contains the polygon P but is contained in the polygon Q.

The inequalities in (1) bracket the desired area $a(R)$. Suppose, for instance, that calculations based on triangular dissections (as in Fig. 5.3.2) yield $a(P) = 7.341$ and $a(Q) = 7.343$. Then the resulting inequality

$$7.341 < a(R) < 7.343$$

implies that $a(R) \approx 7.34$, accurate to two decimal places.

Our primary objective here is to describe a systematic technique by which to approximate the area of an appropriate curvilinear region using easily calculated polygonal areas.

AREAS UNDER GRAPHS

We consider the type of region that is determined by a continuous positive-valued function f defined on a closed interval $[a, b]$. Suppose that we want to calculate the area A of the region R that lies *below* the curve $y = f(x)$ and *above* the interval $[a, b]$ on the x-axis (Fig. 5.3.5). The region R is bounded on the left by the vertical line $x = a$ and on the right by the vertical line $x = b$.

We divide the base interval $[a, b]$ into subintervals, all with the same length. Above each subinterval lies a vertical strip (Fig. 5.3.6), and the area A is the sum of the areas of these strips.

On each of these base subintervals, we erect a rectangle that approximates the corresponding vertical strip. We may choose either an "inscribed" or a "circumscribed" rectangle (both possibilities are illustrated in Fig. 5.3.6) or even a rectangle that is intermediate between the two. These rectangles then make up a polygon that approximates the region R, and therefore the sum of the area of these rectangles *approximates* the desired area A.

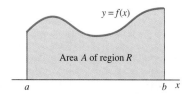

Fig. 5.3.5 The area under the graph of $y = f(x)$ from $x = a$ to $x = b$

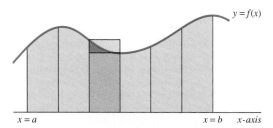

Fig. 5.3.6 Vertical strips determined by a division of $[a, b]$ into equal-length subintervals

For example, suppose that we want to approximate the area A of the region R that lies below the parabola $y = x^2$ above the interval $[0, 3]$. The computer plots in Fig. 5.3.7 show successively

- ❑ 5 inscribed and 5 circumscribed rectangles;
- ❑ 10 inscribed and 10 circumscribed rectangles;
- ❑ 20 inscribed and 20 circumscribed rectangles;
- ❑ 40 inscribed and 40 circumscribed rectangles.

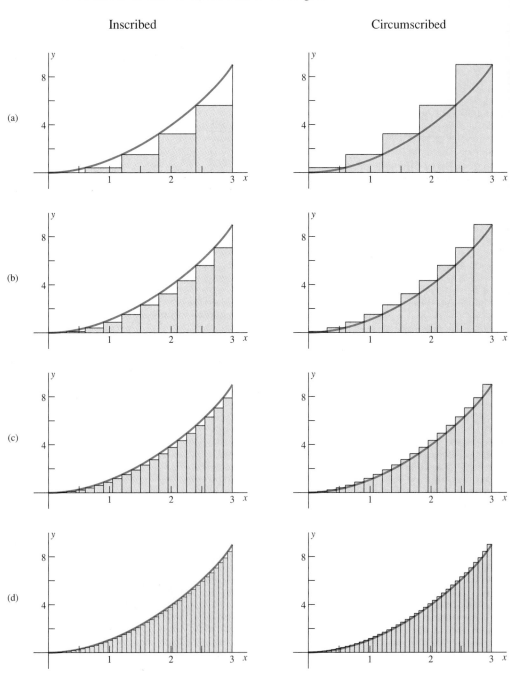

Inscribed   Circumscribed

Fig. 5.3.7 (a) Five inscribed and circumscribed polygons; (b) ten inscribed and circumscribed polygons; (c) twenty inscribed and circumscribed polygons; (d) forty inscribed and circumscribed polygons

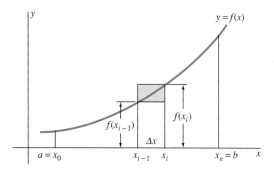

Fig. 5.3.12 Inscribed and circumscribed rectangles on the ith subinterval $[x_{i-1}, x_i]$

respectively. Adding the areas of the inscribed rectangles for $i = 1, 2, 3, \ldots,$ n, we get the underestimate

$$\underline{A}_n = \sum_{i=1}^{n} f(x_{i-1})\, \Delta x \tag{12}$$

of the actual area A. Similarly, the sum of the areas of the circumscribed rectangles is the overestimate

$$\overline{A}_n = \sum_{i=1}^{n} f(x_i)\, \Delta x. \tag{13}$$

The inequality $\underline{A}_n \leqq A \leqq \overline{A}_n$ then yields

$$\sum_{i=1}^{n} f(x_{i-1})\, \Delta x \leqq A \leqq \sum_{i=1}^{n} f(x_i)\, \Delta x. \tag{14}$$

The inequalities in (14) would be reversed if $f(x)$ were decreasing (rather than increasing) on $[a, b]$. (Why?)

An illustration such as Fig. 5.3.7 suggests that if the number n of subintervals is very large, so that Δx is very small, then the areas $\underline{A}_n$ and $\overline{A}_n$ of the inscribed and circumscribed polygons will differ by very little. Hence both will be very close to the actual area A of the region R. Indeed, if f either is increasing or is decreasing on the whole interval $[a, b]$, then the small rectangles in Fig. 5.3.11 (representing the difference between $\overline{A}_n$ and $\underline{A}_n$) can be reassembled in a "stack," as indicated on the right in the figure. It follows that

$$|\overline{A}_n - \underline{A}_n| = |f(b) - f(a)|\, \Delta x. \tag{15}$$

But $\Delta x = (b - a)/n \to 0$ as $n \to \infty$. Thus the difference between the left-hand and right-hand sums in (14) is approaching zero as $n \to \infty$, whereas A does not change as $n \to \infty$. It follows that the area of the region R is given by

$$A = \lim_{n \to \infty} \sum_{i=1}^{n} f(x_{i-1})\, \Delta x = \lim_{n \to \infty} \sum_{i=1}^{n} f(x_i)\, \Delta x. \tag{16}$$

The meaning of these limits is simply that A can be found with any desired accuracy by calculating either sum in Eq. (16) with a sufficiently large number n of subintervals. In applying Eq. (16), recall that

$$\Delta x = \frac{b - a}{n}. \tag{17}$$

Also note that

$$x_i = a + i\,\Delta x \qquad (18)$$

for $i = 0, 1, 2, \ldots, n$, because x_i is i "steps" of length Δx to the right of $x_0 = a$.

EXAMPLE 6 We can now compute exactly the area we approximated in Example 1—the area of the region under the graph of $f(x) = x^2$ over the interval $[0, 3]$. If we divide $[0, 3]$ into n subintervals all of the same length, then Eqs. (17) and (18) give

$$\Delta x = \frac{3}{n} \quad \text{and} \quad x_i = 0 + i \cdot \frac{3}{n} = \frac{3i}{n}$$

for $i = 0, 1, 2, \ldots, n$. Therefore,

$$\sum_{i=1}^{n} f(x_i)\,\Delta x = \sum_{i=1}^{n} (x_i)^2\,\Delta x = \sum_{i=1}^{n} \left(\frac{3i}{n}\right)^2 \left(\frac{3}{n}\right) = \frac{27}{n^3} \sum_{i=1}^{n} i^2.$$

Then Eq. (8) for Σi^2 yields

$$\sum_{i=1}^{n} f(x_i)\,\Delta x = \frac{27}{n^3}\left(\frac{1}{3}n^3 + \frac{1}{2}n^2 + \frac{1}{6}n\right) = 27\left(\frac{1}{3} + \frac{1}{2n} + \frac{1}{6n^2}\right).$$

When we take the limit as $n \to \infty$, Eq. (16) gives

$$A = \lim_{n \to \infty} 27\left(\frac{1}{3} + \frac{1}{2n} + \frac{1}{6n^2}\right) = 9,$$

because the terms $1/(2n)$ and $1/(6n^2)$ approach zero as $n \to \infty$. Thus our earlier inference from the data in Fig. 5.3.10 was correct: $A = 9$ *exactly*.

EXAMPLE 7 Find the area under the graph of $f(x) = 100 - 3x^2$ from $x = 1$ to $x = 5$.

Solution As shown in Fig. 5.3.13, the sum $\Sigma f(x_i)\,\Delta x$ gives the area of the inscribed rectangular polygon. With $a = 1$ and $b = 5$, Eqs. (17) and (18) give

$$\Delta x = \frac{4}{n} \quad \text{and} \quad x_i = 1 + i \cdot \frac{4}{n} = 1 + \frac{4i}{n}.$$

Therefore,

$$\sum_{i=1}^{n} f(x_i)\,\Delta x = \sum_{i=1}^{n} \left[100 - 3 \cdot \left(1 + \frac{4i}{n}\right)^2\right]\left(\frac{4}{n}\right)$$

$$= \sum_{i=1}^{n} \left[97 - \frac{24i}{n} - \frac{48i^2}{n^2}\right]\left(\frac{4}{n}\right)$$

$$= \frac{388}{n} \sum_{i=1}^{n} 1 - \frac{96}{n^2} \sum_{i=1}^{n} i - \frac{192}{n^3} \sum_{i=1}^{n} i^2$$

$$= \frac{388}{n} \cdot n - \frac{96}{n^2}\left(\frac{1}{2}n^2 + \frac{1}{2}n\right) - \frac{192}{n^3}\left(\frac{1}{3}n^3 + \frac{1}{2}n^2 + \frac{1}{6}n\right)$$

$$= 276 - \frac{144}{n} - \frac{32}{n^2}.$$

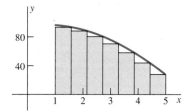

Fig. 5.3.13 The region of Example 7

[We have applied Eqs. (6) through (8)]. Consequently, the second limit in Eq. (16) yields

$$A = \lim_{n \to \infty} \left(276 - \frac{144}{n} - \frac{32}{n^2} \right) = 276$$

for the desired area.

HISTORICAL NOTE—THE NUMBER π

Mathematicians of ancient times tended to employ inscribed and circumscribed triangles rather than rectangles for area approximations. In the third century B.C., Archimedes, the greatest mathematician of antiquity, used such an approach to derive the famous estimate

$$\tfrac{223}{71} = 3\tfrac{10}{71} < \pi < 3\tfrac{1}{7} = \tfrac{22}{7}.$$

Because the area of a circle of radius r is πr^2, the number π may be *defined* to be the area of the unit circle of radius $r = 1$. We will approximate π, then, by approximating the area of the unit circle.

Let P_n and Q_n be n-sided regular polygons, with P_n inscribed in the unit circle and Q_n circumscribed around it (Fig. 5.3.14). Because both polygons are regular, all their sides and angles are equal, so we need to find the area of only *one* of the triangles that we've shown making up P_n and *one* of those making up Q_n.

Let α_n be the central angle subtended by *half* of one of the sides. The angle α_n is the same whether we work with P_n or with Q_n. In degrees,

$$\alpha_n = \frac{360°}{2n} = \frac{180°}{n}.$$

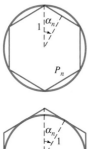

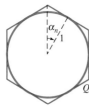

Fig. 5.3.14 Estimating π by using inscribed and circumscribed regular polygons and the unit circle

We can read various dimensions and proportions from Fig. 5.3.14. For example, we see that the area $a(P_n) = \underline{A}_n$ of P_n is given by

$$\underline{A}_n = a(P_n) = n \cdot 2 \cdot \frac{1}{2} \sin \alpha_n \cos \alpha_n = \frac{n}{2} \sin 2\alpha_n = \frac{n}{2} \sin\left(\frac{360°}{n}\right) \quad (19)$$

and that the area of Q_n is

$$\overline{A}_n = a(Q_n) = n \cdot 2 \cdot \frac{1}{2} \tan \alpha_n = n \tan\left(\frac{180°}{n}\right). \quad (20)$$

We substituted selected values of n into Eqs. (19) and (20) to obtain the entries of the table in Fig. 5.3.15. Because $\underline{A}_n \leqq \pi \leqq \overline{A}_n$ for all n, we see that $\pi \approx 3.14159$ to five decimal places. Archimedes' reasoning was *not* circular—he used a direct method for computing the sines and cosines in Eqs. (19) and (20) that does not depend upon *a priori* knowledge of the value of π.*

n	$a(P_n)$	$a(Q_n)$
6	2.598076	3.464102
12	3.000000	3.215390
24	3.105829	3.159660
48	3.132629	3.146086
96	3.139350	3.142715
180	3.140955	3.141912
360	3.141433	3.141672
720	3.141553	3.141613
1440	3.141583	3.141598
2880	3.141590	3.141594
5760	3.141592	3.141593

Fig. 5.3.15 Data for estimating π (rounded to six-place accuracy)

*See Chapter 2 of C. H. Edwards, Jr., *The Historical Development of the Calculus* (New York: Springer-Verlag, 1979).

5.3 Problems

Write the sums in Problems 1 through 8 in summation notation.

1. $1 + 4 + 9 + 16 + 25$

2. $1 - 2 + 3 - 4 + 5 - 6$

3. $1 + \frac{1}{2} + \frac{1}{3} + \frac{1}{4} + \frac{1}{5}$

4. $1 + \frac{1}{4} + \frac{1}{9} + \frac{1}{16} + \frac{1}{25}$

5. $\frac{1}{2} + \frac{1}{4} + \frac{1}{8} + \frac{1}{16} + \frac{1}{32} + \frac{1}{64}$

6. $\frac{1}{3} - \frac{1}{9} + \frac{1}{27} - \frac{1}{81} + \frac{1}{243}$

7. $\frac{2}{3} + \frac{4}{9} + \frac{8}{27} + \frac{16}{81} + \frac{32}{243}$

8. $1 + \sqrt{2} + \sqrt{3} + 2 + \sqrt{5} + \sqrt{6} + \sqrt{7} + 2\sqrt{2} + 3$

Use Eqs. (6) through (9) to find the sums in Problems 9 through 18.

9. $\displaystyle\sum_{i=1}^{10} (4i - 3)$

10. $\displaystyle\sum_{j=1}^{8} (5 - 2j)$

11. $\displaystyle\sum_{i=1}^{10} (3i^2 + 1)$

12. $\displaystyle\sum_{k=1}^{6} (2k - 3k^2)$

13. $\displaystyle\sum_{r=1}^{8} (r - 1)(r + 2)$

14. $\displaystyle\sum_{i=1}^{5} (i^3 - 3i + 2)$

15. $\displaystyle\sum_{i=1}^{6} (i^3 - i^2)$

16. $\displaystyle\sum_{k=1}^{10} (2k - 1)^2$

17. $\displaystyle\sum_{i=1}^{100} i^2$

18. $\displaystyle\sum_{i=1}^{100} i^3$

Use the method of Example 5 to evaluate the limits in Problems 19 and 20.

19. $\displaystyle\lim_{n \to \infty} \frac{1^2 + 2^2 + 3^2 + \cdots + n^2}{n^3}$

20. $\displaystyle\lim_{n \to \infty} \frac{1^3 + 2^3 + 3^3 + \cdots + n^3}{n^4}$

Use Eqs. (6) through (9) to derive concise formulas in terms of n for the sums in Problems 21 and 22.

21. $\displaystyle\sum_{i=1}^{n} (2i - 1)$

22. $\displaystyle\sum_{i=1}^{n} (2i - 1)^2$

In Problems 23 through 32, let R denote the region that lies below the graph of $y = f(x)$ over the interval $[a, b]$ on the x-axis. Use the method of Example 1 to calculate both an underestimate $\underline{A}_n$ and an overestimate $\overline{A}_n$ for the area A of R, based on a division of $[a, b]$ into n subintervals all with the same length $\Delta x = (b - a)/n$.

23. $f(x) = x$ on $[0, 1]$; $\quad n = 5$

24. $f(x) = x$ on $[1, 3]$; $\quad n = 5$

25. $f(x) = 2x + 3$ on $[0, 3]$; $\quad n = 6$

26. $f(x) = 13 - 3x$ on $[0, 3]$; $\quad n = 6$ (Fig. 5.3.16)

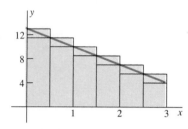

Fig. 5.3.16 Problem 26

27. $f(x) = x^2$ on $[0, 1]$; $\quad n = 5$

28. $f(x) = x^2$ on $[1, 3]$; $\quad n = 5$

29. $f(x) = 9 - x^2$ on $[0, 3]$; $\quad n = 5$ (Fig. 5.3.17)

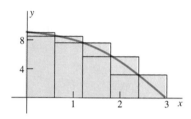

Fig. 5.3.17 Problem 29

30. $f(x) = 9 - x^2$ on $[1, 3]$; $\quad n = 8$

31. $f(x) = x^3$ on $[0, 1]$; $\quad n = 10$

32. $f(x) = \sqrt{x}$ on $[0, 1]$; $\quad n = 10$ (Fig. 5.3.18)

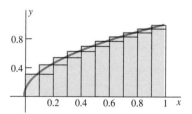

Fig. 5.3.18 Problem 32

33. Derive Eq. (7) by adding the equations

$$\sum_{i=1}^{n} i = 1 + 2 + 3 + \cdots + n$$

and

$$\sum_{i=1}^{n} i = n + (n - 1) + (n - 2) + \cdots + 1.$$

34. Write the n equations obtained by substituting the values $k = 1, 2, 3, \ldots, n$ into the identity

$$(k + 1)^3 - k^3 = 3k^2 + 3k + 1.$$

Add these n equations and use their sum to deduce Eq. (8) from Eq. (7).

In Problems 35 through 40, first calculate (in terms of n) the sum

$$\sum_{i=1}^{n} f(x_i)\, \Delta x$$

to approximate the area A of the region under $y = f(x)$ above the interval $[a, b]$. Then find A exactly (as in Examples 6 and 7) by taking the limit as $n \to \infty$.

35. $f(x) = x$ on $[0, 1]$
36. $f(x) = x^2$ on $[0, 2]$
37. $f(x) = x^3$ on $[0, 3]$
38. $f(x) = x + 2$ on $[0, 2]$
39. $f(x) = 5 - 3x$ on $[0, 1]$
40. $f(x) = 9 - x^2$ on $[0, 3]$

41. As in Fig. 5.3.19, the region under the graph of $f(x) = hx/b$ is a triangle with base b and height h. Use Eq. (7) to verify—with the notation of Eq. (16)—that

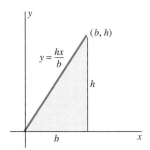

Fig. 5.3.19   Problem 41

$$\lim_{n \to \infty} \sum_{i=1}^{n} f(x_i)\, \Delta x = \tfrac{1}{2} bh,$$

in agreement with the familiar formula for the area of a triangle.

In Problems 42 and 43, let A denote the area and C the circumference of a circle of radius r and let A_n and C_n denote the area and perimeter, respectively, of a regular n-sided polygon inscribed in this circle.

42. Figure 5.3.20 shows one side of the n-sided polygon subtending an angle $2\pi/n$ at the center O of the circle. Show that

$$A_n = nr^2 \sin\!\left(\frac{\pi}{n}\right)\cos\!\left(\frac{\pi}{n}\right) \quad \text{and that} \quad C_n = 2nr \, \sin\!\left(\frac{\pi}{n}\right).$$

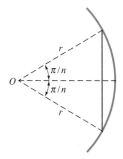

Fig. 5.3.20   Problem 42

43. Deduce that $A = \tfrac{1}{2} rC$ by taking the limit of A_n/C_n as $n \to \infty$. Then, under the assumption that $A = \pi r^2$, deduce that $C = 2\pi r$. Thus the familiar circumference formula for a circle follows from the familiar area formula for a circle.

5.4
Riemann Sums and the Integral

Suppose that f is a positive-valued and increasing function defined on a set of real numbers that includes the interval $[a, b]$. In Section 5.3 we used inscribed and circumscribed rectangles to set up the sums

$$\sum_{i=1}^{n} f(x_{i-1})\, \Delta x \quad \text{and} \quad \sum_{i=1}^{n} f(x_i)\, \Delta x \tag{1}$$

that approximate the area A under the graph of $y = f(x)$ from $x = a$ to $x = b$. Recall that the notation in Eq. (1) is based on a division of the interval $[a, b]$ into n subintervals, all with the same length $\Delta x = (b - a)/n$, and that $[x_{i-1}, x_i]$ denotes the ith subinterval.

The approximating sums in Eq. (1) are both of the form

$$\sum_{i=1}^{n} f(x_i^*)\, \Delta x, \tag{2}$$

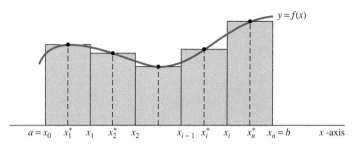

Fig. 5.4.1 The Riemann sum in Eq. (2) as a sum of areas of rectangles

where x_i^* denotes a selected point of the ith subinterval $[x_{i-1}, x_i]$ (Fig. 5.4.1). Sums of the form in (2) appear as approximations in a wide range of applications and form the basis for the definition of the integral. Motivated by our discussion of area in Section 5.3, we want to define the integral of f from a to b as some sort of limit, as $\Delta x \to 0$, of sums such as the one in (2). Our goal is to begin with a fairly general function f and define a computable real number I (the integral of f) that—in the special case when f is continuous and positive-valued on $[a, b]$—will equal the area under the graph of $y = f(x)$.

We begin with a function f defined on $[a, b]$ that is *not* necessarily either continuous or positive-valued. A **partition** P of $[a, b]$ is a collection of subintervals

$$[x_0, x_1], [x_1, x_2], [x_2, x_3], \ldots, [x_{n-1}, x_n]$$

of $[a, b]$ such that

$$a = x_0 < x_1 < x_2 < x_3 < \cdots < x_{n-1} < x_n = b.$$

The **mesh** (or **norm**) of the partition P is the largest of the lengths

$$\Delta x_i = x_i - x_{i-1}$$

of the subintervals in P and is denoted by $|P|$. To get a sum such as the one in (2), we need a point x_i^* in the ith subinterval for each i, $1 \leqq i \leqq n$. A collection of points

$$S = \{x_1^*, x_2^*, x_3^*, \ldots, x_n^*\}$$

with x_i^* in $[x_{i-1}, x_i]$ (for each i) is called a **selection** for the partition P.

Definition *Riemann Sum*

Let f be a function defined on the interval $[a, b]$. If P is a partition of $[a, b]$ and S is a selection for P, then the **Riemann sum** for f determined by P and S is

$$R = \sum_{i=1}^{n} f(x_i^*)\, \Delta x_i. \tag{3}$$

We also say that this Riemann sum is **associated with** the partition P.

The German mathematician G. F. B. Riemann (1826–1866) provided a rigorous definition of the integral. Various special types of "Riemann sums" had appeared in area and volume computations since the time of

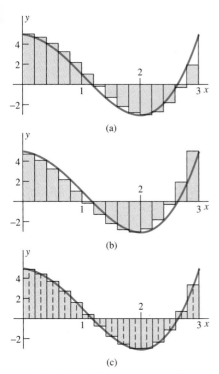

(a)

(b)

(c)

Fig. 5.4.2 Riemann sums for
$f(x) = 2x^3 - 6x^2 + 5$ on $[0, 3]$
(a) Left-endpoint sum
(b) Right-endpoint sum
(c) Midpoint sum

Archimedes, but it was Riemann who framed the preceding definition in its full generality.

The point x_i^* in Eq. (3) is simply a selected point of the ith subinterval $[x_{i-1}, x_i]$. That is, it can be *any* point of this subinterval. But when we compute Riemann sums, we usually choose the points of the selection S in some systematic manner, as illustrated in Fig. 5.4.2. There we show different Riemann sums for the function $f(x) = 2x^3 - 6x^2 + 5$ on the interval $[0, 3]$. Figure 5.4.2(a) shows rectangles associated with the *left-endpoint sum*

$$R_{\text{left}} = \sum_{i=1}^{n} f(x_{i-1})\, \Delta x, \tag{4}$$

in which each x_i^* is selected to be x_{i-1}, the *left endpoint* of the ith subinterval $[x_{i-1}, x_i]$ of length $\Delta x = (b - a)/n$. Figure 5.4.2(b) shows rectangles associated with the *right-endpoint sum*

$$R_{\text{right}} = \sum_{i=1}^{n} f(x_i)\, \Delta x, \tag{5}$$

in which each x_i^* is selected to be x_i, the *right endpoint* of $[x_{i-1}, x_i]$. In each figure, some of the rectangles are inscribed and others are circumscribed.

Figure 5.4.2(c) shows rectangles associated with the *midpoint sum*

$$R_{\text{mid}} = \sum_{i=1}^{n} f(m_i)\, \Delta x, \tag{6}$$

in which

$$x_i^* = m_i = \frac{x_{i-1} + x_i}{2},$$

the *midpoint* of the ith subinterval $[x_{i-1}, x_i]$. The dashed lines in Fig. 5.4.2(c) represent the ordinates of f at these midpoints.

EXAMPLE 1   In Example 1 of Section 5.3 we calculated left- and right-endpoint sums for $f(x) = x^2$ on $[0, 3]$ with $n = 10$ subintervals. We now do this more concisely by using summation notation, and we also calculate the analogous midpoint sum. Figure 5.4.3 shows a typical approximating rectan-

Fig. 5.4.3   Example 1
(a) The case $x_i^* = x_{i-1}$
(b) The case $x_i^* = x_i$
(c) The case $x_i^* = m_i$

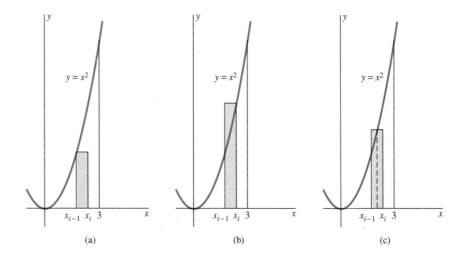

(a)

(b)

(c)

gle for each of these sums. With $a = 0$, $b = 3$, and $\Delta x = (b - a)/n = \frac{3}{10}$, we see that the ith subdivision point is

$$x_i = a + i \cdot \Delta x = \tfrac{3}{10} i.$$

The ith subinterval, as well as its midpoint

$$m_i = \frac{1}{2}(x_{i-1} + x_i) = \frac{1}{2}\left(\frac{3i - 3}{10} + \frac{3}{10}i\right) = \frac{3}{20}(2i - 1),$$

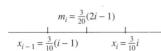

$$m_i = \tfrac{3}{20}(2i - 1)$$
$$x_{i-1} = \tfrac{3}{10}(i - 1) \qquad x_i = \tfrac{3}{10}i$$

Fig. 5.4.4 The ith subinterval of Example 1

are shown in Fig. 5.4.4. With $x_i^* = x_{i-1} = \frac{3}{10}(i - 1)$, we obtain the left-endpoint sum in Eq. (4),

$$R_{\text{left}} = \sum_{i=1}^{n} f(x_{i-1})\,\Delta x = \sum_{i=1}^{10} [\tfrac{3}{10}(i - 1)]^2(\tfrac{3}{10})$$

$$= \tfrac{27}{1000} \cdot (0^2 + 1^2 + 2^2 + \cdots + 9^2)$$

$$= \tfrac{7695}{1000} = 7.695 \quad \text{[using Eq. (8) of Section 5.3].}$$

With $x_i^* = x_i = \frac{3}{10}i$, we get the right-endpoint sum in Eq. (5),

$$R_{\text{right}} = \sum_{i=1}^{n} f(x_i)\,\Delta x = \sum_{i=1}^{10} [\tfrac{3}{10}i]^2(\tfrac{3}{10})$$

$$= \tfrac{27}{1000} \cdot (1^2 + 2^2 + 3^2 + \cdots + 10^2)$$

$$= \tfrac{10{,}395}{1000} = 10.395 \quad \text{[using Eq. (8) of Section 5.3].}$$

Finally, with $x_i^* = m_i = \frac{3}{20}(2i - 1)$, we get the midpoint sum in Eq. (6),

$$R_{\text{mid}} = \sum_{i=1}^{n} f(m_i)\,\Delta x = \sum_{i=1}^{10} [\tfrac{3}{20}(2i - 1)]^2(\tfrac{3}{10})$$

$$= \tfrac{27}{4000} \cdot (1^2 + 3^2 + 5^2 + \cdots + 17^2 + 19^2) = \tfrac{35{,}910}{4000} = 8.9775.$$

The midpoint sum is much closer than either endpoint sum to the actual value 9 (of the area under the graph of $y = x^2$ over $[0, 3]$) that we found in Example 6 of Section 5.3.

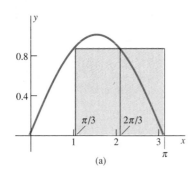

(a)

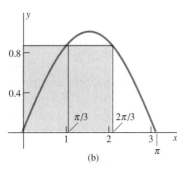

(b)

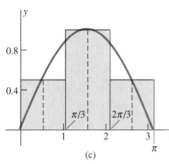

(c)

Fig. 5.4.5 Approximating the area under $y = \sin x$ on $[0, \pi]$ (Example 2)   (a) Left-endpoint sum   (b) Right-endpoint sum   (c) Midpoint sum

EXAMPLE 2 Figure 5.4.5 illustrates Riemann sums for $f(x) = \sin x$ on $[0, \pi]$ based on $n = 3$ subintervals: $[0, \pi/3]$, $[\pi/3, 2\pi/3]$, and $[2\pi/3, \pi]$ of length $\Delta x = \pi/3$ with midpoints $\pi/6$, $\pi/2$, and $5\pi/6$. The left-endpoint sum is

$$R_{\text{left}} = (\Delta x) \cdot \left(\sum_{i=1}^{n} f(x_{i-1})\right) = \frac{\pi}{3} \cdot \left(\sin 0 + \sin \frac{\pi}{3} + \sin \frac{2\pi}{3}\right)$$

$$= \frac{\pi}{3} \cdot \left(0 + \frac{\sqrt{3}}{2} + \frac{\sqrt{3}}{2}\right) = \frac{\pi\sqrt{3}}{3} \approx 1.81.$$

It is clear from the figure that the right-endpoint sum has the same value. The corresponding midpoint sum is

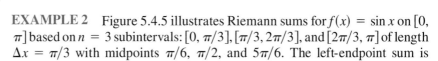

$$R_{\text{mid}} = \frac{\pi}{3} \cdot \left(\sin \frac{\pi}{6} + \sin \frac{\pi}{2} + \sin \frac{5\pi}{6}\right) = \frac{\pi}{3} \cdot \left(\frac{1}{2} + 1 + \frac{1}{2}\right) = \frac{2\pi}{3} \approx 2.09.$$

(It happens that the exact area under one arch of the sine curve is 2.)

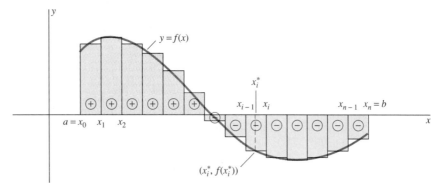

Fig. 5.4.6 A geometric interpretation of the Riemann sum in Eq. (3)

In the case of a function f that has both positive and negative values on $[a, b]$, it is necessary to consider the *signs* indicated in Fig. 5.4.6 when we interpret geometrically the Riemann sum in Eq. (3). On each subinterval $[x_{i-1}, x_i]$, we have a rectangle with width Δx_i and "height" $f(x_i^*)$. If $f(x_i^*) > 0$, then this rectangle stands *above* the x-axis; if $f(x_i^*) < 0$, it lies *below* the x-axis. The Riemann sum R is then the sum of the **signed** areas of these rectangles—that is, the sum of the areas of those rectangles that lie above the x-axis *minus* the sum of the areas of those that lie below the x-axis.

If the widths Δx_i of these rectangles are all very small—that is, if the mesh $|P|$ is small—then it appears that the Riemann sum R will closely approximate the area from a to b under $y = f(x)$ above the x-axis minus the area below the x-axis. This suggests that the integral of f from a to b should be defined by taking the limit of the Riemann sums as the mesh $|P|$ approaches zero:

$$I = \lim_{|P| \to 0} \sum_{i=1}^{n} f(x_i^*)\, \Delta x_i. \tag{7}$$

The formal definition of the integral is obtained by saying precisely what it means for this limit to exist. Briefly, it means that if $|P|$ is sufficiently small, then *all* Riemann sums associated with P are very close to the number I.

Definition *The Definite Integral*

The **definite integral of the function f from a to b** is the number

$$I = \lim_{|P| \to 0} \sum_{i=1}^{n} f(x_i^*)\, \Delta x_i, \tag{8}$$

provided that this limit exists, in which case we say that f is **integrable** on $[a, b]$. Equation (8) means that, for each number $\epsilon > 0$, there exists a number $\delta > 0$ such that

$$\left| I - \sum_{i=1}^{n} f(x_i^*)\, \Delta x_i \right| < \epsilon$$

for every Riemann sum associated with any partition P of $[a, b]$ for which $|P| < \delta$.

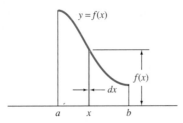

Fig. 5.4.7 Origin of Leibniz's notation for the integral

The customary notation for the integral of f from a to b, due to the German mathematician and philosopher G. W. Leibniz, is

$$I = \int_a^b f(x)\, dx = \lim_{|P| \to 0} \sum_{i=1}^n f(x_i^*)\, \Delta x_i. \tag{9}$$

Considering I to be the area under $y = f(x)$ from a to b, Leibniz first thought of a narrow strip with height $f(x)$ and "infinitesimally small" width dx (as in Fig. 5.4.7), so that its area would be the product $f(x)\, dx$. He regarded the integral as a sum of areas of such strips and denoted this sum by the elongated capital S (for *summa*) that appears as the integral sign in Eq. (9).

We shall see that this integral notation is not only highly suggestive, but also is exceedingly useful in manipulations with integrals. The numbers a and b are called the **lower limit** and **upper limit,** respectively, of the integral; they are the endpoints of the interval of integration. The function $f(x)$ that appears between the integral sign and dx is called the **integrand.** The symbol dx that follows the integrand in Eq. (9) should, for the time being, be thought of as simply an indication of what the independent variable is. Like the index of summation, the independent variable x is a "dummy variable"—it may be replaced by any other variable without affecting the meaning of Eq. (9). Thus if f is integrable on $[a, b]$, we can write

$$\int_a^b f(x)\, dx = \int_a^b f(t)\, dt = \int_a^b f(u)\, du.$$

The definition given of the definite integral applies only if $a < b$, but it is convenient to include the cases $a = b$ and $a > b$ as well. The integral is *defined* in these cases as follows:

$$\int_a^a f(x)\, dx = 0 \tag{10}$$

and

$$\int_a^b f(x)\, dx = -\int_b^a f(x)\, dx, \tag{11}$$

provided that the right-hand integral exists. Thus *interchanging the limits of integration reverses the sign of the integral.*

Just as not all functions are differentiable, not every function is integrable. Suppose that c is a point of $[a, b]$ such that $f(x) \to +\infty$ as $x \to c$. If $[x_{k-1}, x_k]$ is the subinterval of the partition P that contains c, then the Riemann sum in Eq. (3) can be made arbitrarily large by choosing x_k^* to be sufficiently close to c. For our purposes, however, we need to know only that every continuous function is integrable. The following theorem is proved in Appendix E.

Theorem 1   *Existence of the Integral*
If the function f is continuous on $[a, b]$, then f is integrable on $[a, b]$.

Although we omit the details, it is not difficult to show that the definition of the integral can be reformulated in terms of sequences of Riemann sums, as follows.

Theorem 2   *The Integral as a Limit of a Sequence*
The function f is integrable on $[a, b]$ with integral I if and only if

$$\lim_{n \to \infty} R_n = I \tag{12}$$

for every sequence $\{R_n\}_1^\infty$ of Riemann sums associated with a sequence of partitions $\{P_n\}_1^\infty$ of $[a, b]$ such that $|P_n| \to 0$ as $n \to +\infty$.

This reformulation of the definition of the integral is advantageous because it is easier to visualize a specific sequence of Riemann sums than to visualize the vast totality of all possible Riemann sums. In the case of a continuous function f (known to be integrable by Theorem 1), the situation can be simplified even more by using only Riemann sums associated with partitions consisting of subintervals all with the same length

$$\Delta x_1 = \Delta x_2 = \cdots = \Delta x_n = \frac{b - a}{n} = \Delta x.$$

Such a partition of $[a, b]$ into equal-length subintervals is called a **regular partition** of $[a, b]$.

Any Riemann sum associated with a regular partition can be written in the form

$$\sum_{i=1}^{n} f(x_i^*) \, \Delta x, \tag{13}$$

where the absence of a subscript in Δx signifies that the sum is associated with a regular partition. In such a case, the conditions $|P| \to 0$, $\Delta x \to 0$, and $n \to +\infty$ are equivalent, so the integral of a *continuous* function can be defined quite simply:

$$\int_a^b f(x) \, dx = \lim_{n \to \infty} \sum_{i=1}^{n} f(x_i^*) \, \Delta x = \lim_{\Delta x \to 0} \sum_{i=1}^{n} f(x_i^*) \, \Delta x. \tag{14}$$

Therefore, because we are concerned for the most part only with integrals of *continuous* functions, in our subsequent discussions we will employ only regular partitions.

EXAMPLE 3   Use Riemann sums to compute

$$\int_a^b x \, dx,$$

where $a < b$.

Solution   We take $f(x) = x$ and $x_i^* = x_i$, where

$$\Delta x = \frac{b - a}{n} \quad \text{and} \quad x_i = a + i \cdot \Delta x.$$

The Riemann sum in Eq. (13) is then

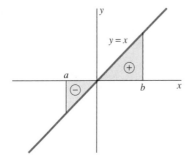

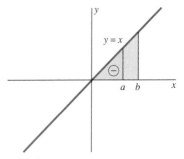

Fig. 5.4.8 Example 3 with $a < 0 < b$ **Fig. 5.4.9** Example 3 with $0 < a < b$

$$\sum_{i=1}^{n} f(x_i) \, \Delta x = \sum_{i=1}^{n} (a + i \cdot \Delta x) \, \Delta x = (a \Delta x) \sum_{i=1}^{n} 1 + (\Delta x)^2 \sum_{i=1}^{n} i$$

$$= a \cdot \frac{b - a}{n} \cdot n + \left(\frac{b - a}{n}\right)^2 \cdot \frac{n}{2} \cdot (n + 1) \quad \text{[using Eqs. (6) and (7) of Section 5.3]}$$

$$= a \cdot (b - a) + (b - a)^2 \cdot \left(\frac{1}{2} + \frac{1}{2n}\right).$$

Because $1/(2n) \to 0$ as $n \to +\infty$, it follows that

$$\int_{a}^{b} x \, dx = a \cdot (b - a) + \tfrac{1}{2} \cdot (b - a)^2$$

$$= (b - a) \cdot (a + \tfrac{1}{2}b - \tfrac{1}{2}a) = \tfrac{1}{2}b^2 - \tfrac{1}{2}a^2. \quad (15)$$

Figures 5.4.8 and 5.4.9 illustrate two of the cases in Example 3. In each case Eq. (15) agrees with the sum of the indicated *signed* areas. The minus sign in Fig. 5.4.8 represents the fact that area beneath the *x*-axis is measured with a negative number.

5.4 Problems

In Problems 1 through 10, express the given limit as a definite integral over the indicated interval [a, b]. Assume that $[x_{i-1}, x_i]$ denotes the ith subinterval of a subdivision of [a, b] into n subintervals, all with the same length $\Delta x = (b - a)/n$, and that $m_i = \frac{1}{2}(x_{i-1} + x_i)$ is the midpoint of the ith subinterval.

1. $\lim_{n \to \infty} \sum_{i=1}^{n} (2x_i - 1) \, \Delta x$ over $[1, 3]$

2. $\lim_{n \to \infty} \sum_{i=1}^{n} (2 - 3x_{i-1}) \, \Delta x$ over $[-3, 2]$

3. $\lim_{n \to \infty} \sum_{i=1}^{n} (x_i^2 + 4) \, \Delta x$ over $[0, 10]$

4. $\lim_{n \to \infty} \sum_{i=1}^{n} (x_i^3 - 3x_i^2 + 1) \, \Delta x$ over $[0, 3]$

5. $\lim_{n \to \infty} \sum_{i=1}^{n} \sqrt{m_i} \, \Delta x$ over $[4, 9]$

6. $\lim_{n \to \infty} \sum_{i=1}^{n} \sqrt{25 - x_i^2} \, \Delta x$ over $[0, 5]$

7. $\lim_{n \to \infty} \sum_{i=1}^{n} \frac{1}{\sqrt{1 + m_i}} \, \Delta x$ over $[3, 8]$

8. $\lim_{n \to \infty} \sum_{i=1}^{n} (\cos 2x_{i-1}) \, \Delta x$ over $[0, \pi/2]$

9. $\lim_{n \to \infty} \sum_{i=1}^{n} (\sin 2\pi m_i) \, \Delta x$ over $[0, 1/2]$

10. $\lim_{n \to \infty} \sum_{i=1}^{n} e^{2x_i} \, \Delta x$ over $[0, 1]$

In Problems 11 through 20, compute the Riemann sum

$$\sum_{i=1}^{n} f(x_i^*) \, \Delta x$$

for the indicated function and a regular partition of the

given interval into n subintervals. Use $x_i^* = x_i$, the right-hand endpoint of the ith subinterval $[x_{i-1}, x_i]$.

11. $f(x) = x^2$ on $[0, 1]$; $n = 5$

12. $f(x) = x^3$ on $[0, 1]$; $n = 5$

13. $f(x) = \dfrac{1}{x}$ on $[1, 6]$; $n = 5$

14. $f(x) = \sqrt{x}$ on $[0, 5]$; $n = 5$

15. $f(x) = 2x + 1$ on $[1, 4]$; $n = 6$

16. $f(x) = x^2 + 2x$ on $[1, 4]$; $n = 6$

17. $f(x) = x^3 - 3x$ on $[1, 4]$; $n = 5$

18. $f(x) = 1 + 2\sqrt{x}$ on $[2, 3]$; $n = 5$

19. $f(x) = \cos x$ on $[0, \pi]$; $n = 6$

20. $f(x) = \sin \pi x$ on $[0, 1]$; $n = 6$

21 through 30. Repeat Problems 11 through 20, except with $x_i^* = x_{i-1}$, the left-hand endpoint.

31 through 40. Repeat Problems 11 through 20, except with $x_i^* = (x_{i-1} + x_i)/2$, the midpoint of the ith subinterval.

41. Work Problem 13 with $x_i^* = (3x_{i-1} + 2x_i)/5$.

42. Work Problem 14 with $x_i^* = (x_{i-1} + 2x_i)/3$.

In Problems 43 through 48, evaluate the given integral by computing

$$\lim_{n \to \infty} \sum_{i=1}^{n} f(x_i) \, \Delta x$$

for a regular partition of the interval of integration.

43. $\displaystyle\int_0^2 x^2 \, dx$ **44.** $\displaystyle\int_0^4 x^3 \, dx$

45. $\displaystyle\int_0^3 (2x + 1) \, dx$ **46.** $\displaystyle\int_1^5 (4 - 3x) \, dx$

47. $\displaystyle\int_0^3 (3x^2 + 1) \, dx$ **48.** $\displaystyle\int_0^4 (x^3 - x) \, dx$

49. Show by the method of Example 3 that

$$\int_0^b x \, dx = \tfrac{1}{2} b^2$$

if $b > 0$.

50. Show by the method of Example 3 that

$$\int_0^b x^3 \, dx = \tfrac{1}{4} b^4$$

if $b > 0$.

51. Let $f(x) = x$, and let $\{x_0, x_1, x_2, \ldots, x_n\}$ be an arbitrary partition of the closed interval $[a, b]$. For each i $(1 \leq i \leq n)$, let $x_i^* = (x_{i-1} + x_i)/2$. Then show that

$$\sum_{i=1}^{n} x_i^* \, \Delta x_i = \tfrac{1}{2} b^2 - \tfrac{1}{2} a^2.$$

Explain why this computation proves that

$$\int_a^b x \, dx = \frac{b^2 - a^2}{2}.$$

52. Suppose that f is a function continuous on $[a, b]$ and that k is a constant. Use Riemann sums to prove that

$$\int_a^b kf(x) \, dx = k \int_a^b f(x) \, dx.$$

53. Suppose that $f(x) \equiv c$, a constant. Use Riemann sums to prove that

$$\int_a^b c \, dx = c(b - a).$$

[*Suggestion:* First consider the case $a < b$.]

5.4 Projects

These projects require the use of a programmable calculator or computer. It is not otherwise practical to calculate Riemann sums with large numbers of subintervals.

The TI calculator and BASIC programs listed in Fig. 5.4.10 can be used to approximate the integral

$$\int_a^b f(x) \, dx$$

by means of a Riemann sum corresponding to a subdivision of the interval $[a, b]$ into n subintervals, each with the same length $h = \Delta x$.

For the TI calculator program, you must define the desired function $f(x)$ (as `Y1`) in the "Y=" menu before the program is executed. For the BASIC version, you define the function in the first line of the program itself. You

TI-81/85	BASIC	Comments
`PRGM1:RIEMANN`	`DEF FN F(X) = X*X`	Define Y1 = $f(x)$
`:Disp "A"`		
`:Input A`	`Input "A"; A`	Left endpoint
`:Disp "B"`		
`:Input B`	`Input "B"; B`	Right endpoint
`:Disp "N"`		
`:Input N`	`Input "N"; N`	Number of subintervals
`:(B—A)/N→H`	`H = (B − A)/N`	Subinterval length
`:0→I`		Initialize counter I
`:0→S`	`S = 0`	Initialize sum S
`:A+H→X`	`X = A + H`	Initial point
`:Lbl 1`	`FOR I = 1 TO N`	Begin loop
`:S+H*Y1→S`	`S = S + H*FNF(X)`	Add next term
`:X+H→X`	`X = X + H`	Update X
`:I+1→I`	`NEXT`	Next I
`:If I<N`		End loop
`:Goto 1`		
`:Disp "N="`	`PRINT "N = ";N`	Display N
`:Disp N`		
`:Disp "SUM="`	`PRINT "Sum = ";S`	Display sum
`:Disp S`		
`:End`	`END`	

Fig. 5.4.10 TI-81/85 and BASIC programs for calculating Riemann sums

enter the lower and upper limits a and b of the integral and the desired number n of subintervals while the program is running.

The "initial point" line of the program determines which type of Riemann sum is calculated:

`A→X`	left-endpoint sum
`A+H→X`	right-endpoint sum (as above)
`A+H/2→X`	midpoint sum

Most computer algebra systems include a SUM function that can be used to calculate Riemann sums directly. Suppose that the integrand function $f(x)$ has already been defined appropriately. Then the single-line commands listed in Fig. 5.4.11 all calculate (in various systems) the right-endpoint sum with n subintervals. The alterations that produce left-endpoint and midpoint sums should be apparent.

Whatever type of calculator or computer you use, reading and comparing the various programs and commands in Figs. 5.4.10 and 5.4.11 may give you additional insight into computational algorithms for approximating integrals.

Computer	Command
Derive	`SUM( f(a + (b − a)*i/n), i, 1, n )*(b − a)/n`
Maple	`sum(f(a + (b − a)*i/n), i= 1..n)*(b − a)/n`
Mathematica	`Sum[ f[x], {x, a + (b − a)/n, b, (b − a)/n} ]*(b − a)/n`
(X)PLORE	`SUM( f(x), x = a + (b − a)/n to b step (b − a)/n)*(b − a)/n`

Fig. 5.4.11 Computer algebra system commands for approximating the integral of f from $x = a$ to $x = b$ with n subintervals

Ch. 5 / The Integral

PROJECT A Verify the approximations to the integral

$$\int_0^3 x^2\, dx$$

that are listed in Fig. 5.3.10. Include also a column of midpoint-sum approximations. Which appears to give a more accurate approximation to this integral for a given number of subintervals: the average of the left- and right-endpoint sums or the midpoint sum?

PROJECT B Use midpoint sums and left- and right-endpoint sum averages to corroborate the claim that

$$\int_0^\pi \sin x\, dx = 2.$$

PROJECT C First explain why Fig. 5.4.12 and the circle area formula $A = \pi r^2$ imply that

$$\int_0^1 4\sqrt{1 - x^2}\, dx = \pi.$$

Then use midpoint-sum approximations to this sum to approximate the number π. Begin with $n = 25$ subintervals, then successively double n. How large must n be for you to obtain the familiar approximation $\pi \approx 3.1416$?

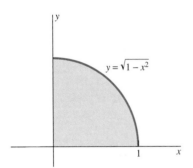

Fig. 5.4.12 Project C

5.5

Evaluation of Integrals

The evaluation of integrals by using Riemann sums, as in Section 5.4, is tedious and time-consuming. Fortunately, we will seldom find it necessary to evaluate an integral in this way. In 1666, Isaac Newton, while still a student at Cambridge University, discovered a much more efficient way to evaluate an integral. A few years later, Gottfried Wilhelm Leibniz, working with a different approach, discovered this method independently.

Newton's key idea was that to evaluate the *number*

$$\int_a^b f(x)\, dx,$$

we should first introduce the *function $A(x)$* defined as follows:

$$A(x) = \int_a^x f(t)\, dt. \tag{1}$$

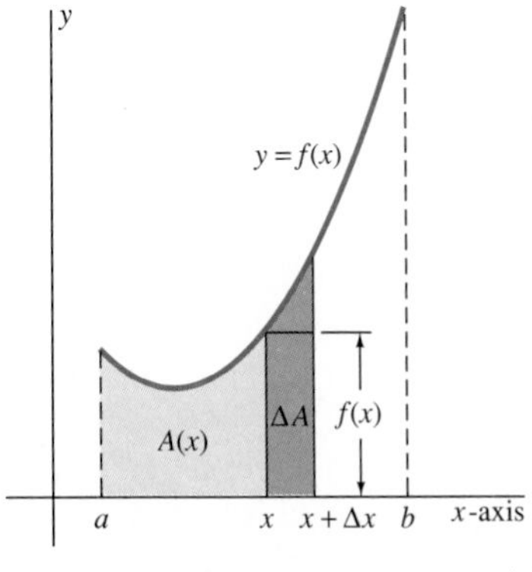

The independent variable x appears in the *upper limit* of the integral in Eq. (1); the dummy variable t is used in the integrand merely to avoid confusion. If f is positive-valued, continuous, and $x > a$, then $A(x)$ is the area below the curve $y = f(x)$ above the interval $[a, x]$ (Fig. 5.5.1).

It is apparent from Fig. 5.5.1 that $A(x)$ increases as x increases. When x increases by Δx, A increases by the area ΔA of the narrow strip in Fig. 5.5.1 with base $[x, x + \Delta x]$. If Δx is very small, the area of this strip is very close to the area $f(x)\,\Delta x$ of the rectangle with base $[x, x + \Delta x]$ and height $f(x)$. Thus

$$\Delta A \approx f(x)\,\Delta x; \qquad \frac{\Delta A}{\Delta x} \approx f(x). \tag{2}$$

Fig. 5.5.1 The area function $A(x)$

Moreover, the figure makes it plausible that we get equality in the limit as $\Delta x \to 0$:

$$\frac{dA}{dx} = \lim_{\Delta x \to 0} \frac{\Delta A}{\Delta x} = f(x).$$

That is,

$$A'(x) = f(x), \tag{3}$$

so *the derivative of the area function $A(x)$ is the curve's height function $f(x)$.* In other words, Eq. (3) implies that $A(x)$ is an *antiderivative* of $f(x)$.

Figure 5.5.2 shows a physical interpretation of Eq. (3). A paint roller is laying down a 1-mm-thick coat of paint to cover the region under the curve $y = f(t)$. The paint roller is of adjustable length—as it rolls with a speed of 1 mm/s from left to right, one end traces the x-axis and the other end traces the curve $y = f(t)$. At any time t, the volume V of paint the roller has laid down equals the area of the region already painted:

$$V = A(t) \qquad (\text{mm}^3).$$

Then Eq. (3) yields

$$\frac{dV}{dt} = A'(t) = f(t).$$

Thus the instantaneous rate at which the roller is depositing paint is equal to the current length of the roller.

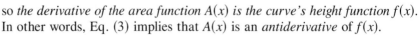

Fig. 5.5.2 The adjustable-length paint roller

THE EVALUATION THEOREM

Equation (3) says that the area function $A(x)$ illustrated in Fig. 5.5.1 is *one* antiderivative of the given function $f(x)$. Now suppose that $G(x)$ is any other antiderivative of $f(x)$—perhaps one found by the methods of Section 5.2. Then

$$A(x) = G(x) + C, \tag{4}$$

because (second corollary to the mean value theorem) two antiderivatives of the same function (on an interval) can differ only by a constant. Also,

$$A(a) = \int_a^a f(t)\, dt = 0 \tag{5}$$

and

$$A(b) = \int_a^b f(t)\, dt = \int_a^b f(x)\, dx \tag{6}$$

by Eq. (1). So it follows that

$$\int_a^b f(x)\, dx = A(b) - A(a) = [G(b) + C] - [G(a) + C],$$

and thus

$$\int_a^b f(x)\, dx = G(b) - G(a). \tag{7}$$

Our intuitive discussion has led us to the statement of Theorem 1.

> **Theorem 1   Evaluation of Integrals**
> If G is an antiderivative of the continuous function f on the interval $[a, b]$, then
> $$\int_a^b f(x)\, dx = G(b) - G(a). \tag{7}$$

In Section 5.6 we will fill in the details of the preceding discussion, thus giving a rigorous proof of Theorem 1 (which is part of the fundamental theorem of calculus). Here we concentrate on the computational applications of this theorem. The difference $G(b) - G(a)$ is customarily abbreviated as $[G(x)]_a^b$, so Theorem 1 implies that

$$\int_a^b f(x)\, dx = \left[G(x) \right]_a^b = G(b) - G(a) \tag{8}$$

if G is any antiderivative of the continuous function f on the interval $[a, b]$. Thus if we can find an antiderivative G of f, we can quickly evaluate the integral *without* having to resort to the paraphernalia of limits of Riemann sums.

If $G'(x) = f(x)$, then (as in Section 5.2) we write

$$\int f(x)\, dx = G(x) + C \tag{9}$$

for the indefinite integral of f. With the indefinite integral $\int f(x)\, dx$ in place of the antiderivative $G(x)$, Eq. (8) takes the form

$$\int_a^b f(x)\, dx = \left[\int f(x)\, dx \right]_a^b. \tag{10}$$

This is the connection between the indefinite integral and the definite integral to which we have alluded in the earlier sections of Chapter 5.

EXAMPLE 1   Because

$$\int x^n\, dx = \frac{x^{n+1}}{n + 1} + C \qquad (\text{if } n \neq -1),$$

it follows that

$$\int_a^b x^n\, dx = \left[\frac{x^{n+1}}{n + 1} \right]_a^b = \frac{b^{n+1} - a^{n+1}}{n + 1}.$$

For instance,

$$\int_0^3 x^2\, dx = \left[\tfrac{1}{3} x^3 \right]_0^3 = \tfrac{1}{3} \cdot 3^3 - \tfrac{1}{3} \cdot 0^3 = 9.$$

Contrast the immediacy of this result with the complexity of the computations of Example 6 in Section 5.3.

EXAMPLE 2   Because

$$\int \cos x\, dx = \sin x + C,$$

it follows that

$$\int_a^b \cos x \, dx = \left[\sin x \right]_a^b = \sin b - \sin a.$$

Similarly,

$$\int_a^b \sin x \, dx = \left[-\cos x \right]_a^b = \cos a - \cos b.$$

For instance, as we mentioned in Example 2 of Section 5.4,

$$\int_0^\pi \sin x \, dx = \left[-\cos x \right]_0^\pi = (-\cos \pi) - (-\cos 0) = (+1) - (-1) = 2.$$

EXAMPLE 3

$$\int_0^2 x^5 \, dx = \left[\tfrac{1}{6} x^6 \right]_0^2 = \tfrac{64}{6} - 0 = \tfrac{32}{3}.$$

$$\int_1^9 (2x - x^{-1/2} - 3) \, dx = \left[x^2 - 2x^{1/2} - 3x \right]_1^9 = 52.$$

$$\int_0^1 (2x + 1)^3 \, dx = \left[\tfrac{1}{8} (2x + 1)^4 \right]_0^1 = \tfrac{1}{8} \cdot (81 - 1) = 10.$$

$$\int_0^{\pi/2} \sin 2x \, dx = \left[-\tfrac{1}{2} \cos 2x \right]_0^{\pi/2} = -\tfrac{1}{2} (\cos \pi - \cos 0) = 1.$$

$$\int_0^1 e^{2x} \, dx = \left[\tfrac{1}{2} e^{2x} \right]_0^1 = \tfrac{1}{2} (e^2 - 1).$$

We have not shown the details of finding the antiderivatives, but you can (and should) check each of these results by showing that the derivative of the function within the brackets on the right is equal to the integrand on the left. In Example 4 we show the details.

EXAMPLE 4   Evaluate $\int_1^5 \sqrt{3x + 1} \, dx.$

Solution   We apply the antiderivative form of the generalized power rule,

$$\int u^k \, du = \frac{u^{k+1}}{k + 1} + C,$$

with

$$u = 3x + 1, \qquad du = 3 \, dx.$$

This gives

$$\int (3x + 1)^{1/2} \, dx = \frac{1}{3} \int (3x + 1)^{1/2} (3 \, dx) = \frac{1}{3} \int u^{1/2} \, du$$

$$= \frac{1}{3} \cdot \frac{u^{3/2}}{\frac{3}{2}} + C = \frac{2}{9} (3x + 1)^{1/2} + C$$

for the indefinite integral, so it follows from Eq. (10) that

$$\int_1^5 \sqrt{3x+1}\, dx = \left[\tfrac{2}{9}(3x+1)^{3/2} \right]_1^5$$

$$= \tfrac{2}{9}(16^{3/2} - 4^{3/2}) = \tfrac{2}{9}(4^3 - 2^3) = \tfrac{112}{9}.$$

If the derivative $F'(x)$ of the function $F(x)$ is continuous, then the evaluation theorem, with $F'(x)$ in place of $f(x)$ and $F(x)$ in place of $G(x)$, yields

$$\int_a^b F'(x)\, dx = \left[F(x) \right]_a^b = F(b) - F(a). \qquad (11)$$

Here is an immediate application.

EXAMPLE 5 Suppose that an animal population $P(t)$ initially numbers $P(0) = 100$ and that its rate of growth after t months is given by

$$P'(t) = 10 + t + (0.06)t^2.$$

What is the population after 10 months?

Solution By Eq. (11), we know that

$$P(10) - P(0) = \int_0^{10} P'(t)\, dt = \int_0^{10} [10 + t + (0.06)t^2]\, dt$$

$$= \left[10t + \tfrac{1}{2}t^2 + (0.02)t^3 \right]_0^{10} = 170.$$

Thus $P(10) = 100 + 170 = 270$ individuals.

EXAMPLE 6 Evaluate

$$\lim_{n \to \infty} \sum_{i=1}^n \frac{2i}{n^2}$$

by recognizing this limit as the value of an integral.

Solution If we write

$$\sum_{i=1}^n \frac{2i}{n^2} = \sum_{i=1}^n \left(\frac{2i}{n} \right)\left(\frac{1}{n} \right),$$

we recognize that we have a Riemann sum for the function $f(x) = 2x$ associated with a partition of the interval $[0, 1]$ into n equal-length subintervals. The ith point of subdivision is $x_i = i/n$, and $\Delta x = 1/n$. Hence it follows from the definition of the integral and from the evaluation theorem that

$$\lim_{n \to \infty} \sum_{i=1}^n \frac{2i}{n^2} = \lim_{n \to \infty} \sum_{i=1}^n 2x_i\, \Delta x = \lim_{n \to \infty} \sum_{i=1}^n f(x_i)\, \Delta x$$

$$= \int_0^1 f(x)\, dx = \int_0^1 2x\, dx.$$

Therefore,

$$\lim_{n \to \infty} \sum_{i=1}^n \frac{2i}{n^2} = \left[x^2 \right]_0^1 = 1.$$

BASIC PROPERTIES OF INTEGRALS

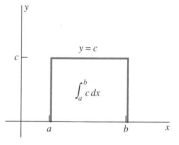

Fig. 5.5.3 The integral of a constant is the area of a rectangle.

Problems 44 through 46 outline elementary proofs of the properties stated next. We assume throughout that f is integrable on $[a, b]$.

Integral of a Constant

$$\int_a^b c \, dx = c(b - a).$$

This property is intuitively obvious because the area represented by the integral is simply a rectangle with base $b - a$ and height c (Fig. 5.5.3).

Constant Multiple Property

$$\int_a^b cf(x) \, dx = c \int_a^b f(x) \, dx.$$

Thus a constant can be "moved across" the integral sign. For example,

$$\int_0^{\pi/2} 2 \sin x \, dx = 2 \int_0^{\pi/2} \sin x \, dx = 2 \left[-\cos x \right]_0^{\pi/2} = 2.$$

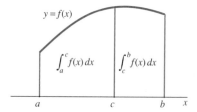

Fig. 5.5.4 The way the interval union property works

Interval Union Property
If $a < c < b$, then

$$\int_a^b f(x) \, dx = \int_a^c f(x) \, dx + \int_c^b f(x) \, dx.$$

Figure 5.5.4 indicates the plausibility of the interval union property.

EXAMPLE 7 If $f(x) = 2|x|$, then

$$f(x) = \begin{cases} -2x & \text{if } x \leqq 0, \\ 2x & \text{if } x \geqq 0. \end{cases}$$

The graph of f is shown in Fig. 5.5.5. An antiderivative of $f(x)$ is not evident, but the interval union property allows us to split the integral into two easily calculated integrals:

$$\int_{-1}^3 2|x| \, dx = \int_{-1}^0 (-2x) \, dx + \int_0^3 (2x) \, dx$$

$$= \left[-x^2 \right]_{-1}^0 + \left[x^2 \right]_0^3 = [0 - (-1)] + [9 - 0] = 10.$$

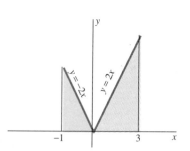

Fig. 5.5.5 The area under the graph of $y = 2|x|$

Does the result agree with Fig. 5.5.5?

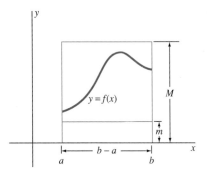

Fig. 5.5.6 Plausibility of the comparison property

> **Comparison Property**
> If $m \leqq f(x) \leqq M$ for all x in $[a, b]$, then
> $$m(b - a) \leqq \int_a^b f(x)\, dx \leqq M(b - a).$$

The plausibility of this property is indicated in Fig. 5.5.6. Note that m and M need not necessarily be the minimum and maximum values of $f(x)$ on $[a, b]$.

EXAMPLE 8 If $2 \leqq x \leqq 3$, then
$$5 \leqq x^2 + 1 \leqq 10,$$
so
$$m = \frac{1}{10} \leqq \frac{1}{x^2 + 1} \leqq \frac{1}{5} = M.$$

Hence the comparison property yields
$$\frac{1}{10} \leqq \int_2^3 \frac{1}{x^2 + 1}\, dx \leqq \frac{1}{5},$$
as illustrated in Fig. 5.5.7.

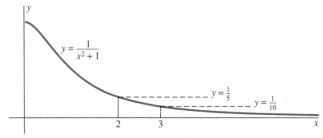

Fig. 5.5.7 The area under $y = \dfrac{1}{x^2 + 1}$ (Example 8)

The properties of integrals stated here are frequently used in computing and will be applied in the proof of the fundamental theorem of calculus in Section 5.6.

5.5 Problems

Apply the evaluation theorem to evaluate the integrals in Problems 1 through 36.

1. $\displaystyle\int_0^1 (3x^2 + 2\sqrt{x} + 3\sqrt[3]{x})\, dx$

2. $\displaystyle\int_1^3 \frac{6}{x^2}\, dx$

3. $\displaystyle\int_0^1 x^3(1 + x)^2\, dx$

4. $\displaystyle\int_{-2}^{-1} \frac{1}{x^4}\, dx$

5. $\displaystyle\int_0^1 (x^4 - x^3)\, dx$

6. $\displaystyle\int_1^2 (x^4 - x^3)\, dx$

7. $\displaystyle\int_{-1}^0 (x + 1)^3\, dx$

8. $\displaystyle\int_1^3 \frac{x^4 + 1}{x^2}\, dx$

9. $\displaystyle\int_0^4 \sqrt{x}\, dx$

10. $\displaystyle\int_1^4 \frac{1}{\sqrt{x}}\, dx$

11. $\displaystyle\int_{-1}^2 (3x^2 + 2x + 4)\, dx$

12. $\displaystyle\int_0^1 x^{99}\, dx$

13. $\displaystyle\int_{-1}^1 x^{99}\, dx$

14. $\displaystyle\int_0^4 (7x^{5/2} - 5x^{3/2})\, dx$

15. $\displaystyle\int_1^3 (x - 1)^5\, dx$

16. $\displaystyle\int_1^2 (x^2 + 1)^3\, dxw$

17. $\displaystyle\int_{-1}^0 (2x + 1)^3\, dx$

18. $\displaystyle\int_1^3 \frac{10}{(2x + 3)^2}\, dx$

19. $\displaystyle\int_1^8 x^{2/3}\,dx$

20. $\displaystyle\int_1^9 (1 + \sqrt{x})^2\,dx$

21. $\displaystyle\int_{-1}^1 (e^x - e^{-x})\,dx$

22. $\displaystyle\int_0^4 \sqrt{3t}\,dt$

23. $\displaystyle\int_0^2 \sqrt{e^{3t}}\,dt$

24. $\displaystyle\int_2^3 \frac{du}{u^2}$ $\left(\text{Note the abbreviation for } \dfrac{1}{u^2}\,du.\right)$

25. $\displaystyle\int_1^2 \frac{1}{t}\,dt$

26. $\displaystyle\int_5^{10} \frac{1}{x}\,dx$

27. $\displaystyle\int_0^1 (e^x - 1)^2\,dx$

28. $\displaystyle\int_0^{\pi/2} \cos 2x\,dx$

29. $\displaystyle\int_0^{\pi/4} \sin x \cos x\,dx$

30. $\displaystyle\int_0^{\pi} \sin^2 x \cos x\,dx$

31. $\displaystyle\int_0^{\pi} \sin 5x\,dx$

32. $\displaystyle\int_0^2 \cos \pi t\,dt$

33. $\displaystyle\int_0^{\pi/2} \cos 3x\,dx$

34. $\displaystyle\int_0^5 \sin \frac{\pi x}{10}\,dx$

35. $\displaystyle\int_0^2 \cos \frac{\pi x}{4}\,dx$

36. $\displaystyle\int_0^{\pi/8} \sec^2 2t\,dt$

In Problems 37 through 42, evaluate the given limit by first recognizing the indicated sum as a Riemann sum associated with a regular partition of [0, 1] and then evaluating the corresponding integral.

37. $\displaystyle\lim_{n\to\infty} \sum_{i=1}^n \left(\frac{2i}{n} - 1\right)\frac{1}{n}$

38. $\displaystyle\lim_{n\to\infty} \sum_{i=1}^n \frac{i^2}{n^3}$

39. $\displaystyle\lim_{n\to\infty} \frac{1 + 2 + 3 + \cdots + n}{n^2}$

40. $\displaystyle\lim_{n\to\infty} \frac{1^3 + 2^3 + 3^3 + \cdots + n^3}{n^4}$

41. $\displaystyle\lim_{n\to\infty} \frac{\sqrt{1} + \sqrt{2} + \sqrt{3} + \cdots + \sqrt{n}}{n\sqrt{n}}$

42. $\displaystyle\lim_{n\to\infty} \sum_{i=1}^n \frac{1}{n} \sin \frac{\pi i}{n}$

43. Evaluate the integral $\displaystyle\int_0^5 \sqrt{25 - x^2}\,dx$ by interpreting it as the area under the graph of a certain function.

44. Use sequences of Riemann sums to establish the constant multiple property.

45. Use sequences of Riemann sums to establish the interval union property of the integral. Note that if R_n' and R_n'' are Riemann sums for f on the intervals $[a, c]$ and $[c, b]$, respectively, then $R_n = R_n' + R_n''$ is a Riemann sum for f on $[a, b]$.

46. Use Riemann sums to establish the comparison property for integrals. Show first that if $m \leq f(x) \leq M$ for all x in $[a, b]$, then $m(b - a) \leq R \leq M(b - a)$ for every Riemann sum R for f on $[a, b]$.

47. Suppose that a tank initially contains 1000 gal of water and that the rate of change of its volume after the tank drains for t min is $V'(t) = (0.8)t - 40$ (in gallons per minute). How much water does the tank contain after it has been draining for a half-hour?

48. Suppose that the population of Juneau in 1970 was 125 (in thousands) and that its rate of growth t years later was $P'(t) = 8 + (0.5)t + (0.03)t^2$ (in thousands per year). What was its population in 1990?

49. Find a lower and an upper bound for
$$\int_1^2 \frac{1}{x}\,dx$$
by using the comparison property and a subdivision of $[1, 2]$ into five equal-length subintervals.

50. Find a lower and an upper bound for
$$\int_0^1 \frac{1}{x^2 + 1}\,dx$$
by using the comparison property and a subdivision of $[0, 1]$ into five equal-length subintervals.

5.6

Average Values and the Fundamental Theorem of Calculus

Newton and Leibniz are generally credited with the invention of calculus in the latter part of the seventeenth century. Actually, others had earlier calculated areas essentially equivalent to integrals and tangent line slopes essentially equivalent to derivatives. The great accomplishments of Newton and Leibniz were the discovery and computational exploitation of the inverse relationship between differentiation and integration. This relationship is embodied in the **fundamental theorem of calculus**. One part of this theorem is the evaluation theorem of Section 5.5: To evaluate

$$\int_a^b f(x)\,dx,$$

it suffices to find an antiderivative of f on $[a, b]$. The other part of the fundamental theorem tells us that doing so is usually possible, at least in theory: Every continuous function has an antiderivative.

THE AVERAGE VALUE OF A FUNCTION

The concept of the *average value* of a function is useful for the proof of the fundamental theorem and has numerous important applications in its own right. The ordinary (arithmetic) **average** of n given numbers $a_1, a_2, \ldots, a_n$ is defined to be

$$\bar{a} = \frac{a_1 + a_2 + \cdots + a_n}{n} = \frac{1}{n} \sum_{i=1}^{n} a_i. \tag{1}$$

But a function f defined on an interval generally has "infinitely many" values $f(x)$, so we cannot simply divide the sum of all these values by their number to find the average value of $f(x)$. We introduce the proper notion by means of a discussion of average temperatures.

EXAMPLE 1 Let the measured temperature T during a particular 24-h day at a certain location be given by the function

$$T = f(t), \qquad 0 \leqq t \leqq 24$$

(with the 24-h clock running from $t = 0$ at one midnight to $t = 24$ at the following midnight). Thus the temperatures $f(1), f(2), \ldots, f(24)$ are recorded at 1-h intervals during the day. We might define the average temperature $\bar{T}$ for the day as the (ordinary arithmetic) average of the hourly temperatures:

$$\bar{T} = \frac{1}{24} \sum_{i=1}^{24} f(i) = \frac{1}{24} \sum_{i=1}^{24} f(t_i),$$

where $t_i = i$. If we divided the day into n equal subintervals rather than into 24 1-h intervals, we would obtain the more general average

$$\bar{T} = \frac{1}{n} \sum_{i=1}^{n} f(t_i).$$

The larger n is, the closer would we expect $\bar{T}$ to be to the "true" average temperature for the entire day. It is therefore plausible to define the true average temperature by letting n increase without bound. This gives

$$\bar{T} = \lim_{n \to \infty} \frac{1}{n} \sum_{i=1}^{n} f(t_i).$$

The right-hand side resembles a Riemann sum, and we can make it a Riemann sum by introducing the factor

$$\Delta t = \frac{b - a}{n},$$

where $a = 0$ and $b = 24$. Then

$$\bar{T} = \lim_{n \to \infty} \frac{1}{n} \cdot \frac{n}{b - a} \sum_{n=1}^{n} f(t_i) \cdot \frac{b - a}{n} = \lim_{n \to \infty} \frac{1}{b - a} \sum_{i=1}^{n} f(t_i) \cdot \frac{b - a}{n}$$

$$= \frac{1}{b - a} \lim_{n \to \infty} \sum_{i=1}^{n} f(t_i) \, \Delta t = \frac{1}{b - a} \int_{a}^{b} f(t) \, dt.$$

Thus

$$\overline{T} = \frac{1}{24} \int_0^{24} f(t)\, dt \qquad (2)$$

under the assumption that f is continuous, so the Riemann sums converge to the integral as $n \to \infty$.

The final result in Eq. (2) is the *integral of the function divided by the length of the interval.* Example 1 motivates the following definition.

> **Definition** *Average Value of a Function*
>
> Suppose that the function f is integrable on $[a, b]$. Then the **average value** $\overline{y}$ of $y = f(x)$ on $[a, b]$ is
>
> $$\overline{y} = \frac{1}{b-a} \int_a^b f(x)\, dx. \qquad (3)$$

We can rewrite Eq. (3) in the form

$$\int_a^b f(x)\, dx = \overline{y} \cdot (b - a). \qquad (4)$$

If f is positive-valued on $[a, b]$, then Eq. (4) implies that the area under $y = f(x)$ over $[a, b]$ is equal to the area of a rectangle with base length $b - a$ and height $\overline{y}$ (Fig. 5.6.1).

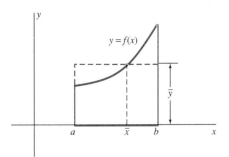

Fig. 5.6.1 Rectangle illustrating the average value of a function

EXAMPLE 2 The average value of $f(x) = x^2$ on $[0, 2]$ is

$$\overline{y} = \frac{1}{2} \int_0^2 x^2\, dx = \frac{1}{2} \left[\frac{1}{3} x^3 \right]_0^2 = \frac{4}{3}.$$

EXAMPLE 3 The mean daily temperature in degrees Fahrenheit in Athens, Georgia, t months after July 15 is closely approximated by

$$T = 61 + 18 \cos \frac{\pi t}{6} = f(t). \qquad (5)$$

Find the average temperature between September 15 ($t = 2$) and December 15 ($t = 5$).

Solution Equation (3) gives

$$\overline{T} = \frac{1}{5-2} \int_2^5 \left(61 + 18 \cos \frac{\pi t}{6} \right) dt = \frac{1}{3} \left[61t + \frac{6 \cdot 18}{\pi} \sin \frac{\pi t}{6} \right]_2^5 \approx 57°\text{F}.$$

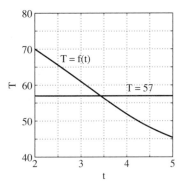

Fig. 5.6.2 The temperature function $T = f(t)$ of Example 3

Figure 5.6.2 shows the graphs of $T = f(t)$ and $T = 57$. Can you see that Eq. (4) implies that the two almost-triangular regions in the figure have equal areas?

Theorem 1 tells us that every continuous function on a closed interval *attains* its average value at some point of the interval.

Theorem 1 *Average Value Theorem*

If f is continuous on $[a, b]$, then

$$f(\bar{x}) = \frac{1}{b - a} \int_a^b f(x)\, dx \qquad (6)$$

for some number $\bar{x}$ in $[a, b]$.

Proof Let $m = f(c)$ be the minimum value of $f(x)$ on $[a, b]$ and let $M = f(d)$ be its maximum value there. Then, by the comparison property of Section 5.5,

$$m = f(c) \leqq \bar{y} = \frac{1}{b - a} \int_a^b f(x)\, dx \leqq f(d) = M.$$

Because f is continuous, we can now apply the intermediate value property. The number $\bar{y}$ is between the two values m and M of f, and consequently, $\bar{y}$ itself must be a value of f. Specifically, $\bar{y} = f(\bar{x})$ for some number $\bar{x}$ between a and b. This yields Eq. (6). ❑

EXAMPLE 4   If $v(t)$ denotes the velocity function of a sports car accelerating during the time interval $a \leqq t \leqq b$, then the car's average velocity is given by

$$\bar{v} = \frac{1}{b - a} \int_a^b v(t)\, dt.$$

The average value theorem implies that $\bar{v} = v(\bar{t})$ for some number $\bar{t}$ in $[a, b]$. Thus $\bar{t}$ is an instant at which the car's instantaneous velocity is equal to its average velocity over the entire time interval.

THE FUNDAMENTAL THEOREM

We state the fundamental theorem of calculus in two parts. The first part is the fact that every function f that is continuous on an interval I has an antiderivative on I. In particular, an antiderivative of f can be obtained by integrating f in a certain way. Intuitively, in the case $f(x) > 0$, we let $F(x)$ denote the area under the graph of f from a fixed point a of I to x, a point of I with $x > a$. We shall prove that $F'(x) = f(x)$. We show the construction of the function F in Fig. 5.6.3. More precisely, we define the function F as follows:

$$F(x) = \int_a^x f(t)\, dt,$$

where we use the dummy variable t in the integrand to avoid confusion with the upper limit x. The proof that $F'(x) = f(x)$ will be independent of the supposition that $x > a$.

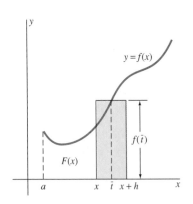

Fig. 5.6.3 The area function F is an antiderivative of f.

Proof of Part 1 By the definition of the derivative,

$$F'(x) = \lim_{h \to 0} \frac{F(x + h) - F(x)}{h} = \lim_{h \to 0} \frac{1}{h}\left(\int_a^{x+h} f(t)\, dt - \int_a^x f(t)\, dt \right).$$

But

$$\int_a^{x+h} f(t)\, dt = \int_a^x f(t)\, dt + \int_x^{x+h} f(t)\, dt$$

by the interval union property of Section 5.5. Thus

$$F'(x) = \lim_{h \to 0} \frac{1}{h} \int_x^{x+h} f(t)\, dt.$$

The average value theorem tells us that

$$\frac{1}{h} \int_x^{x+h} f(t)\, dt = f(\bar{t})$$

for some number $\bar{t}$ in $[x, x + h]$. Finally, we note that $\bar{t} \to x$ as $h \to 0$. Thus, because f is continuous, we see that

$$F'(x) = \lim_{h \to 0} \frac{1}{h} \int_x^{x+h} f(t)\, dt = \lim_{h \to 0} f(\bar{t}) = \lim_{\bar{t} \to x} f(\bar{t}) = f(x).$$

Hence the function F in Eq. (7) is, indeed, an antiderivative of f. ❑

REMARK Figure 5.6.4 indicates why $\bar{t}$ must approach x as $h \to 0$. As the moving washer at $x + h$ approaches the fixed flange at x, the bead $\bar{t}$ between them has nowhere else to go.

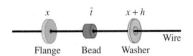

Flange Bead Washer

Fig. 5.6.4 The bead at $\bar{t}$ trapped between the washer at $x + h$ and the flange at x

Proof of Part 2 Here we apply Part 1 to give a proof of the evaluation theorem in Section 5.5. If G is *any* antiderivative of f, then—because it and the function F of Part 1 are both antiderivatives of f on the interval $[a, b]$—we know that

$$G(x) = F(x) + C$$

on $[a, b]$ for some constant C. To evaluate C, we substitute $x = a$ and obtain

$$C = G(a) - F(a) = G(a),$$

because

$$F(a) = \int_a^a f(t) \, dt = 0.$$

Hence $G(x) = F(x) + G(a)$. In other words,

$$F(x) = G(x) - G(a)$$

for all x in $[a, b]$. With $x = b$ this gives

$$G(b) - G(a) = F(b) = \int_a^b f(x) \, dx,$$

which establishes Eq. (8). ❏

Sometimes the fundamental theorem of calculus is interpreted to mean that differentiation and integration are *inverse processes*. Part 1 can be written in the form

$$\frac{d}{dx} \left(\int_a^x f(t) \, dt \right) = f(x) \tag{9}$$

if f is continuous on an open interval containing a and x. That is, if we first integrate the function f (with *variable* upper limit of integration x) and then differentiate with respect to x, the result is the function f again. So differentiation "cancels" the effect of integration of continuous functions.

Moreover, part 2 of the fundamental theorem can be written in the form

$$\int_a^x G'(t) \, dt = G(x) - G(a) \tag{10}$$

if we assume that G' is continuous. If so, this equation means that if we first differentiate the function G and then integrate the result from a to x, the result can differ from the original function G by, at worst, the *constant* $G(a)$. If a is chosen so that $G(a) = 0$, this means that integration "cancels" the effect of differentiation.

COMPUTATIONAL APPLICATIONS

Examples 1 through 4 of Section 5.5 illustrate the use of Part 2 of the fundamental theorem in the evaluation of integrals. Additional examples appear in the end-of-section problems, in this section, and in Section 5.7. Example 5 illustrates the necessity of splitting an integral into a sum of integrals when its integrand has different antiderivative formulas on different intervals.

EXAMPLE 5 Figure 5.6.5 shows the graph of the function f defined by

$$f(x) = \begin{cases} \cos x & \text{if } x \geq 0, \\ 1 - x^2 & \text{if } x \leq 0. \end{cases}$$

Find the area A of the region R bounded above by the graph of $y = f(x)$ and below by the x-axis.

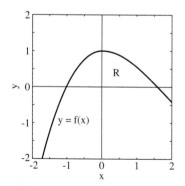

Fig. 5.6.5 The region of Example 5

Solution The x-intercepts shown in the figure are $x = -1$ (where $1 - x^2 = 0$ and $x < 0$) and $x = \pi/2$ (where $\cos x = 0$ and $x > 0$). Hence

$$A = \int_{-1}^{\pi/2} f(x)\,dx = \int_{-1}^{0} (1 - x^2)\,dx + \int_{0}^{\pi/2} \cos x\,dx$$

$$= \left[x - \frac{x^3}{3} \right]_{-1}^{0} + \left[\sin x \right]_{0}^{\pi/2} = \frac{2}{3} + 1 = \frac{5}{3}.$$

EXAMPLE 6 Figure 5.6.6 shows the graph of

$$f(x) = x^3 - x^2 - 6x.$$

Find the area A of the entire region R bounded by the graph of $y = f(x)$ and by the x-axis.

Solution The region R consists of the two regions R_1 and R_2 and extends from $x = -2$ to $x = 3$. The area of R_1 is

$$A_1 = \int_{-2}^{0} (x^3 - x^2 - 6x)\,dx = \left[\tfrac{1}{4}x^4 - \tfrac{1}{3}x^3 - 3x^2 \right]_{-2}^{0} = \tfrac{16}{3}.$$

But on the interval $(0, 3)$, the function $f(x)$ is negative-valued, so to get the (positive) area A_2 of R_2, we must integrate the *negative* of f:

$$A_2 = \int_{0}^{3} (-x^3 + x^2 + 6x)\,dx = \left[-\tfrac{1}{4}x^4 + \tfrac{1}{3}x^3 + 3x^2 \right]_{0}^{3} = \tfrac{63}{4}.$$

Consequently the area of the entire region R is

$$A = A_1 + A_2 = \tfrac{16}{3} + \tfrac{63}{4} = \tfrac{253}{12} \approx 21.08.$$

In effect, we have integrated the *absolute value* of $f(x)$:

$$A = \int_{-2}^{3} |f(x)|\,dx = \int_{-2}^{0} (x^3 - x^2 - 6x)\,dx + \int_{0}^{3} (-x^3 + x^2 + 6x)\,dx = \tfrac{253}{12}.$$

Compare the graph of $y = |f(x)|$ in Fig. 5.6.7 with that of $y = f(x)$ in Fig. 5.6.6.

EXAMPLE 7 Evaluate

$$\int_{-1}^{2} |x^3 - x|\,dx.$$

Solution We note that $x^3 - x \geq 0$ on $[-1, 0]$, that $x^3 - x \leq 0$ on $[0, 1]$, and that $x^3 - x \geq 0$ on $[1, 2]$. So we write

$$\int_{-1}^{2} |x^3 - x|\,dx = \int_{-1}^{0} (x^3 - x)\,dx + \int_{0}^{1} (x - x^3)\,dx + \int_{1}^{2} (x^3 - x)\,dx$$

$$= \left[\tfrac{1}{4}x^4 - \tfrac{1}{2}x^2 \right]_{-1}^{0} + \left[\tfrac{1}{2}x^2 - \tfrac{1}{4}x^4 \right]_{0}^{1} + \left[\tfrac{1}{4}x^4 - \tfrac{1}{2}x^2 \right]_{1}^{2}$$

$$= \tfrac{1}{4} + \tfrac{1}{4} + [2 - (-\tfrac{1}{4})] = \tfrac{11}{4} = 2.75.$$

Part 1 of the fundamental theorem of calculus says that the derivative of an integral with respect to its upper limit is equal to the value of the integrand at the upper limit. For example, if

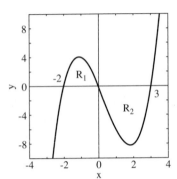

Fig. 5.6.6 The graph $y = x^3 - x^2 - 6x$ of Example 6

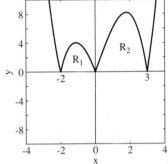

Fig. 5.6.7 The graph $y = |x^3 - x^2 - 6x|$ of Example 6

$$y(x) = \int_0^x t^3 \sin t \, dt,$$

then

$$\frac{dy}{dx} = x^3 \sin x.$$

Example 8 is a bit more complicated in that the upper limit of the integral is a nontrivial function of the independent variable.

EXAMPLE 8   Find $h'(x)$ given

$$h(x) = \int_0^{x^2} t^3 \sin t \, dt.$$

Solution   Let $y = h(x)$ and $u = x^2$. Then

$$y = \int_0^u t^3 \sin t \, dt,$$

so

$$\frac{dy}{du} = u^3 \sin u$$

by the fundamental theorem of calculus. Then the chain rule yields

$$h'(x) = \frac{dy}{dx} = \frac{dy}{du} \cdot \frac{du}{dx} = (u^3 \sin u)(2x) = 2x^7 \sin x^2.$$

INITIAL VALUE PROBLEMS

Note that if

$$y(x) = \int_a^x f(t) \, dt, \tag{11}$$

then $y(a) = 0$. Hence $y(x)$ is a solution of the initial value problem

$$\frac{dy}{dx} = f(x), \qquad y(a) = 0. \tag{12}$$

To get a solution of the initial value problem

$$\frac{dy}{dx} = f(x), \quad y(a) = b, \tag{13}$$

we need only add the desired initial value:

$$y(x) = b + \int_a^x f(t) \, dt. \tag{14}$$

EXAMPLE 9   Express as an integral the solution of the initial value problem

$$\frac{dy}{dx} = \sec x, \qquad y(2) = 3. \tag{15}$$

Solution With $a = 2$ and $b = 3$, Eq. (14) gives

$$y(x) = 3 + \int_2^x \sec t \, dt. \tag{16}$$

With our present knowledge, we cannot antidifferentiate $\sec t$, but for a particular value of x the integral in Eq. (16) can be approximated by using Riemann sums. For instance, with $x = 4$ a calculator with an [INTEGRATE] key gives

$$\int_2^4 \sec t \, dt \approx -2.5121.$$

Hence the value of the solution in Eq. (16) at $x = 4$ is

$$y(4) \approx 3 - 2.5121 = 0.4879.$$

5.6 Problems

In Problems 1 through 12, find the average value of the given function on the specified interval.

1. $f(x) = x^4$; $[0, 2]$ **2.** $g(x) = \sqrt{x}$; $[1, 4]$

3. $h(x) = 3x^2 \sqrt{x^3 + 1}$; $[0, 2]$

4. $f(x) = 8x$; $[0, 4]$ **5.** $g(x) = 8x$; $[-4, 4]$

6. $h(x) = x^2$; $[-4, 4]$ **7.** $f(x) = x^3$; $[0, 5]$

8. $g(x) = x^{-1/2}$; $[1, 4]$

9. $f(x) = \sqrt{x + 1}$; $[0, 3]$

10. $g(x) = \sin 2x$; $[0, \pi/2]$

11. $f(x) = \sin 2x$; $[0, \pi]$

12. $g(t) = e^{2t}$; $[-1, 1]$

Evaluate the integrals in Problems 13 through 28.

13. $\int_{-1}^3 dx$ (Here dx stands for $1 \, dx$.)

14. $\int_1^2 (y^5 - 1) \, dy$ **15.** $\int_1^4 \dfrac{dx}{\sqrt{9x^3}}$

16. $\int_{-1}^1 (x^3 + 2)^2 \, dx$ **17.** $\int_1^3 \dfrac{3t - 5}{t^4} \, dt$

18. $\int_{-2}^{-1} \dfrac{x^2 - x + 3}{\sqrt[3]{x}} \, dx$ **19.** $\int_0^\pi \sin x \cos x \, dx$

20. $\int_{-1}^2 |x| \, dx$ **21.** $\int_1^2 \left(t - \dfrac{1}{2t} \right)^2 dt$

22. $\int_0^1 e^{2x-1} \, dx$ **23.** $\int_0^1 \dfrac{e^{2x} - 1}{e^x} \, dx$

24. $\int_0^2 |x - \sqrt{x}| \, dx$ **25.** $\int_{-2}^2 |x^2 - 1| \, dx$

26. $\int_0^{\pi/3} \sin 3x \, dx$ **27.** $\int_4^8 \dfrac{1}{x} \, dx$

28. $\int_5^9 \dfrac{1}{x - 1} \, dx$

In Problems 29 through 32, the graph of f and the x-axis divide the xy-plane into several regions, some of which are bounded. Find the total area of the bounded regions in each problem.

29. $f(x) = 1 - x^4$ if $x \leq 0$; $f(x) = 1 - x^3$ if $x \geq 0$ (Fig. 5.6.8)

30. $f(x) = (\pi/2)^2 \sin x$ on $[0, \pi/2]$; $f(x) = x(\pi - x)$ on $[\pi/2, \pi]$ (Fig. 5.6.9)

31. $f(x) = x^3 - 9x$ (Fig. 5.6.10)

32. $f(x) = x^3 - 2x^2 - 15x$ (Fig. 5.6.11)

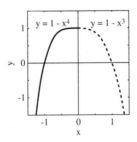

Fig. 5.6.8 Problem 29

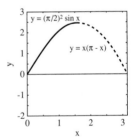

Fig. 5.6.9 Problem 30

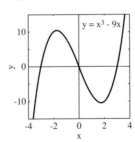

Fig. 5.6.10 Problem 31

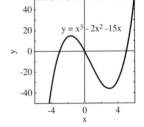

Fig. 5.6.11 Problem 32

33. Rosanne drops a ball from a height of 400 ft. Find the ball's average height and its average velocity between the time it is dropped and the time it strikes the ground.

34. Find the average value of the animal population $P(t) = 100 + 10t + (0.02)t^2$ over the time interval $[0, 10]$.

35. Suppose that a 5000-L water tank takes 10 min to drain and that after t minutes, the amount of water remaining in the tank is $V(t) = 50(10 - t)^2$ liters. What is the average amount of water in the tank during the time it drains?

36. On a certain day the temperature t hours past midnight was

$$T(t) = 80 + 10 \sin\left(\frac{\pi}{12}(t - 10)\right).$$

What was the average temperature between noon and 6 P.M.?

37. Suppose that a heated rod lies along the interval $0 \leq x \leq 10$. If the temperature at points of the rod is given by $T(x) = 4x(10 - x)$, what is the rod's average temperature?

38. Figure 5.6.12 shows a cross section at distance x from the center of a sphere of radius 1. Find the average area of this cross section for $0 \leq x \leq 1$.

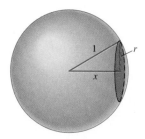

Fig. 5.6.12 The sphere of Problem 38

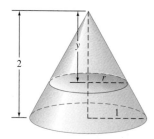

Fig. 5.6.13 The cone of Problem 39

39. Figure 5.6.13 shows a cross section at distance y from the vertex of a cone with base radius 1 and height 2. Find the average area of this cross section for $0 \leq y \leq 2$.

40. A sports car starts from rest ($x = 0, t = 0$) and experiences constant acceleration $x''(t) = a$ for T seconds.

Find, in terms of a and T, (a) its final and average velocities and (b) its final and average positions.

In Problems 41 through 45, apply the fundamental theorem of calculus to find the derivative of the given function.

41. $f(x) = \int_{-1}^{x} (t^2 + 1)^{17} \, dt$

42. $g(t) = \int_{0}^{t} \sqrt{x^2 + 25} \, dx$

43. $h(z) = \int_{2}^{z} \sqrt[3]{u - 1} \, du$　　**44.** $A(x) = \int_{1}^{x} \frac{1}{t} \, dt$

45. $f(x) = \int_{x}^{10} (e^t - e^{-t}) \, dt$

In Problems 46 through 49, $G(x)$ is the integral of the given function $f(t)$ over the specified interval of the form $[a, x]$, $x > a$. Apply Part 1 of the fundamental theorem of calculus to find $G'(x)$.

46. $f(t) = \dfrac{t}{t^2 + 1}$;　$[2, x]$

47. $f(t) = \sqrt{t + 4}$;　$[0, x]$

48. $f(t) = \sin^3 t$;　$[0, x]$

49. $f(t) = \sqrt{t^3 + 1}$;　$[1, x]$

In Problems 50 through 56, differentiate the function by first writing $f(x)$ in the form $g(u)$, where u denotes the upper limit of integration.

50. $f(x) = \int_{0}^{x^2} \sqrt{1 + t^3} \, dt$

51. $f(x) = \int_{2}^{3x} \sin t^2 \, dt$

52. $f(x) = \int_{0}^{\sin x} \sqrt{1 - t^2} \, dt$

53. $f(x) = \int_{0}^{x^2} \sin t \, dt$

54. $f(x) = \int_{1}^{\sin x} (t^2 + 1)^3 \, dt$

55. $f(x) = \int_{1}^{x^2+1} \frac{dt}{t}$

56. $f(x) = \int_{1}^{e^x} \ln (1 + t^2) \, dt$

Use integrals (as in Example 9) to solve the initial value problems in 57 through 60.

57. $\dfrac{dy}{dx} = \dfrac{1}{x}$,　$y(1) = 0$

58. $\dfrac{dy}{dx} = \dfrac{1}{1 + x^2}$,　$y(1) = \dfrac{\pi}{4}$

59. $\dfrac{dy}{dx} = \sqrt{1 + x^2}, \quad y(5) = 10$

60. $\dfrac{dy}{dx} = \tan x, \quad y(1) = 2$

61. The fundamental theorem of calculus *seems* to say that

$$\int_{-1}^{1} \frac{dx}{x^2} = \left[-\frac{1}{x}\right]_{-1}^{1} = -2,$$

in apparent contradiction to the fact that $1/x^2$ is always positive. What's wrong here?

62. Prove that the average rate of change

$$\frac{f(b) - f(a)}{b - a}$$

of the differentiable function f on $[a, b]$ is equal to the average value of its derivative on $[a, b]$.

5.7
Integration by Substitution

The fundamental theorem of calculus in the form

$$\int_a^b f(x)\, dx = \left[\int f(x)\, dx\right]_a^b \tag{1}$$

implies that we can readily evaluate the definite integral on the left if we can find the indefinite integral (that is, antiderivative) on the right. We now discuss a powerful method of antidifferentiation that amounts to "the chain rule in reverse." This method is a generalization of the "generalized power rule in reverse,"

$$\int u^n\, du = \frac{u^{n+1}}{n + 1} + C \qquad (n \neq -1), \tag{2}$$

that we introduced in Section 5.2.

Equation (2) is an abbreviation for the formula

$$\int [g(x)]^n g'(x)\, dx = \frac{[g(x)]^{n+1}}{n + 1} + C \tag{3}$$

that results when we write

$$u = g(x), \qquad du = g'(x)\, dx.$$

So to apply Eq. (2) to a given integral, we must be able to visualize the integrand as a *product* of a power of a differentiable function $g(x)$ and its derivative $g'(x)$.

EXAMPLE 1 With

$$u = 2x + 1, \qquad du = 2\, dx,$$

we see that

$$\int (2x + 1)^5 \cdot 2\, dx = \int u^5\, du = \frac{u^6}{6} + C = \frac{1}{6}(2x + 1)^6 + C.$$

EXAMPLE 2

(a)
$$\int 2x\sqrt{1 + x^2}\, dx = \int (1 + x^2)^{1/2} \cdot 2x\, dx$$

$$= \int u^{1/2}\, du \qquad [u = 1 + x^2,\, du = 2x\, dx]$$

$$= \frac{u^{3/2}}{\frac{3}{2}} + C = \frac{2}{3}(1 + x^2)^{3/2} + C.$$

(b) Similarly, but with $u = 1 + e^{2x}$, $du = 2e^{2x}\,dx$, we get

$$\int 2e^{2x} \sqrt{1 + e^{2x}}\,dx = \int u^{1/2}\,du = \tfrac{2}{3}u^{3/2} + C = \tfrac{2}{3}(1 + e^{2x})^{3/2} + C.$$

Equation (3) is the special case $f(u) = u^n$ of the general integral formula

$$\int f(g(x)) \cdot g'(x)\,dx = \int f(u)\,du. \tag{4}$$

The right-hand side of Eq. (4) results when we make the formal substitution

$$u = g(x), \qquad du = g'(x)\,dx$$

into the left-hand side.

One of the beauties of differential notation is that Eq. (4) is not only plausible but is, in fact, true—with the understanding that u is to be replaced by $g(x)$ after the indefinite integration on the right-hand side of Eq. (4) has been performed. Indeed, Eq. (4) is merely an indefinite integral version of the chain rule. For if $F'(x) = f(x)$, then

$$D_x F(g(x)) = F'(g(x)) \cdot g'(x) = f(g(x)) \cdot g'(x)$$

by the chain rule, so

$$\int f(g(x)) \cdot g'(x)\,dx = \int F'(g(x)) \cdot g'(x)\,dx = F(g(x)) + C$$

$$= F(u) + C \qquad [u = g(x)]$$

$$= \int f(u)\,du.$$

Equation (4) is the basis for the powerful technique of indefinite **integration by substitution.** It may be used whenever the integrand function is recognizable in the form $f'(g(x)) \cdot g'(x)$.

EXAMPLE 3 Find

$$\int x^2 \sqrt{x^3 + 9}\,dx.$$

Solution Note that x^2 is, to within a **constant** factor, the derivative of $x^3 + 9$. We can, therefore, substitute

$$u = x^3 + 9, \qquad du = 3x^2\,dx. \tag{5}$$

The constant factor 3 can be supplied if we compensate by multiplying the integral by $\tfrac{1}{3}$. This gives

$$\int x^2 \sqrt{x^3 + 9}\,dx = \frac{1}{3} \int (x^2 + 9)^{1/2} \cdot 3x^2\,dx = \frac{1}{3} \int u^{1/2}\,du$$

$$= \frac{1}{3} \cdot \frac{u^{3/2}}{\frac{3}{2}} + C = \frac{2}{9}u^{3/2} + C = \frac{2}{9}(x^3 + 9)^{3/2} + C.$$

An alternative way to carry out the substitution in (5) is to solve

$$du = 3x^2\,dx \quad \text{for} \quad x^2\,dx = \tfrac{1}{3}\,du,$$

and then write

$$\int (x^3 + 9)^{1/2}\, dx = \int u^{1/2} \cdot \tfrac{1}{3}\, du = \tfrac{1}{3} \int u^{1/2}\, du,$$

concluding the computation as before.

The following three steps in the solution of Example 3 are worth special mention:

❑ The differential dx along with the rest of the integrand is "transformed," or replaced, in terms of u and du.

❑ Once the integration has been performed, the constant C of integration is added.

❑ A final resubstitution is necessary to write the answer in terms of the original variable x.

SUBSTITUTION IN TRIGONOMETRIC INTEGRALS

By now we know that every differentiation formula yields—upon "reversal"—a corresponding antidifferentiation formula. The familiar formulas for the derivatives of the six trigonometric functions yield the following indefinite-integral formulas:

$$\int \cos u\, du = \sin u + C \tag{6}$$

$$\int \sin u\, du = -\cos u + C \tag{7}$$

$$\int \sec^2 u\, du = \tan u + C \tag{8}$$

$$\int \csc^2 u\, du = -\cot u + C \tag{9}$$

$$\int \sec u \tan u\, du = \sec u + C \tag{10}$$

$$\int \csc u \cot u\, du = -\csc u + C \tag{11}$$

Any of these integrals can appear as the integral $\int f(u)\, du$ that results from an appropriate *u-substitution* in a given integral.

EXAMPLE 4

$$\int \sin(3x + 4)\, dx = \int (\sin u) \cdot \tfrac{1}{3}\, du \quad (u = 3x + 4,\ du = 3\, dx)$$

$$= \tfrac{1}{3} \int \sin u\, du = -\tfrac{1}{3} \cos u\, du + C$$

$$= -\tfrac{1}{3} \cos(3x + 4) + C.$$

EXAMPLE 5

$$\int 3x \cos(x^2)\, dx = 3 \int (\cos x^2) \cdot x\, dx$$

$$= 3 \int (\cos u) \cdot \tfrac{1}{2}\, du \qquad (u = x^2,\ du = 2x\, dx)$$

$$= \tfrac{3}{2} \int \cos u\, du = \tfrac{3}{2} \sin u + C = \tfrac{3}{2} \sin(x^2) + C.$$

EXAMPLE 6

$$\int \sec^2 3x\, dx = \int (\sec^2 u) \cdot \tfrac{1}{3}\, du \qquad (u = 3x,\ du = 3\, dx)$$

$$= \tfrac{1}{3} \tan u + C = \tfrac{1}{3} \tan 3x + C.$$

EXAMPLE 7   Evaluate

$$\int 2 \sin^3 x\, \cos x\, dx.$$

Solution   None of the integrals in Eqs. (6) through (11) appear to "fit," but the substitution

$$u = \sin x, \qquad du = \cos x\, dx$$

yields

$$\int 2 \sin^3 x\, \cos x\, dx = 2 \int u^3 \cdot du = 2 \cdot \frac{u^4}{4} + C = \frac{1}{2} \sin^4 x + C.$$

SUBSTITUTION IN DEFINITE INTEGRALS

The method of integration by substitution can be used with definite integrals as well as with indefinite integrals. Only one additional step is required—evaluation of the final antiderivative at the original limits of integration.

EXAMPLE 8   The substitution of Example 3 gives

$$\int_0^3 x^2 \sqrt{x^3 + 9}\, dx = \int_*^{**} u^{1/2} \cdot \tfrac{1}{3}\, du \qquad (u = x^3 + 9,\ du = 3x^2\, dx)$$

$$= \tfrac{1}{3} \left[\tfrac{2}{3} u^{3/2} \right]_*^{**} = \tfrac{2}{9} \left[(x^3 + 9)^{3/2} \right]_0^3 \qquad \text{(resubstitute)}$$

$$= \tfrac{2}{9}(216 - 27) = 42.$$

The limits on u were left unspecified (we used asterisks) because they weren't calculated—there was no need to know them, because we planned to resubstitute for u in terms of the original variable x before using the original limits of integration.

But sometimes it is more convenient to determine the limits of integration with respect to the new variable u. With the substitution $u = x^3 + 9$, we see that

❑ $u = 9$ when $x = 0$ (lower limit);
❑ $u = 36$ when $x = 3$ (upper limit).

Use of these new limits on u (rather than resubstitution in terms of x) gives

$$\int_0^3 x^2 \sqrt{x^3 + 9} \, dx = \tfrac{1}{3} \int_9^{36} u^{1/2} \, du = \tfrac{1}{3} \left[\tfrac{2}{3} u^{3/2} \right]_9^{36} = 42.$$

Theorem 1 says that the "natural" way of transforming an integral's limits under a u-substitution, like the work just done, is in fact correct.

Theorem 1 *Definite Integral by Substitution*
Suppose that the function g has a continuous derivative on $[a, b]$ and that f is continuous on the set $g([a, b])$. Let $u = g(x)$. Then

$$\int_a^b f(g(x)) \cdot g'(x) \, dx = \int_{g(a)}^{g(b)} f(u) \, du. \tag{12}$$

REMARK   Thus we get the new limits on u by applying the substitution function $u = g(x)$ to the old limits on x. The

❑ New lower limit is $g(a)$, and the
❑ New upper limit is $g(b)$,

whether or not $g(b)$ is greater than $g(a)$.

Proof of Theorem 1   Choose an antiderivative F of f, so $F' = f$. Then, by the chain rule,

$$D_x[F(g(x))] = F'(g(x)) \cdot g'(x) = f(g(x)) \cdot g'(x).$$

Therefore,

$$\int_a^b f(g(x)) \cdot g'(x) \, dx = \left[F(g(x)) \right]_a^b = F(g(b)) - F(g(a))$$

$$= \left[F(u) \right]_{u=g(a)}^{g(b)} = \int_{g(a)}^{g(b)} f(u) \, du.$$

We used the fundamental theorem to obtain the first and last equalities in this argument.   ❑

Whether it is simpler to apply Theorem 1 and transform to new u-limits or to resubstitute $u = g(x)$ and use the old x-limits depends on the specific problem at hand. Examples 9 and 10 illustrate the technique of transforming to new limits.

EXAMPLE 9   Evaluate

$$\int_3^5 \frac{x \, dx}{(30 - x^2)^2}.$$

Solution   Note that $30 - x^2$ is nonzero on $[3, 5]$, so the integrand is continuous there. We substitute

$$u = 30 - x^2, \qquad du = -2x \, dx,$$

and observe that

$$\text{If } x = 3, \quad \text{then } u = 21 \quad \text{(lower limit)};$$
$$\text{If } x = 5, \quad \text{then } u = 5 \quad \text{(upper limit)}.$$

Hence our substitution gives

$$\int_3^5 \frac{x\,dx}{(30 - x^2)^2} = \int_{21}^5 \frac{-\frac{1}{2}\,du}{u^2} = -\frac{1}{2}\left[-\frac{1}{u}\right]_{21}^5 = -\frac{1}{2}\left(-\frac{1}{5} + \frac{1}{21}\right) = \frac{8}{105}.$$

EXAMPLE 10 Evaluate

$$\int_0^{\pi/4} \sin^3 2t \cos 2t\,dt.$$

Solution   We substitute

$$u = \sin 2t, \quad \text{so} \quad du = 2\cos 2t\,dt.$$

Then $u = 0$ when $t = 0$; $u = 1$ when $t = \pi/4$. Hence

$$\int_0^{\pi/4} \sin^3 2t \cos 2t\,dt = \frac{1}{2}\int_0^1 u^3\,du = \frac{1}{2}\left[\frac{1}{4}u^4\right]_0^1 = \frac{1}{8}.$$

5.7 Problems

Evaluate the indefinite integrals in Problems 1 through 30.

1. $\displaystyle\int (x + 1)^6\,dx$

2. $\displaystyle\int (2 - x)^5\,dx$

3. $\displaystyle\int (4 - 3x)^7\,dx$

4. $\displaystyle\int \sqrt{2x + 1}\,dx$

5. $\displaystyle\int \frac{dx}{\sqrt{7x + 5}}$

6. $\displaystyle\int \frac{dx}{(3 - 5x)^2}$

7. $\displaystyle\int \sin(\pi x + 1)\,dx$

8. $\displaystyle\int \cos\frac{\pi t}{3}\,dt$

9. $\displaystyle\int \sec 2\theta \tan 2\theta\,d\theta$

10. $\displaystyle\int \csc^2 5x\,dx$

11. $\displaystyle\int x\sqrt{x^2 - 1}\,dx$

12. $\displaystyle\int 3t(1 - 2t^2)^{10}\,dt$

13. $\displaystyle\int x\sqrt{2 - 3x^2}\,dx$

14. $\displaystyle\int \frac{t\,dt}{\sqrt{2t^2 + 1}}$

15. $\displaystyle\int x^3\sqrt{x^4 + 1}\,dx$

16. $\displaystyle\int \frac{x^2\,dx}{\sqrt[3]{x^3 + 1}}$

17. $\displaystyle\int x^2\cos(2x^3)\,dx$

18. $\displaystyle\int t\sec^2(t^2)\,dt$

19. $\displaystyle\int xe^{-x^2}\,dx$

20. $\displaystyle\int \frac{x}{1 + x^2}\,dx$

21. $\displaystyle\int \cos^3 x \sin x\,dx$

22. $\displaystyle\int \sin^5 3z \cos 3z\,dz$

23. $\displaystyle\int \tan^3\theta \sec^2\theta\,d\theta$

24. $\displaystyle\int \sec^3\theta \tan\theta\,d\theta$

25. $\displaystyle\int \frac{\cos\sqrt{x}}{\sqrt{x}}\,dx$ [*Suggestion:* Try $u = \sqrt{x}$.]

26. $\displaystyle\int \frac{dx}{\sqrt{x}\,(1 + \sqrt{x})^2}$

27. $\displaystyle\int (x^2 + 2x + 1)^4(x + 1)\,dx$

28. $\displaystyle\int \frac{(x + 2)\,dx}{(x^2 + 4x + 3)^3}$

29. $\displaystyle\int \frac{x + 2}{x^2 + 4x + 5}\,dx$

30. $\displaystyle\int \frac{2x + e^x}{(x^2 + e^x + 1)^2}\,dx$

Evaluate the definite integrals in Problems 31 through 44.

31. $\displaystyle\int_1^2 \frac{dt}{(t + 1)^3}$

32. $\displaystyle\int_0^4 \frac{dx}{\sqrt{2x + 1}}$

33. $\displaystyle\int_0^4 x\sqrt{x^2 + 9}\,dx$

34. $\displaystyle\int_1^4 \frac{(1 + \sqrt{x})^4}{\sqrt{x}}\,dx$ [*Suggestion:* Try $u = 1 + \sqrt{x}$.]

35. $\displaystyle\int_0^8 t\sqrt{t + 1}\,dt$ [*Suggestion:* Try $u = t + 1$.]

36. $\displaystyle\int_0^{\pi/2} \sin x \cos x\,dx$

37. $\displaystyle\int_0^{\pi/6} \sin 2x \cos^3 2x\,dx$

38. $\displaystyle\int_0^{\sqrt{\pi}} t\sin\frac{t^2}{2}\,dt$

39. $\displaystyle\int_0^{\pi/2} (1 + 3 \sin \theta)^{3/2} \cos \theta \, d\theta$ [*Suggestion: Try*

$u = 1 + 3 \sin \theta$.]

40. $\displaystyle\int_0^{\pi/2} \sec^2 \frac{x}{2} \, dx$

41. $\displaystyle\int_0^{\pi/2} e^{\sin x} \cos x \, dx$ [*Suggestion: Try $u = \sin x$.*]

42. $\displaystyle\int_1^2 \frac{1 + \ln x}{x} \, dx$ [*Suggestion: Try $u = 1 + \ln x$.*]

43. $\displaystyle\int_1^4 \frac{e^{\sqrt{x}}}{\sqrt{x}} \, dx$ **44.** $\displaystyle\int_{\pi^2/4}^{\pi^2} \frac{\sin \sqrt{x} \cos \sqrt{x}}{\sqrt{x}} \, dx$

Use the half-angle identities

$$\cos^2 \theta = \frac{1 + \cos 2\theta}{2} \quad and \quad \sin^2 \theta = \frac{1 - \cos 2\theta}{2}$$

to evaluate the integrals in Problems 45 through 48.

45. $\displaystyle\int \sin^2 x \, dx$ **46.** $\displaystyle\int \cos^2 x \, dx$

47. $\displaystyle\int_0^\pi \sin^2 3t \, dt$ **48.** $\displaystyle\int_0^1 \cos^2 \pi t \, dt$

Use the identity $1 + \tan^2 \theta = \sec^2 \theta$ to evaluate the integrals in Problems 49 and 50.

49. $\displaystyle\int \tan^2 x \, dx$ **50.** $\displaystyle\int_0^{\pi/12} \tan^2 3t \, dt$

51. Substitute $\sin^3 x = (\sin x)(1 - \cos^2 x)$ to show that

$$\int \sin^3 x \, dx = \tfrac{1}{3} \cos^3 x - \cos x + C.$$

52. Evaluate

$$\int_0^{\pi/2} \cos^3 x \, dx$$

by the method of Problem 51.

53. Substitute first $u = \sin \theta$ and then $u = \cos \theta$ to obtain

$$\int \sin \theta \cos \theta \, d\theta = \tfrac{1}{2} \sin^2 \theta + C_1 = -\tfrac{1}{2} \cos^2 \theta + C_2.$$

Reconcile these results. What is the relation between the constants C_1 and C_2?

54. Substitute first $u = \tan \theta$ and then $u = \sec \theta$ to obtain

$$\int \sec^2 \theta \tan \theta \, d\theta = \tfrac{1}{2} \tan^2 \theta + C_1 = \tfrac{1}{2} \sec^2 \theta + C_2.$$

Reconcile these results. What is the relation between the constants C_1 and C_2?

55. (a) Verify by differentiation that

$$\int \frac{dx}{(1 - x)^2} = \frac{x}{1 - x} + C_1.$$

(b) Substitute $u = 1 - x$ to show that

$$\int \frac{dx}{(1 - x)^2} = \frac{1}{1 - x} + C_2.$$

(c) Reconcile the results of parts (a) and (b).

56. (a) Substitute $u = x^2$ and apply part (a) of Problem 55 to show that

$$\int \frac{x \, dx}{(1 - x^2)^2} = \frac{x^2}{2(1 - x^2)} + C_1.$$

(b) Substitute $u = 1 - x^2$ to show that

$$\int \frac{x \, dx}{(1 - x^2)^2} = \frac{1}{2(1 - x^2)} + C_2.$$

(c) Reconcile the results of parts (a) and (b).

*Problems 57 through 60 deal with even and odd functions. An **even** function f is a function such that*

$$f(-x) = f(x)$$

*for all x. This means that the graph of $y = f(x)$ is symmetric under reflection through the y-axis (Fig. 5.7.1). Examples of even functions include $f(x) = \cos x$, 1, x^2, x^4, and x^6. An **odd** function f is a function such that*

$$f(-x) = -f(x)$$

for all x. This means that the graph of $y = f(x)$ is symmetric under reflections first through the y-axis, then through the x-axis (Fig. 5.7.2). Examples of odd functions are $f(x) = \sin x$, x, x^3, and x^5. Think about the indicated reflections with the (even) cosine function (in Fig. 5.7.3) and the (odd) sine function (in Fig. 5.7.4).

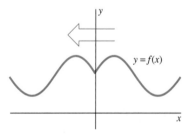

Fig. 5.7.1 The graph of the even function $y = f(x)$ is invariant under reflection through the y-axis.

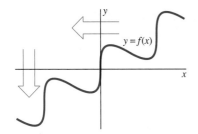

Fig. 5.7.2   The graph of the odd function $y = f(x)$ is invariant under successive reflections through both axes.

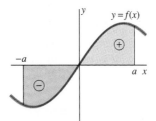

Fig. 5.7.5 f odd: areas cancel (Problem 57)

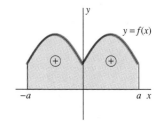

Fig. 5.7.6 f even: areas add (Problem 58)

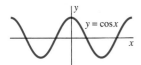

Fig. 5.7.3   The cosine function is even.

Fig. 5.7.4   The sine function is odd.

58. See Fig. 5.7.6. If the continuous function f is even, use the method of Problem 57 to show that

$$\int_{-a}^{a} f(x)\, dx = 2 \int_{0}^{a} f(x)\, dx.$$

59. Explain without extensive computation why it is evident that

$$\int_{-1}^{1} \left[\tan x + \frac{\sqrt[3]{x}}{(1 + x^2)^7} - x^{17} \cos x \right] dx = 0.$$

57. See Fig. 5.7.5. If the continuous function f is odd, substitute $u = -x$ into the integral

$$\int_{-a}^{0} f(x)\, dx \quad \text{to show that} \quad \int_{-a}^{a} f(x)\, dx = 0.$$

60. Explain without extensive computation why it is evident that

$$\int_{-5}^{5} \left(3x^2 - x^{10} \sin x + x^5 \sqrt{1 + x^4} \right) dx = 2 \left[x^3 \right]_{0}^{5} = 250.$$

5.8
Areas of Plane Regions

In Section 5.3 we discussed the area A under the graph of a positive-valued continuous function f on the interval $[a, b]$. This discussion motivated our definition in Section 5.4 of the integral of f from a to b as the limit of Riemann sums. An important result was that

$$A = \int_{a}^{b} f(x)\, dx, \tag{1}$$

by definition.

Here we consider the problem of finding the areas of more general regions in the coordinate plane. Regions such as the ones illustrated in Fig. 5.8.1 may be bounded by the graphs of *two* (or more) different functions.

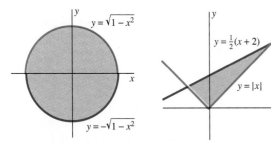
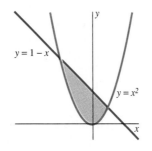
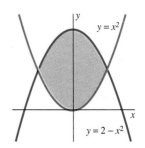

Fig. 5.8.1   Plane regions bounded by pairs of curves

Let f and g be continuous functions such that $f(x) \geqq g(x)$ for all x in the interval $[a, b]$. We are interested in the area A of the region R in Fig. 5.8.2, which lies *between* the graphs of $y = f(x)$ and $y = g(x)$ for x in $[a, b]$. Thus R is bounded by

❑ The curve $y = f(x)$, the upper boundary of R, by
❑ The curve $y = g(x)$, the lower boundary of R,

and by the vertical lines $x = a$ and $x = b$ (if needed).

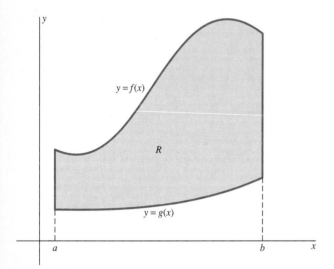

Fig. 5.8.2 A region between two graphs

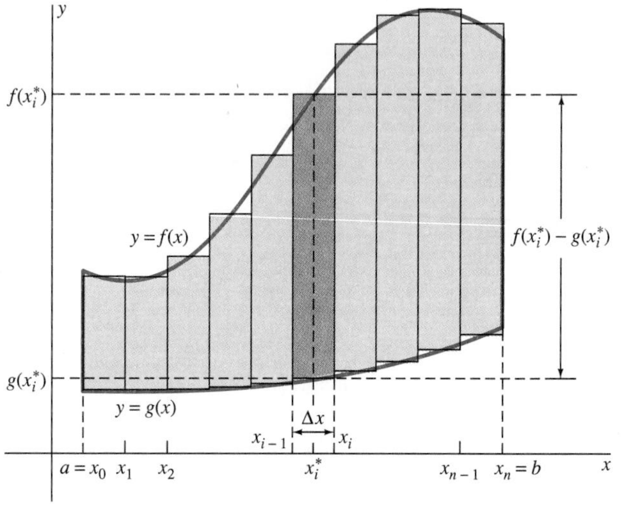

Fig. 5.8.3 A partition of $[a, b]$ divides R into vertical strips that we approximate with rectangular strips.

To approximate A, we consider a partition of $[a, b]$ into n subintervals, all with the same length $\Delta x = (b - a)/n$. If ΔA_i denotes the area of the region between the graphs of f and g over the ith subinterval $[x_{i-1}, x_i]$, and x_i^* is a selected number chosen in that subinterval (all this for $i = 1, 2, 3, \ldots, n$), then ΔA_i is approximately equal to the area of a rectangle with height $f(x_i^*) - g(x_i^*)$ and width Δx (Fig. 5.8.3). Hence

$$\Delta A_i \approx [f(x_i^*) - g(x_i^*)] \, \Delta x;$$

so

$$A = \sum_{i=1}^{n} \Delta A_i \approx \sum_{i=1}^{n} [f(x_i^*) - g(x_i^*)] \, \Delta x.$$

We introduce the *height* function $h(x) = f(x) - g(x)$ and observe that A is approximated by a Riemann sum for $h(x)$ associated with our partition of $[a, b]$:

$$A \approx \sum_{i=1}^{n} h(x_i^*) \, \Delta x.$$

Both intuition and reason suggest that this approximation can be made arbitrarily accurate by choosing n to be sufficiently large (and hence $\Delta x = (b - a)/n$ to be sufficiently small). We therefore conclude that

$$A = \lim_{\Delta x \to 0} \sum_{i=1}^{n} h(x_i^*) \, \Delta x = \int_a^b h(x) \, dx = \int_a^b [f(x) - g(x)] \, dx.$$

Because our discussion is based on an intuitive concept rather than on a precise logical definition of area, it does *not* constitute a proof of this area formula. It does, however, provide justification for the following *definition* of the area in question.

Definition *The Area Between Two Curves*
Let f and g be continuous with $f(x) \geqq g(x)$ for x in $[a, b]$. Then the **area** A of the region bounded by the curves $y = f(x)$ and $y = g(x)$ and by the vertical lines $x = a$ and $x = b$ is

$$A = \int_a^b [f(x) - g(x)] \, dx. \tag{2}$$

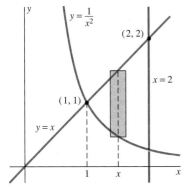

Fig. 5.8.4 The region of Example 1

EXAMPLE 1 Find the area of the region bounded by the lines $y = x$ and $x = 2$ and by the curve $y = 1/x^2$ (Fig. 5.8.4).

Solution Here the top curve is $y = f(x) = x$, the bottom curve is $y = g(x) = 1/x^2$, $a = 1$, and $b = 2$. The vertical line $x = 2$ is "needed" (to provide the right-hand boundary of the region), whereas $x = 1$ is not. Equation (2) yields

$$A = \int_1^2 \left(x - \frac{1}{x^2} \right) dx = \left[\frac{1}{2}x^2 + \frac{1}{x} \right]_1^2 = \left(2 + \frac{1}{2} \right) - \left(\frac{1}{2} + 1 \right) = 1.$$

Equation (1) is the special case of Eq. (2) in which $g(x)$ is identically zero on $[a, b]$. But if $f(x) \equiv 0$ and $g(x) \leqq 0$ on $[a, b]$, then Eq. (2) reduces to

$$A = -\int_a^b g(x) \, dx; \quad \text{that is,} \quad \int_a^b g(x) \, dx = -A.$$

In this case the region R lies beneath the x-axis (Fig. 5.8.5). Thus the integral from a to b of a negative-valued function is the *negative* of the area of the region bounded by its graph, the x-axis, and the vertical lines $x = a$ and $x = b$.

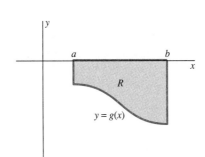

Fig. 5.8.5 The integral gives the negative of the geometric area for a region that lies below the x-axis.

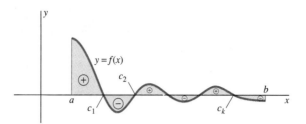

Fig. 5.8.6 The integral computes the area above the x-axis *minus* the area below the x-axis.

More generally, consider a continuous function f with a graph that crosses the x-axis at finitely many points $c_1, c_2, c_3, \ldots, c_k$ between a and b (Fig. 5.8.6). We write

$$\int_a^b f(x)\, dx = \int_a^{c_1} f(x)\, dx + \int_{c_1}^{c_2} f(x)\, dx + \cdots + \int_{c_k}^b f(x)\, dx.$$

Thus we see that

$$\int_a^b f(x)\, dx$$

is equal to the area below $y = f(x)$ and *above* the x-axis *minus* the area above $y = f(x)$ and *below* the x-axis.

The following *heuristic* (suggestive, though nonrigorous) way of setting up integral formulas such as Eq. (2) can be useful. Consider the vertical strip of area that lies above the interval $[x, x + dx]$, shown shaded in Fig. 5.8.7, where we have written

$$y_{\text{top}} = f(x) \quad \text{and} \quad y_{\text{bot}} = g(x)$$

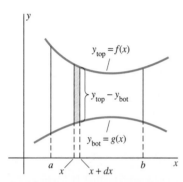

Fig. 5.8.7 Heuristic (suggestive but nonrigorous) approach to setting up area integrals

for the top and bottom boundary curves. We think of the length dx of the interval $[x, x + dx]$ as being so small that we can regard this strip as a rectangle with width dx and height $y_{\text{top}} - y_{\text{bot}}$. Its area is then

$$dA = (y_{\text{top}} - y_{\text{bot}})\, dx.$$

Think now of the region over $[a, b]$ that lies between $y_{\text{top}} = f(x)$ and $y_{\text{bot}} = g(x)$ as being made up of many such vertical strips. We can regard its area as a sum of areas of such rectangular strips. If we write $\int$ for *sum*, we get the formula

$$A = \int dA = \int_a^b (y_{\text{top}} - y_{\text{bot}})\, dx.$$

This heuristic approach bypasses the subscript notation associated with Riemann sums. Nevertheless, it *is not and should not be regarded as a complete* derivation of the last formula. It is best used only as a convenient memory device. For instance, in the figures that accompany examples, we shall often show a strip of width dx as a visual aid in properly setting up the correct integral.

EXAMPLE 2 Find the area A of the region R bounded by the line $y = x$ and by the parabola $y = 6 - x^2$.

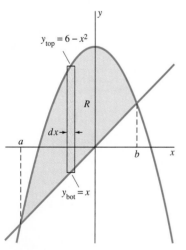

Fig. 5.8.8 The region R of Example 2

Solution The region R is shown in Fig. 5.8.8. We can use Eq. (2) and take $f(x) = 6 - x^2$ and $g(x) = x$. The limits a and b will be the x-coordinates of

328

the two points of intersection of the line and the parabola; our first order of business is to find a and b. To do so, we equate $f(x)$ and $g(x)$ and solve the resulting equation for x:

$$x = 6 - x^2; \quad x^2 + x - 6 = 0; \quad (x - 2)(x + 3) = 0; \quad x = -3, 2.$$

Thus $a = -3$ and $b = 2$, so Eq. (2) gives

$$A = \int_{-3}^{2} (6 - x^2 - x)\,dx = \left[6x - \tfrac{1}{3}x^3 - \tfrac{1}{2}x^2\right]_{-3}^{2}$$

$$= [6 \cdot 2 - \tfrac{1}{3} \cdot 2^3 - \tfrac{1}{2} \cdot 2^2] - [6 \cdot (-3) - \tfrac{1}{3} \cdot (-3)^3 - \tfrac{1}{2} \cdot (-3)^2] = \tfrac{125}{6}.$$

SUBDIVIDING REGIONS BEFORE INTEGRATING

Example 3 shows that it is sometimes necessary to subdivide a region before applying Eq. (2), typically because the formula for either the top or the bottom boundary curve (or both) changes somewhere between $x = a$ and $x = b$.

EXAMPLE 3   Find the area A of the region R bounded by the line $y = \tfrac{1}{2}x$ and by the parabola $y^2 = 8 - x$.

Solution   The region R is shown in Fig. 5.8.9. The points of intersection $(-8, -4)$ and $(4, 2)$ are found by equating $y = \tfrac{1}{2}x$ and $y = \pm\sqrt{8 - x}$ and then solving for x. The lower boundary of R is given by $y_{\text{bot}} = -\sqrt{8 - x}$ on $[-8, 8]$. But the upper boundary of R is given by

$$y_{\text{top}} = \tfrac{1}{2}x \quad \text{on} \quad [-8, 4], \qquad y_{\text{top}} = +\sqrt{8 - x} \quad \text{on} \quad [4, 8].$$

We must therefore divide R into the two regions R_1 and R_2, as indicated in Fig. 5.8.9. Then Eq. (2) gives

$$A = \int_{-8}^{4} \left(\tfrac{1}{2}x + \sqrt{8 - x}\right) dx + \int_{4}^{8} 2\sqrt{8 - x}\,dx$$

$$= \left[\tfrac{1}{4}x^2 - \tfrac{2}{3}(8 - x)^{3/2}\right]_{-8}^{4} + \left[-\tfrac{4}{3}(8 - x)^{3/2}\right]_{4}^{8}$$

$$= \left[(\tfrac{16}{4} - \tfrac{16}{3}) - (\tfrac{64}{4} - \tfrac{128}{3})\right] + \left[0 + \tfrac{32}{3}\right] = 36.$$

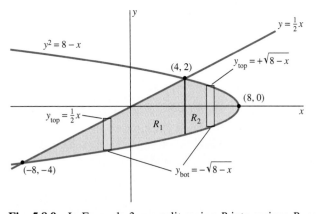

Fig. 5.8.9   In Example 3, we split region R into regions R_1 and R_2.

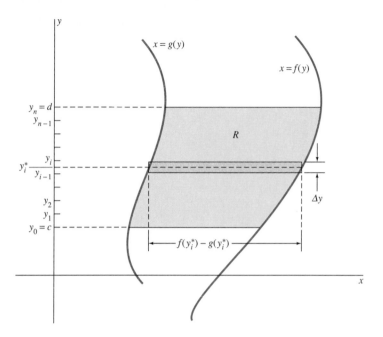

Fig. 5.8.10 Finding areas by using an integral with respect to y

DETERMINING AREA BY INTEGRATING WITH RESPECT TO y

The region in Example 3 appears to be simpler if we consider it to be bounded by graphs of functions of y rather than by graphs of functions of x. Figure 5.8.10 shows a region R bounded by the curves $x = f(y)$ and $x = g(y)$, with $f(y) \geqq g(y)$ for y in $[c, d]$, and by the horizontal lines $y = c$ and $y = d$. To approximate the area A of R, we begin with a partition of $[c, d]$ into n subintervals, all with the same length $\Delta y = (d - c)/n$. We choose a point y_i^* in the ith subinterval $[y_{i-1}, y_i]$ for each i $(1 \leqq i \leqq n)$. The horizontal strip of R lying opposite $[y_{i-1}, y_i]$ is approximated by a rectangle with width Δy (measured vertically) and height $f(y_i^*) - g(y_i^*)$ (measured horizontally). Hence

$$A \approx \sum_{i=1}^{n} [f(y_i^*) - g(y_i^*)] \, \Delta y.$$

Recognition of this sum as a Riemann sum for the integral

$$\int_{c}^{d} [f(y) - g(y)] \, dy$$

motivates the following definition.

Definition *The Area Between Two Curves*

Let f and g be continuous functions of y with $f(y) \geqq g(y)$ for y in $[c, d]$. Then the **area** A of the region bounded by the curves $x = f(y)$ and $x = g(y)$ and by the horizontal lines $y = c$ and $y = d$ is

$$A = \int_{c}^{d} [f(y) - g(y)] \, dy. \qquad (3)$$

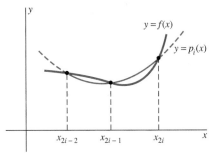

Fig. 5.9.13 The parabolic approximation $y = p_i(x)$ to $y = f(x)$ on $[x_{2i-2}, x_{2i}]$

$$p_i(x) = A_i + B_i x + C_i x^2$$

on $[x_{2i-2}, x_{2i}]$ as follows: We choose the coefficients A_i, B_i, and C_i so that $p_i(x)$ agrees with $f(x)$ at the three points x_{2i-2}, x_{2i-1}, and x_{2i} (Fig. 5.9.13). This can be done by solving the three equations

$$A_i + B_i x_{2i-2} + C_i(x_{2i-2})^2 = f(x_{2i-2}),$$
$$A_i + B_i x_{2i-1} + C_i(x_{2i-1})^2 = f(x_{2i-1}),$$
$$A_i + B_i x_{2i} + C_i(x_{2i})^2 = f(x_{2i})$$

for the three unknowns A_i, B_i, and C_i. A routine (though tedious) algebraic computation—see Problem 48 of Section 5.8—shows that

$$\int_{x_{2i-2}}^{x_{2i}} p_i(x)\, dx = \frac{\Delta x}{3}(y_{2i-2} + 4y_{2i-1} + y_{2i}).$$

We now approximate $\int_a^b f(x)\, dx$ by replacing $f(x)$ with $p_i(x)$ on the interval $[x_{2i-2}, x_{2i}]$ for $i = 1, 2, 3, \ldots, n$. This gives

$$\int_a^b f(x)\, dx = \sum_{i=1}^n \int_{x_{2i-2}}^{x_{2i}} f(x)\, dx \approx \sum_{i=1}^n \int_{x_{2i-2}}^{x_{2i}} p_i(x)\, dx$$

$$= \sum_{i=1}^n \frac{\Delta x}{3}(y_{2i-2} + 4y_{2i-1} + y_{2i})$$

$$= \frac{\Delta x}{3}(y_0 + 4y_1 + 2y_2 + 4y_3 + \cdots + 4y_{2n-3}$$
$$+ 2y_{2n-2} + 4y_{2n-1} + y_{2n}).$$

Thus the parabolic approximation described here results in Simpson's approximation S_{2n} to $\int_a^b f(x)\, dx$.

The numerical methods of this section are especially useful for approximating integrals of functions that are available only in graphical or in tabular form. This is often the case with functions derived from empirical data or from experimental measurements.

EXAMPLE 6 Suppose that the graph in Fig. 5.9.14 shows the velocity $v(t)$ recorded by instruments on board a submarine traveling under the polar ice cap directly toward the North Pole. Use the trapezoidal approximation and Simpson's approximation to estimate the distance $s = \int_a^b v(t)\, dt$ traveled by the submarine during the 10-h period from $t = 0$ to $t = 10$.

Solution We read the following data from the graph.

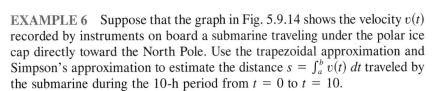

t	0	1	2	3	4	5	6	7	8	9	10	h
v	12	14	17	21	22	21	15	11	11	14	17	mi/h

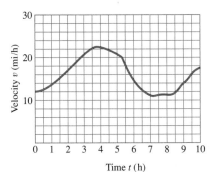

Fig. 5.9.14 Velocity graph for the submarine of Example 6

Using the trapezoidal approximation with $n = 10$ and $\Delta x = 1$, we obtain

$$s = \int_0^{10} v(t)\, dt$$

$$\approx \tfrac{1}{2}[12 + 2\cdot(14 + 17 + 22 + 23 + 21 + 15$$
$$+ 11 + 11 + 14) + 17]$$

$$= 162.5 \quad \text{(mi)}.$$

Using Simpson's approximation with $2n = 10$ and $\Delta x = 1$, we obtain

$$s = \int_0^{10} v(t)\, dt$$

$$\approx \tfrac{1}{3}[12 + 4 \cdot 14 + 2 \cdot 17 + 4 \cdot 22 + 2 \cdot 23 + 4 \cdot 21$$

$$+ 2 \cdot 15 + 4 \cdot 11 + 2 \cdot 11 + 4 \cdot 14 + 17]$$

$$= 163 \quad (\text{mi})$$

as an estimate of the distance traveled by the submarine during this 10-h period.

ERROR ESTIMATES

The trapezoidal approximation and Simpson's approximation are widely used for numerical integration, and there are *error estimates* that can be used to predict the maximum possible error in a particular approximation. The trapezoidal error ET_n and the Simpson's error ES_n are defined by the equations

$$\int_a^b f(x)\, dx = T_n + ET_n \tag{13}$$

and

$$\int_a^b f(x)\, dx = S_n + ES_n \qquad (n \text{ even}), \tag{14}$$

respectively. Thus $|ET_n|$ is the numerical difference between the value of the integral and the trapezoidal approximation with n subintervals, whereas $|ES_n|$ is the numerical difference between the integral and Simpson's approximation.

Theorems 1 and 2 are proved in numerical analysis textbooks.

Theorem 1 *Trapezoidal Error Estimate*
Suppose that the second derivative f'' is continuous on $[a, b]$ and that $|f''(x)| \leqq M_2$ for all x in $[a, b]$. Then

$$|ET_n| \leqq \frac{M_2(b - a)^3}{12n^2}. \tag{15}$$

Theorem 2 *Simpson's Error Estimate*
Suppose that the fourth derivative $f^{(4)}$ is continuous on $[a, b]$ and that $|f^{(4)}(x)| \leqq M_4$ for all x in $[a, b]$. If n is even, then

$$|ES_n| \leqq \frac{M_4(b - a)^5}{180n^4}. \tag{16}$$

REMARK   The factor n^4 in Eq. (16)—compared with the n^2 in Eq. (15)—explains the greater accuracy typical of Simpson's approximation. For instance, if $n = 10$, then $n^2 = 100$ but $n^4 = 10{,}000$, so the denominator in the error formula for Simpson's approximation is much larger.

EXAMPLE 7 In Chapter 7 we will see that the natural logarithm of the number 2 is the value of the integral

$$\ln 2 = \int_1^2 \frac{dx}{x}.$$

Estimate the errors in the trapezoidal approximation and Simpson's approximation to this integral by using $n = 10$ subintervals. (The actual value of $\ln 2$ is approximately 0.693147.)

Solution With $f(x) = 1/x$ we calculate

$$f'(x) = -\frac{1}{x^2}, \quad f''(x) = \frac{2}{x^3},$$

$$f'''(x) = -\frac{6}{x^4}, \quad f^{(4)}(x) = \frac{24}{x^5}.$$

The maximum values of all these derivatives in $[1, 2]$ occur at $x = 1$, so we may take $M_2 = 2$ and $M_4 = 24$ in Eqs. (15) and (16). From Eq. (15) we see that

$$|ET_{10}| \leqq \frac{2 \cdot 1^3}{12 \cdot 10^2} \approx 0.0016667, \tag{17}$$

so we would expect the trapezoidal approximation T_{10} to be accurate to at least two decimal places. From Eq. (16) we see that

$$|ES_{10}| \leqq \frac{24 \cdot 1^5}{180 \cdot 10^4} \approx 0.000013, \tag{18}$$

so we would expect Simpson's approximation to be accurate to at least four decimal places.

It turns out that $T_{10} \approx 0.693771$ and that $S_{10} \approx 0.693150$, so the errors in these approximations (in comparision with the actual value of the integral, about 0.693147) are $ET_{10} \approx 0.000624$ and $ES_{10} \approx 0.000003$. Note that the actual errors are somewhat smaller than their estimates calculated in Eqs. (17) and (18). It is fairly typical of numerical integration that the trapezoidal and Simpson's approximations computed in practice are somewhat more accurate than the "worst-case" estimates provided by Theorems 1 and 2.

5.9 Problems

In Problems 1 through 6, calculate the trapezoidal approximation T_n to the given integral, and compare T_n with the exact value of the integral. Use the indicated number n of subintervals, and round answers to two decimal places.

1. $\int_0^4 x \, dx, \quad n = 4$

2. $\int_1^2 x^2 \, dx, \quad n = 5$

3. $\int_0^1 \sqrt{x} \, dx, \quad n = 5$

4. $\int_1^3 \frac{1}{x^2} \, dx, \quad n = 4$

5. $\int_0^{\pi/2} \cos x \, dx, \quad n = 3$

6. $\int_0^\pi \sin x \, dx, \quad n = 4$

7 through 12. Calculate the midpoint approximations to the integrals in Problems 1 through 6, using the indicated number of subintervals. In each case compare M_n with the exact value of the integral.

In Problems 13 through 20, calculate both the trapezoidal approximation T_n and Simpson's approximation S_n to the given integral. Use the indicated number of subintervals, and round answers to four decimal places. In Problems 13 through 16, also compare these approximations with the exact value of the integral.

13. $\displaystyle\int_1^3 x^2\,dx, \quad n = 4$

14. $\displaystyle\int_1^4 x^3\,dx, \quad n = 4$

15. $\displaystyle\int_2^4 \frac{1}{x^3}\,dx, \quad n = 4$

16. $\displaystyle\int_0^1 \sqrt{1 + x}\,dx, \quad n = 4$

17. $\displaystyle\int_0^2 \sqrt{1 + x^3}\,dx, \quad n = 6$

18. $\displaystyle\int_0^3 \frac{1}{1 + x^4}\,dx, \quad n = 6$

19. $\displaystyle\int_1^5 \sqrt[3]{1 + \ln x}\,dx, \quad n = 8$

20. $\displaystyle\int_0^1 \frac{e^x - 1}{x}\,dx, \quad n = 10$

[*Note:* Make the integrand in Problem 20 continuous by assuming that its value at $x = 0$ is its limit there,

$$\lim_{x \to 0} \frac{e^x - 1}{x} = 1.]$$

In Problems 21 and 22, calculate (a) the trapezoidal approximation and (b) Simpson's approximation to

$$\int_a^b f(x)\,dx,$$

where f is the given tabulated function.

21.

x	$a = 1.00$	1.25	1.50	1.75	2.00	2.25	$2.50 = b$
$f(x)$		3.43	2.17	0.38	1.87	2.65	2.31	1.97

22.

x	$a = 0$	1	2	3	4	5	6	7	8	9	$10 = b$
$f(x)$		23	8	−4	12	35	47	53	50	39	29	5

23. Figure 5.9.15 shows the measured rate of water flow (in liters per minute) into a tank during a 10-min period. Using 10 subintervals in each case, estimate the total amount of water that flows into the tank during this period by using (a) the trapezoidal approximation and (b) Simpson's approximation.

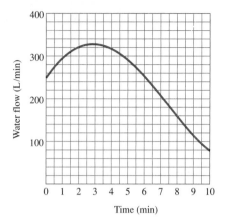

Fig. 5.9.15 Water-flow graph for Problem 23

24. Figure 5.9.16 shows the daily mean temperature recorded during December at Big Frog, California. Using 10 subintervals in each case, estimate the average temperature during that month by using (a) the trapezoidal approximation and (b) Simpson's approximation.

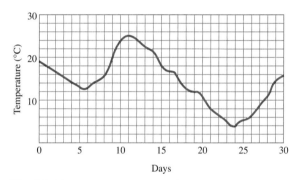

Fig. 5.9.16 Temperature graph for Problem 24

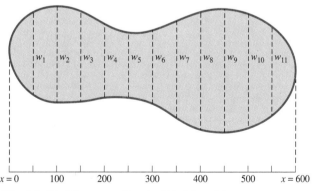

Fig. 5.9.17 The tract of land of Problem 25

25. Figure 5.9.17 shows a tract of land with measurements in feet. A surveyor has measured its width w at 50-ft intervals (the values of x shown in the figure), with the following results.

x	0	50	100	150	200	250	300
w	0	165	192	146	63	42	84

x	350	400	450	500	550	600
w	155	224	270	267	215	0

Use (a) the trapezoidal approximation and (b) Simpson's approximation to estimate the acreage of this tract. [*Note:* An acre is 4840 yd².]

26. The base for natural logarithms is the number e so it follows that

$$\int_1^e \frac{1}{x}\,dx = 1.$$

Approximate the integrals

$$\int_1^{2.7} \frac{1}{x}\,dx \quad \text{and} \quad \int_1^{2.8} \frac{1}{x}\,dx$$

with sufficient accuracy to show that $2.7 < e < 2.8$.

Problems 27 and 28 deal with the integral

$$\ln 2 = \int_1^2 \frac{1}{x}\,dx$$

of Example 7.

27. Use the trapezoidal error estimate to determine how large n must be in order to guarantee that T_n differs from $\ln 2$ by at most 0.0005.

28. Use the Simpson's error estimate to determine how large n must be in order to guarantee that S_n differs from $\ln 2$ by at most 0.000005.

29. Deduce the following from the error estimate for Simpson's approximation: If $p(x)$ is a polynomial of degree at most 3, then Simpson's approximation with $n = 2$ subintervals gives the exact value of the integral

$$\int_a^b p(x)\,dx.$$

30. Use the result of Problem 29 to calculate (without explicit integration) the area of the region shown in Fig. 5.9.18. [*Answer:* 1331/216.]

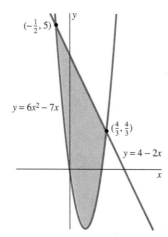

Fig. 5.9.18 The region of Problem 30

5.9 Projects

These projects require the use of a programmable calculator or computer. In the Section 5.4 projects—see Figs. 5.4.10 and 5.4.11—we discussed programs and computer algebra system commands that can be used to calculate the Riemann sums

$$L_n, \quad \text{the left-endpoint approximation,}$$

$$R_n, \quad \text{the right-endpoint approximation, and}$$

$$M_n, \quad \text{the midpoint approximation,}$$

based on a division of $[a, b]$ into n equal-length subintervals, to approximate the integral

$$\int_a^b f(x)\,dx.$$

The Riemann sums L_n, R_n, and M_n suffice, in turn, to calculate the trapezoidal and Simpson approximations of this section. In particular, the trapezoidal approximation is given in Eq. (5) by

$$T_n = \tfrac{1}{2}(L_n + R_n).$$

If n is an *even* integer, then it follows from Eq. (10) that Simpson's approximation with n equal-length subintervals is given by

$$S_n = \tfrac{1}{3}(2M_{n/2} + T_{n/2}) = \tfrac{1}{6}(L_{n/2} + 4M_{n/2} + R_{n/2}).$$

You can use these formulas for T_n and S_n in the following projects.

PROJECT A According to Example 7, the natural logarithm (corresponding to the [LN] or, in some cases, the [LOG] key on your calculator) of the number 2 is the value of the integral

$$\ln 2 = \int_1^2 \frac{dx}{x}.$$

The value of $\ln 2$ correct to 15 decimal places is

$$\ln 2 \approx 0.69314\,71805\,59945.$$

See how many correct decimal places you can obtain in a reasonable period of time by using a Simpson's approximation procedure.

PROJECT B In Chapter 8 we will study the inverse tangent function $y = \arctan x$ (y is the angle between $-\pi/2$ and $\pi/2$ such that $\tan y = x$). There we will discover that the derivative of $y = \arctan x$ is

$$\frac{dy}{dx} = \frac{1}{1 + x^2}.$$

This implies that

$$\int_0^1 \frac{1}{1 + x^2}\,dx = \left[\arctan x\right]_0^1 = \arctan 1 - \arctan 0 = \frac{\pi}{4}.$$

It follows that the number π is the value of the integral

$$\pi = \int_0^1 \frac{4}{1 + x^2}\,dx.$$

The value of π to 15 decimal places is

$$\pi \approx 3.14159\,26535\,89793.$$

See how many correct decimal places you can obtain in a reasonable period of time by using a Simpson's approximation procedure.

Chapter 5 Review: DEFINITIONS, CONCEPTS, RESULTS

Use the following list as a guide to concepts that you may need to review.

1. Antidifferentiation and antiderivatives
2. The most general antiderivative of a function
3. Integral formulas
4. Initial value problems
5. Velocity and acceleration
6. Solution of problems involving constant acceleration

7. Properties of area
8. Summation notation
9. The area under the graph of f from a to b
10. Inscribed and circumscribed rectangular polygons
11. A partition of $[a, b]$
12. The mesh of a partition
13. A Riemann sum associated with a partition
14. The definite integral of f from a to b

15. Existence of the integral of a continuous function

16. The integral as the limit of a sequence of Riemann sums

17. Regular partitions and

$$\lim_{\Delta x \to 0} \sum_{i=1}^{n} f(x_i^*) \, \Delta x$$

18. Riemann sums

19. The constant multiple, interval union, and comparison properties of integrals

20. Evaluation of definite integrals by using antiderivatives

21. The average value of $f(x)$ on the interval $[a, b]$

22. The average value theorem

23. The fundamental theorem of calculus

24. Indefinite integrals

25. The method of integration by substitution

26. Transforming the limits in integration by substitution

27. The area between $y = f(x)$ and $y = g(x)$ by integration with respect to x

28. The area between $x = f(y)$ and $x = g(y)$ by integration with respect to y

29. The right-endpoint and left-endpoint approximations

30. The trapezoidal approximation

31. The midpoint approximation

32. Simpson's approximation

33. Error estimates for the trapezoidal approximation and Simpson's approximation

Chapter 5 Miscellaneous Problems

Find the indefinite integrals in Problems 1 through 24. In Problems 13 through 24, use the indicated substitution.

1. $\displaystyle\int \frac{x^5 - 2x + 5}{x^3} \, dx$

2. $\displaystyle\int \sqrt{x}(1 + \sqrt{x})^3 \, dx$

3. $\displaystyle\int (1 - 3x)^9 \, dx$

4. $\displaystyle\int \frac{7}{(2x + 3)^3} \, dx$

5. $\displaystyle\int \sqrt[3]{9 + 4x} \, dx$

6. $\displaystyle\int \frac{24}{\sqrt{6x + 7}} \, dx$

7. $\displaystyle\int x^3(1 + x^4)^5 \, dx$

8. $\displaystyle\int 3x^2 \sqrt{4 + x^3} \, dx$

9. $\displaystyle\int x \sqrt[3]{1 - x^2} \, dx$

10. $\displaystyle\int \frac{3x}{\sqrt{1 + 3x^2}} \, dx$

11. $\displaystyle\int (7 \cos 5x - 5 \sin 7x) \, dx$

12. $\displaystyle\int (5 \sin^3 4x \cos 4x) \, dx$

13. $\displaystyle\int x^3 \sqrt{1 + x^4} \, dx; \quad u = x^4$

14. $\displaystyle\int \sin^2 x \cos x \, dx; \quad u = \sin x$

15. $\displaystyle\int \frac{1}{\sqrt{x}(1 + \sqrt{x})^2} \, dx; \quad u = 1 + \sqrt{x}$

16. $\displaystyle\int \frac{1}{\sqrt{x}(1 + \sqrt{x})^2} \, dx; \quad u = \sqrt{x}$

17. $\displaystyle\int x^2 \cos 4x^3 \, dx; \quad u = 4x^3$

18. $\displaystyle\int x(x + 1)^{14} \, dx; \quad u = x + 1$

19. $\displaystyle\int x(x^2 + 1)^{14} \, dx; \quad u = x^2 + 1$

20. $\displaystyle\int x^3 \cos x^4 \, dx; \quad u = x^4$

21. $\displaystyle\int x \sqrt{4 - x} \, dx; \quad u = 4 - x$

22. $\displaystyle\int \frac{x + 2x^3}{(x^4 + x^2)^3} \, dx; \quad u = x^4 + x^2$

23. $\displaystyle\int \frac{2x^3}{\sqrt{1 + x^4}} \, dx; \quad u = x^4$

24. $\displaystyle\int \frac{2x + 1}{\sqrt{x^2 + x}} \, dx; \quad u = x^2 + x$

Solve the initial value problems in Problems 25 through 30.

25. $\dfrac{dy}{dx} = 3x^2 + 2x; \quad y(0) = 5$

26. $\dfrac{dy}{dx} = 3\sqrt{x}; \quad y(4) = 20$

27. $\dfrac{dy}{dx} = (2x + 1)^5; \quad y(0) = 2$

28. $\dfrac{dy}{dx} = \dfrac{2}{\sqrt{x + 5}}; \quad y(4) = 3$

29. $\dfrac{dy}{dx} = \dfrac{1}{\sqrt[3]{x}}; \quad y(1) = 1$

30. $\dfrac{dy}{dx} = 1 - \cos x; \quad y(0) = 0$

31. When its brakes are fully applied, a certain automobile has constant deceleration of 22 ft/s². If its initial velocity is 90 mi/h, how long will it take to come to a stop? How many feet will it travel during that time?

32. In Hal Clement's novel *Mission of Gravity*, much of the action takes place in the polar regions of the planet Mesklin, where the acceleration of gravity is 22,500 ft/s². A stone is dropped near the north pole of Mesklin from a height of 450 ft. How long does it remain aloft? With what speed does it strike the ground?

33. An automobile is traveling along the x-axis in the positive direction. At time $t = 0$ its brakes are fully applied, and the car experiences a constant deceleration of 40 ft/s² while skidding. The car skids 180 ft before coming to a stop. What was its initial velocity?

34. If a car starts from rest with an acceleration of 8 ft/s², how far has it traveled by the time it reaches a speed of 60 mi/h?

35. On the planet Zorg, a ball dropped from a height of 20 ft hits the ground in 2 s. If the ball is dropped from the top of a 200-ft building on this planet, how long will it take to reach the ground? With what speed will it hit?

36. Suppose that you can throw a ball from the earth's surface straight upward to a maximum height of 144 ft. (a) How high could you throw it on the planet of Problem 35? (b) How high could you throw it in the polar regions of Mesklin (see Problem 32)?

37. Suppose that a car skids 44 ft if its velocity is 30 mi/h when the brakes are fully applied. Assuming the same constant deceleration, how far will it skid if its velocity is 60 mi/h when the brakes are fully applied?

38. The graph of the velocity of a model rocket fired at time $t = 0$ is shown in Fig. 5.MP.1. (a) At what time was the fuel exhausted? (b) At what time did the parachute open? (c) At what time did the rocket reach its maximum altitude? (d) At what time did the rocket land? (e) How high did the rocket go? (f) How high was the pole on which the rocket landed?

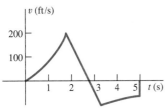

Fig. 5.MP.1 Rocket velocity graph for Problem 38

Find the sums in Problems 39 through 42.

39. $\sum_{i=1}^{100} 17$

40. $\sum_{k=1}^{100} \left(\frac{1}{k} - \frac{1}{k+1} \right)$

41. $\sum_{n=1}^{10} (3n - 2)^2$

42. $\sum_{n=1}^{16} \sin \frac{n\pi}{2}$

In Problems 43 through 45, find the limit of the given Riemann sum associated with a regular partition of the indicated interval [a, b]. First express it as an integral from a to b; then evaluate that integral.

43. $\lim_{n \to \infty} \sum_{i=1}^{n} \frac{\Delta x}{\sqrt{x_i^*}}$;   [1, 2]

44. $\lim_{n \to \infty} \sum_{i=1}^{n} [(x_i^*)^2 - 3x_i^*] \, \Delta x$;   [0, 3]

45. $\lim_{n \to \infty} \sum_{i=1}^{n} 2\pi x_i^* \sqrt{1 + (x_i^*)^2} \, \Delta x$;   [0, 1]

46. Evaluate

$$\lim_{n \to \infty} \frac{1^{10} + 2^{10} + 3^{10} + \cdots + n^{10}}{n^{11}}$$

by expressing this limit as an integral over [0, 1].

47. Use Riemann sums to prove that if $f(x) \equiv c$ (a constant), then

$$\int_a^b f(x) \, dx = c(b - a).$$

48. Use Riemann sums to prove that if f is continuous on $[a, b]$ and $f(x) \geqq 0$ for all x in $[a, b]$, then

$$\int_a^b f(x) \, dx \geqq 0.$$

49. Use the comparison property of integrals (Section 5.5) to prove that

$$\int_a^b f(x) \, dx > 0$$

if f is a continuous function with $f(x) > 0$ on $[a, b]$.

Evaluate the integrals in Problems 50 through 63.

50. $\int_0^1 (1 - x^2)^3 \, dx$

51. $\int \left(\sqrt{2x} - \frac{1}{\sqrt{3x^3}} \right) dx$

52. $\int \frac{(1 - \sqrt[3]{x})^2}{\sqrt{x}} \, dx$

53. $\int \frac{4 - x^3}{2x^2} \, dx$

54. $\int_0^1 \frac{dt}{(3 - 2t)^2}$

55. $\int \sqrt{x} \cos x \sqrt{x} \, dx$

56. $\int_0^2 x^2 \sqrt{9 - x^3} \, dx$

57. $\int \frac{1}{t^2} \sin \frac{1}{t} \, dt$

58. $\int_1^2 \frac{2t + 1}{\sqrt{t^2 + t}} \, dt$

59. $\int \frac{\sqrt[3]{u}}{(1 + u^{4/3})^3} \, du$

60. $\int_0^{\pi/4} \frac{\sin t}{\sqrt{\cos t}} \, dt$

61. $\int_1^4 \frac{(1 + \sqrt{t})^2}{\sqrt{t}} \, dt$

62. $\displaystyle\int \frac{1}{u^2}\sqrt[3]{1 - \frac{1}{u}}\, du$ **63.** $\displaystyle\int \frac{\sqrt{4x^2 - 1}}{x^4}\, dx$

Find the areas of the regions bounded by the curves given in Problems 64 through 70.

64. $y = x^3$, $x = -1$, $y = 1$

65. $y = x^4$, $y = x^5$

66. $y^2 = x$, $3y^2 = x + 6$

67. $y = x^4$, $y = 2 - x^2$

68. $y = x^4$, $y = 2x^2 - 1$

69. $y = (x - 2)^2$, $y = 10 - 5x$

70. $y = x^{2/3}$, $y = 2 - x^2$

71. Evaluate the integral

$$\int_0^2 \sqrt{2x - x^2}\, dx$$

by interpreting it as the area of a region.

72. Evaluate the integral

$$\int_1^5 \sqrt{6x - 5 - x^2}\, dx$$

by interpreting it as the area of a region.

73. Find a function f such that

$$x^2 = 1 + \int_1^x \sqrt{1 + [f(t)]^2}\, dt$$

for all $x > 1$. [*Suggestion:* Differentiate both sides of the equation with the aid of the fundamental theorem of calculus.]

74. Show that $G'(x) = \phi(h(x)) \cdot h'(x)$ if

$$G(x) = \int_a^{h(x)} \phi(t)\, dt.$$

75. Use right-endpoint and left-endpoint approximations to estimate

$$\int_0^1 \sqrt{1 + x^2}\, dx$$

with error not exceeding 0.05.

76. Calculate the trapezoidal approximation and Simpson's approximation to

$$\int_0^\pi \sqrt{1 - \cos x}\, dx$$

with six subintervals. For comparison, use an appropriate half-angle identity to calculate the exact value of this integral.

77. Calculate the midpoint approximation and trapezoidal approximation to

$$\int_1^2 \frac{dx}{x + x^2}$$

with $n = 5$ subintervals. Then explain why the exact value of the integral lies between these two approximations.

In Problems 78 through 80, let $\{x_0, x_1, x_2, \ldots, x_n\}$ be a partition of $[a, b]$, where $a < b$.

78. For $i = 1, 2, 3, \ldots, n$, let x_i^* be given by

$$(x_i^*)^2 = \tfrac{1}{3}[(x_{i-1})^2 + x_{i-1}x_i + (x_i)^2].$$

Show first that $x_{i-1} < x_i^* < x_i$. Then use the algebraic identity

$$(c - d)(c^2 + cd + d^2) = c^3 - d^3$$

to show that

$$\sum_{i=1}^n (x_i^*)^2\, \Delta x_i = \tfrac{1}{3}(b^3 - a^3).$$

Explain why this computation proves that

$$\int_a^b x^2\, dx = \tfrac{1}{3}(b^3 - a^3).$$

79. Let $x_i^* = \sqrt{x_{i-1}x_i}$ for $i = 1, 2, 3, \ldots, n$, and assume that $0 < a < b$. Show that

$$\sum_{i=1}^n \frac{\Delta x_i}{(x_i^*)^2} = \frac{1}{a} - \frac{1}{b}.$$

Then explain why this computation proves that

$$\int_a^b \frac{dx}{x^2} = \frac{1}{a} - \frac{1}{b}.$$

80. Assume that $0 < a < b$. Define x_i^* by means of the equation $(x_i^*)^{1/2}(x_i - x_{i-1}) = \tfrac{2}{3}[(x_i)^{3/2} - (x_{i-1})^{3/2}]$. Show that $x_{i-1} < x_i^* < x_i$. Then use this selection for the given partition to prove that

$$\int_a^b \sqrt{x}\, dx = \tfrac{2}{3}(b^{3/2} - a^{3/2}).$$

81. We derived Eq. (12) for Simpson's approximation in Section 5.9 by giving the midpoint approximation "weight" $\tfrac{2}{3}$ and the trapezoidal approximation weight $\tfrac{1}{3}$. Why not use some *other* weights? Read the discussion in Section 5.9 of parabolic approximations; then work Problem 48 of Section 5.8. This should provide most of the answer.

6

Applications of the Integral

G. F. B. Riemann (1826–1866)

☐ The general concept of integration traces back to the area and volume computations of ancient times, but the integrals used by Newton and Leibniz were not defined with sufficient precision for full understanding. We owe to the German mathematician G. F. Bernhard Riemann the modern definition that uses "Riemann sums." The son of a Protestant minister, Riemann studied theology and philology at Göttingen University (in Germany) until he finally gained his father's permission to concentrate on mathematics. He transferred to Berlin University, where he received his Ph.D. in 1851. The work he did in the next decade justifies his place on everyone's short list of the most profound and creative mathe-

maticians of all time. But in 1862 he was stricken ill. He never fully recovered and in 1866 died prematurely at the age of 39.

☐ Riemann's mathematical investigations were as varied as they were deep, ranging from the basic concepts of functions and integrals to such areas as non-Euclidean (differential) geometry and the distribution of prime numbers. Recall that the integer p is *prime* provided it cannot be factored into smaller integers. In a famous paper of 1859, Riemann analyzed the approximation

$$\pi(x) \approx \int_2^x \frac{dt}{\ln t} = \mathrm{li}(x)$$

to the number $\pi(x)$ of those primes $p \leqq x$ (with $\ln x$ denoting the natural logarithm of x). There is a remarkable correspondence between the values of $\pi(x)$ and the "logarithmic integral" approximation $\mathrm{li}(x)$:

x	1,000,000	10,000,000
$\mathrm{li}(x)$	78,628	664,918
$\pi(x)$	78,498	664,579
error	0.165%	0.051%

x	100,000,000	1,000,000,000
$\mathrm{li}(x)$	5,762,209	50,849,235
$\pi(x)$	5,761,455	50,847,534
error	0.013%	0.003%

Thirty years after Riemann's death, his ideas led ultimately to

a proof that the percentage error in the approximation $\mathrm{li}(x)$ to $\pi(x)$ approaches 0 as $x \to \infty$.

☐ In his 1851 thesis, Riemann introduced a geometric way of visualizing "multi-valued" functions such as the square root function with two values $\pm \sqrt{x}$. The following graph illustrates the *cube root* function. For each complex number $z = x + yi$ in the unit disk $x^2 + y^2 \leqq 1$, the three (complex) cube roots of z are plotted directly above z. Each root is plotted at a height equal to its real part, with color determined by its imaginary part. The result is the "Riemann surface" of the cube root function.

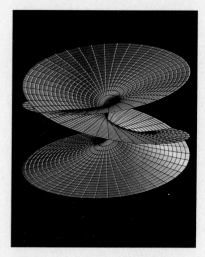

Image created using MATLAB, courtesy of The MathWorks, Inc., Natick, MA.

6.1

Setting Up Integral Formulas

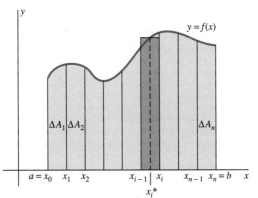

x_1^* x_2^* x_3^* x_i^* x_n^*

$a = x_0$ x_1 x_2 x_3 ··· x_{i-1} x_i ··· x_{n-1} $x_n = b$

Fig. 6.1.1 A division (or partition) of $[a, b]$ into n equal-length subintervals

In Section 5.4 we defined the integral of the function f on the interval $[a, b]$ as a limit of Riemann sums. Specifically, let the interval $[a, b]$ be divided into n subintervals, all with the same length $\Delta x = (b - a)/n$ (Fig. 6.1.1). Then a selection of numbers $x_1^*, x_2^*, \ldots, x_n^*$ in these subintervals (x_i^* being a point of the ith subinterval $[x_{i-1}, x_i]$) produces a Riemann sum

$$\sum_{i=1}^{n} f(x_i^*) \, \Delta x \tag{1}$$

whose value approximates the integral of f on $[a, b]$. The value of the integral is the limiting value (if any) of such sums as the subinterval length Δx approaches zero. That is,

$$\int_a^b f(x) \, dx = \lim_{\Delta x \to 0} \sum_{i=1}^{n} f(x_i^*) \, \Delta x. \tag{2}$$

The wide applicability of the definite integral arises from the fact that many geometric and physical quantities can be approximated arbitrarily closely by Riemann sums. Such approximations lead to integral formulas for the computation of such quantities.

For example, suppose that $f(x)$ is positive-valued on $[a, b]$ and that our goal—as in Section 5.3—is to calculate the area A of the region that lies below the graph of $y = f(x)$ over the interval $[a, b]$. Beginning with the subdivision (or partition) of $[a, b]$ indicated in Fig. 6.1.1, let ΔA_i denote the area of the vertical "strip" that lies under $y = f(x)$ over the ith subinterval $[x_{i-1}, x_i]$. Then, as illustrated in Fig. 6.1.2, the "strip areas"

$$\Delta A_1, \Delta A_2, \ldots, \Delta A_n$$

add up to the total area A:

$$A = \sum_{i=1}^{n} \Delta A_i. \tag{3}$$

But the ith strip is approximated by a rectangle with base $[x_{i-1}, x_i]$ and height $f(x_i^*)$, so its area is given approximately by

$$\Delta A_i \approx f(x_i^*) \, \Delta x. \tag{4}$$

After we substitute Eq. (4) into Eq. (3), it becomes apparent that the total area A under the graph of f is given approximately by

$$A \approx \sum_{i=1}^{n} f(x_i^*) \, \Delta x. \tag{5}$$

Fig. 6.1.2 Approximating an area by means of a Riemann sum

Note that the approximating sum on the right is a Riemann sum for f on $[a, b]$. Moreover,

1. It is intuitively evident that the Riemann sum in (5) approaches the actual area A as $n \to +\infty$ (which forces $\Delta x \to 0$);
2. By the definition of the integral, this Riemann sum approaches $\displaystyle\int_a^b f(x)\,dx$ as $n \to +\infty$.

These observations justify the *definition* of the area A by the formula

$$A = \int_a^b f(x)\,dx. \tag{6}$$

OTHER QUANTITIES AS INTEGRALS

Our justification of the area formula in Eq. (6) illustrates an important general method of setting up integral formulas. Suppose that we want to calculate a certain quantity Q that is associated with an interval $[a, b]$ in such a way that subintervals of $[a, b]$ correspond to specific portions of Q (such as the portion of area lying above a particular subinterval). Then a subdivision of $[a, b]$ into n subintervals produces portions

$$\Delta Q_1, \Delta Q_2, \ldots, \Delta Q_n,$$

which add up to the quantity

$$Q = \sum_{i=1}^n \Delta Q_i. \tag{7}$$

Now suppose that we can find a function f such that the ith portion ΔQ_i is given approximately by

$$\Delta Q_i \approx f(x_i^*)\,\Delta x \tag{8}$$

(for each i, $1 \le i \le n$) for a selected point x_i^* of the ith subinterval $[x_{i-1}, x_i]$ of $[a, b]$. Then substitution of Eq. (8) into Eq. (7) yields the Riemann sum approximation

$$Q \approx \sum_{i=1}^n f(x_i^*)\,\Delta x \tag{9}$$

analogous to the approximation in Eq. (5). The right-hand sum in Eq. (9) is a Riemann sum that approaches the integral

$$\int_a^b f(x)\,dx \qquad \text{as} \qquad n \to +\infty.$$

If it is also evident—for geometric or physical reasons, for example—that this Riemann sum must approach the quantity Q as $n \to +\infty$, then Eq. (9) justifies our setting up the integral formula

$$Q = \int_a^b f(x)\,dx. \tag{10}$$

and

$$\int_{15}^{20} (30 - 2t)\, dt = \left[30t - t^2 \right]_{15}^{20} = -25 \quad \text{(ft)},$$

we see that the particle travels 225 ft forward and then 25 ft backward, for a total distance traveled of 250 ft.

6.1 Problems

In Problems 1 through 10, x_i^* denotes a selected point, and m_i the midpoint, of the ith subinterval $[x_{i-1}, x_i]$ of a partition of the indicated interval $[a, b]$ into n subintervals each of length Δx. Evaluate the given limit by computing the value of the appropriate related integral.

1. $\lim\limits_{n \to \infty} \sum\limits_{i=1}^{n} 2x_i^* \, \Delta x; \quad a = 0, b = 1$

2. $\lim\limits_{n \to \infty} \sum\limits_{i=1}^{n} \dfrac{\Delta x}{(x_i^*)^2}; \quad a = 1, b = 2$

3. $\lim\limits_{n \to \infty} \sum\limits_{i=1}^{n} (\sin \pi x_i^*) \, \Delta x; \quad a = 0, b = 1$

4. $\lim\limits_{n \to \infty} \sum\limits_{i=1}^{n} [3(x_i^*)^2 - 1] \, \Delta x; \quad a = -1, b = 3$

5. $\lim\limits_{n \to \infty} \sum\limits_{i=1}^{n} x_i^* \sqrt{(x_i^*)^2 + 9} \, \Delta x; \quad a = 0, b = 4$

6. $\lim\limits_{n \to \infty} \sum\limits_{i=1}^{n} (x_i)^2 \, \Delta x; \quad a = 2, b = 4$

7. $\lim\limits_{n \to \infty} \sum\limits_{i=1}^{n} (2m_i - 1) \, \Delta x; \quad a = -1, b = 3$

8. $\lim\limits_{n \to \infty} \sum\limits_{i=1}^{n} \sqrt{2m_i + 1} \, \Delta x; \quad a = 0, b = 4$

9. $\lim\limits_{n \to \infty} \sum\limits_{i=1}^{n} \dfrac{m_i}{\sqrt{(m_i)^2 + 16}} \, \Delta x; \quad a = -3, b = 0$

10. $\lim\limits_{n \to \infty} \sum\limits_{i=1}^{n} m_i \cos(m_i)^2 \, \Delta x; \quad a = 0, b = \sqrt{\pi}$

The notation in Problems 11 through 14 is the same as in Problems 1 through 10. Express the given limit as an integral involving the function f.

11. $\lim\limits_{n \to \infty} \sum\limits_{i=1}^{n} 2\pi x_i^* f(x_i^*) \, \Delta x; \quad a = 1, b = 4$

12. $\lim\limits_{n \to \infty} \sum\limits_{i=1}^{n} [f(x_i^*)]^2 \, \Delta x; \quad a = -1, b = 1$

13. $\lim\limits_{n \to \infty} \sum\limits_{i=1}^{n} \sqrt{1 + [f(x_i^*)]^2} \, \Delta x; \quad a = 0, b = 10$

14. $\lim\limits_{n \to \infty} \sum\limits_{i=1}^{n} 2\pi m_i \sqrt{1 + [f(m_i)]^2} \, \Delta x; \quad a = -2, b = 3$

In Problems 15 through 18, a rod coinciding with the interval $[a, b]$ on the x-axis (units in centimeters) has the specified density function $\rho(x)$ that gives its density (in grams per centimeter) at the point x. Find the mass M of the rod.

15. $a = 0, b = 100; \quad \rho(x) = \frac{1}{5}x$

16. $a = 0, b = 25; \quad \rho(x) = 60 - 2x$

17. $a = 0, b = 10; \quad \rho(x) = x(10 - x)$

18. $a = 0, b = 10; \quad \rho(x) = 10 \sin \dfrac{\pi x}{10}$

In Problems 19 through 28, compute both the net distance and the total distance traveled between time $t = a$ and $t = b$ by a particle moving with the given velocity function $v = f(t)$ along a line.

19. $v = -32; \quad a = 0, b = 10$

20. $v = 2t + 10; \quad a = 1, b = 5$

21. $v = 4t - 25; \quad a = 0, b = 10$

22. $v = |2t - 5|; \quad a = 0, b = 5$

23. $v = 4t^3; \quad a = -2, b = 3$

24. $v = t - \dfrac{1}{t^2}; \quad a = 0.1, b = 1$

25. $v = \sin 2t; \quad a = 0, b = \dfrac{\pi}{2}$

26. $v = \cos 2t; \quad a = 0, b = \dfrac{\pi}{2}$

27. $v = 10e^{-t}; \quad a = 0, b = 100$

28. $v = \dfrac{1}{t} - \dfrac{1}{2}; \quad a = 1, b = 10$

29. Suppose that the circular disk of Example 3 has mass density $\rho(x)$ (in grams per square centimeter) at distance x from the origin. Then the annular ring of Figs. 6.1.5 and 6.1.6 has density approximately $\rho(x_i^*)$ at each point. Conclude that the mass M of this disk of radius r is given by

$$M = \int_0^r 2\pi x \rho(x) \, dx.$$

In Problems 30 and 31, use the result of Problem 29 to find the mass of a circular disk with the given radius r and density function ρ.

30. $r = 10; \quad \rho(x) = x$ 31. $r = 5; \quad \rho(x) = 25 - x^2$

32. If a particle is thrown straight upward from the ground with an initial velocity of 160 ft/s, then its velocity after t seconds is $v = -32t + 160$ feet/second, and it attains its maximum height when $t = 5$ s (and $v = 0$). Use Eq. (13) to compute this maximum height. Check your answer by the methods of Section 5.2.

33. Suppose that the rate of water flow into an initially empty tank is $100 - 3t$ gallons/minute at time t (in minutes). How much water flows into the tank during the interval from $t = 10$ to $t = 20$ min?

34. Suppose that the birth rate in Calgary t years after 1970 was $13 + t$ thousands of births per year. Set up and evaluate an appropriate integral to compute the total number of births that occurred between 1970 and 1990.

35. Assume that the city of Problem 34 had a death rate of $5 + (t/2)$ thousands per year t years after 1970. If the population of the city was 125,000 in 1970, what was its population in 1990? Consider both births and deaths.

36. The average daily rainfall in Sioux City is $r(t)$ inches/day at time t (in days), $0 \leqq t \leqq 365$. Begin with a partition of the interval $[0, 365]$ and derive the formula

$$R = \int_0^{365} r(t)\, dt$$

for the average total annual rainfall R.

37. Take the average daily rainfall of Problem 36 to be

$$r(t) = a - b \cos \frac{2\pi t}{365},$$

where a and b are constants to be determined. If the value of $r(t)$ on January 1 ($t = 0$) is 0.1 in. and the value of $r(t)$ on July 1 ($t = 182.5$) is 0.5 in., what is the average total annual rainfall in this locale?

38. Suppose that the rate of water flow into a tank is $r(t)$ liters/minute at time t (in minutes). Use the method of Example 1 to derive the formula

$$Q = \int_a^b r(t)\, dt$$

for the amount of water that flows into the tank between times $t = a$ and $t = b$.

39. Evaluate

$$\lim_{n \to \infty} \frac{\sqrt[3]{1} + \sqrt[3]{2} + \sqrt[3]{3} + \ldots + \sqrt[3]{n}}{n^{4/3}}$$

by first finding a function f such that the limit is equal to

$$\int_0^1 f(x)\, dx.$$

40. In this problem you are to derive the volume formula $V = \frac{4}{3}\pi r^3$ for a spherical ball of radius r, assuming as *known* the formula $S = 4\pi r^2$ for the surface area of a sphere of radius r. Assume it follows that the volume of a thin spherical shell of thickness t (Fig. 6.1.7) is given approximately by $\Delta V \approx S \cdot t = 4\pi r^2 t$. Then divide the spherical ball into concentric spherical shells, analogous to the concentric annular rings of Example 3. Finally, interpret the sum of the volumes of these spherical shells as a Riemann sum.

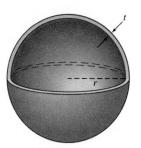

Fig. 6.1.7 A thin spherical shell of thickness t and inner radius r (Problem 40)

41. A spherical ball has radius 1 ft and, at distance x from its center, its density is $100(1 + x)$ lb/ft³. Use Riemann sums to find a function $f(x)$ such that the weight of the ball is

$$W = \int_0^1 f(x)\, dx$$

(in pounds). Then compute W by evaluating this integral. [*Suggestion:* Given a partition $0 = x_0 < x_1 < \ldots < x_n = 1$ of $[0, 1]$, estimate the weight ΔW_i of the spherical shell $x_{i-1} \leqq x \leqq x_i$ of the ball.]

6.2

Volumes by the Method of Cross Sections

Here we use integrals to calculate the volumes of certain solids or regions in space. We begin with an intuitive idea of volume as a measure of solids, analogous to area as a measure of plane regions. In particular, we assume that every simply expressible bounded solid region R has a volume measured by a nonnegative number $v(R)$ such that

❏ If R consists of two nonoverlapping pieces, then $v(R)$ is the sum of *their* volumes;

❏ Two different solids have the same volume if they have the same size and shape.

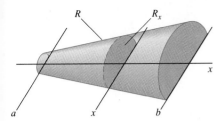

Fig. 6.2.1 R_x is the cross section of R in the plane perpendicular to the x-axis at x.

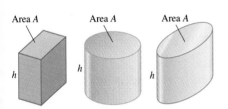

Fig. 6.2.2 Every cylinder of height h and base area A has volume $V = Ah$.

The **method of cross sections** is a way of computing the volume of a solid that is described in terms of its cross sections (or "slices") in planes perpendicular to a fixed *reference line* (such as the x-axis or y-axis). For instance, Fig. 6.2.1 shows a solid R with volume $V = v(R)$ lying alongside the interval $[a, b]$ on the x-axis. That is, a plane perpendicular to the x-axis intersects the solid if and only if this plane meets the x-axis in a point of $[a, b]$. Let R_x denote the intersection of R with the perpendicular plane that meets the x-axis at the point x of $[a, b]$. We call R_x the (plane) **cross section** of the solid R at x.

This situation is especially simple if all the cross sections of R are congruent to one another and are parallel translations of each other. In this case the solid R is called a **cylinder** with **bases** R_a and R_b and **height** $h = b - a$. If R_a and R_b are circular disks, then R is the familiar **circular cylinder**. Recall that the volume formula for a circular cylinder of height h and circular base of radius r and area $A = \pi r^2$ is

$$V = \pi r^2 h = Ah.$$

Figure 6.2.2 shows several (general) cylinders with bases of various shapes. The method of cross sections is based on the fact that the volume V of any cylinder—circular or not—is equal to the product of the cylinder's height h and the area A of its base:

$$V = Ah \qquad \text{(volume of a cylinder)}. \tag{1}$$

The volume of a more general solid, as in Fig. 6.2.1, can be approximated by using cylinders. For each x in $[a, b]$, let $A(x)$ denote the area of the cross section R_x of the solid R:

$$A(x) = a(R_x). \tag{2}$$

We shall assume that the shape of R is sufficiently simple that this **cross-sectional area function** A is continuous (and therefore integrable).

To set up an integral formula for $V = v(R)$, we begin with a division of $[a, b]$ into n subintervals, all with the same length $\Delta x = (b - a)/n$. Let R_i denote the slab or slice of the solid R positioned alongside the ith subinterval $[x_{i-1}, x_i]$ (Fig. 6.2.3). We denote the volume of this ith slice of R by $\Delta V_i = v(R_i)$, so

$$V = \sum_{i=1}^{n} \Delta V_i.$$

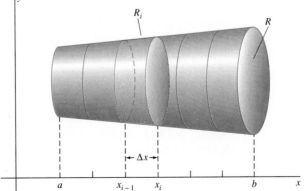

Fig. 6.2.3 Planes through the partition points $x_0, x_1, x_2, \ldots, x_n$ partition the solid R into slabs $R_1, R_2, \ldots, R_n$.

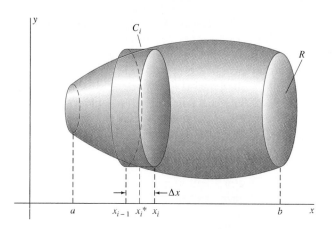

Fig. 6.2.4 The slab R_i is approximated by the cylinder C_i of volume $A(x_i^*)\,\Delta x$.

To approximate ΔV_i, we select a typical point x_i^* in $[x_{i-1},\ x_i]$ and consider the *cylinder* C_i whose height is Δx and whose base is the cross section $R_{x_i^*}$ of R at x_i^*. Figure 6.2.4 suggests that if Δx is small, then $v(C_i)$ is a good approximation to $\Delta V_i = v(R_i)$:

$$\Delta V_i \approx v(C_i) = a(R_{x_i^*})\cdot\Delta x = A(x_i^*)\,\Delta x,$$

a consequence of Eq. (1) with $A = A(x_i^*)$ and $h = \Delta x$.

Then we add the volumes of these approximating cylinders for $i = 1, 2, 3, \ldots, n$. We find that

$$V = \sum_{i=1}^{n} \Delta V_i \approx \sum_{i=1}^{n} A(x_i^*)\,\Delta x.$$

We recognize the approximating sum on the right to be a Riemann sum that approaches $\int_a^b A(x)\,dx$ as $n \to +\infty$. This justifies the following *definition* of the volume of a solid R in terms of its cross-sectional area function $A(x)$.

> **Definition** *Volume by Cross Sections*
> If the solid R lies alongside the interval $[a, b]$ on the x-axis and has continuous cross-sectional area function $A(x)$, then its volume $V = v(R)$ is
>
> $$V = \int_a^b A(x)\,dx. \tag{3}$$

Equation (3) is known as **Cavalieri's principle**, after the Italian mathematician Bonaventura Cavalieri (1598–1647), who systematically exploited the fact that the volume of a solid is determined by the areas of its cross sections perpendicular to a given reference line.

EXAMPLE 1 Figure 6.2.5(a) shows a square-based pyramid oriented so that its height h corresponds to the interval $[0, h]$ on the x-axis. Its *base* is a b-by-b square, and each cross section perpendicular to the x-axis is also a square. To find the area $A(x)$ of the s-by-s cross section at x, we equate height-to-length ratios in the similar triangles of Fig. 6.2.5(b):

$$\frac{s}{x} = \frac{b}{h}, \quad \text{so} \quad s = \frac{b}{h}x.$$

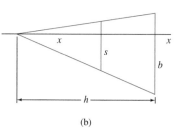

(a)

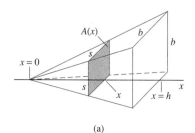

(b)

Fig. 6.2.5 The square-based pyramid of Example 1

Ch. 6 / Applications of the Integral

Therefore,

$$A(x) = s^2 = \frac{b^2}{h^2}x^2,$$

and Eq. (3)—with $[0, h]$ as the interval of integration—gives

$$V = \int_0^h A(x)\,dx = \int_0^h \frac{b^2}{h^2}x^2\,dx = \left[\frac{b^2}{h^2} \cdot \frac{x^3}{3}\right]_{x=0}^{x=h} = \frac{1}{3}b^2h.$$

With $A = b^2$ denoting the area of the base, our result takes the form

$$V = \tfrac{1}{3}Ah$$

for the volume of a pyramid.

CROSS SECTIONS PERPENDICULAR TO THE y-AXIS

In the case of a solid R lying alongside the interval $[c, d]$ on the y-axis, we denote by $A(y)$ the area of the solid's cross section R_y in the plane perpendicular to the y-axis at the point y of $[c, d]$ (Fig. 6.2.6). A similar discussion, beginning with a partition of $[c, d]$, leads to the volume formula

$$V = \int_c^d A(y)\,dy. \tag{4}$$

SOLIDS OF REVOLUTION—DISKS AND WASHERS

An important special case of Eq. (3) gives the volume of a **solid of revolution**. For example, consider the solid R obtained by revolving around the x-axis the region under the graph of $y = f(x)$ over the interval $[a, b]$, where $f(x) \geqq 0$. Such a region and the resulting solid of revolution are shown in Fig. 6.2.7.

Because the solid R is obtained by revolution, each cross section of R at x is a circular *disk* of radius $f(x)$. The cross-sectional area function is then $A(x) = \pi[f(x)]^2$, so Eq. (3) yields

$$V = \int_a^b \pi y^2\,dx = \int_a^b \pi[f(x)]^2\,dx \tag{5}$$

for the **volume of a solid of revolution about the x-axis**.

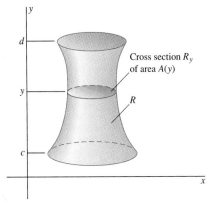

Fig. 6.2.6 $A(y)$ is the area of the cross section R_y in the plane perpendicular to the y-axis at the point y.

Cross section R_y of area $A(y)$

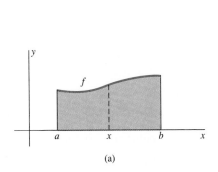

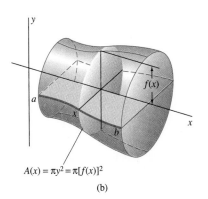

Fig. 6.2.7 **(a)** A region from which we can determine the volume of a **(b)** solid of revolution around the x-axis

$A(x) = \pi y^2 = \pi[f(x)]^2$

(a)

(b)

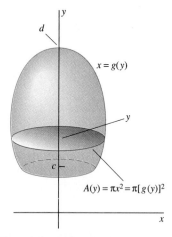

Fig. 6.2.8 Solid of revolution around the *y*-axis

In the expression $\pi y^2\,dx$, the differential dx tells us that the independent variable is x. We *must* express y (and any other dependent variable) in terms of x in order to perform the indicated integration.

By a similar argument, if the region bounded by the curve $x = g(y)$, the *y*-axis, and the horizontal lines $y = c$ and $y = d$ is rotated around the *y*-axis, then the volume of the resulting solid of revolution (Fig. 6.2.8) is

$$V = \int_c^d \pi x^2\,dy = \int_c^d \pi[g(y)]^2\,dy. \qquad (6)$$

We *must* express x (and any other dependent variable) in terms of the independent variable y before we attempt to antidifferentiate in Eq. (6).

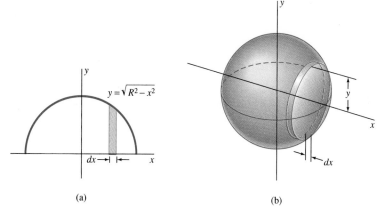

Fig. 6.2.9 (a) A semicircular region that we rotate (b) to generate a sphere (Example 2)

EXAMPLE 2 Use the method of cross sections to verify the familiar formula $V = \frac{4}{3}\pi R^3$ for the volume of a sphere of radius R.

Solution We think of the sphere as the solid of revolution obtained by revolving the semicircular plane region of Fig. 6.2.9 around the *x*-axis. This is the region bounded above by the semicircle $y = \sqrt{R^2 - x^2}$ ($-R \leqq x \leqq R$) and below by the interval $[-R, R]$ on the *x*-axis. To use Eq. (5), we take $f(x) = \sqrt{R^2 - x^2}$, $a = -R$, and $b = R$. This gives

$$V = \int_{-R}^R \pi(\sqrt{R^2 - x^2})^2\,dx = \pi \int_{-R}^R (R^2 - x^2)\,dx$$

$$= \pi\left[R^2 x - \tfrac{1}{3}x^3\right]_{-R}^R = \tfrac{4}{3}\pi R^3.$$

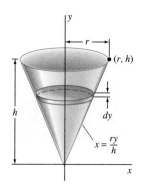

Fig. 6.2.10 Generating a cone by rotation (Example 3)

EXAMPLE 3 Use the method of cross sections to verify the familiar formula $V = \frac{1}{3}\pi r^2 h$ for the volume of a right circular cone with base radius r and height h.

Solution Figure 6.2.10 depicts the cone as the solid of revolution obtained by revolving around the *y*-axis the triangle with vertices $(0, 0)$, $(0, h)$, and (r, h). The similar triangles of Fig. 6.2.11 yield the equation $x/y = r/h$, so the radius of the circular cross section perpendicular to the *y*-axis at the point y is $x = ry/h$. Then Eq. (6), with $g(y) = ry/h$, gives

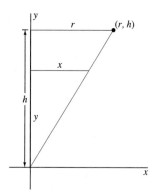

Fig. 6.2.11 Finding the radius x of the circular cross section of Example 3

$$V = \int_a^b A(y)\, dy = \int_a^b \pi x^2\, dy = \int_0^h \pi \left(\frac{ry}{h}\right)^2 dy$$

$$= \frac{\pi r^2}{h^2} \int_0^h y^2\, dy = \frac{1}{3}\pi r^2 h = \frac{1}{3}Ah,$$

where $A = \pi r^2$ is the area of the base of the cone.

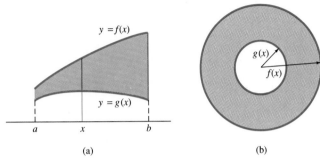

Fig. 6.2.12 **(a)** The region between two positive graphs **(b)** is rotated around the x-axis. Cross sections are annular rings.

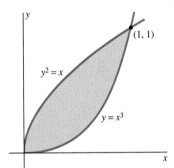

Fig. 6.2.13 The plane region of Example 4

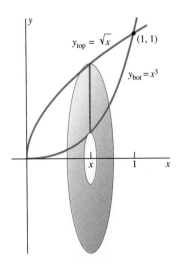

Fig. 6.2.14 Revolution around the x-axis (Example 4)

REVOLVING THE REGION BETWEEN TWO CURVES

Sometimes we need to calculate the volume of a solid generated by revolution of a plane region that lies between two given curves. Suppose that $f(x) > g(x) \geqq 0$ for x in the interval $[a, b]$ and that the solid R is generated by revolving around the x-axis the region between $y = f(x)$ and $y = g(x)$. Then the cross section at x is an **annular ring** (or **washer**) bounded by two circles (Fig. 6.2.12). The ring has inner radius $r_{\text{in}} = g(x)$ and outer radius $r_{\text{out}} = f(x)$, so the formula for the cross-sectional area of R at x is

$$A(x) = \pi (r_{\text{out}})^2 - \pi (r_{\text{in}})^2 = \pi[(y_{\text{top}})^2 - (y_{\text{bot}})^2] = \pi\{[f(x)]^2 - [g(x)]^2\},$$

where we write $y_{\text{top}} = f(x)$ and $y_{\text{bot}} = g(x)$ for the top and bottom curves of the plane region. Therefore, Eq. (3) yields

$$V = \int_a^b \pi[(y_{\text{top}})^2 - (y_{\text{in}})^2]\, dx = \int_a^b \pi\{[f(x)]^2 - [g(x)]^2\}\, dx \qquad (7)$$

for the volume V of the solid.

Similarly, if $f(y) > g(y) \geqq 0$ for $c \leqq y \leqq d$, then the volume of the solid obtained by revolving around the y-axis the region between $x_{\text{right}} = f(y)$ and $x_{\text{left}} = g(y)$ is

$$V = \int_c^d \pi[(x_{\text{right}})^2 - (x_{\text{left}})^2]\, dy = \int_c^d \pi\{[f(y)]^2 - [g(y)]^2\}\, dy. \qquad (8)$$

EXAMPLE 4 Consider the plane region shown in Fig. 6.2.13, bounded by the curves $y^2 = x$ and $y = x^3$, which intersect at the points $(0,0)$ and $(1,1)$. If this region is revolved around the x-axis (Fig. 6.2.14), then Eq. (7) with

$$y_{\text{top}} = \sqrt{x}, \qquad y_{\text{bot}} = x^3$$

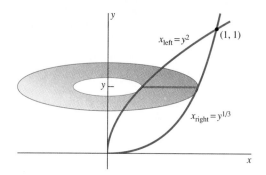

Fig. 6.2.15 Revolution around the y-axis (Example 4)

gives

$$V = \int_0^1 \pi[(\sqrt{x})^2 - (x^3)^2]\,dx = \int_0^1 \pi(x - x^6)\,dx$$

$$= \pi\left[\tfrac{1}{2}x^2 - \tfrac{1}{7}x^7\right]_0^1 = \tfrac{5}{14}\pi$$

for the volume of revolution.

If the same region is revolved around the y-axis (Fig. 6.2.15), then each cross section perpendicular to the y-axis is an annular ring with outer radius $x_{\text{right}} = y^{1/3}$ and inner radius $x_{\text{left}} = y^2$. Hence Eq. (8) gives the volume of revolution generated by this region as

$$V = \int_0^1 \pi[(y^{1/3})^2 - (y^2)^2]\,dy = \int_0^1 \pi(y^{2/3} - y^4)\,dy$$

$$= \pi\left[\tfrac{3}{5}y^{5/3} - \tfrac{1}{5}y^5\right]_0^1 = \tfrac{2}{5}\pi.$$

EXAMPLE 5 Suppose that the plane region of Example 4 (Fig. 6.2.13) is revolved about the vertical line $x = -1$ (Fig. 6.2.16). Then each cross section of the resulting solid is an annular ring with outer radius

$$r_{\text{out}} = 1 + x_{\text{right}} = 1 + y^{1/3}$$

and inner radius

$$r_{\text{in}} = 1 + x_{\text{left}} = 1 + y^2.$$

The area of such a cross section is

$$A(y) = \pi(1 + y^{1/3})^2 - \pi(1 + y^2)^2 = \pi(2y^{1/3} + y^{2/3} - 2y^2 - y^4),$$

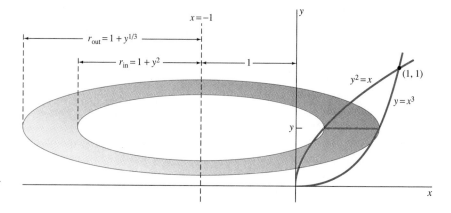

Fig. 6.2.16 The annular ring of Example 5

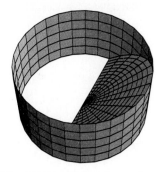

Fig. 6.2.17 The wedge and cylinder of Example 6

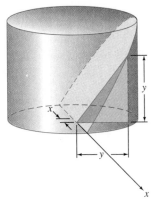

Fig. 6.2.18 A cross section of the wedge—an isosceles triangle (Example 6)

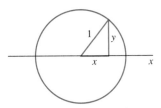

Fig. 6.2.19 The base of the cylinder of Example 6

so the volume of the resulting solid of revolution is

$$V = \int_0^1 \pi(2y^{1/3} + y^{2/3} - 2y^2 - y^4)\, dy$$

$$= \pi \left[\tfrac{3}{2} y^{4/3} + \tfrac{3}{5} y^{5/3} - \tfrac{2}{3} y^3 - \tfrac{1}{5} y^5 \right]_0^1 = \tfrac{37}{30} \pi.$$

EXAMPLE 6 Find the volume of the wedge that is cut from a circular cylinder with unit radius and unit height by a plane that passes through a diameter of the base of the cylinder and through a point on the circumference of its top.

Solution The cylinder and wedge are shown in Fig. 6.2.17. To form such a wedge, fill a cylindrical glass with cider and then drink slowly until half the bottom of the glass is exposed; the remaining cider forms the wedge.

We choose as reference line and x-axis the line through the "edge of the wedge"—the original diameter of the base of the cylinder. We can verify with similar triangles that each cross section of the wedge perpendicular to the diameter is an isosceles right triangle. One of these triangles is shown in Fig. 6.2.18. We denote by y the equal base and height of this triangle.

To determine the cross-sectional area function $A(x)$, we must express y in terms of x. Figure 6.2.19 shows the unit circular base of the original cylinder. We apply the Pythagorean theorem to the right triangle in this figure and find that $y = \sqrt{1 - x^2}$. Hence

$$A(x) = \tfrac{1}{2} y^2 = \tfrac{1}{2}(1 - x^2),$$

so Eq. (3) gives

$$V = \int_{-1}^1 A(x)\, dx = 2 \int_0^1 A(x)\, dx \quad \text{(by symmetry)}$$

$$= 2 \int_0^1 \tfrac{1}{2}(1 - x^2)\, dx = \left[x - \tfrac{1}{3} x^3 \right]_0^1 = \tfrac{2}{3}$$

for the volume of the wedge.

It is a useful habit to check answers for plausibility whenever convenient. For example, we may compare a given solid with one whose volume is already known. Because the volume of the original cylinder in Example 6 is π, we have found that the wedge occupies the fraction

$$\frac{V_{\text{wedge}}}{V_{\text{cyl}}} = \frac{\tfrac{2}{3}}{\pi} \approx 21\%$$

of the volume of the cylinder. A glance at Fig. 6.2.17 indicates that this is plausible. An error in our computations could well have given an unbelievable answer.

The wedge of Example 6 has an ancient history. Its volume was first calculated in the third century B.C. by Archimedes, who also derived the formula $V = \tfrac{4}{3} \pi r^3$ for the volume of a sphere of radius r. His work on the wedge is found in a manuscript that was discovered in 1906 after having been lost for centuries. Archimedes used a method of exhaustion for volume similar to that discussed for areas in Section 5.3.

6.2 Problems

In Problems 1 through 24, find the volume of the solid that is generated by rotating around the indicated axis the plane region bounded by the given curves.

1. $y = x^2$, $y = 0$, $x = 1$; the x-axis

2. $y = \sqrt{x}$, $y = 0$, $x = 4$; the x-axis

3. $y = x^2$, $y = 4$, $x = 0$ (first quadrant only); the y-axis (Fig. 6.2.20)

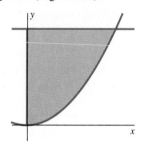 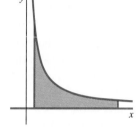

Fig. 6.2.20 Problem 3   **Fig 6.2.21** Problem 4

4. $y = 1/x$, $y = 0$, $x = 0.1$, $x = 1$; the x-axis (Fig. 6.2.21)

5. $y = \sin x$ on $[0, \pi]$, $y = 0$; the x-axis

6. $y = 9 - x^2$, $y = 0$; the x-axis

7. $y = x^2$, $x = y^2$; the x-axis (Fig 6.2.22)

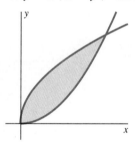 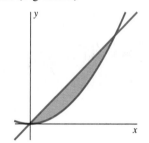

Fig. 6.2.22 Problem 7   **Fig. 6.2.23** Problem 8

8. $y = x^2$, $y = 4x$; the line $x = 5$ (Fig. 6.2.23)

9. $y = \dfrac{1}{\sqrt{x}}$, $x = 1$, $x = 4$; the x-axis

10. $x = y^2$, $x = y + 6$; the y-axis

11. $y = 1 - x^2$, $y = 0$; the x-axis (Fig. 6.2.24)

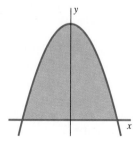

Fig. 6.2.24 Problem 11

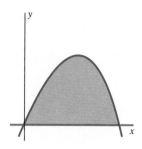

Fig. 6.2.25 Problem 12

12. $y = x - x^3$, $y = 0$ $(0 \le x \le 1)$; the x-axis (Fig. 6.2.25)

13. $y = 1 - x^2$, $y = 0$; the y-axis

14. $y = e^x$, $x = 0$, $y = 0$, $x = 1$; the x-axis

15. $y = 6 - x^2$, $y = 2$; the y-axis (Fig. 6.2.26)

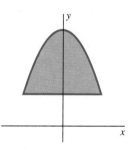

Fig. 6.2.26 Problem 15

16. $y = 1 - x^2$, $y = 0$; the vertical line $x = 2$

17. $y = x - x^3$, $y = 0$ $(0 \le x \le 1)$; the horizontal line $y = -1$

18. $y = e^x$, $y = e^{-x}$, $x = 1$; the x-axis

19. $y = 4$, $x = 0$, $y = x^2$; the y-axis

20. $x = 16 - y^2$, $x = 0$, $y = 0$; the x-axis (Fig. 6.2.27)

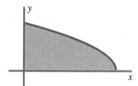

Fig. 6.2.27 Problem 20

21. $y = x^2$, $x = y^2$; the line $y = -2$

22. $y = x^2$, $y = 8 - x^2$; the line $y = -1$

23. $y = x^2$, $x = y^2$; the line $x = 3$

24. $y = 2e^{-x}$, $x = 0$, $y = 1$; the line $y = -1$

25. The region R shown in Fig. 6.2.28 is bounded by the parabolas $y^2 = x$ and $y^2 = 2(x - 3)$. Find the volume of the solid generated by rotating R about the x-axis.

368

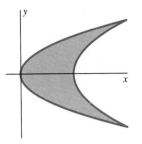

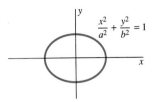

Fig. 6.2.28 The region of Problem 25

Fig. 6.2.29 The ellipse of Problems 26 and 27

26. Find the volume of the ellipsoid generated by rotating around the x-axis the region bounded by the ellipse with equation

$$\left(\frac{x}{a}\right)^2 + \left(\frac{y}{b}\right)^2 = 1$$

(Fig.6.2.29).

27. Repeat Problem 26, except rotate the elliptical region around the y-axis.

28. (a) Find the volume of the unbounded solid generated by rotating the unbounded region of Fig. 6.2.30 around the x-axis. This is the region between the graph of $y = e^{-x}$ and the x-axis for $x \geqq 1$. [*Method:* Compute the volume from $x = 1$ to $x = b$, where $b > 1$. Then find the limit of this volume as $b \to +\infty$.] (b) What happens if $y = 1/\sqrt{x}$ instead?

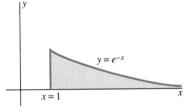

Fig. 6.2.30 The unbounded plane region of Problem 28

29. An observatory (Fig. 6.2.31) is shaped like a solid whose base is a circular disk with diameter AB of length $2a$ (Fig. 6.2.32). Find the volume of this solid if each cross section perpendicular to AB is a square.

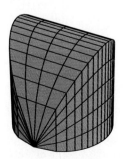

Fig. 6.2.31 The observatory of Problem 29

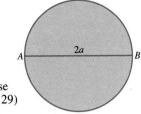

Fig. 6.2.32 The circular base of the observatory (Problem 29)

30. The base of a certain solid is a circular disk with diameter AB of length $2a$. Find the volume of the solid if each cross section perpendicular to AB is a semicircle.

31. The base of a certain solid is a circular disk with diameter AB of length $2a$. Find the volume of the solid if each cross section perpendicular to AB is an equilateral triangle.

32. The base of a solid is the region in the xy-plane bounded by the parabolas $y = x^2$ and $x = y^2$. Find the volume of this solid if every cross section perpendicular to the x-axis is a square with base in the xy-plane.

33. The *paraboloid* generated by rotating about the x-axis the region under the parabola $y^2 = 2px$, $0 \leqq x \leqq h$, is shown in Fig. 6.2.33. Show that the volume of the paraboloid is one-half that of the circumscribed cylinder also shown in the figure.

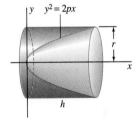

Fig. 6.2.33 The paraboloid and cylinder of Problem 33

34. A pyramid has height h and rectangular base with area A. Show that its volume is $V = \frac{1}{3}Ah$. [*Suggestion:* Note that each cross section parallel to the base is a rectangle.]

35. Repeat Problem 34, except make the base a triangle with area A.

36. Find the volume that remains after a hole of radius 3 is bored through the center of a solid sphere of radius 5 (Fig. 6.2.34).

Fig. 6.2.34 The sphere-with-hole of Problem 36

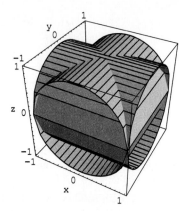

Fig. 6.2.35 The intersecting cylinders of Problem 37

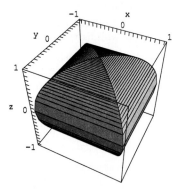

Fig. 6.2.36 The solid of intersection (Problem 37)

37. Two horizontal circular cylinders both have radius a, and their axes intersect at right angles. Find the volume of their solid of intersection (Figs. 6.2.35 and 6.2.36, where $a = 1$). Is it clear to you that each horizontal cross section of the solid is a square?

38. Figure 6.2.37 shows a "spherical segment" of height h that is cut off from a sphere of radius r by a horizontal plane. Show that its volume is

$$V = \tfrac{1}{3}\pi h^2(3r - h).$$

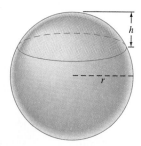

Fig. 6.2.37 A spherical segment (Problem 38)

Fig. 6.2.38 The torus of Problem 39

39. A doughnut-shaped solid, called a *torus* (Fig. 6.2.38), is generated by revolving about the y-axis the circular disk $(x - b)^2 + y^2 \leqq a^2$ centered at the point $(b, 0)$, where $0 < a < b$. Show that the volume of this torus is $V = 2\pi^2 a^2 b$. [*Suggestion:* Note that each cross section perpendicular to the y-axis is an annular ring, and recall that

$$\int_0^a \sqrt{a^2 - y^2}\, dy = \tfrac{1}{4}\pi a^2$$

because the integral represents the area of a quarter-circle of radius a.]

40. The summit of a hill is 100 ft higher than the surrounding level terrain, and each horizontal cross section of the hill is circular. The following table gives the radius r (in feet) for selected values of the height h (in feet) above the surrounding terrain. Use Simpson's approximation to estimate the volume of the hill.

h	0	25	50	75	100
r	60	55	50	35	0

41. *Newton's Wine Barrel* Consider a barrel with the shape of the solid generated by revolving around the x-axis the region under the parabola

$$y = R - kx^2, \qquad -\tfrac{1}{2}h \leqq x \leqq \tfrac{1}{2}h$$

(Fig. 6.2.39). (a) Show that the radius of each end of the barrel is $r = R - \delta$, where $4\delta = kh^2$. (b) Then show that the volume of the barrel is

$$V = \tfrac{1}{3}\pi h(2R^2 + r^2 - \tfrac{2}{5}\delta^2).$$

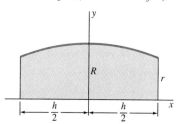

Fig. 6.2.39 The region of Problem 41

42. *The Clepsydra, or Water Clock* Consider a water tank whose side surface is generated by rotating the curve $y = kx^4$ about the y-axis (k is a positive constant). (a) Compute $V(y)$, the volume of water in the tank as a function of the depth y. (b) Suppose that water drains

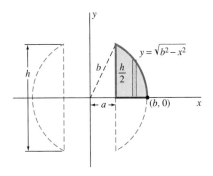

Fig. 6.3.8 Middle cross section of the sphere-with-hole (Example 2)

the region shaded in Fig. 6.3.8. This is the region below the graph of $y = \sqrt{b^2 - x^2}$ (and above the x-axis) from $x = a$ to $x = b$. The volume of the entire sphere-with-hole is then double that of the upper half, and Eq. (2) gives

$$V = 2 \int_a^b 2\pi x (b^2 - x^2)^{1/2} \, dx = 4\pi \left[-\tfrac{1}{3}(b^2 - x^2)^{3/2} \right]_a^b,$$

so

$$V = \tfrac{4}{3}\pi (b^2 - a^2)^{3/2}.$$

A way to check an answer such as this is to test it in some extreme cases. If $a = 0$ and $b = r$, which corresponds to drilling no hole at all through a sphere of radius r, then our result reduces to the volume $V = \tfrac{4}{3}\pi r^3$ of the entire sphere. If $a = b$, which corresponds to using a drill bit as large as the sphere, then $V = 0$; this, too, is correct.

REVOLVING THE REGION BETWEEN TWO CURVES

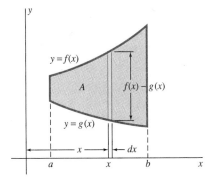

Fig. 6.3.9 The region A between the graphs of f and g over $[a, b]$ is to be rotated about the y-axis.

Now let A denote the region between the curves $y = f(x)$ and $y = g(x)$ over the interval $[a, b]$, where $0 \leq a < b$ and $g(x) \leq f(x)$ for x in $[a, b]$. Such a region is shown in Fig. 6.3.9. When A is rotated around the y-axis, it generates a solid of revolution. Suppose that we want to find the volume V of this solid. A development similar to that of Eq. (2) leads to the approximation

$$V \approx \sum_{i=1}^{n} 2\pi x_i^* [f(x_i^*) - g(x_i^*)] \, \Delta x,$$

from which we may conclude that

$$V = \int_a^b 2\pi x [f(x) - g(x)] \, dx. \qquad (3)$$

Thus

$$V = \int_a^b 2\pi x [y_{\text{top}} - y_{\text{bot}}] \, dx, \qquad (3')$$

where $y_{\text{top}} = f(x)$ and $y_{\text{bot}} = g(x)$.

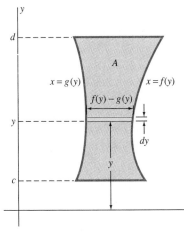

Fig. 6.3.10 The region A is to be rotated about the x-axis.

The method of cylindrical shells is also an effective way to compute volumes of solids of revolution about the x-axis. Figure 6.3.10 shows the region A bounded by the curves $x = f(y)$ and $x = g(y)$ for $c \leq y \leq d$ and by the horizontal lines $y = c$ and $y = d$. Let V be the volume obtained by revolving the region A around the x-axis. To compute V, we begin with a subdivision of $[c, d]$ into n subintervals, all of the same length $\Delta y = (d - c)/n$. Let y_i^* denote the midpoint of the ith subinterval $[y_{i-1}, y_i]$ of the subdivision. Then the volume of the cylindrical shell with average radius y_i^*, height $f(y_i^*) - g(y_i^*)$, and thickness Δy is

$$\Delta V_i = 2\pi y_i^* [f(y_i^*) - g(y_i^*)] \, \Delta y.$$

We add the volume of these cylindrical shells and thus obtain the approximation

$$V \approx \sum_{i=1}^{n} 2\pi y_i^* [f(y_i^*) - g(y_i^*)] \, \Delta y.$$

We recognize the right-hand side to be a Riemann sum for an integral with respect to y from c to d and so conclude that the volume of the solid of revolution is given by

$$V = \int_c^d 2\pi y \, [f(y) - g(y)] \, dy. \tag{4}$$

Thus

$$V = \int_c^d 2\pi y [x_{\text{right}} - x_{\text{left}}] \, dy, \tag{4'}$$

where $x_{\text{right}} = f(y)$ and $x_{\text{left}} = g(y)$.

NOTE   To use Eqs. (3′) and (4′), the integrand must be expressed in terms of the variable of integration specified by the differential.

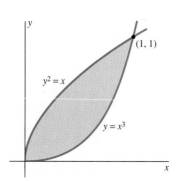

Fig. 6.3.11   The region of Example 3

EXAMPLE 3   Consider the region in the first quadrant bounded by the curves $y^2 = x$ and $y = x^3$ (Fig. 6.3.11). Use the method of cylindrical shells to compute the volume of the solids obtained by revolving this region first around the y-axis and then around the x-axis.

Solution   It is best to use cylindrical shells, as in Figs. 6.3.12 and 6.3.13, rather than memorized formulas, to set up the appropriate integrals. Thus the volume of revolution around the y-axis (Fig. 6.3.12) is given by

$$V = \int_0^1 2\pi x(y_{\text{top}} - y_{\text{bot}}) \, dx = \int_0^1 2\pi x(\sqrt{x} - x^3) \, dx$$

$$= \int_0^1 2\pi(x^{3/2} - x^4) \, dx = 2\pi \left[\tfrac{2}{5} x^{5/2} - \tfrac{1}{5} x^5 \right]_0^1 = \tfrac{2}{5}\pi.$$

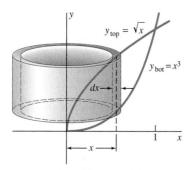

Fig. 6.3.12   Revolution about the y-axis (Example 3)

The volume of revolution around the x-axis (Fig. 6.3.13) is given by

$$V = \int_0^1 2\pi y(x_{\text{right}} - x_{\text{left}}) \, dy = \int_0^1 2\pi y(y^{1/3} - y^2) \, dy$$

$$= \int_0^1 2\pi(y^{4/3} - y^3) \, dy = 2\pi \left[\tfrac{3}{7} y^{7/3} - \tfrac{1}{4} y^4 \right]_0^1 = \tfrac{5}{14}\pi.$$

These answers are the same, of course, as those we obtained by using the method of cross sections in Example 4 of Section 6.2.

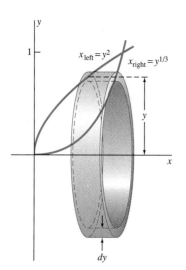

Fig. 6.3.13   Revolution about the x-axis (Example 3)

EXAMPLE 4   Suppose that the region of Example 3 is rotated around the vertical line $x = -1$ (Fig. 6.3.14). Then the area element

$$dA = (y_{\text{top}} - y_{\text{bot}}) \, dx = (\sqrt{x} - x^3) \, dx$$

is revolved through a circle of radius $r = 1 + x$. Hence the volume of the resulting cylindrical shell is

$$dV = 2\pi r \, dA = 2\pi(1 + x)(x^{1/2} - x^3) \, dx$$

$$= 2\pi(x^{1/2} + x^{3/2} - x^3 - x^4) \, dx.$$

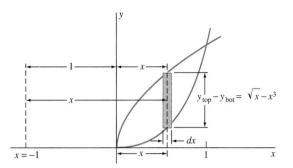

Fig. 6.3.14 Revolution about the line $x = -1$ (Example 4)

The volume of the resulting solid of revolution is then

$$V = \int_0^1 2\pi(x^{1/2} + x^{3/2} - x^3 - x^4)\, dx$$

$$= 2\pi\left[\tfrac{2}{3}x^{3/2} + \tfrac{2}{5}x^{5/2} - \tfrac{1}{4}x^4 - \tfrac{1}{5}x^5\right]_0^1 = \tfrac{37}{30}\,\pi,$$

as we found by using the method of cross sections in Example 5 of Section 6.2.

We may observe finally that the method of cylindrical shells is summarized by the heuristic formula

$$V = \int_*^{**} 2\pi r\, dA,$$

where dA denotes the area of an infinitesimal strip that is revolved through a circle of radius r to generate a thin cylindrical shell. The asterisks indicate limits of integration that you need to find.

6.3 Problems

In Problems 1 through 28, use the method of cylindrical shells to find the volume of the solid generated by revolving around the indicated axis the region bounded by the given curves.

1. $y = x^2$, $y = 0$, $x = 2$;   the y-axis

2. $x = y^2$, $x = 4$;   the y-axis

3. $y = 25 - x^2$, $y = 0$;   the y-axis (Fig. 6.3.15)

4. $y = 2x^2$, $y = 8$;   the y-axis (Fig. 6.3.16)

5. $y = x^2$, $y = 8 - x^2$;   the y-axis

6. $x = 9 - y^2$, $x = 0$;   the x-axis

7. $x = y$, $x + 2y = 3$, $y = 0$;   the x-axis (Fig. 6.3.17)

8. $y = x^2$, $y = 2x$;   the line $y = 5$

9. $y = 2x^2$, $y^2 = 4x$;   the x-axis

10. $y = 3x - x^2$, $y = 0$;   the y-axis

11. $y = 4x - x^3$, $y = 0$;   the y-axis (Fig. 6.3.18)

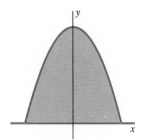

Fig. 6.3.15 Problem 3

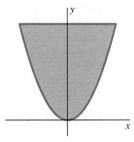

Fig. 6.3.16 Problem 4

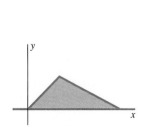

Fig. 6.3.17 Problem 7

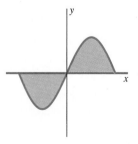

Fig. 6.3.18 Problem 11

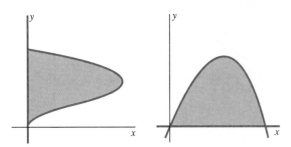

Fig. 6.3.19 Problem 12   **Fig. 6.3.20** Problem 15

12. $x = y^3 - y^4$, $x = 0$; the line $y = -2$ (Fig. 6.3.19)

13. $y = x - x^3$, $y = 0$ $(0 \leqq x \leqq 1)$; the y-axis

14. $x = 16 - y^2$, $x = 0$, $y = 0$ $(0 \leqq y \leqq 4)$; the x-axis

15. $y = x - x^3$, $y = 0$ $(0 \leqq x \leqq 1)$; the line $x = 2$ (Fig. 6.3.20)

16. $y = x^3$, $y = 0$, $x = 2$; the y-axis (Fig. 6.3.21)

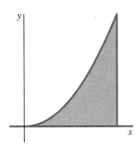   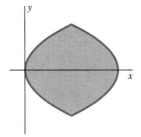

Fig. 6.3.21 Problem 16   **Fig. 6.3.22** Problem 24

17. $y = x^3$, $y = 0$, $x = 2$; the line $x = 3$

18. $y = x^3$, $y = 0$, $x = 2$; the x-axis

19. $y = x^2$, $y = 0$, $x = -1$, $x = 1$; the line $x = 2$

20. $y = x^2$, $y = x$ $(0 \leqq x \leqq 1)$; the y-axis

21. $y = x^2$, $y = x$ $(0 \leqq x \leqq 1)$; the x-axis

22. $y = x^2$, $y = x$ $(0 \leqq x \leqq 1)$; the line $y = 2$

23. $y = x^2$, $y = x$ $(0 \leqq x \leqq 1)$; the line $x = -1$

24. $x = y^2$, $x = 2 - y^2$; the x-axis (Fig. 6.3.22)

25. $x = y^2$, $x = 2 - y^2$; the line $y = 1$

26. $y = e^{-x^2}$, $y = 0$, $x = 0$, $x = 1$; the y-axis

27. $y = \dfrac{1}{x^2}$, $y = 0$, $x = 1$, $x = 2$; the y-axis

28. $y = \dfrac{1}{x^2}$, $y = 0$, $x = 1$, $x = 2$; the line $x = -1$

29. Verify the formula for the volume of a right circular cone by using the method of cylindrical shells. Apply the

method to the figure generated by revolving the triangular region with vertices $(0, 0)$, $(r, 0)$, and $(0, h)$ about the y-axis.

30. Use the method of cylindrical shells to compute the volume of the paraboloid of Problem 33 in Section 6.2.

31. Use the method of cylindrical shells to find the volume of the ellipsoid obtained by revolving the elliptical region bounded by the graph of the equation

$$\left(\frac{x}{a}\right)^2 + \left(\frac{y}{b}\right)^2 = 1$$

about the y-axis.

32. Use the method of cylindrical shells to derive the formula given in Problem 38 of Section 6.2 for the volume of a spherical segment.

33. Use the method of cylindrical shells to compute the volume of the torus of Problem 39 in Section 6.2. [*Suggestion:* Substitute u for $x - b$ in the integral given by the formula in Eq. (2).]

34. (a) Find the volume of the solid generated by revolving the region bounded by the curves $y = x^2$ and $y = x + 2$ around the line $x = -2$. (b) Repeat part (a), but revolve the region around the line $x = 3$.

35. Find the volume of the solid generated by revolving the circular disk $x^2 + y^2 \leqq a^2$ around the vertical line $x = -a$.

36. (a) Verify by differentiation that

$$\int xe^x \, dx = (x - 1)e^x + C.$$

(b) Find the volume of the solid obtained by rotating around the y-axis the area under $y = e^x$ from $x = 0$ to $x = 1$.

37. We found in Example 2 that the volume remaining after a hole of radius a is bored through the center of a sphere of radius $b > a$ is

$$V = \tfrac{4}{3}\pi(b^2 - a^2)^{3/2}.$$

(a) Express the volume V of this formula *without* use of the hole radius a; use instead the hole height h. [*Suggestion:* Use the right triangle in Fig. 6.3.8.] (b) What is remarkable about the answer to part (a)?

38. The plane region R is bounded above and on the right by the graph of $y = 25 - x^2$, on the left by the y-axis, and below by the x-axis. A paraboloid is generated by revolving R around the y-axis. Then a vertical hole of radius 3 that is centered along the y-axis is bored through the paraboloid. Find the volume of the solid that remains by using (a) the method of cross sections and (b) the method of cylindrical shells.

6.3 Project

Fig. 6.3.23 Wedding band

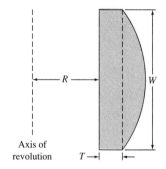

Fig. 6.3.24 Cross section of the wedding band

This project deals with the custom-made gold wedding band pictured in Fig. 6.3.23. Its shape is obtained by revolving the region A shown in Fig. 6.3.24 around the vertical axis shown there. The resulting wedding band has

❑ Inner radius R,
❑ Minimum thickness T, and
❑ Width W.

The curved boundary of the region A is an arc of a circle whose center lies on the axis of revolution. For a typical wedding band, R might be anywhere from 6 to 12 mm, T might be 0.5 to 1.5 mm, and W might be 4 to 10 mm.

If a customer asks the price of a wedding band with given dimensions R, T, and W, the jeweler must first calculate the volume of the desired band to determine how much gold will be required to make it. Use the methods of this section to show that the volume V is given by the formula

$$V = \frac{\pi W}{6}(W^2 + 12RT + 6T^2). \tag{5}$$

If these dimensions are measured in millimeters, then V is given in cubic millimeters. (There are 1000 mm³ in 1 cm³).

Suppose that the jeweler plans to charge the customer $1000 per *troy ounce* of alloy (90% gold, 10% silver) used to make the ring. (The profit on the sale, covering the jeweler's time and overhead in making the ring, is fairly substantial because the price of gold is generally under $400/oz and that of silver, under $5/oz.) The inner radius R of the wedding band is determined by measurement of the customer's finger (in millimeters; there are exactly 25.4 mm per inch). Suppose that the jeweler makes all wedding bands with $T = 1$ (mm). Then, for a given acceptable cost C (in dollars), the customer wants to know the maximum width W of the wedding band he can afford.

PROBLEM   Measure your own ring finger to determine R (you can measure its circumference C with a piece of string and then divide by 2π). Then choose a cost figure C in the $100 to $500 price range. Use Eq. (5) with $T = 1$ to find the width W of a band that costs C dollars (at $1000/oz). You will need to know that the density of the gold-silver alloy is 18.4 gm/cm³ and that 1 lb contains 12 troy ounces and 453.59 gm. Use a graphics calculator or a calculator with a $\boxed{\text{SOLVE}}$ key to solve the resulting cubic equation in W.

6.4
Arc Length and Surface Area of Revolution

If you plan to hike North Carolina's Mountains-to-the-Sea Trail, you will need to know the length of this curved path so you'll know how much food and equipment to take. Here we investigate how to find the length of a curved path and the closely related idea of finding the surface area of a curved surface.

A **smooth arc** is the graph of a smooth function defined on a closed interval; a **smooth function** f on $[a, b]$ is a function whose derivative f' is continuous on $[a, b]$. The continuity of f' rules out the possibility of corner points on the graph of f, points where the direction of the tangent line changes

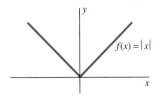

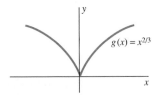

Fig. 6.4.1 Graphs that have corner points

abruptly. The graphs of $f(x) = |x|$ and $g(x) = x^{2/3}$ are shown in Fig. 6.4.1; neither is smooth because each has a corner point at the origin.

To investigate the length of a smooth arc, we begin with the length of a straight line segment, which is simply the distance between its endpoints. Then, given a smooth arc C, we pose the following question: If C were a thin wire and we straightened it without stretching it, how long would the resulting straight wire be? The answer is what we call the *length* of C.

To approximate the length s of the smooth arc C, we can inscribe in C a polygonal arc—one made up of straight line segments—and then calculate the length of this polygonal arc. We proceed in the following way, under the assumption that C is the graph of a smooth function f defined on the closed interval $[a, b]$. Consider a division of $[a, b]$ into n subintervals, all with the same length Δx. Let P_i denote the point $(x_i, f(x_i))$ on the arc C corresponding to the ith subdivision point x_i. Our polygonal arc "inscribed in C" is then the union of the line segments P_0P_1, P_1P_2, P_2P_3, . . . , $P_{n-1}P_n$. So an approximation to the length s of C is

$$s \approx \sum_{i=1}^{n} |P_{i-1}P_i|, \tag{1}$$

the sum of the lengths of these line segments (Fig. 6.4.2).

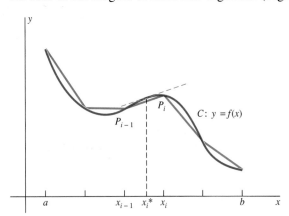

Fig. 6.4.2 A polygonal arc inscribed in the smooth curve C

The length of the typical line segment $P_{i-1}P_i$ is

$$|P_{i-1}P_i| = [(x_i - x_{i-1})^2 + (f(x_i) - f(x_{i-1}))^2]^{1/2}.$$

We apply the mean value theorem to the function f on the interval $[x_{i-1}, x_i]$ and thereby conclude the existence of a point x_i^* in this interval such that

$$f(x_i) - f(x_{i-1}) = f'(x_i^*) \cdot (x_i - x_{i-1}).$$

Hence

$$|P_{i-1}P_i| = \left[1 + \left(\frac{f(x_i) - f(x_{i-1})}{x_i - x_{i-1}}\right)^2\right]^{1/2} \cdot (x_i - x_{i-1})$$

$$= \sqrt{1 + [f'(x_i^*)]^2}\, \Delta x,$$

where $\Delta x = x_i - x_{i-1}$.

We next substitute this expression for $|P_{i-1}P_i|$ into Eq. (1) and get the approximation

$$s \approx \sum_{i=1}^{n} \sqrt{1 + [f'(x_i^*)]^2}\, \Delta x.$$

This sum is a Riemann sum for the function $\sqrt{1 + [f'(x)]^2}$ on $[a, b]$, and therefore—because f' is continuous—such sums approach the integral

$$\int_a^b \sqrt{1 + [f'(x)]^2}\, dx$$

as $\Delta x \to 0$. But our approximation ought to approach, as well, the actual length s as $\Delta x \to 0$. On this basis we *define* the **length** s of the smooth arc C to be

$$s = \int_a^b \sqrt{1 + [f'(x)]^2}\, dx = \int_a^b \sqrt{1 + \left(\frac{dy}{dx}\right)^2}\, dx. \qquad (2)$$

In the case of a smooth arc given as a graph $x = g(y)$ for y on $[c, d]$, a similar discussion beginning with a subdivision of $[c, d]$ leads to the formula

$$s = \int_c^d \sqrt{1 + [g'(y)]^2}\, dy = \int_c^d \sqrt{1 + \left(\frac{dx}{dy}\right)^2}\, dy \qquad (3)$$

for its length. We can compute the length of a more general curve, such as a circle, by dividing it into a finite number of smooth arcs and then applying to each of these arcs whichever of Eqs. (2) and (3) is required.

There is a convenient symbolic device that we can employ to remember both Eqs. (2) and (3) simultaneously. We think of two nearby points $P(x, y)$ and $Q(x + dx, x + dy)$ on the smooth arc C and denote by ds the length of the arc that joins P and Q. Imagine that P and Q are so close together that ds is, for all practical purposes, equal to the length of the straight line segment PQ. Then the Pythagorean theorem applied to the small right triangle in Fig. 6.4.3 gives

$$ds = \sqrt{(dx)^2 + (dy)^2} \qquad (4)$$

$$= \sqrt{1 + \left(\frac{dy}{dx}\right)^2}\, dx \qquad (4')$$

$$= \sqrt{1 + \left(\frac{dx}{dy}\right)^2}\, dy. \qquad (4'')$$

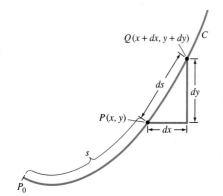

Fig. 6.4.3 Heuristic development of the arc-length formula

Thinking of the entire length s of C as the sum of small pieces such as ds, we write

$$s = \int_*^{**} ds. \qquad (5)$$

Then formal (symbolic) substitution of the expressions in Eqs. (4′) and (4″) for ds in Eq. (5) yields Eqs. (2) and (3); only the limits of integration remain to be determined.

EXAMPLE 1 Find the length of the so-called semicubical parabola (it's not really a parabola) $y = x^{3/2}$ on $[0, 5]$ (Fig. 6.4.4).

Solution We first compute the integrand in Eq. (2):

$$\sqrt{1 + \left(\frac{dy}{dx}\right)^2} = \sqrt{1 + (\tfrac{3}{2}x^{1/2})^2} = \sqrt{1 + \tfrac{9}{4}x} = \tfrac{1}{2}(4 + 9x)^{1/2}.$$

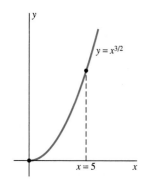

Fig. 6.4.4 The semicubical parabola of Example 1

Hence the length of the arc $y = x^{3/2}$ over the interval $[0, 5]$ is

$$s = \int_0^5 \tfrac{1}{2}(4 + 9x)^{1/2}\, dx = \left[\tfrac{1}{27}(4 + 9x)^{3/2}\right]_0^5 = \tfrac{335}{27} \approx 12.41.$$

As a plausibility check, the endpoints of the arc are $(0, 0)$ and $(5, 5\sqrt{5})$, so the straight line segment connecting these points has length $5\sqrt{6} \approx 12.25$. This is, as it should be, somewhat less than the calculated length of the arc.

EXAMPLE 2 Find the length s of the curve

$$x = \frac{1}{6}y^3 + \frac{1}{2y}, \qquad 1 \leq y \leq 2.$$

Solution Here y is the natural independent variable, so we use the arc-length formula in Eq. (3). First we calculate

$$1 + \left(\frac{dx}{dy}\right)^2 = 1 + \left(\frac{1}{2}y^2 - \frac{1}{2y^2}\right)^2 = 1 + \frac{1}{4}y^4 - \frac{1}{2} + \frac{1}{4y^4}$$

$$= \frac{1}{4}y^4 + \frac{1}{2} + \frac{1}{4y^4} = \left(\frac{1}{2}y^2 + \frac{1}{2y^2}\right)^2.$$

Thus we can "get out from under the radical" in Eq. (3):

$$s = \int_c^d \sqrt{1 + \left(\frac{dx}{dy}\right)^2}\, dy = \int_1^2 \left(\frac{1}{2}y^2 + \frac{1}{2y^2}\right) dy$$

$$= \left[\frac{1}{6}y^3 - \frac{1}{2y}\right]_1^2 = \frac{17}{12}.$$

EXAMPLE 3 A manufacturer needs to make corrugated metal sheets 36 in. wide with cross sections in the shape of the curve

$$y = \tfrac{1}{2}\sin \pi x, \qquad 0 \leq x \leq 36$$

(Fig. 6.4.5). How wide must the original flat sheets be for the manufacturer to produce these corrugated sheets?

Fig. 6.4.5 The corrugated sheet in the shape of $y = \tfrac{1}{2}\sin x$ (Example 3)

Solution If

$$f(x) = \tfrac{1}{2}\sin \pi x, \quad \text{then} \quad f'(x) = \tfrac{1}{2}\pi \cos \pi x.$$

Hence Eq. (2) yields the arc length of the graph of f over $[0, 36]$:

$$s = \int_0^{36} \sqrt{1 + (\tfrac{1}{2}\pi)^2 \cos^2 \pi x}\, dx = 36 \int_0^1 \sqrt{1 + (\tfrac{1}{2}\pi)^2 \cos^2 \pi x}\, dx.$$

These integrals cannot be evaluated in terms of elementary functions. Because of this, we cannot apply the fundamental theorem of calculus. So we estimate their values with the aid of Simpson's approximation (Section 5.9). Both with $n = 6$ and with $n = 12$ subintervals we find that

$$\int_0^1 \sqrt{1 + (\tfrac{1}{2}\pi)^2 \cos^2 \pi x} \; dx \approx 1.46$$

inches. Therefore the manufacturer should use flat sheets of approximate width $36 \cdot 1.46 \approx 52.6$ in.

AREAS OF SURFACES OF REVOLUTION

A **surface of revolution** is a surface obtained by revolving an arc or curve around an axis that lies in the same plane as the arc. The surface of a cylinder or of a sphere and the curved surface of a cone are important as examples of surfaces of revolution.

Our basic approach to finding the area of such a surface is this: First we inscribe a polygonal arc in the curve to be revolved. We then regard the area of the surface generated by revolving the polygonal arc to be an approximation to the surface generated by revolving the original curve. Because a surface generated by revolving a polygonal arc around an axis consists of frusta (sections) of cones, we can calculate its area in a reasonably simple way.

Fig. 6.4.6 A frustum of a cone. The slant height is L.

This approach to surface area originated with Archimedes. For example, he used this method to establish the formula $A = 4\pi r^2$ for the surface area of a sphere of radius r.

We will need the formula

$$A = 2\pi \bar{r} L \tag{6}$$

for the curved surface area of a cone with average radius $\bar{r} = \tfrac{1}{2}(r_1 + r_2)$ and *slant height* L (Fig. 6.4.6). Equation (6) follows from the formula

$$A = \pi r L \tag{7}$$

for the area of a conical surface with base radius r and slant height L (Fig. 6.4.7). It is easy to derive Eq. (7) by "unrolling" the conical surface onto a sector of a circle of radius L, because the area of this sector is

$$A = \frac{2\pi r}{2\pi L} \cdot \pi L^2 = \pi r L.$$

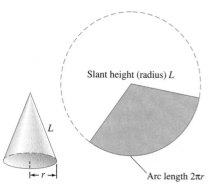

Slant height (radius) L

Arc length $2\pi r$

Fig. 6.4.7 Surface area of a cone: Cut along L; then unroll the cone onto the circular sector.

To derive Eq. (6) from Eq. (7), we think of the frustum as the lower section of a cone with slant height $L_2 = L + L_1$ (Fig. 6.4.8). Then subtraction of the area of the upper conical section from that of the entire cone gives

$$A = \pi r_2 L_2 - \pi r_1 L_1 = \pi r_2(L + L_1) - \pi r_1 L_1 = \pi(r_2 - r_1)L_1 + \pi r_2 L$$

for the area of the frustum. But the similar right triangles in Fig. 6.4.8 yield the proportion

$$\frac{r_1}{L_1} = \frac{r_2}{L_2} = \frac{r_2}{L + L_1},$$

from which we find that $(r_2 - r_1)L_1 = r_1 L$. Hence the area of the frustum is

$$A = \pi r_1 L + \pi r_2 L = 2\pi \bar{r} L,$$

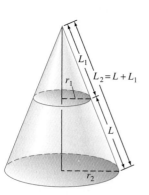

Fig. 6.4.8 Derivation of Eq. (6)

where $\bar{r} = \tfrac{1}{2}(r_1 + r_2)$. So we have verified Eq. (6).

Now suppose that the surface S has area A and is generated by revolving around the x-axis the smooth arc $y = f(x)$, $a \leqq x \leqq b$; suppose also that $f(x)$ is never negative on $[a, b]$. To approximate A we begin with a division of $[a, b]$ into n subintervals, each of length Δx. As in our discussion of arc length leading to Eq. (2), let P_i denote the point $(x_i, f(x_i))$ on the arc. Then, as before, the line segment $P_{i-1}P_i$ has length

$$L_i = |P_{i-1}P_i| = \sqrt{1 + [f'(x_i^*)]^2}\, \Delta x$$

for some point x_i^* in the ith subinterval $[x_{i-1}, x_i]$.

The conical frustum obtained by revolving the segment $P_{i-1}P_i$ around the x-axis has slant height L_i and, as shown in Fig. 6.4.9, average radius

$$\bar{r}_i = \tfrac{1}{2}[f(x_{i-1}) + f(x_i)].$$

Because $\bar{r}_i$ lies between the values $f(x_{i-1})$ and $f(x_i)$, the intermediate value property of continuous functions (Section 2.4) yields a point x_i^{**} in $[x_{i-1}, x_i]$ such that $\bar{r}_i = f(x_i^{**})$. By Eq. (6), the area of this conical frustum is, therefore,

$$2\pi \bar{r}_i L_i = 2\pi f(x_i^{**}) \sqrt{1 + [f'(x_i^*)]^2}\, \Delta x.$$

We add the areas of these conical frusta for $i = 1, 2, 3, \ldots, n$. This gives the approximation

$$A \approx \sum_{i=1}^{n} 2\pi f(x_i^{**}) \sqrt{1 + [f'(x_i^*)]^2}\, \Delta x.$$

If x_i^* and x_i^{**} were the *same* point of the ith subinterval $[x_{i-1}, x_i]$, then this approximation would be a Riemann sum for the integral

$$\int_a^b 2\pi f(x) \sqrt{1 + [f'(x)]^2}\, dx.$$

Even though the numbers x_i^* and x_i^{**} are generally not equal, it still follows (from a result stated in Appendix F) that our approximation approaches the integral above as $\Delta x \to 0$. Intuitively, this is easy to believe: After all, as $\Delta x \to 0$, the difference between x_i^* and x_i^{**} also approaches zero.

We therefore *define* the **area** A of the surface generated by revolving around the x-axis the smooth arc $y = f(x)$, $a \leqq x \leqq b$, by the formula

$$A = \int_a^b 2\pi f(x) \sqrt{1 + [f'(x)]^2}\, dx. \tag{8}$$

If we write y for $f(x)$ and ds for $\sqrt{1 + (dy/dx)^2}\, dx$, as in Eq. (4′), then we can abbreviate Eq. (8) as

$$A = \int_a^b 2\pi y\, ds \qquad (x\text{-axis}). \tag{9}$$

This abbreviated formula is conveniently remembered by thinking of $dA = 2\pi y\, ds$ as the area of the narrow frustum obtained by revolving the tiny arc ds around the x-axis in a circle of radius y (Fig. 6.4.10).

If the smooth arc being revolved around the x-axis is given instead by $x = g(y), c \leqq y \leqq d$, then an approximation based on a subdivision of $[c, d]$ leads to the area formula

$$A = \int_c^d 2\pi y \sqrt{1 + [g'(y)]^2}\, dy. \tag{10}$$

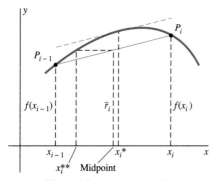

Fig. 6.4.9 Approximation of a surface area of revolution by the surface of a frustum of a cone

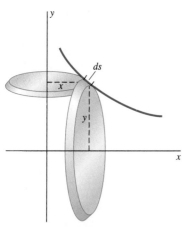

Fig. 6.4.10 The arc ds may be rotated about either the x-axis or the y-axis.

We can obtain Equation (10) by making the formal substitution $ds = \sqrt{1 + (dx/dy)^2}\, dy$ of Eq. (4″) into the abbreviated formula in Eq. (9) for surface area of revolution and then replacing a and b by the correct limits of integration.

Now let us consider the surface generated by revolving a smooth arc around the y-axis rather than around the x-axis. In Fig. 6.4.10 we see that the average radius of the narrow frustum obtained by revolving the tiny arc ds is now x instead of y. This suggests the abbreviated formula

$$A = \int_a^b 2\pi x\, ds \qquad (y\text{-axis}) \qquad (11)$$

for a surface area of revolution about the y-axis. If the smooth arc is given by $y = f(x)$, $a \leq x \leq b$, then the symbolic substitution $ds = \sqrt{1 + (dy/dx)^2}\, dx$ gives

$$A = \int_a^b 2\pi x \sqrt{1 + [f'(x)]^2}\, dx. \qquad (12)$$

But if the smooth arc is presented in the form $x = g(y)$, $c \leq y \leq d$, then the symbolic substitution of $ds = \sqrt{1 + (dx/dy)^2}\, dy$ into Eq. (11) gives

$$A = \int_c^d 2\pi g(y)\sqrt{1 + [g'(y)]^2}\, dy. \qquad (13)$$

Equations (12) and (13) may be verified by using approximations similar to the one leading to Eq. (8).

Thus we have *four* formulas for areas of surfaces of revolution, summarized in the table in Fig. 6.4.11. Which of these formulas is appropriate for computing the area of a given surface depends on two factors:

1. Whether the smooth arc that generates the surface is presented in the form $y = f(x)$ or in the form $x = g(y)$, and

2. Whether this arc is to be revolved around the x-axis or around the y-axis.

Memorizing the four formulas in the table is unnecessary. We suggest that you instead remember the abbreviated formulas in Eqs. (9) and (11) in conjunction with Fig. 6.4.10 and make either the substitution

$$y = f(x), \qquad ds = \sqrt{1 + \left(\frac{dy}{dx}\right)^2}\, dx$$

| | | Axis of revolution | | |
		x-axis		y-axis
	$y = f(x)$, $a \leq x \leq b$	$\displaystyle\int_a^b 2\pi f(x)\sqrt{1 + [f'(x)]^2}\, dx$ (8)		$\displaystyle\int_a^b 2\pi x \sqrt{1 + [f'(x)]^2}\, dx$ (10)
Description of curve C				
	$x = g(y)$, $c \leq y \leq d$	$\displaystyle\int_c^d 2\pi y \sqrt{1 + [g'(y)]^2}\, dy$ (10)		$\displaystyle\int_c^d 2\pi g(y)\sqrt{1 + [g'(y)]^2}\, dy$ (13)

Fig. 6.4.11 Area formulas for surfaces of revolution

or the substitution

$$x = g(y), \qquad ds = \sqrt{1 + \left(\frac{dx}{dy}\right)^2} \, dy,$$

depending on whether the smooth arc is presented as a function of x or as a function of y. It may also be helpful to note that each of these four surface-area formulas is of the form

$$A = \int_{*}^{**} 2\pi r \, ds, \qquad (14)$$

where r denotes the radius of the circle around which the arc length element ds is revolved.

As in earlier sections, we again caution you to identify the independent variable by examining the differential and to express every dependent variable in terms of the independent variable before you antidifferentiate. That is, either express everything, including ds, in terms of x (and dx) or everything in terms of y (and dy).

EXAMPLE 4 Find the area of the paraboloid shown in Fig. 6.4.12, which is obtained by revolving the parabolic arc $y = x^2$, $0 \leq x \leq \sqrt{2}$, around the y-axis.

Solution Following the suggestion that precedes the example, we get

$$A = \int_{*}^{**} 2\pi x \, ds = \int_{a}^{b} 2\pi x \sqrt{1 + \left(\frac{dy}{dx}\right)^2} \, dx$$

$$= \int_{0}^{\sqrt{2}} 2\pi x \sqrt{1 + (2x)^2} \, dx$$

$$= \int_{0}^{\sqrt{2}} \frac{\pi}{4}(1 + 4x^2)^{1/2} \cdot 8x \, dx = \left[\frac{\pi}{6}(1 + 4x^2)^{3/2} \right]_{0}^{\sqrt{2}} = \frac{13}{3} \pi.$$

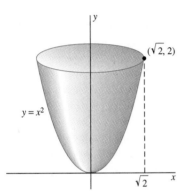

Fig. 6.4.12 The paraboloid of Example 4

The decision of which abbreviated formula—Eq. (9) or Eq. (11)—to use is determined by the axis of revolution. In contrast, the decision of whether the variable of integration should be x or y is made by the way in which the smooth arc is given: as a function of x or as a function of y. In some problems, either x or y may be used as the variable of integration, but the integral is usually much simpler to evaluate if you make the correct choice. Experience is very helpful here. Right now, try Example 4 with independent variable y.

6.4 Problems

In Problems 1 through 10, set up and simplify the integral that gives the length of the given smooth arc. Do not evaluate the integral.

1. $y = x^2$, $\quad 0 \leq x \leq 1$

2. $y = x^{5/2}$, $\quad 1 \leq x \leq 3$

3. $y = 2x^3 - 3x^2$, $\quad 0 \leq x \leq 2$

4. $y = x^{4/3}$, $\quad -1 \leq x \leq 1$

5. $y = 1 - x^2$, $\quad 0 \leq x \leq 100$

6. $x = 4y - y^2$, $\quad 0 \leq y \leq 1$

7. $x = y^4$, $\quad -1 \leq y \leq 2$

8. $x^2 = y$, $\quad 1 \leq y \leq 4$

9. $xy = 1$, $\quad 1 \leq x \leq 2$

10. $x^2 + y^2 = 4$, $\quad 0 \leq x \leq 2$

Ch. 6 / Applications of the Integral

$$dV = -av\, dt = -a\sqrt{2gy}\, dt. \tag{18}$$

But if $A(y)$ denotes the horizontal cross-sectional area of the tank at height y above the hole, then

$$dV = A(y)\, dy, \tag{19}$$

as usual. Comparing Eqs. (18) and (19), we see that $y(t)$ satisfies the differential equation

$$A(y)\frac{dy}{dt} = -a\sqrt{2gy}. \tag{20}$$

In some applications this is a very convenient form of Torricelli's law (see Section 5.2). In other situations you may prefer to work with the differential equation in (18) in the form

$$\frac{dV}{dt} = -a\sqrt{2gy}$$

or, if the area of the bottom hole is unknown, the form

$$\frac{dV}{dt} = -c\sqrt{y},$$

where $c = a\sqrt{2g}$ is a positive constant.

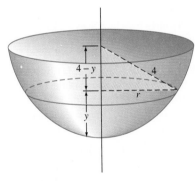

Fig. 6.5.4 Draining a hemispherical tank

EXAMPLE 5 A hemispherical tank has top radius 4 ft and, at time $t = 0$, is full of water. At that moment a circular hole of diameter 1 in. is opened in the bottom of the tank. How long will it take for all the water to drain from the tank?

Solution From the right triangle in Fig. 6.5.4, we see that

$$A(y) = \pi r^2 = \pi[16 - (4-y)^2] = \pi(8y - y^2).$$

With $g = 32$ ft/s², Eq. (20) takes the form

$$\pi(8y - y^2)\frac{dy}{dt} = -\pi\left(\frac{1}{24}\right)^2\sqrt{64y};$$

$$\int (8y^{1/2} - y^{3/2})\, dy = -\int \frac{1}{72}\, dt + C;$$

$$\frac{16}{3}y^{3/2} - \frac{2}{5}y^{5/2} = -\frac{1}{72}t + C.$$

Now $y(0) = 4$, so

$$C = \tfrac{16}{3}\cdot 4^{3/2} - \tfrac{2}{5}\cdot 4^{5/2} = \tfrac{448}{15}.$$

The tank is empty when $y = 0$ — that is, when

$$t = 72\cdot\tfrac{448}{15} \approx 2150 \text{ (s)},$$

or about 35 min 50 s. So it takes slightly less than 36 min for the tank to drain.

6.5 Problems

Find general solutions (implicit if necessary, explicit if possible) of the differential equations in Problems 1 through 10.

1. $\dfrac{dy}{dx} = 2x\sqrt{y}$ **2.** $\dfrac{dy}{dx} = 2xy^2$

3. $\dfrac{dy}{dx} = x^2 y^3$ **4.** $\dfrac{dy}{dx} = (xy)^{3/2}$

5. $\dfrac{dy}{dx} = 2x\sqrt{y-1}$ **6.** $\dfrac{dy}{dx} = 4x^3(y-4)^2$

7. $\dfrac{dy}{dx} = \dfrac{1 + \sqrt{x}}{1 + \sqrt{y}}$ **8.** $\dfrac{dy}{dx} = \dfrac{x + x^3}{y + y^3}$

9. $\dfrac{dy}{dx} = \dfrac{x^2 + 1}{x^2(3y^2 + 1)}$ **10.** $\dfrac{dy}{dx} = \dfrac{(x^3 - 1)y^3}{x^2(2y^3 - 3)}$

Solve the initial value problems in Problems 11 through 20.

11. $\dfrac{dy}{dx} = y^2, \quad y(0) = 1$

12. $\dfrac{dy}{dx} = \sqrt{y}, \quad y(0) = 4$

13. $\dfrac{dy}{dx} = \dfrac{1}{4y^3}, \quad y(0) = 1$

14. $\dfrac{dy}{dx} = \dfrac{1}{x^2 y}, \quad y(1) = 2$

15. $\dfrac{dy}{dx} = \sqrt{xy^3}, \quad y(0) = 4$

16. $\dfrac{dy}{dx} = \dfrac{x}{y}, \quad y(3) = 5$

17. $\dfrac{dy}{dx} = -\dfrac{x}{y}, \quad y(12) = -5$

18. $y^2 \dfrac{dy}{dx} = x^2 + 2x + 1, \quad y(1) = 2$

19. $\dfrac{dy}{dx} = 3x^2 y^2 - y^2, \quad y(0) = 1$

20. $\dfrac{dy}{dx} = 2xy^3(2x^2 + 1), \quad y(1) = 1$

21. Suppose that the fish population $P(t)$ in a lake is attacked by disease at time $t = 0$, with the result that

$$\frac{dP}{dt} = -k\sqrt{P} \quad (k > 0)$$

thereafter. If there were initially 900 fish in the lake and 441 were left after 6 weeks, how long would it take all the fish in the lake to die?

22. Prove that the solution of the initial value problem

$$\frac{dP}{dt} = k\sqrt{P}, \qquad P(0) = P_0,$$

is given by

$$P(t) = \left(\tfrac{1}{2}kt + \sqrt{P_0}\right)^2.$$

23. Suppose that the population of Beaverton satisfies the differential equation of Problem 22. (a) If $P = 100{,}000$ in 1970 and $P = 121{,}000$ in 1980, what will the population be in the year 2000? (b) When will the population be 200,000?

24. Consider a breed of rabbits whose population $P(t)$ satisfies the initial value problem

$$\frac{dP}{dt} = kP^2, \qquad P(0) = P_0,$$

where k is a positive constant. Derive the solution

$$P(t) = \frac{P_0}{1 - kP_0 t}.$$

25. In Problem 24, suppose that $P_0 = 2$ and that there are 4 rabbits after 3 months. What happens in the next 3 months?

26. Suppose that a motorboat is traveling at $v = 40$ ft/s when its motor is cut off at time $t = 0$. Thereafter, its deceleration due to water resistance is given by $dv/dt = -kv^2$, where k is a positive constant. (a) Solve this differential equation to show that the speed of the boat after t seconds is $v = 40/(1 + 40kt)$ feet/second. (b) If the boat's speed after 10 s is 20 ft/s, how long does it take it to slow to 5 ft/s?

27. A tank shaped like a vertical cylinder initially contains water to a depth of 9 ft (Fig. 6.5.5). A bottom plug is pulled at time $t = 0$ (t in hours). After 1 h the depth has dropped to 4 ft. How long will it take all the water to drain from this tank?

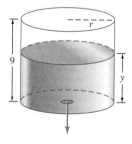

Fig. 6.5.5 The cylindrical tank of Problem 27

28. Suppose that the tank of Problem 27 has a radius of 3 ft and that its bottom hole is circular with radius 1 in.

How long will it take the water, initially 9 ft deep, to drain completely?

29. A water tank is in the shape of a right circular cone with its axis vertical and its vertex at the bottom. The tank is 16 ft high, and the radius of its top is 5 ft. At time $t = 0$, a plug at its vertex is removed and the tank, initially full of water, begins to drain. After 1 h the water in the tank is 9 ft deep. When will the tank be empty (Fig. 6.5.6)?

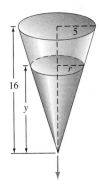

Fig. 6.5.6 The conical tank of Problem 29

30. Suppose that a cylindrical tank (axis vertical) initially containing V_0 liters of water drains through a bottom hole in T minutes. Use Torricelli's law to show that the volume of water in the tank after $t \leqq T$ minutes is $V(t) = V_0[1 - (t/T)]^2$.

31. The shape of a water tank is obtained by revolving the curve $y = x^{4/3}$ around the y-axis (units on the coordinate axes are in feet). A plug at the bottom is removed at 12 noon, when the water depth in the tank is 12 ft. At 1 P.M. the water depth is 6 ft. When will the tank be empty?

32. The shape of a water tank is obtained by revolving the parabola $y = x^2$ around the y-axis (units on the coordinate axes are in feet; see Fig. 6.5.7). The water depth is 4 ft at 12 noon; at that time, the plug in the circular hole at the bottom of the tank is removed. At 1 P.M. the water depth is 1 ft. (a) Find the water depth $y(t)$ after t hours. (b) When will the tank be empty? (c) What is the radius of the circular hole at the bottom?

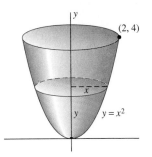

Fig. 6.5.7 The tank of Problem 32

33. A cylindrical tank of length 5 ft and radius 3 ft is situated with its axis horizontal. If a circular bottom hole of radius 1 in. is opened and the tank is initially half full of xylene, how long will it take the liquid to drain completely?

34. A spherical tank of radius 4 ft is full of mercury when a circular bottom hole of radius 1 in. is opened. How long will it be before all of the mercury drains from the tank?

35. *The Clepsydra, or Water Clock*   A 12-h water clock is to be designed with the dimensions shown in Fig. 6.5.8, shaped like the surface obtained by revolving the curve $y = f(x)$ around the y-axis. What equation should this curve have, *and* what radius should the circular bottom hole have, so that the water level will fall at the *constant* rate of 4 in./h?

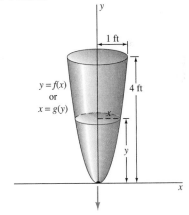

Fig. 6.5.8 The clepsydra of Problem 35

6.6

Force and Work

The concept of *work* is introduced to measure the cumulative effect of a force in moving a body from one position to another. In the simplest case, a particle is moved along a straight line by the action of a *constant* force. The work done by such a force is defined to be the product of the force and the distance through which it acts. Thus if the constant force has magnitude F and the particle is moved through the distance d, then the work done by the force is given by

$$W = F \cdot d. \tag{1}$$

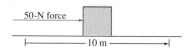

Fig. 6.6.1 A 50–N force does 500 N · m of work in pushing a box 10 m.

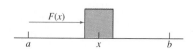

Fig. 6.6.2 A variable force pushing a particle from a to b

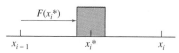

Fig. 6.6.3 The *constant* force $F(x_i^*)$ acting through the ith subinterval

For example, if a constant horizontal force of 50 newtons (N) is applied to a heavy box to push it a distance of 10 m along a rough floor (Fig. 6.6.1), then the work done by the force is

$$W = 50 \cdot 10 = 500$$

newton-meters (N·m). Note the units; because of the definition of work, units of work are always products of force units and distance units. For another example, to lift a weight of 75 lb a vertical distance of 5 ft, a constant force of 75 lb must be applied. The work done by this force is

$$W = 75 \cdot 5 = 375$$

foot-pounds (ft·lb).

Here we use the integral to generalize the definition of work to the case in which a particle is moved along a straight line by a *variable* force. Given a **force function** $F(x)$ defined at each point x of the straight line segment $[a, b]$, we want to define the work W done by this variable force in pushing the particle from the point $x = a$ to the point $x = b$ (Fig. 6.6.2).

We begin with the usual division of the interval $[a, b]$ into n subintervals, all with the same length $\Delta x = (b - a)/n$. For each $i = 1, 2, 3, \ldots, n$, let x_i^* be an arbitrary point of the ith subinterval $[x_{i-1}, x_i]$. The key idea is to approximate the actual work ΔW_i done by the **variable** force $F(x)$ in moving the particle from x_{i-1} to x_i by the work $F(x_i^*) \Delta x$ (force × distance) done in moving a particle the distance Δx from x_{i-1} to x_i (Fig. 6.6.3). Thus

$$\Delta W_i \approx F(x_i^*) \, \Delta x. \tag{2}$$

We approximate the total work W by summing from $i = 1$ to $i = n$:

$$W = \sum_{i=1}^{n} \Delta W_i \approx \sum_{i=1}^{n} F(x_i^*) \, \Delta x. \tag{3}$$

But the final sum in (3) is a Riemann sum for $F(x)$ on the interval $[a, b]$, and as $n \to +\infty$ (and $\Delta x \to 0$), such sums approach the *integral* of $F(x)$ from $x = a$ to $x = b$. We therefore are motivated to *define* the **work** W done by the force $F(x)$ in moving the particle from $x = a$ to $x = b$ to be

$$W = \int_a^b F(x) \, dx. \tag{4}$$

The following heuristic way of setting up Eq. (4) is useful in setting up integrals for work problems. Imagine that dx is so small a number that the value of $F(x)$ does not change appreciably on the tiny interval from x to $x + dx$. Then the work done by the force in moving a particle from x to $x + dx$ should be very close to

$$dW = F(x) \, dx.$$

The natural additive property of work then implies that we could obtain the total work W by adding these tiny elements of work:

$$W = \int_*^{**} dW = \int_a^b F(x) \, dx.$$

396

ELASTIC SPRINGS

Consider a spring whose left end is held fixed and whose right end is free to move along the x-axis. We assume that the right end is at the origin $x = 0$ when the spring has its **natural length**—that is, when the spring is in its rest position, neither compressed nor stretched by outside forces.

According to **Hooke's law** for elastic springs, the force $F(x)$ that must be exerted on the spring to hold its right end at the point x is proportional to the displacement x of the right end from its rest position. That is,

$$F(x) = kx, \tag{5}$$

where k is a positive constant. The constant k, called the **spring constant**, is a characteristic of the particular spring under study.

Figure 6.6.4 shows the arrangement of such a spring along the x-axis. The right end of the spring is held at position x on the x-axis by a force $F(x)$. The figure shows the situation for $x > 0$, so the spring is stretched. The force that the spring exerts on its right-hand end is directed to the left, so—as the figure shows—the external force $F(x)$ must act to the right. The right is the positive direction here, so $F(x)$ must be a positive number. Because x and $F(x)$ have the same sign, k must also be positive. You can check that k is positive as well in the case $x < 0$.

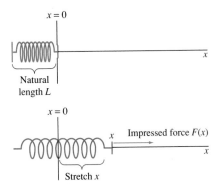

Fig. 6.6.4 The stretch x is proportional to the impressed force F.

EXAMPLE 1 Suppose that a spring has a natural length of 1 ft and that a force of 10 lb is required to hold it compressed to a length of 6 in. How much work is done in stretching the spring from its natural length to a total length of 2 ft?

Solution To move the free end from $x = 0$ (the natural-length position) to $x = 1$ (stretched by 1 ft), we must exert a variable force $F(x)$ determined by Hooke's law. We are given that $F = -10$ (lb) when $x = -0.5$ (ft), so Eq. (5), $F = kx$, implies that the spring constant for this spring is $k = 20$ (lb/ft). Thus $F(x) = 20x$, and so—using Eq. (4)—we find that the work done in stretching this spring in the manner given is

$$W = \int_0^1 20x \, dx = \left[10x^2 \right]_0^1 = 10 \quad \text{(ft·lb)}.$$

*WORK DONE AGAINST GRAVITY

According to Newton's law of gravitation, the force that must be exerted on a body to hold it at a distance r from the center of the earth is inversely proportional to r^2 (if $r \geq R$, the radius of the earth). In other words, if $F(r)$ denotes the holding force, then

$$F(r) = \frac{k}{r^2} \tag{6}$$

Work in the sense of physics is different from work in the sense of physiology. At this moment the weightlifter is doing no work in the physics sense because he is holding the weight still.

for some positive constant k. The value of this force at the surface of the earth, where $r = R \approx 4000$ mi (≈ 6370 km), is called the **weight** of the body.

Given the weight $F(R)$ of a particular body, we can find the corresponding value of k by using Eq. (6):

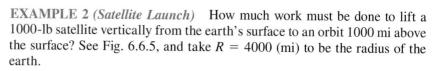

$$k = R^2 \cdot F(R).$$

The work that must be done to lift the body vertically from the surface to a distance $R_1 > R$ from the center of the earth is then

$$W = \int_R^{R_1} \frac{k}{r^2}\, dr. \qquad (7)$$

If distance is measured in miles and force in pounds, then this integral gives the work in mile-pounds. This is a very unconventional unit of work. We shall multiply by 5280 (ft/mi) to convert any such result into foot-pounds.

EXAMPLE 2 *(Satellite Launch)*   How much work must be done to lift a 1000-lb satellite vertically from the earth's surface to an orbit 1000 mi above the surface? See Fig. 6.6.5, and take $R = 4000$ (mi) to be the radius of the earth.

Solution   Because $F = 1000$ (lb) when $r = R = 4000$ (mi), we find from Eq. (6) that

$$k = (4000)^2(1000) = 16 \times 10^9 \ (\text{mi}^2 \cdot \text{lb}).$$

Then by Eq. (7), the work done is

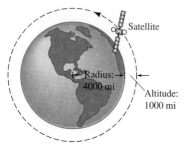

Fig. 6.6.5 A satellite in orbit 1000 mi above the surface of the earth (Example 2)

$$W = \int_{4000}^{5000} \frac{k}{r^2}\, dr = \left[-\frac{k}{r} \right]_{4000}^{5000}$$

$$= (16 \times 10^9)\left(\frac{1}{4000} - \frac{1}{5000} \right) = 8 \times 10^5 \quad (\text{mi} \cdot \text{lb}).$$

We multiply by 5280 (ft/mi) and write the answer as

$$4.224 \times 10^9 = 4{,}224{,}000{,}000 \quad (\text{ft} \cdot \text{lb}).$$

We can instead express the answer to Example 2 in terms of the power that the launch rocket must provide. **Power** is the rate at which work is done. For instance, 1 **horsepower** (hp) is defined to be 33,000 ft·lb/min. If the ascent to orbit takes 15 min and if only 2% of the power generated by the rocket is effective in lifting the satellite (the rest is used to lift the rocket and its fuel), we can convert the answer in Example 2 to horsepower. The *average* power that the rocket engine must product during the 15-min ascent is

$$P = \frac{(50)(4.224 \times 10^9)}{(15)(33{,}000)} \approx 426{,}667 \quad (\text{hp}).$$

The factor of 50 in the numerator comes from the 2% "efficiency" of the rocket: The total power must be multiplied by $1/(0.02) = 50$.

WORK DONE IN FILLING A TANK

Examples 1 and 2 are applications of Eq. (4) for calculating the work done by a variable force in moving a particle a certain distance. Another common type of force-work problem involves the summation of work done by constant forces that act through different distances. For example, consider the problem of pumping a fluid from ground level up into an above-ground tank (Fig. 6.6.6).

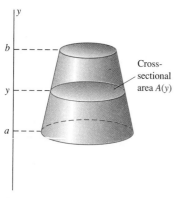

Fig. 6.6.6 An above-ground tank

Ch. 6 / Applications of the Integral

It is convenient to think of the tank as being filled in thin, horizontal layers of fluid, each lifted from the ground to its final position in the tank. No matter how the fluid actually behaves as the tank is filled, this simple way of thinking about the process gives us a way to compute the work done in the filling process. But when we think of filling the tank in this way, we must allow for the fact that different layers of fluid are lifted different distances to reach their final positions in the tank.

Suppose that the bottom of the tank is at height $y = a$ and that its top is at height $y = b > a$. Let $A(y)$ be the cross-sectional area of the tank at height y. Consider a subdivision of $[a, b]$ into n subintervals, all with the same length Δy. Then the volume of the horizontal slice of the tank that corresponds to the ith subinterval $[y_{i-1}, y_i]$ is

$$\Delta V_i = \int_{y_{i-1}}^{y_i} A(y)\, dy = A(y_i^*)\, \Delta y$$

for some number y_i^* in $[y_{i-1}, y_i]$; this is a consequence of the average value theorem for integrals (Section 5.6). If ρ is the density of the fluid (in pounds per cubic foot, for example), then the force required to lift this slice from the ground to its final position in the tank is simply the (constant) weight of the slice:

$$F_i = \rho\, \Delta V_i = \rho A(y_i^*)\, \Delta y.$$

What about the distance through which this force must act? The fluid in question is lifted from ground level to the level of the subinterval $[y_{i-1}, y_i]$, so every particle of the fluid is lifted at least the distance y_{i-1} and at most the distance y_i. Hence the work ΔW_i needed to lift this ith slice of fluid satisfies the inequalities

$$F_i y_{i-1} \leqq \Delta W_i \leqq F_i y_i;$$

that is,

$$\rho y_{i-1} A(y_i^*)\, \Delta y \leqq \Delta W_i \leqq \rho y_i A(y_i^*)\, \Delta y.$$

Now we add these inequalities for $i = 1, 2, 3, \ldots, n$ and find thereby that the total work $W = \Sigma W_i$ satisfies the inequalities

$$\sum_{i=1}^{n} \rho y_{i-1} A(y_i^*)\, \Delta y \leqq W \leqq \sum_{i=1}^{n} \rho y_i A(y_i^*)\, \Delta y.$$

If the three points y_{i-1}, y_i, and y_i^* of $[y_{i-1}, y_i]$ were the same, then both the last two sums would be Riemann sums for the function $f(y) = \rho y A(y)$ on $[a, b]$. Although the three points are not the same, it still follows—from a result stated in Appendix F—that both sums approach

$$\int_a^b \rho y A(y)\, dy \quad \text{as} \quad \Delta y \to 0.$$

The squeeze law of limits therefore gives the formula

$$W = \int_a^b \rho y A(y)\, dy. \tag{8}$$

This is the work W done in pumping fluid of density ρ from the ground into a tank that has horizontal cross-sectional area A(y) and is located between heights y = a and y = b above the ground.

A quick heuristic way to set up Eq. (8), and many variants of it, is to think of a thin, horizontal slice of fluid with volume $dV = A(y)\,dy$ and weight $\rho\,dV = \rho A(y)\,dy$. The work required to lift this slice a distance y is

$$dW = y \cdot \rho\,dV = \rho y A(y)\,dy,$$

so the total work required to fill the tank is

$$W = \int_{*}^{**} dW = \int_{a}^{b} \rho y A(y)\,dy,$$

because the horizontal slices lie between $y = a$ and $y = b$.

EXAMPLE 3 Suppose that it took 20 yr to construct the great pyramid of Khufu at Gizeh, Egypt. This pyramid is 500 ft high and has a square base with edge length 750 ft. Suppose also that the pyramid is made of rock with density $\rho = 120$ lb/ft^3. Finally, suppose that each laborer did 160 ft·lb/h of work in lifting rocks from ground level to their final position in the pyramid and worked 12 h daily for 330 days/yr. How many laborers would have been required to construct the pyramid?

Solution   We assume a constant labor force throughout the 20-yr construction period. We think of the pyramid as being made up of thin, horizontal slabs of rock, each slab lifted (just like a slice of liquid) from ground level to its ultimate height. Hence we can use Eq. (8) to compute the work W required.

Figure 6.6.7 shows a vertical cross section of the pyramid. The horizontal cross section at height y is a square with edge length s. We see from the similar triangles in Fig. 6.6.7 that

$$\frac{s}{750} = \frac{500 - y}{500}, \quad \text{so} \quad s = \frac{3}{2}(500 - y).$$

Hence the cross-sectional area at height y is

$$A(y) = \tfrac{9}{4}(500 - y)^2.$$

Equation (8) therefore gives

$$W = \int_{0}^{500} 120 \cdot y \cdot \tfrac{9}{4}(500 - y)^2\,dy$$

$$= 270 \int_{0}^{500} (250{,}000y - 1000y^2 + y^3)\,dy$$

$$= 270 \left[125{,}000y^2 - \tfrac{1000}{3}y^3 + \tfrac{1}{4}y^4 \right]_{0}^{500},$$

so $W \approx 1.406 \times 10^{12}$ ft·lb.

Because each laborer does

$$160 \cdot 12 \cdot 330 \cdot 20 \approx 1.267 \times 10^7 \quad \text{ft·lb}$$

of work, the construction of the pyramid would—under our assumptions—have required

$$\frac{1.406 \times 10^{12}}{1.267 \times 10^7},$$

or about 111,000 laborers.

The great pyramid of Khufu

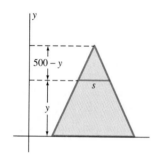

Fig. 6.6.7 Vertical cross section of Khufu's pyramid

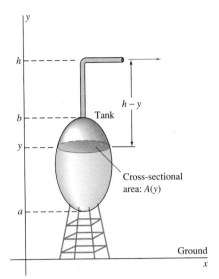

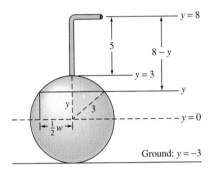

Fig. 6.6.8 Pumping liquid from a tank to a higher level

Fig. 6.6.9 End view of the cylindrical tank of Example 4

EMPTYING A TANK

Suppose now that the tank shown in Fig. 6.6.8 is already filled with a liquid of density ρ lb/ft^3, and we want to pump all this liquid from the tank up to the level $y = h$ above the top of the tank. We imagine a thin, horizontal slice of liquid at height y. If its thickness is dy, then its volume is $dV = A(y)\, dy$, so its weight is $\rho\, dV = \rho A(y)\, dy$. This slice must be lifted the distance $h - y$, so the work done to lift the slice is

$$dW = (h - y)\, \rho\, dV = \rho(h - y)A(y)\, dy.$$

Hence the total amount of work done on all the liquid originally in the tank is

$$W = \int_a^b \rho(h - y)A(y)\, dy. \tag{9}$$

Problem 14 asks you to use Riemann sums to set up this integral.

EXAMPLE 4 A cylindrical tank of radius 3 ft and length 10 ft is lying on its side on horizontal ground. If this tank initially is full of gasoline weighing 40 lb/ft^3, how much work is done in pumping all this gasoline to a point 5 ft above the top of the tank?

Solution Figure 6.6.9 shows an end view of the tank. To exploit circular symmetry, we choose $y = 0$ at the *center* of the circular vertical section, so the tank lies between $y = -3$ and $y = 3$. A horizontal cross section of the tank that meets the y-axis at y is a rectangle of length 10 ft and width w. From the right triangle in Fig. 6.6.9, we see that

$$\tfrac{1}{2}w = (9 - y^2)^{1/2},$$

so the area of this cross section is

$$A(y) = 10w = 20(9 - y^2)^{1/2}.$$

This cross section must be lifted from its initial position y to the final position $5 + 3 = 8$, so it is to be lifted the distance $8 - y$. Thus Eq. (9) with $\rho = 40$, $a = -3$, and $b = 3$ yields

$$W = \int_{-3}^3 40 \cdot (8 - y) \cdot 20 \cdot (9 - y^2)^{1/2}\, dy$$

$$= 6400 \int_{-3}^3 (9 - y^2)^{1/2}\, dy - 800 \int_{-3}^3 y(9 - y^2)^{1/2}\, dy.$$

We attack the two integrals separately. First,

$$\int_{-3}^3 y(9 - y^2)^{1/2}\, dy = \left[-\tfrac{1}{3}(9 - y^2)^{3/2} \right]_{-3}^3 = 0.$$

Second,

$$\int_{-3}^3 (9 - y^2)^{1/2}\, dy = \tfrac{1}{2}\pi \cdot 3^2 = \tfrac{9}{2}\pi,$$

because the integral is simply the area of a semicircle of radius 3. Hence

$$W = 6400 \cdot \tfrac{9}{2}\pi = 28{,}800\pi,$$

approximately 90,478 ft · lb.

As in Example 4, you may use as needed in the problems the integral

$$\int_0^a (a^2 - x^2)^{1/2} \, dx = \tfrac{1}{4}\pi a^2, \qquad (10)$$

which corresponds to the area of a quarter-circle of radius a.

FORCE EXERTED BY A LIQUID

The **pressure** p at depth h in a liquid is the force per unit area exerted by the liquid at that depth. Pressure is given by

$$p = \rho h, \qquad (11)$$

where ρ is the (weight) density of the liquid. For example, at a depth of 10 ft in water, for which $\rho = 62.4$ lb/ft^3, the pressure is $62.4 \cdot 10 = 624$ lb/ft^2. Hence if a thin, flat plate of area 5 ft^2 is suspended in a horizontal position at a depth of 10 ft in water, then the water exerts a downward force of $624 \cdot 5 = 3120$ lb on the top face of the plate and an equal upward force on its bottom face.

It is an important fact that at a given depth in a liquid, the pressure is the same in all directions. But if a flat plate is submerged in a vertical position in the liquid, then the pressure on the face of the plate is *not* constant, because by Eq. (11) the pressure increases with increasing depth. Consequently, the total force exerted on a vertical plate must be computed by integration.

Consider a thin, vertical, flat plate submerged in a liquid of density ρ (Fig. 6.6.10). The surface of the liquid is at the line $y = c$, and the plate lies alongside the interval $a \leqq y \leqq b$. The width of the plate at depth $c - y$ is some function of y, which we denote by $w(y)$.

To compute the total force F exerted by the liquid on either face of this plate, we begin with a division of $[a, b]$ into n subintervals, all with the same length Δy, and denote by y_i^* the midpoint of the subinterval $[y_{i-1}, y_i]$. The horizontal strip of the plate opposite this ith subinterval is approximated by a rectangle of width $w(y_i^*)$ and height Δy, and its average depth in the liquid is $c - y_i^*$. Hence the force ΔF_i exerted by the liquid on this horizontal strip is given approximately by

$$\Delta F_i \approx \rho(c - y_i^*)w(y_i^*) \, \Delta y. \qquad (12)$$

The total force on the entire plate is given approximately by

$$F = \sum_{i=1}^n \Delta F_i \approx \sum_{i=1}^n \rho(c - y_i^*)w(y_i^*) \, \Delta y.$$

We obtain the exact value of F by taking the limit of such Riemann sums as $\Delta y \to 0$:

$$F = \int_a^b \rho(c - y)w(y) \, dy. \qquad (13)$$

EXAMPLE 5 A cylindrical tank 8 ft in diameter is lying on its side and is half full of oil of density $\rho = 75$ lb/ft^3. Find the total force F exerted by the oil on one end of the tank.

Solution We locate the y-axis as indicated in Fig. 6.6.11 so that the surface of the oil is at the level $y = 0$. The oil lies alongside the interval

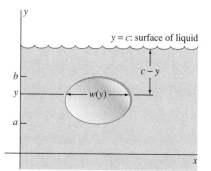

Fig. 6.6.10 A thin plate suspended vertically in a liquid

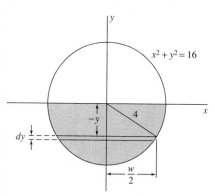

Fig. 6.6.11 View of one end of the tank of Example 5

Ch. 6 / Applications of the Integral

$-4 \leq y \leq 0$. We see from the right triangle in the figure that the width of the oil at depth $-y$ (and thus at location y) is

$$w(y) = 2(16 - y^2)^{1/2}.$$

Hence Eq. (13) gives

$$F = \int_{-4}^{0} 75(-y)[2(16 - y^2)^{1/2}] \, dy = 75\left[\tfrac{2}{3}(16 - y^2)^{3/2}\right]_{-4}^{0} = 3200 \quad \text{(lb)}.$$

6.6 Problems

In Problems 1 through 5, find the work done by the given force $F(x)$ in moving a particle along the x-axis from $x = a$ to $x = b$.

1. $F(x) = 10$; $\quad a = -2, b = 1$

2. $F(x) = 3x - 1$; $\quad a = 1, b = 5$

3. $F(x) = \dfrac{10}{x^2}$; $\quad a = 1, b = 10$

4. $F(x) = -3\sqrt{x}$; $\quad a = 0, b = 4$

5. $F(x) = \sin \pi x$; $\quad a = -1, b = 1$

6. A spring has a natural length of 1 m, and a force of 10 N is required to hold it stretched to a total length of 2 m. How much work is done in compressing this spring from its natural length to a length of 60 cm?

7. A spring has a natural length of 2 ft, and a force of 15 lb is required to hold it compressed to a length of 18 in. How much work is done in stretching this spring from its natural length to a length of 3 ft?

8. Apply Eq. (4) to compute the amount of work done in lifting a 100-lb weight a height of 10 ft, assuming that this work is done against the constant force of gravity.

9. Compute the amount of work (in foot-pounds) done in lifting a 1000-lb weight from an orbit 1000 mi above the earth's surface to one 2000 mi above the earth's surface. Use the value of k given in Example 2.

10. A cylindrical tank of radius 5 ft and height 10 ft is resting on the ground with its axis vertical. Use Eq. (8) to compute the amount of work done in filling this tank with water pumped in from ground level. (Use $\rho = 62.4 \text{ lb/ft}^3$ for the weight density of water.)

11. A conical tank is resting on its base, which is at ground level, and its axis is vertical. The tank has radius 5 ft and height 10 ft (Fig. 6.6.12). Compute the work done in filling this tank with water ($\rho = 62.4 \text{ lb/ft}^3$) pumped in from ground level.

12. Repeat Problem 11, except that now the tank is up-ended: Its vertex is at ground level and its base is 10 ft above the ground (but its axis is still vertical).

13. A tank whose lowest point is 10 ft above the ground has the shape of a cup obtained by rotating the parabola $x^2 = 5y$, $-5 \leq x \leq 5$, around the y-axis (Fig. 6.6.13).

Fig. 6.6.12 The conical tank of Problem 11

Fig. 6.6.13 The cup-shaped tank of Problem 13

The units on the coordinate axes are feet. How much work is done in filling this tank with oil of density 50 lb/ft³ if the oil is pumped in from ground level?

14. Suppose that the tank of Fig. 6.6.8 is filled with fluid of density ρ and that all this fluid must be pumped from the tank to the level $y = h$ above the top of the tank. Use Riemann sums, as in the derivation of Eq. (8), to obtain the formula $W = \int_a^b \rho(h - y)A(y) \, dy$ for the work required to do this.

15. Use the formula of Problem 14 to find the amount of work done in pumping the water in the tank of Problem 10 to a height of 5 ft above the top of the tank.

16. Gasoline at a service station is stored in a cylindrical tank buried on its side, with the highest part of the tank 5 ft below the surface. The tank is 6 ft in diameter and 10 ft long. The density of gasoline is 45 lb/ft³. Assume that the filler cap of each automobile gas tank is 2 ft above the ground (Fig. 6.6.14). (a) How much work is done in

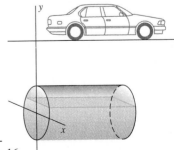

Fig. 6.6.14 The gasoline tank of Problem 16

emptying all the gasoline from this tank, which is initially full, into automobiles? (b) Recall that 1 hp is equivalent to 33,000 ft·lb/min. For electrical conversions, 1 kW (1000 W) is the same as 1.341 hp. The charge for use of electricity generated by a power company is typically about 7.2¢/kWh. Assume that the electrical motor in the gas pump at this station is 30% efficient. How much does it cost to pump all the gasoline from this tank into automobiles?

17. Consider a spherical water tank whose radius is 10 ft and whose center is 50 ft above the ground. How much work is required to fill this tank by pumping water up from ground level? [*Suggestion:* It may simplify your computations to take $y = 0$ at the center of the tank and think of the distance each horizontal slice of water must be lifted.]

18. A hemispherical tank of radius 10 ft is located with its flat side down atop a tower 60 ft high (Fig. 6.6.15). How much work is required to fill this tank with oil of density 50 lb/ft^3 if the oil is to be pumped into the tank from ground level?

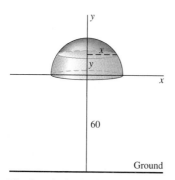

Fig. 6.6.15   The hemispherical tank of Problem 18

19. Water is being drawn from a well 100 ft deep, using a bucket that scoops up 100 lb of water. The bucket is pulled up at the rate of 2 ft/s, but it has a hole in the bottom through which water leaks out at the rate of 0.5 lb/s. How much work is done in pulling the bucket to the top of the well? Neglect the weight of the bucket, the weight of the rope, and the work done in overcoming friction. [*Suggestion:* Take $y = 0$ at the level of water in the well, so that $y = 100$ at ground level. Let $\{y_0, y_1, y_2, \ldots, y_n\}$ be a subdivision of $[0, 100]$ into n equal-length subintervals. Estimate the amount of work ΔW_i required to raise the bucket from y_{i-1} to y_i. Then set up the sum $W = \Sigma \, \Delta W_i$ and proceed to the appropriate integral by letting $n \to +\infty$.]

20. A rope that is 100 ft long and weighs 0.25 lb per linear foot hangs from the edge of a very tall building. How much work is required to pull this rope to the top of the building?

21. Suppose that we plug the hole in the leaky bucket of Problem 19. How much work do we do in lifting the mended bucket, full of water, to the surface, using the rope of Problem 20? Ignore friction and the weight of the bucket, but allow for the weight of the rope.

22. Consider a volume V of gas in a cylinder fitted with a piston at one end, where the pressure p of the gas is a function $p(V)$ of its volume (Fig. 6.6.16). Let A be the area of the face of the piston. Then the force exerted on the piston by gas in the cylinder is $F = pA$. Assume that the gas expands from volume V_1 to volume V_2. Show that the work done by the force F is then given by

$$W = \int_{V_1}^{V_2} p(V) \, dV.$$

[*Suggestion:* If x is the length of the cylinder (from its fixed end to the face of the piston), then $F = A \cdot p(Ax)$. Apply Eq. (4), and substitute $V = Ax$ into the resulting integral.]

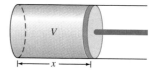

Fig. 6.6.16   A cylinder fitted with a piston (Problem 22)

23. The pressure p and volume V of the steam in a small steam engine satisfy the condition $pV^{1.4} = c$ (where c is a constant). In one cycle, the steam expands from a volume $V_1 = 50$ in.^{3} to $V_2 = 500$ in.^{3} with an initial pressure of 200 lb/in.^{2} Use the formula of Problem 22 to compute the work, in foot-pounds, done by this engine in each such cycle.

24. A tank in the shape of a hemisphere of radius 60 ft is resting on its flat base with the curved surface on top. It is filled with alcohol of density 40 lb/ft^3. How much work is done in pumping all the alcohol to the top of the tank?

25. A tank is in the shape of the surface generated by rotating around the y-axis the graph of $y = x^4$, $0 \leq x \leq 1$. The tank is initially full of oil of density 60 lb/ft^3. The units on the coordinate axes are feet. How much work is done in pumping all the oil to the top of the tank?

26. A cylindrical tank of radius 3 ft and length 20 ft is lying on its side on horizontal ground. Gasoline weighing 40 lb/ft^3 is at ground level and is to be pumped into the tank. Find the work required to fill the tank.

27. The base of a spherical storage tank of radius 12 ft is at ground level. Find the amount of work done in filling the tank with oil of density 50 lb/ft^3 if all the oil is initially at ground level.

28. A 20-lb monkey is attached to a 50-ft chain that weighs 0.5 lb per (linear) foot. The other end of the chain

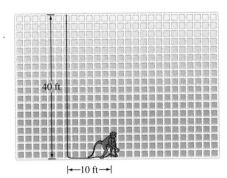

Fig. 6.6.17 The monkey of Problem 28

is attached to the 40-ft-high ceiling of the monkey's cage (Fig. 6.6.17). Find the amount of work the monkey does in climbing up her chain to the ceiling.

29. Tom is flying a kite at a height of 500 ft above the ground. Suppose that the kite string weighs $\frac{1}{16}$ oz per (linear) foot and that its string is stretched in a straight line at a $45°$ angle to the ground. How much work was done by the wind in lifting the string from ground level up to its flying position?

30. The center of a spherical tank of radius R is at a distance $H > R$ above the ground. A liquid of weight density ρ is at ground level. Show that the work required to pump the initially empty tank full of this liquid is the same as that to lift the full tank the distance H.

31. A water trough 10 ft long has a square cross section that is 2 ft wide. If the trough is full ($\rho = 62.4$ lb/ft³), find the force exerted by the water on one end of the trough.

32. Repeat Problem 31 for a trough whose cross section is an equilateral triangle with edges 3 ft long.

33. Repeat Problem 31 for a trough whose cross section is a trapezoid 3 ft high, 2 ft wide at the bottom, and 4 ft wide at the top.

34. Find the force on one end of the cylindrical tank of Example 5 if the tank is filled with oil of density 50 lb/ft³. Remember that

$$\int_0^a (a^2 - y^2)^{1/2} \, dy = \tfrac{1}{4}\pi a^2,$$

because the integral represents the area of a quarter-circle of radius a.

In Problems 35 through 38, a gate in the vertical face of a dam is described. Find the total force of water on this gate if its top is 10 ft beneath the surface of the water.

35. A square of edge length 5 ft whose top is parallel to the water surface.

36. A circle of radius 3 ft.

37. An isosceles triangle 5 ft high and 8 ft wide at the top.

38. A semicircle of radius 4 ft whose top edge is its diameter (also parallel to the water surface).

39. Suppose that the dam of Fig. 6.6.18 is $L = 200$ ft long and $T = 30$ ft thick at its base. Find the force of water on the dam if the water is 100 ft deep and the *slanted* end of the dam faces the water.

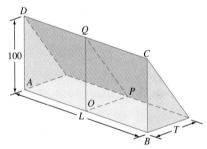

Fig. 6.6.18 View of a model of a dam (Problem 39)

Chapter 6 Review: DEFINITIONS, CONCEPTS, RESULTS

Use the following list as a guide to concepts that you may need to review.

1. The general method of setting up an integral formula for a quantity by approximating it and then recognizing the approximation to be a Riemann sum that corresponds to the desired integral: If the interval $[a, b]$ is divided into n subintervals of equal length $\Delta x = (b - a)/n$ and if x_i^* denotes a point of the ith subinterval, then

$$\lim_{n \to \infty} \sum_{i=1}^{n} f(x_i^*) \, \Delta x = \int_a^b f(x) \, dx.$$

2. Net distance traveled as the integral of velocity:

$$s = \int_a^b v(t) \, dt.$$

3. The method of cross sections for computing volumes:

$$V = \int_a^b A(x) \, dx,$$

where $A(x)$ denotes the area of a slice with infinitesimal thickness dx.

4. Determining the volume of a solid of revolution by the method of cross sections, with the cross sections being either disks or annular rings.

5. Determining the volume of a solid of revolution by the method of cylindrical shells:

$$V = \int_*^{**} 2\pi r\, dA,$$

where r denotes the radius of the circle through which the area element dA is revolved.

6. The arc length of a smooth arc described either in the form $y = f(x)$, $a \leq x \leq b$, or in the form $x = g(y)$, $c \leq y \leq d$:

$$s = \int_*^{**} ds \qquad (ds = \sqrt{(dx)^2 + (dy)^2}),$$

where

$$ds = \sqrt{1 + [f'(x)]^2}\, dx \qquad \text{for } y = f(x);$$
$$ds = \sqrt{1 + [g'(y)]^2}\, dy \qquad \text{for } x = g(y).$$

7. Determining the area of the surface of revolution generated by revolving a smooth arc, given in the form $y = f(x)$ or in the form $x = g(y)$, about either the x-axis or the y-axis:

$$A = \int_*^{**} 2\pi r\, ds,$$

where r denotes the radius of the circle through which the arc length element ds is revolved.

8. The work done by a force function in moving a particle along a straight line segment:

$$W = \int_a^b F(x)\, dx$$

if the force $F(x)$ acts from $x = a$ to $x = b$.

9. Hooke's law and the work done in stretching or compressing an elastic spring.

10. Work done against the varying force of gravity.

11. Work done in filling a tank or in pumping the liquid in a tank to another level:

$$W = \int_a^b \rho h(y) A(y)\, dy,$$

where $h(y)$ denotes the vertical distance that the horizontal fluid slice of volume $dV = A(y)\, dy$, at the height y, must be lifted.

12. The force exerted by a liquid on the face of a submerged vertical plate:

$$F = \int_*^{**} \rho h\, dA,$$

where h denotes the depth of the horizontal area element dA beneath the surface of the fluid of weight density ρ.

13. Solution of the separable differential equation

$$\frac{dy}{dx} = g(x)\phi(y)$$

by integration:

$$\int f(y)\, dy = \int g(x)\, dx + C,$$

where $f(y) = 1/\phi(y)$.

Chapter 6 Miscellaneous Problems

In Problems 1 through 3, find both the net distance and the total distance traveled between times $t = a$ and $t = b$ by a particle moving along a line with the given velocity function $v = f(t)$.

1. $v = t^2 - t - 2; \quad a = 0, b = 3$

2. $v = |t^2 - 4|; \quad a = 1, b = 4$

3. $v = \pi \sin \frac{1}{2}\pi(2t - 1); \quad a = 0, b = 1.5$

In Problems 4 through 8, a solid extends along the x-axis from $x = a$ to $x = b$, and its cross-sectional area at x is $A(x)$. Find its volume.

4. $A(x) = x^3; \quad a = 0, b = 1$

5. $A(x) = \sqrt{x}; \quad a = 1, b = 4$

6. $A(x) = x^3; \quad a = 1, b = 2$

7. $A(x) = \pi(x^2 - x^4); \quad a = 0, b = 1$

8. $A(x) = x^{100}; \quad a = -1, b = 1$

9. Suppose that rainfall begins at time $t = 0$ and that the rate after t hours is $(t + 6)/12$ inches/hour. How many inches of rain falls during the first 12 h?

10. The base of a certain solid is the region in the first quadrant bounded by the curves $y = x^3$ and $y = 2x - x^2$. Find the solid's volume, if each cross section perpendicular to the x-axis is a square with one edge in the base of the solid.

11. Find the volume of the solid generated by revolving around the x-axis the first-quadrant region of Problem 10.

12. Find the volume of the solid generated by revolving the region bounded by $y = 2x^4$ and $y = x^2 + 1$ around (a) the x-axis; (b) the y-axis.

13. A wire made of copper (density 8.5 g/cm³) is shaped like a helix that spirals around the x-axis from $x = 0$ to $x = 20$. Each cross section of this wire perpendicular to the x-axis is a circular disk of radius 0.25 cm. What is the total mass of the wire?

406

14. Derive the formula $V = \frac{1}{3}\pi h(r_1^2 + r_1 r_2 + r_2^2)$ for the volume of a frustum of a cone with height h and base radii r_1 and r_2.

15. Suppose that the point P lies on a line perpendicular to the xy-plane at the origin O, with $|OP| = h$. Consider the "elliptical cone" that consists of all points on line segments from P to points on and within the ellipse with equation

$$\left(\frac{x}{a}\right)^2 + \left(\frac{y}{b}\right)^2 = 1.$$

Show that the volume of this elliptical cone is $V = \frac{1}{3}\pi abh$.

16. Figure 6.MP.1 shows the region R bounded by the ellipse $(x/a)^2 + (y/b)^2 = 1$ and by the line $x = a - h$, where $0 < h < a$. Revolution of R around the x-axis generates a "segment of an ellipsoid" of radius r, height h, and volume V. Show that

$$r^2 = \frac{b^2(2ah - h^2)}{a^2} \quad \text{and that} \quad V = \frac{1}{3}\pi r^2 h \frac{3a - h}{2a - h}.$$

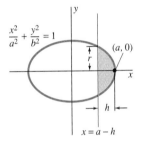

Fig. 6.MP.1 A segment of an ellipse (Problem 16)

Fig. 6.MP.2 The region R of Problem 17

17. Figure 6.MP.2 shows the region R bounded by the hyperbola $(x/a)^2 - (y/b)^2 = 1$ and the line $x = a + h$, where $h > 0$. Revolution of R about the x-axis generates a "segment of a hyperboloid" of radius r, height h, and volume V. Show that

$$r^2 = \frac{b^2(2ah + h^2)}{a^2} \quad \text{and that} \quad V = \frac{1}{3}\pi r^2 h \frac{3a + h}{2a + h}.$$

In Problems 18 through 20, the function $f(x)$ is nonnegative and continuous for $x \geqq 1$. When the region lying under $y = f(x)$ from $x = 1$ to $x = t$ is revolved around the indicated axis, the volume of the resulting solid is $V(t)$. Find the function $f(x)$.

18. $V(t) = \pi\left(1 - \frac{1}{t}\right)$; the x-axis

19. $V(t) = \frac{1}{6}\pi[(1 + 3t)^2 - 16]$; the x-axis

20. $V(t) = \frac{2}{9}\pi[(1 + 3t^2)^{3/2} - 8]$; the y-axis

21. Use the integral formula in Problem 36 of Section 6.3

to find the volume of the solid generated by revolving around the y-axis the first-quadrant region bounded by $y = x$ and $y = \sin(\pi x/2)$.

22. Use the method of cylindrical shells to find the volume of the solid generated by revolving around the line $x = -2$ the region bounded by $y = x^2$ and $y = x + 2$.

23. Find the length of the curve $y = \frac{1}{3}x^{3/2} - x^{1/2}$ from $x = 1$ to $x = 4$.

24. Find the area of the surface generated by revolving the curve of Problem 23 around (a) the x-axis; (b) the y-axis.

25. Find the length of the curve $y = \frac{3}{8}(y^{4/3} - 2y^{2/3})$ from $y = 1$ to $y = 8$.

26. Find the area of the surface generated by revolving the curve of Problem 25 around (a) the x-axis; (b) the y-axis.

27. Find the area of the surface generated by revolving the curve of Problem 23 around the line $x = 1$.

28. If $-r < a < b < r$, then a "spherical zone" of "height" $h = b - a$ is generated by revolving around the x-axis the circular arc $y = \sqrt{r^2 - x^2}$, $a \leqq x \leqq b$. Show that the area of this spherical zone is $A = 2\pi rh$, the same as that of a cylinder of radius r and height h.

29. Apply the result of Problem 28 to show that the surface area of a sphere of radius r is $A = 4\pi r^2$.

30. Let R denote the region bounded by the curves $y = 2x^3$ and $y^2 = 4x$. Find the volumes of the solids obtained by revolving the region R around (a) the x-axis; (b) the y-axis; (c) the line $y = -1$; (d) the line $x = 2$. In each case use both the method of cross sections and the method of cylindrical shells.

In Problems 31 through 42, find the general solution of the given differential equation. If an initial condition is given, find the corresponding particular solution.

31. $\dfrac{dy}{dx} = 2x + \cos x; \quad y(0) = 0$

32. $\dfrac{dy}{dx} = 3\sqrt{x} + \dfrac{1}{\sqrt{x}}; \quad y(1) = 10$

33. $\dfrac{dy}{dx} = (y + 1)^2$ **34.** $\dfrac{dy}{dx} = \sqrt{y + 1}$

35. $\dfrac{dy}{dx} = 3x^2 y^2; \quad y(0) = 1$

36. $\dfrac{dy}{dx} = \sqrt[3]{xy}; \quad y(1) = 1$

37. $x^2 y^2 \dfrac{dy}{dx} = 1$ **38.** $\sqrt{xy}\,\dfrac{dy}{dx} = 1$

39. $\dfrac{dy}{dx} = y^2 \cos x; \quad y(0) = 1$

40. $\dfrac{dy}{dx} = \sqrt{y}\,\sin x; \quad y(0) = 4$

41. $\dfrac{dy}{dx} = \dfrac{y^2(1 - \sqrt{x})}{x^2(1 - \sqrt{y})}$ **42.** $\dfrac{dy}{dx} = \dfrac{\sqrt{y}\,(x + 1)^3}{\sqrt{x}\,(y + 1)^3}$

43. Find the natural length L of a spring if five times as much work is required to stretch it from a length of 2 ft to a length of 5 ft as is required to stretch it from a length of 2 ft to a length of 3 ft.

44. A steel beam weighing 1000 lb hangs from a 50-ft cable that weighs 5 lb per linear foot. How much work is done in winding in 25 ft of the cable with a windlass?

45. A spherical tank of radius R (in feet) is initially full of oil of density ρ lb/ft^3. Find the total work done in pumping all the oil from the sphere to a height of $2R$ above the top of the tank.

46. How much work is done by a colony of ants in building a conical anthill of height and diameter 1 ft, using sand initially at ground level and with a density of 150 lb/ft^3?

47. The gravitational attraction below the earth's surface is directly proportional to the distance from the earth's center. Suppose that a straight cylindrical hole of radius 1 ft is dug from the earth's surface to its center. Assume that the earth has radius 3960 mi and uniform density 350 lb/ft^3. How much work, in foot-pounds, is done in lifting a 1-lb weight from the bottom of this hole to its top?

48. How much work is done in digging the hole of Problem 47—that is, in lifting the material it initially contained to the earth's surface?

49. Suppose that a dam is shaped like a trapezoid of height 100 ft, 300 ft long at the top and 200 ft long at the bottom. When the water level behind the dam is even with its top, what is the total force that the water exerts on the dam?

50. Suppose that a dam has the same top and bottom lengths as the dam of Problem 49 and the same vertical height of 100 ft, but that its face toward the water is slanted at an angle of 30° from the vertical. What is the total force of water pressure on this dam?

51. For $c > 0$, the graphs of $y = c^2 x^2$ and $y = c$ bound a plane region. Revolve this region around the horizontal line $y = -1/c$ to form a solid. For what value of c is the volume of this solid maximal? Minimal?

7

More Exponential and Logarithmic Functions

John Napier (1550–1617)

❏ John Napier was the eighth baron of Merchiston, near Edinburgh, Scotland. As a landowner, he participated vigorously in local affairs and actively managed crops and cattle on his land surrounding Merchiston Castle. He is said to have regarded as his most important and lasting contribution his book *A Plaine Discovery of the Whole Revelation of Saint John* (1593), in which (among other things) he predicted that the world would come to an end in the year 1786. But the world—still here some two centuries later—remembers Napier as the inventor of logarithms. The late sixteenth century was an age of numerical computation, as developments in astronomy and navigation called for increasingly accurate and lengthy trigonometric computations. With his theological work behind him, Napier began in 1594 the 20-year labor that was to revolutionize the practical art of numerical computation. The result of this monumental project was his 1614 book *Mirifici Logarithmorum Canonis Descriptio* (Description of the Wonderful Canon of Logarithms), which contained his logarithmic tables along with a brief introduction and guide to their use.

❏ Students today first learn of the "common" base-10 logarithm $\log_{10} x$ as the power to which 10 must be raised to get the number x. Because

$$\log_{10} xy = \log_{10} x + \log_{10} y,$$

multiplying the two numbers x and y effectively reduces to the simpler operation of adding their logarithms. The "logarithm" $\mathrm{Nlog}(x)$ that Napier used was somewhat different. Whereas $\log_{10} 1 = 0$, Napier's definition implied that $\mathrm{Nlog}(10^7) = 0$ with the previous equation replaced by a rule of the form $\mathrm{Nlog}(xy) = \mathrm{Nlog}(x) + \mathrm{Nlog}(y) - C$. Hence the use of Napier's original table of logarithms would involve continual addition or subtraction of the particular constant $C = 10^7 \times \ln 10^7$ (where ln denotes the modern "natural" logarithm).

❏ In 1615 the English mathematics professor Henry Briggs visited Napier in Scotland. Their discussions led to Briggs' construction of a table of "improved" logarithms, ones for which the logarithm of 1 is 0 and the logarithm of 10 is 1. Within a decade after Napier's death, 10-place tables of these common logarithms had been completed. These tables placed a central role in scientific computation during the next three centuries.

❏ With a slide rule of the type used universally by calculus students prior to the advent of electronic pocket calculators in the 1970s, numbers are inscribed on the scales at distances proportional to their logarithms. Consequently, the logarithms of x on a fixed scale and y on a sliding scale are readily added to determine the product xy.

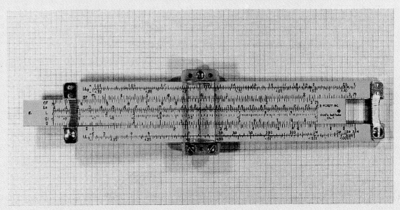

7.1
Introduction

In Section 3.8 we introduced the general exponential function

$$f(x) = a^x \qquad (1)$$

informally as the number $a > 0$ raised to the power x, without dwelling on precisely what this might mean if, for instance, the exponent x is irrational. Applying the definition of the derivative, we concluded that, for a certain carefully chosen value $e \approx 2.71828$ of the base a, the exponential function has an especially simple-looking derivative—namely, itself:

$$D_x e^x = e^x. \qquad (2)$$

We then turned our attention to logarithms and defined the natural logarithm $\ln x$ as the inverse of the natural exponential function e^x. That is, $\ln x$ is the power to which the number e must be raised to get x. Finally, we saw that Eq. (2) and the chain rule imply that

$$D_x \ln x = \frac{1}{x}. \qquad (3)$$

Thus $\ln x$ is an antiderivative of the one power x^{-1} of x that is not the derivative of any other power of x.

Exponential and logarithmic functions are so important in applications of mathematics—ranging from population growth, the spread of epidemics, and the growth of investments to the flow of heat and motion with resistance—that a more detailed analysis of these functions and their applications in this chapter is warranted. But now more of the basic machinery of calculus is available than was the case in Chapter 3. In particular, the definite integral now enables us to put exponentials and logarithms on a firmer footing than intuitive concepts of vaguely defined powers. Somewhat surprisingly, it is easier to begin with the logarithm—the less intuitive of these functions—as we do in Section 7.2. In Section 7.3 we reverse the order of Section 3.8 and define the natural exponential function in terms of the natural logarithm. Later in this chapter we provide a sampling of the diverse real-world applications of these related functions.

7.2
The Natural Logarithm

Section 3.8 was an informal overview of exponential and logarithmic functions. We now present a more systematic development of the properties of these functions.

It is simplest to make the definition of the natural logarithm our starting point. Guided by the results in Section 3.8, we want to define $\ln x$ for $x > 0$ in such a way that

$$\ln 1 = 0 \quad \text{and} \quad D_x \ln x = \frac{1}{x}. \qquad (1)$$

To do so, we recall part 1 of the fundamental theorem of calculus (Section 5.5), according to which

$$D_x \int_a^x f(t)\, dt = f(x)$$

if f is continuous on an interval that contains a and x. In order that $\ln x$ satisfy the equations in (1), we take $a = 1$ and $f(t) = 1/t$.

$$\frac{dy}{dx} = \frac{1}{2}\left(\frac{2x}{x^2+1}\right) - \frac{1}{3}\left(\frac{3x^2}{x^3+1}\right) = \frac{x}{x^2+1} - \frac{x^2}{x^3+1}.$$

LIMITS INVOLVING ln x

Now we establish the limits of $\ln x$ as $x \to 0^+$ and as $x \to +\infty$, as stated in Eqs.(3) and (4). Because $\ln 2$ is the area under $y = 1/x$ from $x = 1$ to $x = 2$, we can inscribe and circumscribe a pair of rectangles (Fig. 7.2.3) and conclude that

$$\tfrac{1}{2} < \ln 2 < 1.$$

Given a large positive number x, let n be the largest integer such that $x > 2^n$. Then, because $\ln x$ is an increasing function, the law of logarithms in Eq. (12) gives

$$\ln x > \ln(2^n) = n \ln 2 > \frac{n}{2}.$$

Because $n \to \infty$ as $x \to \infty$, it follows that we can make $\ln x$ as large as we please simply by choosing x to be sufficiently large. This proves Eq. (4):

$$\lim_{x \to +\infty} \ln x = +\infty.$$

To prove Eq. (3), we use the law of logarithms in Eq. (10) to write

$$\lim_{x \to 0^+} \ln x = -\lim_{x \to 0^+} \ln\left(\frac{1}{x}\right) = -\lim_{y \to +\infty} \ln y = -\infty$$

by taking $y = 1/x$ and applying the result in Eq. (4).

The graphs of the natural logarithm function shown in Figs. 7.2.4 through 7.2.6 indicate that, although $\ln x \to +\infty$ as $x \to +\infty$, the logarithm of x increases *very* slowly. For instance,

$$\ln(1000) \approx 6.908,$$
$$\ln(1{,}000{,}000) \approx 13.812, \quad \text{and} \quad \ln(1{,}000{,}000{,}000) \approx 20.723.$$

Indeed, the function $\ln x$ increases more slowly than any positive integral power of x. By this we mean that

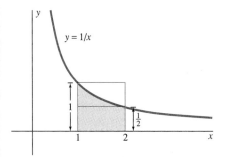

Fig. 7.2.3 Using rectangles to estimate ln 2

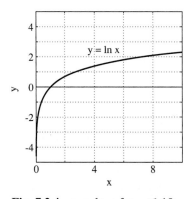

Fig. 7.2.4 $y = \ln x$ for $x \leq 10$

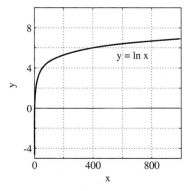

Fig. 7.2.5 $y = \ln x$ for $x \leq 1000$

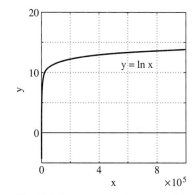

Fig. 7.2.6 $y = \ln x$ for $x \leq 1{,}000{,}000$

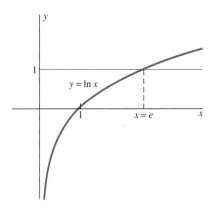

$$\lim_{x \to \infty} \frac{\ln x}{x^n} = 0 \qquad (13)$$

if n is a fixed positive integer. To prove this, note first that

$$\ln x = \int_1^x \frac{dt}{t} \leqq \int_1^x \frac{dt}{t^{1/2}} = 2(x^{1/2} - 1),$$

because $1/t \leqq 1/t^{1/2}$ if $t \geqq 1$. Hence if $x > 1$, then

$$0 < \frac{\ln x}{x^n} \leqq \frac{\ln x}{x} \leqq \frac{2}{x^{1/2}} - \frac{2}{x}.$$

Fig. 7.2.7 The fact that $\ln e = 1$ is expressed graphically here.

The last expression on the right approaches zero as $x \to +\infty$. Equation (13) follows by the squeeze law for limits. Moreover, this argument shows that the positive integer n in Eq. (13) may be replaced by any rational number $k \geqq 1$.

Because $\ln x$ is an increasing function, the intermediate value property implies that the curve $y = \ln x$ crosses the horizontal line $y = 1$ precisely once. The abscissa of the point of intersection is the important number $e \approx 2.71828$ mentioned in Section 3.8 (see Fig. 7.2.7).

> **Definition of e**
> The number e is the unique real number such that
> $$\ln e = 1. \qquad (14)$$

Title page of the book in which Euler introduced the number e.

The letter e has been used to denote the number whose natural logarithm is 1 ever since this number was introduced by the Swiss mathematician Leonhard Euler (1707–1783), who used e for "exponential."

EXAMPLE 8   Sketch the graph of

$$f(x) = \frac{\ln x}{x}, \qquad x > 0.$$

Solution   First we compute the derivative

$$f'(x) = \frac{\frac{1}{x} \cdot x - (\ln x) \cdot 1}{x^2} = \frac{1 - \ln x}{x^2}.$$

Thus the only critical point occurs where $\ln x = 1$—that is, where $x = e$. Because $f'(x) > 0$ if $x < e$ (the same as $\ln x < 1$) and $f'(x) < 0$ if $x > e$ (the same as $\ln x > 1$), we see that f is increasing if $x < e$ and decreasing if $x > e$. Hence f has a local maximum at $x = e$.

The second derivative of f is

$$f''(x) = \frac{-\frac{1}{x} \cdot x^2 - (1 - \ln x) \cdot 2x}{x^4} = \frac{2 \ln x - 3}{x^3},$$

so the only inflection point of the graph is where $\ln x = \frac{3}{2}$—that is, at $x = e^{3/2} \approx 4.48$.

Because

$$\lim_{x \to 0^+} \frac{\ln x}{x} = -\infty \quad \text{and} \quad \lim_{x \to \infty} \frac{\ln x}{x} = 0$$

—consequences of Eqs. (3) and (13), respectively—we conclude that the graph of f looks like Fig. 7.2.8.

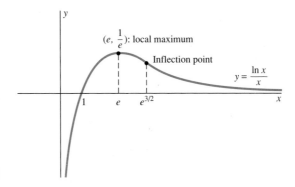

Fig. 7.2.8 The graph of Example 8

LOGARITHMS AND EXPERIMENTAL DATA

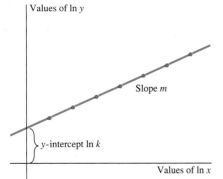

Fig. 7.2.9 Plotting the logarithms of data may reveal a hidden relationship

Certain empirical data can be explained by assuming that the observed dependent variable is a **power** function of the independent variable x. In other words, y is described by a mathematical model of the form

$$y = kx^m,$$

where k and m are constants. If so, the laws of logarithms imply that

$$\ln y = \ln k + m \ln x.$$

An experimenter can then plot values of $\ln y$ against values of $\ln x$. If the power-function model is valid, the resulting data points will lie on a straight line of slope m and y-intercept $\ln k$ (Fig. 7.2.9). This technique makes it easy to see whether or not the data lie on a straight line. And if they do, this technique makes it easy also to measure the slope and y-intercept of the line and thereby to find the values of k and m.

EXAMPLE 9 *(Planetary Motion)* The table in Fig. 7.2.10 gives the period of revolution T and the major semiaxis a of the elliptical orbit of each of the first six planets around the sun, together with the logarithms of these

Planet	T (in days)	a (in 10^6 km)	$\ln T$	$\ln a$
Mercury	87.97	58	4.48	4.06
Venus	224.70	108	5.41	4.68
Earth	365.26	149	5.90	5.00
Mars	686.98	228	6.53	5.43
Jupiter	4332.59	778	8.37	6.66
Saturn	10,759.20	1426	9.28	7.26

Fig. 7.2.10 Data of Example 9

numbers. If we plot $\ln T$ against $\ln a$, it is immediately apparent that the resulting points lie on a straight line of slope $m = \frac{3}{2}$. Hence T and a satisfy an equation of the form $T = ka^{3/2}$, so

$$T^2 = Ca^3.$$

This means that the square of the period T is proportional to the cube of the major semiaxis a. This is Kepler's third law of planetary motion, which Johannes Kepler (1571–1630) discovered empirically in 1619.

7.2 Problems

Differentiate the functions given in Problems 1 through 18.

1. $f(x) = \ln(3x - 1)$

2. $f(x) = \ln(4 - x^2)$

3. $f(x) = \ln\sqrt{1 + 2x}$

4. $f(x) = \ln[(1 + x)^3]$

5. $f(x) = \ln\sqrt[3]{x^3 - x}$

6. $f(x) = \ln(\sin^2 x)$

7. $f(x) = \cos(\ln x)$

8. $f(x) = (\ln x)^3$

9. $f(x) = \dfrac{1}{\ln x}$

10. $f(x) = \ln(\ln x)$

11. $f(x) = \ln(x\sqrt{x^2 + 1})$

12. $g(t) = t^{3/2} \ln(t + 1)$

13. $f(x) = \ln(\cos x)$

14. $f(x) = \ln(2 \sin x)$

15. $f(t) = t^2 \ln(\cos t)$

16. $f(x) = \sin(\ln 2x)$

17. $g(t) = t(\ln t)^2$

18. $g(t) = \sqrt{t}\,[\cos(\ln t)]^2$

In Problems 19 through 28, apply laws of logarithms to simplify the given function; then write its derivative.

19. $f(x) = \ln[(2x + 1)^3(x^2 - 4)^4]$

20. $f(x) = \ln\sqrt{\dfrac{1 - x}{1 + x}}$

21. $f(x) = \ln\sqrt{\dfrac{4 - x^2}{9 + x^2}}$

22. $f(x) = \ln\dfrac{\sqrt{4x - 7}}{(3x - 2)^3}$

23. $f(x) = \ln\dfrac{x + 1}{x - 1}$

24. $f(x) = x^2 \ln\dfrac{1}{2x + 1}$

25. $g(t) = \ln\dfrac{t^2}{t^2 + 1}$

26. $f(x) = \ln\dfrac{\sqrt{x + 1}}{(x - 1)^3}$

27. $f(x) = \ln\dfrac{\sin x}{x}$

28. $f(x) = \ln\dfrac{\sin x}{\cos x}$

In Problems 29 through 32, find dy/dx by implicit differentiation.

29. $y = x \ln y$

30. $y = (\ln x)(\ln y)$

31. $xy = \ln(\sin y)$

32. $xy + x^2(\ln y)^2 = 4$

Evaluate the indefinite integrals in Problems 33 through 50.

33. $\displaystyle\int \dfrac{dx}{2x - 1}$

34. $\displaystyle\int \dfrac{dx}{3x + 5}$

35. $\displaystyle\int \dfrac{x}{1 + 3x^2}\, dx$

36. $\displaystyle\int \dfrac{x^2}{4 - x^3}\, dx$

37. $\displaystyle\int \dfrac{x + 1}{2x^2 + 4x + 1}\, dx$

38. $\displaystyle\int \dfrac{\cos x}{1 + \sin x}\, dx$

39. $\displaystyle\int \dfrac{1}{x}(\ln x)^2\, dx$

40. $\displaystyle\int \dfrac{1}{x \ln x}\, dx$

41. $\displaystyle\int \dfrac{1}{x + 1}\, dx$

42. $\displaystyle\int \dfrac{x}{1 - x^2}\, dx$

43. $\displaystyle\int \dfrac{2x + 1}{x^2 + x + 1}\, dx$

44. $\displaystyle\int \dfrac{x + 1}{x^2 + 2x + 3}\, dx$

45. $\displaystyle\int \dfrac{\ln x}{x}\, dx$

46. $\displaystyle\int \dfrac{\ln(x^3)}{x}\, dx$

47. $\displaystyle\int \dfrac{\sin 2x}{1 - \cos 2x}\, dx$

48. $\displaystyle\int \dfrac{dx}{x(\ln x)^2}$

49. $\displaystyle\int \dfrac{x^2 - 2x}{x^3 - 3x^2 + 1}\, dx$

50. $\displaystyle\int \dfrac{dx}{\sqrt{x}\,(1 + \sqrt{x})}$ [*Suggestion:* Let $u = 1 + \sqrt{x}$.]

Apply Eq. (13) to evaluate the limits in Problems 51 through 56.

51. $\displaystyle\lim_{x \to \infty} \dfrac{\ln \sqrt{x}}{x}$

52. $\displaystyle\lim_{x \to \infty} \dfrac{\ln(x^3)}{x^2}$

53. $\displaystyle\lim_{x \to \infty} \dfrac{\ln x}{\sqrt{x}}$ [*Suggestion:* Substitute $x = u^2$.]

54. $\displaystyle\lim_{x \to 0^+} x \ln x$ [*Suggestion:* Substitute $x = 1/u$.]

55. $\displaystyle\lim_{x \to 0^+} \sqrt{x} \ln x$

56. $\displaystyle\lim_{x \to \infty} \dfrac{(\ln x)^2}{x}$

57. Prove: If $x \geqq 1$, then

$$\ln\left(x + \sqrt{x^2 - 1}\right) = -\ln\left(x - \sqrt{x^2 - 1}\right).$$

58. Find a formula for $f^{(n)}(x)$, given $f(x) = \ln x$.

59. The heart rate R (in beats per minute) and weight W (in pounds) of various mammals were measured, with the

418

results shown in Fig. 7.2.11. Use the method of Example 9 to find a relation between the two of the form $R = kW^m$.

W	25	67	127	175	240	975
R	131	103	88	81	75	53

Fig. 7.2.11   Data of Problem 59

60. During the adiabatic expansion of a certain diatomic gas, its volume V (in liters) and pressure p (in atmospheres) were measured, with the results shown in Fig. 7.2.12. Use the method of Example 9 to find a relation between V and p of the form $p = kV^m$.

V	1.46	2.50	3.51	5.73	7.26
p	28.3	13.3	8.3	4.2	3.0

Fig. 7.2.12   Data of Problem 60

61. Substitute $y = x^p$ and then apply Eq. (13) to show that

$$\lim_{x \to \infty} \frac{\ln x}{x^p} = 0 \quad \text{if } 0 < p < 1.$$

62. Deduce from the result of Problem 61 that

$$\lim_{x \to \infty} \frac{(\ln x)^k}{x} = 0 \quad \text{if } k > 0.$$

63. Substitute $y = 1/x$ and then apply Eq. (13) to show that

$$\lim_{x \to 0^+} x^k \ln x = 0 \quad \text{if } k > 0.$$

Use the limits in Problems 61 through 63 to help you sketch the graphs, for $x > 0$, of the functions given in Problems 64 through 67.

64. $y = x \ln x$ **65.** $y = x^2 \ln x$

66. $y = \sqrt{x} \ln x$ **67.** $y = \dfrac{\ln x}{\sqrt{x}}$

68. Problem 26 of Section 5.9 asks you to show by numerical integration that

$$\int_1^{2.7} \frac{dx}{x} < 1 < \int_1^{2.8} \frac{dx}{x}.$$

Explain carefully why this result proves that $2.7 < e < 2.8$.

69. If n moles of an ideal gas expands at *constant* temperature T, then the pressure and volume satisfy the ideal-gas equation $pV = nRT$ (n and R are constants). With the aid of Problem 22 in Section 6.6, show that the work W done by the gas in expanding from volume V_1 to volume V_2 is

$$W = nRT \ln \frac{V_2}{V_1}.$$

70. "Gabriel's horn" is obtained by revolving around the x-axis the curve $y = 1/x$, $x \geqq 1$ (Fig. 7.2.13). Let A_b denote its surface area from $x = 1$ to $x = b > 1$. Show that $A_b \geqq 2\pi \ln b$, so—as a consequence—$A_b \to +\infty$ as $b \to +\infty$. Thus the surface area of Gabriel's horn is infinite. Is its volume finite or infinite?

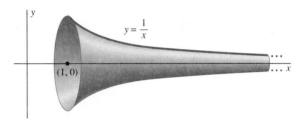

Fig. 7.2.13   Gabriel's horn (Problem 70)

71. According to the prime number theorem, which was conjectured by the great German mathematician Carl Friedrich Gauss in 1792 (when he was 15 yr old) but not proved until 1896 (independently, by Jacques Hadamard and C. J. de la Vallée Poussin), the number of prime numbers between the large positive numbers a and b ($a < b$) is given to a close approximation by the integral

$$\int_a^b \frac{1}{\ln x} \, dx.$$

The midpoint and trapezoidal approximations with $n = 1$ subinterval provide an underestimate and an overestimate of the value of this integral. (Why?) Calculate these estimates with $a = 90{,}000$ and $b = 100{,}000$. The actual number of prime numbers in this range is 879.

7.2 Project

The fact that the number e satisfies (by definition) the equation

$$\int_1^e \frac{dx}{x} = 1 \tag{15}$$

suggests the possibility of using numerical integration to investigate the value of e. You can verify that

$$e \approx 2.71828.$$

The objective of this project is for you to bracket the number e between closer and closer numerical approximations—attempting to determine the upper limit b so that the value of the integral

$$\int_1^b \frac{dx}{x} \qquad (16)$$

is as close as possible to the target value 1.

You may use whatever computational technology is available to you. For example, with $b = 2$ and $n = 50$, a Simpson's approximation program will yield the sum 0.6931; with $b = 3$ and $n = 50$, you'll obtain the sum 1.0986. These results show that $b = 2$ is too small, and that $b = 3$ is too large, to yield 1 for the value of the integral in Eq. (16). Thus $2 < e < 3$.

Interpolation between 0.6931 and 1.0986 suggests that you try the upper limits $b = 2.6$, $b = 2.7$, and perhaps $b = 2.8$. In fact, you should obtain the sums 0.9933 and 1.0296, respectively, when you use $b = 2.7$ and $b = 2.8$ (and $n = 100$). Thus it follows that $2.7 < e < 2.8$.

As you bracket e more closely, you will need to increase the number of decimal places displayed and the number of subintervals used in Simpson's approximation to keep pace with your increasing accuracy. Continue until you have bracketed e between two seven-place approximations that both round to 2.71828.

7.3
The Exponential Function

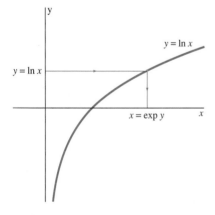

Fig. 7.3.1 To get $x = \exp y$, move straight over from y to the graph $y = \ln x$; then move straight down (or up) to x.

We saw in Section 7.2 that the natural logarithm function $\ln x$ is continuous and increasing for $x > 0$ and that it attains arbitrarily large positive and negative values [because of the limits in Eqs. (3) and (4) of Section 7.2]. It follows that $\ln x$ has an inverse function that is defined for all x. To see this, let y be any (fixed) real number. If a and b are positive numbers such that $\ln a < y < \ln b$, then the intermediate value property gives a number $x > 0$, with x between a and b, such that $\ln x = y$. Because $\ln x$ is an increasing function, there is only *one* such number x such that $\ln x = y$ (Fig. 7.3.1). Because y determines precisely one such value x, we see that x is a *function* of y.

This function x of y is the inverse function of the natural logarithm function, and it is called the *natural exponential function*. It is commonly denoted by exp (for exponential), so

$$x = \exp y \quad \text{provided that} \quad y = \ln x.$$

Interchanging x and y yields the following definition.

Definition *The Natural Exponential Function*
The **natural exponential function** exp is defined for all x as follows:

$$\exp x = y \quad \text{if and only if} \quad \ln y = x. \qquad (1)$$

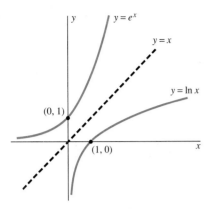

Fig. 7.3.2 The graphs $y = e^x$ and $y = \ln x$ are reflections of each other in the 45° line $y = x$.

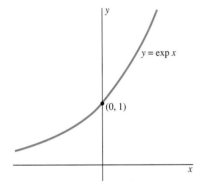

Fig. 7.3.3 The graph of the exponential function, exp

Thus exp x is simply that (positive) number whose natural logarithm is x. It is an immediate consequence of Eq. (1) that

$$\ln(\exp x) = x \qquad \text{for all } x \tag{2}$$

and that

$$\exp(\ln y) = y \qquad \text{for all } y > 0. \tag{3}$$

As in the case of the graphs of $y = a^x$ and $y = \log_a x$ discussed informally in Section 3.8, the fact that exp x and ln x are inverse functions implies that the graphs of $y = \exp x$ and $y = \ln x$ are reflections of each other in the line $y = x$ (Fig. 7.3.2). Therefore, the graph of the exponential function looks like the one shown in Fig. 7.3.3. In particular, exp x is positive-valued for all x, and

$$\exp 0 = 1, \tag{4}$$

$$\lim_{x \to \infty} \exp x = +\infty, \quad \text{and} \tag{5}$$

$$\lim_{x \to -\infty} \exp x = 0. \tag{6}$$

These facts follow from the equation ln $1 = 0$ and from the limits in Eqs. (3) and (4) of Section 7.2.

Recall from Section 7.2 that we defined the number $e \approx 2.71828$ as the number whose natural logarithm is 1. If r is any rational number, it follows that

$$\ln(e^r) = r \ln e = r.$$

But Eq. (1) implies that $\ln(e^r) = r$ if and only if

$$\exp r = e^r.$$

Thus exp x is equal to e^x (e raised to the power x) if x is a rational number. We therefore *define* e^x for irrational as well as rational values of x by

$$e^x = \exp x. \tag{7}$$

This is our first instance of powers with irrational exponents.

Equation (7) is the reason for calling exp the natural exponential function. With this notation, Eqs. (1) through (3) become

$$e^x = y \quad \text{if and only if} \quad \ln y = x, \tag{8}$$

$$\ln(e^x) = x \qquad \text{for all } x, \quad \text{and} \tag{9}$$

$$e^{\ln x} = x \qquad \text{for all } x > 0. \tag{10}$$

To justify Eq. (7), we should show rigorously that powers of e satisfy the laws of exponents. We can do this immediately.

Theorem 1 *Laws of Exponents*

If x and y are real numbers and r is rational, then

$$e^x e^y = e^{x+y}, \tag{11}$$

$$e^{-x} = \frac{1}{e^x}, \quad \text{and} \tag{12}$$

$$(e^x)^r = e^{rx}. \tag{13}$$

Proof The laws of logarithms and Eq. (9) give

$$\ln(e^x e^y) = \ln(e^x) + \ln(e^y) = x + y = \ln(e^{x+y}).$$

Thus Eq. (11) follows from the fact that ln is an increasing function and therefore is one-to-one—if $x_1 \neq x_2$, then $\ln x_1 \neq \ln x_2$. Similarly,

$$\ln([e^x]^r) = r \ln(e^x) = rx = \ln(e^{rx}).$$

So Eq. (13) follows in the same way. The proof of Eq. (12) is almost identical. We will see in Section 7.4 that the restriction that r is rational in Eq. (13) is unnecessary; that is,

$$(e^x)^y = e^{xy}$$

for *all* real numbers x and y. ❑

DERIVATIVES AND INTEGRALS OF EXPONENTIALS

Because e^x is the inverse of the differentiable and increasing function $\ln x$, it follows from Theorem 1 of Section 3.8 that e^x is differentiable and, therefore, also continuous. We may thus differentiate both sides of the equation (actually, the *identity*)

$$\ln(e^x) = x$$

with respect to x. Let $u = e^x$. Then this equation becomes

$$\ln u = x,$$

and the derivatives also must be equal:

$$\frac{1}{u} \cdot \frac{du}{dx} = 1 \qquad \text{(because } u > 0\text{)}.$$

So

$$\frac{du}{dx} = u = e^x;$$

that is,

$$D_x e^x = e^x, \tag{14}$$

as we indicated in Section 3.8.

If u denotes a differentiable function of x, then Eq. (14) in combination with the chain rule gives

$$D_x e^u = e^u \frac{du}{dx}. \tag{15}$$

The corresponding integration formula is

$$\int e^u \, du = e^u + C. \tag{16}$$

The special case of Eq. (15) with $u = kx$ (k constant) is worth noting:

$$D_x e^{kx} = k e^{kx}.$$

For instance, $D_x e^{5x} = 5e^{5x}$.

EXAMPLE 1   Find dy/dx, given $y = e^{\sqrt{x}}$.

Solution   With $u = \sqrt{x}$, Eq. (15) gives

$$\frac{dy}{dx} = e^{\sqrt{x}} D_x(\sqrt{x}) = e^{\sqrt{x}}\left(\frac{1}{2}x^{-1/2}\right) = \frac{e^{\sqrt{x}}}{2\sqrt{x}}.$$

EXAMPLE 2   If $y = x^2 e^{-2x^3}$, then Eq. (15) and the product rule yield

$$\frac{dy}{dx} = (D_x x^2)e^{-2x^3} + x^2 D_x(e^{-2x^3}) = 2xe^{-2x^3} + x^2 e^{-2x^3} D_x(-2x^3)$$

$$= 2xe^{-2x^3} + x^2 e^{-2x^3}(-6x^2).$$

Therefore,

$$\frac{dy}{dx} = (2x - 6x^4)e^{-2x^3}.$$

EXAMPLE 3   Find $\displaystyle\int xe^{-3x^2}\, dx$.

Solution   We substitute $u = -3x^2$, so $du = -6x\, dx$. Then we have $x\, dx = -\frac{1}{6} du$, and hence we obtain

$$\int xe^{-3x^2}\, dx = -\frac{1}{6}\int e^u\, du = -\frac{1}{6}e^u + C = -\frac{1}{6}e^{-3x^2} + C.$$

ORDER OF MAGNITUDE

The exponential function is remarkable for its high rate of increase with increasing x. In fact, e^x increases more rapidly as $x \to +\infty$ than *any* fixed power of x. In the language of limits,

$$\lim_{x \to \infty} \frac{x^k}{e^x} = 0 \qquad \text{for any fixed } k > 0. \tag{17}$$

Alternatively,

$$\lim_{x \to \infty} \frac{e^x}{x^k} = +\infty \qquad \text{for any fixed } k > 0. \tag{17'}$$

Because we have not yet defined x^k for irrational values of k, we prove Eq. (17) for the case in which k is rational. Once we know that (for $x > 1$) the power function x^k is an increasing function of k for all k, then the general case will follow.

We begin by taking logarithms. We find that

$$\ln\left(\frac{e^x}{x^k}\right) = x - k \ln x = \left(\frac{x}{\ln x} - k\right)\ln x.$$

Because we deduce [from Eq. (13) of Section 7.2] that $x/(\ln x) \to +\infty$ as $x \to +\infty$, this makes it clear that

$$\lim_{x \to \infty} \ln\left(\frac{e^x}{x^k}\right) = +\infty.$$

Hence $e^x/x^k \to +\infty$ as $x \to +\infty$, so we have proved both Eqs. (17) and (17′). The table in Fig. 7.3.4 illustrates the limit in Eq. (17) for the case $k = 5$.

x	x^5	e^x	x^5/e^x
10	1.00×10^5	2.20×10^4	4.54×10^0
20	3.20×10^6	4.85×10^8	6.60×10^{-3}
30	2.43×10^7	1.07×10^{13}	2.27×10^{-6}
40	1.02×10^8	2.35×10^{17}	4.35×10^{-10}
50	3.13×10^8	5.18×10^{21}	6.03×10^{-14}
	$\downarrow$	$\downarrow$	$\downarrow$
	∞	∞	0

Fig. 7.3.4 Orders of magnitude of x^5 and e^x

Although both $x^5 \to +\infty$ and $e^x \to +\infty$ as $x \to +\infty$, we see that e^x (the hare) increases so much more rapidly than x^5 (the tortoise) that $x^5/e^x \to 0$.

If we write Eq. (17) in the form

$$\lim_{x \to \infty} \frac{e^{-x}}{1/x^k} = 0,$$

we see that not only does e^{-x} approach zero as $x \to +\infty$ (Fig. 7.3.5), it does so more rapidly than any power reciprocal $1/x^k$.

We can use Eqs. (17) and (17′) to evaluate certain limits with the limit laws. For example,

$$\lim_{x \to \infty} \frac{2e^x - 3x}{3e^x + x^3} = \lim_{x \to \infty} \frac{2 - 3xe^{-x}}{3 + x^3e^{-x}} = \frac{2 - 0}{3 + 0} = \frac{2}{3}.$$

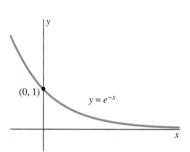

Fig. 7.3.5 The graph of $y = e^{-x}$ has the positive x-axis as an asymptote.

(0, 1)

$y = e^{-x}$

EXAMPLE 4 Sketch the graph of $f(x) = xe^{-x}$.

Solution We see from Eq. (17) that $f(x) \to 0$ as $x \to +\infty$, whereas $f(x) \to -\infty$ as $x \to -\infty$. Because

$$f'(x) = e^{-x} - xe^{-x} = e^{-x}(1 - x),$$

the only critical point of f is $x = 1$, where $y = e^{-1} \approx 0.37$. Moreover,

$$f''(x) = -e^{-x}(1 - x) + e^{-x}(-1) = e^{-x}(x - 2),$$

so that the only inflection point occurs at $x = 2$, where $y = 2e^{-2} \approx 0.27$. Hence the graph of f resembles Fig. 7.3.6.

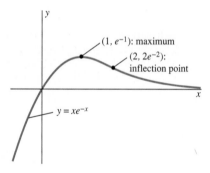

$(1, e^{-1})$: maximum

$(2, 2e^{-2})$: inflection point

$y = xe^{-x}$

Fig. 7.3.6 Graph of the function of Example 4

THE NUMBER e AS A LIMIT

We now establish the following limit expression for the exponential function:

$$e^x = \lim_{n \to \infty} \left(1 + \frac{x}{n}\right)^n. \tag{18}$$

We begin by differentiating $\ln t$, using the definition of the derivative in combination with the fact that we already know that the derivative is $1/t$. Thus

$$\frac{1}{t} = D_t \ln t = \lim_{h \to 0} \frac{\ln(t + h) - \ln t}{h} = \lim_{h \to 0} \frac{1}{h} \ln\left(\frac{t + h}{t}\right)$$

$$= \lim_{h \to 0} \ln\left[\left(1 + \frac{h}{t}\right)^{1/h}\right] \qquad \text{(by laws of logarithms)}$$

$$= \ln\left[\lim_{h \to 0}\left(1 + \frac{h}{t}\right)^{1/h}\right] \qquad \text{(by continuity of the logarithm function)}.$$

The substitution $n = 1/h$ allows us to write

$$\frac{1}{t} = \ln\left[\lim_{n \to \infty}\left(1 + \frac{1}{nt}\right)^n\right].$$

Then the substitution $x = 1/t$ gives

$$x = \ln\left[\lim_{n \to \infty}\left(1 + \frac{x}{n}\right)^n\right].$$

Now Eq. (18) follows, because $x = \ln y$ implies that $e^x = y$. With $x = 1$, we obtain also the following important expression of e as a limit:

$$e = \lim_{n \to \infty}\left(1 + \frac{1}{n}\right)^n. \tag{19}$$

7.3 Problems

Differentiate the functions in Problems 1 through 30.

1. $f(x) = e^{2x}$

2. $f(x) = e^{3x-1}$

3. $f(x) = e^{x^2} = \exp(x^2)$

4. $f(x) = e^{4-x^3}$

5. $f(x) = e^{1/x^2}$

6. $f(x) = x^2 e^{x^3}$

7. $g(t) = t e^{\sqrt{t}}$

8. $g(t) = (e^{2t} + e^{3t})^7$

9. $g(t) = (t^2 - 1)e^{-t}$

10. $g(t) = \sqrt{e^t - e^{-t}}$

11. $g(t) = e^{\cos t}$

12. $f(x) = x e^{\sin x}$

13. $f(x) = \cos(1 - e^{-x})$

14. $f(x) = \sin^2(e^{-x})$

15. $f(x) = \ln(x + e^{-x})$

16. $f(x) = e^x \cos 2x$

17. $f(x) = e^{-2x} \sin 3x$

18. $g(t) = \ln(t e^{t^2})$

19. $g(t) = 3(e^t - \ln t)^5$

20. $g(t) = \sin(e^t) \cos(e^{-t})$

21. $f(x) = \dfrac{2 + 3x}{e^{4x}}$

22. $g(t) = \dfrac{1 + e^t}{1 - e^t}$

23. $g(t) = \dfrac{1 - e^{-t}}{t}$

24. $f(x) = e^{-1/x}$

25. $f(x) = \dfrac{1 - x}{e^x}$

26. $f(x) = e^{\sqrt{x}} + e^{-\sqrt{x}}$

27. $f(x) = e^{(e^x)}$

28. $f(x) = \sqrt{e^{2x} + e^{-2x}}$

29. $f(x) = \sin(2e^x)$

30. $f(x) = \cos(e^x + e^{-x})$

In Problems 31 through 35, find dy/dx by implicit differentiation.

31. $xe^y = y$

32. $\sin(e^{xy}) = x$

33. $e^x + e^y = e^{xy}$

34. $x = ye^y$

35. $e^{x-y} = xy$

Find the antiderivatives indicated in Problems 36 through 53.

36. $\displaystyle\int e^{3x}\, dx$

37. $\displaystyle\int e^{1-2x}\, dx$

38. $\displaystyle\int xe^{x^2}\, dx$

39. $\displaystyle\int x^2 e^{3x^3 - 1}\, dx$

40. $\displaystyle\int \sqrt{x}\, e^{2x\sqrt{x}}\, dx$

41. $\displaystyle\int \frac{e^{2x}}{1 + e^{2x}}\, dx$

42. $\displaystyle\int (\cos x) e^{\sin x}\, dx$

43. $\displaystyle\int (\sin 2x) e^{1 - \cos 2x}\, dx$

44. $\displaystyle\int (e^x + e^{-x})^2\, dx$

45. $\displaystyle\int \frac{x + e^{2x}}{x^2 + e^{2x}}\, dx$

46. $\displaystyle\int e^{2x+3}\, dx$

47. $\displaystyle\int t e^{-t^2/2}\, dt$

48. $\displaystyle\int x^2 e^{1 - x^3}\, dx$

49. $\displaystyle\int \frac{e^{\sqrt{x}}}{\sqrt{x}}\, dx$

50. $\displaystyle\int \frac{e^{1/t}}{t^2}\, dt$

51. $\displaystyle\int \frac{e^x}{1 + e^x}\, dx$

52. $\displaystyle\int \exp(x + e^x)\, dx$

53. $\displaystyle\int \sqrt{x}\, \exp(-\sqrt{x^3})\, dx$

Apply Eq. (18) to evaluate (in terms of the exponential function) the limits in Problems 54 through 58.

54. $\displaystyle\lim_{n \to \infty}\left(1 - \frac{1}{n}\right)^n$

55. $\displaystyle\lim_{n \to \infty}\left(1 + \frac{2}{n}\right)^n$

56. $\displaystyle\lim_{n \to \infty}\left(1 + \frac{2}{3n}\right)^n$

57. $\displaystyle\lim_{h \to 0}(1 + h)^{1/h}$

58. $\lim_{h \to 0} (1 + 2h)^{1/h}$ [*Suggestion:* Substitute $k = 2h$.]

Evaluate the limits in Problems 59 through 62 by applying the fact that $\lim_{x \to \infty} x^k e^{-x} = 0$.

59. $\lim_{x \to \infty} \dfrac{e^x}{x}$ **60.** $\lim_{x \to \infty} \dfrac{e^x}{\sqrt{x}}$

61. $\lim_{x \to \infty} \dfrac{e^{\sqrt{x}}}{x}$ **62.** $\lim_{x \to \infty} x^2 e^{-x}$

In Problems 63 through 65, sketch the graph of the given equation. Show and label all extrema, inflection points, and asymptotes; show the concave structure clearly.

63. $y = x^2 e^{-x}$ **64.** $y = x^3 e^{-x}$

65. $y = \exp(-x^2)$

66. Find the area under the graph of $y = e^x$ from $x = 0$ to $x = 1$.

67. Find the volume generated by revolving the region of Problem 66 around the x-axis.

68. Let R be the plane figure bounded below by the x-axis, above by the graph of $y = \exp(-x^2)$, and on the sides by the vertical lines at $x = 0$ and $x = 1$ (Fig. 7.3.7). Find the volume generated by rotating R around the y-axis.

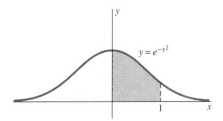

Fig. 7.3.7 The region of Problem 68

69. Find the length of the curve $y = (e^x + e^{-x})/2$ from $x = 0$ to $x = 1$.

70. Find the area of the surface generated by revolving around the x-axis the curve of Problem 69 (Fig. 7.3.8).

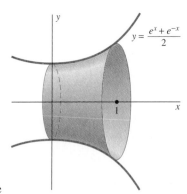

Fig. 7.3.8 The surface of Problem 70

71. Prove that the equation $e^{-x} = x - 1$ has a single solution. (Fig. 7.3.9 may be helpful.) Then use Newton's method to find the solution to three-place accuracy.

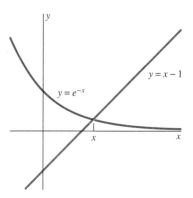

Fig. 7.3.9 The solution of the equation $e^{-x} = x - 1$ (Problem 71)

72. If a chemical plant releases an amount A of pollutant into a canal at time $t = 0$, then the resulting concentration of pollutant at time t in the water at a town on the canal a distance x_0 downstream from the plant is

$$C(t) = \frac{A}{\sqrt{k\pi t}} \exp\left(-\frac{x_0^2}{4kt}\right),$$

where k is a constant. Show that the maximum concentration at the town is

$$C_{\max} = \frac{A}{x_0} \sqrt{\frac{2}{\pi e}}.$$

73. Sketch the graph of $f(x) = x^n e^{-x}$ for $x \geq 0$ (n is a fixed but arbitrary positive integer). In particular, show that the maximum value of $f(x)$ is $f(n) = n^n e^{-n}$.

74. Approximate the number e as follows. First apply Simpson's approximation with $n = 2$ subintervals to the integral

$$\int_0^1 e^x \, dx = e - 1$$

to obtain the approximation $5e - 4\sqrt{e} - 7 \approx 0$. Then solve for e.

75. Suppose that $f(x) = x^n e^{-x}$, where n is a fixed but arbitrary positive integer. Conclude from Problem 73 that the numbers $f(n - 1)$ and $f(n + 1)$ are each less than $f(n) = n^n e^{-n}$. Deduce from this that

$$\left(1 + \frac{1}{n}\right)^n < e < \left(1 - \frac{1}{n}\right)^{-n}.$$

Substitute $n = 1024$ to show that $2.716 < e < 2.720$. Note that $1024 = 2^{10}$, so a^{1024} can be computed easily with almost any calculator by entering a and then squaring ten times in succession.

76. Suppose that the quadratic equation $am^2 + bm + c = 0$ has the two real roots m_1 and m_2, and suppose that C_1 and C_1 are arbitrary constants. Show that the function

$$y = y(x) = C_1 e^{m_1 x} + C_2 e^{m_2 x}$$

satisfies the differential equation $ay'' + by' + c = 0$.

77. Use the result of Problem 76 to find a solution $y = y(x)$ of the differential equation $y'' + y' - 2y = 0$ such that $y(0) = 5$ and $y'(0) = 2$.

7.3 Project

In this brief calculator project you are to use the limit

$$e = \lim_{n \to \infty} \left(1 + \frac{1}{n}\right)^n \tag{20}$$

to investigate numerically the number e. Assuming that this limit exists, you can "accelerate" the convergence to the limit by calculating the quantity $(1 + 1/n)^n$ only for each power $n = 2^k$ of 2 instead of for every positive integer n. That is, consider the sequence $n = 1, 2, 4, 8, \ldots, 2^k, \ldots$ instead of the sequence $n = 1, 2, 3, 4, \ldots$ of all positive integers:

$$e = \lim_{k \to \infty} \left(1 + \frac{1}{2^k}\right)^{2^k}. \tag{21}$$

This method has the advantage that the terms

$$s_k = \left(1 + \frac{1}{2^k}\right)^{2^k} \tag{22}$$

approaching e can be calculated by "successive squaring," because

$$(x^2)^2 = x^4, \qquad (x^4)^2 = x^8, \qquad (x^8)^2 = x^{16},$$

and so forth. Thus to calculate the $(2^k)th$ power $x^{(2^k)}$ of the number x, you need only enter x and then press the ⌨ x^2 key k times in succession. (On some calculators, such as the TI-81, you must first press x^2 and then ENTER each time.)

Consequently, you can calculate the number

$$s_k = \left(1 + \frac{1}{2^k}\right)^{2^k} \tag{22}$$

by the following elementary steps:

1. Calculate 2^k (no squaring here).

2. Calculate the reciprocal $1/2^k$.

3. Add 1 to get the sum $1 + (1/2^k)$.

4. Square the result k times in succession.

Use your calculator to do this with $k = 2, 4, 6, 8, \ldots, 18, 20$. Construct a table showing each result s_k accurate to five decimal places. When you finish you will have verified that $e \approx 2.71828$. Congratulations!

7.4

General Exponential and Logarithmic Functions

The natural exponential function e^x and the natural logarithm function $\ln x$ are often called the exponential and logarithm with *base e*. We now define general exponential and logarithm functions, with the forms a^x and $\log_a x$, whose base is a positive number $a \neq 1$. But it is now convenient to reverse the order of treatment from that of Sections 7.2 and 7.3, so we first consider the general exponential function.

If r is a rational number, then one of the laws of exponents [Eq. (13) of Section 7.3] gives

$$a^r = (e^{\ln a})^r = e^{r \ln a}.$$

We therefore *define* arbitrary powers (rational *and* irrational) of the positive number a in this way:

$$a^x = e^{x \ln a} \tag{1}$$

for all x. Then $f(x) = a^x$ is called the **exponential function with base a.** Note that $a^x > 0$ for all x and that $a^0 = e^0 = 1$ for all $a > 0$.

The *laws of exponents* for general exponentials follow almost immediately from the definition in Eq. (1) and from the laws of exponents for the natural exponential function:

$$a^x a^y = a^{x+y}, \tag{2}$$

$$a^{-x} = \frac{1}{a^x}, \quad \text{and} \tag{3}$$

$$(a^x)^y = a^{xy} \tag{4}$$

for all x and y. To prove Eq. (2), we write

$$a^x a^y = e^{x \ln a} e^{y \ln a} = e^{(x \ln a)+(y \ln a)} = e^{(x+y)\ln a} = a^{x+y}.$$

To derive Eq. (4), note first from Eq. (1) that $\ln a^x = x \ln a$. Then

$$(a^x)^y = e^{y \ln(a^x)} = e^{xy \ln a} = a^{xy}.$$

This follows for all real numbers x and y, so the restriction that r is rational in the formula $(e^x)^r = e^{rx}$ [see Eq. (13) of Section 7.3] has now been removed.

If $a > 1$, so that $\ln a > 0$, then Eqs. (5) and (6) in Section 7.3 immediately give us the results

$$\lim_{x \to \infty} a^x = +\infty \quad \text{and} \quad \lim_{x \to -\infty} a^x = 0. \tag{5}$$

The values of these two limits are interchanged if $0 < a < 1$, because then $\ln a < 0$.

Because

$$D_x a^x = D_x e^{x \ln a} = a^x \ln a \tag{6}$$

is positive for all x if $a > 1$, we see that—in this case—a^x is an increasing function of x. If $a > 1$, then the graph of $y = a^x$ resembles that of $f(x) = e^x$. For example, examine the graph of $y = 2^x$ shown in Fig. 7.4.1. But if $0 < a < 1$, then $\ln a < 0$, and it follows from Eq. (6) that a^x is a decreasing function. In this case, the graph of $y = a^x$ will look like Fig. 7.4.2.

If $u = u(x)$ is a differentiable function of x, then Eq. (6) combined with the chain rule gives

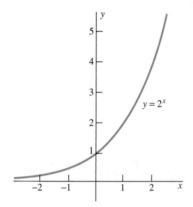

Fig. 7.4.1 The graph of $y = 2^x$

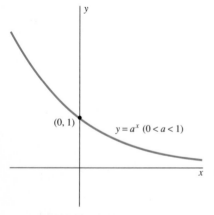

Fig. 7.4.2 The graph of $y = a^x$ is decreasing and concave upward if $0 < a < 1$.

428

$$D_x a^u = (a^u \ln a)\frac{du}{dx}. \tag{7}$$

The corresponding integral formula is

$$\int a^u \, du = \frac{a^u}{\ln a} + C. \tag{8}$$

But rather than using these general formulas, it usually is simpler to rely solely on the definition in Eq. (1), as in Examples 1 and 2.

EXAMPLE 1   To differentiate $f(x) = 3^{x^2}$, we may first write

$$3^{x^2} = (e^{\ln 3})^{x^2} = e^{x^2 \ln 3}.$$

Then

$$D_x 3^{x^2} = D_x e^{x^2 \ln 3} = e^{x^2 \ln 3} D_x(x^2 \ln 3) = 3^{x^2}(\ln 3)(2x).$$

EXAMPLE 2   Find $\displaystyle\int \frac{10^{\sqrt{x}}}{\sqrt{x}} \, dx.$

Solution   We first write $10^{\sqrt{x}} = (e^{\ln 10})^{\sqrt{x}} = e^{\sqrt{x} \ln 10}$. Then

$$\int \frac{10^{\sqrt{x}}}{\sqrt{x}} \, dx = \int \frac{e^{\sqrt{x} \ln 10}}{\sqrt{x}} \, dx$$

$$= \int \frac{2e^u}{\ln 10} \, du \qquad \left(u = \sqrt{x} \ln 10, \, du = \frac{\ln 10}{2\sqrt{x}} \, dx\right)$$

$$= \frac{2e^u}{\ln 10} + C = \frac{2}{\ln 10} 10^{\sqrt{x}} + C.$$

Whether or not the exponent r is rational, the **general power function** $f(x) = x^r$ is now defined for $x > 0$ by

$$x^r = e^{r \ln x}.$$

We may now prove the power rule of differentiation for an *arbitrary* (constant) exponent as follows:

$$D_x x^r = D_x(e^{r \ln x}) = e^{r \ln x} D_x(r \ln x) = x^r \cdot \frac{r}{x} = rx^{r-1}.$$

For example, we now know that

$$D_x x^\pi = \pi x^{\pi - 1} \approx (3.14159) \, x^{2.14159}.$$

If $a > 1$, then the general exponential function a^x is continuous and increasing for all x and attains all positive values [by an argument similar to that in the first paragraph of Section 7.3, using Eq. (6) and the limits in Eq. (5)]. It therefore has an inverse function that is defined for all $x > 0$. This inverse function of a^x is called the **logarithm function with base a** and is denoted by $\log_a x$. Thus

$$y = \log_a x \quad \text{if and only if} \quad x = a^y. \tag{9}$$

The logarithm function with base e is the natural logarithm function: $\log_e x = \ln x$.

The following *laws of logarithms* are easy to derive from the laws of exponents in Eqs. (2) through (4).

$$\log_a xy = \log_a x + \log_a y, \tag{10}$$

$$\log_a\left(\frac{1}{x}\right) = -\log_a x, \tag{11}$$

$$\log_a x^y = y \log_a x. \tag{12}$$

These formulas hold for any positive base $a \neq 1$ and for all positive values of x and y; in Eq. (12), y may be negative or zero as well.

Logarithms with one base are related to logarithms with another base, and the relationship is most easily expressed by the formula

$$(\log_a b)(\log_b c) = \log_a c. \tag{13}$$

This formula holds for all values of a, b, and c for which it makes sense—the bases a and b are positive numbers other than 1 and c is positive. The proof of this formula is outlined in Problem 53. Equation (13) should be easy to remember—it is as if some arcane cancellation law applies.

If we take $c = a$ in Eq. (13), this gives

$$(\log_a b)(\log_b a) = 1, \tag{14}$$

which in turn, with $b = e$, gives

$$\ln a = \frac{1}{\log_a e}. \tag{15}$$

If we replace a with e, b with a, and c with x in Eq. (13),we obtain

$$(\log_e a)(\log_a x) = \log_e x,$$

so

$$\log_a x = \frac{\log_e x}{\log_e a} = \frac{\ln x}{\ln a}. \tag{16}$$

On most calculators, the $\boxed{\texttt{log}}$ key denotes common (base 10) logarithms: $\log x = \log_{10} x$. In contrast, in many programming languages, such as BASIC, and some symbolic algebra programs, such as *Mathematica,* only the natural logarithm appears explicitly—as `LOG(X)` (in BASIC) and as `Log[x]` (in *Mathematica*). To get $\log_{10} x$ we write `LOG(X)/LOG(10)` and `Log[10,x]`, respectively.

Differentiation of both sides of Eq. (16) yields

$$D_x \log_a x = \frac{1}{x \ln a} = \frac{\log_a e}{x}. \tag{17}$$

For example,

$$D_x \log_{10} x = \frac{\log_{10} e}{x} \approx \frac{0.4343}{x}.$$

If we now reason as we did to obtain Eq. (6) of Section 7.2, the chain rule yields the general formula

$$D_x \log_a |u| = \frac{1}{u \ln a} \cdot \frac{du}{dx} = \frac{\log_a e}{u} \cdot \frac{du}{dx} \qquad (u \neq 0) \tag{18}$$

if u is a differentiable function of x. For example,

$$D_x \log_2 \sqrt{x^2 + 1} = \frac{1}{2} D_x \log_2(x^2 + 1) = \frac{1}{2} \cdot \frac{\log_2 e}{x^2 + 1} \cdot 2x \approx \frac{(1.4427)x}{x^2 + 1}.$$

Here we used the fact that $\log_2 e = 1/(\ln 2)$ by Eq. (15).

LOGARITHMIC DIFFERENTIATION

The derivatives of certain functions are most conveniently found by first differentiating their logarithms. This process—called **logarithmic differentiation**—involves the following steps for finding $f'(x)$:

1. Given: $\qquad\qquad\qquad\qquad\qquad\qquad y = f(x).$

2. Take *natural* logarithms; then simplify, using laws of logarithms: $\qquad\qquad \ln y = \ln f(x).$

3. Differentiate with respect to x: $\qquad \dfrac{1}{y} \cdot \dfrac{dy}{dx} = D_x[\ln f(x)].$

4. Multiply both sides by $y = f(x)$ $\qquad \dfrac{dy}{dx} = f(x)\, D_x[\ln f(x)].$

REMARK If $f(x)$ is not positive-valued everywhere, Steps 1 and 2 should be replaced by $y = |f(x)|$ and $\ln y = \ln |f(x)|$, respectively. The differentiation in Step 3 then leads to the result $dy/dx = f(x)\, D_x[\ln |f(x)|]$ in Step 4. In practice, we need not be overly concerned in advance with the sign of $f(x)$, because the appearance of what seems to be the logarithm of a negative quantity will signal the fact that absolute values should be used.

EXAMPLE 3 Find dy/dx, given

$$y = \frac{\sqrt{(x^2 + 1)^3}}{\sqrt[3]{(x^3 + 1)^4}}.$$

Solution The laws of logarithms give

$$\ln y = \ln \frac{(x^2 + 1)^{3/2}}{(x^3 + 1)^{4/3}} = \frac{3}{2} \ln(x^2 + 1) - \frac{4}{3} \ln(x^3 + 1).$$

Then differentiation with respect to x gives

$$\frac{1}{y} \cdot \frac{dy}{dx} = \frac{3}{2} \cdot \frac{2x}{x^2 + 1} - \frac{4}{3} \cdot \frac{3x^2}{x^3 + 1} = \frac{3x}{x^2 + 1} - \frac{4x^2}{x^3 + 1}.$$

Finally, to solve for dy/dx, we multiply both sides by

$$y = \frac{(x^2 + 1)^{3/2}}{(x^3 + 1)^{4/3}},$$

and we obtain

$$\frac{dy}{dx} = \left(\frac{3x}{x^2 + 1} - \frac{4x^2}{x^3 + 1} \right) \frac{(x^2 + 1)^{3/2}}{(x^3 + 1)^{4/3}}.$$

EXAMPLE 4 Find dy/dx, given $y = x^{x+1}$.

Solution   If $y = x^{x+1}$, then

$$\ln y = \ln(x^{x+1}) = (x+1)\ln x,$$

$$\frac{1}{y}\cdot\frac{dy}{dx} = (1)(\ln x) + (x+1)\left(\frac{1}{x}\right) = 1 + \frac{1}{x} + \ln x.$$

Multiplication by $y = x^{x+1}$ gives

$$\frac{dy}{dx} = \left(1 + \frac{1}{x} + \ln x\right)x^{x+1}.$$

7.4 Problems

In Problems 1 through 24, find the derivative of the given function $f(x)$.

1. $f(x) = 10^x$

2. $f(x) = 2^{1/x^2}$

3. $f(x) = \dfrac{3^x}{4^x}$

4. $f(x) = \log_{10}\cos x$

5. $f(x) = 7^{\cos x}$

6. $f(x) = 2^x 3^{x^2}$

7. $f(x) = 2^{x\sqrt{x}}$

8. $f(x) = \log_{100}(10^x)$

9. $f(x) = 2^{\ln x}$

10. $f(x) = 7^{8^x}$

11. $f(x) = 17^x$

12. $f(x) = 2^{\sqrt{x}}$

13. $f(x) = 10^{1/x}$

14. $f(x) = 3^{\sqrt{1-x^2}}$

15. $f(x) = 2^{2^x}$

16. $f(x) = \log_2 x$

17. $f(x) = \log_3\sqrt{x^2 + 4}$

18. $f(x) = \log_{10}(e^x)$

19. $f(x) = \log_3(2^x)$

20. $f(x) = \log_{10}(\log_{10}x)$

21. $f(x) = \log_2(\log_3 x)$

22. $f(x) = \pi^x + x^\pi + \pi^\pi$

23. $f(x) = \exp(\log_{10}x)$

24. $f(x) = \pi^{x^3}$

Evaluate the integrals given in Problems 25 through 32.

25. $\displaystyle\int 3^{2x}\,dx$

26. $\displaystyle\int x\cdot 10^{-x^2}\,dx$

27. $\displaystyle\int \frac{2^{\sqrt{x}}}{\sqrt{x}}\,dx$

28. $\displaystyle\int \frac{10^{1/x}}{x^2}\,dx$

29. $\displaystyle\int x^2 7^{x^3+1}\,dx$

30. $\displaystyle\int \frac{dx}{x\,\log_{10}x}$

31. $\displaystyle\int \frac{\log_2 x}{x}\,dx$

32. $\displaystyle\int (2^x)3^{(2^x)}\,dx$

In Problems 33 through 52, find dy/dx by logarithmic differentiation.

33. $y = \sqrt{(x^2 - 4)\sqrt{2x + 1}}$

34. $y = \dfrac{(3 - x^2)^{1/2}}{(x^4 + 1)^{1/4}}$

35. $y = 2^x$

36. $y = x^x$

37. $y = x^{\ln x}$

38. $y = (1 + x)^{1/x}$

39. $y = \left[\dfrac{(x+1)(x+2)}{(x^2+1)(x^2+2)}\right]^{1/3}$

40. $y = \sqrt{x+1}\,\sqrt[3]{x+2}\,\sqrt[4]{x+3}$

41. $y = (\ln x)^{\sqrt{x}}$

42. $y = (3 + 2^x)^x$

43. $y = \dfrac{(1 + x^2)^{3/2}}{(1 + x^3)^{4/3}}$

44. $y = (x + 1)^x$

45. $y = (x^2 + 1)^{x^2}$

46. $y = \left(1 + \dfrac{1}{x}\right)^x$

47. $y = (\sqrt{x})^{\sqrt{x}}$

48. $y = x^{\sin x}$

49. $y = e^x$

50. $y = (\ln x)^{\ln x}$

51. $y = x^{(e^x)}$

52. $y = (\cos x)^x$

53. Prove Eq. (13). [*Suggestion:* Let $x = \log_a b$, $y = \log_b c$, and $z = \log_a c$. Then show that $a^z = a^{xy}$, and conclude that $z = xy$.]

54. Suppose that u and v are differentiable functions of x. Show by logarithmic differentiation that

$$D_x(u^v) = v(u^{v-1})\frac{du}{dx} + (u^v \ln u)\frac{dv}{dx}.$$

Interpret the two terms on the right in relation to the special cases in which (a) u is a constant; (b) v is a constant.

55. Suppose that $a > 0$. By examining $\ln(a^{1/x})$, show that

$$\lim_{x \to \infty} a^{1/x} = 1.$$

It follows that

$$\lim_{n \to \infty} a^{1/n} = 1.$$

(The inference is that n is a positive integer.) Test this conclusion by entering some positive number in your calculator and then pressing the square root key repeatedly. Make a table of the results of two such experiments.

56. Show that $\lim_{n \to \infty} n^{1/n} = 1$ by showing that $\lim_{x \to \infty} x^{1/x} = 1$. (The inference is that x is an arbitrary positive *real number*, whereas n is a positive *integer*.) Use the method of Problem 55.

57. By examining $\ln(x^x/e^x)$, show that

$$\lim_{x \to \infty} \frac{x^x}{e^x} = +\infty.$$

Thus x^x increases even faster than the exponential function e^x as $x \to +\infty$.

58. Consider the function

$$f(x) = \frac{1}{1 + 2^{1/x}} \qquad \text{for} \quad x \neq 0.$$

Show that both the left-hand and right-hand limits of $f(x)$ at $x = 0$ exist but are unequal.

59. Find dy/dx if $y = \log_x 2$.

60. Suppose that $y = uvw/pqr$, where u, v, w, p, q, and r are nonzero differentiable functions of x. Show by logarithmic differentiation that

$$\frac{dy}{dx} = y \cdot \left(\frac{1}{u} \cdot \frac{du}{dx} + \frac{1}{v} \cdot \frac{dv}{dx} + \frac{1}{w} \cdot \frac{dw}{dx} - \frac{1}{p} \cdot \frac{dp}{dx} - \frac{1}{q} \cdot \frac{dq}{dx} - \frac{1}{r} \cdot \frac{dr}{dx} \right).$$

Is the generalization—for an arbitrary finite number of factors in numerator and denominator—obvious?

7.4 Project

This project investigates the equation

$$2^x = x^{10}. \tag{19}$$

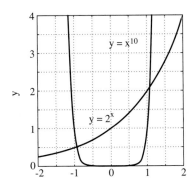

Fig. 7.4.3 $y = 2^x$ and $y = x^{10}$ for $-2 \leq x \leq 2$

1. The graphs of $y = 2^x$ and $y = x^{10}$ (Fig. 7.4.3) *suggest* that Eq. (19) has two solutions—one positive and the other negative. If a graphics calculator or computer is available to you, find these two solutions (accurate to three or four decimal places) by successive magnification (the method of "zooming").

2. Figure 7.4.4 seems to indicated that x^{10} leaves 2^x forever behind as $x \to +\infty$. Show, however, that Eq. (19) has the same positive solutions as the equation

$$\frac{\ln x}{x} = \frac{\ln 2}{10}.$$

Hence conclude from the graph of $y = (\ln x)/x$ (Example 8 of Section 7.2) that Eq. (19) has precisely *two* positive solutions.

3. Tabulate values of 2^x and x^{10} for $x = 10, 20, 30, 40, 50$, and 60 and thereby verify that the missing positive solution is somewhere between $x = 50$ and $x = 60$ (Fig. 7.4.5). If you attempt to locate this solution by successive

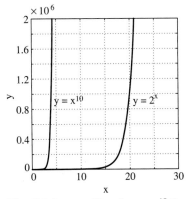

Fig. 7.4.4 $y = 2^x$ and $y = x^{10}$ for $0 \leq x \leq 30$

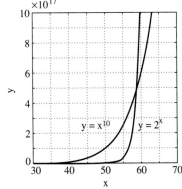

Fig. 7.4.5 $y = 2^x$ and $y = x^{10}$ for $30 \leq x \leq 70$

magnifications, you may well have the feeling of going boldly where no one has gone before!

4. Use Newton's method to approximate (with four-place accuracy) all three solutions of Eq. (19).

7.5
Natural Growth and Decay

In this section we use the natural exponential function to model the "natural growth" of populations. Consider a population that numbers $P(t)$ persons or other animals, bacteria, dollars, molecules—any sort of entity—at time t. We assume that this population has a constant *birth rate* β and a constant *death rate* δ. Roughly speaking, this means that during any 1-yr period, βP births and δP deaths occur.

Because P changes during the course of a year, some allowance must be made for changes in the number of births and the number of deaths. We think of a very brief time interval from t to $t + \Delta t$. For very small values of Δt, the value of $P = P(t)$ will change by such a small amount during the time interval $[t, t + \Delta t]$ that we can regard $P(t)$ as almost constant. We require that the number of births and deaths during this time interval be given with sufficient accuracy by the approximations

$$\text{Number of births:} \quad \text{approximately } \beta P(t) \, \Delta t;$$
$$\text{Number of deaths:} \quad \text{approximately } \delta P(t) \, \Delta t. \tag{1}$$

More precisely, we assume that the ratio to Δt of the error in each of these approximations approaches zero as $\Delta t \to 0$.

On this basis we wish to deduce the form of the function $P(t)$ that describes the population in question. Our strategy begins with finding the **time rate of change** of $P(t)$. Hence we consider the increment

$$\Delta P = P(t + \Delta t) - P(t)$$

of P during the time interval $[t, t + \Delta t]$. Because ΔP is simply the number of births minus the number of deaths, we find from Eq. (1) that

$$\Delta P = P(t + \Delta t) - P(t)$$
$$\approx \beta P(t) \, \Delta t - \delta P(t) \, \Delta t \qquad \text{(the number of births minus the number of deaths).}$$

Therefore,

$$\frac{\Delta P}{\Delta t} = \frac{P(t + \Delta t) - P(t)}{\Delta t} \approx (\beta - \delta)P(t).$$

The quotient on the left-hand side approaches the derivative $P'(t)$ as $\Delta t \to 0$, and, by the assumption following (1), it approaches also the right-hand side, $(\beta - \delta)P(t)$. Hence, when we take the limit as $\Delta t \to 0$, we get the differential equation

$$P'(t) = (\beta - \delta)P(t);$$

that is,

$$\frac{dP}{dt} = kP, \qquad \text{where } k = \beta - \delta. \tag{2}$$

This differential equation may be regarded as a *mathematical model* of the changing population.

THE NATURAL GROWTH EQUATION

With $x(t)$ in place of $P(t)$ in Eq. (2), we have the differential equation

$$\frac{dx}{dt} = kx, \tag{3}$$

which serves as the mathematical model for an extraordinarily wide range of natural phenomena. It is easily solved if we first "separate the variables" and then integrate:

$$\frac{dx}{x} = k\, dt;$$

$$\int \frac{dx}{x} = \int k\, dt;$$

$$\ln x = kt + C.$$

We apply the exponential function to both sides of the last equation to solve for x:

$$x = e^{\ln x} = e^{kt+C} = e^{kt}e^{C} = Ae^{kt}.$$

Here, $A = e^{C}$ is a constant that remains to be determined. But we see that A is simply the value of $x_0 = x(0)$ of $x(t)$ when $t = 0$, and thus $A = x_0$.

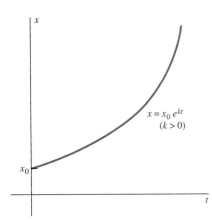

Fig. 7.5.1 Solution of the exponential growth equation for $k > 0$

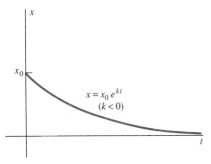

Fig. 7.5.2 Solution of the exponential growth equation—now actually a *decay* equation—for the case $k < 0$

> **Theorem 1** *The Natural Growth Equation*
> The solution of the initial value problem
>
> $$\frac{dx}{dt} = kx, \qquad x(0) = x_0 \tag{4}$$
>
> is
>
> $$x(t) = x_0 e^{kt}. \tag{5}$$

As a consequence, Eq. (3) is often called the **exponential growth equation,** or the **natural growth equation.** We see from Eq. (5) that, with $x_0 > 0$, the solution $x(t)$ is an increasing function if $k > 0$ and a decreasing function if $k < 0$. (The situation $k < 0$ is sometimes called exponential *decay*.) These two cases are illustrated in Figs. 7.5.1 and 7.5.2, respectively. The remainder of this section concerns examples of natural phenomena for which this differential equation serves as a mathematical model.

POPULATION GROWTH

When we compare Eqs. (2), (3), and (5), we see that a population $P(t)$ with constant birth rate β and constant death rate δ is given by

$$P(t) = P_0 e^{kt}, \tag{6}$$

where $P_0 = P(0)$ and $k = \beta - \delta$. If t is measured in years, then k is called the **annual growth rate,** which can be positive, negative, or zero. Its value is often given as a percentage (its decimal value multiplied by 100). If k is close to zero, this value is fairly close to the actual percentage increase (or decrease) of the population each year.

EXAMPLE 1   According to an Associated Press release of March 1987, the world population had then reached 5 billion persons and was increasing at the rate of 380,000 persons each day. Let us assume constant birth and death rates. We want to answer these questions:

1. What is the annual growth rate k?
2. What will the world's population be in the year 2000?
3. How long will it take the world's population to double?
4. When will the world's population be 50 billion (believed by some demographers to be the maximum for which the planet can provide food)?

Solution   We measure the world's population $P(t)$ in billions and measure time t in years. We take $t = 0$ to correspond to 1987, so $P_0 = 5$. The fact that P is increasing by 380,000, or 0.00038 billion, persons per day at time $t = 0$ means that

$$P'(0) = (0.00038)(365.25) \approx 0.1388$$

billion per year. From Eq. (2) we now obtain

$$k = \left[\frac{1}{P} \cdot \frac{dP}{dt}\right]_{t=0} = \frac{P'(0)}{P(0)} = \frac{0.1388}{5} \approx 0.0278.$$

The population of the world is growing rapidly, but how rapidly?

Thus the world's population was growing at the rate of about 2.78% per year in 1987.

We use this value of k to conclude that the world's population at time t should be

$$P(t) = 5e^{(0.0278)t}.$$

For example, $t = 13$ yields

$$P(13) = 5e^{(0.0278)(13)} \approx 7.17 \quad \text{(billion)}$$

for the population in the year 2000.

To find when the population will double to 10 billion, we solve the equation

$$10 = 5e^{(0.0278)t}.$$

We must first take the natural logarithm of each side and then solve for

$$t = \frac{\ln 2}{0.0278} \approx 25 \quad \text{(yr)},$$

which corresponds to the year 2012. Finally, under the assumptions made here, the world's population will reach 50 billion when

$$50 = 5e^{(0.0278)t}; \qquad t = \frac{\ln 10}{0.0278} \approx 83$$

—that is, in the year 2070.

RADIOACTIVE DECAY

Consider a sample of material that contains $N(t)$ atoms of a certain radioactive isotope at time t. Many experiments have confirmed that a constant fraction of these radioactive atoms will spontaneously decay (into atoms of another element or another isotope of the same element) during each given unit of time. Consequently, the sample behaves exactly like a population with a constant death rate but with no births occurring. To write a model for $N(t)$, we use Eq. (2) with N in place of P, $k > 0$ in place of δ, and $\beta = 0$. We thus obtain the differential equation

$$\frac{dN}{dt} = -kN. \tag{7}$$

From the solution in Eq. (5) of Eq. (3), with k replaced by $-k$, we conclude that

$$N(t) = N_0 e^{-kt}, \tag{8}$$

where $N_0 = N(0)$, the number of radioactive atoms of the original isotope present in the sample at time $t = 0$.

The value of the *decay constant k* depends on the particular isotope with which we are dealing. If k is large, then the isotope decays rapidly. If k is near zero, the isotope decays quite slowly and thus may be a relatively persistent factor in its environment. The decay constant k is often specified in terms of another empirical parameter that is more convenient, the *half-life* of the isotope. The **half-life** τ of a sample of a radioactive isotope is the time required for *half* of that sample of decay. To find the relationship between k and τ, we set

$$t = \tau \quad \text{and} \quad N = \tfrac{1}{2} N_0$$

in Eq. (8), so that

$$\tfrac{1}{2} N_0 = N_0 e^{-k\tau}. \tag{9}$$

When we solve for τ, we find that

$$\tau = \frac{\ln 2}{k}. \tag{10}$$

Note that the concept of half-life is meaningful—the value of τ depends *only* on k and thus depends only on the particular isotope involved. It does *not* depend on the amount of that isotope present.

The method of *radiocarbon dating* is based on the fact that the radioactive carbon isotope ^{14}C has a known half-life of about 5700 yr. Living organic matter maintains a constant level of ^{14}C by "breathing" air (or by consuming organic matter that does so). But air contains ^{14}C along with the much more common, stable isotope ^{12}C of carbon, mostly in the gas CO_2. Thus all living organisms maintain the same percentage of ^{14}C as in air, because organic processes seem to make no distinction between the two isotopes. But when an organism dies, it ceases to metabolize carbon, and the process of radioactive decay begins to deplete its ^{14}C content. The fraction of ^{14}C in the air remains roughly constant because new ^{14}C is continuously being generated by the bombardment of nitrogen atoms in the upper atmosphere by cosmic rays, and this generation has long been in a steady-state equilibrium with the loss of ^{14}C through radioactive decay.

EXAMPLE 2 A specimen of charcoal found at Stonehenge contains 63% as much ^{14}C as a sample of present-day charcoal. What is the age of the sample?

Solution We take $t = 0$ (in years) as the time of death of the tree from which the charcoal was made. From Eq. (9) we know that

$$\tfrac{1}{2}N_0 = N_0 e^{-5700k},$$

so

$$k = \frac{\ln 2}{\tau} = \frac{\ln 2}{5700} \approx 0.0001216.$$

We are given that $N = (0.63)N_0$ at present, so we solve the equation

$$(0.63)N_0 = N_0 e^{-kt}$$

with this value of k. We thus find that

$$t = -\frac{\ln(0.63)}{0.0001216} \approx 3800 \quad \text{(yr)}.$$

Therefore, the sample is about 3800 yr old. If it is connected in any way with the builders of Stonehenge, our computations suggest that this observatory, monument, or temple—whichever it may be—dates from almost 1800 B.C.

EXAMPLE 3 According to one cosmological theory, there were equal amounts of the uranium isotopes ^{235}U and ^{238}U at the creation of the universe in the "big bang." At present there are 137.7 ^{238}U atoms for each ^{235}U atom. Using the known half-lives

$$4.51 \text{ billion yr for } {}^{238}\text{U},$$

$$0.71 \text{ billion yr for } {}^{235}\text{U},$$

calculate the age of the universe.

Solution Let $N_8(t)$ and $N_5(t)$ be the numbers of ^{238}U and ^{235}U atoms, respectively, at time t, in billions of years after the creation of the universe. Then

$$N_8(t) = N_0 e^{-kt} \quad \text{and} \quad N_5(t) = N_0 e^{-ct},$$

where N_0 is the initial number of atoms of each isotope. Also,

$$k = \frac{\ln 2}{4.51} \quad \text{and} \quad c = \frac{\ln 2}{0.71},$$

a consequence of Eq. (10). We divide the equation for N_8 by the equation for N_5 and find that when t has the value corresponding to "now,"

$$137.7 = \frac{N_8}{N_5} = e^{(c-k)t}.$$

Finally, we solve this equation for t:

$$t = \frac{\ln(137.7)}{\left(\dfrac{1}{0.71} - \dfrac{1}{4.51}\right)\ln 2} \approx 5.99.$$

Thus we estimate the age of the universe to be about 6 billion years, which is roughly on the same order of magnitude as recent estimates of about 15 billion years.

CONTINUOUSLY COMPOUNDED INTEREST

Consider a savings account that is opened with an initial deposit of A_0 dollars and earns interest at the annual rate r. If there are $A(t)$ dollars in the account at time t and the interest is compounded at time $t + \Delta t$, this means that $rA(t)\,\Delta t$ dollars in interest are added to the account then. So

$$A(t + \Delta t) = A(t) + rA(t)\,\Delta t,$$

and thus

$$\frac{\Delta A}{\Delta t} = \frac{A(t + \Delta t) - A(t)}{\Delta t} = rA(t).$$

Continuous compounding of interest results from taking the limit as $\Delta t \to 0$, so

$$\frac{dA}{dt} = rA. \tag{11}$$

This is an exponential growth equation that has the solution

$$A(t) = A_0 e^{rt}. \tag{12}$$

EXAMPLE 4 If $A_0 = \$1000$ is invested at an annual interest rate of 6% compounded continuously, then $r = 0.06$, and Eq. (12) gives

$$A(1) = 1000e^{(0.06)(1)} = \$1061.84$$

for the value of the investment after one year. Thus the *effective annual interest rate* is 6.184%. The more often interest is compounded, the more rapidly savings grow, but bank advertisements sometimes overemphasize this advantage. For instance, 6% compounded *monthly* multiplies your investment by

$$1 + \frac{0.06}{12} = 1.005$$

at the end of each month, so an initial investment of $1000 would grow in one year to

$$(1000)(1.005)^{12} = \$1061.68,$$

only $0.16 less than would be yielded by continuous compounding.

*DRUG ELIMINATION

The amount $A(t)$ of a certain drug in the human bloodstream, as measured by the excess above the natural level of the drug in the bloodstream, typically declines at a rate proportional to that excess amount. That is,

$$\frac{dA}{dt} = -\lambda A, \quad \text{so} \quad A(t) = A_0 e^{-\lambda t}. \tag{13}$$

The parameter λ is called the *elimination constant* of the drug, and $T = 1/\lambda$ is called the *elimination time*.

EXAMPLE 5 The elimination time for alcohol varies from one person to another. If a person's "sobering time" $T = 1/\lambda$ is 2.5 h, how long will it take the excess bloodstream alcohol concentration to be reduced from 0.10% to 0.02%?

Solution We assume that the normal concentration of alcohol in the blood is zero, so any amount is an excess amount. In this problem, we have $\lambda = 1/2.5 = 0.4$, so Eq. (13) yields

$$0.02 = (0.10)e^{-(0.4)t}.$$

Thus

$$t = -\frac{\ln(0.2)}{0.4} \approx 4.02 \quad \text{(h)}.$$

*SALES DECLINE

According to marketing studies, if advertising for a particular product is halted and other market conditions—such things as number and promotion of competing products, their prices, and so on—remain unchanged, then the sales of the unadvertised product will decline at a rate that is proportional at any time t to the current sales S. That is,

$$\frac{dS}{dt} = -\lambda S, \quad \text{so} \quad S(t) = S_0 e^{-\lambda t}. \tag{14}$$

Here S_0 denotes the initial value of the sales, which we take to be sales in the last month of advertising. If we take months as the units for time t, then $S(t)$ gives the number of sales t months after advertising is halted, and λ might be called the *sales decay constant*.

*LINGUISTICS

Consider a basic list of N_0 words in use in a given language at time $t = 0$. Let $N(t)$ denote the number of these words that are still in use at time t—those that have neither disappeared from the language nor been replaced. According to one theory in linguistics, the rate of decrease of N is proportional to N.

That is,

$$\frac{dN}{dt} = -\lambda N, \quad \text{so} \quad N(t) = N_0 e^{-\lambda t}. \tag{15}$$

If t is measured in millennia (as is standard in linguistics), then $k = e^{-\lambda}$ is the fraction of the words in the original list that survive for 1000 yr.

7.5 Problems

1. *Continuously Compounded Interest* Suppose that $1000 is deposited in a savings account that pays 8% annual interest compounded continuously. At what rate (in $/yr) is it earning interest after 5 yr? After 20 yr?

2. *Population Growth* Coopersville had a population of 25,000 in 1970 and a population of 30,000 in 1980. Assume that its population will continue to grow exponentially at a constant rate. What population can the Coopersville city planners expect in the year 2010?

3. *Population Growth* In a certain culture of bacteria, the number of bacteria increased sixfold in 10 h. Assuming natural growth, how long did it take for their number to double?

4. *Radiocarbon Dating* Carbon extracted from an ancient skull recently unearthed contained only one-sixth as much radioactive ^{14}C as carbon extracted from present-day bone. How old is the skull?

5. *Radiocarbon Dating* Carbon taken from a relic purported to date from A.D. 30 contained 4.6×10^{10} atoms of ^{14}C per gram. Carbon extracted from a present-day specimen of the same substance contained 5.0×10^{10} atoms of ^{14}C per gram. Compute the approximate age of the relic. What is your opinion as to its authenticity?

6. *Continuously Compounded Interest* An amount A of money is invested for t years at an annual interest rate r compounded n times over these years at equal intervals. (a) Explain why the amount accrued after t years is

$$A_{r,n} = A \cdot \left(1 + \frac{rt}{n}\right)^n.$$

(b) Conclude from the limit in Eq. (18) of Section 7.3 that

$$\lim_{n \to \infty} A_{r,n} = A e^{rt},$$

in agreement with Eq. (12) of this section.

7. *Continuously Compounded Interest* If an investment of A_0 dollars returns A_1 dollars after 1 yr, the **effective annual interest rate** r is defined by the equation

$$A_1 = (1 + r)A_0.$$

Banks often advertise that they increase the effective interest rates on their customers' savings accounts by increasing the frequency of compounding. Calculate the effective annual interest rate if a 9% annual interest rate is compounded (a) quarterly; (b) monthly; (c) weekly; (d) daily; (e) continuously.

8. *Continuously Compounded Interest* Upon the birth of their first child, a couple deposited $5000 in a savings account that pays 6% annual interest compounded continuously. The interest payments are allowed to accumulate. How much will the account contain when the child is ready to go to college at age 18?

9. *Continuously Compounded Interest* You discover in your attic an overdue library book on which your great-great-grandfather owed a fine of 30¢ exactly 100 yr ago. If an overdue fine grows exponentially at a 5% annual interest rate compounded continuously, how much would you have to pay if you returned the book today?

10. *Drug Elimination* Suppose that sodium pentobarbitol will anesthetize a dog when its bloodstream contains at least 45 mg of sodium pentobarbitol per kilogram of body weight. Suppose also that sodium pentobarbitol is eliminated exponentially from a dog's bloodstream, with a half-life of 5 h. What single dose should be administered to anesthetize a 50-kg dog for 1 h?

11. *Sales Decline* Moonbeam Motors has discontinued advertising of their minivan. The company plans to resume advertising when sales have declined to 75% of their initial rate. If after 1 week without advertising, sales have declined to 95% of their initial rate, when should the company expect to resume advertising?

12. *Linguistics* The English language evolves in such a way that 77% of all words disappear (or are replaced) every 1000 yr. Of a basic list of words used by Chaucer in A.D. 1400, what percentage should we expect to find still in use today?

13. *Radioactive Decay* The half-life of radioactive cobalt is 5.27 yr. Suppose that a nuclear accident has left the level of cobalt radiation in a certain region at 100 times the level acceptable for human habitation. How long will it be before the region is again habitable? (Ignore the likely presence of other radioactive substances.)

14. *Radioactive Decay*   Suppose that a rare mineral deposit formed in an ancient cataclysm—such as the collision of a meteorite with the earth—originally contained the uranium isotope ^{238}U (which has a half-life of 4.51×10^9 yr) but none of the lead isotope ^{207}Pb, the end product of the radioactive decay of ^{238}U. If the ratio of ^{238}U atoms to ^{207}Pb atoms in the mineral deposit today is 0.9, when did the cataclysm occur?

15. *Radioactive Decay*   A certain moon rock contains equal numbers of potassium atoms and argon atoms. Assume that all the argon is present because of radioactive decay of potassium (its half-life is about 1.28×10^9 yr) and that 1 of every 9 potassium-atom disintegrations yields an argon atom. What is the age of the rock, measured from the time it contained only potassium?

16. If a body is cooling in a medium with constant temperature A, then—according to Newton's law of cooling (Section 7.6)—the rate of change of the body's temperature T is proportional to $T - A$. We want to cool a pitcher of buttermilk initially at 25°C by setting it out on the front porch, where the temperature is 0°C. If the temperature of the buttermilk drops to 15°C after 20 min, when will it be at 5°C?

17. When sugar is dissolved in water, the amount A of sugar that remains undissolved after t minutes satisfies the differential equation $dA/dt = -kA$ $(k > 0)$. If 25% of the sugar dissolves in 1 min, how long does it take for half the sugar to dissolve?

18. The intensity I of light at a depth x meters below the surface of a lake satisfies the differential equation $dI/dx = -(1.4)I$. (a) At what depth is the intensity half the intensity I_0 at the surface (where $x = 0$)? (b) What is the intensity at a depth of 10 m (as a fraction of I_0)? (c) At what depth will the intensity be 1% of its value at the surface?

19. The barometric pressure p (in inches of mercury) at an altitude x miles above sea level satisfies the differential equation $dp/dx = -(0.2)p$; $p(0) = 29.92$. (a) Calculate the barometric pressure at 10,000 ft and again at 30,000 ft. (b) Without prior conditioning, few people can survive when the pressure drops to less than 15 in. of mercury. How high is that?

20. An accident at a nuclear power plant has left the surrounding area polluted with a radioactive element that decays at a rate proportional to its current amount $A(t)$. The initial radiation level is 10 times the maximum amount S that is safe, and 100 days later it is still 7 times that amount. (a) Set up and solve a differential equation to find $A(t)$. (b) How long (to the nearest day after the original accident) will it be before it is safe for people to return to the area?

*7.6

Linear First-Order Differential Equations and Applications

A **first-order differential equation** is one in which only the first derivative (not higher derivatives) of the dependent variable appears. It is a **linear** first-order differential equation if it can be written in the form

$$\frac{dx}{dt} = ax + b, \tag{1}$$

where a and b denote functions of the independent variable t. Here we discuss applications in the special case in which the coefficients a and b are *constants*.

   Equation (1) is separable, so we can separate the variables as in Section 6.5 and immediately integrate. Assuming that $ax + b > 0$, we get

$$\int \frac{a\,dx}{ax + b} = \int a\,dt; \qquad \ln(ax + b) = at + C.$$

Then application of the natural exponential function to both sides gives

$$ax + b = Ke^{at},$$

where $K = e^C$. When we substitute $t = 0$ and denote the resulting value of x by x_0, we find that $K = ax_0 + b$. So

$$ax + b = (ax_0 + b)e^{at}.$$

With $x_0 = 20$, Eq. (2) gives the solution

$$x(t) = 4 + 16e^{-t/16}.$$

We can find the value of t at which $x(t) = 8$ by solving the equation $8 = 4 + 16e^{-t/16}$. This gives

$$t = 16 \ln 4 \approx 22.2 \quad \text{(days)}.$$

7.6 Problems

In Problems 1 through 10, use the method of derivation of Eq. (2) rather than the equation itself to find the solution of the given initial value problem.

1. $\dfrac{dy}{dx} = y + 1; \quad y(0) = 1$

2. $\dfrac{dy}{dx} = 2 - y; \quad y(0) = 3$

3. $\dfrac{dy}{dx} = 2y - 3; \quad y(0) = 2$

4. $\dfrac{dy}{dx} = \dfrac{1}{4} - \dfrac{y}{16}; \quad y(0) = 20$

5. $\dfrac{dx}{dt} = 2(x - 1); \quad x(0) = 0$

6. $\dfrac{dx}{dt} = 2 - 3x; \quad x(0) = 4$

7. $\dfrac{dx}{dt} = 5(x + 2); \quad x(0) = 25$

8. $\dfrac{dx}{dt} = -3 - 4x; \quad x(0) = -5$

9. $\dfrac{dv}{dt} = 10(10 - v); \quad v(0) = 0$

10. $\dfrac{dv}{dt} = -5(10 - v); \quad v(0) = -10$

11. Zembla had a population of 1.5 million in 1990. Assume that this country's population is growing continuously at a 4% annual rate and that Zembla absorbs 50,000 newcomers per year. What will its population be in the year 2010?

12. When a cake is removed from an oven, the temperature of the cake is 210°F. The cake is left to cool at room temperature, which is 70°F. After 30 min the temperature of the cake is 140°F. When will it be 100°F?

13. Payments are made continuously on a mortgage (original loan) of P_0 dollars at the constant rate of c dollars per month. Let $P(t)$ denote the balance (amount still owed) after t months, and let r denote the monthly interest rate paid by the mortgage holder. (For example, $r = 0.06/12 = 0.005$ if the annual interest rate is 6%.) Derive the differential equation

$$\frac{dP}{dt} = rP - c, \qquad P(0) = P_0.$$

14. Your cousin must pay off an auto loan of $3600 continuously over a period of 36 months. Apply the result of Problem 13 to determine the monthly payment required if the annual interest rate is (a) 12%; (b) 18%.

15. A rumor about thiotimoline in the drinking water began to spread one day in a city with a population of 100,000. Within a week, 10,000 people had heard this rumor. Assuming that the rate of increase of the number of people who have heard the rumor is proportional to the number who have not yet heard it, how long will it be until half the population of the city has heard the rumor?

16. A tank contains 1000 L of a solution consisting of 50 kg of salt dissolved in water. Pure water is pumped into the tank at the rate of 5 L/s and the mixture—kept uniform by stirring—is pumped out at the same rate. After how many seconds will only 10 kg of salt remain in the tank?

17. Derive the solution in Eq. (2) of Eq. (1) under the assumption that $ax + b < 0$.

18. Suppose that a body moves through a resisting medium with resistance proportional to its velocity v, so $dv/dt = -kv$. (a) Show that its velocity $v(t)$ and position $x(t)$ at time t are given by

$$v(t) = v_0 e^{-kt} \quad \text{and} \quad x(t) = x_0 + \frac{v_0}{k}(1 - e^{-kt}).$$

(b) Conclude that the body travels only a *finite* distance v_0/k.

19. A motorboat is moving at 40 ft/s when its motor suddenly quits; 10 s later the boat has slowed to 20 ft/s. Assume, as in Problem 18, that the resistance it encounters while it coasts is proportional to its velocity. How far will the motorboat coast in all?

20. The acceleration of a Lamborghini is proportional to the difference between 250 km/h and the velocity of this sports car. If this car can accelerate from rest to 100 km/h in 10 s, how long will it take the car to accelerate from rest to 200 km/h?

21. Consider the linear first-order differential equation

$$\frac{dx}{dt} + p(t)x(t) = q(t)$$

with variable coefficients. Let $P(t)$ be an antiderivative of $p(t)$. Multiply both sides of the given equation by $e^{P(t)}$, and note that the left-hand side of the resulting equation is $D_t[e^{P(t)} x(t)]$. Conclude by antidifferentiation that

$$x(t) = e^{-P(t)}\left[\int e^{P(t)} q(t)\ dt + C\right].$$

22. Use the method of Problem 21 to derive the solution

$$x(t) = x_0 e^{-at} + b\frac{e^{ct} - e^{-at}}{a + c}$$

of the differential equation $dx/dt + ax = be^{ct}$ (under the assumption that $a + c \neq 0$).

23. A 30-yr-old engineer accepts a position with a starting salary of \$30,000/yr. Her salary S increases exponentially, with

$$S(t) = 30e^{(0.05)t}$$

thousand dollars after t years. Meanwhile, 12% of her salary is deposited continuously in a retirement account, which accumulates interest at an annual rate of 6% compounded continuously. (a) Estimate ΔA in terms of Δt to derive this equation for the amount $A(t)$ in her retirement account at time t:

$$\frac{dA}{dt} - (0.06)A = (3.6)e^{(0.05)t}.$$

(b) Apply the result of Problem 22 to compute $A(40)$, the amount available for her retirement at age 70.

24. Pottstown has a fixed population of 10,000 people. On January 1, 1000 people have the flu; on April 1, 2000 people have it. Assume that the rate of increase of the number $N(t)$ who have the flu is proportional to the number who don't have it. How many will have the disease on October 1?

25. Let $x(t)$ denote the number of people in Athens, Georgia, of population 100,000, who have the Tokyo flu. The rate of change of $x(t)$ is proportional to the number of those in Athens who do not yet have the disease. Suppose that 20,000 have the flu on March 1 and that 60,000 have it on March 16. (a) Set up and solve a differential equation to find $x(t)$. (b) On what date will the number of people infected with the disease reach 80,000? (c) What happens in the long run?

Chapter 7 Review: DEFINITIONS, CONCEPTS, RESULTS

Use this list as a guide to concepts that you may need to review.

1. The laws of exponents

2. The laws of logarithms

3. The definition of the natural logarithm function

4. The graph of $y = \ln x$

5. The definition of the number e

6. The definition of the natural exponential function

7. The inverse function relationship between $\ln x$ and e^x

8. The graphs of $y = e^x$ and $y = e^{-x}$

9. Differentiation of $\ln u$ and e^u, where u is a differentiable function of x

10. The order of magnitude of $(\ln x)/x^k$ and x^k/e^x as $x \to +\infty$

11. The number e as a limit

12. The definition of general exponential and logarithm functions

13. Differentiation of a^u and $\log_a u$

14. Logarithmic differentiation

15. Solution of the differential equation $dx/dt = kx$

16. The natural growth equation

17. Radioactive decay and radiocarbon dating

18. Solution of a linear first-order differential equation with constant coefficients

19. Solution of separable first-order differential equations

20. Evaluating the constant of integration in an initial value problem

Chapter 7 Miscellaneous Problems

Differentiate the functions given in Problems 1 through 24.

1. $f(x) = \ln 2\sqrt{x}$

2. $f(x) = e^{-2\sqrt{x}}$

3. $f(x) = \ln(x - e^x)$

4. $f(x) = 10^{\sqrt{x}}$

5. $f(x) = \ln(2^x)$

6. $f(x) = \log_{10}(\sin x)$

7. $f(x) = x^3 e^{-1/x^2}$

8. $f(x) = x(\ln x)^2$

9. $f(x) = (\ln x)[\ln(\ln x)]$

10. $f(x) = \exp(10^x)$

11. $f(x) = 2^{\ln x}$

12. $f(x) = \ln\left(\dfrac{e^x + e^{-x}}{e^x - e^{-x}}\right)$

13. $f(x) = e^{(x+1)/(x-1)}$

14. $f(x) = \ln\left(\sqrt{1 + x}\sqrt[3]{2 + x^2}\right)$

15. $f(x) = \ln\left(\dfrac{x - 1}{3 - 4x^2}\right)^{3/2}$

16. $f(x) = \sin(\ln x)$

17. $f(x) = \exp\left(\sqrt{1 + \sin^2 x}\right)$

18. $f(x) = \dfrac{x}{(\ln x)^2}$

19. $f(x) = \ln(3^x \sin x)$

20. $f(x) = (\ln x)^x$

21. $f(x) = x^{1/x}$

22. $f(x) = x^{\sin x}$

23. $f(x) = (\ln x)^{\ln x}$

24. $f(x) = (\sin x)^{\cos x}$

Evaluate the indefinite integrals in Problems 25 through 36.

25. $\displaystyle\int \frac{dx}{1 - 2x}$

26. $\displaystyle\int \frac{\sqrt{x}}{1 + x^{3/2}} \, dx$

27. $\displaystyle\int \frac{3 - x}{1 + 6x - x^2} \, dx$

28. $\displaystyle\int \frac{e^x - e^{-x}}{e^x + e^{-x}} \, dx$

29. $\displaystyle\int \frac{\sin x}{2 + \cos x} \, dx$

30. $\displaystyle\int \frac{e^{-1/x^2}}{x^3} \, dx$

31. $\displaystyle\int \frac{10^{\sqrt{x}}}{\sqrt{x}} \, dx$

32. $\displaystyle\int \frac{1}{x(\ln x)^2} \, dx$

33. $\displaystyle\int e^x \sqrt{1 + e^x} \, dx$

34. $\displaystyle\int \frac{1}{x}\sqrt{1 + \ln x} \, dx$

35. $\displaystyle\int 2^x 3^x \, dx$

36. $\displaystyle\int \frac{dx}{x^{1/3}(1 + x^{2/3})}$

Solve the initial value problems in Problems 37 through 44.

37. $\dfrac{dx}{dt} = 2t; \quad x(0) = 17$

38. $\dfrac{dx}{dt} = 2x; \quad x(0) = 17$

39. $\dfrac{dx}{dt} = e^t; \quad x(0) = 2$

40. $\dfrac{dx}{dt} = e^x; \quad x(0) = 2$

41. $\dfrac{dx}{dt} = 3x - 2; \quad x(0) = 3$

42. $\dfrac{dx}{dt} = x^2 t^2; \quad x(0) = -1$

43. $\dfrac{dx}{dt} = x \cos t; \quad x(0) = \sqrt{2}$

44. $\dfrac{dx}{dt} = \sqrt{x}; \quad x(1) = 0$

Sketch the graphs of the equations given in Problems 45 through 49.

45. $y = e^{-x}\sqrt{x}$

46. $y = x - \ln x$

47. $y = \sqrt{x} - \ln x$

48. $y = x(\ln x)^2$

49. $y = e^{-1/x}$

50. Find the length of the curve $y = \frac{1}{2}x^2 - \frac{1}{4}\ln x$ from $x = 1$ to $x = e$.

51. A grain warehouse holds B bushels of grain, which is deteriorating in such a way that only $B \cdot 2^{-t/12}$ bushels will be salable after t months. Meanwhile, the grain's market price is increasing linearly: After t months it will be $2 + (t/12)$ dollars per bushel. After how many months should the grain be sold to maximize the revenue obtained?

52. You have borrowed $1000 at 10% annual interest, compounded continuously, to plant timber on a tract of land. Your agreement is to repay the loan, plus interest, when the timber is cut and sold. If the cut timber can be sold after t years for 800 $\exp(\frac{1}{2}\sqrt{t})$ dollars, when should you cut and sell to maximize the profit?

53. Blood samples from 1000 students are to be tested for a certain disease known to occur in 1% of the population. Each tests costs $5, so it would cost $5000 to test the samples individually. Suppose, however, that "lots" made up of x samples each are formed by pooling halves of individual samples, and that these lots are tested first (for $5 each). Only in case a lot tests positive—the probability of this is $1 - (0.99)^x$—will the x samples used to make up this lot be tested individually.
(a) Show that the total expected number of tests is

$$f(x) = \frac{1000}{x}[(1)(0.99)^x + (x + 1)(1 - (0.99)^x)]$$

$$= 1000 + \frac{1000}{x} - 1000 \cdot (0.99)^x \quad \text{if } x \geq 2.$$

(b) Show that the value of x that minimizes $f(x)$ is a root of the equation

$$x = \frac{(0.99)^{-x/2}}{[\ln(100/99)]^{1/2}}.$$

Because the denominator is approximately 0.1, it may be convenient to solve instead the simpler equation $x = 10 \cdot (0.99)^{-x/2}$. (c) From the results in parts (a) and (b), compute the cost of using this batch method to test the original 1000 samples.

54. Deduce from Problem 63 in Section 7.2 that

$$\lim_{x \to 0^+} x^x = 1.$$

55. Show that

$$\lim_{x \to 0} \frac{\ln(1 + x)}{x} = 1$$

by considering the value of $D_x \ln x$ for $x = 1$. Thus show that $\ln(1 + x) \approx x$ if x is very close to zero.

56. (a) Prove that

$$\lim_{h \to 0} \frac{a^h - 1}{h} = \ln a$$

by considering the definition of the derivative of a^x at $x = 0$. (b) Substitute $h = 1/n$ to obtain

$$\ln a = \lim_{n \to \infty} n(a^{1/n} - 1).$$

(c) Approximate $\ln 2$ by taking $n = 1024 = 2^{10}$ and using only the square root key (10 times) on a pocket calculator.

57. Suppose that the fish population $P(t)$ in a lake is attacked by disease at time $t = 0$, with the result that

$$\frac{dP}{dt} = -3\sqrt{P}$$

thereafter. Time t is measured in weeks. Initially, there are $P_0 = 900$ fish in the lake. How long will it take for all the fish to die?

58. A race car sliding along a level surface is decelerated by frictional forces proportional to its speed. Suppose that it decelerates initially at 2 m/s² and travels a total distance of 1800 m. What was its initial velocity? See Problem 18 of Section 7.6.

59. A home mortgage of $120,000 is to be paid off continuously over a period of 25 yr. Apply the result of Problem 13 in Section 7.6 to determine the monthly payment if the annual interest rate, compounded continuously, is (a) 8%; (b) 12%.

60. A powerboat weighs 32,000 lb, and its motor provides a thrust of 5000 lb. Assume that the water resistance is 100 lb for each foot per second of the boat's speed. Then the velocity $v(t)$ (in ft/s) of the boat at time t (in seconds) satisfies the differential equation

$$1000 \frac{dv}{dt} = 5000 - 100v.$$

Find the maximum velocity that the boat can attain if it starts from rest.

61. The temperature inside my freezer is $-16°C$, and the room temperature is a constant $20°C$. At 11 P.M. one evening the power goes off due to an ice storm. At 6 A.M. the next morning I see that the temperature in the freezer has risen to $-10°C$. At what time will the temperature in the freezer reach the critical value of $0°C$ if the power remains off?

62. Suppose that the action of fluorocarbons depletes the ozone in the upper atmosphere by 0.25% annually, so that the amount A of ozone in the upper atmosphere satisfies the differential equation

$$\frac{dA}{dt} = -\frac{1}{400}A \qquad (t \text{ in years}).$$

(a) What percentage of the original amount A_0 of upper-atmospheric ozone will remain 25 yr from now? (b) How long will it take for the amount of upper-atmospheric ozone to be reduced to half its initial amount?

63. A car starts from rest and travels along a straight road. Its engine provides a constant acceleration of a feet per second per second. Air resistance and road friction cause a deceleration of ρ feet per second per second for every foot per second of the car's velocity. (a) Show that the car's velocity after t seconds is

$$v(t) = \frac{a}{\rho}(1 - e^{-\rho t}).$$

(b) If $a = 17.6$ ft/s² and $\rho = 0.1$, find v when $t = 10$ s, and find also the limiting velocity as $t \to +\infty$. Give each answer in *miles per hour* as well as feet per second.

64. Immediately after an accident in a nuclear power plant, the level of radiation there was 10 times the safe limit. After 6 mo it dropped to 9 times the safe limit. Assuming exponential decay, how long (in years) after the accident will the radiation level drop to the safe limit?

65. Figure 7.MP.1 shows the graphs of $f(x) = x^{1/2}$, $g(x) = \ln x$, and $h(x) = x^{1/3}$ plotted on the interval $[0.2, 10]$. You can see that the graph of f remains above the graph of $\ln x$, whereas the graph of h dips below the graph of $\ln x$. But because $\ln x$ increases *less* rapidly than any positive power of x, the graph of h must eventually cross the graph of ln x and rise above it. Finally, you can easily believe that, for a suitable choice of p between 2 and 3, the graph of $j(x) = x^{1/p}$ never dips below the graph of ln x but does drop down just far enough to be tangent to the graph of ln x at a certain point. (a) Show that $f(x) > \ln x$ for all $x > 0$ by finding the global minimum value of $f(x) - \ln x$ on the interval $(0, \infty)$. (b) Use Newton's method to find the value of x at which $h(x)$ crosses the graph of ln x and rises above it—the value of x *not* given in Fig. 7.MP.1. (c) Find the value of p for which the graph of $j(x)$ is tangent to the graph of ln x at the point $(q, \ln q)$.

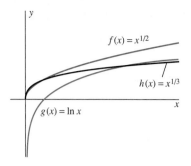

Fig. 7.MP.1   The three functions of Problem 65

Further Calculus of Transcendental Functions

John Bernoulli (1667–1748)

❏ In the eighteenth century the remarkable Swiss family Bernoulli was to mathematics what the Bach family was to music. Eight different Bernoullis were sufficiently prominent that more than two centuries later, they rate entries in the *Dictionary of Scientific Biography*. The brothers James (1654–1705) and John Bernoulli played crucial roles in the early development of Leibniz's version of the calculus based on infinitely small differentials, which in continental European science predominated over Newton's version based more explicitly on limits of ratios. It was James Bernoulli who introduced the word "integral" in suggesting the name *calculus*

integralis (instead of Leibniz's original *calculus summatorius*) for the subject inverse to the *calculus differentialis*.

❏ John Bernoulli first studied mathematics under his older brother James at the university in Basel, Switzerland, but soon they were on an equal footing in mathematical understanding. In 1691, John Bernoulli visited Paris and there met the young Marquis de l'Hôpital (1661–1704), who was anxious to learn the secrets of the new infinitesimal calculus. In return for a generous monthly stipend, Bernoulli agreed to tutor the wealthy Marquis and continued the lessons (and the financial arrangement) by mail after his return to Basel. The result of this correspondence was the first differential calculus textbook, published by l'Hôpital in 1696. This text is remembered mainly for its inclu-

sion of a result of Bernoulli's that is known as "l'Hôpital's rule." This result concerns the limit as $x \to a$ of a quotient $f(x)/g(x)$ whose numerator and denominator both approach 0 as $x \to a$, so mere substitution of the value $x = a$ would give the "indeterminate form" $0/0$. The definition of the derivative involves such a limit, so indeterminate forms pervade the subject of calculus.

❏ Both James and John Bernoulli worked on (and solved) the *catenary problem*, which asks for the shape of a hanging cable suspended between two fixed points, assuming that it is inelastic (unstretchable) but perfectly flexible. The Bernoullis showed that, in terms of hyperbolic functions, such a hanging cable takes the shape of a curve of the form

$$y = a \cosh(x/a).$$

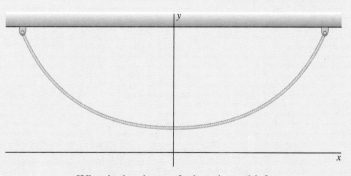

What is the shape of a hanging cable?

8.1
Introduction

The function f is called an **algebraic function** provided that $y = f(x)$ satisfies an equation of the form

$$a_n(x)y^n + a_{n-1}(x)y^{n-1} + \cdots + a_1(x)y + a_0(x) = 0,$$

where the coefficients $a_0(x), a_1(x), \ldots, a_n(x)$ are polynomials in x. For example, because the equation $y^2 - p(x) = 0$ has the form shown above, the square root of the polynomial $p(x)$—that is, $f(x) = \sqrt{p(x)}$—is an algebraic function. The equation $q(x)y - p(x) = 0$ also has the necessary form, so a rational function [a quotient of polynomials; here, $y = p(x)/q(x)$] is also an algebraic function.

A function that is *not* algebraic is said to be **transcendental.** The natural logarithm function $\ln x$ and the natural exponential function e^x are transcendental functions, as are the six familiar trigonometric functions. In this chapter we shall study the remaining transcendental functions of elementary character—the inverse trigonometric functions and the hyperbolic functions. These functions have extensive scientific applications and provide the basis for certain important methods of integration (as discussed in Chapter 9). In Sections 8.3 and 8.4 we also study certain limit expressions ("indeterminate forms") that typically involve transcendental functions.

8.2
Inverse Trigonometric Functions

If the function f is one-to-one on its domain of definition, then it has an inverse function f^{-1}. This inverse function is defined by the fact that

$$f^{-1}(x) = y \quad \text{if and only if} \quad f(y) = x. \tag{1}$$

For example, from Chapter 7 we are familiar with the pair of inverse functions

$$f(x) = e^x \quad \text{and} \quad f^{-1}(x) = \ln x.$$

From a geometric viewpoint, Eq. (1) implies that the graphs $y = f(x)$ and $y = f^{-1}(x)$ are reflections across the 45° line $y = x$, like the familiar graphs $y = e^x$ and $y = \ln x$ in Fig. 8.2.1.

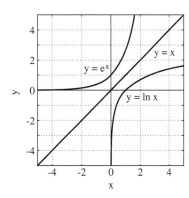

Fig. 8.2.1 The graphs $y = e^x$ and $y = \ln x$ are reflections across the line $y = x$.

THE INVERSE SINE FUNCTION

Here we want to define the inverses of the trigonometric functions, beginning with the inverse sine function. We must, however, confront the fact that the trigonometric functions fail to be one-to-one because the period of each of

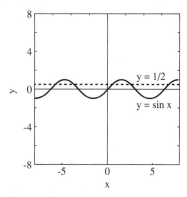

Fig. 8.2.2 Multiple values of x such that $\sin x = \frac{1}{2}$

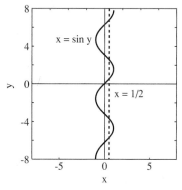

Fig. 8.2.3 There are many possible choices for $y = \sin^{-1}(\frac{1}{2})$.

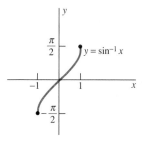

Fig. 8.2.4 The graph of $y = \sin^{-1} x = \arcsin x$

the six is π *or* 2π. For example, $\sin x = \frac{1}{2}$ if x is *either* $\pi/6$ plus any multiple of 2π or $5\pi/6$ plus any multiple of 2π. These many values of x, all with sine equal to $\frac{1}{2}$, correspond to the multiple points of intersection of the graph $y = \sin x$ and the horizontal line $y = \frac{1}{2}$ in Fig. 8.2.2.

Figure 8.2.3 is the reflection of Fig. 8.2.2 across the 45° line $y = x$. The multiple intersections of $x = \sin y$ and the vertical line $x = \frac{1}{2}$ indicate that we must make a choice in order to define $\sin^{-1}(\frac{1}{2})$. That is, we *cannot* define $y = \sin^{-1} x$, the inverse of the sine function, by saying simply that y is the number such that $\sin y = x$.[†] There are *many* such values of y, and we must specify just which particular one of these is to be used.

We do this by suitably restricting the domain of the sine function. Because the function $\sin x$ is increasing on $[-\pi/2, \pi/2]$ and its range of values is $[-1, 1]$, for each x in $[-1, 1]$ there is *one* number y in $[-\pi/2, \pi/2]$ such that $\sin y = x$. This observation leads to the following definition of the *inverse sine* (or *arcsine*) function, denoted by $\sin^{-1} x$ or arcsin x.

Definition *The Inverse Sine Function*
The **inverse sine** (or **arcsine**) **function** is defined as follows:

$$y = \sin^{-1} x \quad \text{if and only if} \quad \sin y = x, \tag{2}$$

where $-1 \leqq x \leqq 1$ and $-\pi/2 \leqq y \leqq \pi/2$.

Thus, if x is between -1 and $+1$ (inclusive), then $\sin^{-1} x$ is that number y between $-\pi/2$ and $\pi/2$ such that $\sin y = x$. Even more briefly, arcsin x is the angle (in radians) nearest zero whose sine is x. For instance,

$$\sin^{-1} 1 = \frac{\pi}{2}, \qquad \sin^{-1} 0 = 0, \qquad \sin^{-1}(-1) = -\frac{\pi}{2},$$

and $\sin^{-1} 2$ does not exist.

Because interchanging x and y in the equation $\sin y = x$ yields $y = \sin x$, it follows from Eq. (2) that the graph of $y = \sin^{-1} x$ is the reflection of the graph of $y = \sin x$, $-\pi/2 \leqq x \leqq \pi/2$, across the line $y = x$ (Fig. 8.2.4).

It also follows from Eq. (2) that

$$\sin(\sin^{-1} x) = x \qquad \text{if } -1 \leqq x \leqq 1 \quad \text{and} \tag{3a}$$

$$\sin^{-1}(\sin x) = x \qquad \text{if } -\pi/2 \leqq x \leqq \pi/2. \tag{3b}$$

Because the derivative of $\sin x$ is positive for $-\pi/2 < x < \pi/2$, it follows from Theorem 1 of Section 7.1 that $\sin^{-1} x$ is differentiable on $(-1, 1)$. We may, therefore, differentiate both sides of the identity in (3a), but we begin by writing it in the form

$$\sin y = x,$$

where $y = \sin^{-1} x$. Then differentiation with respect to x gives

$$(\cos y) \frac{dy}{dx} = 1,$$

[†] The symbol -1 in the notation $\sin^{-1} x$ is not an exponent—it does *not* mean $(\sin x)^{-1}$.

where x is any real number and $0 < y < \pi$. Then differentiation of both sides of the identity $\cot(\cot^{-1} x) = x$ leads, as in the derivation of Eq. (10), to

$$D_x \cot^{-1} x = -\frac{1}{1 + x^2}.$$

If u is a differentiable function of x, then the chain rule gives

$$D_x \cot^{-1} u = -\frac{1}{1 + u^2} \cdot \frac{du}{dx}. \tag{13}$$

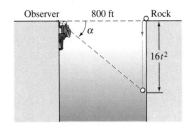

Observer   800 ft   Rock
α
$16t^2$

Fig. 8.2.9 The falling rock of Example 2

EXAMPLE 2 A mountain climber on one edge of a deep canyon 800 ft wide sees a large rock fall from the opposite edge at time $t = 0$. As he watches the rock plummet downward, his eyes first move slowly, then faster, then more slowly again. Let α denote the angle of depression of his line of sight below the horizontal. At what angle α would the rock *seem* to be moving the most rapidly? That is, when would $d\alpha/dt$ be maximal?

Solution From our study of constant acceleration in Section 5.2, we know that the rock will fall $16t^2$ feet in the first t seconds. We refer to Fig. 8.2.9 and see that the value of α at time t will be

$$\alpha = \alpha(t) = \tan^{-1}\left(\frac{16t^2}{800}\right) = \tan^{-1}\left(\frac{t^2}{50}\right).$$

Hence

$$\frac{d\alpha}{dt} = \frac{1}{1 + \left(\frac{t^2}{50}\right)^2} \cdot \frac{2t}{50} = \frac{100t}{t^4 + 2500}.$$

To find when $d\alpha/dt$ is maximal, we find when *its* derivative is zero:

$$\frac{d}{dt}\left(\frac{d\alpha}{dt}\right) = \frac{100(t^4 + 2500) - 100t\,(4t^3)}{(t^4 + 2500)^2} = \frac{100(2500 - 3t^4)}{(t^4 + 2500)^2}.$$

So $d^2\alpha/dt^2$ is zero when $3t^4 = 2500$—that is, when

$$t = \sqrt[4]{\frac{2500}{3}} \approx 5.37 \quad \text{(s)}.$$

This is the value of t when $d\alpha/dt$ is maximal, and at this time we have $t^2 = 50/\sqrt{3}$. So the angle at this time is

$$\alpha = \arctan\left(\frac{1}{50} \cdot \frac{50}{\sqrt{3}}\right) = \arctan\left(\frac{1}{\sqrt{3}}\right) = \frac{\pi}{6}.$$

The *apparent* speed of the falling rock is greatest when the climber's line of sight is 30° below the horizontal. The actual speed of the rock is then $32t$ with $t \approx 5.37$ s and thus is about 172 ft/s.

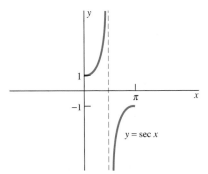

y
1
-1
π
x
$y = \sec x$

Fig. 8.2.10 Restriction of the secant function to the union of the two intervals $[0, \pi/2)$ and $(\pi/2, \pi]$

THE INVERSE SECANT FUNCTION

Figure 8.2.10 shows that the secant function is increasing on each of the intervals $[0, \pi/2)$ and $(\pi/2, \pi]$. On the union of these two intervals, the secant function attains all real values y such that $|y| \geq 1$. We may, therefore, define the inverse secant function, denoted by $\sec^{-1} x$ or by $\operatorname{arcsec} x$, by restricting the secant function to the union of the two intervals $[0, \pi/2)$ and $(\pi/2, \pi]$.

Definition *The Inverse Secant Function*
The **inverse secant** (or **arcsecant**) **function** is defined as follows:

$$y = \sec^{-1} x \quad \text{if and only if} \quad \sec y = x, \tag{14}$$

where $|x| \geqq 1$ and $0 \leqq y \leqq \pi$.

REMARK   Older textbooks offer alternative definitions of the inverse secant based on different intervals of definition of $\sec x$. The definition given here, however, satisfies the condition that

$$\sec^{-1} x = \cos^{-1} \frac{1}{x} \quad \text{(if } x \geqq 1\text{)},$$

which is convenient for calculator-computer calculations (see Problem 57).

The graph of $\sec^{-1} x$ is the reflection of the graph of $y = \sec x$, suitably restricted to the intervals $0 \leqq x < \pi/2$ and $\pi/2 < x \leqq \pi$, across the line $y = x$ (Fig. 8.2.11). It follows from the definition of the inverse secant that

$$\sec(\sec^{-1} x) = x \quad \text{if } |x| \geqq 1, \tag{15a}$$

$$\sec^{-1}(\sec x) = x \quad \text{for } x \text{ in } [0, \pi/2) \cup (\pi/2, \pi]. \tag{15b}$$

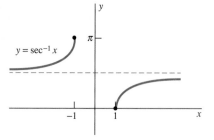

Fig. 8.2.11 The graph of $y = \sec^{-1} x = \operatorname{arcsec} x$

Following the now-familiar pattern, we find $D_x \sec^{-1} x$ by differentiating both sides of Eq. (15a) in the form

$$\sec y = x,$$

where $y = \sec^{-1} x$. This yields

$$(\sec y \tan y) \frac{dy}{dx} = 1,$$

so

$$\frac{dy}{dx} = \frac{1}{\sec y \tan y} = \frac{1}{\pm x \sqrt{x^2 - 1}},$$

because $\tan y = \pm\sqrt{\sec^2 y - 1} = \pm\sqrt{x^2 - 1}$.

To obtain the correct choice of sign here, note what happens in the two cases $x > 1$ and $x < -1$. In the first case, $0 < y < \pi/2$ and $\tan y > 0$, so we choose the plus sign. If $x < -1$, then $\pi/2 < y < \pi$ and $\tan y < 0$, so we take the minus sign. Thus

$$D_x \sec^{-1} x = \frac{1}{|x|\sqrt{x^2 - 1}} \quad (|x| > 1). \tag{16}$$

If u is a differentiable function of x with values that exceed 1 in magnitude, then by the chain rule we have

Ch. 8 / Further Calculus of Transcendental Functions

$$D_x \sec^{-1} u = \frac{1}{|u|\sqrt{u^2 - 1}} \cdot \frac{du}{dx}. \tag{17}$$

EXAMPLE 3 The function $\sec^{-1} e^x$ is defined if $x > 0$, because then $e^x > 1$. Then by Eq. (17),

$$D_x \sec^{-1} e^x = \frac{e^x}{|e^x|\sqrt{e^{2x} - 1}} = \frac{1}{\sqrt{e^{2x} - 1}}$$

because $|e^x| = e^x$ for all x.

The **inverse cosecant** (or **arccosecant**) function is the inverse of the function $y = \csc x$, where x is restricted to the union of the intervals $[-\pi/2, 0)$ and $(0, \pi/2]$. Thus

$$y = \csc^{-1} x \quad \text{if and only if} \quad \csc y = x, \tag{18}$$

where $|x| \geq 1$ and $-\pi/2 < y < \pi/2$. Its derivative formula, which has a derivation similar to that of the inverse secant function, is

$$D_x \csc^{-1} u = -\frac{1}{|u|\sqrt{u^2 - 1}} \cdot \frac{du}{dx}. \tag{19}$$

SUMMARY

The following table summarizes the domains, ranges, and derivatives of the six inverse trigonometric functions.

Function	Domain of definition	Range of values	Derivative
$\sin^{-1} x$	$-1 \leq x \leq 1$	$-\pi/2 \leq y \leq \pi/2$	$\dfrac{1}{\sqrt{1 - x^2}}$
$\cos^{-1} x$	$-1 \leq x \leq 1$	$0 \leq y \leq \pi$	$-\dfrac{1}{\sqrt{1 - x^2}}$
$\tan^{-1} x$	$-\infty < x < +\infty$	$-\pi/2 < y < \pi/2$	$\dfrac{1}{1 + x^2}$
$\cot^{-1} x$	$-\infty < x < +\infty$	$0 < y < \pi$	$-\dfrac{1}{1 + x^2}$
$\sec^{-1} x$	$|x| \geq 1$	$0 \leq y \leq \pi$	$\dfrac{1}{|x|\sqrt{x^2 - 1}}$
$\csc^{-1} x$	$|x| \geq 1$	$-\pi/2 < y < \pi/2$	$-\dfrac{1}{|x|\sqrt{x^2 - 1}}$

It is worth noting that

❑ $\sin^{-1} x$ and $\tan^{-1} x$ share the range $-\pi/2$ to $\pi/2$, whereas

❑ $\cos^{-1} x$ and $\sec^{-1} x$ share the range 0 to π.

Observe also the "difference only in sign" of the derivatives of function/cofunction pairs of inverse functions.

INTEGRALS INVOLVING INVERSE TRIGONOMETRIC FUNCTIONS

The derivatives of the six inverse trigonometric functions are all simple *algebraic* functions. As a consequence, inverse trigonometric functions typically occur when we integrate algebraic functions. Moreover, as mentioned earlier, the derivatives of $\cos^{-1} x$, $\cot^{-1} x$, and $\csc^{-1} x$ differ only in sign from the derivatives of their respective cofunctions. For this reason only the arcsin, arctangent, and arcsecant functions are necessary for integration, and only these three are in common use. That is, you need commit to memory the integral formulas only for the latter three functions. They follow immediately from Eqs. (5), (11), and (17) and may be written in the forms shown next:

$$\int \frac{du}{\sqrt{1 - u^2}} = \sin^{-1} u + C, \tag{20}$$

$$\int \frac{du}{1 + u^2} = \tan^{-1} u + C, \tag{21}$$

$$\int \frac{du}{u\sqrt{u^2 - 1}} = \sec^{-1}|u| + C. \tag{22}$$

It is easy to verify that the absolute value on the right-hand side in Eq. (22) follows from the one in Eq. (17). And remember that because $\sec^{-1}|u|$ is undefined unless $|u| \geqq 1$, the definite integral

$$\int_a^b \frac{du}{u\sqrt{u^2 - 1}}$$

is meaningful only when the limits a and b are both at least 1 or both at most -1.

EXAMPLE 4 It follows immediately from Eq. (21) that

$$\int_0^1 \frac{dx}{1 + x^2} = \left[\tan^{-1} x \right]_0^1 = \tan^{-1} 1 - \tan^{-1} 0 = \frac{\pi}{4}.$$

EXAMPLE 5 The substitution $u = 3x$, $du = 3\, dx$ gives

$$\int \frac{1}{1 + 9x^2} \, dx = \frac{1}{3} \int \frac{3}{1 + (3x)^2} \, dx$$

$$= \frac{1}{3} \int \frac{du}{1 + u^2} = \frac{1}{3} \tan^{-1} u + C = \frac{1}{3} \tan^{-1} 3x + C.$$

EXAMPLE 6 The substitution $u = \frac{1}{2} x$, $du = \frac{1}{2} dx$ gives

$$\int \frac{1}{\sqrt{4 - x^2}} \, dx = \int \frac{1}{2\sqrt{1 - (x/2)^2}} \, dx$$

$$= \int \frac{1}{\sqrt{1 - u^2}} \, du = \arcsin u + C = \arcsin\left(\frac{x}{2}\right) + C.$$

EXAMPLE 7   The substitution $u = x\sqrt{2}$, $du = \sqrt{2}\,dx$ gives

$$\int_1^{\sqrt{2}} \frac{1}{x\sqrt{2x^2 - 1}}\,dx = \int_{\sqrt{2}}^2 \frac{1}{u\sqrt{u^2 - 1}}\,du$$

$$= \left[\sec^{-1}|u|\right]_{\sqrt{2}}^2 = \sec^{-1} 2 - \sec^{-1}\sqrt{2}$$

$$= \frac{\pi}{3} - \frac{\pi}{4} = \frac{\pi}{12}.$$

8.2 Problems

Find the values indicated in each of Problems 1 through 4.

1. (a) $\sin^{-1}(\tfrac{1}{2})$

   (b) $\sin^{-1}(-\tfrac{1}{2})$

   (c) $\sin^{-1}(\tfrac{1}{2}\sqrt{2})$

   (d) $\sin^{-1}(-\tfrac{1}{2}\sqrt{3})$

3. (a) $\tan^{-1} 0$

   (b) $\tan^{-1} 1$

   (c) $\tan^{-1}(-1)$

   (d) $\tan^{-1}\sqrt{3}$

2. (a) $\cos^{-1}(\tfrac{1}{2})$

   (b) $\cos^{-1}(-\tfrac{1}{2})$

   (c) $\cos^{-1}(\tfrac{1}{2}\sqrt{2})$

   (d) $\cos^{-1}(-\tfrac{1}{2}\sqrt{3})$

4. (a) $\sec^{-1} 1$

   (b) $\sec^{-1}(-1)$

   (c) $\sec^{-1} 2$

   (d) $\sec^{-1}(-\sqrt{2})$

Differentiate the functions in Problems 5 through 26.

5. $f(x) = \sin^{-1}(x^{100})$

6. $f(x) = \arctan(e^x)$

7. $f(x) = \sec^{-1}(\ln x)$

8. $f(x) = \ln(\tan^{-1} x)$

9. $f(x) = \arcsin(\tan x)$

10. $f(x) = x \arctan x$

11. $f(x) = \sin^{-1} e^x$

12. $f(x) = \arctan \sqrt{x}$

13. $f(x) = \cos^{-1} x + \sec^{-1}\left(\dfrac{1}{x}\right)$

14. $f(x) = \cot^{-1}\left(\dfrac{1}{x^2}\right)$

15. $f(x) = \csc^{-1} x^2$

16. $f(x) = \arccos\left(\dfrac{1}{\sqrt{x}}\right)$

17. $f(x) = \dfrac{1}{\arctan x}$

18. $f(x) = (\arcsin x)^2$

19. $f(x) = \tan^{-1}(\ln x)$

20. $f(x) = \text{arcsec}\sqrt{x^2 + 1}$

21. $f(x) = \tan^{-1} e^x + \cot^{-1} e^{-x}$

22. $f(x) = \exp(\arcsin x)$

23. $f(x) = \sin(\arctan x)$

24. $f(x) = \sec(\sec^{-1} e^x)$

25. $f(x) = \dfrac{\arctan x}{(1 + x^2)^2}$

26. $f(x) = (\sin^{-1} 2x^2)^{-2}$

In Problems 27 through 30, find dy/dx by implicit differentiation. Then find the line tangent to the graph of the equation at the indicated point P.

27. $\tan^{-1} x + \tan^{-1} y = \dfrac{\pi}{2}$; $P(1, 1)$

28. $\sin^{-1} x + \sin^{-1} y = \dfrac{\pi}{2}$; $P(\tfrac{1}{2}, \tfrac{1}{2}\sqrt{3})$

29. $(\sin^{-1} x)(\sin^{-1} y) = \dfrac{\pi^2}{16}$; $P(\tfrac{1}{2}\sqrt{2}, \tfrac{1}{2}\sqrt{2})$

30. $(\sin^{-1} x)^2 + (\sin^{-1} y)^2 = \dfrac{5\pi^2}{36}$; $P(\tfrac{1}{2}, \tfrac{1}{2}\sqrt{3})$

Evaluate or antidifferentiate, as appropriate, in Problems 31 through 55.

31. $\displaystyle\int_0^1 \frac{dx}{1 + x^2}$

32. $\displaystyle\int_0^{1/2} \frac{dx}{\sqrt{1 - x^2}}$

33. $\displaystyle\int_{\sqrt{2}}^2 \frac{dx}{x\sqrt{x^2 - 1}}$

34. $\displaystyle\int_{-2}^{-2/\sqrt{3}} \frac{dx}{x\sqrt{x^2 - 1}}$

35. $\displaystyle\int_0^3 \frac{dx}{9 + x^2}$

36. $\displaystyle\int_0^{\sqrt{12}} \frac{dx}{\sqrt{16 - x^2}}$

37. $\displaystyle\int \frac{dx}{\sqrt{1 - 4x^2}}$

38. $\displaystyle\int \frac{dx}{9x^2 + 4}$

39. $\displaystyle\int \frac{dx}{x\sqrt{x^2 - 25}}$

40. $\displaystyle\int \frac{dx}{x\sqrt{4x^2 - 9}}$

41. $\displaystyle\int \frac{e^x}{1 + e^{2x}}\,dx$

42. $\displaystyle\int \frac{x^2}{x^6 + 25}\,dx$

43. $\displaystyle\int \frac{dx}{x\sqrt{x^6 - 25}}$

44. $\displaystyle\int \frac{\sqrt{x}}{1 + x^3}\,dx$

45. $\displaystyle\int \frac{dx}{\sqrt{x(1 - x)}}$

46. $\displaystyle\int \frac{\sec x \tan x}{1 + \sec^2 x}\,dx$

47. $\displaystyle\int \frac{x^{49}}{1 + x^{100}}\,dx$

48. $\displaystyle\int \frac{x^4}{\sqrt{1 - x^{10}}}\,dx$

49. $\displaystyle\int \frac{1}{x[1 + (\ln x)^2]}\,dx$

50. $\displaystyle\int \frac{\arctan x}{1 + x^2}\,dx$

51. $\displaystyle\int_0^1 \frac{1}{1 + (2x - 1)^2}\,dx$

52. $\displaystyle\int_0^1 \frac{x^3}{1 + x^4}\,dx$

53. $\displaystyle\int_1^e \frac{dx}{x\sqrt{1 - (\ln x)^2}}$

54. $\displaystyle\int_1^2 \frac{dx}{x\sqrt{x^2 - 1}}$

55. $\displaystyle\int_1^3 \frac{dx}{2\sqrt{x}(1 + x)}$ [*Suggestion:* Let $u = x^{1/2}$.]

56. Conclude from the formula $D_x \cos^{-1} x = -D_x \sin^{-1} x$ that $\sin^{-1} x + \cos^{-1} x = \pi/2$ if $0 \leq x \leq 1$.

57. Show that $D_x \sec^{-1} x = D_x \cos^{-1}(1/x)$ if $x \geq 1$, and conclude that $\sec^{-1} x = \cos^{-1}(1/x)$ if $x \geq 1$. This fact can be used to find arcsecants on a calculator that has the key for the arccosine function, usually written $\boxed{\text{inv}}$ $\boxed{\cos}$ or $\boxed{\cos^{-1}}$.

58. (a) Deduce from the addition formula for tangents (Problem 28 in Appendix A) that

$$\arctan x + \arctan y = \arctan \frac{x + y}{1 - xy}$$

provided that $xy < 1$. (b) Apply part (a) to show that each of the following numbers is equal to $\pi/4$: (i) $\arctan(\frac{1}{2})$ + $\arctan(\frac{1}{3})$; (ii) $2\arctan(\frac{1}{3}) + \arctan(\frac{1}{7})$; (iii) $\arctan(\frac{120}{119})$ − $\arctan(\frac{1}{239})$; (iv) $4\arctan(\frac{1}{5})$ − $\arctan(\frac{1}{239})$.

59. A billboard to be built *parallel* to a highway will be 12 m high, and its bottom will be 4 m above the eye level of the average passing motorist. How far from the highway should the billboard be placed in order to maximize the vertical angle it subtends at the motorist's eyes?

60. Use inverse trigonometric functions to prove that the vertical angle subtended by a rectangular painting on a wall is greatest when the painting is hung with its center at the level of the observer's eyes.

61. Show that the circumference of a circle of radius a is $2\pi a$ by finding the length of the circular arc

$$y = \sqrt{a^2 - x^2}$$

from $x = 0$ to $x = a/\sqrt{2}$ and then multiplying by 8.

62. Find the volume generated by revolving around the x-axis the area under $y = 1/(1 + x^4)$ from $x = 0$ to $x = 1$.

63. The unbounded region R is bounded on the left by the y-axis, below by the x-axis, and above by the graph of $y = 1/(1 + x^2)$. Show that the area of R is finite by evaluating

$$\lim_{a \to \infty} \int_0^a \frac{dx}{1 + x^2}.$$

64. A building 250 ft high is equipped with an external elevator. The elevator starts at the top at time $t = 0$ and descends at the constant rate of 25 ft/s. You are watching the elevator from a window that is 100 ft above the ground and in a building 50 ft from the elevator. At what height does the elevator appear to you to be moving the fastest?

65. Suppose that the function f is defined for all x such that $|x| > 1$ and has the property that

$$f'(x) = \frac{1}{x\sqrt{x^2 - 1}}$$

for all such x. (a) Explain why there exist two constants A and B such that

$$f(x) = \arcsec x + A \qquad \text{if } x > 1;$$
$$f(x) = -\arcsec x + B \qquad \text{if } x < -1.$$

(b) Determine the values of A and B so that $f(2) = 1 = f(-2)$. Then sketch the graph of $y = f(x)$.

66. The arctangent is the only inverse trigonometric function included in some versions of BASIC and FORTRAN, so it is necessary in programming to express $\sin^{-1} x$ and $\sec^{-1} x$ in terms of the arctangent. Show each of the following: (a) If $|x| < 1$, then

$$\sin^{-1} x = \arctan\left(\frac{x}{\sqrt{1 - x^2}}\right).$$

(b) If $x > 1$, then $\sec^{-1} x = \arctan(\sqrt{x^2 - 1})$. (c) If $x < -1$, then $\sec^{-1} x = \pi - \arctan(\sqrt{x^2 - 1})$.

8.3 ▰▰▰

Indeterminate Forms and L'Hôpital's Rule

An *indeterminate form* is a certain type of expression with a limit that is not evident by inspection. There are several types of indeterminate forms. If

$$\lim_{x \to a} f(x) = 0 = \lim_{x \to a} g(x),$$

then we say that the quotient $f(x)/g(x)$ has the **indeterminate form 0/0** at $x = a$. For example, to differentiate the trigonometric functions (Section 3.7), we needed to know that

$$\lim_{x \to 0} \frac{\sin x}{x} = 1. \tag{1}$$

Here, $f(x) = \sin x$ and $g(x) = x$. Because $\sin x$ and x both approach zero as $x \to 0$, the quotient $(\sin x)/x$ has the indeterminate form $0/0$ at $x = 0$. Consequently, we had to use a special geometric argument to find the limit in Eq.

(1); see Section 2.3. Indeed, something of this sort happens whenever we compute a derivative, because the quotient

$$\frac{f(x) - f(a)}{x - a},$$

whose limit as $x \to a$ is the derivative $f'(a)$, has the indeterminate form $0/0$ at $x = a$.

We can sometimes find the limit of an indeterminate form by performing a special algebraic manipulation or construction, as in our earlier computation of derivatives. Often, however, it is more convenient to apply a rule that appeared in the first calculus textbook ever published, by the Marquis de l'Hôpital, in 1696. L'Hôpital was a French nobleman who had hired the Swiss mathematician John Bernoulli as his calculus tutor, and "l'Hôpital's rule" is actually the work of Bernoulli.

Theorem 1 *L'Hôpital's Rule*
Suppose that the functions f and g are differentiable in a deleted neighborhood of the point a and that $g'(x)$ is nonzero in that neighborhood. Suppose also that

$$\lim_{x \to a} f(x) = 0 = \lim_{x \to 0} g(x).$$

Then

$$\lim_{x \to a} \frac{f(x)}{g(x)} = \lim_{x \to a} \frac{f'(x)}{g'(x)}, \tag{2}$$

provided that the limit on the right either exists (as a finite real number) or is $+\infty$ or $-\infty$.

In essence, l'Hôpital's rule says that if $f(x)/g(x)$ has the indeterminate form $0/0$ at $x = a$, then—subject to a few mild restrictions—this quotient has the same limit at $x = a$ as does the quotient $f'(x)/g'(x)$ of *derivatives*. The proof of l'Hôpital's rule is discussed at the end of this section.

EXAMPLE 1 Find $\lim\limits_{x \to 0} \dfrac{e^x - 1}{\sin 2x}$.

Solution   The fraction whose limit we seek has the indeterminate form $0/0$ at $x = 0$. The numerator and denominator are clearly differentiable in some deleted neighborhood of $x = 0$, and the derivative of the denominator is certainly nonzero if the neighborhood is small enough (specifically, if $0 < |x| < \pi/4$). So l'Hôpital's rule applies, and

$$\lim_{x \to 0} \frac{e^x - 1}{\sin 2x} = \lim_{x \to 0} \frac{e^x}{2 \cos 2x} = \frac{1}{2}.$$

If the quotient $f'(x)/g'(x)$ is itself indeterminate, then l'Hôpital's rule may be applied a second (or third, . . .) time, as in Example 2. When the rule is applied repeatedly, however, the conditions for its applicability must be checked at each stage.

EXAMPLE 2 Find $\lim\limits_{x \to 1} \dfrac{1 - x + \ln x}{1 + \cos \pi x}$.

Solution

$$\lim_{x \to 1} \frac{1 - x + \ln x}{1 + \cos \pi x} = \lim_{x \to 1} \frac{-1 + \dfrac{1}{x}}{-\pi \sin \pi x} \qquad \text{(still of the form 0/0)}$$

$$= \lim_{x \to 1} \frac{x - 1}{\pi x \sin \pi x} \qquad \text{(algebraic simplification)}$$

$$= \lim_{x \to 1} \frac{1}{\pi \sin \pi x + \pi^2 x \cos \pi x} \qquad \text{(l'Hôpital's rule again)}$$

$$= -\frac{1}{\pi^2} \qquad \text{(by inspection)}.$$

Because the final limit exists, so do the previous ones; the existence of the final limit in Eq. (2) implies the existence of the first.

When you need to apply l'Hôpital's rule repeatedly in this way, you need only keep differentiating the numerator and denominator separately until at least one of them has a nonzero finite limit. At that point you can recognize the limit of the quotient by inspection, as in the final step in Example 2.

EXAMPLE 3 Find $\lim\limits_{x \to 0} \dfrac{\sin x}{x + x^2}$.

Solution If we simply apply l'Hôpital's rule twice in succession, the result is the *incorrect* computation

$$\lim_{x \to 0} \frac{\sin x}{x + x^2} = \lim_{x \to 0} \frac{\cos x}{1 + 2x}$$

$$= \lim_{x \to 0} \frac{-\sin x}{2} = 0. \qquad \textbf{(Wrong!)}$$

This answer is wrong because $(\cos x)/(1 + 2x)$ is *not* an indeterminate form. Thus l'Hôpital's rule cannot be applied to it. The *correct* computation is

$$\lim_{x \to 0} \frac{\sin x}{x + x^2} = \lim_{x \to 0} \frac{\cos x}{1 + 2x} = \frac{\lim\limits_{x \to 0} \cos x}{\lim\limits_{x \to 0} (1 + 2x)} = \frac{1}{1} = 1.$$

The point of Example 3 is to issue a warning: Verify the hypotheses of l'Hôpital's rule *before* you apply it. It is an oversimplification to say that l'Hôpital's rule works when you need it and doesn't work when you don't, but there is still much truth in this statement.

INDETERMINATE FORMS INVOLVING ∞

L'Hôpital's rule has several variations. In addition to the fact that the limit in Eq. (2) is allowed to be infinite, the real number a in l'Hôpital's rule may be replaced by either $+\infty$ or $-\infty$. For example,

Ch. 8 / Further Calculus of Transcendental Functions

$$\lim_{x \to \infty} \frac{f(x)}{g(x)} = \lim_{x \to \infty} \frac{f'(x)}{g'(x)} \tag{3}$$

provided that the other hypotheses are satisfied. [A deleted neighborhood of $+\infty$ is an interval of the form $(c, +\infty)$.] In particular, to use Eq. (3), we must first verify that

$$\lim_{x \to \infty} f(x) = 0 = \lim_{x \to \infty} g(x)$$

and that the right-hand limit in Eq. (3) exists. The proof of this version of l'Hôpital's rule is outlined in Problem 50.

L'Hôpital's rule may also be used when $f(x)/g(x)$ has the **indeterminate form** ∞/∞. This means that

$$\lim_{x \to a} f(x) \text{ is either } +\infty \text{ or } -\infty$$

and

$$\lim_{x \to a} g(x) \text{ is either } +\infty \text{ or } -\infty.$$

The proof of this extension of the rule is difficult and is omitted here.[†]

EXAMPLE 4 Find $\lim\limits_{x \to \infty} \dfrac{e^x}{x^2 + x}$.

Solution Both quotients $e^x/(x^2 + x)$ and $e^x/(2x + 1)$ have the indeterminate form ∞/∞, so two applications of l'Hôpital's rule yield

$$\lim_{x \to \infty} \frac{e^x}{x^2 + x} = \lim_{x \to \infty} \frac{e^x}{2x + 1} = \lim_{x \to \infty} \frac{e^x}{2} = +\infty.$$

Remember that l'Hôpital's rule also "allows" the final result to be an infinite limit.

EXAMPLE 5

$$\lim_{x \to \infty} \frac{\ln x}{\sqrt{x}} = \lim_{x \to \infty} \frac{\dfrac{1}{x}}{\frac{1}{2} x^{-1/2}} = \lim_{x \to \infty} \frac{2}{\sqrt{x}} = 0.$$

PROOF OF L'HÔPITAL'S RULE

Suppose that the functions f and g of Theorem 1 are not merely differentiable but have continuous derivatives near $x = a$ and that $g'(a) \neq 0$. Then

$$\lim_{x \to a} \frac{f'(x)}{g'(x)} = \frac{\lim\limits_{x \to a} f'(x)}{\lim\limits_{x \to a} g'(x)} = \frac{f'(a)}{g'(a)} \tag{4}$$

by the limit law for quotients. In this case l'Hôpital's rule in Eq. (2) reduces to the limit

$$\lim_{x \to a} \frac{f(x)}{g(x)} = \frac{f'(a)}{g'(a)}, \tag{5}$$

which is a weak form of the rule. It actually is this weak form that is typically applied in single-step applications of l'Hôpital's rule.

[†]For a proof, see (for example) A. E. Taylor and W. R. Mann, *Advanced Calculus,* 3rd ed. (New York: John Wiley, 1983), p. 107.

EXAMPLE 6 In Example 1 we had

$$f(x) = e^x - 1, \qquad g(x) = \sin 2x,$$

so

$$f'(x) = e^x, \qquad g'(x) = 2 \cos 2x,$$

and $g'(0) = 2 \neq 0$. With $a = 0$, Eq. (5) therefore gives

$$\lim_{x \to 0} \frac{e^x - 1}{\sin 2x} = \lim_{x \to 0} \frac{f(x)}{g(x)} = \frac{f'(0)}{g'(0)} = \frac{e^0}{2 \cos 0} = \frac{1}{2}.$$

Theorem 2   L'Hôpital's Rule (weak form)
Suppose that the functions f and g are differentiable at $x = a$, that

$$f(a) = 0 = g(a),$$

and that $g'(a) \neq 0$. Then

$$\lim_{x \to a} \frac{f(x)}{g(x)} = \frac{f'(a)}{g'(a)}. \tag{4}$$

Proof   We begin with the right-hand side of Eq.(4) and work toward the left-hand side:

$$\frac{f'(a)}{g'(a)} = \frac{\displaystyle\lim_{x \to a} \frac{f(x) - f(a)}{x - a}}{\displaystyle\lim_{x \to a} \frac{g(x) - g(a)}{x - a}} \qquad \text{(the definition of the derivative)}$$

$$= \lim_{x \to a} \frac{\dfrac{f(x) - f(a)}{x - a}}{\dfrac{g(x) - g(a)}{x - a}} \qquad \text{(the quotient law of limits)}$$

$$= \lim_{x \to a} \frac{f(x) - f(a)}{g(x) - g(a)} \qquad \text{(algebraic simplification)}$$

$$= \lim_{x \to a} \frac{f(x)}{g(x)} \qquad [\text{because } f(a) = 0 = g(a)]. \qquad \square$$

Appendix G includes a proof of the strong form of l'Hôpital's rule stated in Theorem 1.

8.3 Problems

Find the limits in Problems 1 through 48.

1. $\displaystyle\lim_{x \to 1} \frac{x - 1}{x^2 - 1}$

2. $\displaystyle\lim_{x \to \infty} \frac{3x - 4}{2x - 5}$

3. $\displaystyle\lim_{x \to \infty} \frac{2x^2 - 1}{5x^2 + 3x}$

4. $\displaystyle\lim_{x \to 0} \frac{e^{3x} - 1}{x}$

5. $\displaystyle\lim_{x \to 0} \frac{\sin x^2}{x}$

6. $\displaystyle\lim_{x \to 0^+} \frac{1 - \cos \sqrt{x}}{x}$

7. $\displaystyle\lim_{x \to 1} \frac{x - 1}{\sin x}$

8. $\displaystyle\lim_{x \to 0} \frac{1 - \cos x}{x^3}$

9. $\displaystyle\lim_{x \to 0} \frac{e^x - x - 1}{x^2}$

10. $\displaystyle\lim_{z \to \pi/2} \frac{1 + \cos 2z}{1 - \sin 2z}$

As $x \to 0$, $\cos x \to 1$, and so $\ln \cos x \to 0$; we are now dealing with the indeterminate form $0/0$. Hence two applications of l'Hôpital's rule yield

$$\lim_{x \to 0} \ln y = \lim_{x \to 0} \frac{\ln \cos x}{x^2} = \lim_{x \to 0} \frac{-\tan x}{2x} \qquad (0/0 \text{ form})$$

$$= \lim_{x \to 0} \frac{-\sec^2 x}{2} = -\frac{1}{2}.$$

Consequently,

$$\lim_{x \to 0} (\cos x)^{1/x^2} = e^{-1/2} = \frac{1}{\sqrt{e}}.$$

EXAMPLE 5 Find $\lim_{x \to 0^+} x^{\tan x}$.

Solution This has the indeterminate form 0^0. If $y = x^{\tan x}$, then

$$\ln y = (\tan x)(\ln x) = \frac{\ln x}{\cot x}.$$

Now we have the indeterminate form ∞/∞, and l'Hôpital's rule yields

$$\lim_{x \to 0^+} \ln y = \lim_{x \to 0^+} \frac{\ln x}{\cot x} = \lim_{x \to 0^+} \frac{\dfrac{1}{x}}{-\csc^2 x} = -\lim_{x \to 0^+} \frac{\sin^2 x}{x}$$

$$= -\lim_{x \to 0^+} \left(\frac{\sin x}{x} \right)(\sin x) = (-1) \cdot (0) = 0.$$

Therefore, $\lim_{x \to 0^+} x^{\tan x} = e^0 = 1$.

Although $a^0 = 1$ for any *nonzero* constant a, the form 0^0 is indeterminate—the limit is not necessarily 1 (see Problem 37). But the form 0^∞ is not indeterminate; its limit is zero. For example,

$$\lim_{x \to 0^+} x^{1/x} = 0.$$

8.4 Problems

Find the limits in Problems 1 through 34.

1. $\lim_{x \to 0} x \cot x$

2. $\lim_{x \to 0} \left(\frac{1}{x} - \cot x \right)$

3. $\lim_{x \to 0} \frac{1}{x} \ln\left(\frac{7x + 8}{4x + 8} \right)$

4. $\lim_{x \to 0^+} (\sin x)(\ln \sin x)$

5. $\lim_{x \to 0} x^2 \csc^2 2x$

6. $\lim_{x \to \infty} e^{-x} \ln x$

7. $\lim_{x \to \infty} x(e^{1/x} - 1)$

8. $\lim_{x \to 2} \left(\frac{1}{x - 2} - \frac{1}{\ln(x - 1)} \right)$

9. $\lim_{x \to 0^+} x \ln x$

10. $\lim_{x \to \pi/2} (\tan x)(\cos 3x)$

11. $\lim_{x \to \pi} (x - \pi) \csc x$

12. $\lim_{x \to \infty} e^{-x^2}(x - \sin x)$

13. $\lim_{x \to 0^+} \left(\frac{1}{\sqrt{x}} - \frac{1}{\sin x} \right)$

14. $\lim_{x \to 0} \left(\frac{1}{x} - \frac{1}{e^x - 1} \right)$

15. $\lim_{x \to 1^+} \left(\frac{x}{x^2 + x - 2} - \frac{1}{x - 1} \right)$

16. $\lim_{x \to \infty} (\sqrt{x + 1} - \sqrt{x})$

17. $\lim_{x \to 0} \left(\frac{1}{x} - \frac{1}{\ln(1 + x)} \right)$

18. $\lim_{x \to \infty} (\sqrt{x^2 + x} - \sqrt{x^2 - x})$

19. $\lim_{x \to \infty} (\sqrt[3]{x^3 + 2x + 5} - x)$

20. $\lim_{x \to 0^+} x^x$

21. $\lim_{x \to 0^+} x^{\sin x}$

22. $\lim_{x \to \infty} \left(1 + \frac{1}{x} \right)^x$

23. $\lim_{x \to \infty} (\ln x)^{1/x}$

24. $\lim\limits_{x \to \infty} \left(1 - \dfrac{1}{x^2}\right)^x$

25. $\lim\limits_{x \to 0} \left(\dfrac{\sin x}{x}\right)^{1/x^2}$

26. $\lim\limits_{x \to 0^+} (1 + 2x)^{1/(3x)}$

27. $\lim\limits_{x \to \infty} \left(\cos \dfrac{1}{x^2}\right)^{x^4}$

28. $\lim\limits_{x \to 0^+} (\sin x)^{\sec x}$

29. $\lim\limits_{x \to 0^+} (x + \sin x)^x$

30. $\lim\limits_{x \to \pi/2} (\tan x - \sec x)$

31. $\lim\limits_{x \to 1} x^{1/(1-x)}$

32. $\lim\limits_{x \to 1^+} (x - 1)^{\ln x}$

33. $\lim\limits_{x \to 2^+} \left(\dfrac{1}{\sqrt{x^2 - 4}} - \dfrac{1}{x - 2}\right)$

34. $\lim\limits_{x \to \infty} (\sqrt[5]{x^5 - 3x^4 + 17} - x)$

35. Use l'Hôpital's rule to establish these two limits:

(a) $\lim\limits_{h \to 0} (1 + hx)^{1/h} = e^x$; (b) $\lim\limits_{n \to \infty} \left(1 + \dfrac{x}{n}\right)^n = e^x$.

36. Sketch the graph of $y = x^{1/x}$, $x > 0$.

37. Let $f(x) = \exp(-1/x^2)$ and $g(x) = \cos x - 1$, so that $[f(x)]^{g(x)}$ is indeterminate of the form 0^0 as $x \to 0$. Show that $[f(x)]^{g(x)} \to \sqrt{e}$ as $x \to 0$.

38. Let n be a fixed positive integer and let $p(x)$ be the polynomial

$$p(x) = x^n + a_1 x^{n-1} + a_2 x^{n-2} + \cdots + a_{n-1} x + a_n;$$

the numbers $a_1, a_2, \ldots, a_n$ are fixed real numbers. Prove that

$$\lim\limits_{x \to \infty} ([p(x)]^{1/n} - x) = \dfrac{a_1}{n}.$$

39. As we shall see in Problem 50 of Section 9.6, the surface area of the ellipsoid obtained by revolving the ellipse

$$\dfrac{x^2}{a^2} + \dfrac{y^2}{b^2} = 1 \qquad (a > b > 0)$$

around the x-axis is

$$A = 2\pi ab\left[\dfrac{b}{a} + \dfrac{a}{c} \sin^{-1}\left(\dfrac{c}{a}\right)\right],$$

where $c = \sqrt{a^2 - b^2}$. Use l'Hôpital's rule to show that

$$\lim\limits_{b \to a} A = 4\pi a^2,$$

the surface area of a sphere of radius a.

40. Consider a long, thin rod that has heat diffusivity k and coincides with the x-axis. Suppose that at time $t = 0$ the temperature at x is $A/2\epsilon$ if $-\epsilon \leq x \leq \epsilon$ and is zero if $|x| > \epsilon$. Then it turns out that the temperature $T(x, t)$ of the rod at the point x at time $t > 0$ is given by

$$T(x, t) = \dfrac{A}{\epsilon \sqrt{4\pi kt}} \int_0^t \exp\left(-\dfrac{(x - u)^2}{4kt}\right) du.$$

Use l'Hôpital's rule to show that

$$\lim\limits_{\epsilon \to 0} T(x, t) = \dfrac{A}{\sqrt{4\pi kt}} \exp\left(-\dfrac{x^2}{4kt}\right).$$

This is the temperature resulting from an initial "hot spot" at the origin.

41. Explain why $\lim\limits_{x \to 0^+} (\ln x)^{1/x} \neq 0$.

42. Let α be a fixed real number. (a) Evaluate (in terms of α) the 0^0 indeterminate form

$$\lim\limits_{x \to 0} \left[\exp\left(-\dfrac{1}{x^2}\right)\right]^{\alpha x^2}.$$

(Note that l'Hôpital's rule is not needed.) Thus the indeterminate form 0^0 may have as its limit any positive real number. Explain why. (b) Can its limit be zero, negative, or infinite? Explain.

8.5
Hyperbolic Functions and Inverse Hyperbolic Functions

The **hyperbolic cosine** and the **hyperbolic sine** of the real number x are denoted by $\cosh x$ and $\sinh x$ and are defined to be

$$\cosh x = \dfrac{e^x + e^{-x}}{2}, \qquad \sinh x = \dfrac{e^x - e^{-x}}{2}. \tag{1}$$

These particular combinations of familiar exponentials are useful in certain applications of calculus and are also helpful in evaluating certain integrals. The other four hyperbolic functions—the hyperbolic tangent, cotangent, secant, and cosecant—are defined in terms of $\cosh x$ and $\sinh x$ by analogy with trigonometry:

$$\tanh x = \frac{\sinh x}{\cosh x} = \frac{e^x - e^{-x}}{e^x + e^{-x}},$$

$$\coth x = \frac{\cosh x}{\sinh x} = \frac{e^x + e^{-x}}{e^x - e^{-x}} \qquad (x \neq 0); \qquad (2)$$

$$\operatorname{sech} x = \frac{1}{\cosh x} = \frac{2}{e^x + e^{-x}},$$

$$\operatorname{csch} x = \frac{1}{\sinh x} = \frac{2}{e^x - e^{-x}} \qquad (x \neq 0). \qquad (3)$$

The trigonometric terminology and notation for these hyperbolic functions stem from the fact that these functions satisfy a list of identities that, apart from an occasional difference of sign, much resemble the familiar trigonometric identities:

$$\cosh^2 x - \sinh^2 x = 1; \qquad (4)$$

$$1 - \tanh^2 x = \operatorname{sech}^2 x; \qquad (5)$$

$$\coth^2 x - 1 = \operatorname{csch}^2 x; \qquad (6)$$

$$\sinh(x + y) = \sinh x \cosh y + \cosh x \sinh y; \qquad (7)$$

$$\cosh(x + y) = \cosh x \cosh y + \sinh x \sinh y; \qquad (8)$$

$$\sinh 2x = 2 \sinh x \cosh x; \qquad (9)$$

$$\cosh 2x = \cosh^2 x + \sinh^2 x; \qquad (10)$$

$$\cosh^2 x = \tfrac{1}{2}(\cosh 2x + 1); \qquad (11)$$

$$\sinh^2 x = \tfrac{1}{2}(\cosh 2x - 1). \qquad (12)$$

The identities in Eqs. (4), (7), and (8) follow directly from the definitions of $\cosh x$ and $\sinh x$. For example,

$$\cosh^2 x - \sinh^2 x = \tfrac{1}{4}(e^x + e^{-x})^2 - \tfrac{1}{4}(e^x - e^{-x})^2$$

$$= \tfrac{1}{4}(e^{2x} + 2 + e^{-2x}) - \tfrac{1}{4}(e^{2x} - 2 + e^{-2x}) = 1.$$

The other identities listed here may be derived from Eqs. (4), (7), and (8) in ways that parallel the derivations of the corresponding trigonometric identities.

The trigonometric functions are sometimes called the *circular* functions because the point $(\cos \theta, \sin \theta)$ lies on the circle $x^2 + y^2 = 1$ for all θ. Similarly, the identity in Eq. (4) tells us that the point $(\cosh \theta, \sinh \theta)$ lies on the hyperbola $x^2 - y^2 = 1$, and this is how the name *hyperbolic* function originated (Fig. 8.5.1).

The graphs of $y = \cosh x$ and $y = \sinh x$ are easy to construct. Add (for cosh) or subtract (for sinh) the ordinates of the graphs of $y = \tfrac{1}{2}e^x$ and $y = \tfrac{1}{2}e^{-x}$. The graphs of the other four hyperbolic functions can then be constructed by dividing ordinates. The graphs of all six are shown in Fig. 8.5.2.

These graphs show a striking difference between the hyperbolic functions and the ordinary trigonometric functions: None of the hyperbolic func-

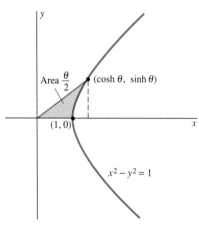

Fig. 8.5.1 Relation of the hyperbolic cosine and hyperbolic sine to the hyperbola $x^2 - y^2 = 1$

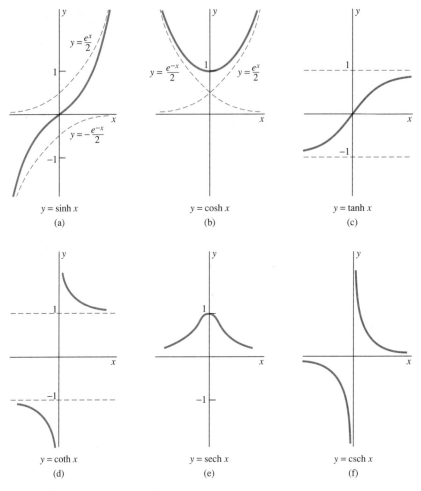

$y = \dfrac{e^x}{2}$

$y = \sinh x$

(a)

$y = \dfrac{e^{-x}}{2}$ $y = \dfrac{e^x}{2}$

$y = -\dfrac{e^{-x}}{2}$

$y = \cosh x$

(b)

$y = \tanh x$

(c)

$y = \coth x$

(d)

$y = \operatorname{sech} x$

(e)

$y = \operatorname{csch} x$

(f)

Fig. 8.5.2 Graphs of the six hyperbolic functions

tions is periodic. They do, however, have even-odd properties, as the circular functions do. The two functions cosh and sech are even, because

$$\cosh(-x) = \cosh x \quad \text{and} \quad \operatorname{sech}(-x) = \operatorname{sech} x$$

for all x. The other four hyperbolic functions are odd:

$$\sinh(-x) = -\sinh x, \quad \tanh(-x) = -\tanh x,$$

and so on.

DERIVATIVES AND INTEGRALS OF HYPERBOLIC FUNCTIONS

The formulas for the derivatives of the hyperbolic functions parallel those for the trigonometric functions, with occasional sign differences. For example,

$$D_x \cosh x = D_x\left(\tfrac{1}{2}e^x + \tfrac{1}{2}e^{-x}\right) = \tfrac{1}{2}e^x - \tfrac{1}{2}e^{-x} = \sinh x.$$

The chain rule then gives

$$D_x \cosh u = (\sinh u)\,\frac{du}{dx} \tag{13}$$

if u is a differentiable function of x. The other five differentiation formulas are

$$D_x \sinh u = (\cosh u)\frac{du}{dx}, \tag{14}$$

$$D_x \tanh u = (\operatorname{sech}^2 u)\frac{du}{dx}, \tag{15}$$

$$D_x \coth u = (-\operatorname{csch}^2 u)\frac{du}{dx}, \tag{16}$$

$$D_x \operatorname{sech} u = (-\operatorname{sech} u \tanh u)\frac{du}{dx}, \tag{17}$$

$$D_x \operatorname{csch} u = (-\operatorname{csch} u \coth u)\frac{du}{dx}. \tag{18}$$

Equation (14) is derived exactly as Eq. (13) is. Then Eqs. (15) through (18) follow from Eqs. (13) and (14) with the aid of the quotient rule and the identities in Eqs. (5) and (6).

As indicated in Example 1, the differentiation of hyperbolic functions using Eqs. (13) through (18) is very similar to the differentiation of trigonometric functions.

EXAMPLE 1   (a) $D_x \cosh 2x = 2 \sinh 2x$.
     (b) $D_x \sinh^2 x = 2 \sinh x \cosh x$.
     (c) $D_x (x \tanh x) = \tanh x + x \operatorname{sech}^2 x$.
     (d) $D_x \operatorname{sech}(x^2) = -2x \operatorname{sech}(x^2) \tanh(x^2)$.

The antiderivative versions of the differentiation formulas in Eqs. (13) through (18) are the following integral formulas:

$$\int \sinh u \, du = \cosh u + C, \tag{19}$$

$$\int \cosh u \, du = \sinh u + C, \tag{20}$$

$$\int \operatorname{sech}^2 u \, du = \tanh u + C, \tag{21}$$

$$\int \operatorname{csch}^2 u \, du = -\coth u + C, \tag{22}$$

$$\int \operatorname{sech} u \tanh u \, du = -\operatorname{sech} u + C, \tag{23}$$

$$\int \operatorname{csch} u \coth u \, du = -\operatorname{csch} u + C. \tag{24}$$

The integrals in Example 2 illustrate the fact that simple hyperbolic integrals may be treated in much the same way as simple trigonometric integrals.

EXAMPLE 2   (a) With $u = 3x$, we have

$$\int \cosh 3x \, dx = \int (\cosh u)(\tfrac{1}{3} \, du) = \tfrac{1}{3} \sinh u + C = \tfrac{1}{3} \sinh 3x + C.$$

(b) With $u = \sinh x$, we have

$$\int \sinh x \cosh x \, dx = \int u \, du = \tfrac{1}{2}u^2 + C = \tfrac{1}{2}\sinh^2 x + C.$$

(c) Using Eq. (12), we find that

$$\int \sinh^2 x \, dx = \int \tfrac{1}{2}(\cosh 2x - 1) \, dx = \tfrac{1}{4}\sinh 2x - \tfrac{1}{2}x + C.$$

(d) Finally, using Eq. (5), we see that

$$\int_0^1 \tanh^2 x \, dx = \int_0^1 (1 - \operatorname{sech}^2 x) \, dx = \left[x - \tanh x \right]_0^1 = 1 - \tanh 1$$

$$= 1 - \frac{e - e^{-1}}{e + e^{-1}} = \frac{2}{e^2 + 1} \approx 0.238406.$$

INVERSE HYPERBOLIC FUNCTIONS

Figure 8.5.2 shows that

❏ The functions $\sinh x$ and $\tanh x$ are increasing for all x;
❏ The functions $\coth x$ and $\operatorname{csch} x$ are decreasing and defined for all $x \neq 0$;
❏ The function $\cosh x$ is increasing on the half-line $x \geq 0$; and
❏ The function $\operatorname{sech} x$ is decreasing on the half-line $x \geq 0$.

It follows that each of the six hyperbolic functions can be "inverted" on the indicated domain where it is either increasing or decreasing. The resulting inverse hyperbolic functions and their domains of definition are listed in the next table and shown in Fig. 8.5.3.

Inverse hyperbolic function:	Defined for:
$\sinh^{-1} x$	All x
$\cosh^{-1} x$	$x \geq 1$
$\tanh^{-1} x$	$|x| < 1$
$\coth^{-1} x$	$|x| > 1$
$\operatorname{sech}^{-1} x$	$0 < x \leq 1$
$\operatorname{csch}^{-1} x$	$x \neq 0$

EXAMPLE 3 Find the numerical value of $\tanh^{-1}(\tfrac{1}{2})$.

Solution If $y = \tanh^{-1}(\tfrac{1}{2})$, then

$$\tanh y = \tfrac{1}{2};$$

$$\frac{e^y - e^{-y}}{e^y + e^{-y}} = \tfrac{1}{2} \qquad \text{[by Eq. (2)]};$$

$$2e^y - 2e^{-y} = e^y + e^{-y};$$

$$e^y = 3e^{-y}; \qquad e^{2y} = 3;$$

$$y = \tfrac{1}{2}\ln 3 \approx 0.5493.$$

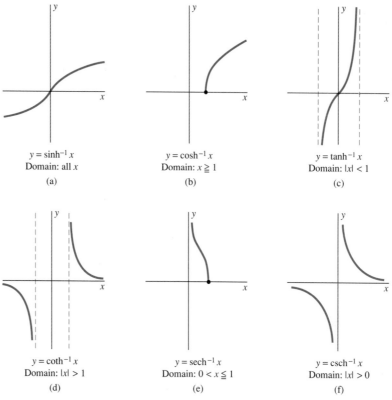

Fig. 8.5.3 The inverse hyperbolic functions

Scientific calculators ordinarily are used to find values of hyperbolic and inverse hyperbolic functions. Many calculators give values only of $\sinh^{-1}$, $\cosh^{-1}$, and $\tanh^{-1}$. Values of the other three inverse hyperbolic functions can then be found by using the identities

$$\text{sech}^{-1} x = \cosh^{-1}\left(\frac{1}{x}\right), \tag{25}$$

$$\text{csch}^{-1} x = \text{sech}^{-1}\left(\frac{1}{x}\right), \quad \text{and} \tag{26}$$

$$\coth^{-1} x = \tanh^{-1}\left(\frac{1}{x}\right). \tag{27}$$

For example,

$$\coth^{-1} 2 = \tanh^{-1}\left(\frac{1}{2}\right) \approx 0.5493.$$

DERIVATIVES OF INVERSE HYPERBOLIC FUNCTIONS

Here are the derivatives of the six inverse hyperbolic functions:

$$D_x \sinh^{-1} x = \frac{1}{\sqrt{x^2 + 1}}, \tag{28}$$

$$D_x \cosh^{-1} x = \frac{1}{\sqrt{x^2 - 1}}, \tag{29}$$

$$D_x \tanh^{-1} x = \frac{1}{1 - x^2}, \tag{30}$$

$$D_x \coth^{-1} x = \frac{1}{1 - x^2}, \tag{31}$$

$$D_x \operatorname{sech}^{-1} x = -\frac{1}{x\sqrt{1 - x^2}}, \tag{32}$$

$$D_x \operatorname{csch}^{-1} x = -\frac{1}{|x|\sqrt{1 - x^2}}. \tag{33}$$

We can derive these formulas by the standard method of finding the derivative of an inverse function, given the derivative of the function itself. The only requirement is that the inverse function is known in advance to be differentiable. For example, to differentiate $\tanh^{-1} x$, we begin with the inverse function relation

$$\tanh(\tanh^{-1} x) = x$$

and substitute $u = \tanh^{-1} x$. Then, because this equation is actually an identity,

$$D_x \tanh u = D_x x = 1,$$

so

$$(\operatorname{sech}^2 u) \frac{du}{dx} = 1.$$

Thus

$$D_x \tanh^{-1} x = \frac{du}{dx} = \frac{1}{\operatorname{sech}^2 u} = \frac{1}{1 - \tanh^2 u}$$

$$= \frac{1}{1 - \tanh^2(\tanh^{-1} x)} = \frac{1}{1 - x^2}.$$

This establishes Eq. (30). We can use similar methods to verify the formulas for the derivatives of the other five inverse hyperbolic functions.

The hyperbolic functions are defined in terms of the natural exponential function e^x, so it's no surprise that their inverses may be expressed in terms of $\ln x$. In fact,

$$\sinh^{-1} x = \ln(x + \sqrt{x^2 + 1}) \qquad \text{for all } x; \tag{34}$$

$$\cosh^{-1} x = \ln(x + \sqrt{x^2 - 1}) \qquad \text{for all } x \geq 1; \tag{35}$$

$$\tanh^{-1} x = \frac{1}{2} \ln\left(\frac{1 + x}{1 - x}\right) \qquad \text{for } |x| < 1; \tag{36}$$

$$\coth^{-1} x = \frac{1}{2} \ln\left(\frac{x + 1}{x - 1}\right) \qquad \text{for } |x| > 1; \tag{37}$$

$$\operatorname{sech}^{-1} x = \ln\left(\frac{1 + \sqrt{1 - x^2}}{x}\right) \qquad \text{if } 0 < x \leq 1; \tag{38}$$

$$\operatorname{csch}^{-1} x = \ln\left(\frac{1}{x} + \frac{\sqrt{1 + x^2}}{|x|}\right) \qquad \text{if } x \neq 0. \tag{39}$$

Each of these identities may be established by showing that each side has the same derivative and also that the two sides agree for at least one value of x in every interval of their respective domains. For example,

$$D_x \ln(x + \sqrt{x^2 + 1}) = \frac{1 + \dfrac{x}{\sqrt{x^2 + 1}}}{x + \sqrt{x^2 + 1}} = \frac{1}{\sqrt{x^2 + 1}} = D_x \sinh^{-1} x.$$

Thus

$$\sinh^{-1} x = \ln(x + \sqrt{x^2 + 1}) + C.$$

But $\sinh^{-1}(0) = 0 = \ln(0 + \sqrt{0 + 1})$. This implies that $C = 0$ and thus establishes Eq. (34). It is not quite so easy to show that $C = 0$ in the proofs of Eqs. (37) and (39); see Problems 64 and 65.

Equations (34) through (39) may be used to calculate the values of inverse hyperbolic functions. This is convenient if you own a calculator whose repertoire does not include the inverse hyperbolic functions or if you are programming in a language such as BASIC, most forms of which do not include these functions.

INTEGRALS INVOLVING INVERSE HYPERBOLIC FUNCTIONS

The principal applications of inverse hyperbolic functions are to the evaluation of algebraic integrals. The differentiation formulas in Eqs. (28) through (33) may, in the usual way, be written as the following integral formulas:

$$\int \frac{du}{\sqrt{u^2 + 1}} = \sinh^{-1} u + C, \tag{40}$$

$$\int \frac{du}{\sqrt{u^2 - 1}} = \cosh^{-1} u + C, \tag{41}$$

$$\int \frac{du}{1 - u^2} = \begin{cases} \tanh^{-1} u + C & \text{if } |u| < 1; \tag{42a} \\ \coth^{-1} u + C & \text{if } |u| > 1; \tag{42b} \end{cases}$$

$$= \frac{1}{2} \ln \left| \frac{1 + u}{1 - u} \right| + C, \tag{42c}$$

$$\int \frac{du}{u\sqrt{1 - u^2}} = -\operatorname{sech}^{-1} |u| + C, \tag{43}$$

$$\int \frac{du}{u\sqrt{1 + u^2}} = -\operatorname{csch}^{-1} |u| + C. \tag{44}$$

The distinction between the two cases $|u| < 1$ and $|u| > 1$ in Eq. (42) results from the fact that the inverse hyperbolic tangent is defined for $|x| < 1$, whereas the inverse hyperbolic cotangent is defined for $|x| > 1$.

EXAMPLE 4 The substitution $u = 2x$, $dx = \frac{1}{2} du$ yields

$$\int \frac{dx}{\sqrt{4x^2 + 1}} = \frac{1}{2} \int \frac{du}{\sqrt{u^2 + 1}} = \frac{1}{2} \sinh^{-1} 2x + C.$$

EXAMPLE 5

$$\int_0^{1/2} \frac{dx}{1 - x^2} = \left[\tanh^{-1} x \right]_0^{1/2}$$

$$= \frac{1}{2} \left[\ln \left| \frac{1 + x}{1 - x} \right| \right]_0^{1/2} = \frac{1}{2} \ln 3 \approx 0.5493.$$

EXAMPLE 6

$$\int_2^5 \frac{dx}{1 - x^2} = \left[\coth^{-1} x \right]_2^5 = \frac{1}{2} \left[\ln \left| \frac{1 + x}{1 - x} \right| \right]_2^5$$

$$= \frac{1}{2} \left[\ln \left(\frac{6}{4} \right) - \ln 3 \right] = -\frac{1}{2} \ln 2 \approx -0.3466.$$

8.5 Problems

Find the derivatives of the functions in Problems 1 through 14.

1. $f(x) = \cosh(3x - 2)$

2. $f(x) = \sinh \sqrt{x}$

3. $f(x) = x^2 \tanh\left(\frac{1}{x} \right)$

4. $f(x) = \text{sech } e^{2x}$

5. $f(x) = \coth^3 4x$

6. $f(x) = \ln \sinh 3x$

7. $f(x) = e^{\text{csch } x}$

8. $f(x) = \cosh \ln x$

9. $f(x) = \sin(\sinh x)$

10. $f(x) = \tan^{-1}(\tanh x)$

11. $f(x) = \sinh x^4$

12. $f(x) = \sinh^4 x$

13. $f(x) = \dfrac{1}{x + \tanh x}$

14. $f(x) = \cosh^2 x - \sinh^2 x$

Evaluate the integrals in Problems 15 through 28.

15. $\displaystyle\int x \sinh x^2 \, dx$

16. $\displaystyle\int \cosh^2 3u \, du$

17. $\displaystyle\int \tanh^2 3x \, dx$

18. $\displaystyle\int \frac{\text{sech}\sqrt{x} \, \tanh \sqrt{x}}{\sqrt{x}} \, dx$

19. $\displaystyle\int \sinh^2 2x \cosh 2x \, dx$

20. $\displaystyle\int \tanh 3x \, dx$

21. $\displaystyle\int \frac{\sinh x}{\cosh^3 x} \, dx$

22. $\displaystyle\int \sinh^4 x \, dx$

23. $\displaystyle\int \coth x \, \text{csch}^2 x \, dx$

24. $\displaystyle\int \text{sech } x \, dx$

25. $\displaystyle\int \frac{\sinh x}{1 + \cosh x} \, dx$

26. $\displaystyle\int \frac{\sinh \ln x}{x} \, dx$

27. $\displaystyle\int \frac{1}{(e^x + e^{-x})^2} \, dx$

28. $\displaystyle\int \frac{e^x + e^{-x}}{e^x - e^{-x}} \, dx$

Find the derivatives of the functions in Problems 29 through 38.

29. $f(x) = \sinh^{-1} 2x$

30. $f(x) = \cosh^{-1}(x^2 + 1)$

31. $f(x) = \tanh^{-1} \sqrt{x}$

32. $f(x) = \coth^{-1} \sqrt{x^2 + 1}$

33. $f(x) = \text{sech}^{-1}\left(\frac{1}{x} \right)$

34. $f(x) = \text{csch}^{-1} e^x$

35. $f(x) = (\sinh^{-1} x)^{3/2}$

36. $f(x) = \sinh^{-1}(\ln x)$

37. $f(x) = \ln(\tanh^{-1} x)$

38. $f(x) = \dfrac{1}{\tanh^{-1} 3x}$

Use inverse hyperbolic functions to evaluate the integrals in Problems 39 through 48.

39. $\displaystyle\int \frac{dx}{\sqrt{x^2 + 9}}$

40. $\displaystyle\int \frac{dy}{\sqrt{4y^2 - 9}}$

41. $\displaystyle\int_{1/2}^1 \frac{dx}{4 - x^2}$

42. $\displaystyle\int_5^{10} \frac{dx}{4 - x^2}$

43. $\displaystyle\int \frac{dx}{x\sqrt{4 - 9x^2}}$

44. $\displaystyle\int \frac{dx}{x\sqrt{x^2 + 25}}$

45. $\displaystyle\int \frac{e^x}{\sqrt{e^{2x} + 1}} \, dx$

46. $\displaystyle\int \frac{x}{\sqrt{x^4 - 1}} \, dx$

47. $\displaystyle\int \frac{1}{\sqrt{1 - e^{2x}}} \, dx$

48. $\displaystyle\int \frac{\cos x}{\sqrt{1 + \sin^2 x}} \, dx$

49. Apply the definitions in Eq. (1) to prove the identity in Eq. (7).

50. Derive the identities in Eqs. (5) and (6) from the identity in Eq. (4).

51. Deduce the identities in Eqs. (10) and (11) from the identity in Eq. (8).

52. Suppose that A and B are constants. Show that the function $x(t) = A \cosh kt + B \sinh kt$ is a solution of the differential equation

$$\frac{d^2x}{dt^2} = k^2 x(t).$$

53. Find the length of the curve $y = \cosh x$ over the interval $[0, a]$.

54. Find the volume of the solid obtained by revolving around the x-axis the area under $y = \sinh x$ from $x = 0$ to $x = \pi$.

55. Show that the area $A(\theta)$ of the shaded sector in Fig. 8.5.1 is $\theta/2$. This corresponds to the fact that the area of the sector of the unit circle between the positive x-axis and the radius to the point $(\cos \theta, \sin \theta)$ is $\theta/2$. [*Suggestion:* Note first that

$$A(\theta) = \tfrac{1}{2} \cosh \theta \, \sinh \theta - \int_1^{\cosh \theta} \sqrt{x^2 - 1} \, dx.$$

Then use the fundamental theorem of calculus to show that $A'(\theta) = \tfrac{1}{2}$ for all θ.]

56. Evaluate the following limits: (a) $\lim\limits_{x \to 0} \dfrac{\sinh x}{x}$;

(b) $\lim\limits_{x \to \infty} \tanh x$; (c) $\lim\limits_{x \to \infty} \dfrac{\cosh x}{e^x}$.

57. Use the method of Example 3 to find the numerical value of $\sinh^{-1} 1$.

58. Apply Eqs. (34) and (39) to verify the identity

$$\operatorname{csch}^{-1} x = \sinh^{-1}\left(\frac{1}{x}\right) \qquad \text{if } x \neq 0.$$

59. Establish the formula for $D_x \sinh^{-1} x$ in Eq. (28).

60. Establish the formula for $D_x \operatorname{sech}^{-1} x$ in Eq. (32).

61. Prove Eq. (36) by differentiating both sides.

62. Establish Eq. (34) by solving the equation

$$x = \sinh y = \frac{e^y - e^{-y}}{2}$$

for y in terms of x.

63. Establish Eq. (37) by solving the equation

$$x = \coth y = \frac{e^y + e^{-y}}{e^y - e^{-y}}$$

for y in terms of x.

64. (a) Differentiate both sides of Eq. (37) to show that they differ by a constant C. (b) Then prove that $C = 0$ by using the definition of $\coth x$ to show that $\coth^{-1} 2 = \tfrac{1}{2} \ln 3$.

65. (a) Differentiate both sides of Eq. (39) to show that they differ by a constant C. (b) Then prove that $C = 0$ by using the definition of $\operatorname{csch} x$ to show that $\operatorname{csch}^{-1} 1 = \ln(1 + \sqrt{2})$.

8.5 Project

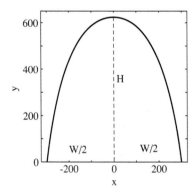

Fig. 8.5.4 The centroid curve of the Gateway Arch

Construction of the Gateway Arch in St. Louis, Missouri, was completed in October 1965. With a hollow stainless-steel design, its *centroid curve* (or centerline) is described very closely by the equation

$$y = 693.86 - (68.767)\cosh \frac{3x}{299}. \tag{1}$$

Here y denotes height above the ground (in feet) and x denotes horizontal distance (in feet) from the arch's vertical axis of symmetry. You might begin by plotting Eq. (1) to verify that the centroid curve has the appearance of Fig. 8.5.4. Problems 1 through 5 involve mathematical aspects of the arch and its centroid curve.

1. Deduce from Eq. (1) that the height H of the centroid curve's highest point is approximately 625 ft and that the width W of the base of the arch is slightly less than 600 ft.

2. Find the arc length of the centroid curve of the Gateway Arch. You may integrate numerically as in the Projects of Sections 5.4 and 5.9.

The arch is a hollow tube with varying cross section. Each cross section normal to the centroid curve is an equilateral triangle. The area (in square feet)

The Gateway Arch, St. Louis, Missouri.

of the cross section that intersects the centroid curve at the point (x, y) is

$$A = 1262.67 - (1.8198)y. \tag{2}$$

3. Show that the edge length l of the arch's triangular cross section varies from about 54 ft at its base to about 17 ft at its top.

4. The perimeter of a triangular cross section of edge length l is $P = 3l$. Explain nonrigorously the formula

$$S = \int_*^{**} P\,ds$$

that gives the surface area S of a tube in terms of its cross-sectional perimeter P and arc length s along its centroid curve. Then apply this formula appropriately to approximate numerically the (outer) surface area of the Gateway Arch. You might begin by expressing the perimeter as a function $P(x)$ and writing

$$ds = \frac{ds}{dx}\,dx = \sqrt{1 + \left(\frac{dy}{dx}\right)^2}\,dx.$$

5. Explain nonrigorously the formula

$$V = \int_*^{**} A\,ds$$

that gives the volume V of a tube in terms of its cross-sectional area A and the arc length s along its centroid curve. Then apply this formula appropriately to find the total volume enclosed by the outer surface of the Gateway Arch.

Chapter 8 Review: DEFINITIONS AND FORMULAS

Use the following list as a guide to concepts that you may need to review.

1. The definitions of the six inverse trigonometric functions

2. The derivatives of the inverse trigonometric functions

3. The integral formulas corresponding to the derivatives of the inverse sine, inverse tangent, and inverse secant functions

4. L'Hôpital's rule and the indeterminate forms $0/0$, ∞/∞, $0 \cdot \infty$, $\infty - \infty$, 0^0, ∞^0, and 1^∞

5. The definitions and derivatives of the hyperbolic functions

6. The definitions and derivatives of the inverse hyperbolic functions

7. The use of inverse hyperbolic functions to evaluate certain algebraic integrals

Chapter 8 Miscellaneous Problems

Differentiate the functions in Problems 1 through 20.

1. $f(x) = \sin^{-1} 3x$

2. $f(x) = \tan^{-1} 7x$

3. $g(t) = \sec^{-1} t^2$

4. $g(t) = \tan^{-1} e^t$

5. $f(x) = \sin^{-1}(\cos x)$

6. $f(x) = \sinh^{-1} 2x$

7. $g(t) = \cosh^{-1} 10t$

8. $h(u) = \tanh^{-1}\left(\dfrac{1}{u}\right)$

9. $f(x) = \sin^{-1}\left(\dfrac{1}{x^2}\right)$

10. $f(x) = \tan^{-1}\left(\dfrac{1}{x}\right)$

11. $f(x) = \arcsin \sqrt{x}$

12. $f(x) = x \sec^{-1} x^2$

13. $f(x) = \tan^{-1}(1 + x^2)$

14. $f(x) = \sin^{-1} \sqrt{1 - x^2}$

15. $f(x) = e^x \sinh e^x$

16. $f(x) = \ln \cosh x$

17. $f(x) = \tanh^2 3x + \operatorname{sech}^2 3x$

18. $f(x) = \sinh^{-1}\sqrt{x^2 - 1}$

19. $f(x) = \cosh^{-1}\sqrt{x^2 + 1}$

20. $f(x) = \tanh^{-1}(1 - x^2)$

Evaluate the integrals in Problems 21 through 40.

21. $\displaystyle\int \frac{dx}{\sqrt{1 - 4x^2}}$

22. $\displaystyle\int \frac{dx}{1 + 4x^2}$

23. $\displaystyle\int \frac{dx}{\sqrt{4 - x^2}}$

24. $\displaystyle\int \frac{dx}{4 + x^2}$

25. $\displaystyle\int \frac{e^x}{\sqrt{1 - e^{2x}}}\, dx$

26. $\displaystyle\int \frac{x}{1 + x^4}\, dx$

27. $\displaystyle\int \frac{1}{\sqrt{9 - 4x^2}}\, dx$

28. $\displaystyle\int \frac{1}{9 + 4x^2}\, dx$

29. $\displaystyle\int \frac{x^2}{1 + x^6}\, dx$

30. $\displaystyle\int \frac{\cos x}{1 + \sin^2 x}\, dx$

31. $\displaystyle\int \frac{1}{x\sqrt{4x^2 - 1}}\, dx$

32. $\displaystyle\int \frac{1}{x\sqrt{x^4 - 1}}\, dx$

33. $\displaystyle\int \frac{1}{\sqrt{e^{2x} - 1}}\, dx$

34. $\displaystyle\int x^2 \cosh x^3 \, dx$

35. $\displaystyle\int \frac{\sinh\sqrt{x}}{\sqrt{x}}\, dx$

36. $\displaystyle\int \operatorname{sech}^2(3x - 2)\, dx$

37. $\displaystyle\int \frac{\arctan x}{1 + x^2}\, dx$

38. $\displaystyle\int \frac{1}{\sqrt{4x^2 - 1}}\, dx$

39. $\displaystyle\int \frac{1}{\sqrt{4x^2 + 9}}\, dx$

40. $\displaystyle\int \frac{x}{\sqrt{x^4 + 1}}\, dx$

Find the limits in Problems 41 through 55.

41. $\displaystyle\lim_{x \to 2} \frac{x - 2}{x^2 - 4}$

42. $\displaystyle\lim_{x \to 0} \frac{\sin 2x}{x}$

43. $\displaystyle\lim_{x \to \pi} \frac{1 + \cos x}{(x - \pi)^2}$

44. $\displaystyle\lim_{x \to 0} \frac{x - \sin x}{x^3}$

45. $\displaystyle\lim_{t \to 0} \frac{\arctan t - \sin t}{t^3}$

46. $\displaystyle\lim_{x \to \infty} \frac{\ln(\ln x)}{\ln x}$

47. $\displaystyle\lim_{x \to 0} (\cot x) \ln(1 + x)$

48. $\displaystyle\lim_{x \to 0^+} (e^{1/x} - 1) \tan x$

49. $\displaystyle\lim_{x \to 0} \left(\frac{1}{x^2} - \frac{1}{1 - \cos x} \right)$

50. $\displaystyle\lim_{x \to \infty} \left(\frac{x^2}{x + 2} - \frac{x^3}{x^2 + 3} \right)$

51. $\displaystyle\lim_{x \to \infty} (\sqrt{x^2 - x + 1} - \sqrt{x})$

52. $\displaystyle\lim_{x \to \infty} x^{1/x}$

53. $\displaystyle\lim_{x \to \infty} (e^{2x} - 2x)^{1/x}$

54. $\displaystyle\lim_{x \to \infty} [1 - \exp(-x^2)]^{1/x^2}$

55. $\displaystyle\lim_{x \to \infty} x\left[\left(1 + \frac{1}{x}\right)^x - e \right]$ [*Suggestion:* Let $u = 1/x$, and take the limit as $u \to 0^+$.]

56. According to Problem 51 of Section 9.6, the surface area of the ellipsoid obtained by revolving around the x-axis the ellipse with equation

$$\left(\frac{x}{a}\right)^2 + \left(\frac{y}{b}\right)^2 = 1 \qquad (0 < a < b)$$

is

$$A = 2\pi ab\left[\frac{b}{a} + \frac{a}{c}\ln\left(\frac{b + c}{a}\right) \right],$$

where $c = \sqrt{b^2 - a^2}$. Use l'Hôpital's rule to show that

$$\lim_{b \to a} A = 4\pi a^2,$$

the surface area of a sphere of radius a.

57. Find the volume generated by revolving around the y-axis the region under $y = \sqrt{1 - x^4}$ from $x = 0$ to $x = 1/\sqrt{2}$.

58. Find the volume generated by revolving around the y-axis the region under $y = 1/\sqrt{x^4 + 1}$ from $x = 0$ to $x = 1$.

59. Use Eqs. (35) through (38) of Section 8.5 to show that

(a) $\coth^{-1}x = \tanh^{-1}\left(\frac{1}{x}\right)$; (b) $\operatorname{sech}^{-1}x = \cosh^{-1}\left(\frac{1}{x}\right)$.

60. Show that $x''(t) = k^2 x(t)$ if

$$x(t) = A \cosh kt + B \sinh kt,$$

where A and B are constants. Determine A and B if (a) $x(0) = 1$, $x'(0) = 0$; (b) $x(0) = 0$, $x'(0) = 1$.

61. Use Newton's method to find the least positive solution of the equation $\cos x \cosh x = 1$. Begin by sketching the graphs of $y = \cos x$ and $y = \operatorname{sech} x$.

62. (a) Verify by differentiation that

$$\int \sec x \, dx = \sinh^{-1}(\tan x) + C.$$

(b) Show similarly that

$$\int \operatorname{sech} x \, dx = \tan^{-1}(\sinh x) + C.$$

Techniques of Integration

Leonhard Euler (1707–1783)

❏ The most prolific mathematician in all history was Leonhard Euler (pronounced "oiler"), who was born in 1707 in Basel, Switzerland, the home of the Bernoulli family of mathematicians. His father preferred a theological career for his son, but young Euler learned mathematics from John Bernoulli and thereby found his true vocation. During his lifetime Euler published more than 500 books and papers. His work continued unabated even after he had lost his sight in 1766. Upon his death in 1783, he left behind more than 300 additional manuscripts whose publication continued in a steady flow for another half century. His collected works fill approximately 75 substantial volumes.

❏ No mathematician of the past more directly affects the modern student of mathematics, because it was largely Euler who shaped the notation and terminology still used today in teaching high school algebra and trigonometry as well as calculus. His *Introductio in Analysin Infinitorum* (Introduction to Infinitesimal Analysis) is the earliest mathematics textbook whose exposition would (in translation from the original Latin) be accessible to a modern student. Here are some now-familiar notations whose use was popularized and standardized by Euler:

e for the base of natural logarithms;

a, b, c for the sides of the triangle *ABC;*

i for the square root of -1;

Σ for the summation symbol;

$f(x)$ for functional notation;

π for the area of the unit circle;

and the trigonometric abbreviations sin, cos, tang, cot, sec, and cosec, which are close to their current forms. It was Euler's *Introductio* that once and for all based calculus squarely on the function concept. His 1755 and 1768 calculus treatises provide the original source for much of the content and methods of modern calculus courses and texts.

❏ Euler originally discovered so many of the standard formulas and identities of mathematics that is customary to attribute a formula to the first mathematician *after* Euler to *re*discover it. But the identity $e^{ix} = \cos x + i \sin x$ relating the exponential and trigonometric functions is still known as Euler's formula. Substitution of $x = \pi$ yields the relation $e^{i\pi} + 1 = 0$, which links five of the most important constants in mathematics.

❏ The photograph—part of a page from Chapter VII of the *Introductio*—shows the first appearance in public print of the number $e \approx 2.71828$. Immediately following its definition as the sum of the infinite series $\Sigma(1/n!)$, Euler gives the numerical value of e accurate to 23 decimal places.

90 DE QUANTITATUM EXPONENTIALIUM

Lib. I. (116) inventam, $a = 1 + \frac{1}{1} + \frac{1}{1.2} + \frac{1}{1.2.3} + \frac{1}{1.2.3.4} + \&c.$, qui termini, si in fractiones decimales convertantur atque actu addantur, præbebunt hunc valorem pro $a =$ 2,718281828459045235360028, cujus ultima adhuc nota veritati est consentanea. Quod si jam ex hac basi Logarithmi construantur, ii vocari solent Logarithmi *naturales* seu *hyperbolici*, quoniam quadratura hyperbolæ per istiusmodi Logarithmos exprimi potest. Ponamus autem brevitatis gratia pro numero hoc 2,718281828459 &c. constanter litteram e, quæ ergo denotabit basin Logarithmorum naturalium seu hyperbolicorum, cui respondet valor litteræ $k = 1$; sive hæc littera e quoque exprimet summam hujus Seriei $1 + \frac{1}{1} + \frac{1}{1.2} + \frac{1}{1.2.3} + \frac{1}{1.2.3.4} + \&c.$ in infinitum.

9.1
Introduction

We have seen in the past three chapters that many geometric and physical quantities can be expressed as definite integrals. The fundamental theorem of calculus reduces the problem of calculating the definite integral

$$\int_a^b f(x)\ dx$$

to that of finding an antiderivative $G(x)$ of $f(x)$. Once this is accomplished, then

$$\int_a^b f(x)\ dx = \left[G(x) \right]_a^b = G(b) - G(a).$$

But as yet we have relied largely on trial-and-error methods for finding the required antiderivative $G(x)$. In some cases a knowledge of elementary derivative formulas, perhaps in combination with a simple substitution, allows us to integrate a given function. This approach can, however, be inefficient and time-consuming, especially in view of the following surprising fact: Some simple-looking integrals, such as

$$\int e^{-x^2}\ dx, \quad \int \frac{\sin x}{x}\ dx, \quad \text{and} \quad \int \sqrt{1 + x^4}\ dx,$$

cannot be evaluated in terms of finite combinations of the familiar algebraic and elementary transcendental functions. For example, the antiderivative

$$H(x) = \int_0^x e^{-t^2}\ dt$$

of e^{-x^2} has no finite expression in terms of elementary functions. Any attempt to find such an expression will, therefore, inevitably be unsuccessful.

The presence of such integrals indicates that we cannot hope to reduce integration to a routine process such as differentiation. In fact, finding antiderivatives is an art that depends on experience and practice. Nevertheless, there are a number of techniques whose systematic use can substantially reduce our dependence on chance and intuition alone. This chapter deals with some of these systematic techniques of integration.

9.2
Integral Tables and Simple Substitutions

Integration would be a simple matter if we had a list of integral formulas, an *integral table,* in which we could locate any integral that we needed to evaluate. But the diversity of integrals that we encounter is too great for such an all-inclusive integral table to be practical. It is more sensible to print or memorize a short table of integrals of the sort seen frequently and to learn techniques by which the range of applicability of this short table can be extended. We begin with the list of integrals in Fig. 9.2.1, which are familiar from earlier chapters. Each formula is equivalent to one of the basic derivative formulas.

A table of 113 integral formulas appears on the inside back cover of this book. Even more extensive integral tables are readily available. For example, the volume of *Standard Mathematical Tables and Formulas,* edited by William H. Beyer and published by the CRC Press, Inc. (Boca Raton, Florida), contains over 700 integral formulas. But even such a lengthy table can be expected to include only a small fraction of the integrals we may need to

$$\int u^n \, du = \frac{u^{n+1}}{n+1} + C \quad [n \neq -1] \quad (1)$$

$$\int \frac{du}{u} = \ln|u| + C \quad (2)$$

$$\int e^u \, du = e^u + C \quad (3)$$

$$\int \cos u \, du = \sin u + C \quad (4)$$

$$\int \sin u \, du = -\cos u + C \quad (5)$$

$$\int \sec^2 u \, du = \tan u + C \quad (6)$$

$$\int \csc^2 u \, du = -\cot u + C \quad (7)$$

$$\int \sec u \tan u \, du = \sec u + C \quad (8)$$

$$\int \csc u \cot u \, du = -\csc u + C \quad (9)$$

$$\int \frac{du}{\sqrt{1 - u^2}} = \sin^{-1} u + C \quad (10)$$

$$\int \frac{du}{1 + u^2} = \tan^{-1} u + C \quad (11)$$

$$\int \frac{du}{u\sqrt{u^2 - 1}} = \sec^{-1}|u| + C \quad (12)$$

Fig. 9.2.1 A short table of integrals

evaluate. Thus it is necessary to learn techniques for deriving new formulas and for transforming a given integral either into one that's already familiar or into one that appears in an accessible table.

The principal such technique is the *method of substitution,* which we first considered in Section 5.7. Recall that if

$$\int f(u) \, du = F(u) + C,$$

then

$$\int f(g(x))g'(x) \, dx = F(g(x)) + C.$$

Thus the substitution

$$u = g(x), \qquad du = g'(x) \, dx$$

transforms the integral $\int f(g(x)) \cdot g'(x) \, dx$ into the simpler integral $\int f(u) \, du$. The key to making this simplification lies in spotting the composition $f(g(x))$ in the given integrand. For this integrand to be converted into a function of u alone, the remaining factor must be a constant multiple of the derivative $g'(x)$ of the "inside function" $g(x)$. In this case we replace $f(g(x))$ by the simpler $f(u)$ and $g'(x) \, dx$ by the simpler du. Chapters 6 through 8 contain numerous illustrations of this method of substitution, and the problems at the end of this section provide an opportunity to review it.

EXAMPLE 1 Find $\int \frac{1}{x}(1 + \ln x)^5 \, dx.$

Solution We need to spot *both* the inner function $g(x)$ and its derivative $g'(x)$. If we choose $g(x) = 1 + \ln x$, then $g'(x) = 1/x$. Hence the given integral is of the form discussed above with $f(u) = u^5$, $u = 1 + \ln x$, and $du = dx/x$. Therefore,

$$\int \frac{1}{x}(1 + \ln x)^5 \, dx = \int u^5 \, du = \frac{1}{6}u^6 + C = \frac{1}{6}(1 + \ln x)^6 + C.$$

EXAMPLE 2 Find $\int \frac{x}{1 + x^4} \, dx.$

Solution Here it is not so clear what the inside function is. But, looking at the integral formula in Eq. (11) (Fig. 9.2.1), we shall try the substitution $u = x^2$, $du = 2x \, dx$. We take advantage of the factor $x \, dx = \frac{1}{2} du$ that is available in the integrand and compute as follows:

$$\int \frac{x}{1 + x^4} \, dx = \frac{1}{2} \int \frac{du}{1 + u^2} = \frac{1}{2} \tan^{-1} u + C = \frac{1}{2} \tan^{-1} x^2 + C.$$

Note that the substitution $u = x^2$ would have been of no use had the integrand been either $1/(1 + x^4)$ or $x^2/(1 + x^4)$.

Example 2 illustrates how to make a substitution that converts a given integral into a familiar one. Often an integral that does not appear in any integral table can be transformed into one that does by using the techniques of this chapter. In Example 3 we employ an appropriate substitution to "reconcile" the given integral with the standard integral formula

$$\int \frac{u^2}{\sqrt{a^2 - u^2}} \, du = \frac{a^2}{2} \sin^{-1}\left(\frac{u}{a}\right) - \frac{u}{2}\sqrt{a^2 - u^2} + C, \qquad (13)$$

which is Formula (56) (inside the back cover).

EXAMPLE 3 Find $\displaystyle\int \frac{x^2}{\sqrt{25 - 16x^2}} \, dx.$

Solution So that $25 - 16x^2$ will be equal to $a^2 - u^2$ in Eq. (13), we take $a = 5$ and $u = 4x$. Then $du = 4 \, dx$, and so $dx = \frac{1}{4} \, du$. This gives

$$\int \frac{x^2}{\sqrt{25 - 16x^2}} \, dx = \int \frac{(\frac{1}{4}u)^2}{\sqrt{25 - u^2}} \cdot \frac{1}{4} \, du = \frac{1}{64} \int \frac{u^2}{\sqrt{25 - u^2}} \, du$$

$$= \frac{1}{64}\left[\frac{25}{2} \sin^{-1}\left(\frac{u}{5}\right) - \frac{u}{2}\sqrt{25 - u^2}\right] + C$$

$$= \frac{25}{128} \sin^{-1}\left(\frac{4x}{5}\right) - \frac{x}{32}\sqrt{25 - 16x^2} + C.$$

In Section 9.6 we will see how to derive integral formulas such as that in Eq. (13).

9.2 Problems

Evaluate the integrals in Problems 1 through 30.

1. $\displaystyle\int (2 - 3x)^4 \, dx$

2. $\displaystyle\int \frac{1}{(1 + 2x)^2} \, dx$

3. $\displaystyle\int x^2\sqrt{2x^3 - 4} \, dx$

4. $\displaystyle\int \frac{5t}{5 + 2t^2} \, dt$

5. $\displaystyle\int \frac{3x}{\sqrt[3]{2x^2 + 3}} \, dx$

6. $\displaystyle\int x \sec^2 x^2 \, dx$

7. $\displaystyle\int \frac{\cot \sqrt{y} \csc \sqrt{y}}{\sqrt{y}} \, dy$

8. $\displaystyle\int \sin \pi(2x + 1) \, dx$

9. $\displaystyle\int (1 + \sin \theta)^5 \cos \theta \, d\theta$

10. $\displaystyle\int \frac{\sin 2x}{4 + \cos 2x} \, dx$

11. $\displaystyle\int e^{-\cot x} \csc^2 x \, dx$

12. $\displaystyle\int \frac{e^{\sqrt{x+4}}}{\sqrt{x + 4}} \, dx$

13. $\displaystyle\int \frac{(\ln t)^{10}}{t} \, dt$

14. $\displaystyle\int \frac{t}{\sqrt{1 - 9t^2}} \, dt$

15. $\displaystyle\int \frac{1}{\sqrt{1 - 9t^2}} \, dt$

16. $\displaystyle\int \frac{e^{2x}}{1 + e^{2x}} \, dx$

17. $\displaystyle\int \frac{e^{2x}}{1 + e^{4x}} \, dx$

18. $\displaystyle\int \frac{e^{\arctan x}}{1 + x^2} \, dx$

19. $\displaystyle\int \frac{3x}{\sqrt{1 - x^4}} \, dx$

20. $\displaystyle\int \sin^3 2x \cos 2x \, dx$

21. $\displaystyle\int \tan^4 3x \sec^2 3x \, dx$

22. $\displaystyle\int \frac{1}{1 + 4t^2} \, dt$

23. $\displaystyle\int \frac{\cos \theta}{1 + \sin^2 \theta} \, d\theta$

24. $\displaystyle\int \frac{\sec^2 \theta}{1 + \tan \theta} \, d\theta$

25. $\displaystyle\int \frac{(1 + \sqrt{x})^4}{\sqrt{x}} \, dx$

26. $\displaystyle\int t^{-1/3}\sqrt{t^{2/3} - 1} \, dt$

27. $\displaystyle\int \frac{1}{(1 + t^2) \arctan t} \, dt$

28. $\displaystyle\int \frac{\sec 2x \tan 2x}{(1 + \sec 2x)^{3/2}} \, dx$

29. $\displaystyle\int \frac{1}{\sqrt{e^{2x} - 1}} \, dx$

30. $\displaystyle\int \frac{x}{\sqrt{\exp(2x^2) - 1}} \, dx$

In Problems 31 through 35, evaluate the given integral by making the indicated substitution.

31. $\displaystyle\int x^2\sqrt{x - 2} \, dx; \quad u = x - 2$

32. $\displaystyle\int \frac{x^2}{\sqrt{x + 3}} \, dx; \quad u = x + 3$

33. $\displaystyle\int \frac{x}{\sqrt{2x+3}}\,dx;\quad u = 2x + 3$

34. $\displaystyle\int x\sqrt[3]{x-1}\,dx;\quad u = x - 1$

35. $\displaystyle\int \frac{x}{\sqrt[3]{x+1}}\,dx;\quad u = x + 1$

In Problems 36 through 50, evaluate the given integral. First make a substitution that transforms it into a standard form. The standard forms with the given formula numbers are on the inside back cover of this book.

36. $\displaystyle\int \frac{1}{100 + 9x^2}\,dx;\quad$ Formula (17)

37. $\displaystyle\int \frac{1}{100 - 9x^2}\,dx;\quad$ Formula (18)

38. $\displaystyle\int \sqrt{9 - 4x^2}\,dx;\quad$ Formula (54)

39. $\displaystyle\int \sqrt{4 + 9x^2}\,dx;\quad$ Formula (44)

40. $\displaystyle\int \frac{1}{\sqrt{16x^2 + 9}}\,dx;\quad$ Formula (45)

41. $\displaystyle\int \frac{x^2}{\sqrt{16x^2 + 9}}\,dx;\quad$ Formula (49)

42. $\displaystyle\int \frac{x^2}{\sqrt{25 + 16x^2}}\,dx;\quad$ Formula (49)

43. $\displaystyle\int x^2\sqrt{25 - 16x^2}\,dx;\quad$ Formula (57)

44. $\displaystyle\int x\sqrt{4 - x^4}\,dx;\quad$ Formula (54)

45. $\displaystyle\int e^x\sqrt{9 + e^{2x}}\,dx;\quad$ Formula (44)

46. $\displaystyle\int \frac{\cos x}{(\sin^2 x)\sqrt{1 + \sin^2 x}}\,dx;\quad$ Formula (50)

47. $\displaystyle\int \frac{\sqrt{x^4 - 1}}{x}\,dx;\quad$ Formula (47)

48. $\displaystyle\int \frac{e^{3x}}{\sqrt{25 + 16e^{2x}}}\,dx;\quad$ Formula (49)

49. $\displaystyle\int \frac{(\ln x)^2}{x}\sqrt{1 + (\ln x)^2}\,dx;\quad$ Formula (48)

50. $\displaystyle\int x^8\sqrt{4x^6 - 1}\,dx;\quad$ Formula (48)

51. The substitution $u = x^2$, $x = \sqrt{u}$, $dx = du/(2\sqrt{u})$ appears to lead to this result:

$$\int_{-1}^{1} x^2\,dx = \tfrac{1}{2}\int_{1}^{1} \sqrt{u}\,du = 0.$$

Do you believe this result? If not, why not?

52. Use the fact that $x^2 + 4x + 5 = (x + 2)^2 + 1$ to evaluate

$$\int \frac{1}{x^2 + 4x + 5}\,dx.$$

53. Use the fact that $1 - (x - 1)^2 = 2x - x^2$ to evaluate

$$\int \frac{1}{\sqrt{2x - x^2}}\,dx.$$

9.2 Project

According to Formula (44) on the inside back cover of this book,

$$\int \sqrt{x^2 + 1}\,dx = G(x) + C, \tag{14}$$

where

$$G(x) = \frac{x}{2}\sqrt{x^2 + 1} + \frac{1}{2}\ln(x + \sqrt{x^2 + 1}). \tag{15}$$

According to Serge Lang's *First Course in Calculus* (5th ed., New York: Springer-Verlag, 1991, p. 376), this same indefinite integral is given by

$$\int \sqrt{x^2 + 1}\,dx = H(x) + C, \tag{16}$$

where

$$H(x) = \tfrac{1}{8}[(x + \sqrt{x^2 + 1})^2 + 4\ln(x + \sqrt{x^2 + 1}) - (x + \sqrt{x^2 + 1})^{-2}]. \tag{17}$$

Your mission in this project is to determine whether the functions $G(x)$ and $H(x)$ in Eqs. (15) and (17) are, in fact, both antiderivatives of $f(x) = \sqrt{x^2 + 1}$. We list possible ways to investigate the relationships among the functions $f(x)$, $G(x)$, and $H(x)$. You should explore several different approaches.

1. If you have a graphics calculator or a computer with a graphing utility, plot both $y = G(x)$ and $y = H(x)$. If the functions $G(x)$ and $H(x)$ are both antiderivatives of $f(x)$, how should their graphs be related?

2. If your calculator or computer can graph the *derivative* of a user-defined function, plot the graphs of $f(x)$ and the derivatives $G'(x)$ and $H'(x)$. Does the visual evidence convince you that $G'(x) = H'(x) = f(x)$?

3. Even if your calculator or computer cannot plot derivatives directly, you can still plot $f(x)$ and the quotients

$$\frac{G(x + h) - G(x - h)}{2h} \quad \text{and} \quad \frac{H(x + h) - H(x - h)}{2h},$$

which—with $h = 0.001$—should closely approximate the derivatives $G'(x)$ and $H'(x)$, respectively. (Why?)

4. With a calculator or computer that can approximate integrals numerically—either with an appropriate function key or by using programs like those in the Section 5.9 projects—you can determine whether

$$\int_a^b f(x)\,dx = G(b) - G(a) = H(b) - H(a), \tag{18}$$

as the fundamental theorem implies if $G' = H' = f$. It should be fairly convincing if you can verify Eq. (18) numerically with several different pairs of limits, such as $a = 1$, $b = 5$ and $a = 7$, $b = 11$.

5. Even without a graphics calculator, you can compute numerical values of $G(x)$, $H(x)$, $G'(x)$, $H'(x)$, and $f(x)$ for several selected values of x. Do the numerical results imply that $G(x) = H(x)$ or that $G'(x) = H'(x) = f(x)$?

6. With a computer algebra system, such as *Derive*, *Maple*, or *Mathematica*, you can calculate the derivatives $G'(x)$ and $H'(x)$ symbolically to determine whether or not both are equal to $f(x)$. Perhaps you can even do this the old-fashioned way—by using nothing but paper and pencil. And you might investigate in the same way whether or not $G(x) = H(x)$.

9.3
Trigonometric Integrals

Here we discuss the evaluation of certain integrals in which the integrand is either a power of a trigonometric function or the product of two such powers.

To evaluate the integrals

$$\int \sin^2 u\,du \quad \text{and} \quad \int \cos^2 u\,du$$

that appear in numerous applications, we use the **half-angle identities**

$$\sin^2 \theta = \tfrac{1}{2}(1 - \cos 2\theta), \tag{1}$$

$$\cos^2 \theta = \tfrac{1}{2}(1 + \cos 2\theta) \tag{2}$$

of Eqs. (11) and (10) in Appendix A.

EXAMPLE 1 Find $\int \sin^2 3x \, dx$.

Solution The identity in Eq. (1)—with $3x$ in place of x—yields

$$\int \sin^2 3x \, dx = \int \tfrac{1}{2}(1 - \cos 6x) \, dx$$

$$= \tfrac{1}{2}(x - \tfrac{1}{6} \sin 6x) + C = \tfrac{1}{12}(6x - \sin 6x) + C.$$

To integrate $\tan^2 x$ and $\cot^2 x$, we use the identities

$$1 + \tan^2 x = \sec^2 x, \qquad 1 + \cot^2 x = \csc^2 x. \tag{3}$$

The first of these follows from the fundamental identity $\sin^2 x + \cos^2 x = 1$ upon division of both sides by $\cos^2 x$. To obtain the second formula in (3), we divide both sides of the fundamental identity by $\sin^2 x$.

EXAMPLE 2 Compute the antiderivative $\int \cot^2 3x \, dx$.

Solution By using the second identity in (3), we obtain

$$\int \cot^2 3x \, dx = \int (\csc^2 3x - 1) \, dx$$

$$= \int (\csc^2 u - 1)(\tfrac{1}{3} du) \qquad (u = 3x)$$

$$= \tfrac{1}{3}(-\cot u - u) + C = -\tfrac{1}{3} \cot 3x - x + C.$$

INTEGRALS OF PRODUCTS OF SINES AND COSINES

The substitution $u = \sin x$, $du = \cos x \, dx$ gives

$$\int \sin^3 x \cos x \, dx = \int u^3 \, du = \tfrac{1}{4}u^4 + C = \tfrac{1}{4} \sin^4 x + C.$$

This substitution, or the similar substitution $u = \cos x$, $du = -\sin x \, dx$, can be used to evaluate an integral of the form

$$\int \sin^m x \cos^n x \, dx \tag{4}$$

in the first of the following two cases:

❏ Case 1: At least one of the two numbers m and n is an *odd positive integer*. If so, then the other may be any real number.

❏ Case 2: Both m and n are *nonnegative even integers*.

Suppose, for example, that $m = 2k + 1$ is an odd positive integer. Then we isolate one $\sin x$ factor and use the identity $\sin^2 x = 1 - \cos^2 x$ to express the remaining $\sin^{m-1} x$ factor in terms of $\cos x$, as follows:

$$\int \sin^m x \cos^n x \, dx = \int \sin^{m-1} x \cos^n x \sin x \, dx = \int (\sin^2 x)^k \cos^n x \sin x \, dx$$

$$= \int (1 - \cos^2 x)^k \cos^n x \sin x \, dx.$$

Now the substitution $u = \cos x$, $du = -\sin x \, dx$ yields

$$\int \sin^m x \cos^n x \, dx = -\int (1 - u^2)^k u^n \, du.$$

The exponent $k = (m - 1)/2$ is a nonnegative integer because m is an odd positive integer. Thus the factor $(1 - u^2)^k$ of the integrand is a polynomial in the variable u, and so its product with u^n is easy to integrate.

In essence, this method consists of peeling off one copy of $\sin x$ (if m is odd) and then converting the remaining sines into cosines. If n is odd, then we can split off one copy of $\cos x$ and convert the remaining cosines into sines.

EXAMPLE 3

(a) $\displaystyle \int \sin^3 x \cos^2 x \, dx = \int (1 - \cos^2 x) \cos^2 x \sin x \, dx$

$$= \int (u^4 - u^2) \, du \qquad (u = \cos x)$$

$$= \tfrac{1}{5} u^5 - \tfrac{1}{3} u^3 + C = \tfrac{1}{5} \cos^5 x - \tfrac{1}{3} \cos^3 x + C.$$

(b) $\displaystyle \int \cos^5 x \, dx = \int (1 - \sin^2 x)^2 \cos x \, dx$

$$= \int (1 - u^2)^2 \, du \qquad (u = \sin x)$$

$$= \int (1 - 2u^2 + u^4) \, du = u - \tfrac{2}{3} u^3 + \tfrac{1}{5} u^5 + C$$

$$= \sin x - \tfrac{2}{3} \sin^3 x + \tfrac{1}{5} \sin^5 x + C.$$

In Case 2 of the sine-cosine integral in Eq. (4), with both m and n nonnegative even integers, we use the half-angle formulas in Eqs. (1) and (2) to halve the even powers of $\sin x$ and $\cos x$. If we repeat this process with the resulting powers of $\cos 2x$ (if necessary), we get integrals involving odd powers, and we have seen how to handle these in Case 1.

EXAMPLE 4   Use of Eqs. (1) and (2) gives

$$\int \sin^2 x \cos^2 x \, dx = \int \tfrac{1}{2}(1 - \cos 2x) \tfrac{1}{2}(1 + \cos 2x) \, dx$$

$$= \tfrac{1}{4} \int (1 - \cos^2 2x) \, dx = \tfrac{1}{4} \int [1 - \tfrac{1}{2}(1 + \cos 4x)] \, dx$$

$$= \tfrac{1}{8} \int (1 - \cos 4x) \, dx = \tfrac{1}{8} x - \tfrac{1}{32} \sin 4x + C.$$

In the third step we have used Eq. (2) with $\theta = 2x$.

EXAMPLE 5 Here we apply Eq. (2), first with $\theta = 3x$ and then with $\theta = 6x$.

$$\int \cos^4 3x \, dx = \int \tfrac{1}{4}(1 + \cos 6x)^2 \, dx$$

$$= \tfrac{1}{4} \int (1 + 2 \cos 6x + \cos^2 6x) \, dx$$

$$= \tfrac{1}{4} \int (\tfrac{3}{2} + 2 \cos 6x + \tfrac{1}{2} \cos 12x) \, dx$$

$$= \tfrac{3}{8}x + \tfrac{1}{12} \sin 6x + \tfrac{1}{96} \sin 12x + C.$$

INTEGRALS OF PRODUCTS OF TANGENTS AND SECANTS

To integrate $\tan x$, the substitution

$$u = \cos x, \qquad du = -\sin x \, dx$$

gives

$$\int \tan x \, dx = \int \frac{\sin x}{\cos x} \, dx = -\int \frac{1}{u} \, du = -\ln|u| + C,$$

and thus

$$\int \tan x \, dx = -\ln|\cos x| + C = \ln|\sec x| + C. \qquad (5)$$

In Eq. (5) we used the fact that $|\sec x| = 1/|\cos x|$.

Similarly,

$$\int \cot x \, dx = \ln|\sin x| + C = -\ln|\csc x| + C. \qquad (6)$$

The first person to integrate $\sec x$ may well have spent much time doing so. Here is one of several methods. First we "prepare" the function for integration:

$$\sec x = \frac{1}{\cos x} = \frac{\cos x}{\cos^2 x} = \frac{\cos x}{1 - \sin^2 x}.$$

Now

$$\frac{1}{1 + z} + \frac{1}{1 - z} = \frac{2}{1 - z^2}.$$

Similarly, working backward, we have

$$\frac{2 \cos x}{1 - \sin^2 x} = \frac{\cos x}{1 + \sin x} + \frac{\cos x}{1 - \sin x}.$$

Therefore,

$$\int \sec x \, dx = \frac{1}{2} \int \left(\frac{\cos x}{1 + \sin x} + \frac{\cos x}{1 - \sin x} \right) dx$$

$$= \tfrac{1}{2} \left(\ln |1 + \sin x| - \ln |1 - \sin x| \right) + C.$$

It's customary to perform some algebraic simplifications of this last result:

$$\int \sec x \, dx = \frac{1}{2} \ln \left| \frac{1 + \sin x}{1 - \sin x} \right| + C = \frac{1}{2} \ln \left| \frac{(1 + \sin x)^2}{1 - \sin^2 x} \right| + C$$

$$= \ln \left| \frac{(1 + \sin x)^2}{\cos^2 x} \right|^{1/2} + C = \ln \left| \frac{1 + \sin x}{\cos x} \right| + C$$

$$= \ln |\sec x + \tan x| + C.$$

After we verify by differentiation that

$$\int \sec x \, dx = \ln |\sec x + \tan x| + C, \tag{7}$$

we could always "derive" this result by using an unmotivated trick:

$$\int \sec x \, dx = \int (\sec x) \frac{\tan x + \sec x}{\sec x + \tan x} \, dx$$

$$= \int \frac{\sec x \tan x + \sec^2 x}{\sec x + \tan x} \, dx = \ln |\sec x + \tan x| + C.$$

A similar technique yields

$$\int \csc x \, dx = -\ln |\csc x + \cot x| + C. \tag{8}$$

EXAMPLE 6   The substitution $u = \frac{1}{2} x$, $du = \frac{1}{2} dx$ gives

$$\int_0^{\pi/2} \sec \frac{x}{2} \, dx = 2 \int_0^{\pi/4} \sec u \, du$$

$$= 2 \left[\ln |\sec u + \tan u| \right]_0^{\pi/4} = 2 \ln(1 + \sqrt{2}) \approx 1.76275.$$

An integral of the form

$$\int \tan^m x \sec^n x \, dx \tag{9}$$

can be routinely evaluated in either of the following two cases:

❑ Case 1: m is an *odd positive integer.*
❑ Case 2: n is an *even positive integer.*

In Case 1, we split off the factor $\sec x \tan x$ to form, along with dx, the differential $\sec x \tan x \, dx$ of $\sec x$. We then use the identity $\tan^2 x = \sec^2 x - 1$ to convert the remaining even power of $\tan x$ into powers of $\sec x$. This prepares the integrand for the substitution $u = \sec x$.

EXAMPLE 7

$$\int \tan^3 x \sec^3 x \, dx = \int (\sec^2 x - 1) \sec^2 x \sec x \tan x \, dx$$

$$= \int (u^4 - u^2) \, du \qquad (u = \sec x)$$

$$= \tfrac{1}{5} u^5 - \tfrac{1}{3} u^3 + C = \tfrac{1}{5} \sec^5 x - \tfrac{1}{3} \sec^3 x + C.$$

To evaluate the integral in Eq. (9) for Case 2, we split off $\sec^2 x$ to form, along with dx, the differential of $\tan x$. We then use the identity $\sec^2 x = 1 + \tan^2 x$ to convert the remaining even power of $\sec x$ into powers of $\tan x$. This prepares the integrand for the substitution $u = \tan x$.

EXAMPLE 8

$$\int \sec^6 2x \, dx = \int (1 + \tan^2 2x)^2 \sec^2 2x \, dx$$

$$= \tfrac{1}{2} \int (1 + \tan^2 2x)^2 (2 \sec^2 2x) \, dx$$

$$= \tfrac{1}{2} \int (1 + u^2)^2 \, du \qquad (u = \tan 2x)$$

$$= \tfrac{1}{2} \int (1 + 2u^2 + u^4) \, du = \tfrac{1}{2} u + \tfrac{1}{3} u^3 + \tfrac{1}{10} u^5 + C$$

$$= \tfrac{1}{2} \tan 2x + \tfrac{1}{3} \tan^3 2x + \tfrac{1}{10} \tan^5 2x + C.$$

Similar methods are effective with integrals of the form

$$\int \csc^m x \cot^n x \, dx.$$

The method of Case 1 succeeds with the integral $\int \tan^n x \, dx$ only when n is an odd positive integer, but there is another approach that works equally well whether n is even *or* odd. We split off the factor $\tan^2 x$ and replace it by $\sec^2 x - 1$:

$$\int \tan^n x \, dx = \int (\tan^{n-2} x)(\sec^2 x - 1) \, dx$$

$$= \int \tan^{n-2} x \sec^2 x \, dx - \int \tan^{n-2} x \, dx.$$

We integrate what we can and find that

$$\int \tan^n x \, dx = \frac{\tan^{n-1} x}{n - 1} - \int \tan^{n-2} x \, dx. \tag{10}$$

Equation (10) is our first example of a **reduction formula**. Its use effectively reduces the original exponent from n to $n - 2$. If we apply Eq. (10) repeatedly, we eventually get either

$$\int \tan^2 x \, dx = \int (\sec^2 x - 1) \, dx = \tan x - x + C$$

or

$$\int \tan x \, dx = \int \frac{\sin x}{\cos x} \, dx = -\ln|\cos x| + C = \ln|\sec x| + C.$$

EXAMPLE 9 Two applications of Eq. (10) give

$$\int \tan^6 x \, dx = \tfrac{1}{5} \tan^5 x - \int \tan^4 x \, dx$$

$$= \tfrac{1}{5} \tan^5 x - \left(\tfrac{1}{3} \tan^3 x - \int \tan^2 x \, dx \right)$$

$$= \tfrac{1}{5} \tan^5 x - \tfrac{1}{3} \tan^3 x + \tan x - x + C.$$

Finally, in the case of an integral involving an unusual mixture of trigonometric functions—tangents and cosecants, for example—an effective procedure is to express the integrand wholly in terms of sines and cosines. Simplification may then yield an expression that's easy to integrate.

9.3 Problems

Evaluate the integrals in Problems 1 through 45.

1. $\displaystyle\int \sin^2 2x \, dx$

2. $\displaystyle\int \cos^2 5x \, dx$

3. $\displaystyle\int \sec^2 \frac{x}{2} \, dx$

4. $\displaystyle\int \tan^2 \frac{x}{2} \, dx$

5. $\displaystyle\int \tan 3x \, dx$

6. $\displaystyle\int \cot 4x \, dx$

7. $\displaystyle\int \sec 3x \, dx$

8. $\displaystyle\int \csc 2x \, dx$

9. $\displaystyle\int \frac{dx}{\csc^2 x}$

10. $\displaystyle\int \sin^2 x \cot^2 x \, dx$

11. $\displaystyle\int \sin^3 x \, dx$

12. $\displaystyle\int \sin^4 x \, dx$

13. $\displaystyle\int \sin^2 \theta \cos^3 \theta \, d\theta$

14. $\displaystyle\int \sin^3 t \cos^3 t \, dt$

15. $\displaystyle\int \cos^5 x \, dx$

16. $\displaystyle\int \frac{\sin t}{\cos^3 t} \, dt$

17. $\displaystyle\int \frac{\sin^3 x}{\sqrt{\cos x}} \, dx$

18. $\displaystyle\int \sin^3 3\phi \cos^4 3\phi \, d\phi$

19. $\displaystyle\int \sin^5 2z \cos^2 2z \, dz$

20. $\displaystyle\int \sin^{3/2} x \cos^3 x \, dx$

21. $\displaystyle\int \frac{\sin^3 4x}{\cos^2 4x} \, dx$

22. $\displaystyle\int \cos^6 4\theta \, d\theta$

23. $\displaystyle\int \sec^4 t \, dt$

24. $\displaystyle\int \tan^3 x \, dx$

25. $\displaystyle\int \cot^3 2x \, dx$

26. $\displaystyle\int \tan \theta \sec^4 \theta \, d\theta$

27. $\displaystyle\int \tan^5 2x \sec^2 2x \, dx$

28. $\displaystyle\int \cot^3 x \csc^2 x \, dx$

29. $\displaystyle\int \csc^6 2t \, dt$

30. $\displaystyle\int \frac{\sec^4 t}{\tan^2 t} \, dt$

31. $\displaystyle\int \frac{\tan^3 \theta}{\sec^4 \theta} \, d\theta$

32. $\displaystyle\int \frac{\cot^3 x}{\csc^2 x} \, dx$

33. $\displaystyle\int \frac{\tan^3 t}{\sqrt{\sec t}} \, dt$

34. $\displaystyle\int \frac{1}{\cos^4 2x} \, dx$

35. $\displaystyle\int \frac{\cot \theta}{\csc^3 \theta} \, d\theta$

36. $\displaystyle\int \sin^2 3\alpha \cos^2 3\alpha \, d\alpha$

37. $\displaystyle\int \cos^3 5t \, dt$

38. $\displaystyle\int \tan^4 x \, dx$

39. $\displaystyle\int \cot^4 3t \, dt$

40. $\displaystyle\int \tan^2 2t \sec^4 2t \, dt$

41. $\displaystyle\int \sin^5 2t \cos^{3/2} 2t \, dt$

42. $\displaystyle\int \cot^3 \xi \csc^{3/2} \xi \, d\xi$

43. $\displaystyle\int \frac{\tan x + \sin x}{\sec x} \, dx$

44. $\displaystyle\int \frac{\cot x + \csc x}{\sin x} \, dx$

45. $\displaystyle\int \frac{\cot x + \csc^2 x}{1 - \cos^2 x} \, dx$

46. Determine a reduction formula, one analogous to that in Eq. (10), for $\int \cot^n x \, dx$.

47. Find $\int \tan x \sec^4 x \, dx$ in two different ways. Then show that your two results are equivalent.

48. Find $\int \cot^3 x \, dx$ in two different ways. Then show that your two results are equivalent.

Problems 49 through 52 are applications of the trigonometric identities

$$\sin A \sin B = \tfrac{1}{2}[\cos(A - B) - \cos(A + B)],$$

$$\sin A \cos B = \tfrac{1}{2}[\sin(A - B) + \sin(A + B)],$$

$$\cos A \cos B = \tfrac{1}{2}[\cos(A - B) + \cos(A + B)].$$

49. Find $\displaystyle\int \sin 3x \cos 5x \, dx$.

50. Find $\displaystyle\int \sin 2x \sin 4x \, dx$.

51. Find $\displaystyle\int \cos x \cos 4x \, dx$.

52. Suppose that m and n are positive integers with $m \neq n$. Show that

(a) $\displaystyle\int_0^{2\pi} \sin mx \sin nx \, dx = 0$;

(b) $\displaystyle\int_0^{2\pi} \cos mx \sin nx \, dx = 0$;

(c) $\displaystyle\int_0^{2\pi} \cos mx \cos nx \, dx = 0$.

53. Substitute $\sec x \csc x = (\sec^2 x)/(\tan x)$ to derive the formula $\int \sec x \csc x \, dx = \ln|\tan x| + C$.

54. Show that

$$\csc x = \frac{1}{2 \sin(x/2) \cos(x/2)},$$

and then apply the result of Problem 53 to derive the formula $\int \csc x \, dx = \ln|\tan(x/2)| + C$.

55. Substitute $x = (\pi/2) - u$ into the integral formula of Problem 54 to show that

$$\int \sec x \, dx = \ln\left|\cot\left(\frac{\pi}{4} - \frac{x}{2}\right)\right| + C.$$

56. Use appropriate trigonometric identities to deduce from the result of Problem 55 that

$$\int \sec x \, dx = \ln|\sec x + \tan x| + C.$$

57. The region between the curves $y = \tan^2 x$ and $y = \sec^2 x$ from $x = 0$ to $x = \pi/4$ is shown in Fig. 9.3.1. Find its area.

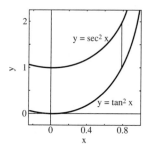

Fig. 9.3.1 The region of Problem 57

58. Let R denote the region between the graph of $y = \sec x$ and the x-axis over the interval $[0, \pi/4]$. Find the volume generated by rotation of R around the x-axis.

59. The region S between the graph of $y = \tan(\pi x^2/4)$ and the x-axis over the interval $[0, 1]$ is shown in Fig. 9.3.2. Find the volume generated by rotation of S around the y-axis.

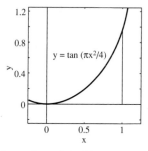

Fig. 9.3.2 The region of Problem 59

60. Find the length of the graph of $y = \ln(\cos x)$ from $x = 0$ to $x = \pi/4$.

9.4
Integration by Parts

One reason for transforming one given integral into another is to produce an integral that is easier to evaluate. There are two general ways to accomplish this. We have seen the first, integration by substitution. The second is *integration by parts*.

The formula for integration by parts is a simple consequence of the product rule for derivatives,

$$D_x(uv) = v\frac{du}{dx} + u\frac{dv}{dx}.$$

If we write this formula in the form

$$u(x)v'(x) = D_x[u(x)v(x)] - v(x)u'(x), \tag{1}$$

then antidifferentiation gives

$$\int u(x)v'(x)\,dx = u(x)v(x) - \int v(x)u'(x)\,dx. \tag{2}$$

This is the formula for **integration by parts**. With $du = u'(x)\,dx$ and $dv = v'(x)\,dx$, Eq. (2) becomes

$$\int u\,dv = uv - \int v\,du. \tag{3}$$

To apply the integration by parts formula to a given integral, we must first factor its integrand into two "parts," u and dv, the latter including the differential dx. We try to choose the parts in accordance with two principles:

1. The antiderivative $v = \int dv$ is easy to find.
2. The new integral $\int v\,du$ is easier to compute than the original integral $\int u\,dv$.

An effective strategy is to choose for dv the most complicated factor that can readily be integrated. Then we differentiate the other part, u, to find du.

We begin with two examples in which we have little flexibility in choosing the parts u and dv.

EXAMPLE 1 Find $\displaystyle\int \ln x\,dx$.

Solution   Here there is little alternative to the natural choice $u = \ln x$ and $dv = dx$. It is helpful to systematize the procedure of integration by parts by writing u, dv, du, and v in a rectangular array like this:

$$\text{Let} \qquad u = \ln x \quad \text{and} \quad dv = dx.$$

$$\text{Then} \quad du = \frac{1}{x}\,dx \quad \text{and} \quad v = x.$$

The first line in the array specifies the choice of u and dv; the second line is computed from the first. Then Eq. (3) gives

$$\int \ln x\,dx = x\ln x - \int dx = x\ln x - x + C.$$

COMMENT 1   The constant of integration appears only at the last step. We know that once we have found one antiderivative, any other may be obtained by adding a constant C to the one we have found.

COMMENT 2   In computing $v = \int dv$, we ordinarily take the constant of integration to be zero. Had we written $v = x + C_1$ in Example 1, the answer would have been

$$\int \ln x \, dx = (x + C_1) \ln x - \int \left(1 + \frac{C_1}{x}\right) dx$$

$$= x \ln x + C_1 \ln x - (x + C_1 \ln x) + C = x \ln x - x + C$$

as before, so introducing the extra constant C_1 has no effect.

EXAMPLE 2 Find $\int \arcsin x \, dx$.

Solution Again, there is only one real choice for u and dv:

$$\text{Let} \quad u = \arcsin x, \quad dv = dx.$$

$$\text{Then} \quad du = \frac{dx}{\sqrt{1 - x^2}}, \quad v = x.$$

Then Eq. (3) gives

$$\int \arcsin x \, dx = x \arcsin x - \int \frac{x}{\sqrt{1 - x^2}} \, dx$$

$$= x \arcsin x + \sqrt{1 - x^2} + C.$$

EXAMPLE 3 Find $\int x e^{-x} \, dx$.

Solution Here we have some flexibility. Suppose that we try

$$u = e^{-x}, \quad dv = x \, dx$$

so that

$$du = -e^{-x} \, dx, \, v = \tfrac{1}{2} x^2.$$

Then integration by parts gives

$$\int x e^{-x} \, dx = \tfrac{1}{2} x^2 e^{-x} + \tfrac{1}{2} \int x^2 e^{-x} \, dx.$$

The new integral on the right looks more troublesome than the one we started with! Let us begin anew:

$$\text{Let} \quad u = x, \quad dv = e^{-x} \, dx.$$

$$\text{Then} \quad du = dx, \quad v = -e^{-x}.$$

Now integration by parts gives

$$\int x e^{-x} \, dx = -x e^{-x} + \int e^{-x} \, dx = -x e^{-x} - e^{-x} + C.$$

Integration by parts can be applied to definite integrals as well as to indefinite integrals. We integrate Eq. (1) from $x = a$ to $x = b$ and apply the fundamental theorem of calculus. This gives

$$\int_a^b u(x)v'(x) \, dx = \int_a^b D_x[u(x)v(x)] \, dx - \int_a^b v(x)u'(x) \, dx$$

$$= \left[u(x)v(x)\right]_a^b - \int_a^b v(x)u'(x) \, dx.$$

In the notation of Eq. (3), this equation would be written

$$\int_{x=a}^{x=b} u \, dv = \left[uv \right]_{x=a}^{x=b} - \int_{x=a}^{x=b} v \, du, \tag{4}$$

although we must not forget that u and v are functions of x. For example, with $u = x$ and $dv = e^{-x} \, dx$, as in Example 3, we obtain

$$\int_0^1 xe^{-x} \, dx = \left[-xe^{-x} \right]_0^1 + \int_0^1 e^{-x} \, dx = -e^{-1} + \left[-e^{-x} \right]_0^1 = 1 - \frac{2}{e}.$$

EXAMPLE 4 Find $\int x^2 e^{-x} \, dx$.

Solution If we choose $u = x^2$, then $du = 2x \, dx$, so we will reduce the exponent of x by this choice:

$$\text{Let} \qquad u = x^2, \qquad dv = e^{-x} \, dx.$$
$$\text{Then} \quad du = 2x \, dx, \qquad v = -e^{-x}.$$

Then integration by parts gives

$$\int x^2 e^{-x} \, dx = -x^2 e^{-x} + 2 \int xe^{-x} \, dx.$$

We apply integration by parts a second time to the right-hand integral and obtain the result

$$\int xe^{-x} \, dx = -xe^{-x} - e^{-x}$$

of Example 3. Substitution then yields

$$\int x^2 e^{-x} \, dx = -x^2 e^{-x} - 2xe^{-x} - 2e^{-x} + C$$

$$= -(x^2 + 2x + 2)e^{-x} + C.$$

In effect, we have annihilated the original factor x^2 by integrating by parts twice in succession.

EXAMPLE 5 Find $\int e^{2x} \sin 3x \, dx$.

Solution This is another example in which repeated integration by parts succeeds, but with a twist:

$$\text{Let} \qquad u = \sin 3x, \qquad dv = e^{2x} \, dx.$$
$$\text{Then} \quad du = 3 \cos 3x \, dx, \qquad v = \tfrac{1}{2} e^{2x}.$$

Therefore,

$$\int e^{2x} \sin 3x \, dx = \tfrac{1}{2} e^{2x} \sin 3x - \tfrac{3}{2} \int e^{2x} \cos 3x \, dx.$$

At first it might appear that little progress has been made, for the integral on the right is as difficult to integrate as the one on the left. We ignore this objection and try again, applying integration by parts to the new integral:

Let $u = \cos 3x,$ $dv = e^{2x}\,dx.$

Then $du = -3 \sin 3x\,dx,$ $v = \tfrac{1}{2}e^{2x}.$

Now we find that

$$\int e^{2x}\cos 3x\,dx = \tfrac{1}{2}e^{2x}\cos 3x + \tfrac{3}{2}\int e^{2x}\sin 3x\,dx.$$

When we substitute this result into the previous equation, we discover that

$$\int e^{2x}\sin 3x\,dx = \tfrac{1}{2}e^{2x}\sin 3x - \tfrac{3}{4}e^{2x}\cos 3x - \tfrac{9}{4}\int e^{2x}\sin 3x\,dx.$$

So we are back where we started. Or *are* we? In fact we are *not*, because we can *solve* this last equation for the desired integral. We add the right-hand integral here to both sides of the last equation. This gives

$$\tfrac{13}{4}\int e^{2x}\sin 3x\,dx = \tfrac{1}{4}e^{2x}(2 \sin 3x - 3 \cos 3x) + C_1,$$

so

$$\int e^{2x}\sin 3x\,dx = \tfrac{1}{13}e^{2x}(2 \sin 3x - 3 \cos 3x) + C.$$

EXAMPLE 6 Find a reduction formula for $\displaystyle\int \sec^n x\,dx.$

Solution The idea is that n is a (large) positive integer and that we want to express the given integral in terms of the integral of a lower power of $\sec x$. The easiest power of $\sec x$ to integrate is $\sec^2 x$, so we proceed as follows:

Let $u = \sec^{n-2} x,$ $dv = \sec^2 x\,dx.$

Then $du = (n - 2) \sec^{n-2} x \tan x\,dx,$ $v = \tan x.$

This gives

$$\int \sec^n x\,dx = \sec^{n-2} x \tan x - (n - 2)\int \sec^{n-2} x \tan^2 x\,dx$$

$$= \sec^{n-2} x \tan x - (n - 2)\int (\sec^{n-2} x)(\sec^2 x - 1)\,dx.$$

Hence

$$\int \sec^n x\,dx = \sec^{n-2} x \tan x - (n - 2)\int \sec^n x\,dx + (n - 2)\int \sec^{n-2} x\,dx.$$

We solve this equation for the desired integral and find that

$$\int \sec^n x\,dx = \frac{\sec^{n-2} x \tan x}{n - 1} + \frac{n - 2}{n - 1}\int \sec^{n-2} x\,dx. \tag{5}$$

This is the desired reduction formula. For example, if we take $n = 3$ in this formula, we find that

$$\int \sec^3 x\,dx = \tfrac{1}{2} \sec x \tan x + \tfrac{1}{2}\int \sec x\,dx$$

$$= \tfrac{1}{2} \sec x \tan x + \tfrac{1}{2} \ln|\sec x + \tan x| + C. \tag{6}$$

500

In the last step we used Eq. (7) of Section 9.3,

$$\int \sec x \, dx = \ln|\sec x + \tan x| + C.$$

The reason for using the reduction formula in Eq. (5) is that repeated application must yield one of the two elementary integrals $\int \sec x \, dx$ and $\int \sec^2 x \, dx$. For instance, with $n = 4$ we get

$$\int \sec^4 x \, dx = \tfrac{1}{3} \sec^2 x \tan x + \tfrac{2}{3} \int \sec^2 x \, dx$$

$$= \tfrac{1}{3} \sec^2 x \tan x + \tfrac{2}{3} \tan x + C, \qquad (7)$$

and with $n = 5$ we get

$$\int \sec^5 x \, dx = \tfrac{1}{4} \sec^3 x \tan x + \tfrac{3}{4} \int \sec^3 x \, dx$$

$$= \tfrac{1}{4} \sec^3 x \tan x + \tfrac{3}{8} \sec x \tan x + \tfrac{3}{8} \ln|\sec x + \tan x| + C, \qquad (8)$$

using Eq. (6) in the last step.

9.4 Problems

Use integration by parts to compute the integrals in Problems 1 through 35.

1. $\int xe^{2x} \, dx$

2. $\int x^2 e^{2x} \, dx$

3. $\int t \sin t \, dt$

4. $\int t^2 \sin t \, dt$

5. $\int x \cos 3x \, dx$

6. $\int x \ln x \, dx$

7. $\int x^3 \ln x \, dx$

8. $\int e^{3z} \cos 3z \, dz$

9. $\int \arctan x \, dx$

10. $\int \dfrac{\ln x}{x^2} \, dx$

11. $\int \sqrt{y} \ln y \, dy$

12. $\int x \sec^2 x \, dx$

13. $\int (\ln t)^2 \, dt$

14. $\int t(\ln t)^2 \, dt$

15. $\int x\sqrt{x + 3} \, dx$

16. $\int x^3\sqrt{1 - x^2} \, dx$

17. $\int x^5\sqrt{x^3 + 1} \, dx$

18. $\int \sin^2 \theta \, d\theta$

19. $\int \csc^3 \theta \, d\theta$

20. $\int \sin(\ln t) \, dt$

21. $\int x^2 \arctan x \, dx$

22. $\int \ln(1 + x^2) \, dx$

23. $\int \sec^{-1}\sqrt{x} \, dx$

24. $\int x \tan^{-1}\sqrt{x} \, dx$

25. $\int \tan^{-1}\sqrt{x} \, dx$

26. $\int x^2 \cos 4x \, dx$

27. $\int x \csc^2 x \, dx$

28. $\int x \arctan x \, dx$

29. $\int x^3 \cos x^2 \, dx$

30. $\int e^{-3x} \sin 4x \, dx$

31. $\int \dfrac{\ln x}{x\sqrt{x}} \, dx$

32. $\int \dfrac{x^7}{(1 + x^4)^{3/2}} \, dx$

33. $\int x \cosh x \, dx$

34. $\int e^x \cosh x \, dx$

35. $\int x^2 \sinh x \, dx$

36. Use the method of cylindrical shells to find the volume generated by rotating around the y-axis the area under the curve $y = \cos x$ for $0 \le x \le \pi/2$.

37. Find the volume of the solid obtained by revolving around the x-axis the area bounded by the graph of $y = \ln x$, by the x-axis, and by the vertical line $x = e$.

38. Use integration by parts to evaluate

$$\int 2x \arctan x \, dx,$$

with $dv = 2x \, dx$, but let $v = x^2 + 1$ rather than $v = x^2$. Is there a reason why v should not be chosen in this way?

39. Use integration by parts to evaluate $\int xe^x \cos x \, dx$.

40. Use integration by parts to evaluate $\int \sin 3x \cos x \, dx$.

Derive the reduction formulas given in Problems 41 through 46.

41. $\displaystyle \int x^n e^x \, dx = x^n e^x - n \int x^{n-1} e^x \, dx$

42. $\displaystyle \int x^n e^{-x^2} \, dx = -\frac{1}{2} x^{n-1} e^{-x^2} + \frac{n-1}{2} \int x^{n-2} e^{-x^2} \, dx$

43. $\displaystyle \int (\ln x)^n \, dx = x(\ln x)^n - n \int (\ln x)^{n-1} \, dx$

44. $\displaystyle \int x^n \cos x \, dx = x^n \sin x - n \int x^{n-1} \sin x \, dx$

45. $\displaystyle \int \sin^n x \, dx = -\frac{\sin^{n-1} x \cos x}{n} + \frac{n-1}{n} \int \sin^{n-2} x \, dx$

46. $\displaystyle \int \cos^n x \, dx = \frac{\cos^{n-1} x \sin x}{n} + \frac{n-1}{n} \int \cos^{n-2} x \, dx$

Use appropriate reduction formulas from the preceding list to evaluate the integrals in Problems 47 through 49.

47. $\displaystyle \int_0^1 x^3 e^x \, dx$ **48.** $\displaystyle \int_0^1 x^5 e^{-x^2} \, dx$ **49.** $\displaystyle \int_1^e (\ln x)^3 \, dx$

50. Apply the reduction formula in Problem 45 to show that for each positive integer n,

$$\int_0^{\pi/2} \sin^{2n} x \, dx = \frac{\pi}{2} \cdot \frac{1}{2} \cdot \frac{3}{4} \cdot \frac{5}{6} \cdots \frac{2n-1}{2n}$$

and

$$\int_0^{\pi/2} \sin^{2n+1} x \, dx = \frac{2}{3} \cdot \frac{4}{5} \cdot \frac{6}{7} \cdot \frac{8}{9} \cdots \frac{2n}{2n+1}.$$

51. Derive the formula

$$\int \ln(x+10) \, dx = (x+10)\ln(x+10) - x + C$$

in three different ways: (a) by substituting $u = x + 10$ and applying the result of Example 1; (b) by integrating by parts with $u = \ln(x+10)$ and $dv = dx$, noting that

$$\frac{x}{x+10} = 1 - \frac{10}{x+10};$$

and (c) by integrating by parts with $u = \ln(x+10)$ and $dv = dx$, but with $v = x + 10$.

52. Derive the formula

$$\int x^3 \tan^{-1} x \, dx = \tfrac{1}{4}(x^4 - 1)\tan^{-1} x - \tfrac{1}{12} x^3 + \tfrac{1}{4} x + C$$

by integrating by parts with $u = \tan^{-1} x$ and $v = \tfrac{1}{4}(x^4 - 1)$.

53. Let $\displaystyle J_n = \int_0^1 x^n e^x \, dx$ for each integer $n \geq 0$. (a) Show that

$$J_0 = 1 - \frac{1}{e} \quad \text{and that} \quad J_n = nJ_{n-1} - \frac{1}{e}$$

for $n \geq 1$. (b) Deduce by mathematical induction that

$$J_n = n! - \frac{n!}{e} \sum_{k=0}^n \frac{1}{k!}$$

for each integer $n \geq 0$. (c) Explain why $J_n \to 0$ as $n \to +\infty$. (d) Conclude that

$$e = \lim_{n \to \infty} \sum_{k=0}^n \frac{1}{k!}.$$

54. Let m and n be positive integers. Derive the reduction formula

$$\int x^m (\ln x)^n \, dx = \frac{x^{m+1}}{m+1} (\ln x)^n - \frac{n}{m+1} \int x^m (\ln x)^{n-1} \, dx.$$

55. A recent advertisement for a symbolic algebra program claimed that an engineer worked for three weeks on the integral

$$\int (k \ln x - 2x^3 + 3x^2 + b)^4 \, dx,$$

which deals with turbulence in an aerospace application. The advertisement said that the engineer never got the same answer twice in the three weeks. Explain how you could use the reduction formula of Problem 54 to find the engineer's integral (but don't actually do it). Can you see any reason why it should have taken three weeks?

56. Figure 9.4.1 shows the region bounded by the x-axis and by the graph of $y = \tfrac{1}{2} x^2 \sin x$, $0 \leq x \leq \pi$. Use Formulas (42) and (43) (inside the back cover)—which are derived by integration by parts—to find (a) the area of this region; (b) the volume obtained by revolving this region around the y-axis.

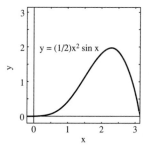

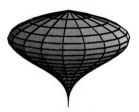

$y = (1/2)x^2 \sin x$

Fig. 9.4.1 The region of Problem 56

Fig. 9.4.2 The top of Problem 57

57. The top shown in Fig. 9.4.2 has the shape of the solid obtained by revolving the region of Problem 56 around the x-axis. Find the volume of this top.

502

9.5
Rational Functions and Partial Fractions

We shall show here how every rational function can be integrated in terms of elementary functions. Recall that a rational function $R(x)$ is a function that can be expressed as the quotient of two polynomials. That is,

$$R(x) = \frac{P(x)}{Q(x)}, \tag{1}$$

where $P(x)$ and $Q(x)$ are polynomials. The **method of partial fractions** is an *algebraic* technique that decomposes $R(x)$ into a sum of terms:

$$R(x) = \frac{P(x)}{Q(x)} = p(x) + F_1(x) + F_2(x) + \cdots + F_k(x), \tag{2}$$

where $p(x)$ is a polynomial and each expression $F_i(x)$ is a fraction that can be integrated with little difficulty.

For example, we can verify (by finding a common denominator on the right) that

$$\frac{x^3 - 1}{x^3 + x} = 1 - \frac{1}{x} + \frac{x - 1}{x^2 + 1}. \tag{3}$$

It follows that

$$\int \frac{x^3 - 1}{x^3 + x}\, dx = \int \left(1 - \frac{1}{x} + \frac{x}{x^2 + 1} - \frac{1}{x^2 + 1} \right) dx$$

$$= x - \ln|x| + \tfrac{1}{2}\ln(x^2 + 1) - \tan^{-1}x + C.$$

The key to this simple integration lies in finding the decomposition given in Eq. (3). The existence of such a decomposition and the technique of finding it are what the method of partial fractions is about.

According to a theorem proved in advanced algebra, every rational function can be written in the form in Eq. (2) with each $F_i(x)$ being a fraction either of the form

$$\frac{A}{(ax + b)^n} \tag{4}$$

or of the form

$$\frac{Bx + C}{(ax^2 + bx + c)^n}. \tag{5}$$

Here the quadratic polynomial $ax^2 + bx + c$ is **irreducible**: It is not a product of linear factors with real coefficients. This is the same as saying that the equation $ax^2 + bx + c = 0$ has no real roots, and the quadratic formula tells us that this is the case exactly when its discriminant is negative: $b^2 - 4ac < 0$.

Fractions of the forms in Eqs. (4) and (5) are called **partial fractions**, and the sum in Eq. (2) is called the **partial fraction decomposition** of $R(x)$. Thus Eq. (3) gives the partial fraction decomposition of $(x^3 - 1)/(x^3 + x)$. A partial fraction of the form in Eq. (4) may be integrated immediately, and we will see in Section 9.7 how to integrate one of the form in Eq. (5).

The first step in finding the partial fraction decomposition of $R(x)$ is to find the polynomial $p(x)$ in Eq. (2). It turns out that $p(x) \equiv 0$ provided that the degree of the numerator $P(x)$ is *less than* that of the denominator $Q(x)$; such a rational fraction $R(x) = P(x)/Q(x)$ is said to be **proper**. If $R(x)$ is not proper, then $p(x)$ may be found dividing $Q(x)$ into $P(x)$, as in Example 1.

EXAMPLE 1 Find $\displaystyle\int \frac{x^3 + x^2 + x - 1}{x^2 + 2x + 2}\,dx.$

Solution Long division of denominator into numerator may be carried out as follows:

$$
\begin{array}{r}
x \ - 1 \quad\longleftarrow\quad p(x) \quad\text{(quotient)} \\
x^2 + 2x + 2\,\overline{)\,x^3 + \ x^2 + \ x - 1\,} \\
\underline{x^3 + 2x^2 + 2x} \\
-x^2 - \ x - 1 \\
\underline{-x^2 - 2x - 2} \\
x + 1 \quad\longleftarrow\quad r(x) \quad\text{(remainder)}
\end{array}
$$

As in simple arithmetic,

$$\text{``fraction} = \text{quotient} + \frac{\text{remainder}}{\text{divisor}}.\text{''}$$

Thus

$$\frac{x^3 + x^2 + x - 1}{x^2 + 2x + 2} = (x - 1) + \frac{x + 1}{x^2 + 2x + 2},$$

and hence

$$\int \frac{x^3 + x^2 + x - 1}{x^2 + 2x + 2}\,dx = \int \left(x - 1 + \frac{x + 1}{x^2 + 2x + 2} \right) dx$$

$$= \tfrac{1}{2}x^2 - x + \tfrac{1}{2}\ln(x^2 + 2x + 2) + C.$$

By using long division as in Example 1, any rational function $R(x)$ can be written as a sum of a polynomial $p(x)$ and a *proper* rational fraction,

$$R(x) = p(x) + \frac{r(x)}{Q(x)}.$$

To see how to integrate an arbitrary rational function, we therefore need only see how to find the partial fraction decomposition of a proper rational fraction.

To obtain such a decomposition, the first step is to factor the denominator $Q(x)$ into a product of linear factors (those of the form $ax + b$) and irreducible quadratic factors (those of the form $ax^2 + bx + c$ with $b^2 - 4ac < 0$). This is always possible in principle but may be difficult in practice. But once we have found the factorization of $Q(x)$, we can obtain the partial fraction decomposition by routine algebraic methods (described next). Each linear or irreducible quadratic factor of $Q(x)$ leads to one or more partial fractions of the forms in Eqs. (4) and (5).

LINEAR FACTORS

Let $R(x) = P(x)/Q(x)$ be a *proper* rational fraction, and suppose that the linear factor $ax + b$ occurs n times in the complete factorization of $Q(x)$. That is, $(ax + b)^n$ is the highest power of $ax + b$ that divides "evenly" into $Q(x)$. In this case we call n the **multiplicity** of the factor $ax + b$.

> **Rule 1** *Linear Factor Partial Fractions*
>
> The part of the partial fraction decomposition of $R(x)$ that corresponds to the linear factor $ax + b$ of multiplicity n is a sum of n partial fractions, specifically
>
> $$\frac{A_1}{ax + b} + \frac{A_2}{(ax + b)^2} + \cdots + \frac{A_n}{(ax + b)^n}, \tag{6}$$
>
> where $A_1, A_2, \ldots, A_n$ are constants.

If *all* the factors of $Q(x)$ are linear, then the partial fraction decomposition of $R(x)$ is a sum of expressions like the one in Eq. (6). The situation is especially simple if each of these linear factors is *nonrepeated*—that is, if each has multiplicity $n = 1$. In this case, the expression in Eq. (6) reduces to its first term, and the partial fraction decomposition of $R(x)$ is a sum of such terms. The solutions in Examples 2 and 3 illustrate how the constant numerators can be determined.

EXAMPLE 2 Find $\displaystyle\int \frac{5}{(2x + 1)(x - 2)}\, dx$.

Solution The linear factors in the denominator are distinct, so we seek a partial fraction decomposition of the form

$$\frac{5}{(2x + 1)(x - 2)} = \frac{A}{2x + 1} + \frac{B}{x - 2}.$$

To find the constants A and B, we multiply both sides of this *identity* by the left-hand (common) denominator $(2x + 1)(x - 2)$. The result is

$$5 = A(x - 2) + B(2x + 1) = (A + 2B)x + (-2A + B).$$

Next we equate coefficients of x and of 1 on the left-hand and right-hand sides of this equation. This yields the simultaneous equations

$$A + 2B = 0,$$

$$-2A + B = 5,$$

which we readily solve for $A = -2$, $B = 1$. Hence

$$\frac{5}{(2x + 1)(x - 2)} = \frac{-2}{2x + 1} + \frac{1}{x - 2},$$

and therefore

$$\int \frac{5}{(2x + 1)(x - 2)}\, dx = -\ln|2x + 1| + \ln|x - 2| + C$$

$$= \ln\left|\frac{x - 2}{2x + 1}\right| + C.$$

EXAMPLE 3 Find $\displaystyle\int \frac{4x^2 - 3x - 4}{x^3 + x^2 - 2x}\, dx$.

Solution   The rational function to be integrated is proper, so we immediately factor its denominator:

$$x^3 + x^2 - 2x = x(x^2 + x - 2) = x(x - 1)(x + 2).$$

We are dealing with three nonrepeated linear factors, so the partial fraction decomposition has the form

$$\frac{4x^2 - 3x - 4}{x^3 + x^2 - 2x} = \frac{A}{x} + \frac{B}{x - 1} + \frac{C}{x + 2}.$$

To find the constants A, B, and C, we multiply both sides of this equation by the common denominator $x(x - 1)(x + 2)$ and find thereby that

$$4x^2 - 3x - 4 = A(x - 1)(x + 2) + Bx(x + 2) + Cx(x - 1). \quad (7)$$

Then we collect coefficients of like powers of x on the right:

$$4x^2 - 3x - 4 = (A + B + C)x^2 + (A + 2B - C)x + (-2A).$$

Because two polynomials are (identically) equal only if the coefficients of corresponding powers of x are the same, we conclude that

$$
\begin{aligned}
A + B + C &= 4, \\
A + 2B - C &= -3, \\
-2A &= -4.
\end{aligned}
$$

We solve these simultaneous equations and thus find that $A = 2$, $B = -1$, and $C = 3$.

There is an alternative way to find A, B, and C that is especially effective in the case of nonrepeated linear factors. Substitute the values of $x = 0$, $x = 1$, and $x = -2$ (the zeros of the linear factors of the denominator) into Eq. (7). Substitution of $x = 0$ into Eq. (7) immediately gives $-4 = -2A$, so $A = 2$. Substitution of $x = 1$ into Eq. (7) gives $-3 = 3B$, so $B = -1$. Substitution of $x = -2$ gives $18 = 6C$, so $C = 3$.

With these values of $A = 2$, $B = -1$, and $C = 3$, however obtained, we find that

$$\int \frac{4x^2 - 3x - 4}{x^3 + x^2 - 2x}\, dx = \int \left(\frac{2}{x} - \frac{1}{x - 1} + \frac{3}{x + 2} \right) dx$$

$$= 2 \ln|x| - \ln|x - 1| + 3 \ln|x + 2| + C.$$

Laws of logarithms allow us to write this antiderivative in the more compact form

$$\int \frac{4x^2 - 3x - 4}{x^3 + x^2 - 2x}\, dx = \ln \left| \frac{x^2(x + 2)^3}{x - 1} \right| + C.$$

EXAMPLE 4   Find $\displaystyle \int \frac{x^3 - 4x - 1}{x(x - 1)^3}\, dx.$

Solution   Here we have a linear factor of multiplicity $n = 3$. According to Rule 1, the partial fraction decomposition of the integrand has the form

$$\frac{x^3 - 4x - 1}{x(x - 1)^3} = \frac{A}{x} + \frac{B}{x - 1} + \frac{C}{(x - 1)^2} + \frac{D}{(x - 1)^3}.$$

To find the constants A, B, C, and D, we multiply both sides of this equation by the least common denominator $x(x - 1)^3$. We find that

$$x^3 - 4x - 1 = A(x - 1)^3 + Bx(x - 1)^2 + Cx(x - 1) + Dx.$$

We expand and then collect coefficients of like powers of x on the right-hand side. This yields

$$x^3 - 4x - 1 = (A + B)x^3 + (-3A - 2B + C)x^2$$
$$+ (3A + B - C + D)x - A.$$

Then we equate coefficients of like powers of x on each side of this equation. We get the four simultaneous equations

$$
\begin{aligned}
A + B &= 1, \\
-3A - 2B + C &= 0, \\
3A + B - C + D &= -4, \\
-A &= -1.
\end{aligned}
$$

The last equation gives $A = 1$, and then the first equation gives $B = 0$. Next, the second equation gives $C = 3$. When we substitute these values into the third equation, we finally get $D = -4$. Hence

$$\int \frac{x^3 - 4x - 1}{x(x - 1)^3}\, dx = \int \left(\frac{1}{x} + \frac{3}{(x - 1)^2} - \frac{4}{(x - 1)^3} \right) dx$$

$$= \ln|x| - \frac{3}{x - 1} + \frac{2}{(x - 1)^2} + C.$$

QUADRATIC FACTORS

Suppose that $R(x) = P(x)/Q(x)$ is a proper rational fraction and that the irreducible quadratic factor $ax^2 + bx + c$ occurs n times in the factorization of $Q(x)$. That is, $(ax^2 + bx + c)^n$ is the highest power of $ax^2 + bx + c$ that divides evenly into $Q(x)$. As before, we call n the *multiplicity* of the quadratic factor $ax^2 + bx + c$.

> **Rule 2 *Quadratic Factor Partial Fractions***
> The part of the partial fraction decomposition of $R(x)$ that corresponds to the irreducible quadratic factor $ax^2 + bx + c$ of multiplicity n is a sum of n partial fractions. It has the form
>
> $$\frac{B_1 x + C_1}{ax^2 + bx + c} + \frac{B_2 x + C_2}{(ax^2 + bx + c)^2} + \cdots + \frac{B_n x + C_n}{(ax^2 + bx + c)^n}, \quad (8)$$
>
> where $B_1, B_2, \ldots, B_n$ and $C_1, C_2, \ldots, C_n$ are constants.

If $Q(x)$ has both linear and irreducible quadratic factors, then the partial fraction decomposition of $R(x)$ is simply the sum of the expressions of the form in Eq. (6) that correspond to the linear factors plus the sum of the expressions of the form in Eq. (8) that correspond to the quadratic factors. In the case of an irreducible quadratic factor of multiplicity $n = 1$, the expression in Eq. (8) reduces to its first term alone.

The most important case is that of a nonrepeated quadratic factor of the sum of squares form $x^2 + k^2$ (where k is a positive constant). The corresponding partial fraction $(Bx + C)/(x^2 + k^2)$ is readily integrated by using the familiar integrals

$$\int \frac{x}{x^2 + k^2} \, dx = \frac{1}{2} \ln(x^2 + k^2) + C,$$

$$\int \frac{1}{x^2 + k^2} \, dx = \frac{1}{k} \arctan \frac{x}{k} + C.$$

We will discuss in Section 9.7 the integration of more general partial fractions involving irreducible quadratic factors.

EXAMPLE 5   Find $\displaystyle\int \frac{5x^3 - 3x^2 + 2x - 1}{x^4 + x^2} \, dx$.

Solution   The denominator $x^4 + x^2 = x^2(x^2 + 1)$ has both a quadratic factor and a repeated linear factor. The partial fraction decomposition takes the form

$$\frac{5x^3 - 3x^2 + 2x - 1}{x^4 + x^2} = \frac{A}{x} + \frac{B}{x^2} + \frac{Cx + D}{x^2 + 1}.$$

We multiply both sides by $x^4 + x^2$ and obtain

$$5x^3 - 3x^2 + 2x - 1 = Ax(x^2 + 1) + B(x^2 + 1) + (Cx + D)x^2$$
$$= (A + C)x^3 + (B + D)x^2 + Ax + B.$$

As before, we equate coefficients of like powers of x. This yields the four simultaneous equations

$$
\begin{array}{rcl}
A \quad\quad + C \quad\quad & = & 5, \\
B \quad\quad + D & = & -3, \\
A \quad\quad\quad\quad\quad & = & 2, \\
B \quad\quad\quad\quad\quad & = & -1.
\end{array}
$$

These equations are easily solved for $A = 2$, $B = -1$, $C = 3$, and $D = -2$. Thus

$$\int \frac{5x^3 - 3x^2 + 2x - 1}{x^4 + x^2} \, dx = \int \left(\frac{2}{x} - \frac{1}{x^2} + \frac{3x - 2}{x^2 + 1} \right) dx$$

$$= 2 \ln|x| + \frac{1}{x} + \frac{3}{2} \int \frac{2x \, dx}{x^2 + 1} - 2 \int \frac{dx}{x^2 + 1}$$

$$= 2 \ln|x| + \frac{1}{x} + \frac{3}{2} \ln(x^2 + 1) - 2 \tan^{-1}x + C.$$

*APPLICATIONS TO DIFFERENTIAL EQUATIONS

Example 6 illustrates the use of partial fractions to solve certain types of separable differential equations.

EXAMPLE 6 Suppose that at time $t = 0$, half of a population of 100,000 people has heard a certain rumor and that the number $P(t)$ of those who have heard it is then increasing at the rate of 1000 people per day. If $P(t)$ satisfies the differential equation

$$\frac{dP}{dt} = kP(100 - P) \tag{9}$$

(with P in thousands of people and t in days), determine how many people will have heard the rumor after $t = 30$ days.

Solution The differential equation in (9) is the model for a simple but widely used assumption: The rate at which the rumor spreads is proportional to the number of contacts between those who already know the rumor and those who have not yet heard it. That is, dP/dt is proportional both to P and to $100 - P$; thus dP/dt is proportional to the product of P and $100 - P$.

To find the constant k of proportionality, we substitute the given values $P(0) = 50$ and $P'(0) = 1$ in Eq. (9). This yields the equation

$$1 = k \cdot 50 \cdot (100 - 50).$$

It follows that $k = 0.0004$, so the differential equation in (9) takes the form

$$\frac{dP}{dt} = (0.0004) \cdot P \cdot (100 - P). \tag{10}$$

Separation of the variables leads to

$$\int \frac{dP}{P(100 - P)} = \int 0.0004 \, dt.$$

Then the partial fraction decomposition

$$\frac{100}{P(100 - P)} = \frac{1}{P} + \frac{1}{100 - P}$$

yields

$$\int \left(\frac{1}{P} + \frac{1}{100 - P} \right) dP = \int 0.04 \, dt;$$

$$\ln P - \ln(100 - P) = (0.04)t + C. \tag{11}$$

Substitution of the initial data $P = 50$ when $t = 0$ now gives $C = 0$, so Eq. (11) takes the form

$$\ln \frac{P}{100 - P} = (0.04)t, \quad \text{so} \quad \frac{P}{100 - P} = e^{(0.04)t}.$$

We readily solve this last equation for the solution

$$P(t) = \frac{100e^{(0.04)t}}{1 + e^{(0.04)t}}. \tag{12}$$

Hence the number of people who have heard the rumor after 30 days is $P(30) \approx 76.85$ thousand people.

The method of Example 6 can be used to solve any differential equation of the form

$$\frac{dx}{dt} = k(x - a)(x - b), \tag{13}$$

where a, b, and k are constants. As Problems 49 through 54 indicate, this differential equation serves as a mathematical model for a wide variety of natural phenomena.

9.5 Problems

Find the integrals in Problems 1 through 35.

1. $\displaystyle\int \frac{x^2}{x + 1}\, dx$

2. $\displaystyle\int \frac{x^3}{2x - 1}\, dx$

3. $\displaystyle\int \frac{1}{x^2 - 3x}\, dx$

4. $\displaystyle\int \frac{x}{x^2 + 4x}\, dx$

5. $\displaystyle\int \frac{1}{x^2 + x - 6}\, dx$

6. $\displaystyle\int \frac{x^3}{x^2 + x - 6}\, dx$

7. $\displaystyle\int \frac{1}{x^3 + 4x}\, dx$

8. $\displaystyle\int \frac{1}{(x + 1)(x^2 + 1)}\, dx$

9. $\displaystyle\int \frac{x^4}{x^2 + 4}\, dx$

10. $\displaystyle\int \frac{1}{(x^2 + 1)(x^2 + 4)}\, dx$

11. $\displaystyle\int \frac{x - 1}{x + 1}\, dx$

12. $\displaystyle\int \frac{2x^3 - 1}{x^2 + 1}\, dx$

13. $\displaystyle\int \frac{x^2 + 2x}{(x + 1)^2}\, dx$

14. $\displaystyle\int \frac{2x - 4}{x^2 - x}\, dx$

15. $\displaystyle\int \frac{1}{x^2 - 4}\, dx$

16. $\displaystyle\int \frac{x^4}{x^2 + 4x + 4}\, dx$

17. $\displaystyle\int \frac{x + 10}{2x^2 + 5x - 3}\, dx$

18. $\displaystyle\int \frac{x + 1}{x^3 - x^2}\, dx$

19. $\displaystyle\int \frac{x^2 + 1}{x^3 + 2x^2 + x}\, dx$

20. $\displaystyle\int \frac{x^2 + x}{x^3 - x^2 - 2x}\, dx$

21. $\displaystyle\int \frac{4x^3 - 7x}{x^4 - 5x^2 + 4}\, dx$

22. $\displaystyle\int \frac{2x^2 + 3}{x^4 - 2x^2 + 1}\, dx$

23. $\displaystyle\int \frac{x^2}{(x + 2)^3}\, dx$

24. $\displaystyle\int \frac{x^2 + x}{(x^2 - 4)(x + 4)}\, dx$

25. $\displaystyle\int \frac{1}{x^3 + x}\, dx$

26. $\displaystyle\int \frac{6x^3 - 18x}{(x^2 - 1)(x^2 - 4)}\, dx$

27. $\displaystyle\int \frac{x + 4}{x^3 + 4x}\, dx$

28. $\displaystyle\int \frac{4x^4 + x + 1}{x^5 + x^4}\, dx$

29. $\displaystyle\int \frac{x}{(x + 1)(x^2 + 1)}\, dx$

30. $\displaystyle\int \frac{x^2 + 2}{(x^2 + 1)^2}\, dx$

31. $\displaystyle\int \frac{x^2 - 10}{2x^4 + 9x^2 + 4}\, dx$

32. $\displaystyle\int \frac{x^2}{x^4 - 1}\, dx$

33. $\displaystyle\int \frac{x^3 + x^2 + 2x + 3}{x^4 + 5x^2 + 6}\, dx$

34. $\displaystyle\int \frac{x^2 + 4}{(x^2 + 1)^2 (x^2 + 2)}\, dx$

35. $\displaystyle\int \frac{x^4 + 3x^2 - 4x + 5}{(x - 1)^2 (x^2 + 1)}\, dx$

In Problems 36 through 39, make a preliminary substitution before using the method of partial fractions.

36. $\displaystyle\int \frac{\cos \theta}{\sin^2 \theta - \sin \theta - 6}\, d\theta$

37. $\displaystyle\int \frac{e^{4t}}{(e^{2t} - 1)^3}\, dt$

38. $\displaystyle\int \frac{\sec^2 t}{\tan^3 t + \tan^2 t}\, dt$

39. $\displaystyle\int \frac{1 + \ln t}{t(3 + 2 \ln t)^2}\, dt$

40. The plane region R shown in Fig. 9.5.1 is bounded by the curve

$$y^2 = \frac{1 - x}{1 + x} x^2, \qquad 0 \leq x \leq 1.$$

Find the volume generated by revolving R around the x-axis.

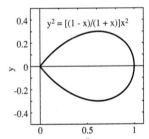

Fig. 9.5.1 The region of Problem 40

41. Figure 9.5.2 shows the region bounded by the curve

$$y^2 = \frac{(1 - x)^2}{(1 + x)^2} x^4, \qquad 0 \leq x \leq 1.$$

Find the volume generated by revolving this region around the x-axis.

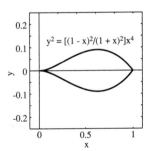

Fig. 9.5.2 The region of Problems 41 and 42

42. Find the volume generated by revolving the region of Problem 41 around the y-axis.

Solve the initial value problems in 43 through 48.

43. $\dfrac{dx}{dt} = x - x^2;\quad x(0) = 2$

44. $\dfrac{dx}{dt} = 10x - x^2;\quad x(0) = 1$

45. $\dfrac{dx}{dt} = 1 - x^2;\quad x(0) = 3$

46. $\dfrac{dx}{dt} = 9 - 4x^2;\quad x(0) = 0$

47. $\dfrac{dx}{dt} = x^2 + 5x + 6;\quad x(0) = 5$

48. $\dfrac{dx}{dt} = 2x^2 + x - 15;\quad x(0) = 10$

49. Suppose that the population $P(t)$ (in millions) of Ruritania satisfies the differential equation

$$\frac{dP}{dt} = k \cdot P \cdot (200 - P) \qquad (k \text{ constant}).$$

Its population in 1940 was 100 million and was then growing at the rate of 1 million per year. Predict this country's population for the year 2000.

50. Suppose that a community contains 15,000 people who are susceptible to Michaud's syndrome, a contagious disease. At time $t = 0$ the number $N(t)$ of people who have caught Michaud's syndrome is 5000 and is increasing then at the rate of 500 per day. Assume that $N'(t)$ is proportional to the product of the numbers of those who have caught the disease and those who have not. How long will it take for another 5000 people to contract Michaud's syndrome?

51. As the salt KNO_3 dissolves in methanol, the number $x(t)$ of grams of the salt in solution after t seconds satisfies the differential equation

$$\frac{dx}{dt} = (0.8)x - (0.004)x^2.$$

(a) If $x = 50$ when $t = 0$, how long will it take an additional 50 g of the salt to dissolve? (b) What is the maximum amount of the salt that will ever dissolve in the methanol?

52. A population $P(t)$ (t in months) of squirrels satisfies the differential equation

$$\frac{dP}{dt} = (0.001)P^2 - kP \qquad (k \text{ constant}).$$

If $P(0) = 100$ and $P'(0) = 8$, how long will it take for this population to double to 200 squirrels?

53. Consider an animal population $P(t)$ (t in years) that satisfies the differential equation

$$\frac{dP}{dt} = kP^2 - (0.01)P \qquad (k \text{ constant}).$$

Suppose also that $P(0) = 200$ and that $P'(0) = 2$. (a) When is $P = 1000$? (b) When will doomsday occur for this population?

54. Suppose that the number $x(t)$ (t in months) of alligators in a swamp satisfies the differential equation

$$\frac{dx}{dt} = (0.0001)x^2 - (0.01)x.$$

(a) If initially there are 25 alligators, solve this equation to determine what happens to this alligator population in the long run. (b) Repeat part (a), but use 150 alligators initially.

9.5 Project

The differential equation in (9) of Example 6 is a **logistic equation**, an equation of the form

$$\frac{dP}{dt} = kP(M - P) \qquad (k, M \text{ constants}). \qquad (14)$$

The logistic equation models many animal (including human) populations more accurately than does the natural growth equation $dP/dt = kP$ that we studied in Section 7.5. For instance, think of an environment that can support a population of at most M individuals. We might then think of $M - P$ as the potential for further expansion when the population is P. The hypothesis that the rate of change dP/dt is therefore proportional to $M - P$ as well as to P itself then yields Eq. (14) with the proportionality constant k. The classic example of such a limited-environment situation is a fruit fly population in a closed container.

The object of this project is to investigate the behavior of populations that can be modeled by the logistic equation.

1. First separate the variables in Eq. (14) and then use partial fractions as in Example 6 to derive the solution

$$P(t) = \frac{MP_0}{P_0 + (M - P_0)e^{-kMt}} \tag{15}$$

that satisfies the initial condition $P(0) = P_0$. If k and M are positive constants, then

$$\lim_{t \to \infty} P(t) = M. \tag{16}$$

Hence M is the *limiting population*.

2. During the period from 1790 to 1930, the U.S. population $P(t)$ (t in years) grew from 3.9 million to 123.2 million. Throughout this period, $P(t)$ remained close to the solution of the initial value problem

$$\frac{dP}{dt} = (0.03135)P - (0.0001589)P^2, \qquad P(0) = 3.9.$$

Has this differential equation continued since 1930 to predict accurately the U.S. population? If so, what is the limiting population of the United States?[†]

3. For your very own logistic equation, choose nonzero integers r and s (for example, the last two nonzero digits of your student ID number), and then take

$$M = 10r, \qquad k = \frac{s}{100M}$$

in Eq. (14). Then plot the corresponding solution in Eq. (15) with several different values of the initial population P_0. What determines whether the graph $P = P(t)$ looks like the upper curve or the lower curve in Fig. 9.5.3?

4. Now, with the same fixed value of the limiting population M as in Problem 3, plot solution curves with both larger and smaller values of k. What appears to be the relationship between the size of k and the rate at which the solution curve approaches its horizontal asymptote $P = M$?

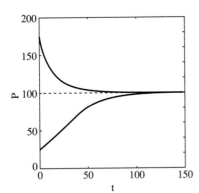

Fig. 9.5.3 Typical solutions of the logistic equation

9.6

Trigonometric Substitution

The method of *trigonometric substitution* can be effective in dealing with integrals when the integrands contain algebraic expressions such as $(a^2 - u^2)^{1/2}$, $(u^2 - a^2)^{3/2}$, and $1/(a^2 + u^2)^2$. There are three basic trigonometric substitutions:

If the integral involves	Then substitute	And use the identity
$a^2 - u^2$	$u = a \sin \theta$	$1 - \sin^2 \theta = \cos^2 \theta$
$a^2 + u^2$	$u = a \tan \theta$	$1 + \tan^2 \theta = \sec^2 \theta$
$u^2 - a^2$	$u = a \sec \theta$	$\sec^2 \theta - 1 = \tan^2 \theta$

[†] This problem is based on a computation by the Belgian demographer Verhulst, who in 1845 used the 1790–1840 U.S. population data to predict accurately the U.S. population through the year 1930.

What we mean by the substitution $u = a \sin \theta$ is, more precisely, the *inverse* trigonometric substitution

$$\theta = \sin^{-1} \frac{u}{a}, \qquad -\frac{\pi}{2} \leq \theta \leq \frac{\pi}{2},$$

where $|u| \leq a$. Suppose, for example, that an integral contains the expression $(a^2 - u^2)^{1/2}$. Then this substitution yields

$$(a^2 - u^2)^{1/2} = (a^2 - a^2 \sin^2 \theta)^{1/2} = (a^2 \cos^2 \theta)^{1/2} = a \cos \theta.$$

We choose the nonnegative square root in the last step because $\cos \theta \geq 0$ for $-\pi/2 \leq \theta \leq \pi/2$. Thus the troublesome factor $(a^2 - u^2)^{1/2}$ becomes $a \cos \theta$ and, meanwhile, $du = a \cos \theta \, d\theta$. If the trigonometric integral that results from this substitution can be evaluated by the methods of Section 9.3, the result will normally involve $\theta = \sin^{-1}(u/a)$ and trigonometric functions of θ. The final step will be to express the answer in terms of the original variable. For this purpose the values of the various trigonometric functions can be read from the right triangle in Fig. 9.6.1, which contains an angle θ such that $\sin \theta = u/a$ (if u is negative, then θ is negative).

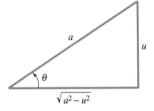

Fig. 9.6.1 The reference triangle for the substitution $u = a \sin \theta$

EXAMPLE 1 Evaluate $\displaystyle\int \frac{x^3}{\sqrt{1 - x^2}} \, dx$, where $|x| < 1$.

Solution Here $a = 1$ and $u = x$, so we substitute

$$x = \sin \theta, \qquad dx = \cos \theta \, d\theta.$$

This gives

$$\int \frac{x^3}{\sqrt{1 - x^2}} \, dx = \int \frac{\sin^3 \theta \cos \theta}{\sqrt{1 - \sin^2 \theta}} \, d\theta$$

$$= \int \sin^3 \theta \, d\theta = \int (\sin \theta)(1 - \cos^2 \theta) \, d\theta$$

$$= \tfrac{1}{3} \cos^3 \theta - \cos \theta + C.$$

Because $\cos \theta = (1 - \sin^2 \theta)^{1/2} = \sqrt{1 - x^2}$, our final answer is

$$\int \frac{x^3}{\sqrt{1 - x^2}} \, dx = \frac{1}{3}(1 - x^2)^{3/2} - \sqrt{1 - x^2} + C.$$

Example 2 illustrates the use of trigonometric substitution to find integrals like those in Formulas (44) through (62) on the inside back cover.

EXAMPLE 2 Find $\displaystyle\int \sqrt{a^2 - u^2} \, du, |u| < a$.

Solution The substitution $u = a \sin \theta$, $du = a \cos \theta \, d\theta$ gives

$$\int \sqrt{a^2 - u^2} \, du = \int \sqrt{a^2 - a^2 \sin^2 \theta} \, (a \cos \theta) \, d\theta$$

$$= \int a^2 \cos^2 \theta \, d\theta = \tfrac{1}{2} a^2 \int (1 + \cos 2\theta) \, d\theta$$

$$= \tfrac{1}{2} a^2 (\theta + \tfrac{1}{2} \sin 2\theta) + C = \tfrac{1}{2} a^2 (\theta + \sin \theta \cos \theta) + C.$$

(We used the identity $\sin 2\theta = 2 \sin \theta \cos \theta$ in the last step.) Now from Fig. 9.6.1 we see that

$$\sin \theta = \frac{u}{a} \quad \text{and} \quad \cos \theta = \frac{\sqrt{a^2 - u^2}}{a}.$$

Hence

$$\int \sqrt{a^2 - u^2} \, du = \frac{1}{2} a^2 \left(\sin^{-1} \frac{u}{a} + \frac{u}{a} \cdot \frac{\sqrt{a^2 - u^2}}{a} \right) + C$$

$$= \frac{u}{2} \sqrt{a^2 - u^2} + \frac{a^2}{2} \sin^{-1} \frac{u}{a} + C.$$

Thus we have obtained Formula (54) of the inside back cover.

What we mean by the substitution $u = a \tan \theta$ in an integral that contains $a^2 + u^2$ is the substitution

$$\theta = \tan^{-1} \frac{u}{a}, \qquad -\frac{\pi}{2} < \theta < \frac{\pi}{2}.$$

In this case

$$\sqrt{a^2 + u^2} = \sqrt{a^2 + a^2 \tan^2 \theta} = \sqrt{a^2 \sec^2 \theta} = a \sec \theta,$$

under the assumption that $a > 0$. We take the positive square root in the last step here, because $\sec \theta > 0$ for $-\pi/2 < \theta < \pi/2$. The values of the various trigonometric functions of θ under this substitution can be read from the right triangle of Fig. 9.6.2, which shows a [positive or negative] acute angle θ such that $\tan \theta = u/a$.

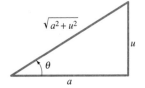

Fig. 9.6.2 The reference triangle for the substitution $u = a \tan \theta$

EXAMPLE 3 Find $\displaystyle\int \frac{1}{(4x^2 + 9)^2} \, dx$.

Solution The factor $4x^2 + 9$ corresponds to $u^2 + a^2$ with $u = 2x$ and $a = 3$. Hence the substitution $u = a \tan \theta$ amounts to

$$2x = 3 \tan \theta, \qquad x = \tfrac{3}{2} \tan \theta, \qquad dx = \tfrac{3}{2} \sec^2 \theta \, d\theta.$$

This gives

$$\int \frac{1}{(4x^2 + 9)^2} \, dx = \int \frac{\tfrac{3}{2} \sec^2 \theta}{(9 \tan^2 \theta + 9)^2} \, d\theta$$

$$= \frac{3}{2} \int \frac{\sec^2 \theta}{(9 \sec^2 \theta)^2} \, d\theta = \frac{1}{54} \int \frac{1}{\sec^2 \theta} \, d\theta$$

$$= \tfrac{1}{54} \int \cos^2 \theta \, d\theta = \tfrac{1}{108} (\theta + \sin \theta \cos \theta) + C.$$

(The integration in the last step is the same as in Example 2.) Now $\theta = \tan^{-1}(2x/3)$, and the triangle of Fig. 9.6.3 gives

$$\sin \theta = \frac{2x}{\sqrt{4x^2 + 9}}, \qquad \cos \theta = \frac{3}{\sqrt{4x^2 + 9}}.$$

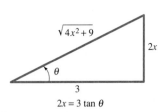

$2x = 3 \tan \theta$

Fig. 9.6.3 The reference triangle for Example 3

Hence

$$\int \frac{1}{(4x^2 + 9)^2}\, dx = \frac{1}{108}\left[\tan^{-1}\left(\frac{2x}{3}\right) + \frac{2x}{\sqrt{4x^2 + 9}} \cdot \frac{3}{\sqrt{4x^2 + 9}} \right] + C$$

$$= \frac{1}{108} \tan^{-1}\left(\frac{2x}{3}\right) + \frac{x}{18(4x^2 + 9)} + C.$$

What we mean by the substitution $u = a \sec \theta$ in an integral that contains $u^2 - a^2$ is the substitution

$$\theta = \sec^{-1}\frac{u}{a}, \qquad 0 \le \theta \le \pi,$$

where $|u| \ge a > 0$ (because of the domain and range of the inverse secant function). Then

$$\sqrt{u^2 - a^2} = \sqrt{a^2 \sec^2 \theta - a^2} = \sqrt{a^2 \tan^2 \theta} = \pm a \tan \theta.$$

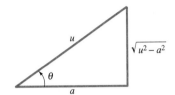

Fig. 9.6.4 The reference triangle for the substitution $u = a \sec \theta$

Here we must take the plus sign if $u > a$, so that $0 < \theta < \pi/2$ and $\tan \theta > 0$. If $u < -a$, so $\pi/2 < \theta < \pi$ and $\tan \theta < 0$, we take the minus sign. In either case the values of the various trigonometric functions of θ can be read from the right triangle in Fig. 9.6.4.

EXAMPLE 4 Find $\displaystyle\int \frac{\sqrt{x^2 - 25}}{x}\, dx, \; x > 5.$

Solution We substitute $x = 5 \sec \theta$, $dx = 5 \sec \theta \tan \theta\, d\theta$. Then

$$\sqrt{x^2 - 25} = \sqrt{25(\sec^2 \theta - 1)} = 5 \tan \theta,$$

because $x > 5$ implies $0 < \theta < \pi/2$, so $\tan \theta > 0$. Hence this substitution gives

$$\int \frac{\sqrt{x^2 - 25}}{x}\, dx = \int \frac{5 \tan \theta}{5 \sec \theta}(5 \sec \theta \tan \theta)\, d\theta$$

$$= 5 \int \tan^2 \theta\, d\theta = 5 \int (\sec^2 \theta - 1)\, d\theta$$

$$= 5 \tan \theta - 5\theta + C = \sqrt{x^2 - 25} - 5 \sec^{-1}\left(\frac{x}{5}\right) + C.$$

Hyperbolic substitutions may be used in a similar way—and with the same effect—as trigonometric substitutions. The three basic hyperbolic substitutions, which are not ordinarily memorized, are listed here for reference.

If the integral involves	Then substitute	And use the identity
$a^2 - u^2$	$u = a \tanh \theta$	$1 - \tanh^2 \theta = \operatorname{sech}^2 \theta$
$a^2 + u^2$	$u = a \sinh \theta$	$1 + \sinh^2 \theta = \cosh^2 \theta$
$u^2 - a^2$	$u = a \cosh \theta$	$\cosh^2 \theta - 1 = \sinh^2 \theta$

EXAMPLE 5 Find $\displaystyle\int \frac{1}{\sqrt{x^2 - 1}}\, dx, \; x > 1.$

Solution For purposes of comparison, we evaluate this integral both by trigonometric substitution and by hyperbolic substitution. The trigonometric substitution

$$x = \sec \theta, \qquad dx = \sec \theta \tan \theta \, d\theta, \qquad \tan \theta = \sqrt{x^2 - 1}$$

gives

$$\int \frac{1}{\sqrt{x^2 - 1}} \, dx = \int \frac{\sec \theta \tan \theta}{\tan \theta} \, d\theta = \int \sec \theta \, d\theta$$

$$= \ln |\sec \theta + \tan \theta| + C \qquad \text{[Eq. (7), Section 9.3]}$$

$$= \ln |x + \sqrt{x^2 - 1}| + C.$$

Using instead the hyperbolic substitution $x = \cosh \theta$, $dx = \sinh \theta \, d\theta$, we have

$$\sqrt{x^2 - 1} = \sqrt{\cosh^2 \theta - 1} = \sinh \theta.$$

We take the positive square root here, because $x > 1$ implies that $\theta = \cosh^{-1} x > 0$ and thus that $\sinh \theta > 0$. Hence

$$\int \frac{1}{\sqrt{x^2 - 1}} \, dx = \int \frac{\sinh \theta}{\sinh \theta} \, d\theta = \int 1 \, d\theta = \theta + C = \cosh^{-1} x + C.$$

The two results appear to differ, but Eq. (35) in Section 8.5 shows that they are equivalent.

9.6 Problems

Use trigonometric substitutions to evaluate the integrals in Problems 1 through 36.

1. $\displaystyle\int \frac{1}{\sqrt{16 - x^2}} \, dx$

2. $\displaystyle\int \frac{1}{\sqrt{4 - 9x^2}} \, dx$

3. $\displaystyle\int \frac{1}{x^2 \sqrt{4 - x^2}} \, dx$

4. $\displaystyle\int \frac{1}{x^2 \sqrt{x^2 - 25}} \, dx$

5. $\displaystyle\int \frac{x^2}{\sqrt{16 - x^2}} \, dx$

6. $\displaystyle\int \frac{x^2}{\sqrt{9 - 4x^2}} \, dx$

7. $\displaystyle\int \frac{1}{(9 - 16x^2)^{3/2}} \, dx$

8. $\displaystyle\int \frac{1}{(25 + 16x^2)^{3/2}} \, dx$

9. $\displaystyle\int \frac{\sqrt{x^2 - 1}}{x^2} \, dx$

10. $\displaystyle\int x^3 \sqrt{4 - x^2} \, dx$

11. $\displaystyle\int x^3 \sqrt{9 + 4x^2} \, dx$

12. $\displaystyle\int \frac{x^3}{\sqrt{x^2 + 25}} \, dx$

13. $\displaystyle\int \frac{\sqrt{1 - 4x^2}}{x} \, dx$

14. $\displaystyle\int \frac{1}{\sqrt{1 + x^2}} \, dx$

15. $\displaystyle\int \frac{1}{\sqrt{9 + 4x^2}} \, dx$

16. $\displaystyle\int \sqrt{1 + 4x^2} \, dx$

17. $\displaystyle\int \frac{x^2}{\sqrt{25 - x^2}} \, dx$

18. $\displaystyle\int \frac{x^3}{\sqrt{25 - x^2}} \, dx$

19. $\displaystyle\int \frac{x^2}{\sqrt{1 + x^2}} \, dx$

20. $\displaystyle\int \frac{x^3}{\sqrt{1 + x^2}} \, dx$

21. $\displaystyle\int \frac{x^2}{\sqrt{4 + 9x^2}} \, dx$

22. $\displaystyle\int (1 - x^2)^{3/2} \, dx$

23. $\displaystyle\int \frac{1}{(1 + x^2)^{3/2}} \, dx$

24. $\displaystyle\int \frac{1}{(4 - x^2)^2} \, dx$

25. $\displaystyle\int \frac{1}{(4 - x^2)^3} \, dx$

26. $\displaystyle\int \frac{1}{(4x^2 + 9)^3} \, dx$

27. $\displaystyle\int \sqrt{9 + 16x^2} \, dx$

28. $\displaystyle\int (9 + 16x^2)^{3/2} \, dx$

29. $\displaystyle\int \frac{\sqrt{x^2 - 25}}{x} \, dx$

30. $\displaystyle\int \frac{\sqrt{9x^2 - 16}}{x} \, dx$

31. $\displaystyle\int x^2 \sqrt{x^2 - 1} \, dx$

32. $\displaystyle\int \frac{x^2}{\sqrt{4x^2 - 9}} \, dx$

33. $\displaystyle\int \frac{1}{(4x^2 - 1)^{3/2}} \, dx$

34. $\displaystyle\int \frac{1}{x^2 \sqrt{4x^2 - 9}} \, dx$

35. $\displaystyle\int \frac{\sqrt{x^2 - 5}}{x^2} \, dx$

36. $\displaystyle\int (4x^2 - 5)^{3/2} \, dx$

Use hyperbolic substitutions to evaluate the integrals in Problems 37 through 41.

37. $\displaystyle\int \frac{1}{\sqrt{25 + x^2}}\, dx$ **38.** $\displaystyle\int \sqrt{1 + x^2}\, dx$

39. $\displaystyle\int \frac{\sqrt{x^2 - 4}}{x^2}\, dx$ **40.** $\displaystyle\int \frac{1}{\sqrt{1 + 9x^2}}\, dx$

41. $\displaystyle\int x^2\sqrt{1 + x^2}\, dx$

42. Compute the arc length of the parabola $y = x^2$ over the interval $[0, 1]$.

43. Compute the area of the surface obtained by revolving around the x-axis the parabolic arc of Problem 42.

44. Show that the length of one arch of the sine curve $y = \sin x$ is equal to half the circumference of the ellipse $x^2 + (y^2/2) = 1$. [*Suggestion:* Substitute $x = \cos\theta$ into the arc-length integral for the ellipse.] See Fig. 9.6.5.

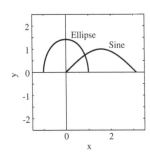

Fig. 9.6.5   Two arcs with the same length (Problem 44)

45. Compute the arc length of the curve $y = \ln x$ over the interval $[1, 2]$.

46. Compute the area of the surface obtained by revolving around the y-axis the curve of Problem 45.

47. A torus (see Fig. 9.6.6) is obtained by revolving around the y-axis the circle

$$(x - b)^2 + y^2 = a^2 \qquad (0 < a \le b).$$

Show that the surface area of the torus is $4\pi^2 ab$.

Fig. 9.6.6   The torus of Problem 47

48. Find the area under the curve $y = \sqrt{9 + x^2}$ over the interval $[0, 4]$.

49. Find the area of the surface obtained by revolving around the x-axis the curve $y = \sin x$, $0 \le x \le \pi$ (see Fig. 9.6.7).

Fig. 9.6.7   The pointed football of Problem 49

50. An ellipsoid of revolution is obtained by revolving the ellipse $x^2/a^2 + y^2/b^2 = 1$ around the x-axis. Suppose that $a > b$. Show that the ellipsoid has surface area

$$A = 2\pi ab\left[\frac{b}{a} + \frac{a}{c}\sin^{-1}\left(\frac{c}{a}\right)\right],$$

where $c = \sqrt{a^2 - b^2}$. Assume that $a \approx b$, so that $c \approx 0$ and $\sin^{-1}(c/a) \approx c/a$. Conclude that $A \approx 4\pi a^2$.

51. Suppose that $b > a$ for the ellipsoid of revolution of Problem 50. Show that its surface area is then

$$A = 2\pi ab\left[\frac{b}{a} + \frac{a}{c}\ln\left(\frac{b + c}{a}\right)\right],$$

where $c = \sqrt{b^2 - a^2}$. Use the fact that $\ln(1 + x) \approx x$ if $x \approx 0$, and thereby conclude that $A \approx 4\pi a^2$ if $a \approx b$.

52. A road is to be built from the point $(2, 1)$ to the point $(5, 3)$, following the path of the parabola

$$y = -1 + 2\sqrt{x - 1}.$$

Calculate the length of this road (the units on the coordinate axes are in miles). [*Suggestion:* Substitute $x = \sec^2\theta$ into the arc-length integral.]

53. Suppose that the cost of the road in Problem 52 is $\sqrt{x}$ million dollars per mile. Calculate the total cost of the road.

54. A kite is flying at a height of 500 ft and at a horizontal distance of 100 ft from the string-holder on the ground. The kite string weighs $1/16$ oz/ft and is hanging in the shape of the parabola $y = x^2/20$ that joins the string-holder at $(0, 0)$ to the kite at $(100, 500)$ (Fig. 9.6.8). Calculate the work (in foot-pounds) done in lifting the kite string from the ground to its present position.

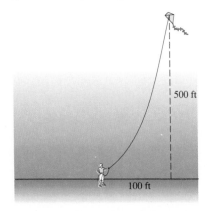

Fig. 9.6.8   The kite string of Problem 54

Many integrals containing a square root or negative power of a quadratic polynomial $ax^2 + bx + c$ can be simplified by the process of *completing the square*. For example,

$$x^2 + 2x + 2 = (x + 1)^2 + 1,$$

and hence the substitution $u = x + 1$, $du = dx$ yields

$$\int \frac{1}{x^2 + 2x + 2}\, dx = \int \frac{1}{u^2 + 1}\, du = \tan^{-1} u + C = \tan^{-1}(x + 1) + C.$$

In general, the objective is to convert $ax^2 + bx + c$ into either a sum or a difference of two squares—either $u^2 \pm a^2$ or $a^2 - u^2$—so that the method of trigonometric substitution can then be used. To see how this works in practice, suppose first that $a = 1$, so that the quadratic in question has the form $x^2 + bx + c$. The sum $x^2 + bx$ of the first two terms can be completed to a perfect square by adding $b^2/4$, the square of half the coefficient of x, and in turn subtracting $b^2/4$ from the constant term c. This gives

$$x^2 + bx + c = \left(x^2 + bx + \frac{b^2}{4}\right) + \left(c - \frac{b^2}{4}\right)$$

$$= \left(x + \frac{b}{2}\right)^2 + \left(c - \frac{b^2}{4}\right).$$

With $u = x + \frac{1}{2}b$, this result is of the form $u^2 + A^2$ or $u^2 - A^2$ (depending on the sign of $c - \frac{1}{4}b^2$). If the coefficient a of x^2 is not 1, we first factor it out and proceed as before:

$$ax^2 + bx + c = a\left(x^2 + \frac{b}{a}x + \frac{c}{a}\right).$$

EXAMPLE 1 Find $\displaystyle \int \frac{1}{9x^2 + 6x + 5}\, dx.$

Solution The first step is to complete the square:

$$9x^2 + 6x + 5 = 9(x^2 + \tfrac{2}{3}x) + 5 = 9(x^2 + \tfrac{2}{3}x + \tfrac{1}{9}) - 1 + 5$$

$$= 9(x + \tfrac{1}{3})^2 + 4 = (3x + 1)^2 + 2^2.$$

Hence

$$\int \frac{1}{9x^2 + 6x + 5}\, dx = \int \frac{1}{(3x + 1)^2 + 4}\, dx$$

$$= \frac{1}{3}\int \frac{1}{u^2 + 4}\, du \qquad (u = 3x + 1)$$

$$= \frac{1}{6}\int \frac{\frac{1}{2}}{(\frac{1}{2}u)^2 + 1}\, du$$

$$= \frac{1}{6}\int \frac{1}{v^2 + 1}\, dv \qquad \left(v = \frac{1}{2}u\right)$$

$$= \frac{1}{6} \tan^{-1} v + C = \frac{1}{6} \tan^{-1}\left(\frac{u}{2}\right) + C$$

$$= \frac{1}{6} \tan^{-1}\left(\frac{3x + 1}{2}\right) + C.$$

EXAMPLE 2   Find $\displaystyle\int \frac{1}{\sqrt{9 + 16x - 4x^2}}\,dx.$

Solution   First we complete the square:

$$9 + 16x - 4x^2 = 9 - 4(x^2 - 4x) = 9 - 4(x^2 - 4x + 4) + 16$$
$$= 25 - 4(x - 2)^2.$$

Hence

$$\int \frac{1}{\sqrt{9 + 16x - 4x^2}}\,dx = \int \frac{1}{\sqrt{25 - 4(x - 2)^2}}\,dx$$

$$= \frac{1}{5} \int \frac{1}{\sqrt{1 - \frac{4}{25}(x - 2)^2}}\,dx$$

$$= \frac{1}{2} \int \frac{1}{\sqrt{1 - u^2}}\,du \qquad \left(u = \frac{2(x - 2)}{5}\right)$$

$$= \frac{1}{2} \sin^{-1} u + C = \frac{1}{2} \sin^{-1}\frac{2(x - 2)}{5} + C.$$

An alternative approach is to make the trigonometric substitution

$$2(x - 2) = 5 \sin \theta, \qquad 2\,dx = 5 \cos \theta\,d\theta$$

immediately after completing the square. This yields

$$\int \frac{1}{\sqrt{9 + 16x - 4x^2}}\,dx = \int \frac{1}{\sqrt{25 - 4(x - 2)^2}}\,dx$$

$$= \int \frac{\frac{5}{2}\cos \theta}{\sqrt{25 - 25 \sin^2 \theta}}\,d\theta$$

$$= \frac{1}{2} \int 1\,d\theta = \frac{1}{2}\theta + C$$

$$= \frac{1}{2} \arcsin\frac{2(x - 2)}{5} + C.$$

Some integrals that contain a quadratic expression can be split into two simpler integrals. Examples 3 and 4 illustrate this technique.

EXAMPLE 3   Find $\displaystyle\int \frac{2x + 3}{9x^2 + 6x + 5}\,dx.$

Solution   Because $D_x(9x^2 + 6x + 5) = 18x + 6$, this would be a simpler integral if the numerator $2x + 3$ were a constant multiple of $18x + 6$. Our strategy is to write

$$2x + 3 = A(18x + 6) + B$$

so that we can split the given integral into a sum of two integrals, one of which has numerator $18x + 6$ in its integrand. By matching coefficients in

$$2x + 3 = 18Ax + (6A + B),$$

we find that $A = \frac{1}{9}$ and $B = \frac{7}{3}$. Hence

$$\int \frac{2x + 3}{9x^2 + 6x + 5} \, dx = \frac{1}{9} \int \frac{18x + 6}{9x^2 + 6x + 5} \, dx + \frac{7}{3} \int \frac{1}{9x^2 + 6x + 5} \, dx.$$

The first integral on the right is a logarithm, and the second is given in Example 1. Thus

$$\int \frac{2x + 3}{9x^2 + 6x + 5} \, dx = \frac{1}{9} \ln(9x^2 + 6x + 5) + \frac{7}{18} \tan^{-1}\left(\frac{3x + 1}{2}\right) + C.$$

Alternatively, we could first complete the square in the denominator. The substitution $u = 3x + 1$, $x = \frac{1}{3}(u - 1)$, $dx = \frac{1}{3} \, du$ then gives

$$\int \frac{2x + 3}{(3x + 1)^2 + 4} \, dx = \int \frac{\frac{2}{3}(u - 1) + 3}{u^2 + 4} \cdot \frac{1}{3} \, du$$

$$= \frac{1}{9} \int \frac{2u}{u^2 + 4} \, du + \frac{7}{9} \int \frac{1}{u^2 + 4} \, du$$

$$= \frac{1}{9} \ln(u^2 + 4) + \frac{7}{18} \tan^{-1}\left(\frac{u}{2}\right) + C$$

$$= \frac{1}{9} \ln(9x^2 + 6x + 5) + \frac{7}{18} \tan^{-1}\left(\frac{3x + 1}{2}\right) + C.$$

EXAMPLE 4   Find $\displaystyle\int \frac{2 + 6x}{(3 + 2x - x^2)^2} \, dx$ given $|x - 1| < 2$.

Solution   Because $D_x(3 + 2x - x^2) = 2 - 2x$, we first write

$$\int \frac{2 + 6x}{(3 + 2x - x^2)^2} \, dx$$

$$= -3 \int \frac{2 - 2x}{(3 + 2x - x^2)^2} \, dx + 8 \int \frac{1}{(3 + 2x - x^2)^2} \, dx.$$

Then let $u = 3 + 2x - x^2$, $du = (2 - 2x) \, dx$ in the first integral to obtain

$$-3 \int \frac{2 - 2x}{(3 + 2x - x^2)^2} \, dx = -3 \int \frac{du}{u^2} = \frac{3}{u} + C_1 = \frac{3}{3 + 2x - x^2} + C_1.$$

Therefore,

$$\int \frac{2 + 6x}{(3 + 2x - x^2)^2} \, dx = \frac{3}{3 + 2x - x^2} + 8 \int \frac{1}{(3 + 2x - x^2)^2} \, dx. \quad (1)$$

(We can drop the constant C_1 because it can be absorbed by the constant C we will obtain when we evaluate the remaining integral.) To evaluate the remaining integral, we complete the square:

$$3 + 2x - x^2 = 4 - (x^2 - 2x + 1) = 4 - (x - 1)^2.$$

Because $|x - 1| < 2$, this suggests the substitution

$$x - 1 = 2 \sin \theta, \qquad dx = 2 \cos \theta \, d\theta,$$

with which

$$3 + 2x - x^2 = 4 - 4 \sin^2 \theta = 4 \cos^2 \theta.$$

This substitution yields

$$8 \int \frac{1}{(3 + 2x - x^2)^2} \, dx = 8 \int \frac{2 \cos \theta}{(4 \cos^2 \theta)^2} \, d\theta = \int \sec^3 \theta \, d\theta$$

$$= \tfrac{1}{2} \sec \theta \tan \theta + \tfrac{1}{2} \int \sec \theta \, d\theta \qquad \begin{array}{l}\text{[by Eq. (6)} \\ \text{in Section 9.4]}\end{array}$$

$$= \tfrac{1}{2} \sec \theta \tan \theta + \tfrac{1}{2} \ln| \sec \theta + \tan \theta| + C$$

$$= \frac{x - 1}{3 + 2x - x^2} + \frac{1}{2} \ln \left| \frac{x + 1}{\sqrt{3 + 2x - x^2}} \right| + C. \quad (2)$$

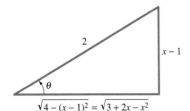

$$\sqrt{4 - (x - 1)^2} = \sqrt{3 + 2x - x^2}$$

Fig. 9.7.1 The reference triangle for Example 4

In the last step we read the values of $\sec \theta$ and $\tan \theta$ from the right triangle in Fig. 9.7.1. When we substitute Eq. (2) into Eq. (1), we finally obtain the result

$$\int \frac{2 + 6x}{(3 + 2x - x^2)^2} \, dx = \frac{x + 2}{3 + 2x - x^2} + \frac{1}{2} \ln \left| \frac{x + 1}{\sqrt{3 + 2x - x^2}} \right| + C.$$

The method of Example 4 can be used to evaluate a general integral of the form

$$\int \frac{Ax + B}{(ax^2 + bx + c)^n} \, dx, \quad (3)$$

where n is a positive integer. By splitting such an integral into two simpler ones and by completing the square in the quadratic expression in the denominator, the problem of evaluating the integral in Eq. (3) can be reduced to that of computing

$$\int \frac{1}{(a^2 \pm u^2)^n} \, du. \quad (4)$$

If the sign in the denominator of Eq. (4) is the plus sign, then the substitution $u = a \tan \theta$ transforms the integral into the form

$$\int \cos^m \theta \, d\theta$$

(see Problem 35). This integral can be handled by the methods of Section 9.3 or by using the reduction formula

$$\int \cos^k \theta \, d\theta = \frac{1}{k} \cos^{k-1} \theta \sin \theta + \frac{k - 1}{k} \int \cos^{k-2} \theta \, d\theta$$

of Problem 46 in Section 9.4.

If the sign in the denominator of Eq. (4) is the minus sign, then the substitution $u = a \sin \theta$ transforms the integral into the form

$$\int \sec^m \theta \; d\theta$$

(see Problem 36). This integral may be evaluated with the aid of the reduction formula

$$\int \sec^k \theta \; d\theta = \frac{1}{k-1} \sec^{k-2} \theta \tan \theta + \frac{k-2}{k-1} \int \sec^{k-2} \theta \; d\theta$$

[Eq. (5) of Section 9.4].

9.7 Problems

Evaluate the antiderivatives in Problems 1 through 34.

1. $\displaystyle\int \frac{1}{x^2 + 4x + 5} \, dx$ **2.** $\displaystyle\int \frac{2x + 5}{x^2 + 4x + 5} \, dx$

3. $\displaystyle\int \frac{5 - 3x}{x^2 + 4x + 5} \, dx$ **4.** $\displaystyle\int \frac{x + 1}{(x^2 + 4x + 5)^2} \, dx$

5. $\displaystyle\int \frac{1}{\sqrt{3 - 2x - x^2}} \, dx$ **6.** $\displaystyle\int \frac{x + 3}{\sqrt{3 - 2x - x^2}} \, dx$

7. $\displaystyle\int x\sqrt{3 - 2x - x^2} \, dx$ **8.** $\displaystyle\int \frac{1}{4x^2 + 4x - 3} \, dx$

9. $\displaystyle\int \frac{3x + 2}{4x^2 + 4x - 3} \, dx$ **10.** $\displaystyle\int \sqrt{4x^2 + 4x - 3} \, dx$

11. $\displaystyle\int \frac{1}{x^2 + 4x + 13} \, dx$ **12.** $\displaystyle\int \frac{1}{\sqrt{2x - x^2}} \, dx$

13. $\displaystyle\int \frac{1}{3 + 2x - x^2} \, dx$ **14.** $\displaystyle\int x\sqrt{8 + 2x - x^2} \, dx$

15. $\displaystyle\int \frac{2x - 5}{x^2 + 2x + 2} \, dx$ **16.** $\displaystyle\int \frac{2x - 1}{4x^2 + 4x - 15} \, dx$

17. $\displaystyle\int \frac{x}{\sqrt{5 + 12x - 9x^2}} \, dx$

18. $\displaystyle\int (3x - 2)\sqrt{9x^2 + 12x + 8} \, dx$

19. $\displaystyle\int (7 - 2x)\sqrt{9 + 16x - 4x^2} \, dx$

20. $\displaystyle\int \frac{2x + 3}{\sqrt{x^2 + 2x + 5}} \, dx$

21. $\displaystyle\int \frac{x + 4}{(6x - x^2)^{3/2}} \, dx$ **22.** $\displaystyle\int \frac{x - 1}{(x^2 + 1)^2} \, dx$

23. $\displaystyle\int \frac{2x + 3}{(4x^2 + 12x + 13)^2} \, dx$ **24.** $\displaystyle\int \frac{x^3}{(1 - x^2)^4} \, dx$

25. $\displaystyle\int \frac{3x - 1}{x^2 + x + 1} \, dx$ **26.** $\displaystyle\int \frac{3x - 1}{(x^2 + x + 1)^2} \, dx$

27. $\displaystyle\int \frac{1}{(x^2 - 4)^2} \, dx$ **28.** $\displaystyle\int (x - x^2)^{3/2} \, dx$

29. $\displaystyle\int \frac{x^2 + 1}{x^3 + x^2 + x} \, dx$ **30.** $\displaystyle\int \frac{x^2 + 2}{(x^2 + 1)^2} \, dx$

31. $\displaystyle\int \frac{2x^2 + 3}{x^4 - 2x^2 + 1} \, dx$

32. $\displaystyle\int \frac{x^2 + 4}{(x^2 + 1)^2(x^2 + 2)} \, dx$

33. $\displaystyle\int \frac{3x + 1}{(x^2 + 2x + 5)^2} \, dx$ **34.** $\displaystyle\int \frac{x^3 - 2x}{x^2 + 2x + 2} \, dx$

35. Show that the substitution $u = a \tan \theta$ gives

$$\int \frac{1}{(a^2 + u^2)^n} \, du = \frac{1}{a^{2n-1}} \int \cos^{2n-2} \theta \; d\theta.$$

36. Show that the substitution $u = a \sin \theta$ gives

$$\int \frac{1}{(a^2 - u^2)^n} \, du = \frac{1}{a^{2n-1}} \int \sec^{2n-2} \theta \; d\theta.$$

37. Your task is to build a road that joins the points $(0, 0)$ and $(3, 2)$ and follows the path of the circle with equation $(4x + 4)^2 + (4y - 19)^2 = 377$. Find the length of this road. (Units on the coordinate axes are measured in miles.)

38. Suppose that the road of Problem 37 costs $10/(1 + x)$ million dollars per mile. (a) Calculate its total cost. (b) With the same cost per mile, calculate the total cost of a straight line road from $(0, 0)$ to $(3, 2)$. You should find that it is *more* expensive than the *longer* circular road!

In Problems 39 through 41, factor the denominator by first noting by inspection a root r of the denominator and then employing long division by x − r. Finally, use the method of partial fractions to aid in finding the indicated antiderivative.

39. $\displaystyle\int \frac{3x + 2}{x^3 + x^2 - 2} \, dx$ **40.** $\displaystyle\int \frac{1}{x^3 + 8} \, dx$

41. $\displaystyle\int \frac{x^4 + 2x^2}{x^3 - 1}\, dx$

42. (a) Find constants a and b such that

$$x^4 + 1 = (x^2 + ax + 1)(x^2 + bx + 1).$$

(b) Prove that

$$\int_0^1 \frac{x^2 + 1}{x^4 + 1}\, dx = \frac{\pi}{2\sqrt{2}}.$$

[*Suggestion:* If u and v are positive numbers and $uv = 1$, then

$$\arctan u + \arctan v = \frac{\pi}{2}.]$$

43. Factor $x^4 + x^2 + 1$ with the aid of ideas suggested in Problem 42. Then evaluate

$$\int \frac{2x^3 + 3x}{x^4 + x^2 + 1}\, dx.$$

9.8

Improper Integrals

To show the existence of the definite integral, we have relied until now on the existence theorem stated in Section 5.4. This is the theorem that guarantees the existence of the definite integral $\int_a^b f(x)\, dx$ provided that the function f is *continuous* on the closed and bounded interval $[a, b]$. Certain applications in calculus, however, lead naturally to the formulation of integrals in which either

1. The interval of integration is not bounded; it has one of the forms

$$[a, +\infty), \qquad (-\infty, a], \quad \text{or} \quad (-\infty, +\infty); \text{ or}$$

2. The integrand has an infinite discontinuity at some point c:

$$\lim_{x \to c} f(x) = \pm\infty.$$

An example of Case 1 is the integral

$$\int_1^\infty \frac{1}{x^2}\, dx.$$

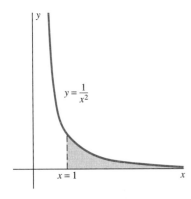

A geometric interpretation of this integral is the area of the unbounded region (shaded in Fig. 9.8.1) that lies between the curve $y = 1/x^2$ and the x-axis and to the right of the vertical line $x = 1$. An example of Case 2 is the integral

$$\int_0^1 \frac{1}{\sqrt{x}}\, dx.$$

Fig. 9.8.1 The shaded area cannot be measured by using our earlier techniques.

This integral may be interpreted to be the area of the unbounded region (shaded in Fig. 9.8.2) that lies under the curve $y = 1/\sqrt{x}$ from $x = 0$ to $x = 1$.

Such integrals are called **improper integrals.** The natural interpretation of an improper integral is the area of an unbounded region. It is perhaps surprising that such an area can nevertheless be finite, and here we shall show how to find such areas—that is, how to evaluate improper integrals.

To see why improper integrals require special care, let us consider the integral

$$\int_{-1}^1 \frac{1}{x^2}\, dx.$$

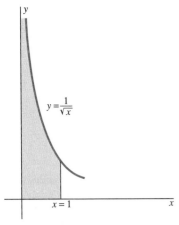

Fig. 9.8.2 Another area that must be measured with an improper integral

This integral is improper because its integrand $f(x) = 1/x^2$ is unbounded as $x \to 0$, and thus f is not continuous at $x = 0$. If we thoughtlessly applied the fundamental theorem of calculus, we would obtain

$$\int_{-1}^{1} \frac{1}{x^2}\, dx = \left[-\frac{1}{x}\right]_{-1}^{1} = (-1) - (+1) = -2. \qquad \textbf{(Wrong!)}$$

The negative answer is obviously incorrect, because the area shown in Fig. 9.8.3 lies above the x-axis and hence cannot be negative. This simple example emphasizes that we cannot ignore the hypotheses—*continuous* function and *bounded closed* interval—of the fundamental theorem of calculus.

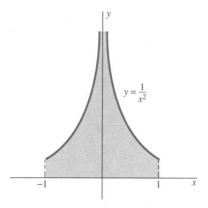

Fig. 9.8.3 The area under $y = 1/x^2$, $-1 \leq x \leq 1$

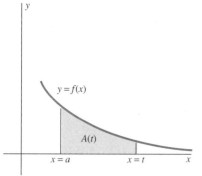

Fig. 9.8.4 The shaded area $A(t)$ exists provided that f is continuous.

INFINITE LIMITS OF INTEGRATION

Suppose that the function f is continuous and nonnegative on the unbounded interval $[a, +\infty)$. Then, for any fixed $t > a$, the area $A(t)$ of the region under $y = f(x)$ from $x = a$ to $x = t$ (shaded in Fig. 9.8.4) is given by the (ordinary) definite integral

$$A(t) = \int_{a}^{t} f(x)\, dx.$$

Suppose now that we let $t \to +\infty$ and find that the limit of $A(t)$ exists. Then we may regard this limit as the area of the unbounded region that lies under $y = f(x)$ and over $[a, +\infty)$. For f continuous on $[a, +\infty)$, we therefore *define*

$$\int_{a}^{\infty} f(x)\, dx = \lim_{t \to \infty} \int_{a}^{t} f(x)\, dx \qquad (1)$$

provided that this limit exists (as a finite number). If this limit exists, we say that the improper integral on the left **converges;** if the limit does not exist, we say that the improper integral **diverges.** If $f(x)$ is nonnegative on $[a, +\infty)$, then the limit in Eq. (1) either exists or is infinite, and in the latter case we write

$$\int_{a}^{\infty} f(x)\, dx = +\infty$$

and say that the improper integral **diverges to infinity.**

If the function f has both positive and negative values on $[a, +\infty)$, then the improper integral can diverge *by oscillation*—that is, without diverging to infinity. This occurs with $\int_{0}^{\infty} \sin x\, dx$, because it is easy to verify that $\int_{0}^{t} \sin x\, dx$ is zero if t is an even multiple of π but is 2 if t is an odd multiple of π. Thus $\int_{0}^{t} \sin x\, dx$ oscillates between 0 and 2 as $t \to +\infty$, and so the limit in Eq. (1) does not exist.

We handle an infinite lower limit of integration similarly: We define

$$\int_{-\infty}^{b} f(x)\, dx = \lim_{t \to -\infty} \int_{t}^{b} f(x)\, dx \qquad (2)$$

provided that the limit exists. If the function f is continuous on the whole real line, we define

$$\int_{-\infty}^{\infty} f(x)\, dx = \int_{-\infty}^{c} f(x)\, dx + \int_{c}^{\infty} f(x)\, dx \qquad (3)$$

Ch. 9 / Techniques of Integration

for any convenient choice of c, provided that both improper integrals on the right-hand side converge. Note that $\int_{-\infty}^{\infty} f(x)\,dx$ is *not* necessarily equal to

$$\lim_{t \to \infty} \int_{-t}^{t} f(x)\,dx$$

(see Problem 28).

It makes no difference what value of c is used in Eq. (3), because if $c < d$, then

$$\int_{-\infty}^{c} f(x)\,dx + \int_{c}^{\infty} f(x)\,dx = \int_{-\infty}^{c} f(x)\,dx + \int_{c}^{d} f(x)\,dx + \int_{d}^{\infty} f(x)\,dx$$

$$= \int_{-\infty}^{d} f(x)\,dx + \int_{d}^{\infty} f(x)\,dx,$$

under the assumption that the limits involved all exist.

EXAMPLE 1 Investigate the improper integrals (a) $\int_{1}^{\infty} \dfrac{1}{x^2}\,dx$ and (b) $\int_{-\infty}^{0} \dfrac{1}{\sqrt{1-x}}\,dx.$

Solution

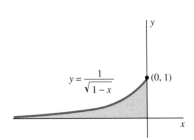

$y = \dfrac{1}{\sqrt{1-x}}$ (0, 1)

(a) $\displaystyle\int_{1}^{\infty} \frac{1}{x^2}\,dx = \lim_{t \to \infty} \int_{1}^{t} \frac{1}{x^2}\,dx = \lim_{t \to \infty} \left[-\frac{1}{x}\right]_{1}^{t} = \lim_{t \to \infty} \left(-\frac{1}{t} + 1\right) = 1.$

Thus this improper integral converges, and this is the area of the region shaded in Fig. 9.8.1.

(b) $\displaystyle\int_{-\infty}^{0} \frac{1}{\sqrt{1-x}}\,dx = \lim_{t \to -\infty} \int_{t}^{0} \frac{1}{\sqrt{1-x}}\,dx = \lim_{t \to -\infty} \left[-2\sqrt{1-x}\right]_{t}^{0}$

$$= \lim_{t \to -\infty} (2\sqrt{1-t} - 2) = +\infty.$$

Fig. 9.8.5 The unbounded region represented by the improper integral of Example 1(b)

Thus the second improper integral of the example diverges to $+\infty$ (Fig. 9.8.5).

EXAMPLE 2 Investigate the improper integral $\displaystyle\int_{-\infty}^{\infty} \frac{1}{1+x^2}\,dx.$

Solution The choice $c = 0$ in Eq. (3) gives

$$\int_{-\infty}^{\infty} \frac{1}{1+x^2}\,dx = \int_{-\infty}^{0} \frac{1}{1+x^2}\,dx + \int_{0}^{\infty} \frac{1}{1+x^2}\,dx$$

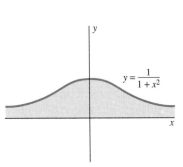

$y = \dfrac{1}{1+x^2}$

$$= \lim_{s \to -\infty} \int_{s}^{0} \frac{1}{1+x^2}\,dx + \lim_{t \to \infty} \int_{0}^{t} \frac{1}{1+x^2}\,dx$$

$$= \lim_{s \to -\infty} \left[\tan^{-1} x\right]_{s}^{0} + \lim_{t \to \infty} \left[\tan^{-1} x\right]_{0}^{t}$$

$$= \lim_{s \to -\infty} (-\tan^{-1} s) + \lim_{t \to \infty} (\tan^{-1} t) = \frac{\pi}{2} + \frac{\pi}{2} = \pi.$$

Fig. 9.8.6 The area measured by the integral of Example 2

The shaded region in Fig. 9.8.6 is a geometric interpretation of the integral of Example 2.

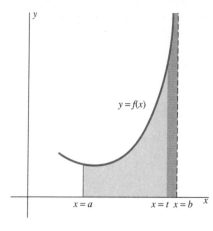

Fig. 9.8.7 An improper integral of the second type: $f(x) \to \infty$ as $x \to b^-$.

INFINITE INTEGRANDS

Suppose that the function f is continuous and nonnegative on $[a, b]$ but that $f(x) \to +\infty$ as $x \to b^-$. The graph of such a function appears in Fig. 9.8.7. The area $A(t)$ of the region lying under $y = f(x)$ from $x = a$ to $x = t < b$ is the value of the (ordinary) definite integral

$$A(t) = \int_a^t f(x)\, dx.$$

If the limit of $A(t)$ exists as $t \to b^-$, then this limit may be regarded as the area of the (unbounded) region under $y = f(x)$ from $x = a$ to $x = b$. For f continuous on $[a, b)$, we therefore *define*

$$\int_a^b f(x)\, dx = \lim_{t \to b^-} \int_a^t f(x)\, dx, \qquad (4)$$

provided that this limit exists (as a finite number), in which case we say that the improper integral on the left **converges**; if the limit does not exist, we say that it **diverges**. If

$$\int_a^b f(x)\, dx = \lim_{t \to b^-} \int_a^t f(x)\, dx = \infty,$$

then we say that the improper integral **diverges to infinity**.

If f is continuous on $(a, b]$ but the limit of $f(x)$ as $x \to a^+$ is infinite, then we *define*

$$\int_a^b f(x)\, dx = \lim_{t \to a^+} \int_t^b f(x)\, dx, \qquad (5)$$

provided that the limit exists. If f is continuous at every point of $[a, b]$ except for the point c in (a, b) and one or both one-sided limits of f at c are infinite, then we *define*

$$\int_a^b f(x)\, dx = \int_a^c f(x)\, dx + \int_c^b f(x)\, dx \qquad (6)$$

provided that both improper integrals on the right converge.

526

EXAMPLE 3 Investigate the improper integrals

(a) $\int_0^1 \frac{1}{\sqrt{x}}\, dx$ and (b) $\int_1^2 \frac{1}{(x-2)^2}\, dx.$

Solution

(a) The integrand $1/\sqrt{x}$ becomes infinite as $x \to 0^+$, so

$$\int_0^1 \frac{1}{\sqrt{x}}\, dx = \lim_{t \to 0^+} \int_t^1 \frac{1}{\sqrt{x}}\, dx$$

$$= \lim_{t \to 0^+} \left[2\sqrt{x} \right]_t^1 = \lim_{t \to 0^+} 2(1 - \sqrt{t}) = 2.$$

Thus the area of the unbounded region shown in Fig. 9.8.2 is 2.

(b) Here the integrand becomes infinite as x approaches the right-hand endpoint, so

$$\int_1^2 \frac{1}{(x-2)^2}\, dx = \lim_{t \to 2^-} \int_1^t \frac{1}{(x-2)^2}\, dx$$

$$= \lim_{t \to 2^-} \left[-\frac{1}{x-2} \right]_1^t = \lim_{t \to 2^-} \left(-1 - \frac{1}{t-2} \right) = +\infty.$$

Hence this improper integral diverges to infinity (Fig. 9.8.8). It follows that the improper integral

$$\int_1^3 \frac{1}{(x-2)^2}\, dx = \int_1^2 \frac{1}{(x-2)^2}\, dx + \int_2^3 \frac{1}{(x-2)^2}\, dx$$

also diverges, because not both of the right-hand improper integrals converge. (You can verify that the second one also diverges to $+\infty$.)

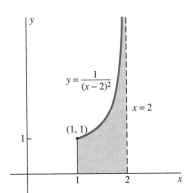

Fig. 9.8.8 The unbounded region represented by the improper integral of Example 3(b)

EXAMPLE 4 Investigate the improper integral

$$\int_0^2 \frac{1}{(2x-1)^{2/3}}\, dx.$$

Solution This improper integral corresponds to the region shaded in Fig. 9.8.9. The integrand has an infinite discontinuity at the point $c = \frac{1}{2}$ within the interval of integration, so we write

$$\int_0^2 \frac{1}{(2x-1)^{2/3}}\, dx = \int_0^{1/2} \frac{1}{(2x-1)^{2/3}}\, dx + \int_{1/2}^2 \frac{1}{(2x-1)^{2/3}}\, dx$$

and investigate separately the two improper integrals on the right. We find that

$$\int_0^{1/2} \frac{1}{(2x-1)^{2/3}}\, dx = \lim_{t \to (1/2)^-} \int_0^t \frac{1}{(2x-1)^{2/3}}\, dx$$

$$= \lim_{t \to (1/2)^-} \left[\frac{3}{2}(2x-1)^{1/3} \right]_0^t$$

$$= \lim_{t \to (1/2)^-} \frac{3}{2} \left[(2t-1)^{1/3} - (-1)^{1/3} \right] = \frac{3}{2},$$

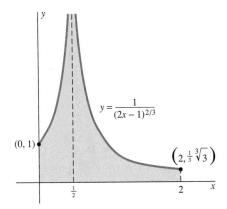

Fig. 9.8.9 The region of Example 4

and

$$\int_{1/2}^{2} \frac{1}{(2x-1)^{2/3}}\, dx = \lim_{t \to (1/2)^+} \int_{t}^{2} \frac{1}{(2x-1)^{2/3}}\, dx$$

$$= \lim_{t \to (1/2)^+} \left[\frac{3}{2}(2x-1)^{1/3} \right]_{t}^{2}$$

$$= \lim_{t \to (1/2)^+} \frac{3}{2}[3^{1/3} - (2t-1)^{1/3}] = \frac{3}{2}\sqrt[3]{3}.$$

Therefore,

$$\int_{0}^{2} \frac{1}{(2x-1)^{2/3}}\, dx = \frac{3}{2}(1 + \sqrt[3]{3}).$$

Special functions in advanced mathematics frequently are defined by means of improper integrals. An important example is the **gamma function** $\Gamma(t)$ that the prolific Swiss mathematician Leonhard Euler (1707–1783) introduced to "interpolate" the factorial function $n!$. The gamma function is defined for all real numbers $t > 0$ as follows:

$$\Gamma(t) = \int_{0}^{\infty} x^{t-1} e^{-x}\, dx. \tag{7}$$

This definition gives a continuous function of t such that

$$\Gamma(n+1) = n! \tag{8}$$

if n is a nonnegative integer (see Problems 29 and 30).

EXAMPLE 5 Find the volume V of the unbounded solid obtained by revolving around the y-axis the region under the curve $y = \exp(-x^2)$, $x \geqq 0$.

Solution The region in question is shown in Fig. 9.8.10. Let V_t denote the volume generated by revolving the shaded part of this region between $x = 0$ and $x = t > 0$. Then the method of cylindrical shells gives

$$V_t = \int_{0}^{t} 2\pi x \exp(-x^2)\, dx.$$

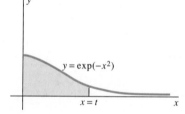

Fig. 9.8.10 The *whole* region under the graph is rotated around the y-axis; this produces an unbounded solid (Example 5).

We get the whole volume V by letting $t \to +\infty$:

$$V = \lim_{t \to \infty} V_t = \int_{0}^{\infty} 2\pi x \exp(-x^2)\, dx = \lim_{t \to \infty} \left[-\pi \exp(-x^2) \right]_{0}^{t} = \pi.$$

*ESCAPE VELOCITY

We saw in Section 6.6 how to compute the work W_r required to lift a body of mass m from the surface of a planet of mass M and radius R to a distance $r > R$ from the center of the planet. According to Eq. (7) there, the answer is

$$W_r = \int_{R}^{r} \frac{GMm}{x^2}\, dx.$$

So the work required to move the mass m "infinitely far" from the planet is

$$W = \lim_{r \to \infty} W_r = \int_{R}^{\infty} \frac{GMm}{x^2}\, dx = \lim_{r \to \infty} \left[-\frac{GMm}{x} \right]_{R}^{r} = \frac{GMm}{R}.$$

Suppose that the mass is projected with initial velocity v straight upward from the planet's surface, as in Jules Verne's novel *From the Earth to the Moon* (1865), in which a spacecraft was fired from an immense cannon. Then the initial kinetic energy $\frac{1}{2}mv^2$ is available to supply this work—by conversion into potential energy. From the equation

$$\frac{1}{2}mv^2 = \frac{GMm}{R},$$

we find that

$$v = \sqrt{\frac{2GM}{R}}.$$

Substitution of appropriate numerical values for the constants G, M, and R yields the value $v \approx 11{,}175$ m/s $\approx 25{,}000$ mi/h for the *escape velocity* from the earth.

*PRESENT VALUE OF A PERPETUITY

Consider a perpetual annuity, under which you and your heirs (and theirs, ad infinitum) will be paid A dollars annually. The question we pose is this: What is the fair market value of such an annuity? What should you pay to purchase it?

If the annual interest rate r is continuously compounded, then a dollar deposited in a savings account would grow to e^{rt} dollars in t years. Hence e^{-rt} dollars deposited now would yield \$1 after t years. Consequently, the **present value** of the amount you (and your heirs) will receive between time $t = 0$ (the present) and time $t = T > 0$ is defined to be

$$P_T = \int_0^T A e^{-rt}\, dt.$$

Hence the total present value of the perpetual annuity is

$$P = \lim_{T \to \infty} P_T = \int_0^\infty A e^{-rt}\, dt = \lim_{T \to \infty} \left[-\frac{A}{r} e^{-rt} \right]_0^T = \frac{A}{r}.$$

Thus $A = rP$. For instance, at an annual interest rate of 8% ($r = 0.08$), you should be able to purchase for $P = (\$50{,}000)/(0.08) = \$625{,}000$ a perpetuity that pays you (and your heirs) an annual sum of \$50,000.

9.8 Problems

Determine whether or not the improper integrals in Problems 1 through 24 converge. Evaluate those that converge.

1. $\displaystyle\int_4^\infty \frac{1}{x\sqrt{x}}\, dx$

2. $\displaystyle\int_1^\infty \frac{1}{x^{2/3}}\, dx$

3. $\displaystyle\int_0^4 \frac{1}{x\sqrt{x}}\, dx$

4. $\displaystyle\int_0^8 \frac{1}{x^{2/3}}\, dx$

5. $\displaystyle\int_1^\infty \frac{1}{x+1}\, dx$

6. $\displaystyle\int_3^\infty \frac{1}{\sqrt{x+1}}\, dx$

7. $\displaystyle\int_5^\infty \frac{1}{(x-1)^{3/2}}\, dx$

8. $\displaystyle\int_0^4 \frac{1}{\sqrt{4-x}}\, dx$

9. $\displaystyle\int_0^9 \frac{1}{(9-x)^{3/2}}\, dx$

10. $\displaystyle\int_0^3 \frac{1}{(x-3)^2}\, dx$

11. $\displaystyle\int_{-\infty}^{-2} \frac{1}{(x+1)^3}\, dx$

12. $\displaystyle\int_{-\infty}^0 \frac{1}{\sqrt{4-x}}\, dx$

13. $\displaystyle\int_{-1}^8 \frac{1}{\sqrt[3]{x}}\, dx$

14. $\displaystyle\int_{-4}^4 \frac{1}{(x+4)^{2/3}}\, dx$

15. $\displaystyle\int_2^\infty \frac{1}{\sqrt[3]{x-1}}\,dx$

16. $\displaystyle\int_{-\infty}^\infty \frac{x}{(x^2+4)^{3/2}}\,dx$

17. $\displaystyle\int_{-\infty}^\infty \frac{x}{x^2+4}\,dx$

18. $\displaystyle\int_0^\infty e^{-(x+1)}\,dx$

19. $\displaystyle\int_0^1 \frac{e^{\sqrt{x}}}{\sqrt{x}}\,dx$

20. $\displaystyle\int_0^2 \frac{x}{x^2-1}\,dx$

21. $\displaystyle\int_1^\infty \frac{1}{x\ln x}\,dx$

22. $\displaystyle\int_0^\infty \sin^2 x\,dx$

23. $\displaystyle\int_0^\infty xe^{-2x}\,dx$

24. $\displaystyle\int_0^\infty e^{-x}\sin x\,dx$

Problems 25 through 27 deal with Gabriel's horn, the surface obtained by revolving the curve $y = 1/x$, $x \geq 1$, around the x-axis (Fig. 9.8.11).

Fig. 9.8.11 Gabriel's horn (Problems 25 through 27)

25. Show that the area under the curve $y = 1/x$, $x \geq 1$, is infinite.

26. Show that the volume of revolution enclosed by Gabriel's horn is finite, and compute it.

27. Show that the surface area of Gabriel's horn is infinite. [*Suggestion:* Let A_t denote the surface area from $x = 1$ to $x = t > 1$. Prove that $A_t > 2\pi \ln t$.] In any case, the implication is that we could fill Gabriel's horn with a finite amount of paint (Problem 26), but no finite amount suffices to paint its surface.

28. Show that

$$\int_{-\infty}^\infty \frac{1+x}{1+x^2}\,dx$$

diverges, but that

$$\lim_{t\to\infty} \int_{-t}^t \frac{1+x}{1+x^2}\,dx = \pi.$$

29. Let x be a fixed positive number. Begin with the integral in Eq. (7), the one where the gamma function is defined, and integrate by parts to prove that $\Gamma(x+1) = x\Gamma(x)$.

30. (a) Prove that $\Gamma(1) = 1$. (b) Use the results of part (a) and Problem 29 to prove by mathematical induction that $\Gamma(n+1) = n!$ if n is a positive integer.

31. Use the substitution $x = e^{-u}$ and the fact that $\Gamma(n+1) = n!$ to prove that if m and n are fixed but arbitrary positive integers, then

$$\int_0^1 x^m(\ln x)^n\,dx = \frac{n!\,(-1)^n}{(m+1)^{n+1}}.$$

32. Consider a perpetual annuity under which you and your heirs will be paid at the rate of $10 + t$ thousand dollars per year t years hence. Thus you will receive $20,000 10 years from now, your heir will receive $110,000 100 years from now, and so on. Assuming a constant annual interest rate of 10%, show that the present value of this perpetuity is

$$P = \int_0^\infty (10+t)e^{-t/10}\,dt,$$

and then evaluate this improper integral.

33. A "semi-infinite" uniform rod occupies the nonnegative x-axis ($x \geq 0$) and has linear density ρ; that is, a segment of length dx has mass $\rho\,dx$. Show that the force of gravitational attraction that the rod exerts on a point mass m at $(-a, 0)$ is

$$F = \int_0^\infty \frac{Gm\rho}{(a+x)^2}\,dx = \frac{Gm\rho}{a}.$$

34. A rod of linear density ρ occupies the entire y-axis. A point mass m is located at $(a, 0)$ on the x-axis, as indicated in Fig. 9.8.12. Show that the total (horizontal) gravitational attraction that the rod exerts on m is

$$F = \int_{-\infty}^\infty \frac{Gm\rho \cos\theta}{r^2}\,dy = \frac{2Gm\rho}{a},$$

where $r^2 = a^2 + y^2$ and $\cos\theta = a/r$.

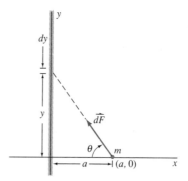

Fig. 9.8.12 Gravitational attraction exerted on a point mass by an infinite rod (Problem 34)

35. Given:

$$\int_0^\infty e^{-x^2}\,dx = \tfrac{1}{2}\sqrt{\pi}.$$

Deduce that $\Gamma(\tfrac{1}{2}) = \sqrt{\pi}$.

36. Recall that $\Gamma(x + 1) = x\Gamma(x)$ if $x > 0$. Suppose that n is a positive integer. Use Problem 35 to establish that

$$\Gamma\left(n + \frac{1}{2}\right) = \frac{1 \cdot 3 \cdot 5 \cdots (2n - 1)}{2^n}\sqrt{\pi}.$$

37. (a) Suppose that $k > 1$. Use integration by parts to show that

$$\int_0^\infty x^k \exp(-x^2)\, dx = \frac{k - 1}{2}\int_0^\infty x^{k-2}\exp(-x^2)\, dx.$$

(b) Suppose that n is a positive integer. Prove that

$$\int_0^\infty x^{n-1}\exp(-x^2)\, dx = \frac{1}{2}\Gamma\left(\frac{n}{2}\right).$$

38. Suppose that you win the Florida lottery and decide to use part of your winnings to purchase a perpetual annuity that will pay you and your heirs $10,000 per year (forever). Assuming an annual interest rate of 6%, what is a fair price for an insurance company to charge you for such an annuity?

9.8 Project

The improper integral $\int_0^\infty e^{-x^2}\, dx$ is important in applications ranging from probability and statistics (political polling, for example) to traffic flow and the theory of heat. Because the function $f(x) = e^{-x^2}$ has no elementary antiderivative as a finite combination of familiar functions, a simple and direct evaluation of the limit

$$\int_0^\infty e^{-x^2}\, dx = \lim_{b\to\infty}\int_0^b e^{-x^2}\, dx \qquad (9)$$

by using only the fundamental theorem of calculus is not feasible. But the fact that $e^{-x^2} \leqq e^{-x}$ for $x \geqq 1$ implies that the improper integral in Eq. (9) converges rather than diverges to infinity. (Can you explain why?)

We will see in Section 15.4 that the exact value of the integral in Eq. (9) is given by

$$\int_0^\infty e^{-x^2}\, dx = \frac{\sqrt{\pi}}{2} \approx 0.88622\,693. \qquad (10)$$

To verify this value numerically, we used the [integrate] key on an HP-48SX calculator to compute values of the integral $\int_0^b e^{-x^2}\, dx$ with successively larger values of the upper limit b. The results appear in Fig. 9.8.13.

As an alternative to the [integrate] key on a scientific calculator, you could use the numerical integration methods described in the project of Section 5.4. If you use a method such as Simpson's approximation, it is a good idea to increase the accuracy by doubling, say, the number of subintervals with each increase in the value b of the upper limit.

Use one of the techniques indicated here to verify numerically the values specified in Problems 1 through 5.

b	$\int_0^b e^{-x^2}\, dx$
1	0.74682 413
2	0.88208 139
3	0.88620 735
4	0.88622 691
5	0.88622 693
10	0.88622 693
100	0.88622 693

Fig. 9.8.13 Verifying the numerical value in Eq. (10)

1. $\displaystyle\int_0^\infty x^5 e^{-x}\, dx = 120$

2. $\displaystyle\int_0^\infty \frac{\sin x}{x}\, dx = \frac{\pi}{2}$

3. $\displaystyle\int_0^\infty \frac{1}{x^2 + 2}\, dx = \frac{\pi}{2\sqrt{2}}$

4. $\displaystyle\int_0^\infty e^{-x^2}\cos 2x\, dx = \frac{\pi}{2e}$

5. $\displaystyle\int_0^\infty \frac{1 - e^{-3x}}{x}\, dx = \frac{\ln 10}{2}$

When you confront the problem of evaluating a particular integral, you must first decide which of the several methods of this chapter to try. There are only two *general* methods of integration:

❑ Integration by substitution (Section 9.2), and

❑ Integration by parts (Section 9.4).

These are the analogues for integration of the chain rule and product rule, respectively, for differentiation.

Look first at the given integral to see if you can spot a substitution that would transform it into an elementary or familiar integral or one likely to be found in an integral table. In the case of an integral $\int f(x) q(x)\, dx$, whose integrand is an unfamiliar product of two functions, one of which is easily differentiated and the other easily integrated, then an attempt to integrate by parts is indicated.

Beyond these two general methods, the chapter deals with a number of *special* methods. In the case of an integral that is obviously trigonometric, $\int \mathrm{trig}(x)\, dx$, the simple "split-off" methods of Section 9.3 may succeed. Remember that reduction formulas [such as Eq. (5) and Problems 45 and 46 of Section 9.4] are available for integrating an integral power of a single trigonometric function.

Any integral of a rational function—that is, an integral of the form $\int (p(x)/q(x))\, dx$, where the integrand is a quotient of polynomials—can be evaluated by the method of partial fractions (Section 9.5). If the degree of the numerator is not less than that of the denominator—that is, if the rational function is not proper—first use long division to express it as the sum of a polynomial (easily integrated) and a proper rational fraction. Then decompose the latter into partial fractions. Partial fractions corresponding to linear factors are easily integrated, and those corresponding to irreducible quadratic factors can be integrated by completing the square and making (if necessary) a trigonometric substitution. As we explained in Section 9.7, the trigonometric integrals that result can always be evaluated.

In the case of an integral involving $\sqrt{ax^2 + bx + c}$, first complete the square (Section 9.7) and then rationalize the integral by making an appropriate trigonometric substitution (Section 9.6). This will leave you with a trigonometric integral.

Some additional special substitutions are introduced in the Miscellaneous Problems that follow. Notable among these is the substitution

$$u = \tan\frac{\theta}{2},$$

which transforms any integral $\int R(\sin\theta, \cos\theta)\, d\theta$ of a rational function of $\sin\theta$ and $\cos\theta$ into an integral of a rational function of u. The latter integral can then be evaluated by the method of partial fractions.

A final comment: Computer algebra systems are increasingly used for the evaluation of integrals such as those studied in this chapter. Nevertheless, the availability of these systems is no panacea. For instance, such computer systems are likely to be stumped by the integral

$$\int (1 + \ln x)\sqrt{1 + (x \ln x)^2} \, dx.$$

But you probably notice that the substitution

$$u = x \ln x, \qquad du = (1 + \ln x) \, dx$$

transforms this integral into the integral $\int \sqrt{1 + u^2} \, du$, which is amenable to trigonometric substitution (and can be found in any integral table). Thus the human factor remains—thankfully—essential.

Chapter 9 Miscellaneous Problems

Evaluate the integrals in Problems 1 through 100.

1. $\int \dfrac{1}{\sqrt{x}\,(1 + x)} \, dx$ [*Suggestion:* Let $x = u^2$.]

2. $\int \dfrac{\sec^2 t}{1 + \tan t} \, dt$

3. $\int \sin x \sec x \, dx$

4. $\int \dfrac{\csc x \cot x}{1 + \csc^2 x} \, dx$

5. $\int \dfrac{\tan \theta}{\cos^2 \theta} \, d\theta$

6. $\int \csc^4 x \, dx$

7. $\int x \tan^2 x \, dx$

8. $\int x^2 \cos^2 x \, dx$

9. $\int x^5 \sqrt{2 - x^3} \, dx$

10. $\int \dfrac{1}{\sqrt{x^2 + 4}} \, dx$

11. $\int \dfrac{x^2}{\sqrt{25 + x^2}} \, dx$

12. $\int (\cos x)\sqrt{4 - \sin^2 x} \, dx$

13. $\int \dfrac{1}{x^2 - x + 1} \, dx$

14. $\int \sqrt{x^2 + x + 1} \, dx$

15. $\int \dfrac{5x + 31}{3x^2 - 4x + 11} \, dx$

16. $\int \dfrac{x^4 + 1}{x^2 + 2} \, dx$

17. $\int \sqrt{x^4 + x^7} \, dx$

18. $\int \dfrac{\sqrt{x}}{1 + x} \, dx$ [*Suggestion:* Let $x = u^2$.]

19. $\int \dfrac{\cos x}{\sqrt{4 - \sin^2 x}} \, dx$

20. $\int \dfrac{\cos 2x}{\cos x} \, dx$

21. $\int \dfrac{\tan x}{\ln(\cos x)} \, dx$

22. $\int \dfrac{x^7}{\sqrt{1 - x^4}} \, dx$

23. $\int \ln(1 + x) \, dx$

24. $\int x \sec^{-1} x \, dx$

25. $\int \sqrt{x^2 + 9} \, dx$

26. $\int \dfrac{x^2}{\sqrt{4 - x^2}} \, dx$

27. $\int \sqrt{2x - x^2} \, dx$

28. $\int \dfrac{4x - 2}{x^3 - x} \, dx$

29. $\int \dfrac{x^4}{x^2 - 2} \, dx$

30. $\int \dfrac{\sec x \tan x}{\sec x + \sec^2 x} \, dx$

31. $\int \dfrac{x}{(x^2 + 2x + 2)^2} \, dx$

32. $\int \dfrac{x^{1/3}}{x^{1/2} + x^{1/4}} \, dx$ [*Suggestion:* Let $x = u^{12}$.]

33. $\int \dfrac{1}{1 + \cos 2\theta} \, d\theta$

34. $\int \dfrac{\sec x}{\tan x} \, dx$

35. $\int \sec^3 x \, \tan^3 x \, dx$

36. $\int x^2 \tan^{-1} x \, dx$

37. $\int x(\ln x)^3 \, dx$

38. $\int \dfrac{1}{x\sqrt{1 + x^2}} \, dx$

39. $\int e^x \sqrt{1 + e^{2x}} \, dx$

40. $\int \dfrac{x}{\sqrt{4x - x^2}} \, dx$

41. $\int \dfrac{1}{x^3 \sqrt{x^2 - 9}} \, dx$

42. $\int \dfrac{x}{(7x + 1)^{17}} \, dx$

43. $\int \dfrac{4x^2 + x + 1}{4x^3 + x} \, dx$

44. $\int \dfrac{4x^3 - x + 1}{x^3 + 1} \, dx$

45. $\int \tan^2 x \sec x \, dx$

46. $\int \dfrac{x^2 + 2x + 2}{(x + 1)^3} \, dx$

47. $\int \dfrac{x^4 + 2x + 2}{x^5 + x^4} \, dx$

48. $\int \dfrac{8x^2 - 4x + 7}{(x^2 + 1)(4x + 1)} \, dx$

49. $\int \dfrac{3x^5 - x^4 + 2x^3 - 12x^2 - 2x + 1}{(x^3 - 1)^2} \, dx$

50. $\int \dfrac{x}{x^4 + 4x^2 + 8} \, dx$

51. $\int (\ln x)^6 \, dx$

52. $\int \dfrac{(1 + x^{2/3})^{3/2}}{x^{1/3}} \, dx$ [*Suggestion:* Let $x = u^3$.]

53. $\int \dfrac{(\arcsin x)^2}{\sqrt{1 - x^2}} \, dx$

54. $\int \dfrac{1}{x^{3/2}(1 + x^{1/3})} \, dx$ [*Suggestion:* Let $x = u^6$.]

55. $\int \tan^3 z \, dz$

56. $\int \sin^2 \omega \cos^4 \omega \, d\omega$

57. $\int \dfrac{xe^{x^2}}{1 + e^{2x^2}} \, dx$

58. $\int \dfrac{\cos^3 x}{\sqrt{\sin x}} \, dx$

59. $\displaystyle\int x^3 e^{-x^2}\, dx$

60. $\displaystyle\int \sin\sqrt{x}\, dx$

61. $\displaystyle\int \frac{\arcsin x}{x^2}\, dx$

62. $\displaystyle\int \sqrt{x^2 - 9}\, dx$

63. $\displaystyle\int x^2\sqrt{1 - x^2}\, dx$

64. $\displaystyle\int x\sqrt{2x - x^2}\, dx$

65. $\displaystyle\int \frac{x - 2}{4x^2 + 4x + 1}\, dx$

66. $\displaystyle\int \frac{2x^2 - 5x - 1}{x^3 - 2x^2 - x + 2}\, dx$

67. $\displaystyle\int \frac{e^{2x}}{e^{2x} - 1}\, dx$

68. $\displaystyle\int \frac{\cos x}{\sin^2 x - 3\sin x + 2}\, dx$

69. $\displaystyle\int \frac{2x^3 + 3x^2 + 4}{(x + 1)^4}\, dx$

70. $\displaystyle\int \frac{\sec^2 x}{\tan^2 x + 2\tan x + 2}\, dx$

71. $\displaystyle\int \frac{x^3 + x^2 + 2x + 1}{x^4 + 2x^2 + 1}\, dx$

72. $\displaystyle\int \sin x \cos 3x\, dx$

73. $\displaystyle\int x^5\sqrt{x^3 - 1}\, dx$

74. $\displaystyle\int \ln(x^2 + 2x)\, dx$

75. $\displaystyle\int \frac{\sqrt{1 + \sin x}}{\sec x}\, dx$

76. $\displaystyle\int \frac{1}{x^{2/3}(1 + x^{2/3})}\, dx$

77. $\displaystyle\int \frac{\sin x}{\sin 2x}\, dx$

78. $\displaystyle\int \sqrt{1 + \cos t}\, dt$

79. $\displaystyle\int \sqrt{1 + \sin t}\, dt$

80. $\displaystyle\int \frac{\sec^2 t}{1 - \tan^2 t}\, dt$

81. $\displaystyle\int \ln(x^2 + x + 1)\, dx$

82. $\displaystyle\int e^x \sin^{-1}(e^x)\, dx$

83. $\displaystyle\int \frac{\arctan x}{x^2}\, dx$

84. $\displaystyle\int \frac{x^2}{\sqrt{x^2 - 25}}\, dx$

85. $\displaystyle\int \frac{x^3}{(x^2 + 1)^2}\, dx$

86. $\displaystyle\int \frac{1}{x\sqrt{6x - x^2}}\, dx$

87. $\displaystyle\int \frac{3x + 2}{(x^2 + 4)^{3/2}}\, dx$

88. $\displaystyle\int x^{3/2}\ln x\, dx$

89. $\displaystyle\int \frac{\sqrt{1 + \sin^2 x}}{\sec x \csc x}\, dx$

90. $\displaystyle\int \frac{e^{\sqrt{\sin x}}}{(\sec x)\sqrt{\sin x}}\, dx$

91. $\displaystyle\int xe^x \sin x\, dx$

92. $\displaystyle\int x^2 e^{x^{3/2}}\, dx$

93. $\displaystyle\int \frac{\arctan x}{(x - 1)^3}\, dx$

94. $\displaystyle\int \ln\left(1 + \sqrt{x}\right) dx$

95. $\displaystyle\int \frac{2x + 3}{\sqrt{3 + 6x - 9x^2}}\, dx$

96. $\displaystyle\int \frac{1}{\sqrt{e^{2x} - 1}}\, dx$

97. $\displaystyle\int \frac{x^4}{(x - 1)^2}\, dx$ [*Suggestion:* Let $u = x - 1$.]

98. $\displaystyle\int x^{3/2}\tan^{-1}\sqrt{x}\, dx$

99. $\displaystyle\int \operatorname{arcsec}\sqrt{x}\, dx$

100. $\displaystyle\int x\sqrt{\frac{1 - x^2}{1 + x^2}}\, dx$

101. Find the area of the surface generated by revolving the curve $y = \cosh x$, $0 \le x \le 1$, around the x-axis.

102. Find the length of the curve $y = e^{-x}$, $0 \le x \le 1$.

103. (a) Find the area A_t of the surface generated by revolving the curve $y = e^{-x}$, $0 \le x \le t$, around the x-axis. (b) Find $\lim_{t\to\infty} A_t$.

104. (a) Find the area A_t of the surface generated by revolving the curve $y = 1/x$, $1 \le x \le t$, around the x-axis. (b) Find $\lim_{t\to\infty} A_t$.

105. Find the area of the surface generated by revolving the curve $y = \sqrt{x^2 - 1}$, $1 \le x \le 2$, around the x-axis.

106. (a) Derive the reduction formula

$$\int x^m (\ln x)^n\, dx$$

$$= \frac{1}{m + 1} x^{m+1}(\ln x)^n - \frac{n}{m + 1}\int x^m(\ln x)^{n-1}\, dx.$$

(b) Evaluate $\displaystyle\int_1^e x^3(\ln x)^3\, dx$.

107. Derive the reduction formula

$$\int \sin^m x \cos^n x\, dx = -\frac{1}{m + n}\sin^{m-1}x \cos^{n+1}x$$

$$+ \frac{m - 1}{m + n}\int \sin^{m-2}x \cos^n x\, dx.$$

108. Use the reduction formulas of Problem 107 here and Problem 46 of Section 9.4 to evaluate

$$\int_0^{\pi/2} \sin^6 x \cos^5 x\, dx.$$

109. Find the area bounded by the curve $y^2 = x^5(2 - x)$, $0 \le x \le 2$. [*Suggestion:* Substitute $x = 2\sin^2\theta$; then use the result of Problem 50 of Section 9.4.]

110. Show that

$$0 < \int_0^1 \frac{t^4(1 - t)^4}{1 + t^2}\, dt$$

and that

$$\int_0^1 \frac{t^4(1 - t)^4}{1 + t^2}\, dt = \frac{22}{7} - \pi.$$

534

111. Evaluate $\displaystyle\int_0^1 t^4(1-t)^4\,dt$; then apply the results of Problem 110 to conclude that

$$\tfrac{22}{7} - \tfrac{1}{630} < \pi < \tfrac{22}{7} - \tfrac{1}{1260}.$$

Thus $3.1412 < \pi < 3.1421$.

112. Find the length of the curve $y = \frac{4}{5}x^{5/4},\ 0 \le x \le 1$.

113. Find the length of the curve $y = \frac{4}{3}x^{3/4},\ 1 \le x \le 4$.

114. An initially empty water tank is shaped like a cone whose axis is vertical. Its vertex is at the bottom; the cone is 9 ft deep and has a top radius of 4.5 ft. Beginning at time $t = 0$, water is poured into this tank at 50 ft^3/min. Meanwhile, water leaks out a hole at the bottom at the rate of $10\sqrt{y}$ cubic feet per minute, where y is the depth of water in the tank. (This is consistent with Torricelli's law of draining.) How long does it take to fill the tank?

115. (a) Evaluate $\displaystyle\int \frac{1}{1 + e^x + e^{-x}}\,dx$.

(b) Explain why your substitution in part (a) suffices to integrate any rational function of e^x.

116. (a) The equation $x^3 + x + 1 = 0$ has one real root r. Use Newton's method to find it, accurate to at least two places. (b) Use long division to find (approximately) the irreducible quadratic factor of $x^3 + x + 1$. (c) Use the factorization obtained in part (b) to evaluate (approximately)

$$\int_0^1 \frac{1}{x^3 + x + 1}\,dx.$$

117. Evaluate $\displaystyle\int \frac{1}{1 + e^x}\,dx$.

118. The integral

$$\int \frac{1 + 2x^2}{x^5(1 + x^2)^3}\,dx = \int \frac{x + 2x^3}{(x^4 + x^2)^3}\,dx$$

would require you to solve 11 equations in 11 unknowns if you were to use the method of partial fractions to evaluate it. Use the substitution $u = x^4 + x^2$ to evaluate it much more simply.

119. Evaluate $\displaystyle\int \sqrt{\tan\theta}\,d\theta$. [*Suggestion:* First substitute

$u = \tan\theta$. Then substitute $u = x^2$. Finally, use the method of partial fractions; see Problem 42 of Section 9.7.]

120. Prove that if $p(x)$ is a polynomial, then the substitution $u^n = (ax + b)/(cx + d)$ transforms the integral

$$\int p(x)\left(\frac{ax + b}{cx + d}\right)^{1/n} dx$$

into the integral of a rational function of u. (The substitution indicated here is called a *rationalizing substitution;* its name comes from the fact that it converts the integrand into a *rational* function of u.)

In Problems 121 through 129, use the rationalizing substitution indicated in Problem 120.

121. $\displaystyle\int x^3\sqrt{3x - 2}\,dx$

122. $\displaystyle\int x^3\sqrt[3]{x^2 + 1}\,dx$

123. $\displaystyle\int \frac{x^3}{(x^2 - 1)^{4/3}}\,dx$

124. $\displaystyle\int x^2(x - 1)^{3/2}\,dx$

125. $\displaystyle\int \frac{x^5}{\sqrt{x^3 + 1}}\,dx$

126. $\displaystyle\int x^7\sqrt[3]{x^4 + 1}\,dx$

127. $\displaystyle\int \sqrt{\frac{1 + x}{1 - x}}\,dx$

128. $\displaystyle\int \frac{x}{\sqrt{x + 1}}\,dx$

129. $\displaystyle\int \frac{\sqrt[3]{x + 1}}{x}\,dx$

130. Substitute $x = u^2$ to find $\displaystyle\int \sqrt{1 + \sqrt{x}}\,dx$.

131. Substitute $u^2 = 1 + e^{2x}$ to find $\displaystyle\int \sqrt{1 + e^{2x}}\,dx$.

132. Find the area A of the surface obtained by revolving the curve $y = \frac{2}{3}x^{3/2},\ 3 \le x \le 8$, around the x-axis. [*Suggestion:* Substitute $x = u^2$ into the surface area integral. *Note:* $A \approx 732.39$.]

133. Find the area bounded by the loop of the curve

$$y^2 = x^2(1 - x), \qquad 0 \le x \le 1.$$

134. Find the area bounded by the loop of the curve

$$y^2 = x^2\left(\frac{1 - x}{1 + x}\right), \qquad 0 \le x \le 1.$$

MORE GENERAL TRIGONOMETRIC INTEGRALS

As a last resort, any trigonometric integral can be transformed into an integral

$$\int R(\sin\theta, \cos\theta)\,d\theta \tag{1}$$

of sines and cosines. If the integrand in Eq. (1) is a quotient of polynomials in the variables $\sin\theta$ and $\cos\theta$, then the special substitution

$$u = \tan\frac{\theta}{2} \tag{2}$$

suffices for its evaluation.

To carry out the substitution indicated in Eq. (2), we must express $\sin \theta$, $\cos \theta$, and $d\theta$ in terms of u and du. Note first that

$$\theta = 2 \tan^{-1} u, \quad \text{so} \quad d\theta = \frac{2\,du}{1 + u^2}. \tag{3}$$

We see from the triangle in Fig. 9.MP.1 that

$$\sin \frac{\theta}{2} = \frac{u}{\sqrt{1 + u^2}}, \qquad \cos \frac{\theta}{2} = \frac{1}{\sqrt{1 + u^2}}.$$

Hence

$$\sin \theta = 2 \sin \frac{\theta}{2} \cos \frac{\theta}{2} = \frac{2u}{1 + u^2}, \tag{4}$$

$$\cos \theta = \cos^2 \frac{\theta}{2} - \sin^2 \frac{\theta}{2} = \frac{1 - u^2}{1 + u^2}. \tag{5}$$

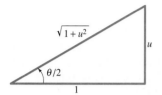

Fig. 9.MP.1 The special rationalizing substitution $u = \tan \dfrac{\theta}{2}$

These substitutions will convert the integral in Eq. (1) into an integral of a rational function of u. The latter can then be evaluated by the methods of Section 9.5.

EXAMPLE 1

$$\int \frac{1}{5 + 3 \cos \theta}\, d\theta = \int \frac{1}{5 + 3 \cdot \dfrac{1 - u^2}{1 + u^2}} \cdot \frac{2\,du}{1 + u^2}$$

$$= \int \frac{2}{8 + 2u^2}\, du = \int \frac{1}{4 + u^2}\, du$$

$$= \frac{1}{2} \tan^{-1} \frac{u}{2} + C = \frac{1}{2} \tan^{-1}\left(\frac{1}{2} \tan \frac{\theta}{2}\right) + C.$$

In Problems 135 through 142, use the rationalizing substitution given in Eqs. (2) through (5).

135. $\displaystyle\int \frac{1}{1 + \cos \theta}\, d\theta$ **136.** $\displaystyle\int \frac{1}{5 + 4 \cos \theta}\, d\theta$

137. $\displaystyle\int \frac{1}{1 + \sin \theta}\, d\theta$ **138.** $\displaystyle\int \frac{1}{(1 - \cos \theta)^2}\, d\theta$

139. $\displaystyle\int \frac{1}{\sin \theta + \cos \theta}\, d\theta$

140. $\displaystyle\int \frac{1}{2 + \sin \phi + \cos \phi}\, d\phi$

141. $\displaystyle\int \frac{\sin \theta}{2 + \cos \theta}\, d\theta$ **142.** $\displaystyle\int \frac{\sin \theta - \cos \theta}{\sin \theta + \cos \theta}\, d\theta$

143. (a) Substitute $u = \tan(\theta/2)$ to show that

$$\int \sec \theta\, d\theta = \ln\left| \frac{1 + \tan \dfrac{\theta}{2}}{1 - \tan \dfrac{\theta}{2}} \right| + C.$$

(b) Use the trigonometric identity

$$\tan \frac{\theta}{2} = \sqrt{\frac{1 - \cos \theta}{1 + \cos \theta}}$$

to derive our earlier formula

$$\int \sec \theta\, d\theta = \ln|\sec \theta + \tan \theta| + C$$

from the solution in part (a).

144. (a) Use the method of Problem 143 to show that

$$\int \csc \theta\, d\theta = \ln\left| \tan \frac{\theta}{2} \right| + C.$$

(b) Use trigonometric identities to derive the formula

$$\int \csc \theta\, d\theta = \ln|\csc \theta - \cot \theta| + C$$

from part (a).

Polar Coordinates and Conic Sections

Pierre de Fermat (1601–1665)

❑ Pierre de Fermat exemplifies the distinguished tradition of great amateurs in mathematics. Like his contemporary René Descartes, he was educated as a lawyer. But Fermat actually practiced law as his profession and served in the local regional parliament. His ample leisure time, however, was devoted to mathematics and to other intellectual pursuits, such as the study of ancient Greek manuscripts.

❑ In a margin of one such manuscript (by the Greek mathematician Diophantus) was found a handwritten note that has remained an enigma ever since. Fermat asserts that for *no* integer $n > 2$ do positive integers $x, y,$ and z exist such that $x^n + y^n = z^n$. For instance, although $3^2 + 4^2 = 5^2$, the sum of two (integer) cubes cannot be a cube. "I have found an admirable proof of this," he wrote, "but this margin is too narrow to contain it." Despite the publi-

cation of many incorrect proofs, "Fermat's last theorem" has remained unproved for three and one–half centuries (although it has been verified for all $n < 1{,}000{,}000$). However, in a June 1993 lecture, the British mathematician Andrew Wiles of Princeton University announced a long and complex proof of Fermat's last theorem. This is not the first time such a proof has been announced, but the proposed proof is based on several decades of systematic development of the tools that experts in the field believe are needed for a complete and valid proof.

❑ Descartes and Fermat shared in the discovery of analytic geometry. But whereas Descartes typically used geometrical methods to solve algebraic equations (see the Chapter 1 opening), Fermat concentrated on the investigation of geometric curves defined by algebraic equations. For instance, he introduced the translation and rotation methods of this chapter to show that the graph of an equation of the form $Ax^2 + Bxy + Cy^2 + Dx + Ey + F = 0$ is generally a conic

section. Most of his mathematical work remained unpublished during his lifetime, but it contains numerous tangent line (derivative) and area (integral) computations.

❑ The brilliantly colored left-hand photograph is a twentieth-century example of a geometric object defined by means of algebraic operations. Starting with the point $P(a, b)$ in the xy-plane, we interpret P as the complex number $c = a + bi$ and define the sequence $\{z_n\}$ of points of the complex plane iteratively (as in Section 3.10) by the equations

$$z_0 = c, \quad z_{n+1} = z_n^2 + c$$

(for $n > 0$).

If this sequence of points remains inside the circle $x^2 + y^2 = 4$ for all n, then the original point $P(a, b)$ is colored black. Otherwise, the color assigned to P is determined by the speed with which this sequence "escapes" the circular disk. The set of all black points is the famous *Mandelbrot set*, discovered in 1980 by the French mathematician Benoit Mandelbrot.

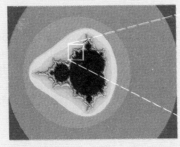 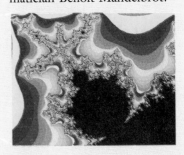

The object in the right–hand figure is a subset of that in the left–hand figure.

Plane analytic geometry, the main topic of this chapter, is the use of algebra and calculus to study the properties of curves in the xy-plane. The ancient Greeks used deductive reasoning and the methods of axiomatic Euclidean geometry to study lines, circles, and the **conic sections** (parabolas, ellipses, and hyperbolas). The properties of conic sections have played an important role in diverse scientific applications since the seventeenth century, when Kepler discovered—and Newton explained—the fact that the orbits of planets and other bodies in the Solar System are conic sections.

The French mathematicians Descartes and Fermat, working almost independently of one another, initiated analytic geometry in 1637. The central idea of analytic geometry is the correspondence between an equation $F(x, y) = 0$ and its **locus** (typically, a curve), the set of all those points (x, y) in the plane with coordinates that satisfy this equation.

A central idea of analytic geometry is this: Given a geometric locus or curve, its properties can be derived algebraically or analytically from its defining equation $F(x, y) = 0$. For example, suppose that the equation of a given curve turns out to be the linear equation

$$Ax + By = C, \tag{1}$$

where A, B, and C are constants with $B \neq 0$. This equation may be written in the form

$$y = mx + b, \tag{2}$$

where $m = -A/B$ and $b = C/B$. But Eq. (2) is the slope-intercept equation of the straight line with slope m and y-intercept b. Hence the given curve is this straight line. We use this approach in Example 1 to show that a specific geometrically described locus is a particular straight line.

EXAMPLE 1 Prove that the set of all points equidistant from the points $(1, 1)$ and $(5, 3)$ is the perpendicular bisector of the line segment that joins these two points.

Solution The typical point $P(x, y)$ in Fig. 10.1.1 is equidistant from $(1, 1)$ and $(5, 3)$ if and only if

$$(x - 1)^2 + (y - 1)^2 = (x - 5)^2 + (y - 3)^2;$$
$$x^2 - 2x + 1 + y^2 - 2y + 1 = x^2 - 10x + 25 + y^2 - 6y + 9;$$
$$2x + y = 8;$$
$$y = -2x + 8. \tag{3}$$

Thus the given locus is the straight line in Eq. (3) whose slope is -2. The straight line through $(1, 1)$ and $(5, 3)$ has equation

$$y - 1 = \tfrac{1}{2}(x - 1) \tag{4}$$

and thus has slope $\frac{1}{2}$. Because the product of the slopes of these two lines is -1, it follows (from Theorem 2 in Section 1.2) that these lines are perpendicular. If we solve Eqs. (3) and (4) simultaneously, we find that the intersection of these lines is, indeed, the midpoint $(3, 2)$ of the given line segment. Thus the locus described is the perpendicular bisector of this line segment.

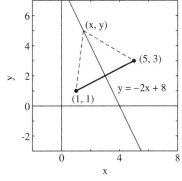

Fig. 10.1.1 The perpendicular bisector of Example 1

θ	2θ	$-4\sin 2\theta$
$0 < \theta < \dfrac{\pi}{2}$	$0 < 2\theta < \pi$	Negative
$\dfrac{\pi}{2} < \theta < \pi$	$\pi < 2\theta < 2\pi$	Positive
$\pi < \theta < \dfrac{3\pi}{2}$	$2\pi < 2\theta < 3\pi$	Negative
$\dfrac{3\pi}{2} < \theta < 2\pi$	$3\pi < 2\theta < 4\pi$	Positive

When θ lies in the first or the third quadrant, the quantity $-4\sin 2\theta$ is negative, so the equation $r^2 = -4\sin 2\theta$ cannot be satisfied for any real value of r.

Example 6 illustrates a peculiarity of graphs of polar equations, caused by the fact that a single point has multiple representations in polar coordinates. The point with polar coordinates $(2, \pi/2)$ clearly lies on the four-leaved rose, but these coordinates do *not* satisfy the equation $r = 2\cos 2\theta$. This means that a point may have one pair of polar coordinates that satisfy a given equation and others that do not. Hence we must be careful to understand this: The graph of a polar equation consists of all those points with *at least one* polar-coordinates representation that satisfies the given equation.

Another result of the multiplicity of polar coordinates is that the simultaneous solution of two polar equations does not always give all the points of intersection of their graphs. For instance, consider the circles $r = 2\sin \theta$ and $r = 2\cos \theta$ shown in Fig. 10.2.8. The origin is clearly a point of intersection of these two circles. Its polar representation $(0, \pi)$ satisfies the equation $r = 2\sin \theta$, and its representation $(0, \pi/2)$ satisfies the other equation, $r = 2\cos \theta$. But the origin has no *single* polar representation that satisfies both equations simultaneously! If we think of θ as increasing uniformly with time, then the corresponding moving points on the two circle pass through the origin at different times. Hence the origin cannot be discovered as a point of intersection of the two circles by solving their equations simultaneously.

As a consequence of the phenomenon illustrated by this example, the only way we can be certain of finding *all* points of intersection of two curves in polar coordinates is to graph both curves.

EXAMPLE 8 Find all points of intersection of the graphs of the equations $r = 1 + \sin \theta$ and $r^2 = 4\sin \theta$.

Solution The graph of $r = 1 + \sin \theta$ is a scaled-down version of the cardioid of Example 5. In Problem 52 we ask you to show that the graph of $r^2 = 4\sin \theta$ is the figure-eight curve shown with the cardioid in Fig. 10.2.13. The figure shows four points of intersection: A, B, C, and O. Can we find all four with algebra?

Given the two equations, we begin by eliminating r. Because

$$(1 + \sin \theta)^2 = r^2 = 4\sin \theta,$$

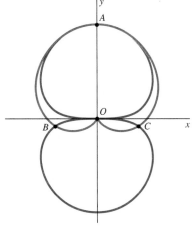

Fig. 10.2.13 The cardioid $r = 1 + \sin \theta$ and the figure eight $r^2 = 4\sin \theta$ meet in four points (Example 8).

it follows that

$$\sin^2 \theta - 2 \sin \theta + 1 = 0;$$

$$(\sin \theta - 1)^2 = 0;$$

and thus that $\sin \theta = 1$. So θ must be an angle of the form $\pi/2 + 2n\pi$, where n is an integer. All points on the cardioid and all points on the figure-eight curve are produced by letting θ increase from 0 to 2π, so $\theta = \pi/2$ will produce all the solutions that we can possibly obtain by algebraic methods. The only such point is $A(2, \pi/2)$, and the other three points of intersection are detected only when the two equations are graphed.

10.2 Problems

1. Plot the points with the given polar coordinates, and then find the rectangular coordinates of each.

(a) $(1, \pi/4)$ (b)$(-2, 2\pi/3)$

(c) $(1, -\pi/3)$ (d)$(3, 3\pi/2)$

(e) $(2, 9\pi/4)$ (f)$(-2, -7\pi/6)$

(g) $(2, 5\pi/6)$

2. Find two polar-coordinates representations, one with $r > 0$ and the other with $r < 0$, for the points with the given rectangular coordinates.

(a) $(-1, -1)$ (b) $(\sqrt{3}, -1)$

(c) $(2, 2)$ (d) $(-1, \sqrt{3})$

(e) $(\sqrt{2}, -\sqrt{2})$ (f) $(-3, \sqrt{3})$

In Problems 3 through 10, express the given rectangular equations in polar form.

3. $x = 4$ **4.** $y = 6$

5. $x = 3y$ **6.** $x^2 + y^2 = 25$

7. $xy = 1$ **8.** $x^2 - y^2 = 1$

9. $y = x^2$ **10.** $x + y = 4$

In each of Problems 11 through 18, express the given polar equation in rectangular coordinates.

11. $r = 3$ **12.** $\theta = 3\pi/4$

13. $r = -5 \cos \theta$ **14.** $r = \sin 2\theta$

15. $r = 1 - \cos 2\theta$ **16.** $r = 2 + \sin \theta$

17. $r = 3 \sec \theta$ **18.** $r^2 = \cos 2\theta$

For the curves described in Problems 19 through 28, write equations in both rectangular and polar coordinates.

19. The vertical line through $(2, 0)$

20. The horizontal line through $(1, 3)$

21. The line with slope -1 through $(2, -1)$

22. The line with slope 1 through $(4, 2)$

23. The line through the points $(1, 3)$ and $(3, 5)$

24. The circle with center $(3, 0)$ that passes through the origin

25. The circle with center $(0, -4)$ that passes through the origin

26. The circle with center $(3, 4)$ and radius 5

27. The circle with center $(1, 1)$ that passes through the origin

28. The circle with center $(5, -2)$ that passes through the point $(1, 1)$

In Problems 29 through 38, match the given polar-coordinates equation with its graph, using Figs. 10.2.14 through 10.2.23.

29. $r = 5 \sin \theta$ **30.** $r = 5 \cos \theta$

31. $r = -5 \sin \theta$ **32.** $r = -5 \cos \theta$

33. $r = 2 + 2 \sin \theta$ **34.** $r = 2 + 2 \cos \theta$

35. $r = 3 + 2 \sin \theta$ **36.** $r = 3 + 2 \cos \theta$

37. $r = 2 + 3 \sin \theta$ **38.** $r = 2 + 3 \cos \theta$

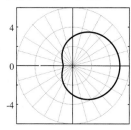

Fig. 10.2.14

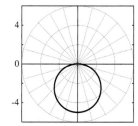

Fig. 10.2.15

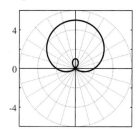

Fig. 10.2.16

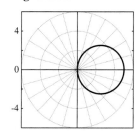

Fig. 10.2.17

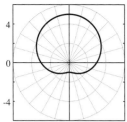

Fig. 10.2.18

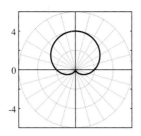

Fig. 10.2.19

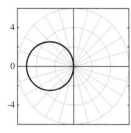

Fig. 10.2.20

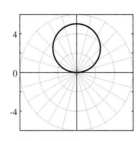

Fig. 10.2.21

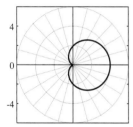

Fig. 10.2.22

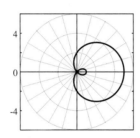

Fig. 10.2.23

Sketch the graphs of the polar equations in Problems 39 through 52. Indicate any symmetries about either axis or the origin.

39. $r = 2 \cos \theta$ (circle)

40. $r = 2 \sin \theta + 2 \cos \theta$ (circle)

41. $r = 1 + \cos \theta$ (cardioid)

42. $r = 1 - \sin \theta$ (cardioid)

43. $r = 2 + 4 \cos \theta$ (limaçon)

44. $r = 4 + 2 \cos \theta$ (limaçon)

45. $r^2 = 4 \sin 2\theta$ (lemniscate)

46. $r^2 = 4 \cos 2\theta$ (lemniscate)

47. $r = 2 \sin 2\theta$ (four-leaved rose)

48. $r = 3 \sin 3\theta$ (three-leaved rose)

49. $r = 3 \cos 3\theta$ (three-leaved rose)

50. $r = 3\theta$ (spiral of Archimedes)

51. $r = 2 \sin 5\theta$ (five-leaved rose)

52. $r^2 = 4 \sin \theta$ (figure eight)

In Problems 53 through 58, find all points of intersection of the curves with the given polar equations.

53. $r = 2$, $r = \cos \theta$

54. $r = \sin \theta$, $r^2 = 3 \cos^2 \theta$

55. $r = \sin \theta$, $r = \cos 2\theta$

56. $r = 1 + \cos \theta$, $r = 1 - \sin \theta$

57. $r = 1 - \cos \theta$, $r^2 = 4 \cos \theta$

58. $r^2 = 4 \sin \theta$, $r^2 = 4 \cos \theta$

59. (a) The straight line L passes through the point with polar coordinates (p, α) and is perpendicular to the line segment joining the pole and the point (p, α). Write the polar-coordinates equation of L. (b) Show that the rectangular-coordinates equation of L is

$$x \cos \alpha + y \sin \alpha = p.$$

60. Show that the graph of the rectangular-coordinates equation $x^2 + y^2 = (x + x^2 + y^2)^2$ is the cardioid with the polar equation $r = 1 - \cos \theta$.

61. Did you wonder if the cusp shown in the cardioid of Example 5 is really there? After all, it's conceivable that the cardioid has a *horizontal* tangent at the origin. Here is one way to be sure about the actual shape of a polar graph. Begin with the equation $r = 2 + 2 \sin \theta$ of the cardioid of Example 5. Use the equations in (2) to write

$$x = r \cos \theta = 2 \cos \theta + 2 \sin \theta \cos \theta$$

and

$$y = r \sin \theta = 2 \sin \theta + 2 \sin^2 \theta.$$

Compute $dx/d\theta$, $dy/d\theta$, and

$$\frac{dx}{dy} = \frac{\dfrac{dx}{d\theta}}{\dfrac{dy}{d\theta}}$$

as functions of θ. Compute the limit of dx/dy as θ approaches $3\pi/2$ by using l'Hôpital's rule (Section 8.3). You will obtain the limit zero, and thus the cardioid has a vertical tangent at the origin; Fig. 10.2.10 is correctly drawn.

62. Plot the polar equations

$$r = 1 + \cos \theta \quad \text{and} \quad r = -1 + \cos \theta$$

on the same coordinate plane. Comment on the results.

The graph of the polar-coordinates equation $r = f(\theta)$ can be plotted in rectangular coordinates by using the equations

$$x = r \cos \theta = f(\theta) \cos \theta, \tag{5}$$

$$y = r \sin \theta = f(\theta) \sin \theta. \tag{6}$$

Then, as θ ranges from 0 to 2π (or, in some cases, through a much larger domain), the point (x, y) traces the polar graph $r = f(\theta)$.

For instance, with $r = 1 + \cos \theta$, the equations

$$x = (1 + \cos \theta) \cos \theta,$$

$$y = (1 + \cos \theta) \sin \theta$$

yield the cardioid shown in Fig. 10.2.24. To plot this cardioid with a TI graphics calculator (for instance), we define the variables

```
X₁T=(1+cos T) cos T
Y₁T=(1+cos T) sin T
```

in PARAMETRIC mode before graphing (with T in place of θ). The table in Fig. 10.2.25 shows corresponding commands in several common computer systems for plotting the polar-coordinates curve $r = f(\theta)$.

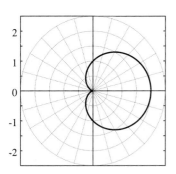

Fig. 10.2.24 The cardioid $r = 1 + \cos \theta$

System	Command
Derive	Plot the expression [f(t)*cos(t), f(t)*sin(t)]
Maple	plot([f(t)*cos(t), f(t)*sin(t), 0..2*Pi])
Mathematica	ParametricPlot[{f[t] Cos[t], f[t] Sin[t]}, {t, 0, 2 Pi}]
X(PLORE)	paramg(f(t)*cos(t), f(t)*sin(t), t=0 to 2*pi)

Fig. 10.2.25 Computer system commands to plot the polar curve $r = f(\theta)$

PROJECT A Plot the polar-coordinates curve

$$r = (a + b \cos m\theta)(c + d \sin n\theta)$$

with various values of the coefficients a, b, c, d, and the positive integers m and n. You might begin with the special case $a = 1, b = 0$, or the special case $c = 1, d = 0$, or the special case $a = c = 0$. Figures 10.2.26 through 10.2.29 illustrate just a few of the possibilities.

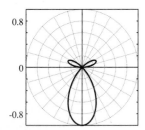

Fig. 10.2.26 $r = \sin \theta \cos 2\theta$

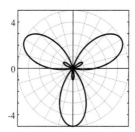

Fig. 10.2.27 $r = 2 + 3 \sin 3\theta$

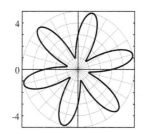

Fig. 10.2.28 $r = 3 + 2 \sin 6\theta$

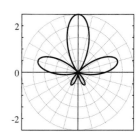

Fig. 10.2.29 $r = (1 + \sin \theta) \cos 2\theta$

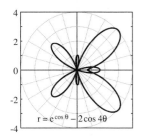

Fig. 10.2.30 The butterfly curve

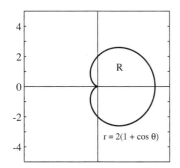

Fig. 10.3.1 What is the area of the region R bounded by the cardioid $r = 2(1 + \cos \theta)$?

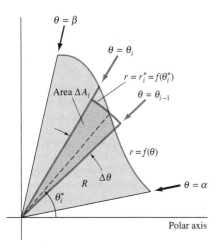

Fig. 10.3.2 We obtain the area formula from Riemann sums.

PROJECT B The simple "butterfly" shown in Fig. 10.2.30 is the graph of the polar-coordinates equation

$$r = e^{\cos\theta} - 2\cos 4\theta.$$

Now plot the polar-coordinates equation

$$r = e^{\cos\theta} - 2\cos 4\theta + \sin^5 \frac{\theta}{12}$$

for $0 \leq \theta \leq 24\pi$. The incredibly beautiful curve that results was discovered by Temple H. Fay.[†]

10.3

Area Computations in Polar Coordinates

The graph of the polar-coordinates equation $r = f(\theta)$ may bound an area, as does the cardioid $r = 2(1 + \cos \theta)$ (Fig. 10.3.1). To calculate the area of this region, we may find it convenient to work directly in polar coordinates rather than to change to rectangular coordinates.

To see how to set up an area integral in polar coordinates, we consider the region R of Fig. 10.3.2. This region is bounded by the two radial lines $\theta = \alpha$ and $\theta = \beta$ and by the curve $r = f(\theta)$, $\alpha \leq \theta \leq \beta$. To approximate the area A of R, we begin with a partition

$$\alpha = \theta_0 < \theta_1 < \theta_2 < \cdots < \theta_n = \beta$$

of the interval $[\alpha, \beta]$ into n subintervals, all with the same length $\Delta\theta = (\beta - \alpha)/n$. We select a point θ_i^* in the ith subinterval $[\theta_{i-1}, \theta_i]$ for $i = 1, 2, \ldots, n$.

Let ΔA_i denote the area of the sector bounded by the lines $\theta = \theta_{i-1}$ and $\theta = \theta_i$ and by the curve $r = f(\theta)$. We see from Fig. 10.3.2 that for small values of $\Delta\theta$, ΔA_i is approximately equal to the area of the *circular* sector that has radius $r_i^* = f(\theta_i^*)$ and is bounded by the same lines. That is,

$$\Delta A_i \approx \tfrac{1}{2}(r_i^*)^2 \, \Delta\theta = \tfrac{1}{2}[f(\theta_i^*)]^2 \, \Delta\theta.$$

We add the areas of these sectors for $i = 1, 2, \ldots, n$ and thereby find that

$$A = \sum_{i=1}^{n} \Delta A_i \approx \sum_{i=1}^{n} \tfrac{1}{2}[f(\theta_i^*)]^2 \, \Delta\theta.$$

The right-hand sum is a Riemann sum for the integral

$$\int_{\alpha}^{\beta} \tfrac{1}{2}[f(\theta)]^2 \, d\theta.$$

Hence, if f is continuous, the value of this integral is the limit, as $\Delta\theta \to 0$, of the preceding sum. We therefore conclude that the *area A of the region R bounded by the lines $\theta = \alpha$ and $\theta = \beta$ and by the curve $r = f(\theta)$ is*

$$A = \int_{\alpha}^{\beta} \tfrac{1}{2}[f(\theta)]^2 \, d\theta. \qquad (1)$$

[†] Fay's article "The Butterfly Curve" (*American Mathematical Monthly,* May 1989, p. 442) is well worth a trip to the library.

The infinitesimal sector shown in Fig. 10.3.3, with radius r, central angle $d\theta$, and area $dA = \frac{1}{2}r^2\,d\theta$, serves as a useful device for remembering Eq. (1) in the abbreviated form

$$A = \int_\alpha^\beta \frac{1}{2} r^2\,d\theta. \tag{2}$$

EXAMPLE 1 Find the area of the region bounded by the limaçon with equation $r = 3 + 2\cos\theta$, $0 \le \theta \le 2\pi$ (Fig. 10.3.4).

Solution We could apply Eq. (2) with $\alpha = 0$ and $\beta = 2\pi$. Here, instead, we will make use of symmetry. We will calculate the area of the upper half of the region and then double the result. Note that the infinitesimal sector shown in Fig. 10.3.4 sweeps out the upper half of the limaçon as θ increases from 0 to π (Fig. 10.3.5). Hence

$$A = 2\int_\alpha^\beta \frac{1}{2}r^2\,d\theta = \int_0^\pi (3 + 2\cos\theta)^2\,d\theta$$

$$= \int_0^\pi (9 + 12\cos\theta + 4\cos^2\theta)\,d\theta.$$

Because

$$4\cos^2\theta = 4\cdot\frac{1 + \cos 2\theta}{2} = 2 + 2\cos 2\theta,$$

we now get

$$A = \int_0^\pi (11 + 12\cos\theta + 2\cos 2\theta)\,d\theta$$

$$= \left[11\theta + 12\sin\theta + \sin 2\theta \right]_0^\pi = 11\pi.$$

EXAMPLE 2 Find the area bounded by each loop of the limaçon with equation $r = 1 + 2\cos\theta$ (Fig. 10.3.6).

Solution The equation $1 + 2\cos\theta = 0$ has two solutions for θ in the interval $[0, 2\pi]$: $\theta = 2\pi/3$ and $\theta = 4\pi/3$. The upper half of the outer loop of the limaçon corresponds to values of θ between 0 and $2\pi/3$, where r is positive. Because the curve is symmetric about the x-axis, we can find the

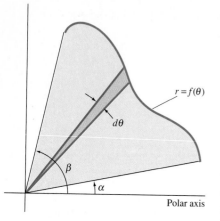

Fig. 10.3.3 Nonrigorous derivation of the area formula in polar coordinates

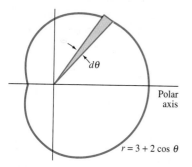

Fig. 10.3.4 The limaçon of Example 1

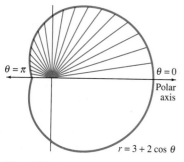

Fig. 10.3.5 Infinitesimal sectors from $\theta = 0$ to $\theta = \pi$ (Example 1)

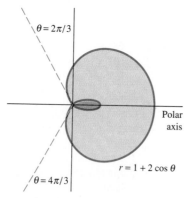

Fig. 10.3.6 The limaçon of Example 2

Ch. 10 / Polar Coordinates and Conic Sections

total area A_1 bounded by the outer loop by integrating from 0 to $2\pi/3$ and then doubling. Thus

$$A_1 = 2 \int_0^{2\pi/3} \tfrac{1}{2}(1 + 2\cos\theta)^2 \, d\theta = \int_0^{2\pi/3} (1 + 4\cos\theta + 4\cos^2\theta) \, d\theta$$

$$= \int_0^{2\pi/3} (3 + 4\cos\theta + 2\cos 2\theta) \, d\theta$$

$$= \left[3\theta + 4\sin\theta + \sin 2\theta \right]_0^{2\pi/3} = 2\pi + \tfrac{3}{2}\sqrt{3}.$$

The inner loop of the limaçon corresponds to values of θ between $2\pi/3$ and $4\pi/3$, where r is negative. Hence the area bounded by the inner loop is

$$A_2 = \int_{2\pi/3}^{4\pi/3} \tfrac{1}{2}(1 + 2\cos\theta)^2 \, d\theta$$

$$= \tfrac{1}{2} \left[3\theta + 4\sin\theta + \sin 2\theta \right]_{2\pi/3}^{4\pi/3} = \pi - \tfrac{3}{2}\sqrt{3}.$$

The area of the region lying *between* the two loops of the limaçon is then

$$A = A_1 - A_2 = 2\pi + \tfrac{3}{2}\sqrt{3} - (\pi - \tfrac{3}{2}\sqrt{3}) = \pi + 3\sqrt{3}.$$

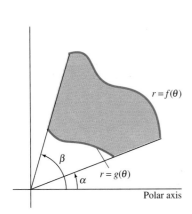

Fig. 10.3.7 The area between the graphs of f and g

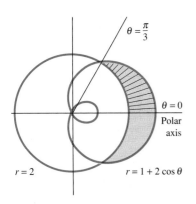

Fig. 10.3.8 The radial line segment illustrates the radii r_{inner} and r_{outer} of Eq. (4).

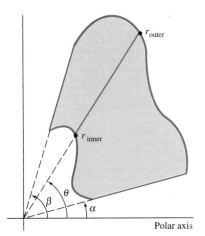

Fig. 10.3.9 The region of Example 3

THE AREA BETWEEN TWO POLAR CURVES

Now consider two curves $r = f(\theta)$ and $r = g(\theta)$, with $f(\theta) \geq g(\theta) \geq 0$ for $\alpha \leq \theta \leq \beta$. Then we can find the area of the region bounded by these curves and by the rays (or radial lines) $\theta = \alpha$ and $\theta = \beta$ (Fig. 10.3.7) by subtracting the area bounded by the inner curve from that bounded by the outer curve. That is, the area A between the two curves is given by

$$A = \int_\alpha^\beta \tfrac{1}{2}[f(\theta)]^2 \, d\theta - \int_\alpha^\beta \tfrac{1}{2}[g(\theta)]^2 \, d\theta$$

$$= \tfrac{1}{2} \int_\alpha^\beta \{[f(\theta)]^2 - [g(\theta)]^2\} \, d\theta. \qquad (3)$$

With r_{outer} for the outer curve and r_{inner} for the inner curve, we get the abbreviated formula

$$A = \tfrac{1}{2} \int_\alpha^\beta [(r_{\text{outer}})^2 - (r_{\text{inner}})^2] \, d\theta \qquad (4)$$

for the area of the region shown in Fig. 10.3.8.

EXAMPLE 3 Find the area A of the region that lies within the limaçon $r = 1 + 2\cos\theta$ and outside the circle $r = 2$.

Solution The circle and limaçon are shown in Fig. 10.3.9, with the area A between them shaded. The points of intersection of the circle and limaçon are given by

$$1 + 2\cos\theta = 2, \quad \text{so} \quad \cos\theta = \tfrac{1}{2},$$

and the figure shows that we should choose the solutions $\theta = \pm\pi/3$. These two values of θ are the needed limits of integration. When we use Eq. (3), we find that

$$A = \tfrac{1}{2} \int_{-\pi/3}^{\pi/3} \left[(1 + 2\cos\theta)^2 - 2^2\right] d\theta$$

$$= \int_{0}^{\pi/3} (4\cos\theta + 4\cos^2\theta - 3)\, d\theta \qquad \text{(because of symmetry about the polar axis)}$$

$$= \int_{0}^{\pi/3} (4\cos\theta + 2\cos 2\theta - 1)\, d\theta$$

$$= \left[4\sin\theta + \sin 2\theta - \theta\right]_{0}^{\pi/3} = \frac{15\sqrt{3} - 2\pi}{6}.$$

10.3 Problems

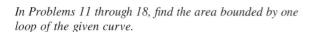

In Problems 1 through 10, find the area bounded by the given curve.

1. $r = 2\cos\theta$ **2.** $r = 4\sin\theta$

3. $r = 1 + \cos\theta$

4. $r = 2 - 2\sin\theta$ (Fig. 10.3.10)

5. $r = 2 - \cos\theta$

6. $r = 3 + 2\sin\theta$ (Fig. 10.3.11)

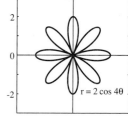

Fig. 10.3.12 The three-leaved rose of Problem 12

Fig. 10.3.13 The eight-leaved rose of Problem 13

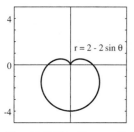

Fig. 10.3.10 The cardioid of Problem 4

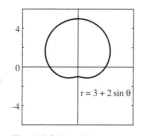

Fig. 10.3.11 The limaçon of Problem 6

7. $r = -4\cos\theta$ **8.** $r = 5(1 + \sin\theta)$

9. $r = 3 - \cos\theta$ **10.** $r = 2 + \sin\theta + \cos\theta$

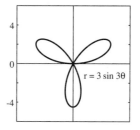

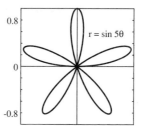

Fig. 10.3.14 The five-leaved rose of Problem 14

Fig. 10.3.15 The lemniscate of Problem 16

In Problems 11 through 18, find the area bounded by one loop of the given curve.

11. $r = 2\cos 2\theta$

12. $r = 3\sin 3\theta$ (Fig. 10.3.12)

13. $r = 2\cos 4\theta$ (Fig. 10.3.13)

14. $r = \sin 5\theta$ (Fig. 10.3.14)

15. $r^2 = 4\sin 2\theta$

16. $r^2 = 4\cos 2\theta$ (Fig. 10.3.15)

17. $r^2 = 4\sin\theta$ **18.** $r = 6\cos 6\theta$

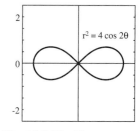

In Problems 19 through 30, find the area of the region described.

19. Inside $r = 2\sin\theta$ and outside $r = 1$

20. Inside both $r = 4\cos\theta$ and $r = 2$

21. Inside both $r = \cos\theta$ and $r = \sqrt{3}\sin\theta$

22. Inside $r = 2 + \cos\theta$ and outside $r = 2$

23. Inside $r = 3 + 2\cos\theta$ and outside $r = 4$

24. Inside $r^2 = 2\cos 2\theta$ and outside $r = 1$

25. Inside $r^2 = \cos 2\theta$ and $r^2 = \sin 2\theta$ (Fig. 10.3.16)

554

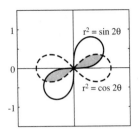

Fig. 10.3.16 Problem 25

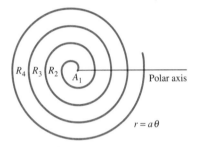

Fig. 10.3.17 Problem 26

26. Inside the large loop and outside the small loop of $r = 1 - 2 \sin \theta$ (Fig. 10.3.17)

27. Inside $r = 2(1 + \cos \theta)$ and outside $r = 1$

28. Inside the figure-eight curve $r^2 = 4 \cos \theta$ and outside $r = 1 - \cos \theta$

29. Inside both $r = 2 \cos \theta$ and $r = 2 \sin \theta$

30. Inside $r = 2 + 2 \sin \theta$ and outside $r = 3$

31. Find the area of the circle $r = \sin \theta + \cos \theta$ by integration in polar coordinates (Fig. 10.3.18). Check your answer by writing the equation of the circle in rectangular coordinates, finding its radius, and then using the familiar formula for the area of a circle.

32. Find the area of the region that lies interior to all three circles $r = 1$, $r = 2 \cos \theta$, and $r = 2 \sin \theta$.

33. The *spiral of Archimedes,* shown in Fig. 10.3.19, has the simple equation $r = a\theta$ (a is a constant). Let A_n denote the area bounded by the nth turn of the spiral, where $2(n - 1)\pi \leqq \theta \leqq 2n\pi$, and by the portion of the polar axis joining its endpoints. For each $n \geqq 2$, let $R_n = A_n - A_{n-1}$ denote the area between the $(n - 1)$th and the nth turns. Then derive the following results of Archimedes:

(a) $A_1 = \frac{1}{3}\pi(2\pi a)^2$; (b) $A_2 = \frac{7}{12}\pi(4\pi a)^2$;

(c) $R_2 = 6A_1$; (d) $R_{n+1} = nR_2$ for $n \geqq 2$.

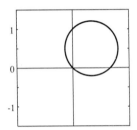

Fig. 10.3.18 The circle $r = \sin \theta + \cos \theta$ (Problem 31)

Fig. 10.3.19 The spiral of Archimedes (Problem 33)

34. Two circles both have radius a, and each circle passes through the center of the other. Find the area of the region that lies within both circles.

10.3 Project

We can use midpoint sums corresponding to the polar-coordinates area integral

$$A = \int_{\alpha}^{\beta} \tfrac{1}{2}[f(\theta)]^2 \, d\theta \tag{5}$$

to approximate numerically the area of the region bounded by the curve $r = f(\theta)$, $\alpha \leqq \theta \leqq \beta$ (Fig. 10.3.2). Let θ_i^* denote the *midpoint* of the ith subinterval $[\theta_{i-1}, \theta_i]$ of a subdivision of $[\alpha, \beta]$ into n subintervals, each of length $\Delta\theta = (\beta - \alpha)/n$. Then the corresponding midpoint approximation to the area in Eq. (5) is

$$A_n = \sum_{i=1}^{n} \tfrac{1}{2}[f(\theta_i^*)]^2 \, \Delta\theta. \tag{6}$$

We can evaluate such sums numerically by using calculator or computer programs or with commands like those listed in Figs. 5.4.10 and 5.4.11 (of the Section 5.4 project).

In Problems 1 through 7, first approximate the area of the given region

R by using Eq. (6) with $n = 10$ and with $n = 20$. Then compare your approximations to the exact area calculated from Eq. (5).

1. *R* is the unit circle bounded by $r = 1$. Do you get the same results with $n = 10$ as with $n = 20$? Why?

2. *R* is the unit circle with boundary $r = 2 \cos \theta$.

3. *R* is bounded by the cardioid $r = 2 - 2 \sin \theta$ (Fig. 10.3.10).

4. *R* is bounded by the limaçon $r = 3 + 2 \sin \theta$ (Fig. 10.3.11).

5. *R* is bounded by one loop of the three-leaved rose $r = 3 \sin 3\theta$ (Fig. 10.3.12).

6. *R* is bounded by one loop of the eight-leaved rose $r = 2 \cos 4\theta$ (Fig. 10.3.13).

7. *R* is bounded by one loop of the curve $r^2 = 4 \cos 2\theta$ (Fig. 10.3.15).

10.4
The Parabola

The case $e = 1$ of Example 3 in Section 10.1 is motivation for this formal definition.

> **Definition** *The Parabola*
> A **parabola** is the set of all points *P* in the plane that are equidistant from a fixed point *F* (called the **focus** of the parabola) and from a fixed line *L* (called the parabola's **directrix**) not containing *F*.

If the focus of the parabola is $F(p, 0)$ and its directrix is the vertical line $x = -p$, $p > 0$, then it follows from Eq. (12) of Section 10.1 that the equation of this parabola is

$$y^2 = 4px. \tag{1}$$

When we replace *x* by $-x$ both in the equation and in the discussion that precedes it, we get the equation of the parabola whose focus is $(-p, 0)$ and whose directrix is the vertical line $x = p$. The new parabola has equation

$$y^2 = -4px. \tag{2}$$

The old and new parabolas appear in Fig. 10.4.1.

We could also interchange *x* and *y* in Eq. (1). This would give the equation of a parabola whose focus is $(0, p)$ and whose directrix is the horizontal line $y = -p$. This parabola opens upward, as in Fig. 10.4.2(a); its equation is

$$x^2 = 4py. \tag{3}$$

Finally, we replace *y* with $-y$ in Eq. (3). This gives the equation

$$x^2 = -4py \tag{4}$$

of a parabola opening downward, with focus $(0, -p)$ and with directrix $y = p$, as in Fig. 10.4.2(b).

Each of the parabolas discussed so far is symmetric about one of the coordinate axes. The line about which a parabola is symmetric is called the parabola's **axis.** The point of a parabola midway between its focus and its

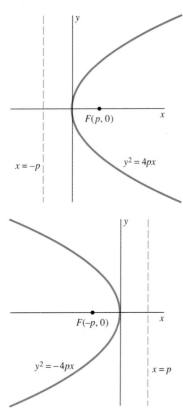

Fig. 10.4.1 Two parabolas with vertical directrices

Ch. 10 / Polar Coordinates and Conic Sections

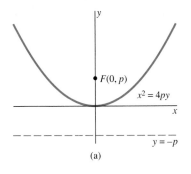

$F(0, p)$

$x^2 = 4py$

$y = -p$

(a)

$y = p$

$F(0, -p)$

$x^2 = -4py$

(b)

Fig. 10.4.2 Two parabolas with horizontal directrices: (a) opening upward; (b) opening downward

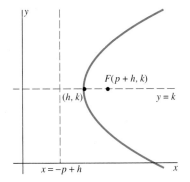

$F(p + h, k)$

(h, k)

$y = k$

$x = -p + h$

Fig. 10.4.3 A translation of the parabola $y^2 = 4px$

directrix is called the parabola's **vertex.** The vertex of each parabola that we discussed in connections with Eqs. (1) through (4) is the origin $(0, 0)$.

EXAMPLE 1 Determine the focus, directrix, axis, and vertex of the parabola $x^2 = 12y$.

Solution We write the given equation as $x^2 = 4 \cdot (3y)$. In this form it matches Eq. (3) with $p = 3$. Hence the focus of the given parabola is $(0, 3)$, and the parabola's directrix is the horizontal line $y = -3$. The y-axis is its axis of symmetry, and the parabola opens upward from its vertex at the origin.

Suppose that we begin with the parabola of Eq. (1) and translate it in such a way that its vertex moves to the point (h, k). Then the translated parabola has equation

$$(y - k)^2 = 4p(x - h). \tag{1a}$$

The new parabola has focus $F(p + h, k)$, and its directrix is the vertical line $x = -p + h$ (Fig. 10.4.3). Its axis is the horizontal line $y = k$.

We can obtain the translates of the other three parabolas in Eqs. (2) through (4) in the same way. If the vertex is moved from the origin to the point (h, k), then the three equations take these forms:

$$(y - k)^2 = -4p(x - h), \tag{2a}$$

$$(x - h)^2 = 4p(y - k), \quad \text{and} \tag{3a}$$

$$(x - h)^2 = -4p(y - k). \tag{4a}$$

Equations (1a) and (2a) both take the general form

$$y^2 + Ax + By + C = 0 \qquad (A \neq 0), \tag{5}$$

whereas Eqs. (3a) and (4a) both take the general form

$$x^2 + Ax + By + C = 0 \qquad (B \neq 0). \tag{6}$$

What is significant about Eqs. (5) and (6) is what they have in common: Both are linear in one of the coordinate variables and quadratic in the other. In fact, we can reduce *any* such equation to one of the standard forms in Eqs. (1a) through (4a) by completing the square in the coordinate variable that appears quadratically. This means that the graph of any equation of the form of either Eq. (5) or (6) is a parabola. The features of the parabola can be read from the standard form of its equation, as in Example 2.

EXAMPLE 2 Determine the graph of the equation

$$4y^2 - 8x - 12y + 1 = 0.$$

Solution This equation is linear in x and quadratic in y. We divide through by the coefficient of y^2 and then collect on one side of the equation all terms that include y:

$$y^2 - 3y = 2x - \tfrac{1}{4}.$$

Then we complete the square in the variable y and thus find that

$$y^2 - 3y + \tfrac{9}{4} = 2x - \tfrac{1}{4} + \tfrac{9}{4} = 2x + 2 = 2(x + 1).$$

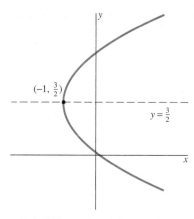

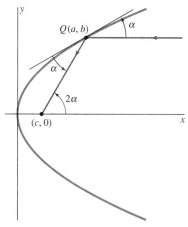

Fig. 10.4.4 The parabola of Example 2

Fig. 10.4.5 The reflection property of the parabola: $\alpha = \beta$

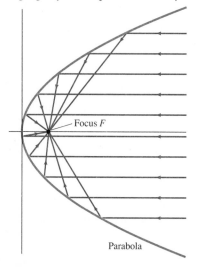

Fig. 10.4.6 Incident rays parallel to the axis reflect through the focus.

The final step is to write in the form $4p(x - h)$ the terms on the right-hand side that include x:

$$(y - \tfrac{3}{2})^2 = 4 \cdot \tfrac{1}{2} \cdot (x + 1).$$

This equation has the form of Eq. (1a), with $p = \tfrac{1}{2}, h = -1$, and $k = \tfrac{3}{2}$. Thus the graph is a parabola that opens to the right from the vertex at $(-1, \tfrac{3}{2})$. Its focus is at $(-\tfrac{1}{2}, \tfrac{3}{2})$, its directrix is the vertical line $x = -\tfrac{3}{2}$, and its axis is the horizontal line $y = \tfrac{3}{2}$ (Fig. 10.4.4).

APPLICATIONS OF PARABOLAS

The parabola $y^2 = 4px$ $(p > 0)$ is shown in Fig. 10.4.5 along with an incoming ray of light traveling to the left and parallel to the x-axis. This light ray strikes the parabola at the point $Q(a, b)$ and is reflected toward the x-axis, which it meets at the point $(c, 0)$. The light ray's angle of reflection must equal its angle of incidence, which is why both of these angles—measured with respect to the tangent line L at Q—are labeled α in the figure. The angle vertical to the angle of incidence is also equal to α. Hence, because the incoming ray is parallel to the x-axis, the angle the reflected ray makes with the x-axis at $(c, 0)$ is 2α.

Using the points Q and $(c, 0)$ to compute the slope of the reflected light ray, we find that

$$\frac{b}{a - c} = \tan 2\alpha = \frac{2 \tan \alpha}{1 - \tan^2 \alpha}.$$

(The second equality follows from a trigonometric identity in Problem 58 of Section 8.2). But the angle α is related to the slope of the tangent line L at Q. To find that slope, we begin with

$$y = 2\sqrt{px} = 2(px)^{1/2}$$

and compute

$$\frac{dy}{dx} = (p/x)^{1/2}.$$

Hence the slope of L is both $\tan \alpha$ and dy/dx evalauted at (a, b); that is,

$$\tan \alpha = (p/a)^{1/2}.$$

Therefore,

$$\frac{b}{a - c} = \frac{2 \tan \alpha}{1 - \tan^2 \alpha} = \frac{2\sqrt{p/a}}{1 - \dfrac{p}{a}} = \frac{2\sqrt{pa}}{a - p} = \frac{b}{a - p},$$

because $b = 2\sqrt{pa}$. Hence $c = p$. The surprise is that c is independent of a and b and depends only on the equation $y^2 = 4px$ of the parabola. Therefore *all* incoming light rays parallel to the x-axis will be reflected to the single point $F(p, 0)$. This is why F is called the focus of the parabola.

This **reflection property** of the parabola is exploited in the design of parabolic mirrors. Such a mirror has the shape of the surface obtained by revolving a parabola around its axis of symmetry. Thus a beam of incoming light rays parallel to the axis will be focused at F, as shown in Fig. 10.4.6. The

An application of the reflection properties of conic sections.

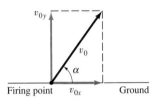

Fig. 10.4.7 Resolution of the initial velocity v_0 into horizontal and vertical components.

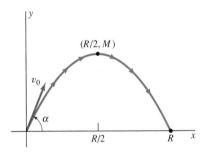

Figure 10.4.8 The trajectory of the projectile, showing its maximum altitude and its range.

reflection property can also be used in reverse—rays emitted at the focus are reflected in a beam parallel to the axis, thus keeping the light beam intense. Moreover, applications are not limited to light rays alone; parabolic mirrors are used in visual and radio telescopes, radar antennas, searchlights, automobile headlights, microphone systems, satellite ground stations, and solar heating devices.

Galileo discovered early in the seventeenth century that the trajectory of a projectile fired from a gun is a parabola (under the assumptions that air resistance can be ignored and that the gravitational acceleration remains constant). Suppose that a projectile is fired with initial velocity v_0 at time $t = 0$ from the origin and at an angle α of inclination from the horizontal x-axis. Then the initial velocity of the projectile splits into the components

$$v_{0x} = v_0 \cos \alpha \quad \text{and} \quad v_{0y} = v_0 \sin \alpha,$$

as indicated in Fig. 10.4.7. The fact that the projectile continues to move horizontally with *constant* speed v_{0x}, together with Eq. (34) of Section 5.2, implies that its x- and y-coordinates after t seconds are

$$x = (v_0 \cos \alpha)t, \tag{7}$$

$$y = -\tfrac{1}{2}gt^2 + (v_0 \sin \alpha)t. \tag{8}$$

By substituting $t = x/(v_0 \cos \alpha)$ from Eq. (7) into Eq. (8) and then completing the square, we can derive (as in Problem 24) an equation of the form

$$y - M = -4p(x - \tfrac{1}{2}R)^2. \tag{9}$$

Here,

$$M = \frac{v_0^2 \sin^2 \alpha}{2g} \tag{10}$$

is the maximum height attained by the projectile, and

$$R = \frac{v_0^2 \sin 2\alpha}{g} \tag{11}$$

is the **range,** or horizontal distance, the projectile will travel before it returns to the ground. Thus its trajectory is the parabola shown in Fig. 10.4.8.

10.4 Problems

In Problems 1 through 5, find the equation and sketch the graph of the parabola with vertex V and focus F.

1. $V(0, 0)$, $F(3, 0)$ **2.** $V(0, 0)$, $F(0, -2)$

3. $V(2, 3)$, $F(2, 1)$ **4.** $V(-1, -1)$, $F(-3, -1)$

5. $V(2, 3)$, $F(0, 3)$

In Problems 6 through 10, find the equation and sketch the graph of the parabola with the given focus and directrix.

6. $F(1, 2)$, $x = -1$ **7.** $F(0, -3)$, $y = 0$

8. $F(1, -1)$, $x = 3$ **9.** $F(0, 0)$, $y = -2$

10. $F(-2, 1)$, $x = -4$

In Problems 11 through 18, sketch the parabola with the given equation. Show and label its vertex, focus, axis, and directrix.

11. $y^2 = 12x$ **12.** $x^2 = -8y$

13. $y^2 = -6x$ **14.** $x^2 = 7y$

15. $x^2 - 4x - 4y = 0$

16. $y^2 - 2x + 6y + 15 = 0$

17. $4x^2 + 4x + 4y + 13 = 0$

18. $4y^2 - 12y + 9x = 0$

19. Prove that the point of the parabola $y^2 = 4px$ closest to its vertex is its focus.

20. Find the equation of the parabola that has a vertical axis and passes through the points (2, 3), (4, 3), and (6, −5).

21. Show that the equation of the line tangent to the parabola $y^2 = 4px$ at the point (x_0, y_0) is

$$2px - y_0 y + 2px_0 = 0.$$

Conclude that the tangent line intersects the x-axis at the point $(-x_0, 0)$. This fact provides a quick method for constructing a line tangent to a parabola at a given point.

22. A comet's orbit is a parabola with the sun at its focus. When the comet is $100\sqrt{2}$ million miles from the sun, the line from the sun to the comet makes an angle of $45°$ with the axis of the parabola (Fig. 10.4.9). What will be the minimum distance between the comet and the sun? [*Suggestion:* Write the equation of the parabola with the origin at the focus, and then use the result of Problem 19.]

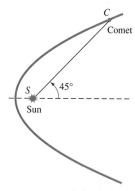

Fig. 10.4.9 The comet of Problem 22 in parabolic orbit around the sun

23. Suppose that the angle of Problem 22 increases from $45°$ to $90°$ in 3 days. How much longer will be required for

the comet to reach its point of closest approach to the sun? Assume that the line segment from the sun to the comet sweeps out area at a constant rate (Kepler's second law).

24. Use Eqs. (7) and (8) to derive Eq. (9) with the values of M and R given in Eqs. (10) and (11).

25. Deduce from Eq. (11) that, given a fixed initial velocity v_0, the maximum range of the projectile is $R_{\max} = v_0^2/g$ and is attained when $\alpha = 45°$.

In Problems 26 through 28, assume that a projectile is fired with initial velocity $v_0 = 50$ m/s from the origin and at an angle of inclination α. Use $g = 9.8$ m/s^2.

26. If $\alpha = 45°$, find the range of the projectile and the maximum height it attains.

27. For what value or values of α is the range $R = 125$ m?

28. Find the range R of the projectile and the length of time it remains above the ground if (a) $\alpha = 30°$; (b) $\alpha = 60°$.

29. The book *Elements of Differential and Integral Calculus* by William Granville, Percey Smith, and William Longley (Ginn and Company: Boston, 1929) lists a number of "curves for reference"; the curve with equation $\sqrt{x} + \sqrt{y} = \sqrt{a}$ is called a parabola. Verify that the curve in question actually is a parabola, or show that it is not.

30. The 1992 edition of the study guide for the national actuarial examinations has a problem similar to this one: Every point on the plane curve K is equally distant from the point $(-1, -1)$ and from the line $x + y = 1$, and K has equation $x^2 + Bxy + Cy^2 + Dx + Ey + F = 0$. Which is the value of D: (a) -2, (b) 2, (c) 4, (d) 6, or (e) 8?

10.5
The Ellipse

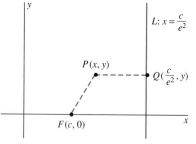

Fig. 10.5.1 Ellipse: focus F, directrix L, eccentricity e

An ellipse is a conic section with eccentricity e less than 1, as in Example 3 of Section 10.1.

> **Definition** *The Ellipse*
> Suppose that $e < 1$, and let F be a fixed point and L a fixed line not containing F. The **ellipse** with **eccentricity** e, **focus** F, and **directrix** L is the set of all points P such that the distance $|PF|$ is e times the (perpendicular) distance from P to the line L.

The equation of the ellipse is especially simple if F is the point $(c, 0)$ on the x-axis and L is the vertical line $x = c/e^2$. The case $c > 0$ is shown in Fig. 10.5.1. If Q is the point $(c/e^2, y)$, then PQ is the perpendicular from $P(x, y)$ to L. The condition $|PF| = e|PQ|$ then gives

$$(x - c)^2 + y^2 = e^2\left(x - \frac{c}{e^2}\right)^2;$$

$$x^2 - 2cx + c^2 + y^2 = e^2x^2 - 2cx + \frac{c^2}{e^2};$$

$$x^2(1 - e^2) + y^2 = c^2\left(\frac{1}{e^2} - 1\right) = \frac{c^2}{e^2}(1 - e^2).$$

Thus

$$x^2(1 - e^2) + y^2 = a^2(1 - e^2),$$

where

$$a = \frac{c}{e}. \tag{1}$$

We divide both sides of the next-to-last equation by $a^2(1 - e^2)$ and get

$$\frac{x^2}{a^2} + \frac{y^2}{a^2(1 - e^2)} = 1.$$

Finally, with the aid of the fact that $e < 1$, we may let

$$b^2 = a^2(1 - e^2) = a^2 - c^2. \tag{2}$$

Then the equation of the ellipse with focus $(c, 0)$ and directrix $x = c/e^2 = a/e$ takes the simple form

$$\frac{x^2}{a^2} + \frac{y^2}{b^2} = 1. \tag{3}$$

We see from Eq. (3) that this ellipse is symmetric about both coordinate axes. Its x-intercepts are $(\pm a, 0)$, and its y-intercepts are $(0, \pm b)$. The points $(\pm a, 0)$ are called the **vertices** of the ellipse, and the line segment joining them is called its **major axis**. The line segment joining $(0, b)$ and $(0, -b)$ is called the **minor axis** [note from Eq. (2) that $b < a$]. The alternative form

$$a^2 = b^2 + c^2 \tag{4}$$

of Eq. (2) is the Pythagorean relation for the right triangle of Fig. 10.5.2. Indeed, visualization of this triangle is an excellent way to remember Eq. (4). The numbers a and b are the lengths of the major and minor **semiaxes**, respectively.

Because $a = c/e$, the directrix of the ellipse in Eq. (3) is $x = a/e$. If we had begun instead with the focus $(-c, 0)$ and directrix $x = -a/e$, we would still have obtained Eq. (3), because only the squares of a and c are involved in its derivation. Thus the ellipse in Eq. (3) has *two* foci, $(c, 0)$ and $(-c, 0)$, and *two* directrices, $x = a/e$ and $x = -a/e$ (Fig. 10.5.3).

The larger the eccentricity $e < 1$, the more elongated the ellipse. (Remember that $e = 1$ is the eccentricity of a parabola.) But if $e = 0$, then Eq. (2) gives $b = a$, so Eq. (3) reduces to the equation of a circle of radius a. Thus a circle is an ellipse of eccentricity zero. Compare the three cases shown in Fig. 10.5.4.

EXAMPLE 1 Find the equation of the ellipse with foci $(\pm 3, 0)$ and vertices $(\pm 5, 0)$.

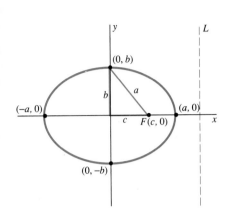

Fig. 10.5.2 The parts of an ellipse

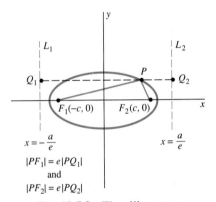

Fig. 10.5.3 The ellipse as a conic section: two foci, two directrices

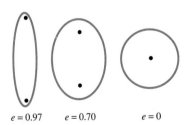

$e = 0.97$ $e = 0.70$ $e = 0$

Fig. 10.5.4 The relation between the eccentricity of an ellipse and its shape

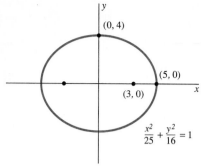

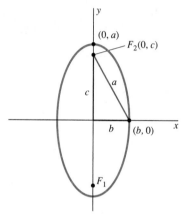

Fig. 10.5.5 The ellipse of Example 1

Fig. 10.5.6 An ellipse whose major axis is vertical

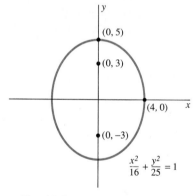

Fig. 10.5.7 The ellipse of Example 2

Solution We are given $c = 3$ and $a = 5$, so Eq. (2) gives $b = 4$. Thus Eq. (3) gives

$$\frac{x^2}{25} + \frac{y^2}{16} = 1$$

for the desired equation. This ellipse is shown in Fig. 10.5.5.

If the two foci of an ellipse are on the y-axis, such as $F_1(0, c)$ and $F_2(0, -c)$, then the equation of the ellipse is

$$\frac{x^2}{b^2} + \frac{y^2}{a^2} = 1,\tag{5}$$

and it is still true that $a^2 = b^2 + c^2$, as in Eq. (4). But now the major axis of length $2a$ is vertical, and the minor axis of length $2b$ is horizontal. The derivation of Eq. (5) is similar to that of Eq. (3); see Problem 23. Figure 10.5.6 shows the case of an ellipse whose major axis is vertical. The vertices of such an ellipse are at $(0, \pm a)$; they are always the endpoints of the major axis.

In practice there is little chance of confusing Eqs. (3) and (5). The equation or the given data will make it clear whether the major axis of the ellipse is horizontal or vertical. Just use the equation to read the ellipse's intercepts. The two intercepts that are farthest from the origin are the endpoints of the major axis, and the other two are the endpoints of the minor axis. The two foci lie on the major axis, each at distance c from the center of the ellipse—which will be the origin if the equation of the ellipse has the form of either Eq. (3) or Eq. (5).

EXAMPLE 2 Sketch the graph of the equation

$$\frac{x^2}{16} + \frac{y^2}{25} = 1.$$

Solution The x-intercepts are $(\pm 4, 0)$; the y- intercepts are $(0, \pm 5)$. So the major axis is vertical. We take $a = 5$ and $b = 4$ in Eq. (4) and find that $c = 3$. The foci are thus at $(0, \pm 3)$. Hence this ellipse has the appearance of the one shown in Fig. 10.5.7.

Any equation of the form

$$Ax^2 + Cy^2 + Dx + Ey + F = 0,\tag{6}$$

in which the coefficients A and C of the squared terms are *both nonzero* and *have the same sign,* may be reduced to the form

$$A(x - h)^2 + C(y - k)^2 = G$$

by completing the square in x and y. We may assume that A and C are both positive. Then if $G < 0$, there are no points that satisfy Eq. (6), and the graph is the empty set. If $G = 0$, then there is exactly one point on the locus—the single point (h, k). And if $G > 0$, we can divide both sides of the last equation by G and get an equation that resembles one of these two:

$$\frac{(x - h)^2}{a^2} + \frac{(y - k)^2}{b^2} = 1,\tag{7a}$$

$$\frac{(x-h)^2}{b^2} + \frac{(y-k)^2}{a^2} = 1. \tag{7b}$$

Which equation should you choose? Select the one that is consistent with the condition $a \geqq b > 0$. Finally, note that either of the equations in (7) is the equation of a translated ellipse. Thus, apart from the exceptional cases already noted, the graph of Eq. (6) is an ellipse if $AC > 0$.

EXAMPLE 3 Determine the graph of the equation

$$3x^2 + 5y^2 - 12x + 30y + 42 = 0.$$

Solution We collect terms containing x and terms containing y and complete the square in each variable. This gives

$$3(x^2 - 4x) + 5(y^2 + 6y) = -42;$$

$$3(x^2 - 4x + 4) + 5(y^2 + 6y + 9) = 15;$$

$$\frac{(x-2)^2}{5} + \frac{(y+3)^2}{3} = 1.$$

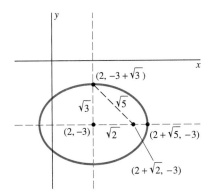

Fig. 10.5.8 The ellipse of Example 3

Thus the given equation is that of a translated ellipse. Its center is at $(2, -3)$. The ellipse's horizontal major semiaxis has length $a = \sqrt{5}$, and its minor semiaxis has length $b = \sqrt{3}$ (Fig. 10.5.8). The distance from the center to each focus is $c = \sqrt{2}$, and the eccentricity is $e = c/a = \sqrt{\frac{2}{5}}$.

APPLICATIONS OF ELLIPSES

EXAMPLE 4 The orbit of the earth is an ellipse with the sun at one focus. The planet's maximum distance from the sun is 94.56 million miles, and its minimum distance is 91.44 million miles. What are the major and minor semiaxes of the earth's orbit, and what is its eccentricity?

Solution As Fig. 10.5.9 shows, we have

$$a + c = 94.56 \quad \text{and} \quad a - c = 91.44,$$

with units in millions of miles. We conclude from these equations that $a = 93.00$, that $c = 1.56$, and then that

$$b = \sqrt{(93.00)^2 - (1.56)^2} \approx 92.99$$

million miles. Finally,

$$e = \frac{1.56}{93.00} \approx 0.017,$$

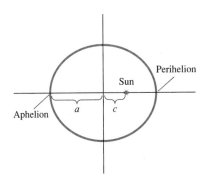

Fig. 10.5.9 The orbit of the earth with its eccentricity exaggerated (Example 4)

a number relatively close to zero. This means that the earth's orbit is nearly circular. Indeed, the major and minor semiaxes are so nearly equal that, on any usual scale, the earth's orbit would appear to be a perfect circle. But the difference between uniform circular motion and the earth's actual motion has some important aspects, including the facts that the sun is 1.56 million miles off center and—as we shall see in Chapter 13—that the orbital speed of the earth is not constant.

EXAMPLE 5 One of the most famous comets is Halley's comet, named for Edmund Halley (1656–1742), a disciple of Newton. By studying the records of the paths of earlier-sighted comets, Halley deduced that the comet of 1682 was the same one that had been sighted in 1607, in 1531, in 1456, and in 1066 (an omen at the Battle of Hastings). In 1682 he predicted that this comet would return in 1759, in 1835, and in 1910; he was correct each time. The period of Halley's comet is about 76 years—it can vary a couple of years in either direction because of perturbations of its orbit by the planet Jupiter. The orbit of Halley's comet is an ellipse with the sun at one focus. In terms of astronomical units (1 AU is the mean distance from the earth to the sun), the major and minor semiaxes of this elliptical orbit are 18.09 AU and 4.56 AU, respectively. What are the maximum and minimum distances from the sun of Halley's comet?

Solution We are given that $a = 18.09$ (all measurements are in astronomical units) and that $b = 4.56$, so

$$c = \sqrt{(18.09)^2 - (4.56)^2} \approx 17.51.$$

Hence its maximum distance from the sun is $a + c \approx 35.60$ AU, and its minimum distance is $a - c \approx 0.58$ AU. The eccentricity of its orbit is

$$e = \frac{17.51}{18.09} \approx 0.97,$$

a very eccentric orbit.

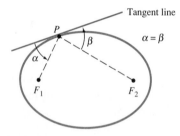

Fig. 10.5.10 The reflection property: $\alpha = \beta$

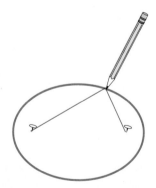

Fig. 10.5.11 One way to draw an ellipse

The *reflection property* of the ellipse states that the tangent line at a point P of an ellipse makes equal angles with the two lines PF_1 and PF_2 from P to the two foci of the ellipse (Fig. 10.5.10). This property is the basis of the "whispering gallery" phenomenon, which has been observed in the so-called whispering gallery of the U.S. Senate. Suppose that the ceiling of a large room is shaped like half an ellipsoid obtained by revolving an ellipse about its major axis. Sound waves, like light waves, are reflected at equal angles of incidence and reflection. Thus if two diplomats are holding a quiet conversation near one focus of the ellipsoidal surface, a reporter standing near the other focus—perhaps 50 ft away—would be able to eavesdrop on their conversation even if the conversation were inaudible to others in the same room.

Some billiard tables are manufactured in the shape of an ellipse. The foci of such tables are plainly marked for the convenience of enthusiasts of this unusual game.

A more serious application of the reflection property of ellipses is the nonsurgical kidney-stone treatment called *shockwave lithotripsy*. An elliptical reflector with a transducer (an energy transmitter) at one focus is positioned outside the patient's body so that the offending kidney stone is located at the other focus. The stone then is pulverized by reflected shock waves emanating from the transducer.[†]

An alternative definition of the ellipse with foci F_1 and F_2 and major axis of length $2a$ is this: It is the locus of a point P such that the sum of the distances $|PF_1|$ and $|PF_2|$ is the constant $2a$ (see Problem 26). This fact gives us a convenient way to draw the ellipse by using two tacks placed at F_1 and F_2, a string of length $2a$, and a pencil (Fig. 10.5.11).

[†] For further details, see the COMAP *Newsletter* **20**, November 1986.

10.5 Problems

In Problems 1 through 15, find an equation of the ellipse specified.

1. Vertices $(\pm 4, 0)$ and $(0, \pm 5)$

2. Foci $(\pm 5, 0)$, major semiaxis 13

3. Foci $(0, \pm 8)$, major semiaxis 17

4. Center $(0, 0)$, vertical major axis 12, minor axis 8

5. Foci $(\pm 3, 0)$, eccentricity $\frac{3}{4}$

6. Foci $(0, \pm 4)$, eccentricity $\frac{2}{3}$

7. Center $(0, 0)$, horizontal major axis 20, eccentricity $\frac{1}{2}$

8. Center $(0, 0)$, horizontal minor axis 10, eccentricity $\frac{1}{2}$

9. Foci $(\pm 2, 0)$, directrices $x = \pm 8$

10. Foci $(0, \pm 4)$, directrices $y = \pm 9$

11. Center $(2, 3)$, horizontal axis 8, vertical axis 4

12. Center $(1, -2)$, horizontal major axis 8, eccentricity $\frac{3}{4}$

13. Foci $(-2, 1)$ and $(4, 1)$, major axis 10

14. Foci $(-3, 0)$ and $(-3, 4)$, minor axis 6

15. Foci $(-2, 2)$ and $(4, 2)$, eccentricity $\frac{1}{3}$

Sketch the graphs of the equations in Problems 16 through 20. Indicate centers, foci, and lengths of axes.

16. $4x^2 + y^2 = 16$

17. $4x^2 + 9y^2 = 144$

18. $4x^2 + 9y^2 = 24x$

19. $9x^2 + 4y^2 - 32y + 28 = 0$

20. $2x^2 + 3y^2 + 12x - 24y + 60 = 0$

21. The orbit of the comet Kahoutek is an ellipse of extreme eccentricity $e = 0.999925$; the sun is at one focus of this ellipse. The minimum distance between the sun and Kahoutek is 0.13 AU. What is the maximum distance between Kahoutek and the sun?

22. The orbit of the planet Mercury is an ellipse of eccentricity $e = 0.206$. Its maximum and minimum distances from the sun are 0.467 and 0.307 AU, respectively. What are the major and minor semiaxes of the orbit of Mercury? Does "nearly circular" accurately describe the orbit of Mercury?

23. Derive Eq. (5) for an ellipse whose foci lie on the y-axis.

24. Show that the line tangent to the ellipse

$$\frac{x^2}{a^2} + \frac{y^2}{b^2} = 1$$

at the point $P(x_0, y_0)$ of that ellipse has equation

$$\frac{x_0 x}{a^2} + \frac{y_0 y}{b^2} = 1.$$

25. Use the result of Problem 24 to establish the reflection property of the ellipse. [*Suggestion:* Let m be the slope of the line normal to the ellipse at $P(x_0, y_0)$, and let m_1 and m_2 be the slopes of the lines PF_1 and PF_2, respectively, from P to the two foci F_1 and F_2 of the ellipse. Show that

$$\frac{m - m_1}{1 + m_1 m} = \frac{m_2 - m}{1 + m_2 m};$$

then use the identity for $\tan(A - B)$.]

26. Given $F_1(-c, 0)$ and $F_2(c, 0)$ and $a > c > 0$, prove that the ellipse $x^2/a^2 + y^2/b^2 = 1$ (with $b^2 = a^2 - c^2$) is the locus of a point P such that $|PF_1| + |PF_2| = 2a$.

27. Find the equation of the ellipse with horizontal and vertical axes that passes through the points $(-1, 0)$, $(3, 0)$, $(0, 2)$, and $(0, -2)$.

28. Derive an equation for the ellipse with foci $(3, -3)$ and $(-3, 3)$ and major axis of length 10. Note: The foci of this ellipse lie on neither a vertical line nor a horizontal line.

10.6
The Hyperbola

A hyperbola is a conic section defined in the same way as is an ellipse, except that the eccentricity e of a hyperbola is greater than 1.

Definition *The Hyperbola*

Suppose that $e > 1$, and let F be a fixed point and L a fixed line not containing F. Then the **hyperbola** with **eccentricity** e, **focus** F, and **directrix** L is the set of all points P such that the distance $|PF|$ is e times the (perpendicular) distance from P to the line L.

As with the ellipse, the equation of a hyperbola is simplest if F is the point $(c, 0)$ on the x-axis and L is the vertical line $x = c/e^2$. The case $c > 0$

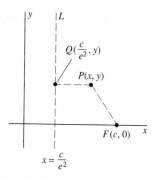

Fig. 10.6.1 The definition of the hyperbola

is shown in Fig. 10.6.1. If Q is the point $(c/e^2, y)$, then PQ is the perpendicular from $P(x, y)$ to L. The condition $|PF| = e|PQ|$ gives

$$(x - c)^2 + y^2 = e^2\left(x - \frac{c}{e^2}\right)^2;$$

$$x^2 - 2cx + c^2 + y^2 = e^2 x^2 - 2cx + \frac{c^2}{e^2};$$

$$(e^2 - 1)x^2 - y^2 = c^2\left(1 - \frac{1}{e^2}\right) = \frac{c^2}{e^2}(e^2 - 1).$$

Thus

$$(e^2 - 1)x^2 - y^2 = a^2(e^2 - 1),$$

where

$$a = \frac{c}{e}. \tag{1}$$

If we divide both sides of the next-to-last equation by $a^2(e^2 - 1)$, we get

$$\frac{x^2}{a^2} - \frac{y^2}{a^2(e^2 - 1)} = 1.$$

To simplify this equation, we let

$$b^2 = a^2(e^2 - 1) = c^2 - a^2. \tag{2}$$

This is permissible because $e > 1$. So the equation of the hyperbola with focus $(c, 0)$ and directrix $x = c/e^2 = a/e$ takes the form

$$\frac{x^2}{a^2} - \frac{y^2}{b^2} = 1. \tag{3}$$

The minus sign on the left-hand side is the only difference between the equation of a hyperbola and that of an ellipse. Of course, Eq. (2) differs from the relation

$$b^2 = a^2(1 - e^2) = a^2 - c^2$$

for the case of the ellipse.

The hyperbola of Eq. (3) is clearly symmetric about both coordinate axes and has x-intercepts $(\pm a, 0)$. But it has no y-intercept. If we rewrite Eq. (3) in the form

$$y = \pm\frac{b}{a}\sqrt{x^2 - a^2}, \tag{4}$$

then we see that there are points on the graph only if $|x| \geqq a$. Hence the hyperbola has two **branches**, as shown in Fig. 10.6.2. We also see from Eq. (4) that $|y| \to \infty$ as $|x| \to \infty$.

The x-intercepts $V_1(-a, 0)$ and $V_2(a, 0)$ are the **vertices** of the hyperbola, and the line segment joining them is its **transverse axis** (Fig. 10.6.3). The line segment joining $W_1(0, -b)$ and $W_2(0, b)$ is its **conjugate axis**. The alternative form

$$c^2 = a^2 + b^2 \tag{5}$$

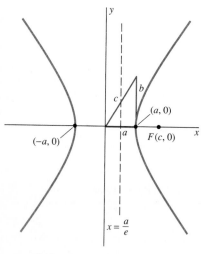

Fig. 10.6.2 A hyperbola has two *branches.*

of Eq. (2) is the Pythagorean relation for the right triangle shown in Fig. 10.6.2.

The lines $y = \pm bx/a$ that pass through the **center** $(0, 0)$ and the opposite vertices of the rectangle in Fig. 10.6.3 are **asymptotes** of the two branches of the hyperbola in both directions. That is, if

$$y_1 = \frac{bx}{a} \quad \text{and} \quad y_2 = \frac{b}{a}\sqrt{x^2 - a^2},$$

then

$$\lim_{x \to \infty} (y_1 - y_2) = 0 = \lim_{x \to -\infty} (y_1 - (-y_2)). \tag{6}$$

To verify the first limit, note that

$$\lim_{x \to \infty} \frac{b}{a}(x - \sqrt{x^2 - a^2}) = \lim_{x \to \infty} \frac{b}{a} \cdot \frac{(x - \sqrt{x^2 - a^2})(x + \sqrt{x^2 - a^2})}{x + \sqrt{x^2 - a^2}}$$

$$= \lim_{x \to \infty} \frac{b}{a} \cdot \frac{a^2}{x + \sqrt{x^2 - a^2}} = 0.$$

Just as in the case of the ellipse, the hyperbola with focus $(c, 0)$ and directrix $x = a/e$ also has focus $(-c, 0)$ and directrix $x = -a/e$ (Figs. 10.6.3 and 10.6.4). Because $c = ae$ by Eq. (1), the foci $(\pm ae, 0)$ and the directrices $x = \pm a/e$ take the same forms in terms of a and e for both the hyperbola $(e > 1)$ and the ellipse $(e < 1)$.

If we interchange x and y in Eq. (3), we obtain

$$\frac{y^2}{a^2} - \frac{x^2}{b^2} = 1. \tag{7}$$

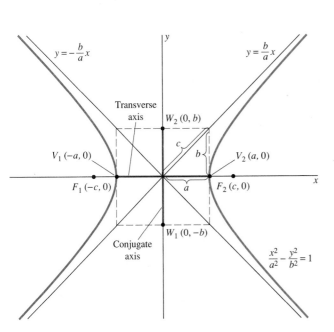

Fig. 10.6.3 The parts of a hyperbola

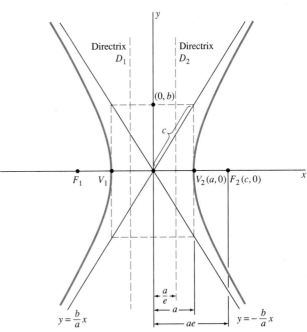

Fig. 10.6.4 The relations between the parts of a hyperbola

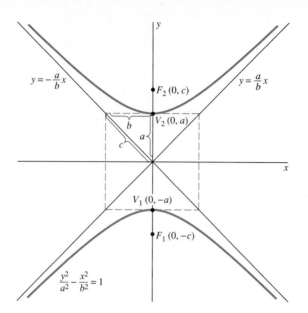

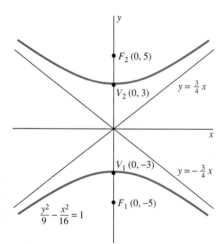

Fig. 10.6.5 The hyperbola of Eq. (7) has horizontal directrices.

This hyperbola has foci at $(0, \pm c)$. The foci as well as this hyperbola's transverse axis lie on the y-axis. Its asymptotes are $y = \pm ax/b$, and its graph generally resembles the one in Fig. 10.6.5.

When we studied the ellipse, we saw that its orientation—whether the major axis is horizontal or vertical—is determined by the relative sizes of a and b. In the case of the hyperbola, the situation is quite different, because the relative sizes of a and b make no such difference: They affect only the slopes of the asymptotes. The direction in which the hyperbola opens—horizontal as in Fig. 10.6.4 or vertical as in Fig. 10.6.5—is determined by the signs of the terms that contain x^2 and y^2.

EXAMPLE 1 Sketch the graph of the hyperbola with equation

$$\frac{y^2}{9} - \frac{x^2}{16} = 1.$$

Solution This is an equation of the form in Eq. (7), so the hyperbola opens vertically. Because $a = 3$ and $b = 4$, we find that $c = 5$ by using Eq. (5): $c^2 = a^2 + b^2$. Thus the vertices are $(0, \pm 3)$, the foci are the two points $(0, \pm 5)$, and the asymptotes are the two lines $y = \pm 3x/4$. This hyperbola appears in Fig. 10.6.6.

Fig. 10.6.6 The hyperbola of Example 1

EXAMPLE 2 Find the equation of the hyperbola with foci $(\pm 10, 0)$ and asymptotes $y = \pm 4x/3$.

Solution Because $c = 10$, we have

$$a^2 + b^2 = 100 \quad \text{and} \quad \frac{b}{a} = \frac{4}{3}.$$

Thus $b = 8$ and $a = 6$, and the equation of the hyperbola is

$$\frac{x^2}{36} - \frac{y^2}{64} = 1.$$

As we noted in Section 10.5, any equation of the form

$$Ax^2 + Cy^2 + Dx + Ey + F = 0 \tag{8}$$

with A and C nonzero can be reduced to the form

$$A(x - h)^2 + C(y - k)^2 = G$$

by completing the square in x and y. Now suppose that the coefficients A and C of the quadratic terms have *opposite signs*. For example, suppose that $A = p^2$ and that $C = -q^2$. The last equation then becomes

$$p^2(x - h)^2 - q^2(y - k)^2 = G. \qquad (9)$$

If $G = 0$, then factorization of the difference of squares on the left-hand side yields the equations

$$p(x - h) + q(y - k) = 0 \quad \text{and} \quad p(x - h) - q(y - k) = 0$$

of two straight lines through (h, k) with slopes $\pm p/q$. If $G \ne 0$, then division of Eq. (9) by G gives an equation that looks either like

$$\frac{(x - h)^2}{a^2} - \frac{(y - k)^2}{b^2} = 1 \qquad \text{(if } G > 0\text{)}$$

or like

$$\frac{(y - k)^2}{a^2} - \frac{(x - h)^2}{b^2} = 1 \qquad \text{(if } G < 0\text{)}.$$

Thus if $AC < 0$ in Eq. (8), the graph is either a pair of intersecting straight lines or a hyperbola.

EXAMPLE 3 Determine the graph of the equation

$$9x^2 - 4y^2 - 36x + 8y = 4.$$

Solution We collect the terms that contain x and those that contain y, and we then complete the square in each variable. We find that

$$9(x - 2)^2 - 4(y - 1)^2 = 36,$$

so

$$\frac{(x - 2)^2}{4} - \frac{(y - 1)^2}{9} = 1.$$

Hence the graph is a hyperbola with a horizontal transverse axis and center $(2, 1)$. Because $a = 2$ and $b = 3$, we find that $c = \sqrt{13}$. The vertices of the hyperbola are $(0, 1)$ and $(4, 1)$, and its foci are the two points $(2 \pm \sqrt{13}, 1)$. Its asymptotes are the two lines

$$y - 1 = \pm\tfrac{3}{2}(x - 2),$$

translates of the asymptotes $y = \pm 3x/2$ of the hyperbola $x^2/4 - y^2/9 = 1$. Figure 10.6.7 shows the graph of the translated hyperbola.

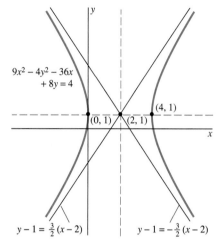

$9x^2 - 4y^2 - 36x + 8y = 4$

$(4, 1)$

$(0, 1)$ $(2, 1)$

$y - 1 = \frac{3}{2}(x - 2)$ $y - 1 = -\frac{3}{2}(x - 2)$

Fig. 10.6.7 The hyperbola of Example 3, a translate of the hyperbola $x^2/4 - y^2/9 = 1$

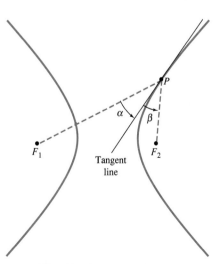

Fig. 10.6.8 The reflection property of the hyperbola

APPLICATIONS OF HYPERBOLAS

The *reflection property* of the hyperbola takes the same form as that for the ellipse. If P is a point on a hyperbola, then the two lines PF_1 and PF_2 from P to the two foci make equal angles with the tangent line at P. In Fig. 10.6.8 this means that $\alpha = \beta$.

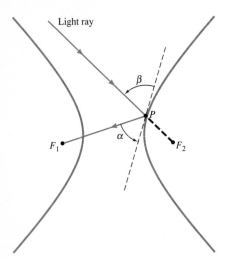

Fig. 10.6.9 How a hyperbolic mirror reflects a ray aimed at one focus: $\alpha = \beta$ again

For an important application of this reflection property, consider a mirror that is shaped like one branch of a hyperbola and is reflective on its outer (convex) surface. An incoming light ray aimed toward one focus will be reflected toward the other focus (Fig. 10.6.9). Figure 10.6.10 indicates the design of a reflecting telescope that makes use of the reflection properties of the parabola and the hyperbola. The parallel incoming light rays first are reflected by the parabola toward its focus at F. Then they are intercepted by an auxiliary hyperbolic mirror with foci at E and F and reflected into the eyepiece located at E.

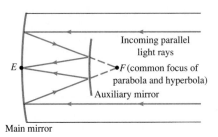

Fig. 10.6.10 One type of reflecting telescope: main mirror parabolic, auxiliary mirror hyperbolic

Example 4 illustrates how hyperbolas are used to determine the position of ships at sea.

EXAMPLE 4 A ship lies in the Labrador Sea due east of Wesleyville, point A, on the long north-south coastline of Newfoundland. Simultaneous radio signals are transmitted by radio stations at A and at St. John's, point B, which is on the coast 200 km due south of A. The ship receives the signal from A 500 microseconds (μs) before it receives the signal from B. Assume that the speed of radio signals is 300 m/μs. How far out at sea is the ship?

Solution The situation is diagramed in Fig. 10.6.11. The difference between the distances of the ship at S from A and B is

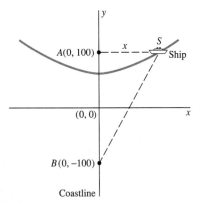

Fig. 10.6.11 A navigation problem (Example 4)

$$|SB| - |SA| = 500 \cdot 300 = 150{,}000$$

meters; that is, 150 km. Thus (by Problem 24) the ship lies on a hyperbola with foci A and B. From Fig. 10.6.5 we see that $c = 100$, so $a = \frac{1}{2} \cdot 150 = 75$, and thus

$$b = \sqrt{c^2 - a^2} = \sqrt{100^2 - 75^2} = 25\sqrt{7}.$$

In the coordinate system of Fig. 10.6.11, the hyperbola has equation

$$\frac{y^2}{75^2} - \frac{x^2}{7 \cdot 25^2} = 1.$$

We substitute $y = 100$ because the ship is due east of A. Thus we find that the ship's distance from the coastline is $x = \frac{175}{3} \approx 58.3$ km.

10.6 Problems

In Problems 1 through 14, find an equation of the hyperbola described there.

1. Foci (± 4, 0),   vertices (± 1, 0)
2. Foci (0, ± 3),   vertices (0, ± 2)
3. Foci (± 5, 0),   asymptotes $y = \pm 3x/4$
4. Vertices (± 3, 0),   asymptotes $y = \pm 3x/4$
5. Vertices (0, ± 5),   asymptotes $y = \pm x$
6. Vertices (± 3, 0),   eccentricity $e = \frac{5}{3}$
7. Foci (0, ± 6),   eccentricity $e = 2$
8. Vertices (± 4, 0) and passing through (8, 3)
9. Foci (± 4, 0),   directrices $x = \pm 1$
10. Foci (0, ± 9),   directrices $y = \pm 4$
11. Center (2, 2),   horizontal transverse axis of length 6, eccentricity $e = 2$
12. Center (-1, 3),   vertices (-4, 3) and (2, 3),   foci (-6, 3) and (4, 3)
13. Center (1, -2),   vertices (1, 1) and (1, -5), asymptotes $3x - 2y = 7$ and $3x + 2y = -1$
14. Focus   (8, -1),   asymptotes $3x - 4y = 13$ and $3x + 4y = 5$

Sketch the graphs of the equations given in Problems 15 through 20; indicate centers, foci, and asymptotes.

15. $x^2 - y^2 - 2x + 4y = 4$
16. $x^2 - 2y^2 + 4x = 0$
17. $y^2 - 3x^2 - 6y = 0$
18. $x^2 - y^2 - 2x + 6y = 9$
19. $9x^2 - 4y^2 + 18x + 8y = 31$
20. $4y^2 - 9x^2 - 18x - 8y = 41$
21. Show that the graph of the equation

$$\frac{x^2}{15 - c} - \frac{y^2}{c - 6} = 1$$

is (a) a hyperbola with foci (± 3, 0) if $6 < c < 15$ and (b) an ellipse if $c < 6$. (c) Identify the graph in the case $c > 15$.

22. Establish that the line tangent to the hyperbola

$$\frac{x^2}{a^2} - \frac{y^2}{b^2} = 1$$

at the point $P(x_0, y_0)$ has equation

$$\frac{x_0 x}{a^2} - \frac{y_0 y}{b^2} = 1.$$

23. Use the result of Problem 22 to establish the reflection property of the hyperbola. (See the suggestion for Problem 25 of Section 10.5.)

24. Suppose that $0 < a < c$, and let $b = \sqrt{c^2 - a^2}$. Show that the hyperbola $x^2/a^2 - y^2/b^2 = 1$ is the locus of a point P such that the *difference* between the distances $|PF_1|$ and $|PF_2|$ is equal to $2a$ (F_1 and F_2 are the foci of the hyperbola).

25. Derive an equation for the hyperbola with the foci (± 5, ± 5) and vertices ($\pm 3/\sqrt{2}$, $\pm 3/\sqrt{2}$). Use the difference definition of a hyperbola implied by Problem 24.

26. Two radio signaling stations at A and B lie on an east-west line, with A 100 mi west of B. A plane is flying west on a line 50 mi north of the line AB. Radio signals are sent (traveling at 980 ft/μs) simultaneously from A and B, and the one sent from B arrives at the plane 400 μs before the one sent from A. Where is the plane?

27. Two radio signaling stations are located as in Problem 26 and transmit radio signals that travel at the same speed. But now we know only that the plane is generally somewhere north of the line AB, that the signal from B arrives 400 μs before the one sent from A, and that the signal sent from A and reflected by the plane takes a total of 600 μs to reach B. Where is the plane?

10.7

Rotation of Axes and Second-Degree Curves

In the preceding three sections we studied the second-degree equation

$$Ax^2 + Cy^2 + Dx + Ey + F = 0, \tag{1}$$

which contains no xy-term. We found that its graph is always a conic section, apart from exceptional cases of these types:

$$
\begin{array}{ll}
2x^2 + 3y^2 = -1 & \text{(no locus),} \\
2x^2 + 3y^2 = 0 & \text{(a single point),} \\
(2x - 1)^2 = 0 & \text{(a line),} \\
(2x - 1)^2 = 1 & \text{(two parallel lines),} \\
x^2 - y^2 = 0 & \text{(two intersecting lines).}
\end{array}
$$

We may, therefore, say that the graph of Eq. (1) is a conic section, possibly **degenerate** (any of the exceptional cases just listed). If either A or C (but not both) is zero, then the graph is a parabola. It is an ellipse if $AC > 0$ [by the discussion of Eq.(6) in Section 10.5] but a hyperbola if $AC < 0$ [by the discussion of Eq. (8) in Section 10.6.]

Let us assume that $AC \neq 0$. Then we can determine by completing squares the particular conic section represented by Eq. (1). That is, we write Eq. (1) in the form

$$A(x - h)^2 + C(y - k)^2 = G. \tag{2}$$

This equation can be simplified further by a **translation of coordinates** to a new $\overline{x}\,\overline{y}$-coordinate system centered at the point (h, k) in the old xy-system. The geometry of this change in coordinates is shown in Fig. 10.7.1. The relation between the old and the new coordinates is

$$\left.\begin{aligned} \overline{x} &= x - h, \\ \overline{y} &= y - k \end{aligned}\right\} \quad \text{or} \quad \begin{cases} x = \overline{x} + h, \\ y = \overline{y} + k. \end{cases} \tag{3}$$

In the new $\overline{x}\,\overline{y}$-coordinate system, Eq. (2) takes the simpler form

$$A\overline{x}^2 + C\overline{y}^2 = G, \tag{2'}$$

from which it is clear whether we have an ellipse, a hyperbola, or a degenerate case.

We now turn to the general second-degree equation

$$Ax^2 + Bxy + Cy^2 + Dx + Ey + F = 0. \tag{4}$$

Note the presence of the "cross-product" term, the xy-term. In order to recognize the graph of Eq. (4), we need to change to a new $x'y'$-coordinate system obtained by a **rotation of axes**.

We get the $x'y'$-axis from the xy-axes by a rotation through an angle α in the counterclockwise direction. In the notation of Fig. 10.7.2, we have

$$x = |OQ| = |OP| \cos(\phi + \alpha) \quad \text{and} \quad y = |PQ| = |OP| \sin(\phi + \alpha). \tag{5}$$

Similarly,

$$x' = |OR| = |OP| \cos \phi \quad \text{and} \quad y' = |PR| = |OP| \sin \phi. \tag{6}$$

Recall the addition formulas

$$\cos(\phi + \alpha) = \cos \phi \cos \alpha - \sin \phi \sin \alpha,$$
$$\sin(\phi + \alpha) = \sin \phi \cos \alpha + \cos \phi \sin \alpha.$$

With the aid of these identities and the substitution of the equations in (6) into the equations in (5), we obtain this result.

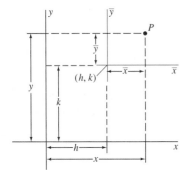

Fig. 10.7.1 A translation of coordinates

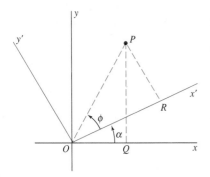

Fig. 10.7.2 A rotation of coordinates through the angle α

> **Equations for Rotation of Axes**
>
> $$x = x' \cos \alpha - y' \sin \alpha,$$
> $$y = x' \sin \alpha + y' \cos \alpha. \tag{7}$$

These equations express the old xy-coordinates of the point P in terms of its new $x'y'$-coordinates and the rotation angle α. Example 1 illustrates how the equations in (7) may be used to transform the equation of a curve from xy-coordinates into the rotated $x'y'$-coordinates.

EXAMPLE 1 The xy-axes are rotated through an angle of $\alpha = 45°$. Find the equation of the curve $2xy = 1$ in the new $x'y'$-coordinate system.

Solution Because $\cos 45° = \sin 45° = 1/\sqrt{2}$, the equations in (7) yield

$$x = \frac{x' - y'}{\sqrt{2}} \quad \text{and} \quad y = \frac{x' + y'}{\sqrt{2}}.$$

The original equation $2xy = 1$ thus becomes

$$(x')^2 - (y')^2 = 1.$$

So, in the $x'y'$-coordinate system, we have a hyperbola for which $a = b = 1$ and $c = \sqrt{2}$ and with the foci $(\pm\sqrt{2}, 0)$. In the original xy-coordinate system, its foci are $(1, 1)$ and $(-1, -1)$ and its asymptotes are the x- and y-axes. This hyperbola is shown in Fig. 10.7.3. (A hyperbola of this form, one with the equation $xy = k$, is called a **rectangular** hyperbola because its asymptotes are perpendicular.)

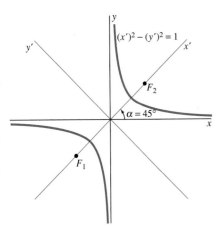

Fig. 10.7.3 The graph of $2xy = 1$ is a hyperbola (Example 1).

Example 1 strongly suggests that the cross-product term Bxy of Eq. (4) may disappear if we properly rotate the coordinate axes. We can, indeed, always choose an appropriate angle of rotation so that, in the new coordinate system, there is no cross-product term.

To determine the appropriate rotation angle, we substitute the equations in (7) for x and y into the general second-degree equation in (4). We obtain the following new second-degree equation:

$$A'(x')^2 + B'x'y' + C'(y')^2 + D'x' + E'y' + F' = 0. \tag{8}$$

The new coefficients are given in terms of the old ones and the angle α by the following equations:

$$A' = A\cos^2\alpha + B\cos\alpha\sin\alpha + C\sin^2\alpha,$$

$$B' = B(\cos^2\alpha - \sin^2\alpha) + 2(C - A)\sin\alpha\cos\alpha,$$

$$C' = A\sin^2\alpha - B\sin\alpha\cos\alpha + C\cos^2\alpha,$$

$$D' = D\cos\alpha + E\sin\alpha,$$

$$E' = -D\sin\alpha + E\cos\alpha, \quad \text{and}$$

$$F' = F.$$

$$\tag{9}$$

Now suppose that an equation of the form in (4) is given, with $B \neq 0$. We simply choose α so that $B' = 0$ in the list of new coefficients in (9). Then Eq. (8) will have no cross-product term, and we can identify and sketch the curve with little trouble in the $x'y'$-coordinate system. But is it really easy to choose such an angle α?

It is. We recall that

$$\cos 2\alpha = \cos^2\alpha - \sin^2\alpha \quad \text{and} \quad \sin 2\alpha = 2\sin\alpha\cos\alpha.$$

So the equation for B' in (9) may be written

$$B' = B\cos 2\alpha + (C - A)\sin 2\alpha.$$

We can cause B' to be zero by choosing as α that (unique) acute angle such that

$$\cot 2\alpha = \frac{A - C}{B}. \tag{10}$$

If we plan to use the equations in (9) to calculate the coefficients in the transformed Eq. (8), we will need to know the values of $\sin \alpha$ and $\cos \alpha$ that follow from Eq. (10). It is sometimes convenient to calculate these values directly from $\cot 2\alpha$, as follows. From the right triangle in Fig. 10.7.4, we can read the numerical value of $\cos 2\alpha$. Because the cosine and cotangent are both positive in the first quadrant and both negative in the second quadrant, we give $\cos 2\alpha$ the same sign as $\cot 2\alpha$. Then we use the half-angle formulas to get $\sin \alpha$ and $\cos \alpha$:

$$\sin \alpha = \sqrt{\frac{1 - \cos 2\alpha}{2}}, \qquad \cos \alpha = \sqrt{\frac{1 + \cos 2\alpha}{2}}. \tag{11}$$

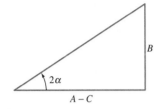

Fig. 10.7.4 Finding $\sin \alpha$ and $\cos \alpha$, given $\cot 2\alpha = (A - C)/B$

Once we have the values of $\sin \alpha$ and $\cos \alpha$, we can compute the coefficients in the resulting Eq. (8) by means of the equations in (9). Alternatively, it's frequently simpler to get Eq. (8) directly by substituting the equations in (7), with the numerical values of $\sin \alpha$ and $\cos \alpha$ obtained as before, in Eq. (4). If we are using a calculator or a computer that has an inverse tangent function, then we can calculate α more briefly by using the formula

$$\alpha = \frac{\pi}{4} - \frac{1}{2} \tan^{-1}\left(\frac{A - C}{B}\right), \tag{12}$$

which follows from Eq. (10) and from the observation that $\cot^{-1} x = (\pi/2) - \tan^{-1} x$.

EXAMPLE 2 Determine the graph of the equation

$$73x^2 - 72xy + 52y^2 - 30x - 40y - 75 = 0.$$

Solution We begin with Eq. (10) and find that $\cot 2\alpha = -\frac{7}{24}$, so $\cos 2\alpha = -\frac{7}{25}$. Thus

$$\sin \alpha = \sqrt{\frac{1 - (-\frac{7}{25})}{2}} = \frac{4}{5}, \qquad \cos \alpha = \sqrt{\frac{1 + (-\frac{7}{25})}{2}} = \frac{3}{5}.$$

Then, with $A = 73$, $B = -72$, $C = 52$, $D = -30$, $E = -40$, and $F = -75$, the equations in (9) yield

$A' = 25,$	$D' = -50,$
$B' = 0$ (this was our goal),	$E' = 0,$
$C' = 100,$	$F' = -75.$

Consequently the equation in the new $x'y'$-coordinate system, obtained by rotation through an angle $\alpha = \arcsin\left(\frac{4}{5}\right)$ (approximately 53.13°), is

$$25(x')^2 + 100(y')^2 - 50x' = 75.$$

Alternatively, we could have obtained this equation by substituting

$$x = \tfrac{3}{5}x' - \tfrac{4}{5}y', \qquad y = \tfrac{4}{5}x' + \tfrac{3}{5}y'$$

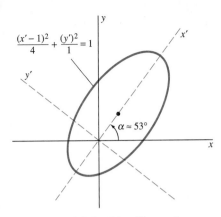

$$\frac{(x'-1)^2}{4} + \frac{(y')^2}{1} = 1$$

$\alpha \approx 53°$

Fig. 10.7.5 The ellipse of Example 2

into the original equation.

By completing the square in x' we finally obtain

$$25(x' - 1)^2 + 100(y')^2 = 100,$$

which we put into the standard form

$$\frac{(x' - 1)^2}{4} + \frac{(y')^2}{1} = 1.$$

Thus the original curve is an ellipse with major semiaxis 2, minor semiaxis 1, and center $(1, 0)$ in the $x'y'$-coordinate system (Fig. 10.7.5).

CLASSIFICATION OF CONICS

Example 2 illustrates the general procedure for finding the graph of a second-degree equation. First, if there is a cross-product term, rotate the axes to eliminate it. Then translate axes (if necessary) to reduce the equation to the standard form of a parabola, ellipse, or hyperbola (or a degenerate case).

There *is* a test by which the nature of the curve may be discovered without actually carrying out these transformations. This test depends on the fact that, whatever the angle α of rotation, the equations in (9) imply that

$$(B')^2 - 4A'C' = B^2 - 4AC. \tag{13}$$

Thus the **discriminant** $B^2 - 4AC$ is an *invariant* under any rotation of axes. If α is chosen so that $B' = 0$, then the left-hand side in Eq. (13) is simply $-4A'C'$. Because A' and C' are the coefficients of the squared terms, our earlier discussion of Eq. (1) now applies. It follows that the graph will be

❑ A *parabola* if $B^2 - 4AC = 0$,
❑ An *ellipse* if $B^2 - 4AC < 0$,
❑ A *hyperbola* if $B^2 - 4AC > 0$.

Of course, degenerate cases may occur.
Here are some examples:

1. $x^2 + 2xy + y^2 = 1$ is a (degenerate) parabola.
2. $x^2 + xy + y^2 = 1$ is an ellipse.
3. $x^2 + 3xy + y^2 = 1$ is a hyperbola.

10.7 Problems

In Problems 1 through 6, the graph of the given equation is a translated conic section, possibly degenerate. Give its equation in standard position in the appropriate $x'y'$-coordinate system, and give the origin (in xy-coordinates) of the new coordinate system.

1. $2x^2 + y^2 - 8x - 6y + 13 = 0$
2. $x^2 + 3y^2 + 6x + 12y + 18 = 0$
3. $9x^2 - 16y^2 - 18x - 32y - 151 = 0$
4. $y^2 - 4x - 4y - 8 = 0$
5. $2x^2 + 3y^2 - 8x - 18y + 35 = 0$
6. $4x^2 - y^2 - 8x - 4y = 0$

In Problems 7 through 20, the graph of the given equation is a rotated conic section (possibly degenerate). Identify it; give the counterclockwise angle α of rotation that yields a rotated $x'y'$-system in which the graph is in stan-

dard position, and give the transformed equation in $x'y'$-coordinates.

7. $3x^2 + 2xy + 3y^2 = 1$ **8.** $x^2 + 6xy + y^2 = 1$

9. $4x^2 + 4xy + y^2 = 20$ **10.** $9x^2 + 4xy + 6y^2 = 40$

11. $4x^2 + 6xy - 4y^2 = 5$

12. $19x^2 + 6xy + 11y^2 = 40$

13. $22x^2 + 12xy + 17y^2 = 26$

14. $9x^2 + 12xy + 4y^2 = 13$

15. $52x^2 + 72xy + 73y^2 = 100$

16. $9x^2 + 24xy + 16y^2 = 0$

17. $33x^2 + 8xy + 18y^2 = 68$

18. $40x^2 + 36xy + 25y^2 = 52$

19. $119x^2 + 240xy - 119y^2 = 0$

20. $313x^2 + 120xy + 194y^2 = 0$

In Problems 21 through 26, identify the graph of the given equation by carrying out first a rotation and then a translation. Give the angle of rotation, the translated center of coordinates in the rotated coordinate system, and the final transformed equation.

21. $34x^2 - 24xy + 41y^2 - 40x - 30y - 25 = 0$

22. $41x^2 - 24xy + 34y^2 + 20x - 140y + 125 = 0$

23. $23x^2 - 72xy + 2y^2 + 140x + 20y - 75 = 0$

24. $9x^2 + 24xy + 16y^2 - 170x - 60y + 245 = 0$

25. $161x^2 + 480xy - 161y^2 - 510x - 272y = 0$

26. $144x^2 - 120xy + 25y^2 - 65x - 156y - 169 = 0$

27. Solve the equations in (7) to show that the rotated coordinates (x', y') are given in terms of the original coordinates by

$$x' = x \cos \alpha + y \sin \alpha, \qquad y' = -x \sin \alpha + y \cos \alpha.$$

28. Use the equations in (9) to verify Eq. (13).

29. Prove that the sum $A + C$ of the coefficients of x^2 and y^2 in Eq. (4) is invariant under rotation. That is, show that $A' + C' = A + C$ for any rotation through an angle α.

30. Use the equations in (9) to prove that any rotation of axes transforms the equation $x^2 + y^2 = r^2$ into the equation $(x')^2 + (y')^2 = r^2$.

31. Consider the equation

$$x^2 + Bxy - y^2 + Dx + Ey + F = 0.$$

Prove that there is a rotation of axes such that $A' = 0 = C'$ in the resulting equation. [*Suggestion:* Find the angle α for which $A' = 0$. Then apply the result of Problem 29. What can you conclude about the graph of the given equation?]

32. Suppose that $B^2 - 4AC < 0$, so that the equation

$$Ax^2 + Bxy + Cy^2 = 1$$

represents an ellipse. Prove that its area is

$$\pi ab = \frac{2\pi}{\sqrt{4AC - B^2}},$$

where a and b are the lengths of its semiaxes.

33. Show that the equation $27x^2 + 37xy + 17y^2 = 1$ represents an ellipse, and then find the points of the ellipse that are closest to and farthest from the origin.

34. Show that the equation $x^2 + 14xy + 49y^2 = 100$ represents a parabola (possibly degenerate), and then find the point of this parabola that is closest to the origin.

Chapter 10 Review: PROPERTIES OF CONIC SECTIONS

The parabola with focus $(p, 0)$ and directrix $x = -p$ has eccentricity $e = 1$ and equation $y^2 = 4px$. The accompanying table compares the properties of an ellipse and a hyperbola, each with foci $(\pm c, 0)$ and major axis of length $2a$.

	Ellipse	Hyperbola
Eccentricity	$e = \dfrac{c}{a} < 1$	$e = \dfrac{c}{a} > 1$
a, b, c relation	$a^2 = b^2 + c^2$	$c^2 = a^2 + b^2$
Equation	$\dfrac{x^2}{a^2} + \dfrac{y^2}{b^2} = 1$	$\dfrac{x^2}{a^2} - \dfrac{y^2}{b^2} = 1$
Vertices	$(\pm a, 0)$	$(\pm a, 0)$
y-intercepts	$(0, \pm b)$	None
Directrices	$x = \pm \dfrac{a}{e}$	$x = \pm \dfrac{a}{e}$
Asymptotes	None	$y = \pm \dfrac{bx}{a}$

Use the following list as a guide to additional concepts that you may need to review.

1. Conic sections
2. The relationship between rectangular and polar coordinates
3. The graph of an equation in polar coordinates
4. The area formula in polar coordinates
5. Translation of coordinates
6. Equations and the procedure for rotation of axes
7. Use of the discriminant to classify the graph of a second-degree equation

Chapter 10 Miscellaneous Problems

Sketch the graphs of the equations in Problems 1 through 30. In Problems 1 through 18, if the graph is a conic section, label its center, foci, and vertices.

1. $x^2 + y^2 - 2x - 2y = 2$
2. $x^2 + y^2 = x + y$
3. $x^2 + y^2 - 6x + 2y + 9 = 0$
4. $y^2 = 4(x + y)$
5. $x^2 = 8x - 2y - 20$
6. $x^2 + 2y^2 - 2x + 8y + 8 = 0$
7. $9x^2 + 4y^2 = 36x$
8. $x^2 - y^2 = 2x - 2y - 1$
9. $y^2 - 2x^2 = 4x + 2y + 3$
10. $9y^2 - 4x^2 = 8x + 18y + 31$
11. $x^2 + 2y^2 = 4x + 4y - 12$
12. $x^2 + 2xy + y^2 + 1 = 0$
13. $x^2 + 2xy - y^2 = 7$
14. $xy + 8 = 0$
15. $3x^2 - 22xy + 3y^2 = 4$
16. $x^2 - 6xy + y^2 = 4$
17. $9x^2 - 24xy + 16y^2 = 20x + 15y$
18. $7x^2 + 48xy - 7y^2 + 25 = 0$
19. $r = -2 \cos \theta$
20. $\cos \theta + \sin \theta = 0$
21. $r = \dfrac{1}{\sin \theta - \cos \theta}$
22. $r \sin^2 \theta = \cos \theta$
23. $r = 3 \csc \theta$
24. $r = 2(\cos \theta - 1)$
25. $r^2 = 4 \cos \theta$
26. $r\theta = 1$
27. $r = 3 - 2 \sin \theta$
28. $r = \dfrac{1}{1 + \cos \theta}$
29. $r = \dfrac{4}{2 + \cos \theta}$
30. $r = \dfrac{4}{1 - 2 \cos \theta}$

In Problems 31 through 38, find the area of the region described.

31. Inside both $r = 2 \sin \theta$ and $r = 2 \cos \theta$
32. Inside $r^2 = 4 \cos \theta$
33. Inside $r = 3 - 2 \sin \theta$ and outside $r = 4$
34. Inside $r^2 = 2 \sin 2\theta$ and outside $r = 2 \sin \theta$
35. Inside $r = 2 \sin 2\theta$ and outside $r = \sqrt{2}$
36. Inside $r = 3 \cos \theta$ and outside $r = 1 + \cos \theta$
37. Inside $r = 1 + \cos \theta$ and outside $r = \cos \theta$
38. Between the loops of $r = 1 - 2 \sin \theta$

39. Find a polar-coordinate equation of the circle that passes through the origin and is centered at the point with polar coordinates (p, α).

40. Find a simple equation of the parabola whose focus is the origin and whose directrix is the line $y = x + 4$. Recall from Miscellaneous Problem 71 of Chapter 3 that the distance from the point (x_0, y_0) to the line with equation $Ax + By + C = 0$ is

$$\frac{|Ax_0 + By_0 + C|}{\sqrt{A^2 + B^2}}.$$

41. A *diameter* of an ellipse is a chord through its center. Find the maximum and minimum lengths of diameters of the ellipse with equation

$$\frac{x^2}{a^2} + \frac{y^2}{b^2} = 1.$$

42. Use calculus to prove that the ellipse of Problem 41 is normal to the coordinate axes at each of its four vertices.

43. The parabolic arch of a bridge has base width b and height h at its center. Write its equation, choosing the origin on the ground at the left end of the arch.

44. Use methods of calculus to find the points on the ellipse

$$\frac{x^2}{a^2} + \frac{y^2}{b^2} = 1$$

that are nearest to and farthest from (a) the center $(0, 0)$; (b) the focus $(c, 0)$.

45. Consider a line segment QR that contains a point P such that $|QP| = a$ and $|PR| = b$. Suppose that Q is

constrained to move on the y-axis, whereas R must remain on the x-axis. Prove that the locus of P is an ellipse.

46. Suppose that $a > 0$ and that F_1 and F_2 are two fixed points in the plane with $|F_1 F_2| > 2a$. Imagine a point P that moves in such a way that $|PF_2| = 2a + |PF_1|$. Prove that the locus of P is one branch of a hyperbola with foci F_1 and F_2. Then—as a consequence—explain how to construct points on a hyperbola by drawing appropriate circles centered at its foci.

47. Let Q_1 and Q_2 be two points on the parabola $y^2 = 4px$. Let P be the point of the parabola at which the tangent line is parallel to $Q_1 Q_2$. Prove that the horizontal line through P bisects the segment $Q_1 Q_2$.

48. Determine the locus of a point P such that the product of its distances from the two fixed points $F_1(-a, 0)$ and $F_2(a, 0)$ is a^2.

49. Find the eccentricity of the conic section with equation $3x^2 - y^2 + 12x + 9 = 0$.

50. Find the area bounded by the loop of the *strophoid*

$$r = \sec \theta - 2 \cos \theta$$

shown in Fig. 10.MP.1.

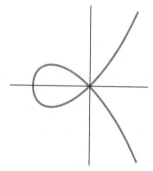

Fig. 10.MP.1 The strophoid of Problem 50

51. Find the area bounded by the loop of the *folium of Descartes* with equation $x^3 + y^3 = 3xy$ shown in Fig. 10.MP.2. [*Suggestion:* Change to polar coordinates and then substitute $u = \tan \theta$ to evaluate the area integral.]

52. Use the method of Problem 51 to find the area bounded by the first-quadrant loop (similar to the folium of Problem 51) of the curve $x^5 + y^5 = 5x^2y^2$.

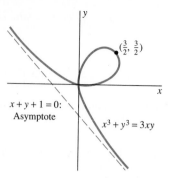

Fig. 10.MP.2 The folium of Descartes $x^3 + y^3 = 3xy$ (Problem 51)

53. Show that the graph of the equation

$$2929x^2 - 3456xy + 1921y^2 - 9000x$$
$$- 12{,}000y - 15{,}625 = 0$$

is the ellipse shown in Fig. 10.MP.3. The vertices of this ellipse are at $(15, 20)$ and $(-0.6, -0.8)$, and its major axis makes the angle $\arctan(\frac{4}{3})$ with the horizontal.

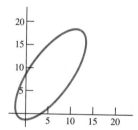

Fig. 10.MP.3 The ellipse of Problem 53

54. Suppose that $0 < c < \pi$. Prove that the graph of the equation

$$\sin^{-1} x + \sin^{-1} y = c$$

is an ellipse rotated (from standard position) through an angle of $45°$. [*Suggestion:* If $u = \sin^{-1} x$ and $v = \sin^{-1} y$, then $u + v = c$. Expand the left-hand side of the equation $\cos(u + v) = \cos c = k$.]

55. The graph of a conic section in the xy-plane has intercepts at $(5, 0)$, $(-5, 0)$, $(0, 4)$, and $(0, -4)$. Deduce all the information you can about this conic. Can you determine whether it is a parabola, a hyperbola, or an ellipse? What if you know that the graph of this conic is normal to the y-axis at $(0, 4)$?

Infinite Series

Srinivasa Ramanujan (1887–1920)

❑ On a January day in 1913, the eminent Cambridge mathematics professor G. H. Hardy received a lettter from an unknown 25-year-old clerk in the accounting department of a government office in Madras, India. Its author, Srinivasa Ramanujan, had no university education, he admitted—he had flunked out—but "after leaving school I have employed the spare time at my disposal to work at Mathematics. . . . I have not trodden through the conventional regular course . . . but am striking out a new path for myself." The 10 pages that followed listed in neat handwritten script approximately 50 formulas, most dealing with integrals and infinite series that Ramanujan had discovered, and asked Hardy's advice whether they contained anything of value. The formulas were of such exotic and unlikely appearance that Hardy at first suspected a hoax,

but he and his colleague J. E. Littlewood soon realized that they were looking at the work of an extraordinary mathematical genius.

❑ Thus began one of the most romantic episodes in the history of mathematics. In April 1914 Ramanujan arrived in England a poor, self-taught Indian mathematical amateur called to collaborate as an equal with the most sophisticated professional mathematicians of the day. For the next three years a steady stream of remarkable discoveries poured forth from his pen. But in 1917 he fell seriously ill, apparently with tuberculosis. The following year he returned to India to attempt to regain his health but never recovered, and he died in 1920 at the age of 32. Up to the very end he worked feverishly to record his final discoveries. He left behind notebooks outlining work whose completion has occupied prominent mathematicians throughout the twentieth century.

❑ With the possible exception of Euler, no one before or since has exhibited Ramanujan's virtuosity with infinite series. An example of his discoveries is the infinite series

$$\frac{1}{\pi} = \frac{\sqrt{8}}{9801} \sum_{n=0}^{\infty} \frac{(4n)!}{(n!)^4} \frac{(1103 + 26{,}390n)}{396^{4n}},$$

whose first term yields the familiar approximation $\pi \approx 3.14159$, with each additional term giving π to roughly eight more decimal places of accuracy. For instance, just four terms of Ramanujan's

series are needed to calculate the 30-place approximation

$$\pi \approx 3.14159\ 26535\ 89793$$
$$23846\ 26433\ 83279$$

that suffices for virtually any imaginable "practical" application—if the universe were a sphere with a radius of 10 billion light years, then this value of π would give its circumference accurate to the nearest hundredth of an inch. But in recent years Ramanujan's ideas have been used to calculate the value of π accurate to a *billion* decimal places! Indeed, such gargantuan computations of π are commonly used to check the accuracy of new supercomputers.

A typical page of Ramanujan's letter to Hardy, listing formulas Ramanujan had discovered, but with no hint of proof or derivation.

11.1
Introduction

Fig. 11.1.1 Subdivision of an interval to illustrate Zeno's paradox

In the fifth century B.C., the Greek philosopher Zeno proposed the following paradox: In order for a runner to travel a given distance, the runner must first travel halfway, then half the remaining distance, then half the distance that yet remains, and so on ad infinitum. But, Zeno argued, it is clearly impossible for a runner to take infinitely many steps in a finite period of time, so motion from one point to another is impossible.

Zeno's paradox suggests the infinite subdivision of [0, 1] indicated in Fig. 11.1.1. There is one subinterval of length $1/2^n$ for each integer $n = 1, 2, 3, \ldots$. If the length of the interval is the sum of the lengths of the subintervals into which it is divided, then it would appear that

$$1 = \frac{1}{2} + \frac{1}{4} + \frac{1}{8} + \frac{1}{16} + \cdots + \frac{1}{2^n} + \cdots,$$

with infinitely many terms somehow adding up to 1. But the formal infinite sum

$$1 + 2 + 3 + \cdots + n + \cdots$$

of all the positive integers seems meaningless—it does not appear to add up to *any* (finite) value.

The question is this: What, if anything, do we mean by the sum of an *infinite* collection of numbers? This chapter explores conditions under which an *infinite* sum

$$a_1 + a_2 + a_3 + \cdots + a_n + \cdots,$$

known as an *infinite series,* is meaningful. We discuss methods for computing the sum of an infinite series and applications of the algebra and calculus of infinite series. Infinite series are important in science and mathematics because many functions either arise most naturally in the form of infinite series or have infinite series representations (such as the Taylor series of Section 11.4) that are useful for numerical computations.

11.2
Infinite Sequences

An **infinite sequence** of real numbers is a function whose domain of definition is the set of all positive integers. Thus if s is a sequence, then to each positive integer n there corresponds a real number $s(n)$. Ordinarily, a sequence is most conveniently described by listing its values in order, beginning with $s(1)$:

$$s(1), s(2), s(3), \ldots, s(n), \ldots.$$

With subscript notation rather than function notation, we usually write

$$s_1, s_2, s_3, \ldots, s_n, \ldots \tag{1}$$

for this list of values. The values in this list are the **terms** of the sequence; s_1 is the first term, s_2 the second term, s_n the **nth term.**

We use the notation $\{s_n\}_{n=1}^{\infty}$, or simply $\{s_n\}$, as an abbreviation for the **ordered** list in Eq. (1), and we may refer to the sequence by saying simply "the sequence $\{s_n\}$." When a particular sequence is so described, the nth term s_n is generally (though not always) given by a formula in terms of its subscript n. In this case, listing the first few terms of the sequence often helps us to see it more concretely.

EXAMPLE 1 The following table lists explicitly the first four terms of several sequences.

$\{s_n\}_1^\infty$	$s_1, s_2, s_3, s_4, \ldots$
$\left\{\dfrac{1}{n}\right\}_1^\infty$	$1, \dfrac{1}{2}, \dfrac{1}{3}, \dfrac{1}{4}, \ldots$
$\left\{\dfrac{1}{10^n}\right\}_1^\infty$	$0.1, 0.01, 0.001, 0.0001, \ldots$
$\left\{\dfrac{1}{n!}\right\}_1^\infty$	$1, \dfrac{1}{2}, \dfrac{1}{6}, \dfrac{1}{24}, \ldots$
$\left\{\sin\dfrac{n\pi}{2}\right\}_1^\infty$	$1, 0, -1, 0, \ldots$
$\{1 + (-1)^n\}_1^\infty$	$0, 2, 0, 2, \ldots$

EXAMPLE 2 The **Fibonacci sequence** $\{F_n\}$ may be defined as follows:

$$F_1 = 1, \qquad F_2 = 1, \quad \text{and} \quad F_{n+1} = F_n + F_{n-1} \qquad \text{for } n \geqq 2.$$

The first ten terms of the Fibonacci sequence are

$$1, 1, 2, 3, 5, 8, 13, 21, 34, 55.$$

This is a *recursively defined sequence* —after the initial terms are given, each term is defined in terms of its predecessors. Such sequences are particularly important in computer science; a computer's state at each tick of its internal clock typically depends on its state at the previous tick.

The limit of a sequence is defined in much the same way as the limit of an ordinary function (Section 2.2).

Definition *Limit of a Sequence*

We say that the sequence $\{s_n\}$ **converges** to the real number L, or has the **limit** L, and we write

$$\lim_{n \to \infty} s_n = L, \tag{2}$$

provided that s_n can be made as close to L as we please merely by choosing n to be sufficiently large. That is, given any number $\epsilon > 0$, there exists an integer N such that

$$|s_n - L| < \epsilon \qquad \text{for all } n \geqq N. \tag{3}$$

If the sequence $\{s_n\}$ does *not* converge, then we say that $\{s_n\}$ **diverges.**

Figure 11.2.1 illustrates geometrically the definition of the limit of a sequence. Because

$$|s_n - L| < \epsilon \quad \text{means that} \quad L - \epsilon < s_n < L + \epsilon,$$

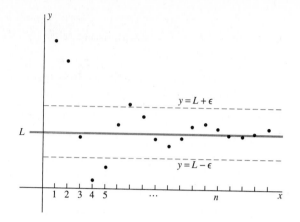

Fig. 11.2.1 The point (n, s_n) approaches the line $y = L$ as $n \to \infty$.

the condition in (3) means that if $n \geq N$, then the point (n, s_n) lies between the horizontal lines $y = L - \epsilon$ and $y = L + \epsilon$.

EXAMPLE 3 Prove that $\lim\limits_{n \to \infty} \dfrac{1}{n} = 0$.

Proof We need to show this: To each positive real number ϵ, there corresponds an integer N such that, for all $n \geq N$,

$$\left| \frac{1}{n} - 0 \right| = \frac{1}{n} < \epsilon.$$

It suffices to choose any fixed integer $N > 1/\epsilon$. For example, let N denote the *smallest* integer that is greater than the real number $1/\epsilon$ (Fig. 11.2.2). Then $n \geq N$ implies that

$$\frac{1}{n} \leq \frac{1}{N} < \epsilon,$$

as desired. ❑

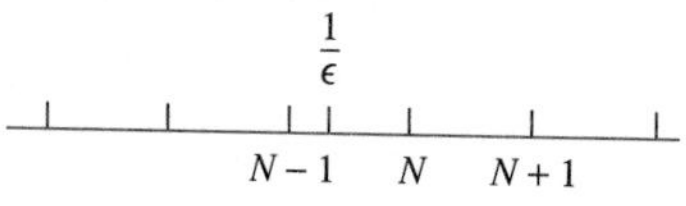

Fig. 11.2.2 The integer N of Example 3

EXAMPLE 4 (a) The sequence $\{(-1)^n\}$ diverges because its successive terms "oscillate" between the two values $+1$ and -1. Hence $(-1)^n$ cannot approach any single value as $n \to \infty$. (b) The terms of the sequence $\{n^2\}$ increase without bound as $n \to \infty$. Thus $\{n^2\}$ diverges. In this case, we might also say that $\{n^2\}$ diverges to *infinity*.

USING LIMIT LAWS

The limit laws in Section 2.2 for limits of functions have natural analogues for limits of sequences. Their proofs are based on techniques similar to those used in Appendix B.

Theorem 1   Limit Laws for Sequences

If the limits

$$\lim_{n \to \infty} a_n = A \quad \text{and} \quad \lim_{n \to \infty} b_n = B$$

exist (so A and B are real numbers), then

1. $\lim\limits_{n \to \infty} ca_n = cA$ (c any real number);

2. $\lim\limits_{n \to \infty} (a_n + b_n) = A + B$;

3. $\lim\limits_{n \to \infty} a_n b_n = AB$;

4. $\lim\limits_{n \to \infty} \dfrac{a_n}{b_n} = \dfrac{A}{B}$.

In part 4 we must assume that $B \neq 0$ and that $b_n \neq 0$ for all sufficiently large values of n.

Theorem 2   Substitution Law for Sequences

If $\lim\limits_{n \to \infty} a_n = A$ and the function f is continuous at $x = A$, then

$$\lim_{n \to \infty} f(a_n) = f(A).$$

Theorem 3   Squeeze Law for Sequences

If $a_n \leqq b_n \leqq c_n$ for all n and

$$\lim_{n \to \infty} a_n = L = \lim_{n \to \infty} c_n,$$

then $\lim\limits_{n \to \infty} b_n = L$ as well.

These theorems can be used to compute limits of many sequences formally, without recourse to the definition. For example, if k is a positive integer and c is a constant, then Example 3 and the product law (Theorem 1, part 3) give

$$\lim_{n \to \infty} \frac{c}{n^k} = c \cdot 0 \cdot 0 \cdots 0 = 0.$$

EXAMPLE 5   Show that $\lim\limits_{n \to \infty} \dfrac{(-1)^n \cos n}{n^2} = 0$.

Solution   This result follows from the squeeze law and from the fact that $1/n^2 \to 0$ as $n \to \infty$, because

$$-\frac{1}{n^2} \leqq \frac{(-1)^n \cos n}{n^2} \leqq \frac{1}{n^2}.$$

EXAMPLE 6   Show that if $a > 0$, then $\lim\limits_{n \to \infty} \sqrt[n]{a} = 1$.

Solution   We apply the substitution law with $f(x) = a^x$ and $A = 0$. Because $1/n \to 0$ as $n \to \infty$ and f is continuous at $x = 0$, this gives

$$\lim_{n \to \infty} a^{1/n} = a^0 = 1.$$

EXAMPLE 7 The limit laws and the continuity of $f(x) = \sqrt{x}$ at $x = 4$ yield

$$\lim_{n \to \infty} \sqrt{\frac{4n - 1}{n + 1}} = \sqrt{\lim_{n \to \infty} \frac{4 - \dfrac{1}{n}}{1 + \dfrac{1}{n}}} = \sqrt{4} = 2.$$

EXAMPLE 8 Show that if $|r| < 1$, then $\lim_{n \to \infty} r^n = 0$.

Solution Because $|r^n| = |(-r)^n|$, we may assume that $0 < r < 1$. Then $1/r = 1 + a$ for some number $a > 0$, so the binomial formula yields

$$\frac{1}{r^n} = (1 + a)^n = 1 + na + \{\text{positive terms}\} > 1 + na;$$

$$0 < r^n < \frac{1}{1 + na}.$$

Now $1/(1 + na) \to 0$ as $n \to \infty$. Therefore, the squeeze law implies that $r^n \to 0$ as $n \to \infty$.

Figure 11.2.3 shows the graph of a function f such that $\lim_{x \to \infty} f(x) = L$. If the sequence $\{a_n\}$ is defined by the formula $a_n = f(n)$ for each positive integer n, then all the points $(n, f(n))$ lie on the graph of $y = f(x)$. It therefore follows from the definition of the limit of a function that $\lim_{n \to \infty} a_n = L$ as well.

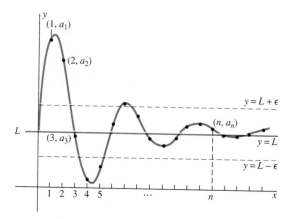

Fig. 11.2.3 The limit of the sequence is the limit of the function.

Theorem 4 Limits of Functions and Sequences
If $a_n = f(n)$ for each positive integer n, then

$$\lim_{x \to \infty} f(x) = L \quad \text{implies} \quad \lim_{n \to \infty} a_n = L. \tag{4}$$

The converse of the statement in (4) is generally false. For example, take $f(x) = \sin \pi x$ and, for each positive integer n, let $a_n = f(n) = \sin n\pi$. Then

$$\lim_{n \to \infty} a_n = \lim_{n \to \infty} \sin n\pi = 0, \quad \text{but}$$

$$\lim_{n \to \infty} f(x) = \lim_{n \to \infty} \sin nx \quad \text{does not exist.}$$

584

Because of (4) we can use **l'Hôpital's rule for sequences:** If $a_n = f(n)$, $b_n = g(n)$, and $f(x)/g(x)$ has the indeterminate form ∞/∞ as $x \to \infty$, then

$$\lim_{n \to \infty} \frac{a_n}{b_n} = \lim_{x \to \infty} \frac{f(x)}{g(x)} = \lim_{x \to \infty} \frac{f'(x)}{g'(x)}, \tag{5}$$

provided that f and g satisfy the other hypotheses of l'Hôpital's rule, including the assumption that the right-hand limit exists.

EXAMPLE 9   Show that $\lim\limits_{n \to \infty} \dfrac{\ln n}{n} = 0$.

Solution   The function $(\ln x)/x$ is defined for all $x \geq 1$ and agrees with the given sequence $\{(\ln n)/n\}$ when $x = n$, a positive integer. Because $(\ln x)/x$ has the indeterminate form ∞/∞ as $x \to \infty$, l'Hôpital's rule gives

$$\lim_{n \to \infty} \frac{\ln n}{n} = \lim_{x \to \infty} \frac{\ln x}{x} = \lim_{x \to \infty} \frac{\dfrac{1}{x}}{1} = 0.$$

EXAMPLE 10   Show that $\lim\limits_{n \to \infty} \sqrt[n]{n} = 1$.

Solution   First we note that

$$\ln \sqrt[n]{n} = \ln n^{1/n} = \frac{\ln n}{n} \to 0 \qquad \text{as } n \to \infty,$$

by Example 9. By the substitution law with $f(x) = e^x$, this gives

$$\lim_{n \to \infty} n^{1/n} = \lim_{n \to \infty} \exp(\ln n^{1/n}) = e^0 = 1.$$

EXAMPLE 11   Find $\lim\limits_{n \to \infty} \dfrac{3n^3}{e^{2n}}$.

Solution   We apply l'Hôpital's rule repeatedly, although we must be careful at each intermediate step to verify that we still have an indeterminate form. Thus we find that

$$\lim_{n \to \infty} \frac{3n^3}{e^{2n}} = \lim_{x \to \infty} \frac{3x^3}{e^{2x}} = \lim_{x \to \infty} \frac{9x^2}{2e^{2x}} = \lim_{x \to \infty} \frac{18x}{4e^{2x}} = \lim_{x \to \infty} \frac{18}{8e^{2x}} = 0.$$

BOUNDED MONOTONIC SEQUENCES

The set of all *rational* numbers has by itself all the most familiar elementary algebraic properties of the entire real number system. To guarantee the existence of irrational numbers, we must assume in addition a "completeness property" of the real numbers. Otherwise, the real line might have "holes" where the irrational numbers ought to be. One way of stating this completeness property is in terms of the convergence of an important type of sequence, a bounded monotonic sequence.

The sequence $\{a_n\}_1^\infty$ is said to be **increasing** if

$$a_1 \leq a_2 \leq a_3 \leq \cdots \leq a_n \leq \cdots$$

and **decreasing** if

$$a_1 \geq a_2 \geq a_3 \geq \cdots \geq a_n \geq \cdots.$$

The sequence $\{a_n\}$ is **monotonic** if it is *either* increasing *or* decreasing. The sequence $\{a_n\}$ is **bounded** if there is a number M such that $|a_n| \leq M$ for all n. The following assertion may be taken to be an axiom for the real number system.

> **Bounded Monotonic Sequence Property**
> Every bounded monotonic infinite sequence converges—that is, has a finite limit.

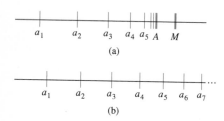

(a)

(b)

Fig. 11.2.4 (a) A bounded increasing sequence; (b) an increasing sequence that is not bounded above

Suppose, for example, that the increasing sequence $\{a_n\}_1^\infty$ is bounded above by a number M, meaning that $a_n \leq M$ for all n. Because it is also bounded below (by a_1, for instance), the bounded monotonic sequence property implies that

$$\lim_{n \to \infty} a_n = A \qquad \text{for some real number } A \leq M,$$

as in Fig. 11.2.4(a). If the increasing sequence $\{a_n\}$ is *not* bounded above (by any number), then it follows that

$$\lim_{n \to \infty} a_n = +\infty,$$

as in Fig. 11.2.4(b) (see Problem 38).

EXAMPLE 12 Investigate the sequence $\{a_n\}$ that is defined recursively by

$$a_1 = \sqrt{2}, \qquad a_{n+1} = \sqrt{2 + a_n} \qquad \text{for } n \geq 1. \tag{6}$$

Solution The first four terms of $\{a_n\}$ are

$$\sqrt{2}, \quad \sqrt{2 + \sqrt{2}}, \quad \sqrt{2 + \sqrt{2 + \sqrt{2}}}, \quad \sqrt{2 + \sqrt{2 + \sqrt{2 + \sqrt{2}}}}.$$

If the sequence $\{a_n\}$ has a limit A, then A would seem to be the natural interpretation of the value of the infinite expression

$$\sqrt{2 + \sqrt{2 + \sqrt{2 + \sqrt{2 + \cdots}}}}.$$

The sequence $\{a_n\}$ defined in (6) is an increasing sequence with $a_n < 2$ for all n, as Problem 43 asks you to verify. Therefore, the bounded monotonic sequence property implies that the sequence $\{a_n\}$ has a limit A. It does not tell us what the number A is. But now that we know that the limit A of the sequence $\{a_n\}$ exists, we can write

$$A = \lim_{n \to \infty} a_{n+1} = \lim_{n \to \infty} \sqrt{2 + a_n} = \sqrt{2 + A},$$

and thus

$$A^2 = 2 + A.$$

The roots of this equation are -1 and 2. It is clear that $A > 0$, so we conclude that

$$\lim_{n \to \infty} a_n = 2.$$

Thus

$$\sqrt{2 + \sqrt{2 + \sqrt{2 + \sqrt{2 + \cdots}}}} = 2.$$

To indicate what the bounded monotonic sequence property has to do with the "completeness property" of the real numbers, in Problem 42 we outline a proof, using this property, of the existence of the number $\sqrt{2}$. In Problems 45 and 46, we outline a proof of the equivalence of the bounded monotonic sequence property and another common statement of the completeness of the real numbers—the *least upper bound property*.

11.2 Problems

In Problems 1 through 35, determine whether or not the sequence $\{a_n\}$ converges, and find its limit if it does converge.

1. $a_n = \dfrac{2n}{5n - 3}$

2. $a_n = \dfrac{1 - n^2}{2 + 3n^2}$

3. $a_n = \dfrac{n^2 - n + 7}{2n^3 + n^2}$

4. $a_n = \dfrac{n^3}{10n^2 + 1}$

5. $a_n = 1 + \left(\frac{9}{10}\right)^n$

6. $a_n = 2 - \left(-\frac{1}{2}\right)^n$

7. $a_n = 1 + (-1)^n$

8. $a_n = \dfrac{1 + (-1)^n}{\sqrt{n}}$

9. $a_n = \dfrac{1 + (-1)^n \sqrt{n}}{\left(\frac{3}{2}\right)^n}$

10. $a_n = \dfrac{\sin n}{3^n}$

11. $a_n = \dfrac{\sin^2 n}{\sqrt{n}}$

12. $a_n = \sqrt{\dfrac{2 + \cos n}{n}}$

13. $a_n = n \sin \pi n$

14. $a_n = n \cos \pi n$

15. $a_n = \pi^{-(\sin n)/n}$

16. $a_n = 2^{\cos n\pi}$

17. $a_n = \dfrac{\ln n}{\sqrt{n}}$

18. $a_n = \dfrac{\ln 2n}{\ln 3n}$

19. $a_n = \dfrac{(\ln n)^2}{n}$

20. $a_n = n \sin\left(\dfrac{1}{n}\right)$

21. $a_n = \dfrac{\tan^{-1} n}{n}$

22. $a_n = \dfrac{n^3}{e^{n/10}}$

23. $a_n = \dfrac{2^n + 1}{e^n}$

24. $a_n = \dfrac{\sinh n}{\cosh n}$

25. $a_n = \left(1 + \dfrac{1}{n}\right)^n$

26. $a_n = (2n + 5)^{1/n}$

27. $a_n = \left(\dfrac{n - 1}{n + 1}\right)^n$

28. $a_n = (0.001)^{-1/n}$

29. $a_n = \sqrt[n]{2^{n+1}}$

30. $a_n = \left(1 - \dfrac{2}{n^2}\right)^n$

31. $a_n = \left(\dfrac{2}{n}\right)^{3/n}$

32. $a_n = (-1)^n (n^2 + 1)^{1/n}$

33. $a_n = \left(\dfrac{2 - n^2}{3 + n^2}\right)^n$

34. $a_n = \dfrac{\left(\frac{2}{3}\right)^n}{1 - \sqrt[n]{n}}$

35. $a_n = \dfrac{\left(\frac{2}{3}\right)^n}{\left(\frac{1}{2}\right)^n + \left(\frac{9}{10}\right)^n}$

36. Suppose that $\lim\limits_{n \to \infty} a_n = A$. Prove that $\lim\limits_{n \to \infty} |a_n| = |A|$.

37. Prove that $\{(-1)^n a_n\}$ diverges if $\lim\limits_{n \to \infty} a_n = A \neq 0$.

38. Suppose that $\{a_n\}$ is an increasing sequence that is not bounded. Prove that $\lim\limits_{n \to \infty} a_n = +\infty$.

39. Suppose that $A > 0$. Given x_1 arbitrary, define the sequence $\{x_n\}$ recursively as follows:

$$x_{n+1} = \frac{1}{2}\left(x_n + \frac{A}{x_n}\right) \qquad \text{if } n \geq 1.$$

Prove that if $L = \lim\limits_{n \to \infty} x_n$ exists, then $L = \pm\sqrt{A}$.

40. Let $\{F_n\}$ be the Fibonacci sequence of Example 2. Assume that

$$\tau = \lim_{n \to \infty} \frac{F_{n+1}}{F_n}$$

exists, and prove that $\tau = \frac{1}{2}(1 + \sqrt{5})$.

41. Let the sequence $\{a_n\}$ be defined recursively by

$$a_1 = 2; \qquad a_{n+1} = \frac{1}{2}(a_n + 4) \qquad \text{for } n \geq 1.$$

(a) Prove by induction on n that $a_n < 4$ for each n and that $\{a_n\}$ is an increasing sequence. (b) Find the limit of this sequence.

42. For each positive integer n, let a_n be the largest integral multiple of $1/10^n$ such that $a_n^2 \leq 2$. (a) Prove that $\{a_n\}$ is a bounded increasing sequence, so $A = \lim_{n \to \infty} a_n$ exists. (b) Prove that if $A^2 > 2$, then $a_n^2 > 2$ for n sufficiently large. (c) Prove that if $A^2 < 2$, then $a_n^2 < B$ for some number $B < 2$ and all sufficiently large n. (d) Conclude that $A^2 = 2$.

43. (a) Square both sides of the defining equation $a_{n+1} = \sqrt{2 + a_n}$ of the sequence in Example 12 to show that if $a_n < 2$, then it follows that $a_{n+1} < 2$. Why does this fact imply that $a_n < 2$ for all n? (b) Compute $(a_{n+1})^2 - a_n^2$ to show that $a_{n+1} > a_n$ for all n.

44. Use the method of Example 12 to show that

(a) $\sqrt{6 + \sqrt{6 + \sqrt{6 + \cdots}}} = 3$;

(b) $\sqrt{20 + \sqrt{20 + \sqrt{20 + \cdots}}} = 5$.

Problems 45 and 46 deal with the least upper bound *property of the real numbers: If the nonempty set S of real numbers has an upper bound, then S has a least upper bound. The number M is an* **upper bound** *for the set S if* $x \leqq M$ *for all x in S. The upper bound L of S is a* ***least upper bound*** *for S if no number smaller than L is a least upper bound for S. You can show easily that if the set S has least upper bounds L_1 and L_2, then $L_1 = L_2$; in other words, if a least upper bound for a set exists, then it is unique.*

45. Prove that the least upper bound property implies the bounded monotonic sequence property. [*Suggestion:* If $\{a_n\}$ is a bounded increasing sequence and A is the least upper bound of the set $\{a_n : n \geqq 1\}$ of terms of the sequence, you can prove that $A = \lim\limits_{n \to \infty} a_n$.]

46. Prove that the bounded monotonic sequence property implies the least upper bound property. [*Suggestion:* For each positive integer n, let a_n be the least integral multiple of $1/10^n$ that is an upper bound of the set S. Prove that $\{a_n\}$ is a bounded decreasing sequence and then that $A = \lim\limits_{n \to \infty} a_n$ is a least upper bound for S.]

11.3

Infinite Series and Convergence

An **infinite series** is an expression of the form

$$\sum_{n=1}^{\infty} a_n = a_1 + a_2 + a_3 + \cdots + a_n + \cdots, \qquad (1)$$

where $\{a_n\}$ is an infinite sequence of real numbers. The number a_n is called the **nth term** of the series. The symbol $\sum_{n=1}^{\infty} a_n$ is simply an abbreviation for the right-hand side of Eq. (1). An example of an infinite series is the series

$$\sum_{n=1}^{\infty} \frac{1}{2^n} = \frac{1}{2} + \frac{1}{4} + \frac{1}{8} + \frac{1}{16} + \cdots + \frac{1}{2^n} + \cdots$$

that we mentioned in Section 11.1. The nth term of this particular infinite series is $a_n = 1/2^n$.

To say what such an infinite sum means, we introduce the *partial sums* of the infinite series in Eq. (1). The **nth partial sum S_n** of the series is the sum of its first n terms:

$$S_n = a_1 + a_2 + a_3 + \cdots + a_n. \qquad (2)$$

Thus each infinite series is associated with an infinite **sequence of partial sums**

$$S_1, S_2, S_3, \ldots, S_n, \ldots.$$

We define the *sum* of the infinite series to be the limit of its sequence of partial sums, provided that this limit exists.

Definition   *The Sum of an Infinite Series*
We say that the infinite series

$$\sum_{n=1}^{\infty} a_n \quad \textbf{converges (or is convergent)}$$

with **sum** S provided that the limit of its sequence of partial sums,

$$S = \lim_{n \to \infty} S_n, \qquad (3)$$

exists (and is finite). Otherwise we say that the series **diverges** (or is **divergent**). If a series diverges, then it has no sum.

Thus an infinite series is a limit of finite sums,

$$S = \sum_{n=1}^{\infty} a_n = \lim_{N \to \infty} \sum_{n=1}^{N} a_n,$$

provided that this limit exists.

EXAMPLE 1 Show that the series

$$\sum_{n=1}^{\infty} \left(\tfrac{1}{2}\right)^n = \tfrac{1}{2} + \tfrac{1}{4} + \tfrac{1}{8} + \tfrac{1}{16} + \cdots$$

converges, and find its sum.

Solution The first four partial sums are

$$S_1 = \tfrac{1}{2}, \qquad S_2 = \tfrac{3}{4}, \qquad S_3 = \tfrac{7}{8}, \quad \text{and} \quad S_4 = \tfrac{15}{16}.$$

It seems likely that $S_n = (2^n - 1)/2^n$, and indeed this follows easily by induction on n, because

$$S_{n+1} = S_n + \frac{1}{2^{n+1}} = \frac{2^n - 1}{2^n} + \frac{1}{2^{n+1}} = \frac{2^{n+1} - 2 + 1}{2^{n+1}} = \frac{2^{n+1} - 1}{2^{n+1}}.$$

Hence the sum of the given series is

$$S = \lim_{n \to \infty} S_n = \lim_{n \to \infty} \frac{2^n - 1}{2^n} = \lim_{n \to \infty} \left(1 - \frac{1}{2^n}\right) = 1.$$

EXAMPLE 2 Show that the series

$$\sum_{n=1}^{\infty} (-1)^{n+1} = 1 - 1 + 1 - 1 + \cdots$$

diverges.

Solution The sequence of partial sums of this series is

$$1, 0, 1, 0, 1, \ldots,$$

which has no limit. Therefore the series diverges.

EXAMPLE 3 Show that the infinite series

$$\sum_{n=1}^{\infty} \frac{1}{n(n + 1)}$$

converges, and find its sum.

Solution We need a formula for the nth partial sum S_n so that we can evaluate its limit as $n \to \infty$. To find such a formula, we begin with the observation that the nth term of the series is

$$a_n = \frac{1}{n(n + 1)} = \frac{1}{n} - \frac{1}{n + 1}.$$

(In more complicated cases, such as those in Problems 36 through 40, such a decomposition can be obtained by the method of partial fractions.) It follows that the sum of the first n terms of the given series is

$$S_n = \left(1 - \frac{1}{2}\right) + \left(\frac{1}{2} - \frac{1}{3}\right) + \left(\frac{1}{3} - \frac{1}{4}\right)$$
$$+ \left(\frac{1}{4} - \frac{1}{5}\right) + \cdots + \left(\frac{1}{n} - \frac{1}{n+1}\right)$$
$$= 1 - \frac{1}{n+1} = \frac{n}{n+1}.$$

Hence

$$\sum_{n=1}^{\infty} \frac{1}{n(n+1)} = \lim_{n \to \infty} \frac{n}{n+1} = 1.$$

The sum for S_n in Example 3, called a *telescoping* sum, provides us with a way to find the sums of certain series. The series in Examples 1 and 2 are examples of a more common and more important type of series, the *geometric series*.

Definition   *Geometric Series*
The series $\sum_{n=0}^{\infty} a_n$ is said to be a **geometric series** if each term after the first is a fixed multiple of the term immediately before it. That is, there is a number r, called the **ratio** of the series, such that

$$a_{n+1} = ra_n \qquad \text{for all } n \geq 0.$$

Thus every geometric series takes the form

$$a_0 + ra_0 + r^2 a_0 + r^3 a_0 + \cdots = \sum_{n=0}^{\infty} r^n a_0. \qquad (4)$$

It is convenient to begin the summation at $n = 0$, and thus we regard the sum
$$S_n = a_0(1 + r + r^2 + \cdots + r^n)$$
of the first $n + 1$ terms to be the nth partial sum of the series.

EXAMPLE 4   The infinite series
$$\sum_{n=0}^{\infty} \frac{2}{3^n} = 2 + \frac{2}{3} + \frac{2}{9} + \cdots + \frac{2}{3^n} + \cdots$$

is a geometric series whose first term is $a_0 = 2$ and whose ratio is $r = \frac{1}{3}$.

Theorem 1   *Sum of a Geometric Series*
If $|r| < 1$, then the geometric series in Eq. (4) converges, and its sum is

$$S = \sum_{n=0}^{\infty} r^n a_0 = \frac{a_0}{1 - r}. \qquad (5)$$

If $|r| \geq 1$ and $a_0 \neq 0$, then the geometric series diverges.

Proof If $r = 1$, then $S_n = (n + 1)a_0$, so the series certainly diverges if $a_0 \ne 0$. If $r = -1$ and $a_0 \ne 0$, then the series diverges by an argument like the one in Example 2. So we may suppose that $|r| \ne 1$. Then the elementary identity

$$1 + r + r^2 + \cdots + r^n = \frac{1 - r^{n+1}}{1 - r}$$

follows if we multiply each side by $1 - r$. Hence the nth partial sum of the geometric series is

$$S_n = a_0(1 + r + r^2 + \cdots + r^n) = a_0\left(\frac{1}{1 - r} - \frac{r^{n+1}}{1 - r}\right).$$

If $|r| < 1$, then $r^{n+1} \to 0$ as $n \to \infty$, by Example 7 in Section 11.2. So in this case the geometric series converges to

$$S = \lim_{n \to \infty} a_0\left(\frac{1}{1 - r} - \frac{r^{n+1}}{1 - r}\right) = \frac{a_0}{1 - r}.$$

But if $|r| > 1$, then $\lim_{n \to \infty} r^{n+1}$ does not exist, so $\lim_{n \to \infty} S_n$ does not exist. This establishes the theorem. ❑

EXAMPLE 5 With $a_0 = 1$ and $r = -\frac{1}{2}$, we find that

$$1 - \frac{1}{2} + \frac{1}{4} - \frac{1}{8} + \cdots = \sum_{n=0}^{\infty} \left(-\frac{1}{2}\right)^n = \frac{1}{1 - (-\frac{1}{2})} = \frac{2}{3}.$$

Theorem 2 implies that the operations of addition and of multiplication by a constant can be carried out term by term in the case of *convergent* series. Because the sum of an infinite series is the limit of its sequence of partial sums, this theorem follows immediately from the limit laws for sequences (Theorem 1 of Section 11.2).

Theorem 2 *Termwise Addition and Multiplication*
If the series $A = \Sigma a_n$ and $B = \Sigma b_n$ converge to the indicated sums and c is a constant, then the series $\Sigma (a_n + b_n)$ and Σca_n also converge, with sums

1. $\sum (a_n + b_n) = A + B$;
2. $\sum ca_n = cA$.

The geometric series in Eq. (5) may be used to find the rational number represented by a given infinite repeating decimal.

EXAMPLE 6

$$0.55555\ldots = \frac{5}{10} + \frac{5}{100} + \frac{5}{1000} + \cdots$$

$$= \sum_{n=0}^{\infty} \frac{5}{10}\left(\frac{1}{10}\right)^n = \frac{\frac{5}{10}}{1 - \frac{1}{10}} = \frac{5}{10} \cdot \frac{10}{9} = \frac{5}{9}.$$

In a more complicated situation, we may need to use the termwise algebra of Theorem 2:

$$0.72828\,28\ldots = \frac{7}{10} + \frac{28}{10^3} + \frac{28}{10^5} + \frac{28}{10^7} + \cdots$$

$$= \frac{7}{10} + \frac{28}{10^3}\left(1 + \frac{1}{10^2} + \frac{1}{10^4} + \cdots\right)$$

$$= \frac{7}{10} + \frac{28}{1000}\sum_{n=0}^{\infty}\left(\frac{1}{100}\right)^n = \frac{7}{10} + \frac{28}{1000}\left(\frac{1}{1 - \frac{1}{100}}\right)$$

$$= \frac{7}{10} + \frac{28}{1000}\cdot\frac{100}{99} = \frac{7}{10} + \frac{28}{990} = \frac{721}{990}.$$

This technique can be used to show that every repeating infinite decimal represents a rational number. Consequently, the decimal expansions of irrational numbers such as π, e, and $\sqrt{2}$ must be nonrepeating as well as infinite. Conversely, if p and q are integers with $q \neq 0$, then long division of q into p yields a repeating decimal expansion for the rational number p/q, because such a division can yield at each stage only q possible different remainders.

EXAMPLE 7   Suppose that Paul and Mary toss a fair six-sided die in turn until one of them wins by getting the first "six." If Paul tosses first, calculate the probability that he will win the game.

Solution   Because the die is fair, the probability that Paul gets a "six" on the first round is $\frac{1}{6}$. The probability that he gets the game's first "six" on the second round is $(\frac{5}{6})^2(\frac{1}{6})$—the product of the probability $(\frac{5}{6})^2$ that neither Paul nor Mary rolls a "six" in the first round and the probability $\frac{1}{6}$ that Paul rolls a "six" in the second round. Paul's probability p of getting the first "six" in the game is the *sum* of his probabilities of getting it in the first round, in the second round, in the third round, and so on. Hence

$$p = \frac{1}{6} + \left(\frac{5}{6}\right)^2\left(\frac{1}{6}\right) + \left(\frac{5}{6}\right)^2\left(\frac{5}{6}\right)^2\left(\frac{1}{6}\right) + \cdots = \frac{1}{6}\left[1 + \left(\frac{5}{6}\right)^2 + \left(\frac{5}{6}\right)^4 + \cdots\right]$$

$$= \frac{1}{6}\cdot\frac{1}{1 - (\frac{5}{6})^2} = \frac{1}{6}\cdot\frac{36}{11} = \frac{6}{11}.$$

Because he has the advantage of tossing first, Paul has more than the fair probability $\frac{1}{2}$ of getting the first "six" and thus winning the game.

Theorem 3 is often useful in showing that a given series does *not* converge.

> **Theorem 3   The *n*th-Term Test for Divergence**
> If either
>
> $$\lim_{n\to\infty} a_n \neq 0$$
>
> or this limit does not exist, then the infinite series $\sum a_n$ diverges.

Proof We want to show under the stated hypothesis that the series $\Sigma\, a_n$ diverges. It suffices to show that *if* the series $\Sigma\, a_n$ does converge, then $\lim_{n\to\infty} a_n = 0$. So suppose that $\Sigma\, a_n$ converges with sum $S = \lim_{n\to\infty} S_n$, where

$$S_n = a_1 + a_2 + a_3 + \cdots + a_n$$

is the *n*th partial sum of the series. Because $a_n = S_n - S_{n-1}$,

$$\lim_{n\to\infty} a_n = \lim_{n\to\infty} (S_n - S_{n-1}) = \lim_{n\to\infty} S_n - \lim_{n\to\infty} S_{n-1} = S - S = 0.$$

Consequently, if $\lim_{n\to\infty} a_n \neq 0$, then the series $\Sigma\, a_n$ diverges. ❑

EXAMPLE 8 The series

$$\sum_{n=1}^{\infty} (-1)^{n-1} n^2 = 1 - 4 + 9 - 16 + 25 - \cdots$$

diverges because $\lim_{n\to\infty} a_n$ does not exist, whereas the series

$$\sum_{n=1}^{\infty} \frac{n}{3n+1} = \frac{1}{4} + \frac{2}{7} + \frac{3}{10} + \frac{4}{13} + \cdots$$

diverges because

$$\lim_{n\to\infty} \frac{n}{3n+1} = \frac{1}{3} \neq 0.$$

WARNING The converse of Theorem 3 is *false*! The condition

$$\lim_{n\to\infty} a_n = 0$$

is necessary *but not sufficient* to guarantee convergence of the series

$$\sum_{n=1}^{\infty} a_n.$$

That is, a series may satisfy the condition $a_n \to 0$ as $n \to \infty$ and yet diverge. An important example of a divergent series with terms that approach zero is the **harmonic series**

$$\sum_{n=1}^{\infty} \frac{1}{n} = 1 + \frac{1}{2} + \frac{1}{3} + \frac{1}{4} + \frac{1}{5} + \cdots. \tag{6}$$

Theorem 4
The harmonic series diverges.

Proof Because each term of the harmonic series is positive, its sequence of partial sums $\{S_n\}$ is increasing. We shall prove that

$$\lim_{n\to\infty} S_n = +\infty,$$

and thus that the harmonic series diverges, by showing that there are arbitrarily large partial sums. Consider the closed interval $[0, k]$ on the *x*-axis, where k is a positive integer. On its subinterval $[0, 1]$, imagine a rectangle of height

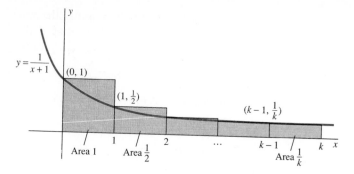

Fig. 11.3.1 Idea of the proof of Theorem 4

1 with that subinterval as its base. On the subinterval $[1, 2]$, imagine a rectangle of height $\frac{1}{2}$ whose base is that subinterval. Build a rectangle of height $\frac{1}{3}$ over the subinterval $[2, 3]$, a rectangle of height $\frac{1}{4}$ over the subinterval $[3, 4]$, and so on. The last rectangle will have the subinterval $[k - 1, k]$ as its base and will have height $1/k$. These rectangles are shown in Fig. 11.3.1.

Next, note that the total area of all k of these rectangles is the kth partial sum of the harmonic series:

$$S_k = 1 + \frac{1}{2} + \frac{1}{3} + \frac{1}{4} + \cdots + \frac{1}{k}.$$

Finally, the graph of the function

$$f(x) = \frac{1}{x + 1}$$

—also shown in Fig. 11.3.1—passes through the upper left-hand corner point of each rectangle. Because f is decreasing for all $x \geqq 0$, the graph of f is never above the top of any rectangle on the corresponding subinterval. So the sum of the areas of the k rectangles is greater than the area under the graph of f over $[0, k]$. That is,

$$S_k \geqq \int_0^k \frac{1}{x + 1}\, dx = \left[\ln(x + 1) \right]_0^k = \ln(k + 1).$$

But

$$\lim_{k \to \infty} \ln(k + 1) = +\infty,$$

so $\ln(k + 1)$ takes on arbitrarily large positive values with increasing k. Consequently, because $S_k \geqq \ln(k + 1)$ for all $k \geqq 1$, S_k also takes on arbitrarily large positive values. Therefore $\lim_{k \to \infty} S_k = +\infty$, and hence the harmonic series diverges. ❑

If the sequence of partial sums of the series Σa_n diverges to infinity, then we say that the series **diverges to infinity,** and we write

$$\sum_{n=1}^{\infty} a_n = \infty.$$

The series $\Sigma(-1)^{n+1}$ of Example 2 is a series that diverges but does not diverge to infinity. In the nineteenth century it was common to say that such a series was divergent by oscillation; today we say merely that it diverges.

594

Our proof of Theorem 4 shows that

$$\sum_{n=1}^{\infty} \frac{1}{n} = \infty.$$

But the partial sums of the harmonic series diverge to infinity very slowly. If N_A denotes the smallest integer such that

$$\sum_{n=1}^{N_A} \frac{1}{n} \geqq A,$$

then it is known that

$$N_5 = 83,$$

(This can be verified with the aid of a programmable calculator.)

$$N_{10} = 12{,}367,$$

$$N_{20} = 272{,}400{,}600,$$

$$N_{100} \approx 1.5 \times 10^{43}, \quad \text{and}$$

$$N_{1000} \approx 1.1 \times 10^{434}.$$

Thus you would need to add more than a quarter of a billion terms of the harmonic series to get a partial sum that exceeds 20. At this point each of the next few terms would be approximately $0.000000004 = 4 \times 10^{-9}$. The number of terms you'd have to add to reach 1000 is far larger than the estimated number of elementary particles in the entire universe (10^{80}).[†]

Theorem 5 says that if two infinite series have the same terms from some point on, then either both series converge or both series diverge. The proof is left for Problem 43.

> **Theorem 5** *Series that Are Eventually the Same*
> If there exists a positive integer k such that $a_n = b_n$ for all $n > k$, then the series $\Sigma \, a_n$ and $\Sigma \, b_n$ either both converge or both diverge.

It follows that a *finite* number of terms can be changed, deleted from, or adjoined to an infinite series without altering its convergence or divergence (although the *sum* of a convergent series will generally be changed by such alterations). In particular, taking $b_n = 0$ for $n \leqq k$ and $b_n = a_n$ for $n > k$, we see that the series

$$\sum_{n=1}^{\infty} a_n$$

and the series

$$\sum_{n=k+1}^{\infty} a_n$$

that is obtained by deleting its first k terms either both converge or both diverge.

[†] If you enjoy such large numbers, see the article "Partial sums of inifinite series, and how they grow," by R. P. Boas, Jr., in *American Mathematical Monthly* **84** (1977): 237–248.

11.3 Problems

In Problems 1 through 29, determine whether the given infinite series converges or diverges. If it converges, find its sum.

1. $1 + \dfrac{1}{3} + \dfrac{1}{9} + \cdots + \dfrac{1}{3^n} + \cdots$

2. $1 + e^{-1} + e^{-2} + \cdots + e^{-n} + \cdots$

3. $1 + 3 + 5 + 7 + \cdots + (2n - 1) + \cdots$

4. $\dfrac{1}{2} + \dfrac{1}{\sqrt{2}} + \dfrac{1}{\sqrt[3]{2}} + \cdots + \dfrac{1}{\sqrt[n]{2}} + \cdots$

5. $1 - 2 + 4 - 8 + \cdots + (-2)^n + \cdots$

6. $1 - \frac{1}{4} + \frac{1}{16} - \cdots + (-\frac{1}{4})^n + \cdots$

7. $4 + \dfrac{4}{3} + \dfrac{4}{9} + \dfrac{4}{27} + \cdots + \dfrac{4}{3^n} + \cdots$

8. $\dfrac{1}{3} + \dfrac{2}{9} + \dfrac{4}{27} + \cdots + \dfrac{2^{n-1}}{3^n} + \cdots$

9. $1 + (1.01) + (1.01)^2 + (1.01)^3 + \cdots + (1.01)^n + \cdots$

10. $1 + \dfrac{1}{\sqrt{2}} + \dfrac{1}{\sqrt[3]{3}} + \cdots + \dfrac{1}{\sqrt[n]{n}} + \cdots$

11. $\displaystyle\sum_{n=0}^{\infty} \dfrac{(-1)^n n}{n + 1}$

12. $\displaystyle\sum_{n=1}^{\infty} \left(\dfrac{e}{10}\right)^n$

13. $\displaystyle\sum_{n=0}^{\infty} (-1)^n \left(\dfrac{3}{e}\right)^n$

14. $\displaystyle\sum_{n=0}^{\infty} \dfrac{3^n - 2^n}{4^n}$

15. $\displaystyle\sum_{n=1}^{\infty} (\sqrt{2})^{1-n}$

16. $\displaystyle\sum_{n=1}^{\infty} \left(\dfrac{2}{n} - \dfrac{1}{2^n}\right)$

17. $\displaystyle\sum_{n=1}^{\infty} \dfrac{n}{10n + 17}$

18. $\displaystyle\sum_{n=1}^{\infty} \dfrac{\sqrt{n}}{\ln(n + 1)}$

19. $\displaystyle\sum_{n=1}^{\infty} (5^{-n} - 7^{-n})$

20. $\displaystyle\sum_{n=0}^{\infty} \dfrac{1}{1 + (\frac{9}{10})^n}$

21. $\displaystyle\sum_{n=1}^{\infty} \left(\dfrac{e}{\pi}\right)^n$

22. $\displaystyle\sum_{n=1}^{\infty} \left(\dfrac{\pi}{e}\right)^n$

23. $\displaystyle\sum_{n=0}^{\infty} (\frac{100}{99})^n$

24. $\displaystyle\sum_{n=0}^{\infty} (\frac{99}{100})^n$

25. $\displaystyle\sum_{n=0}^{\infty} \dfrac{1 + 2^n + 3^n}{5^n}$

26. $\displaystyle\sum_{n=0}^{\infty} \dfrac{1 + 2^n + 5^n}{3^n}$

27. $\displaystyle\sum_{n=0}^{\infty} \dfrac{7 \cdot 5^n + 3 \cdot 11^n}{13^n}$

28. $\displaystyle\sum_{n=1}^{\infty} \sqrt[n]{2}$

29. $\displaystyle\sum_{n=1}^{\infty} [(\frac{7}{11})^n - (\frac{3}{5})^n]$

30. Use the method of Example 6 to verify that

(a) $0.666\,666\,666\ldots = \frac{2}{3}$; (b) $0.111\,111\,111\ldots = \frac{1}{9}$;

(c) $0.249\,999\,999\ldots = \frac{1}{4}$; (d) $0.999\,999\,999\ldots = 1$.

In Problems 31 through 35, find the rational number represented by the given repeating decimal.

31. $0.4747\,4747\ldots$ **32.** $0.2525\,2525\ldots$

33. $0.123\,123\,123\ldots$ **34.** $0.3377\,3377\,3377\ldots$

35. $3.14159\,14159\,14159\ldots$

In Problems 36 through 40, use the method of Example 3 to find a formula for the nth partial sum S_n, and then compute the sum of the infinite series.

36. $\displaystyle\sum_{n=1}^{\infty} \dfrac{1}{4n^2 - 1}$

37. $\displaystyle\sum_{n=1}^{\infty} \ln\left(\dfrac{n + 1}{n}\right)$

38. $\displaystyle\sum_{n=0}^{\infty} \dfrac{4}{16n^2 - 8n - 3}$

39. $\displaystyle\sum_{n=2}^{\infty} \dfrac{2}{n^2 - 1}$

40. $\dfrac{1}{1 \cdot 3} + \dfrac{1}{2 \cdot 4} + \dfrac{1}{3 \cdot 5} + \dfrac{1}{4 \cdot 6} + \cdots$

41. Prove: If $\Sigma\, a_n$ diverges and c is a nonzero constant, then $\Sigma\, c a_n$ diverges.

42. Suppose that $\Sigma\, a_n$ converges and that $\Sigma\, b_n$ diverges. Prove that $\Sigma\, (a_n + b_n)$ diverges.

43. Let S_n and T_n denote the nth partial sums of $\Sigma\, a_n$ and $\Sigma\, b_n$, respectively. Suppose that $a_n = b_n$ for all $n > k$. Show that $S_n - T_n = S_k - T_k$ if $n > k$. Hence prove Theorem 5.

44. Suppose that $0 < x \le 1$. Integrate both sides of the identity

$$\dfrac{1}{1 + t} = 1 - t + t^2 - t^3 + \cdots$$

$$+ (-1)^n t^n + \dfrac{(-1)^{n+1} t^{n+1}}{1 + t}$$

from $t = 0$ to $t = x$ to show that

$$\ln(1 + x) = x - \dfrac{x^2}{2} + \dfrac{x^3}{3} - \cdots + (-1)^n \dfrac{x^{n+1}}{n + 1} + R_n,$$

where $\lim_{n \to \infty} R_n = 0$. Hence conclude that

$$\ln(1 + x) = \sum_{n=1}^{\infty} (-1)^{n+1} \dfrac{x^n}{n}$$

if $0 < x \le 1$.

45. Criticize the following "proof" that $2 = 1$. Substitution of $x = 1$ into the result of Problem 44 gives the fact that

$$\ln 2 = 1 - \tfrac{1}{2} + \tfrac{1}{3} - \tfrac{1}{4} + \cdots.$$

If

$$S = 1 + \tfrac{1}{2} + \tfrac{1}{3} + \tfrac{1}{4} + \cdots,$$

then

$$\ln 2 = S - 2 \cdot (\tfrac{1}{2} + \tfrac{1}{4} + \tfrac{1}{6} + \tfrac{1}{8} + \cdots) = S - S = 0.$$

Hence $2 = e^{\ln 2} = e^0 = 1$.

46. A ball has *bounce coefficient* $r < 1$ if, when it is dropped from a height h, it bounces back to a height of rh (Fig. 11.3.2). Suppose that such a ball is dropped from the initial height a and subsequently bounces infinitely many times. Use a geometric series to show that the total up-and-down distance it travels in all its bouncing is

$$D = a \frac{1 + r}{1 - r}.$$

Note that D is *finite*.

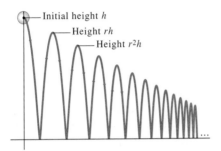

Fig. 11.3.2   Successive bounces of the ball of Problems 46 and 47

47. A ball with bounce coefficient $r = 0.64$ (see Problem 46) is dropped from an initial height of $a = 4$ ft. Use a geometric series to compute the total time required for it to complete its infinitely many bounces. The time required for a ball to drop h feet (from rest) is $\sqrt{2h/g}$ seconds where $g = 32$ ft/s^2.

48. Suppose that the government spends $1 billion and that each recipient of a fraction of this wealth spends 90% of the dollars that he or she receives. In turn, the secondary recipients spend 90% of the dollars they receive, and so on. How much total spending results from the original injection of $1 billion into the economy?

49. A tank initially contains a mass M_0 of air. Each stroke of a vacuum pump removes 5% of the air in the container.

Compute (a) the mass M_n of air remaining in the tank after n strokes of the pump; (b) $\lim\limits_{n \to \infty} M_n$.

50. Paul and Mary toss a fair coin in turn until one of them wins the game by getting the first "head." Calculate for each the probability that he or she wins the game.

51. Peter, Paul, and Mary toss a fair coin in turn until one of them wins by getting the first "head." Calculate for each the probability that he or she wins the game. Check your answer by verifying that the sum of the three probabilities is 1.

52. Peter, Paul, and Mary roll a fair die in turn until one of them wins the game by getting the first "six." Calculate for each the probability that he or she wins the game. Check your answer by verifying that the sum of the three probabilities is 1.

53. A pane of a certain type of glass reflects half the incident light, absorbs one-fourth, and transmits one-fourth. A window is made of two panes of this glass separated by a small space (Fig. 11.3.3). What fraction of the incident light I is transmitted by the double window?

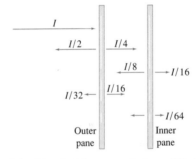

Fig. 11.3.3   The double-pane window of Problem 53

54. Criticize the following evaluation of the sum of an infinite series:

Let $x = 1 - 2 + 4 - 8 + 16 - 32 + 64 - \cdots$.
Then $2x = 2 - 4 + 8 - 16 + 32 - 64 + \cdots$.
Add the equations to obtain $3x = 1$. Thus $x = \tfrac{1}{3}$, and "therefore"

$$1 - 2 + 4 - 8 + 16 - 32 + 64 - \cdots = \tfrac{1}{3}.$$

11.3 Projects

These projects require the use of a programmable calculator or computer. It is not otherwise practical to calculate partial sums with large numbers of terms.

The TI-81/85 and BASIC programs listed in Fig. 11.3.4 can be used to approximate the sum of the convergent infinite series

$$\sum_{n=1}^{\infty} a_n = a_1 + a_2 + a_3 + \cdots$$

TI-81/85	BASIC	Comments
PRGM1:SERIES :Disp "K"	DEF FN A(N) = (1/2)^N	Define Y1 = $a(n)$
:Input K	INPUT "K"; K	Number of terms to sum
:0->S	S = 0	Initialize sum
:1->N		Initialize index
:Lbl 1	FOR N = 1 TO K	Begin loop
:Y1->T	T = FN A(N)	Compute next term
:S+T->S	S = S + T	Add next term
:N+1->N	NEXT	Next N
:If N ≦ K		End loop
:Goto 1		
:Disp "SUM="	PRINT "Sum = "; S	Display sum
:Disp S		
:End	END	

Fig. 11.3.4 TI calculator and BASIC programs for calculating partial sums of Σa_n

by calculating the kth partial sum

$$S_k = \sum_{n=1}^{k} a_n = a_1 + a_2 + a_3 + \cdots + a_k.$$

In the case of an infinite series $\Sigma_{n=0}^{\infty} a_n$ beginning at $n = 0$, the initial term a_0 must be added manually or the program altered appropriately.

Even if you don't program in BASIC or in the TI calculator language, you may find it helpful to compare these alternative descriptions of the same computational algorithm. For the TI calculator program, the formula $a_n = a(n)$ that gives the nth term as a function of n must be defined (as Y1) in the Y = menu before the program is executed. (For instance, enter Y1=(1/2)∧N to sum the geometric series $\Sigma_{n=1}^{\infty} (1/2)^n$.) In the BASIC version the nth term function is defined in the first line of the program itself. The desired number k of terms to sum is entered while the program is running.

Most computer algebra systems include a SUM function that can be used to calculate partial sums directly. Suppose that the nth term function $a(n)$ (expressing a_n as a a function of n) has already been defined appropriately. Then all the single-line commands listed in Fig. 11.3.5 calculate (in various systems) the kth partial sum $\Sigma_{n=1}^{k} a_n$. The similarities among these commands in different systems are striking.

Program	Command
Derive	SUM(a(n), n, 1, k)
HP-48SX	'Σ (n=1,k,a(n))'
Maple	sum(a(n), n=1..k)
Mathematica	Sum[a[n], {n, 1, k}]
TI-85	sum seq(a(n), n, 1, k, 1)
(X)PLORE	sum(a(n), n=1 to k)

Fig. 11.3.5 Commands for calculating the kth partial sum of the infinite series Σa_n

PROJECT A Calculate partial sums of the geometric series $\Sigma_{n=0}^{\infty} r^n$ with $r = 0.2, 0.5, 0.75, 0.9, 0.95,$ and 0.99. For each value of r, calculate k-term partial sums with $k = 10, 20, 30, \ldots$, continuing until two successive results agree to four or five decimal places. (For $r = 0.95$ and 0.99, you may decide to use $k = 100, 200, 300, \ldots$.) How does the apparent rate of convergence—as measured by the number of terms required for the desired accuracy—depend on the value of r?

PROJECT B Calculate partial sums of the harmonic series $\Sigma\, 1/n$ with $k = 100, 200, 300, \ldots$ terms (or with $k = 1000, 2000, 3000, \ldots$ if you have a powerful microcomputer). Interpret your results in light of our discussion of the harmonic series.

PROJECT C In Section 11.4 we will see that the famous number e is given by

$$e = 1 + \sum_{n=1}^{\infty} \frac{1}{n!} = 1 + \frac{1}{1!} + \frac{1}{2!} + \frac{1}{3!} + \cdots,$$

where the *factorial* $n! = 1 \cdot 2 \cdot 3 \cdots n$ denotes the product of the first n positive integers. Sum enough terms of this rapidly convergent infinite series to convince yourself that $e = 2.7\,1828\,1828$ (accurate to nine decimal places).

11.4
Taylor Series and Taylor Polynomials

The infinite series we studied in Section 11.3 have *constant* terms, and the sum of such a series (assuming it converges) is a *number*. In contrast, much of the practical importance of infinite series derives from the fact that many functions have useful representations as infinite series with *variable* terms.

EXAMPLE 1 If we write $r = x$ for the ratio in a geometric series, Theorem 1 in Section 11.3 gives the infinite series representation

$$\frac{1}{1 - x} = \sum_{n=0}^{\infty} x^n = 1 + x + x^2 + x^3 + \cdots \qquad (1)$$

of the function $f(x) = 1/(1 - x)$. The infinite series in Eq. (1) converges to $1/(1 - x)$ for every number x in the interval $(-1, 1)$. Each partial sum

$$1 + x + x^2 + x^3 + \cdots + x^n \qquad (2)$$

of the geometric series in Eq. (1) is a polynomial approximation to the function $f(x) = 1/(1 - x)$. If $|x| < 1$, then the convergence of the series in Eq. (1) implies that the approximation

$$\frac{1}{1 - x} \approx 1 + x + x^2 + x^3 + \cdots + x^n \qquad (3)$$

is accurate if n is sufficiently large.

REMARK The approximation in (3) could be used to calculate numerical quotients with a calculator that has only $+$, $-$, $\times$ keys (but no $\div$ key). For instance,

$$\frac{329}{73} = \frac{3.29}{0.73} = 3.29 \times \frac{1}{1 - 0.27}$$

$$\approx (3.29)[1 + (0.27) + (0.27)^2 + \cdots + (0.27)^{10}]$$

$$\approx (3.29)(1.36986); \quad \text{thus}$$

$$\frac{329}{73} \approx 4.5068,$$

599

accurate to four decimal places. This is a simple illustration of the use of polynomial approximation for numerical computation.

The definitions of the various elementary transcendental functions leave it unclear how to compute their values precisely, except at a few isolated points. For example,

$$\ln x = \int_1^x \frac{1}{t}\, dt \qquad (x > 0)$$

by definition, so obviously $\ln 1 = 0$, but no other value of $\ln x$ is obvious. The natural exponential function is the inverse of $\ln x$, so it is clear that $e^0 = 1$, but it is not at all clear how to compute e^x for $x \neq 0$. Indeed, even such an innocent-looking expression as $\sqrt{x}$ is not computable (precisely and in a finite number of steps) unless x happens to be the square of a rational number.

But *any* value of a polynomial

$$P(x) = c_0 + c_1 x + c_2 x^2 + \cdots + c_n x^n$$

with known coefficients $c_0, c_1, c_2, \ldots, c_n$ is easy to calculate—as in the preceding remark, only addition and multiplication are required. One goal of this section is to use the fact that polynomial values are so readily computable to help us calculate approximate values of functions such as $\ln x$ and e^x.

POLYNOMIAL APPROXIMATIONS AND TAYLOR'S FORMULA

Suppose that we want to calculate (or, at least, closely approximate) a specific value $f(x_0)$ of a given function f. It would suffice to find a polynomial $P(x)$ with a graph that is very close to that of f on some interval containing x_0. For then we could use the value $P(x_0)$ as an approximation to the actual value of $f(x_0)$. Once we know how to find such an approximating polynomial $P(x)$, the next question would be how accurately $P(x_0)$ approximates the desired value $f(x_0)$.

The simplest example of polynomial approximation is the linear approximation

$$f(x) \approx f(a) + f'(a)(x - a)$$

obtained by writing $\Delta x = x - a$ in the linear approximation formula, Eq. (3) of Section 4.2. The graph of the first-degree polynomial

$$P_1(x) = f(a) + f'(a)(x - a) \tag{4}$$

is the line tangent to the curve $y = f(x)$ at the point $(a, f(a))$; see Fig. 11.4.1. This first-degree polynomial agrees with f and with its first derivative at $x = a$. That is,

$$P_1(a) = f(a) \quad \text{and} \quad P_1'(a) = f'(a).$$

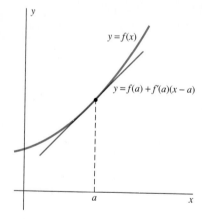

Fig. 11.4.1 The tangent line at $(a, f(a))$ is the best linear approximation to $y = f(x)$ near a.

EXAMPLE 2 Suppose that $f(x) = \ln x$ and that $a = 1$. Then $f(1) = 0$ and $f'(1) = 1$, so $P_1(x) = x - 1$. Hence we expect that $\ln x \approx x - 1$ for x near 1. With $x = 1.1$, we find that

$$P_1(1.1) = 0.1000, \quad \text{whereas} \quad \ln(1.1) \approx 0.0953.$$

600

The error in this approximation is about 5%.

To better approximate $\ln x$ near $x = 1$, let us look for a second-degree polynomial

$$P_2(x) = c_0 + c_1 x + c_2 x^2$$

that has not only the same value and the same first derivative as does f at $x = 1$, but has also the same second derivative there: $P_2''(1) = f''(1) = -1$. To satisfy these conditions, we must have

$$P_2(1) = c_2 + c_1 + c_0 = 0,$$

$$P_2'(1) = 2c_2 + c_1 = 1,$$

$$P_2''(1) = 2c_2 = -1.$$

When we solve these equations, we find that $c_0 = -\frac{3}{2}$, $c_1 = 2$, and $c_2 = -\frac{1}{2}$, so

$$P_2(x) = -\tfrac{1}{2}x^2 + 2x - \tfrac{3}{2}.$$

With $x = 1.1$ we find that $P_2(1.1) = 0.0950$, which is accurate to three decimal places because $\ln(1.1) \approx 0.0953$. The graph of $y = -\frac{1}{2}x^2 + 2x - \frac{3}{2}$ is a parabola through $(1, 0)$ with the same value, slope, *and curvature* there as $y = \ln x$ (Fig. 11.4.2).

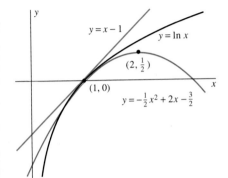

Fig. 11.4.2 The linear and parabolic approximations to $y = \ln x$ near the point $(1, 0)$ (Example 2)

The tangent line and the parabola used in the computations of Example 2 illustrate one general approach to polynomial approximation. To approximate the function $f(x)$ near $x = a$, we look for an nth-degree polynomial

$$P_n(x) = c_0 + c_1 x + c_2 x^2 + \cdots + c_n x^n$$

such that its value at a and the values of its first n derivatives at a agree with the corresponding values of f. That is, we require that

$$\begin{aligned}
P_n(a) &= f(a), \\
P_n'(a) &= f'(a), \\
P_n''(a) &= f''(a), \\
&\;\;\vdots \\
P_n^{(n)}(a) &= f^{(n)}(a).
\end{aligned} \tag{5}$$

We can use these $n + 1$ conditions to evaluate the values of the $n + 1$ coefficients $c_0, c_1, \ldots, c_n$.

The algebra involved is much simpler, however, if we begin with $P_n(x)$ expressed as an nth-degree polynomial in powers of $x - a$ rather than in powers of x:

$$P_n(x) = b_0 + b_1(x - a) + b_2(x - a)^2 + \cdots + b_n(x - a)^n. \tag{6}$$

Then substitution of $x = a$ into Eq. (3) yields

$$b_0 = P_n(a) = f(a)$$

by the first condition in Eq. (5). Substitution of $x = a$ into

$$P_n'(x) = b_1 + 2b_2(x - a) + 3b_3(x - a)^2 + \cdots + nb_n(x - a)^{n-1}$$

yields

$$b_1 = P_n'(a) = f'(a)$$

by the second condition in Eq. (5). Next, substitution of $x = a$ into

$$P_n''(x) = 2b_2 + 3 \cdot 2b_3(x - a) + \cdots + n(n - 1)b_n(x - a)^{n-2}$$

yields $2b_2 = P_n''(a) = f''(a)$, so

$$b_2 = \tfrac{1}{2} f''(a).$$

We continue this process to find $b_3, b_4, \ldots, b_n$. In general, the constant term in the kth derivative $P_n^{(k)}(x)$ is $k!b_k$, because it is the kth derivative of the kth-degree term $b_k(x - a)^k$ in $P_n(x)$:

$$P_n^{(k)}(x) = k!b_k + \{\text{powers of } x - a\}.$$

(Recall that $k! = 1 \cdot 2 \cdot 3 \cdots (k - 1) \cdot k$ denotes the *factorial* of the positive integer k, read "k factorial.") So when we substitute $x = a$ into $P_n^{(k)}(x)$, we find that

$$k!b_k = P_n^{(k)}(a) = f^{(k)}(a)$$

and thus that

$$b_k = \frac{f^{(k)}(a)}{k!} \tag{7}$$

for $k = 1, 2, 3, \ldots, n$.

Indeed, Eq. (7) holds also for $k = 0$ if we use the universal convention that $0! = 1$ and agree that the zeroth derivative $g^{(0)}$ of the function g is just g itself. With such conventions, our computations establish the following theorem.

Theorem 1   The nth-Degree Taylor Polynomial
Suppose that the first n derivatives of the function $f(x)$ exist at $x = a$. Let $P_n(x)$ be the nth-degree polynomial

$$P_n(x) = \sum_{k=0}^{n} \frac{f^{(k)}(a)}{k!}(x - a)^k$$

$$= f(a) + f'(a)(x - a) + \frac{f''(a)}{2!}(x - a)^2$$

$$+ \cdots + \frac{f^{(n)}(a)}{n!}(x - a)^n. \tag{8}$$

Then the values of $P_n(x)$ and its first n derivatives agree, at $x = a$, with the values of f and its first n derivatives there. That is, the equations in (5) all hold.

The polynomial in Eq. (8) is called the **nth-degree Taylor polynomial of the function f at the point $x = a$**. Note that $P_n(x)$ is a polynomial in powers of $x - a$ rather than in powers of x. To use $P_n(x)$ effectively for the approximation of $f(x)$ near a, we must be able to compute the value $f(a)$ and the values of its derivatives $f'(a), f''(a)$, and so on, all the way to $f^{(n)}(a)$.

The line $y = P_1(x)$ is simply the line tangent to the curve $y = f(x)$ at the point $(a, f(a))$. Thus $y = f(x)$ and $y = P_1(x)$ have the same slope at this point. Now recall from Section 4.6 that the second derivative measures the way the curve $y = f(x)$ is bending as it passes through $(a, f(a))$. Therefore, let us call $f''(a)$ the "concavity" of $y = f(x)$ at $(a, f(a))$. Then, because $P_2''(a) = f''(a)$, it follows that $y = P_2(x)$ has the same value, the same slope, *and* the same concavity at $(a, f(a))$ as does $y = f(x)$. Moreover, $P_3(x)$ and $f(x)$ will also have the same rate of change of concavity at $(a, f(a))$. Such observations suggest that the larger n is, the more closely the nth-degree Taylor polynomial will approximate $f(x)$ for x near a.

EXAMPLE 3 Find the nth-degree Taylor polynomial of $f(x) = \ln x$ at $a = 1$.

Solution The first few derivatives of $f(x) = \ln x$ are

$$f'(x) = \frac{1}{x}, \quad f''(x) = -\frac{1}{x^2}, \quad f^{(3)}(x) = \frac{2}{x^3}, \quad f^{(4)}(x) = -\frac{3!}{x^4}, \quad f^{(5)}(x) = \frac{4!}{x^5}.$$

The pattern is clear:

$$f^{(k)}(x) = (-1)^{k-1}\frac{(k-1)!}{x^k} \qquad \text{for } k \geqq 1.$$

Hence $f^{(k)}(1) = (-1)^{k-1}(k-1)!$, so Eq. (8) gives

$$P_n(x) = (x - 1) - \frac{1}{2}(x - 1)^2 + \frac{1}{3}(x - 1)^3$$

$$- \frac{1}{4}(x - 1)^4 + \cdots + \frac{(-1)^{n-1}}{n}(x - 1)^n.$$

With $n = 2$ we obtain the quadratic polynomial

$$P_2(x) = (x - 1) - \tfrac{1}{2}(x - 1)^2 = -\tfrac{1}{2}x^2 + 2x - \tfrac{3}{2},$$

the same as in Example 2. With the third-degree Taylor polynomial

$$P_3(x) = (x - 1) - \tfrac{1}{2}(x - 1)^2 + \tfrac{1}{3}(x - 1)^3,$$

we can go a step further in approximating $\ln(1.1)$: The value $P_3(1.1) = 0.095333\ldots$ is correct to four decimal places. See Fig. 11.4.3 for the graphs of $\ln x$, $P_1(x)$, $P_2(x)$, and $P_3(x)$.

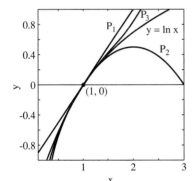

Fig. 11.4.3 The graphs of $f(x) = \ln x$ and its first three Taylor polynomial approximations at $x = 1$ (Example 3)

In the common case $a = 0$, the nth-degree Taylor polynomial in Eq. (8) reduces to

$$P_n(x) = f(0) + f'(0) \cdot x + \frac{f''(0)}{2!}x^2 + \cdots + \frac{f^{(n)}(0)}{n!}x^n. \qquad (9)$$

EXAMPLE 4 Find the nth-degree Taylor polynomial for $f(x) = e^x$ at $a = 0$.

Solution This is the easiest of all Taylor polynomials to compute, because $f^{(k)}(x) = e^x$ for all $k \geqq 0$. Hence $f^{(k)}(0) = 1$ for all $k \geqq 0$, so Eq. (9) yields

$$P_n(x) = 1 + x + \frac{x^2}{2!} + \frac{x^3}{3!} + \cdots + \frac{x^n}{n!}.$$

The first few Taylor polynomials of the natural exponential function at $a = 0$ are, therefore,

$$P_0(x) = 1,$$

$$P_1(x) = 1 + x,$$

$$P_2(x) = 1 + x + \tfrac{1}{2}x^2,$$

$$P_3(x) = 1 + x + \tfrac{1}{2}x^2 + \tfrac{1}{6}x^3,$$

$$P_4(x) = 1 + x + \tfrac{1}{2}x^2 + \tfrac{1}{6}x^3 + \tfrac{1}{24}x^4,$$

$$P_5(x) = 1 + x + \tfrac{1}{2}x^2 + \tfrac{1}{6}x^3 + \tfrac{1}{24}x^4 + \tfrac{1}{120}x^5.$$

The table in Fig. 11.4.4 shows how these polynomials approximate $f(x) = e^x$ for $x = 0.1$ and for $x = 0.5$. At least for these two values of x, the closer x is to $a = 0$, the more rapidly $P_n(x)$ appears to approach $f(x)$ as n increases.

$$x = 0.1$$

n	$P_n(X)$	e^x	$e^x - P_n(x)$
0	1.00000	1.10517	0.10517
1	1.10000	1.10517	0.00517
2	1.10500	1.10517	0.00017
3	1.10517	1.10517	0.00000
4	1.10517	1.10517	0.00000

$$x = 0.5$$

n	$P_n(X)$	e^x	$e^x - P_n(x)$
0	1.00000	1.64872	0.64872
1	1.50000	1.64872	0.14872
2	1.62500	1.64872	0.02372
3	1.64583	1.64872	0.00289
4	1.64844	1.64872	0.00028
5	1.64879	1.64872	0.00002

Fig. 11.4.4 Approximating $y = e^x$ with Taylor polynomials at $a = 0$

The closeness with which $P_n(x)$ approximates $f(x)$ is measured by the difference

$$R_n(x) = f(x) - P_n(x),$$

for which

$$f(x) = P_n(x) + R_n(x). \tag{10}$$

This difference $R_n(x)$ is called the **nth-degree remainder for $f(x)$ at $x = a$**. It is the *error* made if the value $f(x)$ is replaced by the approximation $P_n(x)$.

The theorem that lets us estimate the error, or remainder, $R_n(x)$ is called **Taylor's formula**, after Brook Taylor (1685–1731), a follower of Newton who introduced Taylor polynomials in an article published in 1715. The particular expression for $R_n(x)$ that we give next is called the *Lagrange form* for the remainder because it first appeared in 1797 in a book written by the French mathematician Joseph Louis Lagrange (1736–1813).

Theorem 2 Taylor's Formula

Suppose that the $(n + 1)$th derivative of the function f exists on an interval containing the points a and b. Then

$$f(b) = f(a) + f'(a)(b - a) + \frac{f''(a)}{2!}(b - a)^2$$

$$+ \frac{f^{(3)}(a)}{3!}(b - a)^3 + \cdots + \frac{f^{(n)}(a)}{n!}(b - a)^n$$

$$+ \frac{f^{(n+1)}(z)}{(n + 1)!}(b - a)^{n+1} \tag{11}$$

for some number z between a and b.

REMARK   With $n = 0$, Eq. (11) reduces to the equation

$$f(b) = f(a) + f'(z) \cdot (b - a)$$

in the conclusion of the mean value theorem (Section 4.3). Thus Taylor's formula is a far-reaching generalization of the mean value theorem of differential calculus.

The proof of Taylor's formula is given in Appendix H. If we replace b by x in Eq. (11), we get the **nth-degree Taylor formula with remainder at $x = a$,**

$$f(x) = f(a) + f'(a)(x - a) + \frac{f''(a)}{2!}(x - a)^2 + \frac{f^{(3)}(a)}{3!}(x - a)^3$$

$$+ \cdots + \frac{f^{(n)}(a)}{n!}(x - a)^n + \frac{f^{(n+1)}(z)}{(n + 1)!}(x - a)^{n+1}, \tag{12}$$

where z is some number between a and x. Thus the nth-degree remainder term is

$$R_n(x) = \frac{f^{(n+1)}(z)}{(n + 1)!}(x - a)^{n+1}, \tag{13}$$

which is easy to remember—it's the same as the *last* term of $P_{n+1}(x)$, except that $f^{(n+1)}(a)$ is replaced by $f^{(n+1)}(z)$.

EXAMPLE 3 continued   To estimate the accuracy of the approximation

$$\ln(1.1) \approx 0.095333,$$

we substitute $x = 1$ into the formula

$$f^{(k)}(x) = (-1)^{k-1}\frac{(k-1)!}{x^k}$$

for the kth derivative of $f(x) = \ln x$ and get

$$f^{(k)}(1) = (-1)^{k-1}(k-1)!.$$

Hence the third-degree Taylor formula *with remainder* at $a = 1$ is

$$\ln x = (x - 1) - \frac{1}{2}(x - 1)^2 + \frac{1}{3}(x - 1)^3 - \frac{3!}{4! \, z^4}(x - 1)^4$$

with z between $a = 1$ and x. With $x = 1.1$ this gives

$$\ln(1.1) = 0.095333 \ldots - \frac{(0.1)^4}{4z^4}$$

with $1 < z < 1.1$. The value $z = 1$ gives the largest possible value $(0.1)^4/4 = 0.000025$ of the remainder term. It follows that

$$0.0953083 < \ln(1.1) < 0.0953334,$$

so we can conclude that $\ln(1.1) = 0.0953$ to four-place accuracy.

TAYLOR SERIES

If the function f has derivatives of all orders, then we can write Taylor's formula [Eq. (11)] with any degree n that we please. Ordinarily, the exact value of z in the Taylor remainder term in Eq. (13) is unknown. Nevertheless, we can sometimes use Eq. (13) to show that the remainder approaches zero as $n \to \infty$:

$$\lim_{n \to \infty} R_n(x) = 0 \tag{14}$$

for some particular *fixed* value of x. Then Eq. (10) gives

$$f(x) = \lim_{n \to \infty} [P_n(x) + R_n(x)] = \lim_{n \to \infty} P_n(x) = \lim_{n \to \infty} \sum_{k=0}^{n} \frac{f^{(k)}(a)}{k!}(x - a)^k;$$

that is,

$$f(x) = \sum_{k=0}^{\infty} \frac{f^{(k)}(a)}{k!}(x - a)^k. \tag{15}$$

The infinite series

$$\sum_{n=0}^{\infty} \frac{f^{(n)}(a)}{n!}(x - a)^n = f(a) + f'(a)(x - a) + \frac{f''(a)}{2}(x - a)^2$$

$$+ \cdots + \frac{f^{(n)}(a)}{n!}(x - a)^n + \cdots \tag{16}$$

is called the **Taylor series** of the function f at $x = a$. Its partial sums are the successive Taylor polynomials of f at $x = a$.

We can write the Taylor series of a function f without knowing that it converges. But if the limit in Eq. (14) can be established, then it follows as in Eq. (15) that the Taylor series in Eq. (16) actually converges to $f(x)$. If so, then we can approximate the value of $f(x)$ accurately by calculating the value of a Taylor polynomial of f of sufficiently high degree.

EXAMPLE 5   In Example 4 we noted that if $f(x) = e^x$, then $f^{(k)}(x) = e^x$ for all $k \geq 0$. Hence the Taylor formula

$$f(x) = f(0) + f'(0) \cdot x + \frac{f''(0)}{2!}x^2 + \cdots + \frac{f^{(n)}(0)}{n!}x^n + \frac{f^{(n+1)}(z)}{(n + 1)!}x^{n+1}$$

at $a = 0$ gives

$$e^x = 1 + x + \frac{x^2}{2!} + \frac{x^3}{3!} + \cdots + \frac{x^n}{n!} + \frac{e^z x^{n+1}}{(n+1)!} \tag{17}$$

for some z between 0 and x. Thus the remainder term $R_n(x)$ satisfies the inequalities

$$0 < |R_n(x)| < \frac{|x|^{n+1}}{(n+1)!} \qquad \text{if } x < 0,$$

$$0 < |R_n(x)| < \frac{e^x x^{n+1}}{(n+1)!} \qquad \text{if } x > 0.$$

Therefore, the fact that

$$\lim_{n \to \infty} \frac{x^n}{n!} = 0 \tag{18}$$

for all x (see Problem 49) implies that $\lim_{n \to \infty} R_n(x) = 0$ for all x. This means that the Taylor series for e^x converges to e^x for all x, and we may write

$$e^x = \sum_{n=0}^{\infty} \frac{x^n}{n!} = 1 + x + \frac{x^2}{2!} + \frac{x^3}{3!} + \frac{x^4}{4!} + \cdots. \tag{19}$$

The series in Eq. (19) is the most famous and most important of all Taylor series.

With $x = 1$, Eq. (19) yields a numerical series

$$e = \sum_{n=0}^{\infty} \frac{1}{n!} = 1 + \frac{1}{1!} + \frac{1}{2!} + \frac{1}{3!} + \frac{1}{4!} + \cdots \tag{20}$$

for the number e itself. The 10th and 20th partial sums of this series gives the approximations

$$e \approx 1 + \frac{1}{1!} + \frac{1}{2!} + \cdots + \frac{1}{10!} \approx 2.71828\ 18$$

and

$$e \approx 1 + \frac{1}{1!} + \frac{1}{2!} + \cdots + \frac{1}{20!} \approx 2.71828\ 18284\ 59045\ 235,$$

both of which are accurate to the number of decimal places shown.

EXAMPLE 6   To find the Taylor series at $a = 0$ for $f(x) = \cos x$, we first calculate the derivatives

$$f(x) = \cos x, \qquad\qquad f'(x) = -\sin x,$$

$$f''(x) = -\cos x, \qquad\qquad f^{(3)}(x) = \sin x,$$

$$f^{(4)}(x) = \cos x, \qquad\qquad f^{(5)}(x) = -\sin x,$$

$$\vdots \qquad\qquad\qquad\qquad \vdots$$

$$f^{(2n)}(x) = (-1)^n \cos x, \qquad f^{(2n+1)}(x) = (-1)^{n+1} \sin x.$$

It follows that

$$f^{(2n)}(0) = (-1)^n \quad \text{but} \quad f^{(2n+1)}(0) = 0,$$

so the Taylor polynomials and Taylor series of $f(x) = \cos x$ include only terms of *even* degree. The Taylor formula of degree $2n$ for $\cos x$ at $a = 0$ is

$$\cos x = 1 - \frac{x^2}{2!} + \frac{x^4}{4!} - \cdots + (-1)^n \frac{x^{2n}}{(2n)!} + (-1)^{n+1} \frac{\cos z}{(2n+2)!} x^{2n+2},$$

where z is between 0 and x. Because $|\cos z| \leq 1$ for all z, it follows from Eq. (18) that the remainder term approaches zero as $n \to \infty$ *for all* x. Hence the desired Taylor series of $f(x) = \cos x$ at $a = 0$ is

$$\cos x = \sum_{n=0}^{\infty} \frac{(-1)^n x^{2n}}{(2n)!} = 1 - \frac{x^2}{2!} + \frac{x^4}{4!} - \frac{x^6}{6!} + \cdots. \qquad (21)$$

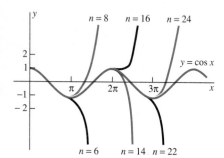

Fig. 11.4.5 Approximating $\cos x$ with nth-degree Taylor polynomials

In Problem 43 we ask you to show similarly that the Taylor series at $a = 0$ of $f(x) = \sin x$ is

$$\sin x = \sum_{n=0}^{\infty} \frac{(-1)^n x^{2n+1}}{(2n+1)!} = x - \frac{x^3}{3!} + \frac{x^5}{5!} - \frac{x^7}{7!} + \cdots. \qquad (22)$$

Figures 11.4.5 and 11.4.6 illustrate the increasingly better approximations to $\cos x$ and $\sin x$ that we get by using more and more terms of the series in Eqs. (21) and (22).

The case $a = 0$ of Taylor's series is called the **Maclaurin series** of the function $f(x)$,

$$\sum_{n=0}^{\infty} \frac{f^{(n)}(0)}{n!} x^n = f(0) + f'(0) \cdot x + \frac{f''(0)}{2!} x^2 + \frac{f^{(3)}(0)}{3!} x^3 + \cdots. \qquad (23)$$

Colin Maclaurin (1698–1746) was a Scottish mathematician who used this series as a basic tool in a calculus book he published in 1742. The three Maclaurin series

$$e^x = \sum_{n=0}^{\infty} \frac{x^n}{n!} = 1 + x + \frac{x^2}{2!} + \frac{x^3}{3!} + \frac{x^4}{4!} + \cdots, \qquad (19)$$

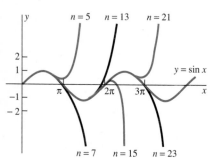

Fig. 11.4.6 Approximating $\sin x$ with nth-degree Taylor polynomials

$$\cos x = \sum_{n=0}^{\infty} \frac{(-1)^n x^{2n}}{(2n)!} = 1 - \frac{x^2}{2!} + \frac{x^4}{4!} - \frac{x^6}{6!} + \cdots, \quad \text{and} \qquad (21)$$

$$\sin x = \sum_{n=0}^{\infty} \frac{(-1)^n x^{2n+1}}{(2n+1)!} = x - \frac{x^3}{3!} + \frac{x^5}{5!} - \frac{x^7}{7!} + \cdots \qquad (22)$$

(which actually were discovered by Newton) bear careful examination and comparison. Observe that

❑ The terms in the *even* cosine series are the *even*-degree terms in the exponential series but with alternating signs.

❑ The terms in the *odd* sine series are the *odd*-degree terms in the exponential series but with alternating signs.

These series are *identities* that hold for all values of x. Consequently, new series can be derived by substitution, as in Examples 7 and 8.

EXAMPLE 7 The substitution $x = -t^2$ into Eq. (19) yields

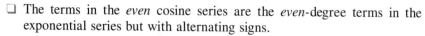

$$e^{-t^2} = 1 - t^2 + \frac{t^4}{2!} - \frac{t^6}{3!} + \cdots + (-1)^n \frac{t^{2n}}{n!} + \cdots.$$

EXAMPLE 8 The substitution $x = 2t$ into Eq. (22) gives

$$\sin 2t = 2t - \tfrac{4}{3}t^3 + \tfrac{4}{15}t^5 - \tfrac{8}{315}t^7 + \cdots.$$

*THE NUMBER π

In Section 5.3 we described how Archimedes used polygons inscribed in and circumscribed about the unit circle to show that $3\tfrac{10}{71} < \pi < 3\tfrac{1}{7}$. With the aid of electronic computers, π has been calculated to well over a *billion* decimal places. We describe now some of the methods that have been used for such computations.[†]
  We begin with the elementary algebraic identity

$$\frac{1}{1+x} = 1 - x + x^2 - x^3 + \cdots + (-1)^{k-1}x^{k-1} + \frac{(-1)^k x^k}{1+x}, \quad (24)$$

which can be verified by multiplying both sides by $1 + x$. We substitute t^2 for x and $n + 1$ for k and thus find that

$$\frac{1}{1+t^2} = 1 - t^2 + t^4 - t^6 + \cdots + (-1)^n t^{2n} + \frac{(-1)^{n+1} t^{2n+2}}{1+t^2}.$$

Because $D_t \tan^{-1} t = 1/(1 + t^2)$, integration of both sides of this last equation from $t = 0$ to $t = x$ gives

$$\tan^{-1}x = x - \frac{x^3}{3} + \frac{x^5}{5} - \frac{x^7}{7} + \cdots + (-1)^n \frac{x^{2n+1}}{2n+1} + R_{2n+1}, \quad (25)$$

where

$$|R_{2n+1}| = \left| \int_0^x \frac{t^{2n+2}}{1+t^2}\,dt \right| \leqq \left| \int_0^x t^{2n+2}\,dt \right| = \frac{|x|^{2n+3}}{2n+3}. \quad (26)$$

This estimate of the error makes it clear that

$$\lim_{n \to \infty} R_n = 0$$

if $|x| \leqq 1$. Hence we obtain the Taylor series for the inverse tangent function:

$$\tan^{-1}x = \sum_{n=0}^{\infty} (-1)^n \frac{x^{2n+1}}{2n+1} = x - \frac{x^3}{3} + \frac{x^5}{5} - \frac{x^7}{7} + \cdots, \quad (27)$$

valid for $-1 \leqq x \leqq 1$.
  If we substitute $x = 1$ into Eq. (27), we obtain *Leibniz's series*

$$\frac{\pi}{4} = 1 - \frac{1}{3} + \frac{1}{5} - \frac{1}{7} + \cdots.$$

Although this is a beautiful series, it is not an effective way to compute π. But the error estimate in Eq. (26) shows that we can use Eq. (25) to calculate $\tan^{-1}x$ if $|x|$ is small. For example, if $x = \tfrac{1}{5}$, then the fact that

$$\frac{1}{9 \cdot 5^9} \approx 0.00000\,0057 < 0.0000001$$

[†] For a chronicle of humanity's perennial fascination with the number π, see Howard Eves, *An Introduction to the History of Mathematics*, (Boston: Allyn and Bacon, 4th ed., 1976), pp. 96–102.

implies that the approximation

$$\tan^{-1}(\tfrac{1}{5}) \approx \tfrac{1}{5} - \tfrac{1}{3}(\tfrac{1}{5})^3 + \tfrac{1}{5}(\tfrac{1}{5})^5 - \tfrac{1}{7}(\tfrac{1}{5})^7$$

is accurate to six decimal places.

Accurate inverse tangent calculations lead to accurate computations of the number π. For example, we can use the addition formula for the tangent function to show (Problem 46) that

$$\frac{\pi}{4} = 4 \tan^{-1}\left(\frac{1}{5}\right) - \tan^{-1}\left(\frac{1}{239}\right). \tag{28}$$

In 1706, John Machin (?–1751) used Eq. (28) to calculate the first 100 decimal places of π. (In Problem 48 we ask you to use it to show that $\pi = 3.14159$ to five decimal places.) In 1844 the lightning-fast mental calculator Zacharias Dase (1824–1861) of Germany computed the first 200 decimal places of π, using the related formula

$$\frac{\pi}{4} = \tan^{-1}\left(\frac{1}{2}\right) + \tan^{-1}\left(\frac{1}{5}\right) + \tan^{-1}\left(\frac{1}{8}\right). \tag{29}$$

You might enjoy verifying this formula (see Problem 47). A recent computation of 1 million decimal places of π used the formula

$$\frac{\pi}{4} = 12 \tan^{-1}\left(\frac{1}{18}\right) + 8 \tan^{-1}\left(\frac{1}{57}\right) - 5 \tan^{-1}\left(\frac{1}{239}\right).$$

For derivations of this formula and others like it, with further discussion of the computations of the number π, see the article "An algorithm for the calculation of π" by George Miel in the *American Mathematical Monthly* **86** (1979), pp. 694–697. Although no practical application is likely to require more than ten or twelve decimal places of π, these computations provide dramatic evidence of the power of Taylor's formula. Moreover, the number π continues to serve as a challenge both to human ingenuity and to the accuracy and efficiency of modern electronic computers. For an account of how investigations of the Indian mathematical genius Srinivasa Ramanujan (1887–1920) have led recently to the computation of over a billion decimal places of π, see the article "Ramanujan and pi," Jonathan M. Borwein and Peter B. Borwein, *Scientific American* (Feb. 1988), pp. 112–117.

11.4 Problems

In Problems 1 through 10, find Taylor's formula for the given function f at a = 0. Find both the Taylor polynomial $P_n(x)$ of the indicated degree n and the remainder term $R_n(x)$.

1. $f(x) = e^{-x}, \quad n = 5$

2. $f(x) = \sin x, \quad n = 4$

3. $f(x) = \cos x, \quad n = 4$

4. $f(x) = \dfrac{1}{1 - x}, \quad n = 4$

5. $f(x) = \sqrt{1 + x}, \quad n = 3$

6. $f(x) = \ln(1 + x), \quad n = 4$

7. $f(x) = \tan x, \quad n = 3$

8. $f(x) = \arctan x, \quad n = 2$

9. $f(x) = \sin^{-1}x, \quad n = 2$

10. $f(x) = x^3 - 3x^2 + 5x - 7, \quad n = 4$

In Problems 11 through 20, find the Taylor polynomial with remainder by using the given values of a and n.

11. $f(x) = e^x; \quad a = 1, n = 4$

12. $f(x) = \cos x; \quad a = \pi/4, n = 3$

13. $f(x) = \sin x$; $a = \pi/6$, $n = 3$

14. $f(x) = \sqrt{x}$; $a = 100$, $n = 3$

15. $f(x) = \dfrac{1}{(x-4)^2}$; $a = 5$, $n = 5$

16. $f(x) = \tan x$; $a = \pi/4$, $n = 4$

17. $f(x) = \cos x$; $a = \pi$, $n = 4$

18. $f(x) = \sin x$; $a = \pi/2$, $n = 4$

19. $f(x) = x^{3/2}$; $a = 1$, $n = 4$

20. $f(x) = \dfrac{1}{\sqrt{1-x}}$; $a = 0$, $n = 4$

In Problems 21 through 28, find the Maclaurin series of the given function f by substitution in one of the known series in Eqs. (19), (21), and (22).

21. $f(x) = e^{-x}$ **22.** $f(x) = e^{2x}$

23. $f(x) = e^{-3x}$ **24.** $f(x) = \exp(x^3)$

25. $f(x) = \cos 2x$ **26.** $f(x) = \sin \dfrac{x}{2}$

27. $f(x) = \sin x^2$ **28.** $f(x) = \cos \sqrt{x}$

In Problems 29 through 42, find the Taylor series [Eq. (16)] of the given function at the indicated point a.

29. $f(x) = \ln(1 + x)$, $a = 0$

30. $f(x) = \dfrac{1}{1-x}$, $a = 0$

31. $f(x) = e^{-x}$, $a = 0$

32. $f(x) = \cosh x$, $a = 0$

33. $f(x) = \ln x$, $a = 1$

34. $f(x) = \sin x$, $a = \pi/2$

35. $f(x) = \cos x$, $a = \pi/4$

36. $f(x) = e^{2x}$, $a = 0$

37. $f(x) = \sinh x$, $a = 0$

38. $f(x) = \dfrac{1}{(1-x)^2}$, $a = 0$

39. $f(x) = \dfrac{1}{x}$, $a = 1$

40. $f(x) = \cos x$, $a = \pi/2$

41. $f(x) = \sin x$, $a = \pi/4$

42. $f(x) = \sqrt{1+x}$, $a = 0$

43. Derive, as in Example 5, the Taylor series in Eq. (22) of $f(x) = \sin x$ at $a = 0$.

44. Granted that it is valid to differentiate the sine and cosine Taylor series in a term-by-term manner, use these series to verify that $D_x \cos x = -\sin x$ and $D_x \sin x = \cos x$.

45. Use the Taylor series for the cosine and sine functions to verify that $\cos(-x) = \cos x$ and $\sin(-x) = -\sin x$ for all x.

46. Beginning with $\alpha = \tan^{-1}(\frac{1}{5})$, use the addition formula

$$\tan(A + B) = \frac{\tan A + \tan B}{1 - \tan A \tan B}$$

to show in turn that: (a) $\tan 2\alpha = \frac{5}{12}$; (b) $\tan 4\alpha = \frac{120}{119}$; (c) $\tan(\pi/4 - 4\alpha) = -\frac{1}{239}$. Finally, show that (c) implies Eq. (28).

47. Apply the addition formula for the tangent function to verify Eq. (29).

48. Use Eqs. (25), (26), and (28) to show that $\pi = 3.14159$ to five decimal places. [*Suggestion:* Compute $\arctan(\frac{1}{5})$ and $\arctan(\frac{1}{239})$, both with error less than 5×10^{-8}. You can do this by applying Eq. (25) with $n = 4$ and with $n = 1$. Carry your computations to seven decimal places, and keep track of errors.]

49. Prove that

$$\lim_{n \to \infty} \frac{x^n}{n!} = 0$$

if x is a real number. [*Suggestion:* Choose an integer k such that $k > |2x|$, and let $L = |x|^k/k!$. Then show that

$$\frac{|x|^n}{n!} < \frac{L}{2^{n-k}}$$

if $n > k$.]

50. (a) Substitute $-x$ for x in Eq. (24). (b) Now suppose that t is a positive real number but is otherwise arbitrary. Integrate both sides in the result of part (a) from $x = 0$ to $x = t$ to show that

$$\ln(1 + t) = t - \frac{t^2}{2} + \frac{t^3}{3} - \cdots + (-1)^{k-1}\frac{t^k}{k} + R_k,$$

where

$$|R_k| \le \frac{t^{k+1}}{k+1}.$$

(c) Conclude that

$$\ln(1 + t) = t - \frac{t^2}{2} + \frac{t^3}{3} - \cdots = \sum_{k=1}^{\infty} (-1)^{k-1}\frac{t^k}{k}$$

if $0 \le t \le 1$.

51. Use the formula $\ln 2 = \ln(\frac{5}{4}) + 2\ln(\frac{6}{5}) + \ln(\frac{10}{9})$ to calculate $\ln 2$ accurate to three decimal places. See part (b) of Problem 50 for the Taylor formula for $\ln(1 + x)$.

11.4 Project

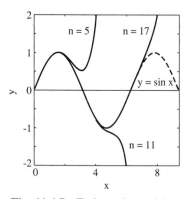

Fig. 11.4.7 Taylor polynomial approximations to sin x

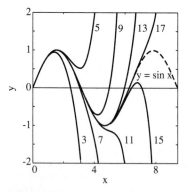

Fig. 11.4.8 Taylor polynomial approximations to sin x of degrees $n = 3, 5, 7, \ldots, 17$

This project involves the use of a graphics calculator or computer graphing program to plot Taylor polynomials. By plotting several successive Taylor polynomials on the same set of coordinate axes, we get a visual sense of the way in which a function is approximated by partial sums of its Taylor series.

Figure 11.4.7 shows the graph of sin x and its Taylor polynomials of degrees $n = 5, 11,$ and 17. Using a graphics calculator with a Y = menu, you can generate this figure by plotting the functions defined by

```
Y1=X − X^3/3! + X^5/5!
Y2=Y1 − X^7/7! + X^9/9! − X^11/11!
Y3=Y2 + X^13/13! − X^15/15! + X^17/17!
Y4=sin X
```

With a calculator or computer system that has a [sum] function (Fig. 11.3.5), we can define high-degree Taylor polynomials by commands such as the *X(PLORE)* command

```
SUM( (x^n)/n!, n=0 to k )
```

which defines the kth partial sum of the exponential series in Eq. (19) of this section. With a calculator or computer that can plot a number of graphs simultaneously, we can generate a picture like Fig. 11.4.8, which shows the graphs of the Taylor polynomials of sin x of degrees $n = 3, 5, 7, \ldots, 17$.

For every function given in Problems 1 through 7, generate several pictures, each of which shows the function's graph and several Taylor polynomial approximations.

1. $f(x) = e^{-x}$ **2.** $f(x) = \sin x$

3. $f(x) = \cos x$ **4.** $f(x) = \dfrac{1}{1 + x}$

5. $f(x) = \ln(1 + x)$ **6.** $f(x) = \dfrac{1}{1 + x^2}$

7. $f(x) = \tan^{-1} x$

11.5
The Integral Test

A Taylor series (as in Section 11.4) is a special type of infinite series with *variable* terms. We saw that Taylor's formula can sometimes be used—as in the case of the exponential, sine, and cosine series—to establish the convergence of such a series.

But given an infinite series $\Sigma\, a_n$ with *constant* terms, it is the exception rather than the rule when a simple formula for the nth partial sum of that series can be found and used directly to determine whether the series converges or diverges. There are, however, several *convergence tests* that use the *terms* of an infinite series rather than its partial sums. Such a test, when successful, will tell us whether or not the series converges. Once we know that the series $\Sigma\, a_n$ does converge, it is then a separate matter to *find* its sum S. It may be necessary to approximate S by adding sufficiently many terms; in this case we shall need to know how many terms are required for the desired accuracy.

Here and in Section 11.6, we concentrate our attention on **positive-term series**—that is, series with terms that are all positive. If $a_n > 0$ for all n, then

$$S_1 < S_2 < S_3 < \cdots < S_n < \cdots,$$

so the sequence $\{S_n\}$ of partial sums of the series is increasing. Hence there are just two possibilities. If the sequence $\{S_n\}$ is *bounded*—there exists a number M such that $S_n \leq M$ for all n—then the bounded monotonic sequence property (Section 11.2) implies that $S = \lim_{n \to \infty} S_n$ exists, so the series Σa_n *converges*. Otherwise, it diverges to infinity (by Problem 38 in Section 11.2).

A similar alternative holds for improper integrals. Suppose that the function f is continuous and positive-valued for $x \geq 1$. Then it follows (from Problem 35) that the improper integral

$$\int_1^\infty f(x)\ dx = \lim_{b \to \infty} \int_1^b f(x)\ dx \tag{1}$$

either converges (the limit is a real number) or diverges to infinity (the limit is $+\infty$). This analogy between positive-term series and improper integrals of positive functions is the key to the **integral test**. We compare the behavior of the series Σa_n with that of the improper integral in Eq. (1), where f is an appropriately chosen function. [Among other things, we require that $f(n) = a_n$ for all n.]

Theorem 1 *The Integral Test*

Suppose that Σa_n is a positive-term series and that f is a positive-valued, decreasing, continuous function for $x \geq 1$. If $f(n) = a_n$ for all integers $n \geq 1$, then the series and the improper integral

$$\sum_{n=1}^\infty a_n \quad \text{and} \quad \int_1^\infty f(x)\ dx$$

either both converge or both diverge.

Proof Because f is a decreasing function, the rectangular polygon with area

$$S_n = a_1 + a_2 + a_3 + \cdots + a_n$$

shown in Fig. 11.5.1 contains the region under $y = f(x)$ from $x = 1$ to $x = n + 1$. Hence

$$\int_1^{n+1} f(x)\ dx \leq S_n. \tag{2}$$

Similarly, the rectangular polygon with area

$$S_n - a_1 = a_2 + a_3 + a_4 + \cdots + a_n$$

shown in Fig. 11.5.2 is contained in the region under $y = f(x)$ from $x = 1$ to $x = n$. Hence

$$S_n - a_1 \leq \int_1^n f(x)\ dx. \tag{3}$$

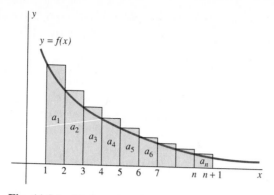

Fig. 11.5.1 Underestimating the partial sums with an integral

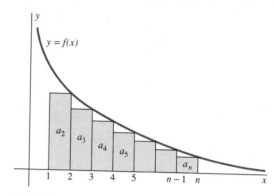

Fig. 11.5.2 Overestimating the partial sums with an integral

Suppose first that the improper integral $\int_1^\infty f(x)\,dx$ diverges (necessarily to $+\infty$). Then

$$\lim_{n\to\infty}\int_1^{n+1} f(x)\,dx = +\infty,$$

so it follows from Eq. (2) that $\lim_{n\to\infty} S_n = +\infty$ as well, and hence the infinite series $\Sigma\,a_n$ likewise diverges.

Now suppose instead that the improper integral $\int_1^\infty f(x)\,dx$ converges and has the (finite) value I. Then (3) implies that

$$S_n \leqq a_1 + \int_1^n f(x)\,dx \leqq a_1 + I,$$

so the increasing sequence $\{S_n\}$ is bounded. Thus the infinite series

$$\sum_{n=1}^{\infty} a_n = \lim_{n\to\infty} S_n$$

converges as well. Hence we have shown that the infinite series and the improper integral either both converge or both diverge. ❑

EXAMPLE 1 We used a version of the integral test to prove in Section 11.3 that the harmonic series

$$\sum_{n=1}^{\infty} \frac{1}{n} = 1 + \frac{1}{2} + \frac{1}{3} + \frac{1}{4} + \cdots$$

diverges. Using the test as stated in Theorem 1 is a little simpler: We note that $f(x) = 1/x$ is positive, continuous, and decreasing for $x \geqq 1$ and that $f(n) = 1/n$ for each positive integer n. Now

$$\int_1^\infty \frac{1}{x}\,dx = \lim_{b\to\infty}\int_1^b \frac{1}{x}\,dx = \lim_{b\to\infty}\left[\ln x\right]_1^b = \lim_{b\to\infty}(\ln b - \ln 1) = +\infty.$$

Thus the improper integral diverges and, therefore, so does the harmonic series.

The harmonic series is the case $p = 1$ of the **p-series**

$$\sum_{n=1}^{\infty} \frac{1}{n^p} = 1 + \frac{1}{2^p} + \frac{1}{3^p} + \cdots + \frac{1}{n^p} + \cdots. \tag{4}$$

Whether the p-series converges or diverges depends on the value of p.

614

EXAMPLE 2 Show that the p-series converges if $p > 1$ but diverges if $0 < p \le 1$.

Solution The case $p = 1$ has already been settled in Example 1. If $p > 0$ but $p \ne 1$, then the function $f(x) = 1/x^p$ satisfies the conditions of the integral test, and

$$\int_1^\infty \frac{1}{x^p}\,dx = \lim_{b \to \infty} \int_1^b \frac{1}{x^p}\,dx = \lim_{b \to \infty}\left[-\frac{1}{(p-1)x^{p-1}} \right]_1^b$$

$$= \lim_{b \to \infty} \frac{1}{p-1}\left(1 - \frac{1}{b^{p-1}} \right).$$

If $p > 1$, then

$$\int_1^\infty \frac{1}{x^p}\,dx = \frac{1}{p-1} < \infty,$$

so the integral and the series both converge. But if $0 < p < 1$, then

$$\int_1^\infty \frac{1}{x^p}\,dx = \lim_{b \to \infty} \frac{1}{1-p}(b^{1-p} - 1) = \infty,$$

aı ᴄ in this case the integral and the series both diverge.

As specific examples, the series

$$\sum_{n=1}^\infty \frac{1}{n^2} = 1 + \frac{1}{2^2} + \frac{1}{3^2} + \cdots + \frac{1}{n^2} + \cdots$$

converges ($p = 2$), whereas the series

$$\sum_{n=1}^\infty \frac{1}{\sqrt{n}} = 1 + \frac{1}{\sqrt{2}} + \frac{1}{\sqrt{3}} + \cdots + \frac{1}{\sqrt{n}} + \cdots$$

diverges ($p = \frac{1}{2}$).

Now suppose that the positive-term series $\Sigma\, a_n$ converges by the integral test and that we wish to approximate its sum by adding sufficiently many of its initial terms. The difference between the sum S and the nth partial sum S_n is the **remainder**

$$R_n = S - S_n = a_{n+1} + a_{n+2} + a_{n+3} + \cdots. \tag{5}$$

This remainder is the error made when the sum S is estimated by using in its place the partial sum S_n.

Theorem 2 *The Integral Test Remainder Estimate*
Suppose that the infinite series and improper integral

$$\sum_{n=1}^\infty a_n \quad \text{and} \quad \int_1^\infty f(x)\,dx$$

satisfy the hypotheses of the integral test, and suppose in addition that both converge. Then

$$\int_{n+1}^\infty f(x)\,dx \le R_n \le \int_n^\infty f(x)\,dx, \tag{6}$$

where R_n is the remainder given in Eq. (5).

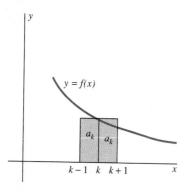

Fig. 11.5.3 Establishing the integral test remainder estimate

Proof We see from Figure 11.5.3 that

$$\int_k^{k+1} f(x) \, dx \le a_k \le \int_{k-1}^k f(x) \, dx$$

for $k = n + 1, n + 2, \ldots$. We add these inequalities for all such values of k, and the result is the inequality in (6), because

$$\sum_{k=n+1}^{\infty} \int_k^{k+1} f(x) \, dx = \int_{n+1}^{\infty} f(x) \, dx$$

and

$$\sum_{k=n+1}^{\infty} \int_{k-1}^k f(x) \, dx = \int_n^{\infty} f(x) \, dx. \qquad \square$$

If we substitute $R_n = S - S_n$, then it follows from (6) that the sum S of the series satisfies the inequality

$$S_n + \int_{n+1}^{\infty} f(x) \, dx \le S \le S_n + \int_n^{\infty} f(x) \, dx. \qquad (7)$$

If the nth partial sum is known and the difference

$$\int_n^{n+1} f(x) \, dx$$

between the two integrals is small, then (7) provides an accurate estimate of the sum S.

EXAMPLE 3 We will see in Section 11.8 that the exact sum of the p-series with $p = 2$ is $\pi^2/6$, thus giving the beautiful formula

$$\frac{\pi^2}{6} = 1 + \frac{1}{2^2} + \frac{1}{3^2} + \frac{1}{4^2} + \cdots. \qquad (8)$$

Approximate the number π by applying to the series in Eq. (8) the integral test remainder estimate with $n = 50$.

Solution The sum of the first 50 terms in Eq. (8) is, to the accuracy shown,

$$\sum_{n=1}^{50} \frac{1}{n^2} = 1.62513\,273.$$

You can add these 50 terms one by one on a pocket calculator in a few minutes, but this is precisely the sort of thing for which a computer or programmable calculator is effective.

Because

$$\int_a^{\infty} \frac{1}{x^2} \, dx = \lim_{b \to \infty} \left[-\frac{1}{x} \right]_a^b = \frac{1}{a},$$

the integral test remainder estimate with $f(x) = 1/x^2$ gives $\frac{1}{51} \le R_{50} \le \frac{1}{50}$, so (7) yields

$$\frac{1}{51} + 1.6251327 \le \frac{\pi^2}{6} \le \frac{1}{50} + 1.62513\,28.$$

We multiply by 6, extract the square root, and round to four-place accuracy. The result is that

$$3.1414 < \pi < 3.1418.$$

With $n = 200$, a programmable calculator or computer gives

$$\sum_{n=1}^{200} \frac{1}{n^2} = 1.63994\,655,$$

so

$$\frac{1}{201} + 1.63994\,65 \leq \frac{\pi^2}{6} \leq \frac{1}{200} + 1.63994\,66.$$

This leads to the inequality $3.14158 < \pi < 3.14161$, and it follows that $\pi = 3.1416$ rounded to four decimal places.

EXAMPLE 4 Show that the series

$$\sum_{n=2}^{\infty} \frac{1}{n(\ln n)^2}$$

converges, and find how many terms you would need to add to find its sum accurate to within 0.01.

Solution We begin the sum at $n = 2$ because $\ln 1 = 0$. Let $f(x) = x(\ln x)^2$. Then

$$\int_a^\infty \frac{1}{x(\ln x)^2}\,dx = \lim_{b \to \infty}\left[-\frac{1}{\ln x}\right]_a^b = \frac{1}{\ln a}.$$

With $a = 2$, the series converges by the integral test. With $a = n$, the right-hand inequality in (6) gives $R_n \leq 1/(\ln n)$, so we need

$$\frac{1}{\ln n} \leq 0.01, \qquad \ln n \geq 100, \qquad n \geq e^{100} \approx 2.7 \times 10^{43}.$$

This is a far larger number of terms than any conceivable computer could add within the expected lifetime of the universe. But accuracy to within 0.05 would require only that $n \geq 4.85 \times 10^8$, fewer than half a billion terms—well within the range of a modern computer.

11.5 Problems

In Problems 1 through 28, use the integral test to test the given series for convergence.

1. $\displaystyle\sum_{n=1}^{\infty} \frac{n}{n^2 + 1}$

2. $\displaystyle\sum_{n=1}^{\infty} \frac{n}{e^{n^2}}$

3. $\displaystyle\sum_{n=1}^{\infty} \frac{1}{\sqrt{n + 1}}$

4. $\displaystyle\sum_{n=1}^{\infty} \frac{1}{(n + 1)^{4/3}}$

5. $\displaystyle\sum_{n=1}^{\infty} \frac{1}{n^2 + 1}$

6. $\displaystyle\sum_{n=1}^{\infty} \frac{1}{n(n + 1)}$

7. $\displaystyle\sum_{n=2}^{\infty} \frac{1}{n \ln n}$

8. $\displaystyle\sum_{n=1}^{\infty} \frac{\ln n}{n}$

9. $\displaystyle\sum_{n=1}^{\infty} \frac{1}{2^n}$

10. $\displaystyle\sum_{n=1}^{\infty} \frac{n}{e^n}$

11. $\displaystyle\sum_{n=1}^{\infty} \frac{n^2}{e^n}$

12. $\displaystyle\sum_{n=1}^{\infty} \frac{1}{17n - 13}$

13. $\displaystyle\sum_{n=1}^{\infty} \frac{\ln n}{n^2}$

14. $\displaystyle\sum_{n=1}^{\infty} \frac{n + 1}{n^2}$

15. $\sum_{n=1}^{\infty} \dfrac{n}{n^4 + 1}$

16. $\sum_{n=1}^{\infty} \dfrac{1}{n^3 + n}$

17. $\sum_{n=1}^{\infty} \dfrac{2n + 5}{n^2 + 5n + 7}$

18. $\sum_{n=1}^{\infty} \ln\!\left(\dfrac{n + 1}{n}\right)$

19. $\sum_{n=1}^{\infty} \ln\!\left(1 + \dfrac{1}{n^2}\right)$

20. $\sum_{n=1}^{\infty} \dfrac{2^{1/n}}{n^2}$

21. $\sum_{n=1}^{\infty} \dfrac{n}{4n^2 + 5}$

22. $\sum_{n=1}^{\infty} \dfrac{n}{(4n^2 + 5)^{3/2}}$

23. $\sum_{n=2}^{\infty} \dfrac{1}{n\sqrt{\ln n}}$

24. $\sum_{n=2}^{\infty} \dfrac{1}{n(\ln n)^3}$

25. $\sum_{n=1}^{\infty} \dfrac{1}{4n^2 + 9}$

26. $\sum_{n=1}^{\infty} \dfrac{n + 1}{n + 100}$

27. $\sum_{n=1}^{\infty} \dfrac{n}{n^4 + 2n^2 + 1}$

28. $\sum_{n=1}^{\infty} \dfrac{1}{(n + 1)^3}$

In Problems 29 through 31, tell why the integral test does not apply to the given infinite series.

29. $\sum_{n=1}^{\infty} \dfrac{(-1)^n}{n}$

30. $\sum_{n=1}^{\infty} e^{-n} \sin n$

31. $\sum_{n=1}^{\infty} \dfrac{2 + \sin n}{n^2}$

In Problems 32 through 34, find the least positive integer n such that the remainder R_n in Theorem 2 is less than E.

32. $\sum_{n=1}^{\infty} \dfrac{1}{n^2}$; $\quad E = 0.0001$

33. $\sum_{n=1}^{\infty} \dfrac{1}{n^2}$; $\quad E = 0.00005$

34. $\sum_{n=1}^{\infty} \dfrac{1}{n^6}$; $\quad E = 2 \times 10^{-11}$

35. Suppose that the function f is continuous and positive-valued for $x \geq 1$. Let $b_n = \int_1^n f(x)\, dx$ for $n = 1, 2, 3,$ .... (a) Suppose that the increasing sequence $\{b_n\}$ is bounded, so $B = \lim_{n \to \infty} b_n$ exists. Prove that

$$\int_1^{\infty} f(x)\, dx = B.$$

(b) Prove that if the sequence $\{b_n\}$ is not bounded, then

$$\int_1^{\infty} f(x)\, dx = +\infty.$$

36. Show that the series $\sum_{n=2}^{\infty} \dfrac{1}{n(\ln n)^p}$ converges if $p > 1$ and diverges if $p \leq 1$.

37. Use the integral test remainder estimate to find

$$\sum_{n=1}^{\infty} \dfrac{1}{n^5}$$

with three digits to the right of the decimal correct or correctly rounded.

38. Using the integral test remainder estimate, how many terms are needed to approximate

$$\sum_{n=1}^{\infty} \dfrac{1}{n^{3/2}}$$

with two digits to the right of the decimal correct?

39. Deduce from the inequalities in (2) and (3) with the function $f(x) = 1/x$ that

$$\ln n \leq 1 + \dfrac{1}{2} + \dfrac{1}{3} + \cdots + \dfrac{1}{n} \leq 1 + \ln n$$

for $n = 1, 2, 3,$ .... If a computer adds 1 million terms of the harmonic series per second, how long will it take for the partial sum to reach 50?

40. (a) Let

$$c_n = 1 + \dfrac{1}{2} + \dfrac{1}{3} + \cdots + \dfrac{1}{n} - \ln n$$

for $n = 1, 2, 3,$ .... Deduce from Problem 39 that $0 \leq c_n \leq 1$ for all n. (b) Note that

$$\int_n^{n+1} \dfrac{1}{x}\, dx \geq \dfrac{1}{n + 1}.$$

Conclude that the sequence $\{c_n\}$ is decreasing. Therefore, the sequence $\{c_n\}$ converges. The number

$$\gamma = \lim_{n \to \infty} \left(1 + \dfrac{1}{2} + \dfrac{1}{3} + \cdots + \dfrac{1}{n} - \ln n\right) \approx 0.57722$$

is known as **Euler's constant**.

41. It is known that

$$\sum_{n=1}^{\infty} \dfrac{1}{n^4} = \dfrac{\pi^4}{90}.$$

Use the integral test remainder estimate and the first 10 terms of this series to show that $\pi = 3.1416$ rounded to four decimal places.

11.5 Project

You can readily use the programs listed in Fig. 11.3.4 to calculate partial sums of the *p*-series

$$\zeta(p) = \sum_{n=1}^{\infty} \dfrac{1}{n^p} = 1 + \dfrac{1}{2^p} + \dfrac{1}{3^p} + \cdots, \tag{9}$$

where the notation $\zeta(p)$—read "zeta of p."—is a standard abbreviation for this p-series. Given the value of p to be used, you need only define `Y1=1/N^P` before you execute the TI calculator program, or define `A(N)=1/N^P` in the first line of the BASIC program.

Alternatively, with an appropriate calculator or computer system you can use a SUM function such as the *Maple* command

```
sum( 1/n^2, n=1..1000 )
```

to calculate the sum of the first thousand terms of $\zeta(2) = \Sigma\, 1/n^2$.

In Problems 1 through 5, use the listed known value of $\zeta(p)$ to see how accurately you can approximate the value

$$\pi = 3.14159\,26535\,89793\ldots.$$

For instance, in Problem 2 you might calculate the kth partial sum s_k of $\Sigma 1/n^4$ for $k = 10, 20, 40, 80, \ldots$ and observe how accurately $\sqrt[4]{90s_k}$ approximates π.

1. $\zeta(2) = \dfrac{\pi^2}{6}$ **2.** $\zeta(4) = \dfrac{\pi^4}{90}$

3. $\zeta(6) = \dfrac{\pi^6}{945}$ **4.** $\zeta(8) = \dfrac{\pi^8}{9450}$

5. $\zeta(10) = \dfrac{\pi^{10}}{93555}$

6. Calculate partial sums of the harmonic series $\Sigma 1/n$ to approximate Euler's constant

$$\gamma = \lim_{n\to\infty} \left(1 + \frac{1}{2} + \frac{1}{3} + \cdots + \frac{1}{n} - \ln n\right) \approx 0.57722$$

(see Problem 40 of this section). Unless you use a very powerful microcomputer, you will probably have to be content with accuracy to two or three decimal places only.

11.6
Comparison Tests for Positive-Term Series

With the integral test we attempt to determine whether or not an infinite series converges by comparing it with an improper integral. The methods of this section involve comparing the terms of the *positive-term* series Σa_n with those of another positive-term series Σb_n whose convergence or divergence is known. We have already developed two families of *reference series* for the role of the known series Σb_n; these are the geometric series of Section 11.3 and the p-series of Section 11.5. They are well adapted for our new purposes because their convergence or divergence is quite easy to determine. Recall that the geometric series Σr^n converges if $|r| < 1$ and diverges if $|r| \geq 1$, and the p-series $\Sigma 1/n^p$ converges if $p > 1$ and diverges if $0 < p \leq 1$.

Let Σa_n and Σb_n be positive-term series. Then we say that the series Σb_n **dominates** the series Σa_n provided that $a_n \leq b_n$ for all n. Theorem 1 says that the positive-term series Σa_n converges if it is dominated by a convergent series and diverges if it dominates a positive-term divergent series.

> **Theorem 1** *Comparison Test*
>
> Suppose that $\Sigma\, a_n$ and $\Sigma\, b_n$ are positive-term series. Then
>
> **1.** $\Sigma\, a_n$ converges if $\Sigma\, b_n$ converges and $a_n \leqq b_n$ for all n;
> **2.** $\Sigma\, a_n$ diverges if $\Sigma\, b_n$ diverges and $a_n \geqq b_n$ for all n.

Proof Denote the nth partial sums of the series $\Sigma\, a_n$ and $\Sigma\, b_n$ by S_n and T_n, respectively. Then $\{S_n\}$ and $\{T_n\}$ are increasing sequences. To prove part (1), suppose that $\Sigma\, b_n$ converges, so $T = \lim_{n\to\infty} T_n$ exists (T is a real number). Then the fact that $a_n \leqq b_n$ for all n implies that $S_n \leqq T_n \leqq T$ for all n. Thus the sequence $\{S_n\}$ of partial sums of $\Sigma\, a_n$ is bounded and increasing and therefore converges. Thus $\Sigma\, a_n$ converges.

Part (2) is merely a restatement of part (1). If the series $\Sigma\, a_n$ converged, then the fact that $\Sigma\, a_n$ dominates $\Sigma\, b_n$ would imply—by part 1, with a_n and b_n interchanged—that $\Sigma\, b_n$ converged. But $\Sigma\, b_n$ diverges, so it follows that $\Sigma\, a_n$ must also diverge. $\square$

We know by Theorem 5 of Section 11.3 that the convergence or divergence of an infinite series is not affected by the addition or deletion of a finite number of terms. Consequently, the conditions $a_n \leqq b_n$ and $a_n \geqq b_n$ in the two parts of the comparison test really need to hold only for all $n \geqq k$, where k is some fixed positive integer. Thus we can say that the positive-term series $\Sigma\, a_n$ converges if it is "eventually dominated" by the convergent positive-term series $\Sigma\, b_n$.

EXAMPLE 1 Because

$$\frac{1}{n(n + 1)(n + 2)} < \frac{1}{n^3}$$

for all $n \geqq 1$, the series

$$\sum_{n=1}^{\infty} \frac{1}{n(n + 1)(n + 2)} = \frac{1}{1\cdot 2\cdot 3} + \frac{1}{2\cdot 3\cdot 4} + \frac{1}{3\cdot 4\cdot 5} + \cdots$$

is dominated by the series $\Sigma 1/n^3$, which is a convergent p-series with $p = 3$. Both are positive-term series, and hence the series $\Sigma 1/[n(n + 1)(n + 2)]$ converges by part (1) of the comparison test.

EXAMPLE 2 Because

$$\frac{1}{\sqrt{2n - 1}} > \frac{1}{\sqrt{2n}}$$

for all $n \geqq 1$, the positive-term series

$$\sum_{n=1}^{\infty} \frac{1}{\sqrt{2n - 1}} = 1 + \frac{1}{\sqrt{3}} + \frac{1}{\sqrt{5}} + \frac{1}{\sqrt{7}} + \cdots$$

dominates the series

$$\sum_{n=1}^{\infty} \frac{1}{\sqrt{2n}} = \frac{1}{\sqrt{2}} \sum_{n=1}^{\infty} \frac{1}{n^{1/2}}.$$

620

But $\Sigma 1/n^{1/2}$ is a divergent p-series with $p = \frac{1}{2}$, and a constant nonzero multiple of a divergent series diverges. So part (2) of the comparison test shows that the series $\Sigma 1/\sqrt{2n - 1}$ also diverges.

EXAMPLE 3 Test the series

$$\sum_{n=0}^{\infty} \frac{1}{n!} = 1 + \frac{1}{1!} + \frac{1}{2!} + \frac{1}{3!} + \cdots$$

for convergence.

Solution We note first that if $n \geqq 1$, then

$$n! = n(n - 1)(n - 2) \cdots 3 \cdot 2 \cdot 1$$

$$\geqq 2 \cdot 2 \cdot 2 \cdots 2 \cdot 2 \cdot 1; \qquad \text{(the same number of factors)}$$

that is, $n! \geqq 2^{n-1}$ for $n \geqq 1$. Thus

$$\frac{1}{n!} \leqq \frac{1}{2^{n-1}} \text{ for } n \geqq 1,$$

so the series

$$\sum_{n=0}^{\infty} \frac{1}{n!} \quad \text{is dominated by the series} \quad 1 + \sum_{n=1}^{\infty} \frac{1}{2^{n-1}} = 1 + \sum_{n=0}^{\infty} \frac{1}{2^n},$$

which is a convergent geometric series (after the first term). Both are positive-term series, and so by the comparison test the series converges. We saw in Section 11.4 that the sum of this series is the number e, so

$$e = 1 + \frac{1}{1!} + \frac{1}{2!} + \frac{1}{3!} + \cdots + \frac{1}{n!} + \cdots.$$

Indeed, this series provides perhaps the simplest way of showing that

$$e \approx 2.71828\,18284\,59045\,23536.$$

Suppose that Σa_n is a positive-term series such that $a_n \to 0$ as $n \to \infty$. Then, in connection with the nth-term divergence test of Section 11.3, the series Σa_n has at least a *chance* of converging. How do we choose an appropriate positive-term series Σb_n with which to compare it? A good idea is to pick b_n as a *simple* function of n, simpler than a_n but such that a_n and b_n approach zero at the same rate as $n \to \infty$. If the formula for a_n is a fraction, we can try discarding all but the terms of largest magnitude in its numerator and denominator to form b_n. For example, if

$$a_n = \frac{3n^2 + n}{n^4 + \sqrt{n}},$$

then we reason that n is small in comparison with $3n^2$ and $\sqrt{n}$ is small in comparison with n^4 when n is quite large. This suggests that we choose $b_n = 3n^2/n^4 = 3/n^2$. The series $\Sigma 3/n^2$ converges ($p = 2$), but when we attempt to compare Σa_n and Σb_n, we find that $a_n \geqq b_n$ (rather than $a_n \leqq b_n$). Consequently, the comparison test does not apply immediately—the fact that Σa_n dominates a convergent series does *not* imply that Σa_n itself converges. Theorem 2 provides a convenient way of handling such a situation.

> **Theorem 2** *Limit Comparison Test*
> Suppose that $\Sigma\, a_n$ and $\Sigma\, b_n$ are positive-term series. If the limit
>
> $$L = \lim_{n\to\infty} \frac{a_n}{b_n}$$
>
> exists and $0 < L < +\infty$, then either both series converge or both series diverge.

Proof Choose two fixed positive numbers P and Q such that $P < L < Q$. Then $P < a_n/b_n < Q$ for n sufficiently large, and so

$$Pb_n < a_n < Qb_n$$

for all sufficiently large values of n. If $\Sigma\, b_n$ converges, then $\Sigma\, a_n$ is eventually dominated by the convergent series $\Sigma\, Qb_n = Q\Sigma\, b_n$, so part (1) of the comparison test implies that $\Sigma\, a_n$ also converges. If $\Sigma\, b_n$ diverges, then $\Sigma\, a_n$ eventually dominates the divergent series $\Sigma\, Pb_n = P\Sigma\, b_n$, so part (2) of the comparison test implies that $\Sigma\, a_n$ also diverges. Thus the convergence of either series implies the convergence of the other. ❑

EXAMPLE 4 With

$$a_n = \frac{3n^2 + n}{n^4 + \sqrt{n}} \quad\text{and}\quad b_n = \frac{1}{n^2}$$

(motivated by the discussion preceding Theorem 2), we find that

$$\lim_{n\to\infty} \frac{a_n}{b_n} = \lim_{n\to\infty} \frac{3n^4 + n^3}{n^4 + \sqrt{n}} = \lim_{n\to\infty} \frac{3 + \dfrac{1}{n}}{1 + \dfrac{1}{n^{7/2}}} = 3.$$

Because $\Sigma\, 1/n^2$ is a convergent p-series ($p = 2$), the limit comparison test tells us that the series

$$\sum_{n=1}^{\infty} \frac{3n^2 + n}{n^4 + \sqrt{n}}$$

also converges.

EXAMPLE 5 Test for convergence: $\displaystyle\sum_{n=1}^{\infty} \frac{1}{2n + \ln n}$.

Solution Because $\lim_{n\to\infty} (\ln n)/n = 0$ by l'Hôpital's rule, $\ln n$ is very small in comparison with $2n$ when n is large. We therefore take $a_n = 1/(2n + \ln n)$ and, ignoring the constant coefficient 2, we take $b_n = 1/n$. Then we find that

$$\lim_{n\to\infty} \frac{a_n}{b_n} = \lim_{n\to\infty} \frac{n}{2n + \ln n} = \lim_{n\to\infty} \frac{1}{2 + \dfrac{\ln n}{n}} = \frac{1}{2}.$$

Because the harmonic series $\Sigma\, 1/n = \Sigma\, b_n$ diverges, it follows that the given series $\Sigma\, a_n$ also diverges.

It is important to realize that if $L = \lim(a_n/b_n)$ is either zero or ∞, then the limit comparison test does not apply. (See Problem 42 for a discussion of what conclusions may sometimes be drawn in these cases.) Note, for example, that if $a_n = 1/n^2$ and $b_n = 1/n$, then $\lim(a_n/b_n) = 0$. But in this case $\Sigma\, a_n$ converges, whereas $\Sigma\, b_n$ diverges.

We close our discussion of positive-term series with the observation that the sum of a convergent *positive*-term series is not altered by grouping or rearranging its terms. For example, let $\Sigma\, a_n$ be a convergent positive-term series and consider

$$\sum_{n=1}^{\infty} b_n = (a_1 + a_2 + a_3) + a_4 + (a_5 + a_6) + \cdots.$$

That is, the new series has terms $b_1 = a_1 + a_2 + a_3$, $b_2 = a_4$, $b_3 = a_5 + a_6$, and so on. Then every partial sum T_n of $\Sigma\, b_n$ is equal to some partial sum $S_{n'}$ of $\Sigma\, a_n$. Because $\{S_n\}$ is an increasing sequence with limit $S = \Sigma\, a_n$, it follows easily that $\{T_n\}$ is an increasing sequence with the same limit. Thus $\Sigma\, b_n = S$ as well. The argument is more subtle if terms of $\Sigma\, a_n$ are moved "out of place," as in

$$\sum_{n=1}^{\infty} b_n = a_2 + a_1 + a_4 + a_3 + a_6 + a_5 + \cdots,$$

but the same conclusion holds: Any rearrangement of a convergent *positive*-term series also converges, and it converges to the same sum.

Similarly, it is easy to prove that any grouping or rearrangement of a divergent positive-term series also diverges. But these observations all fail in the case of an infinite series with both positive and negative terms. For example, the series $\Sigma\, (-1)^n$ diverges, but it has the convergent grouping

$$(-1 + 1) + (-1 + 1) + (-1 + 1) + \cdots = 0 + 0 + 0 + \cdots = 0.$$

It follows from Problem 44 of Section 11.3 that

$$\ln 2 = 1 - \frac{1}{2} + \frac{1}{3} - \frac{1}{4} + \frac{1}{5} - \cdots,$$

but the rearrangement

$$1 + \frac{1}{3} - \frac{1}{2} + \frac{1}{5} + \frac{1}{7} - \frac{1}{4} + \frac{1}{9} + \frac{1}{11} - \frac{1}{6} + \cdots$$

converges instead to $\frac{3}{2} \ln 2$. This series for $\ln 2$ even has rearrangements that converge to zero and others that diverge to $+\infty$. One could write a book on this fascinating aspect of infinite series.

11.6 Problems

Use comparison tests to determine whether the infinite series in Problems 1 through 35 converge or diverge.

1. $\displaystyle\sum_{n=1}^{\infty} \frac{1}{n^2 + n + 1}$

2. $\displaystyle\sum_{n=1}^{\infty} \frac{n^3 + 1}{n^4 + 2}$

3. $\displaystyle\sum_{n=1}^{\infty} \frac{1}{n + \sqrt{n}}$

4. $\displaystyle\sum_{n=1}^{\infty} \frac{1}{n + n^{3/2}}$

5. $\displaystyle\sum_{n=1}^{\infty} \frac{1}{1 + 3^n}$

6. $\displaystyle\sum_{n=1}^{\infty} \frac{10n^2}{n^4 + 1}$

7. $\displaystyle\sum_{n=2}^{\infty} \frac{10n^2}{n^3 - 1}$

8. $\displaystyle\sum_{n=1}^{\infty} \frac{n^2 - n}{n^4 + 2}$

9. $\displaystyle\sum_{n=1}^{\infty} \frac{1}{\sqrt{37n^3 + 3}}$

10. $\displaystyle\sum_{n=1}^{\infty} \frac{1}{\sqrt{n^2 + 1}}$

11. $\displaystyle\sum_{n=1}^{\infty} \frac{\sqrt{n}}{n^2 + n}$

12. $\displaystyle\sum_{n=1}^{\infty} \frac{1}{3 + 5^n}$

13. $\displaystyle\sum_{n=2}^{\infty} \frac{1}{\ln n}$

14. $\displaystyle\sum_{n=1}^{\infty} \frac{1}{n - \ln n}$

15. $\displaystyle\sum_{n=1}^{\infty} \frac{\sin^2 n}{n^2 + 1}$

16. $\displaystyle\sum_{n=1}^{\infty} \frac{\cos^2 n}{3^n}$

17. $\displaystyle\sum_{n=1}^{\infty} \frac{n + 2^n}{n + 3^n}$

18. $\displaystyle\sum_{n=1}^{\infty} \frac{1}{2^n + 3^n}$

19. $\displaystyle\sum_{n=2}^{\infty} \frac{1}{n^2 \ln n}$

20. $\displaystyle\sum_{n=1}^{\infty} \frac{1}{n^{1+\sqrt{n}}}$

21. $\displaystyle\sum_{n=1}^{\infty} \frac{\ln n}{n^2}$

22. $\displaystyle\sum_{n=1}^{\infty} \frac{\arctan n}{n}$

23. $\displaystyle\sum_{n=1}^{\infty} \frac{\sin^2(1/n)}{n^2}$

24. $\displaystyle\sum_{n=1}^{\infty} \frac{e^{1/n}}{n}$

25. $\displaystyle\sum_{n=1}^{\infty} \frac{\ln n}{e^n}$

26. $\displaystyle\sum_{n=1}^{\infty} \frac{n^2 + 2}{n^3 + 3n}$

27. $\displaystyle\sum_{n=1}^{\infty} \frac{n^{3/2}}{n^2 + 4}$

28. $\displaystyle\sum_{n=1}^{\infty} \frac{1}{n \cdot 2^n}$

29. $\displaystyle\sum_{n=1}^{\infty} \frac{3}{4 + \sqrt{n}}$

30. $\displaystyle\sum_{n=1}^{\infty} \frac{n^2 + 1}{e^n(n + 1)^2}$

31. $\displaystyle\sum_{n=1}^{\infty} \frac{2n^2 - 1}{n^2 \cdot 3^n}$

32. $\displaystyle\sum_{n=1}^{\infty} \frac{1}{\sqrt[3]{2n^4 + 1}}$

33. $\displaystyle\sum_{n=1}^{\infty} \frac{2 + \sin n}{n^2}$

34. $\displaystyle\sum_{n=1}^{\infty} \frac{\ln n}{n^3}$

35. $\displaystyle\sum_{n=1}^{\infty} \frac{(n + 1)^n}{n^{n+1}}$ [*Suggestion:* $\displaystyle\lim_{n\to\infty}\left(1 + \frac{1}{n}\right)^n = e$.]

36. (a) Prove that $\ln n < n^{1/8}$ for all sufficiently large values of n. (b) Explain why part (a) shows that the series $\Sigma \, 1/(\ln n)^8$ diverges.

37. Prove that if $\Sigma\, a_n$ is a convergent positive-term series, then $\Sigma\, (a_n/n)$ converges.

38. Suppose that $\Sigma\, a_n$ is a convergent positive-term series and that $\{c_n\}$ is a sequence of positive numbers with limit zero. Prove that $\Sigma\, a_n c_n$ converges.

39. Use the result of Problem 38 to prove that if $\Sigma\, a_n$ and $\Sigma\, b_n$ are convergent positive-term series, then $\Sigma\, a_n b_n$ converges.

40. Prove that the series

$$\sum_{n=1}^{\infty} \frac{1}{1 + 2 + 3 + \cdots + n}$$

converges.

41. Use the result of Problem 40 in Section 11.5 to prove that the series

$$\sum_{n=1}^{\infty} \frac{1}{1 + \dfrac{1}{2} + \dfrac{1}{3} + \cdots + \dfrac{1}{n}}$$

diverges.

42. Adapt the proof of the limit comparison test to prove the following two results. (a) Suppose that $\Sigma\, a_n$ and $\Sigma\, b_n$ are positive-term series and that $\Sigma\, b_n$ converges. If

$$L = \lim_{n\to\infty} \frac{a_n}{b_n} = 0,$$

then $\Sigma\, a_n$ converges. (b) Suppose that $\Sigma\, a_n$ and $\Sigma\, b_n$ are positive-term series and that $\Sigma\, b_n$ diverges. If

$$L = \lim_{n\to\infty} \frac{a_n}{b_n} = +\infty,$$

then $\Sigma\, a_n$ diverges.

11.7
Alternating Series and Absolute Convergence

In Sections 11.5 and 11.6 we concentrated on positive-term series. Now we discuss infinite series that have both positive terms and negative terms. An important example is a series with terms that are alternately positive and negative. An **alternating series** is an infinite series of the form

$$\sum_{n=1}^{\infty} (-1)^{n+1} a_n = a_1 - a_2 + a_3 - a_4 + a_5 - \cdots \qquad (1)$$

or of the form $\displaystyle\sum_{n=1}^{\infty} (-1)^n a_n$, where $a_n > 0$ for all n. For example, the series

$$\sum_{n=1}^{\infty} \frac{(-1)^{n+1}}{n} = 1 - \frac{1}{2} + \frac{1}{3} - \frac{1}{4} + \frac{1}{5} - \cdots$$

is an alternating series. Theorem 1 shows that this series converges because the sequence of absolute values of its terms is decreasing and has limit zero.

> **Theorem 1**   **Alternating Series Test**
> If $a_n > a_{n+1} > 0$ for all n and $\lim\limits_{n\to\infty} a_n = 0$, then the alternating series in Eq. (1) converges.

Proof   We first consider the even-numbered partial sums $S_2, S_4, S_6, \ldots, S_{2n}$. We may write

$$S_{2n} = (a_1 - a_2) + (a_3 - a_4) + \cdots + (a_{2n-1} - a_{2n}).$$

Because $a_k - a_{k+1} \geqq 0$ for all k, the sequence $\{S_{2n}\}$ is increasing. Also, because

$$S_{2n} = a_1 - (a_2 - a_3) - \cdots - (a_{2n-2} - a_{2n-1}) - a_{2n},$$

$S_{2n} \leqq a_1$ for all n, so the increasing sequence $\{S_{2n}\}$ is bounded above. Hence the limit

$$S = \lim_{n\to\infty} S_{2n}$$

exists by the bounded monotonic sequence property of Section 11.2. It remains only for us to verify that the odd-numbered partial sums $S_1, S_3, S_5, \ldots$ also converge to S. But $S_{2n+1} = S_{2n} + a_{2n+1}$ and $\lim\limits_{n\to\infty} a_{2n+1} = 0$, so

$$\lim_{n\to\infty} S_{2n+1} = \lim_{n\to\infty} S_{2n} + \lim_{n\to\infty} a_{2n+1} = S.$$

Thus $\lim\limits_{n\to\infty} S_n = S$, and therefore the series converges. $\square$

EXAMPLE 1   The series

$$\sum_{n=1}^{\infty} \frac{(-1)^{n+1}}{2n-1} = 1 - \frac{1}{3} + \frac{1}{5} - \frac{1}{7} + \frac{1}{9} - \cdots$$

satisfies the conditions of Theorem 1 and therefore converges. The alternating series test does not tell us the sum of this series, but we saw in Section 11.4 that its sum is $\pi/4$.

EXAMPLE 2   The series

$$\sum_{n=1}^{\infty} \frac{(-1)^{n+1}n}{2n-1} = 1 - \frac{2}{3} + \frac{3}{5} - \frac{4}{7} + \frac{5}{9} - \cdots$$

is an alternating series, and it is easy to verify that

$$a_n = \frac{n}{2n-1} > \frac{n+1}{2n+1} = a_{n+1}$$

for all $n \geqq 1$. But

$$\lim_{n\to\infty} a_n = \frac{1}{2} \neq 0,$$

so the alternating series test *does not apply*. (This fact alone does not imply that the series in question diverges—many series in Sections 11.5 and 11.6 converge even though the alternating series test does not apply. But the series of this example diverges by the nth-term divergence test.)

If a series converges by the alternating series test, then Theorem 2 shows how to approximate its sum with any desired degree of accuracy—*if* you have a computer fast enough to add a large number of its terms.

> **Theorem 2** *Alternating Series Error Estimate*
> Suppose that the series $\Sigma\,(-1)^{n+1}a_n$ satisfies the conditions of the alternating series test and therefore converges. Let S denote the sum of the series. Denote by $R_n = S - S_n$ the error made in replacing S by the nth partial sum S_n of the series. Then R_n has the same sign as the next term $(-1)^{n+2}a_{n+1}$ of the series, and
>
> $$0 < |R_n| < a_{n+1}. \tag{2}$$

In particular, the *sum S of a convergent alternating series lies between any two consecutive partial sums*. This follows from the proof of Theorem 1, where we saw that $\{S_{2n}\}$ is an increasing sequence, whereas $\{S_{2n+1}\}$ is a decreasing sequence, each converging to S. The resulting inequalities

$$S_{2n-1} > S > S_{2n} = S_{2n-1} - a_{2n}$$

and

$$S_{2n} < S < S_{2n+1} = S_{2n} + a_{2n+1}$$

imply the inequality in (2).

EXAMPLE 3   We saw in Section 11.4 that

$$e^x = \sum_{n=0}^{\infty} \frac{x^n}{n!}$$

for all x and thus that

$$\frac{1}{e} = e^{-1} = 1 - 1 + \frac{1}{2!} - \frac{1}{3!} + \frac{1}{4!} - \cdots.$$

Use this alternating series to compute e^{-1} accurate to four decimal places.

Solution   We want $|R_n| < 1/(n + 1)! \leqq 0.00005$. The least value of n for which this inequality holds is $n = 7$. Then

$$e^{-1} = 1 - 1 + \frac{1}{2!} - \frac{1}{3!} + \frac{1}{4!} - \frac{1}{5!} + \frac{1}{6!} - \frac{1}{7!} + R_7 = 0.367857 + R_7.$$

(We are carrying six decimal places because we want four-place accuracy in the final answer.) Now the inequality in (2) gives

$$0 < R_7 < \frac{1}{8!} < 0.000025.$$

Thus

$$0.367857 < e^{-1} < 0.367882.$$

Thus $e^{-1} = 0.3679$ rounded to four places, and $e = 2.718$ rounded to three places.

ABSOLUTE CONVERGENCE

The series

$$\sum_{n=1}^{\infty} \frac{(-1)^{n+1}}{n} = 1 - \frac{1}{2} + \frac{1}{3} - \frac{1}{4} + \frac{1}{5} - \cdots$$

converges, but if we simply replace each term with its absolute value, we get the *divergent* series

$$1 + \frac{1}{2} + \frac{1}{3} + \frac{1}{4} + \frac{1}{5} + \cdots .$$

In contrast, the *convergent* series

$$\sum_{n=1}^{\infty} \frac{(-1)^{n}}{2^{n}} = 1 - \frac{1}{2} + \frac{1}{4} - \frac{1}{8} + \cdots = \frac{2}{3}$$

has the property that the associated positive-term series

$$1 + \frac{1}{2} + \frac{1}{4} + \frac{1}{8} + \cdots = 2$$

also converges. Theorem 3 tells us that if a series of *positive* terms converges, then we may insert minus signs in front of any of the terms—every other one, for instance—and the resulting series will also converge.

Theorem 3 *Absolute Convergence Implies Convergence*
If the series $\sum |a_n|$ converges, then so does the series $\sum a_n$.

Proof   Suppose that the series $\sum |a_n|$ converges. Note that

$$0 \leqq a_n + |a_n| \leqq 2|a_n|$$

for all n. Let $b_n = a_n + |a_n|$. It then follows from the comparison test that the positive-term series $\sum b_n$ converges, because it is dominated by the convergent series $\sum 2|a_n|$. It is easy to verify, too, that the termwise difference of two convergent series also converges. Hence we now see that the series

$$\sum a_n = \sum (b_n - |a_n|) = \sum b_n - \sum |a_n|$$

converges. ❏

Thus we have another convergence test, one not limited to positive-term or to alternating series: Given the series $\sum a_n$, test the series $\sum |a_n|$ for convergence. If the latter converges, then so does the former. (But the converse is *not* true!) This phenomenon motivates us to make the following definition.

Definition *Absolute Convergence*
The series $\sum a_n$ is said to **converge absolutely** (and is called **absolutely convergent**) provided that the series

$$\sum |a_n| = |a_1| + |a_2| + |a_3| + \cdots + |a_n| + \cdots$$

converges.

Thus we have explained the title of Theorem 3, and we can rephrase the theorem as follows: *If a series converges absolutely, then it converges.* The two examples preceding Theorem 3 show that a convergent series may either converge absolutely or fail to do so:

$$1 - \tfrac{1}{2} + \tfrac{1}{4} - \tfrac{1}{8} + \cdots$$

is an absolutely convergent series because

$$1 + \tfrac{1}{2} + \tfrac{1}{4} + \tfrac{1}{8} + \cdots$$

converges, whereas

$$1 - \tfrac{1}{2} + \tfrac{1}{3} - \tfrac{1}{4} + \tfrac{1}{5} - \cdots$$

is a series that, though convergent, is *not* absolutely convergent. A series that converges but does not converge absolutely is said to be **conditionally convergent.** Consequently, the terms *absolutely convergent, conditionally convergent,* and *divergent* are simultaneously all-inclusive and mutually exclusive. (There's no such thing as "absolute divergence.")

There is some advantage in the application of Theorem 3, because to apply it we test the *positive*-term series $\Sigma |a_n|$ for convergence—and we have a variety of tests, such as comparison tests or the integral test, designed for use on positive-term series.

Note also that absolute convergence of the series Σa_n means that *another* series $\Sigma |a_n|$ converges, and the two sums will generally differ. For example, with $a_n = (-\tfrac{1}{3})^n$, the formula for the sum of a geometric series gives

$$\sum_{n=0}^{\infty} a_n = \sum_{n=0}^{\infty} \left(-\frac{1}{3}\right)^n = \frac{1}{1 - (-\frac{1}{3})} = \frac{3}{4},$$

whereas

$$\sum_{n=0}^{\infty} |a_n| = \sum_{n=0}^{\infty} \left(\frac{1}{3}\right)^n = \frac{1}{1 - \frac{1}{3}} = \frac{3}{2}.$$

EXAMPLE 4   Discuss the convergence of the series

$$\sum_{n=1}^{\infty} \frac{\cos n}{n^2} = \cos 1 + \frac{\cos 2}{4} + \frac{\cos 3}{9} + \cdots.$$

Solution   Let $a_n = (\cos n)/n^2$. Then

$$|a_n| = \frac{|\cos n|}{n^2} \leqq \frac{1}{n^2}$$

for all $n \geqq 1$. Hence the positive-term series $\Sigma |a_n|$ converges by the comparison test, because it is dominated by the convergent p-series $\Sigma (1/n^2)$. Thus the given series is absolutely convergent, and it therefore converges by Theorem 3.

One reason for the importance of absolute convergence is the fact (proved in advanced calculus) that the terms of an absolutely convergent series may be regrouped or rearranged without changing the sum of the series. As we suggested at the end of Section 11.6, this is *not* true of conditionally convergent series.

THE RATIO TEST AND THE ROOT TEST

Our next two convergence tests involve a way of measuring the rate of growth or decrease of the sequence $\{a_n\}$ of terms of a series to determine whether $\Sigma\, a_n$ converges absolutely or diverges.

Theorem 4   The Ratio Test

Suppose that the limit

$$\rho = \lim_{n \to \infty} \left| \frac{a_{n+1}}{a_n} \right| \tag{3}$$

either exists or is infinite. Then the infinite series $\Sigma\, a_n$ of nonzero terms

1. Converges absolutely if $\rho < 1$;
2. Diverges if $\rho > 1$.

If $\rho = 1$, the ratio test is inconclusive.

Proof   If $\rho < 1$, choose a (fixed) number r with $\rho < r < 1$. Then Eq. (3) implies that there exists an integer N such that $|a_{n+1}| \leqq r|a_n|$ for all $n \geqq N$. It follows that

$$|a_{N+1}| \leqq r|a_N|,$$

$$|a_{N+2}| \leqq r|a_{N+1}| \leqq r^2|a_N|,$$

$$|a_{N+3}| \leqq r|a_{N+2}| \leqq r^3|a_N|,$$

and in general that

$$|a_{N+k}| \leqq r^k|a_N| \qquad \text{for } k \geqq 0.$$

Hence the series

$$|a_N| + |a_{N+1}| + |a_{N+2}| + \cdots$$

is dominated by the geometric series

$$|a_N|(1 + r + r^2 + r^3 + \cdots),$$

and the latter converges because $r < 1$. Thus the series $\Sigma\, |a_n|$ converges, so the series $\Sigma\, a_n$ converges absolutely.

   If $\rho > 1$, then Eq. (3) implies that there exists a positive integer N such that $|a_{n+1}| > |a_n|$ for all $n \geqq N$. It follows that $|a_n| > |a_N| > 0$ for all $n > N$. Thus the sequence $\{a_n\}$ cannot approach zero as $n \to \infty$, and consequently, by the nth-term divergence test, the series $\Sigma\, a_n$ diverges. ◻

   To see that $\Sigma\, a_n$ may either converge or diverge if $\rho = 1$, consider the divergent series $\Sigma\,(1/n)$ and the convergent series $\Sigma\,(1/n^2)$. You should verify that, for both series, the value of the ratio ρ is 1.

EXAMPLE 5   Consider the series

$$\sum_{n=1}^{\infty} \frac{(-1)^n 2^n}{n!} = -2 + \frac{4}{2!} - \frac{8}{3!} + \frac{16}{4!} - \cdots.$$

Then

$$\rho = \lim_{n \to \infty} \left| \frac{a_{n+1}}{a_n} \right| = \lim_{n \to \infty} \left| \frac{\dfrac{(-1)^{n+1}2^{n+1}}{(n+1)!}}{\dfrac{(-1)^n 2^n}{n!}} \right| = \lim_{n \to \infty} \frac{2}{n+1} = 0.$$

Because $\rho < 1$, the series converges absolutely.

EXAMPLE 6 Test for convergence: $\displaystyle\sum_{n=1}^{\infty} \frac{n}{2^n}$.

Solution We have

$$\rho = \lim_{n \to \infty} \left| \frac{a_{n+1}}{a_n} \right| = \lim_{n \to \infty} \frac{\dfrac{n+1}{2^{n+1}}}{\dfrac{n}{2^n}} = \lim_{n \to \infty} \frac{n+1}{2n} = \frac{1}{2}.$$

Because $\rho < 1$, this series converges (absolutely).

EXAMPLE 7 Test for convergence: $\displaystyle\sum_{n=1}^{\infty} \frac{3^n}{n^2}$.

Solution Here we have

$$\rho = \lim_{n \to \infty} \left| \frac{a_{n+1}}{a_n} \right| = \lim_{n \to \infty} \frac{\dfrac{3^{n+1}}{(n+1)^2}}{\dfrac{3^n}{n^2}} = \lim_{n \to \infty} \frac{3n^2}{(n+1)^2} = 3.$$

In this case $\rho > 1$, so the given series diverges.

Theorem 5   The Root Test

Suppose that the limit

$$\rho = \lim_{n \to \infty} \sqrt[n]{|a_n|} \tag{4}$$

exists or is infinite. Then the infinite series $\Sigma\, a_n$

1. Converges absolutely if $\rho < 1$;
2. Diverges if $\rho > 1$.

If $\rho = 1$, the test is inconclusive.

Proof If $\rho < 1$, choose a (fixed) number r such that $\rho < r < 1$. Then $|a_n|^{1/n} < r$, and hence $|a_n| < r^n$, for n sufficiently large. Thus the series $\Sigma |a_n|$ is eventually dominated by the convergent geometric series Σr^n. Therefore $\Sigma |a_n|$ converges, and so the series $\Sigma\, a_n$ converges absolutely.

If $\rho > 1$, then $|a_n|^{1/n} > 1$, and hence $|a_n| > 1$, for n sufficiently large. Therefore the nth-term test for divergence implies that the series $\Sigma\, a_n$ diverges. ❑

The ratio test is generally simpler to apply than the root test, and therefore it is ordinarily the one to try first. But there are certain series for which the root test succeeds and the ratio test fails, as in Example 8.

EXAMPLE 8 Consider the series

$$\sum_{n=1}^{\infty} \frac{1}{2^{n+(-1)^n}} = \frac{1}{2} + \frac{1}{1} + \frac{1}{8} + \frac{1}{4} + \frac{1}{32} + \frac{1}{16} + \cdots.$$

Then $a_{n+1}/a_n = 2$ if n is even, whereas $a_{n+1}/a_n = \frac{1}{8}$ if n is odd. So the limit required for the ratio test does not exist. But

$$\lim_{n\to\infty} |a_n|^{1/n} = \lim_{n\to\infty} \left| \frac{1}{2^{n+(-1)^n}} \right|^{1/n} = \lim_{n\to\infty} \frac{1}{2} \left| \frac{1}{2^{(-1)^n/n}} \right| = \frac{1}{2},$$

so the given series converges by the root test. (Its convergence also follows from the fact that it is a rearrangement of the positive-term convergent geometric series $\Sigma\, 1/2^n$.)

11.7 Problems

Determine whether or not the alternating series in Problems 1 through 10 converge.

1. $\displaystyle\sum_{n=1}^{\infty} \frac{(-1)^{n+1}}{n^2}$

2. $\displaystyle\sum_{n=1}^{\infty} \frac{(-1)^{n+1} n}{3n + 2}$

3. $\displaystyle\sum_{n=1}^{\infty} \frac{(-1)^{n+1} n}{n^2 + 1}$

4. $\displaystyle\sum_{n=2}^{\infty} \frac{(-1)^n n}{\ln n}$

5. $\displaystyle\sum_{n=2}^{\infty} \frac{(-1)^n}{\ln n}$

6. $\displaystyle\sum_{n=1}^{\infty} \frac{(-1)^{n+1}}{\sqrt{2n + 1}}$

7. $\displaystyle\sum_{n=0}^{\infty} \frac{(-1)^n}{n!}$

8. $\displaystyle\sum_{n=1}^{\infty} \frac{(-1)^{n+1}(1.01)^n}{n^4 + 1}$

9. $\displaystyle\sum_{n=1}^{\infty} \frac{(-1)^{n+1}}{\sqrt[n]{n}}$

10. $\displaystyle\sum_{n=1}^{\infty} \frac{(-1)^{n+1} n!}{(2n)!}$

Determine whether the series in Problems 11 through 32 converge absolutely, converge conditionally, or diverge.

11. $\displaystyle\sum_{n=1}^{\infty} \frac{(-1)^{n+1}}{2^n}$

12. $\displaystyle\sum_{n=1}^{\infty} \frac{1}{n^2 + 1}$

13. $\displaystyle\sum_{n=1}^{\infty} \frac{(-1)^{n+1} \ln n}{n}$

14. $\displaystyle\sum_{n=1}^{\infty} \frac{1}{n^n}$

15. $\displaystyle\sum_{n=1}^{\infty} \left(\frac{10}{n}\right)^n$

16. $\displaystyle\sum_{n=1}^{\infty} \frac{3^n}{n!n}$

17. $\displaystyle\sum_{n=0}^{\infty} \frac{(-10)^n}{n!}$

18. $\displaystyle\sum_{n=1}^{\infty} \frac{(-1)^{n+1} n!}{n^n}$

19. $\displaystyle\sum_{n=1}^{\infty} (-1)^{n+1}\left(\frac{n}{n + 1}\right)^n$

20. $\displaystyle\sum_{n=1}^{\infty} \frac{n!\, n^2}{(2n)!}$

21. $\displaystyle\sum_{n=1}^{\infty} \left(\frac{\ln n}{n}\right)^n$

22. $\displaystyle\sum_{n=0}^{\infty} \frac{(-1)^n 2^{3n}}{7^n}$

23. $\displaystyle\sum_{n=0}^{\infty} (-1)^n(\sqrt{n + 1} - \sqrt{n})$

24. $\displaystyle\sum_{n=1}^{\infty} n(\tfrac{3}{4})^n$

25. $\displaystyle\sum_{n=1}^{\infty} \left[\ln\!\left(\frac{1}{n}\right) \right]^n$

26. $\displaystyle\sum_{n=0}^{\infty} \frac{(n!)^2}{(2n)!}$

27. $\displaystyle\sum_{n=1}^{\infty} \frac{(-1)^{n+1} 3^n}{n(2^n + 1)}$

28. $\displaystyle\sum_{n=1}^{\infty} \frac{(-1)^{n+1} \arctan n}{n}$

29. $\displaystyle\sum_{n=1}^{\infty} \frac{(-1)^{n+1} n!}{1\cdot 3\cdot 5 \cdots (2n - 1)}$

30. $\displaystyle\sum_{n=1}^{\infty} (-1)^{n+1}\frac{1\cdot 3\cdot 5 \cdots (2n - 1)}{1\cdot 4\cdot 7 \cdots (3n - 2)}$

31. $\displaystyle\sum_{n=1}^{\infty} \frac{(n + 2)!}{3^n (n!)^2}$

32. $\displaystyle\sum_{n=1}^{\infty} \frac{(-1)^{n+1} n^n}{3^{n^2}}$

In Problems 33 through 36, find the least positive integer n such that $|R_n| = |S - S_n| < 0.0005$, so the nth partial sum S_n of the given alternating series approximates its sum S accurate to three decimal places.

33. $\displaystyle\sum_{n=1}^{\infty} \frac{(-1)^{n+1}}{n}$

34. $\displaystyle\sum_{n=1}^{\infty} \frac{(-1)^{n+1}}{n^2}$

35. $\displaystyle\sum_{n=0}^{\infty} \frac{(-1)^n}{3^n}$

36. $\displaystyle\sum_{n=1}^{\infty} \frac{(-1)^{n+1}}{n^n}$

In Problems 37 through 40, approximate the sum of the given series accurate to three decimal places.

37. $\displaystyle e^{-1/2} = \sum_{n=0}^{\infty} \frac{(-1)^n}{n!\, 2^n}$

38. $\displaystyle \cos 1 = \sum_{n=0}^{\infty} \frac{(-1)^n}{(2n)!}$

39. $\displaystyle \ln(1.1) = \sum_{n=1}^{\infty} \frac{(-1)^{n+1}(0.1)^n}{n}$

40. $\displaystyle\sum_{n=1}^{\infty} \frac{(-1)^{n+1}}{n^5}$

41. Approximate the sum of the series

$$\frac{\pi^2}{12} = \sum_{n=1}^{\infty} \frac{(-1)^{n+1}}{n^2}$$

with error less than 0.01. Use the corresponding partial sum and error estimate to verify that $3.13 < \pi < 3.15$.

42. Prove that $\Sigma |a_n|$ diverges if the series Σa_n diverges.

43. Give an example of a pair of convergent series Σa_n and Σb_n such that $\Sigma a_n b_n$ diverges.

44. (a) Suppose that r is a (fixed) number such that $|r| < 1$. Use the ratio test to prove that the series

$$\sum_{n=0}^{\infty} nr^n$$

converges. Let S denote its sum. (b) Show that

$$(1 - r)S = \sum_{n=1}^{\infty} r^n.$$

Show how to conclude that

$$\sum_{n=0}^{\infty} nr^n = \frac{r}{(1 - r)^2}.$$

11.8

Power Series

The most important infinite series representations of functions are those whose terms are constant multiples of (successive) integral powers of the independent variable x—that is, series that resemble "infinite polynomials." For example, we discussed in Section 11.4 the geometric series

$$\frac{1}{1 - x} = 1 + x + x^2 + x^3 + \cdots \qquad (|x| < 1) \tag{1}$$

and the Taylor series

$$e^x = \sum_{n=0}^{\infty} \frac{x^n}{n!} = 1 + x + \frac{x^2}{2!} + \frac{x^3}{3!} + \frac{x^4}{4!} + \cdots, \tag{2}$$

$$\cos x = \sum_{n=0}^{\infty} (-1)^n \frac{x^{2n}}{(2n)!} = 1 - \frac{x^2}{2!} + \frac{x^4}{4!} - \frac{x^6}{6!} + \cdots, \quad \text{and} \tag{3}$$

$$\sin x = \sum_{n=0}^{\infty} (-1)^n \frac{x^{2n+1}}{(2n + 1)!} = x - \frac{x^3}{3!} + \frac{x^5}{5!} - \frac{x^7}{7!} + \cdots. \tag{4}$$

There we used Taylor's formula to show that the series in Eqs. (2) through (4) converge, for all x, to the functions e^x, $\cos x$, and $\sin x$, respectively. Here we investigate the convergence of a "power series" without knowing in advance the function (if any) to which it converges.

All the infinite series in Eqs. (1) through (4) have the form

$$\sum_{n=0}^{\infty} a_n x^n = a_0 + a_1 x + a_2 x^2 + \cdots + a_n x^n + \cdots \tag{5}$$

with the constant *coefficients* $a_0, a_1, a_2, \ldots$. An infinite series of this form is called a **power series** in (powers of) x. In order that the initial terms of the two sides of Eq. (5) agree, we adopt here the convention that $x^0 = 1$ even if $x = 0$.

CONVERGENCE OF POWER SERIES

The power series in Eq. (5) obviously converges when $x = 0$. In general, it will converge for some nonzero values of x and diverge for others. Because of the way in which powers of x are involved, the ratio test is particularly effective in determining the values of x for which a power series converges.

Assume that the limit

$$\rho = \lim_{n \to \infty} \left| \frac{a_{n+1}}{a_n} \right| \qquad (6)$$

exists. This is the limit that we need if we want to apply the ratio test to the series Σa_n of constants. To apply the ratio test to the power series in Eq. (5), we write $u_n = a_n x^n$ and compute the limit

$$\lim_{n \to \infty} \left| \frac{u_{n+1}}{u_n} \right| = \lim_{n \to \infty} \left| \frac{a_{n+1} x^{n+1}}{a_n x^n} \right| = \rho |x|. \qquad (7)$$

If $\rho = 0$, then $\Sigma a_n x^n$ converges absolutely for all x. If $\rho = +\infty$, then $\Sigma a_n x^n$ diverges for all $x \neq 0$. If ρ is a positive real number, we see from Eq. (7) that $\Sigma a_n x^n$ converges absolutely for all x such that $\rho |x| < 1$—that is, when

$$|x| < R = \frac{1}{\rho} = \lim_{n \to \infty} \left| \frac{a_n}{a_{n+1}} \right|. \qquad (8)$$

In this case the ratio test also implies that $\Sigma a_n x^n$ diverges if $|x| > R$ but is inconclusive when $x = \pm R$. We have therefore proved Theorem 1, under the additional hypothesis that the limit in Eq. (6) exists. In Problems 43 and 44 we outline a proof that does not require this additional hypothesis.

Theorem 1 Convergence of Power Series
If $\Sigma a_n x^n$ is a power series, then either

1. The series converges absolutely for all x, or
2. The series converges only when $x = 0$, or
3. There exists a number $R > 0$ such that $\Sigma a_n x^n$ converges absolutely if $|x| < R$ and diverges if $|x| > R$.

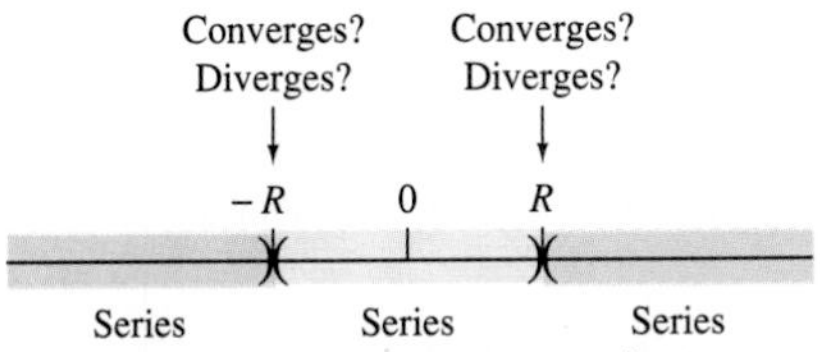

The number R of case 3 is called the **radius of convergence** of the power series $\Sigma a_n x^n$. We shall write $R = \infty$ in case 1 and $R = 0$ in case 2. The set of all real numbers x for which the series converges is called its **interval of convergence** (Fig. 11.8.1); note that this set *is* an interval. If $0 < R < \infty$, then the interval of convergence is one of the intervals

$$(-R, R), \qquad (-R, R], \qquad [-R, R), \quad \text{or} \quad [-R, R].$$

When we substitute either of the endpoints $x = \pm R$ into the series $\Sigma a_n x^n$, we obtain an infinite series with constant terms whose convergence must be determined separately. Because these will be numerical series, the earlier tests of this chapter are appropriate.

Converges? Converges?
Diverges? Diverges?

$-R$ 0 R

Series diverges   Series converges   Series diverges

Fig. 11.8.1 The interval of convergence if

$$0 < R = \lim_{n \to \infty} \left| \frac{a_n}{a_{n+1}} \right| < \infty$$

EXAMPLE 1 Find the interval of convergence of the series

$$\sum_{n=1}^{\infty} \frac{x^n}{n \cdot 3^n}.$$

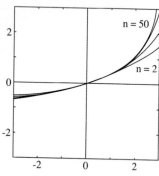

Fig. 11.8.2 Graphs of partial sums of the series of Example 1, with degrees $n = 2, 4, 10, 50$

Solution   With $u_n = x_n/(n \cdot 3^n)$ we find that

$$\lim_{n \to \infty} \left| \frac{u_{n+1}}{u_n} \right| = \lim_{n \to \infty} \left| \frac{\dfrac{x^{n+1}}{(n+1) \cdot 3^{n+1}}}{\dfrac{x^n}{n \cdot 3^n}} \right| = \lim_{n \to \infty} \frac{n|x|}{3(n+1)} = \frac{|x|}{3}.$$

Now $|x|/3 < 1$ provided that $|x| < 3$, so the ratio test implies that the given series converges absolutely if $|x| < 3$ and diverges if $|x| > 3$. When $x = 3$, we have the divergent harmonic series $\Sigma\,(1/n)$, and when $x = -3$ we have the convergent alternating series $\Sigma\,(-1)^n/n$. Thus the interval of convergence of the given power series is $[-3, 3)$. We see graphically in Fig. 11.8.2 the difference between convergence at $x = -3$ and divergence at $x = +3$.

EXAMPLE 2   Find the interval of convergence of $\displaystyle\sum_{n=0}^{\infty} \frac{2^n x^n}{n!}$.

Solution   With $u_n = 2^n x^n/n!$, we find that

$$\lim_{n \to \infty} \left| \frac{u_{n+1}}{u_n} \right| = \lim_{n \to \infty} \left| \frac{\dfrac{2^{n+1} x^{n+1}}{(n+1)!}}{\dfrac{2^n x^n}{n!}} \right| = \lim_{n \to \infty} \frac{2|x|}{n+1} = 0$$

for all x. Hence the ratio test implies that the power series converges for all x, and its interval of convergence is $(-\infty, \infty)$, the entire real line.

EXAMPLE 3   Find the interval of convergence of the series $\displaystyle\sum_{n=0}^{\infty} n^n x^n$.

Solution   With $u_n = n^n x^n$ we find that

$$\lim_{n \to \infty} \left| \frac{u_{n+1}}{u_n} \right| = \lim_{n \to \infty} \left| \frac{(n+1)^{n+1} x^{n+1}}{n^n x^n} \right| = \lim_{n \to \infty} (n+1)\left(1 + \frac{1}{n}\right)^n |x| = +\infty$$

for all $x \neq 0$, because

$$\lim_{n \to \infty} \left(1 + \frac{1}{n}\right)^n = e.$$

Thus the given series diverges for all $x \neq 0$, and its interval of convergence consists of the single point $x = 0$.

EXAMPLE 4   Use the ratio test to verify that the Taylor series for $\cos x$ in Eq. (3) converges for all x.

Solution   With $u_n = (-1)^n x^{2n}/(2n)!$ we find that

$$\lim_{n \to \infty} \left| \frac{u_{n+1}}{u_n} \right| = \lim_{n \to \infty} \left| \frac{\dfrac{(-1)^{n+1} x^{2n+2}}{(2n+2)!}}{\dfrac{(-1)^n x^{2n}}{(2n)!}} \right| = \lim_{n \to \infty} \frac{x^2}{(2n+1)(2n+2)} = 0$$

for all x, so the series converges for all x.

IMPORTANT In Example 4, the ratio test tells us only that the series for cos x converges to *some* number, *not* necessarily the particular number cos x. The argument of Section 11.4, using Taylor's formula with remainder, is required to establish that the sum of the series is actually cos x.

An infinite series of the form

$$\sum_{n=0}^{\infty} a_n(x - c)^n = a_0 + a_1(x - c) + a_2(x - c)^2 + \cdots, \qquad (9)$$

where c is a constant, is called a **power series in** (powers of) $x - c$. By the same reasoning that led us to Theorem 1, with x^n replaced by $(x - c)^n$ throughout, we conclude that either

1. The series in Eq. (9) converges absolutely for all x, or
2. That series converges only when $x - c = 0$—that is, when $x = c$—or
3. There exists a number $R > 0$ such that the series in Eq. (9) converges absolutely if $|x - c| < R$ and diverges if $|x - c| > R$.

As in the case of a power series with $c = 0$, the number R is called the **radius of convergence** of the series, and the **interval of convergence** of the series $\sum a_n(x - c)^n$ is the set of all numbers x for which it converges (Fig. 11.8.3). As before, when $0 < R < \infty$, the convergence of the series at the endpoints $x = c - R$ and $x = c + R$ of its interval of convergence must be checked separately.

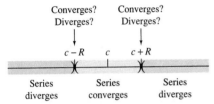

Fig. 11.8.3 The interval of convergence of

$$\sum_{n=0}^{\infty} a_n(x - c)^n$$

EXAMPLE 5 Determine the interval of convergence of the series

$$\sum_{n=1}^{\infty} \frac{(-1)^n(x - 2)^n}{n \cdot 4^n}.$$

Solution We let $u_n = (-1)^n(x - 2)^n/(n \cdot 4^n)$. Then

$$\lim_{n \to \infty} \left| \frac{u_{n+1}}{u_n} \right| = \lim_{n \to \infty} \left| \frac{\dfrac{(-1)^{n+1}(x - 2)^{n+1}}{(n + 1) \cdot 4^{n+1}}}{\dfrac{(-1)^n(x - 2)^n}{n \cdot 4^n}} \right|$$

$$= \lim_{n \to \infty} \frac{|x - 2|}{4} \cdot \frac{n}{n + 1} = \frac{|x - 2|}{4}.$$

Hence the given series converges when $|x - 2|/4 < 1$—that is, when $|x - 2| < 4$—so the radius of convergence is $R = 4$. Because $c = 2$, the series converges when $-2 < x < 6$ and diverges if either $x < -2$ or $x > 6$. When $x = -2$, the series reduces to the divergent harmonic series, and when $x = 6$, it reduces to the convergent alternating series $\sum (-1)^n/n$. Thus the interval of convergence of the given power series is $(-2, 6]$.

POWER SERIES REPRESENTATIONS OF FUNCTIONS

Power series are important tools for computing (or approximating) values of functions. Suppose that the series $\sum a_n x^n$ converges to the value $f(x)$; that is,

$$f(x) = a_0 + a_1 x + a_2 x^2 + \cdots + a_n x^n + \cdots$$

for each x in the interval of convergence of the power series. Then we call $\Sigma a_n x^n$ a **power series representation** of $f(x)$. For example, the geometric series Σx^n in Eq. (1) is a power series representation of the function $f(x) = 1/(1 - x)$ on the interval $(-1, 1)$.

We saw in Section 11.4 how Taylor's formula with remainder can often be used to find a power series representation of a given function. Recall that the nth-degree Taylor's formula for $f(x)$ at $x = a$ is

$$f(x) = f(a) + f'(a)(x - a) + \frac{f''(a)}{2!}(x - a)^2 + \frac{f^{(3)}(a)}{3!}(x - a)^3$$

$$+ \cdots + \frac{f^{(n)}(a)}{n!}(x - a)^n + R_n(x). \tag{10}$$

The remainder $R_n(x)$ is given by

$$R_n(x) = \frac{f^{(n+1)}(z)}{(n + 1)!}(x - a)^{n+1},$$

where z is some number between a and x. If we let $n \to \infty$ in Eq. (10), we obtain Theorem 2.

Theorem 2 *Taylor Series Representations*

Suppose that the function f has derivatives of all orders on some interval containing a and also that

$$\lim_{n \to \infty} R_n(x) = 0 \tag{11}$$

for each x in that interval. Then

$$f(x) = \sum_{n=0}^{\infty} \frac{f^{(n)}(a)}{n!}(x - a)^n \tag{12}$$

for each x in the interval.

The power series in Eq. (12) is the **Taylor series** of the function f **at** $x = a$ (or *in powers of $x - a$*, or *with center a*). If $a = 0$, we obtain the power series

$$f(x) = \sum_{n=0}^{\infty} \frac{f^{(n)}(0)}{n!}x^n = f(0) + f'(0)x + \frac{f''(0)}{2!}x^2 + \cdots, \tag{13}$$

commonly called the **Maclaurin series** of f. Thus the power series in Eqs. (2) through (4) are the Maclaurin series of the functions e^x, $\cos x$, and $\sin x$, respectively.

Upon replacing x by $-x$ in the Maclaurin series for e^x, we obtain

$$e^{-x} = 1 - x + \frac{x^2}{2!} - \frac{x^3}{3!} + \cdots + (-1)^n \frac{x^n}{n!} + \cdots.$$

Let us add the series for e^x and e^{-x} and divide by 2. This gives

$$\cosh x = \frac{e^x + e^{-x}}{2} = \frac{1}{2}\left(1 + x + \frac{x^2}{2!} + \frac{x^3}{3!} + \frac{x^4}{4!} + \cdots\right)$$

$$+ \frac{1}{2}\left(1 - x + \frac{x^2}{2!} - \frac{x^3}{3!} + \frac{x^4}{4!} - \cdots\right),$$

636

so

$$\cosh x = 1 + \frac{x^2}{2!} + \frac{x^4}{4!} + \frac{x^6}{6!} + \cdots.$$

Similarly,

$$\sinh x = x + \frac{x^3}{3!} + \frac{x^5}{5!} + \frac{x^7}{7!} + \cdots.$$

Note the strong resemblance to Eqs. (3) and (4), the series for $\cos x$ and $\sin x$, respectively.

Upon replacing x by $-x^2$ in the series for e^x, we obtain

$$e^{-x^2} = \sum_{n=0}^{\infty} (-1)^n \frac{x^{2n}}{n!} = 1 - x^2 + \frac{x^4}{2!} - \frac{x^6}{3!} + \cdots.$$

Because this power series converges to $\exp(-x^2)$ for all x, it must be the Maclaurin series for $\exp(-x^2)$ (see Problem 40). Think how tedious it would be to compute the derivatives of $\exp(-x^2)$ needed to write its Maclaurin series directly from Eq. (13).

THE BINOMIAL SERIES

Example 6 gives one of the most famous and useful of all series, the *binomial series*, which was discovered by Newton in the 1660s. It is the infinite series generalization of the (finite) binomial formula of elementary algebra.

EXAMPLE 6 Suppose that α is a nonzero real number. Show that the Maclaurin series of $f(x) = (1 + x)^\alpha$ is

$$(1 + x)^\alpha = 1 + \sum_{n=1}^{\infty} \frac{\alpha(\alpha - 1)(\alpha - 2) \cdots (\alpha - n + 1)}{n!} x^n$$

$$= 1 + \alpha x + \frac{\alpha(\alpha - 1)}{2!} x^2 + \frac{\alpha(\alpha - 1)(\alpha - 2)}{3!} x^3 + \cdots. \quad (14)$$

Also determine the interval of convergence of this **binomial series**.

Solution To derive the series itself, we simply list all the derivatives of $f(x) = (1 + x)^\alpha$, including its "zeroth" derivative:

$$f(x) = (1 + x)^\alpha,$$
$$f'(x) = \alpha(1 + x)^{\alpha-1},$$
$$f''(x) = \alpha(\alpha - 1)(1 + x)^{\alpha-2},$$
$$f^{(3)}(x) = \alpha(\alpha - 1)(\alpha - 2)(1 + x)^{\alpha-3},$$
$$\vdots$$
$$f^{(n)}(x) = \alpha(\alpha - 1)(\alpha - 2) \cdots (\alpha - n + 1)(1 + x)^{\alpha-n}.$$

Thus

$$f^{(n)}(0) = \alpha(\alpha - 1)(\alpha - 2) \cdots (\alpha - n + 1).$$

If we substitute this value of $f^{(n)}(0)$ into the Maclaurin series formula in Eq. (13), we get the binomial series in Eq. (14).

To determine the interval of convergence of the binomial series, we let

$$u_n = \frac{\alpha(\alpha - 1)(\alpha - 2) \cdots (\alpha - n + 1)}{n!} x^n.$$

We find that

$$\lim_{n \to \infty} \left| \frac{u_{n+1}}{u_n} \right| = \lim_{n \to \infty} \left| \frac{\dfrac{\alpha(\alpha - 1)(\alpha - 2) \cdots (\alpha - n)x^{n+1}}{(n + 1)!}}{\dfrac{\alpha(\alpha - 1)(\alpha - 2) \cdots (\alpha - n + 1)x^n}{n!}} \right|$$

$$= \lim_{n \to \infty} \left| \frac{(\alpha - n)x}{n + 1} \right| = |x|.$$

Hence the ratio test shows that the binomial series converges absolutely if $|x| < 1$ and diverges if $|x| > 1$. Its convergence at the endpoints $x = \pm 1$ depends on the value of α; we shall not pursue this problem. Problem 41 outlines a proof that the sum of the binomial series actually is $(1 + x)^\alpha$ if $|x| < 1$.

If $\alpha = k$, a positive integer, then the coefficient of x^n is zero for $n > k$, and the binomial series reduces to the binomial formula

$$(1 + x)^k = \sum_{n=0}^{k} \frac{k!}{n!(k - n)!} x^n.$$

Otherwise Eq. (14) is an infinite series. For example, with $\alpha = \frac{1}{2}$, we obtain

$$\sqrt{1 + x} = 1 + \frac{\frac{1}{2}}{1!}x + \frac{(\frac{1}{2})(-\frac{1}{2})}{2!}x^2 + \frac{(\frac{1}{2})(-\frac{1}{2})(-\frac{3}{2})}{3!}x^3$$

$$+ \frac{(\frac{1}{2})(-\frac{1}{2})(-\frac{3}{2})(-\frac{5}{2})}{4!}x^4 + \cdots$$

$$= 1 + \tfrac{1}{2}x - \tfrac{1}{8}x^2 + \tfrac{1}{16}x^3 - \tfrac{5}{128}x^4 + \cdots. \qquad (15)$$

If we replace x by $-x$ and take $\alpha = -\frac{1}{2}$, we get the series

$$\frac{1}{\sqrt{1 - x}} = 1 + \frac{-\frac{1}{2}}{1!}(-x) + \frac{(-\frac{1}{2})(-\frac{3}{2})}{2!}(-x)^2$$

$$+ \cdots + \frac{1 \cdot 3 \cdot 5 \cdots (2n - 1)}{n! \cdot 2^n}x^n + \cdots,$$

which in summation notation takes the form

$$\frac{1}{\sqrt{1 - x}} = 1 + \sum_{n=1}^{\infty} \frac{1 \cdot 3 \cdot 5 \cdots (2n - 1)}{2 \cdot 4 \cdot 6 \cdots (2n)}x^n. \qquad (16)$$

We will find this series quite useful in Example 10 and in Problem 42.

DIFFERENTIATION AND INTEGRATION OF POWER SERIES

Sometimes it is inconvenient to compute the repeated derivatives of a function in order to find its Taylor series. An alternative method of finding new power series is by the differentiation and integration of known power series.

Suppose that a power series representation of the function $f(x)$ is known. Then Theorem 3 (we leave its proof to advanced calculus) implies that the function $f(x)$ may be differentiated by separately differentiating the individual terms in its power series. That is, the power series obtained by termwise differentiation converges to the derivative $f'(x)$. Similarly, a function can be integrated by termwise integration of its power series.

Theorem 3 *Termwise Differentiation and Integration*

Suppose that the function f has a power series representation

$$f(x) = \sum_{n=0}^{\infty} a_n x^n = a_0 + a_1 x + a_2 x^2 + a_3 x^3 + \cdots$$

with nonzero radius of convergence R. Then f is differentiable on $(-R, R)$ and

$$f'(x) = \sum_{n=1}^{\infty} n a_n x^{n-1} = a_1 + 2a_2 x + 3a_3 x^2 + 4a_4 x^3 + \cdots. \quad (17)$$

Also,

$$\int_0^x f(t)\, dt = \sum_{n=0}^{\infty} \frac{a_n x^{n+1}}{n+1} = a_0 x + \tfrac{1}{2} a_1 x^2 + \tfrac{1}{3} a_2 x^3 + \cdots \quad (18)$$

for each x in $(-R, R)$. Moreover, the power series in Eqs. (17) and (18) have the same radius of convergence R.

REMARK   Although we omit the proof of Theorem 3, we observe that the radius of convergence of the series in Eq. (17) is

$$R = \lim_{n \to \infty} \left| \frac{n a_n}{(n+1) a_{n+1}} \right| = \left(\lim_{n \to \infty} \frac{n}{n+1} \right) \left(\lim_{n \to \infty} \left| \frac{a_n}{a_{n+1}} \right| \right) = \lim_{n \to \infty} \left| \frac{a_n}{a_{n+1}} \right|.$$

Thus, by Eq. (8), the power series for $f(x)$ and the power series for $f'(x)$ have the same radius of convergence (under the assumption that the preceding limit exists).

EXAMPLE 7   Termwise differentiation of the geometric series for

$$f(x) = \frac{1}{1 - x}$$

yields

$$\frac{1}{(1 - x)^2} = D_x \left(\frac{1}{1 - x} \right) = D_x (1 + x + x^2 + x^3 + \cdots)$$

$$= 1 + 2x + 3x^2 + 4x^3 + \cdots.$$

Thus

$$\frac{1}{(1 - x)^2} = \sum_{n=1}^{\infty} n x^{n-1} = \sum_{n=0}^{\infty} (n + 1)\, x^n.$$

The series converges to $1/(1 - x)^2$ if $-1 < x < 1$.

EXAMPLE 8 Replacement of x by $-t$ in the geometric series of Example 7 gives

$$\frac{1}{1+t} = 1 - t + t^2 - t^3 + \cdots + (-1)^n t^n + \cdots.$$

Because $D_t \ln(1+t) = 1/(1+t)$, termwise integration from $t = 0$ to $t = x$ now gives

$$\ln(1+x) = \int_0^x \frac{1}{1+t}\, dt$$

$$= \int_0^x (1 - t + t^2 - \cdots + (-1)^n t^n + \cdots)\, dt;$$

$$\ln(1+x) = x - \frac{1}{2}x^2 + \frac{1}{3}x^3 - \frac{1}{4}x^4 + \cdots + \frac{(-1)^{n-1}}{n}x^n + \cdots \quad (19)$$

if $|x| < 1$.

EXAMPLE 9 Find a power series representation for the arctangent function.

Solution Because $D_t \tan^{-1} t = 1/(1+t^2)$, termwise integration of the series

$$\frac{1}{1+t^2} = 1 - t^2 + t^4 - t^6 + t^8 - \cdots$$

gives

$$\tan^{-1} x = \int_0^x \frac{1}{1+t^2}\, dt = \int_0^x (1 - t^2 + t^4 - t^6 + t^8 - \cdots)\, dt.$$

Therefore,

$$\tan^{-1} x = x - \tfrac{1}{3}x^3 + \tfrac{1}{5}x^5 - \tfrac{1}{7}x^7 + \tfrac{1}{9}x^9 - \cdots \quad (20)$$

if $-1 < x < 1$.

EXAMPLE 10 Find a power series representation for the arcsine function.

Solution First we substitute t^2 for x in Eq. (16). This yields

$$\frac{1}{\sqrt{1-t^2}} = 1 + \sum_{n=1}^{\infty} \frac{1 \cdot 3 \cdot 5 \cdots (2n-1)}{2 \cdot 4 \cdot 6 \cdots (2n)} t^{2n}$$

if $|t| < 1$. Because $D_t \sin^{-1} t = 1/\sqrt{1-t^2}$, termwise integration of this series from $t = 0$ to $t = x$ gives

$$\sin^{-1} x = \int_0^x \frac{1}{\sqrt{1-t^2}}\, dt = x + \sum_{n=1}^{\infty} \frac{1 \cdot 3 \cdot 5 \cdots (2n-1)}{2 \cdot 4 \cdot 6 \cdots (2n)} \cdot \frac{x^{2n+1}}{2n+1} \quad (21)$$

if $|x| < 1$. Problem 42 shows how to use this series to derive the series

$$\frac{\pi^2}{6} = 1 + \frac{1}{2^2} + \frac{1}{3^2} + \frac{1}{4^2} + \cdots + \frac{1}{n^2} + \cdots,$$

which we used in Example 3 of Section 11.5 to approximate the number π.

Theorem 3 has this important consequence: If both power series $\Sigma\, a_n x^n$ and $\Sigma\, b_n x^n$ converge and, for all x with $|\,x\,| < R$ $(R > 0)$, $\Sigma\, a_n x^n = \Sigma\, b_n x^n$, then $a_n = b_n$ for all n. In particular, the Taylor series of a function is its unique power series representation (if any). See Problem 40.

11.8 Problems

Find the interval of convergence of the power series in Problems 1 through 20.

1. $\displaystyle\sum_{n=1}^{\infty} \frac{1}{n} x^n$

2. $\displaystyle\sum_{n=0}^{\infty} \frac{(-1)^n}{n^2 + 1} x^n$

3. $\displaystyle\sum_{n=1}^{\infty} (-1)^{n+1} n^2 x^n$

4. $\displaystyle\sum_{n=1}^{\infty} n! x^n$

5. $\displaystyle\sum_{n=1}^{\infty} \frac{(-1)^{n+1} x^{2n}}{2n - 1}$

6. $\displaystyle\sum_{n=1}^{\infty} \frac{n x^n}{5^n}$

7. $\displaystyle\sum_{n=0}^{\infty} (5x - 3)^n$

8. $\displaystyle\sum_{n=1}^{\infty} \frac{(2x - 1)^n}{n^4 + 16}$

9. $\displaystyle\sum_{n=1}^{\infty} \frac{2^n (x - 3)^n}{n^2}$

10. $\displaystyle\sum_{n=1}^{\infty} \frac{n!}{n^n} x^n$ (Do not test the endpoints; the series diverges at each.)

11. $\displaystyle\sum_{n=1}^{\infty} \frac{(2n)!}{n!} x^n$

12. $\displaystyle\sum_{n=1}^{\infty} \frac{1 \cdot 3 \cdot 5 \cdots (2n + 1)}{n!} x^n$ (Do not test the endpoints; the series diverges at each.)

13. $\displaystyle\sum_{n=1}^{\infty} \frac{n^3 (x + 1)^n}{3^n}$

14. $\displaystyle\sum_{n=1}^{\infty} \frac{(-1)^{n+1} (x - 2)^n}{n^2}$

15. $\displaystyle\sum_{n=1}^{\infty} \frac{(3 - x)^n}{n^3}$

16. $\displaystyle\sum_{n=1}^{\infty} \frac{(-1)^{n+1} 10^n}{n!} (x - 10)^n$

17. $\displaystyle\sum_{n=1}^{\infty} \frac{n!}{2^n} (x - 5)^n$

18. $\displaystyle\sum_{n=1}^{\infty} \frac{(-1)^{n+1}}{n \cdot 10^n} (x - 2)^n$

19. $\displaystyle\sum_{n=0}^{\infty} x^{(2^n)}$

20. $\displaystyle\sum_{n=0}^{\infty} \left(\frac{x^2 + 1}{5} \right)^n$

In Problems 21 through 30, use power series established in this section to find a power series representation of the given function. Then determine the radius of convergence of the resulting series.

21. $f(x) = x^2 e^{-3x}$

22. $f(x) = \dfrac{1}{10 + x}$

23. $f(x) = \sin x^2$

24. $f(x) = \cos^2 x$ [*Suggestion:* $\cos^2 x = \frac{1}{2}(1 + \cos 2x)$.]

25. $f(x) = \sqrt[3]{1 - x}$

26. $f(x) = (1 + x^2)^{3/2}$

27. $f(x) = (1 + x)^{-3}$

28. $f(x) = \dfrac{1}{\sqrt{9 + x^3}}$

29. $f(x) = \dfrac{\ln(1 + x)}{x}$

30. $f(x) = \dfrac{x - \arctan x}{x^3}$

In Problems 31 through 36, find a power series representation for the given function $f(x)$ by using termwise integration.

31. $f(x) = \displaystyle\int_0^x \sin t^3 \, dt$

32. $f(x) = \displaystyle\int_0^x \frac{\sin t}{t} \, dt$

33. $f(x) = \displaystyle\int_0^x \exp(-t^3) \, dt$

34. $f(x) = \displaystyle\int_0^x \frac{\arctan t}{t} \, dt$

35. $f(x) = \displaystyle\int_0^x \frac{1 - \exp(-t^2)}{t^2} \, dt$

36. $\tanh^{-1} x = \displaystyle\int_0^x \frac{1}{1 - t^2} \, dt$

37. Deduce from the arctangent series (Example 9) that

$$\pi = \frac{6}{\sqrt{3}} \sum_{n=0}^{\infty} \frac{(-1)^n}{2n + 1} \left(\frac{1}{3} \right)^n.$$

Then use this alternating series to show that $\pi = 3.14$ accurate to two decimal places.

38. Substitute the Maclaurin series for $\sin x$, and then assume the validity of termwise integration of the resulting series, to derive the formula

$$\int_0^{\infty} e^{-t} \sin xt \, dt = \frac{x}{1 + x^2} \qquad (|x| < 1).$$

Use the fact that

$$\int_0^{\infty} t^n e^{-t} \, dt = \Gamma(n + 1) = n!$$

(Section 9.8).

39. (a) Deduce from the Maclaurin series for e^t that

$$\frac{1}{x^x} = \sum_{n=0}^{\infty} \frac{(-1)^n}{n!} (x \ln x)^n.$$

(b) Assuming the validity of termwise integration of the series in part (a), use the integral formula of Problem 31 in Section 9.8 to conclude that

$$\int_0^1 \frac{1}{x^x} \, dx = \sum_{n=1}^{\infty} \frac{1}{n^n}.$$

40. Suppose that $f(x)$ is represented by the power series $\sum_{n=0}^{\infty} a_n x^n$ for all x in some open interval centered at $x = 0$. Show by repeated termwise differentiation of the series, substituting $x = 0$ each time, that $a_n = f^{(n)}(0)/n!$ for all n. Thus the only power series in x that represents a function at and near $x = 0$ is its Maclaurin series.

41. (a) Consider the binomial series

$$f(x) = \sum_{n=0}^{\infty} \frac{\alpha(\alpha - 1)(\alpha - 2) \cdots (\alpha - n + 1)}{n!} x^n,$$

which converges (to *something*) if $|x| < 1$. Compute the derivative $f'(x)$ by termwise differentiation, and show that it satisfies the differential equation $(1 + x)f'(x) = \alpha f(x)$. (b) Solve the differential equation in part (a) to obtain $f(x) = C(1 + x)^{\alpha}$ for some constant C. Finally, show that $C = 1$. Thus the binomial series converges to $(1 + x)^{\alpha}$ if $|x| < 1$.

42. (a) Show by direct integration that

$$\int_0^1 \frac{\arcsin x}{\sqrt{1 - x^2}} \, dx = \frac{\pi^2}{8}.$$

(b) Use the result of Problem 50 in Section 9.4 to show that

$$\int_0^1 \frac{x^{2n+1}}{\sqrt{1 - x^2}} \, dx = \frac{2 \cdot 4 \cdot 6 \cdots (2n)}{1 \cdot 3 \cdot 5 \cdots (2n + 1)}.$$

(c) Substitute the series of Example 10 for $\arcsin x$ into the integral of part (a); then use the integral of part (b) to

integrate termwise. Conclude that

$$\int_0^1 \frac{\arcsin x}{\sqrt{1 - x^2}} \, dx = 1 + \frac{1}{3^2} + \frac{1}{5^2} + \cdots.$$

(d) Note that

$$\sum_{n=1}^{\infty} \frac{1}{n^2} = \sum_{n=1}^{\infty} \frac{1}{(2n - 1)^2} + \sum_{n=1}^{\infty} \frac{1}{(2n)^2}.$$

Use this information and parts (a) and (c) to show that

$$\sum_{n=1}^{\infty} \frac{1}{n^2} = \frac{\pi^2}{6}.$$

43. Prove that if the power series $\sum a_n x^n$ converges for some $x = x_0 \neq 0$, then it converges absolutely for all x such that $|x| < |x_0|$. [*Suggestion:* Conclude from the fact that $\lim_{n \to \infty} a_n x_0^n = 0$ that $|a_n x^n| \leq |x/x_0|^n$ for n sufficiently large. Thus the series $\sum |a_n x^n|$ is eventually dominated by the geometric series $\sum |x/x_0|^n$, which converges if $|x| < |x_0|$.]

44. Suppose that the power series $\sum a_n x^n$ converges for some but not all nonzero values of x. Let S be the set of real numbers for which the series converges absolutely. (a) Conclude from Problem 43 that the set S is bounded above. (b) Let λ be the least upper bound of the set S (see Problem 46 of Section 11.2). Then show that $\sum a_n x^n$ converges absolutely if $|x| < \lambda$ and diverges if $|x| > \lambda$. Explain why this proves Theorem 1 without the additional hypothesis that $\lim_{n \to \infty} |a_{n+1}/a_n|$ exists.

11.8 Project

The dashed curve in Fig. 11.8.4 is the graph of the function

$$\frac{\sin x}{x} = 1 - \frac{x^2}{3!} + \frac{x^4}{5!} - \frac{x^6}{7!} + \cdots, \tag{22}$$

and the other graphs are those of eight of the curve's Taylor polynomial approximations. The dashed curve in Fig. 11.8.5 is the graph of the antiderivative

$$f(x) = \int_0^x \frac{\sin t}{t} \, dt = x - \frac{x^3}{3! \, 3} + \frac{x^5}{5! \, 5} - \frac{x^7}{7! \, 7} + \cdots \tag{23}$$

of the function in Eq. (22); the other graphs are those of eight of the curve's Taylor polynomial approximations.

First verify the Taylor series in Eq. (22) and (23). Then use them to produce your own versions of Figs. 11.8.4 and 11.8.5. You may want to plot a smaller or large number of Taylor polynomials, depending on the type of calculator or computer graphing program you have available. You may find useful the series summation techniques described in the projects of Sections 11.3 and 11.4.

Then do the same with one of the other function-antiderivative pairs of Problems 31 through 36 of this section.

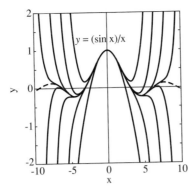

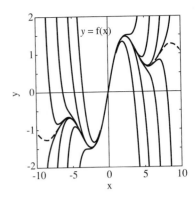

Fig. 11.8.4 Approximating the function $(\sin x)/x$ with its Taylor polynomials of degrees $n = 2, 4, 6, 8, \ldots, 16$

Fig. 11.8.5 Approximating the function $f(x) = \int_0^x (\sin t)/t \, dt$ with its Taylor polynomials of degrees $n = 3, 5, 7, \ldots, 17$

11.9
Power Series Computations

Power series often are used to approximate numerical values of functions and integrals. *Alternating* power series are especially useful. Recall the alternating series remainder (or "error") estimate of Section 11.7 (Theorem 2). It may be applied to a convergent alternating series $\Sigma(-1)^{n+1}a_n$ whose terms are decreasing in magnitude:

$$a_1 > a_2 > a_3 > \cdots > a_n > a_{n+1} > \cdots.$$

The remainder estimate says that when such a convergent alternating series is *truncated*—the terms that follow a_n are simply chopped off and discarded—the error E made has the same sign as the first term a_k omitted and $|E| < |a_k|$: The error is less in magnitude than the first omitted term in the truncation.

EXAMPLE 1 Use the binomial series

$$\sqrt{1 + x} = 1 + \tfrac{1}{2}x - \tfrac{1}{8}x^2 + \tfrac{1}{16}x^3 - \tfrac{5}{128}x^4 + \cdots$$

to approximate $\sqrt{105}$ and estimate the accuracy.

Solution The binomial series shown here is, after the first term, an alternating series. Hence

$$\sqrt{105} = \sqrt{100 + 5} = 10\sqrt{1 + 0.05}$$

$$= 10[1 + \tfrac{1}{2}(0.05) - \tfrac{1}{8}(0.05)^2 + \tfrac{1}{16}(0.05)^3 + E]$$

$$\approx 10 \cdot (1.02469\,531 + E) = 10.24695\,31 + 10E,$$

where the error E is negative and

$$|10E| < \tfrac{50}{128}(0.05)^4 < 0.000003.$$

Consequently,

$$10.246950 < \sqrt{105} < 10.246953,$$

so $\sqrt{105} = 10.24695$ to five decimal places.

Suppose that we had been asked in advance to approximate $\sqrt{105}$ accurate to five decimal places. A convenient way to do this is to continue writing terms of the series until it is clear that they have become too small in magnitude to affect the fifth decimal place. A good rule of thumb is to use two more decimal places in the computations than are required in the final answer. Thus we use seven decimal places in this case and get

$$\sqrt{105} = 10(1 + 0.05)^{1/2}$$

$$\approx 10(1 + 0.025 - 0.0003\,125 + 0.0000\,078 - 0.0000\,002 + \cdots)$$

$$\approx 10.246951 \approx 10.24695.$$

EXAMPLE 2 Approximate

$$\int_0^1 \frac{1 - \cos x}{x^2}\,dx$$

accurate to five places.

Solution We replace $\cos x$ by its Maclaurin series and get

$$\int_0^1 \frac{1 - \cos x}{x^2}\,dx = \int_0^1 \frac{1}{x^2}\left(\frac{x^2}{2!} - \frac{x^4}{4!} + \frac{x^6}{6!} - \cdots\right)dx$$

$$= \int_0^1 \left(\frac{1}{2!} - \frac{x^2}{4!} + \frac{x^4}{6!} - \frac{x^6}{8!} + \cdots\right)dx$$

$$= \frac{1}{2!} - \frac{1}{4!\,3} + \frac{1}{6!\,5} - \frac{1}{8!\,7} + \cdots.$$

Because this last series is a convergent alternating series, it follows that

$$\int_0^1 \frac{1 - \cos x}{x^2}\,dx = \frac{1}{2!} - \frac{1}{4!\,3} + \frac{1}{6!\,5} + E \approx 0.48638\,89 + E,$$

where E is negative and

$$|E| < \frac{1}{8!\,7} < 0.0000036.$$

Therefore,

$$\int_0^1 \frac{1 - \cos x}{x^2}\,dx \approx 0.48639$$

rounded to five decimal places.

EXAMPLE 3 The binomial series with $\alpha = \frac{1}{3}$ gives

$$(1 + x^2)^{1/3} = 1 + \tfrac{1}{3}x^2 - \tfrac{1}{9}x^4 + \tfrac{5}{81}x^6 - \tfrac{10}{243}x^8 + \cdots,$$

which alternates after its first term. Use the first five terms of this series to estimate the value of

$$\int_0^{1/2} \sqrt[3]{1 + x^2}\,dx.$$

Solution Termwise integration of the binomial series gives

$$\int_0^{1/2} \sqrt[3]{1 + x^2}\, dx = \int_0^{1/2} \left(1 + \tfrac{1}{3}x^2 - \tfrac{1}{9}x^4 + \tfrac{5}{81}x^6 - \tfrac{10}{243}x^8 + \cdots\right) dx$$

$$= \left[x + \tfrac{1}{9}x^3 - \tfrac{1}{45}x^5 + \tfrac{5}{567}x^7 - \tfrac{10}{2187}x^9 + \cdots \right]_0^{1/2}$$

$$= \tfrac{1}{2} + \tfrac{1}{9}\left(\tfrac{1}{2}\right)^3 - \tfrac{1}{45}\left(\tfrac{1}{2}\right)^5 + \tfrac{5}{567}\left(\tfrac{1}{2}\right)^7 - \tfrac{10}{2187}\left(\tfrac{1}{2}\right)^9 + \cdots.$$

Because this last series is a convergent alternating series, it follows that

$$\int_0^{1/2} \sqrt[3]{1 + x^2}\, dx = \tfrac{1}{2} + \tfrac{1}{9}\left(\tfrac{1}{2}\right)^3 - \tfrac{1}{45}\left(\tfrac{1}{2}\right)^5 + \tfrac{5}{567}\left(\tfrac{1}{2}\right)^7 + E \approx 0.513263 + E,$$

where E is negative and

$$|E| < \tfrac{10}{2187}\left(\tfrac{1}{2}\right)^9 < 0.000009.$$

Therefore,

$$0.513254 < \int_0^{1/2} \sqrt[3]{1 + x^2}\, dx < 0.513264,$$

so

$$\int_0^{1/2} \sqrt[3]{1 + x^2}\, dx = 0.51326 \pm 0.00001.$$

THE ALGEBRA OF POWER SERIES

Theorem 1, which we state without proof, implies that power series may be added and multiplied much like polynomials. The guiding principle is that of collecting coefficients of like powers of x.

Theorem 1 *Adding and Multiplying Power Series*

Let $\Sigma\, a_n x^n$ and $\Sigma\, b_n x^n$ be power series with nonzero radii of convergence. Then

$$\sum_{n=0}^{\infty} a_n x^n + \sum_{n=0}^{\infty} b_n x^n = \sum_{n=0}^{\infty} (a_n + b_n)x^n \qquad (1)$$

and

$$\left(\sum_{n=0}^{\infty} a_n x^n\right)\left(\sum_{n=0}^{\infty} b_n x^n\right) = \sum_{n=0}^{\infty} c_n x^n$$

$$= a_0 b_0 + (a_0 b_1 + a_1 b_0)x + (a_0 b_2 + a_1 b_1 + a_2 b_0)x^2 + \cdots, \qquad (2)$$

where

$$c_n = a_0 b_n + a_1 b_{n-1} + \cdots a_{n-1} b_1 + a_n b_0. \qquad (3)$$

The series in Eqs. (1) and (2) converge for any x that lies interior to the intervals of convergence of both $\Sigma\, a_n x^n$ and $\Sigma\, b_n x^n$.

Thus if $\Sigma\, a_n x^n$ and $\Sigma\, b_n x^n$ are power series representations of the functions $f(x)$ and $g(x)$, respectively, then the product power series $\Sigma\, c_n x^n$ found by "ordinary multiplication" and collection of terms is a power series representation of the product function $f(x)g(x)$. This fact can also be used to divide one power series by another, *provided* that the quotient is known to have a power series representation.

EXAMPLE 4 Assume that the tangent function has a power series representation $\tan x = \Sigma\, a_n x^n$. Use the Maclaurin series for $\sin x$ and $\cos x$ to find a_0, a_1, a_2, and a_3.

Solution We multiply series to obtain

$$\sin x = \tan x \cos x$$

$$= (a_0 + a_1 x + a_2 x^2 + a_3 x^3 + \cdots)\left(1 - \frac{x^2}{2} + \frac{x^4}{24} - \cdots\right)$$

$$= a_0 + a_1 x + \left(a_2 - \frac{1}{2}a_0\right)x^2 + \left(a_3 - \frac{1}{2}a_1\right)x^3 + \cdots.$$

But because

$$\sin x = x - \tfrac{1}{6}x^3 + \tfrac{1}{120}x^5 - \cdots,$$

comparison of coefficients gives the equations

$$
\begin{aligned}
a_0 & & & = 0, \\
& a_1 & & = 1, \\
-\tfrac{1}{2}a_0 & & + a_2 & = 0, \\
& -\tfrac{1}{2}a_1 & + a_3 & = -\tfrac{1}{6}.
\end{aligned}
$$

Then we find that $a_0 = 0$, $a_1 = 1$, $a_2 = 0$, and $a_3 = \tfrac{1}{3}$. So

$$\tan x = x + \tfrac{1}{3}x^3 + \cdots.$$

Things are not always as they first appear. The continuation of the tangent series is

$$\tan x = x + \tfrac{1}{3}x^3 + \tfrac{2}{15}x^5 + \tfrac{17}{315}x^7 + \cdots.$$

For the general form of the nth coefficient, see K. Knopp's *Theory and Application of Infinite Series* (New York: Hafner Press, 1971), p. 204. You may check that the first few terms agree with the result of ordinary division:

$$
\begin{array}{r}
x + \tfrac{1}{3}x^3 + \tfrac{2}{15}x^5 + \cdots \\[4pt]
\hline
1 - \tfrac{1}{2}x^2 + \tfrac{1}{24}x^4 - \cdots \overline{)\, x - \tfrac{1}{3}x^3 + \tfrac{1}{120}x^5 + \cdots}.
\end{array}
$$

POWER SERIES AND INDETERMINATE FORMS

According to Theorem 3 of Section 11.8, a power series is differentiable and therefore continuous within its interval of convergence. It follows that

$$\lim_{x \to c} \sum_{n=0}^{\infty} a_n(x - c)^n = a_0. \tag{4}$$

Examples 5 and 6 illustrate the use of this simple observation to find the limit of the indeterminate form $f(x)/g(x)$. The technique is first to substitute power series representations for $f(x)$ and $g(x)$.

EXAMPLE 5 Find $\lim\limits_{x \to 0} \dfrac{\sin x \; - \; \arctan x}{x^2 \ln(1 + x)}$.

Solution The power series of Eqs. (4), (19), and (20) in Section 11.8 give

$$\sin x - \arctan x = \left(x - \tfrac{1}{6}x^3 + \tfrac{1}{120}x^5 - \cdots\right) - \left(x - \tfrac{1}{3}x^3 + \tfrac{1}{5}x^5 - \cdots\right)$$

$$= \tfrac{1}{6}x^3 - \tfrac{23}{120}x^5 + \cdots$$

and

$$x^2 \ln(1 + x) = x^2\left(x - \tfrac{1}{2}x^2 + \tfrac{1}{3}x^3 - \cdots\right) = x^3 - \tfrac{1}{2}x^4 + \tfrac{1}{3}x^5 - \cdots.$$

Hence

$$\lim_{x \to 0} \frac{\sin x - \arctan x}{x^2 \ln(1 + x)} = \lim_{x \to 0} \frac{\tfrac{1}{6}x^3 - \tfrac{23}{120}x^5 + \cdots}{x^3 - \tfrac{1}{2}x^4 + \cdots}$$

$$= \lim_{x \to 0} \frac{\tfrac{1}{6} - \tfrac{23}{120}x^2 + \cdots}{1 - \tfrac{1}{2}x + \cdots} = \frac{1}{6}.$$

EXAMPLE 6 Find $\lim\limits_{x \to 1} \dfrac{\ln x}{x - 1}$.

Solution We first replace x by $x - 1$ in the power series for $\ln(1 + x)$ used in Example 5. This gives us

$$\ln x = (x - 1) - \tfrac{1}{2}(x - 1)^2 + \tfrac{1}{3}(x - 1)^3 - \cdots.$$

Hence

$$\lim_{x \to 1} \frac{\ln x}{x - 1} = \lim_{x \to 1} \frac{(x - 1) - \tfrac{1}{2}(x - 1)^2 + \tfrac{1}{3}(x - 1)^3 - \cdots}{x - 1}$$

$$= \lim_{x \to 1} \left[1 - \frac{1}{2}(x - 1) + \frac{1}{3}(x - 1)^2 - \cdots \right] = 1.$$

The method of Examples 5 and 6 provides a useful alternative to l'Hôpital's rule, especially when repeated differentiation of numerator and denominator is inconvenient or too time-consuming.

11.9 Problems

In Problems 1 through 10, use an infinite series to approximate the indicated number accurate to three decimal places.

1. $\sqrt[3]{65}$ **2.** $\sqrt[4]{630}$ **3.** $\sin(0.5)$ **4.** $e^{-0.2}$

5. $\tan^{-1}(0.5)$ **6.** $\ln(1.1)$ **7.** $\sin\left(\dfrac{\pi}{10}\right)$

8. $\cos\left(\dfrac{\pi}{20}\right)$ **9.** $\sin 10°$ **10.** $\cos 5°$

In Problems 11 through 20, use an infinite series to approximate the value of the given integral accurate to three decimal places.

11. $\displaystyle\int_0^1 \frac{\sin x}{x} \, dx$ **12.** $\displaystyle\int_0^1 \frac{\sin x}{\sqrt{x}} \, dx$

13. $\displaystyle\int_0^{0.5} \frac{\arctan x}{x} \, dx$ **14.** $\displaystyle\int_0^1 \sin x^2 \, dx$

15. $\displaystyle\int_0^{0.1} \frac{\ln(1 + x)}{x} \, dx$ **16.** $\displaystyle\int_0^{0.5} \frac{1}{\sqrt{1 + x^4}} \, dx$

17. $\displaystyle\int_0^{0.5} \frac{1 - e^{-x}}{x} \, dx$ **18.** $\displaystyle\int_0^{0.5} \sqrt{1 + x^3} \, dx$

19. $\displaystyle\int_0^1 e^{-x^2} \, dx$ **20.** $\displaystyle\int_0^{0.5} \frac{1}{1 + x^5} \, dx$

In Problems 21 through 26, use power series rather than l'Hôpital's rule to evaluate the given limit.

21. $\lim\limits_{x \to 0} \dfrac{1 + x - e^x}{x^2}$ **22.** $\lim\limits_{x \to 0} \dfrac{x - \sin x}{x^3 \cos x}$

23. $\lim\limits_{x \to 0} \dfrac{1 - \cos x}{x(e^x - 1)}$

24. $\lim\limits_{x \to 0} \dfrac{e^x - e^{-x} - 2x}{x - \arctan x}$

25. $\lim\limits_{x \to 0} \left(\dfrac{1}{x} - \dfrac{1}{\sin x} \right)$

26. $\lim\limits_{x \to 1} \dfrac{\ln(x^2)}{x - 1}$

27. Derive the geometric series by long division of $1 - x$ into 1.

28. Derive the series for $\tan x$ listed in Example 4 by long division of the Taylor series of $\cos x$ into the Taylor series of $\sin x$.

29. Derive the geometric series representation of $1/(1 - x)$ by finding $a_0, a_1, a_2, \ldots$ such that

$$(1 - x)(a_0 + a_1 x + a_2 x^2 + a_3 x^3 + \cdots) = 1.$$

30. Derive the first five coefficients in the binomial series for $\sqrt{1 + x}$ by finding a_0, a_1, a_2, a_3, and a_4 such that

$$(a_0 + a_1 x + a_2 x^2 + a_3 x^3 + a_4 x^4 + \cdots)^2 = 1 + x.$$

31. Use the method of Example 4 to find the coefficients a_0, a_1, a_2, a_3, and a_4 in the series

$$\sec x = \dfrac{1}{\cos x} = \sum_{n=0}^{\infty} a_n x^n.$$

32. Multiply the geometric series for $1/(1 - x)$ and the series for $\ln(1 - x)$ to show that if $|x| < 1$, then

$$\dfrac{1}{1 - x} \ln(1 - x) = x + (1 + \tfrac{1}{2})x^2 + (1 + \tfrac{1}{2} + \tfrac{1}{3})x^3$$
$$+ (1 + \tfrac{1}{2} + \tfrac{1}{3} + \tfrac{1}{4})x^4 + \cdots.$$

33. Take as known the logarithmic series

$$\ln(1 + x) = x - \tfrac{1}{2}x^2 + \tfrac{1}{3}x^3 - \tfrac{1}{4}x^4 + \cdots.$$

Find the first four coefficients in the series for e^x by finding a_0, a_1, a_2, and a_3 such that

$$1 + x = e^{\ln(1 + x)} = \sum_{n=0}^{\infty} a_n(x - \tfrac{1}{2}x^2 + \tfrac{1}{3}x^3 - \tfrac{1}{4}x^4 + \cdots)^n.$$

This is exactly how the power series for e^x was first discovered (by Newton)!

34. Use the method of Example 4 to show that

$$\dfrac{x}{\sin x} = 1 + \dfrac{1}{6}x^2 + \dfrac{7}{360}x^4 + \cdots.$$

35. Show that long division of power series gives

$$\dfrac{2 + x}{1 + x + x^2} = 2 - x - x^2 + 2x^3 - x^4 - x^5 + 2x^6$$
$$- x^7 - x^8 + 2x^9 - x^{10} - x^{11} + \cdots.$$

Show also that the radius of convergence of this series is $R = 1$.

11.9 Project

This project explores the use of known power series such as

$$\sin x = x - \dfrac{x^3}{6} + \dfrac{x^5}{120} - \dfrac{x^7}{5040} + \cdots, \tag{5}$$

$$\tan x = x + \dfrac{x^3}{3} + \dfrac{2x^5}{15} + \dfrac{17x^7}{315} + \cdots, \tag{6}$$

$$\sin^{-1} x = x + \dfrac{x^3}{6} + \dfrac{3x^5}{40} + \dfrac{5x^7}{112} + \cdots, \tag{7}$$

$$\tan^{-1} x = x - \dfrac{x^3}{3} + \dfrac{x^5}{5} - \dfrac{x^7}{7} + \cdots \tag{8}$$

to evaluate certain indeterminate forms for which the use of l'Hôpital's rule would be inconvenient or impractical.

1. Evaluate

$$\lim_{x \to 0} \dfrac{\sin x - \tan x}{\sin^{-1} x - \tan^{-1} x}. \tag{9}$$

When you substitute Eqs. (5) and (6) into the numerator and Eqs. (7) and (8) into the denominator, you will find that both the numerator series and the denominator series have leading terms that are multiples of x^3. Hence division of each term by x^3 leads quickly to the desired value of the indeterminate form. Can you explain why this implies that l'Hôpital's rule would have to be applied three times in succession? Would you like to calculate (by hand) the third derivative of the denominator in (9)?

2. Evaluate

$$\lim_{x \to 0} \frac{\sin(\tan x) - \tan(\sin x)}{\sin^{-1}(\tan^{-1} x) - \tan^{-1}(\sin^{-1} x)}. \tag{10}$$

This is a much more substantial problem, and the use of a computer algebra system is highly desirable. The table in Fig. 11.9.1 lists commands in common systems that generate the Taylor series

$$\sin(\tan x) = x + \frac{x^3}{6} - \frac{x^5}{40} - \frac{55x^7}{1008} + \cdots. \tag{11}$$

Program	Command
Derive	`taylor( sin(tan(x)), x,0,7 )`
Maple	`taylor( sin(tan(x)), x=0,8 )`
Mathematica	`Series[ Sin[Tan[x]], {x,0,7} ]`

Fig. 11.9.1 Computer algebra system commands that generate the series in Eq. (11)

If one of these systems is available to you, generate similarly the Taylor series

$$\tan(\sin x) = x + \frac{x^3}{6} - \frac{x^5}{40} - \frac{107x^7}{5040} + \cdots, \tag{12}$$

$$\sin^{-1}(\tan^{-1} x) = x - \frac{x^3}{6} + \frac{13x^5}{120} - \frac{341x^7}{5040} + \cdots, \quad \text{and} \tag{13}$$

$$\tan^{-1}(\sin^{-1} x) = x - \frac{x^3}{6} + \frac{13x^5}{120} - \frac{173x^7}{5040} + \cdots. \tag{14}$$

Then show that when you substitute Eqs. (11) and (12) into the numerator of Eq. (10) and Eqs. (13) and (14) into the denominator, both resulting series have leading terms that are multiples of x^7. Hence division of each series by x^7 leads quickly to the desired evaluation. It follows that seven successive applications of l'Hôpital's rule would be required to evaluate the indeterminate form in Eq. (10). (Why?) Can you even conceive of calculating the seventh derivative of either the numerator or denominator in (10)? The seventh derivative of $\sin(\tan x)$ is a sum of sixteen terms, a typical one of which is $3696 \cos(\tan x) \sec^2 x \tan^2 x$.

The fact that the power series for both

$$f(x) = \sin(\tan x) - \tan(\sin x) \tag{15}$$

and

$$g(x) = \sin^{-1}(\tan^{-1} x) - \tan^{-1}(\sin^{-1} x) \tag{16}$$

begin with the term involving x^7 means, geometrically, that the graphs of both are quite "flat" near the origin (Figs. 11.9.2 and 11.9.3).

The graph of $y = f(x)$ in Fig. 11.9.2 is especially exotic. Can you explain the conspicuous oscillations that appear in certain sections of the graph?

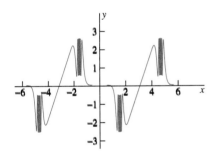

Fig. 11.9.2 The graph $y = \sin(\tan x) - \tan(\sin x)$

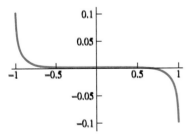

Fig. 11.9.3 The graph $y = \sin^{-1}(\tan^{-1} x) - \tan^{-1}(\sin^{-1} x)$

Chapter 11 Review: DEFINITIONS, CONCEPTS, RESULTS

Use the following list as a guide to concepts that you may need to review.

1. Definition of the limit of a sequence
2. The limit laws for sequences
3. The bounded monotonic sequence property
4. Definition of the sum of an infinite series
5. Formula for the sum of a geometric series
6. The nth-term test for divergence
7. Divergence of the harmonic series
8. The nth-degree Taylor polynomial of the function f at the point $x = a$
9. Taylor's formula with remainder
10. The Taylor series of the elementary transcendental functions
11. The integral test
12. Convergence of p-series
13. The comparison and limit comparison tests
14. The alternating series test
15. Absolute convergence: definition *and* the fact that it implies convergence
16. The ratio test
17. The root test
18. Power series; the radius and interval of convergence
19. The binomial series
20. Termwise differentiation and integration of power series
21. The use of power series to approximate values of functions and integrals
22. The product of two power series
23. The use of power series to evaluate indeterminate forms

Chapter 11 Miscellaneous Problems

In Problems 1 through 15, determine whether or not the sequence $\{a_n\}$ converges, and find its limit if it does converge.

1. $a_n = \dfrac{n^2 + 1}{n^2 + 4}$

2. $a_n = \dfrac{8n - 7}{7n - 8}$

3. $a_n = 10 - (0.99)^n$

4. $a_n = n \sin \pi n$

5. $a_n = \dfrac{1 + (-1)^n \sqrt{n}}{n + 1}$

6. $a_n = \sqrt{\dfrac{1 + (-0.5)^n}{n + 1}}$

7. $a_n = \dfrac{\sin 2n}{n}$

8. $a_n = 2^{-(\ln n)/n}$

9. $a_n = (-1)^{\sin(n\pi/2)}$

10. $a_n = \dfrac{(\ln n)^3}{n^2}$

11. $a_n = \dfrac{1}{n} \sin \dfrac{1}{n}$

12. $a_n = \dfrac{n - e^n}{n + e^n}$

13. $a_n = \dfrac{\sinh n}{n}$

14. $a_n = \left(1 + \dfrac{2}{n}\right)^{2n}$

15. $a_n = (2n^2 + 1)^{1/n}$

Determine whether each infinite series in Problems 16 through 30 converges or diverges.

16. $\displaystyle\sum_{n=1}^{\infty} \dfrac{(n^2)!}{n^n}$

17. $\displaystyle\sum_{n=1}^{\infty} \dfrac{(-1)^{n+1} \ln n}{n}$

18. $\displaystyle\sum_{n=0}^{\infty} \dfrac{3^n}{2^n + 4^n}$

19. $\displaystyle\sum_{n=0}^{\infty} \dfrac{n!}{e^{n^2}}$

20. $\displaystyle\sum_{n=1}^{\infty} \dfrac{1}{n^{3/2}} \sin \dfrac{1}{n}$

21. $\displaystyle\sum_{n=0}^{\infty} \dfrac{(-2)^n}{3^n + 1}$

22. $\displaystyle\sum_{n=1}^{\infty} 2^{-(2/n^2)}$

23. $\displaystyle\sum_{n=2}^{\infty} \dfrac{(-1)^n n}{(\ln n)^3}$

24. $\displaystyle\sum_{n=1}^{\infty} \dfrac{(-1)^n}{10^{1/n}}$

25. $\displaystyle\sum_{n=1}^{\infty} \dfrac{\sqrt{n} + \sqrt[3]{n}}{n^2 + n^3}$

26. $\displaystyle\sum_{n=1}^{\infty} \dfrac{(-1)^{n+1}}{n^{[1+(1/n)]}}$

27. $\displaystyle\sum_{n=1}^{\infty} \dfrac{(-1)^{n+1} \arctan n}{\sqrt{n}}$

28. $\displaystyle\sum_{n=1}^{\infty} n \sin \dfrac{1}{n}$

29. $\displaystyle\sum_{n=3}^{\infty} \dfrac{1}{n(\ln n)(\ln \ln n)}$

30. $\displaystyle\sum_{n=3}^{\infty} \dfrac{1}{n(\ln n)(\ln \ln n)^2}$

Find the interval of convergence of the power series in Problems 31 through 40.

31. $\displaystyle\sum_{n=0}^{\infty} \dfrac{2^n x^n}{n!}$

32. $\displaystyle\sum_{n=0}^{\infty} \dfrac{(3x)^n}{2^{n+1}}$

33. $\displaystyle\sum_{n=1}^{\infty} \dfrac{(x - 1)^n}{n \cdot 3^n}$

34. $\displaystyle\sum_{n=0}^{\infty} \dfrac{(2x - 3)^n}{4^n}$

35. $\displaystyle\sum_{n=1}^{\infty} \frac{(-1)^n x^n}{4n^2 - 1}$ **36.** $\displaystyle\sum_{n=0}^{\infty} \frac{(2x - 1)^n}{n^2 + 1}$

37. $\displaystyle\sum_{n=0}^{\infty} \frac{n! \, x^{2n}}{10^n}$ **38.** $\displaystyle\sum_{n=2}^{\infty} \frac{x^n}{\ln n}$

39. $\displaystyle\sum_{n=0}^{\infty} \frac{1 + (-1)^n}{2(n!)} x^n$ **40.** $\displaystyle\sum_{n=1}^{\infty} \left(1 + \frac{1}{n}\right)^n (x - 1)^n$

Find the set of values of x for which each series in Problems 41 through 43 converges.

41. $\displaystyle\sum_{n=1}^{\infty} (x - n)^n$ **42.** $\displaystyle\sum_{n=1}^{\infty} (\ln x)^n$ **43.** $\displaystyle\sum_{n=0}^{\infty} \frac{e^{nx}}{n!}$

44. Find the rational number that has repeated decimal expansion $2.7\,1828\,1828\,1828\ldots$.

45. Give an example of two convergent numerical series $\Sigma \, a_n$ and $\Sigma \, b_n$ such that the series $\Sigma \, a_n b_n$ diverges.

46. Prove that if $\Sigma \, a_n$ is a convergent positive-term series, then $\Sigma \, a_n^2$ converges.

47. Let the sequence $\{a_n\}$ be defined recursively by

$$a_1 = 1, \qquad a_{n+1} = 1 + \frac{1}{1 + a_n} \qquad \text{if } n \geq 1.$$

The limit of the sequence $\{a_n\}$ is the value of the *continued fraction*

$$1 + \cfrac{1}{2 + \cfrac{1}{2 + \cfrac{1}{2 + \cfrac{1}{2 + \cdots}}}}.$$

Assuming that $A = \lim_{n \to \infty} a_n$ exists, prove that $A = \sqrt{2}$.

48. Let $\{F_n\}_1^{\infty}$ be the Fibonacci sequence of Example 2 in Section 11.2. (a) Prove that $0 < F_n \leq 2^n$ for all $n \geq 1$, and hence conclude that the power series $F(x) = \sum_{n=1}^{\infty} F_n x^n$ converges if $|x| < \frac{1}{2}$. (b) Show that $(1 - x - x^2)F(x) = x$, so

$$F(x) = \frac{x}{1 - x - x^2}.$$

49. We say that the *infinite product* indicated by

$$\prod_{n=1}^{\infty} (1 + a_n) = (1 + a_1)(1 + a_2)(1 + a_3) \cdots$$

converges provided that the infinite series

$$S = \sum_{n=1}^{\infty} \ln(1 + a_n)$$

converges, in which case the value of the infinite product is e^S. Use the integral test to prove that

$$\prod_{n=1}^{\infty} \left(1 + \frac{1}{n}\right)$$

diverges.

50. Prove that the infinite product (see Problem 49)

$$\prod_{n=1}^{\infty} \left(1 + \frac{1}{n^2}\right)$$

converges, and use the integral test remainder estimate to approximate its value. The actual value of this infinite product is known to be $(\sinh \pi)/\pi \approx 3.67607\,791$.

In Problems 51 through 55, use infinite series to approximate the indicated number accurate to three decimal places.

51. $\sqrt[5]{1.5}$ **52.** $\ln(1.2)$ **53.** $\displaystyle\int_0^{0.5} e^{-x^2} \, dx$

54. $\displaystyle\int_0^{0.5} \sqrt[3]{1 + x^4} \, dx$ **55.** $\displaystyle\int_0^1 \frac{1 - e^{-x}}{x} \, dx$

56. Substitute the Maclaurin series for $\sin x$ into that for e^x to obtain

$$e^{\sin x} = 1 + x + \tfrac{1}{2}x^2 - \tfrac{1}{8}x^4 + \cdots.$$

57. Substitute the Maclaurin series for the cosine and then integrate termwise to derive the formula

$$\int_0^{\infty} e^{-t^2} \cos 2xt \, dt = \frac{\sqrt{\pi}}{2} e^{-x^2}.$$

Use the reduction formula

$$\int_0^{\infty} t^{2n} e^{-t^2} \, dt = \frac{2n - 1}{2} \int_0^{\infty} t^{2n-2} e^{-t^2} \, dt$$

derived in Problem 42 of Section 9.4. The validity of this improper termwise integration is subject to verification.

58. Prove that

$$\tanh^{-1} x = \int_0^x \frac{1}{1 - t^2} \, dt = \sum_{n=0}^{\infty} \frac{x^{2n+1}}{2n + 1}$$

if $|x| < 1$.

59. Prove that

$$\sinh^{-1} x = \int_0^x \frac{1}{\sqrt{1 + t^2}} \, dt$$

$$= \sum_{n=0}^{\infty} (-1)^n \frac{1 \cdot 3 \cdot 5 \cdots (2n - 1)}{2 \cdot 4 \cdot 6 \cdots (2n)} \cdot \frac{x^{2n+1}}{2n + 1}$$

if $|x| < 1$.

60. Suppose that $\tan y = \Sigma \, a_n y^n$. Determine a_0, a_1, a_2, and a_3 by substituting the inverse tangent series [Eq.(27) of Section 11.4] into the equation

$$x = \tan(\tan^{-1} x) = \sum_{n=0}^{\infty} a_n (\tan^{-1} x)^n.$$

61. According to *Stirling's series*, the value of $n!$ for large n is given to a close approximation by

$$n! = \sqrt{2\pi n}\left(\frac{n}{e}\right)^n e^{\mu(n)},$$

where

$$\mu(n) = \frac{1}{12n} - \frac{1}{360n^3} + \frac{1}{1260n^5}.$$

Substitute $\mu(n)$ into Maclaurin's series for e^x to show that

$$e^{\mu(n)} = 1 + \frac{1}{12n} + \frac{1}{288n^2} - \frac{139}{51,840n^3} + \cdots.$$

Can you show that the next term in the last series is $-571/(2,488,320n^4)$?

62. Define

$$T(n) = \int_0^{\pi/4} \tan^n x \, dx$$

for $n \geq 0$. (a) Show by "reduction" of the integral that

$$T(n + 2) = \frac{1}{n + 1} - T(n)$$

for $n \geq 0$. (b) Conclude that $T(n) \to 0$ as $n \to \infty$. (c) Show that $T_0 = \pi/4$ and that $T_1 = \frac{1}{2}\ln 2$. (d) Prove by induction on n that

$$T(2n) = (-1)^{n+1}\left(1 - \frac{1}{3} + \frac{1}{5} - \cdots \pm \frac{1}{2n - 1} - \frac{\pi}{4}\right).$$

(e) Conclude from parts (b) and (d) that

$$1 - \frac{1}{3} + \frac{1}{5} - \frac{1}{7} + \cdots = \frac{\pi}{4}.$$

(f) Prove by induction on n that

$$T(2n + 1) = \frac{1}{2}(-1)^n\left(1 - \frac{1}{2} + \frac{1}{3} - \cdots \pm \frac{1}{n} - \ln 2\right).$$

(g) Conclude from parts (b) and (f) that

$$1 - \frac{1}{2} + \frac{1}{3} - \frac{1}{4} + \cdots = \ln 2.$$

63. Prove as follows that the number e is irrational. First suppose to the contrary that $e = p/q$, where p and q are positive integers. Note that $q > 1$. Write

$$\frac{p}{q} = e = \frac{1}{0!} + \frac{1}{1!} + \frac{1}{2!} + \frac{1}{3!} + \cdots + \frac{1}{q!} + R_q,$$

where $0 < R_q < 3/(q + 1)!$. (Why?) Then show that multiplication of both sides of this equation by $q!$ would lead to the contradiction that one side of the result is an integer but the other side is not.

64. Evaluate the infinite product (see Problem 49)

$$\prod_{n=2}^{\infty} \frac{n^2}{n^2 - 1}$$

by finding an explicit formula for

$$\prod_{n=2}^{k} \frac{n^2}{n^2 - 1} \qquad (k \geq 2)$$

and then taking the limit as $k \to \infty$.

Appendices

Appendix A
Review of Trigonometry

In elementary trigonometry, the six basic trigonometric functions of an acute angle θ in a right triangle are defined as ratios between pairs of sides of the triangle. As in Fig. A.1, where "adj" stands for "adjacent," "opp" for "opposite," and "hyp" for "hypotenuse,"

$$\cos \theta = \frac{\text{adj}}{\text{hyp}}, \qquad \sin \theta = \frac{\text{opp}}{\text{hyp}}, \qquad \tan \theta = \frac{\text{opp}}{\text{adj}},$$

$$\sec \theta = \frac{\text{hyp}}{\text{adj}}, \qquad \csc \theta = \frac{\text{hyp}}{\text{opp}}, \qquad \cot \theta = \frac{\text{adj}}{\text{opp}}. \tag{1}$$

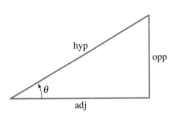

Fig. A.1 The sides and angle θ of a right triangle

We generalize these definitions to *directed* angles of arbitrary size in the following way. Suppose that the initial side of the angle θ is the positive x-axis, so its vertex is at the origin. The angle is **directed** if a direction of rotation from its initial side to its terminal side is specified. We call θ a **positive angle** if this rotation is counterclockwise and a **negative angle** if it is clockwise.

Let $P(x, y)$ be the point at which the terminal side of θ intersects the *unit circle* $x^2 + y^2 = 1$. Then we define

$$\cos \theta = x, \qquad \sin \theta = y, \qquad \tan \theta = \frac{y}{x},$$

$$\sec \theta = \frac{1}{x}, \qquad \csc \theta = \frac{1}{y}, \qquad \cot \theta = \frac{x}{y}. \tag{2}$$

We assume that $x \neq 0$ in the case of $\tan \theta$ and $\sec \theta$ and that $y \neq 0$ in the case of $\cot \theta$ and $\csc \theta$. If the angle θ is positive and acute, then it is clear from Fig. A.2 that the definitions in Eqs. (2) agree with the right triangle definitions in Eqs. (1) in terms of the coordinates of P. A glance at the figure also shows which of the functions are positive for angles in each of the four quadrants. Figure A.3 summarizes this information.

Here we discuss primarily the two most basic trigonometric functions, the sine and the cosine. From Eqs. (2) we see immediately that the other four trigonometric functions are defined in terms of $\sin \theta$ and $\cos \theta$ by

$$\tan \theta = \frac{\sin \theta}{\cos \theta}, \qquad \sec \theta = \frac{1}{\cos \theta},$$

$$\cot \theta = \frac{\cos \theta}{\sin \theta}, \qquad \csc \theta = \frac{1}{\sin \theta}. \tag{3}$$

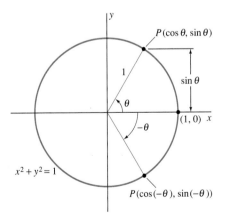

Fig. A.2 Using the unit circle to define the trigonometric functions

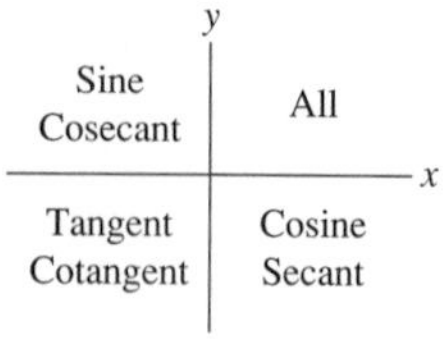

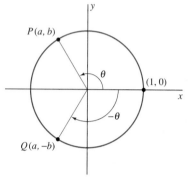

Sine Cosecant	All
Tangent Cotangent	Cosine Secant

Positive in quadrants shown

Fig. A.3 The signs of the trigonometric functions

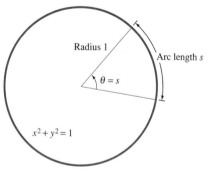

Fig. A.4 The effect of replacing θ by $-\theta$ in sine and cosine functions

Next, we compare the angles θ and $-\theta$ in Fig. A.4. We see that

$$\cos(-\theta) = \cos\theta \quad \text{and} \quad \sin(-\theta) = -\sin\theta. \tag{4}$$

Because $x = \cos\theta$ and $y = \sin\theta$ in Eqs. (2), the equation $x^2 + y^2 = 1$ of the unit circle translates immediately into the **fundamental identity of trigonometry,**

$$\cos^2\theta + \sin^2\theta = 1. \tag{5}$$

Dividing each term of this fundamental identity by $\cos^2\theta$ gives the identity

$$1 + \tan^2\theta = \sec^2\theta. \tag{5'}$$

Similarly, dividing each term in Eq. (5) by $\sin^2\theta$ yields the identity

$$1 + \cot^2\theta = \csc^2\theta. \tag{5''}$$

(See Problem 9 of this appendix.)

In Problems 15 and 16 we outline derivations of the **addition formulas**

$$\sin(\alpha + \beta) = \sin\alpha\cos\beta + \cos\alpha\sin\beta, \tag{6}$$

$$\cos(\alpha + \beta) = \cos\alpha\cos\beta - \sin\alpha\sin\beta. \tag{7}$$

With $\alpha = \theta = \beta$ in Eqs. (6) and (7), we get the **double-angle formulas**

$$\sin 2\theta = 2\sin\theta\cos\theta, \tag{8}$$

$$\cos 2\theta = \cos^2\theta - \sin^2\theta \tag{9}$$

$$= 2\cos^2\theta - 1 \tag{9a}$$

$$= 1 - 2\sin^2\theta, \tag{9b}$$

where Eqs. (9a) and (9b) are obtained from Eq. (9) by use of the fundamental identity in Eq. (5).

If we solve Eq. (9a) for $\cos^2\theta$ and Eq. (9b) for $\sin^2\theta$, we get the **half-angle formulas**

$$\cos^2\theta = \tfrac{1}{2}(1 + \cos 2\theta), \tag{10}$$

$$\sin^2\theta = \tfrac{1}{2}(1 - \cos 2\theta). \tag{11}$$

Equations (10) and (11) are especially important in integral calculus.

RADIAN MEASURE

In elementary mathematics, angles frequently are measured in *degrees,* with $360°$ in one complete revolution. In calculus it is more convenient—and is often essential—to measure angles in *radians.* The **radian measure** of an angle is the length of the arc it subtends in (that is, the arc it cuts out of) the unit circle when the vertex of the angle is at the center of the circle (Fig. A.5).

Recall that the area A and circumference C of a circle of radius r are given by the formulas

$$A = \pi r^2 \quad \text{and} \quad C = 2\pi r,$$

where the irrational number π is approximately 3.14159. Because the circumference of the unit circle is 2π and its central angle is $360°$, it follows that

$$2\pi \text{ rad} = 360°; \qquad 180° = \pi \text{ rad} \approx 3.14159 \text{ rad}. \tag{12}$$

Fig. A.5 The radian measure of an angle

Radians	Degrees
0	0
$\pi/6$	30
$\pi/4$	45
$\pi/3$	60
$\pi/2$	90
$2\pi/3$	120
$3\pi/4$	135
$5\pi/6$	150
π	180
$3\pi/2$	270
2π	360
4π	720

Fig. A.6 Some radian-degree conversions

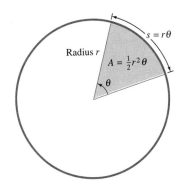

Fig. A.7 The area of a sector and arc length of a circle

Using Eq. (12) we can easily convert back and forth between radians and degrees:

$$1 \text{ rad} = \frac{180°}{\pi} \approx 57°\,17'\,44.8'', \tag{12a}$$

$$1° = \frac{\pi}{180} \text{ raq} \approx 0.01745 \text{ rad.} \tag{12b}$$

Figure A.6 shows radian-degree conversioins for some common angles.

Now consider an angle of θ radians at the center of a circle of radius r (Fig. A.7). Denote by s the length of the arc subtended by θ; denote by A the area of the sector of the circle bounded by this angle. Then the proportions

$$\frac{s}{2\pi r} = \frac{A}{\pi r^2} = \frac{\theta}{2\pi}$$

give the formulas

$$s = r\theta \qquad (\theta \text{ in radians}) \tag{13}$$

and

$$A = \tfrac{1}{2} r^2 \theta \qquad (\theta \text{ in radians}). \tag{14}$$

The definitions in Eqs. (2) refer to trigonometric functions of *angles* rather than trigonometric functions of *numbers*. Suppose that t is a real number. Then the number $\sin t$ is, *by definition,* the sine of an angle of t radians—recall that a positive angle is directed counterclockwise from the positive x-axis, whereas a negative angle is directed clockwise. Briefly, $\sin t$ is the sine of an angle of t *radians.* The other trigonometric functions of the number t have similar definitions. Hence, when we write $\sin t$, $\cos t$, and so on, with t a real number, it is *always* in reference to an angle of t *radians.*

When we need to refer to the sine of an angle of t *degrees,* we will henceforth write $\sin t°$. The point is that $\sin t$ and $\sin t°$ are quite different functions of the variable t. For example, you would get

$$\sin 1° \approx 0.0175 \quad \text{and} \quad \sin 30° = 0.5000$$

on a calculator set in degree mode. But in radian mode, a calculator would give

$$\sin 1 \approx 0.8415 \quad \text{and} \quad \sin 30 \approx -0.9880.$$

The relationship between the functions $\sin t$ and $\sin t°$ is

$$\sin t° = \sin\!\left(\frac{\pi t}{180}\right). \tag{15}$$

The distinction extends even to programming languages. In FORTRAN, the function `SIN` is the radian sine function, and you must write $\sin t°$ in the form `SIND(T)`. In BASIC you must write `SIN(PI*T/180)` to get the correct value of the sine of an angle of t degrees.

An angle of 2π rad corresponds to one revolution around the unit circle. This implies that the sine and cosine functions have **period** 2π, meaning that

$$\sin(t + 2\pi) = \sin t,$$
$$\cos(t + 2\pi) = \cos t. \tag{16}$$

It follows from Eqs. (16) that

Fig. A.8 Periodicity of the sine and cosine functions

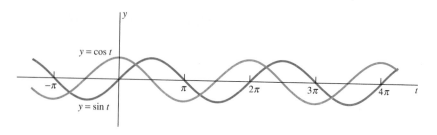

$$\sin(t + 2n\pi) = \sin t \quad \text{and} \quad \cos(t + 2n\pi) = \cos t \qquad (17)$$

for any integer n. This periodicity of the sine and cosine functions is evident in their graphs (Fig. A.8). From Eqs. (3), the other four trigonometric functions also must be periodic, as their graphs in Figs. A.9 and A.10 show.

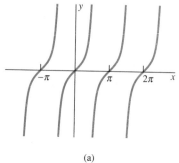

(a)

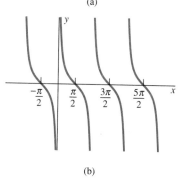

(b)

Fig. A.9 The graphs of (a) the tangent function and (b) the cotangent function

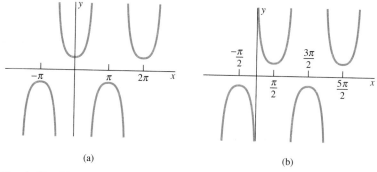

(a)   (b)

Fig. A.10 The graphs of (a) the secant function and (b) the cosecant function

We see from Eqs. (2) that

$$\sin 0 = 0, \quad \sin \frac{\pi}{2} = 1, \quad \sin \pi = 0,$$

$$\cos 0 = 1, \quad \cos \frac{\pi}{2} = 0, \quad \cos \pi = -1. \qquad (18)$$

The trigonometric functions of $\pi/6$, $\pi/4$, and $\pi/3$ (the radian equivalents of $30°$, $45°$, and $60°$, respectively) are easy to read from the well-known triangles of Fig. A.11. For instance,

$$\sin \frac{\pi}{6} = \cos \frac{\pi}{3} = \frac{1}{2} = \frac{\sqrt{1}}{2},$$

$$\sin \frac{\pi}{4} = \cos \frac{\pi}{4} = \frac{1}{\sqrt{2}} = \frac{\sqrt{2}}{2}, \quad \text{and} \qquad (19)$$

$$\sin \frac{\pi}{3} = \cos \frac{\pi}{6} = \frac{\sqrt{3}}{2}.$$

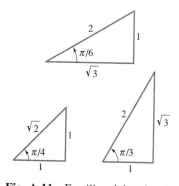

Fig. A.11 Familiar right triangles

To find the values of trigonometric functions of angles larger than $\pi/2$, we can use their periodicity and the identities

$$\sin(\pi \pm \theta) = \mp\sin \theta,$$

$$\cos(\pi \pm \theta) = -\cos \theta, \quad \text{and} \qquad (20)$$

$$\tan(\pi \pm \theta) = \pm\tan \theta$$

A-4

(Problem 14) as well as similar identities for the cosecant, secant, and cotangent functions.

EXAMPLE 1

$$\sin \frac{5\pi}{4} = \sin\left(\pi + \frac{\pi}{4}\right) = -\sin \frac{\pi}{4} = -\frac{\sqrt{2}}{2};$$

$$\cos \frac{2\pi}{3} = \cos\left(\pi - \frac{\pi}{3}\right) = -\cos \frac{\pi}{3} = -\frac{1}{2};$$

$$\tan \frac{3\pi}{4} = \tan\left(\pi - \frac{\pi}{4}\right) = -\tan \frac{\pi}{4} = -1;$$

$$\sin \frac{7\pi}{6} = \sin\left(\pi + \frac{\pi}{6}\right) = -\sin \frac{\pi}{6} = -\frac{1}{2};$$

$$\cos \frac{5\pi}{3} = \cos\left(2\pi - \frac{\pi}{3}\right) = \cos\left(-\frac{\pi}{3}\right) = \cos \frac{\pi}{3} = \frac{1}{2};$$

$$\sin \frac{17\pi}{6} = \sin\left(2\pi + \frac{5\pi}{6}\right) = \sin \frac{5\pi}{6}$$

$$= \sin\left(\pi - \frac{\pi}{6}\right) = \sin \frac{\pi}{6} = \frac{1}{2}.$$

EXAMPLE 2 Find the solutions (if any) of the equation

$$\sin^2 x - 3\cos^2 x + 2 = 0$$

that lie in the interval $[0, \pi]$.

Solution Using the fundamental identity in Eq. (5), we substitute $\cos^2 x = 1 - \sin^2 x$ into the given equation to obtain

$$\sin^2 x - 3(1 - \sin^2 x) + 2 = 0;$$

$$4\sin^2 x - 1 = 0;$$

$$\sin x = \pm\frac{1}{2}.$$

Because $\sin x \geqq 0$ for x in $[0, \pi]$, $\sin x = -\frac{1}{2}$ is impossible. But $\sin x = \frac{1}{2}$ for $x = \pi/6$ and for $x = \pi - \pi/6 = 5\pi/6$. These are the solutions of the given equation in $[0, \pi]$.

Appendix A Problems

ANSWERS TO APPENDIX PROBLEMS APPEAR AT THE END OF THE ANSWERS TO ODD-NUMBERED PROBLEMS.

Express in radian measure the angles in Problems 1 through 5.

1. 40°   **2.** −270°   **3.** 315°   **4.** 210°   **5.** −150°

In Problems 6 through 10, express in degrees the angles given in radian measure.

6. $\dfrac{\pi}{10}$ **7.** $\dfrac{2\pi}{5}$ **8.** 3π **9.** $\dfrac{15\pi}{4}$ **10.** $\dfrac{23\pi}{60}$

In Problems 11 through 14, evaluate the six trigonometric functions of x at the given values.

11. $x = -\dfrac{\pi}{3}$ **12.** $x = \dfrac{3\pi}{4}$

13. $x = \dfrac{7\pi}{6}$ **14.** $x = \dfrac{5\pi}{3}$

Find all solutions x of each equation in Problems 15 through 23.

15. $\sin x = 0$ **16.** $\sin x = 1$ **17.** $\sin x = -1$

18. $\cos x = 0$ **19.** $\cos x = 1$ **20.** $\cos x = -1$

21. $\tan x = 0$ **22.** $\tan x = 1$ **23.** $\tan x = -1$

24. Suppose that $\tan x = \frac{3}{4}$ and that $\sin x < 0$. Find the values of the other five trigonometric functions of x.

25. Suppose that $\csc x = -\frac{5}{3}$ and that $\cos x > 0$. Find the values of the other five trigonometric functions of x.

Deduce the identities in Problems 26 and 27 from the fundamental identity

$$\cos^2 \theta + \sin^2 \theta = 1$$

and from the definitions of the other four trigonometric functions.

26. $1 + \tan^2 \theta = \sec^2 \theta$ **27.** $1 + \cot^2 \theta = \csc^2 \theta$

28. Deduce from the addition formulas for the sine and cosine the addition formula for the tangent:

$$\tan(x + y) = \frac{\tan x + \tan y}{1 - \tan x \tan y}.$$

In Problems 29 through 36, use the method of Example 1 to find the indicated values.

29. $\sin \dfrac{5\pi}{6}$ **30.** $\cos \dfrac{7\pi}{6}$ **31.** $\sin \dfrac{11\pi}{6}$

32. $\cos \dfrac{19\pi}{6}$ **33.** $\sin \dfrac{2\pi}{3}$ **34.** $\cos \dfrac{4\pi}{3}$

35. $\sin \dfrac{5\pi}{3}$ **36.** $\cos \dfrac{10\pi}{3}$

37. Apply the addition formulas for the sine, cosine, and tangent functions (the latter from Problem 28) to show that if $0 < \theta < \pi/2$, then

(a) $\cos\left(\dfrac{\pi}{2} - \theta\right) = \sin \theta$;

(b) $\sin\left(\dfrac{\pi}{2} - \theta\right) = \cos \theta$;

(c) $\cot\left(\dfrac{\pi}{2} - \theta\right) = \tan \theta$.

The prefix *co-* is an abbreviation for the adjective *complementary*, which describes two angles whose sum is $\pi/2$. For example, $\pi/6$ and $\pi/3$ are complementary angles, so (a) implies that $\cos \pi/6 = \sin \pi/3$.

Suppose that $0 < \theta < \pi/2$. Derive the identities in Problems 38 through 40.

38. $\sin(\pi \pm \theta) = \mp\sin \theta$

39. $\cos(\pi \pm \theta) = -\cos \theta$

40. $\tan(\pi \pm \theta) = \pm\tan \theta$

41. The points $A(\cos \theta, -\sin \theta)$, $B(1, 0)$, $C(\cos \phi, \sin \phi)$, and $D(\cos(\theta + \phi), \sin(\theta + \phi))$ are shown in Fig. A.12; all are points on the unit circle. Deduce from the fact that the line segments AC and BD have the same length (because they are subtended by the same angle $\theta + \phi$) that

$$\cos(\theta + \phi) = \cos \theta \cos \phi - \sin \theta \sin \phi.$$

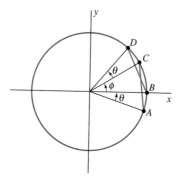

Fig. A.12 Deriving the cosine addition formula (Problem 41)

42. (a) Use the triangles shown in Fig. A.13 to deduce that

$$\sin\left(\theta + \frac{\pi}{2}\right) = \cos \theta \quad \text{and} \quad \cos\left(\theta + \frac{\pi}{2}\right) = -\sin \theta.$$

(b) Use the results of Problem 41 and part (a) to derive the addition formula for the sine function.

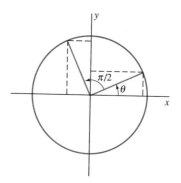

Fig. A.13 Deriving the identities of Problem 42

In Problems 43 through 48, find all solutions of the given equation that lie in the interval $[0, \pi]$.

43. $3 \sin^2 x - \cos^2 x = 2$ **44.** $\sin^2 x = \cos^2 x$

45. $2 \cos^2 x + 3 \sin^2 x = 3$

46. $2 \sin^2 x + \cos x = 2$

47. $8 \sin^2 x \cos^2 x = 1$

48. $\cos 2\theta - 3 \cos \theta = -2$

Appendix B
Proofs of the Limit Laws

Recall the definition of the limit:

$$\lim_{x \to a} F(x) = L$$

provided that, given $\epsilon > 0$, there exists a number $\delta > 0$ such that

$$0 < |x - a| < \delta \quad \text{implies that} \quad |F(x) - L| < \epsilon. \tag{1}$$

Note that the number ϵ comes *first*. Then a value of $\delta > 0$ must be found so that the implication in (1) holds. To prove that $F(x) \to L$ as $x \to a$, you must, in effect, be able to stop the next person you see and ask him or her to pick a positive number ϵ at random. Then you must *always* be ready to respond with a positive number δ. This number δ must have the property that the implication in (1) holds for your number δ and the given number ϵ. The *only* restriction on x is that

$$0 < |x - a| < \delta,$$

as given in (1).

To do all this, you will ordinarily need to give an explicit method—a recipe or formula—for producing a value of δ that works for each value of ϵ. As Examples 1 through 3 show, the method will depend on the particular function F under study as well as the values of a and L.

EXAMPLE 1 Prove that $\lim_{x \to 3} (2x - 1) = 5$.

Solution Given $\epsilon > 0$, we must find $\delta > 0$ such that

$$|(2x - 1) - 5| < \epsilon \quad \text{if} \quad 0 < |x - 3| < \delta.$$

Now

$$|(2x - 1) - 5| = |2x - 6| = 2|x - 3|,$$

so

$$0 < |x - 3| < \frac{\epsilon}{2} \quad \text{implies that} \quad |(2x - 1) - 5| < 2 \cdot \frac{\epsilon}{2} = \epsilon.$$

Hence, given $\epsilon > 0$, it suffices to choose $\delta = \epsilon/2$. This illustrates the observation that the required number δ is generally a function of the given number ϵ.

EXAMPLE 2 Prove that $\lim_{x \to 2} (3x^2 + 5) = 17$.

Solution Given $\epsilon > 0$, we must find $\delta > 0$ such that

$$0 < |x - 2| < \delta \quad \text{implies that} \quad |(3x^2 + 5) - 17| < \epsilon.$$

Now

$$|(3x^2 + 5) - 17| = |3x^2 - 12| = 3 \cdot |x + 2| \cdot |x - 2|.$$

Our problem, therefore, is to show that $|x + 2| \cdot |x - 2|$ can be made as small as we please by choosing $x - 2$ sufficiently small. The idea is that $|x + 2|$ cannot be too large if $|x - 2|$ is fairly small. For example, if $|x - 2| < 1$, then

$$|x + 2| = |(x - 2) + 4| \leq |x - 2| + 4 < 5.$$

Therefore,

$$0 < |x - 2| < 1 \quad \text{implies that} \quad |(3x^2 + 5) - 17| < 15 \cdot |x - 2|.$$

Consequently, let us choose δ to be the minimum of the two numbers 1 and $\epsilon/15$. Then

$$0 < |x - 2| < \delta \quad \text{implies that} \quad |(3x^2 + 5) - 17| < 15 \cdot \frac{\epsilon}{15} = \epsilon,$$

as desired.

EXAMPLE 3 Prove that

$$\lim_{x \to a} \frac{1}{x} = \frac{1}{a} \quad \text{if} \quad a \neq 0.$$

Solution For simplicity, we will consider only the case in which $a > 0$ (the case $a < 0$ is similar).

Suppose that $\epsilon > 0$ is given. We must find a number δ such that

$$0 < |x - a| < \delta \quad \text{implies that} \quad \left| \frac{1}{x} - \frac{1}{a} \right| < \epsilon.$$

Now

$$\left| \frac{1}{x} - \frac{1}{a} \right| = \left| \frac{a - x}{ax} \right| = \frac{|x - a|}{a|x|}.$$

The idea is that $1/|x|$ cannot be too large if $|x - a|$ is fairly small. For example, if $|x - a| < a/2$, then $a/2 < x < 3a/2$. Therefore,

$$|x| > \frac{a}{2}, \quad \text{so} \quad \frac{1}{|x|} < \frac{2}{a}.$$

In this case it would follow that

$$\left| \frac{1}{x} - \frac{1}{a} \right| < \frac{2}{a^2} \cdot |x - a|$$

if $|x - a| < a/2$. Thus, if we choose δ to be the minimum of the two numbers $a/2$ and $a^2\epsilon/2$, then

$$0 < |x - a| < \delta \quad \text{implies that} \quad \left| \frac{1}{x} - \frac{1}{a} \right| < \frac{2}{a^2} \cdot \frac{a^2\epsilon}{2} = \epsilon.$$

Therefore,

$$\lim_{x \to a} \frac{1}{x} = \frac{1}{a} \quad \text{if} \quad a \neq 0,$$

as desired.

We are now ready to give proofs of the limit laws stated in Section 2.2.

Constant Law

If $f(x) \equiv C$, a constant, then

$$\lim_{x \to a} f(x) = \lim_{x \to a} C = C.$$

Proof Because $|C - C| = 0$, we merely choose $\delta = 1$, regardless of the previously given value of $\epsilon > 0$. Then, if $0 < |x - a| < \delta$, it is automatic that $|C - C| < \epsilon$. ❑

Addition Law

If $\lim\limits_{x \to a} F(x) = L$ and $\lim\limits_{x \to a} G(x) = M$, then
$$\lim\limits_{x \to a} [F(x) + G(x)] = L + M.$$

Proof Let $\epsilon > 0$ be given. Because L is the limit of $F(x)$ as $x \to a$, there exists a number $\delta_1 > 0$ such that
$$0 < |x - a| < \delta_1 \quad \text{implies that} \quad |F(x) - L| < \frac{\epsilon}{2}.$$

Because M is the limit of $G(x)$ as $x \to a$, there exists a number $\delta_2 > 0$ such that
$$0 < |x - a| < \delta_2 \quad \text{implies that} \quad |G(x) - M| < \frac{\epsilon}{2}.$$

Let $\delta = \min\{\delta_1, \delta_2\}$. Then
$$0 < |x - a| < \delta \quad \text{implies that} \quad |(F(x) - G(x)) - (L + M)|$$
$$\leqq |F(x) - L| + |G(x) - M| < \frac{\epsilon}{2} + \frac{\epsilon}{2} = \epsilon.$$

Therefore,
$$\lim\limits_{x \to a}[F(x) + G(x)] = L + M,$$

as desired. ❑

Product Law

If $\lim\limits_{x \to a} F(x) = L$ and $\lim\limits_{x \to a} G(x) = M$, then
$$\lim\limits_{x \to a} [F(x) \cdot G(x)] = L \cdot M.$$

Proof Given $\epsilon > 0$, we must find a number $\delta > 0$ such that
$$0 < |x - a| < \delta \quad \text{implies that} \quad |F(x) \cdot G(x) - L \cdot M| < \epsilon.$$

But first, the triangle inequality gives the result
$$|F(x) \cdot G(x) - L \cdot M| = |F(x) \cdot G(x) - L \cdot G(x) + L \cdot G(x) - L \cdot M|$$
$$\leqq |G(x)| \cdot |F(x) - L| + |L| \cdot |G(x) - M|. \quad (2)$$

Because $\lim\limits_{x \to a} F(x) = L$, there exists $\delta_1 > 0$ such that
$$0 < |x - a| < \delta_1 \quad \text{implies that} \quad |F(x) - L| < \frac{\epsilon}{2(|M| + 1)}. \quad (3)$$

And because $\lim_{x \to a} G(x) = M$, there exists $\delta_2 > 0$ such that

$$0 < |x - a| < \delta_2 \quad \text{implies that} \quad |G(x) - M| < \frac{\epsilon}{2(|L| + 1)}. \quad (4)$$

Moreover, there is a *third* number $\delta_3 > 0$ such that

$$0 < |x - a| < \delta_3 \quad \text{implies that} \quad |G(x) - M| < 1,$$

which in turn implies that

$$|G(x)| < |M| + 1. \quad (5)$$

We now choose $\delta = \min\{\delta_1, \delta_2, \delta_3\}$. Then we substitute (3), (4), and (5) into (2) and, finally, see that $0 < |x - a| < \delta$ implies

$$|F(x) \cdot G(x) - L \cdot M| < (|M| + 1) \cdot \frac{\epsilon}{2(|M| + 1)} + |L| \cdot \frac{\epsilon}{2(|L| + 1)}$$

$$< \frac{\epsilon}{2} + \frac{\epsilon}{2} = \epsilon,$$

as desired. The use of $|M| + 1$ and $|L| + 1$ in the denominators avoids the technical difficulty that arises should either L or M be zero. ❑

Substitution Law

If $\lim_{x \to a} g(x) = L$ and $\lim_{x \to L} f(x) = f(L)$, then

$$\lim_{x \to a} f(g(x)) = f(L).$$

Proof Let $\epsilon > 0$ be given. We must find a number $\delta > 0$ such that

$$0 < |x - a| < \delta \quad \text{implies that} \quad |f(g(x)) - f(L)| < \epsilon.$$

Because $\lim_{y \to L} f(y) = f(L)$, there exists $\delta_1 > 0$ such that

$$0 < |y - L| < \delta_1 \quad \text{implies that} \quad |f(y) - f(L)| < \epsilon. \quad (6)$$

Also, because $\lim_{x \to a} g(x) = L$, we can find $\delta > 0$ such that

$$0 < |x - a| < \delta \quad \text{implies that} \quad |g(x) - L| < \delta_1,$$

that is, such that

$$|y - L| < \delta_1,$$

where $y = g(x)$. From (6) we see that

$$0 < |x - a| < \delta \text{ implies that } |f(g(x)) - f(L)| = |f(y) - f(L)| < \epsilon,$$

as desired. ❑

Reciprocal Law

If $\lim_{x \to a} g(x) = L$ and $L \neq 0$, then

$$\lim_{x \to a} \frac{1}{g(x)} = \frac{1}{L}.$$

Proof   Let $f(x) = 1/x$. Then, as we saw in Example 3,

$$\lim_{x \to a} f(x) = \lim_{x \to a} \frac{1}{x} = \frac{1}{L} = f(L).$$

Hence the substitution law gives the result

$$\lim_{x \to a} \frac{1}{g(x)} = \lim_{x \to a} f(g(x)) = f(L) = \frac{1}{L},$$

as desired.   ❑

Quotient Law

If $\lim\limits_{x \to a} F(x) = L$ and $\lim\limits_{x \to a} G(x) = M \neq 0$, then

$$\lim_{x \to a} \frac{F(x)}{G(x)} = \frac{L}{M}.$$

Proof   It follows immediately from the product and reciprocal laws that

$$\lim_{x \to a} \frac{F(x)}{G(x)} = \lim_{x \to a} F(x) \cdot \frac{1}{G(x)} = \left(\lim_{x \to a} F(x) \right) \left(\lim_{x \to a} \frac{1}{G(x)} \right) = L \cdot \frac{1}{M} = \frac{L}{M},$$

as desired.   ❑

Squeeze Law

Suppose that $f(x) \leq g(x) \leq h(x)$ in some deleted neighborhood of a and that

$$\lim_{x \to a} f(x) = L = \lim_{x \to a} h(x).$$

Then

$$\lim_{x \to a} g(x) = L.$$

Proof   Given $\epsilon > 0$, we choose $\delta_1 > 0$ and $\delta_2 > 0$ such that

$$0 < |x - a| < \delta_1 \quad \text{implies that} \quad |f(x) - L| < \epsilon$$

and

$$0 < |x - a| < \delta_2 \quad \text{implies that} \quad |h(x) - L| < \epsilon.$$

Let $\delta = \min\{\delta_1, \delta_2\}$. Then $\delta > 0$. Moreover, if $0 < |x - a| < \delta$, then both $f(x)$ and $h(x)$ are points of the open interval $(L - \epsilon, L + \epsilon)$. So

$$L - \epsilon < f(x) \leq g(x) \leq h(x) < L + \epsilon.$$

Thus

$$0 < |x - a| < \delta \quad \text{implies that} \quad |g(x) - L| < \epsilon,$$

as desired.   ❑

Appendix B Problems

In Problems 1 through 10, apply the definition of the limit to establish the given equality.

1. $\lim\limits_{x \to a} x = a$

2. $\lim\limits_{x \to 2} 3x = 6$

3. $\lim\limits_{x \to 2} (x + 3) = 5$

4. $\lim\limits_{x \to -3} (2x + 1) = -5$

5. $\lim\limits_{x \to 1} x^2 = 1$

6. $\lim\limits_{x \to a} x^2 = a^2$

7. $\lim\limits_{x \to -1} (2x^2 - 1) = 1$

8. $\lim\limits_{x \to a} \dfrac{1}{x^2} = \dfrac{1}{a^2}$ if $a \neq 0$

9. $\lim\limits_{x \to a} \dfrac{1}{x^2 + 1} = \dfrac{1}{a^2 + 1}$

10. $\lim\limits_{x \to a} \dfrac{1}{\sqrt{x}} = \dfrac{1}{\sqrt{a}}$ if $a > 0$

11. Suppose that $\lim_{x \to a} f(x) = L$ and that $\lim_{x \to a} f(x) = M$. Apply the definition of the limit to prove that $L = M$. Thus a limit of a function is unique if it exists.

12. Suppose that C is a constant and that $\lim_{x \to a} f(x) = L$. Apply the definition of the limit to prove that

$$\lim_{x \to a} C \cdot f(x) = C \cdot L.$$

13. Suppose that $L \neq 0$ and that $\lim_{x \to a} f(x) = L$. Use the method of Example 3 and the definition of the limit to

show directly that

$$\lim_{x \to a} \frac{1}{f(x)} = \frac{1}{L}.$$

14. Use the algebraic identity

$$x^n - a^n$$
$$= (x - a)(x^{n-1} + x^{n-2}a + \cdots + xa^{n-2} + a^{n-1})$$

to show directly from the definition of the limit that $\lim\limits_{x \to a} x^n = a^n$ if n is a positive integer.

15. Apply the identity

$$|\sqrt{x} - \sqrt{a}| = \frac{|x - a|}{\sqrt{x} + \sqrt{a}}$$

to show directly from the definition of the limit that $\lim\limits_{x \to a} \sqrt{x} = \sqrt{a}$ if $a > 0$.

16. Suppose that $\lim_{x \to a} f(x) = f(a) > 0$. Prove that there exists a neighborhood of a on which $f(x) > 0$; that is, prove that there exists $\delta > 0$ such that

$$|x - a| < \delta \quad \text{implies that} \quad f(x) > 0.$$

Appendix C
The Completeness of the Real Number System

Here we present a self-contained treatment of those consequences of the completeness of the real number system that are relevant to this text. Our principal objective is to prove the intermediate value theorem and the maximum value theorem. We begin with the least upper bound property of the real numbers, which we take to be an axiom.

Definition *Upper Bound and Lower Bound*

The set S of real numbers is said to be **bounded above** if there is a number b such that $x \leqq b$ for every number x in S, and the number b is called an **upper bound** for S. Similarly, if there is a number a such that $x \geqq a$ for every number x in S, then S is said to be **bounded below,** and a is called a **lower bound** for S.

Definition *Least Upper Bound and Greatest Lower Bound*

The number λ is said to be a **least upper bound** for the set S of real numbers provided that

1. λ is an upper bound for S, and
2. If b is an upper bound for S, then $\lambda \leqq b$.

Similarly, the number γ is said to be a **greatest lower bound** for S if γ is a lower bound for S and $\gamma \geqq a$ for every lower bound a of S.

A-12

EXERCISE   Prove that if a set S has a least upper bound λ, then it is unique. That is, prove that if λ and μ are both least upper bounds for S, then $\lambda = \mu$.

It is easy to show that the greatest lower bound γ of a set S, if any, is also unique. At this point you should construct examples to illustrate that a set with a least upper bound λ may or may not contain λ and that a similar statement is true of the set's greatest lower bound.

We now state the *completeness axiom* of the real number system.

> **Least Upper Bound Axiom**
>
> If the nonempty set S of real numbers has an upper bound, then it has a least upper bound.

By working with the set T consisting of the numbers $-x$, where x is in S, it is not difficult to show the following consequence of the least upper bound axiom: If the nonempty set S of real numbers is bounded below, then S has a greatest lower bound. Because of this symmetry, we need only one axiom, not two; results for least upper bounds also hold for greatest lower bounds, provided that some attention is paid to the direction of the inequalities.

The restriction that S be nonempty is annoying but necessary. If S is the "empty" set of real numbers, then 15 is an upper bound for S, but S has no least upper bound because $14, 13, 12, \ldots, 0, -1, -2, \ldots$ are also upper bounds for S.

> **Definition**   *Increasing, Decreasing, and Monotonic Sequences*
>
> The infinite sequence $x_1, x_2, x_3, \ldots, x_k, \ldots$ is said to be **nondecreasing** if $x_n \leqq x_{n+1}$ for every $n \geqq 1$. This sequence is said to be **nonincreasing** if $x_n \geqq x_{n+1}$ for every $n \geqq 1$. If the sequence $\{x_n\}$ is either nonincreasing or nondecreasing, then it is said to be **monotonic.**

Theorem 1 gives the **bounded monotonic sequence property** of the set of real numbers. (Recall that a set S of real numbers is said to be **bounded** if it is contained in an interval of the form $[a, b]$.)

> **Theorem 1**   *Bounded Monotonic Sequences*
>
> Every bounded monotonic sequence of real numbers converges.

Proof   Suppose that the sequence

$$S = \{x_n\} = \{x_1, x_2, x_3, \ldots, x_k, \ldots\}$$

is bounded and nondecreasing. By the least upper bound axiom, S has a least upper bound λ. We claim that λ is the limit of the sequence $\{x_n\}$. Consider an open interval centered at λ—that is, an interval of the form $I = (\lambda - \epsilon, \lambda + \epsilon)$, where $\epsilon > 0$. Some terms of the sequence must lie within I, or else $\lambda - \epsilon$ would be an upper bound for S that is less than its least upper bound λ. But if x_N is within I, then—because we are dealing with a nondecreasing sequence—x_k must also lie in I for all $k \geq N$. Because ϵ is an arbitrary positive number, λ is by definition (Problem 39 of Section 11.2) the limit of the sequence $\{x_n\}$. That is, a bounded nondecreasing sequence converges. A similar proof can be constructed for nonincreasing sequences by working with the greatest lower bound. ❏

Therefore, the least upper bound axiom implies the bounded monotonic sequence property of the real numbers. With just a little effort, you can prove that the two are logically equivalent: If you take the bounded monotonic sequence property as an axiom, then the least upper bound property follows as a theorem. The **nested interval property** of Theorem 2 is also equivalent to the least upper bound property, but we shall prove only that it follows from the least upper bound property, because we have chosen the latter as the fundamental completeness axiom for the real number system.

Theorem 2 *Nested Interval Property of the Real Numbers*

Suppose that $I_1, I_2, I_3, \ldots, I_n, \ldots$ is a sequence of closed intervals (so I_n is of the form $[a_n, b_n]$ for each positive integer n) such that

1. I_n contains I_{n+1} for each $n \geq 1$, and
2. $\lim\limits_{n \to \infty} (b_n - a_n) = 0$.

Then there exists exactly one real number c such that c belongs to I_n for each n. Thus

$$\{c\} = I_1 \cap I_2 \cap I_3 \cap \ldots.$$

Proof It is clear from hypothesis (2) of Theorem 2 that there is at most one such number c. The sequence $\{a_n\}$ of the left-hand endpoints of the intervals is a bounded (by b_1) nondecreasing sequence and thus has a limit a by the bounded monotonic sequence property. Similarly, the sequence $\{b_n\}$ has a limit b. Because $a_n \leq b_n$ for all n, it follows easily that $a \leq b$. It is clear that $a_n \leq a \leq b_n$ for all $n \geq 1$, so a belongs to every interval I_n; so does b, by a similar argument. But then property (2) of Theorem 2 implies that $a = b$, and clearly this common value—call it c—is the number satisfying the conclusion of Theorem 2. ❏

We can now use these results to prove several important theorems used in the text.

Theorem 3 *Intermediate Value Property of Continuous Functions*

If the function f is continuous on the interval $[a, b]$ and $f(a) < K < f(b)$, then $K = f(c)$ for some number c in (a, b).

Proof Let $I_1 = [a, b]$. Suppose that I_n has been defined for $n \geq 1$. We describe (inductively) how to define I_{n+1}, and this shows in particular how to define I_2, I_3, and so forth. Let a_n be the left-hand endpoint of I_n, b_n be its right-hand endpoint, and m_n be its midpoint. If $f(m_n) > K$, then $f(a_n) < K < f(m_n)$; in this case, let $a_{n+1} = a_n$, $b_{n+1} = m_n$, and $I_{n+1} = [a_{n+1}, b_{n+1}]$. If $f(m_n) < K$, then let $a_{n+1} = m_n$ and $b_{n+1} = b_n$. Thus at each stage we bisect I_n and let I_{n+1} be the half of I_n on which f takes on values both above and below K. Note that if $f(m_n)$ is ever actually equal to K, we simply let $c = m_n$ and stop.

It is easy to show that the sequence $\{I_n\}$ of intervals satisfies the hypotheses of Theorem 2. Let c be the (unique) real number common to all the intervals I_n. We will show that $f(c) = K$, and this will conclude the proof.

The sequence $\{b_n\}$ has limit c, so by the continuity of f, the sequence $\{f(b_n)\}$ has limit $f(c)$. But $f(b_n) > K$ for all n, so the limit of $\{f(b_n)\}$ can be no less than K; that is, $f(c) \geqq K$. By considering the sequence $\{a_n\}$, it follows that $f(c) \leqq K$. Therefore, $f(c) = K$. ❑

Lemma 1

If f is continuous on the closed interval $[a, b]$, then f is bounded there.

Proof　Suppose by way of contradiction that f is not bounded on $I_1 = [a, b]$. Bisect I_1 and let I_2 be either half on which f is unbounded—if f is unbounded on both halves, then let I_2 be the left half of I_1. In general, let I_{n+1} be a half of I_n on which f is unbounded.

Again it is easy to show that the sequence $\{I_n\}$ of closed intervals satisfies the hypotheses of Theorem 2. Let c be the number common to them all. Because f is continuous, there is a number $\epsilon > 0$ such that f is bounded on the interval $(c - \epsilon, c + \epsilon)$. But for sufficiently large values of n, I_n is a subset of $(c - \epsilon, c + \epsilon)$. This contradiction shows that f must be bounded on $[a, b]$. ❑

Theorem 4　*Maximum Value Property of Continuous Functions*

If the function f is continuous on the closed and bounded interval $[a, b]$, then there exists a number c in $[a, b]$ such that $f(x) \leqq f(c)$ for all x in $[a, b]$.

Proof　Consider the set $S = \{f(x) \mid a \leqq x \leqq b\}$. By Lemma 1, this set is bounded; let λ be its least upper bound. Our goal is to show that λ is a value $f(c)$ of f.

With $I_1 = [a, b]$, bisect I_1 as before. Note that λ is the least upper bound of the values of f on at least one of the two halves of I_1; let I_2 be that half. Having defined I_n, let I_{n+1} be the half of I_n on which λ is the least upper bound of the values of f. Let c be the number common to all these intervals. It then follows from the continuity of f, much as in the proof of Theorem 3, that $f(c) = \lambda$. And it is clear that $f(x) \leqq \lambda$ for all x in $[a, b]$. ❑

The technique we are using in these proofs is called the *method of bisection*. We now use it once again to establish the *Bolzano-Weierstrass property* of the real number system.

> **Definition** *Limit Point*
> Let S be a set of real numbers. The number p is said to be a **limit point** of S if every open interval containing p also contains points of S other than p.

> **Theorem 5** *Bolzano-Weierstrass Theorem*
> Every bounded infinite set of real numbers has a limit point.

Proof   Let I_0 be a closed interval containing the bounded infinite set S of real numbers. Let I_1 be one of the closed half-intervals of I_0 that contains infinitely many points of S. If I_n has been chosen, let I_{n+1} be one of the closed half-intervals of I_n containing infinitely many points of S. An application of Theorem 2 yields a number p common to all the intervals I_n. If J is an open interval containing p, then J contains I_n for some sufficiently large value of n and thus contains infinitely many points of S. Therefore, p is a limit point of S. ❑

Our final goal is in sight: We can now prove that a sequence of real numbers converges if and only if it is a Cauchy sequence.

> **Definition** *Cauchy Sequence*
> The sequence $\{a_n\}_1^\infty$ is said to be a **Cauchy sequence** if, for every $\epsilon > 0$, there exists an integer N such that
> $$|a_m - a_n| < \epsilon$$
> for all $m, n \geqq N$.

> **Lemma 2** *Convergent Subsequences*
> Every bounded sequence of real numbers has a convergent subsequence.

Proof   If $\{a_n\}$ has only a finite number of values, then the conclusion of Lemma 2 follows easily. We therefore focus our attention on the case in which $\{a_n\}$ is an infinite set. It is easy to show that this set is also bounded, and thus we may apply the Bolzano-Weierstrass theorem to obtain a limit point p of $\{a_n\}$. For each integer $k \geqq 1$, let $a_{n(k)}$ be a term of the sequence $\{a_n\}$ such that

1. $n(k + 1) > n(k)$ for all $k \geqq 1$, and

2. $|a_{n(k)} - p| < \dfrac{1}{k}$.

It is then easy to show that $\{a_{n(k)}\}$ is a convergent (to p) subsequence of $\{a_n\}$. ❑

> **Theorem 6 Convergence of Cauchy Sequences**
> A sequence of real numbers converges if and only if it is a Cauchy sequence.

Proof It follows immediately from the triangle inequality that every convergent sequence is a Cauchy sequence. Thus suppose that the sequence $\{a_n\}$ is a Cauchy sequence.

Choose N such that

$$|a_m - a_n| < 1$$

if $m, n \geqq N$. It follows that if $n \geqq N$, then a_n lies in the closed interval $[a_N - 1, a_N + 1]$. This implies that the sequence $\{a_n\}$ is bounded, and thus by Lemma 2 it has a convergent subsequence $\{a_{n(k)}\}$. Let p be the limit of this subsequence.

We claim that $\{a_n\}$ itself converges to p. Given $\epsilon > 0$, choose M such that

$$|a_m - a_n| < \frac{\epsilon}{2}$$

if $m, n \geqq N$. Next choose K such that $n(K) \geqq M$ and

$$|a_{n(K)} - p| < \frac{\epsilon}{2}.$$

Then if $n \geqq M$,

$$|a_n - p| \leqq |a_n - a_{n(K)}| + |a_{n(K)} - p| < \epsilon.$$

Therefore, $\{a_n\}$ converges to p by definition. ☐

Appendix D
Proof of the Chain Rule

To prove the chain rule, we need to show that if f is differentiable at a and g is differentiable at $f(a)$, then

$$\lim_{h \to 0} \frac{g(f(a + h)) - g(f(a))}{h} = g'(f(a)) \cdot f'(a). \tag{1}$$

If the quantities h and

$$k(h) = f(a + h) - f(a) \tag{2}$$

are nonzero, then we can write the difference quotient on the left-hand side of Eq. (1) as

$$\frac{g(f(a + h)) - g(f(a))}{h} = \frac{g(f(a) + k(h)) - g(f(a))}{k(h)} \cdot \frac{k(h)}{h}. \tag{3}$$

To investigate the first factor on the right-hand side of Eq. (3), we define a new function ϕ as follows:

$$\phi(k) = \begin{cases} \dfrac{g(f(a) + k) - g(f(a))}{k} & \text{if } k \neq 0; \\ g'(f(a)) & \text{if } k = 0. \end{cases} \tag{4}$$

By the definition of the derivative of g, we see from Eq. (4) that ϕ is continuous at $k = 0$; that is,

$$\lim_{k \to 0} \phi(k) = g'(f(a)). \tag{5}$$

Next,

$$\lim_{h \to 0} k(h) = \lim_{h \to 0} [f(a + h) - f(a)] = 0 \tag{6}$$

because f is continuous at $x = a$, and $\phi(0) = g'(f(a))$. It therefore follows from Eq. (5) that

$$\lim_{h \to 0} \phi(k(h)) = g'(f(a)). \tag{7}$$

We are now ready to assemble all this information. By Eq. (3), if $h \neq 0$, then

$$\frac{g(f(a + h)) - g(f(a))}{h} = \phi(k(h)) \cdot \frac{f(a + h) - f(a)}{h} \tag{8}$$

even if $k(h) = 0$, because in this case both sides of Eq. (8) are zero. Hence the product rule for limits yields

$$\lim_{h \to 0} \frac{g(f(a + h)) - g(f(a))}{h} = \lim_{h \to 0} \phi(k(h)) \cdot \frac{f(a + h) - f(a)}{h}$$

$$= g'(f(a)) \cdot f'(a),$$

a consequence of Eq. (7) and the definition of the derivative of the function f. We have therefore established the chain rule in the form of Eq. (1). ❏

Appendix E ▰
Existence of the Integral

When the basic computational algorithms of the calculus were discovered by Newton and Leibniz in the latter half of the seventeenth century, the logical rigor that had been a feature of the Greek method of exhaustion was largely abandoned. When computing the area A under the curve $y = f(x)$, for example, Newton took it as intuitively obvious that the area function existed, and he proceeded to compute it as the antiderivative of the height function $f(x)$. Leibniz regarded A as an infinite sum of infinitesimal area elements, each of the form $dA = f(x)\, dx$, but in practice computed the area

$$A = \int_a^b f(x)\, dx$$

by antidifferentiation just as Newton did—that is, by computing

$$A = \left[D^{-1} f(x) \right]_a^b.$$

The question of the *existence* of the area function—one of the conditions that a function f must satisfy in order for its integral to exist—did not at first seem to be of much importance. Eighteenth-century mathematicians were mainly occupied (and satisfied) with the impressive applications of calculus to the solution of real-world problems and did not concentrate on the logical foundations of the subject.

The first attempt at a precise definition of the integral and a proof of its existence for continuous functions was that of the French mathematician

Augustin Louis Cauchy (1789–1857). Curiously enough, Cauchy was trained as an engineer, and much of his research in mathematics was in fields that we today regard as applications-oriented: hydrodynamics, waves in elastic media, vibrations of elastic membranes, polarization of light, and the like. But he was a prolific researcher, and his writings cover the entire spectrum of mathematics, with occasional essays into almost unrelated fields.

Around 1824, Cauchy defined the integral of a continuous function in a way that is familiar to us, as a limit of left-endpoint approximations:

$$\int_a^b f(x)\,dx = \lim_{\Delta x \to 0} \sum_{i=1}^{n} f(x_{i-1})\,\Delta x.$$

This is a much more complicated sort of limit than the ones we discussed in Chapter 2. Cauchy was not entirely clear about the nature of the limit process involved in this equation, nor was he clear about the precise role that the hypothesis of the continuity of f played in proving that the limit exists.

A complete definition of the integral, as we gave in Section 5.4, was finally produced in the 1850s by the German mathematician Georg Bernhard Riemann. Riemann was a student of Gauss; he met Gauss upon his arrival at Göttingen, Germany, for the purpose of studying theology, when he was about 20 years old and Gauss was about 70. Riemann soon decided to study mathematics and became known as one of the truly great mathematicians of the nineteenth century. Like Cauchy, he was particularly interested in applications of mathematics to the real world; his research particularly emphasized electricity, heat, light, acoustics, fluid dynamics, and—as you might infer from the fact that Wilhelm Weber (for whom the unit of magnetic flux, the weber, was named) was a major influence on Riemann's education—magnetism. Riemann also made significant contributions to mathematics itself, particularly in the field of complex analysis. A major conjecture of his, involving the zeta function

$$\zeta(s) = \sum_{n=1}^{\infty} \frac{1}{n^s}, \tag{1}$$

remains unsolved to this day and has important consequences in the theory of the distribution of prime numbers because

$$\zeta(k) = \prod \left(1 - \frac{1}{p^k}\right)^{-1},$$

where the product $\prod$ is taken over all primes p. [The zeta function is defined in Eq. (1) for complex numbers s to the right of the vertical line at $x = 1$ and is extended to other complex numbers by the requirement that it be differentiable.] Riemann died of tuberculosis shortly before his fortieth birthday.

Here we give a proof of the existence of the integral of a continuous function. We will follow Riemann's approach. Specifically, suppose that the function f is continuous on the closed and bounded interval $[a, b]$. We will prove that the definite integral

$$\int_a^b f(x)\,dx$$

exists. That is, we will demonstrate the existence of a number I that satisfies the following condition: For every $\epsilon > 0$ there exists $\delta > 0$ such that, for *every* Riemann sum R associated with *any* partition P with $|P| < \delta$,

$$|I - R| < \epsilon.$$

(Recall that the mesh $|P|$ of the partition P is the length of the longest subinterval in the partition.) In other words, every Riemann sum associated with every sufficiently "fine" partition is close to the number I. If this happens, then the definite integral

$$\int_a^b f(x)\, dx$$

is said to **exist,** and I is its **value.**

Now we begin the proof. Suppose throughout that f is a function continuous on the closed interval $[a, b]$. Given $\epsilon > 0$, we need to show the existence of a number $\delta > 0$ such that

$$\left| I - \sum_{i=1}^{n} f(x_i^*)\, \Delta x_i \right| < \epsilon \tag{2}$$

for every Riemann sum associated with any partition P of $[a, b]$ with $|P| < \delta$.

Given a partition P of $[a, b]$ into n subintervals that are *not necessarily of equal length*, let p_i be a point in the subinterval $[x_{i-1}, x_i]$ at which f attains its minimum value $f(p_i)$. Similarly, let $f(q_i)$ be its maximum value there. These numbers exist for $i = 1, 2, 3, \ldots, n$ because of the maximum value property of continuous functions (Theorem 4 of Appendix C).

In what follows we will denote the resulting lower and upper Riemann sums associated with P by

$$L(P) = \sum_{i=1}^{n} f(p_i)\, \Delta x_i \tag{3a}$$

and

$$U(P) = \sum_{i=1}^{n} f(q_i)\, \Delta x_i, \tag{3b}$$

respectively. Then Lemma 1 is obvious.

Lemma 1
For any partition P of $[a, b]$, $L(P) \leqq U(P)$.

Now we need a definition. The partition P' is called a *refinement* of the partition P if each subinterval of P' is contained in some subinterval of P. That is, P' is obtained from P by adding more points of subdivision to P.

Lemma 2
Suppose that P' is a refinement of P. Then

$$L(P) \leqq L(P') \leqq U(P') \leqq U(P). \tag{4}$$

Proof The inequality $L(P') \leq U(P')$ is a consequence of Lemma 1. We will show that $L(P) \leq L(P')$; the proof that $U(P') \leq U(P)$ is similar.

The refinement P' is obtained from P by adding one or more points of subdivision to P. So all we need show is that the Riemann sum $L(P)$ cannot be decreased by adding a single point of subdivision. Thus we will suppose that the partition P' is obtained from P by dividing the kth subinterval $[x_{k-1}, x_k]$ of P into two subintervals $[x_{k-1}, z]$ and $[z, x_k]$ by means of the new point z.

The only resulting effect on the corresponding Riemann sum is to replace the term

$$f(p_k) \cdot (x_k - x_{k-1})$$

in $L(P)$ by the two-term sum

$$f(u) \cdot (z - x_{k-1}) + f(v) \cdot (x_k - z),$$

where $f(u)$ is the minimum of f on $[x_{k-1}, z]$ and $f(v)$ is the minimum of f on $[z, x_k]$. But

$$f(p_k) \leq f(u) \quad \text{and} \quad f(p_k) \leq f(v).$$

Hence

$$f(u) \cdot (z - x_{k-1}) + f(v) \cdot (x_k - z) \geq f(p_k) \cdot (z - x_{k-1}) + f(p_k) \cdot (x_k - z)$$
$$= f(p_k) \cdot (z - x_{k-1} + x_k - z)$$
$$= f(p_k) \cdot (x_k - x_{k-1}).$$

So the replacement of $f(p_k) \cdot (x_k - x_{k-1})$ cannot decrease the sum $L(P)$ in question, and therefore $L(P) \leq L(P')$. Because this is all we needed to show, we have completed the proof of Lemma 2. ❑

To prove that all the Riemann sums for sufficiently fine partitions are close to some number I, we must first give a construction of I. This is accomplished through Lemma 3.

Lemma 3

Let P_n denote the regular partition of $[a, b]$ into 2^n subintervals of equal length. Then the (sequential) limit

$$I = \lim_{n \to \infty} L(P_n) \tag{5}$$

exists.

Proof We begin with the observation that each partition P_{n+1} is a refinement of P_n, so (by Lemma 2)

$$L(P_1) \leq L(P_2) \leq \cdots \leq L(P_n) \leq \cdots.$$

Therefore, $\{L(P_n)\}$ is a nondecreasing sequence of real numbers. Moreover,

$$L(P_n) = \sum_{i=1}^{2^n} f(p_i) \, \Delta x_i \leq M \sum_{i=1}^{2^n} \Delta x_i = M(b - a),$$

where M is the maximum value of f on $[a, b]$.

Theorem 1 of Appendix C guarantees that a bounded monotonic sequence of real numbers must converge. Thus the number

$$I = \lim_{n \to \infty} L(P_n)$$

exists. This establishes Eq. (5), and the proof of Lemma 3 is complete. ❑

It is proved in advanced calculus that if f is continuous on $[a, b]$, then—for every number $\epsilon > 0$—there exists a number $\delta > 0$ such that

$$|f(u) - f(v)| < \epsilon$$

for every two points u and v of $[a, b]$ such that

$$|u - v| < \delta.$$

This property of a function is called **uniform continuity** of f on the interval $[a, b]$. Thus the theorem from advanced calculus that we need to use states that every continuous function on a closed and bounded interval is uniformly continuous there.

NOTE   The fact that f is continuous on $[a, b]$ means that for each number u in the interval and each $\epsilon > 0$, there exists $\delta > 0$ such that if v is a number in the interval with $|u - v| < \delta$, then $|f(u) - f(v)| < \epsilon$. But *uniform* continuity is a more stringent condition. It means that given $\epsilon > 0$, you can find not only a value δ_1 that "works" for u_1, a value δ_2 that works for u_2, and so on, but more: You can find a universal value of δ that works for *all* values of u in the interval. This should not be obvious when you notice the possibility that $\delta_1 = 1$, $\delta_2 = \frac{1}{2}$, $\delta_3 = \frac{1}{3}$, and so on. In any case, it is clear that uniform continuity of f on an interval implies its continuity there.

Remember that throughout we have a continuous function f defined on the closed interval $[a, b]$.

Lemma 4

Suppose that $\epsilon > 0$ is given. Then there exists a number $\delta > 0$ such that if P is a partition of $[a, b]$ with $|P| < \delta$ and P' is a refinement of P, then

$$|R(P) - R(P')| < \frac{\epsilon}{3} \tag{6}$$

for any two Riemann sums $R(P)$ associated with P and $R(P')$ associated with P'.

Proof   Because f must be uniformly continuous on $[a, b]$, there exists a number $\delta > 0$ such that if

$$|u - v| < \delta, \quad \text{then } |f(u) - f(v)| < \frac{\epsilon}{3(b - a)}.$$

Suppose now that P is a partition of $[a, b]$ with $|P| < \delta$. Then

$$|U(P) - L(P)| = \sum_{i=1}^{n} |f(q_i) - f(p_i)| \, \Delta x_i < \frac{\epsilon}{3(b - a)} \sum_{i=1}^{n} \Delta x_i = \frac{\epsilon}{3}.$$

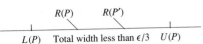

Fig. E.1 Part of the proof of Lemma 4

This is valid because $|p_i - q_i| < \delta$, for both p_i and q_i belong to the same subinterval $[x_{i-1}, x_i]$ of P, and $|P| < \delta$.

Now, as shown in Fig. E.1, we know that $L(P)$ and $U(P)$ differ by less than $\epsilon/3$. We know also that

$$L(P) \leqq R(P) \leqq U(P)$$

for every Riemann sum $R(P)$ associated with P. But

$$L(P) \leqq L(P') \leqq U(P') \leqq U(P)$$

by Lemma 2, because P' is a refinement of P; moreover,

$$L(P') \leqq R(P') \leqq U(P')$$

for every Riemann sum $R(P')$ associated with P'.

As Fig. E.1 shows, both the numbers $R(P)$ and $R(P')$ belong to the interval $[L(P), U(P)]$ of length less than $\epsilon/3$, so Eq. (6) follows, as desired. This concludes the proof of Lemma 4. ❑

Theorem 1   *Existence of the Integral*
If f is continuous on the closed and bounded interval $[a, b]$, then the integral

$$\int_a^b f(x)\, dx$$

exists.

Proof   Suppose that $\epsilon > 0$ is given. We must show the existence of a number $\delta > 0$ such that, for every partition P of $[a, b]$ with $|P| < \delta$, we have

$$|I - R(P)| < \epsilon,$$

where I is the number given in Lemma 3 and $R(P)$ is an arbitrary Riemann sum for f associated with P.

We choose the number δ provided by Lemma 4 such that

$$|R(P) - R(P')| < \frac{\epsilon}{3}$$

if $|P| < \delta$ and P' is a refinement of P.

By Lemma 3, we can choose an integer N so large that

$$|P_N| < \delta \quad \text{and} \quad |L(P_N) - I| < \frac{\epsilon}{3}. \tag{7}$$

Given an arbitrary partition P such that $|P| < \delta$, let P' be a common refinement of both P and P_N. You can obtain such a partition P', for example, by using all the points of subdivision of both P and P_N to form the subintervals of $[a, b]$ that constitute P'.

Because P' is a refinement of both P and P_N and both the latter partitions have mesh less than δ, Lemma 4 implies that

$$|R(P) - R(P')| < \frac{\epsilon}{3} \quad \text{and} \quad |L(P_N) - R(P')| < \frac{\epsilon}{3}. \tag{8}$$

Here $R(P)$ and $R(P')$ are (arbitrary) Riemann sums associated with P and P', respectively.

Given an arbitrary Riemann sum $R(P)$ associated with the partition P with mesh less than δ, we see that

$$|I - R(P)| = |I - L(P_N) + L(P_N) - R(P') + R(P') - R(P)|$$

$$\leq |I - L(P_N)| + |L(P_N) - R(P')| + |R(P') - R(P)|.$$

In the last sum, both of the last two terms are less than $\epsilon/3$ by virtue of the inequalities in (8). We also know, by (7), that the first term is less than $\epsilon/3$. Consequently,

$$|I - R(P)| < \epsilon.$$

This establishes Theorem 1. ❑

We close with an example that shows that some hypothesis of continuity is required for integrability.

EXAMPLE 1 Suppose that f is defined for $0 \leq x \leq 1$ as follows:

$$f(x) = \begin{cases} 1 & \text{if } x \text{ is irrational;} \\ 0 & \text{if } x \text{ is rational.} \end{cases}$$

Then f is not continuous anywhere. (Why?) Given a partition P of $[0, 1]$, let p_i be a rational point and q_i an irrational point of the ith subinterval of P for $i = 1, 2, 3, \ldots, n$. As before, f attains its minimum value 0 at each p_i and its maximum value 1 at each q_i. Also

$$L(P) = \sum_{i=1}^{n} f(p_i) \, \Delta x_i = 0, \quad \text{whereas} \quad U(P) = \sum_{i=1}^{n} f(q_i) \, \Delta x_i = 1.$$

Thus if we choose $\epsilon = \frac{1}{2}$, then there is *no* number I that can lie within ϵ of both $L(P)$ and $U(P)$, no matter how small the mesh of P. It follows that f is *not* Riemann integrable on $[0, 1]$.

Appendix F ▱
Approximations and Riemann Sums

Several times in Chapter 6 our attempt to compute some quantity Q led to the following situation. Beginning with a regular partition of an appropriate interval $[a, b]$ into n subintervals, each of length Δx, we find an approximation A_n to Q of the form

$$A_n = \sum_{i=1}^{n} g(u_i)h(v_i) \, \Delta x, \tag{1}$$

where u_i and v_i are two (generally different) points of the ith subinterval $[x_{i-1}, x_i]$. For example, in our discussion of surface area of revolution that precedes Eq. (8) of Section 6.4, we found the approximation

$$\sum_{i=1}^{n} 2\pi f(u_i)\sqrt{1 + [f'(v_i)]^2} \, \Delta x \tag{2}$$

to the area of the surface generated by revolving the curve $y = f(x)$, $a \leq x \leq b$, around the x-axis. (In Section 6.4 we wrote x_i^{**} for u_i and x_i^* for

A-24

v_i.) Note that the expression in (2) is the same as the right-hand side of Eq. (1); take $g(x) = 2\pi f(x)$ and $h(x) = \sqrt{1 + [f'(x)]^2}$.

In such a situation we observe that if u_i and v_i were the *same* point x_i^* of $[x_{i-1}, x_i]$ for each i ($i = 1, 2, 3, \ldots, n$), then the approximation in Eq. (1) would be a Riemann sum for the function $g(x)h(x)$ on $[a, b]$. This leads us to suspect that

$$\lim_{\Delta x \to 0} \sum_{i=1}^{n} g(u_i)h(v_i)\, \Delta x = \int_a^b g(x)h(x)\, dx. \tag{3}$$

In Section 6.4, we assumed the validity of Eq. (3) and concluded from the approximation in (2) that the surface area of revolution ought to be defined to be

$$A = \lim_{\Delta x \to 0} \sum_{i=1}^{n} 2\pi f(u_i)\sqrt{1 + [f'(v_i)]^2}\, \Delta x = \int_a^b 2\pi f(x)\sqrt{1 + [f'(x)]^2}\, dx.$$

Theorem 1 guarantees that Eq. (3) holds under mild restrictions on the functions g and h.

Theorem 1   *A Generalization of Riemann Sums*
Suppose that h and g' are continuous on $[a, b]$. Then

$$\lim_{\Delta x \to 0} \sum_{i=1}^{n} g(u_i)h(v_i)\, \Delta x = \int_a^b g(x)h(x)\, dx, \tag{3}$$

where u_i and v_i are arbitrary points of the ith subinterval of a regular partition of $[a, b]$ into n subintervals, each of length Δx.

Proof   Let M_1 and M_2 denote the maximum values on $[a, b]$ of $|g'(x)|$ and $|h(x)|$, respectively. Note that

$$\sum_{i=1}^{n} g(u_i)h(v_i)\, \Delta x = R_n + S_n, \quad \text{where} \quad R_n = \sum_{i=1}^{n} g(v_i)h(v_i)\, \Delta x$$

is a Riemann sum approaching $\int_a^b g(x)h(x)\, dx$ as $\Delta x \to 0$, and

$$S_n = \sum_{i=1}^{n} [g(u_i) - g(v_i)]\, h(v_i)\, \Delta x.$$

To prove Eq. (3) it is sufficient to show that $S_n \to 0$ as $\Delta x \to 0$. The mean value theorem gives

$$|g(u_i) - g(v_i)| = |g'(\bar{x}_i)| \cdot |u_i - v_i| \qquad [\bar{x}_i \text{ in } (u_i, v_i)]$$

$$\leq M_1\, \Delta x,$$

because both u_i and v_i are points of the interval $[x_{i-1}, x_i]$ of length Δx. Then

$$|S_n| \leq \sum_{i=1}^{n} |g(u_i) - g(v_i)| \cdot |h(v_i)|\, \Delta x \leq \sum_{i=1}^{n} (M_1\, \Delta x) \cdot (M_2\, \Delta x)$$

$$= (M_1 M_2\, \Delta x) \sum_{i=1}^{n} \Delta x = M_1 M_2 (b - a)\, \Delta x,$$

from which it follows that $S_n \to 0$ as $\Delta x \to 0$, as desired. $\square$

As an application of Theorem 1, let us give a rigorous derivation of Eq. (2) of Section 6.3,

$$V = \int_a^b 2\pi x f(x) \, dx, \tag{4}$$

for the volume of the solid generated by revolving around the y-axis the region lying below $y = f(x)$, $a \leqq x \leqq b$. Beginning with the usual regular partition of $[a, b]$, let $f(x_i^\flat)$ and $f(x_i^\sharp)$ denote the minimum and maximum values of f on the ith subinterval $[x_{i-1}, x_i]$. Denote by x_i^* the midpoint of this subinterval. From Fig. F.1, we see that the part of the solid generated by revolving the region below $y = f(x)$, $x_{i-1} \leqq x \leqq x_i$, contains a cylindrical shell with average radius x_i^*, thickness Δx, and height $f(x_i^\flat)$ and is contained in another cylindrical shell with the same average radius and thickness but with height $f(x_i^\sharp)$. Hence the volume ΔV_i of this part of the solid satisfies the inequalities

$$2\pi x_i^* f(x_i^\flat) \, \Delta x \leqq \Delta V_i \leqq 2\pi x_i^* f(x_i^\sharp) \, \Delta x.$$

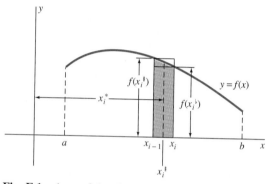

Fig. F.1 A careful estimate of the volume of a solid of revolution around the y-axis

We add these inequalities for $i = 1, 2, 3, \ldots, n$ and find that

$$\sum_{i=1}^n 2\pi x_i^* f(x_i^\flat) \, \Delta x \leqq V \leqq \sum_{i=1}^n 2\pi x_i^* f(x_i^\sharp) \, \Delta x.$$

Because Theorem 1 implies that both of the last two sums approach $\int_a^b f(x) \, dx$, the squeeze law of limits now implies Eq. (4).

We will occasionally need a generalization of Theorem 1 that involves the notion of a continuous function $F(x, y)$ of two variables. We say that F is *continuous* at the point (x_0, y_0) provided that the value $F(x, y)$ can be made arbitrarily close to $F(x_0, y_0)$ merely be choosing the point (x, y) sufficiently close to (x_0, y_0). We discuss continuity of functions of two variables in Chapter 14. Here it will suffice to accept the following facts: If $g(x)$ and $h(y)$ are continuous functions of the single variables x and y, respectively, then simple combinations such as

$$g(x) \pm h(y), \qquad g(x)h(y), \quad \text{and} \quad \sqrt{[g(x)]^2 + [h(y)]^2}$$

are continuous functions of the two variables x and y.

Now consider a regular partition of $[a, b]$ into n subintervals, each of length Δx, and let u_i and v_i denote arbitrary points of the ith subinterval $[x_{i-1}, x_i]$. Theorem 2—we omit the proof—tells us how to find the limit as $\Delta x \to 0$ of a sum such as

$$\sum_{i=1}^{n} F(u_i, v_i) \, \Delta x.$$

Theorem 2   A Further Generalization

Let $F(x, y)$ be continuous for x and y both in the interval $[a, b]$. Then, in the notation of the preceding paragraph,

$$\lim_{\Delta x \to 0} \sum_{i=1}^{n} F(u_i, v_i) \, \Delta x = \int_{a}^{b} F(x, x) \, dx. \tag{5}$$

Theorem 1 is the special case $F(x, y) = g(x)h(y)$ of Theorem 2. Moreover, the integrand $F(x, x)$ on the right in Eq. (5) is merely an ordinary function of the single variable x. As a formal matter, the integral corresponding to the sum in Eq. (5) is obtained by replacing the summation symbol with an integral sign, changing both u_i and v_i to x, replacing Δx by dx, and inserting the correct limits of integration. For example, if the interval $[a, b]$ is $[0, 4]$, then

$$\lim_{\Delta x \to 0} \sum_{i=1}^{n} \sqrt{9u_i^2 + v_i^4} \, \Delta x = \int_{0}^{4} \sqrt{9x^2 + x^4} \, dx$$

$$= \int_{0}^{4} x(9 + x^2)^{1/2} \, dx = \left[\frac{1}{3}(9 + x^2)^{3/2} \right]_{0}^{4}$$

$$= \tfrac{1}{3}[(25)^{3/2} - (9)^{3/2}] = \tfrac{98}{3}.$$

Appendix F Problems

In Problems 1 through 7, u_i and v_i are arbitrary points of the ith subinterval of a regular partition of $[a, b]$ into n subintervals, each of length Δx. Express the given limit as an integral from a to b, then compute the value of this integral.

1. $\lim\limits_{\Delta x \to 0} \sum\limits_{i=1}^{n} u_i v_i \, \Delta x; \quad a = 0, b = 1$

2. $\lim\limits_{\Delta x \to 0} \sum\limits_{j=1}^{n} (3u_j + 5v_j) \, \Delta x; \quad a = -1, b = 3$

3. $\lim\limits_{\Delta x \to 0} \sum\limits_{i=1}^{n} u_i \sqrt{4 - v_i^2} \, \Delta x; \quad a = 0, b = 2$

4. $\lim\limits_{\Delta x \to 0} \sum\limits_{i=1}^{n} \dfrac{u_i \, \Delta x}{\sqrt{16 + v_i^2}}; \quad a = 0, b = 3$

5. $\lim\limits_{\Delta x \to 0} \sum\limits_{i=1}^{n} \sin u_i \cos v_i \, \Delta x; \quad a = 0, b = \dfrac{\pi}{2}$

6. $\lim\limits_{\Delta x \to 0} \sum\limits_{i=1}^{n} \sqrt{\sin^2 u_i + \cos^2 v_i} \, \Delta x; \quad a = 0, b = \pi$

7. $\lim\limits_{\Delta x \to 0} \sum\limits_{k=1}^{n} \sqrt{u_k^4 + v_k^7} \, \Delta x; \quad a = 0, b = 2$

8. Explain how Theorem 1 applies to show that Eq. (8) of Section 6.4 follows from the discussion that precedes it in that section.

9. Use Theorem 1 to derive Eq. (10) of Section 6.4.

Appendix G
L'Hôpital's Rule and Cauchy's Mean Value Theorem

Here we give a proof of l'Hôpital's rule,

$$\lim_{x \to a} \frac{f(x)}{g(x)} = \lim_{x \to a} \frac{f'(x)}{g'(x)}, \tag{1}$$

under the hypotheses of Theorem 1 in Section 8.3. The proof is based on a generalization of the mean value theorem due to the French mathematician Augustin Louis Cauchy. Cauchy used this generalization in the early nineteenth century to give rigorous proofs of several calculus results not previously established firmly.

Cauchy's Mean Value Theorem

Suppose that the functions f and g are continuous on the closed and bounded interval $[a, b]$ and differentiable on (a, b). Then there exists a number c in (a, b) such that

$$[f(b) - f(a)]g'(c) = [g(b) - g(a)]f'(c). \tag{2}$$

REMARK 1   To see that this theorem is indeed a generalization of the (ordinary) mean value theorem, we take $g(x) = x$. Then $g'(x) \equiv 1$, and the conclusion in Eq. (2) reduces to the fact that

$$f(b) - f(a) = (b - a)f'(c)$$

for some number c in (a, b).

REMARK 2   Equation (2) has a geometric interpretation like that of the ordinary mean value theorem. Let us think of the equations $x = g(t), y = f(t)$ as describing the motion of a point $P(x, y)$ moving along a curve C in the xy-plane as t increases from a to b (Fig. G.1). That is, $P(x, y) = P(g(t), f(t))$ is the location of the point P at time t. Under the assumption that $g(b) \neq g(a)$, the slope of the line L connecting the endpoints of the curve C is

$$m = \frac{f(b) - f(a)}{g(b) - g(a)}. \tag{3}$$

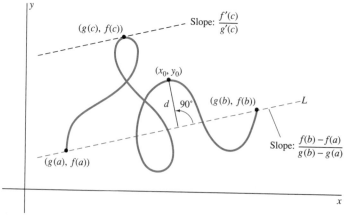

Fig. G.1   The idea of Cauchy's mean value theorem

But if $g'(c) \neq 0$, then the chain rule gives

$$\frac{dy}{dx} = \frac{dy/dt}{dx/dt} = \frac{f'(c)}{g'(c)} \tag{4}$$

for the slope of the line tangent to the curve C at the point $(g(c), f(c))$. But if $g(b) \neq g(a)$ and $g'(c) \neq 0$, then Eq. (2) may be written as

$$\frac{f(b) - f(a)}{g(b) - g(a)} = \frac{f'(c)}{g'(c)}, \tag{5}$$

so the two slopes in Eqs. (3) and (4) are equal. Thus Cauchy's mean value theorem implies that (under our assumptions) there is a point on the curve C where the tangent line is *parallel* to the line joining the endpoints of C. This is exactly what the (ordinary) mean value theorem says for an explicitly defined curve $y = f(x)$. This geometric interpretation motivates the following proof of Cauchy's mean value theorem.

Proof The line L through the endpoints in Fig. G.1 has point-slope equation

$$y - f(a) = \frac{f(b) - f(a)}{g(b) - g(a)}[x - g(a)],$$

which can be rewritten in the form $Ax + By + C = 0$ with

$$A = f(b) - f(a), \qquad B = -[g(b) - g(a)], \quad \text{and}$$
$$C = f(a)[g(b) - g(a)] - g(a)[f(b) - f(a)]. \tag{6}$$

According to Miscellaneous Problem 71 at the end of Chapter 3, the (perpendicular) distance from the point (x_0, y_0) to the line L is

$$d = \frac{|Ax_0 + By_0 + C|}{\sqrt{A^2 + B^2}}.$$

Figure G.1 suggests that the point $(g(c), f(c))$ will maximize this distance d for points on the curve C.

We are motivated, therefore, to define the auxiliary function

$$\phi(t) = Ag(t) + Bf(t) + C, \tag{7}$$

with the constants A, B, and C as defined in Eq. (6). Thus $\phi(t)$ is essentially a constant multiple of the distance from $(g(t), f(t))$ to the line L in Fig. G.1.

Now $\phi(a) = 0 = \phi(b)$ (why?), so Rolle's theorem (Section 4.3) implies the existence of a number c in (a, b) such that

$$\phi'(c) = Ag'(c) + Bf'(c) = 0. \tag{8}$$

We substitute the values of A and B from Eq. (6) into (8) and obtain the equation

$$[f(b) - f(a)]g'(c) - [g(b) - g(a)]f'(c) = 0.$$

This is the same as Eq. (2) in the conclusion of Cauchy's mean value theorem, and the proof is complete. ❏

NOTE Whereas the assumptions that $g(b) \neq g(a)$ and $g'(c) \neq 0$ were needed for our geometric interpretation of the theorem, they were not used in its proof—only in the motivation for the method of proof.

PROOF OF L'HÔPITAL'S RULE

Suppose that $f(x)/g(x)$ has the indeterminate form $0/0$ at $x = a$. We may invoke continuity of f and g to allow the assumption that $f(a) = 0 = g(a)$. That is, we simply define $f(a)$ and $g(a)$ to be zero in case their values at $x = a$ are not originally given.

Now we restrict our attention to values of x in a fixed deleted neighborhood of a on which both f and g are differentiable. Choose one such value of x, and hold it temporarily constant. Then apply Cauchy's mean value theorem on the interval $[a, x]$. (If $x < a$, use the interval $[x, a]$.) We find that there is a number z between a and x that behaves as c does in Eq. (2). Hence, by virtue of Eq. (2), we obtain the equation

$$\frac{f(x)}{g(x)} = \frac{f(x) - f(a)}{g(x) - g(a)} = \frac{f'(z)}{g'(z)}.$$

Now z depends on x, but z is trapped between x and a, so z is forced to approach a as $x \to a$. We conclude that

$$\lim_{x \to a} \frac{f(x)}{g(x)} = \lim_{z \to a} \frac{f'(z)}{g'(z)} = \lim_{x \to a} \frac{f'(x)}{g'(x)},$$

under the assumption that the right-hand limit exists. Thus we have verified l'Hôpital's rule in the form of Eq. (1). ❑

Appendix H
Proof of Taylor's Formula

Several different proofs of Taylor's formula (Theorem 2 of Section 11.4) are known, but none of them seems very well motivated—each requires some "trick" to begin the proof. The trick we employ here (suggested by C. R. MacCluer) is to begin by introducing an auxiliary function $F(x)$, defined as follows:

$$F(x) = f(b) - f(x) - f'(x)(b - x) - \frac{f''(x)}{2!}(b - x)^2$$

$$- \cdots - \frac{f^{(n)}(x)}{n!}(b - x)^n - K(b - x)^{n+1}, \tag{1}$$

where the *constant* K is chosen so that $F(a) = 0$. To see that there *is* such a value of K, we could substitute $x = a$ on the right and $F(x) = F(a) = 0$ on the left in Eq. (1) and then solve routinely for K, but we have no need to do this explicitly.

Equation (1) makes it quite obvious that $F(b) = 0$ as well. Therefore, Rolle's theorem (Section 4.3) implies that

$$F'(z) = 0 \tag{2}$$

for some point z of the open interval (a, b) (under the assumption that $a < b$).

A-30

To see what Eq. (2) means, we differentiate both sides of Eq. (1) and find that

$$F'(x) = -f'(x) + [f'(x) - f''(x)(b - x)]$$

$$+ \left[f''(x)(b - x) - \frac{1}{2!} f^{(3)}(x)(b - x)^2 \right]$$

$$+ \left[\frac{1}{2!} f^{(3)}(x)(b - x)^2 - \frac{1}{3!} f^{(4)}(x)(b - x)^3 \right]$$

$$+ \cdots + \left[\frac{1}{(n - 1)!} f^{(n)}(x)(b - x)^{n-1} - \frac{1}{n!} f^{(n+1)}(x)(b - x)^n \right]$$

$$+ (n + 1)K(b - x)^n.$$

Upon careful inspection of this result, we see that all terms except the final two cancel in pairs. Thus the sum "telescopes" to give

$$F'(x) = (n + 1)K(b - x)^n - \frac{f^{(n+1)}(x)}{n!} (b - x)^n. \qquad (3)$$

Hence Eq. (2) means that

$$(n + 1)K(b - z)^n - \frac{f^{(n+1)}(z)}{n!} (b - z)^n = 0.$$

Consequently we can cancel $(b - z)^n$ and solve for

$$K = \frac{f^{(n+1)}(z)}{(n + 1)!}. \qquad (4)$$

Finally, we return to Eq. (1) and substitute $x = a$, $F(x) = 0$, and the value of K given in Eq. (4). The result is the equation

$$0 = f(b) - f(a) - f'(a)(b - a) - \frac{f''(a)}{2!}(b - a)^2$$

$$- \cdots - \frac{f^{(n)}(a)}{n!}(b - a)^n - \frac{f^{(n+1)}(z)}{(n + 1)!}(b - a)^{n+1},$$

which is equivalent to the desired Taylor's formula, Eq. (11) of Section 11.4. ❑

Appendix I ▰
Units of Measurement and Conversion Factors

MKS SCIENTIFIC UNITS

❑ *Length* in meters (m), *mass* in kilograms (kg), *time* in seconds (s)
❑ *Force* in newtons (N); a force of 1 N imparts an acceleration of 1 m/s² to a mass of 1 kg.
❑ *Work* in joules (J); 1 J is the work done by a force of 1 N acting through a distance of 1 m.
❑ *Power* in watts (W); 1 W is 1 J/s.

BRITISH ENGINEERING UNITS (fps)

❑ *Length* in feet (ft), *force* in pounds (lb), *time* in seconds (s)
❑ *Mass* in slugs; 1 lb of force imparts an acceleration of 1 ft/s² to a mass of

1 slug. A mass of m slugs at the surface of the earth has a *weight* of $w = mg$ pounds (lb), where $g \approx 32.17$ ft/s².

❏ *Work* in ft·lb, *power* in ft·lb/s.

CONVERSION FACTORS

$$1 \text{ in.} = 2.54 \text{ cm} = 0.0254 \text{ m}, \quad 1 \text{ m} \approx 3.2808 \text{ ft}$$

$$1 \text{ mi} = 5280 \text{ ft}; \quad 60 \text{ mi/h} = 88 \text{ ft/s}$$

$$1 \text{ lb} \approx 4.4482 \text{ N}; \quad 1 \text{ slug} \approx 14.594 \text{ kg}$$

$$1 \text{ hp} = 550 \text{ ft·lb/s} \approx 745.7 \text{ W}$$

❏ *Gravitational acceleration:* $g \approx 32.17$ ft/s² ≈ 9.807 m/s²

❏ *Atmospheric pressure:* 1 atm is the pressure exerted by a column of mercury 76 cm high; 1 atm ≈ 14.70 lb/in.² $\approx 1.013 \times 10^5$ N/m²

❏ *Heat energy:* 1 Btu ≈ 778 ft·lb ≈ 252 cal, 1 cal ≈ 4.184 J

Appendix J
Formulas from Algebra, Geometry, and Trigonometry

LAWS OF EXPONENTS

$$a^m a^n = a^{m+n}, \quad (a^m)^n = a^{mn}, \quad (ab)^n = a^n b^n, \quad a^{m/n} = \sqrt[n]{a^m};$$

in particular,

$$a^{1/2} = \sqrt{a}.$$

If $a \neq 0$, then

$$a^{m-n} = \frac{a^m}{a^n}, \quad a^{-n} = \frac{1}{a^n}, \quad \text{and} \quad a^0 = 1.$$

QUADRATIC FORMULA

The quadratic equation

$$ax^2 + bx + c = 0 \quad (a \neq 0)$$

has solutions

$$x = \frac{-b \pm \sqrt{b^2 - 4ac}}{2a}.$$

FACTORING

$$a^2 - b^2 = (a - b)(a + b)$$
$$a^3 - b^3 = (a - b)(a^2 + ab + b^2)$$
$$a^4 - b^4 = (a - b)(a^3 + a^2b + ab^2 + b^3)$$
$$\qquad = (a - b)(a + b)(a^2 + b^2)$$
$$a^5 - b^5 = (a - b)(a^4 + a^3b + a^2b^2 + ab^3 + b^4)$$

A-32

(The pattern continues.)

$$a^3 + b^3 = (a + b)(a^2 - ab + b^2)$$

$$a^5 + b^5 = (a + b)(a^4 - a^3b + a^2b^2 - ab^3 + b^4)$$

(The pattern continues for odd exponents.)

BINOMIAL FORMULA

$$(a + b)^n = a^n + na^{n-1}b + \frac{n(n - 1)}{1 \cdot 2} a^{n-2}b^2$$

$$+ \frac{n(n - 1)(n - 2)}{1 \cdot 2 \cdot 3} a^{n-3}b^3 + \cdots + nab^{n-1} + b^n$$

if n is a positive integer.

AREA AND VOLUME

In Fig. J.1, the symbols have the following meanings.

A:	area	b: length of base	r: radius
B:	area of base	C: circumference	V: volume
h:	height	ℓ: length	w: width

 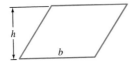 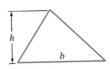 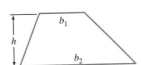

Rectangle: $A = bh$ Parallelogram: $A = bh$ Triangle: $A = \frac{1}{2}bh$ Trapezoid: $A = \frac{1}{2}(b_1 + b_2)h$ Circle: $C = 2\pi r$ and $A = \pi r^2$

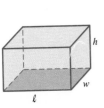 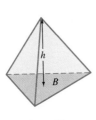  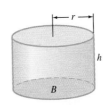

Rectangular parallelepiped:
$V = \ell w h$

Pyramid:
$V = \frac{1}{3}Bh$

Right circular cone:
$V = \frac{1}{3}\pi r^2 h = \frac{1}{3}Bh$

Right circular cylinder:
$V = \pi r^2 h = Bh$

Sphere:
$V = \frac{4}{3}\pi r^3$ and $A = 4\pi r^2$

Fig. J.1 The basic geometric shapes

PYTHAGOREAN THEOREM

In a right triangle with legs a and b and hypotenuse c,

$$a^2 + b^2 = c^2.$$

FORMULAS FROM TRIGONOMETRY

$$\sin(-\theta) = -\sin\theta, \qquad \cos(-\theta) = \cos\theta$$

$$\sin^2\theta + \cos^2\theta = 1$$

$$\sin 2\theta = 2\sin\theta\cos\theta$$

$$\cos 2\theta = \cos^2\theta - \sin^2\theta$$

$$\sin(\alpha + \beta) = \sin\alpha\cos\beta + \cos\alpha\sin\beta$$

$$\cos(\alpha + \beta) = \cos\alpha\cos\beta - \sin\alpha\sin\beta$$

$$\tan(\alpha + \beta) = \frac{\tan\alpha + \tan\beta}{1 - \tan\alpha\tan\beta}$$

$$\sin^2\frac{\theta}{2} = \frac{1}{2}(1 - \cos\theta)$$

$$\cos^2\frac{\theta}{2} = \frac{1}{2}(1 + \cos\theta)$$

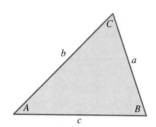

Fig. J.2 An arbitrary triangle

For an arbitrary triangle (Fig J.2):

Law of cosines: $c^2 = a^2 + b^2 - 2ab\cos C.$

Law of sines: $\dfrac{\sin A}{a} = \dfrac{\sin B}{b} = \dfrac{\sin C}{c}.$

Appendix K
The Greek Alphabet

A	α	alpha	I	ι	iota	P	ρ	rho
B	β	beta	K	κ	kappa	Σ	σ	sigma
Γ	γ	gamma	Λ	λ	lambda	T	τ	tau
Δ	δ	delta	M	μ	mu	Υ	υ	upsilon
E	ϵ	epsilon	N	ν	nu	Φ	ϕ	phi
Z	ζ	zeta	Ξ	ξ	xi	X	χ	chi
H	η	eta	O	o	omicron	Ψ	ψ	psi
Θ	θ	theta	Π	π	pi	Ω	ω	omega

A-34

Answers to Odd-Numbered Problems

Section 1.1 (page 11)

1. 14 **3.** 0.5 **5.** 25 **7.** 27 **9.** $\frac{22}{7} - \pi$

11. (a) $-\frac{1}{a}$; (b) a; (c) $\frac{1}{\sqrt{a}}$; (d) $\frac{1}{a^2}$

13. (a) $\frac{1}{a^2 + 5}$; (b) $\frac{a^2}{1 + 5a^2}$; (c) $\frac{1}{a + 5}$; (d) $\frac{1}{a^4 + 5}$

15. $\frac{1}{3}$ **17.** ± 3 **19.** 100 **21.** $3h$

23. $2ah + h^2$ **25.** $-\dfrac{h}{a(a + h)}$ **27.** $\{-1, 0, 1\}$

29. $\{-1, 1\}$ **31.** R **33.** R **35.** $[\frac{5}{3}, \infty)$

37. $t \leqq \frac{1}{2}$ **39.** All real numbers other than 3

41. R **43.** $[0, 16]$

45. All real numbers other than 0

47. $C(A) = 2\sqrt{\pi A}, A \geqq 0$ (or $A > 0$)

49. $C(F) = \frac{5}{9}(F - 32), F > -459.67$

51. $A(x) = x\sqrt{16 - x^2}, 0 \leqq x \leqq 4$ (or $0 < x < 4$)

53. $C(x) = 3x^2 + \dfrac{1296}{x}, x > 0$

55. $A(r) = 2\pi r^2 + \dfrac{2000}{r}, r > 0$

57. $V(x) = x(50 - 2x)^2, 0 \leqq x \leqq 25$ (or $0 < x < 25$)

59. Drill 10 new wells **61.** 0.38 **63.** 1.24

65. 0.72 **67.** 3.21 **69.** 1.62

Section 1.2 (page 22)

1. AB and BC have slope 1.

3. AB has slope -2, but BC has slope $-\frac{4}{3}$.

5. AB and CD have slope $-\frac{1}{2}$; BC and DA have slope 2.

7. AB has slope 2, and AC has slope $-\frac{1}{2}$.

9. $m = \frac{2}{3}, b = 0$ **11.** $m = 2, b = 3$

13. $m = -\frac{2}{5}, b = \frac{3}{5}$ **15.** $y = -5$

17. $y - 3 = 2(x - 5)$ **19.** $y - 2 = 4 - x$

21. $y - 5 = -2(x - 1)$ **23.** $x + 2y = 13$

25. $\frac{4}{13}\sqrt{26} \approx 1.568929$

33. $K = \dfrac{125F + 57,461}{225}; F = -459.688$ when $K = 0$.

35. 1136 gal/week **37.** $x = -2.75, y = 3.5$

39. $x = \frac{37}{6}, y = -\frac{1}{2}$ **41.** $x = \frac{22}{5}, y = -\frac{1}{5}$

43. $x = -\frac{7}{4}, y = \frac{33}{8}$ **45.** $x = \frac{119}{12}, y = -\frac{19}{4}$

Section 1.3 (page 30)

1. Center $(2, 0)$, radius 2

3. Center $(-1, -1)$, radius 2

5. Center $(-\frac{1}{2}, \frac{1}{2})$, radius 1

7. Opens upward, vertex at $(3, 0)$

9. Opens upward, vertex at $(-1, 3)$

11. Opens upward, vertex at $(-2, 3)$

13. Circle, center $(3, -4)$, radius 5

15. There are no points on this graph.

17. The graph is the straight line segment joining and including the two points $(-1, 7)$ and $(1, -3)$.

19. Parabola, opening downward, vertex at $(0, 10)$

21. **23.**

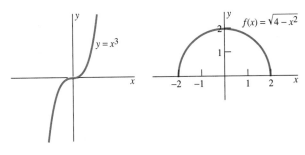

25. The domain of f consists of those numbers x such that $|x| \geq 3$.

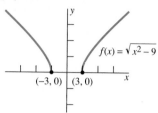

$f(x) = \sqrt{x^2 - 9}$

$(-3, 0)$ $(3, 0)$

27.

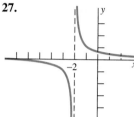

29.

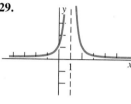

31.

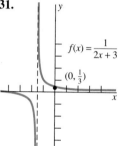

$f(x) = \dfrac{1}{2x + 3}$

$(0, \frac{1}{3})$

33.

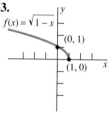

$f(x) = \sqrt{1 - x}$

$(0, 1)$

$(1, 0)$

35.

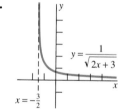

$y = \dfrac{1}{\sqrt{2x + 3}}$

$x = -\dfrac{3}{2}$

37.

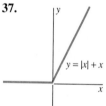

$y = |x| + x$

39.

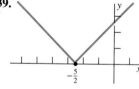

$-\dfrac{5}{2}$

41.

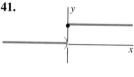

43.

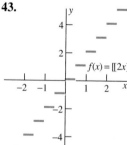

$f(x) = [\![2x]\!]$

45.

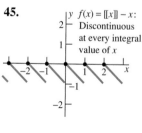

$f(x) = [\![x]\!] - x$:
Discontinuous at every integral value of x

47. $(1.5, 2.5)$ **49.** $(2.25, 1.75)$ **51.** $(2.25, 8.5)$
53. $\left(-\frac{4}{3}, \frac{25}{3}\right)$ **55.** 144 ft **57.** 625

Section 1.4 (page 40)

1. $(f + g)(x) = x^2 + 3x - 2$, domain R;
$(f \cdot g)(x) = x^3 + 3x^2 - x - 3$, domain R; and
$(f/g)(x) = \dfrac{x + 1}{x^2 + 2x - 3}$, domain $x \neq 1, -3$

3. $(f + g)(x) = \sqrt{x} + \sqrt{x - 2}$, domain $x \geq 2$;
$(f \cdot g)(x) = \sqrt{x^2 - 2x}$, domain $x \geq 2$;
$(f/g)(x) = \sqrt{\dfrac{x}{x - 2}}$, domain $x > 2$

5. $(f + g)(x) = \sqrt{x^2 + 1} + \dfrac{1}{\sqrt{4 - x^2}}$,
$(f \cdot g)(x) = \dfrac{\sqrt{x^2 + 1}}{\sqrt{4 - x^2}}$, $(f/g)(x) = \sqrt{4 + 3x^2 - x^4}$,
each with domain $-2 < x < 2$

7. $(f + g)(x) = x + \sin x$, domain R;
$(f \cdot g)(x) = x \sin x$, domain R; $(f/g)(x) = \dfrac{x}{\sin x}$,
domain all real numbers not integral multiples of π

9. $(f + g)(x) = \sqrt{x^2 + 1} + \tan x$, $(f \cdot g)(x) = (x^2 + 1)^{1/2} \tan x$, both with domain all real numbers x such that x is not an odd integral multiple of $\pi/2$;
$(f/g)(x) = \dfrac{\sqrt{x^2 + 1}}{\tan x}$, domain all real numbers x such that x is not an integral multiple of $\pi/2$

11. Fig. 1.4.20 **13.** Fig. 1.4.21 **15.** Fig. 1.4.24
17. Fig. 1.4.23 **19.** Fig. 1.4.28 **21.** Fig. 1.4.29
23. Fig. 1.4.30 **25.** 3 **27.** 1 **29.** 0
31. 1 **33.** 5 **35.** 3

Chapter 1 Miscellaneous Problems (page 47)

1. $x \geq 4$ **3.** $x \neq \pm 3$ **5.** $x \geq 0$ **7.** $x \leq \frac{2}{3}$
9. R **11.** $4 \leq p \leq 8$ **13.** $2 < I < 4$
15. $V(S) = (S/6)^{3/2}$, $0 < S < \infty$
17. $A = A(P) = \dfrac{\sqrt{3}}{36} P^2$, $0 < P < \infty$
19. $y - 5 = 2(x + 3)$ **21.** $2y = x - 10$
23. $x + 2y = 11$ **25.**

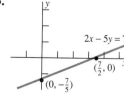

$2x - 5y = 7$

$\left(\frac{7}{2}, 0\right)$

$\left(0, -\frac{7}{5}\right)$

27.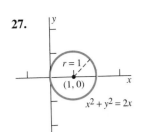

29.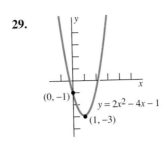

31.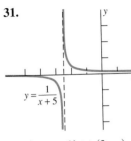

33.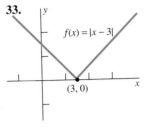

37. $(-\infty, -2) \cup (3, \infty)$

41. $-1.140, 6.140$

45. $-5.021, 0.896$

49. $(\frac{7}{4}, -\frac{5}{4})$

53. $x \approx 0.4505$

57. 3

39. $(-\infty, -2) \cup (4, \infty)$

43. $1.191, 2.309$

47. $(\frac{5}{2}, \frac{3}{4})$

51. $(-\frac{33}{16}, \frac{31}{32})$

55. 3

59. 3

Section 2.1 (page 57)

1. 0   **3.** $2x$ **5.** 4   **7.** $4x - 3$ **9.** $4x + 6$

11. $2 - \dfrac{x}{50}$ **13.** $8x$ **15.** $(0, 10)$ **17.** $(1, 0)$

19. $(50, 25)$ **21.** $f'(x) = 3; y - 5 = 3(x - 2)$

23. $f'(x) = 4x - 3; y - 7 = 5(x - 2)$

25. $f'(x) = 2x - 2; y - 1 = 2(x - 2)$

27. $f'(x) = -\dfrac{1}{x^2}; x + 4y = 4$

29. $f'(x) = -\dfrac{2}{x^3}; x + 4y = 3$

31. $f'(x) = -\dfrac{2}{(x - 1)^2}; y - 2 = -2(x - 2)$

33. $\dfrac{dy}{dx} = 2x; y - 4 = -4(x + 2), y - 4 = \frac{1}{4}(x + 2)$

35. $\dfrac{dy}{dx} = 4x + 3; y - 9 = 11(x - 2),$
$y - 9 = -\frac{1}{11}(x - 2)$

37. $y(3) = 144$ (ft)   **39.** 625   **41.** $y = 12(x - 3)$

43. $(1, 1)$ **45.** 12   **47.** 0.5   **49.** -1

Section 2.2 (page 69)

1. -12 **3.** -1 **5.** 128   **7.** $-\frac{1}{3}$ **9.** 6

11. $16\sqrt{2}$ **13.** 1   **15.** $\frac{1}{4}$ **17.** $-\frac{1}{54}$ **19.** 2

21. 4   **23.** $-\frac{3}{2}$ **25.** -32 **27.** 1   **29.** 0

31. 9   **33.** 2   **35.** 2   **37.** 13

39. Does not exist   **41.** $-\frac{1}{2} x^{-3/2}$

43. $\dfrac{1}{(2x + 1)^2}$ **45.** $\dfrac{x^2 + 2x}{(x + 1)^2}$

49. The limit exists exactly when the real number a is not an integer.

51. $g(x) \to 0$ as $x \to 3$, but $g(3) = 1$.

53. $h(\frac{1}{2}) = 0$, but $h(1/\pi) < 0$.

Section 2.3 (page 80)

1. 0   **3.** $\frac{1}{2}$ **5.** $-\infty$ (or "does not exist")

7. 5   **9.** Does not exist   **11.** $\frac{1}{3}$ **13.** 0

15. 1   **17.** $\frac{1}{2}$ **19.** 1   **21.** $\frac{1}{3}$

23. $\frac{1}{4}$ **25.** $\frac{1}{4}$ **27.** 0   **29.** 3

31. Does not exist   **33.** 0   **35.** 0

37. $+\infty$ (or "does not exist")   **39.** -1 **41.** 1

43. -1 **45.** 2   **47.** -1

49. $f(x) \to +\infty$ as $x \to 1^+, f(x) \to -\infty$ as $x \to 1^-$.

51. $f(x) \to +\infty$ as $x \to -1^-, f(x) \to -\infty$ as $x \to -1^+$.

53. $f(x) \to +\infty$ as $x \to -2^-, f(x) \to -\infty$ as $x \to -2^+$.

55. $f(x) \to +\infty$ as $x \to 1$.

57. $f(x) \to -\infty$ as $x \to -2^+, f(x) \to +\infty$ as $x \to -2^-$, and $f(x) \to -\frac{1}{4}$ as $x \to 2$.

59. If n is an integer, then $f(x) \to 2$ as $x \to n$.

61. If n is an even integer, then the right-hand limit at n is 1 and the left-hand limit at n is -1. If n is an odd integer, then the right-hand limit at n is -1 and the left-hand limit at n is 1.

63. The right-hand limit at the integer n is 1; the left-hand limit at n is zero.

Section 2.4 (page 90)

1. $f(g(x)) = -4x^2 - 12x - 8; g(f(x)) = -2x^2 + 5$

3. $f(g(x)) = \sqrt{x^4 + 6x^2 + 6}, g(f(x)) = x^2$

5. $f(g(x)) = g(f(x)) = x$

7. $f(g(x)) = \sin x^3; g(f(x)) = \sin^3 x$

9. $f(g(x)) = 1 + \tan^2 x; g(f(x)) = \tan(1 + x^2)$

Problems 11 through 19 have many correct answers. We give only the most natural.

11. $k = 2, g(x) = 2 + 3x$ **13.** $k = \frac{1}{2}, g(x) = 2x - x^2$

15. $k = \frac{3}{2}, g(x) = 5 - x^2$ **17.** $k = -1, g(x) = x + 1$

19. $k = -\frac{1}{2}, g(x) = x + 10$

21. R **23.** $(-\infty, -3) \cup (-3, +\infty)$

25. R **27.** $(-\infty, 5) \cup (5, +\infty)$

29. $(-\infty, 2) \cup (2, +\infty)$ **31.** $(-\infty, 1) \cup (1, +\infty)$

33. $(-\infty, 0) \cup (0, 1) \cup (1, +\infty)$ **35.** $(-2, 2)$

37. $(-\infty, 0) \cup (0, +\infty)$

39. At every real number other than the integral multiples of $\pi/2$

41. R

43. Nonremovable discontinuity at $x = -3$

45. Removable discontinuity at $x = 2$; let $f(2) = \frac{1}{4}$
Nonremovable discontinuity at $x = -2$

47. Nonremovable discontinuities at $x = \pm 1$

49. Nonremovable discontinuity at $x = 17$

51. Removable discontinuity at $x = 0$; let $f(0) = 0$

59. Given: $f(x) = x^3 - 4x + 1$. Values of $f(x)$:

x	-3	-2	-1	0	1	2	3
$f(x)$	-14	1	4	1	-2	1	16

63. Discontinuous at $x = 3n$ for each integer n

Chapter 2 Miscellaneous Problems   (page 92)

1. 4   **3.** 0   **5.** $-\frac{5}{3}$ **7.** -2 **9.** 0

11. 4   **13.** 8   **15.** $\frac{1}{6}$ **17.** $-\frac{1}{54}$ **19.** -1

21. 1   **23.** Does not exist   **25.** $+\infty$ (or "does
not exist")   **27.** $+\infty$ **29.** $-\infty$ **31.** 3   **33.** $\frac{3}{2}$

35. 0   **37.** $\frac{9}{4}$ **39.** 2

41. $f'(x) = 4x$; $y - 5 = 4(x - 1)$

43. $f'(x) = 6x + 4$; $y - 2 = 10(x - 1)$

45. $f'(x) = 4x - 3$; $y = x - 1$

47. $4x + 3$ **49.** $\dfrac{1}{(3 - x)^2}$ **51.** $1 + \dfrac{1}{x^2}$

53. $-\dfrac{2}{(x - 1)^2}$ **55.** $a = 3 \pm \sqrt{5}$

57. $g(x) = x - x^2$ or $g(x) = x^2 - x$

59. $g(x) = x + 1$ **61.** $g(x) = \sqrt{x^4 + 1}$

63. Nonremovable discontinuity at $x = -1$; removable
discontinuity at $x = 1$; let $f(1) = \frac{1}{2}$.

65. Nonremovable discontinuity at $x = -3$; removable
discontinuity at $x = 1$; let $f(1) = \frac{3}{4}$.

Section 3.1   (page 105)

1. $f'(x) = 4$ **3.** $h'(z) = 25 - 2z$

5. $\dfrac{dy}{dx} = 4x + 3$ **7.** $\dfrac{dz}{du} = 10u - 3$

9. $\dfrac{dx}{dy} = -10y + 17$ **11.** $f'(x) = 2$

13. $f'(x) = 2x$ **15.** $f'(x) = \dfrac{-2}{(2x + 1)^2}$

17. $f'(x) = \dfrac{1}{\sqrt{2x + 1}}$ **19.** $f'(x) = \dfrac{1}{(1 - 2x)^2}$

21. $x(0) = 100$

23. $x(2.5) = 99$ **25.** $x(-2) = 120$

27. $y(2) = 64$ (ft)   **29.** $y(3) = 194$ (ft)

31. $\dfrac{dA}{dC} = \dfrac{C}{2\pi}$ **33.** 500 ft; 10 s

35. (a) 2.5 months; (b) 50 chipmunks per month

37. *Very* roughly: $v(20) = 50$ mi/h; $v(40) = 62$ mi/h

41. $V'(30) = -\dfrac{25\pi}{12}$ in.³/s; that is, air is leaking out at
about 6.545 in.³/s then.

43. (a) $V'(6) = -144\pi$ cm³/h; (b) -156π cm³/h

45. At $t = 2$ s, $v = 0$ m/s

Section 3.2   (page 115)

1. $f'(x) = 6x - 1$

3. $f'(x) = 2(3x - 2) + 3(2x + 3)$

5. $h'(x) = 3x^2 + 6x + 3$ **7.** $f'(y) = 12y^2 - 1$

9. $g'(x) = \dfrac{1}{(x - 1)^2} - \dfrac{1}{(x + 1)^2}$

11. $h'(x) = -\dfrac{3(2x + 1)}{(x^2 + x + 1)^2}$

13. $g'(t) = (t^2 + 1)(3t^2 + 2t) + (t^3 + t^2 + 1)(2t)$

15. $g'(z) = -\dfrac{1}{2z^2} + \dfrac{2}{3z^3}$

17. $g'(y) = 30y^4 + 48y^3 + 48y^2 - 8y - 6$

19. $g'(t) = \dfrac{3 - t}{(t + 1)^3}$ **21.** $v'(t) = -\dfrac{3}{(t - 1)^4}$

23. $g'(x) = -\dfrac{6x^3 + 15}{(x^3 + 7x - 5)^2}$

25. $g'(x) = \dfrac{4x^3 - 13x^2 + 12x}{(2x - 3)^2}$

27. $\dfrac{dy}{dx} = 3x^2 - 30x^4 - 6x^{-5}$

29. $\dfrac{dy}{dx} = \dfrac{2x^5 + 4x^2 - 15}{x^4}$

31. $\dfrac{dy}{dx} = 3 + \dfrac{1}{2x^3}$ **33.** $\dfrac{dy}{dx} = \dfrac{2x - 1 - 4x^2}{3x^2(x - 1)^2}$

35. $\dfrac{dy}{dx} = \dfrac{x^4 + 31x^2 - 10x - 36}{(x^2 + 9)^2}$

37. $\dfrac{dy}{dx} = \dfrac{30x^5(5x^5 - 8)}{(15x^5 - 4)^2}$ **39.** $\dfrac{dy}{dx} = \dfrac{x^2 + 2x}{(x + 1)^2}$

41. $12x - y = 16$ **43.** $x + y = 3$

45. $5x - y = 10$ **47.** $18x - y = -25$

49. $3x + y = 0$

51. (a) It contracts; (b) -0.06427 cm³/°C

53. $14{,}400\pi \approx 45{,}239$ cm³/cm

55. $y = 3x + 2$

57. Suppose that some line is tangent at both (a, a^2) and
(b, b^2). Use the derivative to show that $a = b$.

59. $x = \dfrac{n-1}{n} x_0$

65. $g'(x) = 17(x^3 - 17x + 35)^{16}(3x^2 - 17)$

71. $0, \pm\sqrt{3}$

Section 3.3 (page 124)

1. $\dfrac{dy}{dx} = 15(3x + 4)^4$ **3.** $\dfrac{dy}{dx} = -3(3x - 2)^{-2}$

5. $\dfrac{dy}{dx} = 3(x^2 + 3x + 4)^2(2x + 3)$

7. $\dfrac{dy}{dx} = -4(2 - x)^3(3 + x)^7 + 7(2 - x)^4(3 + x)^6$

9. $\dfrac{dy}{dx} = -\dfrac{6x + 22}{(3x - 4)^4}$

11. $\dfrac{dy}{dx} = 12[1 + (1 + x)^3]^3(1 + x)^2$

13. $\dfrac{dy}{dx} = -\dfrac{6}{x^3}\left(\dfrac{1}{x^2} + 1\right)^2$

15. $\dfrac{dy}{dx} = 48[1 + (4x - 1)^4]^2(4x - 1)^3$

17. $\dfrac{dy}{dx} = \dfrac{12(1 - x^{-4})^2}{x^9} - \dfrac{4(1 - x^{-4})^3}{x^5}$

$= -\dfrac{4(x^{12} - 6x^8 + 9x^4 - 4)}{x^{17}}$

19. $\dfrac{dy}{dx} = -4x^{-5}(x^{-2} - x^{-8})^3$

$+ 3x^{-4}(8x^{-9} - 2x^{-3})(x^{-2} - x^{-8})^2$

$= -\dfrac{2(x^3 - 1)^2(x^3 + 1)^2(5x^6 - 14)}{x^{29}}$

21. $u(x) = 2x - x^2, n = 3;$
$f'(x) = 3(2x - x^2)^2(2 - 2x)$

23. $u(x) = 1 - x^2, n = -4; f'(x) = 8x(1 - x^2)^{-5}$

25. $u(x) = \dfrac{x + 1}{x - 1}, n = 7; f'(x) = -\dfrac{14(x + 1)^6}{(x - 1)^8}$

27. $g'(y) = 1 + 10(2y - 3)^4$

29. $F'(s) = 3(s - s^{-2})^2(1 + 2s^{-3})$

31. $f'(u) = 8u(u + 1)^3(u^2 + 1)^3 + 3(u + 1)^2(u^2 + 1)^4$

33. $h'(v) = 2(v - 1)(v^2 - 2v + 2)(v^{-3})(2 - v)^{-3}$

35. $F'(z) = 10(4 - 25z^4)(3 - 4z + 5z^5)^{-11}$

37. $\dfrac{dy}{dx} = 4(x^3)^3 \cdot 3x^2 = 12x^{11}$

39. $\dfrac{dy}{dx} = 2(x^2 - 1)(2x) = 4x^3 - 4x$

41. $\dfrac{dy}{dx} = 4(x + 1)^3 = 4x^3 + 12x^2 + 12x + 4$

43. $\dfrac{dy}{dx} = -\dfrac{2x}{(x^2 + 1)^2}$ **45.** $f'(x) = 3x^2 \cos x^3$

47. $g'(z) = 6(\sin 2z)^2 \cos 2z$ **49.** 40π in.^{2}/s

51. 40 in.^{2}/s   **53.** 600 in.^{3}/h   **55.** -18

57. $400\pi \approx 1256.64$ cm^3/s   **59.** 5 cm

61. Total melting time: $2/(2 - 4^{1/3}) \approx 4.85$ h; all melted by about 2:50:50 P.M. that day.

Section 3.4 (page 129)

1. $f'(x) = 10x^{3/2} - x^{-3/2}$ **3.** $f'(x) = (2x + 1)^{-1/2}$

5. $f'(x) = -3x^{-3/2} - \frac{3}{2}x^{1/2}$ **7.** $f'(x) = 3(2x + 3)^{1/2}$

9. $f'(x) = 6x(3 - 2x^2)^{-5/2}$ **11.** $f'(x) = \dfrac{3x^2}{2\sqrt{x^3 + 1}}$

13. $f'(x) = 2x(2x^2 + 1)^{-1/2}$

15. $f'(t) = 3t^2(2t^3)^{-1/2} = \frac{3}{2}\sqrt{2t}$

17. $f'(x) = \frac{3}{2}(2x^2 - x + 7)^{1/2}(4x - 1)$

19. $g'(x) = -\frac{4}{3}(x - 2x^3)^{-7/3}(1 - 6x^2)$

21. $f'(x) = (1 - x^2)^{1/2} - x^2(1 - x^2)^{-1/2}$
$= (1 - 2x^2)(1 - x^2)^{-1/2}$

23. $f'(t) = \dfrac{1}{2}\left(\dfrac{t^2 + 1}{t^2 - 1}\right)^{-1/2} \cdot \dfrac{(t^2 - 1)(2t) - (2t)(t^2 + 1)}{(t^2 - 1)^2}$

$= -2t(t^2 + 1)^{-1/2}(t^2 - 1)^{-3/2}$

25. $f'(x) = 3\left(x - \dfrac{1}{x}\right)^2\left(1 + \dfrac{1}{x^2}\right)$

27. $f'(v) = -\dfrac{v + 2}{2v^2\sqrt{v + 1}}$

29. $f'(x) = \frac{1}{3}(1 - x^2)^{-2/3}(-2x)$

31. $f'(x) = (3 - 4x)^{1/2} - 2x(3 - 4x)^{-1/2}$

33. $f'(x) = (-2x)(2x + 4)^{4/3} + \frac{8}{3}(1 - x^2)(2x + 4)^{1/3}$

35. $g'(t) = -2t^{-2}(1 + t^{-1})(3t^2 + 1)^{1/2}$
$+ 3t(1 + t^{-1})^2(3t^2 + 1)^{-1/2} = \dfrac{3t^4 - 3t^2 - 2t - 2}{t^3\sqrt{3t^2 + 1}}$

37. $f'(x) = \dfrac{2(3x + 4)^5 - 15(3x + 4)^4(2x - 1)}{(3x + 4)^{10}}$

$= \dfrac{23 - 24x}{(3x + 4)^6}$

39. $f'(x) =$

$\dfrac{(3x + 4)^{1/3}(2x + 1)^{-1/2} - (3x + 4)^{-2/3}(2x + 1)^{1/2}}{(3x + 4)^{2/3}}$

$= \dfrac{x + 3}{(3x + 4)^{4/3}(2x + 1)^{1/2}}$

Answers to Odd-Numbered Problems

A-39

41. $h'(y) = \dfrac{(1 + y)^{-1/2} - (1 - y)^{-1/2}}{2y^{5/3}}$

$\qquad - \dfrac{5[(1 + y)^{1/2} + (1 - y)^{1/2}]}{3y^{8/3}}$

$\qquad = \dfrac{(7y - 10)\sqrt{1 + y} - (7y + 10)\sqrt{1 - y}}{6y^{8/3}\sqrt{1 - y^2}}$

43. $g'(t) = \frac{1}{2}[t + (t + t^{1/2})^{1/2}]^{-1/2} \times$
$\qquad [1 + \frac{1}{2}(t + t^{1/2})^{-1/2}(1 + \frac{1}{2}t^{-1/2})]$

$\qquad = \dfrac{1 + \dfrac{1 + \dfrac{1}{2\sqrt{t}}}{2\sqrt{t + \sqrt{t}}}}{2\sqrt{t + \sqrt{t + \sqrt{t}}}}$

45. No horizontal tangents; vertical tangent line at $(0, 0)$

47. Horizontal tangent where $x = \frac{1}{3}$ and $y = \frac{2}{9}\sqrt{3}$; vertical tangent line at $(0, 0)$

49. No horizontal or vertical tangents

51. $\pi^2/32 \approx 0.3084$ (s/ft)

53. $(2/\sqrt{5}, 1/\sqrt{5})$ and $(-2/\sqrt{5}, -1/\sqrt{5})$

55. $x + 4y = 18$

57. $3x + 2y = 5$ and $3x - 2y = -5$

59. Equation (3) is an *identity,* and if two functions have identical graphs on an interval, then their derivatives are also identically equal on that interval.

Section 3.5   (page 138)

1. Max.: 2; no min.   **3.** No max.; min.: 0

5. Max.: 2; min.: 0   **7.** Max.: 2; min.: 0

9. Max.: $-\frac{1}{6}$; min.: $-\frac{1}{2}$ **11.** Max.: 7; min.: -8

13. Max.: 3; min.: -5 **15.** Max.: 9; min.: 0

17. Max.: 52; min.: -2 **19.** Max.: 5; min.: 4

21. Max.: 5; min.: 1   **23.** Max.: 9; min.: -16

25. Max.: 10; min.: -22 **27.** Max.: 56; min.: -56

29. Max.: 13; min.: 5   **31.** Max.: 17; min.: 0

33. Max.: $\frac{3}{4}$; min.: 0
35. Max.: $\frac{1}{2}$ (at $x = -1$); min.: $-\frac{1}{6}$ (at $x = 3$)
37. Max.: $f(1/\sqrt{2}) = \frac{1}{2}$; min.: $f(-1/\sqrt{2}) = -\frac{1}{2}$

39. Max.: $f(\frac{3}{2}) = 3 \cdot 2^{-4/3}$; min.: $f(3) = -3$

41. Consider the cases $A = 0$ and $A \neq 0$.

47. (c)   **49.** (d)   **51.** (a)

Section 3.6   (page 149)

1. 25 and 25   **3.** 1250   **5.** 500 in.³

7. 1152   **9.** 250   **11.** 11,250 yd²   **13.** 128

15. Approximately 3.9665°C   **17.** 1000 cm³

19. 0.25 m³ (all cubes, no open-top boxes)

21. Two equal pieces yield minimum total area 200 in.²; no cut yields one square of maximum area 400 in.²

23. 30.000 m²   **25.** Approximately 9259.26 in.³

27. Five presses

29. The minimizing value of x is $-2 + \frac{10}{3}\sqrt{6}$ in. To the nearest integer, use $x = 6$ in. of insulation for an annual saving of $285.

31. Either $1.10 or $1.15   **33.** Radius $\frac{2}{3}R$, height $\frac{1}{3}H$

35. Let R denote the radius of the circle, and remember that R is constant.

37. $\dfrac{2000\pi\sqrt{3}}{27}$ **39.** Max.: 4, min.: $\sqrt[3]{16}$ **41.** $\frac{1}{2}\sqrt{3}$

43. Each plank has width $\dfrac{-3\sqrt{2} + \sqrt{34}}{8} \approx 0.198539$,

height $\dfrac{\sqrt{7 - \sqrt{17}}}{2} \approx 0.848071$, and area

$\dfrac{\sqrt{7 - \sqrt{17}}\,(-3\sqrt{2} + \sqrt{34})}{4} \approx 0.673500$.

45. $\frac{2}{3}\sqrt{3} \approx 1.1547$ km from the point nearest the island

47. $\frac{1}{3}\sqrt{3}$ **49.** Actual $x \approx 3.45246$.

51. To minimize the sum, choose the radius of the sphere to be $5[10/(\pi + 6)]^{1/2}$ and the edge length of the cube to be $10[10/(\pi + 6)]^{1/2}$. To maximize the sum, choose the edge length of the cube to be zero.

Section 3.7   (page 161)

1. $f'(x) = 6 \sin x \cos x$ **3.** $f'(x) = \cos x - x \sin x$

5. $f'(x) = \dfrac{x \cos x - \sin x}{x^2}$

7. $f'(x) = \cos^3 x - 2 \sin^2 x \cos x$

9. $g'(t) = 4(1 + \sin t)^3 \cos t$

11. $g'(t) = \dfrac{\sin t - \cos t}{(\sin t + \cos t)^2}$

13. $f'(x) = 2 \sin x + 2x \cos x - 6x \cos x + 3x^2 \sin x$

15. $f'(x) = 3 \cos 2x \cos 3x - 2 \sin 2x \sin 3x$

17. $g'(t) = 3t^2 \sin^2 2t + 4t^3 \sin 2t \cos 2t$

19. $g'(t) = -\frac{5}{2}(\cos 3t + \cos 5t)^{3/2}(3 \sin 3t + 5 \sin 5t)$

21. $\dfrac{dy}{dx} = \dfrac{1}{\sqrt{x}} \sin \sqrt{x} \cos \sqrt{x}$

23. $\dfrac{dy}{dx} = 2x \cos(3x^2 - 1) - 6x^3 \sin(3x^2 - 1)$

25. $\dfrac{dy}{dx} = 2 \cos 2x \cos 3x - 3 \sin 2x \sin 3x$

27. $\dfrac{dy}{dx} = -\dfrac{3 \sin 5x \sin 3x + 5 \cos 5x \cos 3x}{\sin^2 5x}$

29. $\dfrac{dy}{dx} = 4x \sin x^2 \cos x^2$

31. $\dfrac{dy}{dx} = \dfrac{\cos 2\sqrt{x}}{\sqrt{x}}$

33. $\dfrac{dy}{dx} = \sin x^2 + 2x^2 \cos x^2$

35. $\dfrac{dy}{dx} = \frac{1}{2} x^{-1/2} \sin x^{1/2} + \frac{1}{2} \cos x^{1/2}$

37. $\dfrac{dy}{dx} = \frac{1}{2} x^{-1/2}(x - \cos x)^3$
$\qquad\qquad + 3x^{1/2}(x - \cos x)^2(1 + \sin x)$

39. $\dfrac{dy}{dx} = -2x[\sin(\sin x^2)] \cos x^2$

41. $\dfrac{dy}{dx} = 7x^6 \sec^2 x^7$ **43.** $\dfrac{dy}{dx} = 7 \sec^2 x \tan^6 x$

45. $\dfrac{dy}{dx} = 5x^7 \sec^2 5x + 7x^6 \tan 5x$

47. $\dfrac{dy}{dx} = \dfrac{\sec \sqrt{x} + \sqrt{x} \sec \sqrt{x} \tan \sqrt{x}}{2\sqrt{x}}$

49. $\dfrac{dy}{dx} = \dfrac{2 \cot \dfrac{1}{x^2} \csc \dfrac{1}{x^2}}{x^3}$

51. $\dfrac{dy}{dx} = \dfrac{5 \tan 3x \sec 5x \tan 5x - 3 \sec 5x \sec^2 3x}{\tan^2 3x}$
$\qquad = 5 \cot 3x \sec 5x \tan 5x - 3 \csc^2 3x \sec 5x$

53. $\dfrac{dy}{dx} = \sec x \csc x + x \sec x \tan x \csc x$
$\qquad\quad - x \sec x \csc x \cot x$
$\qquad = x \sec^2 x + \sec x \csc x - x \csc^2 x$

55. $\dfrac{dy}{dx} = [\sec(\sin x) \tan(\sin x)] \cos x$

57. $\dfrac{dy}{dx} = \dfrac{\sec x \cos x - \sin x \sec x \tan x}{\sec^2 x}$
$\qquad = \cos^2 x - \sin^2 x$

59. $\dfrac{dy}{dx} = -\dfrac{5 \csc^2 5x}{2\sqrt{1 + \cot 5x}}$ **63.** $\pi/4$

65. $\dfrac{\pi}{18} \sec^2 \dfrac{5\pi}{18} \approx 0.4224$ mi/s (about 1521 mi/h)

67. $\dfrac{2000\pi}{27}$ ft/s (about 158.67 mi/h)   **69.** $\pi/3$

71. $\frac{8}{3}\pi R^3$, twice the volume of the sphere!

73. $\dfrac{3\sqrt{3}}{4}$

75. *Suggestion:* $A(\theta) = \dfrac{s^2(\theta - \sin\theta)}{2\theta^2}$

Section 3.8 (page 173)

1. 128   **3.** 64   **5.** 1   **7.** 16
9. 16   **11.** 4   **13.** 3   **15.** 3

17. $3 \ln 2$ **19.** $\ln 2 + \ln 3$
21. $3 \ln 2 + 2 \ln 3$ **23.** $3 \ln 2 - 3 \ln 3$
25. $3 \ln 3 - 3 \ln 2 - \ln 5$ **27.** $2^{81} > 2^{12}$

29. There are three solutions of the equation. The one *not* obvious by inspection is approximately -0.7666647.

31. 6   **33.** -2 **35.** 1, 2   **37.** $81 = 3^4$
39. 0   **41.** $(x + 1)e^x$ **43.** $(x^{1/2} + \frac{1}{2} x^{-1/2})e^x$
45. $x^{-3}(x - 2)e^x$ **47.** $1 + \ln x$
49. $x^{-1/2}(1 + \frac{1}{2} \ln x)$ **51.** $(1 - x)e^{-x}$
53. $3/x$ **57.** $f'(x) = \frac{1}{10} e^{x/10}$
61. $P'(t) = 3^t \ln 3$ **63.** $P'(t) = -(2^{-t} \ln 2)$
65. (a) $P'(0) = \ln 3 \approx 1.09861$ (millions per hour);
(b) $P'(4) = 3^4 \cdot \ln 3 \approx 88.9876$ (millions per hour)

Section 3.9 (page 180)

1. $\dfrac{dy}{dx} = \dfrac{x}{y}$ **3.** $\dfrac{dy}{dx} = -\dfrac{16x}{25y}$ **5.** $\dfrac{dy}{dx} = -\sqrt{\dfrac{y}{x}}$

7. $\dfrac{dy}{dx} = -\left(\dfrac{y}{x}\right)^{1/3}$ **9.** $\dfrac{dy}{dx} = \dfrac{3x^2 - 2xy - y^2}{3y^2 + 2xy + x^2}$

11. $\dfrac{dy}{dx} = -\dfrac{x}{y}$; $3x - 4y = 25$

13. $\dfrac{dy}{dx} = \dfrac{1 - 2xy}{x^2}$; $3x + 4y = 10$

15. $\dfrac{dy}{dx} = -\dfrac{2xy + y^2}{2xy + x^2}$; $y = -2$

17. $\dfrac{dy}{dx} = \dfrac{25y - 24x}{24y - 25x}$; $4x = 3y$

19. $\dfrac{dy}{dx} = -\dfrac{y^4}{x^4}$; $x + y = 2$

21. $\dfrac{dy}{dx} = \dfrac{5x^4y^2 - y^3}{3xy^2 - 2x^5y}$,   so if $y \neq 0$,
$\dfrac{dy}{dx} = \dfrac{5x^4y - y^2}{3xy - 2x^5}$; the slope at $(1, 2)$ is $\frac{3}{2}$.

23. $\dfrac{dy}{dx} = \dfrac{y - x^2}{y^2 - x}$, but there are no horizontal tangents (see Problem 63).

25. $(2, 2 \pm 2\sqrt{2})$ **27.** $y = 2(x - 3)$, $y = 2(x + 3)$

29. Horizontal tangents at all four points where $|x| = \frac{1}{4}\sqrt{6}$ and $|y| = \frac{1}{4}\sqrt{2}$; vertical tangents at the two points $(-1, 0)$ and $(1, 0)$

31. $\dfrac{4}{5\pi} \approx 0.25645$ ft/s   **33.** $\dfrac{32\pi}{125} \approx 0.80425$ m/h

35. 20 cm²/s   **37.** 0.25 cm/s   **39.** 6 ft/s
41. 384 mi/h
43. (a) About 0.047 ft/min; (b) about 0.083 ft/min

A-41

45. $\frac{400}{9} \approx 44.44$ ft/s   **47.** Increasing at 16π cm³/s

49. 6000 mi/h

51. (a) $\frac{11}{15}\sqrt{21} \approx 3.36$ ft/s downward; (b) $\frac{242}{15}\sqrt{119}$ ft/s downward

53. At $t = 12$ min; $32\sqrt{13} \approx 115.38$ mi.

55. $-\dfrac{50}{81\pi} \approx -0.1965$ ft/s

57. $-\dfrac{10}{81\pi} \approx -0.0393$ in./min

59. $300\sqrt{2} \approx 424.26$ mi/h   **61.** $\frac{1}{30}$ ft/s

63. $x^3 + y^3 - 3xy + 1$
$= \frac{1}{2}(x + y + 1)[(x - y)^2 + (x - 1)^2 + (y - 1)^2]$.

Section 3.10 (page 192)

Note: In this section your results may differ from these answers in the last one or two decimal places due to differences in calculators or in methods of solving the equations.

1. 2.2361   **3.** 2.5119   **5.** 0.3028   **7.** -0.7402

9. 0.7391   **11.** 1.2361   **13.** 2.3393   **15.** 2.0288

17. 0.5671   **19.** 0.4429   **21.** (b) 1.25992

23. 0.45018   **27.** 0.755 (0.75487766624669276)

29. -1.8955, 0, and 1.8955

31. $dy/dx = e^x \cos x - e^x \sin x$

33. $dy/dx = -3e^x(2 + 3e^x)[1 + (2 + 3e^x)^{-3/2}]^{-1/3}$

35. $dy/dx = -x^{-1}\sin(1 + \ln x)$ **37.** 0.2261

39. $\alpha_1 \approx 2.029 \approx 1.29\,\dfrac{\pi}{2}$, $\alpha_2 \approx 4.913 \approx 3.13\,\dfrac{\pi}{2}$

(also, $\alpha_3 \approx 5.08\,\dfrac{\pi}{2}$, $\alpha_4 \approx 7.06\,\dfrac{\pi}{2}$)

Chapter 3 Miscellaneous Problems (page 197)

1. $\dfrac{dy}{dx} = 2x - \dfrac{6}{x^3}$ **3.** $\dfrac{dy}{dx} = \dfrac{1}{2\sqrt{x}} - \dfrac{1}{3x^{4/3}}$

5. $\dfrac{dy}{dx} = 7(x - 1)^6(3x + 2)^9 + 27(x - 1)^7(3x + 2)^8$

7. $\dfrac{dy}{dx} = 4(3x - \frac{1}{2}x^{-2})^3(3 + x^{-3})$

9. $\dfrac{dy}{dx} = -\dfrac{y}{x} = -\dfrac{9}{x^2}$

11. $\dfrac{dy}{dx} = -\frac{3}{2}(x^3 - x)^{-5/2}(3x^2 - 1)$

13. $\dfrac{dy}{dx} = \dfrac{-2(1 + x^2)^3}{(x^4 + 2x^2 + 2)^2} \cdot \dfrac{-2x}{(1 + x^2)^2}$
$= \dfrac{4x(1 + x^2)}{(x^4 + 2x^2 + 2)^2}$

15. $\dfrac{dy}{dx} = \frac{7}{3}[x^{1/2} + (2x)^{1/3}]^{4/3}[\frac{1}{2}x^{-1/2} + \frac{2}{3}(2x)^{-2/3}]$

17. $\dfrac{dy}{dx} = \dfrac{dy}{du} \cdot \dfrac{du}{dx} = \dfrac{(u - 1) - (u + 1)}{(u - 1)^2} \cdot \dfrac{1}{2}(x + 1)^{-1/2}$
$= -\dfrac{1}{\sqrt{x + 1}\,(\sqrt{x + 1} - 1)^2}$

19. $\dfrac{dy}{dx} = \dfrac{1 - 2xy^2}{2x^2y - 1}$

21. $\dfrac{dy}{dx} = \dfrac{1 + \dfrac{2 + \dfrac{\sqrt{3}}{2\sqrt{x}}}{2\sqrt{\sqrt{3x} + 2x}}}{2\sqrt{x + \sqrt{\sqrt{3x} + 2x}}}$

23. $\dfrac{dy}{dx} = -\left(\dfrac{y}{x}\right)^{2/3}$

25. $\dfrac{dy}{dx} = -\dfrac{18(x^3 + 3x^2 + 3x + 3)^2}{(x + 1)^{10}}$

27. $\dfrac{dy}{dx} = \dfrac{(2\cos x + \cos^2 x + 1)\sin x}{2(1 + \cos x)^2\sqrt{\dfrac{\sin^2 x}{1 + \cos x}}}$

29. $\dfrac{dy}{dx} = -\dfrac{3\cos 2x \cos 3x + 4\sin 2x \sin 3x}{2(\sin 3x)^{3/2}}$

31. $\dfrac{dy}{dx} = (\sin^3 2x)(2)(\cos 3x)(-\sin 3x)(3)$
$+ (\cos^2 3x)(3\sin^2 2x)(2\cos 2x)$
$= 6\cos 3x \cos 5x \sin^2 2x$

33. $\dfrac{dy}{dx} = 5\left[\sin^4\!\left(x + \dfrac{1}{x}\right)\right]\left[\cos\!\left(x + \dfrac{1}{x}\right)\right]\left(1 - \dfrac{1}{x^2}\right)$

35. $\dfrac{dy}{dx}$
$= [\cos^2(x^4 + 1)^{1/3}][-\sin(x^4 + 1)^{1/3}](x^4 + 1)^{-2/3}(4x^3)$

37. $x = 1$ **39.** $x = 0$ **41.** 0.5 ft/min

43. $\frac{1}{3}$ **45.** $\frac{1}{4}$ **47.** 0

49. $h'(x) = -x(x^2 + 25)^{-3/2}$ **51.** $h'(x) = \frac{5}{3}(x - 1)^{2/3}$

53. $h'(x) = -2x\sin(x^2 + 1)$

55. $\dfrac{dV}{dS} = \frac{1}{4}\sqrt{\dfrac{S}{\pi}}$

57. $\dfrac{2}{\cos^2 50°} \cdot \dfrac{\pi}{36} \approx 0.4224$ mi/s, about 1521 mi/h

59. R^2

61. Minimum area: $(36\pi V^2)^{1/3}$, obtained by making *one* sphere of radius $(3V/4\pi)^{1/3}$; maximum area: $(72\pi V^2)^{1/3}$, obtained by making two equal spheres of radius $\frac{1}{2}(3V/\pi)^{1/3}$

63. $32\pi R^3/81$ **65.** $M/2$ **67.** 36 ft³   **69.** $3\sqrt{3}$

73. 2 mi from the shore point nearest the first town

75. (a) $\dfrac{m^2v^2}{64(m^2+1)}$; (b) when $m = 1$, thus when $\alpha = \pi/4$

77. 2.6458   **79.** 2.3714   **81.** -0.3473

83. 0.7402   **85.** -0.7391 **87.** -1.2361

89. Approximately 1.54785 ft

91. -2.7225, 0.8013, 2.3100   **97.** 4 in.^{2}/s

99. $-50/(9\pi) \approx -1.7684$ ft/min   **101.** 1 in./min

Section 4.2   (page 208)

1. $(6x + 8x^{-3})\,dx$ **3.** $[1 + \frac{3}{2}x^2(4 - x^3)^{-1/2}]\,dx$

5. $[6x(x - 3)^{3/2} + \frac{9}{2}x^2(x - 3)^{1/2}]\,dx$

7. $[(x^2 + 25)^{1/4} + \frac{1}{2}x^2(x^2 + 25)^{-3/4}]\,dx$

9. $-\frac{1}{2}x^{-1/2}\sin x^{1/2}\,dx$

11. $(2\cos^2 2x - 2\sin^2 2x)\,dx$

13. $(\frac{2}{3}x^{-1}\cos 2x - \frac{1}{3}x^{-2}\sin 2x)\,dx$

15. $(\sin x + x\cos x)(1 - x\sin x)^{-2}\,dx$

17. $f(x) \approx 1 + x$ **19.** $f(x) \approx 1 + 2x$

21. $f(x) \approx 1 - 3x$ **23.** $f(x) \approx x$

25. $3 - \frac{2}{27} \approx 2.926$ **27.** $2 - \frac{1}{32} \approx 1.969$

29. $\frac{95}{1536} \approx 0.06185$

31. $\dfrac{1 + \dfrac{\pi}{90}}{\sqrt{2}} \approx 0.7318$ **33.** $\sin\dfrac{\pi}{2} - \dfrac{\pi}{90}\cos\dfrac{\pi}{2} = 1.000$

35. $\dfrac{dy}{dx} = -\dfrac{x}{y}$ **37.** $\dfrac{dy}{dx} = \dfrac{y - x^2}{y^2 - x}$

41. -4 in.^{2} **43.** $-405\pi/2$ cm^3

45. 10 ft   **47.** 6 W

49. $25\pi \approx 78.54$ in.^{3} **51.** $4\pi \approx 12.57$ m^2

Section 4.3   (page 218)

1. Increasing for $x < 0$, decreasing for $x > 0$; (c)

3. Decreasing for $x < -2$, increasing for $x > -2$; (f)

5. Increasing for $x < -1$ and for $x > 2$, decreasing on $(-1, 2)$; (d)

7. $f(x) = 2x^2 + 5$ **9.** $f(x) = 2 - \dfrac{1}{x}$

11. Increasing on R

13. Increasing for $x < 0$, decreasing for $x > 0$

15. Increasing for $x < \frac{3}{2}$, decreasing for $x > \frac{3}{2}$

17. Increasing on $(-1, 0)$ and for $x > 1$, decreasing for $x < -1$ and on $(0, 1)$

19. Increasing on $(-2, 0)$ and for $x > 1$, decreasing for $x < -2$ and on $(0, 1)$

21. Increasing for $x < 2$, decreasing for $x > 2$

23. Increasing for $x < -\sqrt{3}$, for $-\sqrt{3} < x < 1$, and for $x > 3$; decreasing for $1 < x < \sqrt{3}$ and for $\sqrt{3} < x < 3$

25. $f(0) = 0 = f(2)$, $f'(x) = 2x - 2$; $c = 1$

27. $f(-1) = 0 = f(1)$, $f'(x) = -\dfrac{4x}{(1 + x^2)^2}$; $c = 0$

29. $f'(0)$ does not exist.

31. $f(0) \neq f(1)$ **33.** $c = -\frac{1}{2}$ **35.** $c = \frac{35}{27}$

37. The average slope is $\frac{1}{3}$, but $|f'(x)| = 1$ where $f'(x)$ exists.

39. The average slope is 1, but $f'(x) = 0$ wherever it exists.

41. If $g(x) = x^5 + 2x - 3$, then $g'(x) > 0$ for all x in $[0, 1]$ and $g(1) = 0$. So $x = 1$ is the only root of the equation in the given interval.

43. If $g(x) = x^4 - 3x - 20$, then $g(2) = -10$ and $g(3) = 52$. If x is in $[2, 3]$, then

$$g'(x) = 4x^3 - 3 \geqq 4 \cdot 2^3 - 3 = 29 > 0,$$

so g is an increasing function on $[2, 3]$. Hence $g(x)$ can have at most one zero in $[2, 3]$. It has at least one solution because $g(2) < 0 < g(3)$ and g is continuous.

45. Note that $f'(x) = \frac{3}{2}(-1 + \sqrt{x + 1})$.

47. Assume that $f'(x)$ has the form

$$a_0 + a_1x + \cdots + a_{n-1}x^{n-1}.$$

Construct a polynomial such that $p'(x) = f'(x)$. Conclude that $f(x) = p(x) + C$ on $[a, b]$.

Section 4.4   (page 228)

1. Global min. at $x = 2$

3. Local max. at $x = 0$, local min. at $x = 2$

5. No extremum at $x = 1$

7. Local min. at $x = -2$, local max. at $x = 5$

9. Global min. at $x = \pm 1$, local max. at $x = 0$

11. Local max. at $x = -1$, local min. at $x = 1$

13. Local min. at $x = 1$

15. Global min. at $(0, 0)$, local max. at $(2, 4e^{-2})$

17. Global max. at $(\pi/2, 1)$

19. Global max. at $(\pi/2, 1)$, global min. at $(-\pi/2, -1)$

21. Global min. at $(0, 0)$

23. Global max. at (π, π), global min. at $(-\pi, -\pi)$

25. Global max. at $(e^{1/2}, 1/(2e^2))$ **27.** -10 and 10

29. $(1, 1)$ **31.** 9 in. wide, 18 in. long, 6 in. high

33. Radius $5\pi^{-1/3}$ cm, height $10\pi^{-1/3}$ cm

37. Base 5 in. by 5 in., height 2.5 in.

39. Radius $(25/\pi)^{1/3} \approx 1.9965$ in., height 4 times that radius

41. $(\frac{1}{2}\sqrt{6}, \frac{3}{2})$ and $(-\frac{1}{2}\sqrt{6}, \frac{3}{2})$; $(0, 0)$ is *not* the nearest point.

43. 8 cm

45. $L = \sqrt{20 + 12\sqrt[3]{4} + 24\sqrt[3]{2}} \approx 8.324$ m

49. Height $(6V)^{1/3}$, base edge $(\frac{9}{2}V^2)^{1/6}$

Section 4.5 (page 237)

1. (c)   **3.** (d)

5. Parabola opening upward; global min. at $(1, 2)$

7. Increasing if $|x| > 2$, decreasing on $(-2, 2)$; local max. at $(-2, 16)$, local min. at $(2, -16)$

9. Increasing for $x < 1$ and for $x > 3$, decreasing on $(1, 3)$; local max. at $(1, 4)$, local min. at $(3, 0)$

11. Increasing for all x, no extrema

13. Decreasing for $x < -2$ and on $(-0.5, 1)$, increasing on $(-2, -0.5)$ and for $x > 1$; global min. at $(-2, 0)$ and $(1, 0)$, local max. at $(-0.5, 5.0625)$.

15. Increasing for $0 < x < 1$, decreasing for $x > 1$; global max. at $(1, 2)$, no graph for $x < 0$

17. Increasing for $|x| > 1$, decreasing for $|x| < 1$, but with a horizontal tangent at $(0, 0)$; local max. at $(-1, 2)$, local min. at $(1, -2)$

19. Decreasing for $x < -2$ and on $(0, 2)$, increasing for $x > 2$ and on $(-2, 0)$; global min. at $(-2, -9)$ and at $(2, -9)$, local max. at $(0, 7)$

21. Decreasing for $x < \frac{3}{4}$, increasing for $x > \frac{3}{4}$

23. Decreasing on $(-2, 1)$, increasing for $x < -2$ and for $x > 1$; local max. at $(-2, 20)$, local min. at $(1, -7)$

25. Increasing for $x < 0.6$ and for $x > 0.8$, decreasing on $(0.6, 0.8)$; local max. at $(0.6, 16.2)$, local min. at $(0.8, 16.0)$

27. Increasing for $x > 2$ and on $(-1, 0)$, decreasing for $x < -1$ and on $(0, 2)$; local max. at $(0, 8)$, local min. at $(-1, 3)$, global min. at $(2, -24)$

29. Increasing for $|x| > 2$, decreasing for $|x| < 2$; local max. at $(-2, 64)$, local min. at $(2, -64)$

31. Increasing everywhere, no extrema; graph passes through origin; minimum slope $\frac{9}{2}$ occurs at $x = -\frac{1}{2}$

33. Increasing for $x < -\sqrt{2}$ and on $(0, \sqrt{2})$, decreasing for $x > \sqrt{2}$ and on $(-\sqrt{2}, 0)$; global max. 16 occurs where $x = \pm\sqrt{2}$, local min. at $(0, 0)$

35. Increasing for $x < 1$, decreasing for $x > 1$; global max. at $(1, 3)$, vertical tangent at $(0, 0)$

37. Decreasing on $(0.6, 1)$, increasing for $x < 0.6$ and for $x > 1$; local min. and a cusp at $(1, 0)$, local max. at $(0.6, 0.3527)$ (ordinate approximate)

39.
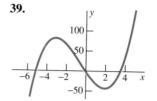

41.
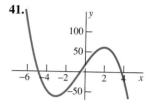

43.
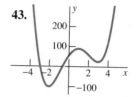

45. (b) $x^3 - 3x + 3 \approx (x + 2.1038)(x^2 - 2.1038x + 1.42599)$; (c) $x \approx 1.0519 \pm 0.5652i$

Section 4.6 (page 250)

1. $8x^3 - 9x^2 + 6$, $24x^2 - 18x$, $48x - 18$

3. $-8(2x - 1)^{-3}$, $48(2x - 1)^{-4}$, $-384(2x - 1)^{-5}$

5. $4(3t - 2)^{1/3}$, $4(3t - 2)^{-2/3}$, $-8(3t - 2)^{-5/3}$

7. $(y + 1)^{-2}$, $-2(y + 1)^{-3}$, $6(y + 1)^{-4}$

9. $g'(t) = t + 2t \ln t$, $g''(t) = 3 + 2 \ln t$, $g'''(t) = 2/t$

11. $3 \cos 3x$, $-9 \sin 3x$, $-27 \cos 3x$

13. $\cos^2 x - \sin^2 x$, $-4 \sin x \cos x$, $4 \sin^2 x - 4 \cos^2 x$

15. $\dfrac{x \cos x - \sin x}{x^2}$, $\dfrac{2 \sin x - 2x \cos x - x^2 \sin x}{x^3}$, $\dfrac{3x^2 \sin x - x^3 \cos x + 6x \cos x - 6 \sin x}{x^4}$

17. $-\dfrac{2x + y}{x + 2y}$, $-\dfrac{18}{(x + 2y)^3}$

19. $-\dfrac{1 + 2x}{3y^2}$, $\dfrac{2y^3 - 42}{9y^5}$

21. $\dfrac{y}{\cos y - x}$, $\dfrac{2y \cos y - 2xy + y^2 \sin y}{(\cos y - x)^3}$

23. $(-3, 81)$, $(5, -175)$, $(1, -47)$

25. $(\frac{9}{2}, -\frac{941}{2})$, $(-\frac{7}{2}, \frac{1107}{2})$, $(\frac{1}{2}, \frac{83}{2})$

27. $(3\sqrt{3}, -492)$, $(-3\sqrt{3}, -492)$, $(0, 237)$, $(3, -168)$, $(-3, -168)$

29. $(\frac{16}{3}, -\frac{181,144}{81})$, $(0, 1000)$, $(4, -1048)$

31. Local min. at $(2, -1)$; no inflection points

33. Local max. at $(-1, 3)$, local min. at $(1, -1)$; inflection point at $(0, 1)$

35. No local extrema; inflection point at $(0, 0)$

37. No local extrema; inflection point at $(0, 0)$

39. Local max. at $(\frac{1}{2}, \frac{1}{16})$, local min. at $(0, 0)$ and $(1, 0)$; the abscissas of the inflection points are the roots of $6x^2 - 6x + 1 = 0$, and the inflection points are at (approximately) $(0.79, 0.03)$ and $(1.21, 0.03)$.

41. Global max. at $(\pi/2, 1)$, global min. at $(3\pi/2, -1)$; inflection point at $(\pi, 0)$

43. Inflection point at $(0, 0)$

45. Global max. at $(0, 1)$ and $(\pi, 1)$, global min. at $(\pi/2, 0)$; inflection points at $(-\pi/4, \frac{1}{2})$, $(\pi/4, \frac{1}{2})$, $(3\pi/4, \frac{1}{2})$, and $(5\pi/4, \frac{1}{2})$

47. Local max. at $(1, e^{-1})$, inflection point at $(2, 2e^{-2})$

49. Local max. at (e, e^{-1}), inflection point at $(e^{3/2}, \frac{3}{2}e^{-3/2})$

63. Increasing for $x < -1$ and for $x > 2$, decreasing on $(-1, 2)$, local max. at $(-1, 10)$, local min. at $(2, -17)$, inflection point at $(\frac{1}{2}, -\frac{7}{2})$

65. Increasing for $x < -2$ and on $(0, 2)$, decreasing for $x > 2$ and on $(-2, 0)$, global max. at $(\pm 2, 22)$, local min. at $(0, 6)$, inflection points where $x^2 = \frac{4}{3}$ (and $y = \frac{134}{9}$)

67. Decreasing for $x < -1$ and on $(0, 2)$, increasing for $x > 2$ and on $(-1, 0)$, local min. at $(-1, -6)$, local max. at $(0, -1)$, global min. at $(2, -33)$, inflection points with approximate coordinates $(1.22, -19.36)$ and $(-0.55, -3.68)$

69. Local max. at $(\frac{3}{7}, \frac{6912}{823,543}) \approx (0.43, 0.0084)$, local min. at $(1, 0)$, inflection points at $(0, 0)$ and at the two solutions of $7x^2 - 6x + 1 = 0$: approximately $(0.22, 0.0042)$ and $(0.63, 0.0047)$. See graph.

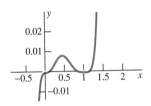

71. The graph is increasing everywhere, with a vertical tangent and inflection point at $(0, 1)$; the other intercept is $(-1, 0)$, and there are no extrema.

73. Global min. at $(0, 0)$, increasing for all $x > 0$, inflection point at $(1, 4)$, concave up for $x > 1$, vertical tangent at the origin

75. Increasing for $x < 1$, decreasing for $x > 1$, vertical tangent and inflection point at $(0, 0)$, another inflection point at $(-2, -7.56)$ (ordinate approximate), global max. at $(1, 3)$

77. (c)   **79.** (b)   **81.** (d)

89. $a = 3pV^2 \approx 3{,}583{,}858.8$, $b = \dfrac{V}{3} = 42.7$,

$$R = \frac{8pV}{3T} \approx 81.80421$$

Section 4.7   (page 261)

1. 1   **3.** 3   **5.** 2   **7.** 1   **9.** 4

11. 0   **13.** 2

15. $+\infty$ (or "does not exist")   **17.** (g)

19. (a)   **21.** (f)   **23.** (j)   **25.** (l)   **27.** (k)

29. No critical points or inflection points, vertical asymptote $x = 3$, horizontal asymptote $y = 0$, sole intercept at $(0, -\frac{2}{3})$

31. No critical points or inflection points, vertical asymptote $x = -2$, horizontal asymptote $y = 0$, sole intercept at $(0, -\frac{3}{4})$

33. No critical points or inflection points, vertical asymptote $x = \frac{3}{2}$, horizontal asymptote $y = 0$

35. Global min. at $(0, 0)$, inflection points where $3x^2 = 1$ (and $y = \frac{1}{4}$), horizontal asymptote $y = 1$

37. Local max. at $(0, -\frac{1}{9})$, no inflection points, vertical asymptotes $x = \pm 3$, horizontal asymptote $y = 0$

39. Local max. at $(-\frac{1}{2}, -\frac{4}{25})$, horizontal asymptote $y = 0$, vertical asymptotes $x = -3$ and $x = 2$, no inflection points

41. Local min. at $(1, 2)$, local max. at $(-1, -2)$, no inflection points, vertical asymptote $x = 0$; the line $y = x$ is also an asymptote.

43. Local min. at $(2, 4)$, local max. at $(0, 0)$, no inflection points, asymptotes $x = 1$ and $y = x + 1$

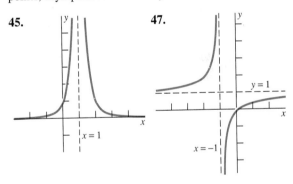

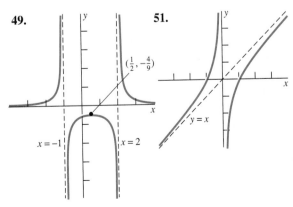

53.

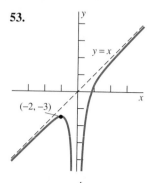

55. Local min. at $(1, 3)$, inflection point at $(\sqrt[3]{-2}, 0)$, vertical asymptote $x = 0$

Chapter 4 Miscellaneous Problems (page 263)

1. $dy = 3(4x - x^2)^{1/2}(2 - x)\, dx$

3. $dy = -2(x - 1)^{-2}\, dx$

5. $dy = (2x \cos x^{1/2} - \frac{1}{2}x^{3/2} \sin x^{1/2})\, dx$

7. $\frac{12{,}801}{160} = 80.00625$ (true value: about 80.00624 975588)

9. $\frac{128{,}192}{125} = 1025.536$ (true value: about 1025.537037)

11. $\frac{601}{60} \approx 10.016667$ (true value: about 10.016639)

13. 132.5 (true value: about 132.574507)

15. 2.03125 (true value: about 2.03054 3185)

17. 7.5 (in.^{3}) **19.** $10\pi \approx 31.416$ (cm^3)

21. $\pi/96 \approx 0.0327$ s **23.** $c = \sqrt{3}$ **25.** $c = 1$

27. $c = \sqrt[4]{2.2}$

29. Decreasing for $x < 3$, increasing for $x > 3$, global min. at $(3, -5)$; concave upward everywhere

31. Increasing everywhere, no extrema

33. Increasing for $x < \frac{1}{4}$, decreasing for $x > \frac{1}{4}$, vertical tangent at $(0, 0)$, global max. at $x = \frac{1}{4}$

35. $3x^2 - 2,\ 6x,\ 6$

37. $-(t^{-2}) + 2(2t + 1)^{-2},\ 2t^{-3} - 8(2t + 1)^{-3},$ $-6t^{-4} + 48(2t + 1)^{-4}$

39. $3t^{1/2} - 4t^{1/3},\ \frac{3}{2}t^{-1/2} - \frac{4}{3}t^{-2/3},\ -\frac{3}{4}t^{-3/2} + \frac{8}{9}t^{-5/3}$

41. $-4(t - 2)^{-2},\ 8(t - 2)^{-3},\ -24(t - 2)^{-4}$

43. $-\frac{4}{3}(5 - 4x)^{-2/3},\ -\frac{32}{9}(5 - 4x)^{-5/3},$ $-\frac{640}{27}(5 - 4x)^{-8/3}$

45. $\dfrac{dy}{dx} = -\left(\dfrac{y}{x}\right)^{2/3},\ \dfrac{d^2y}{dx^2} = \dfrac{2y^{1/3}}{3x^{5/3}}$

47. $\dfrac{dy}{dx} = \dfrac{1}{2\sqrt{x}\,(5y^4 - 4)}$,

$\dfrac{d^2y}{dx^2} = -\dfrac{20y^3\sqrt{x} + (5y^4 - 4)^2}{4x\sqrt{x}\,(5y^4 - 4)^3}$

49. $\dfrac{dy}{dx} = \dfrac{5y - 2x}{2y - 5x}$

51. $\dfrac{dy}{dx} = \dfrac{2xy}{3y^2 - x^2 - 1}$

53. Global min. at $(2, -48)$, x-intercepts 0 and (approximately) 3.1748, no inflection points or asymptotes; concave upward everywhere

55. Local max. at $(0, 0)$, global min. where $x^2 = \frac{4}{3}$ (and $y = -\frac{32}{27}$), inflection points where $x^2 = \frac{4}{5}$ (and $y = -\frac{96}{125}$), no asymptotes

57. Global max. at $(3, 3)$, inflection points where $x = 6$ and $(4, 0)$ (and a vertical tangent at the latter point), no asymptotes

59. Local max. at $(0, -\frac{1}{4})$, no inflection points, horizontal asymptote $y = 1$, vertical asymptotes $x = \pm 2$

61. The inflection point has abscissa the only real solution of $x^3 + 6x^2 + 4 = 0$, approximately -6.10724.

63.   **65.**

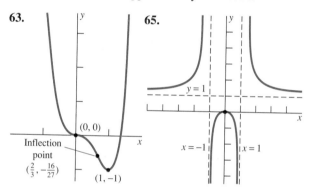

67.   **69.**

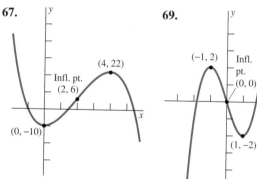

71.

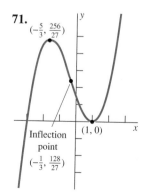

73. Maximum value 1 at $x = -1$

75. Base 15×30, height 10

77. Base 5×10, height 8

79. $100 \cdot (\frac{2}{9})^{2/5} \approx 54.79$ mi/h

81. Two horizontal tangents, where $x \approx 0.42$, $y \approx \pm 0.62$; intercepts at $(0, 0)$, $(1, 0)$, and $(2, 0)$; vertical tangent at each intercept; inflection points corresponding to the only positive solution of $3x^2(x - 2)^2 = 4$—that is, $x \approx 2.46789$, $y \approx \pm 1.30191$; no asymptotes

83. 240 ft   **85.** $2\sqrt{2A(n + 2)}$ ft

87. There is neither a maximum nor a minimum.

89. 288 in.^{2} **91.** 270 cm^2

93. In both cases, $m = 1$ and $b = -\frac{2}{3}$.

Section 5.2   (page 278)

1. $x^3 + x^2 + x + C$ **3.** $x - \frac{2}{3}x^3 + \frac{3}{4}x^4 + C$

5. $-\frac{3}{2}x^{-2} + \frac{4}{5}x^{5/2} - x + C$ **7.** $t^{3/2} + 7t + C$

9. $\frac{3}{5}x^{5/3} - 16x^{-1/4} + C$ **11.** $x^4 - 2x^2 + 6x + C$

13. $49e^{x/7} + C$ **15.** $\frac{1}{5}(x + 1)^5 + C$

17. $-\frac{1}{6}(x - 10)^{-6} + C$

19. $\frac{2}{3}x^{3/2} - \frac{4}{5}x^{5/2} + \frac{2}{7}x^{7/2} + C$

21. $\frac{2}{21}x^3 - \frac{3}{14}x^2 - \frac{5}{7}x^{-1} + C$ **23.** $\frac{1}{54}(9t + 11)^6 + C$

25. $\frac{1}{2}e^{2x} - \frac{1}{2}e^{-2x} + C$

27. $\frac{1}{2}\sin 10x + 2\cos 5x + C$

29. $\dfrac{3\sin \pi t}{\pi} + \dfrac{\sin 3\pi t}{3\pi} + C$

33. $\frac{1}{2}(x - \sin x \cos x) + C$; $\frac{1}{2}(x + \sin x \cos x) + C$

35. $y = f(x) = x^2 + x + 3$

37. $y = f(x) = \frac{2}{3}(x^{3/2} - 8)$

39. $y = f(x) = 2\sqrt{x + 2} - 5$

41. $y = f(x) = \frac{3}{4}x^4 - 2x^{-1} + \frac{9}{4}$

43. $y = f(x) = \frac{1}{4}(x - 1)^4 + \frac{7}{4}$

45. $y = f(x) = 2\sqrt{x - 13} - 2$

47. 144 ft; 6 s   **49.** 144 ft   **51.** 5 s; 112 ft/s

53. $\sqrt{60} \approx 7.75$ s; $32\sqrt{60} \approx 247.87$ ft/s

55. 120 ft/s   **57.** 5 s; -160 ft/s   **59.** 400 ft

61. $\frac{1}{4}(-5 + 2\sqrt{145}) \approx 4.77$ s; $16\sqrt{145} \approx 192.6655$ ft/s

63. $\frac{544}{3}$ ft/s   **65.** 22 ft/s^2

67. Approximately 886.154 ft

69. $5\sqrt{210} \approx 72.5$ mi/h

Section 5.3   (page 290)

1. $\displaystyle\sum_{i=1}^{5} i^2$ **3.** $\displaystyle\sum_{n=1}^{5} \frac{1}{n}$ **5.** $\displaystyle\sum_{j=1}^{6} \frac{1}{2^j}$ **7.** $\displaystyle\sum_{k=1}^{5} \left(\frac{2}{3}\right)^k$

9. 190   **11.** 1165   **13.** 224   **15.** 350

17. 338,350   **19.** $\frac{1}{3}$ **21.** n^2 **23.** $\frac{2}{5}$, $\frac{3}{5}$

25. $\frac{33}{2}$, $\frac{39}{2}$ **27.** $\frac{6}{25}$, $\frac{11}{25}$ **29.** $\frac{378}{25}$, $\frac{513}{25}$ **31.** $\frac{81}{400}$, $\frac{121}{400}$

35. $\dfrac{n(n + 1)}{2n^2} \to \dfrac{1}{2}$ as $n \to \infty$

37. $\dfrac{81n^2(n + 1)^2}{4n^4} \to \dfrac{81}{4}$ as $n \to \infty$

39. $5n \cdot \dfrac{1}{n} - \dfrac{3n(n + 1)}{2n^2} \to \dfrac{7}{2}$ as $n \to \infty$

Section 5.4   (page 298)

1. $\displaystyle\int_{1}^{3} (2x - 1)\, dx$ **3.** $\displaystyle\int_{0}^{10} (x^2 + 4)\, dx$

5. $\displaystyle\int_{4}^{9} \sqrt{x}\, dx$ **7.** $\displaystyle\int_{3}^{8} \frac{1}{\sqrt{1 + x}}\, dx$

9. $\displaystyle\int_{0}^{1/2} \sin 2\pi x\, dx$ **11.** 0.44   **13.** 1.45

15. 19.5   **17.** 58.8   **19.** $-\pi/6$ **21.** 0.24

23. $\frac{137}{60} \approx 2.28333$ **25.** 16.5   **27.** 26.4   **29.** $\pi/6$

31. 0.33   **33.** $\frac{6086}{3465} \approx 1.75462$ **35.** 18   **37.** 40.575

39. 0   **41.** 1.83753   **43.** $\frac{8}{3}$ **45.** 12   **47.** 30

Section 5.5   (page 307)

1. $\frac{55}{12}$ **3.** $\frac{49}{60}$ **5.** $-\frac{1}{20}$ **7.** $\frac{1}{4}$ **9.** $\frac{16}{3}$ **11.** 24

13. 0   **15.** $\frac{32}{3}$ **17.** 0   **19.** $\frac{93}{5}$ **21.** 0

23. $\frac{28}{3}$ **25.** $\ln 2$ **27.** $\frac{1}{2}e^2 - 2e + \frac{5}{2}$ **29.** $\frac{1}{4}$

31. $\frac{2}{5}$ **33.** $-\frac{1}{3}$ **35.** $4/\pi$ **37.** 0   **39.** $\frac{1}{2}$

41. $\frac{2}{3}$ **43.** $25\pi/4 \approx 19.635$

47. $1000 + \displaystyle\int_{0}^{30} V'(t)\, dt = 160$ (gal)

49. Let $Q = \left(\dfrac{1}{1.2} + \dfrac{1}{1.4} + \dfrac{1}{1.6} + \dfrac{1}{1.8}\right)(0.2) \approx 0.5456$

and $I = \displaystyle\int_{1}^{2} \frac{1}{x}\, dx$. Then $Q + 0.1 \leqq I \leqq Q + 0.2$.

Hence $0.64 < I < 0.75$.

Section 5.6   (page 316)

1. $\frac{16}{5}$ **3.** $\frac{26}{3}$ **5.** 0   **7.** $\frac{125}{4}$ **9.** $\frac{14}{9}$ **11.** 0

13. 4   **15.** $\frac{1}{3}$ **17.** $-\frac{22}{81} \approx -0.271605$ **19.** 0

21. $\frac{35}{24} \approx 1.458333$ **23.** $e + e^{-1} - 2$ **25.** 4

27. $\ln 2$ **29.** $\frac{31}{20}$ **31.** $\frac{81}{4} + \frac{81}{4} = \frac{81}{2}$

33. Average height: $\frac{800}{3} \approx 266.67$ ft; average velocity: -80 ft/s

35. $\frac{5000}{3} \approx 1666.67$ (L)   **37.** $\frac{200}{3}$ **39.** $\pi/3$

41. $f'(x) = (x^2 + 1)^{17}$ **43.** $h'(z) = (z - 1)^{1/3}$

45. $f'(x) = -(e^x - e^{-x})$ **47.** $G'(x) = (x + 4)^{1/2}$

49. $G'(x) = (x^3 + 1)^{1/2}$ **51.** $f'(x) = 3\sin 9x^2$

53. $f'(x) = 2x \sin x^2$ **55.** $f'(x) = \dfrac{2x}{x^2 + 1}$

57. $y(x) = \displaystyle\int_1^x \dfrac{1}{t}\, dt$ **59.** $y(x) = 10 + \displaystyle\int_5^x \sqrt{1 + t^2}\, dt$

61. The integral does not exist.

Section 5.7 (page 323)

1. $\frac{1}{7}(x + 1)^7 + C$ **3.** $-\frac{1}{24}(4 - 3x)^8 + C$

5. $\frac{2}{7}\sqrt{7x + 5} + C$ **7.** $-\dfrac{1}{\pi}\cos(\pi x + 1) + C$

9. $\frac{1}{2}\sec 2\theta + C$ **11.** $\frac{1}{3}(x^2 - 1)^{3/2} + C$

13. $-\frac{1}{9}(2 - 3x^2)^{3/2} + C$ **15.** $\frac{1}{6}(x^4 + 1)^{3/2} + C$

17. $\frac{1}{6}\sin 2x^3 + C$ **19.** $-\frac{1}{2}e^{-x^2} + C$

21. $-\frac{1}{4}\cos^4 x + C$ **23.** $\frac{1}{4}\tan^4 \theta + C$

25. $2\sin\sqrt{x} + C$

27. $\frac{1}{10}(x + 1)^{10} + C = \frac{1}{10}(x^2 + 2x + 1)^5 + C$

29. $\frac{1}{2}\ln(x^2 + 4x + 5) + C$

31. $\frac{5}{72}$ **33.** $\frac{98}{3}$ **35.** $\frac{1192}{15}$ **37.** $\frac{15}{128}$ **39.** $\frac{62}{15}$

41. $e - 1$ **43.** $2e^2 - 2e$ **45.** $\frac{1}{2}x - \frac{1}{4}\sin 2x + C$

47. $\dfrac{\pi}{2}$ **49.** $-x + \tan x + C$

Section 5.8 (page 332)

1. $\frac{2}{3}$ **3.** $\frac{16}{3}$ **5.** $+\frac{1}{4}$ **7.** $\frac{1}{4}$ **9.** $+\frac{16}{3}$ **11.** $\frac{1}{4}$

13. $e + e^{-1} - 2$ **15.** $\frac{1}{2}(1 - e^{-4})$

17. $\frac{32}{3}$ **19.** $\frac{128}{3}\sqrt{2}$ **21.** $\frac{500}{3}$ **23.** $\frac{64}{3}$ **25.** $\frac{500}{3}$

27. $\frac{4}{3}$ **29.** $\frac{1}{12}$ **31.** $\frac{16}{3}$ **33.** $\frac{16}{3}$ **35.** $\frac{27}{2}$

37. $\frac{1}{2}(e^2 - e^{-2})$ **39.** $\frac{1}{2}(e - 1)$ **41.** $\sqrt{2} - 1$

47. $\frac{253}{6}$ **49.** $f(x) = 5x^4$ **51.** $f(x) = \frac{1}{2}x$

Section 5.9 (page 345)

1. $T_4 = 8$; true value: 8   **3.** $T_5 \approx 0.65$; true value: $\frac{2}{3}$

5. $T_3 = 0.98$; true value: 1   **7.** $M_4 = 8$; true value: 8

9. $M_5 \approx 0.67$; true value: $\frac{2}{3}$

11. $M_3 \approx 1.01$; true value: 1

13. $T_4 = 8.75$; $S_4 \approx 8.6667$; true value: $\frac{26}{3}$

15. $T_4 \approx 0.0973$; $S_4 \approx 0.0940$; true value: 0.09375

17. $T_6 \approx 3.2599$; $S_6 \approx 3.2411$; true value: approximately 3.2413

19. $T_8 \approx 5.0140$, $S_8 \approx 5.0197$

21. (a) 3.0200; (b) 3.0717   **23.** (a) 2519; (b) 2521

25. (a) 2.09; (b) 2.15   **27.** 19

Chapter 5 Miscellaneous Problems (page 349)

1. $\frac{1}{3}x^3 + 2x^{-1} - \frac{5}{2}x^{-2} + C$

3. $-\frac{1}{30}(1 - 3x)^{10} + C$ **5.** $\frac{3}{16}(4x + 9)^{4/3} + C$

7. $\frac{1}{24}(x^4 + 1)^6 + C$ **9.** $-\frac{3}{8}(1 - x^2)^{4/3} + C$

11. $\frac{7}{5}\sin 5x + \frac{5}{7}\cos 7x + C$

13. $\frac{1}{6}(1 + x^4)^{3/2} + C$ **15.** $C - \dfrac{2}{1 + \sqrt{x}}$

17. $\frac{1}{12}\sin 4x^3 + C$ **19.** $\frac{1}{30}(x^2 + 1)^{15} + C$

21. $\frac{2}{5}(4 - x)^{5/2} - \frac{8}{3}(4 - x)^{3/2} + C$

23. $\sqrt{1 + x^4} + C$ **25.** $y(x) = x^3 + x^2 + 5$

27. $y(x) = \frac{1}{12}(2x + 1)^6 + \frac{23}{12}$ **29.** $y(x) = \frac{3}{2}x^{2/3} - \frac{1}{2}$

31. 6 s; 396 ft   **33.** 120 ft/s

35. Impact time: $t = 2\sqrt{10} \approx 6.32$ s; impact speed: $20\sqrt{10} \approx 63.25$ ft/s

37. 176 ft   **39.** 1700   **41.** 2845

43. $2(\sqrt{2} - 1) \approx 0.82843$

45. $\frac{2}{3}\pi(2\sqrt{2} - 1) \approx 3.82945$

47. Show that every Riemann sum is equal to $c(b - a)$.

51. $\dfrac{2x\sqrt{2x}}{3} + \dfrac{2}{\sqrt{3x}} + C$ **53.** $-2x^{-1} - \frac{1}{4}x^2 + C$

55. $\frac{2}{3}\sin x^{3/2} + C$ **57.** $\cos(t^{-1}) + C$

59. $-\frac{3}{8}(1 + u^{4/3})^{-2} + C$ **61.** $\frac{38}{3}$

63. $\frac{1}{3}x^{-3}(4x^2 - 1)^{3/2} + C$ (use $u = 1/x$)

65. $\frac{1}{30}$ **67.** $\frac{44}{15}$ **69.** $\frac{125}{6}$

71. Semicircle, center $(1, 0)$, radius 1: area $\pi/2$

73. $f(x) = \sqrt{4x^2 - 1}$

75. Use $n \geq 9$. $L_{10} \approx 1.12767$, $R_{10} \approx 1.16909$. The integrand is increasing on $[0, 1]$, so the average of these approximations yields 1.1483 ± 0.05.

77. $M_5 \approx 0.28667$, $T_5 \approx 0.28971$. They bound the true value of the integral because the second derivative of the integrand is positive on $[1, 2]$.

Section 6.1 (page 359)

1. 1   **3.** $2/\pi$ **5.** $\frac{98}{3}$ **7.** 4   **9.** -1

11. $\displaystyle\int_1^4 2\pi x f(x)\, dx$ **13.** $\displaystyle\int_0^{10} \sqrt{1 + [f'(x)]^2}\, dx$

15. 1000   **17.** $\frac{500}{3}$ **19.** -320; 320

21. -50; 106.25   **23.** 65; 97   **25.** 1; 1

27. Both answers: $10(1 - e^{-100}) \approx 10$

31. $625\pi/2$ **33.** 550 gal

35. 385,000   **37.** 109.5 in.   **39.** $\frac{3}{4}$

41. $f(x) = 400\pi x^2(1 + x)$; $700\pi/3$ lb

Section 6.2 (page 368)

1. $\pi/5$ **3.** 8π **5.** $\pi^2/2$ **7.** $3\pi/10$

9. $2\pi \ln 2$ **11.** $16\pi/15$ **13.** $\pi/2$ **15.** 8π

17. $121\pi/210$ **19.** 8π **21.** $49\pi/30$

23. $17\pi/10$ **25.** 9π **27.** $\frac{4}{3}\pi a^2 b$ **29.** $\frac{16}{3}a^3$
31. $\frac{4}{3}a^3\sqrt{3}$ **37.** $\frac{16}{3}a^3$ **43.** (a) \$3000; (b) \$3000

Section 6.3   (page 377)

1. 8π **3.** $625\pi/2$ **5.** 16π **7.** π **9.** $6\pi/5$
11. $256\pi/15$ **13.** $4\pi/15$ **15.** $11\pi/15$
17. $56\pi/5$ **19.** $8\pi/3$ **21.** $2\pi/15$ **23.** $\pi/2$
25. $16\pi/3$ **27.** $2\pi \ln 2$ **31.** $\frac{4}{3}\pi a^2 b$
33. $V = 2\pi^2 a^2 b$ **35.** $V = 2\pi^2 a^3$ **37.** (a) $V = \frac{1}{6}\pi h^3$

Section 6.4   (page 386)

Note: In 1 through 19, the integrand is given, followed by the interval of integration.

1. $\sqrt{1 + 4x^2}$; $[0,1]$
3. $\sqrt{1 + 36(x^4 - 2x^3 + x^2)}$; $[0,2]$
5. $\sqrt{1 + 4x^2}$; $[0,100]$ **7.** $\sqrt{1 + 16y^6}$; $[-1,2]$
9. $\dfrac{\sqrt{x^4 + 1}}{x^2}$; $[1,2]$ **11.** $2\pi x^2 \sqrt{1 + 4x^2}$; $[0,4]$
13. $2\pi(x - x^2)\sqrt{2 - 4x + 4x^2}$; $[0,1]$
15. $2\pi(2 - x)\sqrt{1 + 4x^2}$; $[0,1]$
17. $\pi\sqrt{4x + 1}$; $[1,4]$
19. $\pi(x + 1)\sqrt{4 + 9x}$; $[1,4]$ **21.** $\frac{22}{3}$ **23.** $\frac{14}{3}$
25. $\frac{123}{32} = 3.84375$
27. $\frac{1}{27}(104\sqrt{13} - 125) \approx 9.25842$
29. $\frac{1}{6}\pi(5\sqrt{5} - 1) \approx 5.3304$ **31.** $\frac{339}{16}\pi \approx 66.5625$
33. $\frac{1}{9}\pi(82\sqrt{82} - 1) \approx 258.8468$ **35.** 4π
37. 3.8194 (true value: approximately 3.820197789)
41. Avoid the problem when $x = 0$ as follows:

$$L = 8 \int_{1/(2\sqrt{2})}^{1} x^{-1/3}\, dx = 6.$$

Section 6.5   (page 394)

1. $y(x) = (\frac{1}{2}x^2 + C)^2$ **3.** $3y^{-2} + 2x^3 = C$
5. $y(x) = 1 + (\frac{1}{2}x^2 + C)^2$
7. $3y + 2y^{3/2} = 3x + 2x^{3/2} + C$
9. $y^3 + y = x - x^{-1} + C$ **11.** $y(x) = (1 - x)^{-1}$
13. $y(x) = (x + 1)^{1/4}$ **15.** $y(x) = (\frac{1}{2} - \frac{1}{3}x^{3/2})^{-2}$
17. $x^2 + y^2 = 169$ **19.** $y(x) = (1 + x - x^3)^{-1}$
21. 20 weeks **23.** (a) 169,000; (b) early in 2011
25. $P(t) \to +\infty$ as $t \to 6^-$. **27.** 3 h
29. 1 h 18 min 40 s after the plug is pulled
31. 1:20 P.M. **33.** About 6 min 2.9 s
35. $f(x) = (\pi/86,400)^2 x^4$; the radius of the hole should be $\dfrac{1}{240\sqrt{3}}$ ft, about 0.02887 in.

Section 6.6   (page 403)

1. 30 **3.** 9 **5.** 0 **7.** 15 ft·lb
9. 2.816×10^9 ft·lb (with $R = 4000$ mi, $g = 32$ ft/s²)
11. $13,000\pi \approx 40,841$ ft·lb
13. $125,000\pi/3 \approx 130,900$ ft·lb
15. $156,000\pi \approx 490,088$ ft·lb
17. $4,160,000\pi \approx 13,069,025$ ft·lb
19. 8750 ft·lb **21.** 11,250 ft·lb
23. $25,000 \cdot [1 - (0.1)^{0.4}]$ in.·lb ≈ 1254 ft·lb
25. 16π ft·lb **27.** $1,382,400\pi$ ft·lb
29. About 690.53 ft·lb **31.** 249.6 lb
33. 748.8 lb **35.** 19,500 lb
37. $700\rho/3 \approx 14,560$ lb **39.** About 32,574 tons

Chapter 6 Miscellaneous Problems   (page 406)

1. $-\frac{3}{2}$; $\frac{31}{6}$ **3.** 1; 3 **5.** $\frac{14}{3}$ **7.** $2\pi/15$ **9.** 12 in.
11. $41\pi/105$ **13.** $10.625\pi \approx 33.379$ g
19. $f(x) = \sqrt{1 + 3x}$ **21.** $\dfrac{24 - 2\pi^2}{3\pi} \approx 0.4521$
23. $\frac{10}{3}$ **25.** $\frac{63}{8}$ **27.** $52\pi/5$
31. $y(x) = x^2 + \sin x$ **33.** $y(x) = (C - x)^{-1} - 1$
35. $y(x) = (1 - x^3)^{-1}$ **37.** $y(x) = (4 - 3x^{-1})^{1/3}$
39. $y(x) = (1 - \sin x)^{-1}$
41. $(2y^{1/2} - 1)x = (Cx + 2x^{1/2} - 1)y$ **43.** 1 ft
45. $W = 4\pi R^4 \rho$ **47.** 10,454,400 ft·lb
49. 36,400 tons
51. There is no maximum volume; $c = \frac{1}{3}\sqrt{5}$ minimizes the volume V.

Section 7.2   (page 418)

1. $f'(x) = \dfrac{3}{3x - 1}$ **3.** $f'(x) = \dfrac{1}{1 + 2x}$
5. $f'(x) = \dfrac{3x^2 - 1}{3x^3 - 3x}$ **7.** $f'(x) = -\dfrac{1}{x}\sin(\ln x)$
9. $f'(x) = -\dfrac{1}{x(\ln x)^2}$ **11.** $f'(x) = \dfrac{1}{x} + \dfrac{x}{x^2 + 1}$
13. $f'(x) = -\tan x$
15. $f'(t) = 2t \ln(\cos t) - t^2 \tan t$
17. $g'(t) = (\ln t)^2 + 2\ln t$
19. $f'(x) = 6(2x + 1)^{-1} + 8x(x^2 - 4)^{-1}$
21. $f'(x) = -x(4 - x^2)^{-1} - x(9 + x^2)^{-1}$
23. $f'(x) = \dfrac{1}{x + 1} - \dfrac{1}{x - 1}$
25. $g'(t) = 2t^{-1} - 2t(t^2 + 1)^{-1}$
27. $f'(x) = -x^{-1} + \cot x$

29. $\dfrac{y \ln y}{y - x}$ **31.** $\dfrac{y}{-x + \cot y}$

33. $\frac{1}{2} \ln|2x - 1| + C$ **35.** $\frac{1}{6} \ln(1 + 3x^2) + C$

37. $\frac{1}{4} \ln|2x^2 + 4x + 1| + C$

39. $\frac{1}{3}(\ln x)^3 + C$ **41.** $\ln|x + 1| + C$

43. $\ln(x^2 + x + 1) + C$

45. $\frac{1}{2}(\ln x)^2 + C$ **47.** $\frac{1}{2} \ln(1 - \cos 2x) + C$

49. $\frac{1}{3} \ln|x^3 - 3x^2 + 1| + C$

51. 0 **53.** 0 **55.** 0

59. $m \approx -0.2479$, $k \approx 291.7616$

65. $y \to 0$ as $x \to 0^+$; $dy/dx \to 0$ as $x \to 0^+$. The point $(0,0)$ is not on the graph. Intercept at $(1, 0)$; global min. where $x = e^{-1/2}$—the coordinates are approximately $(0.61, -0.18)$; inflection point where $x = e^{-3/2}$—the coordinates are approximately $(0.22, -0.07)$.

67. $y \to -\infty$ as $x \to 0^+$; $y \to 0$ as $x \to +\infty$. Global max. at $(e^2, 2/e)$; only intercept is $(1, 0)$; inflection point where $x = e^{8/3}$. The x-axis is a horizontal asymptote, and the y-axis is a vertical asymptote.

71. Midpoint estimate: approximately 872.47; trapezoidal estimate: approximately 872.60; true value of integral: approximately 872.5174.

Section 7.3 (page 425)

1. $f'(x) = 2e^{2x}$

3. $f'(x) = 2xe^{(x^2)} = 2xe^{x^2} = 2x \exp(x^2)$

5. $f(x) = -2x^{-3} \exp(x^{-2})$

7. $g'(t) = \left(1 + \dfrac{1}{2}\sqrt{t}\right)e^{\sqrt{t}}$

9. $g'(t) = (1 + 2t - t^2)e^{-t}$

11. $g'(t) = -e^{\cos t} \sin t$

13. $f'(x) = -e^{-x} \sin(1 - e^{-x})$

15. $f'(x) = \dfrac{1 - e^{-x}}{x + e^{-x}}$

17. $f'(x) = e^{-2x}(3 \cos 3x - 2 \sin 3x)$

19. $g'(t) = 15(e^t - t^{-1})(e^t - \ln t)^4$

21. $f'(x) = -(5 + 12x)e^{-4x}$

23. $g'(t) = (e^{-t} + te^{-t} - 1)t^{-2}$

25. $f'(x) = (x - 2)e^{-x}$

27. $f'(x) = e^x \exp(e^x) = e^x e^{e^x}$

29. $f'(x) = 2e^x \cos 2e^x$

31. $\dfrac{e^y}{1 - xe^y} = \dfrac{e^y}{1 - y}$ **33.** $\dfrac{e^x - ye^{xy}}{xe^{xy} - e^y}$

35. $\dfrac{e^{x-y} - y}{e^{x-y} + x} = \dfrac{xy - y}{xy + x}$ **37.** $-\frac{1}{2}e^{1-2x} + C$

39. $\frac{1}{9} \exp(3x^3 - 1) + C$ **41.** $\frac{1}{2} \ln(1 + e^{2x}) + C$

43. $\frac{1}{2} \exp(1 - \cos 2x) + C$ **45.** $\frac{1}{2} \ln(x^2 + e^{2x}) + C$

47. $-\exp(-\frac{1}{2}t^2) + C$ **49.** $2e^{\sqrt{x}} + C$

51. $\ln(1 + e^x) + C$ **53.** $-\frac{2}{3} \exp(-x^{3/2}) + C$

55. e^2 **57.** e **59.** $+\infty$ **61.** $+\infty$

63. Global min. and intercept at $(0, 0)$, local max. at $(2, 4e^{-2})$; inflection points where $x = 2 \pm \sqrt{2}$. The x-axis is an asymptote.

65. Global max. at $(0, 1)$, the only intercept; inflection points where $x = \pm\sqrt{2}$. The x-axis is the only asymptote.

67. $\frac{1}{2}\pi(e^2 - 1) \approx 10.0359$

69. $\frac{1}{2}(e - e^{-1}) \approx 1.1752$

71. The solution is approximately 1.278. Note that if $f(x) = e^{-x} - x + 1$, then $f'(x) < 0$ for all x.

73. $f'(x) = 0$ when $x = 0$ and when $x = n$; f is increasing on $(0, n)$ and decreasing for $x > n$. Thus $x = n$ yields the absolute max. value of $f(x)$ for $x \geqq 0$. The x-axis is a horizontal asymptote, and there are inflection points where $x = n \pm \sqrt{n}$.

77. $y(x) = e^{-2x} + 4e^x$

Section 7.4 (page 432)

1. $f'(x) = 10^x \ln 10$

3. $f'(x) = 3^x 4^{-x} \ln 3 - 3^x 4^{-x} \ln 4 = \left(\frac{3}{4}\right)^x \ln\left(\frac{3}{4}\right)$

5. $f'(x) = -(7^{\cos x})(\ln 7)(\sin x)$

7. $f'(x) = 2^{x\sqrt{x}}\left(\frac{3}{2} \ln 2\right)\sqrt{x}$

9. $f'(x) = x^{-1} 2^{\ln x} \ln 2$

11. $f'(x) = 17^x \ln 17$

13. $f'(x) = -x^{-2} 10^{1/x} \ln 10$

15. $f'(x) = (2^{2^x} \ln 2)(2^x \ln 2)$

17. $f'(x) = \dfrac{1}{\ln 3} \cdot \dfrac{x}{x^2 + 4}$ **19.** $f'(x) = \dfrac{\ln 2}{\ln 3} = \log_3 2$

21. $f'(x) = \dfrac{1}{x(\ln 2)(\ln x)}$ **23.** $f'(x) = \dfrac{\exp(\log_{10} x)}{x \ln 10}$

25. $\dfrac{3^{2x}}{2 \ln 3} + C$ **27.** $\dfrac{2 \cdot 2^{\sqrt{x}}}{\ln 2} + C$

29. $\dfrac{7^{x^3+1}}{3 \ln 7} + C$ **31.** $\dfrac{(\ln x)^2}{2 \ln 2} + C$

33. $[x(x^2 - 4)^{-1} + (4x + 2)^{-1}](x^2 - 4)^{1/2}(2x + 1)^{1/4}$

35. $2^x \ln 2$ **37.** $\dfrac{(x^{\ln x})(2 \ln x)}{x}$

39. $\dfrac{y}{3}\left(\dfrac{1}{x + 1} + \dfrac{1}{x + 2} - \dfrac{2x}{x^2 + 1} - \dfrac{2x}{x^2 + 2}\right)$

41. $(\ln x)^{\sqrt{x}}[\frac{1}{2}x^{-1/2} \ln(\ln x) + x^{-1/2}(\ln x)^{-1}]$

43. $\left(\dfrac{3x}{1 + x^2} - \dfrac{4x^2}{1 + x^3}\right)(1 + x^2)^{3/2}(1 + x^3)^{-4/3}$

A-50

45. $\left[\dfrac{2x^3}{x^2 + 1} + 2x \ln(x^2 + 1) \right](x^2 + 1)^{(x^2)}$

47. $\frac{1}{4} x^{-1/2}(2 + \ln x)(\sqrt{x})^{\sqrt{x}}$ **49.** e^x

51. $x^{\exp(x)} e^x (x^{-1} + \ln x)$

57. Note that $\ln \dfrac{x^x}{e^x} = x \ln \dfrac{x}{e}$.

59. $\dfrac{dy}{dx} = -\dfrac{\ln 2}{x(\ln x)^3}$

Section 7.5 (page 441)

1. \$119.35; \$396.24   **3.** About 3.8685 h

5. About 686 yr old

7. (a) 9.308%; (b) 9.381%; (c) 9.409%; (d) 9.416%; (e) 9.417%

9. \$44.52   **11.** After an additional 32.26 days

13. About 35 yr   **15.** About 4.2521×10^9 yr old

17. 2.40942 min

19. (a) 20.486 in.; 9.604 in. (b) 3.4524 mi, about 18,230 ft

Section 7.6 (page 447)

1. $y(x) = -1 + 2e^x$ **3.** $y(x) = \frac{1}{2}(e^{2x} + 3)$

5. $x(t) = 1 - e^{2t}$ **7.** $x(t) = 27e^{5t} - 2$

9. $v(t) = 10(1 - e^{-10t})$ **11.** 4,870,238

15. About 46 days after the rumor starts

19. $\dfrac{400}{\ln 2} \approx 577$ ft   **23.** (b) \$1,308,283

25. (a) $x(t) = 100{,}000 - 80{,}000e^{-kt}$, where $k = \frac{1}{14} \ln 2$; (b) on March 29; (c) everyone gets the flu.

Chapter 7 Miscellaneous Problems (page 449)

1. $f'(x) = \dfrac{1}{2x}$ **3.** $f'(x) = \dfrac{1 - e^x}{x - e^x}$

5. $f'(x) \equiv \ln 2$ **7.** $f'(x) = (2 + 3x^2)e^{-1/x^2}$

9. $f'(x) = \dfrac{1 + \ln \ln x}{x}$ **11.** $f'(x) = x^{-1} 2^{\ln x} \ln 2$

13. $f'(x) = -\dfrac{2}{(x - 1)^2} \exp\left(\dfrac{x + 1}{x - 1}\right)$

15. $f'(x) = \dfrac{3}{2}\left(\dfrac{1}{x - 1} + \dfrac{8x}{3 - 4x^2}\right)$

17. $f'(x) = \dfrac{(\sin x \cos x) \exp(\sqrt{1 + \sin^2 x})}{\sqrt{1 + \sin^2 x}}$

19. $f'(x) = \cot x + \ln 3$ **21.** $f'(x) = \dfrac{x^{1/x}(1 - \ln x)}{x^2}$

23. $f'(x) = \left(\dfrac{1 + \ln \ln x}{x}\right)(\ln x)^{\ln x}$

25. $-\frac{1}{2} \ln|1 - 2x| + C$

27. $\frac{1}{2} \ln|1 + 6x - x^2| + C$

29. $-\ln(2 + \cos x) + C$ **31.** $\dfrac{2 \cdot 10^{\sqrt{x}}}{\ln 10} + C$

33. $\frac{2}{3}(1 + e^x)^{3/2} + C$ **35.** $\dfrac{6^x}{\ln 6} + C$

37. $x(t) = t^2 + 17$ **39.** $x(t) = 1 + e^t$

41. $x(t) = \frac{1}{3}(2 + 7e^{3t})$ **43.** $x(t) = \sqrt{2}\, e^{\sin t}$

45. Horizontal asymptote: the x-axis; global max. where $x = \frac{1}{2}$; inflection point where $x = (1 + \sqrt{2})/2$—approximately $(1.21, 0.33)$; global min. and intercept at $(0,0)$, with a vertical tangent there as well

47. Global min. at $(4, 2 - \ln 4)$, inflection point at $(16, 1.23)$ (ordinate approximate). The y-axis is a vertical asymptote; there is no horizontal asymptote. (The graph continues to rise for large increasing x.)

49. Inflection point at $(0.5, e^{-2})$. The horizontal line $y = 1$ and the x-axis are asymptotes. The point $(0, 0)$ is *not* on the graph. As $x \to 0^+$, $y \to 0$; as $x \to 0^-$, $y \to +\infty$. As $|x| \to +\infty$, $y \to 1$.

51. Sell immediately!

53. (b) The minimizing value is about 10.516. But because the batch size must be an integer, 11 (rather than 10) minimizes $f(x)$. (c) \$977.85

57. 20 weeks

59. (a) \$925.20; (b) \$1262.88

61. About 22.567 h after the power failure; that is, at about 9:34 P.M. the following evening

63. (b) $v(10) = 176(1 - e^{-1}) \approx 111.2532$ ft/s, about 75.85 mi/h. The limiting velocity is $a/\rho = 176$ ft/s, exactly 120 mi/h.

65. (a) The minimum of $f(x) - g(x)$ occurs when $x = 4$ and is $2(1 - \ln 2) > 0$. Hence $f(x) > g(x)$ for all $x > 0$. (b) The (large) solution of $g(x) = h(x)$ is (by Newton's method) approximately 93.354460835. (c) $p = e$

Section 8.2 (page 461)

1. (a) $\pi/6$, (b) $-\pi/6$, (c) $\pi/4$, (d) $-\pi/3$

3. (a) 0, (b) $\pi/4$, (c) $-\pi/4$, (d) $\pi/3$

5. $f'(x) = 100x^{99}(1 - x^{-200})^{-1/2}$

7. $f'(x) = \dfrac{1}{x |\ln x| \sqrt{(\ln x)^2 - 1}}$

9. $f'(x) = \dfrac{\sec^2 x}{\sqrt{1 - \tan^2 x}}$

11. $f'(x) = \dfrac{e^x}{\sqrt{1 - e^{2x}}}$

13. $f'(x) = -\dfrac{2}{\sqrt{1 - x^2}}$

15. $f'(x) = -\dfrac{2}{x\sqrt{x^4 - 1}}$

17. $f'(x) = -\dfrac{1}{(1 + x^2)(\arctan x)^2}$

19. $f'(x) = \dfrac{1}{x[1 + (\ln x)^2]}$

21. $f'(x) = \dfrac{2e^x}{1 + e^{2x}}$

23. $f'(x) = \dfrac{\cos(\arctan x)}{1 + x^2}$

25. $f'(x) = \dfrac{1 - 4x \arctan x}{(1 + x^2)^3}$

27. $\dfrac{dy}{dx} = -\dfrac{1 + y^2}{1 + x^2}; \ x + y = 2$

29. $\dfrac{dy}{dx} = -\dfrac{\sqrt{1 - y^2} \arcsin y}{\sqrt{1 - x^2} \arcsin x}; \ x + y = \sqrt{2}$

31. $\pi/4$　　**33.** $\pi/12$　　**35.** $\pi/12$

37. $\frac{1}{2} \arcsin 2x + C$　　**39.** $\frac{1}{5} \operatorname{arcsec} |x/5| + C$

41. $\arctan(e^x) + C$　　**43.** $\frac{1}{15} \operatorname{arcsec} |x^3/5| + C$

45. Both $\arcsin(2x - 1) + C$ and $2 \arcsin x^{1/2} + C$ are correct.

47. $\frac{1}{50} \arctan(x^{50}) + C$　　**49.** $\arctan(\ln x) + C$

51. $\pi/4$　　**53.** $\pi/2$　　**55.** $\pi/12$　　**59.** 8 m

63. $\pi/2$　　**65.** (b) $A = 1 - \frac{1}{3}\pi, \ B = 1 + \frac{2}{3}\pi$

Section 8.3　(page 466)

1. $\frac{1}{2}$　　**3.** $\frac{2}{5}$　　**5.** 0　　**7.** 0　　**9.** $\frac{1}{2}$

11. 2　　**13.** 0　　**15.** 1　　**17.** 1　　**19.** $\frac{3}{5}$

21. $\frac{3}{2}$　　**23.** $\frac{1}{3}$　　**25.** $\dfrac{\ln 2}{\ln 3}$　　**27.** $\frac{1}{2}$　　**29.** 1

31. $\frac{1}{3}$　　**33.** $-\frac{1}{2}$　　**35.** 1　　**37.** $\frac{1}{4}$　　**39.** $\frac{2}{3}$

41. 6　　**43.** $\frac{4}{3}$　　**45.** $\frac{2}{3}$　　**47.** 0

Section 8.4　(page 471)

1. 1　　**3.** $\frac{3}{8}$　　**5.** $\frac{1}{4}$　　**7.** 1　　**9.** 0

11. -1　　**13.** $-\infty$　　**15.** $-\infty$　　**17.** $-\frac{1}{2}$　　**19.** 0

21. 1　　**23.** 1　　**25.** $e^{-1/6}$　　**27.** $e^{-1/2}$　　**29.** 1

31. e^{-1}　　**33.** $-\infty$

Section 8.5　(page 480)

1. $f'(x) = 3 \sinh(3x - 2)$

3. $f'(x) = 2x \tanh(1/x) - \operatorname{sech}^2(1/x)$

5. $f'(x) = -12 \coth^2 4x \operatorname{csch}^2 4x$

7. $f'(x) -e^{\operatorname{csch} x} \operatorname{csch} x \coth x$

9. $f'(x) = (\cosh x) \cos(\sinh x)$

11. $f'(x) = 4x^3 \cosh x^4$　　**13.** $f'(x) = -\dfrac{1 + \operatorname{sech}^2 x}{(x + \tanh x)^2}$

15. $\frac{1}{2} \cosh x^2 + C$　　**17.** $x - \frac{1}{3} \tanh 3x + C$

19. $\frac{1}{6} \sinh^3 2x + C$　　**21.** $-\frac{1}{2} \operatorname{sech}^2 x + C$

23. $-\frac{1}{2} \operatorname{csch}^2 x + C$　　**25.** $\ln(1 + \cosh x) + C$

27. $\frac{1}{4} \tanh x + C$　　**29.** $f'(x) = 2(4x^2 + 1)^{-1/2}$

31. $f'(x) = \frac{1}{2} x^{-1/2}(1 - x)^{-1}$

33. $f'(x) = (x^2 - 1)^{-1/2}$

35. $f'(x) = \frac{3}{2}(\sinh^{-1} x)^{1/2}(x^2 + 1)^{-1/2}$

37. $f'(x) = (1 - x^2)^{-1}(\tanh^{-1} x)^{-1}$

39. $\operatorname{arcsinh}(x/3) + C$　　**41.** $\frac{1}{4} \ln \frac{9}{5} \approx 0.14695$

43. $-\frac{1}{2} \operatorname{sech}^{-1}|3x/2| + C$　　**45.** $\sinh^{-1}(e^x) + C$

47. $-\operatorname{sech}^{-1}(e^x) + C$　　**53.** $\sinh a$

57. $\ln(1 + \sqrt{2}) \approx 0.8813735887$

Chapter 8 Miscellaneous Problems　(page 482)

1. $f'(x) = 3(1 - 9x^2)^{-1/2}$

3. $g'(t) = 2t^{-1}(t^4 - 1)^{-1/2}$

5. $f'(x) = -(\sin x)(1 - \cos^2 x)^{-1/2} = -\dfrac{\sin x}{|\sin x|}$

7. $g'(t) = 10(100t^2 - 1)^{-1/2}$

9. $f'(x) = -2x^{-1}(x^4 - 1)^{-1/2}$

11. $f'(x) = \frac{1}{2} x^{-1/2}(1 - x)^{-1/2}$

13. $f'(x) = \dfrac{2x}{x^4 + 2x^2 + 2}$

15. $f'(x) = e^x \sinh e^x + e^{2x} \cosh e^x$

17. $f'(x) \equiv 0$

19. $f'(x) = \dfrac{x}{|x|\sqrt{x^2 + 1}}$

21. $\frac{1}{2} \arcsin 2x + C$

23. $\arcsin(x/2) + C$

25. $\sin^{-1}(e^x) + C$

27. $\frac{1}{2} \arcsin(2x/3) + C$

29. $\frac{1}{3} \arctan(x^3) + C$

31. $\sec^{-1}|2x| + C$

33. $\operatorname{arcsec}(e^x) + C$

35. $2 \cosh \sqrt{x} + C$

37. $\frac{1}{2}(\arctan x)^2 + C$

39. $\frac{1}{2} \sinh^{-1}(2x/3) + C$

41. $\frac{1}{4}$　　**43.** $\frac{1}{2}$　　**45.** $-\frac{1}{6}$　　**47.** 1　　**49.** $-\infty$

51. $+\infty$　　**53.** e^2　　**55.** $-\dfrac{e}{2}$　　**57.** $\dfrac{\pi^2}{6}$

61. $x \approx 4.730041$

Section 9.2　(page 487)

1. $-\frac{1}{15}(2 - 3x)^5 + C$　　**3.** $\frac{1}{9}(2x^3 - 4)^{3/2} + C$

5. $\frac{9}{8}(2x^2 + 3)^{2/3} + C$　　**7.** $-2 \csc \sqrt{y} + C$

9. $\frac{1}{6}(1 + \sin \theta)^6 + C$ **11.** $e^{-\cot x} + C$

13. $\frac{1}{11}(\ln t)^{11} + C$ **15.** $\frac{1}{3}\arcsin 3t + C$

17. $\frac{1}{2}\arctan(e^{2x}) + C$ **19.** $\frac{3}{2}\arcsin(x^2) + C$

21. $\frac{1}{15}\tan^5 3x + C$ **23.** $\tan^{-1}(\sin \theta) + C$

25. $\frac{2}{5}(1 + \sqrt{x})^5 + C$ **27.** $\ln|\arctan t| + C$

29. $\sec^{-1} e^x + C$

31. $\frac{2}{7}(x - 2)^{7/2} + \frac{8}{5}(x - 2)^{5/2} + \frac{8}{3}(x - 2)^{3/2} + C$

33. $\frac{1}{3}(2x + 3)^{1/2}(x - 3) + C$

35. $\frac{3}{10}(x + 1)^{2/3}(2x - 3) + C$

37. $\dfrac{1}{60}\ln\left|\dfrac{3x + 10}{3x - 10}\right| + C$

39. $\frac{1}{2}x(4 + 9x^2)^{1/2} + \frac{2}{3}\ln|3x + (4 + 9x^2)^{1/2}| + C$

41. $\frac{1}{32}x(16x^2 + 9)^{1/2} - \frac{9}{128}\ln|4x + (16x^2 + 9)^{1/2}| + C$

43. $\frac{1}{128}x(32x^2 - 25)(25 - 16x^2)^{1/2} + \frac{625}{512}\arcsin\frac{4}{5}x + C$

45. The substitution $u = e^x$ leads to an integral in the form of Formula (44) in the Table of Integrals inside the back cover. The answer is

$$\frac{1}{2}e^x(9 + e^{2x})^{1/2} + \frac{9}{2}\ln[e^x + (9 + e^{2x})^{1/2}] + C.$$

47. With $u = x^2$ and (47) in the Table of Integrals, $\frac{1}{2}((x^4 - 1)^{1/2} - \text{arcsec } x^2) + C$

49. With $u = \ln x$ and (48) in the Table of Integrals, $\frac{1}{8}\{(\ln x)[2(\ln x)^2 + 1][(\ln x)^2 + 1]^{1/2} - \ln|(\ln x) + [(\ln x)^2 + 1]^{1/2}|\} + C$

53. $\sin^{-1}(x - 1) + C$

Section 9.3   (page 495)

1. $\frac{1}{4}(2x - \sin 2x \cos 2x) + C$

3. $2 \tan \dfrac{x}{2} + C$

5. $\frac{1}{3}\ln|\sec 3x| + C$

7. $\frac{1}{3}\ln|\sec 3x + \tan 3x| + C$

9. $\frac{1}{2}(x - \sin x \cos x) + C$

11. $\frac{1}{3}\cos^3 x - \cos x + C$

13. $\frac{1}{3}\sin^3 \theta - \frac{1}{5}\sin^5 \theta + C$

15. $\frac{1}{5}\sin^5 x - \frac{2}{3}\sin^3 x + \sin x + C$

17. $\frac{2}{5}(\cos x)^{5/2} - 2(\cos x)^{1/2} + C$

19. $-\frac{1}{14}\cos^7 2z + \frac{1}{5}\cos^5 2z - \frac{1}{6}\cos^3 2z + C$

21. $\frac{1}{4}(\sec 4x + \cos 4x) + C$

23. $\frac{1}{3}\tan^3 t + \tan t + C$

25. $-\frac{1}{4}\csc^2 2x - \frac{1}{2}\ln|\sin 2x| + C$

27. $\frac{1}{12}\tan^6 2x + C$

29. $-\frac{1}{10}\cot^5 2t - \frac{1}{3}\cot^3 2t - \frac{1}{2}\cot 2t + C$

31. $\frac{1}{4}\cos^4 \theta - \frac{1}{2}\cos^2 \theta + C_1 = \frac{1}{4}\sin^4 \theta + C_2$

33. $\frac{2}{3}(\sec t)^{3/2} + 2(\sec t)^{-1/2} + C$

35. $\frac{1}{3}\sin^3 \theta + C$

37. $\frac{1}{5}\sin 5t - \frac{1}{15}\sin^3 5t + C$

39. $t + \frac{1}{3}\cot 3t - \frac{1}{9}\cot^3 3t + C$

41. $-\frac{1}{5}\cos^{5/2} 2t + \frac{2}{9}\cos^{9/2} 2t - \frac{1}{13}\cos^{13/2} 2t + C$

43. $\frac{1}{2}\sin^2 x - \cos x + C$

45. $-\cot x - \frac{1}{3}\cot^3 x - \frac{1}{2}\csc^2 x + C$

49. $\frac{1}{4}\cos 2x - \frac{1}{16}\cos 8x + C$

51. $\frac{1}{6}\sin 3x + \frac{1}{10}\sin 5x + C$

57. $\pi/4$

59. $2\ln 2 \approx 1.38629\,43611\,2$

Section 9.4   (page 501)

1. $\frac{1}{2}xe^{2x} - \frac{1}{4}e^{2x} + C$

3. $-t \cos t + \sin t + C$

5. $\frac{1}{3}x \sin 3x + \frac{1}{9}\cos 3x + C$

7. $\frac{1}{4}x^4 \ln x - \frac{1}{16}x^4 + C$

9. $x \arctan x - \frac{1}{2}\ln(1 + x^2) + C$

11. $\frac{2}{3}y^{3/2}\ln y - \frac{4}{9}y^{3/2} + C$

13. $t(\ln t)^2 - 2t \ln t + 2t + C$

15. $\frac{2}{3}x(x + 3)^{3/2} - \frac{4}{15}(x + 3)^{5/2} + C$

17. $\frac{2}{9}x^3(x^3 + 1)^{3/2} - \frac{4}{45}(x^3 + 1)^{5/2} + C$

19. $-\frac{1}{2}(\csc \theta \cot \theta + \ln|\csc \theta + \cot \theta|) + C$

21. $\frac{1}{3}x^3 \arctan x - \frac{1}{6}x^2 + \frac{1}{6}\ln(1 + x^2) + C$

23. $x \text{ arcsec } x^{1/2} - (x - 1)^{1/2} + C$

25. $(x + 1)\arctan x^{1/2} - x^{1/2} + C$

27. $-x \cot x + \ln|\sin x| + C$

29. $\frac{1}{2}x^2 \sin x^2 + \frac{1}{2}\cos x^2 + C$

31. $-2x^{-1/2}(2 + \ln x) + C$

33. $x \sinh x - \cosh x + C$

35. $x^2 \cosh x - 2x \sinh x + 2 \cosh x + C$

37. $\pi(e - 2) \approx 2.25655$

39. $\frac{1}{2}(x - 1)e^x \sin x + \frac{1}{2}xe^x \cos x + C$

47. $6 - 2e \approx 0.563436$

49. $6 - 2e$

57. $\frac{1}{80}(2\pi^6 - 10\pi^4 + 15\pi^2) \approx 13.709144$

Section 9.5   (page 510)

1. $\frac{1}{2}x^2 - x + \ln|x + 1| + C$

3. $\dfrac{1}{3}\ln\left|\dfrac{x - 3}{x}\right| + C$

5. $\dfrac{1}{5}\ln\left|\dfrac{x - 2}{x + 3}\right| + C$

7. $\frac{1}{4}\ln|x| - \frac{1}{8}\ln(x^2 + 4) + C$

9. $\frac{1}{3}x^3 - 4x + 8 \arctan \frac{1}{2}x + C$

11. $x - 2\ln|x + 1| + C$

13. $x + (x + 1)^{-1} + C$

15. $\frac{1}{4} \ln \left| \dfrac{x - 2}{x + 2} \right| + C$

17. $\frac{3}{2} \ln |2x - 1| - \ln |x + 3| + C$

19. $\ln |x| + 2(x + 1)^{-1} + C$

21. $\frac{3}{2} \ln |x^2 - 4| + \frac{1}{2} \ln |x^2 - 1| + C$

23. $\ln |x + 2| + 4(x + 2)^{-1} - 2(x + 2)^{-2} + C$

25. $\dfrac{1}{2} \ln \left(\dfrac{x^2}{x^2 + 1} \right) + C$

27. $\dfrac{1}{2} \ln \left(\dfrac{x^2}{x^2 + 4} \right) + \dfrac{1}{2} \arctan \dfrac{x}{2} + C$

29. $-\frac{1}{2} \ln |x + 1| + \frac{1}{4} \ln(x^2 + 1) + \frac{1}{2} \arctan x + C$

31. $\arctan \frac{1}{2} x - \frac{3}{2} \sqrt{2} \arctan x \sqrt{2} + C$

33. $\dfrac{1}{\sqrt{2}} \arctan \dfrac{x}{\sqrt{2}} + \dfrac{1}{2} \ln(x^2 + 3) + C$

35. $x + \frac{1}{2} \ln |x - 1| - 5(2x - 2)^{-1}$
$\quad + \frac{3}{4} \ln(x^2 + 1) + 2 \arctan x + C$

37. $\frac{1}{4} (1 - 2e^{2t})(e^{2t} - 1)^{-2} + C$

39. $\frac{1}{4} \ln |3 + 2 \ln t| + \frac{1}{4} (3 + 2 \ln t)^{-1} + C$

41. $\frac{4}{15} \pi(52 - 75 \ln 2) \approx 0.0116963324$

43. $x(t) = \dfrac{2e^t}{2e^t - 1}$

45. $x(t) = \dfrac{2e^{2t} + 1}{2e^{2t} - 1}$

47. $x(t) = \dfrac{21e^t - 16}{8 - 7e^t}$

49. About 153,700,000

51. (a) 1.37 s; (b) 200 g

53. $P(t) = \dfrac{200}{2 - e^{t/100}}$;

(a) $t = 100 \ln 1.8 \approx 58.8$ (days);
(b) $t = 100 \ln 2 \approx 69.3$ (days)

Section 9.6 (page 516)

1. $\arcsin \frac{1}{4} x + C$

3. $-\dfrac{\sqrt{4 - x^2}}{4x} + C$

5. $8 \arcsin \dfrac{x}{4} - \dfrac{x\sqrt{16 - x^2}}{2} + C$

7. $\dfrac{x}{9\sqrt{9 - 16x^2}} + C$

9. $\ln |x + \sqrt{x^2 - 1}| - \dfrac{\sqrt{x^2 - 1}}{x} + C$

11. $\frac{1}{80} [(9 + 4x^2)^{5/2} - 15(9 + 4x^2)^{3/2}] + C$

13. $\sqrt{1 - 4x^2} - \ln \left| \dfrac{1 + \sqrt{1 - 4x^2}}{2x} \right| + C$

15. $\frac{1}{2} \ln |2x + \sqrt{9 + 4x^2}| + C$

17. $\frac{25}{2} \arcsin \frac{1}{5} x - \frac{1}{2} x \sqrt{25 - x^2} + C$

19. $\frac{1}{2} x \sqrt{x^2 + 1} - \frac{1}{2} \ln |x + \sqrt{1 + x^2}| + C$

21. $\frac{1}{18} x \sqrt{4 + 9x^2} - \frac{2}{27} \ln |3x + \sqrt{4 + 9x^2}| + C$

23. $x(1 + x^2)^{-1/2} + C$

25. $\dfrac{1}{512} \left[3 \ln \left| \dfrac{x + 2}{x - 2} \right| - \dfrac{12x}{x^2 - 4} + \dfrac{32x}{(x^2 - 4)^2} \right] + C$

27. $\frac{1}{2} x \sqrt{9 + 16x^2} + \frac{9}{8} \ln |4x + \sqrt{9 + 16x^2}| + C$

29. $\sqrt{x^2 - 25} - 5 \operatorname{arcsec} \frac{1}{5} x + C$

31. $\frac{1}{8} x(2x^2 - 1)(x^2 - 1)^{1/2} - \frac{1}{8} \ln |x + (x^2 - 1)^{1/2}| + C$

33. $-x(4x^2 - 1)^{-1/2} + C$

35. $-x^{-1}(x^2 - 5)^{1/2} + \ln |x + (x^2 - 5)^{1/2}| + C$

37. $\operatorname{arcsinh} \frac{1}{5} x + C$

39. $\operatorname{arccosh} \frac{1}{2} x - x^{-1}(x^2 - 4)^{1/2} + C$

41. $\frac{1}{8} x(1 + 2x^2)(1 + x^2)^{1/2} - \frac{1}{8} \operatorname{arcsinh} x + C$

43. $\frac{1}{32} \pi[18\sqrt{5} - \ln(2 + \sqrt{5})] \approx 3.8097$

45. $\sqrt{5} - \sqrt{2} + \ln \left(\dfrac{2 + 2\sqrt{2}}{1 + \sqrt{5}} \right) \approx 1.222016$

49. $2\pi[\sqrt{2} + \ln(1 + \sqrt{2})] \approx 14.4236$

53. $\$6\frac{2}{3}$ million

Section 9.7 (page 522)

1. $\arctan(x + 2) + C$

3. $11 \arctan(x + 2) - \frac{3}{2} \ln(x^2 + 4x + 5) + C$

5. $\arcsin \frac{1}{2}(x + 1) + C$

7. $-2 \arcsin \frac{1}{2}(x + 1) - \frac{1}{2}(x + 1)(3 - 2x - x^2)^{1/2}$
$\quad - \frac{1}{3}(3 - 2x - x^2)^{3/2} + C$

9. $\frac{5}{16} \ln |2x + 3| + \frac{7}{16} \ln |2x - 1| + C$

11. $\frac{1}{3} \arctan \frac{1}{3}(x + 2) + C$

13. $\dfrac{1}{4} \ln \left| \dfrac{1 + x}{3 - x} \right| + C$

15. $\ln(x^2 + 2x + 2) - 7 \arctan(x + 1) + C$

17. $\frac{2}{9} \arcsin(x - \frac{2}{3}) - \frac{1}{9}(5 + 12x - 9x^2)^{1/2} + C$

19. $\frac{75}{4} \arcsin \frac{2}{5}(x - 2) + \frac{3}{2}(x - 2)(9 + 16x - 4x^2)^{1/2}$
$\quad + \frac{1}{6}(9 + 16x - 4x^2)^{3/2} + C$

21. $\frac{1}{9}(7x - 12)(6x - x^2)^{-1/2} + C$

23. $-(16x^2 + 48x + 52)^{-1} + C$

25. $\frac{3}{2} \ln(x^2 + x + 1) - \frac{5}{3} \sqrt{3} \arctan(\frac{1}{3}\sqrt{3}[2x + 1]) + C$

27. $\dfrac{1}{32} \ln \left| \dfrac{x + 2}{x - 2} \right| - \dfrac{x}{8(x^2 - 4)} + C$

29. $\ln |x| - \frac{2}{3} \sqrt{3} \arctan(\frac{1}{3}\sqrt{3}[2x + 1]) + C$

31. $-\frac{5}{4}(x - 1)^{-1} - \frac{1}{4} \ln |x - 1| - \frac{5}{4}(x + 1)^{-1}$
$\quad + \frac{1}{4} \ln |x + 1| + C$

33. $-\frac{1}{4}(x + 7)(x^2 + 2x + 5)^{-1} - \frac{1}{8} \arctan(\frac{1}{2}[x + 1]) + C$

37. About 3.69 mi

39. $\ln|x - 1| - \frac{1}{2}\ln(x^2 + 2x + 2) + \arctan(x + 1) + C$

41. $\frac{1}{2}x^2 + \ln|x - 1| + \frac{1}{2}\ln(x^2 + x + 1)$
$\quad + \frac{1}{3}\sqrt{3}\arctan(\frac{1}{3}\sqrt{3}\,[2x + 1]) + C$

43. $\frac{1}{2}\ln(x^4 + x^2 + 1) - \frac{2\sqrt{3}}{3}\arctan\left(\frac{\sqrt{3}}{2x^2 + 1}\right) + C$

Section 9.8 (page 529)

1. 1   **3.** $+\infty$ **5.** $+\infty$ **7.** 1

9. $+\infty$ **11.** $-\frac{1}{2}$ **13.** $\frac{9}{2}$ **15.** $+\infty$

17. Does not exist   **19.** $2(e - 1)$

21. $+\infty$ **23.** $\frac{1}{4}$

Chapter 9 Miscellaneous Problems   (page 533)

Note: Different techniques of integration may produce answers that appear to differ from these. If both are correct, they must differ by only a constant.

1. $2\arctan\sqrt{x} + C$ **3.** $\ln|\sec x| + C$

5. $\frac{1}{2}\sec^2\theta + C$

7. $x\tan x - \frac{1}{2}x^2 + \ln|\cos x| + C$

9. $\frac{2}{15}(2 - x^3)^{5/2} - \frac{4}{9}(2 - x^3)^{3/2} + C$

11. $\frac{1}{2}x(25 + x^2)^{1/2} - \frac{25}{2}\ln\left|x + (25 + x^2)^{1/2}\right| + C$

13. $\frac{2}{3}\sqrt{3}\arctan(\frac{1}{3}\sqrt{3}\,[2x - 1]) + C$

15. $\frac{103}{87}\sqrt{29}\arctan(\frac{1}{29}\sqrt{29}\,[3x - 2])$
$\quad + \frac{5}{6}\ln(9x^2 - 12x + 33) + C$

17. $\frac{2}{9}(1 + x^3)^{3/2} + C$ **19.** $\arcsin(\frac{1}{2}\sin x) + C$

21. $-\ln|\ln\cos x| + C$

23. $(x + 1)\ln(x + 1) - x + C$

25. $\frac{1}{2}x\sqrt{x^2 + 9} + \frac{9}{2}\ln\left|x + \sqrt{x^2 + 9}\right| + C$

27. $\frac{1}{2}(x - 1)(2x - x^2)^{1/2} + \frac{1}{2}\arcsin(x - 1) + C$

29. $\frac{1}{3}x^3 + 2x - \sqrt{2}\ln\left|\frac{x + \sqrt{2}}{x - \sqrt{2}}\right| + C$

31. $\frac{1}{2}(x^2 + x)(x^2 + 2x + 2)^{-1} - \frac{1}{2}\arctan(x + 1) + C$

33. $\dfrac{\sin 2\theta}{2(1 + \cos 2\theta)} + C$ **35.** $\frac{1}{5}\sec^5 x - \frac{1}{3}\sec^3 x + C$

37. $\frac{1}{8}x^2[4(\ln x)^3 - 6(\ln x)^2 + 6(\ln x) - 3] + C$

39. $\frac{1}{2}e^x(1 + e^{2x})^{1/2} + \frac{1}{2}\ln[e^x + (1 + e^{2x})^{1/2}] + C$

41. $\frac{1}{54}\text{arcsec}\left|\frac{1}{3}x\right| + \frac{1}{18}x^{-2}(x^2 - 9)^{1/2} + C$

43. $\ln|x| + \frac{1}{2}\arctan 2x + C$

45. $\frac{1}{2}(\sec x\tan x - \ln|\sec x + \tan x|) + C$

47. $\ln|x + 1| - \frac{2}{3}x^{-3} + C$

49. $\ln|x - 1| + \ln(x^2 + x + 1) + (x - 1)^{-1}$
$\quad - 2(x^2 + x + 1)^{-1} + C$

51. $x[(\ln x)^6 - 6(\ln x)^5 + 30(\ln x)^4 - 120(\ln x)^3$
$\quad + 360(\ln x)^2 - 720\ln x + 720] + C$

53. $\frac{1}{3}(\arcsin x)^3 + C$ **55.** $\frac{1}{2}\sec^2 z + \ln|\cos z| + C$

57. $\frac{1}{2}\arctan(\exp(x^2)) + C$

59. $-\frac{1}{2}(x^2 + 1)\exp(-x^2) + C$

61. $-\frac{1}{x}\arcsin x - \ln\left|\frac{1 + \sqrt{1 - x^2}}{x}\right| + C$

63. $\frac{1}{8}\arcsin x + \frac{1}{8}x(2x^2 - 1)(1 - x^2)^{1/2} + C$

65. $\frac{1}{4}\ln|2x + 1| + \frac{5}{4}(2x + 1)^{-1} + C$

67. $\frac{1}{2}\ln|e^{2x} - 1| + C$

69. $2\ln|x + 1| + 3(x + 1)^{-1} - \frac{5}{3}(x + 1)^{-3} + C$

71. $\frac{1}{2}\ln(x^2 + 1) + \arctan x - \frac{1}{2}(x^2 + 1)^{-1} + C$

73. $\frac{1}{45}(x^3 + 1)^{3/2}(6x^3 + 4) + C$

75. $\frac{2}{3}(1 + \sin x)^{3/2} + C$

77. $\frac{1}{2}\ln|\sec x + \tan x| + C$

79. $-2(1 - \sin t)^{1/2} + C$

81. $-2x + \sqrt{3}\arctan(\frac{1}{3}\sqrt{3}\,[2x + 1])$
$\quad + \frac{1}{2}(2x + 1)\ln(x^2 + x + 1) + C$

83. $-x^{-1}\arctan x + \ln\left|x(1 + x^2)^{-1/2}\right| + C$

85. $\frac{1}{2}\ln(x^2 + 1) + \frac{1}{2}(x^2 + 1)^{-1} + C$

87. $\frac{1}{2}(x - 6)(x^2 + 4)^{-1/2} + C$

89. $\frac{1}{3}(1 + \sin^2 x)^{3/2} + C$

91. $\frac{1}{2}e^x(x\sin x - x\cos x + \cos x) + C$

93. $-\frac{1}{2}(x - 1)^{-2}\arctan x + \frac{1}{8}\ln(x^2 + 1)$
$\quad - \frac{1}{4}\ln|x - 1| - \frac{1}{4}(x - 1)^{-1} + C$

95. $\frac{11}{9}\arcsin\frac{1}{2}(3x - 1) - \frac{2}{9}(3 + 6x - 9x^2)^{1/2} + C$

97. $\frac{1}{3}x^3 + x^2 + 3x + 4\ln|x - 1| - (x - 1)^{-1} + C$

99. $x\,\text{arcsec}\,x^{1/2} - (x - 1)^{1/2} + C$

101. $\frac{1}{4}\pi(e^2 - e^{-2} + 4)$

103. (a) $A_t = \pi\Big(\sqrt{2} - e^{-t}(1 + e^{-2t})^{1/2}$
$\quad + \ln\left[\dfrac{1 + \sqrt{2}}{e^{-t} + (1 + e^{-2t})^{1/2}}\right]\Big);$
(b) $\pi[\sqrt{2} + \ln(1 + \sqrt{2})] \approx 7.2118$

105. $\dfrac{\pi\sqrt{2}}{2}\left[2\sqrt{14} - \sqrt{2} + \ln\left(\dfrac{1 + \sqrt{2}}{2\sqrt{2} + \sqrt{7}}\right)\right]$
$\quad \approx 11.66353$

109. $\frac{5}{4}\pi \approx 3.92699$

111. The value of the integral is $\frac{1}{630}$.

113. $\dfrac{5\sqrt{6} - 3\sqrt{2}}{2} + \frac{1}{2}\ln\left(\dfrac{1 + \sqrt{2}}{\sqrt{3} + \sqrt{2}}\right) \approx 3.869983$

115. The substitution is $u = e^x$.
(a) $\frac{2}{3}\sqrt{3}\arctan(\frac{1}{3}\sqrt{3}\,[1 + 2e^x]) + C$

119. $\frac{1}{4}\sqrt{2}\ln\left|\dfrac{1 + \tan\theta - \sqrt{2}\tan\theta}{1 + \tan\theta + \sqrt{2}\tan\theta}\right|$
$\quad - \frac{1}{2}\sqrt{2}\arctan\sqrt{2\cot\theta} + C$

121. $\frac{2}{25,515}(3x - 2)^{3/2}(945x^3 + 540x^2 + 288x + 128) + C$

123. $\frac{3}{4}(x^2 - 1)^{-1/3}(x^2 - 3) + C$

125. $\frac{2}{9}(x^3 + 1)^{1/2}(x^3 - 2) + C$

127. $2\arctan\sqrt{\dfrac{1 + x}{1 - x}} - (1 - x)\sqrt{\dfrac{1 + x}{1 - x}} + C$

129. $3(x + 1)^{1/3} - \sqrt{3}\arctan\dfrac{1 + 2(x + 1)^{1/3}}{\sqrt{3}}$
$\quad + \ln\left|(x + 1)^{1/3} - 1\right|$
$\quad - \frac{1}{2}\ln\left|(x + 1)^{2/3} + (x + 1)^{1/3} + 1\right| + C$

131. $\sqrt{1 + e^{2x}} + \dfrac{1}{2}\ln\left(\dfrac{\sqrt{1 + e^{2x}} - 1}{\sqrt{1 + e^{2x}} + 1}\right) + C$

133. $\frac{8}{15}$

135. $\tan\dfrac{\theta}{2} + C = \dfrac{1 - \cos\theta}{\sin\theta} + C = \dfrac{\sin\theta}{1 + \cos\theta} + C$

137. $-\dfrac{2\sin\theta}{1 + \sin\theta - \cos\theta} + C$

139. $\dfrac{\sqrt{2}}{2}\ln\left|\dfrac{1 - \cos\theta + (\sqrt{2} - 1)\sin\theta}{1 - \cos\theta - (\sqrt{2} + 1)\sin\theta}\right| + C$

141. $\ln\dfrac{1 - \cos\theta}{1 - \cos\theta + \sin^2\theta} + C$

Section 10.1 (page 542)

1. $x + 2y + 3 = 0$
3. $4y + 25 = 3x$
5. $x + y = 1$
7. Center $(-1, 0)$, radius $\sqrt{5}$
9. Center $(2, -3)$, radius 4
11. Center $(0.5, 0)$, radius 1
13. Center $(0.5, -1.5)$, radius 3
15. Center $(-\frac{1}{3}, \frac{4}{3})$, radius 2
17. The point $(3, 2)$
19. No points
21. $(x + 1)^2 + (y + 2)^2 = 34$
23. $(x - 6)^2 + (y - 6)^2 = 0.8$
25. $2x + y = 13$
27. $(x - 6)^2 + (y - 11)^2 = 18$
29. $(x/5)^2 + (y/3)^2 = 1$
31. $y - 7 + 4\sqrt{3} = (4 - 2\sqrt{3})(x - 2 + \sqrt{3})$,
$\quad y - 7 - 4\sqrt{3} = (4 + 2\sqrt{3})(x - 2 - \sqrt{3})$
33. $y - 1 = 4(x - 4)$ and $y + 1 = 4(x + 4)$
35. $a = |h^2 - p^2|^{1/2}$, $b = a(e^2 - 1)^{1/2}$, $h = \dfrac{p(e^2 + 1)}{1 - e^2}$

Section 10.2 (page 548)

1. (a) $(\frac{1}{2}\sqrt{2}, \frac{1}{2}\sqrt{2})$ (b) $(1, -\sqrt{3})$
  (c) $(\frac{1}{2}, -\frac{1}{2}\sqrt{3})$ (d) $(0, -3)$
  (e) $(\sqrt{2}, \sqrt{2})$ (f) $(\sqrt{3}, -1)$
  (g) $(-\sqrt{3}, 1)$

3. $r\cos\theta = 4$
5. $\theta = \arctan\frac{1}{3}$
7. $r^2\cos\theta\sin\theta = 1$
9. $r = \tan\theta\sec\theta$
11. $x^2 + y^2 = 9$
13. $x^2 + 5x + y^2 = 0$
15. $(x^2 + y^2)^3 = 4y^4$
17. $x = 3$
19. $x = 2$; $r = 2\sec\theta$
21. $x + y = 1$; $r = \dfrac{1}{\cos\theta + \sin\theta}$
23. $y = x + 2$; $r = \dfrac{2}{\sin\theta - \cos\theta}$
25. $x^2 + y^2 + 8y = 0$; $r = -8\sin\theta$
27. $x^2 + y^2 = 2x + 2y$; $r = 2(\cos\theta + \sin\theta)$
29. Fig. 10.2.21
31. Fig. 10.2.15
33. Fig. 10.2.19
35. Fig. 10.2.18
37. Fig. 10.2.16
39. Symmetric about the x-axis
41. Symmetric about the x-axis
43. Symmetric about the x-axis
45. Symmetric about the origin
47. Symmetric about both axes and the origin
49. Symmetric about the x-axis
51. Symmetric about the y-axis
53. No points of intersection
55. $(0, 0)$, $(\frac{1}{2}, \pi/6)$, $(\frac{1}{2}, 5\pi/6)$, $(1, \pi/2)$
57. The pole, the point $(r, \theta) = (2, \pi)$, and the two points $r = 2(\sqrt{2} - 1)$, $|\theta| = \arccos(3 - 2\sqrt{2})$
59. (a) $r\cos(\theta - \alpha) = p$

Section 10.3 (page 554)

1. π **3.** $3\pi/2$ **5.** $9\pi/2$
7. 4π **9.** $19\pi/2$
11. $\pi/2$ (one of *four* loops)
13. $\pi/4$ (one of *eight* loops)
15. 2 (one of *two* loops)   **17.** 4 (one of *two* loops)
19. $\frac{1}{6}(2\pi + 3\sqrt{3})$ **21.** $\frac{1}{24}(5\pi - 6\sqrt{3})$
23. $\frac{1}{6}(39\sqrt{3} - 10\pi)$ **25.** $\frac{1}{2}(2 - \sqrt{2})$
27. $\frac{1}{6}(20\pi + 21\sqrt{3})$ **29.** $\frac{1}{2}(2 + \pi)$ **31.** $\pi/2$

Section 10.4 (page 559)

1. $y^2 = 12x$ **3.** $(x - 2)^2 = -8(y - 3)$
5. $(y - 3)^2 = -8(x - 2)$
7. $x^2 = -6(y + \frac{3}{2})$ **9.** $x^2 = 4(y + 1)$
11. $y^2 = 12x$; vertex $(0, 0)$, axis the x-axis
13. $y^2 = -6x$; vertex $(0\ 0)$, axis the x-axis
15. $x^2 - 4x - 4y = 0$; vertex $(2, -1)$, axis the line $x = 2$

17. $4y = -12 - (2x + 1)^2$; vertex $(-0.5, -3)$, axis the line $x = -0.5$

23. About 0.693 days; that is, about 16 h 38 min

27. $\alpha = \frac{1}{2}\arcsin(0.49) \approx 0.256045$ (14° 40′ 13″), $\alpha = \frac{1}{2}(\pi - \arcsin(0.49)) \approx 1.314751$ (75° 19′ 47″)

29. *Suggestion:* $x^2 - 2xy + y^2 - 2ax - 2ay + a^2 = 0$

Section 10.5 (page 565)

1. $\left(\dfrac{x}{4}\right)^2 + \left(\dfrac{y}{5}\right)^2 = 1$ **3.** $\left(\dfrac{x}{15}\right)^2 + \left(\dfrac{y}{17}\right)^2 = 1$

5. $\dfrac{x^2}{16} + \dfrac{y^2}{7} = 1$ **7.** $\dfrac{x^2}{100} + \dfrac{y^2}{75} = 1$

9. $\dfrac{x^2}{16} + \dfrac{y^2}{12} = 1$

11. $\dfrac{(x-2)^2}{16} + \dfrac{(y-3)^2}{4} = 1$

13. $\dfrac{(x-1)^2}{25} + \dfrac{(y-1)^2}{16} = 1$

15. $\dfrac{(x-1)^2}{81} + \dfrac{(y-2)^2}{72} = 1$

17. Center $(0, 0)$, foci $(\pm 2\sqrt{5}, 0)$, major axis 12, minor axis 8

19. Center $(0, 4)$, foci $(0, 4 \pm \sqrt{5})$, major axis 6, minor axis 4

21. About 3466.36 AU—that is, about 3.22×10^{11} mi, or about 20 light-days

27. $\dfrac{(x-1)^2}{4} + \dfrac{y^2}{16/3} = 1$

Section 10.6 (page 571)

1. $\dfrac{x^2}{1} - \dfrac{y^2}{15} = 1$ **3.** $\dfrac{x^2}{16} - \dfrac{y^2}{9} = 1$

5. $\dfrac{y^2}{25} - \dfrac{x^2}{25} = 1$ **7.** $\dfrac{y^2}{9} - \dfrac{x^2}{27} = 1$

9. $\dfrac{x^2}{4} - \dfrac{y^2}{12} = 1$ **11.** $\dfrac{(x-2)^2}{9} - \dfrac{(y-2)^2}{27} = 1$

13. $\dfrac{(y+2)^2}{9} - \dfrac{(x-1)^2}{4} = 1$

15. Center $(1, 2)$, foci $(1 \pm \sqrt{2}, 2)$, asymptotes $y - 2 = \pm(x - 1)$

17. Center $(0, 3)$, foci $(0, 3 \pm 2\sqrt{3})$, asymptotes $y = 3 \pm x\sqrt{3}$

19. Center $(-1, 1)$, foci $(-1 \pm \sqrt{13}, 1)$, asymptotes $y = \frac{1}{2}(3x + 5)$, $y = -\frac{1}{2}(3x + 1)$

21. There are no points on the graph if $c > 15$.

25. $16x^2 + 50xy + 16y^2 = 369$

27. About 16.42 mi north of B and 8.66 mi west of B; that is, about 18.56 mi from B at a bearing of 27°48′ west of north.

Section 10.7 (page 575)

1. $2(x')^2 + (y')^2 = 4$: ellipse; origin $(2, 3)$

3. $9(x')^2 - 16(y')^2 = 144$: hyperbola; origin $(1, -1)$

5. The single point $(2, 3)$

7. Ellipse; 45°, $4(x')^2 + 2(y')^2 = 1$

9. The two parallel lines (a "degenerate parabola") $(x')^2 = 4$; $\tan^{-1}(\frac{1}{2}) \approx 26.57°$

11. Hyperbola; $\tan^{-1}(\frac{1}{3}) \approx 18.43°$, $(x')^2 - (y')^2 = 1$

13. Ellipse; $\tan^{-1}(\frac{2}{3}) \approx 33.69°$, $2(x')^2 + (y')^2 = 2$

15. Ellipse; $\tan^{-1}(\frac{4}{3}) \approx 53.13°$, $4(x')^2 + (y')^2 = 4$

17. Ellipse; $\tan^{-1}(\frac{1}{4}) \approx 14.04°$, $2(x')^2 + (y')^2 = 4$

19. The two perpendicular lines $y' = x'$ and $y' = -x'$ (a "degenerate hyperbola"); $\tan^{-1}(\frac{5}{12}) \approx 22.62°$

21. Ellipse, $\tan^{-1}(\frac{4}{3}) \approx 53.13°$, $25(x' - 1)^2 + 50(y')^2 = 50$

23. Hyperbola, $\tan^{-1}(\frac{4}{3}) \approx 53.13°$, $2(y' - 1)^2 - (x' - 2)^2 = 1$

25. Hyperbola, $\tan^{-1}(\frac{8}{15}) \approx 28.07°$, $(x' - 1)^2 - (y')^2 = 1$

Chapter 10 Miscellaneous Problems (page 577)

1. Circle, center $(1, 1)$, radius 2

3. Circle, center $(3, -1)$, radius 1

5. Parabola, vertex $(4, -2)$, opening downward

7. Ellipse, center $(2, 0)$, major axis 6, minor axis 4

9. Hyperbola, center $(-1, 1)$, vertical axis, foci at $(-1, 1 \pm \sqrt{3})$

11. There are no points on the graph.

13. Hyperbola, axis inclined at 22.5° from the horizontal

15. Ellipse, center at the origin, major axis $2\sqrt{2}$, minor axis 1, rotated through the angle $\alpha = \pi/4$

17. Parabola, vertex $(0, 0)$, opening to the "northeast," axis at angle $\alpha = \tan^{-1}(\frac{3}{4})$ from the horizontal

19. Circle, center $(1, 0)$, radius 1

21. Straight line $y = x + 1$

23. Horizontal line $y = 3$

25. Two ovals tangent to each other and to the y-axis at $(0, 0)$

27. Apple-shaped curve, symmetric about the y-axis

29. Ellipse, one focus at $(0, 0)$, directrix $x = 4$, eccentricity $e = 0.5$

31. $\frac{1}{2}(\pi - 2)$ **33.** $\frac{1}{6}(39\sqrt{3} - 10\pi) \approx 6.02234$

35. 2   **37.** $5\pi/4$ **39.** $r = 2p \cos(\theta - \alpha)$

41. If $a > b$, the maximum is $2a$ and the minimum is $2b$.

43. $b^2 y = 4hx(b - x)$; alternatively,

$$r = b \sec\theta - \frac{b^2}{4h} \sec\theta \tan\theta$$

45. *Suggestion:* Let θ be the angle that QR makes with the

$$r = b \sec\theta - \frac{b^2}{4h} \sec\theta \tan\theta$$

x-axis.

49. The curve is a hyperbola with one focus at the origin, directrix $x = -\frac{3}{2}$, and eccentricity $e = 2$.

51. $\frac{3}{2}$

Section 11.2   (page 587)

1. $\frac{2}{5}$ **3.** 0   **5.** 1   **7.** Does not converge

9. 0   **11.** 0   **13.** 0   **15.** 1

17. 0   **19.** 0   **21.** 0   **23.** 0

25. e **27.** e^{-2} **29.** 2   **31.** 1

33. Does not converge

35. 0   **41.** (b) 4

Section 11.3   (page 596)

1. $\frac{3}{2}$ **3.** Diverges   **5.** Diverges

7. 6   **9.** Diverges   **11.** Diverges

13. Diverges   **15.** $\dfrac{\sqrt{2}}{\sqrt{2}-1} = 2 + \sqrt{2}$

17. Diverges   **19.** $\frac{1}{12}$ **21.** $\dfrac{e}{\pi - e}$

23. Diverges   **25.** $\frac{65}{12}$ **27.** $\frac{247}{8}$ **29.** $\frac{1}{4}$

31. $\frac{47}{99}$ **33.** $\frac{41}{333}$ **35.** $\frac{314,156}{99,999}$

37. $S_n = \ln(n + 1)$; diverges

39. $S_n = \dfrac{3}{2} - \dfrac{1}{n} - \dfrac{1}{n+1}$; the sum is $\frac{3}{2}$.

45. Computations with S are meaningless because S is not a number.

47. 4.5 s   **49.** (a) $M_n = (0.95)^n M_0$; (b) 0

51. Peter $\frac{4}{7}$, Paul $\frac{2}{7}$, Mary $\frac{1}{7}$ **53.** $\frac{1}{12}$

Section 11.4   (page 610)

1. $e^{-x} = 1 - \dfrac{x}{1!} + \dfrac{x^2}{2!} - \dfrac{x^3}{3!} + \dfrac{x^4}{4!} - \dfrac{x^5}{5!} + \left(\dfrac{e^{-z}}{6!}\right)x^6$ for some z between 0 and x.

3. $\cos x = 1 - \dfrac{x^2}{2!} + \dfrac{x^4}{4!} - \left(\dfrac{\sin z}{5!}\right)x^5$ for some z between 0 and x.

5. $\sqrt{1 + x} = 1 + \dfrac{x}{1!2} - \dfrac{x^2}{2!4} + \dfrac{3x^3}{3!8} - \dfrac{5x^4}{128}(1 + z)^{-7/2}$ for some z between 0 and x.

7. $\tan x = \dfrac{x}{1!} + \dfrac{2x^3}{3!}$

$+ \dfrac{16 \sec^4 z \tan z + 8 \sec^2 z \tan^3 z}{4!}x^4$ for some z between 0 and x.

9. $\sin^{-1} z = \dfrac{x}{1!} + \dfrac{x^3}{3!}\cdot\dfrac{1 + 2z^2}{(1 - z^2)^{5/2}}$ for some z between 0 and x.

11. $e^x = e + \dfrac{e}{1!}(x - 1) + \dfrac{e}{2!}(x - 1)^2 + \dfrac{e}{3!}(x - 1)^3$

$+ \dfrac{e}{4!}(x - 1)^4 + \dfrac{e^z}{5!}(x - 1)^5$ for some z between 1 and x.

13. $\sin x = \dfrac{1}{2} + \dfrac{\sqrt{3}}{1!2}\left(x - \dfrac{\pi}{6}\right) - \dfrac{1}{2!2}\left(x - \dfrac{\pi}{6}\right)^2$

$- \dfrac{\sqrt{3}}{3!2}\left(x - \dfrac{\pi}{6}\right)^3 + \dfrac{\sin z}{4!2}\left(x - \dfrac{\pi}{6}\right)^4$ for some z between $\pi/6$ and x.

15. $\dfrac{1}{(x - 4)^2} = 1 - 2(x - 5) + 3(x - 5)^2 - 4(x - 5)^3 + 5(x - 5)^4 - 6(x - 5)^5 + \dfrac{7}{(z - 4)^8}(x - 5)^6$ for some z between 5 and x.

17. $\cos x = -1 + \dfrac{(x - \pi)^2}{2!} - \dfrac{(x - \pi)^4}{4!} - \dfrac{(x - \pi)^5}{5!}\sin z$ for some z between π and x.

19. $x^{3/2} = 1 + \dfrac{3(x - 1)}{2} + \dfrac{3(x - 1)^2}{8} - \dfrac{(x - 1)^3}{16}$

$+ \dfrac{3(x - 1)^4}{128} - \dfrac{3(x - 1)^5}{256 z^{7/2}}$ for some z between 1 and x.

21. $e^{-x} = 1 - x + \dfrac{x^2}{2!} - \dfrac{x^3}{3!} + \dfrac{x^4}{4!} - \cdots$

23. $e^{-3x} = 1 - 3x + \dfrac{9x^2}{2} - \dfrac{27x^3}{3!} + \dfrac{81x^4}{4!} - \cdots$

25. $\cos 2x = 1 - \dfrac{4x^2}{2!} + \dfrac{16x^4}{4!} - \dfrac{64x^6}{6!} + \dfrac{256x^8}{8!} - \cdots$

27. $\sin x^2 = x^2 - \dfrac{x^6}{3!} + \dfrac{x^{10}}{5!} - \dfrac{x^{14}}{7!} + \dfrac{x^{18}}{9!} - \cdots$

29. $\ln(1 + x) = x - \frac{1}{2}x^2 + \frac{1}{3}x^3 - \frac{1}{4}x^4 + \frac{1}{5}x^5 - \cdots$

31. $e^{-x} = 1 - x + \dfrac{x^2}{2!} - \dfrac{x^3}{3!} + \dfrac{x^4}{4!} - \cdots$

A-58

33. $\ln x = (x - 1) - \frac{1}{2}(x - 1)^2 + \frac{1}{3}(x - 1)^3$
$- \frac{1}{4}(x - 1)^4 + \cdots$

35. $\cos x = \frac{\sqrt{2}}{2} - \frac{\sqrt{2}}{1!\,2}\left(x - \frac{\pi}{2}\right) - \frac{\sqrt{2}}{2!\,2}\left(x - \frac{\pi}{2}\right)^2$
$+ \frac{\sqrt{2}}{3!\,2}\left(x - \frac{\pi}{2}\right)^3 + \frac{\sqrt{2}}{4!\,2}\left(x - \frac{\pi}{2}\right)^4 - \frac{\sqrt{2}}{5!\,2}\left(x - \frac{\pi}{2}\right)^5$
$- \frac{\sqrt{2}}{6!\,2}\left(x - \frac{\pi}{2}\right)^6 + \frac{\sqrt{2}}{7!\,2}\left(x - \frac{\pi}{2}\right)^7 + \cdots$

37. $\sinh x = x + \frac{x^3}{3!} + \frac{x^5}{5!} + \frac{x^7}{7!} + \frac{x^9}{9!} + \cdots$

39. $\frac{1}{x} = 1 - (x - 1) + (x - 1)^2 - (x - 1)^3$
$+ (x - 1)^4 - \cdots$

41. $\sin x = \frac{\sqrt{2}}{2}\left[1 + \left(x - \frac{\pi}{4}\right) - \frac{(x - \pi/4)^2}{2!}\right.$
$\left. - \frac{(x - \pi/4)^3}{3!} + \frac{(x - \pi/4)^4}{4!} + \cdots \right]$

Section 11.5 (page 617)

1. Diverges **3.** Diverges **5.** Converges
7. Diverges **9.** Converges **11.** Converges
13. Converges **15.** Converges **17.** Diverges
19. Converges **21.** Diverges **23.** Diverges
25. Converges **27.** Converges
29. The terms are not nonnegative.
31. The terms are not decreasing.
33. $n = 100$
37. With $n = 6$, $1.0368 < S < 1.0370$. The true value of the sum is not known exactly but is $1.03692\,77551\,4337$ to the accuracy shown.
39. About a million centuries
41. With $n = 10$, $S_{10} \approx 1.08203\,658$ and $3.141566 < \pi < 3.141627$.

Section 11.6 (page 623)

1. Converges **3.** Diverges **5.** Converges
7. Diverges **9.** Converges **11.** Converges
13. Diverges **15.** Converges **17.** Converges
19. Converges **21.** Converges **23.** Converges
25. Converges **27.** Diverges **29.** Diverges
31. Converges **33.** Converges **35.** Diverges

Section 11.7 (page 631)

1. Converges **3.** Converges **5.** Converges
7. Converges **9.** Diverges

11. Converges absolutely
13. Converges conditionally
15. Converges absolutely **17.** Converges absolutely
19. Diverges **21.** Converges absolutely
23. Converges conditionally **25.** Diverges
27. Diverges **29.** Converges absolutely
31. Converges absolutely
33. $n = 1999$ **35.** $n = 6$
37. The first six terms give the estimate 0.6065.
39. The first three terms give the estimate 0.0953.
41. Using the first ten terms: $S_{10} \approx 0.81796\,22$ and $3.1329 < \pi < 3.1488$.

Section 11.8 (page 641)

1. $[-1, 1)$ **3.** $(-1, 1)$ **5.** $[-1, 1]$
7. $(0.4, 0.8)$ **9.** $[2.5, 3.5]$
11. Converges only for $x = 0$
13. $(-4, 2)$ **15.** $[2, 4]$
17. Converges only for $x = 5$ **19.** $(-1, 1)$
21. $f(x) = x^2 - 3x^3 + \frac{9x^4}{2!} - \frac{27x^5}{3!} + \frac{81x^6}{4!} - \cdots$;
$R = +\infty$.
23. $f(x) = x^2 - \frac{x^6}{3!} + \frac{x^{10}}{5!} - \frac{x^{14}}{7!} + \frac{x^{18}}{9!} - \cdots$;
$R = +\infty$
25. $(1 - x)^{1/3} = 1 - \frac{x}{3} - \frac{2x^2}{2!\,3^2} - \frac{2 \cdot 5x^3}{3!\,3^3} - \frac{2 \cdot 5 \cdot 8x^4}{4!\,3^4}$
$- \cdots$; $R = 1$
27. $f(x) = 1 - 3x + 6x^2 - 10x^3 + 15x^4 - \cdots$;
$R = 1$
29. $f(x) = 1 - \frac{1}{2}x + \frac{1}{3}x^2 - \frac{1}{4}x^3 + \frac{1}{5}x^4 - \cdots$; $R = 1$
31. $f(x) = \frac{x^4}{4} - \frac{x^{10}}{3!\,10} + \frac{x^{16}}{5!\,16} - \frac{x^{22}}{7!\,22} + \cdots$
$= \sum_{n=0}^{\infty} \frac{(-1)^n}{(2n + 1)!\,(6n + 4)} x^{6n+4}$
33. $f(x) = x - \frac{x^4}{4} + \frac{x^7}{2!\,7} - \frac{x^{10}}{3!\,10} + \frac{x^{13}}{4!\,13} - \cdots$
$= \sum_{n=0}^{\infty} \frac{(-1)^n}{n!\,(3n + 1)} x^{3n+1}$
35. $f(x) = x - \frac{x^3}{2!\,3} + \frac{x^5}{3!\,5} - \frac{x^7}{4!\,7} + \frac{x^9}{5!\,9} - \cdots$
$= \sum_{n=1}^{\infty} \frac{(-1)^{n+1}}{n!\,(2n - 1)} x^{2n-1}$
37. Using six terms: $3.14130\,878 < \pi < 3.14167\,44$

Section 11.9 (page 647)

1. $65^{1/3} = (4 + \frac{1}{64})^{1/3}$. The first four terms of the binomial series give $65^{1/3} \approx 4.020726$; answer: 4.021.

3. Three terms of the usual sine series give 0.479427 with error less than 0.000002; answer: 0.479.

5. Five terms of the usual arctangent series give 0.463684 with error less than 0.000045; answer: 0.464.

7. 0.309   **9.** 0.174   **11.** 0.946   **13.** 0.487

15. 0.0976   **17.** 0.444   **19.** 0.747   **21.** −0.5

23. 0.5   **25.** 0

31. The first five coefficients are $1, 0, \frac{1}{2}, 0$, and $\frac{5}{24}$.

Chapter 11 Miscellaneous Problems (page 650)

1. 1   **3.** 10   **5.** 0   **7.** 0   **9.** No limit

11. 0   **13.** $+\infty$ (or "no limit")   **15.** 1

17. Converges   **19.** Converges   **21.** Converges

23. Diverges   **25.** Converges   **27.** Converges

29. Diverges   **31.** $(-\infty, +\infty)$

33. $[-2, 4)$ **35.** $[-1, 1]$

37. Converges only for $x = 0$ **39.** $(-\infty, +\infty)$

41. Converges for *no x*   **43.** Converges for all x

51. Seven terms of the binomial series give 1.084.

53. 0.461   **55.** 0.797

Section 12.1 (page 659)

1. $y = 2x - 3$ **3.** $y^2 = x^3$

5. $y = 2x^2 - 5x + 2$ **7.** $y = 4x^2, x > 0$

9. $\left(\frac{x}{5}\right)^2 + \left(\frac{y}{3}\right)^2 = 1$ **11.** $x^2 - y^2 = 1$

13. (a) $y - 5 = \frac{9}{4}(x - 3)$; (b) $\frac{d^2y}{dx^2} = \frac{9}{16}t^{-1}$, concave upward at $t = 1$

15. (a) $y = -\frac{1}{2}\pi(x - \frac{1}{2}\pi)$; (b) concave downward

17. (a) $x + y = 3$; (b) concave downward

19. $\psi = \pi/6$ (constant)   **21.** $\psi = \pi/2$

25. $x = \frac{p}{m^2}, y = \frac{2p}{m}$

Section 12.2 (page 666)

1. $\frac{22}{5}$ **3.** $\frac{4}{3}$ **5.** $\frac{1}{2}(1 + e^\pi) \approx 12.0703$

7. $\frac{358}{35}\pi \approx 32.13400$ **9.** $\frac{16}{15}\pi \approx 3.35103$

11. $\frac{74}{3}$ **13.** $\frac{1}{4}\pi\sqrt{2} \approx 1.11072$

15. $(e^{2\pi} - 1)\sqrt{5} \approx 1195.1597$

17. $\frac{8}{3}\pi(5\sqrt{5} - 2\sqrt{2}) \approx 69.96882$

19. $\frac{2}{27}\pi(13\sqrt{13} - 8) \approx 9.04596$

21. $16\pi^2 \approx 157.91367$ **23.** $5\pi^2 a^3$

25. (a) $A = \pi ab$; (b) $V = \frac{4}{3}\pi ab^2$

27. $\pi\sqrt{1 + 4\pi^2} + \frac{1}{2}\ln(2\pi + \sqrt{1 + 4\pi^2}) \approx 21.25629$

29. $\frac{3}{8}\pi a^2$ **31.** $\frac{12}{5}\pi a^2$ **33.** $\frac{216}{5}\sqrt{3} \approx 74.8246$

35. $\frac{243}{4}\pi\sqrt{3} \approx 330.5649$ **39.** $6\pi^3 a^3$

Section 12.3 (page 675)

1. $\sqrt{5}, 2\sqrt{13}, 4\sqrt{2}, \langle -2, 0 \rangle, \langle 9, -10 \rangle$; no

3. $2\sqrt{2}, 10, \sqrt{5}, \langle -5, -6 \rangle, \langle 0, 2 \rangle$; no

5. $\sqrt{10}, 2\sqrt{29}, \sqrt{65}; 3\mathbf{i} - 2\mathbf{j}; -\mathbf{i} + 19\mathbf{j}$; no

7. $4, 14, \sqrt{65}, 4\mathbf{i} - 7\mathbf{j}, 12\mathbf{i} + 14\mathbf{j}$; yes

9. $\mathbf{u} = \langle -\frac{3}{5}, -\frac{4}{5} \rangle; \mathbf{v} = \langle \frac{3}{5}, \frac{4}{5} \rangle$

11. $\mathbf{u} = \frac{8}{17}\mathbf{i} + \frac{15}{17}\mathbf{j}; \mathbf{v} = -\frac{8}{17}\mathbf{i} - \frac{15}{17}\mathbf{j}$

13. $-4\mathbf{j}$ **15.** $8\mathbf{i} - 14\mathbf{j}$ **17.** Yes   **19.** No

21. (a) $15\mathbf{i} - 21\mathbf{j}$; (b) $\frac{5}{3}\mathbf{i} - \frac{7}{3}\mathbf{j}$

23. (a) $\frac{35}{58}\mathbf{i}\sqrt{58} - \frac{15}{58}\mathbf{j}\sqrt{58}$; (b) $-\frac{40}{89}\mathbf{i}\sqrt{89} - \frac{25}{89}\mathbf{j}\sqrt{89}$

25. $c = 0$ **33.** $\mathbf{v}_a = (500 + 25\sqrt{2})\mathbf{i} + 25\mathbf{j}\sqrt{2}$

35. $\mathbf{v}_a = -225\mathbf{i}\sqrt{2} + 275\mathbf{j}\sqrt{2}$

41. The question makes no sense.

Section 12.4 (page 682)

1. $0, 0$ **3.** $2\mathbf{i} - \mathbf{j}, 4\mathbf{i} + \mathbf{j}$ **5.** $6\pi\mathbf{i}, 12\pi^2\mathbf{j}$

7. $\mathbf{j}, \mathbf{i}$ **9.** $\frac{1}{2}(2 - \sqrt{2})\mathbf{i} + \mathbf{j}\sqrt{2}$ **11.** $\frac{484}{15}\mathbf{i}$

13. 11   **15.** 0   **17.** $\mathbf{i} + 2t\mathbf{j}, t\mathbf{i} + t^2\mathbf{j}$

19. $\frac{1}{2}t^2\mathbf{i} + \frac{1}{3}t^3\mathbf{j}; (1 + \frac{1}{6}t^3)\mathbf{i} + \frac{1}{12}t^4\mathbf{j}$

21. $v_0 \approx 411.047$ ft/s

25. (a) 100 ft, $400\sqrt{3}$ ft; (b) 200, 800; (c) 300, $400\sqrt{3}$

27. $140\sqrt{5} \approx 313$ (m/s)

29. Inclination angle
$$\alpha = \arctan\left(\frac{8049 - 280\sqrt{10}}{8000}\right) \approx 0.730293 \text{ (about}$$
$41° 50'34''$); initial velocity
$$v_0 = \frac{5600}{(20\sqrt{10} - 7)\cos\alpha} \approx 133.64595 \text{ m/s}$$

35. Begin with $\frac{d}{dt}(\mathbf{v}\cdot\mathbf{v}) = 0$.

37. A repulsive force acting directly away from the origin, with magnitude proportional to distance from the origin

Section 12.5 (page 689)

1. $\mathbf{v} = a\mathbf{u}_\theta, \mathbf{a} = -a\mathbf{u}_r$

3. $\mathbf{v} = \mathbf{u}_r + t\mathbf{u}_\theta, \mathbf{a} = -t\mathbf{u}_r + 2\mathbf{u}_\theta$

5. $\mathbf{v} = (12\cos 4t)\mathbf{u}_r + (6\sin 4t)\mathbf{u}_\theta$,
$\mathbf{a} = (-60\sin 4t)\mathbf{u}_r + (48\cos 4t)\mathbf{u}_\theta$

9. 36.65 mi/s, 24.13 mi/s

11. 0.672 mi/s, 0.602 mi/s

13. About -795 mi—thus it's not possible.

15. About 1.962 h

Chapter 12 Miscellaneous Problems   (page 690)

1. The straight line $y = x + 2$

3. The circle $(x - 2)^2 + (y - 1)^2 = 1$

5. Equation: $y^2 = (x - 1)^3$

7. $y - 2\sqrt{2} = -\frac{4}{3}(x - \frac{3}{2}\sqrt{2})$ **9.** $2\pi y + 4x = \pi^2$

11. 24   **13.** 3π **15.** $\frac{1}{27}(13\sqrt{13} - 8) \approx 1.4397$

17. $\frac{43}{6}$ **19.** $\frac{1}{8}(4\pi - 3\sqrt{3}) \approx 0.92128$

21. $\frac{471,295}{1024}\pi \approx 1445.915$

23. $\frac{1}{2}\pi\sqrt{5}(e^\pi + 1) \approx 84.7919$

25. $x = a\theta - b\sin\theta,\ y = a - b\cos\theta$

27. *Suggestion:* Compute $r^2 = x^2 + y^2$.   **29.** $6\pi^3 a^3$

35. Two solutions: $\alpha \approx 0.033364$ rad (about $1°54'53''$) and $\alpha \approx 1.29116$ rad (about $73°58'40''$)

Section 13.1   (page 700)

1. (a) $\langle 5, 8, -11\rangle$; (b) $\langle 2, 23, 0\rangle$; (c) 4; (d) $\sqrt{51}$; (e) $\frac{1}{15}\sqrt{5}\langle 2, 5, -4\rangle$

3. (a) $2\mathbf{i} + 3\mathbf{j} + \mathbf{k}$; (b) $3\mathbf{i} - \mathbf{j} + 7\mathbf{k}$; (c) 0; (d) $\sqrt{5}$; (e) $\frac{1}{3}\sqrt{3}\,(\mathbf{i} + \mathbf{j} + \mathbf{k})$

5. (a) $4\mathbf{i} - \mathbf{j} - 3\mathbf{k}$; (b) $6\mathbf{i} - 7\mathbf{j} + 12\mathbf{k}$; (c) -1; (d) $\sqrt{17}$; (e) $\frac{1}{5}\sqrt{5}(2\mathbf{i} - \mathbf{j})$

7. $\theta = \cos^{-1}(-\frac{13}{50}\sqrt{10}) \approx 2.536$

9. $\theta = \cos^{-1}\left(-\dfrac{34}{\sqrt{3154}}\right) \approx 2.221$

11. $\text{comp}_b\,\mathbf{a} = \frac{2}{7}\sqrt{14},\ \text{comp}_a\,\mathbf{b} = \frac{4}{15}\sqrt{5}$

13. $\text{comp}_b\,\mathbf{a} = 0,\ \text{comp}_a\,\mathbf{b} = 0$

15. $\text{comp}_b\,\mathbf{a} = -\frac{1}{10}\sqrt{10},\ \text{comp}_a\,\mathbf{b} = -\frac{1}{5}\sqrt{5}$

17. $(x + 2)^2 + (y - 1)^2 + (z + 5)^2 = 7$

19. $(x - 4)^2 + (y - 5)^2 + (z + 2)^2 = 38$

21. Center $(-2, 3, 0)$, radius $\sqrt{13}$

23. Center $(0, 0, 3)$, radius 5

25. A plane perpendicular to the z-axis at $z = 10$

27. All points in the three coordinate planes

29. The point $(1, 0, 0)$

31. $\alpha = \cos^{-1}(\frac{1}{9}\sqrt{6}) \approx 74.21°,\ \beta = \gamma = \cos^{-1}(\frac{5}{18}\sqrt{6}) \approx 47.12°$

33. $\alpha = \cos^{-1}(\frac{3}{10}\sqrt{2}) \approx 64.90°,\ \beta = \cos^{-1}(\frac{2}{5}\sqrt{2}) \approx 55.55°,\ \gamma = \cos^{-1}(\frac{1}{2}\sqrt{2}) = 45°$

35. 48

37. If there's no friction, the work done is *mgh*.

41. $A = \frac{3}{2}\sqrt{69} \approx 12.46$

43. $\cos^{-1}(\frac{1}{3}\sqrt{3}) \approx 0.9553$, about $54.7356°$

49. $2x + 9y - 5z = 23$; the plane through the midpoint of the segment AB perpendicular to AB

51. $60°$

Section 13.2   (page 708)

1. $\langle 0, -14, 7\rangle$ **3.** $\langle -10, -7, 1\rangle$

7. $(\mathbf{a} \times \mathbf{b}) \times \mathbf{c} = \langle -1, 1, 0\rangle,\ \mathbf{a} \times (\mathbf{b} \times \mathbf{c}) = \langle 0, 0, -1\rangle$

11. $A = \frac{1}{2}\sqrt{2546} \approx 25.229$

13. (a) 55; (b) $\frac{55}{6}$ **15.** 4395.657 (m²)

17. 31,271.643 (ft²)   **21.** (b) $\frac{1}{38}\sqrt{9842} \approx 2.6107$

Section 13.3   (page 715)

1. $x = t,\ y = 2t,\ z = 3t$

3. $x = 4 + 2t,\ y = 13,\ z = -3 - 3t$

5. $x = 3 + 3t,\ y = 5 - 13t,\ z = 7 + 3t$

7. $x + 2y + 3z = 0$

9. $x - z + 8 = 0$

11. $x = 2 + t,\ y = 3 - t,\ z = -4 - 2t$; $x - 2 = -y + 3 = \dfrac{-z - 4}{2}$

13. $x = 1,\ y = 1,\ z = 1 + t$; $x - 1 = 0 = y - 1$, z arbitrary

15. $x = 2 + 2t,\ y = -3 - t,\ z = 4 + 3t$; $\dfrac{x - 2}{2} = -y - 3 = \dfrac{z - 4}{3}$

17. $y = 7$ **19.** $7x + 11y = 114$

21. $3x + 4y - z = 0$ **23.** $2x - y - z = 0$

25. $\theta = \cos^{-1}\left(\dfrac{1}{\sqrt{3}}\right) \approx 54.736°$

27. The planes are parallel: $\theta = 0$.

29. $\dfrac{x - 3}{2} = y - 3 = \dfrac{-z + 1}{5}$

31. $3x + 2y + z = 6$ **33.** $7x - 5y - 2z = 9$

35. $x - 2y + 4z = 3$ **37.** $\frac{10}{3}\sqrt{3}$

41. (b) $\dfrac{133}{\sqrt{501}} \approx 5.942$

Section 13.4   (page 720)

1. $\mathbf{v} = 5\mathbf{k},\ \mathbf{a} = \mathbf{0},\ v = 5$

3. $\mathbf{v} = \mathbf{i} + 2t\mathbf{j} + 3t^2\mathbf{k},\ \mathbf{a} = 2\mathbf{j} + 6t\mathbf{k},\ v = \sqrt{1 + 4t^2 + 9t^4}$

5. $\mathbf{v} = \mathbf{i} + 3e^t\mathbf{j} + 4e^t\mathbf{k},\ \mathbf{a} = 3e^t\mathbf{j} + 4e^t\mathbf{k},\ v = \sqrt{1 + 25e^t}$

7. $\mathbf{v} = (-3 \sin t)\mathbf{i} + (3 \cos t)\mathbf{j} - 4\mathbf{k}$, $\mathbf{a} = (-3 \cos t)\mathbf{i} - (3 \sin t)\mathbf{j}$, $v = 5$

9. $\mathbf{r}(t) = t^2\mathbf{i} + 10t\mathbf{j} - 2t^2\mathbf{k}$

11. $\mathbf{r}(t) = 2\mathbf{i} + t^2\mathbf{j} + (5t - t^3)\mathbf{k}$

13. $\mathbf{r}(t) = (10 + \frac{1}{6}t^3)\mathbf{i} + (10t + \frac{1}{12}t^4)\mathbf{j} + \frac{1}{20}t^5\mathbf{k}$

15. $\mathbf{r}(t) = (1 - t - \cos t)\mathbf{i} + (1 + t - \sin t)\mathbf{j} + 5t\mathbf{k}$

17. $\mathbf{v} = 3\sqrt{2}(\mathbf{i} + \mathbf{j}) + 8\mathbf{k}$, $v = 10$, $\mathbf{a} = 6\sqrt{2}(-\mathbf{i} + \mathbf{j})$

21. *Suggestion:* Compute $\dfrac{d}{dt}(\mathbf{r} \cdot \mathbf{r})$.

23. 100 ft, $\sqrt{6425} \approx 80.16$ ft/s

27. $9 \cdot (2 - \sqrt{3})$ ft

Section 13.5 (page 731)

1. 10π **3.** $19(e - 1) \approx 32.647$

5. $2 + \frac{9}{10} \ln 3 \approx 2.9888$ **7.** 0   **9.** 1

11. $\dfrac{40\sqrt{2}}{41\sqrt{41}} \approx 0.2155$ **13.** At $(-\frac{1}{2} \ln 2, \frac{1}{2}\sqrt{2})$

15. Maximum curvature $\frac{5}{9}$ at $(\pm 5, 0)$, minimum curvature $\frac{3}{25}$ at $(0, \pm 3)$

17. $\mathbf{T} = \frac{1}{10}\sqrt{10}(\mathbf{i} + 3\mathbf{j})$, $\mathbf{N} = \frac{1}{10}\sqrt{10}(3\mathbf{i} - \mathbf{j})$

19. $\mathbf{T} = \frac{1}{57}\sqrt{57}(3\mathbf{i} - 4\mathbf{j}\sqrt{3})$, $\mathbf{N} = \frac{1}{57}\sqrt{57}(4\mathbf{i}\sqrt{3} + 3\mathbf{j})$

21. $\mathbf{T} = -\frac{1}{2}\sqrt{2}(\mathbf{i} + \mathbf{j})$, $\mathbf{N} = \frac{1}{2}\sqrt{2}(\mathbf{i} - \mathbf{j})$

23. $a_T = 18t(9t^2 + 1)^{-1/2}$, $a_N = 6(9t^2 + 1)^{-1/2}$

25. $a_T = t(1 + t^2)^{-1/2}$, $a_N = (2 + t^2)(1 + t^2)^{-1/2}$

27. $1/a$ **29.** $x^2 + (y - \frac{1}{2})^2 = \frac{1}{4}$

31. $(x - \frac{3}{2})^2 + (y - \frac{3}{2})^2 = 2$ **33.** $\frac{1}{2}$

35. $\frac{1}{3}e^{-t}\sqrt{2}$ **37.** $a_T = 0 = a_N$

39. $a_T = (4t + 18t^3)(1 + 4t^2 + 9t^4)^{-1/2}$, $a_N = 2(1 + 9t^2 + 9t^4)^{1/2}(1 + 4t^2 + 9t^4)^{-1/2}$

41. $a_T = t(t^2 + 2)^{-1/2}$, $a_N = (t^4 + 5t^2 + 8)^{1/2}(t^2 + 2)^{-1/2}$

43. $\mathbf{T} = \frac{1}{2}\sqrt{2}\langle 1, \cos t, -\sin t\rangle$, $\mathbf{N} = \langle 0, -\sin t, -\cos t\rangle$; at $(0, 0, 1)$, $\mathbf{T} = \frac{1}{2}\sqrt{2}\langle 1, 1, 0\rangle$, $\mathbf{N} = \langle 0, 0, -1\rangle$

45. $\mathbf{T} = \frac{1}{3}\sqrt{3}(\mathbf{i} + \mathbf{j} + \mathbf{k})$, $\mathbf{N} = \frac{1}{2}\sqrt{2}(-\mathbf{i} + \mathbf{j})$

47. $x = 2 + \frac{4}{13}s$, $y = 1 - \frac{12}{13}s$, $z = 3 + \frac{3}{13}s$

49. $x(s) = 3 \cos \frac{1}{5}s$, $y(s) = 3 \sin \frac{1}{5}s$, $z(s) = \frac{4}{5}s$

51. Begin with $\dfrac{d}{dt}(\mathbf{v} \cdot \mathbf{v})$.

53. $|t|^{-1}$

55. $A = 3$, $B = -8$, $C = 6$, $D = 0$, $E = 0$, $F = 0$

Section 13.6 (page 741)

1. A plane with intercepts $(\frac{20}{3}, 0, 0)$, $(0, 10, 0)$, and $(0, 0, 2)$

3. A vertical circular cylinder with radius 3

5. A vertical cylinder intersecting the xy-plane in the rectangular hyperbola $xy = 4$

7. An elliptical paraboloid opening upward from its vertex at the origin

9. A circular paraboloid opening downward from its vertex at $(0, 0, 4)$

11. A paraboloid opening upward, vertex at the origin, axis the z-axis

13. A cone, vertex the origin, axis the z-axis (both nappes)

15. A parabolic cylinder perpendicular to the xz-plane, its trace there the parabola opening upward with axis the z-axis and vertex at $(x, z) = (0, -2)$

17. An elliptical cylinder perpendicular to the xy-plane, its trace there the ellipse with center $(0, 0)$ and intercepts $(\pm 1, 0)$ and $(0, \pm 2)$

19. An elliptical cone, vertex $(0, 0, 0)$, axis the x-axis

21. A paraboloid, opening downward, vertex at the origin, axis the z-axis

23. A hyperbolic paraboloid, saddle point at the origin, meeting the xz-plane in a parabola with vertex the origin and opening downward, meeting the xy-plane in a parabola with vertex the origin and opening upward, meeting each plane parallel to the yz-plane in a hyperbola with directrices parallel to the y-axis

25. A hyperboloid of one sheet, axis the z-axis, trace in the xy-plane the circle with center $(0, 0)$ and radius 3, traces in parallel planes larger circles, and traces in planes parallel to the z-axis hyperbolas

27. An elliptic paraboloid, axis the y-axis, vertex at the origin

29. A hyperboloid of two sheets, axis the y-axis

31. A paraboloid, axis the x-axis, vertex at the origin, equation $x = 2(y^2 + z^2)$

33. Hyperboloid of one sheet, equation $x^2 + y^2 - z^2 = 1$

35. A paraboloid, vertex at the origin, axis the x-axis, equation $y^2 + z^2 = 4x$

37. The surface resembles a rug covering a turtle: highest point $(0, 0, 1)$; $z \to 0$ from above as $|x|$ or $|y|$ (or both) increase without bound, equation $z = \exp(-x^2 - y^2)$.

39. A circular cone with axis of symmetry the z-axis

41. Ellipses with semiaxes 2 and 1

43. Circles   **45.** Parabolas opening downward

47. Parabolas opening upward if $k > 0$, downward if $k < 0$

51. The projection of the intersection has equation $x^2 + y^2 = 2y$; it is the circle with center $(0, 1)$ and radius 1.

A-62

53. Equation: $5x^2 + 8xy + 8y^2 - 8x - 8y = 0$. Because the discriminant is negative, it is an ellipse. In a uv-plane rotated approximately $55°16'41''$ from the xy-plane, the ellipse has center $(u, v) = (0.517, -0.453)$, minor axis 0.352 in the u-direction, and major axis 0.774 in the v-direction.

Section 13.7 (page 747)

1. $(0, 0, 5)_{cyl}$, $(5, 0, 0)_{sph}$

3. $\left(\sqrt{2}, \frac{\pi}{4}, 0\right)_{cyl}$, $\left(\sqrt{2}, \frac{\pi}{2}, \frac{\pi}{4}\right)_{sph}$

5. $\left(\sqrt{2}, \frac{\pi}{4}, 1\right)_{cyl}$, $\left(\sqrt{3}, \tan^{-1}\sqrt{2}, \frac{\pi}{4}\right)_{sph}$

7. $(\sqrt{5}, \tan^{-1}\frac{1}{2}, -2)_{cyl}$,
$\left(3, \frac{\pi}{2} + \tan^{-1}(\frac{1}{2}\sqrt{5}), \tan^{-1}\frac{1}{2}\right)_{sph}$

9. $(5, \tan^{-1}\frac{4}{3}, 12)_{cyl}$, $(13, \tan^{-1}\frac{5}{12}, \tan^{-1}\frac{4}{3})_{sph}$

11. A cylinder, radius 5, axis the z-axis

13. The *plane* $y = x$

15. The upper nappe of the cone $x^2 + y^2 = 3z^2$

17. The xy-plane

19. A cylinder, axis the vertical line $x = 0$, $y = 1$; its trace in the xy-plane is the circle $x^2 + (y - 1)^2 = 1$.

21. The vertical plane with trace the line $y = -x$ in the xy-plane

23. The horizontal plane $z = 1$

25. $r^2 + z^2 = 25$, $\rho = 5$

27. $r(\cos\theta + \sin\theta) + z = 1$,
$\rho(\sin\phi\cos\theta + \sin\phi\sin\theta + \cos\phi) = 1$

29. $r^2 + z^2 = r(\cos\theta + \sin\theta) + z$,
$\rho = \sin\phi\cos\theta + \sin\phi\sin\theta + \cos\phi$

31. $z = r^2$

33. (a) $1 \leq r^2 \leq 4 - z^2$; (b) $\csc\phi \leq \rho \leq 2$
(and, as a consequence, $\frac{\pi}{6} \leq \phi \leq \frac{5\pi}{6}$)

35. About 3821 mi $\approx$ 6149 km

37. Just under 50 km $\approx$ 31 mi

39. $0 \leq \rho \leq H \sec\phi$, $0 \leq \phi \leq \arctan\left(\frac{R}{H}\right)$, θ arbitrary

Chapter 13 Miscellaneous Problems (page 750)

7. $x = 1 + 2t$, $y = -1 + 3t$, $z = 2 - 3t$;
$\dfrac{x - 1}{2} = \dfrac{y + 1}{3} = \dfrac{2 - z}{3}$

9. $-13x + 22y + 6z = -23$

11. $x - y + 2z = 3$

15. 3

17. $\frac{1}{9}$; $a_T = 2$, $a_N = 1$

21. $3x - 3y + z = 1$

27. $\rho = 2\cos\phi$

29. $\rho^2 = 2\cos 2\phi$; shaped like an hourglass with rounded ends

37. The curvature is zero when x is an integral multiple of π and reaches the maximum value 1 when x is an odd integral multiple of $\pi/2$.

39. $\mathbf{N} = -\dfrac{2}{\sqrt{\pi^2 + 4}}\,\mathbf{i} - \dfrac{\pi}{\sqrt{\pi^2 + 4}}\,\mathbf{j}$,

$\mathbf{T} = -\dfrac{\pi}{\sqrt{\pi^2 + 4}}\,\mathbf{i} + \dfrac{2}{\sqrt{\pi^2 + 4}}\,\mathbf{j}$

43. $A = \frac{15}{8}$, $B = -\frac{5}{4}$, $C = \frac{3}{8}$

Section 14.2 (page 761)

1. All (x, y)

3. Except on the line $x = y$

5. Except on the coordinate axes $x = 0$ and $y = 0$

7. Except on the lines $y = \pm x$

9. Except at the origin $(0, 0, 0)$

11. The horizontal plane with equation $z = 10$

13. A plane that makes a $45°$ angle with the xy-plane, intersecting it in the line $x + y = 0$

15. A circular paraboloid opening upward from its vertex at the origin

17. The upper hemispherical surface of radius 2 centered at the origin

19. A circular cone opening downward from its vertex at $(0, 0, 10)$

21. Straight lines of slope 1

23. Ellipses centered at $(0, 0)$, each with major axis twice the minor axis and lying on the x-axis

25. Vertical (y-direction) translates of the curve $y = x^3$

27. Circles centered at the point $(2, 0)$

29. Circles centered at the origin

31. Circular paraboloids opening upward, each with its vertex on the z-axis

33. Spheres centered at $(2, -1, 3)$

35. Elliptical cylinders, each with axis the vertical line through $(2, 1, 0)$ and with the length of the x-semiaxis twice that of the y-semiaxis

37.

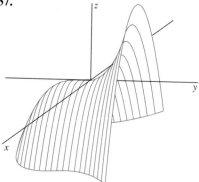

39.

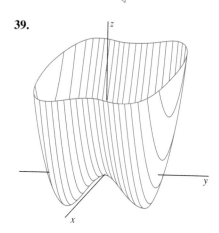

41. Corresponding figures are 14.2.27 and 14.2.35; 14.2.28 and 14.2.36; 14.2.29 and 14.2.33; 14.2.30 and 14.2.34; 14.2.31 and 14.2.38; 14.2.32 and 14.2.37.

Section 14.3 (page 767)

1. 7   **3.** e **5.** $\frac{5}{3}$ **7.** 0

9. 1   **11.** $-\frac{3}{2}$ **13.** 1   **15.** -4

17. y, x **19.** $y^2, 2xy$

29. Note that $f(x, y) = (x - \frac{1}{2})^2 + (y + 1)^2 - \frac{1}{4} \geqq -\frac{1}{4}$ for all x.

Section 14.4 (page 774)

1. $\dfrac{\partial f}{\partial x} = 4x^3 - 3x^2 y + 2xy^2 - y^3$,

$\dfrac{\partial f}{\partial y} = -x^3 + 2x^2 y - 3xy^2 + 4y^3$

3. $\dfrac{\partial f}{\partial x} = e^x(\cos y - \sin y), \dfrac{\partial f}{\partial y} = -e^x(\cos y + \sin y)$

5. $\dfrac{\partial f}{\partial x} = -\dfrac{2y}{(x - y)^2}, \dfrac{\partial f}{\partial y} = \dfrac{2x}{(x - y)^2}$

7. $\dfrac{\partial f}{\partial x} = \dfrac{2x}{x^2 + y^2}, \dfrac{\partial f}{\partial y} = \dfrac{2y}{x^2 + y^2}$

9. $\dfrac{\partial f}{\partial x} = yx^{y-1}, \dfrac{\partial f}{\partial y} = x^y \ln x$

11. $\dfrac{\partial f}{\partial x} = 2xy^3 z^4, \dfrac{\partial f}{\partial y} = 3x^2 y^2 z^4, \dfrac{\partial f}{\partial z} = 4x^2 y^3 z^3$

13. $\dfrac{\partial f}{\partial x} = yze^{xyz}, \dfrac{\partial f}{\partial y} = xze^{xyz}, \dfrac{\partial f}{\partial z} = xye^{xyz}$

15. $\dfrac{\partial f}{\partial x} = 2xe^y \ln z, \dfrac{\partial f}{\partial y} = x^2 e^y \ln z, \dfrac{\partial f}{\partial z} = \dfrac{x^2 e^y}{z}$

17. $\dfrac{\partial f}{\partial r} = \dfrac{4rs^2}{(r^2 + s^2)^2}, \dfrac{\partial f}{\partial s} = -\dfrac{4r^2 s}{(r^2 + s^2)^2}$

19. $\dfrac{\partial f}{\partial u} = e^v + we^u, \dfrac{\partial f}{\partial v} = e^w + ue^v, \dfrac{\partial f}{\partial w} = e^u + ve^w$

21. $z_{xy} = z_{yx} = -4$

23. $z_{xy} = z_{yx} = -4xy \exp(-y^2)$

25. $z_{xy} = z_{yx} = \dfrac{1}{(x + y)^2}$

27. $z_{xy} = z_{yx} = 3e^{-3x} \sin y$

29. $z_{xy} = z_{yx} = -4xy^{-3} \sinh y^{-2}$

31. $6x + 8y - z = 25$ **33.** $z \equiv -1$

35. $27x - 12y - z = 38$ **37.** $x - y + z = 1$

39. $10x - 16y - z = 9$

43. $f_{xyz}(x, y, z) = (x^2 y^2 z^2 + 3xyz + 1)e^{xyz}$

51. $(10, -7, -58)$

53. (a) A decrease of about 2750 cm³; (b) an increase of about 82.5 cm³

Section 14.5 (page 784)

1. None   **3.** $(0, 0, 5)$ **5.** $(3, -1, -5)$

7. $(-2, 0, -4)$ **9.** $(-2, 0, -7)$ and $(-2, 1, -9)$

11. $(0, 0, 0)$, $(1, 0, 2/e)$, $(-1, 0, 2/e)$, $(0, 1, 3/e)$, and $(0, -1, 3/e)$

13. Min. value: $f(1, 1) = 1$

15. Max. value: $f(1, \pm1) = 2$

17. Min. value: $f(4, -2) = f(-4, 2) = -16$

19. Max. value: $f(1, -2) = e^5$

21. Max. value: $f(1, 1) = 3$, min. value: $f(-1, -1) = -3$

23. Max value: $f(0, 2) = 4$, min. value: $f(1, 0) = -1$

25. Let $t = 1/\sqrt{2}$. Max. value: $f(t, t) = f(-t, -t) = 1$, min. value: $f(t, -t) = f(-t, t) = -1$

27. $10 \times 10 \times 10$ **29.** $10 \times 10 \times 10$ cm

31. $\frac{4}{3}$; $x = 2, y = 1, z = \frac{2}{3}$

33. Height 10 ft, front and back 40 ft wide, sides 20 ft deep

35. Base 2 m × 2 m, height 3 m

37. 11,664 in.³   **39.** $\frac{1}{2}$

A-64

41. Max. area: 900 (one square), min. area: 300 (three equal squares)

45. Local min. at $(1, 1)$, saddle point at $(-\frac{1}{3}, -\frac{1}{3})$

47. Global max. 1 at $(\pm 1, 1)$, global min. -1 at $(\pm 1, -1)$, local max. at *every* point $(0, y)$ for which $-1 \leq y < 0$, local min. at every point $(0, y)$ for which $0 < y \leq 1$, a sort of half-saddle at $(0, 0)$

49. 1 m^3 **51.** $abc/27$

53. Base 20×20 in., height 60 in.   **55.** $(\frac{2}{3}, \frac{1}{3})$

57. The base b of each triangular end should equal its height; the depth of the A-frame should be $b\sqrt{2}$.

Section 14.6   (page 793)

1. $dw = (6x + 4y)\, dx + (4x - 6y^2)\, dy$

3. $dw = (x\, dx + y\, dy)/\sqrt{1 + x^2 + y^2}$

5. $dw = (-y\, dx + x\, dy)/(x^2 + y^2)$

7. $dw = (2x\, dx + 2y\, dy + 2z\, dz)/(x^2 + y^2 + z^2)$

9. $dw = (\tan yz)\, dx + (xz \sec^2 yz)\, dy$
$\qquad + (xy \sec^2 yz)\, dz$

11. $dw = -e^{-xyz}(yz\, dx + xz\, dy + xy\, dz)$

13. $dw = \exp(-v^2)(2u\, du - 2u^2 v\, dv)$

15. $dw = (x\, dx + y\, dy + z\, dz)/\sqrt{x^2 + y^2 + z^2}$

17. $\Delta f \approx 0.014$ (true value: about $0.01422\,975$)

19. $\Delta f \approx -0.0007$ **21.** $\Delta f \approx \frac{53}{1300} \approx 0.04077$

23. $\Delta f \approx 0.06$ **25.** 191.1   **27.** 1.4

29. $x \approx 1.95$ **31.** 8.18 in.^{3} **33.** 0.022 acres

35. The period increases by about 0.0278 s.

37. About 303.8 ft

Section 14.7   (page 800)

1. $-(2t + 1) \exp(-t^2 - t)$ **3.** $6t^5 \cos t^6$

5. $\dfrac{\partial w}{\partial s} = \dfrac{\partial w}{\partial t} = \dfrac{2}{s + t}$ **7.** $\dfrac{\partial w}{\partial s} = 0, \dfrac{\partial w}{\partial t} = 5e^t$

9. $\partial r/\partial x = (y + z) \exp(yz + xy + xz)$,
$\qquad \partial r/\partial y = (x + z) \exp(yz + xy + xz)$,
$\qquad \partial r/\partial z = (x + y) \exp(yz + xy + xz)$

11. $\dfrac{\partial z}{\partial x} = -\left(\dfrac{z}{x}\right)^{1/3}, \dfrac{\partial z}{\partial y} = -\left(\dfrac{z}{y}\right)^{1/3}$

13. $\dfrac{\partial z}{\partial x} = -\dfrac{yz(e^{xy} + e^{xz}) + (xy + 1)e^{xy}}{e^{xy} + xye^{yz}}$,

$\qquad \dfrac{\partial z}{\partial y} = -\dfrac{x(x + z)\, e^{xy} + e^{xz}}{xye^{xz} + e^{xy}}$

15. $\dfrac{\partial z}{\partial x} = -\dfrac{c^2 x}{a^2 z}, \dfrac{\partial z}{\partial y} = -\dfrac{c^2 y}{b^2 z}$

17. $\dfrac{\partial w}{\partial x} = \dfrac{(2x^2 - y^2)y^{1/2}}{2x^{1/2}(x^2 - y^2)^{3/4}}$,

$\qquad \dfrac{\partial w}{\partial y} = \dfrac{(x^2 - 2y^2)x^{1/2}}{2y^{1/2}(x^2 - y^2)^{3/4}}$

19. $\dfrac{\partial w}{\partial x} = (y^3 - 3x^2 y)(x^2 + y^2)^{-3} - y$,

$\qquad \dfrac{\partial w}{\partial y} = (x^3 - 3xy^2)(x^2 + y^2)^{-3} - x$

21. $x + y - 2z = 7$

23. $5x + 5y + 11z = 31$

27. $\partial w/\partial x = f'(u)(\partial u/\partial x) = f'(u)$, and so on.

29. Show that $w_u = w_x + w_y$. Then note that

$$w_{uv} = \frac{\partial}{\partial v} w_u = \frac{\partial w_u}{\partial x} \cdot \frac{\partial x}{\partial v} + \frac{\partial w_u}{\partial y} \cdot \frac{\partial y}{\partial v}.$$

Section 14.8   (page 809)

1. $\langle 3, -7 \rangle$ **3.** $\langle 0, 0 \rangle$ **5.** $\langle 0, 6, -4 \rangle$

7. $\langle 1, 1, 1 \rangle$ **9.** $\langle 2, -\frac{3}{2}, -2 \rangle$ **11.** $8\sqrt{2}$

13. $\frac{12}{13}\sqrt{13}$ **15.** $-\frac{13}{20}$ **17.** $-\frac{1}{6}$ **19.** $-6\sqrt{2}$

21. Max.: $\sqrt{170}$; direction: $\langle 7, 11 \rangle$

23. Max.: $14\sqrt{2}$; direction: $\langle 3, 5, -8 \rangle$

25. Max.: $2\sqrt{14}$; direction: $\langle 1, 2, 3 \rangle$

27. $29(x - 2) - 4(y + 3) = 0$

29. $x + y + z = 1$

35. (a) $\frac{34}{3}$°C/ft; (b) 13°C/ft and $\langle 4, 3, 12 \rangle$

37. (a) $z = \frac{3}{10}x + \frac{1}{5}y - \frac{2}{5}$ (b) 0.44 (true value: 0.448)

39. $x - 2y + z + 10 = 0$

43. Compass heading about 36.87°; climbing at 45°

45. Compass heading about 203.2°; climbing at about 75.29°

Section 14.9   (page 818)

1. Max. 4 at $(\pm 2, 0)$, min. -4 at $(0, \pm 2)$

3. Max. 3 at $(\frac{3}{2}\sqrt{2}, \sqrt{2})$ and $(-\frac{3}{2}\sqrt{2}, -\sqrt{2})$, min. -3 at $(-\frac{3}{2}\sqrt{2}, \sqrt{2})$ and $(\frac{3}{2}\sqrt{2}, -\sqrt{2})$

5. Min. $\frac{18}{7}$ at $(\frac{9}{7}, \frac{6}{7}, \frac{3}{7})$, no max.

7. Max. 7 at $(\frac{36}{7}, \frac{9}{7}, \frac{4}{7})$, min. -7 at $(-\frac{36}{7}, -\frac{9}{7}, -\frac{4}{7})$

9. Min. $\frac{25}{3}$ at $(-\frac{5}{3}, \frac{1}{3}, \frac{7}{3})$. There is no max. because, in effect, we seek the extrema of the square of the distance between the origin and a point (x, y, z) on an unbounded straight line.

21. $(2, -2, 1)$ and $(-2, 2, 1)$

25. $(2, 3)$ and $(-2, -3)$

27. Highest: $(\frac{2}{5}\sqrt{5}, \frac{1}{5}\sqrt{5}, \sqrt{5} - 4)$; lowest: $(-\frac{2}{5}\sqrt{5}, -\frac{1}{5}\sqrt{5}, -\sqrt{5} - 4)$

29. Farthest: $x = -\frac{1}{20}(15 + 9\sqrt{5})$, $y = 2x$,
$z = \frac{1}{4}(9 + 3\sqrt{5})$; nearest: $x = -\frac{1}{20}(15 - 9\sqrt{5})$,
$y = 2x$, $z = \frac{1}{4}(9 - 3\sqrt{5})$

33. $\dfrac{3 - 2\sqrt{2}}{4}p^2$

Section 14.10 (page 827)

1. Min. $(-1, 2, -1)$, no other extrema

3. Saddle point $(-\frac{1}{2}, -\frac{1}{2}, \frac{29}{4})$, no extrema

5. Min. $(-3, 4, -9)$, no other extrema

7. Saddle point $(0, 0, 3)$, local max. $(-1, -1, 4)$

9. No extrema

11. Saddle point $(0, 0, 0)$, local min. $(-1, -1, -2)$ and $(1, 1, -2)$

13. Saddle point $(-1, 1, 5)$, local min. $(3, -3, -27)$

15. Saddle point $(0, -2, 32)$, local min. $(-5, 3, -93)$

17. Saddle point $(0, 0, 0)$, local max. $(1, 2, 2)$ and $(-1, -2, 2)$

19. Saddle point $(-1, 0, 17)$, local min. $(2, 0, -10)$

21. Saddle point $(0, 0, 0)$, local (actually, global) max.
$\left(\dfrac{\sqrt{2}}{2}, \dfrac{\sqrt{2}}{2}, \dfrac{1}{2e}\right)$ and $\left(-\dfrac{\sqrt{2}}{2}, -\dfrac{\sqrt{2}}{2}, \dfrac{1}{2e}\right)$, local
(actually, global) min. $\left(-\dfrac{\sqrt{2}}{2}, \dfrac{\sqrt{2}}{2}, -\dfrac{1}{2e}\right)$ and
$\left(\dfrac{\sqrt{2}}{2}, -\dfrac{\sqrt{2}}{2}, -\dfrac{1}{2e}\right)$

23. Local (actually, global) min.

25. Local (actually, global) max.

27. Min. value 3 at $(1, 1)$ and at $(-1, -1)$

29. See Problem 41 of Section 14.5 and its answer

31. The critical points are of the form (m, n), where both m and n are either even integers or odd integers. The critical point (m, n) is a saddle point if both m and n are even, but a local maximum if both m and n are of the form $4k + 1$ or of the form $4k + 3$. It is a local minimum in the remaining cases.

35. Local min. at $(-1.8794, 0)$ and $(1.5321, 0)$, saddle point at $(0.3473, 0)$

37. Local min. at $(-1.8794, 1.8794)$ and $(1.5321, -1.5321)$, saddle point at $(0.3473, -0.3473)$

39. Local min. at $(3.6247, 3.9842)$ and $(3.6247, -3.9842)$, saddle point at $(0, 0)$

Chapter 14 Miscellaneous Problems (page 830)

3. On the line $y = x$, $g(x, y) \equiv \frac{1}{2}$, except that $g(0, 0) = 0$.

5. $f(x, y) = x^2y^3 + e^x \sin y + y + C$

7. All points of the form $(a, b, \frac{1}{2})$ (so $a^2 + b^2 = \frac{1}{2}$) together with $(0, 0, 0)$

9. The normal to the cone at $(a, b, \sqrt{a^2 + b^2})$ meets the z-axis at $(0, 0, 2\sqrt{a^2 + b^2})$.

15. Base $2\sqrt[3]{3} \times 2\sqrt[3]{3}$ ft, height $5\sqrt[3]{3}$ ft

17. $200 \pm 2 \, \Omega$

19. 3%

21. $(\pm 4, 0, 0)$, $(0, \pm 2, 0)$, and $(0, 0, \pm\frac{4}{3})$

25. Parallel to the vector $\langle 4, -3 \rangle$; that is, at an approximate bearing of either $126.87°$ or $306.87°$

27. 1

31. Semiaxes 1 and 2

33. There is no such triangle of minimum perimeter, unless we consider as a triangle the figure with all sides of length zero—a single point on the circumference of the circle. The triangle of *maximum* perimeter is equilateral, with perimeter $3\sqrt{3}$.

35. Closest: $(\frac{1}{3}\sqrt{6}, \frac{1}{6}\sqrt{6})$, farthest: $(-\frac{1}{3}\sqrt{6}, -\frac{1}{6}\sqrt{6})$

39. Max. 1, min. $-\frac{1}{2}$

41. Local min. -1 at $(1, 1)$ and at $(-1, -1)$; horizontal tangent plane (but no extrema) at $(0, 0, 0)$, $(\sqrt{3}, 0, 0)$, and $(-\sqrt{3}, 0, 0)$

43. Local min. -8 at $(2, 2)$, horizontal tangent plane at $(0, 0, 0)$

45. Local max. $\frac{1}{432}$ at $(\frac{1}{2}, \frac{1}{3})$. All points on the intervals $(-\infty, 0)$ and $(1, +\infty)$ on the x-axis are local min. (value 0), and all points on the interval $(0, 1)$ on the x-axis are local max. (value 0); saddle point at $(0, 1)$. Horizontal tangent plane at the origin; it really isn't a saddle point.

47. Saddle point at $(0, 0)$; each point on the hyperbola $xy = \ln 2$ yields a global min.

49. No extrema; saddle points at $(1, 1)$ and $(-1, -1)$

Section 15.1 (page 839)

1. 80   **3.** -78 **5.** $\frac{513}{4}$

7. $-\frac{9}{2}$ **9.** 1   **11.** $\frac{1}{2}(e - 1)$

13. $2(e - 1)$ **15.** $2\pi + \frac{1}{4}\pi^4$ **17.** 1

19. $2 \ln 2$ **21.** -32 **23.** $\frac{4}{15}(9\sqrt{3} - 8\sqrt{2} + 1)$

Section 15.2 (page 845)

1. $\frac{5}{6}$ **3.** $\frac{1}{2}$ **5.** $\frac{1}{12}$ **7.** $\frac{1}{20}$

9. $-\frac{1}{18}$ **11.** $\frac{1}{2}(e - 2)$ **13.** $\frac{61}{3}$

15. $\displaystyle\int_0^4 \int_{-\sqrt{y}}^{\sqrt{y}} x^2y \, dx \, dy = \frac{512}{21}$

17. $\displaystyle\int_0^1 \int_{-\sqrt{y}}^{\sqrt{y}} x \, dx \, dy + \int_1^9 \int_{(y-3)/2}^{\sqrt{y}} x \, dx \, dy = \frac{32}{3}$

A-66

19. $\int_0^4 \int_{2-\sqrt{4-y}}^{y/2} 1 \, dx \, dy = \frac{4}{3}$ **21.** $\int_0^\pi \int_0^y \frac{\sin y}{y} \, dx \, dy = 2$

23. $\int_0^1 \int_0^x \frac{1}{1 + x^4} \, dy \, dx = \frac{\pi}{8}$

Section 15.3   (page 850)

1. $\frac{1}{6}$ **3.** $\frac{32}{3}$ **5.** $\frac{5}{6}$ **7.** $\frac{32}{3}$

9. $2 \ln 2$ **11.** 2 **13.** $2e$ **15.** $\frac{1}{3}$

17. $\frac{41}{60}$ **19.** $\frac{4}{15}$ **21.** $\frac{10}{3}$ **23.** 19

25. $\frac{4}{3}$ **27.** $\frac{1}{6} abc$ **29.** $\frac{2}{3}$

33. $\frac{625}{2} \pi \approx 981.748$

35. $\frac{1}{6}(2\pi + 3\sqrt{3})R^3 \approx (1.913)R^3$ **37.** $\frac{256}{15}$

Section 15.4   (page 858)

3. $\frac{3}{2}\pi$ **5.** $\frac{1}{6}(4\pi - 3\sqrt{3}) \approx 1.22837$

7. $\frac{1}{2}(2\pi - 3\sqrt{3}) \approx 0.5435$ **9.** $\frac{16}{3}\pi$

11. $\frac{23}{8}\pi$ **13.** $\frac{1}{4}\pi \ln 2$ **15.** $\frac{16}{5}\pi$

17. $\frac{1}{4}\pi(1 - \cos 1) \approx 0.36105$

19. 2π **21.** 4π **27.** 2π

29. $\frac{1}{3}\pi a^3(2 - \sqrt{2}) \approx (0.6134)a^3$

31. $\frac{1}{4}\pi$ **35.** $2\pi^2 a^2 b$

Section 15.5   (page 868)

1. $(2, 3)$ **3.** $(1, 1)$ **5.** $(\frac{4}{3}, \frac{2}{3})$

7. $(\frac{3}{2}, \frac{6}{5})$ **9.** $(0, -\frac{8}{5})$ **11.** $\frac{1}{24}, (\frac{2}{5}, \frac{2}{5})$

13. $\frac{256}{15}, (0, \frac{16}{7})$ **15.** $\frac{1}{12}, (\frac{9}{14}, \frac{9}{14})$ **17.** $\frac{1}{3}, (0, \frac{22}{35})$

19. $2, (\frac{1}{2}\pi, \frac{1}{8}\pi)$ **21.** $a^3, (\frac{7}{12}a, \frac{7}{12}a)$ **23.** $\frac{128}{5}, (0, \frac{20}{7})$

25. $\pi; \bar{x} = \frac{\pi^2 - 4}{\pi} \approx 1.87, \bar{y} = \frac{\pi}{8} \approx 0.39$

27. $\frac{1}{3}\pi a^3; \bar{x} = 0, \bar{y} = \frac{3a}{2\pi}$

29. $\frac{2}{3}\pi + \frac{1}{4}\sqrt{3}; \bar{x} = 0, \bar{y} = \frac{36\pi + 33\sqrt{3}}{32\pi + 12\sqrt{3}} \approx 1.4034$

31. $\frac{2\pi a^{n+4}}{n + 4}$ **33.** $\frac{3}{2}\pi k$ **35.** $\frac{1}{9}$

37. $\hat{x} = \frac{2}{21}\sqrt{105}, \hat{y} = \frac{4}{3}\sqrt{5}$

39. $\hat{x} = \hat{y} = \frac{1}{10}a\sqrt{30}$

41. $(4r/3\pi, 4r/3\pi)$ **43.** $(2r/\pi, 2r/\pi)$

51. (a) $\bar{x} = 0, \bar{y} = \frac{4a^2 + 3\pi ab + 6b^2}{12b + 3\pi a}$;

(b) $\frac{1}{3}\pi a(4a^2 + 2\pi ab + 6b^2)$

53. $(1, \frac{1}{4})$ **55.** $\frac{484}{3} k$

Section 15.6   (page 876)

1. 18 **3.** 128 **5.** $\frac{1}{60}$ **7.** 0 **9.** 12

11. $V = \int_0^3 \int_0^{3-(2x/3)} \int_0^{6-2x-2y} 1 \, dz \, dy \, dx = 6$

13. $\frac{128}{5}$ **15.** $\frac{332}{105}$ **17.** $\frac{256}{15}$

19. $\frac{11}{30}$ **21.** $(0, \frac{20}{7}, \frac{20}{7})$ **23.** $(0, \frac{8}{7}, \frac{12}{7})$

25. $\bar{x} = 0, \bar{y} = \frac{44 - 9\pi}{72 - 9\pi}, \bar{z} = \frac{9\pi - 16}{72 - 9\pi}$

27. $\frac{8}{7}$ **29.** $\frac{1}{30}$ **33.** $\frac{2}{3}a^5$

35. $\frac{38}{45} ka^7$ **37.** $\frac{1}{3}k$ **39.** $(\frac{9}{64}\pi, \frac{9}{64}\pi, \frac{3}{8})$

41. 24π **43.** $\frac{1}{6}\pi$

Section 15.7   (page 884)

1. 8π **5.** $\frac{4}{3}\pi(8 - 3\sqrt{3})$ **7.** $\frac{1}{2}\pi a^2 h^2$

9. $\frac{1}{4}\pi a^4 h^2$ **11.** $\frac{81}{2}\pi, (0, 0, 3)$ **13.** 24π

15. $\frac{1}{6}\pi(8\sqrt{2} - 7)$ **17.** $\frac{1}{12}\pi\delta a^2 h(3a^2 + 4h^2)$

19. $\frac{1}{3}\pi$ **21.** $(0,0,3a/8)$ **23.** $\frac{1}{3}\pi$

25. $\frac{1}{3}\pi a^3(2 - \sqrt{2}); \bar{x} = 0 = \bar{y}, \bar{z} = \frac{3}{16}(2 + \sqrt{2})a \approx (0.6402)a$

27. $\frac{7}{5}ma^2$

29. The surface obtained by rotating the circle in the xz-plane with center $(a, 0)$ and radius a around the z-axis—a doughnut with an infinitesimal hole; $2\pi^2 a^3$

31. $\frac{2}{15}(128 - 51\sqrt{3})\pi a^5$ **33.** $\frac{37}{48}\pi a^4, \bar{z} = \frac{105}{74}a$

Section 15.8   (page 891)

1. $6\pi\sqrt{11}$ **3.** $\frac{1}{6}\pi(17\sqrt{17} - 1)$

5. $3\sqrt{2} + \frac{1}{2}\ln(3 + 2\sqrt{2}) \approx 5.124$ **7.** $3\sqrt{14}$

9. $\frac{2}{3}\pi(2\sqrt{2} - 1) \approx 3.829$ **11.** $\frac{1}{6}\pi(65\sqrt{65} - 1)$

15. $8a^2$

Section 15.9   (page 899)

1. $x = \frac{1}{2}(u + v), y = \frac{1}{2}(u - v); J = -\frac{1}{2}$

3. $x = \sqrt{u/v}, y = \sqrt{uv}; J = (2v)^{-1}$

5. $x = \frac{1}{2}(u + v), y = \frac{1}{2}\sqrt{u - v}; J = -\frac{1}{4}(u - v)^{-1/2}$

7. $\frac{3}{5}$ **9.** $\ln 2$ **11.** $\frac{1}{8}(2 - \sqrt{2})$

13. $\frac{39}{2}\pi$ **15.** 8

17. S is the region $3u^2 + v^2 \leq 3$; the value of the integral is $\dfrac{2\pi\sqrt{3}(e^3 - 1)}{3e^3}$.

Chapter 15 Miscellaneous Problems   (page 901)

1. $\frac{1}{3}(2 - \sqrt{2})$ **3.** $\frac{e - 1}{2e}$ **5.** $\frac{1}{4}(e^4 - 1)$

7. $\frac{4}{3}$ **9.** $9\pi, \bar{z} = \frac{9}{16}$ **11.** 4π

13. 4π **15.** $\frac{1}{16}(\pi - 2)$ **17.** $\frac{128}{15}, (\frac{32}{7}, 0)$

19. $k\pi$, $(1, 0)$ **21.** $\bar{y} = 4b/3\pi$ **23.** $(0, \frac{8}{5})$

25. $\frac{10}{3}\pi(\sqrt{5} - 2) \approx 2.4721$

27. $\frac{3}{10}Ma^2$ **29.** $\frac{1}{5}M(b^2 + c^2)$

31. $\frac{128}{225}\delta(15\pi - 26) \approx (12.017)\delta$, where δ is the (constant) density

33. $\frac{8}{3}\pi$ **41.** $\frac{18}{7}$

43. $\frac{1}{6}\pi(37\sqrt{37} - 17\sqrt{17}) \approx 81.1418$

47. $4\sqrt{2}$ **48.** Approximately 3.49608

51. 3δ **53.** $\frac{8}{15}\pi abc$

Section 16.1 (page 909)

1.

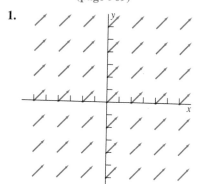

3.

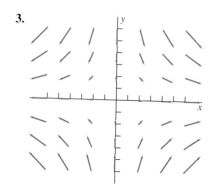

5.

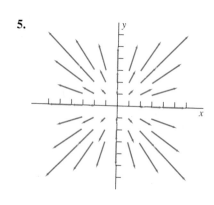

7.

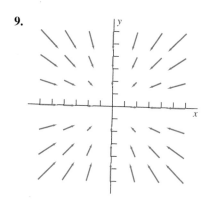

9.

11. 3, **0**

13. 0, **0**

15. $x^2 + y^2 + z^2$, $\langle -2yz, -2xz, -2xy \rangle$

17. 0, $\langle 2y - 2z, 2z - 2x, 2x - 2y \rangle$

19. 3, $\langle x \cos xy - x \cos xz, y \cos yz - y \cos xy, z \cos xz - z \cos yz \rangle$

Section 16.2 (page 918)

1. $\frac{310}{3}$, $\frac{248}{3}$, 62 **3.** $3\sqrt{2}$, 3, 3

5. $\frac{49}{24}$, $\frac{3}{2}$, $\frac{4}{3}$ **7.** $\frac{6}{5}$

9. 315 **11.** $\frac{19}{60}$

13. $\pi + 2\pi^2$ **15.** 28

17. $\frac{1}{6}(14\sqrt{14} - 1) \approx 8.563867$

19. $(0, 2a/\pi)$

21. $10k\pi$, $(0, 0, 4\pi)$

23. $\frac{1}{2}ka^3$, $(\frac{2}{3}a, \frac{2}{3}a)$, $I_x = I_y = \frac{1}{4}ka^5$

25. (a) $\frac{1}{2}k \ln 2$; (b) $-\frac{1}{2}k \ln 2$

Section 16.3 (page 925)

1. $f(x, y) = x^2 + 3xy + y^2$

3. $f(x, y) = x^3 + 2xy^2 + 2y^3$

5. $f(x, y) = \frac{1}{4}x^4 + \frac{1}{3}y^3 + y \ln x$

7. $f(x, y) = \sin x + x \ln y + e^y$

9. $f(x, y) = x^3y^3 + xy^4 + \frac{1}{5}y^5$

11. $f(x, y) = x^2y^{-1} + y^2x^{-3} + 2y^{1/2}$

17. 6

19. $1/e$

21. $-\pi$

23. $f(x, y, z) = xyz$

25. $f(x, y, z) = xy \cos z - yze^x$

27. **F** is not conservative on any region containing $(0, 0)$.

29. $Q_z = 0 \neq 2y = R_y$

Section 16.4 (page 933)

1. 0 **3.** 3 **5.** $\frac{3}{10}$ **7.** 2 **9.** 0

11. $\frac{16}{105}$ **13.** πa^2 **15.** $\frac{3}{8}\pi$ **25.** $\frac{3}{2}$ **31.** 2π

Section 16.5 (page 941)

1. $\frac{1}{120}\sqrt{3}$ **3.** $\frac{1}{60}\pi(1 + 391\sqrt{17}) \approx 84.4635$

5. $\frac{16}{3}\pi$ **7.** 24π **9.** 0

11. $(\frac{1}{2}a, \frac{1}{2}a, \frac{1}{2}a)$

13. $\overline{z} = \dfrac{1 + (24a^4 + 2a^2 - 1)\sqrt{1 + 4a^2}}{10[(1 + 4a^2)^{3/2} - 1]}$,

$I_z = \frac{1}{60}\pi\delta[1 + (24a^4 + 2a^2 - 1)\sqrt{1 + 4a^2}]$

15. $\overline{x} = \dfrac{4}{3\pi - 6} \approx 1.16796$, $\overline{y} = 0$,

$\overline{z} = \dfrac{\pi}{2\pi - 4} \approx 1.13797$

17. Net flux: 0

19. Net flux: 1458π

Section 16.6 (page 949)

1. Both values: 4π **3.** Both values: 24

5. Both values: $\frac{1}{2}$ **7.** $\frac{2385}{2}\pi$ **9.** $\frac{1}{4}$

11. $\frac{703,125}{4}\pi$ **13.** 16π

23. $\frac{1}{48}\pi(482,620 + 29,403 \ln 11) \approx 36,201.967$

Section 16.7 (page 957)

1. -20π **3.** 0 **5.** -52π **7.** -8π **9.** -2

11. $\phi(x, y, z) = yz - 2xz + 3xy$

13. $\phi(x, y, z) = 3xe^z + 5y \cos x + 17$

Chapter 16 Miscellaneous Problems (page 959)

1. $\frac{125}{3}$

3. $\frac{69}{8}$ (Use the fact that the integral of $\mathbf{F} \cdot \mathbf{T}$ is independent of the path.)

5. $\frac{2148}{5}$

9. $\frac{1}{3}(5\sqrt{5} - 1) \approx 3.3934$; $I_y = \frac{1}{15}(2 + 50\sqrt{5}) \approx 7.5869$

11. $\frac{2816}{7}$

17. $\frac{371}{30}\pi$

19. 72π

29. (a) $\phi'(r)(\mathbf{r}/r)$; (b) $3\phi(r) + r\phi'(r)$; (c) $\mathbf{0}$

Appendix A

1. $2\pi/9$ **3.** $7\pi/4$ **5.** $-5\pi/6$ **7.** $72°$

9. $675°$

11. $\sin x = -\frac{1}{2}\sqrt{3}$, $\cos x = \frac{1}{2}$, $\tan x = -\sqrt{3}$, $\csc x = -2/\sqrt{3}$, $\sec x = 2$, $\cot x = -1/\sqrt{3}$

13. $\sin x = -\frac{1}{2}$, $\cos x = -\frac{1}{2}\sqrt{3}$, $\cot x = \sqrt{3}$, $\csc x = -2$, $\sec x = -2/\sqrt{3}$, $\tan x = 1/\sqrt{3}$

15. $x = n\pi$ (n any integer)

17. $x = \frac{3}{2}\pi + 2n\pi$ (n any integer)

19. $x = 2n\pi$ (n any integer)

21. $x = n\pi$ (n any integer)

23. $x = \frac{3}{4}\pi + n\pi$ (n any integer)

25. $\sin x = -\frac{3}{5}$, $\cos x = \frac{4}{5}$, $\tan x = -\frac{3}{4}$, $\sec x = \frac{5}{4}$, $\cot x = -\frac{4}{3}$

29. $\frac{1}{2}$ **31.** $-\frac{1}{2}$ **33.** $\frac{1}{2}\sqrt{3}$

35. $-\frac{1}{2}\sqrt{3}$ **43.** $\pi/3, 2\pi/3$ **45.** $\pi/2$

47. $\pi/8, 3\pi/8, 5\pi/8, 7\pi/8$

Appendix F

1. $\displaystyle\int_0^1 x^2\, dx = \frac{1}{3}$

3. $\displaystyle\int_0^2 x\sqrt{4 - x^2}\, dx = \frac{8}{3}$

5. $\displaystyle\int_0^{\pi/2} \sin x \cos x\, dx = \frac{1}{2}$

7. $\displaystyle\int_0^2 \sqrt{x^4 + x^7}\, dx = \frac{52}{9}$

References for Further Study

References 2, 3, 7, and 10 may be consulted for historical topics pertinent to calculus. Reference 14 provides a more theoretical treatment of single-variable calculus topics than ours. References 4, 5, 8, and 15 include advanced topics in multivariable calculus. Reference 11 is a standard work on infinite series. References 1, 9, and 13 are differential equations textbooks. Reference 6 discusses topics in calculus together with computing and programming in BASIC. Those who would like to pursue the topic of fractals should look at Reference 12.

1. BOYCE, WILLIAM E., and RICHARD C. DiPRIMA, *Elementary Differential Equations* (5th ed.). New York: John Wiley, 1991.
2. BOYER, CARL B., *A History of Mathematics* (2nd ed.). New York: John Wiley, 1991.
3. BOYER, CARL B., *The History of the Calculus and Its Conceptual Development.* New York: Dover Publications, 1959.
4. BUCK, R. CREIGHTON, *Advanced Calculus* (3rd ed.). New York: McGraw-Hill, 1977.
5. COURANT, RICHARD, and FRITZ JOHN, *Introduction to Calculus and Analysis.* Vols. I and II. New York: Springer-Verlag, 1989.
6. EDWARDS, C. H., JR., *Calculus and the Personal Computer.* Englewood Cliffs, N.J.: Prentice Hall, 1986.
7. EDWARDS, C. H., JR., *The Historical Development of the Calculus.* New York: Springer-Verlag, 1982.
8. EDWARDS, C. H., JR., *Advanced Calculus of Several Variables.* New York: Academic Press, 1973.
9. EDWARDS, C. H., JR., and DAVID E. PENNEY, *Elementary Differential Equations with Boundary Value Problems* (3rd ed.). Englewood Cliffs, N.J.: Prentice Hall, 1993.
10. KLINE, MORRIS, *Mathematical Thought from Ancient to Modern Times.* Vols. I, II, and III. New York: Oxford University Press, 1990.
11. KNOPP, KONRAD, *Theory and Application of Infinite Series* (2nd ed.). New York: Hafner Press, 1990.
12. PEITGEN, H. O. and P. H. RICHTER, *The Beauty of Fractals.* New York: Springer-Verlag, 1986.
13. SIMMONS, GEORGE F., *Differential Equations with Applications and Historical Notes.* New York: McGraw-Hill, 1972.
14. SPIVAK, MICHAEL E., *Calculus* (2nd ed.). Berkeley: Publish or Perish, 1980.
15. TAYLOR, ANGUS E., and W. ROBERT MANN, *Advanced Calculus* (3rd ed.). New York: John Wiley, 1983.

Index

Boldface type indicates page on which a term is defined.